ESSENTIALS OF GEOLOGY

FIFTH EDITION

Plate Tectonics

Objective: The objective of this laboratory exercise is to:
1. Examine plate interactions on a global scale, and
2. Identify geologic processes and features associated with tectonic plate boundaries.

After completing this laboratory, you should be able to:
- identify major Earth tectonic plates.
- identify plate interactions at major tectonic plate boundaries
- describe the relationship among earthquakes, volcanoes, and plate tectonic boundaries
- describe how mountain ranges associated with convergent boundaries form.

Reading: Marshak, Chapter 2, The Way the Earth Works: Plate Tectonics (pp. 42-81)

Before Lab Assignment: On each diagram, label the type of plate interaction and write a brief description of the geological processes at this location.

I. Locating Tectonic Plates & Plate Interactions: This portion of the lab is designed to familiarize you with the world's major tectonic plates and the interactions that take place at plate boundaries. Each lab group will be assigned a specific tectonic plate:
- North American Plate
- South American Plate
- Australian-Indian Plate
- Eurasian Plate
- Caribbean Plate
- African Plate

The information below is designed to guide your lab group as you create a brief (approx. 5 minute) summary of your tectonic plate. Divide your presentation among group members, with each person presenting at least one topic.
Information you should include in your presentation and record on your answer sheet:
- Location and Motion:
  - Location of plate (use map pp. 60 in your textbook)
  - Identify type of plate activity at each margin
  - Rate and direction of plate motion (map, p. 76).
- Geographic Features Associated With Plate Boundaries
  - Use Google Earth to investigate the underwater boundaries of your plate, and describe the features you see.
  - Plate boundaries may be associated with volcanoes and earthquakes.
    - Historic Earthquake and Volcanic activity: pp. 60 and 154 in Marshak, 5th edition
    - What activity has been associated with your plate in the last two weeks?
      - Earthquakes: Go to: http://www.iris.edu/seismon/
      - Volcanoes: http://www.volcano.si.edu/weekly_report.cfm
      - While most earthquakes and volcanoes occur at plate boundaries, some do not. Has there been any recent (two weeks) activity in the interior in "your" plate?

Presentations: You will want to take notes during the presentations…you will need this information for the synthesis portion of the lab section.

Synthesis: Using the information from each of the presentations, create a series of brief statements that address the following:
- What features are associated with the underwater portion of each type of plate boundary?
- Describe the relationship of earthquakes and volcanoes to specific types of tectonic plate boundaries.

FIFTH EDITION

ESSENTIALS OF GEOLOGY

STEPHEN MARSHAK
UNIVERSITY OF ILLINOIS

W. W. NORTON & COMPANY
NEW YORK | LONDON

W. W. Norton & Company has been independent since its founding in 1923, when William Warder Norton and Mary D. Herter Norton first published lectures delivered at the People's Institute, the adult education division of New York City's Cooper Union. The firm soon expanded its program beyond the Institute, publishing books by celebrated academics from America and abroad. By mid-century, the two major pillars of Norton's publishing program—trade books and college texts—were firmly established. In the 1950s, the Norton family transferred control of the company to its employees, and today—with a staff of four hundred and a comparable number of trade, college, and professional titles published each year—W. W. Norton & Company stands as the largest and oldest publishing house owned wholly by its employees.

Printed in the United States of America.
Fifth Edition

EDITOR **ERIC SVENDSEN**
SENIOR PROJECT EDITOR **THOMAS FOLEY**
ASSOCIATE PRODUCTION DIRECTOR **BENJAMIN REYNOLDS**
COPY EDITOR **CONNIE PARKS**
MANAGING EDITOR, COLLEGE **MARIAN JOHNSON**
MANAGING EDITOR, COLLEGE DIGITAL MEDIA **KIM YI**
MEDIA EDITOR **ROB BELLINGER**
ASSOCIATE MEDIA EDITOR **CAILIN BARRETT-BRESSACK**
MEDIA PROJECT EDITOR **MARCUS VAN HARPEN**
MEDIA EDITORIAL ASSISTANT **LIZ VOGT**
MARKETING MANAGER, GEOLOGY **JAKE SCHINDEL**
DESIGN DIRECTOR **RUBINA YEH**
DESIGNER **JILLIAN BURR**
PHOTO EDITOR **STEPHANIE ROMEO**
PERMISSIONS MANAGER **MEGAN JACKSON**
EDITORIAL ASSISTANT **RACHEL GOODMAN**

COMPOSITION AND PAGE LAYOUT BY PRECISION GRAPHICS / LACHINA
ILLUSTRATIONS BY PRECISION GRAPHICS / LACHINA

SENIOR ARTIST AT PRECISION GRAPHICS / LACHINA: STAN MADDOCK
PROJECT MANAGER AT PRECISION GRAPHICS / LACHINA: REBECCA MARSHALL

MANUFACTURING BY R. R. DONNELLEY—KENDALLVILLE, IN

978-0-393-26339-8

W. W. NORTON & COMPANY, INC., 500 FIFTH AVENUE, NEW YORK, NY 10110
WWNORTON.COM
W. W. NORTON & COMPANY LTD., CASTLE HOUSE, 75/76 WELLS STREET, LONDON W1T 3QT

1 2 3 4 5 6 7 8 9 0

Title page photo: In 1770, Captain James Cook named these 500 m-high hills, here seen at sunset on the east coast of Australia, the Glass House Mountains. Cook thought that they resembled glass furnaces in Yorkshire. In fact, they are remnants of geologic furnaces. Their shape developed when erosion of volcanoes that had erupted 26 million years ago left behind hard rock that solidified from melt within and just below the volcanoes. (Photo courtesy of Stephen Marshak.)

DEDICATION

—In memory of Jack Repcheck—

friend, mentor, and editor,

whose perpetual enthusiasm

permeates the pages of this book.

Brief Contents

Contents

BLACK SEA
Eurasian Plate
MEDITERRANEAN SEA
African Plate

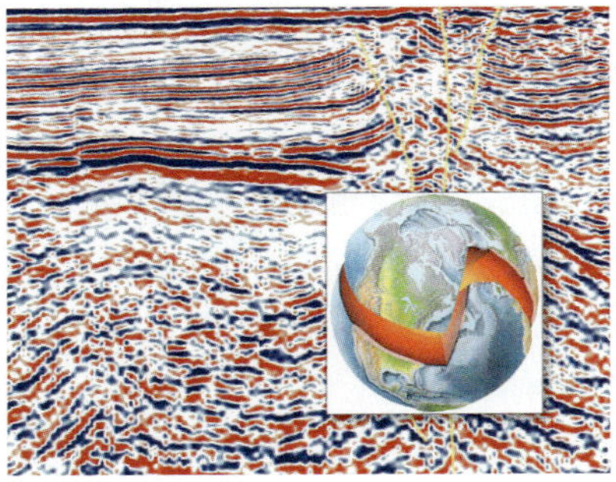

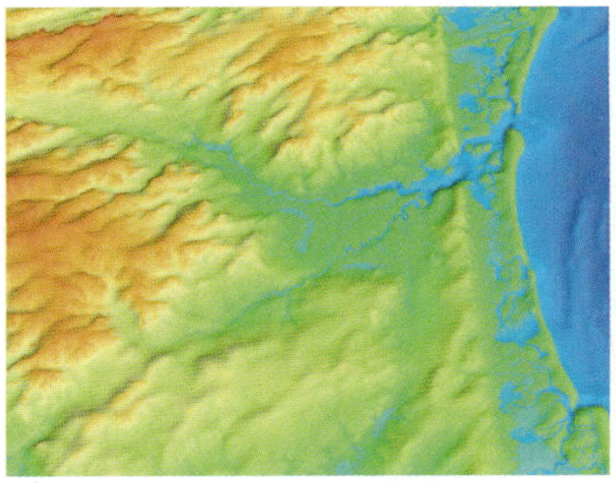

CHAPTER 14

CHAPTER 15

CHAPTER 16

CHAPTER 17
Dry Regions: The Geology of Deserts • 549

CHAPTER 18
Amazing Ice: Glaciers and Ice Ages • 569

CHAPTER 19
Global Change in the Earth System • 603

Preface

NARRATIVE THEMES

Why do earthquakes, volcanoes, floods, and landslides happen? What causes mountains to rise? How do beautiful landscapes develop? Do climate and life change through time? When did the Earth form, and by what process? Where do we dig to find valuable metals, and where do we drill to find oil? Does sea level change? Can continents move? The study of geology addresses these important questions and many more. But from the birth of the discipline in the late-18th century until the mid-20th century, geologists considered each question largely in isolation, without pondering its relation to the others. This approach changed, beginning in the 1960s, in response to the formulation of two "paradigm-shifting" ideas that have unified thinking about the Earth and its features. The first idea, called the *theory of plate tectonics,* states that the Earth's outer shell consists of discrete plates that slowly move relative to each other so that the map of our planet continuously changes. Plate interactions cause earthquakes and volcanoes, build mountains, provide gases that make up the atmosphere, and affect the distribution of life on Earth. The second idea, called the *Earth System concept,* emphasizes that our planet's water, land, atmosphere, and living inhabitants are dynamically interconnected. In the Earth System, materials constantly cycle among various living and nonliving reservoirs on, above, and within the planet. We have come to realize that the history of life is intimately linked to the history of the physical Earth.

Essentials of Geology, Fifth Edition, offers an introduction to the study of our planet that employs both the theory of plate tectonics and the concept of the Earth System throughout the book, to weave together a number of narrative themes, including:

1. The solid Earth, the oceans, the atmosphere, and life interact in complex ways, yielding a unique planet, unlike any other in the Solar System.
2. Most geologic processes reflect the interactions among plates.
3. The Earth is a planet, formed like other planets from dust and gas. But, in contrast to other planets, the Earth is a dynamic place where new geologic features continue to form and old features continue to be destroyed.
4. The Earth is very old—about 4.57 billion years have passed since its birth. During this time the surface, subsurface, and atmosphere of the planet have changed, and life has evolved.
5. Internal processes (driven by the Earth's internal heat) and external processes (driven by heat from the Sun) interact at the Earth's surface to produce complex landscapes.
6. Geologic knowledge can help society understand natural hazards such as earthquakes, volcanoes, landslides, and floods, and in some cases can reduce the danger that these hazards pose.
7. Energy and mineral resources come from the Earth and form in response to geologic phenomena. Geologic study can help locate these resources and mitigate the consequences of their use.
8. Physical features of the Earth are linked to life processes, and vice versa.
9. Science comes from observation; people make scientific discoveries.
10. Geology utilizes ideas from physics, chemistry, and biology, so the study of geology provides an excellent means to improve science literacy.

These narrative themes serve as the *take-home message* of this book, a message that students will remember long after finishing an introductory geology course. In effect, these themes provide a mental framework on which students can organize and connect ideas, and develop a modern, coherent image of our planet.

PEDAGOGICAL APPROACH

Students learn best from textbooks when they can actively engage with a combination of narrative text and narrative art. Some students respond more to words, which help them to organize information, provide answers to questions, fill in the essential steps that link ideas together, and develop a personal context for understanding information. Some students respond more to narrative art—art designed to tell a story—for visual images help students comprehend and remember processes. And some respond to question-and-answer-based active learning, an approach through which students can in effect "practice" their knowledge in real time. *Essentials of Geology*, Fifth Edition, supports all three of these learning approaches.

The text and ancillary educational package of this book have been crafted to engage students in its narrative style, the art has been configured to tell a story, the chapters are laid out to help students internalize key principles, and the online videos, animations, and activities have been designed to increase student interest and involvement, as well as to provide them with active feedback. For example, *Did You Ever Wonder* panels prompt students to connect new information to their existing knowledge base by asking geology-related questions that they have probably already thought about. A *Take-Home Message* panel at the end of each section helps students solidify key themes before proceeding to the next section. Questions at the end of each chapter not only test basic knowledge, but also stimulate critical thinking. Animations and videos support each chapter and are designed to work hand in hand with the text. New student assessment tools that work in your school's learning management system (LMS), including Smartwork5 online homework, help students prepare for class with dynamic visual questions. Finally, the *See for Yourself* and *Geotour* features guide students on virtual field trips, via *Google Earth*™, to locales around the globe where they can apply their newly acquired knowledge to the interpretation of real-world geologic features.

ORGANIZATION

Topics covered in this book have been arranged so that students can build their knowledge of geology on a foundation of overarching principles. The book starts by considering how the Earth formed, and how it is structured, overall, from its surface to its center. With this basic background, students can delve into plate tectonics, the grand unifying theory of geology. Plate tectonics appears early in the book, so that students can use the theory as a foundation from which they can interpret and link ideas presented in subsequent chapters. Knowledge of plate tectonics, for example, helps students understand the suite of chapters on minerals, rocks, and the rock cycle. Knowledge of plate tectonics and rocks together, in turn, provides a basis for studying volcanoes, earthquakes, and mountains. And with this background, students are prepared to see how the map of the Earth has changed through the vast expanse of geologic time, and how energy and mineral resources have developed. The book's final chapters address processes and problems occurring at or near the Earth's surface, from the unstable slopes of hills, down the course of rivers, to the shores of the sea, and beyond. This section concludes with a topic of growing concern in society—global change, particularly climate change.

Although we have arranged the sequence of chapters to provide a pedagogically consistent narrative, each chapter is self-contained, and we reiterate relevant material where necessary. As a result, instructors can choose their own strategies for teaching geology, and they can resequence chapters if useful for their own course's design.

SPECIAL AND UPDATED FEATURES FOR THIS EDITION

Narrative Art and *What a Geologist Sees*

To help students visualize topics, we have lavishly illustrated this book by creating figures and selecting photographs that provide a realistic context for interpreting geologic features without overwhelming students with extraneous detail. In this edition, many drawings and photographs have been integrated into *narrative art* that has been laid out, labeled, and annotated to tell a story—the figures are drawn to teach! Subcaptions are positioned adjacent to the relevant parts of a figure, labels point out key features, and balloons provide additional detail. Where relevant, we've arranged subparts of figures to convey time progression. Color schemes in drawings tie directly to those of relevant photos, so that students can easily visualize the relationships among drawings and photos. In some examples, annotated sketches labeled *What a Geologist Sees* accompany photographs. These "WAGS" help students confirm that they actually see the specific features that the photograph was intended to show.

Narrative Art Videos, Animations, and Simulations

The Fifth Edition of *Essentials of Geology* provides a rich collection of over fifty new animations and videos to illustrate geologic processes and course concepts. Animations and simulations utilize a consistent style applying a new 3-D perspective. Many allow students to control and simulate aspects of a geologic process. In a brand-new set of *narrative art videos*, the author enhances explanations of core concepts in the text by describing the processes displayed in animated versions of the book's figures. We provide these resources to students, at the book's student website, through the Marshak YouTube channel, or through LMS coursepacks. They are available to instructors through an easy-to-use USB-compatible flash drive, and at the instructor resource center.

Featured Paintings: *Geology at a Glance*

In addition to individual figures, renowned British artist Gary Hincks has created spectacular two-page annotated paintings, called *Geology at a Glance*. The Fifth Edition features a brand-new painting that captures the history of the Earth, displaying stages in the evolution of land and life over time, to scale. All Hincks paintings integrate key concepts introduced in the chapters and visually emphasize the relationships among components of the Earth System. And, they provide students with a way to review a subject . . . *at a glance*.

New Coverage of Current Topics

To ensure that *Essentials of Geology*, Fifth Edition, reflects the latest research discoveries and helps students understand geologic events that have been featured in recent news headlines, we have updated many topics throughout the book. For example, the extensively revised resources chapter explains how new technologies—such as directional drilling and hydraulic fracturing—have been game changers in the interpretation of global energy reserves, by making so-called "unconventional reserves" accessible. The book utilizes new discoveries from EarthScope projects, as well as updates from the latest IPCC report on climate change. *Essentials of Geology*, Fifth Edition, also discusses lessons learned from natural disasters such as Hurricane Sandy, Typhoon Haiyan, the Oso landslide, and the Tōhoku tsunami.

Assessment That Works in Your School's LMS

Monitoring whether students are prepared for class and understand the chapter material is a constant challenge. We have designed two tools that help you and that work directly in your school's learning management system—free Norton Coursepacks and the upcoming update of Norton's automated system, Smartwork5. Developed with substantial input from instructors and students, these tools allow students to receive the coaching they need in order to work through their assignments, while instructors get real-time assessment of student progress with automatic grading and item analysis. The questions in these products were developed with the help of the author, Stephen Marshak, in order to work seamlessly with the text.

Built-In Review Features

Each chapter begins with *Learning Objectives* that frame the chapter's major concepts, and every section ends with a *Take-Home Message*, a brief summary to help readers identify and remember the highlights of that section before moving on to the next. *Did You Ever Wonder* questions occur throughout to help show how the book's narrative and figures may address questions that you've already asked about the Earth. Each chapter then concludes with an integrated *Chapter Review*, designed to pull together summary points, key terms, basic review questions, critical thinking questions (*On Further Thought*), and highlights of available free student media, into a visually compact, two-page, easy-to-follow layout.

Interludes

The book contains several Interludes. These "mini-chapters" focus on self-contained key topics that are not broad enough to require an entire chapter. Arranging some content into Interludes not only keeps chapters reasonable in length, but also provides instructors with additional flexibility to sequence topics within a course. Interludes in the Fifth Edition contain dedicated summary sections, structured like those in the numbered chapters.

Discussion of Societal Issues

We address geology's practical applications in several chapters, providing students with an opportunity to learn about energy resources, mineral resources, global change, and natural disasters. *Science and Society* features, provided for free on the open student website and in Norton's LMS coursepacks, challenge students to apply material that they learn in *Essentials of Geology*, Fifth Edition to the interpretation of news articles and publicly available geologic data.

Student Website

Free for students, this site contains all the animations, simulations, and videos developed for this text and highlighted in the chapter summaries. Students will also find vocabulary flashcards, *Science and Society* features that challenge students to use course concepts in analyzing news articles, real-time geologic data, and information on how best to utilize the *Google Earth*™ materials provided for this book. This support includes a video designed to help with start-up, all the book's *See For Yourself* sites in one downloadable file, instructions to enter latitude and longitude coordinates for each site, and links to Google tutorials and information on updates. Students can access the student site at *digital.wwnorton.com/essgeo5*.

MEDIA AND ANCILLARY MATERIALS FOR INSTRUCTORS

Assessment That Works in Your School's LMS

Coursepacks. Available at no cost to professors or students, Norton Coursepacks bring high-quality Norton digital media into a new or existing online course. Working within your school's LMS, and without additional login, Norton Coursepacks offer a selection of rich visual and media-based questions, as well as updated Reading Quizzes for each chapter, featuring art from the text. Instructors can also provide on-line access to all the videos and animations developed for *Essentials of Geology*, Fifth Edition. Additional content includes *Science and Society* features and questions, *Geotour* questions, and links to the ebook (for students) and our test bank (for instructors). Coursepacks were prepared with the consultation of the author and other instructors, especially Cindy Liutkus-Pierce and Brian Zimmer of Appalachian State.

See for Yourself

USING *GOOGLE EARTH*™

Visiting Field Sites Identified in the Text

There's no better way to appreciate geology than to see it firsthand in the field, but the great variety of geologic features that we discuss in this book can't be visited from any one locality. So, even if your class takes geology field trips during the semester, you'll probably see examples of only a few geologic settings. Fortunately, *Google Earth*™ makes it possible for you to fly to spectacular geologic field sites anywhere in the world in a matter of seconds—you can take a virtual field trip electronically. Throughout the chapter, you will find *See For Yourself* panels identifying geologic sites that you can explore on your own personal computer (Mac or PC) using *Google Earth*™ software, or on your Apple/Android smartphone or tablet with the appropriate *Google Earth*™ app.

TO GET STARTED, FOLLOW THESE THREE SIMPLE STEPS:

1. Check to see whether *Google Earth*™ is installed on your personal computer, smartphone, or tablet. If not, please download the software from **earth.google.com** or the app from the Apple or Android app store.
2. Find a site in the *See For Yourself* panel that you're interested in visiting. In addition to a thumbnail photo and very brief description of the site (highlighting what you will see at the site), we provide the latitude and longitude of the site, in degrees, minutes, and seconds.
3. Open *Google Earth*™, and enter the coordinates of the site in the search window. As an example, let's find Mt. Fuji, a beautiful volcano in Japan. We specify the coordinates in the book as follows:

LATITUDE	35°21′41.78″N
LONGITUDE	138°43′50.74″E

Type these coordinates into the search window as:

35 21 41.78N, 138 43 50.74E

Note that the degree (°), minute (′), and second (″) symbols are left as blank spaces.

When you click ENTER or RETURN, your device will bring you to the viewpoint right above Mt. Fuji illustrated by the thumbnail on the left. Note that you can use the tools built into *Google Earth*™ to vary the elevation, tilt, orientation, and position of your viewpoint. The thumbnail on the right shows the view you'll see of the same location if you tilt your viewing direction and look north.

View Looking Down.

View Looking North.

NEED MORE HELP?

If you have any trouble operating *Google Earth*™, please visit digital.wwnorton.com/essgeo5 and navigate to the student site to find a video showing you how to download and install *Google Earth*™, as well as more detailed instructions on how to find the *See for Yourself* sites, links to *Google Earth*™ videos describing basic functions, and links to any hardware and software requirements. Notes addressing important *Google Earth*™ updates will be available at this site.

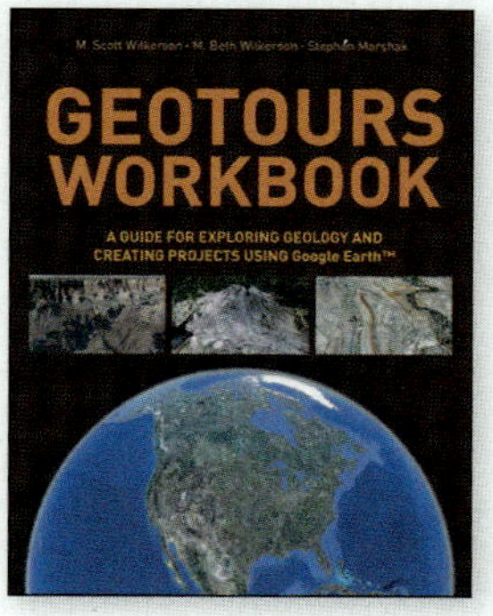

To provide additional opportunities to study geology using *Google Earth*™, we also offer a separate book—the *Geotours Workbook* (ISBN 978-0-393-91891-5)—that identifies even more interesting geologic sites to visit. The *Geotours Workbook* provides active-learning exercises linked to the sites and explains how you can create your own virtual field trips.

New Smartwork5 Online Tutorial and Assessment System. The new Smartwork5 online assessment available for use with *Essentials of Geology,* Fifth Edition features visual assignments developed with the eye of the author, with focused feedback. Because students learn best when they can interact with art as well as with text, Smartwork5 includes drag-and-drop figure-based questions, animation- and video-based questions, and *What a Geologist Sees* photo interpretations. Smartwork5 also provides questions based on real field examples, via the *Geotours Workbook*, and helps students check their knowledge as they go by working with reading-based questions and pre-made and easy-to-assign reading quizzes. Designed to be intuitive and easy to use for both students and instructors, Smartwork5 makes it a snap to assign, assess, and report on student performance, and to keep the class on track. Smartwork5 now works with tablet and mobile environments, and also has single sign-on capability with your institution's learning management system

Smartwork5 is available for free with most newly purchased print or electronic versions of the text. Immediate online access can also be purchased at digital.wwnorton.com/essgeo5. Select the option to buy a registration code before you create your account. Instructors can request their own Smartwork5 course at wwnorton.com/instructors.

Art Files and PowerPoints

The following visual resources are available from wwnorton.com/instructors, or through this book's instructor USB drive.

- *Enhanced Art Lecture PowerPoints*—Designed for instant classroom use, these slides utilize photographs and line art from the book in a format that has been optimized for use in the PowerPoint environment. The art has been relabeled and resized for projection formats. *Enhanced Art PowerPoints* also include supplemental photographs courtesy of Ron Parker of Fronterra Geosciences. Author Brian Zimmer of Appalachian State has also included new *Think-Pair-Share* slides to encourage classroom discussion, as well as *Takeaway Points* slides to summarize key concepts.
- *Labeled and Unlabeled Art PowerPoints*—These PowerPoints include all art from the book. We offer one set from which all labeling has been stripped and one set in which labeling remains.
- *Art JPEGs*—We provide a complete file of individual JPEGs for art and photographs used in the book.
- *Clicker questions*—Designed for use with any classroom signaling device, these sets also contain *Think-Pair-Share* questions designed for work in small groups.
- *Update PowerPoints*—W. W. Norton offers a semester-by-semester update service that provides new PowerPoint slides, with instructor support, covering recent geologic events.

New Animations, Simulations, and Videos. As described earlier, this extensive set of material has been completely revised and updated for quality and consistency working with the text's author and Alex Glass of Duke University. These materials are free and available at the student site, on the Marshak YouTube channel, in Norton Coursepacks, and through instructor formats including the new USB drive. In addition, working with Melissa Hudley of the University of North Carolina, Chapel Hill; Heather Lehto of Angelo State University; and Meghan Lindsey of the University of South Florida, our DVD, coursepacks and instructor support website contain a selection of streaming videos of geologic processes including content developed by IRIS. All animations and videos are ready to go and perfect for classroom or online use.

Norton Instructor USB Flash Drive. The Instructor USB drive offers a wealth of easy-to-use multimedia resources, all structured around the text. Resources include all PowerPoint art files, animations, and videos described earlier in the Preface, and electronic versions of the Instructor's Manual, Test Bank, and ExamView test-generation software. Further resources include clicker questions, as well as supporting files for using *Google Earth*™.

Instructor's Manual and Test Bank. The Instructor's Manual, prepared by John Werner of Seminole State College of Florida, is designed to help instructors prepare lectures and exams. It contains detailed *Learning Objectives, Chapter Summaries*, and complete answers to end-of-chapter Review and *On Further Thought* questions for every chapter and interlude. New to this edition are animation and video descriptions for these Norton resources, including suggestions for classroom implementation and two discussion questions per multimedia asset.

The Test Bank, revised by Heather Lehto of Angelo State University and carefully reviewed by Dylan Blumentritt of SUNY Potsdam, has been rewritten not only to correlate to the new edition of the text, but to provide greater, more rounded assessment than ever before. During the course of this revision, at least four geologists have reviewed and improved on each question, making sure that it is scientifically reliable and truly tests students' understanding of the most important topics in each chapter. Each chapter now features new short-answer questions that test student critical thinking and knowledge-application skills. These supplements are available on the Instructor USB drive and are also downloadable from wwnorton.com/instructors.

Instructor's Website—wwnorton.com/instructors. Online access is available for a rich array of resources: Test Bank, Instructor's Manual, PowerPoints, JPEGs, *Google Earth*™ file of sites from the text, art from the text, animations, simulations, videos, and WebCT- and Blackboard-ready content.

The Google Earth™ Geotours Workbook. Created by Scott Wilkerson, Beth Wilkerson, and Stephen Marshak, *Geotours* are active-learning opportunities that take students on virtual field trips to see outstanding examples of geology at localities around the world. They are available as Worksheets, both in print format (these come free with the book and include complete user instructions and advanced instruction) and electronically with auto-grading through Smartwork5 or your campus LMS. Request a sample copy to preview each worksheet.

See for Yourself Google Earth™ Sample Site File. Users who simply want access to sample field sites for classroom presentations or distribution to students can download the sites from the book's *See for Yourself* panels. This single download is available at the Norton instructor download site as well as from the student site.

Ebook. The *Essentials of Geology,* Fifth Edition ebook is available at digital.wwnorton.com/essgeo5. It contains live links to animations, simulations, and videos, as well as *Google Earth™* sites. When a student purchases ebooks at this site, Smartwork5 is available for free. The Norton ebook reader works on all computers and mobile devices and includes intuitive highlighting, note taking, and bookmarking features.

ACKNOWLEDGMENTS

Essentials of Geology, and its parent book, *Earth: Portrait of a Planet,* would not have come into existence without the inspiration and strong support of Jack Repcheck. As acquisitions editor for the first three editions, Jack proposed a number of the features that attracted readers to the book in the first place, and was a constant source of encouragement and support throughout the long, long process of writing, illustrating, and revising. Sadly, Jack passed away in October 2015. He will be greatly missed by all authors who had the good fortune to work with him over his immensely productive career.

I am very grateful for the assistance of many people in bringing *Essentials of Geology* from the concept stage to the shelf, and for helping to provide the momentum needed to bring this revision to completion. First and foremost, I wish to thank my wife, Kathy, who served as in-home project manager for this book. Kathy helped to construct the manuscript based on prior editions of this book and its parent, *Earth: Portrait of a Planet*; coordinated manuscript and proof traffic between our household and the publisher; maintained schedule, standards, and consistency; and served as an invaluable extra set of eyes. This book would not have appeared without Kathy. I also wish to thank my daughter, Emma, and my son, David, for their willingness to adopt "the book" as a member of our household when they were growing up, and to endure the overabundance of geo-photo stops on family trips. Emma helped develop the concept of narrative art used in the book, provided feedback about how the book works from a student's perspective, and provided several of the book's photos.

I am very grateful to all the staff of W. W. Norton & Company for their incredible efforts during the development of my books over the past two decades. It has been a privilege to work with an employee-owned company that is willing to work so closely with its authors. Many thanks to senior editor Eric Svendsen, who injected new enthusiasm and ideas into the project. Eric's experience and skill have guided the book in new directions and have helped connect the project to new trends in science pedagogy and book design. Thom Foley, as always, has been an incredible project manager. Thom handles the complicated process of assembling chapter proofs and figures into the final book, and puts in extra hours to meet deadline after deadline, all while remaining incredibly calm. Thom invested untold hours in sorting out composition and design issues—it's thanks to Thom that this edition made it to the shelf on schedule. I also greatly appreciate the efforts of Rob Bellinger for his innovative approach to ancillary development and for overseeing the development of the coursepack and Smartwork5 supplements; Sunny Hwang for her expertise and work as the new developmental editor for this project; Stephanie Romeo for her expert and thoughtful editing of the photo collection and permissions; Ben Reynolds for coordinating the back-and-forth between the publisher and various suppliers; Marcus Van Harpen and Kim Yi for their dedicated and creative production work on the media components, including the ebook and Smartwork5; Jake Schindel, who has so ably assumed the mantle of marketing manager for the book; Caitlin Barrett-Bressack and Tori Reuter for creatively handling the ancillaries and Smartwork5; Rachel Goodman for expertly performing all the tasks that come with a project of this complexity; Liz Vogt for assisting Rob with everything emedia; and Connie Parks for her excellent copyediting work. I also wish to thank Susan Gaustad, the outstanding developmental editor of the First Edition, who helped refine the prose style of the book.

Production of the illustrations has involved many people over many years. I am particularly indebted to Stan Maddock, an amazingly talented artist who helped to create the style of the figures from the very first, and who has kept up with constantly evolving graphics technology so the book's illustrations remain at the cutting edge. I am grateful to all the artists and compositors involved in this project.

It has been great fun to interact with Gary Hincks, who painted the incredible two-page spreads, in part using his own designs and geologic insights. Some of Gary's paintings originally appeared in *Earth Story* (BBC Worldwide, 1998) and were based on illustrations conceived with Simon Lamb and Felicity Maxwell. Others were developed specifically for *Earth: Portrait of a Planet* and *Essentials of Geology.* Some of the chapter-opening quotes were found in *Language of the Earth,* compiled by F. T. Rhodes and R. O. Stone (Pergamon, 1981).

The five editions of this book and of its parent, *Earth: Portrait of a Planet,* have benefited greatly from input by expert reviewers for specific chapters, by general reviewers of the entire book, and by comments from faculty and students who have used the book and were kind enough to contact me or the publisher. In particular, in this edition I would like to thank Michael Rygel of SUNY-Potsdam and Geoffrey and Heather Cook, who greatly helped with the process of updating Smartwork5. Kurt Wilkie of Washington State University continues to offer extremely helpful input on the book, Smartwork5, and the coursepack.

Reviewers who have provided helpful feedback for this and previous editions include:

Jack C. Allen, *Bucknell University*
David W. Anderson, *San Jose State University*
Sytle Antao, *University of Calgary*
Martin Appold, *University of Missouri, Columbia*
Philip Astwood, *University of South Carolina*
Eric Baer, *Highline University*
Victor Baker, *University of Arizona*
Julie Baldwin, *University of Montana*
Sandra Barr, *Acadia University*
Miriam Barquero-Molina, *University of Missouri*
Keith Bell, *Carleton University*
Mary Lou Bevier, *University of British Columbia*
Jim Black, *Tarrant County College*
Daniel Blake, *University of Illinois*
Ted Bornhorst, *Michigan Technological University*
Michael Bradley, *Eastern Michigan University*
Mike Branney, *University of Leicester, UK*
Sam Browning, *Massachusetts Institute of Technology*
Bill Buhay, *University of Winnipeg*
Michael Bunds, *Utah Valley University*
Rachel Burks, *Towson University*
Peter Burns, *University of Notre Dame*
Matthew Campbell, *Charleston Southern University*
Katherine Cashman, *University of Oregon*
Christian Maloney Cicimurri, *University of South Carolina*
Christine Clark, *Eastern Michigan University*
George S. Clark, *University of Manitoba*
Kevin Cole, *Grand Valley State University*
Patrick M. Colgan, *Northeastern University*
Geoffrey Cook, *University of California, San Diego*
Heather Cook, *California State University San Marcos*
Peter Copeland, *University of Houston*
Winton Cornell, *University of Tulsa*
John W. Creasy, *Bates College*
Dyanna Czeck, *University of Wisconsin, Milwaukee*
Norbert Cygan, *Chevron Oil, retired*
Michael Dalman, *Blinn College*
Peter DeCelles, *University of Arizona*
Carlos Dengo, *ExxonMobil Exploration Company*
John Dewey, *University of California, Davis*
Charles Dimmick, *Central Connecticut State University*
Robert T. Dodd, *Stony Brook University*
Missy Eppes, *University of North Carolina, Charlotte*
Eric Essene, *University of Michigan*
James E. Evans, *Bowling Green State University*
Susan Everett, *University of Michigan, Dearborn*
Dori Farthing, *State University of New York, Geneseo*
Mark Feigenson, *Rutgers University*
Grant Ferguson, *St. Francis Xavier University*
Eric Ferré, *Southern Illinois University*
Leon Follmer, *Illinois Geological Survey*
Nels Forman, *University of North Dakota*
Bruce Fouke, *University of Illinois*
Frederic Marton, *Bergen Community College*
David Furbish, *Vanderbilt University*
Steve Gao, *University of Missouri*
Yongli Gao, *University of Texas at San Antonio*
Grant Garvin, *John Hopkins University*
Christopher Geiss, *Trinity College, Connecticut*
Richard Gibson, *Texas A & M University*
Gayle Gleason, *State University of New York, Cortland*
Cyrena Goodrich, *Kingsborough Community College*
William D. Gosnold, *University of North Dakota*
Todd Greene, *California State University, Chico*
Lisa Greer, *William & Mary College*
Alessandro Grippo, *Santa Monica College*
Steve Guggenheim, *University of Illinois, Chicago*
Henry Halls, *University of Toronto, Mississuaga*
Jacquelyn Hams, *Los Angeles Valley College*
Bryce M. Hand, *Syracuse University*
Kathleen Harper, *University of Montana*
Bruce Herbert, *Texas A & M University*
Anders Hellstrom, *Stockholm University*
Tom Henyey, *University of South Carolina*
James Hinthorne, *University of Texas, Pan American*
Paul Hoffman, *Harvard University*
Curtis Hollabaugh, *University of West Georgia*
Bernie Housen, *Western Washington University*
Mary Hubbard, *Kansas State University*
Paul Hudak, *University of North Texas*
Melissa Hudley, *University of North Carolina, Chapel Hill*
Warren Huff, *University of Cincinnati*
Robin Humphreys, *College of Charleston*
John Huntley, *University of Missouri*
Neal Iverson, *Iowa State University*
Duncan Johannessen, *University of Victoria*
Charles Jones, *University of Pittsburgh*
Donna M. Jurdy, *Northwestern University*
Thomas Juster, *University of Southern Florida*
H. Karlsson, *Texas Tech*
Daniel Karner, *Sonoma State University*
Dennis Kent, *Lamont-Doherty Earth Observatory/Rutgers University*
Charles Kerton, *Iowa State University*
Susan Kieffer, *University of Illinois*
Jeffrey Knott, *California State University, Fullerton*
Ulrich Kruse, *University of Illinois*
Robert S. Kuhlman, *Montgomery County Community College*
Lee Kump, *Pennsylvania State University*
David R. Lageson, *Montana State University*
Robert Lawrence, *Oregon State University*
Karen Layou, *Reynolds Community College*
Heather Lehto, *Angelo State University*
René A. Shroat-Lewis, *University of Arkansas, Little Rock*

Scott Lockert, *Bluefield Holdings*
Leland Timothy Long, *Georgia Tech*
Craig Lundstrom, *University of Illinois*
John A. Madsen, *University of Delaware*
Jerry Magloughlin, *Colorado State University*
Fred Marton, *Bergen Community College*
Jennifer McGuire, *Texas A&M University*
Judy McIlrath, *University of South Florida*
Paul Meijer, *Utrecht University, Netherlands*
Charles Merguerian, *Hofstra University*
Jamie Dustin Mitchem, *California University of Pennsylvania*
Alan Mix, *Oregon State University*
Stephen Moysey, *Clemson University*
Otto Muller, *Alfred University*
Martha Murphy, *Sonoma State University*
Kathy Nagy, *University of Illinois, Chicago*
Donald W. Neal, *East Carolina University*
Pamela Nelson, *Glendale Community College*
Roger L. Nielsen, *Oregon State University*
Robert Nowack, *Purdue University*
Charlie Onasch, *Bowling Green State University*
David Osleger, *University of California, Davis*
William P. Patterson, *University of Saskatchewan*
Eric Peterson, *Illinois State University*
Ginny Peterson, *Grand Valley State University*
Stephen Piercey, *Laurentian University*
Adrian Pittari, *University of Waikato, New Zealand*
Lisa M. Pratt, *Indiana University*
Mark Ragan, *University of Iowa*
Kent Ratajeski, *University of Kentucky*
Sara Rathburn, *Colorado State University, Fort Collins*
Robert Rauber, *University of Illinois*
Robert S. Reece, *Texas A&M University*
Bob Reynolds, *Central Oregon Community College*
Joshua J. Roering, *University of Oregon*
Michael Rygel, *State University of New York, Potsdam*
Eric Sandvol, *University of Missouri*
William E. Sanford, *Colorado State University*
Jeffrey Schaffer, *Napa Valley Community College*
Roy A. Schiesser, *Chandler Gilbert Community College*
Roy Schlische, *Rutgers University*
Sahlemedhin Sertsu, *Bowie State University*
Doug Shakel, *Pima Community College*
Anne Sheehan, *University of Colorado*
Roger D. Shew, *University of North Carolina, Wilmington*
Virginia Sisson, *University of Houston*
Norma Small-Warren, *Howard University*
Donny Smoak, *University of South Florida*
David Sparks, *Texas A&M University*
Angela Speck, *University of Missouri*
Tim Stark, *University of Illinois*
Seth Stein, *Northwestern University*
David Stetty, *Jacksonville State University*
Kevin G. Stewart, *University of North Carolina, Chapel Hill*
Michael Stewart, *University of Illinois*
Don Stierman, *University of Toledo*
Gina Marie Seegers Szablewski, *University of Wisconsin, Milwaukee*
Barbara Tewksbury, *Hamilton College*
Thomas M. Tharp, *Purdue University*
Kathryn Thornbjarnarson, *San Diego State University*
Basil Tikoff, *University of Wisconsin*
Spencer Titley, *University of Arizona*
Robert T. Todd, *Stony Brook University*
Torbjörn Törnqvist, *University of Illinois, Chicago*
Jon Tso, *Radford University*
Stacey Verardo, *George Mason University*
Barry Weaver, *University of Oklahoma*
John Werner, *Seminole State College of Florida*
Scott White, *University of South Carolina*
Alan Whittington, *University of Missouri*
John Wickham, *University of Texas, Arlington*
Lorraine Wolf, *Auburn University*
Christopher J. Woltemade, *Shippensburg University*
Adam Woods, *California State University, Fullerton*

I apologize if I inadvertently left anyone off this list.

THANKS!

I am very grateful to the students who engaged so energetically with earlier editions of this book, and to the instructors who have selected this book for their classes. I welcome your comments and corrections and can be reached at *smarshak@illinois.edu*.

Stephen Marshak

Geology, perhaps more than any other department of natural philosophy, is a science of contemplation. It demands only an enquiring mind and senses alive to the facts almost everywhere presented in nature.

—Sir Humphry Davy (British scientist, 1778–1829)

ABOUT THE AUTHOR

STEPHEN MARSHAK is a professor of geology at the University of Illinois, Urbana-Champaign, where he is also the director of the School of Earth, Society, and Environment. He holds an A.B. from Cornell University, an M.S. from the University of Arizona, and a Ph.D. from Columbia University. Steve's research interests in structural geology and tectonics have taken him into the field on several continents. He loves teaching and has won his college's and university's highest teaching awards, as well as the Neil Miner Award of the National Association of Geoscience Teachers "for exceptional contributions to the stimulation of interest in the earth sciences." In addition to research papers and *Essentials of Geology*, Steve has authored *Earth: Portrait of a Planet* and has co-authored *Laboratory Manual for Introductory Geology*; *Earth Structure: An Introduction to Structural Geology and Tectonics*; and *Basic Methods of Structural Geology*.

ESSENTIALS OF

GEOLOGY

FIFTH EDITION

LEARNING OBJECTIVES

By the end of this chapter, you should understand...

1. the scope and applications of geology.
2. the foundational themes of modern geologic study.
3. how geologists employ the scientific method.

▲ Students exploring a small canyon in Illinois. Geology is everywhere!

PRELUDE

And Just What Is Geology?

P.1 In Search of Ideas

We arrived in the late-night darkness, at a campsite in western Arizona. Here in the desert, so little rain falls over the course of a year that hardly any plants can survive, and rocks crop out on many hills. Under the dry sky, there's no need for tents, so we could sleep under the stars with our sleeping bags on a bed of sand. At dawn, the red rays of the first sunlight made the slope of the steep-sided hill near our campsite start to glow, and we could see our target, a prominent ledge of rusty-brown rock that formed a shelf at the top of the hill. To reach it, though, we'd have to climb a steep slope littered with jagged boulders.

After a quick breakfast, we loaded our day packs with water bottles and granola bars, slathered on a layer of sunscreen, and set off toward the slope. It was the breezeless morning of what was going to be a truly hot day, and we wanted to gain elevation before the sun rose too high in the sky. After a tiring hour finding our way through the boulder obstacle course, we reached the base of the ledge and decided to take a break before ascending the final cliff. But just as we leaned to rest our backs against a rock, we heard an unnerving vibration. Somewhere nearby, too close for comfort, a

> *Civilization exists by geological consent, subject to change without notice.*
>
> WILL DURANT
>
> (American writer, historian, and philosopher, 1885–1981)

rattlesnake shook an urgent warning with its tail. Rest would have to wait, and we scrambled up the ledge. It was the right choice, for the view from the top of the surrounding landscape was amazing (**Fig. P.1a**). But the rocks beneath our feet were even more amazing. Close up, we could see curving ribbons of light and dark layers, cut by stripes of white quartz. The ledge preserved the story of a distant age in our planet's past when the rock we now stood on was kilometers below ground level and was able to flow like soft plastic, but ever so slowly (**Fig. P.1b**). We now set to the task of figuring out what it all meant.

Geologists—scientists who study the Earth—explore many areas, including remote deserts, high mountains, damp rainforests, frigid glaciers, and deep canyons (**Fig. P.2**). Such efforts can strike people in other professions as a strange way to make a living. This sentiment underlies the Scottish poet Walter Scott's (1771–1832) description of geologists at work: "Some rin uphill and down dale, knappin' the chucky stones to pieces like sa' many roadmakers run daft. They say it is to see how the warld was made!" Indeed—to see how the world was made, to see how it continues to evolve, to find its resources, to protect against its natural hazards, and to predict what its future may bring. These are the questions that have driven geologists to explore the Earth, on all continents and in all oceans, from the equator to the poles, and everything in between.

Geologic discovery continues today, and involves a variety of techniques. While some geologists continue to work in the field with hammers and hand lenses, others have moved into laboratories where they employ sophisticated electronic instruments to analyze microscopic quantities of Earth materials or detect the configuration of layers underground, use satellites to detect the motions of continents or the stability of volcanoes, and use high-speed computers to locate earthquakes or analyze the flow of underground water. For over two centuries, geologists have pored over the Earth in search of ideas to explain the processes that form and change our planet. In this prelude, we look at the questions geologists ask and have tried to answer. You'll see that many of these answers are not just of academic interest but have implications for society as a whole.

P.2 The Nature of Geology

Geology, or geoscience, is the study of the Earth. Not only do geologists address fundamental questions such as the formation and composition of our planet, the causes of earthquakes and ice ages, and the evolution of life, but they also address practical problems such as how to keep pollution out of groundwater, how to find oil and minerals, and how to avoid landslides.

FIGURE P.1 Geologic exploration provides beautiful views, and mysteries to solve.

(a) A view of the western Arizona desert is not just beautiful—it holds clues to the Earth's past and to the changes taking place today.

(b) The contortions of the rock layers speak of a time when the rock flowed, like soft plastic.

FIGURE P.2 Field study in many environments.

(a) A desert cliff in Utah.

(b) A rainforest in Peru.

(c) Mountains in Alaska.

(d) The shore in Massachusetts.

The fascination of geology attracts many to careers in this science. Tens of thousands of geologists work for energy, mining, water, engineering, and environmental companies, while a smaller number work in universities, government geological surveys, and research laboratories. Nevertheless, since most of the students reading this book will not become professional geologists, it's fair to ask the question, "Why should people, in general, study geology?"

Did you ever wonder...
will an earthquake happen near where you live?

First, geology may be one of the most practical subjects you can learn. When a news report begins, "Scientists say," and then continues, "an earthquake occurred today off Japan," or "landslides will threaten the city," or "chemicals from a toxic waste dump will ruin the town's water supply," or "there's only a limited supply of oil left," or "the floods of the last few days are the worst on record," the scientists that the report refers to are geologists. In fact, ask yourself the following questions, and you'll realize that geologic phenomena and materials play major roles in daily life:

- Do you live in a region threatened by landslides, volcanoes, earthquakes, or floods (**Fig. P.3**)?
- Are you worried about the price of energy or about whether there will be a war in an oil-supplying country?

FIGURE P.3 Human-made cities cannot withstand the vibrations of a large eathquake. These apartment buildings collapsed during an earthquake in Turkey.

- Do you ever wonder about where the copper in your home's wires comes from? Or the lithium in the battery of your cell phone?
- Have you seen fields of green crops surrounded by desert and wondered where the irrigation water comes from?
- Would you like to buy a dream house on a beach or near a river?
- Are you following news stories about how toxic waste can migrate underground into your town's water supply?

Clearly, all citizens of the 21st century, not just professional geologists, need to make decisions and understand news reports addressing Earth-related issues. A basic understanding of geology will help you do so.

Second, the study of geology gives you an awareness of the planet that no other field can. As you will see, the Earth is a complicated entity, where living organisms, oceans, atmosphere, and solid rock interact with one another in a great variety of ways. Geologic study reveals Earth's antiquity and demonstrates how the planet has changed profoundly during its existence. What our ancestors considered to be the center of the Universe has become, with the development of geologic perspective, our "island in space" today. And what was believed to be an unchanging orb originating at the same time as humanity has become a dynamic planet that existed long before people did and continues to evolve.

Third, the study of geology puts the accomplishments and consequences of human civilization in a broader context. View the aftermath of a large earthquake, flood, or hurricane, and it's clear that the might of natural geologic phenomena greatly exceeds the strength of human-made structures. But watch a bulldozer clear a swath of forest, a dynamite explosion remove the top of a hill, or a prairie field evolve into a housing development, and it's clear that people can change the face of the Earth at rates often exceeding those of natural geologic processes.

Finally, when you finish reading this book, your view of the world may be forever colored by geologic curiosity. If you walk in the mountains, you will think of the many forces that shape and reshape the Earth's surface. If you hear about a natural disaster, you will have insight into the processes that brought it about. And if you go on a road trip, the rock exposures along the highway will no longer be gray, faceless cliffs, but will present complex puzzles of texture and color telling a story of the Earth's long history.

P.3 Themes of This Book

A number of narrative themes appear (and reappear) throughout this text. These themes, listed below, can be viewed as this book's overall take-home message.

- *Studying geology helps you understand physical science in general:* Geology incorporates many of the basic concepts of physics and chemistry because Earth materials are a form of matter, and energy drives geologic processes. Thus, studying geology can help you develop a better grasp of key ideas in physical science. Those readers who pursue teaching careers in elementary or secondary education will find that geological examples can help students develop STEM (science-technology-engineering-math) learning skills.
- *The Earth has an internal structure:* The Earth is not a homogeneous ball, but rather it consists of concentric layers. From center to surface, Earth has a **core**, **mantle**, and **crust**. We live on the surface of the crust, where it meets the atmosphere and the oceans.
- *The outer layer of the Earth consists of moving plates:* In the 1960s, geologists recognized that the crust, together with the uppermost part of the underlying mantle, forms a 100- to 150-km-thick semi-rigid shell called the **lithosphere**. Large cracks separate this shell into discrete pieces, called **plates**, which move very slowly relative to one another (**Fig. P.4**). The theory that describes this movement and its consequences is called the **theory of plate tectonics**, and it serves as the foundation for understanding most geologic phenomena. Plate movements yield earthquakes, volcanoes, and mountain ranges, and cause the map of Earth's surface to change very slowly over time.
- *We can picture the Earth as a complex system with many interconnected realms:* The Earth is not static, but rather

FIGURE P.4 A simplified map of the Earth's plates. The arrows indicate the direction each plate is moving, and the length of the arrow indicates plate velocity (the longer the arrow, the faster the motion).

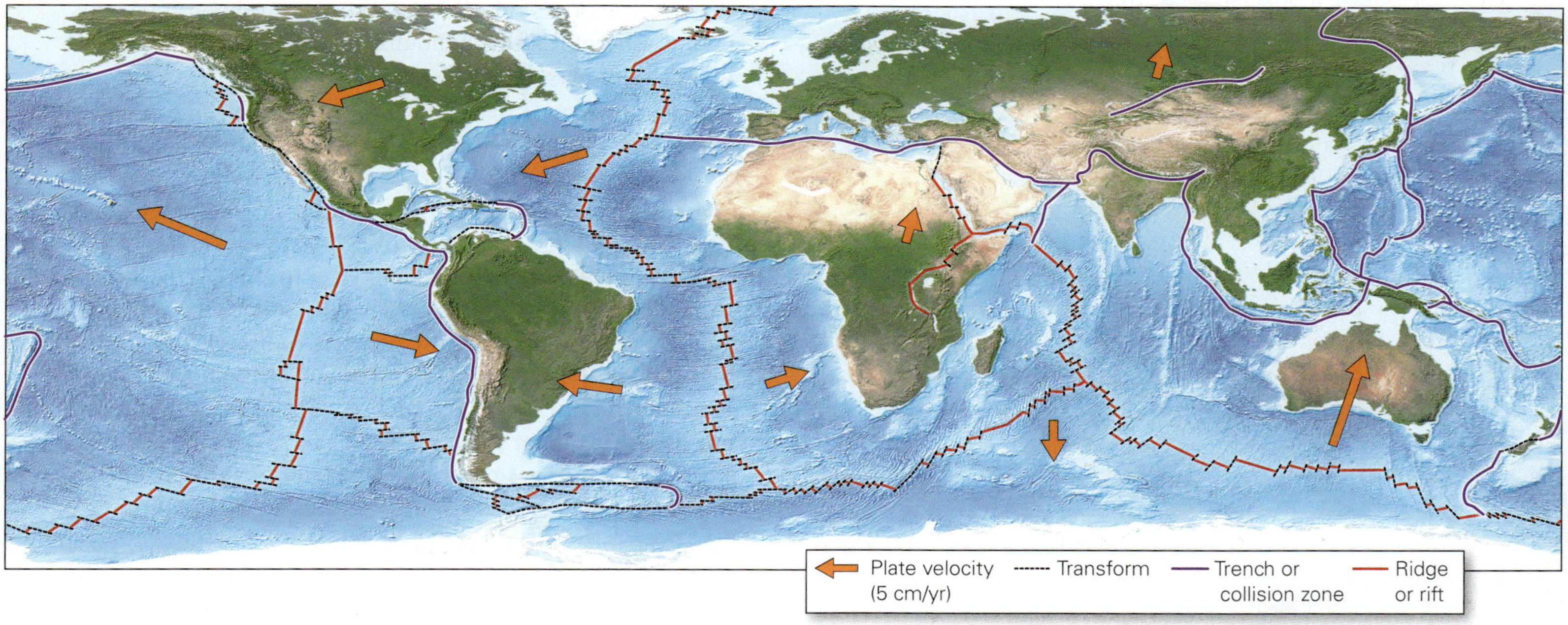

it is a dynamic entity whose components can move and change over time. Our planet's interior, solid surface, oceans, atmosphere, and life all interact with one another in many ways to yield the land, oceans, and air in which we and other species of organisms can live. Geologists refer to this interconnected web of interacting realms of materials and processes as the **Earth System**. Within the Earth System, certain materials cycle among rock, sea, and air, and among all of these entities and life. Over time, the distribution of these materials among various components of the Earth System can change.

- *The Earth is a planet:* Despite the uniqueness of the Earth System, the Earth is a planet, formed like the other planets of the Solar System. But because of the way the Earth System operates, our planet differs from others by having plate tectonics, an oxygen-rich atmosphere, liquid-water ocean, and abundant life.
- *The Earth is very old:* Geologic data indicate that the Earth formed 4.54 billion years ago—plenty of time for geologic processes to build mountains and grind them down many times over, for life forms to evolve and go extinct, and for the map of the planet to change. Even plate movement at rates of only a few centimeters per year can move a continent thousands of kilometers if those movements continue for hundreds of millions of years.

Did you ever wonder...
if a map of the Earth's surface today looks like a map of the surface 200 million years ago?

The Earth has a history, and it extends far into the past. **Geologic time** represents the duration of this history.

- *The geologic time scale divides Earth's history into intervals:* To refer to specific portions of geologic time, geologists developed the **geologic time scale** (**Fig. P.5**). The last 541 million years comprise the *Phanerozoic Eon*, and all time before that makes up the *Precambrian*. The Precambrian can be further divided into three main intervals named, from oldest to youngest: the *Hadean*, the *Archean*, and the *Proterozoic Eons*. The Phanerozoic Eon, in turn, can be divided into three main intervals named, from oldest to youngest: the *Paleozoic*, the *Mesozoic*, and the *Cenozoic Eras*.
- *Internal and external processes drive geologic phenomena: Internal processes* are driven by heat from inside the Earth. Plate movement is an example. Because plate movements cause mountain building, earthquakes, and volcanoes, we consider all of these phenomena to be manifestations of internal processes. *External processes* are driven by energy coming to the Earth from the Sun. The heat produced by this energy drives the movement of air and water, which grinds and sculpts the Earth's surface and transports the debris to new locations, where it accumulates. The interaction between internal and external processes forms and shapes the mountains, canyons, beaches, and plains of our planet. As we'll see, **gravity**—the pull that one mass exerts on another—plays an important role in both internal and external processes.

FIGURE P.5 The geologic time scale.

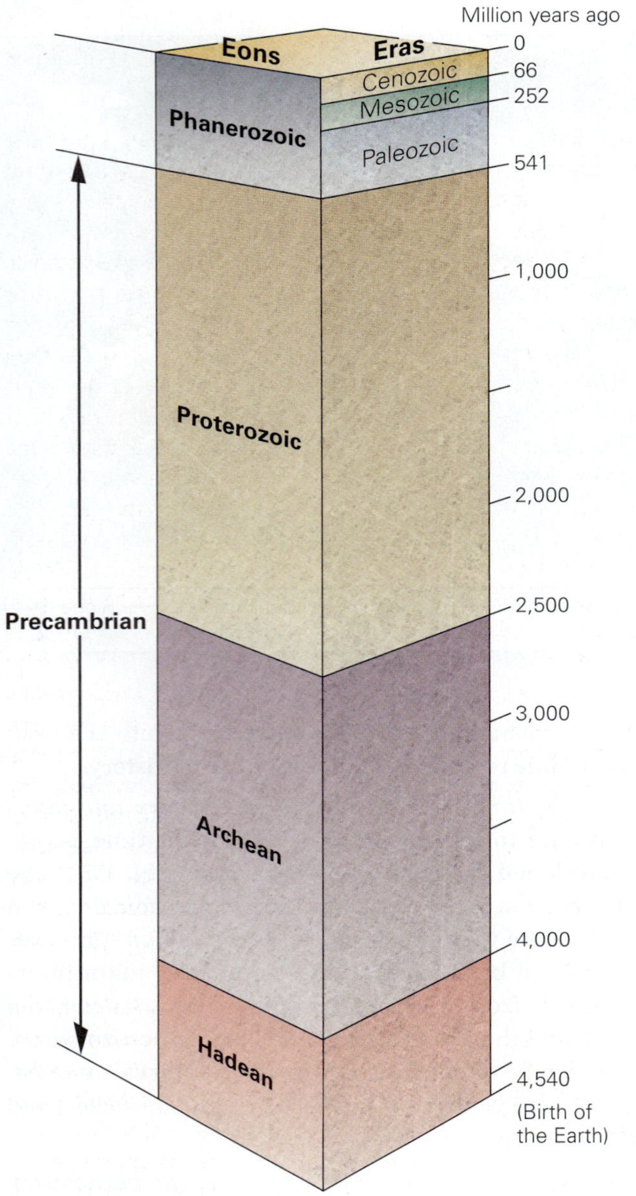

(a) The scale has been divided into eons and eras.

One thousand years ago = 1 **Ka**
(Ka stands for kilo-annum)

One million years ago = 1 **Ma**
(Ma stands for mega-annum)

One billion years ago = 1 **Ga**
(Ga stands for giga-annum)

(b) Abbreviations for time units.

- *Geologic phenomena affect society:* Volcanoes, earthquakes, landslides, floods, groundwater, energy sources, and mineral reserves are of vital interest to every inhabitant of this planet. Therefore, throughout this book we emphasize the linkages among geology, the environment, and society.
- *Physical aspects of the Earth System link to life processes:* All life on this planet depends on such physical features as the composition of soil; the temperature, humidity, and composition of the atmosphere; and the flow of surface and subsurface water. And life in turn affects and alters physical features. For example, the oxygen in the Earth's atmosphere comes from photosynthesis, a life activity in plants. This oxygen in turn permits complex animals to survive and controls chemical reactions among air, water, and rock. Without the physical Earth, life could not exist; but without life, this planet's surface might have become a frozen wasteland, like that of Mars, or a cloud-enshrouded oven, like that of Venus.
- *The Earth has changed dramatically in many ways over geologic time and continues to change:* The landscape that you see outside your window today is not what you would have seen a thousand, a million, or a billion years ago. Over Earth history, the planet's surface, composition of the atmosphere, and sea level have all changed. Also, continents move relative to one another. Change continues today, and aspects of the Earth System are changing faster than ever before because of human activity.
- *Most of the resources that we use come from geologic materials:* Modern society uses vast quantities of oil, gas, coal, metal, concrete, clay, fertilizer, and other materials. Most of these come from the solid Earth (**Fig. P.6**).

FIGURE P.6 Workers excavate limestone in a quarry near Chicago. This rock commonly consists of shells and shell fragments, and can be used in the production of concrete.

BOX P.1 CONSIDER THIS...

The Scientific Method

Sometime during the past 200 million years, a large block of rock or metal, which had been orbiting the Sun, slammed into our planet. It made contact at a site in what is now the central United States, a landscape of flat cornfields. The impact of this block, a *meteorite*, released more energy than a nuclear bomb—a cloud of shattered rock and dust blasted skyward, and once-horizontal layers of rock from deep below the ground sprang upward and tilted on end beneath the gaping crater left by the impact. When the event was over, the land surface looked radically different—a layer of debris surrounded and partially filled the crater at the impact site. Later in Earth history, running water and blowing wind wore down this jagged scar and carried away the debris. Then, about 15,000 years ago, sand, gravel, and mud carried by a vast sheet of ice, a glacier, buried what remained, hiding it entirely from view (**Fig. BxP.1**). Wow! So much history beneath a cornfield. How do we know this? It takes scientific investigation.

The movies often portray science as a dangerous tool, capable of creating Frankenstein's monster, and scientists as nerdy characters with thick glasses and poor taste in clothes. In reality, **science** is simply the use of observation, experiment, and calculation to explain how nature operates, and scientists are people who study and try to understand natural phenomena. Scientists guide their work using the **scientific method**, an idealized thought process for systematically analyzing scientific problems in a way that leads to verifiable results. Let's see how geologists employed the scientific method to come up with the meteorite-impact story.

- *Recognizing the problem:* Any scientific project, like any detective story, begins by identifying a mystery. The cornfield mystery came to light when water drillers discovered that limestone, a rock typically made of shell fragments, lies just below the 15,000-year-old glacial sediment. In surrounding regions, the rock beneath the glacial sediment consists instead of sandstone, a rock made of cemented-together sand grains. Since limestone can be used to build roads, make cement, and produce the agricultural lime used in treating soil, workers stripped off the glacial sediment and dug a quarry to excavate the limestone. They were amazed to find that rock layers exposed in the quarry were tilted steeply and had been shattered by large cracks. In the surrounding regions, all rock layers are horizontal like the layers in a birthday cake, the limestone layer lies underneath the sandstone, and the rocks contain relatively few cracks. When curious geologists came to investigate, they soon realized that the geologic features of the land just beneath the cornfield presented a problem to be solved. What phenomena had brought limestone up close to the Earth's surface, had tilted the layering in the rocks, and had shattered the rocks?
- *Collecting data:* The scientific method proceeds with the collection of observations or clues that point to an answer. Geologists studied the quarry and determined the age of its rocks, measured the orientation of the rock layers, and documented (made a written or photographic record of) the fractures that broke up the rocks.
- *Proposing hypotheses.* A scientific **hypothesis** is merely a possible explanation, involving only natural processes, that can explain a set of observations. Scientists propose hypotheses during or after their initial data collection. In this example, the geologists working in the quarry came up with two alternative hypotheses: either the features in this region resulted from a volcanic explosion, or they were caused by a meteorite impact.
- *Testing hypotheses.* Since a hypothesis is just an idea that can be either right or wrong, scientists try to put hypotheses through a series of tests to see if they work. The geologists at the quarry compared their field observations with published observations made at other sites of volcanic explosions and meteorite impacts, and they studied the results of experiments designed to simulate such events. If the geologic features visible in the quarry were the result of volcanism, the quarry should contain rocks formed by the freezing of molten rock erupted by a volcano. But no such rocks were found. If, however, the features were produced by an impact, the rocks should contain **shatter cones**, tiny cracks that fan out from a point. Shatter cones can be overlooked, so the geologists returned to the quarry specifically to search for them and found them in abundance. The impact hypothesis passed the test!

Note that we said "idealized" thought process. Sometimes serendipity works into the process, and scientists make discoveries by chance. Also, because we can't travel through time, we can't always completely test all geologic hypotheses.

Theories are scientific ideas supported by an abundance of evidence; they have passed many tests and have failed none. Scientists are much more confident in the correctness of a theory than of a hypothesis. Continued study in the quarry eventually yielded so much evidence for impact that the impact hypothesis came to be viewed as a theory. Scientists continue to test theories over a long time. Successful theories withstand these tests and are supported by so many observations that

(continued)

FIGURE BxP.1 An ancient meteorite impact excavates a crater and permanently changes rock beneath the surface.

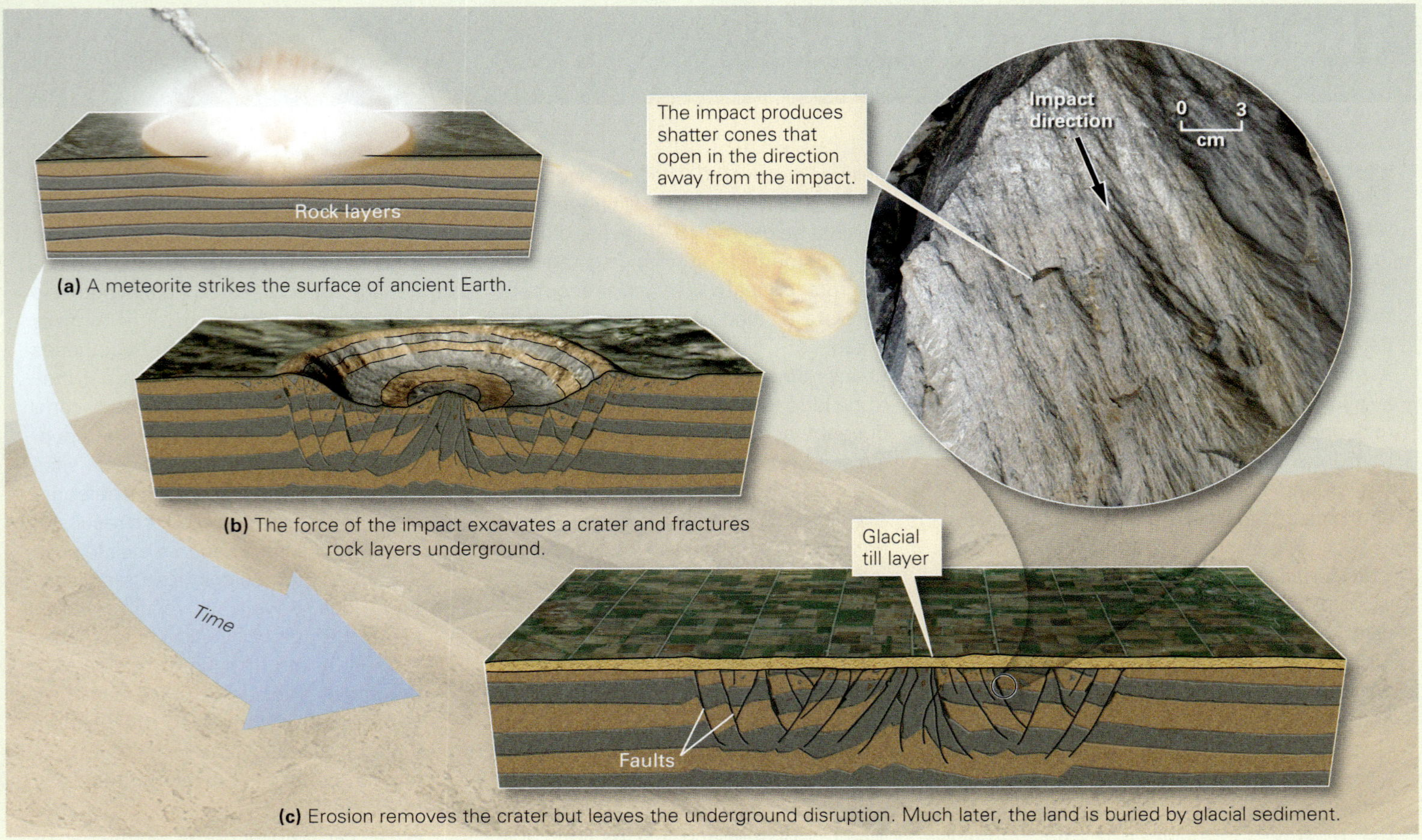

(a) A meteorite strikes the surface of ancient Earth.

(b) The force of the impact excavates a crater and fractures rock layers underground.

(c) Erosion removes the crater but leaves the underground disruption. Much later, the land is buried by glacial sediment.

they become part of a discipline's foundation. However, some theories may eventually be disproved and replaced by better ones.

In a few cases, scientists have been able to devise concise statements that completely describe a specific relationship or phenomenon. Such statements are called **scientific laws**. Scientific laws are not the same as scientific theories, as the former do not provide an explanation for a phenomenon, while the latter do. For example, the *law* of gravity, a simple equation that explains how objects accelerate (speed up) when subjected to gravitational force, does not explain why gravity exists. In contrast, the *theory of evolution* does explain why evolution occurs, by providing a testable mechanism for why the assemblages of life on Earth change over time.

> *Science comes from observation, and people make scientific discoveries:* Science does not consist of subjective guesses or arbitrary dogmas, but rather of a consistent set of objective statements resulting from the application of the scientific method (**Box P.1**). Every scientific idea must be tested thoroughly and should be used only when supported by documented observations. Further, scientific ideas do not appear out of nowhere; they are the result of human efforts. Wherever possible, this book shows where geologic ideas came from, and tries to answer the question, "How do we know that?"

As you read this book, please keep these themes in mind. Don't view geology as a list of words to memorize, but rather as an interconnected set of concepts to digest. Most of all, enjoy yourself as you learn about what may be the most fascinating planet in the Universe. It certainly is to humans!

Prelude Review

Prelude Summary

- Geologists are scientists who study the Earth. They search for the answers to the mysteries of our home planet, from why volcanoes explode to where we can find diamonds.
- Geologic study can involve field exploration, laboratory experiments, high-tech measurements, and calculations with computers.
- Geologic research not only provides answers to academic questions such as how the Earth formed, but also addresses practical problems like how to find resources and to avoid landslides. Many people pursue careers as geologists.
- A set of themes underlies geologic thinking. Key concepts are that the Earth's outer shell consists of moving plates whose interactions produce earthquakes, volcanoes, and mountains; that the Earth is very old; and that interacting realms of material on the planet comprise the "Earth System."

Guide Terms

core (p. 4)
crust (p. 4)
Earth System (p. 5)
geologic time scale (p. 5)
geologist (p. 2)
geology (p. 2)
gravity (p. 5)
hypothesis (p. 7)
lithosphere (p. 4)
mantle (p. 4)
plate (p. 4)
science (p. 7)
scientific law (p. 8)
scientific method (p. 7)
shatter cone (p. 7)
theory (p. 7)
theory of plate tectonics (p. 4)

Review Questions

1. What are some of the practical applications of geology?
2. Explain the difference between internal processes and external processes.
3. How would the Earth's atmosphere differ if life didn't exist?
4. Explain the difference between a hypothesis and a theory, in the context of science.
5. What are the sources of data that geologists can use to understand the Earth?
6. What are the major subdivisions of geologic time? Which time unit is longer, the Precambrian or the Paleozoic?

LEARNING OBJECTIVES

By the end of this chapter, you should understand...

1. modern concepts concerning the basic architecture of our Universe and its components.
2. the character of our Solar System.
3. scientific explanations for the formation of the Universe and the Earth.
4. the overall character of the Earth's magnetic field, atmosphere, and surface.
5. the variety and composition of materials that make up our planet.
6. the nature of the Earth's internal layering.

▲ The Hubble Space Telescope captured this image of the Orion Nebula, a cloud of gas and dust 24 light years across, in which new stars are forming.

CHAPTER 1

The Earth in Context

1.1 Introduction

Sometime in the distant past, perhaps more than 50,000 generations ago, our ancestors developed the capacity for complex, conscious thought. This amazing ability, which distinguishes our species from all others, brought with it the gift of curiosity, an innate desire to understand and explain the workings of all our surroundings—of our *Universe*. Questions that we ask about the Universe differ little from questions a child asks of a new friend: Where do you come from? How old are you? Such musings first spawned legends in which heroes used supernatural powers to mold the planets and sculpt the landscape. Eventually, researchers began to apply scientific principles to **cosmology**, the study of the overall structure and history of the Universe, and a different picture has emerged.

In this chapter, we begin with a brief introduction to the principles of scientific cosmology. We explain the basic architecture of the Universe, introduce the Big Bang theory for the formation of the Universe, and discuss the nebular theory for the birth of the Solar System. Finally, we characterize our home planet by building an image of its surroundings, surface, and interior. This introductory high-speed tour of the Earth provides a context for the remainder of this book.

> *I believe everyone should have a broad picture of how the Universe operates and our place in it. It is a basic human desire. And it also puts our worries in perspective.*
>
> STEPHEN HAWKING (British cosmologist, born 1942)

FIGURE 1.1 Contrasting views of the Universe, as drawn by artists hundreds of years ago.

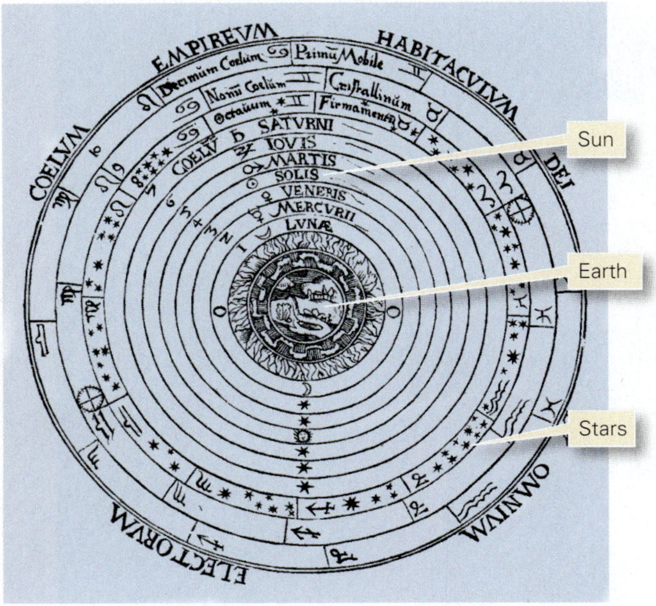

(a) The geocentric image of the Universe shows the Earth at the center, surrounded by air, fire, and the other planets, all contained within the globe of the stars.

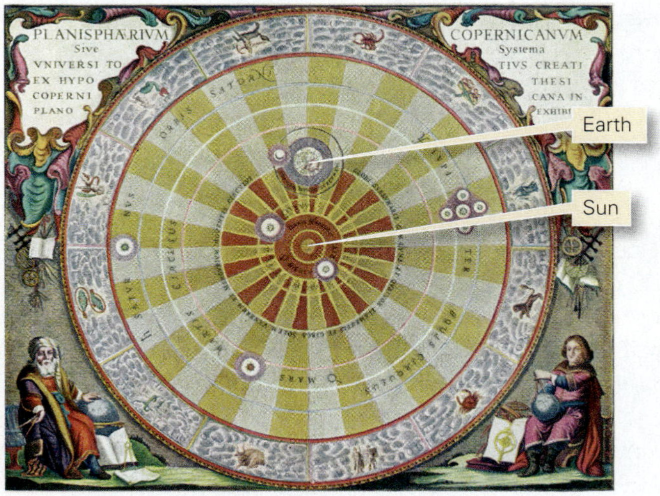

(b) The heliocentric image of the Universe shows the Sun at the center, as envisioned by Copernicus.

1.2 An Image of Our Universe

Stars, Galaxies, and Beyond

Think about the mysterious spectacle of a clear night sky. What objects are up there? How far away are they? How do they move? How are they arranged? In addressing such questions, ancient philosophers learned to distinguish *stars* (points of light whose locations are fixed, relative to each other) from *planets* (tiny spots of light that move relative to the backdrop of stars). Over the centuries, two schools of thought developed concerning how to explain the configuration of stars and planets, and their relationships to the Earth, Sun, and Moon. The first school advocated a **geocentric model** (**Fig. 1.1a**), in which the Earth sits motionless at the center of the Universe while the Moon and the planets whirl around it, and everything lies inside a revolving globe of stars. The second school advocated a **heliocentric model** (**Fig. 1.1b**), in which the Sun lies at the center of the Universe while the Earth and other planets orbit around it.

For millennia, the geocentric model garnered the most followers, due to the influence of an Egyptian mathematician, Ptolemy (100–170 C.E.), who developed equations that appeared to predict the wanderings of the planets in the context of the model, and thus seemed to prove the model. During the Middle Ages (ca. 476–1400 C.E.), church leaders in Europe adopted the geocentric model as dogma, because it justified the comforting thought that humanity's home occupies the most important place in the Universe, the center. Anyone who disagreed with this view risked charges of heresy.

Then came the Renaissance. In 15th-century Europe, bold thinkers spawned a new age of exploration and scientific discovery. Thanks to the efforts of Nicolaus Copernicus (1473–1543) and Galileo Galilei (1564–1642), people gradually came to realize that the Earth and planets did indeed orbit the Sun, so the Earth could not possibly sit at the center of the Universe, and that orbits are elliptical, not cylindrical. As the Renaissance progressed, Isaac Newton (1642–1727) established the foundations of physics and optics, and others began to understand fundamental phenomena of chemistry. In the light of this work, the language and laws of modern science came into focus. Using this language, we now define the **Universe** as all of space and the matter and energy within it. **Box 1.1** provides a brief refresher on the meanings of these terms, and a few related ones.

As telescopes and other instruments improved so that astronomers could see and measure features progressively farther into space, the interpretation of stars evolved. Although it looks like a point of light, a **star** is actually an

FIGURE 1.2 A galaxy may contain about 300 billion stars.

(a) The Milky Way on a clear night. The "haze" actually consists of millions of faraway stars.

(b) A spiral galaxy that looks like the Milky Way, as viewed from the top.

(c) A Hubble Space Telescope view of deep space showing some of the billions of galaxies in the Universe.

immense ball of plasma (incandescent gas) that emits intense heat and light. Stars are not randomly scattered through the Universe. Rather, gravity holds them together in immense groups called **galaxies**. Our Sun, together with over 300 billion other stars, makes up the *Milky Way galaxy*. From our vantage point on Earth, the Milky Way looks like a hazy band (**Fig. 1.2a**), but if we could view the Milky Way from a great distance, it would resemble a flattened spiral with great curving arms slowly swirling around a glowing, disk-like center (**Fig. 1.2b**). Presently, our Sun lies near the outer edge of one of these arms, orbiting the center of the Milky Way once every 250 million years. Astronomers estimate that more than 100 billion galaxies constitute the visible Universe (**Fig. 1.2c**). Clearly, human understanding of Earth's place in the Universe has evolved radically over the past few centuries. Today, we realize that neither the Earth, nor the Sun, nor even the Milky Way occupies the center of the Universe—and everything is in motion.

Welcome to the Neighborhood: Our Solar System

Our Sun's gravitational pull holds on to many objects which, together with the Sun, comprise the **Solar System** (**Fig. 1.3a**). Most of the mass of the Solar System—99.8%, to be exact—resides in the Sun itself. The remaining 0.2% includes a great variety of objects, the largest of which are planets. Astronomers define a **planet** as an object that orbits a star, is roughly spherical, and has "cleared its neighborhood of other objects." The last phrase in this definition sounds a bit strange at first, but it merely implies that a planet's gravity has pulled in all particles of matter in its orbit. According to this definition, formalized in 2005, our Solar System includes eight planets—Mercury, Venus, Earth, Mars, Jupiter, Saturn, Uranus, and Neptune. Until 2005, astronomers considered one more object, Pluto, to be a planet. But Pluto—first photographed close up in 2015—does not fit the modern definition of a planet because it has not cleared its orbit of other objects, so it has been dropped from the roster. The eight planets orbit the Sun in the same direction and more or less in the same plane, called the **ecliptic** (**Fig. 1.3b**). Our Solar System is not alone in hosting planets. Land-based instruments, as well as those of the Kepler Space Telescope, have allowed astronomers as of 2015 to locate over 2,000 *exoplanets*, planets that orbit stars other than our Sun. Some of these are similar in size to the Earth.

Planets in our Solar System differ radically from one another in both size and composition. The inner planets (Mercury, Venus, Earth, and Mars), the ones closer to the Sun, are relatively small. These are called the **terrestrial planets** because, like Earth, they consist of a shell of rock surrounding a ball of metal. The outer planets (Jupiter, Saturn, Uranus, and Neptune) are known as the **giant planets**, or Jovian planets. These planets are indeed huge—Jupiter, for example, contains 318 times as much mass as the Earth, and accounts for about 71% of the non-solar mass in the Solar

FIGURE 1.3 The relative sizes and positions of planets in the Solar System.

(a) Relative sizes of the planets. All are much smaller than the Sun, but the gas-giant planets are much larger than the terrestrial planets. Jupiter's diameter is about 11.2 times greater than that of Earth.

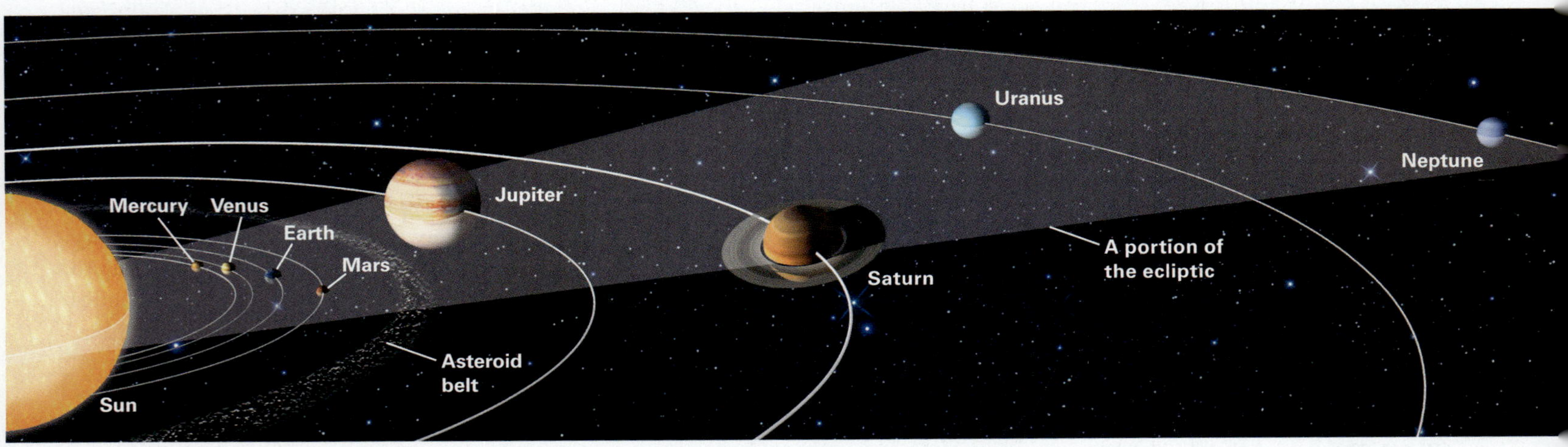

(b) Relative positions of the planets. This figure is not to scale. If the Sun in this figure were the size of a large orange, the Earth would be the size of a sesame seed 15 meters (49 feet) away. Note that all planetary orbits lie roughly in the same plane, called the ecliptic.

System. The overall composition of giant planets differs markedly from that of terrestrial planets. Specifically, most of the mass of Jupiter and Saturn consists of hydrogen and helium, in gas or liquid form—these planets are known as the *gas giants*. While Neptune and Uranus do contain significant quantities of hydrogen and helium, they also contain solid water, ammonia (NH_3), and methane, so these planets are known as the *ice giants*. Note that researchers use the word "ice" to mean not only frozen water but also other solid materials that could exist in gas form under conditions found at the Earth's surface.

In addition to the planets, the Solar System contains a great many smaller objects. A **moon** is a sizable body locked in orbit around a planet. All but two planets (Mercury and Venus) have moons in varying numbers—Earth has 1, Mars has 2, and Jupiter has at least 67. Some moons, such as the Earth's Moon, are large and spherical, but many are small and have irregular shapes. **Asteroids** are rocky and/or metallic objects, with diameters ranging from 100 m to about 930 km. Millions of asteroids occupy a belt between the orbits of Mars and Jupiter. About a trillion bodies of ice lie outside the orbit of Neptune. Most of these icy bodies are tiny, but a few (including Pluto) are spheres with diameters of over 1,500 km and are known as *dwarf planets*. The gravitational pull of planets has sent some of the icy objects on paths that take them into the inner part of the Solar System, where they heat up and release long tails of gas and dust—these objects are **comets**.

TAKE-HOME MESSAGE

People once thought the Earth lay at the center of the Universe. Now it's clear that it is one of eight spinning planets orbiting our Sun, one of 300 billion stars of the revolving, spiral-shaped Milky Way galaxy. Hundreds of billions of galaxies speckle the immense visible Universe.

QUICK QUESTION How do the inner planets differ from the outer planets?

1.3 Forming the Universe

A Key Clue: Discovering That the Universe Expands

We stand on a planet, in orbit around a star, speeding through space on the arm of a galaxy. Beyond our galaxy lie hundreds of billions of other galaxies. Where did all this "stuff"—the matter of the Universe—come from, and when did it first form? For most of human history, a scientific solution to these questions seemed intractable. But in the 1920s, unexpected observations opened a new door of understanding. Astronomers such as Edwin Hubble (1889-1953), after whom the Hubble Space Telescope was named, braved many a frosty night beneath the open dome of a mountaintop observatory in order to aim telescopes into deep space. These researchers were searching for distant galaxies. At first, they documented only the location and shape of newly discovered galaxies, but eventually they also began to document the motion of the galaxies, which they could deduce by studying characteristics of the light emitted by the galaxies. (An astronomy book can provide the details of the method.) The results yielded a surprise that would forever change humanity's perception of the Universe. They found that all distant galaxies are moving away from the Earth, no matter which direction they looked. Furthermore, the farther away the galaxy lies, the faster it's moving away.

Did you ever wonder... if galaxies move?

Around 1929, Hubble finally recognized what these observations meant—the volume of the whole Universe must be increasing! This idea came to be known as the **expanding Universe theory**. To picture the expanding Universe, imagine a ball of bread dough with raisins scattered throughout. As the dough bakes and expands into a loaf, each raisin moves away from its neighbors, in every direction (**Fig. 1.4a**). Note how raisins that were farther apart to start with have moved more, relative to each other, than raisins that were closer together to

FIGURE 1.4 The concept of the expanding Universe and the Big Bang.

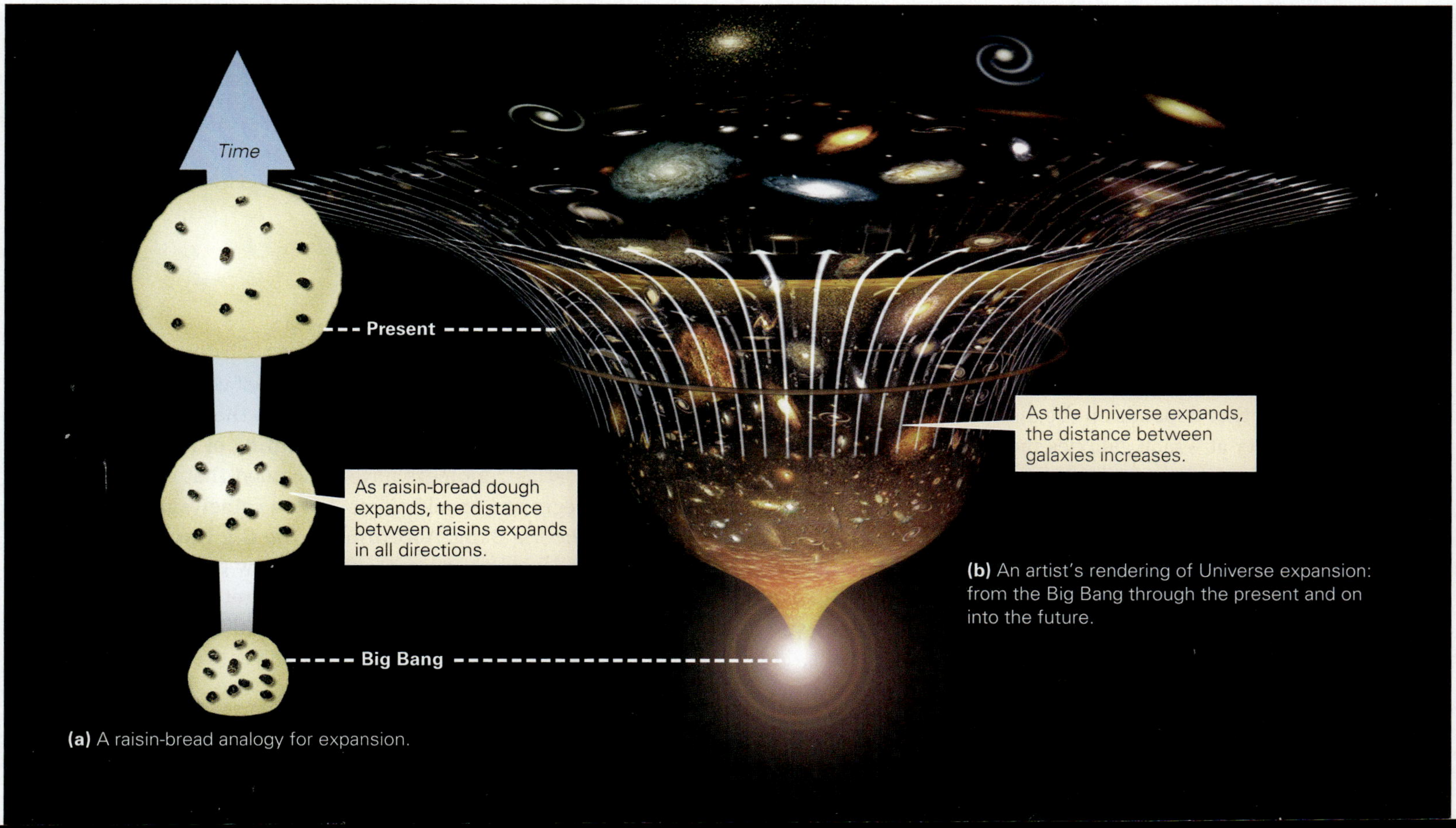
Time
Present
As raisin-bread dough expands, the distance between raisins expands in all directions.
Big Bang
As the Universe expands, the distance between galaxies increases.

(a) A raisin-bread analogy for expansion.

(b) An artist's rendering of Universe expansion: from the Big Bang through the present and on into the future.

BOX 1.1 Consider This...

The Basics of Matter, Force, and Energy

Matter, Atoms, and Molecules

The material substance of the Universe consists of *matter*—it takes up space and you can feel it. We refer to the amount of matter in an object as its *mass*, so an object with greater mass contains more matter. *Density* refers to the amount of mass occupying a given volume of space—a cubic centimeter of a denser material contains more mass than does a cubic centimeter of a less dense material.

An *element* is matter that cannot be subdivided into other components with different properties. If you were to keep subdividing an element into smaller and smaller pieces that have the same properties, you would end up with an *atom*, the smallest piece of an element that has the properties of the element. To picture how tiny atoms are, keep in mind that about 5 to 10 trillion atoms fit within the area of the period at the end of this sentence. We refer to elements by abbreviations, called *chemical symbols*, such as H (hydrogen), O (oxygen), Si (silicon), and Fe (iron).

Atoms themselves can be subdivided. Specifically, a single atom can contain three types of *subatomic particles*: *protons* that have a positive charge, *neutrons* that have a neutral charge, and *electrons* that have a negative charge. (Simplistically, the term "charge" refers to the electrical behavior of the particle.) Protons and neutrons, which are about the same size, stick together in a dense ball, the *nucleus*, at the center of an atom—the "glue" that holds particles together in a nucleus is called a *nuclear bond*. Electrons are much smaller than protons or neutrons, and swirl around the nucleus in an *electron cloud*—it's the outer edge of the electron cloud that defines the surface of an atom (**Fig. Bx1.1a**).

An atom of one element differs from an atom of another in terms of the number of protons in its nucleus. We specify this quantity as the *atomic number* of the element. Hydrogen atoms have an atomic number of 1, helium has an atomic number of 2, and iron has an atomic number of 26. With the exception of the most common form of hydrogen, all nuclei contain neutrons. The sum of the number of protons plus the number of neutrons roughly equals the *atomic mass* of an atom. All atoms, except for hydrogen, have neutrons, so for all atoms except hydrogen, the atomic mass is more than the atomic number.

Atoms tend not to exist in isolation, but rather attach to other atoms, connected by an invisible "glue" called a *chemical bond*. Scientists use the word *molecule* for a particle composed of a combination of two or more atoms bonded together. Some molecules contain atoms of only the same element, whereas others contain atoms of two or more different elements. A substance in which molecules consist of two or more elements is a *compound*. We can specify the composition of a molecule by providing its *chemical formula*, a recipe which uses symbols to represent names and proportions of elements in a compound. For example, hydrogen gas consists of H_2 molecules, oxygen gas of O_2 molecules, water of H_2O molecules, and methane of CH_4 molecules. A molecule of methane contains one carbon atom and four hydrogen atoms. Note that a molecule is the smallest piece of a compound that has the characteristics of the compound, but not all molecules are pieces of compounds.

In everyday life, we see matter in four forms, called *states of matter* (**Fig. Bx1.1b**). In a *gas*, atoms and/or molecules are free to move with respect to one another, so a gas can expand to fill a volume that it occupies. In a *liquid*, atoms and/or molecules are loosely connected, so a liquid can flow and assume the shape of its container, but it will not change its volume as it flows. In a *solid*, atoms and/or molecules are strongly connected, so a solid can retain its shape for a long time. Under very high temperatures, plasma, a fourth state of matter forms—*plasma* is a gas in which the electrons have been stripped from atoms.

start with. Therefore, the velocity at which raisins move relative to each other depends on the distance between them, just as the velocity at which galaxies move away from each other depends on the distance between them.

The Big Bang

Hubble's idea marked a revolution in cosmological thinking. Now we picture the Universe as an expanding bubble, in which galaxies race away from each other at incredible speeds. This image immediately triggers a key question of cosmology: Did the expansion begin at some specific time in the past? If it did, then that instant would mark the beginning of the physical Universe. Most astronomers have concluded that expansion did indeed begin at a specific time, with a cataclysmic explosion called the Big Bang. According to the **Big Bang theory**, all matter and energy—everything that now constitutes the Universe—was initially packed into an infinitesimally small point. The point "exploded" and the Universe began, according to current estimates, about 13.8 billion years ago (Ga).

Force and Energy

Matter in the Universe does not sit still. Objects spin, they move from one place to another, and they pull on or push against each other. Isaac Newton showed that one mass can exert a *force* on another mass, and by doing so, change the velocity (speed and/or direction) of the mass. A change in velocity is called an *acceleration*, so Newton wrote this relationship as a scientific law: $F = ma$. In this equation, F is force, m is mass, and a is acceleration.

Forces can be applied in two ways. A *contact force* happens when one mass touches another—for example, you apply a contact force when you strike a ball with a bat and send the ball careening away. A *field force* happens when an object applies a force to another object without touching it. Examples of field forces include *gravity*, which is the attraction that one mass exerts on another, and *magnetism*, which is the attraction or repulsion that a magnet or electric current exerts on magnetic or electrically charged materials.

Simplistically, *work*, in the jargon of physics, is the consequence of a force acting on an object over a distance. For example, when you lift a weight by 1 meter, against the pull of gravity, you have done work. If you lift the same weight by 2 meters, you have done twice as much work. You also do work when you heat a pot of water and the water turns to steam that rises into the air. In other words, work happens when a mass has changed its position or state.

FIGURE Bx1.1 The nature of atoms and matter.

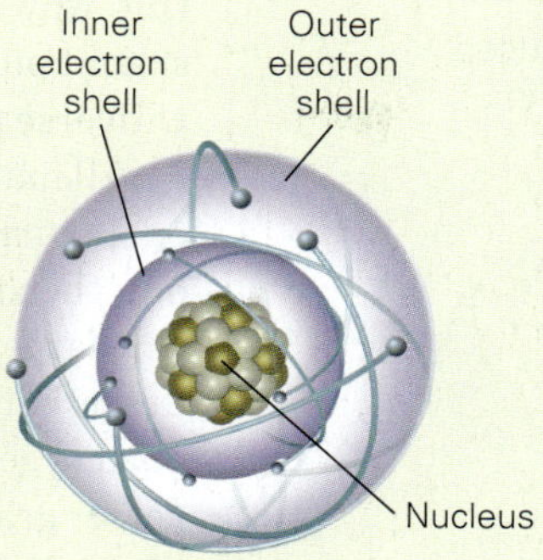

(a) An image of an atom with a nucleus orbited by electrons.

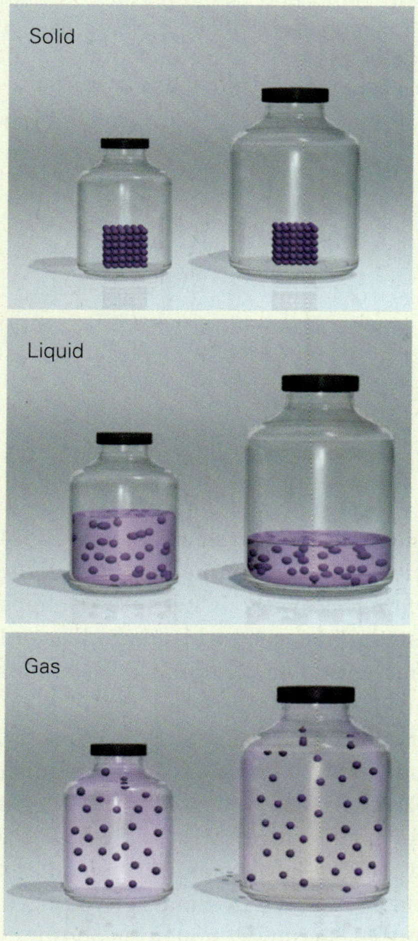

(b) The three common states of matter. Solids keep their shape, liquids conform to the shape of their container without changing density, and gases expand to fill their container.

Physicists refer to the ability or capacity of a system (a specified region of space, of any size, and all it contains) to do work as energy. A moving object, light, magnetism, and gravity all can cause matter to change in some way, and thus can do work, so they provide energy. We can distinguish between *kinetic energy*, the energy of motion, and *potential energy*, the energy stored in a mass. A flying ball has kinetic energy, which can do work when it strikes a window and shatters the glass. A boulder sitting on a hillslope has potential energy, because gravitational force is pulling on it, so it could move (bounce downhill), if the dirt beneath it gives way. Similarly, a liter of gasoline contains potential energy, in that if it burns, it can cause a car to move.

Of course, no one was present at the instant of the Big Bang, so no one actually saw it happen. But by combining clever calculations with careful observations, researchers have developed a consistent model of how the Universe evolved, beginning an instant after the Big Bang (**Fig. 1.4b**). According to this model, profound change happened at a fast and furious rate at the outset. During the first instant of existence, the Universe was so small, so dense, and so hot that it consisted entirely of energy—atoms, or even the smallest subatomic particles that make up atoms (see Box 1.1), could not even exist. Within a few seconds, however, subatomic particles (neutrons, protons, and electrons) came into existence. And by the time the Universe reached an age of 20 minutes, all hydrogen atoms that would form had formed (including normal hydrogen, which consists only of a proton, and two other versions that include neutrons), and enough hydrogen atoms had joined together to form helium atoms so that the proportion of hydrogen to helium in the newborn Universe was about 3 to 1. The process by which the nuclei of two smaller atoms collide, stick together, and form a new atom with a larger

FIGURE 1.5 Fusion happens when colliding particles fuse to form a new, larger nucleus. Deuterium and tritium are versions of hydrogen in which a proton fused to one or two neutrons, respectively. These nuclei, in turn, can collide to form a helium nucleus.

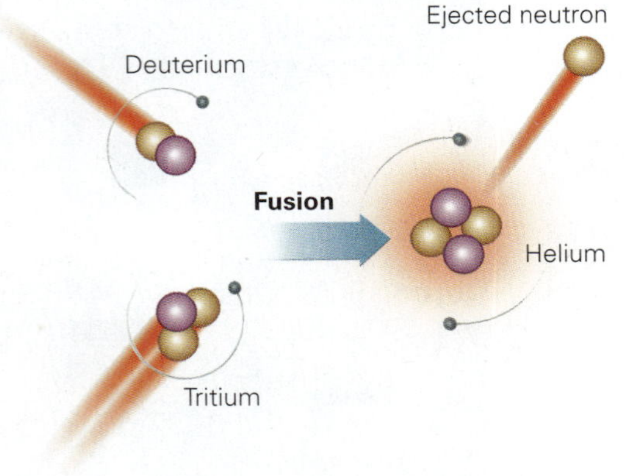

nucleus is called **nuclear fusion** (**Fig. 1.5**). Formation of new nuclei during and immediately after the Big Bang is called *Big Bang nucleosynthesis* because it happened before any stars existed. This process could produce only small atoms, that is, atoms with a small number of protons (atomic number less than 5).

Eventually, the Universe became cool enough for chemical bonds to bind atoms of certain elements together to produce molecules (see Box 1.1). Most notably, hydrogen atoms joined to form molecules of H_2. As the Universe continued to expand and cool further, atoms and molecules slowed down and accumulated into patchy clouds of hydrogen and helium gas called **nebulae** (see chapter opening photo).

Birth of the First Stars: The Nebular Theory, Part 1

When the Universe reached its 200 millionth birthday, it contained immense, slowly swirling nebulae of hydrogen and helium gas separated by vast voids of empty space. Empty space is a **vacuum**, a region in which the density of matter is extremely low. In the regions between nebulae, the vacuum of space may contain less than 1 atom or molecule per cubic meter—by comparison, the air you breathe at sea level contains on the order of 300,000,000,000,000,000,000,000 molecules per cubic meter. The Universe could not remain this way forever, though, because of the invisible but persistent pull of gravity. Eventually, gravity began to remold the Universe pervasively and permanently.

All matter exerts gravitational attraction—a type of force—on its surroundings (see Box 1.1). As Isaac Newton first pointed out, the amount of attraction depends on the amount of mass (the larger the mass, the greater its attraction) and on the distance between the masses (gravitational force increases as objects get closer to each other). Somewhere in the young Universe, gravity produced by an initially more massive region of a nebula began to suck in surrounding gas and, in a grand example of the rich getting richer, grew in mass and, therefore, density. As progressively more gas drew into this denser region, the dense region became even denser, and the initial swirling movement of gas transformed into a rotation around an axis. As gas continued to move inward, cramming into a progressively smaller volume, the rotation rate became faster and faster—a similar phenomenon happens when a spinning ice skater pulls her arms inward. Because of its increased rotation, the nebula evolved into a bulbous plate-like shape called an **accretionary disk** (**Fig. 1.6**). Eventually, gravity caused the inner portion of the accretionary disk to collapse inward to form a dense ball. When gas in this ball squeezed into a smaller and smaller space, the gas's temperature increased dramatically (**Box 1.2**), so the ball became hot enough to glow. At this time, it became a **protostar**. The remaining mass of the accretionary disk, as we will see, eventually clumped into smaller spheres, the planets. The overall model for explaining the sequence of stages during which stars and planets form from a nebula is called the **nebular theory**.

A protostar continues to grow, by pulling in successively more mass, until its core becomes so dense, and its temperature so high (about 10,000,000°C), that hydrogen nuclei within it move incredibly fast and slam together so forcefully that fusion takes place and new helium nuclei form. Fusion reactions produce huge amounts of energy, so when they begin, the protostar "ignites" to becomes a nuclear furnace, and the ball becomes a true star. The first true star may have formed about 800 million years after the Big Bang, and when this happened, the first starlight pierced the newborn Universe. Star formation happened again and again in the young

FIGURE 1.6 Nebular theory for planet formation.

Universe, and soon many first-generation stars had come into existence.

First-generation stars tended to be very massive, having perhaps 100 times the mass of the Sun. Astronomers have shown that the larger the star, the hotter it burns and the faster it runs out of fuel and dies. A huge star may survive only a few million years to a few tens of millions of years before it runs out of fuel, collapses, explodes, and becomes a **supernova**. This name comes from the impression, when their light reaches the Earth, that a new but temporary star has appeared in the night sky. The giant explosion blasts much of the star's matter back into space. Thus, not long after the first generation of stars formed, the Universe began to be peppered with the first generation of supernovas.

TAKE-HOME MESSAGE

All distant galaxies are moving away from us, an observation that requires the Universe to be expanding. According to the Big Bang theory, this expansion, and thus the beginning of our Universe, began with a cataclysmic explosion at about 13.8 Ga. According to the nebular theory, atoms formed during the Big Bang collected into nebulae which, due to gravity, collapsed into dense balls that began to produce energy by nuclear fusion reactions. These objects were the first stars.

QUICK QUESTION What elements made up most of the nebulae in the very young Universe?

1.4 Forming the Solar System, Including the Earth

We Are All Made of Stardust

Nebulae from which the first-generation stars formed consisted entirely of the lightest atoms (atoms with a very small nucleus), because only these atoms were generated by Big Bang nucleosynthesis. In contrast, the Universe of today contains 92 naturally occurring elements. Where did the other 87 elements come from? In other words, how did heavier elements with larger atomic numbers (such as carbon, sulfur, silicon, iron, gold, and uranium) form? Physicists have shown that intermediate-weight elements form by fusion reactions during the life cycle of stars, a process called **stellar nucleosynthesis**. Because of stellar nucleosynthesis, we can consider stars to be "element factories," constantly fashioning larger atoms out of smaller atoms. Most heavy atoms, those with atomic numbers greater than that of iron, require even hotter environments to form than generally occurs within a star. In fact, most very heavy atoms form during the inconceivably hot conditions of supernova explosion, a process called *supernova nucleosynthesis*.

How do the atoms formed in stars get into space? Some escape during the star's lifetime, simply by moving fast enough to overcome the star's gravitational pull. The stream of atoms emitted from a star during its lifetime is a **stellar wind** (**Fig. 1.7a**). Some escape only when a star dies. Large

FIGURE 1.7 Element factories in space.

(a) A photo of solar (stellar) wind streaming into space.

(b) This expanding cloud of gas, ejected into space from a supernova explosion whose light reached the Earth in 1054 C.E., is called the Crab Nebula.

BOX 1.2 Science Toolbox...

Heat, Temperature, and Heat Transfer

The atoms and molecules that make up an object do not stay rigidly fixed in place, but rather jiggle and jostle with respect to one another. This movement creates *thermal energy*—the faster the atoms move, the greater the thermal energy and the "hotter" the object. Put another way, the thermal energy in a substance represents the sum of the kinetic energy (energy of motion) of all the substance's atoms in a given volume. This includes the back-and-forth displacements that an atom makes as it vibrates, as well as the movement of an atom from one place to another. Because of the relation between kinetic energy and thermal energy, kinetic energy can be transformed into thermal energy. You can demonstrate this with a simple home experiment. Start to hammer a large nail into a block of hardwood, and then after a few strong blows, touch the end of the nail. You'll feel that it has gotten warmer. The reason is that although some of the kinetic energy of the moving hammer head has gone into pushing the nail into the wood, most has gone into agitating the atoms in the nail. Because the atoms have started moving around more quickly, the thermal energy in the nail has increased.

When we say that one object is hotter or colder than another, we are describing its temperature. **Temperature** is a measure of warmth relative to a known reference value, and represents the average kinetic energy of atoms in the material. When we compress a gas, the molecules collide more frequently and so the average kinetic energy in the gas increases, so the temperature goes up. To specify temperature, meaning to indicate relative hotness or coldness, scientists use the *centigrade*, or Celsius, *scale*, whose unit is 1 degree (1°C). The scale is calibrated so that the boiling point of water at sea level is 100°C, and the freezing point is 0°C. An increment of 1°C, therefore, represents 1/100 of the temperature difference between the boiling and freezing points of water. (Note that the English system, commonly used in weather reports in the United States, uses the *Fahrenheit scale* for temperature. On this scale water boils at 212°F and freezes at 32°F.)

Heat is the thermal energy transferred from one object to another. We measure heat in *calories*. One thousand calories can heat 1 kilogram of water by 1°C. The transfer of heat from one place to another can occur in four distinct ways:

- *Radiation:* When an object glows, it sends electromagnetic waves (such as visible light, infrared light, and ultraviolet light) into the space around it. This energy, called *radiation*, can cause an object at a distance from the source to heat up, even if there is a vacuum between the source and the object. Radiation traveling from the Sun is responsible for heating the Earth's surface (**Fig. Bx1.2a**).
- *Conduction:* If you stick the end of an iron bar in a fire, *conduction* takes place and heat moves up the bar. This happens because atoms in the bar at the hot end start to vibrate more energetically, and this motion incites atoms farther up the bar to start jiggling. Though heat flows along the bar, the atoms do not actually move from one locality to another (**Fig. Bx1.2b**). Heat always conducts from hotter to cooler material.
- *Convection:* When you place a pot on a stove and heat it, water at the base of the pot gets hotter and expands, so its density decreases. The hot, less-dense water becomes buoyant relative to the colder, denser water above it. In a gravitational field, buoyant material rises if

stars, as we've noted, blast an immense amount of matter into space during supernova explosions (**Fig. 1.7b**). Small or medium-sized stars (like our Sun) do not explode, but rather balloon to become a *red giant*, when they start to run out of fuel. Eventually, when the star runs out of fuel entirely, its interior collapses, leaving its outer shells of gas to drift out into space. Once ejected into space, atoms from stars, collapsed red giants, and supernova explosions form new nebulae or mix back into existing nebulae.

When the first generation of stars died, they left a legacy of new, heavier elements that mixed with residual hydrogen and helium from the Big Bang. A second generation of stars and associated planets formed out of these compositionally more diverse nebulae. Second-generation stars lived and died, and contributed heavier elements to third-generation stars. Succeeding generations contain a progressively greater proportion of heavier elements. Because not all stars live for the same duration of time, at any given moment the Universe contains multiple generations of stars. Our Sun may be a second- or third-generation star, so the mix of elements we find on Earth includes relicts of primordial gas from the Big Bang as well as the disgorged guts of dead stars. Think of it—the atoms that make up your body once resided inside a star!

Did you ever wonder...
where the atoms in your body first formed?

the material above it is weak enough to flow out of the way. Since liquid water flows easily, the hot water can rise. When this happens, cold water sinks to take its place (**Fig. Bx1.2c**). The resulting circulation, during which flow of the material itself carries heat, is called *convection*; the path of flow defines *convection cells*.

- *Advection:* When a hot fluid flows into cracks and pores within a solid, it heats up the solid. Transfer of heat carried by a hot fluid passing through a solid is called *advection*. The hotness of a metal water pipe's surface, for example, comes from the hot water in the pipe. Advection occurs in the Earth when molten rock rises into the crust (**Fig. Bx1.2d**).

FIGURE Bx1.2 The four processes of heat transfer.

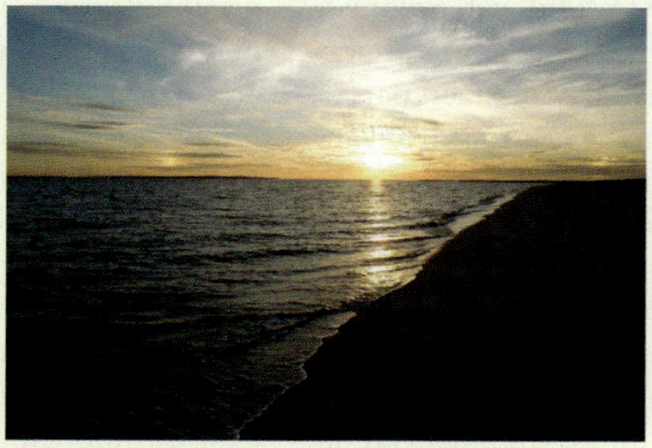

(a) Radiation from sunlight warms the Earth.

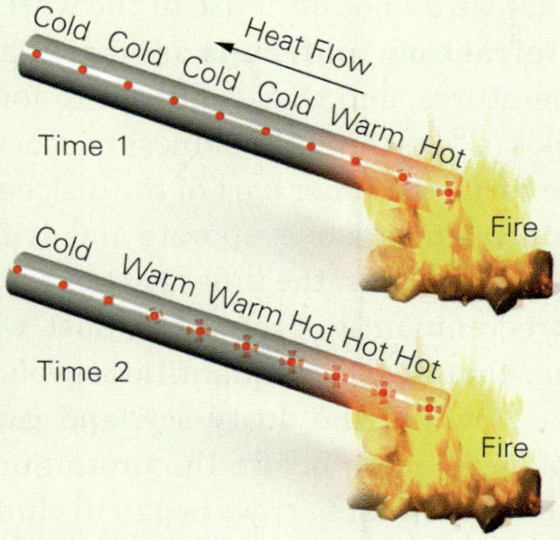

(b) Conduction occurs when you heat the end of an iron bar in a flame. Heat flows from the hot region toward the cold region as vibrating atoms cause their neighbors to vibrate.

(c) Convection takes place when moving fluid carries heat with it. Hot fluid rises while cool fluid sinks, setting up a convection cell.

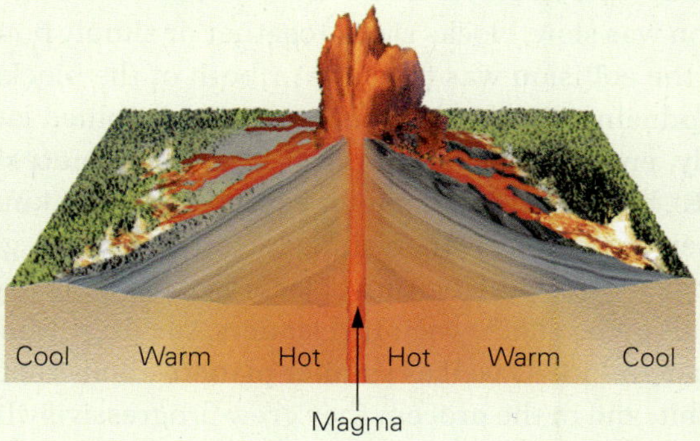

(d) During advection, a hot liquid (such as molten rock) rises into cooler material, and heat then conducts from the hot liquid into the cooler material.

Forming the Solar System: The Nebular Theory, Part 2

Earlier in this chapter, we introduced the nebular theory as a scientific concept that explains how stars form from nebulae. But we delayed our discussion of how the planets and other objects in our Solar System originated until we had discussed the production of heavier atoms such as carbon, silicon, iron, and uranium, because terrestrial planets consist predominantly of these elements. Now that we've discussed stars as element factories, we return to the early history of the Solar System and extend the nebular theory to explain how planets, moons, asteroids, and comets form. According to the nebular theory, these objects came from matter in the flattened outer part of the accretionary disk, the material that did not become incorporated into the star. This outer part is called the *protoplanetary disk*.

The protoplanetary disk from which our Solar System formed contained all 92 elements, some as isolated atoms and some bonded to others in molecules. Geologists divide the material formed from these atoms and molecules into two classes. **Volatile materials**—such as hydrogen, helium, methane, ammonia, water, and carbon dioxide (CO_2)—can exist as gas at the Earth's surface. In the pressure and temperature conditions of space, all volatile materials remain in a gaseous state closer to the Sun. But beyond a distance called the *frost line*, some volatiles can freeze into ice. (Recall that we do not limit use of the word *ice* to solid water alone.) **Refractory materials** are those that melt only at high temperatures, and they condense to form solid soot-sized particles of "dust" in the coldness of space. As the proto-Sun began to form, the inner part of the disk became hotter, causing volatile elements to evaporate and drift to the outer portions of the disk. Thus, the inner part of the disk ended up consisting predominantly of refractory dust, whereas the outer portions accumulated large quantities of volatile materials and ice.

How did the dusty, icy, and gassy rings transform into planets? Even before the proto-Sun ignited, the material of the surrounding rings began to clump and bind together, due to gravity and electrical attraction, forming a series of concentric rings around the proto-Sun. In these rings, matter was denser, and particles got close enough to each other to attract. First, soot-sized particles merged to form sand- to marble-sized grains. These then stuck together to form grainy basketball-sized blocks, which in turn collided. If the collision was slow, blocks stuck together or simply bounced apart. If the collision was fast, one or both of the blocks shattered, producing smaller fragments that recombined later. Eventually, enough blocks coalesced to form **planetesimals** (**Fig. 1.8**), bodies whose diameter exceeded about 1 km. Because of their mass, the planetesimals exerted enough gravitational attraction to pull in other objects that were nearby. Planetesimals behaved like vacuum cleaners, sucking in small pieces of dust and ice as well as smaller planetesimals that lay in their orbit, and in the process they grew progressively larger. Eventually, victors in the competition to attract mass grew into **protoplanets**, bodies approaching the size of today's planets. Once a protoplanet succeeded in incorporating virtually all the debris within its orbit, it became a full-fledged planet.

FIGURE 1.8 The grainy texture of this meteorite fragment may resemble the texture of a small planetesimal.

Early stages in the planet-forming process probably occurred very quickly—some computer models suggest that it may have taken less than a million years to go from the dust and gas stage to the large planetesimal stage. Planets may have grown from planetesimals in 10 to 200 million years. In the inner orbits, where the protoplanetary disk consisted mostly of refractory dust, small terrestrial planets (such as the Earth) composed of rock and metal formed, whereas in the outer part of the Solar System, where significant amounts of volatiles remained, the rocky and metallic protoplanets became surrounded by thick shells of gas and/or ice, and evolved into the giant planets. Fragments of materials that were not incorporated in planets remain today as asteroids and comets.

When did the Solar System form? Using techniques introduced in Chapter 10, geologists have found that special types of meteorites thought to be leftover planetesimals formed at 4.57 Ga, and thus consider that date to be the birth date of the Solar System. If the date is correct, it means that the Solar System formed about 9 billion years after the Big Bang, and thus is only about a third as old as the Universe.

Why Are Planets Round?

Small planetesimals were jagged or irregular in shape, and asteroids today have irregular shapes. Planets and larger moons, on the other hand, are more or less spherical. Why? The rock composing a small planetesimal is cool and strong enough so that the force of gravity cannot cause the rock inside the object to flow, so irregularities like bumps and dimples can survive. But once a planetesimal grows beyond a certain critical size (about 500 km in diameter), its interior becomes warm and soft enough to flow in response to the pull of gravity. As a consequence, protrusions are pulled inward toward the center, and the planetesimal takes on a special shape—a sphere—for in a sphere, mass is evenly distributed around the center so the force of gravity can be nearly the same at all points on its surface (**Geology at a Glance**, pp. 24–25).

Differentiation of the Earth

When planetesimals started to form, they had a fairly homogeneous distribution of material throughout, because the smaller pieces from which they formed all had much the same composition and collected together in no particular order. But large planetesimals did not stay homogeneous for long, because they began to heat up. The heat came primarily from three sources: the heat produced during collisions (similar to the phenomenon that happens when you bang on a nail with a hammer and they both get warm), the heat produced when

FIGURE 1.9 During nuclear fission, a uranium atom splits to form two smaller atoms. The process releases neutrons.

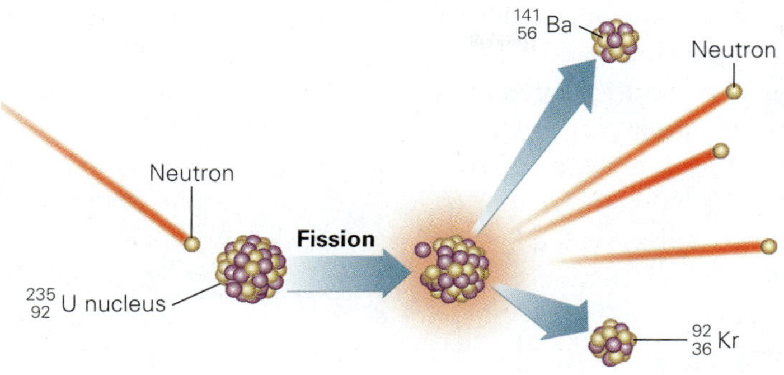

matter is squeezed into a smaller volume (see Box 1.2), and the heat produced from the decay of radioactive elements. A *radioactive element* is one whose nuclei spontaneously emits subatomic particles, or undergoes **nuclear fission** by breaking into two smaller nuclei (**Fig. 1.9**). Such "decay" breaks nuclear bonds and releases energy. In bodies whose temperature rose sufficiently to cause internal melting, denser iron alloy separated out and sank to the center of the body, whereas lighter rocky materials remained in a shell surrounding the center (**Fig. 1.10**). By this process, called **differentiation**, protoplanets and large planetesimals developed internal layering early in their history. As we will see later in this chapter, the Earth's central ball of iron alloy constitutes the body's *core* and the shell surrounding it is its *mantle*. Geologists have concluded that differentiation of protoplanets had occurred by 4.54 Ga, based on the ages of meteorites that represent fragments of differentiated protoplanets that later broke apart, and they consider this age to be the age of the Earth.

TAKE-HOME MESSAGE

Heavier elements formed in stars and supernovae added to gases in nebulae from which new generations of stars formed. Planets formed from rings of dust and ice orbiting the stars, so we are all formed of stardust. When they became large enough, planets differentiated, with denser materials sinking to the center.

QUICK QUESTION Why are the gas-giant planets farther from the Sun than are the terrestrial planets?

1.5 Earth's Moon and Magnetic Field

The Earth does not orbit the Sun all on its own. Rather, the near space around the Earth encompasses two important entities—the Moon and the magnetic field.

The Moon and Its Formation

As seen from Earth, the Moon glows an eerie white. Its radius of 1,737 km is about a quarter of the Earth's. Even with the

FIGURE 1.10 Differentiation of the Earth's interior.

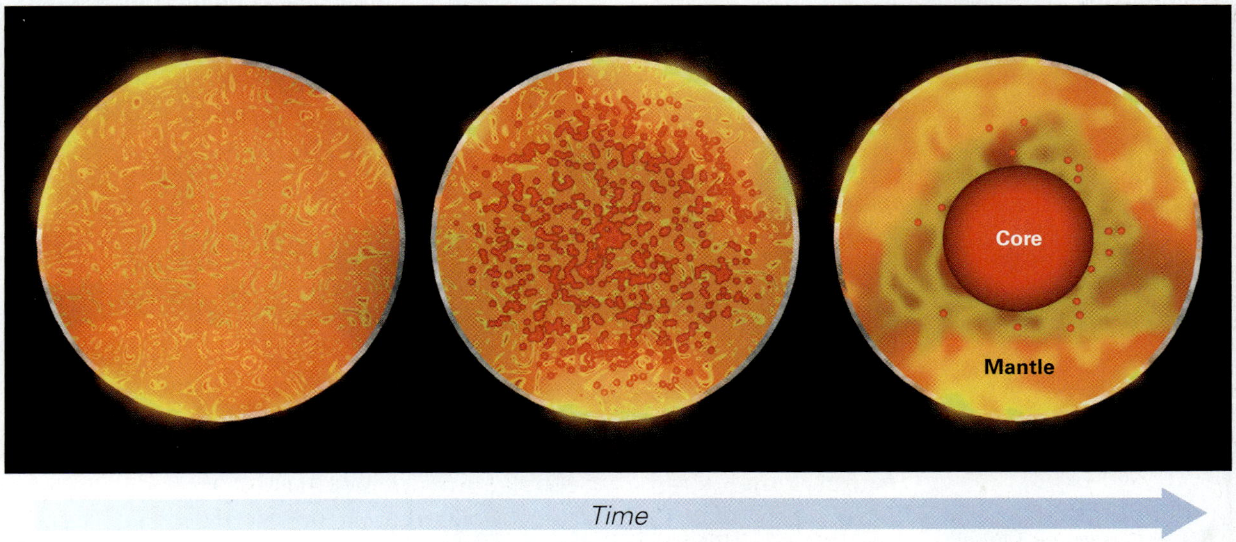

(a) Early on, the Earth was fairly homogeneous inside.

(b) When the temperature got hot enough, iron began to melt.

(c) The iron accumulated at the center of the planet to form a metallic core.

Forming the Planets and the Earth-Moon System

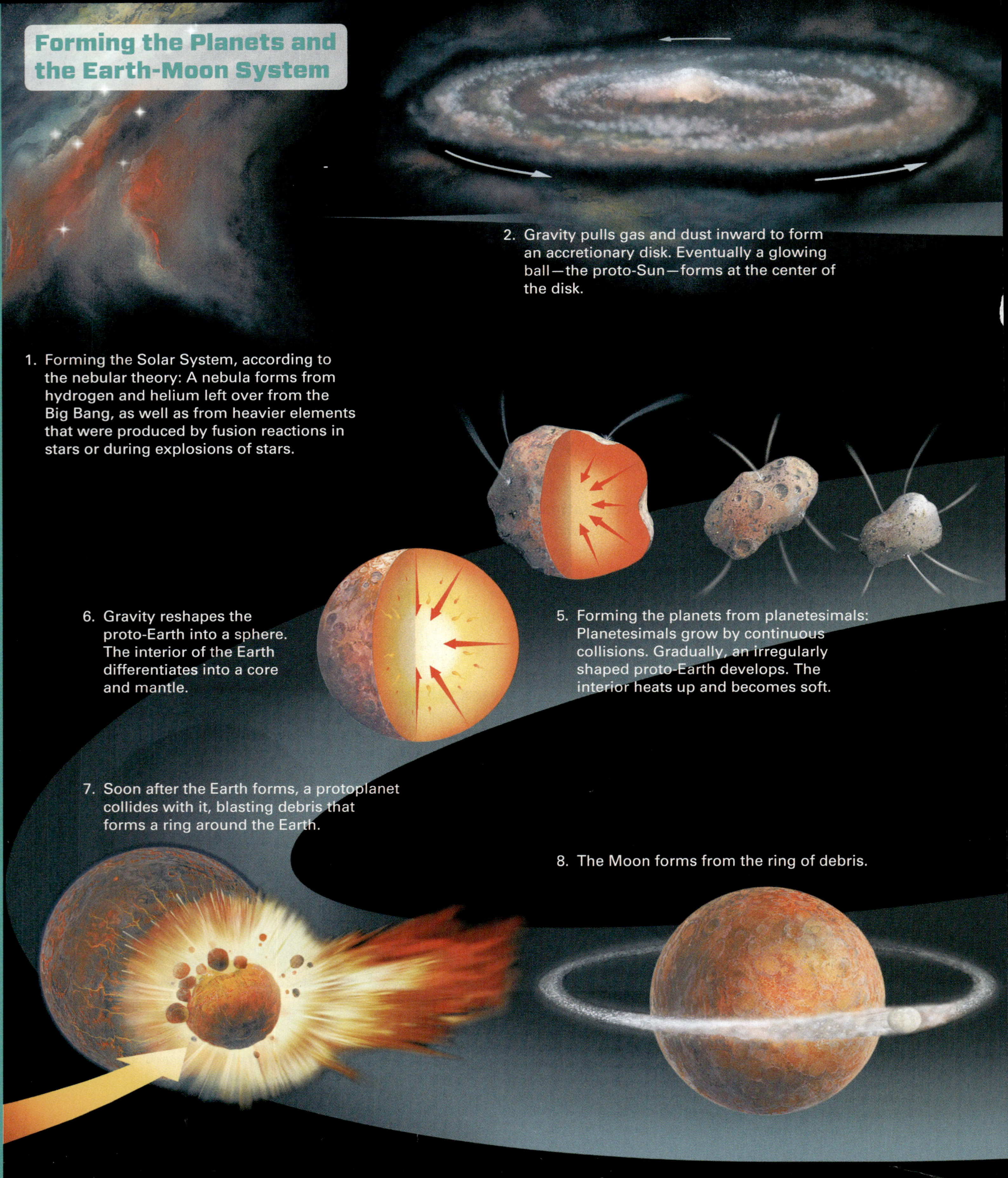
1. Forming the Solar System, according to the nebular theory: A nebula forms from hydrogen and helium left over from the Big Bang, as well as from heavier elements that were produced by fusion reactions in stars or during explosions of stars.
2. Gravity pulls gas and dust inward to form an accretionary disk. Eventually a glowing ball—the proto-Sun—forms at the center of the disk.
5. Forming the planets from planetesimals: Planetesimals grow by continuous collisions. Gradually, an irregularly shaped proto-Earth develops. The interior heats up and becomes soft.
6. Gravity reshapes the proto-Earth into a sphere. The interior of the Earth differentiates into a core and mantle.
7. Soon after the Earth forms, a protoplanet collides with it, blasting debris that forms a ring around the Earth.
8. The Moon forms from the ring of debris.

3. "Dust" (particles of refractory materials) concentrates in the inner rings, while "ice" (particles of volatile materials) concentrates in the outer rings. Eventually, the dense ball of gas at the center of the disk becomes hot enough for fusion reactions to begin. When it ignites, it becomes the Sun.

4. Dust and ice particles collide and stick together, forming planetesimals.

9. Eventually, the atmosphere develops from volcanic gases. When the Earth becomes cool enough, moisture condenses and rains to produce the oceans. Some gases may be added by passing comets.

naked eye, you can see that its surface is variable in character—the whiter areas, the lunar highlands, have been pockmarked by countless craters, whereas the darker areas, the maria (singular: mare), are smoother and less cratered (**Fig. 1.11**). Gravitational interaction between the Earth and the Moon keeps the same side of the Moon facing the Earth, so humans did not know what the far side of the Moon looked like until satellites and astronauts sent back pictures.

How did the Moon form? The answer remains a subject of scientific debate. In the currently most popular hypothesis, the Moon resulted when a protoplanet, perhaps comparable in size to Mars, collided with the newborn Earth. In the process, the colliding body disintegrated and melted or even vaporized, along with a large part of the Earth's mantle (see Geology at a Glance, pp. 24–25). A ring of debris formed around the remaining, probably molten Earth. Gravity quickly caused this material to coalesce into the Moon. Not all moons in the Solar System necessarily formed in this manner. Some may have been independent protoplanets or comets that were captured by a larger planet's gravity. Based on analysis and the dating of Moon rocks, most geologists have concluded that the Moon-forming event happened at about 4.53 Ga. Most of the craters on the Moon formed between the time of the Moon's birth and about 3.0 Ga. In fact, around 3.9 Ga, heavy bombardment by meteorites may have pulverized the surfaces of planets. The Earth must have been hit as frequently as the Moon, but on the Earth, geologic processes have erased most of these craters.

Did you ever wonder...
if the Moon is as old as the Earth?

FIGURE 1.11 The Earth's moon has distinct maria (dark areas), and lighter-colored, intensively cratered highlands.

Earth's Magnetic Field

A **magnetic field**, in a general sense, is the region affected by the force emanating from a magnet. Magnetic force, which grows progressively stronger as you approach the magnet, can attract or repel another magnet and can cause charged particles to move. Earth has a magnetic field, which is largely a **dipole**, meaning that it has two different ends—a north pole and a south pole—just like the magnetic field around a familiar bar magnet (**Fig. 1.12a, b**). When you bring two bar magnets close to one another, their opposite poles attract and their like poles repel. By convention, we represent the orientation of a magnetic dipole by an arrow that points from the south pole to the north pole, and we represent the magnetic field of a magnet by a set of invisible magnetic field lines that curve through the space around the magnet and enter the magnet at its poles. Arrowheads along these lines point in a direction to complete a loop. Magnetized needles, such as iron filings or compass needles, when placed in a field, align with the magnetic field lines.

If we represent the Earth's magnetic field as emanating from an imaginary bar magnet in the planet's interior, the north pole of this bar lies near the south geographic pole of the Earth, whereas the south pole of the bar lies near the north geographic pole. (The *geographic poles* are the places where the spin axis of the Earth intersects the planet's surface.) Nevertheless, by convention geologists and geographers refer to the magnetic pole closer to the north geographic pole as the *north magnetic pole*, and the magnetic pole closer to the south geographic pole is called the *south magnetic pole*. By using this convention, the north-seeking end of a compass points toward the north geographic pole, since opposite ends of a magnet attract.

The Sun's stellar wind, known as the *solar wind*, interacts with Earth's magnetic field, distorting it into a huge teardrop pointing away from the Sun. The solar wind consists of dangerous, high-velocity charged particles. Fortunately, the magnetic field deflects most (but not all) of the particles, so that they do not reach Earth's surface. In this way, the magnetic field acts like a shield against the solar wind; the region inside this magnetic shield is called the **magnetosphere**

FIGURE 1.12 A magnetic field permeates the space around the Earth. It can be symbolized by a bar magnet.

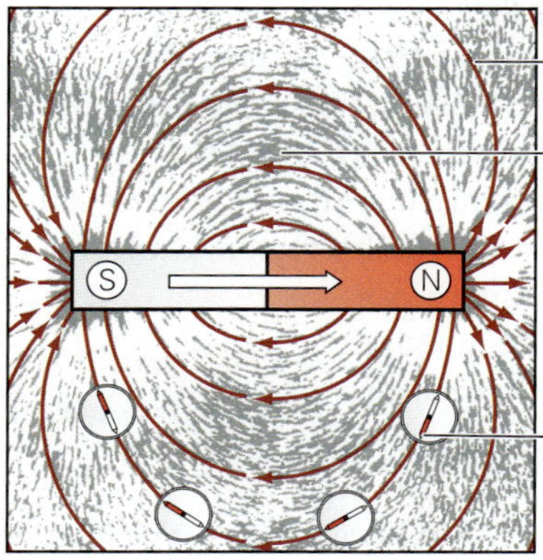

(a) Magnetic field lines produced by a bar magnet point into the magnet's south pole and out from its north pole.

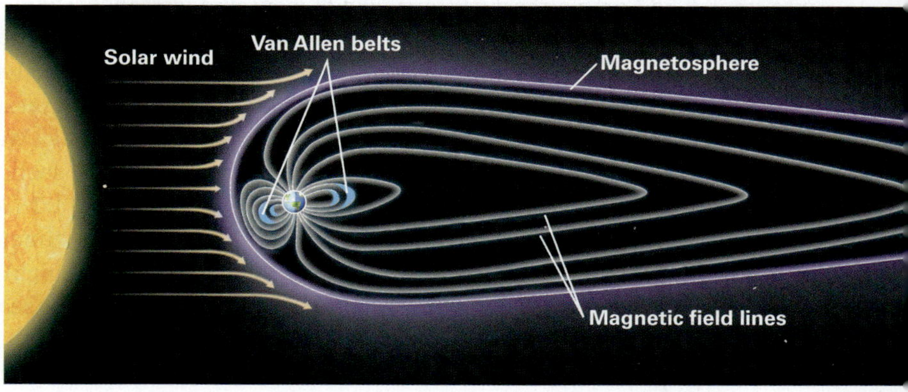

(c) Earth behaves like a magnetic dipole, but the field lines are distorted by the solar wind. The Van Allen radiation belts trap charged particles.

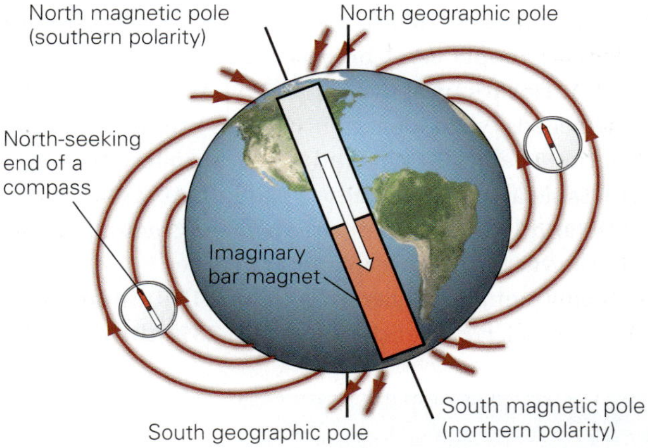

(b) We can represent the Earth's magnetic field by an imaginary bar magnet inside.

(**Fig. 1.12c**). At distances of about 3,000 km and 10,500 km out from the Earth, the *Van Allen radiation belts,* named for American physicist, who first recognized them in 1959, have developed. These belts, which lie well within the magnetosphere, trap solar wind particles as well as cosmic rays (nuclei of atoms emitted from supernova explosions) that were moving so fast they were able to penetrate the weaker outer part of the magnetic field. Some charged particles make it past the Van Allen belts and are channeled along magnetic field lines to the polar regions of Earth.

TAKE-HOME MESSAGE

A collision of a protoplanet with the Earth soon after formation of the Earth may have yielded the material from which the Moon formed. The Earth produces a magnetic field that deflects solar wind.

QUICK QUESTION What is a magnetic pole?

1.6 Introducing the Earth System

So far in this chapter, we've described scientific ideas about how the Universe, and then the Solar System, formed. Now let's focus on our home planet and develop an image of the Earth's overall architecture. To develop this image, pretend that we are explorers from space rocketing toward the Earth for a first visit. As our spacecraft approaches the Earth, we admire its beautiful bluish glow (**Fig. 1.13**). At an elevation of about 10,000 km

FIGURE 1.13 The Earth, as seen from space, has an overall bluish glow. You can distinguish land, sea, clouds, ice, and vegetation.

(6,200 miles) above the surface, we slow the spacecraft and go into orbit. The Earth almost fills our field of view, and we can see that its surface and surroundings consist of several distinct realms or domains, which we begin to explore. We'll see that all these realms interact with one another and exchange materials, over time. Geologists refer to all the realms, and the interactions among them, as the **Earth System**.

The Atmosphere

At an elevation of 10,000 km, our instruments detect a slight increase in the density of gas outside. This change marks the feather edge of the Earth's **atmosphere**, the envelope of gas that surrounds the planet (**Fig. 1.14a**). The weight of overlying atmosphere pushes down on the atmosphere below, so both the density of air and the *air pressure* (the amount of push that the air exerts on material beneath it) increases progressively closer to the surface (**Fig. 1.14b**).

We can specify pressure in units of force per unit area. One such unit, an *atmosphere* (abbreviated atm), has a value of 1.03 kilograms per square centimeter (14.7 pounds per square inch), the average air pressure at sea level. In scientific discussions, another unit, the bar, is commonly used for defining pressure—an atmosphere and a bar are almost the same, for 1 atm = 1.01 bar. Air pressure (and therefore, density) decreases by half for every 5.6 km that you rise above sea level. Thus, at the peak of Mt. Everest, the highest point of the planet (8.85 km above sea level), air pressure is only 0.3 atm. People can't breathe in enough gas to survive for long at elevations above about 5.5 km. Because of the decrease in density with elevation, 99% of atmospheric gas lies at elevations below 50 km, and the atmosphere becomes barely detectable at elevations above 120 km. At elevations of between 700 and 10,000 km, the atmosphere gradually transitions into the vacuum of interplanetary space.

Did you ever wonder...
how thick our atmosphere is?

By analyzing samples of gas from the Earth's atmosphere, we find that overall, it currently consists of 78% nitrogen (N_2)

FIGURE 1.14 Characteristics of the atmosphere that envelops the Earth.

(a) An orbiting astronaut's photograph shows the haze of the atmosphere fading up into the blackness of space.

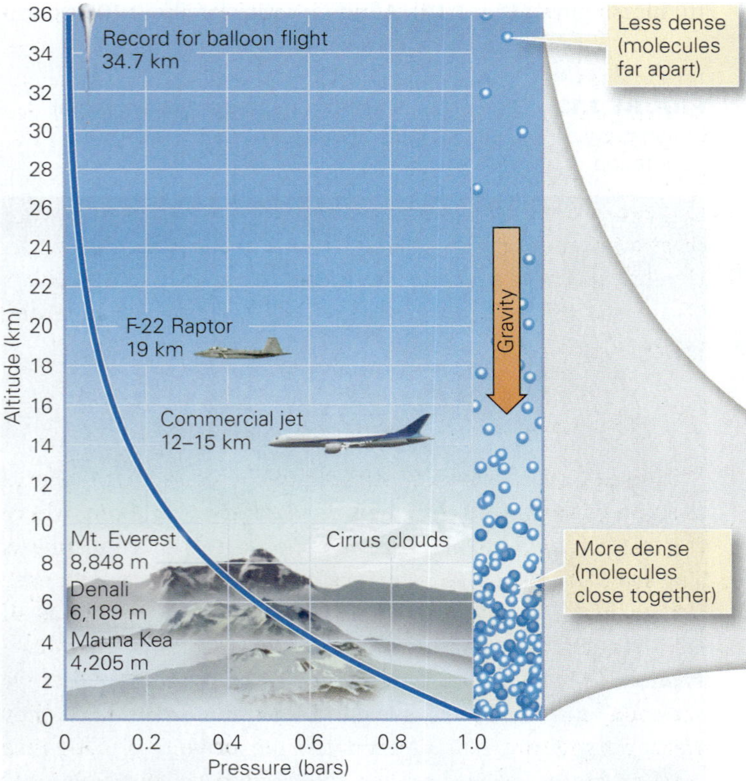

(b) Molecules pack together more tightly at the base of the atmosphere, so atmospheric pressure changes with elevation.

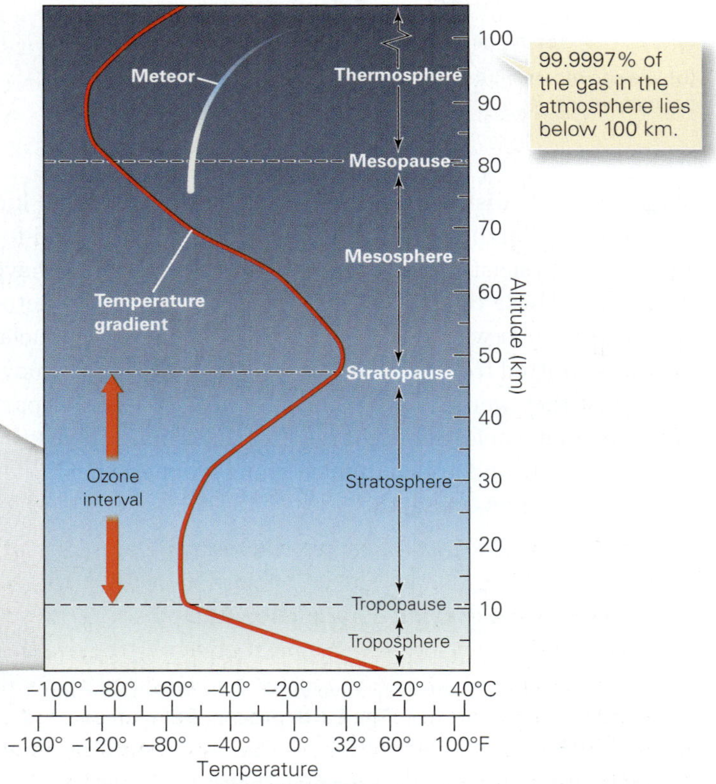

(c) The atmosphere can be divided into several distinct layers. We live in the troposphere.

and 21% oxygen (O_2). The remaining 1% consists of a variety of trace gases, including argon (Ar), carbon dioxide (CO_2), neon (Ne), methane (CH_4), ozone (O_3), carbon monoxide (CO), and sulfur dioxide (SO_2). We will see, in Chapter 19, that the composition of the atmosphere has evolved over time. Early in the Earth's history, the atmosphere consisted mostly of CO_2 and H_2O. Breathable quantities of O_2 have only existed since about the dawn of the Phanerozoic.

The temperature of the Earth's atmosphere changes with increasing distance from the Earth's surface, in some intervals getting hotter with increasing elevation and in some intervals getting cooler. Boundaries at which temperature stops decreasing and starts increasing, or vice versa, divide the atmosphere into layers. The layers of the atmosphere are named, in sequence from base to top: the *troposphere*, the *stratosphere*, the *mesosphere*, and the *thermosphere* (**Fig. 1.14c**). The term *exosphere* applies to the transition from air to space above 700 km. Boundaries between layers are named for the underlying layer. For example, the boundary between the troposphere and the overlying stratosphere is the *tropopause*.

As we circle the Earth, we can see the swirls and streaks of *clouds*, mists of tiny water droplets or hazes of tiny ice crystals, that float in the air (see Fig. 1.13). The complex distribution and shapes of clouds emphasize that the troposphere is in constant motion, with winds that transport air horizontally, and updrafts and downdrafts that transport air vertically. (Most winds and clouds develop only in the troposphere, where convection takes place rapidly; see Box 1.2 for a discussion of convection.) The distribution of clouds emphasizes that the water content of the atmosphere varies with location and over time. We also see that the atmosphere interacts with the magnetosphere. High-energy particles stream along the Earth's magnetic field lines toward the poles and interact with gases in the atmosphere (**Fig. 1.15a**). This causes the gases to glow, creating spectacular *aurorae*, pulsating sheets of color, in the atmosphere (**Fig. 1.15b**). Taking all of the characteristics of the Earth's atmosphere into account, we see that it is radically different in composition and density than the atmospheres of other terrestrial planets.

The Hydrosphere and Cryosphere

The surface of the Earth beneath the atmosphere presents a variety of different textures and colors. Our first task, as explorers, is to make a map of the planet. What features should go on this map? Starting with the most obvious, we note that water covers about 70% of the surface and land covers about 30% of the surface (see Fig. 1.13). About 98% of the water resides in the *oceans*, regions where the water contains about 3.5% dissolved salt and has an average depth of about 4.5 km. A relatively minor amount of water occurs as "fresh" (less

FIGURE 1.15 The aurorae are an atmospheric phenomenon.

(a) Charged particles flow toward Earth's magnetic poles.

(b) Particles cause gases in the atmosphere to glow, forming colorful aurorae in polar skies.

than 0.05% salt) surface water on land, in the form of lakes, streams, and swamps. A larger amount exists as **groundwater**, water filling tiny holes and cracks underground, down to a depth of several kilometers. Geologists refer to the realm including oceans, surface water on land, and groundwater as the **hydrosphere**.

In polar regions, and at high elevations where temperatures are colder, water occurs in solid form, as ice. Much of this ice has built into thick sheets or streams, called *glaciers*, that last all year and slowly flow in response to gravity. Ice also forms a relatively thin sheet over the surface of the ocean in polar regions, and near-surface groundwater remains frozen all year in the form of permanently frozen ground, or *permafrost*, adjacent to many glacier-covered areas. The frozen portion of the Earth's hydrosphere constitutes the **cryosphere**. The prefix *cryo-* comes from the Greek word *kryo*, meaning very cold or ice-like.

The Geosphere

The solid Earth, from the surface to the center, constitutes the **geosphere**. As we've seen, *land* refers to regions where the surface of the geosphere is exposed to the atmosphere. The *seafloor* refers to the surface portion submerged beneath the oceans. Today's Earth includes several large areas of land—called continents—and thousands of smaller ones—islands. As we'll see later in the book, the continents that we see today have not always existed. At times in the past, they were merged into vast supercontinents, or were divided into smaller pieces. Also, because sea level rises and falls over Earth history, some regions of what is currently dry land were once submerged by shallow seas, and some regions that are now submerged were once dry. Similarly, since glaciers grow and shrink over Earth history, some areas of dry land today were covered by glaciers in the past.

The surface of the geosphere is not perfectly smooth (**Fig. 1.16**). From our spacecraft in orbit, we can see that the land surface elevation ranges from a few hundred meters below sea level (the shore of the Dead Sea) to 8.85 km above sea level (the peak of Mt. Everest). We call this variation **topography**. Our instruments also allow us to study **bathymetry**, variations in the depth of the seafloor. We find, along the edges of many continents, *continental shelves*, regions where water depths are generally less than 200 m. The seafloor descends from these down to broad flat regions, called *abyssal plains*, at depths of between 4 and 5 km. Elongate submarine mountain belts, known as *mid-ocean ridges*, where water decreases to less than 2.5 km deep, serve to separate abyssal plains from one another. Here and there, individual submarine mountains, *seamounts*, protrude from the seafloor, and some of these rise above the surface as *sea islands*. The deepest parts of the oceans are in elongate troughs, called *trenches*. In the deepest trenches, water can be as deep as 11 km. Note that the total vertical distance between the deepest point in the sea and the highest point on land is 19.8 km, only about 0.3% of the Earth's radius (6,371 km).

What does the surface of the geosphere consist of? The exploratory probes that we send down to the surface make a preliminary analysis and find the following **Earth materials**:

- *Organic chemicals:* A carbon-containing compound that either occurs in living organisms or has characteristics that resemble compounds in living organisms is an *organic chemical*.
- *Minerals:* A solid, natural substance in which atoms are arranged in an orderly pattern is a **mineral**. A coherent sample of a mineral that grew to its present shape is a **crystal**. Most minerals are *inorganic* chemicals.
- *Glasses:* A solid in which atoms are not arranged in an orderly pattern is called *glass*.
- *Rocks:* An aggregate of mineral crystals or grains, or a mass of natural glass, is called a **rock**. As we'll discuss in Interlude A and succeeding chapters, a rock belongs to one of three main classes: (1) *igneous rocks* develop when hot molten (liquid) rock cools and freezes solid; (2) *sedimentary rocks* form from fragments that break off preexisting rock and become cemented together, or from minerals that precipitate out of a water solution; and (3) *metamorphic rocks* form when preexisting rocks undergo change in response to heat and pressure, to develop new minerals and/or arrangements of minerals.

FIGURE 1.16 This map shows the Earth's topography (variations in land elevation) and bathymetry (variations of ocean-floor depth), and shows where glaciers cover the land (white areas).

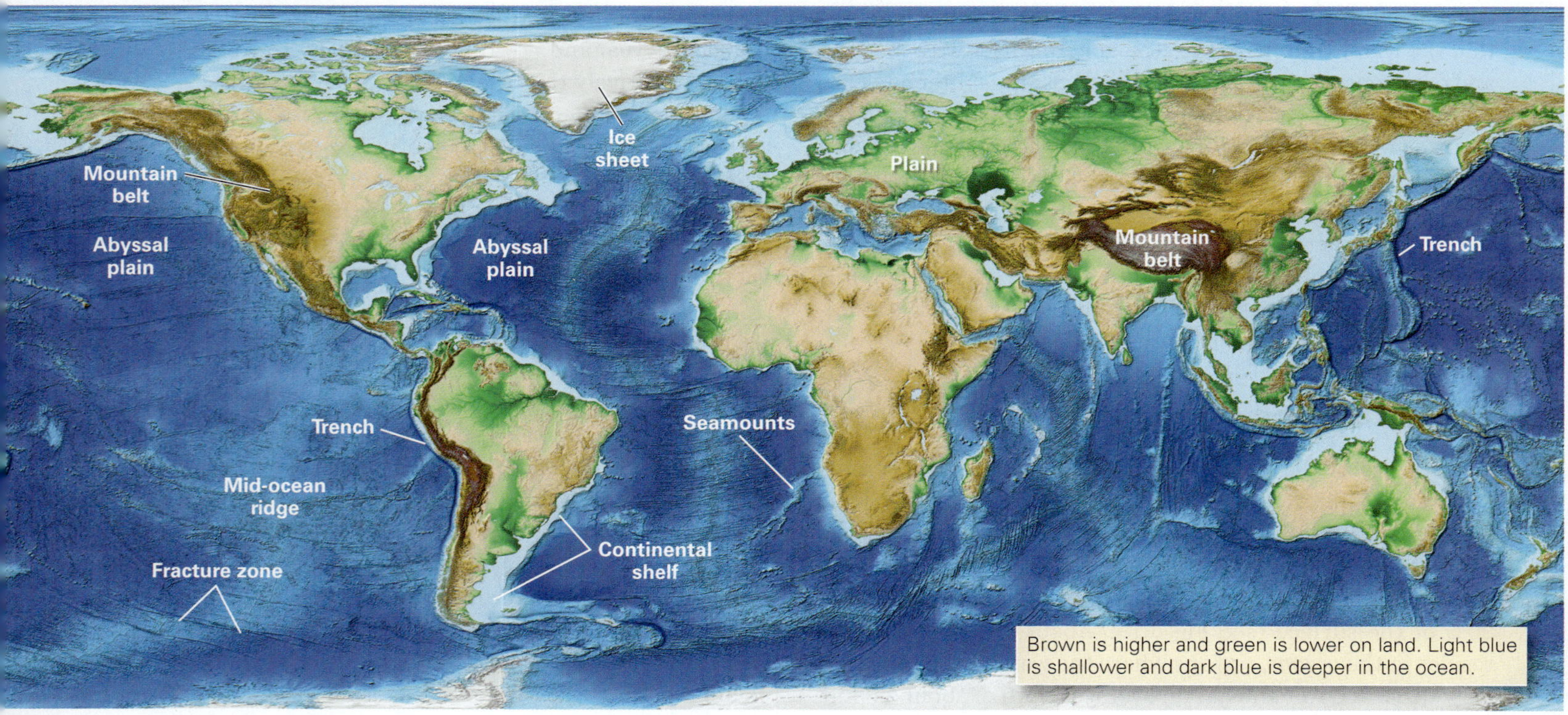

BOX 1.3 Consider This...

Journey to the Center of the Earth?

Since ancient times, people have speculated about what's inside our planet. What is the source of incandescent lavas that spew from volcanoes, of precious gems and metals found in mines, of sparkling mineral waters that bubble from springs, and of the mysterious forces that shake the ground and topple buildings? The cultures of ancient Greece and Rome considered the subsurface to be the underworld, Hades, home of the dead, a region of fire and sulfurous fumes. Perhaps this image was inspired by the molten rock and smoke emitted by the volcanoes of the Mediterranean region. In the 18th and 19th centuries, European writers thought the Earth's interior resembled a sponge containing open caverns variously filled with molten rock, water, or air. In fact, in the popular 1864 novel *Journey to the Center of the Earth*, by the French author Jules Verne, three explorers hike through interconnected caverns down to the Earth's center. Such a journey is impossible, as there are no caverns below a depth of several kilometers. Even with modern technology, we can't dig or drill down very far. Indeed, the deepest mine penetrates only about 3.5 km beneath the surface of South Africa, and the deepest drill hole probes only 12 km below the surface of northern Russia. Compared with the 6,371 km radius of the Earth, this hole goes less than 0.2% of the way to the center and is nothing more than a pinprick.

portions of the Earth's interior soft enough to flow, allowing pieces of its relatively rigid shell to move relative to one another, and these movements, in turn, lead to mountain building.

External energy is energy reaching the Earth's surface from outside. Most of this energy comes in the form of radiation from the Sun. This energy warms the atmosphere, oceans, and the land surface. External energy, acting together with gravity, causes convection of the atmosphere and hydrosphere, resulting in winds, waves, and currents, and therefore, it also causes transfer of water from the oceans to the land surface, feeding rivers and glaciers that flow over and grind away or **erode** the land surface.

TAKE-HOME MESSAGE

Geologists use the term *Earth System* to refer to the variety of interacting realms in, on, and around the planet. Thirty percent of the surface is land, while 70% is sea. Both the land surface and seafloor display notable variations in elevation and depth, respectively, but the difference between the highest point and the lowest is only about 0.3% of Earth's radius. The solid Earth consists of many materials, the most common of which is silicate rock.

QUICK QUESTION How do we distinguish among types of silicate rocks?

1.7 Looking Inward—Discovering the Earth's Interior

Having mapped the Earth's surface from our spacecraft, we've now got a clear understanding that the planet is a variable and dynamic place. Our final task is to determine what's inside. We begin by describing basic characterizations of the Earth's interior that can be determined from outside the planet (**Box 1.3**). Then, we refine the image by placing instruments on the Earth's surface.

Clues from Examining the Earth's Shape and Density

By making measurements of the gravitational pull of the Earth, and by determining the Earth's volume, we can calculate its *average density*. A quick comparison shows that this average density exceeds the density of common rocks found on the Earth's surface, so the interior of the Earth must contain denser material than its outermost layer, without open space. In fact, we find that the overall mass of the Earth is so great that the planet must contain a large amount of metal. And since the Earth is close to being a sphere, the metal must be concentrated near the center—otherwise, centrifugal force due to the spin of the Earth on its axis would pull the equator

FIGURE 1.20 An early image of Earth's internal layers.

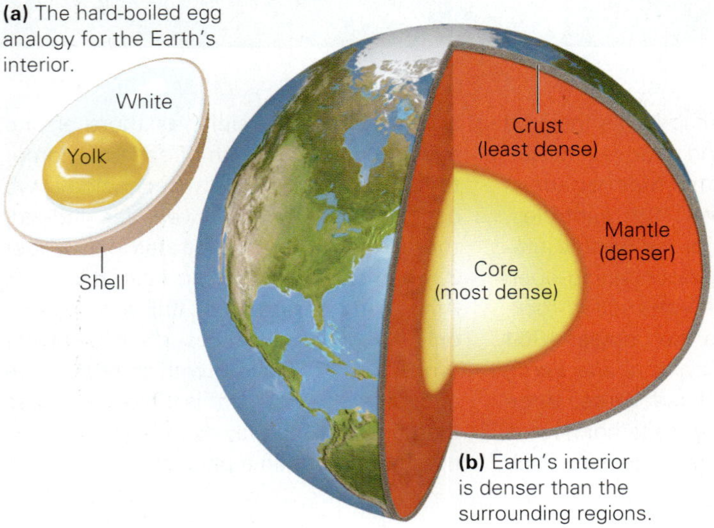

(a) The hard-boiled egg analogy for the Earth's interior.

(b) Earth's interior is denser than the surrounding regions.

outward and the planet would become somewhat disk-shaped. (To picture why, consider that when you swing a hammer, your hand feels more force if you hold the end of the light wooden shaft, rather than the heavy metal head.) Further, the interior must be mostly solid, because if it weren't, the land surface would rise and fall due to tidal forces much more than it does, as the Moon orbits the planet.

With these key observations on hand, we conclude that the Earth resembles a hard-boiled egg, in that it has three principal layers: a not-so-dense *crust* (like an eggshell), a denser solid *mantle* in the middle (like an egg white), and a very dense *core* (the egg yolk) (**Fig. 1.20**). We can determine the composition of the crust by sampling. But to determine the character of materials make up the mantle and core, we must study the composition of meteorites (**Box 1.4**), for they provide samples from the differentiated interiors of planetesimals, or protoplanets, that likely resemble the interior of the Earth.

SEE FOR YOURSELF...

METEOR CRATER

Latitude
35°1'37.18" N

Longitude
111°1'20.17" W

Zoom to an elevation of 5.7 km (19,000 feet) and look straight down.

You are seeing Meteor Crater, Arizona, which is 1.2 km across and 180 m deep. It formed in response to the impact of a 40- to 50-m-wide asteroid fragment, less than 50,000 years ago.

Clues from Studying Seismic Waves: Refining the Image

Clearly, many questions remain after our initial survey. Exactly how thick are the Earth's internal layers, and are the boundaries between layers sharp or gradational? We can gain insight into these questions by studying *seismic waves*, vibrations or shocks that pass through the Earth from a location where rock suddenly breaks and slips along a fracture called a *fault* (**Fig. 1.21a**). You can simulate this process of generating vibrations, at a small scale, when you break a stick and feel the snap in your hands (**Fig. 1.21b**). Where seismic waves reach the surface of the Earth, and cause it to shake, people feel an **earthquake**. Study of seismic waves reveals the character of the Earth's insides much as ultrasound measurements help doctors study the insides of a patient, for seismic waves travel at different velocities through different materials, and bend or reflect when they reach boundaries between different materials. To study these waves, we need to place *seismometers*, instruments that can detect seismic waves, at various sites on the Earth's surface. Using methods described in Chapter 8 and Interlude D, records made by our

FIGURE 1.21 Faulting and the generation of seismic waves.

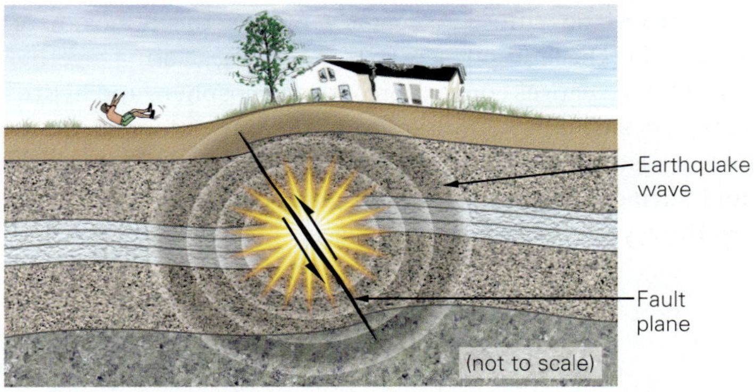

(a) When the rock inside the Earth suddenly breaks and slips on a a fault, it generates seismic waves that pass through the Earth. When they reach the surface, we feel an earthquake.

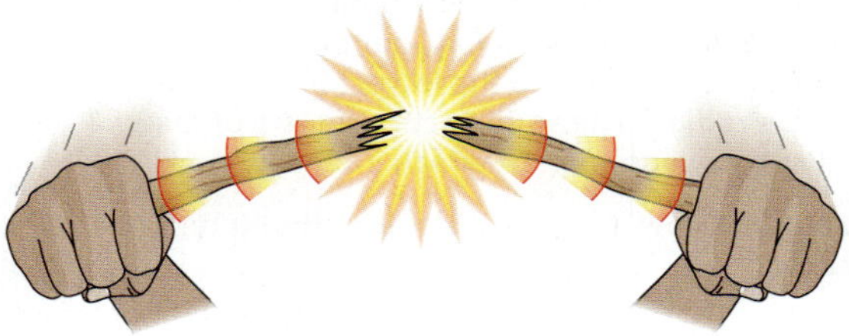

(b) You are feeling seismic waves, in effect, when you snap a stick. The vibrations pass through the stick to your hands.

BOX 1.4 Consider This...

Meteors and Meteorites

Collisions of objects from space with the Earth happened frequently during the early history of the Earth. But even today collisions with space objects continue, and an average of over 1,000 tons of material (rock, metal, dust, and ice) fall to Earth every year. Occasionally, a much bigger object strikes the Earth. The vast majority of this colliding material consists of fragments derived from comets and asteroids sent careening into the path of the Earth after billiard-ball-like collisions with each other out in space, or because the gravitational pull of a passing planet deflected their orbit. Some of the material, however, consists of chips of the Moon or Mars, ejected into space when large objects collided with those bodies.

Astronomers refer to any object from space that enters the Earth's atmosphere as a *meteoroid*. Meteoroids move at speeds of 20 to 75 km/s (over 45,000 mph), so fast that when they reach an altitude of about 150 km, friction with the atmosphere causes them to heat up and vaporize, leaving a streak of bright, glowing gas. The glowing streak, an atmospheric phenomenon, is a **meteor**, also known colloquially, though incorrectly, as a falling star (Fig. Bx1.4a). Most visible meteors completely vaporize by an altitude of about 30 km. But dust-sized ones may slow down sufficiently to float to Earth, and larger ones (fist-sized or bigger) can survive the heat of entry to reach the surface of the planet. In some cases, meteoroids explode in brilliant fireballs.

Objects that strike the Earth are called **meteorites**. Although almost all meteorites are small and have not caused notable damage on Earth during human history, a very few have smashed through houses, dented cars, and bruised people. During the longer term of Earth history, however, some catastrophic collisions have left huge craters (Fig. Bx1.4b).

Researchers recognize three basic classes of meteorites: *iron* (made of iron-nickel alloy), *stony* (made of rock), and *stony iron* (rock embedded in a matrix of metal). Of all known meteorites, about 93% are stony and 6% are iron (Fig. Bx1.4c). Researchers have concluded that most stony meteorites and all iron meteorites are fragments of planetesimals that had differentiated into a metallic core and a rocky mantle before being shattered by collisions with other planetesimals. Most of these appear to be about 4.54 Ga. A few stony meteorites were derived from planetesimals that never differentiated into core and mantle sectors, and these are as old as 4.57 Ga—the oldest Solar System materials ever measured.

FIGURE Bx1.4 Meteors and meteorites.

(a) A shower of meteors over Hong Kong in 2001.

(c) Examples of stony meteorites (left) and iron meteorites (right).

The meteorite that formed the crater was 50 m across.

(b) The Barringer meteor crater in Arizona. It formed about 50,000 years ago and is 1.1 km in diameter.

seismometers allow us to pinpoint the depth of boundaries between layers, to identify sublayers within the major layers, and to determine whether layers are solid or liquid.

Pressure and Temperature Inside the Earth

The mass of the overlying rock layer increases with depth, so the pressure within the Earth also increases progressively with depth. In solid rock, the pressure at a depth of 1 km is about 300 atm, so a mine tunnel requires an appropriate support structure to prevent collapse. At the Earth's center, pressure probably reaches about 3,600,000 atm.

> ***Did you ever wonder...***
> how hot it gets at the center of this planet?

Temperature also increases with depth in the Earth. Even on a cool winter's day, if we could descend down into the crust, we would find that at almost 4 km below the surface, the depth of the deepest gold mine on Earth, the surrounding rock is 60°C (140°F). (At these temperatures, humans can survive no more than 10 minutes, without air-conditioning.) We refer to the rate of change in temperature with depth as the **geothermal gradient**. In the upper part of the crust, the geothermal gradient averages between 20° and 30°C per kilometer. At greater depths, the gradient decreases to 10°C/km or less. Thus, 35 km below the surface of a continent, the temperature reaches 400°C to 700°C, and the mantle-core boundary is about 3,500°C. No one will ever be able to measure the temperature at the Earth's center, but calculations suggest it may exceed 4,700°C, not much cooler than the Sun's surface temperature of 5,500°C.

> ***TAKE-HOME MESSAGE***
>
> Measurements of the Earth's mass, shape, and tidal response led to the conclusion that the Earth has three principal internal layers—the crust, the mantle, and the core. Study of earthquake waves passing through the interior has refined this image.
>
> **QUICK QUESTION** What can we learn about the Earth by studying meteorites?

1.8 Basic Characteristics of the Earth's Layers

At this point, we leave our fantasy spacecraft and describe the basic properties of Earth's interior, based on studies that geologists have undertaken during the past 150 years of research. We will be using this information in the next chapter's discussion of plate tectonics, geology's grand unifying theory. Later, after we have thoroughly discussed earthquakes and seismic waves, we will add further details to our image of the interior in Interlude D.

The Crust

When you look at the surface of the Earth, you are standing on top of its outermost layer, the **crust** (**Fig. 1.22a**). This layer of the Earth overlaps with the biosphere, and is the source of all of humanity's resources. How thick is this all-important crust? Or, in other words, what is the depth to the crust-mantle boundary? An answer came from the studies of Andrija Mohorovičić (1857-1936), a researcher working in Zagreb, Croatia. In 1909, he discovered that the velocity of earthquake waves suddenly increased at a depth of tens of kilometers beneath the Earth's surface, and he suggested that this increase was caused by an abrupt change in the properties of rock (see Interlude D for further details). Later studies showed that this change can be found most everywhere around our planet, though it occurs at different depths in different locations. Specifically, the area of change is deeper beneath continents than beneath oceans. Geologists now consider the change to define the base of the crust, and they refer to it as the **Moho** in Mohorovičić's honor. The relatively shallow depth of the Moho (7 to 70 km, depending on location), as compared to the radius of the Earth (6,371 km), emphasizes that the crust is very thin indeed. In fact, the crust is only 0.1% to 1.0% of the Earth's radius, so if the Earth were the size of a balloon, the crust would be about the thickness of the balloon's skin.

The crust is not simply cooled mantle, like the skin on chocolate pudding, but rather consists of a variety of rocks that differ in composition (chemical makeup) from mantle rock. Oceanic crust, which underlies the seafloor, is only 7 to 10 km thick—at highway speeds (100 km per hour), you could drive a distance equal to the thickness of the oceanic crust in about 5 minutes. The top of the oceanic crust is composed of a blanket, generally less than 1 km thick, of sediment—clay and tiny shells that settled out of the sea like snow. Beneath this blanket, the oceanic crust consists of a layer of basalt and, below that, a layer of gabbro. Continental crust, in contrast, is generally 35 to 40 km thick (four to five times the thickness of oceanic crust), but its thickness varies significantly. Where continental crust has been stretched, it can be as thin as 25 km, and where it has been squeezed horizontally, it can be up to 70 km thick. Continental crust contains a great variety of rock types, ranging from mafic to felsic in composition. On average, upper continental crust has a felsic (granite-like) to intermediate composition—so continental crust overall is less dense than oceanic crust.

FIGURE 1.22 A modern view of Earth's interior layers.

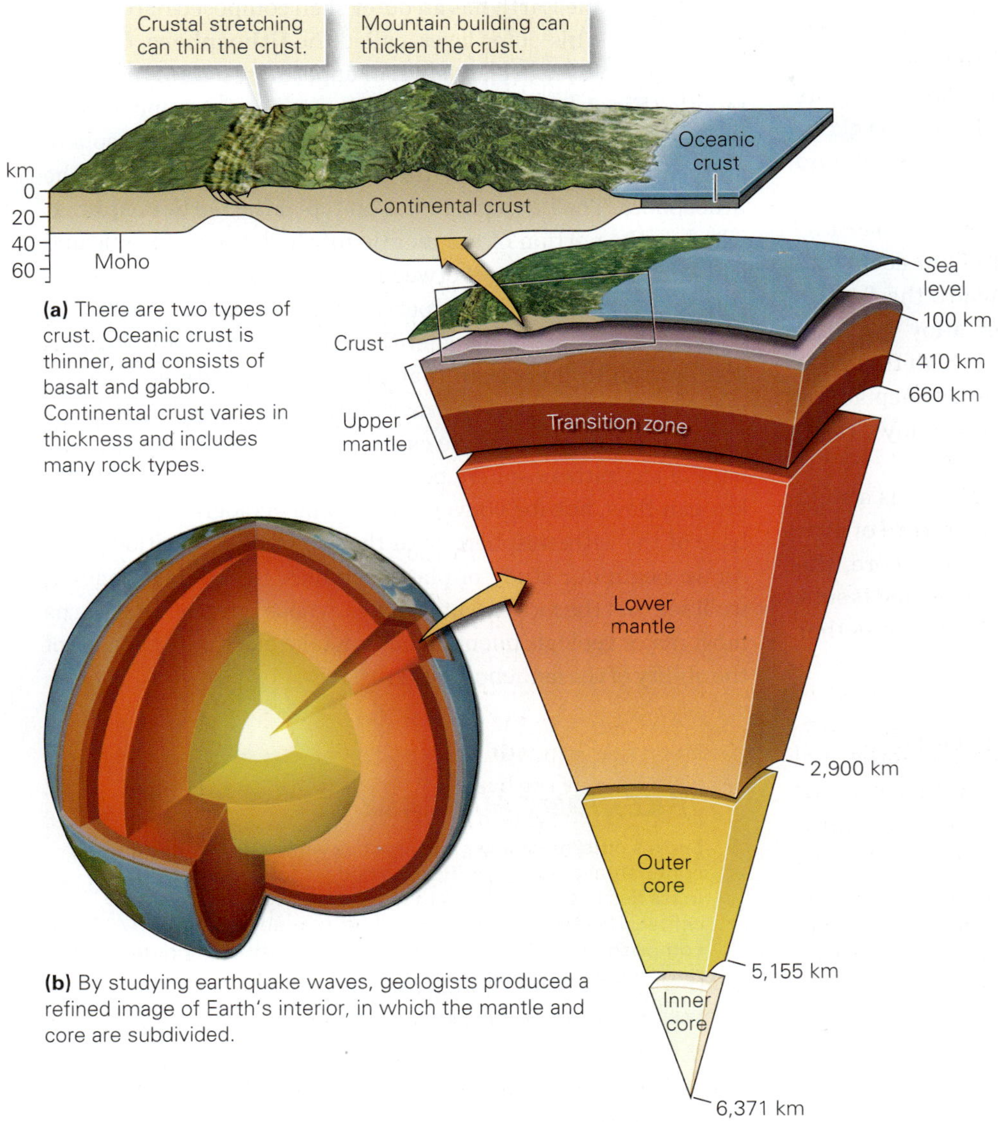

(a) There are two types of crust. Oceanic crust is thinner, and consists of basalt and gabbro. Continental crust varies in thickness and includes many rock types.

(b) By studying earthquake waves, geologists produced a refined image of Earth's interior, in which the mantle and core are subdivided.

TABLE 1.1 Average Relative Abundances of Elements in the Earth's Crust

Element	Symbol	% by weight	% by volume	% by atoms
Oxygen	O	46.3	93.8	60.5
Silicon	Si	28.0	0.9	20.5
Aluminum	Al	8.1	0.8	6.2
Iron	Fe	5.5	0.5	1.9
Calcium	Ca	3.4	1.0	1.9
Magnesium	Mg	2.8	0.3	1.4
Sodium	Na	2.4	1.2	2.5
Potassium	K	2.3	1.5	1.8
All others	—	1.2	>0.1	3.3

Notably, oxygen is the most abundant element in the crust, with silicon a close runner-up, because most rocks contain abundant silica (**Table 1.1**).

The Mantle

The *mantle* of the Earth forms a 2,885-km-thick layer surrounding the core. In terms of volume, it is the largest part of the Earth (**Fig. 1.22b**). In contrast to the crust, the mantle consists entirely of an ultramafic (dark and dense) rock called peridotite. This means that peridotite, though rare at the Earth's surface, is actually the most abundant rock in our planet! Seismic-wave velocities change at a depth of 410 km and again at a depth of 660 km in the mantle. Based on this observation, geologists divide the mantle into two sublayers: the *upper mantle,* down to a depth of 660 km, and the *lower mantle,* from 660 km down to 2,900 km. The lower part of the upper mantle, the interval between 410 and 660 km deep, is called the *transition zone*—in this zone, seismic-wave velocities increase in a series of steps, due to abrupt changes in the structure of minerals making up mantle rock.

Almost all of the mantle is solid rock—melt, which occurs in films or bubbles between grains, accounts for just a few percent of its volume, and generally occurs at a depth of 100 to 200 km, only beneath the ocean floor. But even though it's mostly solid, mantle rock below a depth of 100 to 150 km is so hot that it's soft enough to flow, because rock, like wax, softens as it warms up. This flow, however, takes place extremely slowly, at a rate of less than 15 cm a year. So "soft" here does not mean liquid—it simply means that over long periods of time mantle rock can change shape without breaking, a phenomenon known as *plasticity.* Because of its softness, the mantle can undergo convection very slowly. During this convection, warmer mantle rock rises, while cooler mantle rock sinks.

The Core

Early calculations suggested that the core had the same density as gold, so for many years people held the fanciful hope that vast riches lay at the heart of our planet. Alas, geologists eventually concluded that the core consists of a far less glamorous material—iron alloy (iron mixed with 4% nickel, and up to 10% oxygen and/or silicon or sulfur). Studies of seismic waves led geologists to divide the core into two parts (see Fig. 1.22), the *outer core* (between 2,900 and 5,155 km deep) and the *inner core* (from a depth of 5,155 km down to the Earth's center at 6,371 km). The outer core's iron alloy is liquid because the temperature in the outer core is so high that even the great pressures squeezing the region cannot keep atoms locked into a solid framework. The liquid iron alloy's rapid flow generates Earth's magnetic field.

The inner core, with a radius of about 1,220 km, is a solid iron alloy (see Fig. 1.18) that may reach a temperature of over 4,700°C. Even though it is hotter than the outer core, the inner core's alloy remains solid because it is deeper and is subjected to greater pressure, which keeps atoms locked together tightly in solid crystals.

The Lithosphere and the Asthenosphere

So far, we have identified three major layers (crust, mantle, and core) inside the Earth whose compositions differ from each other. Seismic waves travel at different velocities through these layers. An alternative way of thinking about Earth layers comes from studying the degree to which the material of each layer can flow. In this context, we distinguish between *rigid* materials, which can bend or break but cannot flow, and *plastic* materials, which are relatively soft and can flow without breaking.

The outer 100 to 150 km of the Earth is relatively rigid. In other words, the Earth has an outer shell composed of rock that cannot flow. This outer layer, called the **lithosphere**, consists of the crust plus the relatively cool uppermost part of the upper mantle (**Fig. 1.23**). We refer to this portion of the upper mantle as the *lithospheric mantle*. Note that the terms lithosphere and crust are not the same—the crust is only the top layer of the lithosphere. The lithosphere lies on top of the **asthenosphere**, which is the portion of the mantle in which rock is plastic and can flow. The boundary between the lithosphere and asthenosphere occurs where the temperature reaches about 1,280°C, for when temperatures are higher than this, mantle rock becomes soft enough to flow. Note that the asthenosphere is made up entirely of a section of the mantle that generally lies below a depth of 100 to 150 km. We can't assign a specific depth to the base of the asthenosphere because all of the mantle below 150 km can flow, but for convenience, some geologists equate the base of the asthenosphere with the top of the transition zone. As we begin our study of plate tectonics, in the next chapter, we'll see that the movement of plates, whose interaction governs most major geologic phenomena, can take place only because of the ability of the asthenosphere to flow.

TAKE-HOME MESSAGE

Earth's outermost layer, the crust, is very thin; oceanic and continental crust differ in composition. Most of Earth's mass lies in the mantle. A metallic core lies at this planet's center. The crust and the outermost mantle together comprise the rigid lithosphere, while the asthenosphere is the inner mantle whose material can flow.

QUICK QUESTION Is the entire core solid?

FIGURE 1.23 A block diagram of the lithosphere emphasizing the difference between continental and oceanic lithosphere.

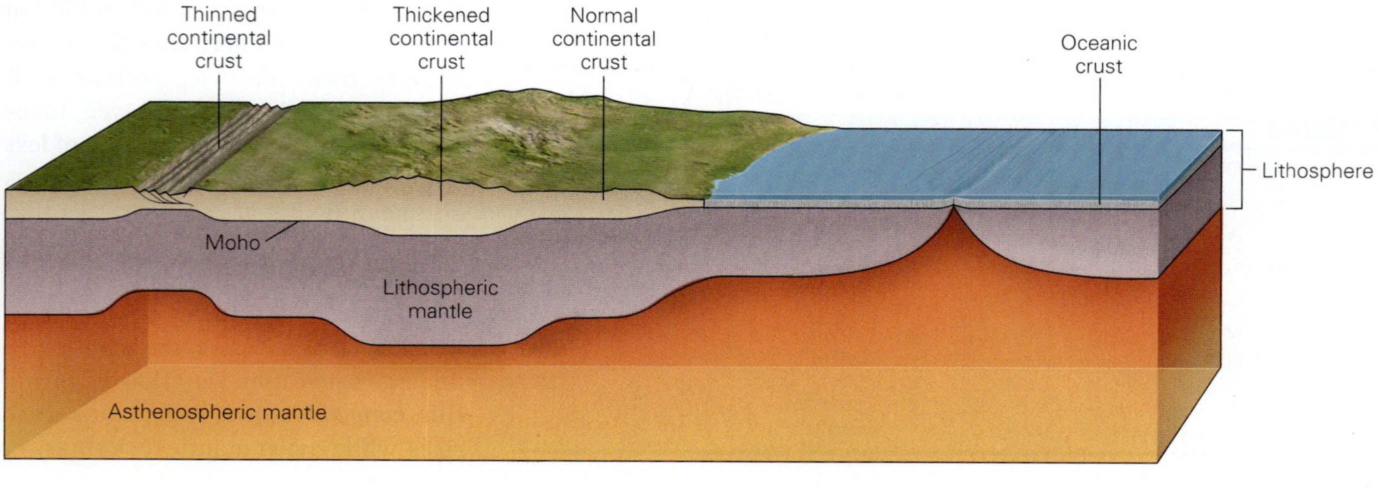

Another View Astronomers are discovering many planetary systems around other nearby stars in our galaxy. This is an artist's conception of what the nearest planetary system, Epsilon Eridani, might look like. This example has an asteroid belt and comets. Their tails follow the currents of the stellar wind.

Chapter 1 Review

Chapter Summary

- The geocentric model of the Universe placed the Earth at the center of the Universe. The heliocentric model placed the Sun at the center.
- The Earth is one of eight planets orbiting the Sun. The Solar System lies on the outer edge of the Milky Way galaxy. The Universe contains hundreds of billions of galaxies.
- Most astronomers agree that the Universe has been expanding since it began at the Big Bang, about 13.8 billion years ago.
- The first atoms (hydrogen and helium) of the Universe developed within minutes of the Big Bang. These atoms formed vast gas clouds, called nebulae.
- Only very small atoms formed during Big Bang nucleosynthesis. The Earth, and the life forms on it, contain elements that could have been produced only during the life cycle of stars. Thus, we are all made of stardust.
- Gravity causes clumps of gas in the nebulae to coalesce into flattened disks with bulbous centers. As the central ball of this accretionary disk collapsed inward, it became a warm protostar. Eventually, the ball became so hot and dense that fusion reactions began and it became a true star.
- Planets developed from the rings of gas and dust surrounding newborn stars. These condensed into planetesimals that then clumped together to form protoplanets, and finally true planets.
- The Moon formed from debris ejected when a protoplanet collided with the Earth in the very young Solar System.
- When a protoplanet grows large enough, it eventually becomes warm enough inside to differentiate into a core and mantle, and to assumes a near-spherical shape.
- The Earth has a magnetic field that shields it from solar wind and cosmic rays.
- A layer of gas surrounds the Earth. This atmosphere, which consists of 78% N_2, 21% O_2, and 1% other gases, can be subdivided into layers. Air pressure decreases with elevation.
- The surface of the Earth can be divided into land (30%) and ocean (70%).
- Earth materials include organic chemicals, minerals, glasses, rocks, metals, melts, and volatiles. Most rocks on Earth contain silica (SiO_2). We distinguish among various major rock types based on the proportion of silica.
- The Earth's interior can be divided into three distinct layers: the very thin crust, the rocky mantle, and the metallic core.
- Pressure and temperature both increase with depth in the Earth. The rate at which temperature increases as depth increases is the geothermal gradient.
- The crust is a thin skin that varies in thickness from 7–10 km (beneath the oceans) to 25–70 km (beneath the continents). Oceanic crust is mafic in composition, whereas average upper continental crust is felsic to intermediate. The mantle is composed of ultramafic rock. The core is made of iron alloy.
- Studies of earthquake waves reveal that the mantle can be subdivided into an upper mantle (including the transition zone) and a lower mantle. The core can be subdivided into the liquid outer core and a solid inner core. Circulation of the outer core produces the Earth's magnetic field.
- The crust plus the upper part of the mantle constitute the lithosphere, a rigid shell. The lithosphere lies over the asthenosphere, mantle that can flow.

Guide Terms

accretionary disk (p. 18)
asteroid (p. 14)
asthenosphere (p. 38)
atmosphere (p. 28)
bathymetry (p. 30)
Big Bang theory (p. 16)
biosphere (p. 31)
comet (p. 14)
cosmology (p. 11)
crust (p. 36)
cryosphere (p. 29)
crystal (p. 30)
differentiation (p. 23)
dipole (p. 26)
earthquake (p. 34)
Earth materials (p. 30)
Earth System (p. 28)
ecliptic (p. 13)
erosion (p. 33)
expanding Universe theory (p. 15)
external energy (p. 33)
galaxy (p. 13)
geocentric model (p. 12)
geosphere (p. 29)
geothermal gradient (p. 36)
giant planet (p. 13)
groundwater (p. 29)
heat (p. 20)
heliocentric model (p. 12)
hydrosphere (p. 29)
internal energy (p. 32)
lithosphere (p. 38)
magnetic field (p. 26)
magnetosphere (p. 26)
melt (p. 31)

GEOTOURS *THIS CHAPTER'S GEOTOUR EXERCISE (A) FEATURES:*

- Nebular Supernova
- Spiral Galaxy
- Meteorite Impacts

metal (p. 31)
meteor (p. 35)
meteorite (p 35)
mineral (p. 30)
Moho (p. 36)
moon (p. 14)
nebula (p. 18)
nebular theory (p. 18)
nuclear fission (p. 23)
nuclear fusion (p. 18)
planet (p. 13)
planetesimal (p. 22)
protoplanet (p. 22)
protostar (p. 18)
refractory materials (p. 22)
rock (p. 30)
silicate rock (p. 31)
Solar System (p. 13)
star (p. 12)
stellar nucleosynthesis (p. 19)
stellar wind (p. 19)
supernova (p. 19)
temperature (p. 20)
terrestrial planet (p. 13)
topography (p. 30)
Universe (p. 12)
vacuum (p. 18)
volatile materials (pp. 21, 31)

Review Questions

1. Contrast the geocentric and heliocentric Universe concepts.
2. What is the ecliptic, and why are the orbits of the planets within the ecliptic? Why is Pluto no longer considered to be a planet?
3. Explain the expanding Universe theory.
4. What is the Big Bang, and when did it occur?
5. Describe the steps in the formation of the Solar System according to the nebular theory.
6. Why isn't the Earth homogeneous?
7. Describe how the Moon was formed.
8. Why is the Earth round?
9. What is the Earth's magnetic field? Draw a representation of the field on a piece of paper. What causes aurorae?
10. What is the Earth's atmosphere composed of? Why would you die of suffocation if you were to eject from a fighter plane at an elevation of 12 km without an oxygen tank?
11. What is the proportion of land area to sea area on Earth?
12. Describe the major categories of materials constituting the Earth. On what basis do geologists distinguish among different kinds of silicate rocks?
13. What are the principal layers of the Earth?
14. How do temperature and pressure change with increasing depth in the Earth?
15. What are meteorites, and how does the study of them provide insight into the character of the Earth's interior?
16. What is the Moho? Describe the differences between continental crust and oceanic crust.
17. What is the mantle composed of? Is there any melt in it?
18. What is the core composed of? How do the inner and outer cores differ? Which produces the magnetic field?
19. What is the difference between the lithosphere and asthenosphere? At what depth does the lithosphere-asthenosphere boundary occur? Is this above or below the Moho? Is the asthenosphere entirely liquid?

On Further Thought

20. Are all stars that we see today considered to be first-generation stars? What is the evidence for your answer?
21. Why are the giant planets, which contain abundant gas and ice, farther from the Sun?
22. Recent observations suggest that the Moon has a very small, solid core that is less than 3% of its mass. In comparison, Earth's core is about 33% of its mass. Explain why this difference might exist.
23. The Moon has virtually no magnetosphere. Why?
24. Popular media sometimes imply that the crust floats on a "sea of magma." Is this a correct image of the mantle just below the Moho? Explain your answer.

Online Resources

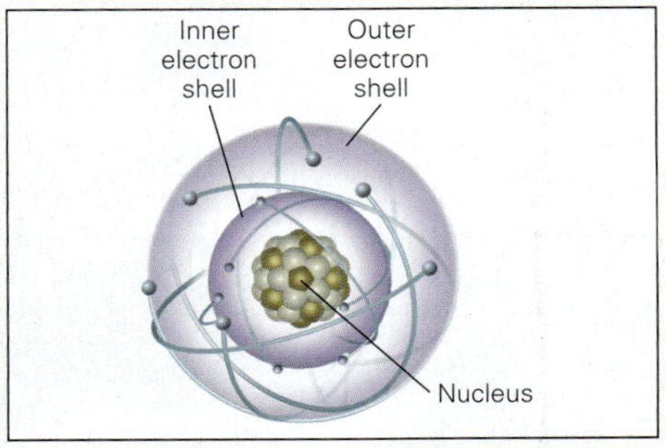

Videos

This chapter features videos showing how our Solar System and the Earth formed. Other topics covered include nebular theory, proto-Earth, the formation of the Moon, and the beginning of the atmosphere.

Assessment

This chapter features visual labeling exercises on topics such as atomic structures and our Solar System.

LEARNING OBJECTIVES

By the end of this chapter, you should understand...

1. Wegener's evidence for continental drift.
2. how the study of paleomagnetism proves that continents move.
3. how seafloor spreading works, and how geologists can prove that it takes place.
4. that the Earth's lithosphere is divided into about 20 plates that move relative to one another.
5. the three kinds of plate boundaries and the basis for recognizing them.
6. how fast plates move, and how we can measure the rate of movement.

▲ Measurements with GPS indicate that Turkey is moving west relative to Asia. Such measurements allow us to "see" plate movements. Each yellow arrow indicates the velocity of the point at the end of the arrow. Red arrows give overall plate-movement direction.

CHAPTER 2

The Way the Earth Works: Plate Tectonics

2.1 Introduction

In September 1930, a group of 15 explorers led by a German meteorologist, Alfred Wegener, set out across the endless snowfields of Greenland to resupply two weather observers stranded at a remote camp. The observers had been planning to spend the long polar night recording wind speeds and temperatures on Greenland's polar plateau. Wegener was well known, not only to researchers studying climate but also to geologists. Some 15 years earlier, he had published a small book, *The Origin of the Continents and Oceans,* in which he had dared to challenge geologists' long-held assumption that the continents had remained fixed in position through all of Earth history. Wegener thought, instead, that the continents had once fit together like pieces of a giant jigsaw puzzle, to make one vast supercontinent. This supercontinent, which he named *Pangaea* (pronounced pan-JEE-ah; Greek for all land), later fragmented into separate continents that "drifted" apart, moving slowly to their present positions (**Fig. 2.1**). This phenomenon came to be known as *continental drift.*

Wegener presented many observations that he believed proved that continental drift had occurred, but he met with strong resistance. At a widely publicized 1926 geology

> *It is only by combing the information furnished by all the earth sciences that we can hope to determine "truth" here.*
>
> ALFRED WEGENER
> (German geophysicist and meteorologist, 1880–1930)

conference in New York City, a crowd of celebrated American professors scoffed, "What force could possibly be great enough to move the immense mass of a continent?" Wegener's writings didn't provide a good answer, so despite all the supporting observations he had provided, most of the meeting's participants rejected continental drift. Four years later, Wegener faced his greatest challenge—survival. Sadly, he lost. On October 30, 1930, he reached the observers and dropped off enough supplies to last the winter. Wegener and one companion set out on the return trip the next day, but they never made it home.

Had Wegener survived to old age, he would have seen his hypothesis become the foundation of a scientific revolution. Today geologists accept Wegener's basic conclusion and take for granted that the map of the Earth constantly changes as continents waltz around this planet's surface, variously combining and breaking apart through geologic time. The revolution began in 1960, when an American geologist, Harry Hess (1906-1969), proposed that as continents drift apart, new ocean floor forms between them by a process that his contemporary, Robert Dietz (1914-1995), named *seafloor spreading.* Hess and others suggested that continents move toward each other when the old ocean floor between bends down along the edge of a continent and sinks back down into the Earth's interior, a process now called *subduction.* By 1968, geologists had developed a fairly complete model encompassing continental drift, seafloor spreading, and subduction. In this model, Earth's lithosphere, its outer, relatively rigid shell, consists of about 20 distinct pieces, or **plates**, that slowly move relative to each other. Because we can confirm this model using many observations, it has gained the status of a theory, now called the *theory of plate tectonics*, or simply **plate tectonics**, from the Greek word *tekton,* which means builder; plate movements "build" regional geologic features. Geologists view plate tectonics as the grand unifying theory of geology, because it can successfully explain a great many geologic phenomena.

In this chapter, we introduce the observations that led Wegener to propose his continental drift hypothesis. Then we look at paleomagnetism, the record of Earth's magnetic field in the past, because it provides a key proof of continental drift. Next, we learn how observations about the seafloor, made by geologists during the mid-20th century, led Harry Hess to propose the concept of seafloor spreading. We conclude by describing the many facets of modern plate tectonics theory.

FIGURE 2.1 Alfred Wegener and his model of continental drift.

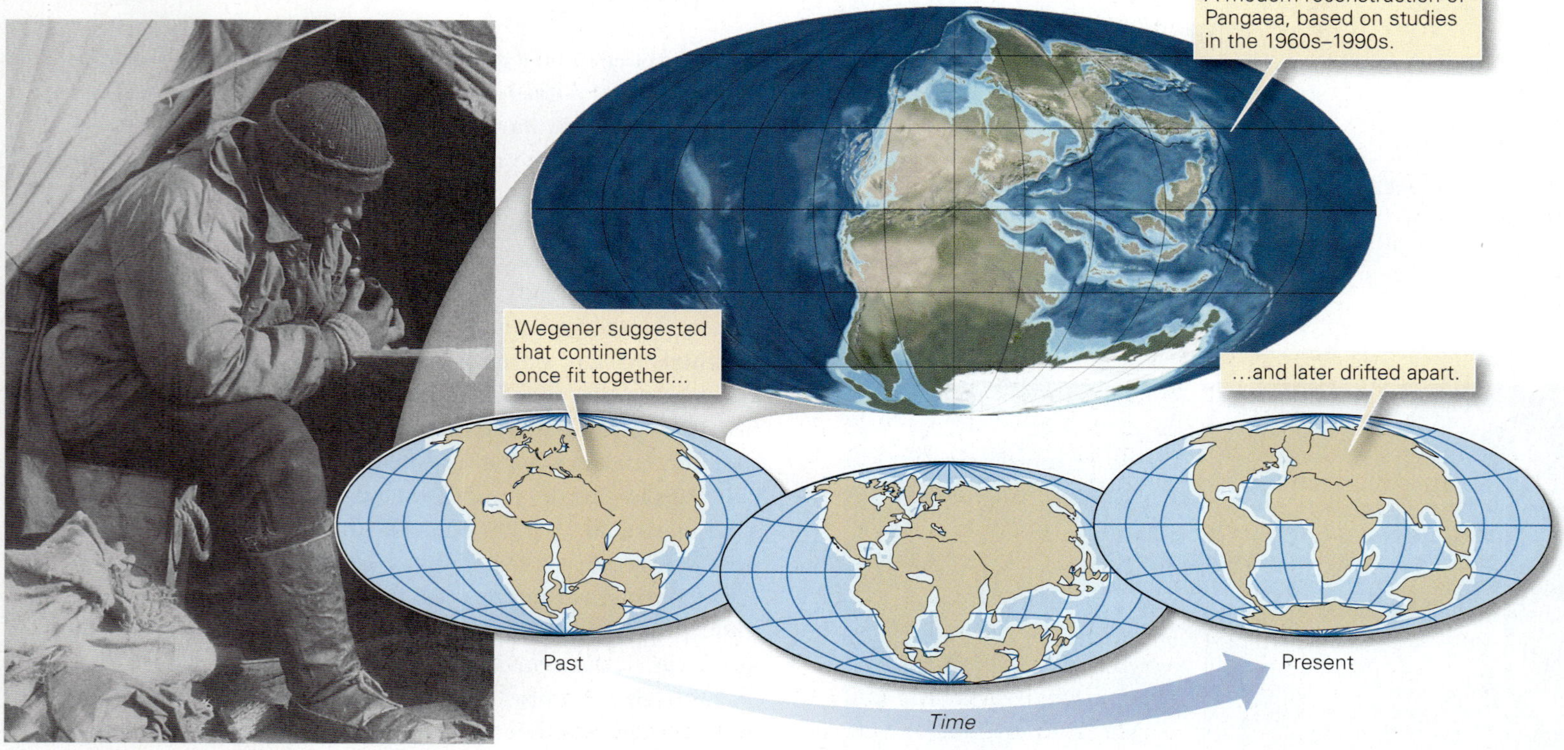
A modern reconstruction of Pangaea, based on studies in the 1960s–1990s.
Wegener suggested that continents once fit together...
...and later drifted apart.
Past
Present
Time

(a) Wegener in Greenland.

(b) Wegener's maps illustrating continental drift.

2.2 Wegener's Evidence for Continental Drift

Wegener suggested that a vast supercontinent, Pangaea, existed until near the end of the Mesozoic Era (the interval of geologic time that lasted from 252 to 66 million years ago). He suggested that Pangaea then broke apart, and the landmasses moved away from each other to form the continents we see today. Let's look at some of Wegener's arguments and see what led him to formulate this hypothesis of continental drift.

The Fit of the Continents

Almost as soon as maps of the Atlantic coastlines became available in the 1500s, scholars noticed the fit of the continents (see Fig. 2.1). The northwestern coast of Africa looks like it could tuck in snugly against the eastern coast of North America, and the bulge of eastern South America could nestle tightly into the indentation of southwestern Africa. Australia, Antarctica, and India could all connect to the southeast of Africa, while Greenland, Europe, and Asia could pack against the northeastern margin of North America. In fact, all the continents could be joined, with remarkably few overlaps or gaps, to produce Pangaea. Wegener concluded that the fit was too good to be coincidence and thus that the continents of today did fit together in the past.

Did you ever wonder...
why opposite coasts of the Atlantic look like they fit together?

SEE FOR YOURSELF...

THE FIT OF CONTINENTS

Latitude
33°31′25.77″ N

Longitude
39°26′23.32″ W

Zoom to an elevation of 14,800 km (9,200 miles) and look straight down.

Note how northwest Africa could fit snuggly along eastern North America. Wegener used this fit as evidence for Pangaea.

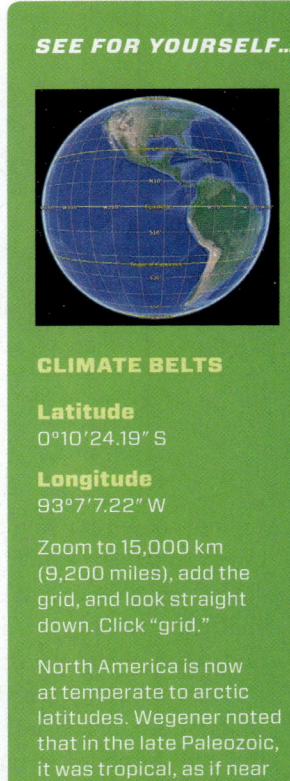

Locations of Past Glaciations

Glaciers are rivers or sheets of ice that flow across the land surface. As a glacier flows, it carries sediment grains of all sizes (clay, silt, sand, pebbles, and boulders). Grains protruding from the base of the moving ice carve scratches, called *striations*, into the substrate. When the ice melts, sediment stays behind in a deposit called *till*, which may bury striations. Thus, the till and striations at a particular location serve as evidence that the region was covered by a glacier in the past. By studying the age of glacial till deposits, geologists have determined that large areas of land were covered by glaciers during time intervals of Earth history called *ice ages*. One of these ice ages occurred from about 280 to 260 Ma (million years ago), near the end of the Paleozoic Era.

Wegener was an Arctic climate scientist by training, so it's no surprise that he had a strong interest in glaciers. He knew that glaciers form at high (polar) latitudes today, so he was bothered by the observation that sediments indicative of late Paleozoic glaciation occurred in southern South America, southern Africa, southern India, Antarctica, and southern Australia. These places are now widely separated and, with the exception of Antarctica, do not currently lie in cold polar regions (**Fig. 2.2a**). To Wegener's amazement, all late Paleozoic glaciated areas lie adjacent to each other on his map of Pangaea. Furthermore, when he plotted the orientation of glacial striations, they all pointed roughly outward from a location in southeastern Africa. In other words, Wegener determined that the distribution of glaciations at the end of the Paleozoic could easily be explained if the continents had been united in Pangaea, with the southern part of Pangaea lying beneath the center of a huge ice cap. This distribution of glaciation could not be explained if the continents had always been in their present positions.

The Distribution of Climatic Belts

If the southern part of Pangaea had straddled the South Pole at the end of the Paleozoic, then during this same time interval, southern North America, southern Europe, and northwestern Africa would have straddled the equator and would have had tropical or subtropical climates. Wegener searched for evidence that this was so by studying descriptions of sedimentary rocks that were formed at this time, for the material making up these rocks can reveal clues to the past climate. For example, in the swamps and jungles of tropical regions, thick deposits of plant material accumulate, and when deeply buried, this material transforms into coal (see Chapter 12). And, in the clear, shallow seas of tropical regions, large reefs develop. Finally, subtropical

SEE FOR YOURSELF...

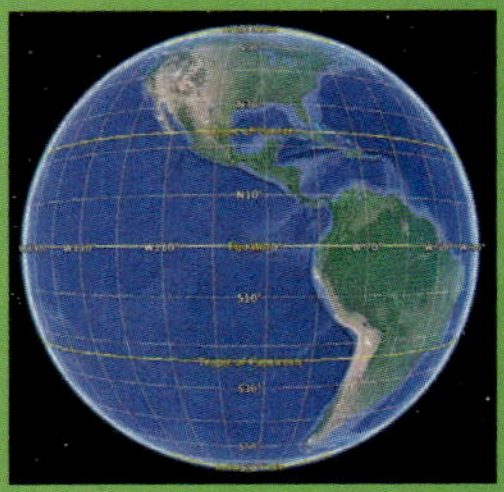

CLIMATE BELTS

Latitude
0°10′24.19″ S

Longitude
93°7′7.22″ W

Zoom to 15,000 km (9,200 miles), add the grid, and look straight down. Click "grid."

North America is now at temperate to arctic latitudes. Wegener noted that in the late Paleozoic, it was tropical, as if near the equator.

FIGURE 2.2 Wegener's evidence for continental drift came from analyzing the geologic record.

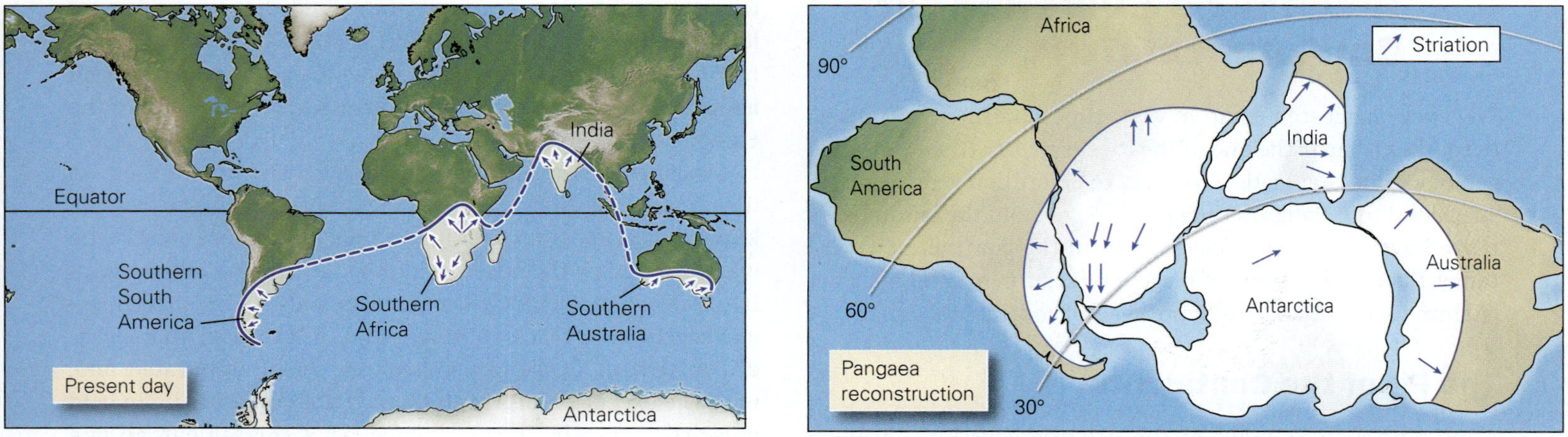

(a) The distribution of late Paleozoic glacial deposits and striations on present-day Earth (left) are hard to explain. But on Pangaea (right), areas with glacial deposits fit together within a southern polar cap.

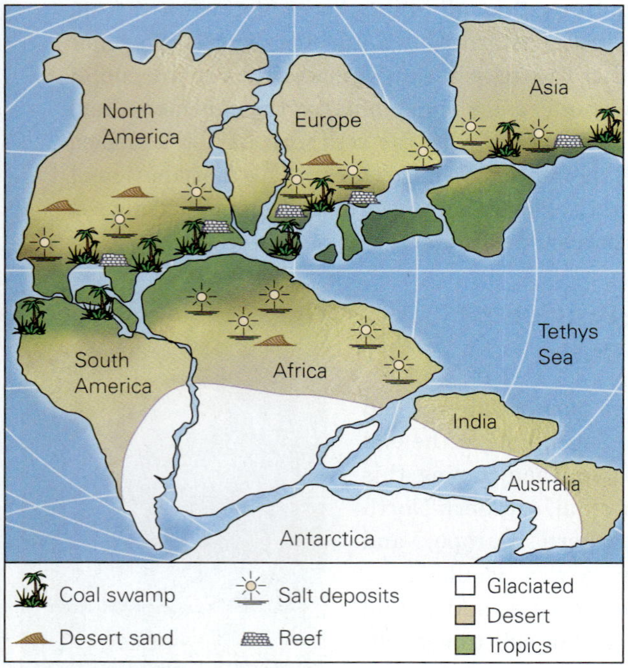

(b) The distribution of late Paleozoic rock types plots sensibly in Pangaea's climate belts.

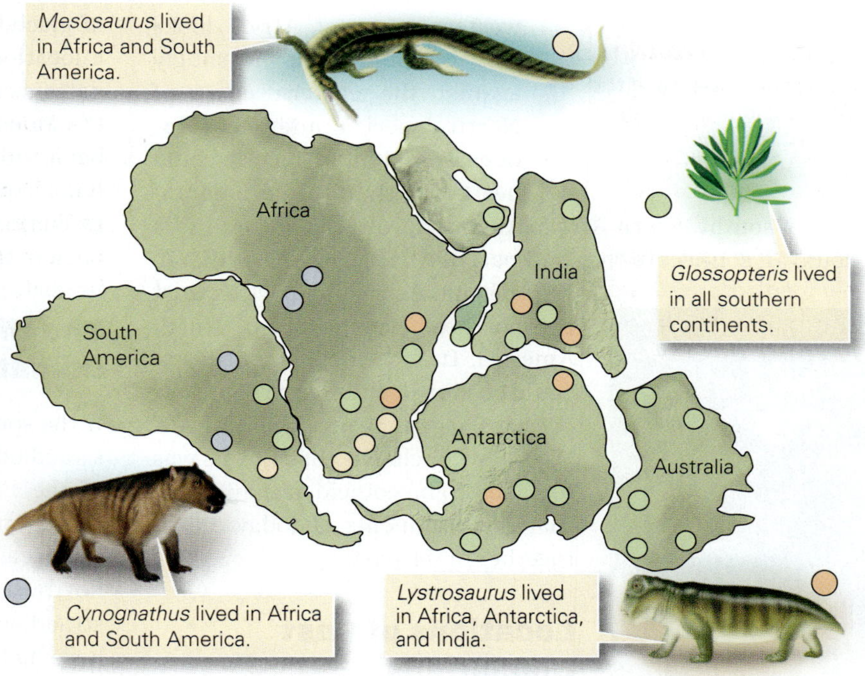

(c) A plot of fossil localities shows that Mesozoic land-dwelling organisms occured on multiple continents. This would be hard to explain if continents were separated.

regions, on either side of the tropical belt, contain deserts, an environment in which sand dunes form and salt from evaporating seawater or salt lakes accumulates. Wegener speculated that the distribution of late Paleozoic coal, reef, sand-dune, and salt deposits could define climate belts on Pangaea.

Sure enough, in the belt of Pangaea that Wegener expected to be equatorial, late Paleozoic sedimentary rock layers include abundant coal and the relicts of reefs. In the portions of Pangaea that Wegener predicted would be subtropical, late Paleozoic sedimentary rock layers include relicts of desert dunes and deposits of salt (**Fig. 2.2b**). On a present-day map of our planet, exposures of these ancient rock layers scatter around the globe at a variety of latitudes. On Wegener's Pangaea, the exposures align in continuous bands that occupy appropriate latitudes.

The Distribution of Fossils

Today, different continents provide homes for different species. Kangaroos, for example, live in the wild only in Australia.

Similarly, many kinds of plants grow only on one continent and not on others. Why? Because land-dwelling species of animals and plants cannot swim across vast oceans, and thus evolved independently on different continents. During a period of Earth history when all continents were in contact, however, land animals and plants could have migrated among many continents.

With this concept in mind, Wegener searched for records of fossil occurrences indicating where land-dwelling species lived during the late Paleozoic and early Mesozoic (between about 300 and 210 million years ago), and he found that these species had indeed existed on several continents (**Fig. 2.2c**). Wegener argued that the distribution of fossil species required the continents to have been adjacent to one another in the late Paleozoic and early Mesozoic.

Matching Geologic Units

Art historians can recognize a Picasso painting, architects know what makes a building look "Victorian," and geoscientists can identify distinctive assemblages of rocks. Wegener found that the same distinctive Precambrian rock assemblages occurred on the eastern coast of South America and the western coast of Africa, regions now separated by an ocean (**Fig. 2.3a**). If the continents had been joined to produce Pangaea in the past, then these matching rock groups would have been adjacent to each other, and thus could have composed continuous blocks or belts. Wegener also noted that features of the Appalachian Mountains of the United States and Canada closely resemble mountain belts in southern Greenland, Great Britain, Scandinavia, and northwestern Africa (**Fig. 2.3b, c**), regions that would have lain adjacent to each other in Pangaea. Wegener thus demonstrated that not only did the coastlines of continents match, so too did the rocks adjacent to the coastlines.

Criticism of Wegener's Ideas

Wegener's model of a supercontinent that later broke apart explained the distribution of ancient glaciers, coal, sand dunes, rock assemblages, and fossils. Clearly, he had compiled a strong circumstantial case for continental drift. But as noted earlier, he could not adequately explain how or why continents drifted. He left for Greenland having failed to convince his peers, and he died with no clue that his ideas, after lying dormant for decades, would be reborn as the basis of the broader theory of plate tectonics.

In effect, Wegener was ahead of his time. It would take an additional 30 years of research before geologists obtained sufficient data to test his hypotheses properly. Collecting this data required instruments and techniques that did not exist in Wegener's day. Of the many geologic discoveries that ultimately opened the door to plate tectonics, perhaps the most important came from the discovery of a phenomenon called paleomagnetism, which we discuss next.

FIGURE 2.3 Further evidence of continental drift: rocks on different sides of the ocean match.

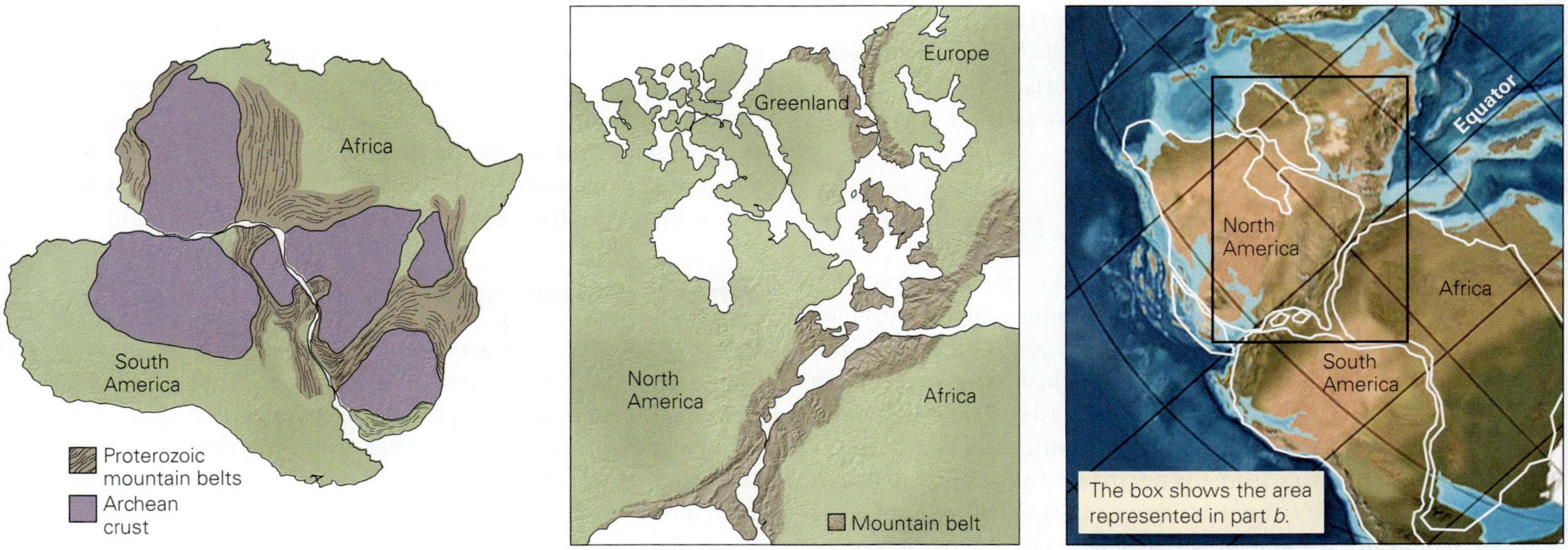

(a) Without the Atlantic, distinctive belts of rock in South America would align with similar ones in Africa.

(b) If the Atlantic didn't exist, Paleozoic mountain belts on both coasts would be adjacent.

(c) A modern reconstruction showing the positions of mountain belts in Pangaea. Modern continents are outlined in white.

TAKE-HOME MESSAGE

In the early 20th century, Alfred Wegener argued that the continents had once been connected in a supercontinent, Pangaea, that later broke up to produce smaller continents that drifted apart. He showed that the matching shapes of coastlines, as well as the distribution of ancient glaciers, climate belts, fossils, and rock units, make better sense if Pangaea existed. But Wegener couldn't convince his peers.

QUICK QUESTION Why were Wegener's peers skeptical of continental drift?

2.3 Paleomagnetism—Proving That Continents Move

More than 1,500 years ago, Chinese sailors discovered that a piece of lodestone, when suspended from a thread, points in a northerly direction and can help guide a voyage. Lodestone exhibits this behavior because it consists of magnetite, an iron-rich mineral that, like a compass needle, aligns with Earth's magnetic field lines. Several other rock types contain tiny crystals of magnetite or other magnetic minerals, and thus, while not as magnetic as lodestone, they behave overall like weak magnets. In this section, we explain how the study of such magnetic behavior led to the realization that rocks preserve *paleomagnetism*, a record of Earth's magnetic field in the past. An understanding of paleomagnetism provided proof of continental drift and, as we'll see later in this chapter, contributed to the development of plate tectonics theory. As a foundation for introducing paleomagnetism, we first provide additional detail about the basic nature of the Earth's magnetic field.

Did you ever wonder... why compasses always point to the north?

Details of Earth's Magnetic Field

As we mentioned in Chapter 1, circulation of liquid iron alloy in the outer core of the Earth generates a magnetic field. Earth's magnetic field resembles the field produced by a bar magnet, in that it is a *dipole*. A dipole has two ends, a north pole and a south pole—opposite poles attract, and like poles repel. We can represent Earth's magnetic field by an imaginary arrow pointing from the north pole to the south pole (see Fig. 1.12b). Earth's dipole intersects the surface of the planet at two points, known as the *magnetic poles* (**Fig. 2.4a**). By convention, the north magnetic pole lies near the north geographic pole and the south magnetic pole lies near the south geographic pole. (*Geographic poles* are the points where the Earth's spin axis intersects the planet's surface.) We use this convention so that the north-seeking (red) end of a compass needle is attracted to, and therefore points to, the north magnetic pole.

Earth's magnetic poles move constantly, and thus generally do not coincide exactly with the geographic poles (**Fig. 2.4b**). At present, the magnetic poles lie hundreds of kilometers away from the geographic poles, so the magnetic dipole tilts at about 12° relative to the Earth's spin axis. Because of this difference, a compass today does not point exactly to geographic north; the angle between the direction that a compass needle points and a line of longitude at a given location is the **magnetic declination**. Measurements indicate that for about the past century, the north magnetic pole has been moving across the Arctic Ocean toward Russia at 50 to 60 km per year (**Fig. 2.4c**). Significantly, poles don't seem to stray farther than about 2,000 km (about 20° of latitude) from the geographic poles. Because of their overall fairly random movements, however, geologists assume that averaged over thousands of years, the locations of the magnetic poles do roughly coincide with Earth's geographic poles. This relationship may exist because the Earth's spin may cause the flow in the outer core to organize into patterns resembling spring-like spirals that align with the axis (Fig. 2.4a).

Invisible field lines curve through space between the magnetic poles. In a cross-sectional view, these lines lie parallel to the surface of the Earth (that is, are horizontal) at the equator, tilt at an angle to the surface in midlatitudes, and plunge perpendicular to the surface (are vertical) at the magnetic poles (**Fig. 2.4d**). The angle between a magnetic field line and the surface of the Earth, at a given location, is called the **magnetic inclination**. If you place a magnetic needle on a horizontal axis so that it can pivot up and down, and then carry it from the magnetic equator to the magnetic pole, you'll see that the inclination varies with latitude—it is 0° at the magnetic equator and 90° at the magnetic poles. (Note that the compass you may carry with you on a hike does not show inclination because it has been balanced to remain horizontal.)

What Is Paleomagnetism?

In the early 20th century, researchers developed instruments that could measure the weak magnetic field produced by rocks and made a surprising discovery. In a rock that formed millions of years ago, the orientation of the dipole representing the magnetic field of the rock is not the same as that of present-day Earth (**Fig. 2.5a**). To understand this statement, imagine traveling to a location near the coast on the equator in South America where the inclination and declination today are presently 0°. If you measure the weak

FIGURE 2.4 Features of Earth's magnetic field.

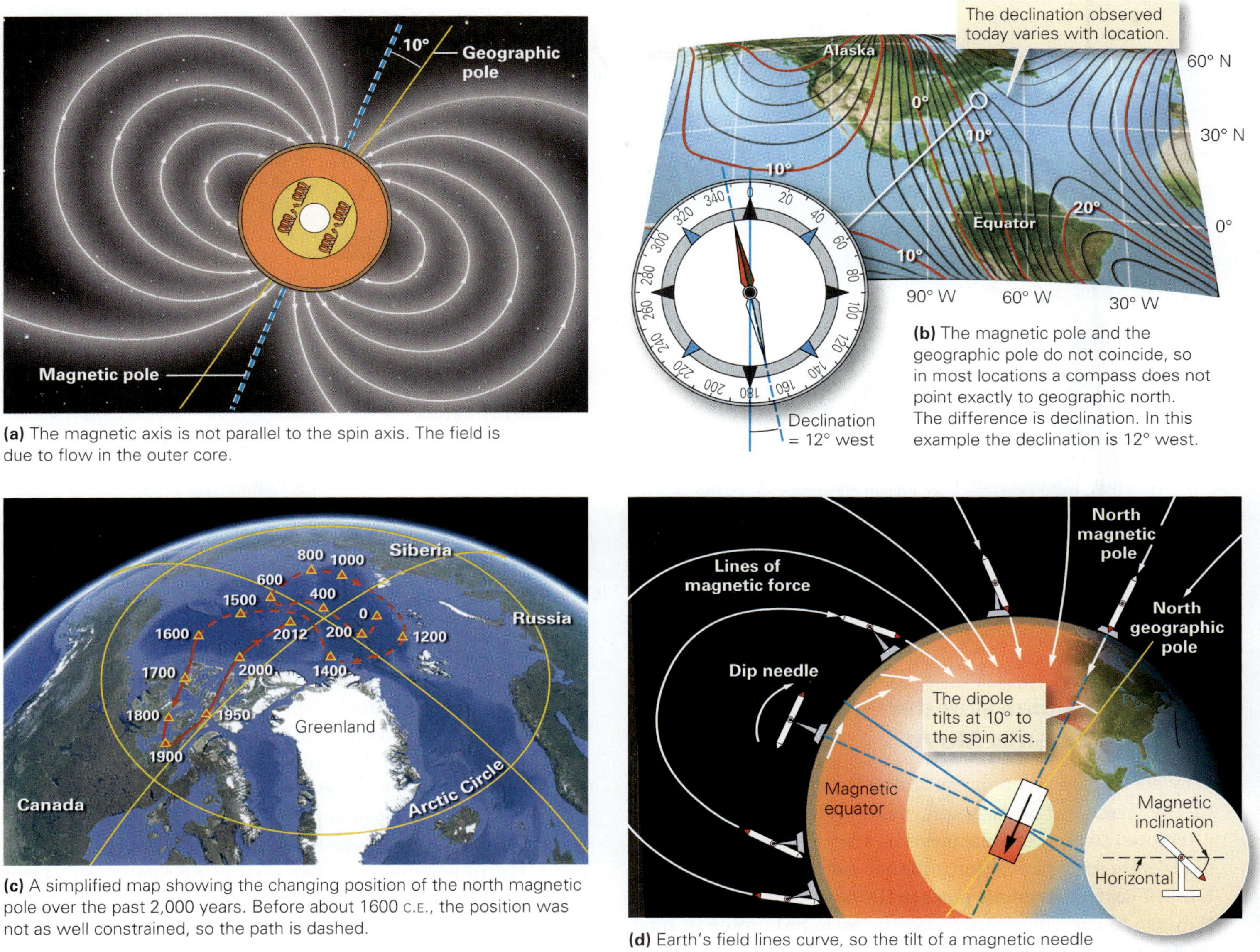

(a) The magnetic axis is not parallel to the spin axis. The field is due to flow in the outer core.

(b) The magnetic pole and the geographic pole do not coincide, so in most locations a compass does not point exactly to geographic north. The difference is declination. In this example the declination is 12° west.

(c) A simplified map showing the changing position of the north magnetic pole over the past 2,000 years. Before about 1600 C.E., the position was not as well constrained, so the path is dashed.

(d) Earth's field lines curve, so the tilt of a magnetic needle changes with latitude. This tilt is the magnetic inclination.

magnetic field produced by, say, a 90-million-year-old rock, and represent the orientation of this field by an imaginary bar magnet, you'll find that this imaginary bar magnet does not point to the present-day north magnetic pole, and you'll find that its inclination is not 0°. The reason for this difference is that the magnetic fields of ancient rocks indicate the orientation of the magnetic field, relative to the rock, at the time the rock formed. This record, preserved in rock, is **paleomagnetism**.

Paleomagnetism can develop in many different ways. For example, when lava, molten rock containing no crystals, starts to cool and solidify into rock, tiny magnetite crystals begin to grow (**Fig. 2.5b**). At first, thermal energy causes the tiny magnetic dipole associated with each crystal to wobble and tumble chaotically. Thus, at any given instant, the dipoles of the magnetite specks are randomly oriented and the magnetic forces they produce cancel each other out. Eventually, however, the rock cools sufficiently that the dipoles slow down and, like tiny compass needles, align with the Earth's magnetic field. As the rock cools still more, these tiny compass needles lock into permanent parallelism with the Earth's magnetic field at the time the cooling takes place. Since the magnetic dipoles of all the grains point in the same direction, they add together and produce a measurable field.

FIGURE 2.5 Paleomagnetism and how it can form during the solidification and cooling of lava.

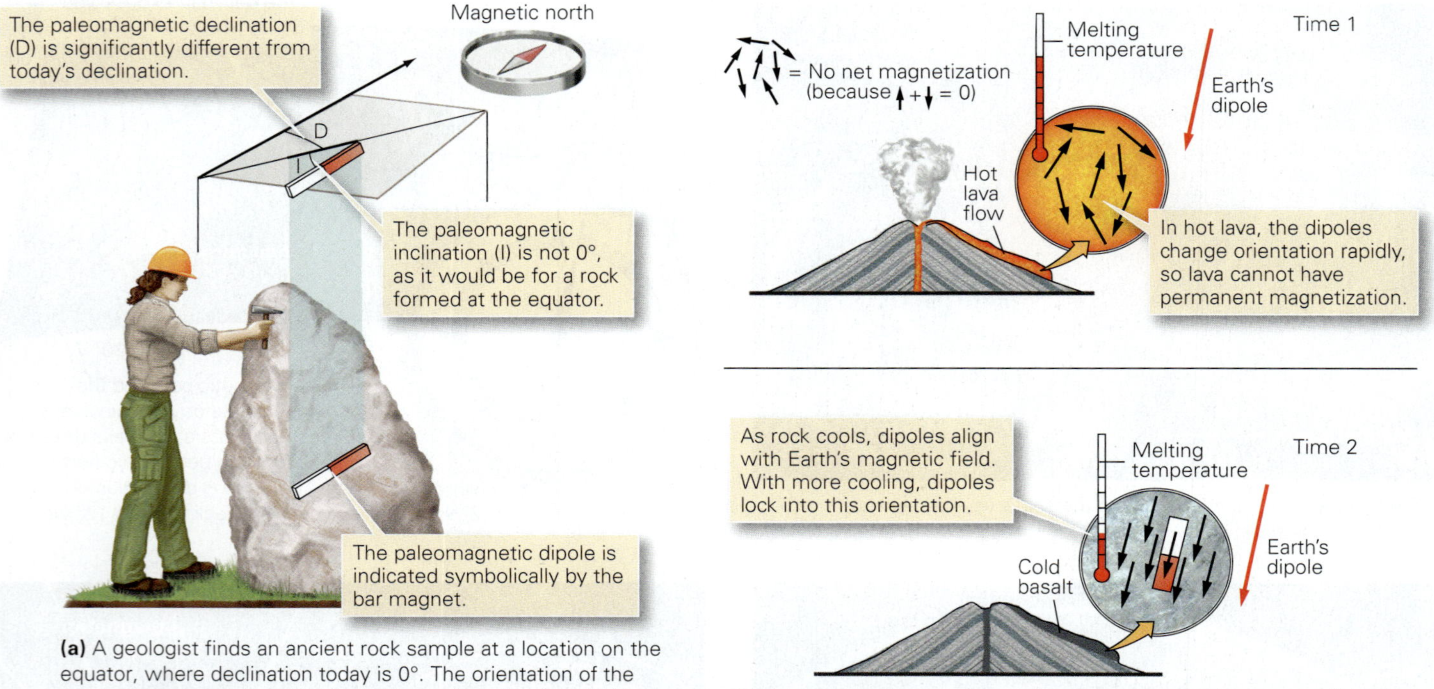

The paleomagnetic declination (D) is significantly different from today's declination.
Magnetic north
D
I
The paleomagnetic inclination (I) is not 0°, as it would be for a rock formed at the equator.
The paleomagnetic dipole is indicated symbolically by the bar magnet.
= No net magnetization (because ↑+↓ = 0)
Melting temperature
Time 1
Earth's dipole
Hot lava flow
In hot lava, the dipoles change orientation rapidly, so lava cannot have permanent magnetization.
As rock cools, dipoles align with Earth's magnetic field. With more cooling, dipoles lock into this orientation.
Melting temperature
Time 2
Earth's dipole
Cold basalt

(a) A geologist finds an ancient rock sample at a location on the equator, where declination today is 0°. The orientation of the rock's paleomagnetism is different from that of today's field.

(b) Paleomagnetism can form when lava cools and becomes solid rock.

Apparent Polar Wander—A Proof That Continents Move

Why doesn't the paleomagnetic dipole in ancient rocks point to the present-day magnetic field? When geologists first attempted to answer this question, they assumed that continents were fixed in position and thus concluded that the positions of Earth's magnetic poles in the past were different than they are today. They introduced the term **paleopole** to refer to the supposed position of the Earth's magnetic north pole in the past. With this concept in mind, they set out to track what they thought was the change in position of the paleopole over time. To do this, they measured the paleomagnetism in a succession of rocks of different ages from the same general location on a continent, and they plotted the position of the associated succession of paleopole positions on a map (**Fig. 2.6a**). The successive positions of dated paleopoles trace out a curving line that came to be known as an **apparent polar-wander path**.

At first, geologists assumed that the apparent polar-wander path actually represented how the position of Earth's magnetic pole migrated through time. But were they in for a surprise! When they obtained polar-wander paths from many different continents, they found that each continent has a different apparent polar-wander path. The hypothesis that continents are fixed in position cannot explain this observation, for if the magnetic pole moved while all the continents stayed fixed, measurements from all continents should produce the same apparent polar-wander paths. Geologists suddenly realized that they were looking at apparent polar-wander paths in the wrong way. It's not the pole that moves relative to fixed continents, but rather the continents that move relative to a fixed pole (**Fig. 2.6b**). Since each continent has its own unique polar-wander path (**Fig. 2.6c**), the continents must move with respect to each other. The discovery proved that Wegener was right all along—continents do move!

TAKE-HOME MESSAGE

A rock can contain a record of the position of the Earth's magnetic poles, relative to the rock, at the time the rock formed. Study of such paleomagnetism indicates that the continents have moved relative to the Earth's magnetic poles. Each continent has a different apparent polar-wander path, which is possible only if the continents move relative to one another.

QUICK QUESTION How does comparison of apparent polar-wander paths for different continents prove that continents move relative to each other?

FIGURE 2.6 Apparent polar-wander paths and their interpretation.

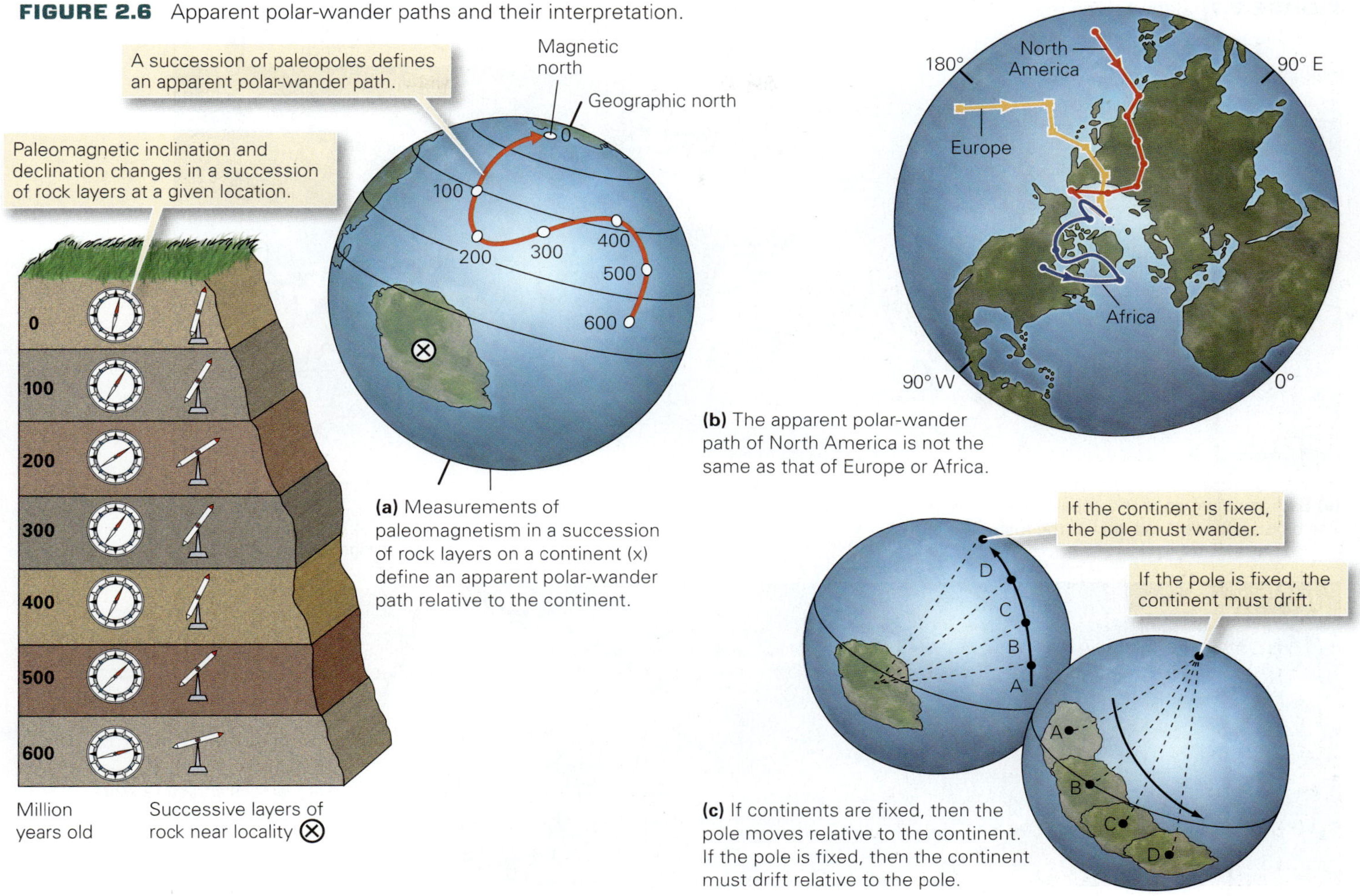

(a) Measurements of paleomagnetism in a succession of rock layers on a continent (x) define an apparent polar-wander path relative to the continent.

(b) The apparent polar-wander path of North America is not the same as that of Europe or Africa.

(c) If continents are fixed, then the pole moves relative to the continent. If the pole is fixed, then the continent must drift relative to the pole.

2.4 The Discovery of Seafloor Spreading

New Images of Seafloor Bathymetry

Military needs during World War II gave a boost to seafloor exploration, for as submarine fleets grew, navies required detailed information about bathymetry, or depth variations. The invention of sonar (echo sounding) permitted such information to be gathered quickly. Echo sounding works on the same principle that a bat uses to navigate and find insects. A ship emits a sound pulse that travels down through the water, bounces off the seafloor, and returns up as an echo through the water to a receiver on the ship. Since sound waves travel at a known velocity, the time between the sound emission and the echo detection indicates the distance between the ship and the seafloor. (Recall that Velocity = distance/time, so Distance = velocity × time.) As the ship travels, observers can obtain a continuous record of the depth of the seafloor. The resulting cross section showing depth plotted as a function of location is called a *bathymetric profile*. By cruising back and forth across the ocean many times, investigators obtained a series of bathymetric profiles and from these constructed *bathymetric maps* of the seafloor. (Geologists can now produce such maps rapidly using satellite data.) Bathymetric maps reveal several important features (**Fig. 2.7a**).

- *Mid-ocean ridges:* The floor beneath all major oceans includes *abyssal plains,* which are broad, relatively flat regions of the ocean that lie at a depth of about 4 to 5 km below sea level, and **mid-ocean ridges**, submarine mountain ranges whose peaks lie only about 2 to 2.5 km below sea level (**Fig. 2.7b**). Geologists call the crest of a mid-ocean ridge the *ridge axis.* Mid-ocean ridges are roughly symmetrical—bathymetry on one side of the axis is nearly a mirror image of bathymetry on the other side.

FIGURE 2.7 Bathymetric features of the ocean floor.

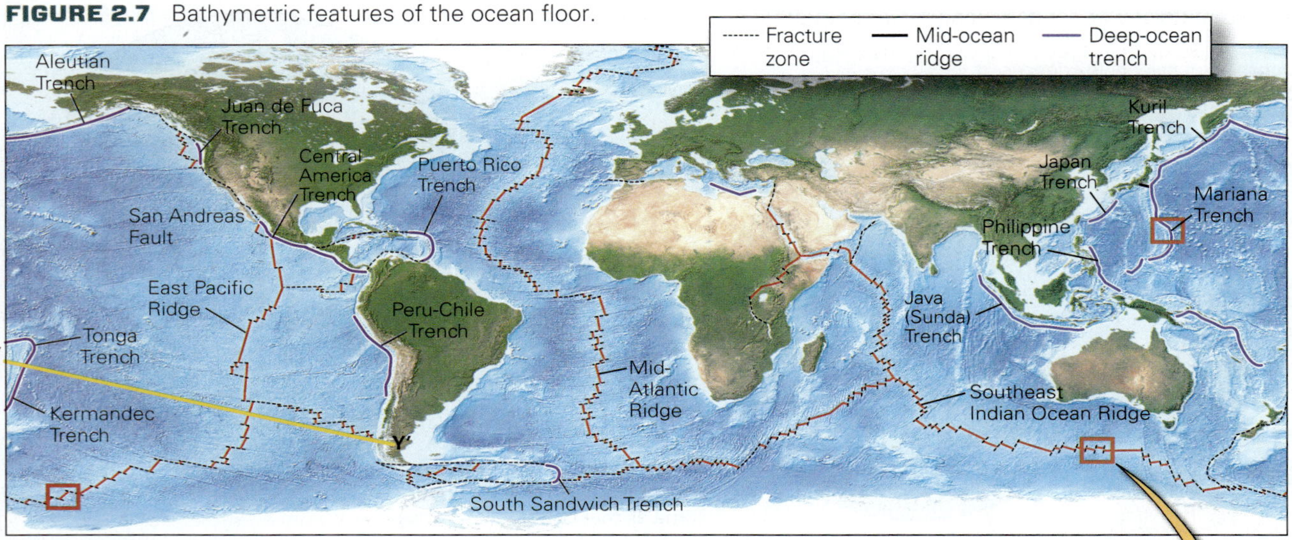

(a) Bathymetric map of the ocean floor, highlighting key bathymetric features. The boxes show the location of the maps in b and c.

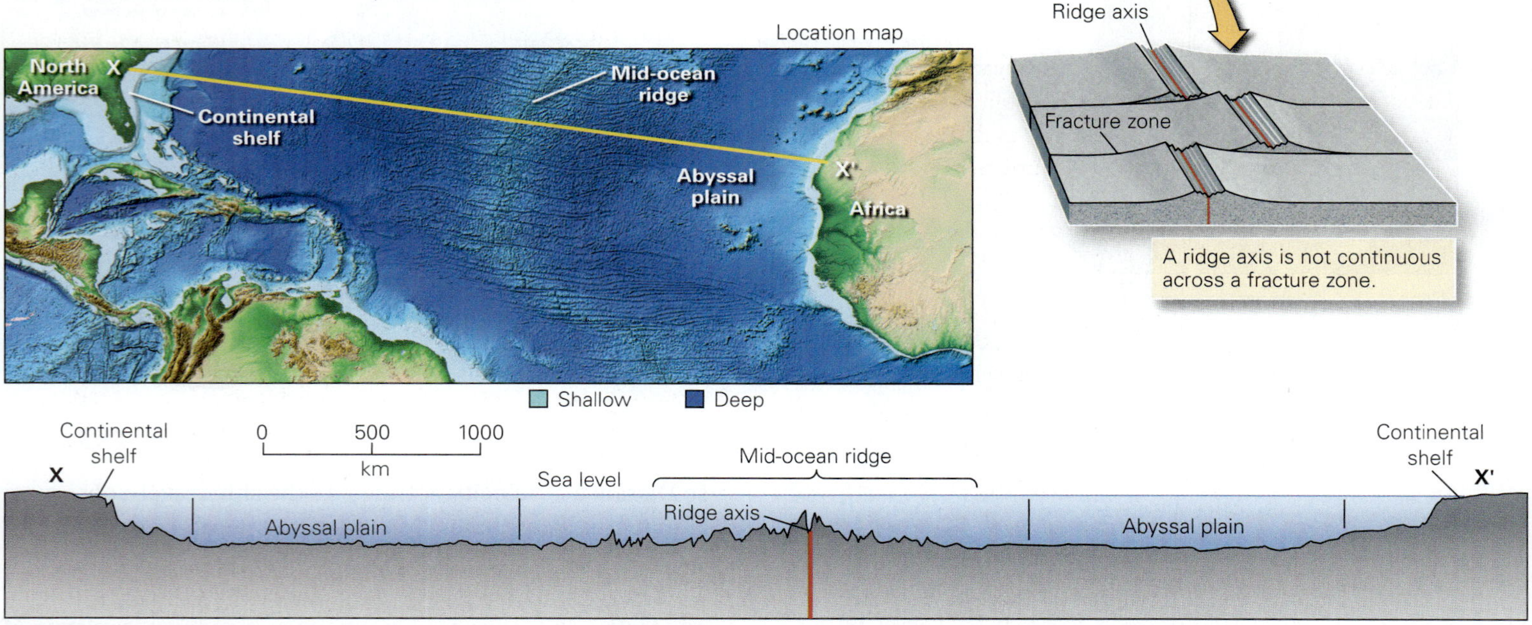

(b) A bathymetric profile along line XX'. The location map shows the position of XX', and the inset shows a fracture zone intersecting a ridge axis, in 3-D.

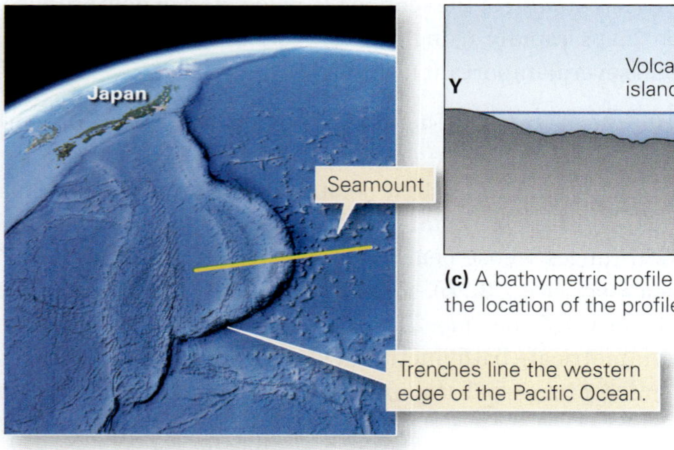

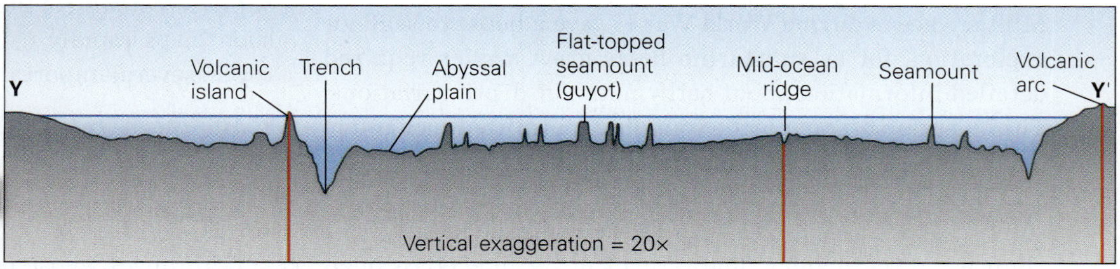

(c) A bathymetric profile of a trench, and nearby seamounts. The yellow line on the map shows the location of the profile.

- *Fracture zones:* Surveys reveal that narrow bands of vertical cracks and broken-up rock cut the ocean floor of mid-ocean ridges. These **fracture zones** lie roughly at right angles to mid-ocean ridges, and interrupt the continuity of the ridge axis (see the inset in Fig. 2.7a).
- *Deep-ocean trenches:* Along much of the perimeter of the Pacific Ocean, and in a few additional localities, the ocean floor reaches depths greater than 5 km. These deep areas define elongate troughs that are referred to as **trenches** (**Fig. 2.7c**). Trenches border **volcanic arcs**, curving chains of active volcanoes.
- *Seamount chains:* Numerous volcanic islands poke up from the ocean floor; for example, the Hawaiian Islands lie in the middle of the Pacific. In addition to islands that rise above sea level, sonar has detected many **seamounts**, isolated submarine mountains, which were once volcanoes but no longer erupt. Volcanic islands and seamounts typically occur in chains, but in contrast to the volcanic arcs that border deep-ocean trenches, only the one island at the end of a seamount and island chain remains capable of erupting today.

Did you ever wonder…
why Hawaii rises above the middle of the ocean?

New Observations on the Nature of Oceanic Crust

By the mid-20th century, geologists had discovered many important characteristics of the seafloor crust. These discoveries led them to realize that oceanic crust differs from continental crust, and that bathymetric features of the ocean floor provide clues to the origin of the crust. Specifically:

- A layer of sediment composed of clay and the tiny shells of dead plankton covers much of the ocean floor. This layer becomes progressively thicker away from the mid-ocean ridge axis. But even at its thickest, the sediment layer is too thin to have been accumulating for the entirety of Earth history.
- The composition of the oceanic crust is fundamentally different from that of continental crust. Beneath its sediment cover, oceanic crust bedrock consists primarily of basalt—it does not contain the great variety of rock types found on continents.
- *Heat flow,* the rate at which heat rises from the Earth's interior up through the crust, is not the same everywhere in the oceans. Rather, more heat rises beneath mid-ocean ridges than elsewhere. This observation led researchers to speculate that hot magma rises into the crust just below a mid-ocean ridge axis.
- When maps showing the distribution of earthquakes in oceanic regions became available in the years after World War II, it became clear that earthquakes occur not randomly, but rather in distinct belts (**Fig. 2.8**). Some belts follow trenches, some follow mid-ocean ridge axes, and others lie along portions of fracture zones. Since earthquakes define locations where rocks break and move, geologists realized that these bathymetric features are places where motion is taking place.
- The ridge axis of some mid-ocean ridges is marked by a narrow (a few kilometers wide), elongate trough hundreds of meters deeper than its borders. In this regard, the bathymetry of a mid-ocean ridge resembles the topography of the East African rift valley, a place where the crust of Africa is stretching and breaking apart.

SEE FOR YOURSELF…

TRENCHES

Latitude
35°37′59.86″ N

Longitude
145°36′12.78″ E

Fly to an elevation of 5,500 km (3,500 miles) and look straight down. Zoom in and use the elevation tool to measure trench depth.

Trenches of the western Pacific Ocean, near Japan.

FIGURE 2.8 A 1953 map showing the distribution of earthquake locations in the ocean basins. Note that earthquakes occur in belts.

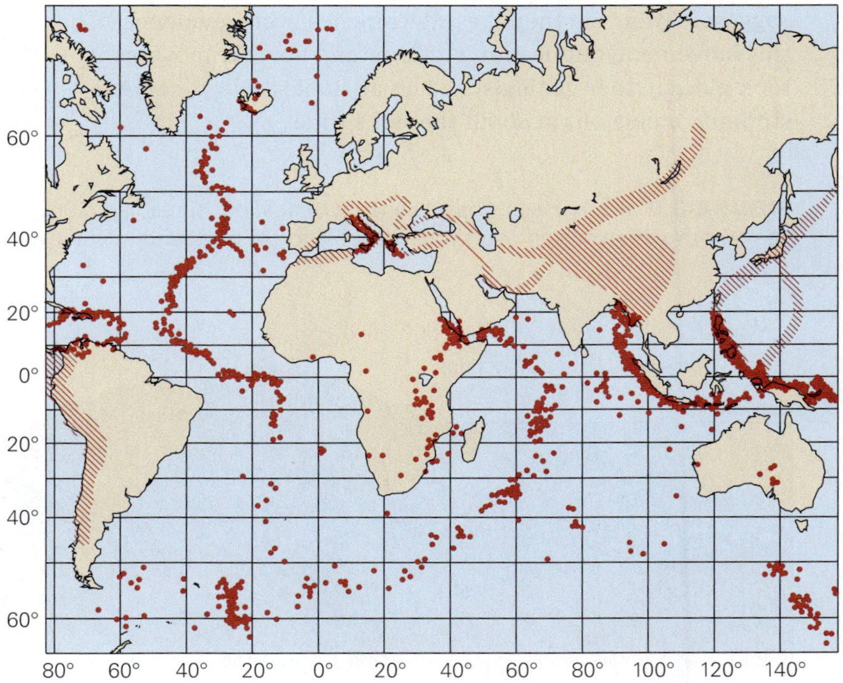

Harry Hess and His "Essay in Geopoetry"

In the late 1950s, Harry Hess studied the observations described above and made an important deduction. He realized that, because the sediment layer on the ocean floor was thin overall, the ocean floor might be much younger than the continents. Also, because sediment thickens progressively away from a mid-ocean ridge axis, the ridges themselves likely were younger than the deeper parts of the ocean floor. If this deduction were correct, then somehow new ocean floor must be forming at the ridges, so an ocean basin could be getting wider with time. But how? The association of earthquakes with mid-ocean ridges suggested to Hess that the seafloor was cracking and splitting apart at the ridge, and the similarity in form to the East African rift valley suggested that the region was stretching. The discovery of high heat flow along mid-ocean ridge axes provided the final piece of the puzzle, for it suggested the presence of molten rock beneath the ridges.

In 1960, Hess brought these ideas together and suggested that magma rose upward at mid-ocean ridges, and that this material solidified to form the basalt of oceanic crust (**Fig. 2.9**). The new seafloor then moved away from the ridge axis, leading to the widening of the ocean basin, a process we now call **seafloor spreading**. Hess realized that old ocean floor must be consumed somewhere, otherwise the Earth would be expanding, so he suggested that deep-ocean trenches might be places where the seafloor sank back into the mantle, a process now known as *subduction*. Hess suggested that earthquakes at trenches were evidence of this movement, but he didn't understand how the movement took place. Other geologists, such as Robert Dietz, came to similar conclusions at about the same time.

Did you ever wonder...
if the distance between New York and Paris changes?

Hess and his contemporaries realized that the seafloor-spreading hypothesis instantly provided the long-sought explanation of how continental drift occurs. Continents passively move apart as the seafloor between them spreads at mid-ocean ridges, and they passively move together as the seafloor between them sinks back into the mantle, or subducts, at trenches. (As we will see later, geologists now realize that it is the whole lithosphere that moves, not just the crust.) Understanding seafloor spreading and subduction proved to be an important step on the route to plate tectonics—the ideas seemed so good that Hess referred to his paper describing them as an "essay in geopoetry." But first, the idea needed to be tested, and other key discoveries would have to take place before the whole theory of plate tectonics could come together.

TAKE-HOME MESSAGE

New observations about seafloor bathymetry, sediment cover, heat flow, and seismicity led to Hess's proposal of seafloor spreading—new seafloor forms at mid-ocean ridges and then moves away from the ridge axis, so ocean basins can get wider with time. As this happens, old ocean floor sinks back into the mantle by subduction.

QUICK QUESTION How do seafloor spreading and subduction provide an explanation for how continents move?

2.5 Evidence for Seafloor Spreading

For a hypothesis to become a theory (see Box P.1), researchers must demonstrate that the idea really works. During the 1960s, geologists found that the seafloor-spreading

FIGURE 2.9 Harry Hess's basic concept of seafloor spreading (1962). Hess implied, incorrectly, that only the crust moved. We will see that this sketch is an oversimplification and contains errors, but it was an important step in the development of plate tectonics.

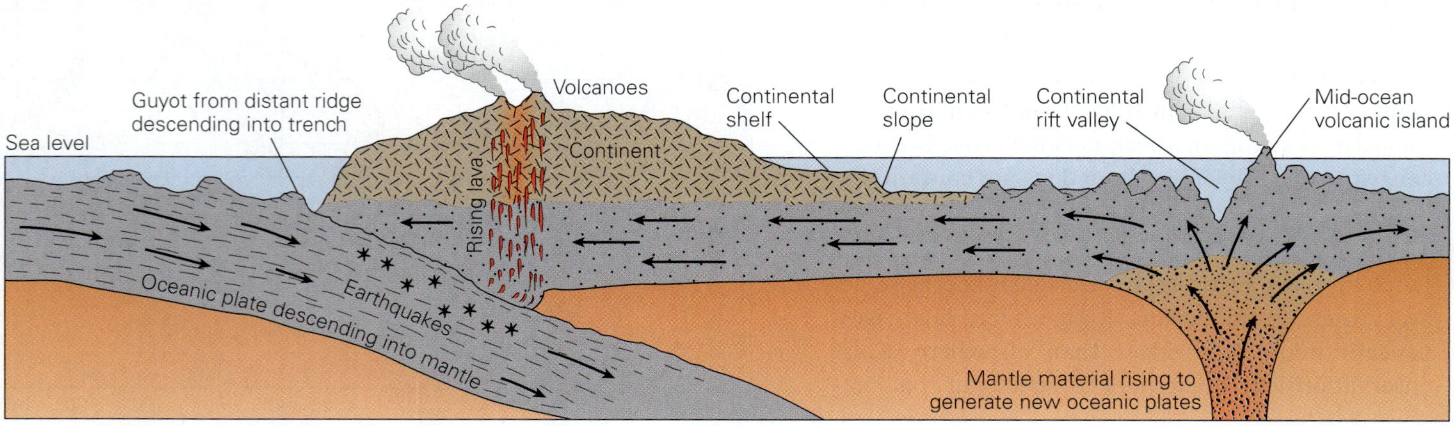

hypothesis successfully explains several previously baffling observations. Here we discuss two: (1) the existence of orderly variations in the strength of the measured magnetic field over the seafloor, producing a pattern of stripes called marine magnetic anomalies; and (2) the age of the oldest deep-sea sediment, relative to the position of the ridge axis.

Marine Magnetic Anomalies

Recognizing Anomalies Geologists can measure the strength of Earth's magnetic field with an instrument called a *magnetometer*. At any given location on the surface of the Earth, the magnetic field that you measure includes two parts: one produced by the main dipole of the Earth generated by circulation of molten iron in the outer core, and another produced by the magnetism of near-surface rock. A **magnetic anomaly** is the difference between the expected strength of the Earth's main dipole field at a certain location and the actual measured strength of the magnetic field at that location. Places where the field strength is stronger than expected are *positive anomalies*, and places where the field strength is weaker than expected are *negative anomalies*.

Geologists towed magnetometers back and forth across the ocean to map variations in magnetic field strength (**Fig. 2.10a**). As a ship cruised along its course, the magnetometer's gauge might first detect an interval of strong signal (a positive anomaly) and then an interval of weak signal (a negative anomaly). A graph of signal strength versus distance along the traverse, therefore, has a sawtooth shape (**Fig. 2.10b**). When geologists compiled data from many cruises on a map, these **marine magnetic anomalies** defined distinctive, alternating bands. If we color positive anomalies dark and negative anomalies light, the pattern made by the anomalies resembles the stripes on a candy cane (**Fig. 2.10c**). The origin of the marine magnetic anomaly pattern, however, remained an unsolved mystery until geologists recognized the existence of magnetic reversals.

Magnetic Reversals Recall that Earth's magnetic field can be represented by an arrow, representing the dipole, that presently points from the north magnetic pole to the south magnetic pole. When researchers measured the paleomagnetism of a succession of rock layers that had accumulated over a long period, they found that the *polarity* (which end of a magnet points north and which end points south) of the paleomagnetic field preserved in some layers was the same as that of Earth's present magnetic field, whereas in other layers it was the opposite (**Fig. 2.11a**).

At first, reversed polarity was thought to be the result of lightning strikes or of local chemical reactions between rock and water. But when repeated measurements from around the world revealed a systematic pattern of alternating normal

FIGURE 2.10 The discovery of marine magnetic anomalies.

(a) A ship towing a magnetometer detects changes in the strength of the magnetic field.

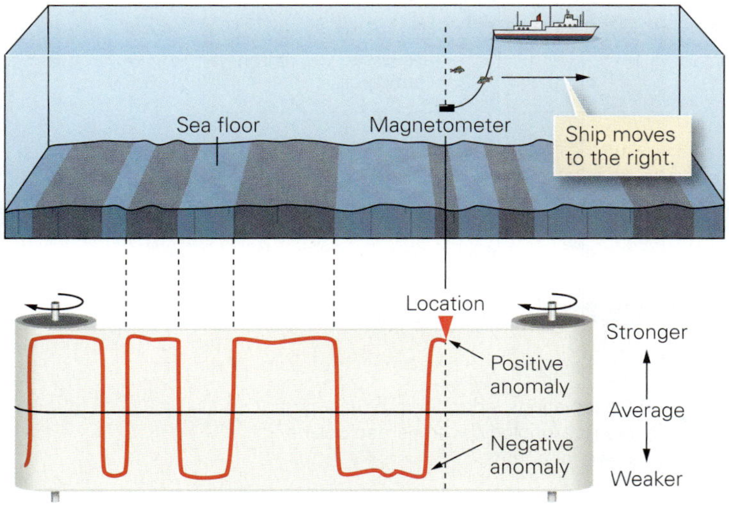

(b) On a paper record, intervals of stronger magnetism (positive anomalies) alternate with intervals of weaker magnetism (negative anomalies).

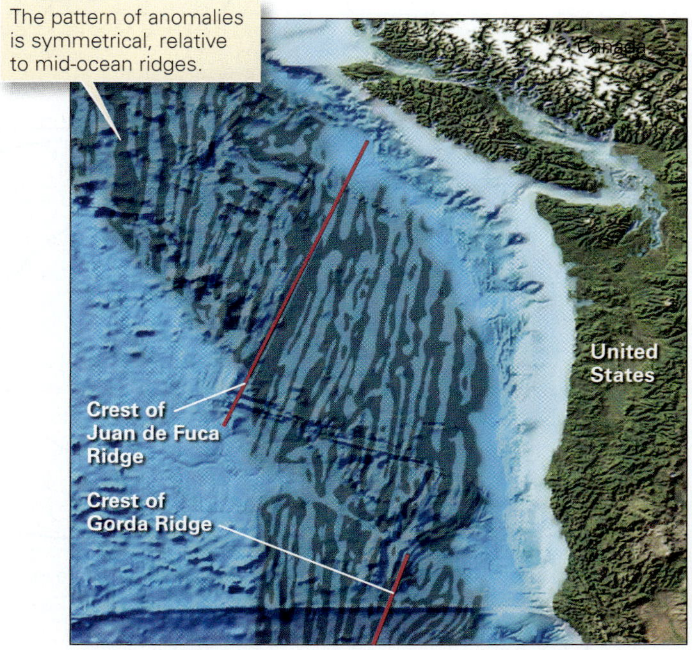

(c) A map showing areas of positive anomalies (dark) and negative anomalies (light) off the west coast of North America. The pattern of anomalies resembles candy-cane stripes.

FIGURE 2.11 Magnetic polarity reversals and the chronology of reversals.

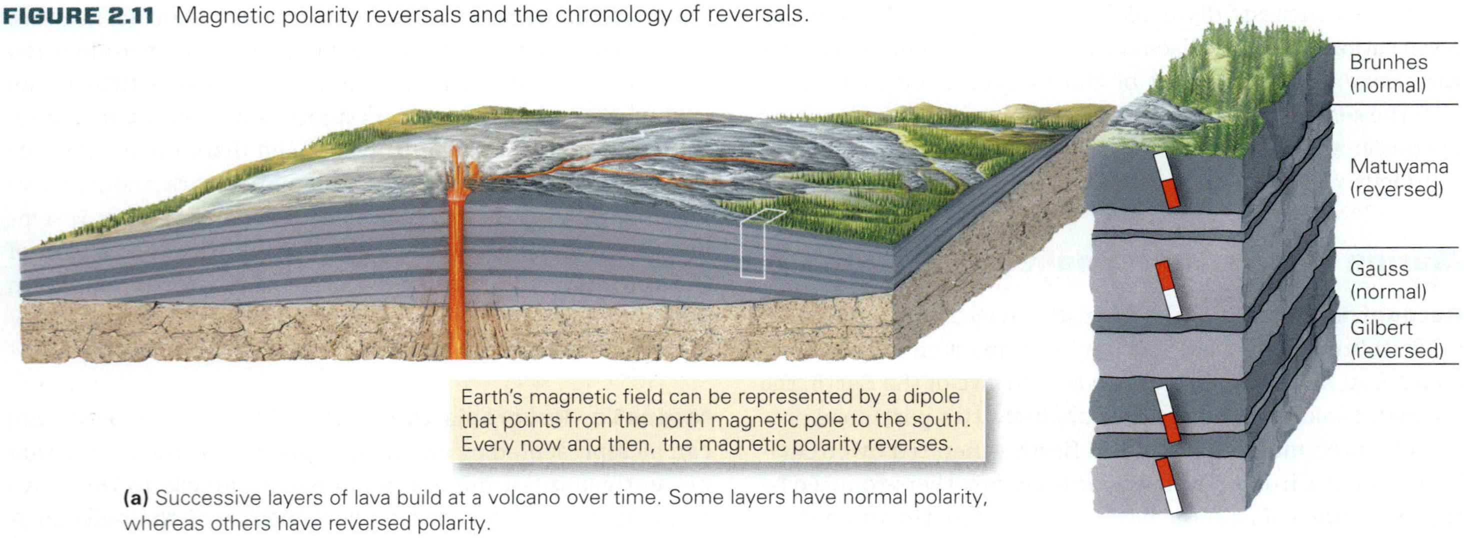

(a) Successive layers of lava build at a volcano over time. Some layers have normal polarity, whereas others have reversed polarity.

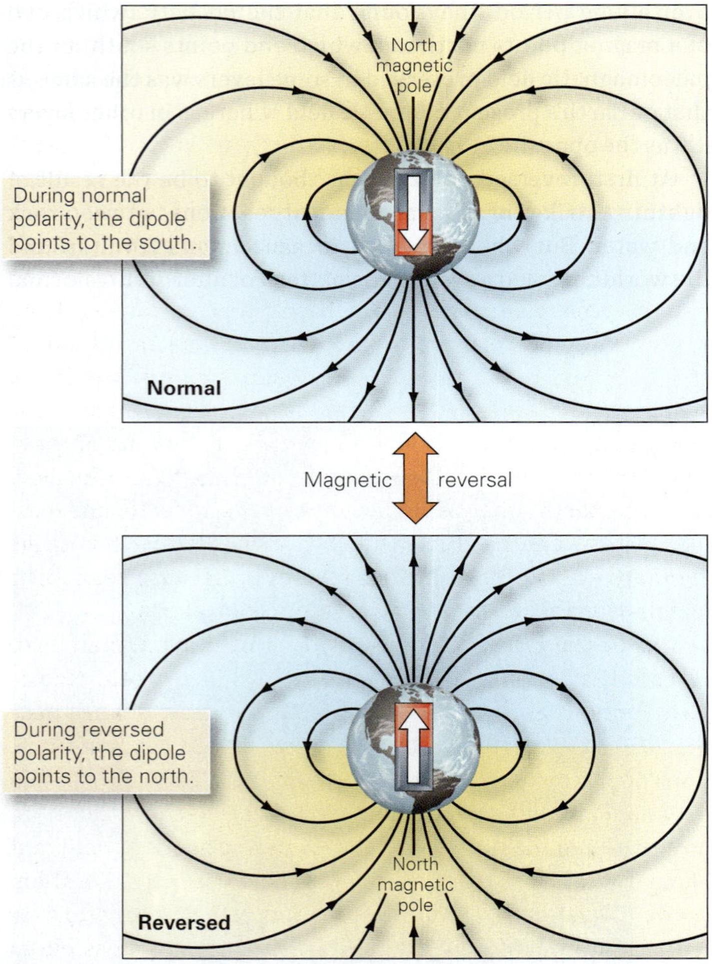

(b) The polarity of the Earth's magnetic field reverses every now and then.

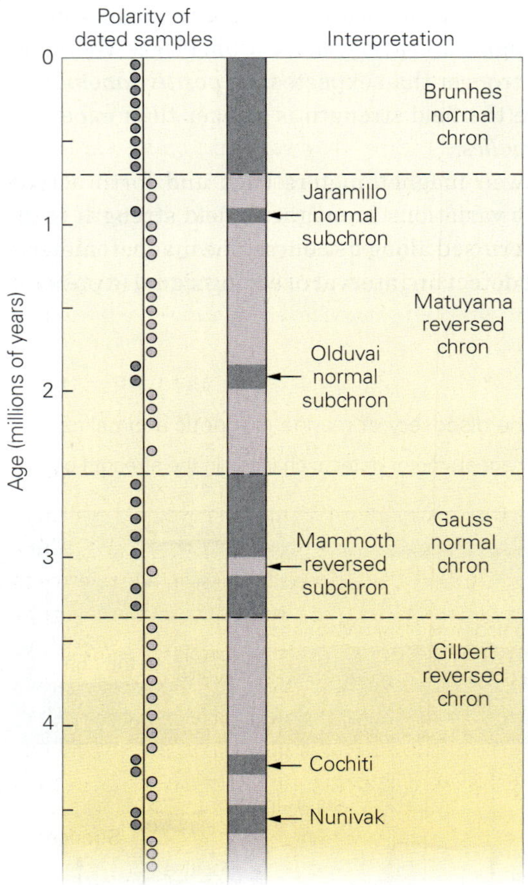

(c) Observations led to the production of a reversal chronology, with named polarity intervals, or chrons.

and reversed polarity in rock layers, geologists realized that reversals were a worldwide, not a local, phenomenon. They reached the unavoidable conclusion that, at various times during Earth history, the polarity of Earth's magnetic field has suddenly flipped! In other words, sometimes the Earth has *normal polarity*, as it does today, and sometimes it has *reversed polarity* (**Fig. 2.11b**). A time when the Earth's field flips from normal to reversed polarity, or vice versa, is called a **magnetic reversal**. When the Earth has reversed polarity, the south magnetic pole lies near the north geographic pole, and the north magnetic pole lies near the south geographic pole. Thus, if you were to use a compass during periods when the Earth's magnetic field was reversed, the north-seeking end of the needle would point to the south geographic pole. Note that the Earth itself doesn't turn upside down during a reversal—it is just the magnetic field that reverses.

In the 1950s, about the same time researchers discovered polarity reversals, they developed a technique to measure the age of a rock in years. The technique, called *isotopic dating*, will be discussed in detail in Chapter 10. Geologists applied the technique to determine the ages of rock layers in which they obtained their paleomagnetic measurements, and thus determined when the magnetic field of the Earth reversed. With this information, they constructed a history of magnetic reversals for the past 4.5 million years; this history is now called the **magnetic-reversal chronology**, and the time interval between successive reversals is called a *polarity chron*.

A diagram representing the Earth's magnetic-reversal chronology (**Fig. 2.11c**) shows that reversals do not occur regularly, so the lengths of different polarity chrons are different. For example, we have had a normal-polarity chron for the last 700,000 years. Before that, a reversed-polarity chron occurred. The youngest four polarity chrons (Brunhes, Matuyama, Gauss, and Gilbert) were named after scientists who made important contributions to the study of magnetism. As more measurements became available, investigators realized that some short-duration reversals (less than 200,000 years long) took place within the chrons, and they called these shorter durations *polarity subchrons*. Using isotopic dating, it was possible to determine the age of chrons back to 4.5 Ma.

Interpreting Marine Magnetic Anomalies Why do marine magnetic anomalies exist? In 1963, researchers in Britain and Canada proposed a solution to this riddle. Simply put, a positive anomaly occurs over areas of the seafloor where underlying basalt has normal polarity. In these areas, the weak magnetic force produced by the magnetite grains in basalt adds to the force produced by the Earth's dipole—the sum of these forces yields a stronger magnetic signal than expected due to the dipole alone (**Fig. 2.12a**). A negative anomaly occurs over regions of the seafloor where the underlying basalt has a reversed polarity. In these regions, the magnetic force of the basalt subtracts from the force produced by the Earth's dipole, so the measured magnetic signal is weaker than expected.

The seafloor-spreading model easily explains not only why positive and negative magnetic anomalies exist over the seafloor, but also why they define stripes that trend parallel to the mid-ocean ridge and why the pattern of stripes on one side of the ridge is the mirror image of the pattern on the other side (**Fig. 2.12b**). To see why, let's examine stages in the process of seafloor spreading (**Fig. 2.12c**). Imagine that at Time 1 in the past, the Earth's magnetic field has normal polarity. As the basalt rising at the mid-ocean ridge during this time interval cools and solidifies, the tiny magnetic grains in basalt align with the Earth's field, and thus the rock as a whole has a normal polarity. Seafloor formed during Time 1 will therefore generate a positive anomaly and appear as a dark stripe on an anomaly map. As it forms, the rock of this stripe moves away from the ridge axis, so half goes to the right and half to the left. Now imagine that later, at Time 2, Earth's field has reversed polarity. Seafloor basalt formed during Time 2, therefore, has reversed polarity and will appear as a light stripe on an anomaly map. As it forms, this reversed-polarity stripe moves away from the ridge axis, and even younger crust forms along the axis. The basalt in each new stripe of crust preserves the polarity that was present at the time it formed, so as the Earth's magnetic field flips back and forth, alternating positive and negative anomaly stripes form. A positive anomaly exists over the ridge axis today because seafloor is forming during the present normal-polarity chron.

Closer examination of a seafloor magnetic-anomaly map reveals that anomalies are not all the same width. Geologists found that the relative widths of anomaly stripes near the Mid-Atlantic Ridge are the same as the relative durations of paleomagnetic chrons (**Fig. 2.12d**). This relationship between anomaly-stripe width and polarity-chron duration indicates that the rate of seafloor spreading has been constant along the Mid-Atlantic Ridge for at least the last 4.5 million years. If you assume that the spreading rate was constant for tens to hundreds of millions of years, then it is possible to estimate the age of stripes right up to the edge of the ocean.

FIGURE 2.12 The progressive development of magnetic anomalies and the long-term reversal chronology.

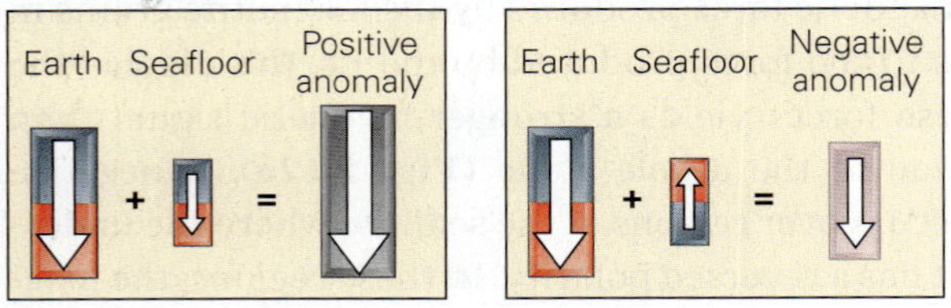

(a) Positive anomalies form when seafloor rock has the same polarity as the present magnetic field. Negative anomalies form when seafloor rock has polarity that is opposite to the present field.

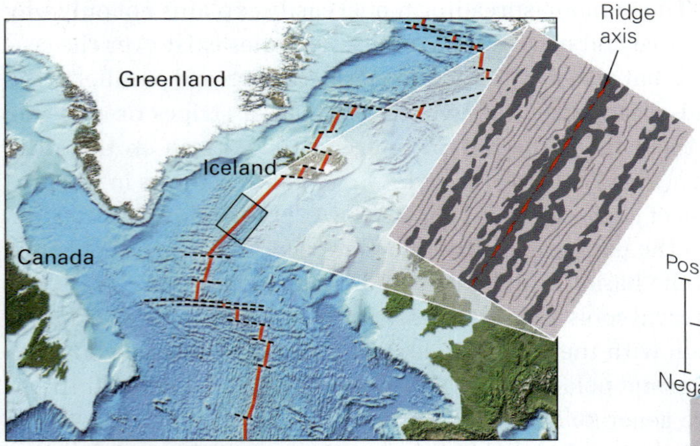

(b) The seafloor-spreading model predicts that magnetic anomalies are symmetrical relative to the mid-ocean ridge.

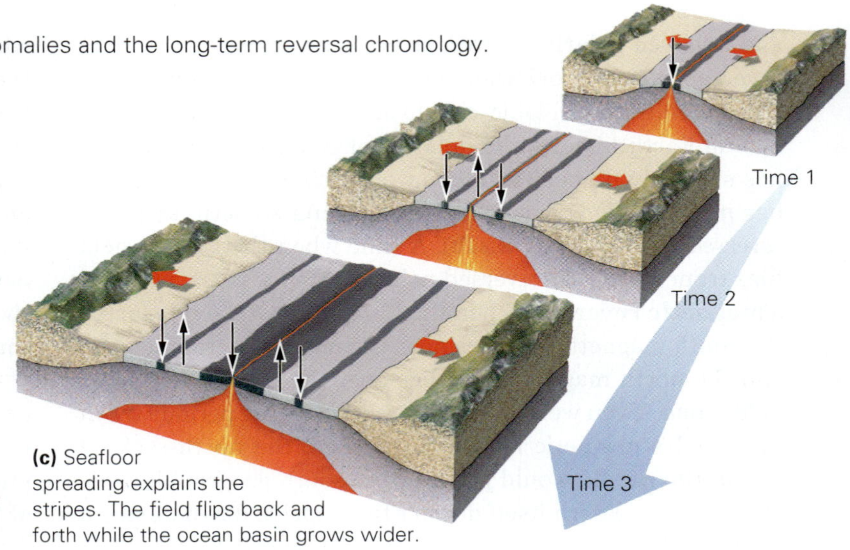

(c) Seafloor spreading explains the stripes. The field flips back and forth while the ocean basin grows wider.

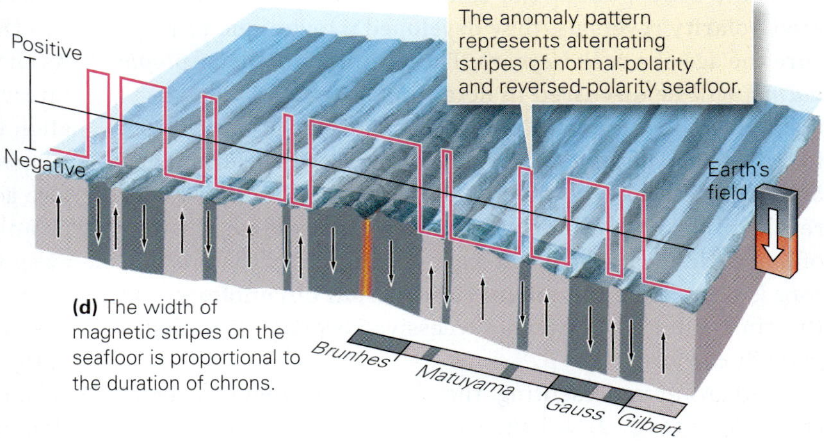

(d) The width of magnetic stripes on the seafloor is proportional to the duration of chrons.

Evidence from Deep-Sea Drilling

In the late 1960s, a research drilling ship called the *Glomar Challenger* set out to sail around the ocean drilling holes into the seafloor. This amazing ship could lower enough drill pipe to drill in 5-km-deep water and could continue to drill until the hole reached a depth of about 1.7 km (1.1 miles) below the seafloor. Drillers brought up cores of rock and sediment that geoscientists then studied on board.

To test the seafloor-spreading hypothesis, researchers proposed that the *Glomar Challenger* drill a series of holes, spaced at progressively greater distances from the axis of the Mid-Atlantic Ridge, through seafloor sediment to the basalt layer. If the model of seafloor spreading was correct, then not only should the sediment layer be progressively thicker away from the axis, but the age of the oldest sediment just above the basalt should be progressively older away from the axis. When the drilling and analyses were complete, this prediction was confirmed (**Fig. 2.13**). Thus, studies of both marine magnetic anomalies and the age of the seafloor supported the seafloor-spreading model.

FIGURE 2.13 Drilling into the sediment layer of the ocean floor confirmed that the age of the oldest sediment in contact with ocean-crust basalt gets older the farther away it is from the ridge. For example, Point A is older than Point B.

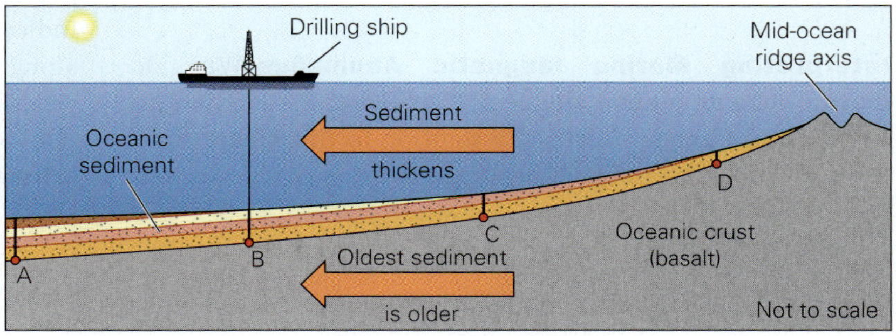

TAKE-HOME MESSAGE

The Earth's magnetic polarity flips every now and then. As a result, different stripes of ocean floor formed at mid-ocean ridges preserve different polarities, causing marine magnetic anomalies. The discovery of these anomalies, as well as documentation that the seafloor gets older away from the ridge axis, proved the seafloor-spreading hypothesis was correct.

QUICK QUESTION What determines the widths of marine magnetic anomalies?

2.6 What Do We Mean by Plate Tectonics?

The paleomagnetic proof of continental drift and the discovery of seafloor spreading set off a scientific revolution in geology in the 1960s and 1970s. Geologists realized that many of their existing interpretations of global geology, based on the premise that the positions of continents and oceans remain fixed in position through time, were simply wrong! Researchers dropped what they were doing and turned their attention to studying the broader implications of continental drift and seafloor spreading. It became clear that these phenomena required that the outer shell of the Earth be divided into rigid plates that moved relative to one another. New studies clarified the meaning of a plate, defined the types of plate boundaries, constrained plate motions, related plate motions to earthquakes and volcanoes, showed how plate interactions can explain mountain belts and seamount chains, and outlined the history of past plate motions. From these, the modern theory of plate tectonics evolved. Below, we first describe lithosphere plates and their boundaries, and then outline the basic principles of plate tectonics theory.

The Concept of a Lithosphere Plate

We learned earlier that geoscientists divide the outer part of the Earth into two layers. The **lithosphere** consists of the crust plus the top (cooler) part of the upper mantle. It behaves relatively rigidly, meaning that when a force pushes or pulls on it, it does not flow but rather bends or breaks (**Fig. 2.14a**). The lithosphere floats on a relatively soft, or "plastic," layer called the **asthenosphere**, composed of warmer (>1,280°C) mantle that can flow very slowly when acted on by a force.

FIGURE 2.14 The nature and behavior of the lithosphere.

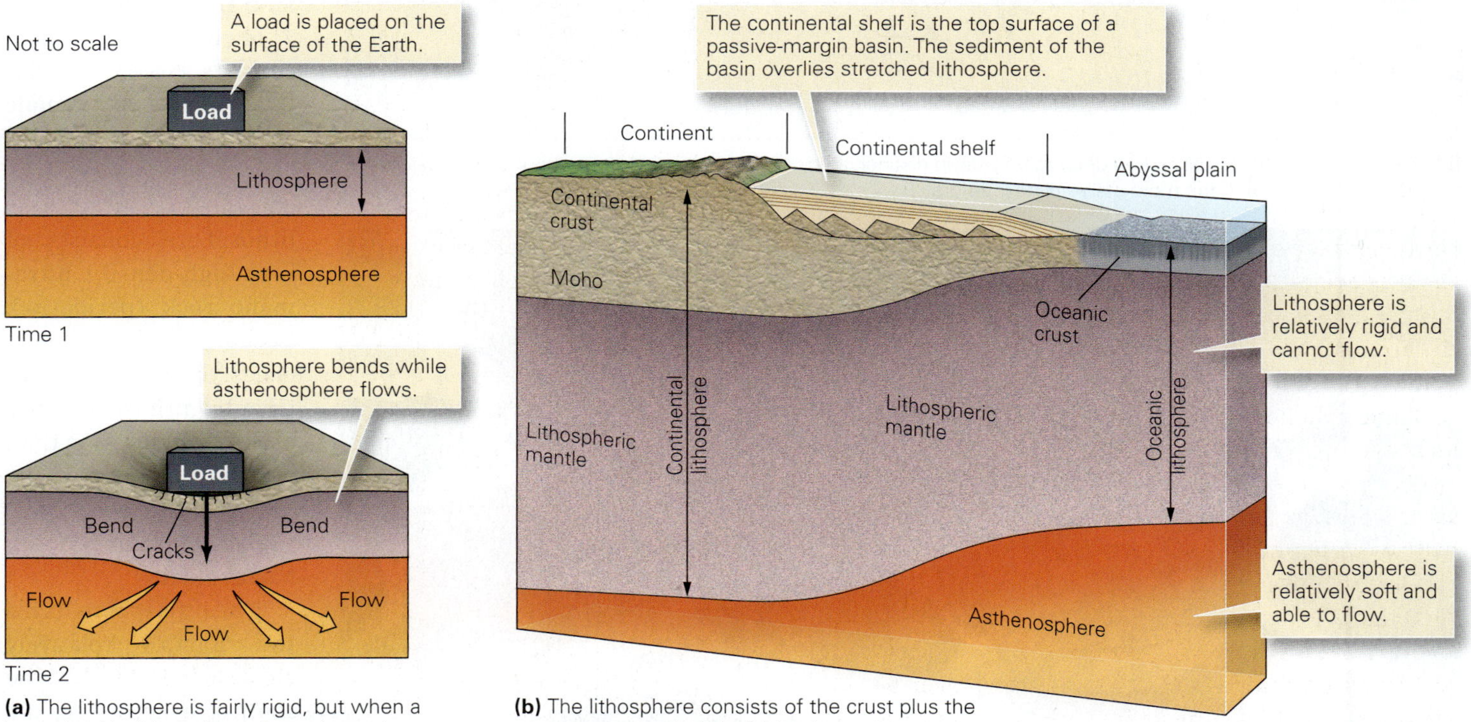

(a) The lithosphere is fairly rigid, but when a heavy load builds on its surface, the surface bends down. This can happen because the "plastic" asthenosphere can flow out of the way.

(b) The lithosphere consists of the crust plus the uppermost mantle. It is thicker beneath continents than beneath oceans.

FIGURE 2.15 The locations of plate boundaries and the distribution of earthquakes.

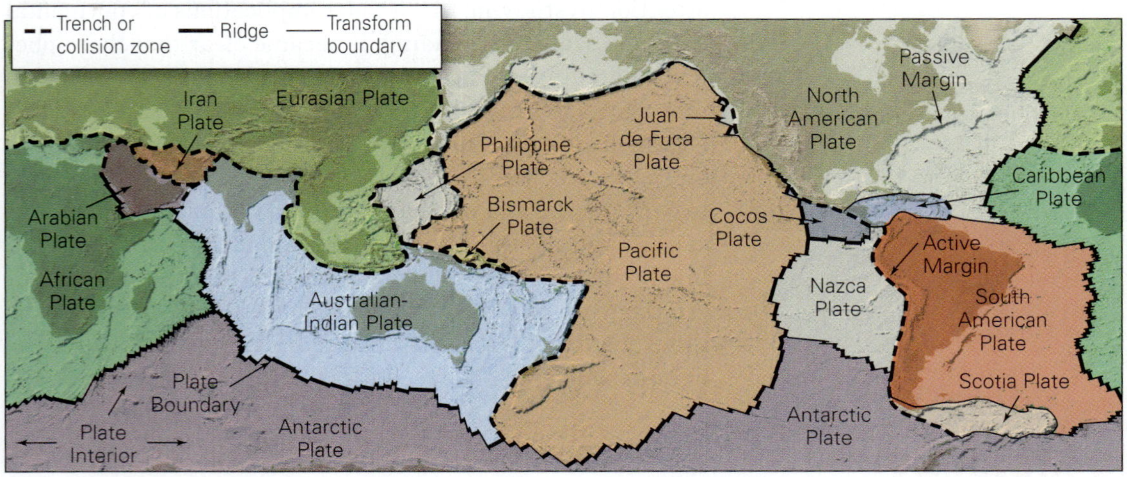

(a) A map of major plates shows that some consist entirely of oceanic lithosphere, whereas others consist of both continental and oceanic lithosphere. Active continental margins lie along plate boundaries; passive margins do not.

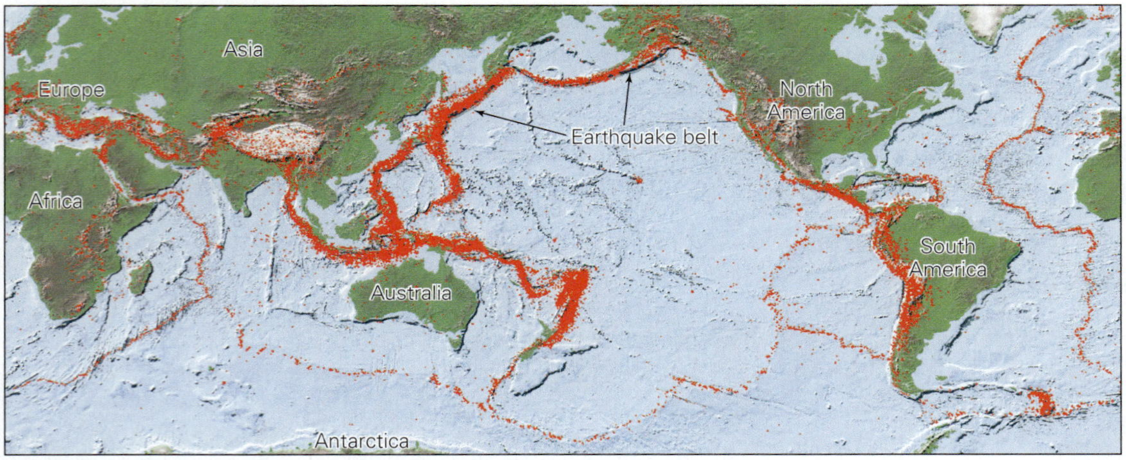

(b) The locations of earthquakes (red dots) mostly fall in distinct bands that correspond to plate boundaries. Relatively few earthquakes occur in the more stable plate interiors.

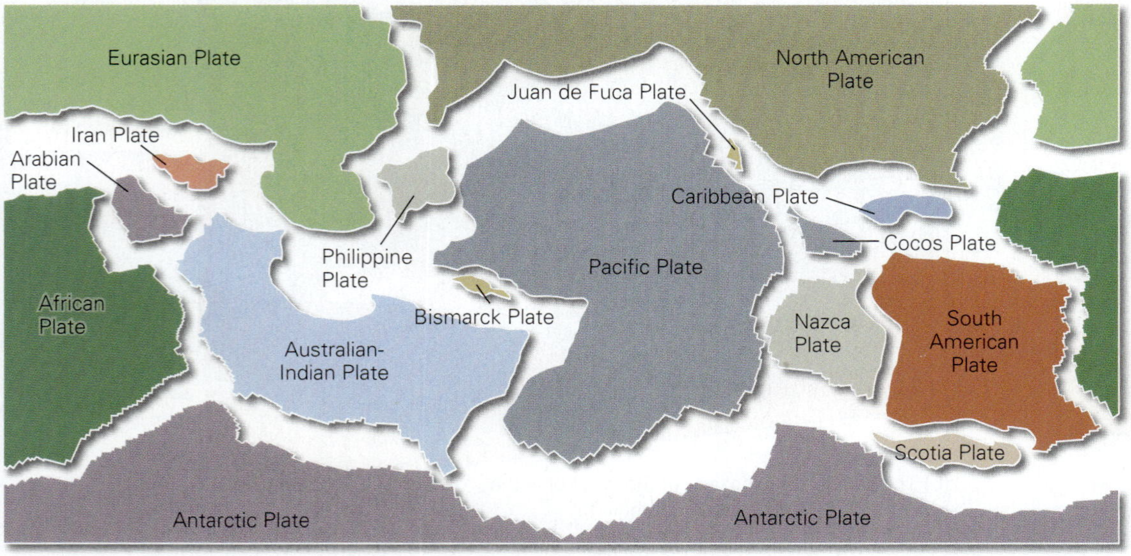

(c) An exploded view of the plates emphasizes the variation in shape and size of the plates.

As a result, the asthenosphere convects, like water in a pot, though much more slowly.

Continental lithosphere and oceanic lithosphere differ markedly in their thicknesses. On average, continental lithosphere has a thickness of 150 km, whereas old oceanic lithosphere has a thickness of about 100 km (**Fig. 2.14b**). (For reasons discussed later in this chapter, new oceanic lithosphere at a mid-ocean ridge is much thinner.) Recall that the crustal part of continental lithosphere ranges from 25 to 70 km thick and consists largely of low-density felsic and intermediate rock (see Chapter 1). In contrast, the crustal part of oceanic lithosphere is only 7 to 10 km thick and consists largely of relatively high-density mafic rock (basalt and gabbro). The mantle part of both continental and oceanic lithosphere consists of very-high-density ultramafic rock (peridotite). Because of these differences, the surface of continental lithosphere lies at a higher elevation than does the surface of the oceanic lithosphere—that's why distinct ocean basins exist.

The lithosphere forms the Earth's relatively rigid shell. But unlike the shell of a hen's egg, the lithospheric shell contains a number of major breaks, which separate it into distinct pieces. As noted earlier, we call the pieces *lithosphere plates*, or simply *plates*. The breaks between plates are known as **plate boundaries** (**Fig. 2.15a**). Geoscientists distinguish 12 major plates and several microplates.

The Basic Principles of Plate Tectonics Theory

With the background provided above, we can restate plate tectonics theory concisely as follows. The Earth's lithosphere is divided into plates that move relative to each other. As a plate moves, its internal area remains mostly, but not perfectly, rigid and intact. But rock along plate boundaries undergoes intense deformation (cracking, sliding, bending, stretching, and squashing) as the plate grinds or scrapes against its neighbors or pulls away from its neighbors. As plates move, so do the continents that form part of the plates. Because of plate tectonics, the map of Earth's surface constantly changes.

Identifying Plate Boundaries

How do we recognize the location of a plate boundary? The answer becomes clear from looking at a map showing the locations of earthquakes (**Fig. 2.15b**). Recall from Chapter 1 that earthquakes are vibrations caused by shock waves that are generated where rock breaks and suddenly slips along a fault. The *epicenter* marks the point on the Earth's surface directly above the *focus*, the location at which rock broke or slipped to cause the earthquake. Earthquake epicenters do not speckle the globe randomly, like buckshot on a target. Rather, the majority occur in relatively narrow, distinct belts. These earthquake belts, or *seismic belts*, define the position of plate boundaries because the fracturing and slipping that occurs along plate boundaries generates earthquakes. *Plate interiors*, regions away from the plate boundaries, remain relatively earthquake-free because they do not accommodate as much movement. While earthquakes serve as the most definitive indicator of a plate boundary, other prominent geologic features also develop along plate boundaries, as you will learn by the end of this chapter.

Did you ever wonder...
why earthquakes don't occur everywhere?

Note that some plates consist entirely of oceanic lithosphere, whereas some plates consist of both oceanic and continental lithosphere. Also, note that not all plates are the same size (**Fig. 2.15c**), and that some plate boundaries follow continental margins, the boundary between a continent and an ocean, but others do not. For this reason, we distinguish between **active margins**, which are plate boundaries, and **passive margins**, which are not plate boundaries. Earthquakes are common at active margins, but not at passive margins. Along passive margins, continental crust is thinner than in continental interiors. Thick (10 to 15 km) accumulations of sediment cover this thinned crust. The surface of this sediment layer is a broad, shallow (less than 500 m deep) region called the *continental shelf*, home to the major fisheries of the world.

Geologists define three types of plate boundaries, based simply on the relative motions of the plates on either side of the boundary (**Fig. 2.16**). A boundary at which two plates move apart from each other is a **divergent boundary**. A boundary at which two plates move toward each other so that one plate sinks beneath the other is a **convergent boundary**. And a boundary at which two plates slide sideways past each other is a **transform boundary**. Each type of boundary looks and behaves differently from the others, as we will see in the next three sections.

FIGURE 2.16 The three types of plate boundaries differ based on the nature of relative movement.

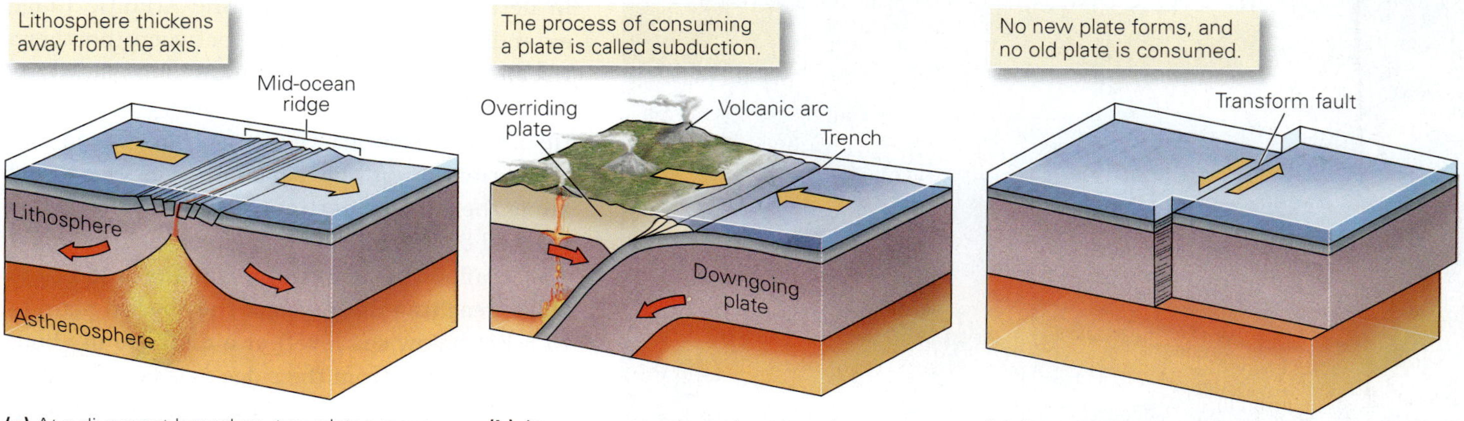

(a) At a divergent boundary, two plates move away from the axis of a mid-ocean ridge. New oceanic lithosphere forms.

(b) At a convergent boundary, two plates move toward each other; the downgoing plate sinks beneath the overriding plate.

(c) At a transform boundary, two plates slide past each other on a vertical fault surface.

FIGURE 2.17 Divergent-plate boundaries are delineated by mid-ocean ridges, where seafloor spreading occurs.

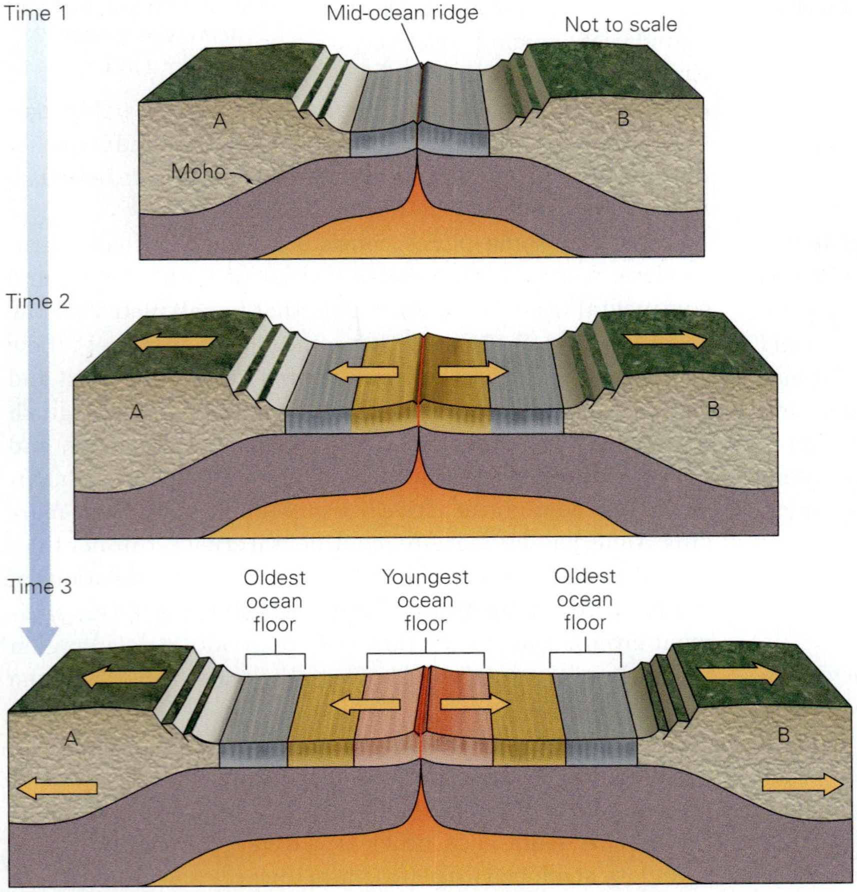

(a) During seafloor spreading, the ocean floor gets wider, and continents on either side move apart. New oceanic crust forms at the ridge axis. Rising asthenosphere melts beneath the axis.

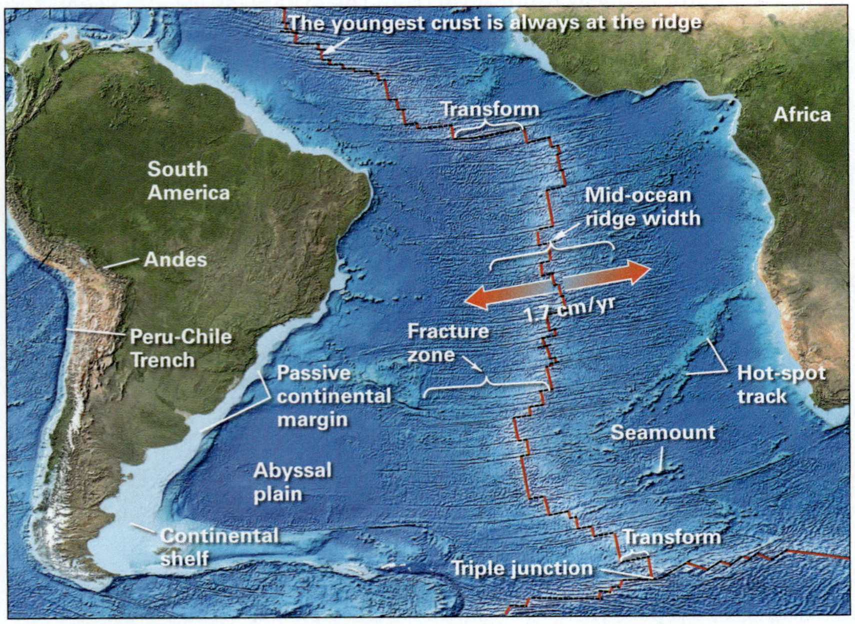

(b) The Mid-Atlantic Ridge in the South Atlantic Ocean. The lighter shades of blue are shallower water depths.

TAKE-HOME MESSAGE

Earth's rigid shell, its lithosphere, is divided into about 20 plates that move relative to one another. The motion of one plate relative to its neighbor is accommodated by fracturing and sliding along plate boundaries, generating earthquakes. Seismic belts, therefore, define plate boundaries.

QUICK QUESTION Can a plate include both continental and oceanic lithosphere? Are all continental margins seismically active?

2.7 Divergent Plate Boundaries and Seafloor Spreading

At a divergent boundary, or spreading boundary, two oceanic plates move apart by the process of seafloor spreading. Note that an open space does not develop between diverging plates. Rather, as the plates move apart, new oceanic lithosphere forms continually along the divergent boundary (**Fig. 2.17a**). This process takes place at submarine mountain ranges, called mid-ocean ridges, that rise 2 km above the adjacent abyssal plains of the ocean. Thus, geologists commonly refer to a divergent boundary as a *mid-ocean ridge*, or simply as a *ridge*. Water depth above ridges averages about 2.5 km.

To characterize a divergent boundary more completely, let's look at one mid-ocean ridge in detail (**Fig. 2.17b**). The Mid-Atlantic Ridge extends from the waters between northern Greenland and northern Scandinavia southward across the equator to the latitude of the southern tip of South America. Geologists have found that the formation of new seafloor takes place only along the axis (centerline) of the ridge, which is marked by an elongate trough, or *median valley*. The seafloor slopes away from the mid-ocean ridge, reaching the depth of the abyssal plain (4 to 5 km below sea level) at a distance of about 500 to 800 km from the ridge axis (see Fig. 2.7). Roughly speaking, the Mid-Atlantic Ridge is symmetrical—its eastern half looks like a mirror image of its western half. The ridge consists,

FIGURE 2.18 Geologic activity at a mid-ocean ridge.

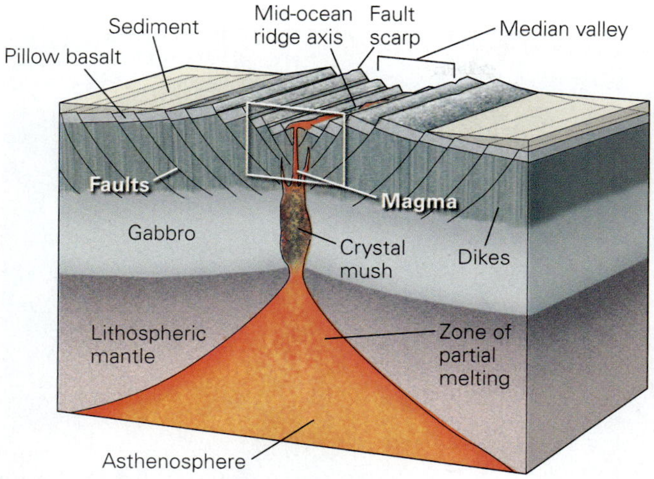

(a) Beneath a mid-ocean ridge, there is a magma chamber. Gabbro forms on the side of the magma chamber. Basalt dikes protrude upward.

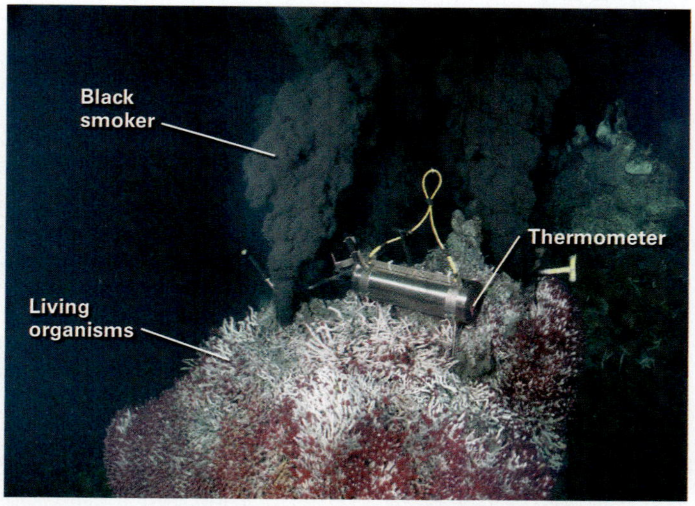

(b) Fine mineral grains precipitate as a cloud from a column of superhot water gushing from a black smoker. Bacteria, shrimp, and worms live around the vent.

along its length, of short segments (tens to hundreds of kilometers long) that step over at breaks that, as we noted earlier, are called fracture zones. Later, we will see that these zones contain transform boundaries.

How Does Oceanic Crust Form at a Mid-Ocean Ridge?

As seafloor spreading takes place, hot asthenosphere rises beneath the ridge and begins to melt, and molten rock, or magma, forms. (We will explain why this magma forms in Chapter 4.) Magma has a lower density than solid rock, so it behaves buoyantly and rises, as oil rises above vinegar in salad dressing. A mush of magma and solid crystals accumulates in the crust below the ridge axis, filling a region called a *magma chamber*. Some of the magma mush cools and solidifies completely along the side of the chamber to make the coarse-grained, mafic igneous rock called gabbro (**Fig. 2.18a**). (Igneous rocks, in general, form when a melt cools, so molecules lock together to produce solid crystals or glass.) Some of the liquid magma rises still higher to fill vertical cracks, where it solidifies and forms wall-like sheets, or dikes, of basalt. Some magma makes it all the way to the surface of the seafloor at the ridge axis and spills out of small submarine volcanoes as lava, which cools in a layer of blob-like shapes called pillow basalt. Observers in research submarines have found places along ridge axes where hot, mineralized water spews out from small chimneys, called **black smokers** because the rising water looks like a cloud of dark smoke. The dark color comes from a suspension of tiny mineral grains that precipitate in the water the instant that the hot water cools (**Fig. 2.18b**). As they accumulate, these minerals build the chimney of a black smoker.

As soon as it forms, new oceanic crust moves away from the ridge axis, and when this happens, more magma rises from below, so still more crust forms. In other words, like a vast, continuously moving conveyor belt, magma from the mantle rises to the Earth's surface at the ridge, solidifies to form oceanic crust, and then moves laterally away from the ridge. Because all seafloor forms at mid-ocean ridges, the youngest seafloor occurs on either side of the ridge axis, and seafloor becomes progressively older away from the ridge. In the Atlantic Ocean, the oldest seafloor, therefore, lies adjacent to the passive continental margins on either side of the ocean (**Fig. 2.19**). The oldest ocean floor on our planet underlies the western Pacific Ocean; this crust formed about 200 Ma (million years ago).

SEE FOR YOURSELF...

THE MID-ATLANTIC RIDGE

Latitude
2°34′26.28″ S

Longitude
15°37′59.86″ W

Fly to an elevation of 11,300 km (6,800 miles) and look straight down to see a view of the Mid-Atlantic Ridge in the equatorial Atlantic Ocean. Zoom in to see the fracture zone.

The tension (stretching force) applied to newly formed solid crust as spreading takes place breaks the crust, resulting in the formation of faults. Slip on the faults causes divergent-boundary earthquakes and produces numerous cliffs, or scarps, that trend parallel to the ridge axis.

FIGURE 2.19 This map of the world shows the age of the seafloor. Note how the seafloor grows older with increasing distance from the ridge axis. (Ma = million years ago.)

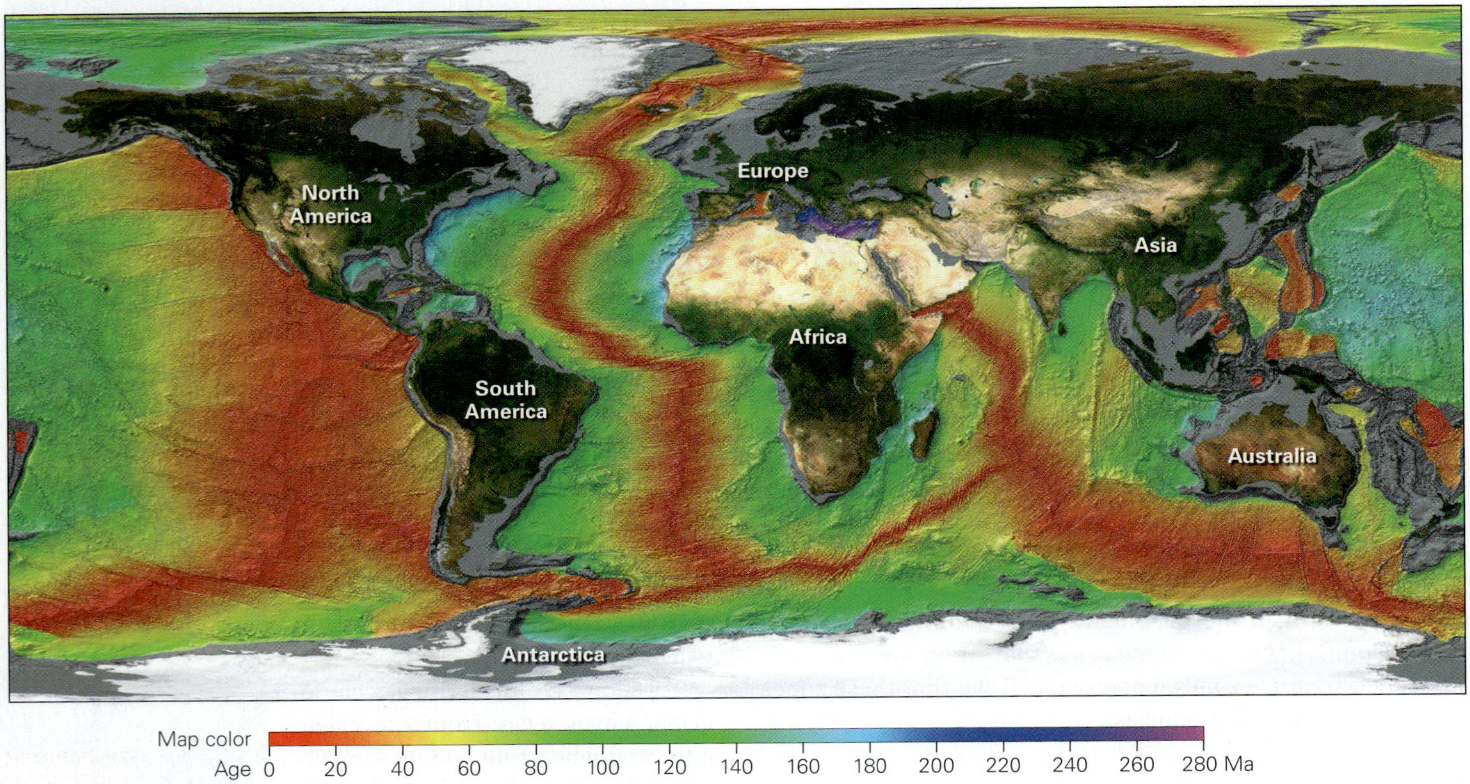

FIGURE 2.20 Changes accompanying the aging of lithosphere.

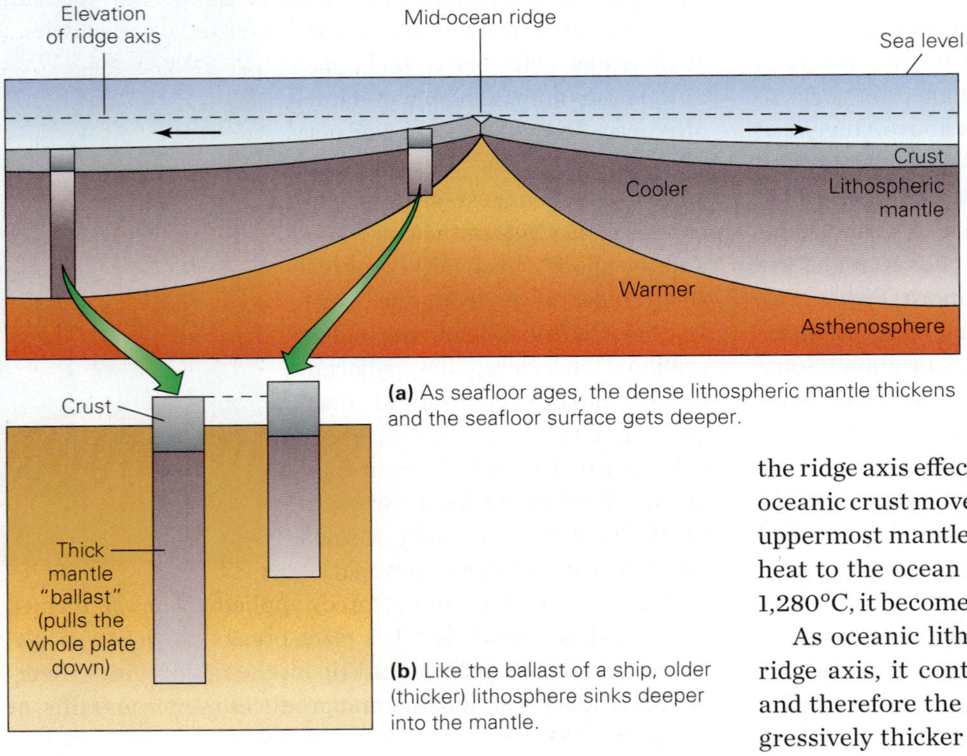

(a) As seafloor ages, the dense lithospheric mantle thickens and the seafloor surface gets deeper.

(b) Like the ballast of a ship, older (thicker) lithosphere sinks deeper into the mantle.

How Does the Lithospheric Mantle Form at a Mid-Ocean Ridge?

So far, we've seen how oceanic crust forms at mid-ocean ridges. How does the mantle part of the oceanic lithosphere form? This part consists of the cooler uppermost layer of the mantle, in which temperatures are less than about 1,280°C. At the ridge axis, such temperatures occur almost at the base of the crust, because of the presence of rising hot asthenosphere and hot magma, so lithospheric mantle beneath the ridge axis effectively doesn't exist. But as the newly formed oceanic crust moves away from the ridge axis, the crust and the uppermost mantle directly beneath it gradually cool by losing heat to the ocean above. As soon as mantle rock cools below 1,280°C, it becomes, by definition, part of the lithosphere.

As oceanic lithosphere continues to move away from the ridge axis, it continues to cool, so the lithospheric mantle, and therefore the oceanic lithosphere as a whole, grows progressively thicker (**Fig. 2.20**). Note that this process doesn't change the thickness of the oceanic crust, for the crust formed entirely at the ridge axis. The rate at which cooling and lithospheric thickening occur decreases progressively with increasing distance from the ridge axis. In fact, by the time the lithosphere is about 80 million years old, it has just about reached its maximum thickness. As lithosphere thickens and gets cooler and denser, it sinks down into the asthenosphere, like a ship taking on ballast. Thus, the deeper abyssal plains lie over older ocean floor, whereas the mid-ocean ridges lie over younger ocean floor.

TAKE-HOME MESSAGE

Seafloor spreading occurs at divergent boundaries, defined by mid-ocean ridges. New oceanic crust solidifies from basaltic magma along the ridge axis. As plates move away from the axis, they cool, and the lithospheric mantle forms and thickens.

QUICK QUESTION What are black smokers, and why do they form?

2.8 Convergent Plate Boundaries and Subduction

At convergent boundaries, two plates, at least one of which is oceanic, move toward one another. But rather than butting each other like angry rams, one oceanic plate bends and sinks down into the asthenosphere beneath the other plate. Geologists refer to the sinking process as **subduction**, so convergent boundaries are also known as *subduction zones*. Because subduction at a convergent boundary "consumes" old ocean lithosphere, geologists also refer to convergent boundaries as *consuming boundaries*, and because they are delineated by deep-ocean trenches, they are sometimes simply called trenches (**Fig. 2.21a**). The amount of oceanic plate consumption worldwide, averaged over time, equals the amount of seafloor spreading worldwide, so the surface area of the Earth remains constant through time.

Subduction occurs for a simple reason: oceanic lithosphere, once it has aged at least 10 million years, is denser than the underlying asthenosphere and thus can sink through the asthenosphere if given an opportunity. Where it lies flat on the surface of the asthenosphere, oceanic lithosphere can't sink. However, once the end of the convergent plate bends down and slips into the mantle, it continues downward like an anchor falling to the bottom of a lake (**Fig. 2.21b**). As the lithosphere sinks, asthenosphere flows out of its way, just as water flows out of the way of a sinking anchor. But unlike water, the asthenosphere can flow only very slowly, so oceanic lithosphere can sink only very slowly, at a rate of less than about 15 cm per year. To visualize the difference, imagine how much faster a coin can sink through water than it can through honey. Note that it's only because the asthenosphere can flow that plate tectonics can take place—if the asthenosphere were too rigid to flow, plates could not subduct and asthenosphere wouldn't be able to rise beneath mid-ocean ridges.

While the *downgoing plate*, the plate that has been subducted, must be composed of oceanic lithosphere, the *overriding plate*, which does not sink, can consist of either oceanic or continental lithosphere. Continental crust cannot be subducted because it is too buoyant; the low-density rocks of continental crust act like a life preserver keeping the continent afloat. If continental crust moves into a convergent margin, subduction eventually stops. Because of subduction, all ocean floor on the planet is less than about 200 million years old. Because continental crust cannot subduct, some continental crust has persisted at the surface of the Earth for over 3.8 billion years.

Earthquakes and the Fate of Subducted Plates

At convergent boundaries, the downgoing plate grinds along the base of the overriding plate, a process that generates large earthquakes. These earthquakes occur fairly close to the Earth's surface, so some of them cause massive destruction in coastal cities. But earthquakes also happen in downgoing plates at greater depths. In fact, geologists have detected earthquakes within downgoing plates to a depth of 660 km. The band of earthquakes in a downgoing plate is called a **Wadati-Benioff zone**, after its two discoverers (**Fig. 2.21c**).

At depths greater than 660 km, conditions leading to earthquakes in subducted lithosphere evidently do not occur. Recent observations, however, indicate that some downgoing plates do continue to sink below a depth of 660 km—they just do so without generating earthquakes. In fact, the lower mantle may be a graveyard for old subducted plates.

Geologic Features of a Convergent Boundary

To become familiar with the various geologic features that occur along a convergent boundary, let's look at an example, the boundary between the western coast of the South American Plate and the eastern edge of the Nazca Plate (a portion of the Pacific Ocean floor). A deep-ocean trench, the Peru-Chile Trench, delineates this boundary (**Fig. 2.22a, b**). Such trenches form where the plate bends as it starts to sink into the asthenosphere.

FIGURE 2.21 During the process of subduction, oceanic lithosphere sinks back into the deeper mantle.

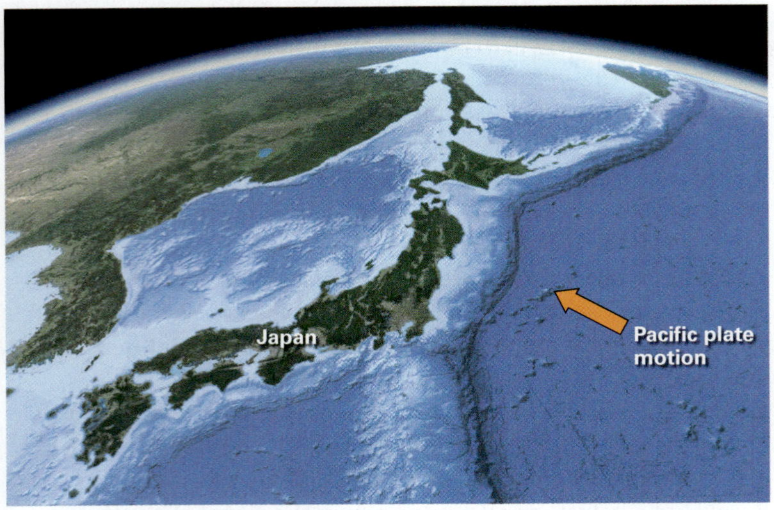

(a) A trench delineates the place where the Pacific Plate is being subducted.

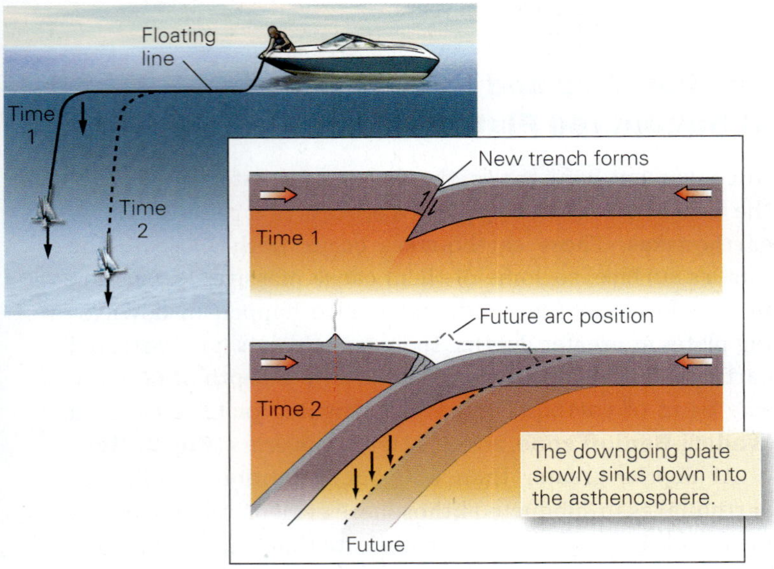

(b) Sinking of the downgoing plate resembles sinking of an anchor attached to a line.

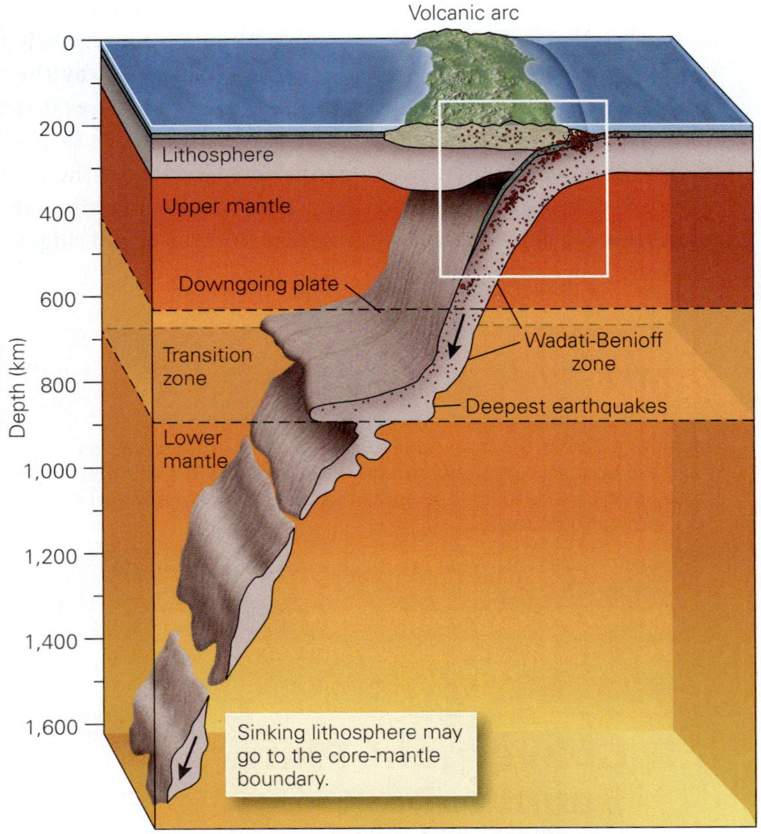

(c) A belt of earthquakes (dots) defines the position of the downgoing plate in the region above a depth of about 660 km. Sometimes plates "pile up" at a depth of 660 km.

In the Peru-Chile Trench, as the downgoing plate slides under the overriding plate, sediment (clay and plankton) that had settled on the surface of the downgoing plate, as well as sand that fell into the trench from the shores of South America, gets scraped up and incorporated in a wedge-shaped mass known as an **accretionary prism** (**Fig. 2.22c**). An accretionary prism forms in basically the same way as a pile of snow or sand gathers in front of a plow, and like snow, the sediment tends to be squashed and contorted.

A chain of volcanoes known as a volcanic arc develops on the edge of the overriding plate behind the accretionary prism. As we will see in Chapter 4, the magma that feeds these volcanoes forms just above the surface of the downgoing plate where the plate reaches a depth of about 150 km below the Earth's surface. If the volcanic arc forms where an oceanic plate subducts beneath continental lithosphere, the resulting chain of volcanoes grows on the continent and forms a *continental volcanic arc*. (In some cases, the plates squeeze together across a continental arc, causing a belt of faults to form behind the arc.) If, however, the volcanic arc grows where one oceanic plate subducts beneath another oceanic plate, the resulting volcanoes form a chain of islands known as a *volcanic island arc* (**Fig. 2.22d**). A *back-arc basin* exists either where subduction happens to begin offshore, trapping ocean lithosphere behind the arc, or where stretching of the lithosphere behind the arc leads to the formation of a small spreading ridge behind the arc (**Fig. 2.22e**).

FIGURE 2.22 The nature of a convergent-plate margin varies with location.

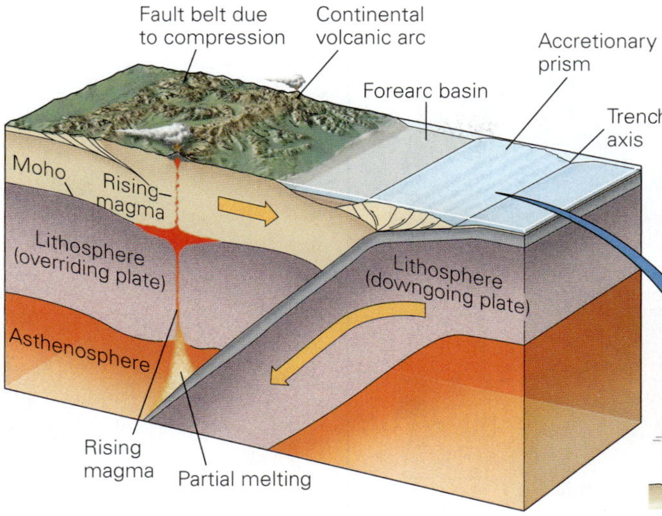

(a) A subduction zone along the edge of a continent. Here compression caused faulting behind the arc.

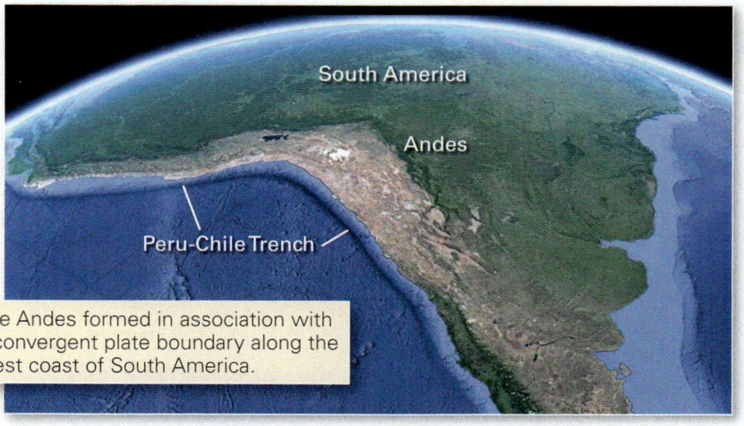

(b) The Peru-Chile Trench, as seen from space.

(c) The overriding plate acts like a bulldozer, scraping up sediment off the downgoing plate to build an accretionary prism.

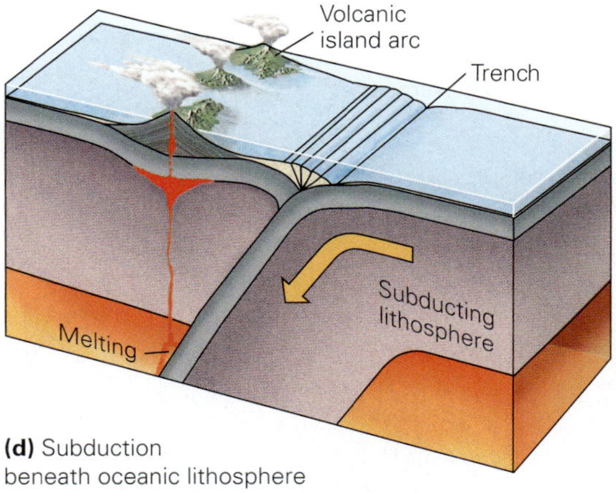

(d) Subduction beneath oceanic lithosphere produces an island arc.

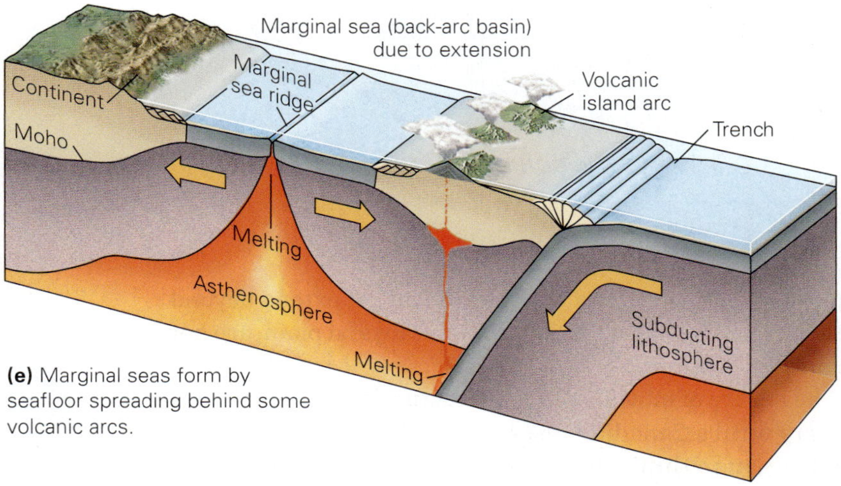

(e) Marginal seas form by seafloor spreading behind some volcanic arcs.

TAKE-HOME MESSAGE

At a convergent boundary, an oceanic plate sinks into the mantle beneath the edge of another plate. A volcanic arc and a trench delineate such plate boundaries, and earthquakes happen along the contact between the two plates as well as in the downgoing slab. Volcanic arcs can form on the edge of a continent or as a chain of islands in the sea.

QUICK QUESTION Can continents be completely subducted?

2.9 Transform Plate Boundaries

When researchers began to explore the bathymetry of mid-ocean ridges in detail, they discovered that mid-ocean ridges are not long, uninterrupted lines, but rather consist of short segments that appear to be offset laterally from each other (**Fig. 2.23a**) by narrow belts of broken and irregular seafloor. These belts, or **fracture zones**, lie roughly at right angles to the ridge segments, intersect the ends of the segments, and extend beyond the ends of the segments. Originally, researchers incorrectly assumed that the entire length of each fracture zone was a fault, and that slip on a fracture zone had displaced segments of the mid-ocean ridge sideways, relative to each other. In other words, they imagined that a mid-ocean ridge initiated as a continuous, fence-like line that only later was broken up by faulting. But when information about the distribution of earthquakes along mid-ocean ridges became available, it was clear that this model could not be correct. Earthquakes, and therefore active fault slip, occur only on the segment of a fracture zone that lies between two ridge

FIGURE 2.23 The concept of transform faulting.

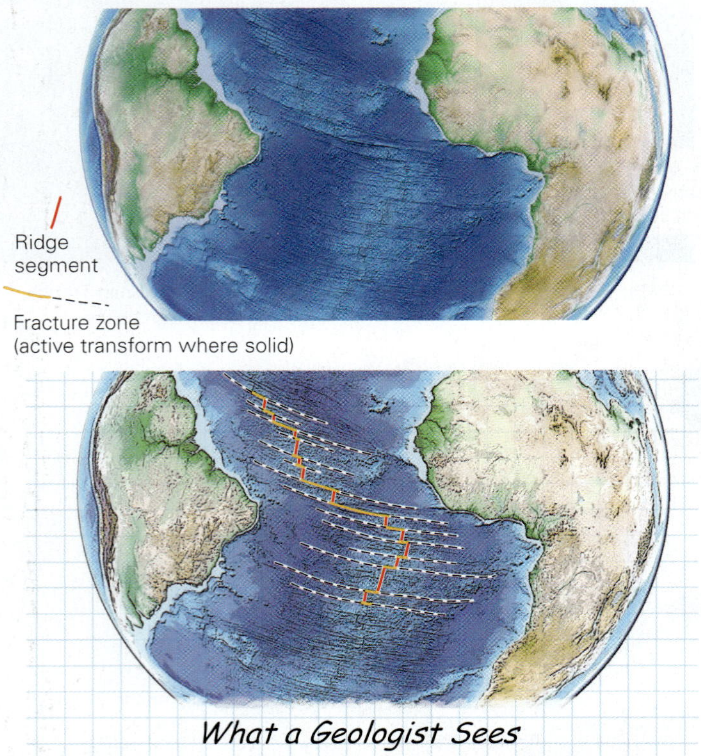

(a) Numerous transform faults segment the Mid-Atlantic Ridge.

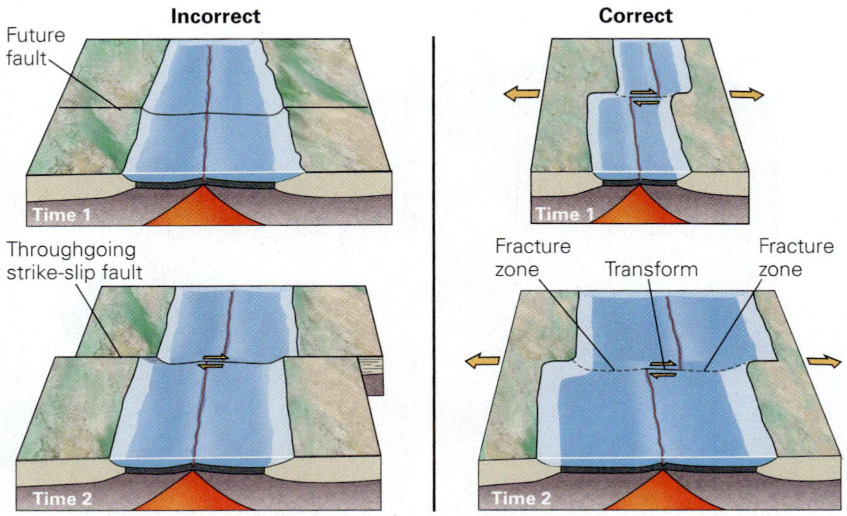

(b) A comparison of the old model of transform faults with the new model required by the seafloor-spreading hypothesis.

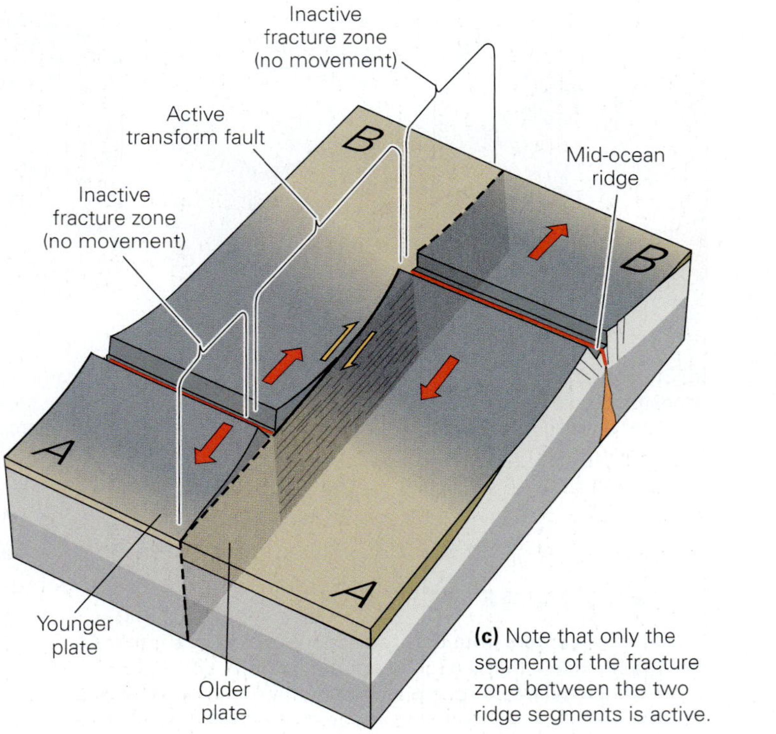

(c) Note that only the segment of the fracture zone between the two ridge segments is active.

segments. The portions of fracture zones that extend beyond the edges of ridge segments, out into the abyssal plain, are not seismically active.

The nature of movement along fracture zones remained a mystery until a Canadian researcher, J. Tuzo Wilson (1908–1993), began to think about fracture zones in the context of the seafloor-spreading concept. Wilson proposed that fracture zones formed at the same time as the ridge axis itself, and thus the ridge consisted of separate segments to start with. These segments were linked (not offset) by fracture zones. With this idea in mind, he drew a sketch map showing two ridge-axis segments linked by a fracture zone, and he drew arrows to indicate the direction that ocean floor was moving, relative to the ridge axis, as a result of seafloor spreading (**Fig. 2.23b**). Look at the arrows in Figure 2.23b. Clearly, the movement direction on the active portion of the fracture zone must be opposite to the movement direction that researchers originally thought occurred across the structure. Further, in Wilson's model, slip occurs only along the segment of the fracture zone between the two ridge segments (**Fig. 2.23c**). Plates on opposite sides of the inactive part of a fracture zone move together, as one plate.

Wilson introduced the term *transform boundary,* or *transform fault*, for the actively slipping segment of a fracture zone between two ridge segments, and he pointed out that these are a third type of plate boundary. At a transform boundary, one plate slides sideways past another, but no new plate forms and no old plate is consumed. Transform boundaries are, therefore, defined by a vertical fault on which the slip direction parallels the Earth's surface. The slip breaks up the crust and forms a set of steep fractures.

So far we've discussed only transforms along mid-ocean ridges. Not all transforms link ridge segments. Some, such as the Alpine fault of New Zealand, link trenches, while others link a trench to a ridge segment. Further, not all transform

FIGURE 2.24 The San Andreas fault—a continental transform boundary.

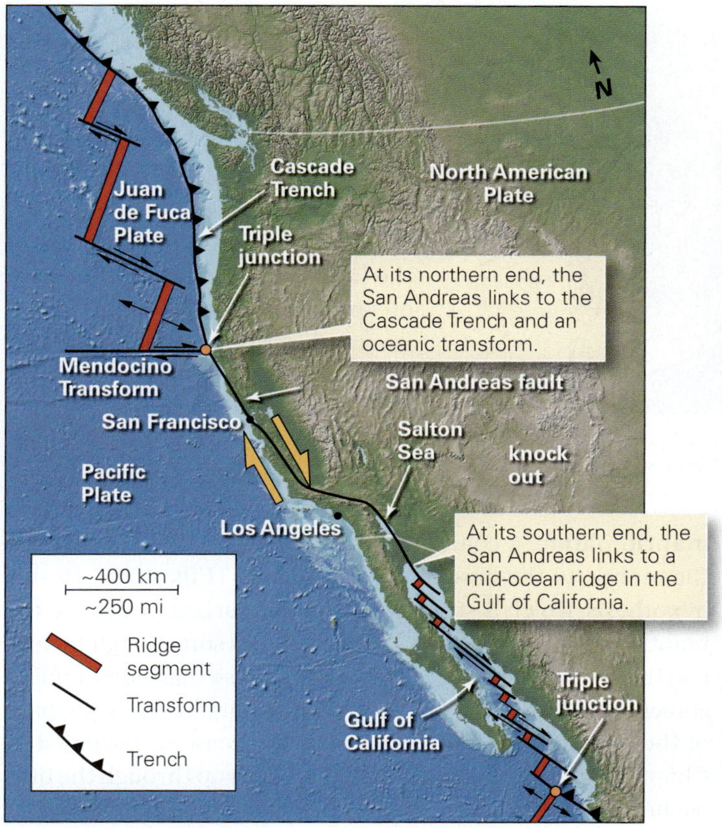

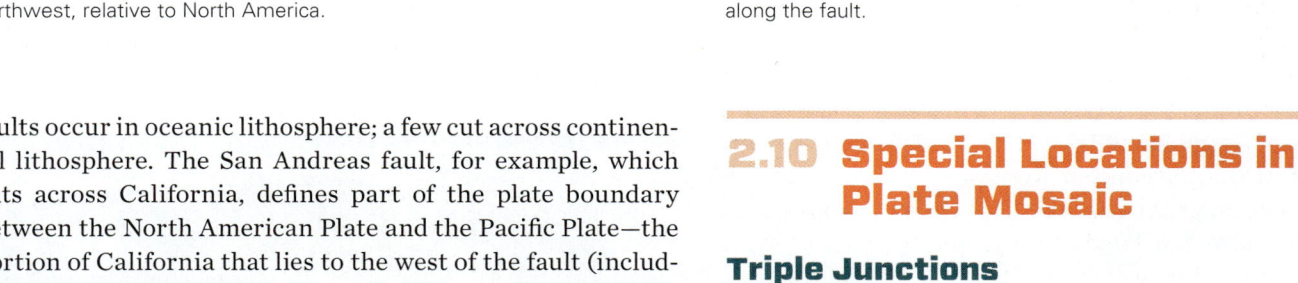

(a) The San Andreas fault is a transform plate boundary between the North American and Pacific Plates. The Pacific is moving northwest, relative to North America.

(b) In southern California, the San Andreas fault cuts a dry landscape. The fault trace is in the narrow valley. The land has been pushed up slightly along the fault.

faults occur in oceanic lithosphere; a few cut across continental lithosphere. The San Andreas fault, for example, which cuts across California, defines part of the plate boundary between the North American Plate and the Pacific Plate—the portion of California that lies to the west of the fault (including Los Angeles) is part of the Pacific Plate, while the portion that lies to the east of the fault is part of the North American Plate (**Fig. 2.24**).

TAKE-HOME MESSAGE

At transform boundaries, one plate slips sideways past another along a vertical fault. Thus, there is no production or destruction of lithosphere at a transform boundary. Most transform boundaries link segments of mid-ocean ridges, but some, such as the San Andreas fault, cut across continental crust.

QUICK QUESTION Why do earthquakes only occur on the portion of a fracture zone that links mid-ocean ridge segments?

2.10 Special Locations in the Plate Mosaic

Triple Junctions

Geologists refer to a point where three plate boundaries intersect as a **triple junction**. We can name triple junctions after the types of boundaries that intersect. For example, the triple junction formed where the Southwest Indian Ocean Ridge intersects two arms of the Mid–Indian Ocean Ridge is a ridge-ridge-ridge triple junction (**Fig. 2.25a**), whereas the triple junction north of San Francisco, where the Cascadia trench, the San Andreas fault, and the Mendocino fracture zone intersect, is a trench-transform-transform triple junction (**Fig. 2.25b**).

Hot Spots

The volcanoes of both volcanic arcs and mid-ocean ridges are *plate-boundary volcanoes*, in that they formed as a

FIGURE 2.25 Examples of triple junctions. The triple junctions are marked by dots.

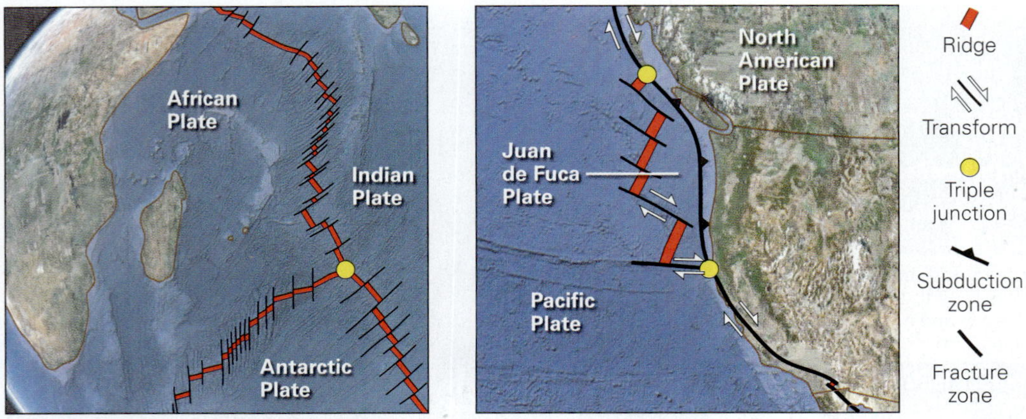

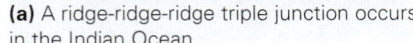

(a) A ridge-ridge-ridge triple junction occurs in the Indian Ocean.

(b) A trench-transform-transform triple junction occurs at the north end of the San Andreas fault.

consequence of movement along the boundary. Not all volcanoes on Earth, however, are plate-boundary volcanoes. Worldwide, geoscientists have identified about 100 volcanoes that exist as isolated points. These are called hot-spot volcanoes, or simply **hot spots** (**Fig. 2.26**).

What causes hot-spot volcanoes? In the early 1960s, J. Tuzo Wilson noted that active hot-spot volcanoes, examples that are erupting or may erupt in the future, occur at the end of a chain of extinct volcanoes, meaning volcanoes that will never erupt again. This configuration is different from that of volcanic arcs along convergent boundaries—at volcanic arcs, all of the volcanoes are active. With this image in mind, Wilson suggested that the position of the heat source causing a hot-spot volcano is fixed, relative to the moving plate. In Wilson's model, the active volcano represents the present-day location of the heat source, whereas the chain of extinct volcanic islands represents locations on the plate that were once over the heat source but progressively moved off.

A few years after Wilson's proposal, researchers suggested that the heat source for hot spots is a **mantle plume**, a column of very hot rock rising up through the mantle to the base of the lithosphere (**Fig. 2.27**). In this hypothesis, which is still debated, plumes originate deep in the mantle. Rock in the plume, though solid, is soft enough to flow, and it rises buoyantly because it is less dense than surrounding cooler rock. When the hot rock of the plume reaches the base of the lithosphere, it begins to melt (for reasons discussed in Chapter 4) and produces magma that seeps up through the lithosphere to the Earth's surface. The chain of extinct volcanoes delineates a **hot-spot track**, and it forms when the overlying

FIGURE 2.26 Locations of selected hot-spot volcanoes and hot-spot tracks. The most recent volcano (dot) is at one end of the track (red lines). Some of these volcanoes are extinct. Some hot spots are fairly recent and do not have tracks. Dashed tracks were broken by seafloor spreading.

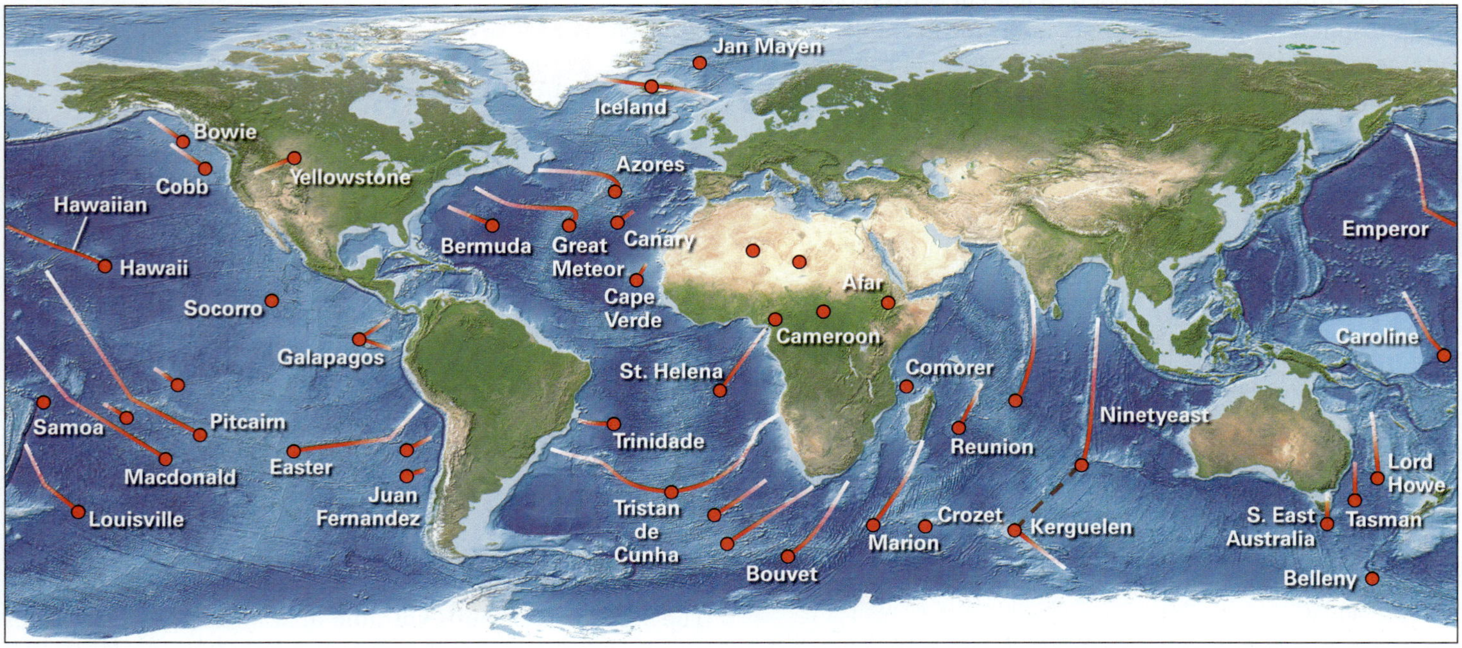

FIGURE 2.27 The deep mantle plume hypothesis for the formation of hot-spot tracks.

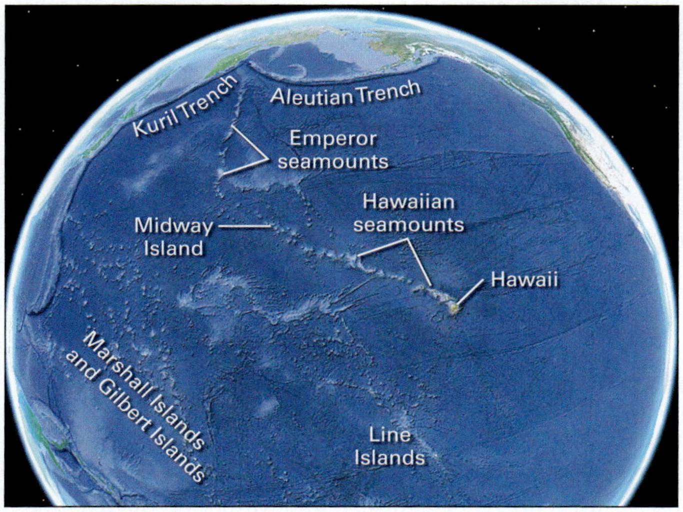

(a) A bathymetric map showing the hot-spot tracks of the Pacific Ocean.

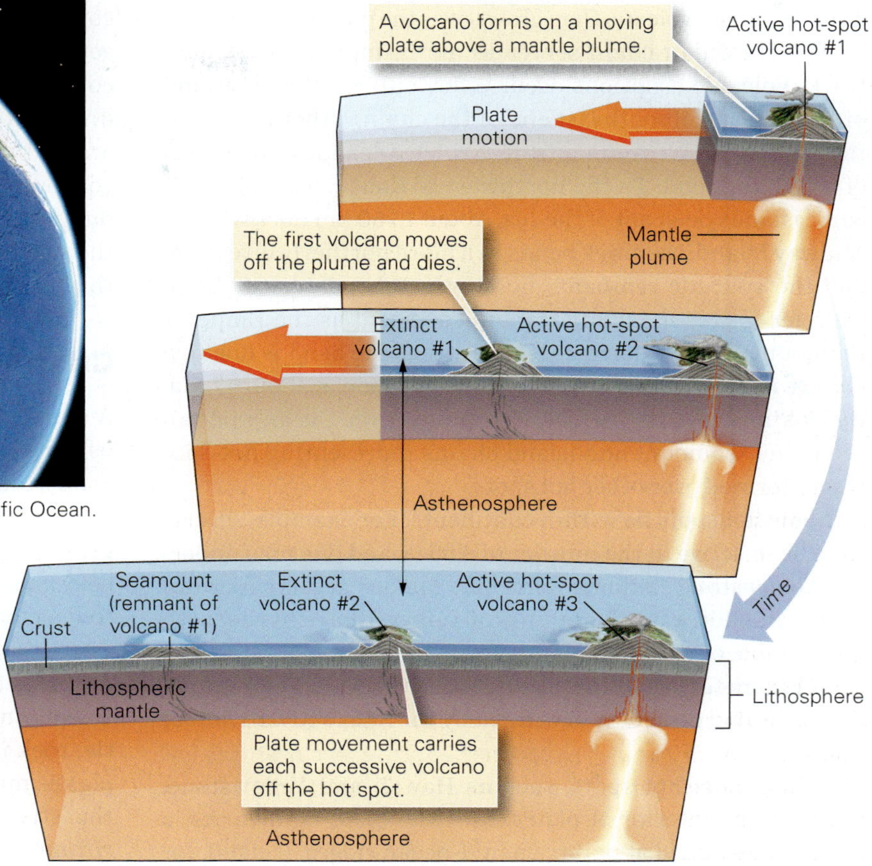

(b) Progressive stages in the development of a hot-spot track, according to the plume model.

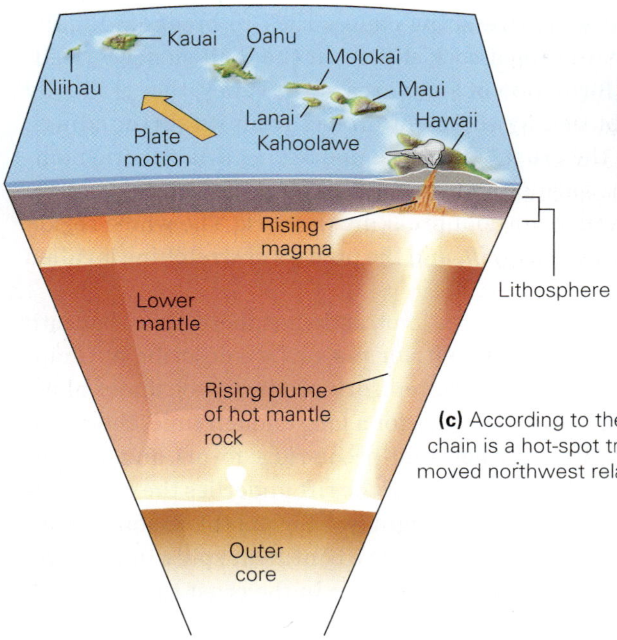

(c) According to the plume model, the Hawaiian island chain is a hot-spot track that formed as the Pacific Plate moved northwest relative to a plume.

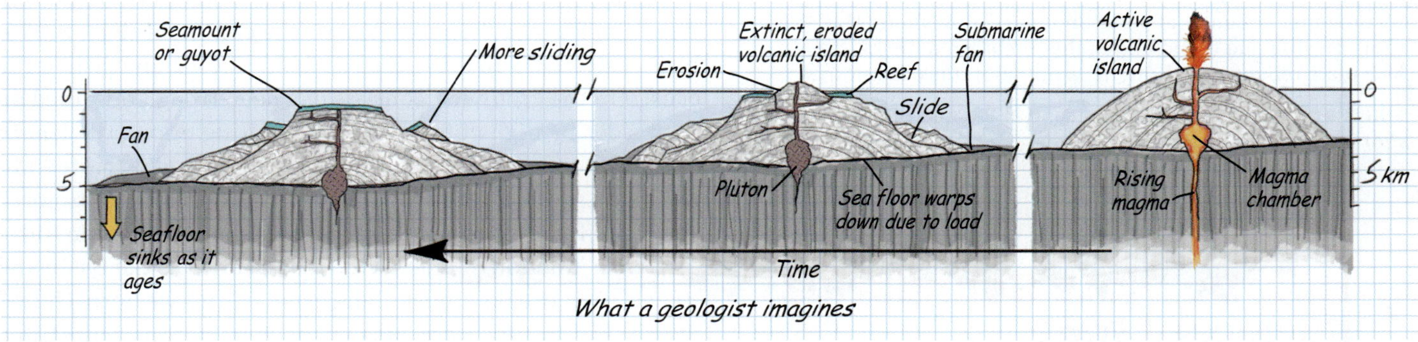

(d) As a volcano moves off the hot spot, it gradually sinks below sea level due to sinking of the plate, erosion, and submarine landslides.

plate moves over a fixed plume. This movement slowly carries the volcano off the top of the plume, so that it becomes extinct. A new, younger volcano then grows over the plume.

The Hawaiian chain serves as an example of a hot-spot track. Volcanic eruptions occur today only on the big island of Hawaii, at the southeast end of the chain. Other islands to the northwest are remnants of extinct volcanoes, the oldest of which is Kauai. To the northwest of Kauai, still older volcanic islands peak out of the sea. About 1,750 km northwest of Midway Island, the track bends in a more northerly direction, and the volcanic remnants no longer poke above sea level; we refer to this northerly trending segment as the Emperor seamount chain. Geologists suggest that the bend is due to a change in the direction of Pacific Plate motion at about 47 Ma. By looking at a bathymetric map of the Pacific Ocean, you can see many seamount and island chains representing hot-spot tracks formed over other hot spots.

Some hot spots lie within continents. For example, several have been active in the interior of Africa, and one now underlies Yellowstone National Park. The famous geysers (natural steam and hot-water fountains) of Yellowstone exist because hot magma, or still hot recently formed igneous rock, lies just a few kilometers below the landscape of the park. Yellowstone lies at the end of a hot spot defined by the Snake River Plain of Idaho.

While most hot spots, such as Hawaii and Yellowstone, occur in the interior of plates, far from plate boundaries, a few are positioned at points on mid-ocean ridges. Iceland, for example, is the product of hot-spot volcanism on the axis of the Mid-Atlantic Ridge. We recognize Iceland as a hot spot because its volcanic activity yields vastly more molten rock than do typical localities along the Mid-Atlantic Ridge. The additional magma production built Iceland into plateau that has risen almost 3 km above normal ridge-axis depths.

TAKE-HOME MESSAGE

A triple junction marks the point where three plate boundaries join. A hot spot is a point where volcanism occurs independently of plate-boundary movement. Hot spots may be due to melting at the top of a mantle plume. As a plate moves over a plume, a hot-spot track develops.

QUICK QUESTION Why is the volcano at the end of a hot-spot track the only one to be active?

2.11 How Do Plate Boundaries Form and Die?

The configuration of plates and plate boundaries visible on our planet today has not existed for all of geologic history, and it will not exist indefinitely into the future. Because of plate motion, oceanic plates form and are later consumed, while continents merge and later split apart. How does a new divergent boundary come into existence, and how does an existing convergent boundary eventually cease to exist? Most new divergent boundaries form when a continent splits and separates into two continents. We call this process **rifting**. A convergent boundary ceases to exist when a piece of relatively buoyant crust, such as a continent or an island arc, moves into the subduction zone and, in effect, jams up the system. We call this process **collision**.

Continental Rifting

A **continental rift** is a linear belt in which continental lithosphere pulls apart (**Fig. 2.28**). During the process, the lithosphere stretches horizontally and thins vertically, much like a piece of chewing gum that you pull between your fingers. Nearer the surface of the continent, where the crust is cold and brittle, stretching causes rock to break and faults to develop. Blocks of rock slip down the fault surfaces, leading to the formation of a low area, a rift valley, that gradually becomes buried by sediment eroded from its surroundings. Deeper in the crust, and in the underlying lithospheric mantle, rock is warmer and softer, so stretching takes place in a plastic manner, without breaking the rock. The whole region that stretches is the *rift,* and the process of stretching is called *rifting.*

As rifting takes place, hot asthenosphere rises beneath the rift and starts to melt, for reasons we describe in Chapter 4. Eruption of the molten rock produces volcanoes along the rift. If rifting continues for a long enough time, the continent breaks in two, a new mid-ocean ridge forms, and seafloor spreading begins. The relict of the rift underlies the continental shelf along a passive margin (see Fig. 2.14b). In some cases, however, rifting stops before the continent splits in two, and a rift remains as a permanent scar in the crust that fills with sediment.

A major rift, the East African Rift, extends in a north-south direction for over 3,500 km across the continent (**Fig. 2.29a, b**). To astronauts in orbit, the rift looks like a giant gash in the crust. On the ground, it consists of a deep trough bordered on both sides by high cliffs formed by faulting. Along the length of the rift, several major volcanoes, such as snow-crested Mt. Kilimanjaro, smoke and fume above the savannah. At its north end, the rift joins the Red Sea Ridge and the Gulf of Aden Ridge at a triple junction. A much wider rift, known as the Basin and Range Rift, breaks up the landscape of the western United States (**Fig. 2.29c**). Here, movement on numerous faults tilted blocks of crust to form narrow mountain ranges, while sediment that eroded from the blocks filled the adjacent basins (the low areas between the ranges).

FIGURE 2.28 During rifting, continental lithosphere stretches and thins. If rifting succeeds, a new mid-ocean ridge forms.

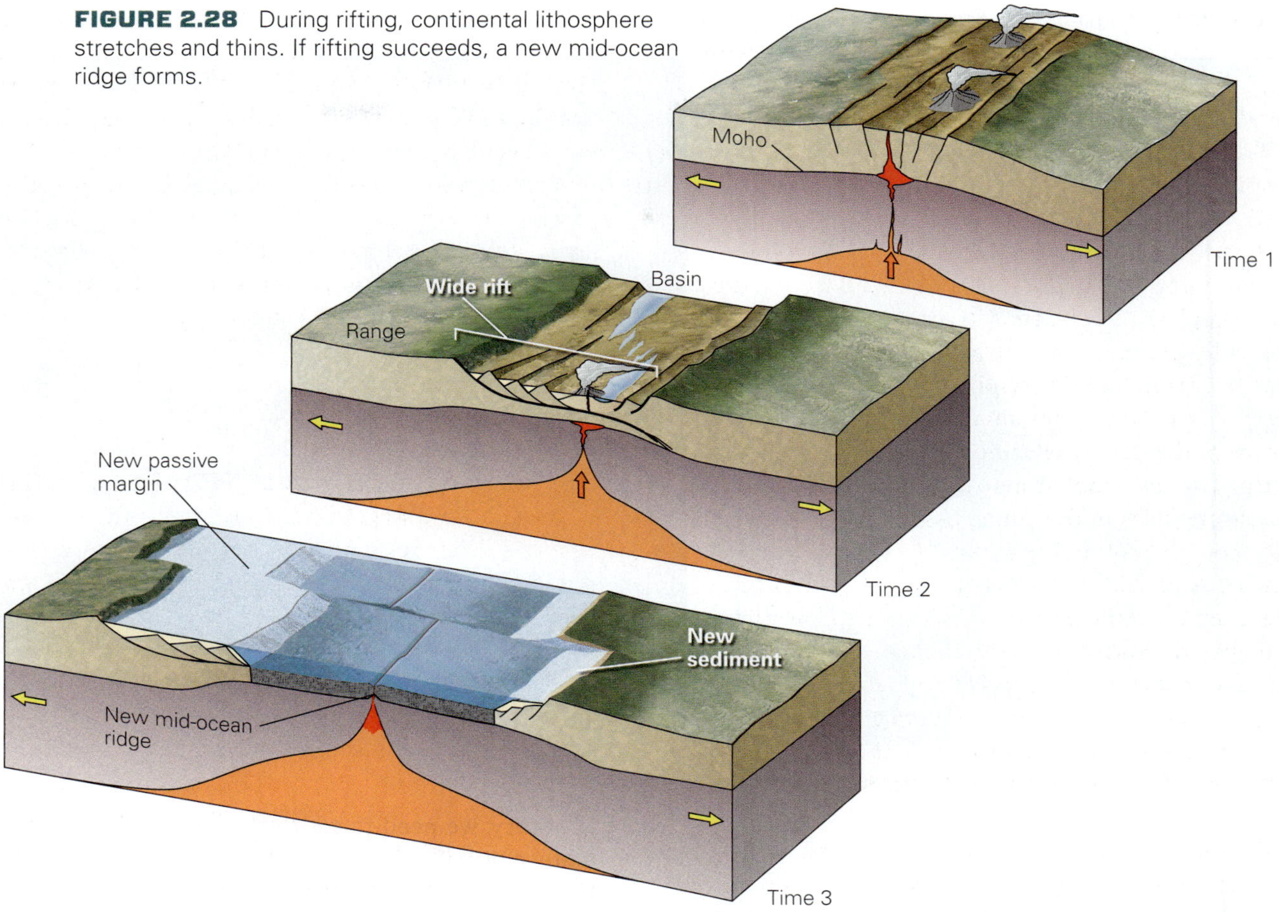

Collision

After the breakup of Pangaea, India was a small, separate continent that lay far to the south of Asia. As time passed, subduction consumed the ocean between India and Asia, and India moved northward, finally slamming into the southern margin of Asia about 40 to 50 Ma. Continental crust, unlike oceanic crust, is too buoyant to subduct. So when India collided with Asia, the attached oceanic plate broke off and sank down into the deep mantle while India pushed hard into and partly under Asia, squeezing the rocks and sediment that once lay between the two continents into the 8-km-high welt that we now know as the Himalayan Mountains. During this process, not only did the surface of the Earth rise, but the crust became thicker. In fact, the crust beneath a collisional mountain range can be up to 60 to 70 km thick, about twice the thickness of normal continental crust. The boundary between what was once two separate continents is called a *suture*; slivers of ocean crust may be trapped along a suture.

Geoscientists refer to the process during which two buoyant pieces of lithosphere converge and squeeze together as collision (**Fig. 2.30**). Some collisions involve two continents, whereas some involve continents and an island arc. When a collision is complete, the convergent boundary that once existed between the two colliding pieces ceases to exist. Collisions yield some of the most spectacular mountains on the planet, such as the Himalayas and the Alps. They also yielded major mountain ranges in the past, which subsequently eroded away so that today we see only their relicts. For example, the Appalachian Mountains in the eastern United States formed as a consequence of three collisions. After the last one, a collision between Africa and North America around 280 Ma, North America became part of the Pangaea supercontinent.

TAKE-HOME MESSAGE

Rifting can split a continent in two and can lead to the formation of a new divergent boundary. When two buoyant crustal blocks, such as continents and island arcs, collide, a mountain belt forms and subduction ceases.

QUICK QUESTION Do rifting and collision affect crustal thickness? If so, how?

FIGURE 2.29 Examples of present-day crustal rifting.

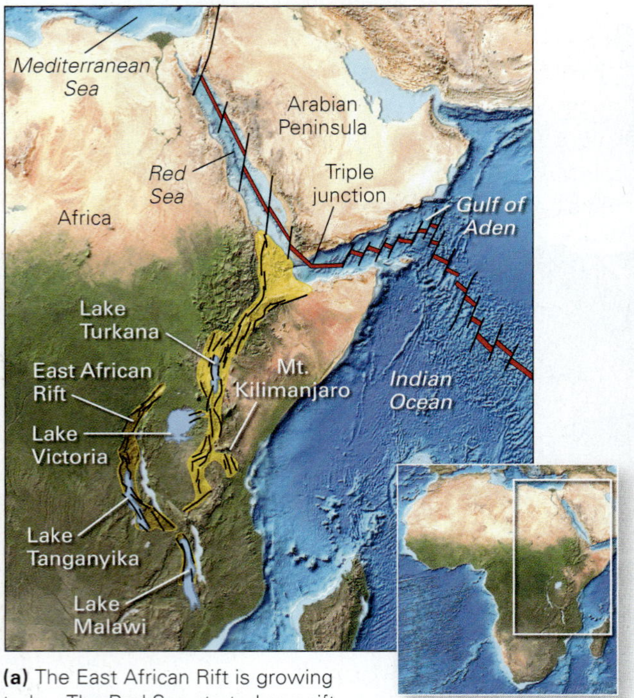

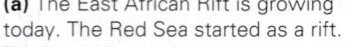

(a) The East African Rift is growing today. The Red Sea started as a rift. The inset shows the map location relative to the rest of Africa.

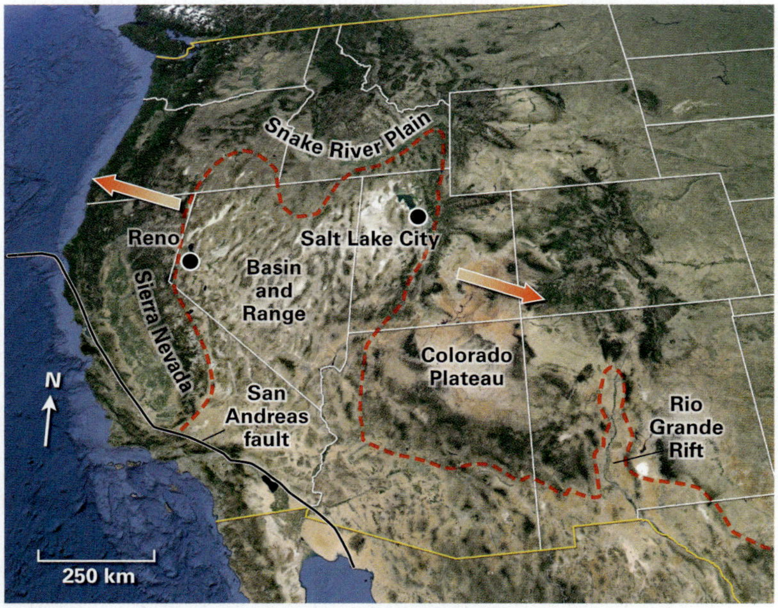

(c) The Basin and Range Province is a rift. Faulting bounds the narrow north-south-trending mountains, separated by basins. The arrows indicate the direction of stretching.

(b) An air photo of the northern end of the East African Rift, showing faults and volcanoes.

2.12 Moving Plates

Forces Acting on Plates

We've now discussed the many facets of plate tectonics theory (see **Geology at a Glance**, pp. 78–79). But to complete the story, we need to address a major question: What drives plate motion? When geoscientists first proposed plate tectonics, they thought the process occurred simply because convective flow in the asthenosphere actively dragged plates along, as if the plates were simply rafts on a flowing river. Thus, early images depicting plate motion showed simple convection cells—elliptical flow paths—in the asthenosphere. At first glance, this hypothesis looked pretty good. But, on closer examination it became clear that a model of simple convection cells carrying plates on their backs can't explain the complex geometry of plate boundaries and the great variety of plate motions that we observe on the Earth, so it is considered to be wrong (**Fig. 2.31a**). Researchers now prefer a model in which convection, along with two other forces, ridge push and slab pull, all contribute to driving plates. Let's look at each of these in turn.

Convection is involved in plate motions in two ways. Recall that, at a mid-ocean ridge, hot asthenosphere rises and then cools to form oceanic lithosphere which slowly moves away from the ridge until, eventually, it sinks back into the mantle at a trench. Since the material forming the plate starts out hot, cools, and then sinks, we can view the plate itself as the top of a convection cell and plate motion as a form of convection. But in this view, convection is effectively a consequence of plate motion, not the cause. Can convection actively cause plates to move? The answer may come from studies which demonstrate

FIGURE 2.30 Continental collision (not to scale).

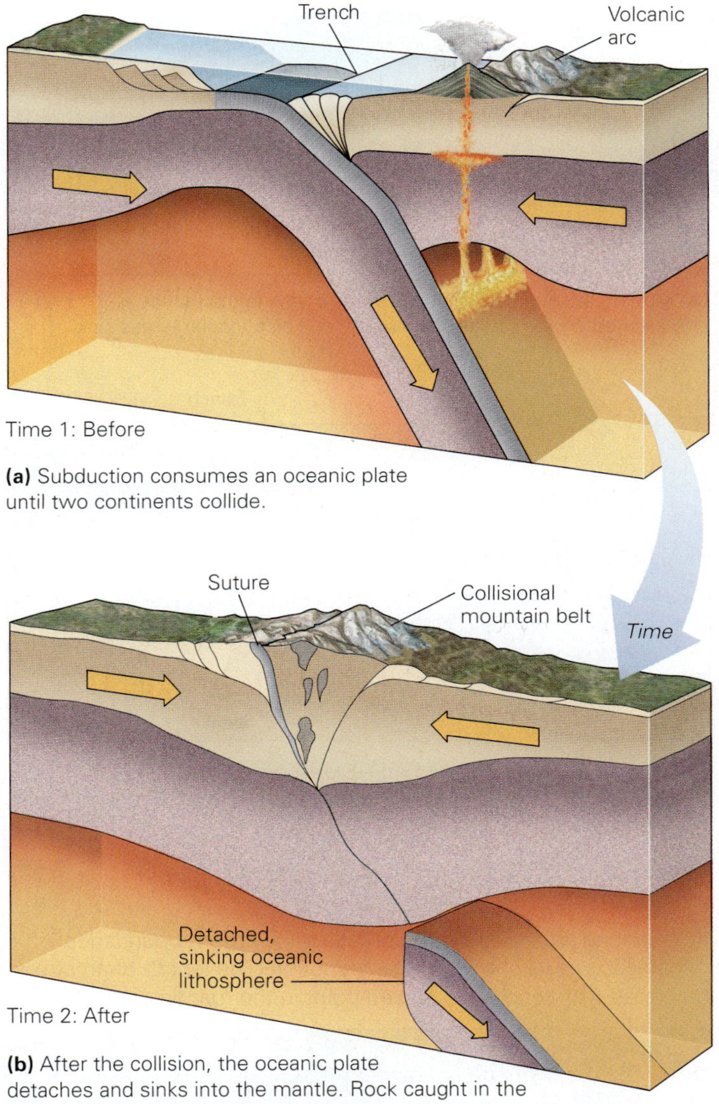

(a) Subduction consumes an oceanic plate until two continents collide.

(b) After the collision, the oceanic plate detaches and sinks into the mantle. Rock caught in the collision zone gets broken, bent, and squashed and forms a mountain range.

that the interior of the mantle, beneath the plates, is indeed convecting on a very broad scale (see Interlude D). Specifically, geologists have found that in some places deeper, hotter asthenosphere is rising or *upwelling*, and in some places shallower, colder asthenosphere is sinking or *downwelling* (**Fig. 2.31b**). Such asthenospheric flow probably does exert a force on the base of plates. But the pattern of upwelling and downwelling on a global scale does not match the pattern of plate boundaries exactly. So, conceivably, asthenospheric flow may either speed up or slow down plates depending on the orientation of the flow direction relative to the movement direction of the overlying plate.

Ridge-push force develops simply because the lithosphere of mid-ocean ridges lies at a higher elevation than that of the adjacent abyssal plains (**Fig. 2.31c**). To understand ridge-push force, imagine you have a glass containing a layer of water over a layer of honey. By tilting the glass momentarily and then returning it to its upright position, you can create a temporary slope in the boundary between these substances. While the boundary has this slope, gravity causes the weight of elevated honey to push against the glass adjacent to the side where the honey surface lies at lower elevation. The geometry of a mid-ocean ridge resembles this situation, for the seafloor of a mid-ocean ridge is higher than the seafloor of abyssal plains. Gravity causes the elevated lithosphere at the ridge axis to push on the lithosphere that lies farther from the axis, making it move away. As lithosphere moves away from the ridge axis, during seafloor spreading, new hot asthenosphere rises to fill the gap. Note that the local upward movement of asthenosphere beneath a mid-ocean ridge is a consequence of seafloor spreading, not the cause.

SEE FOR YOURSELF...

HIMALAYAN COLLISION

Latitude
27°59′17.95″ N

Longitude
86°55′30.71″ E

Zoom to an elevation of 7000 km (4300 miles) and look straight down.

You're seeing the Himalayas. Zoom closer, and you'll be at the peak of Mt. Everest.

Slab-pull force, the force that subducting, downgoing plates apply to oceanic lithosphere at a convergent margin, arises simply because lithosphere that was formed more than 10 million years ago is denser than asthenosphere, so it can sink into the asthenosphere (**Fig. 2.31d**). Thus, once an oceanic plate starts to sink, it gradually pulls the rest of the plate along behind it, like an anchor pulling down the anchor line. This "pull" is the slab-pull force.

The Velocity of Plate Motions

How fast do plates move? It depends on your frame of reference. To illustrate this concept, imagine two cars speeding in the same direction down the highway. From the viewpoint of a tree along the side of the road, Car A zips by at 100 km an hour, while Car B moves at 80 km an hour. But relative to Car B, Car A moves at only 20 km an hour. Geologists use two different frames of reference for describing plate velocity. If we describe the movement of Plate A with respect to Plate B, then we are speaking about **relative plate velocity**. But if we

FIGURE 2.31 Forces driving plate motions.

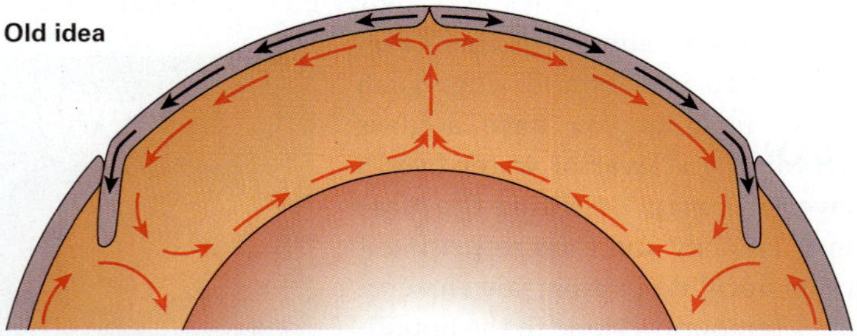

(a) The old, incorrect, image of simple convection cells. In this image, the plates were viewed as passive rafts carried along by convecting mantle.

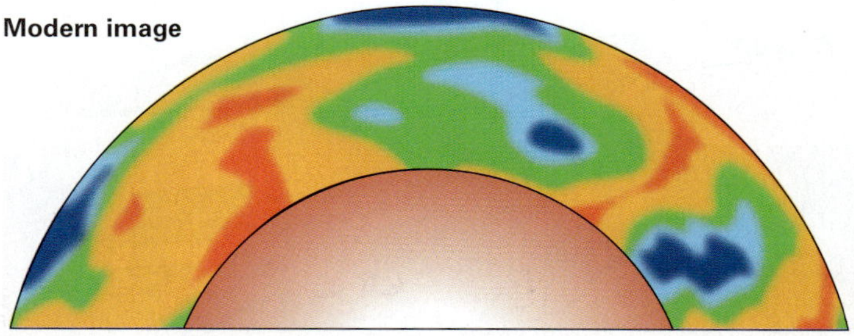

(b) Modern studies show that convective flow in the mantle is complicated. Warm (redder) areas rise, and blue (cooler) areas sink.

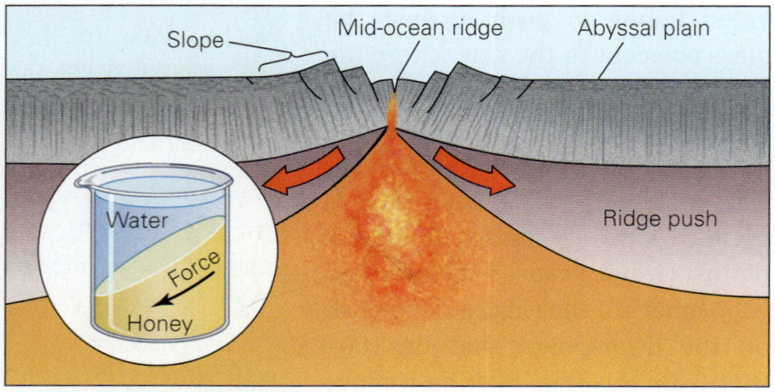

(c) Ridge push develops because the region of a rift is elevated. Like a wedge of honey with a sloping surface, the mass of the ridge pushes sideways.

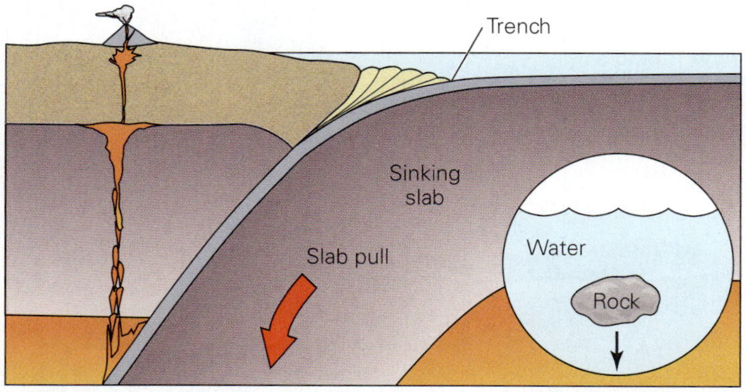

(d) Slab pull develops because lithosphere is denser than the underlying asthenosphere and sinks like a stone in water (though much more slowly).

FIGURE 2.32 The black arrows show relative plate velocities. Outward-pointing arrows indicate spreading (divergent boundaries), inward-pointing arrows indicate subduction (convergent boundaries), and parallel arrows show transform motion. The length of an arrow represents the velocity. The red arrows show the absolute velocity of the plates with respect to a fixed point in the mantle.

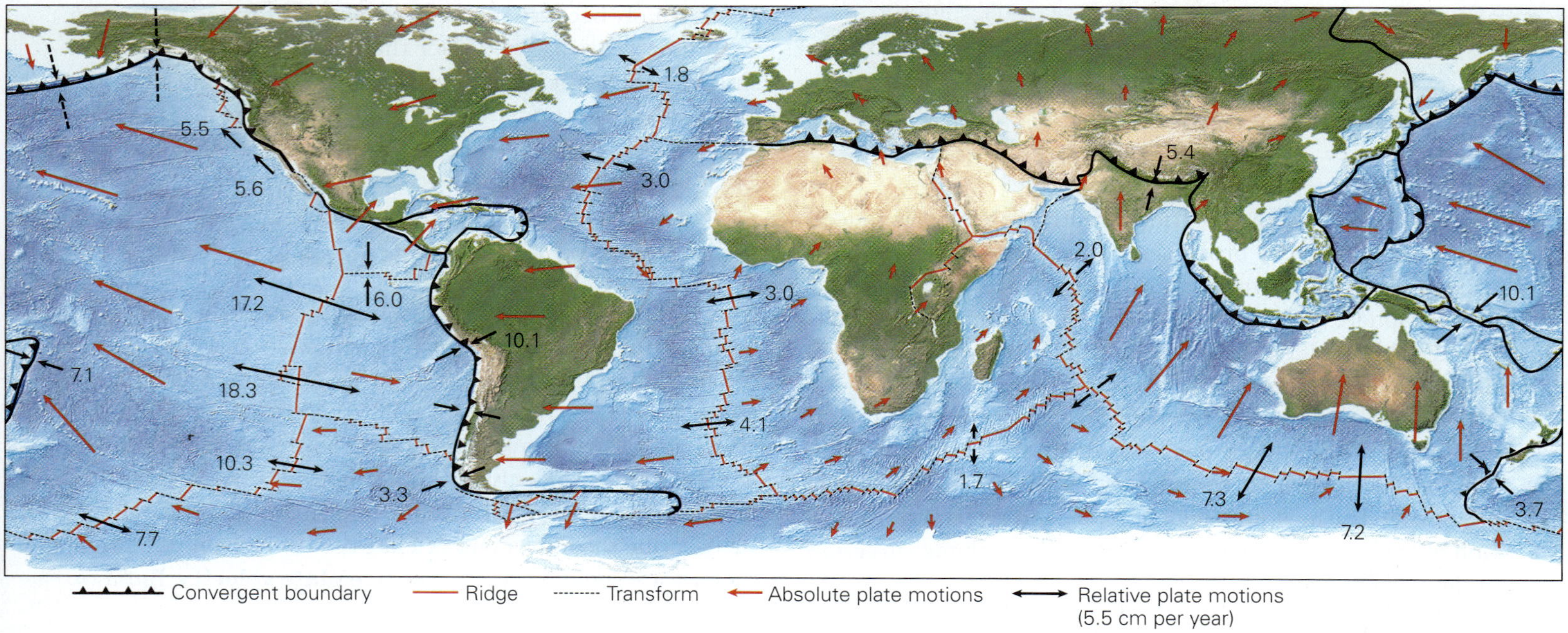

FIGURE 2.33 Due to plate tectonics, the map of Earth's surface slowly changes. Here we see the assembly, and later the breakup, of Pangaea during the past 400 million years.

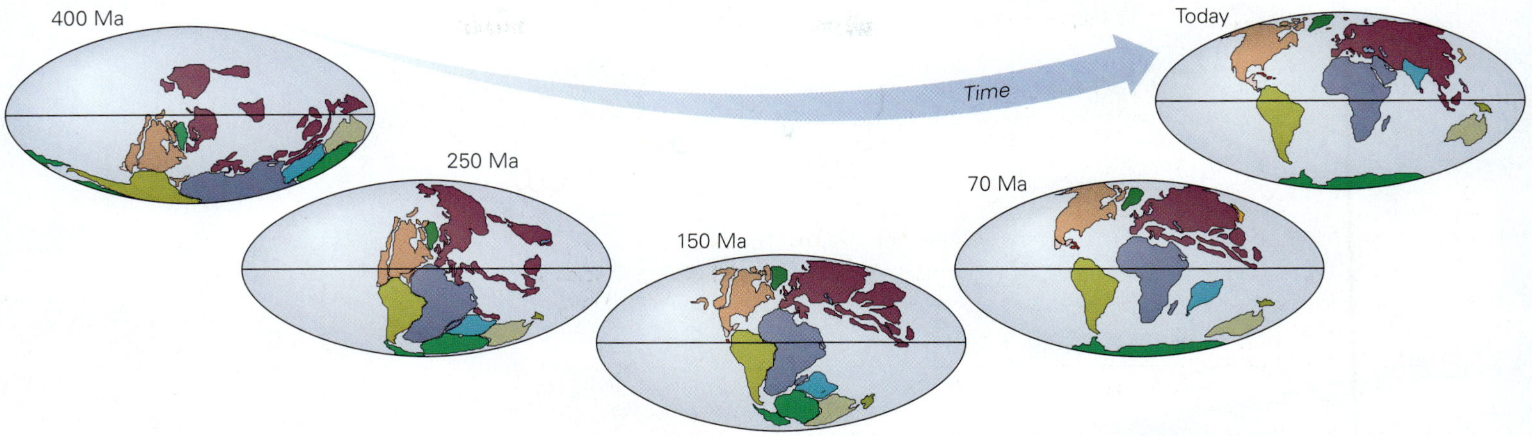

describe the movement of both plates relative to a fixed location in the mantle below the plates, then we are speaking of **absolute plate velocity** (**Fig. 2.32**).

To determine relative plate velocity, geoscientists measure the distance between the ridge axis and a location containing ocean floor of a specified age, and then apply this equation: plate velocity equals the distance from the location to the ridge axis, divided by the age of the seafloor at the location. The velocity of the plate on one side of the ridge relative to the plate on the other is twice this value. (Remember that velocity, by definition, is distance divided by time.) As an example, image a location where the seafloor is 1 million years old, and which lies 10 km from the ridge axis. Given that 10 km is 1,000,000 cm, the plate is moving 1 cm per year away from the ridge axis. The spreading rate across the ridge is therefore 2 cm/year.

Did you ever wonder...
whether we can really "see" continents drift?

To estimate absolute plate motions, we can assume that the position of a mantle plume does not change much for a long time. If this is so, then the track of hot-spot volcanoes on the plate moving over the plume provides a record of the plate's absolute velocity and indicates the direction of movement. (In reality, plumes are not completely fixed; geologists use other, more complex methods to calculate absolute plate motions.)

Working from the calculations described above, geologists have determined that plate motions on Earth today occur at rates of 1 to 15 cm per year—about the rate that your fingernails grow. These rates, though small, can yield large displacements given the immensity of geologic time. At a rate of 10 cm per year, a plate can move 100 km in a million years! Can we detect such slow rates? Yes, by using the *global positioning system* (*GPS*), the same technology that automobile drivers can use to find their destinations. By setting up a fixed GPS receiver that collects data over many years, geologists can detect displacements as small as about 2 mm per year (see chapter-opening photo). Since plates move at 5 to 75 times this rate, we indeed can see the plates move—this observation serves as the ultimate proof of plate tectonics.

Taking into account many data sources that define the motion of plates, geologists have greatly refined the image of continental drift that Wegener tried so hard to prove nearly a century ago. We can now see how the map of our planet's surface has evolved radically during the past 400 million years (**Fig. 2.33**), and even before.

TAKE-HOME MESSAGE

Plate motion takes place because plates are acted on by ridge push, slab pull, and convective shear. This motion takes place at rates of 1 to 15 cm per year. Relative motion specifies the rate that a plate moves relative to its neighbor, whereas absolute motion specifies the rate that a plate moves relative to a fixed point beneath the plate. GPS measurements can now detect relative plate motions directly.

QUICK QUESTION What causes the map-view bend in the Hawaiian-Emperor seamount chain?

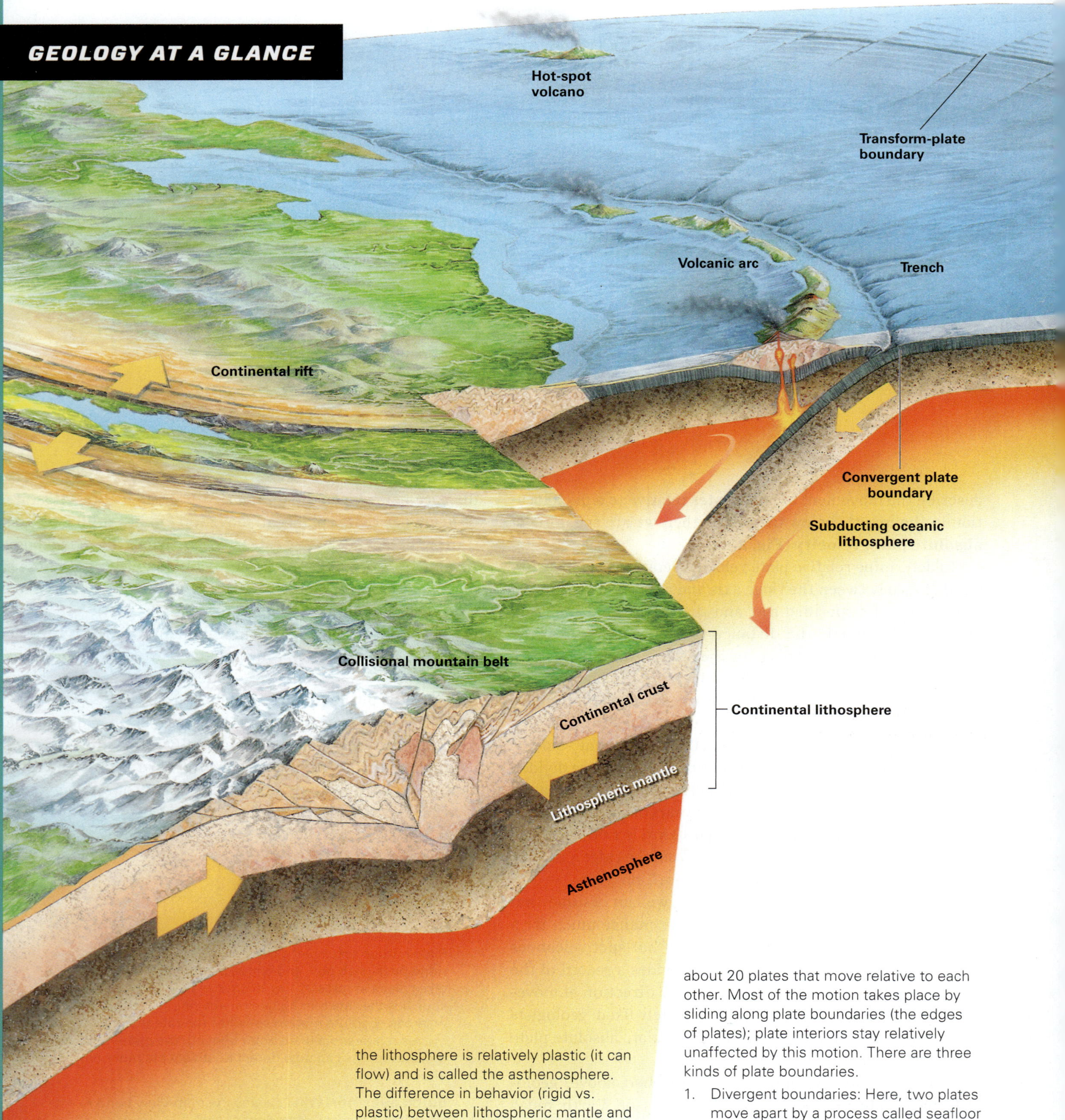

The Theory of Plate Tectonics

The outer portion of the Earth is a relatively rigid layer called the lithosphere. It consists of the crust (oceanic or continental) and the uppermost mantle. The mantle below the lithosphere is relatively plastic (it can flow) and is called the asthenosphere. The difference in behavior (rigid vs. plastic) between lithospheric mantle and asthenospheric mantle is a consequence of temperature—the former is cooler than the latter. Continental lithosphere is typically about 150 km thick, while oceanic lithosphere is about 100 km thick. (Note: These are not drawn to scale in this image.)

According to the theory of plate tectonics, the lithosphere is broken into about 20 plates that move relative to each other. Most of the motion takes place by sliding along plate boundaries (the edges of plates); plate interiors stay relatively unaffected by this motion. There are three kinds of plate boundaries.

1. Divergent boundaries: Here, two plates move apart by a process called seafloor spreading. A mid-ocean ridge delineates a divergent boundary. Asthenospheric mantle rises beneath a mid-ocean ridge and melts, forming magma. The magma rises and solidifies to become new oceanic crust. The lithospheric mantle thickens progressively away from the ridge axis as the plate cools.

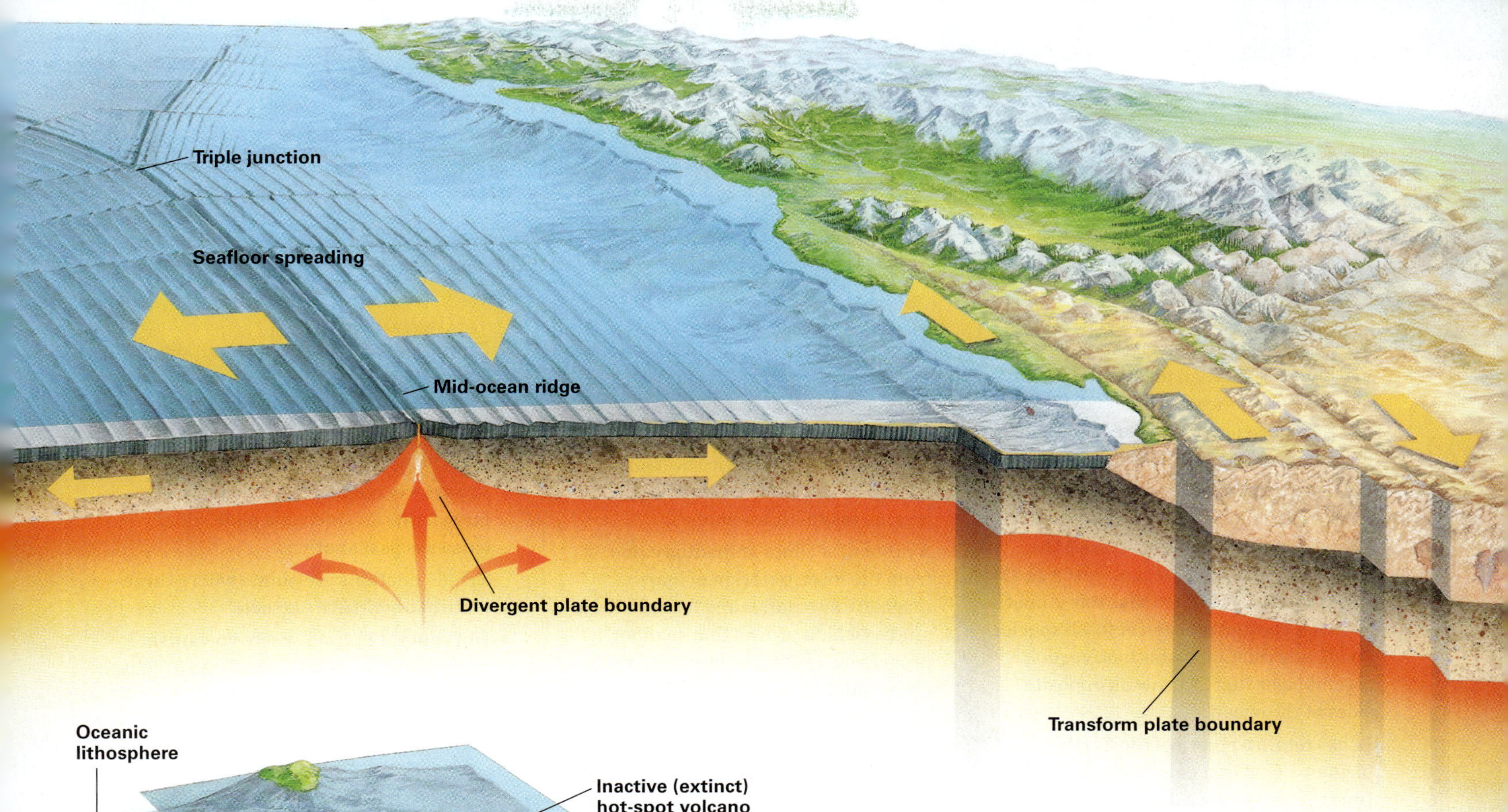

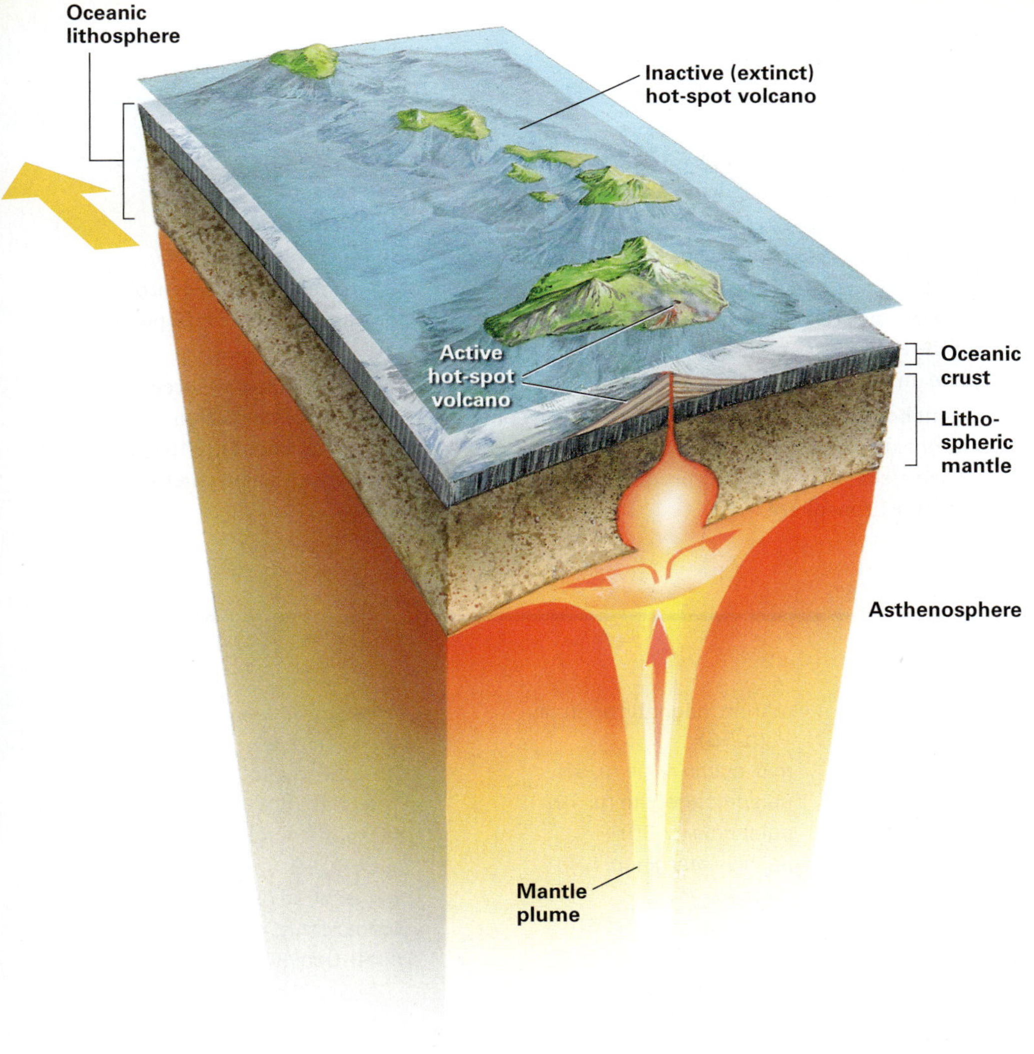

2. Convergent boundaries: Here, two plates move together, and one plate subducts beneath another (it sinks down into the mantle). Only oceanic lithosphere can subduct. At the Earth's surface, the boundary between the two plates is marked by a deep-ocean trench. During subduction, melting above the downgoing plate produces magma that rises to form a volcanic arc.
3. Transform boundaries: Here, one plate slides sideways past another, without the production of a new plate or the subduction of an old one. The boundary is marked by a large fault, a fracture on which sliding occurs. Transform boundaries link segments of mid-ocean ridges. They may also cut through continental lithosphere.

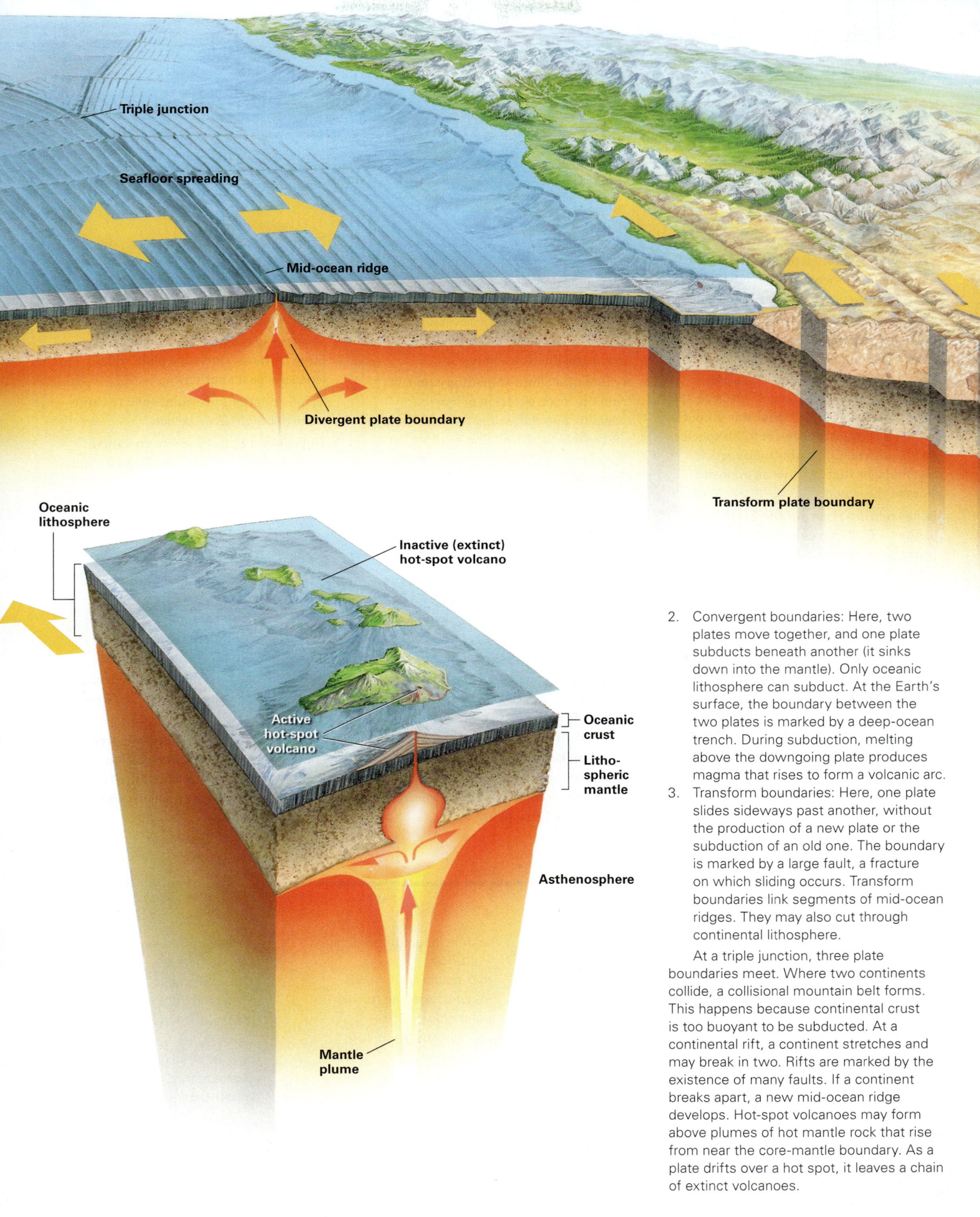

At a triple junction, three plate boundaries meet. Where two continents collide, a collisional mountain belt forms. This happens because continental crust is too buoyant to be subducted. At a continental rift, a continent stretches and may break in two. Rifts are marked by the existence of many faults. If a continent breaks apart, a new mid-ocean ridge develops. Hot-spot volcanoes may form above plumes of hot mantle rock that rise from near the core-mantle boundary. As a plate drifts over a hot spot, it leaves a chain of extinct volcanoes.

Chapter 2 Review

Chapter Summary

- Alfred Wegener proposed that continents had once been joined together to form a single huge supercontinent (Pangaea) and had subsequently drifted apart. This idea is the continental-drift hypothesis.
- Wegener drew from several different sources of data to support his hypothesis: (1) the matching of coastlines; (2) the distribution of late Paleozoic glaciers; (3) the distribution of late Paleozoic climatic belts; (4) the distribution of fossil species; and (5) correlation of distinctive rock assemblages that are now on opposite sides of the ocean.
- Rocks retain a record of the Earth's magnetic field that existed at the time the rocks formed. This record is called paleomagnetism. By measuring paleomagnetism in successively older rocks, geologists discovered apparent polar-wander paths.
- The contrasts among apparent polar-wander paths for different continents serve as a proof of continental drift.
- Around 1960, Harry Hess proposed the hypothesis of seafloor spreading. According to this hypothesis, new seafloor forms at mid-ocean ridges, then spreads symmetrically away from the ridge axis. Eventually, the ocean floor sinks back into the mantle at deep-ocean trenches.
- Geologists documented that the Earth's magnetic field reverses polarity every now and then. Magnetic reversals explain the existence of stripe-like marine magnetic anomalies.
- A proof of seafloor spreading came from the interpretation of marine magnetic anomalies and from drilling studies which proved that seafloor gets progressively older away from a mid-ocean ridge.
- The lithosphere is broken into discrete plates that move relative to each other. Continental drift and seafloor spreading are manifestations of plate movement.
- Most earthquakes and volcanoes occur along plate boundaries; the interiors of plates remain relatively rigid and intact.
- There are three types of plate boundaries—divergent, convergent, and transform—distinguished from each other by the movement the plate on one side of the boundary makes relative to the plate on the other side.
- Divergent boundaries are marked by mid-ocean ridges. At divergent boundaries, seafloor spreading produces new oceanic lithosphere.
- Convergent boundaries are marked by deep-ocean trenches and volcanic arcs. At convergent boundaries, oceanic lithosphere subducts beneath an overriding plate.
- Transform boundaries are marked by large faults at which one plate slides sideways past another.
- Triple junctions are points where three plate boundaries intersect.
- Hot spots are places where volcanism occurs at an isolated volcano. As a plate moves over the hot spot, the volcano moves off and dies, and a new volcano forms over the hot spot. Hot spots may be caused by mantle plumes.
- During rifting, continental lithosphere stretches and thins. If it finally breaks apart, a new mid-ocean ridge forms and seafloor spreading begins.
- Convergent boundaries cease to exist when a buoyant piece of crust (a continent or an island arc) moves into the subduction zone. When that happens, collision occurs.
- Ridge-push force and slab-pull force contribute to driving plate motions. Plates move at rates of about 1 to 15 cm per year. GPS satellite measurements can detect these motions.

Guide Terms

GEOTOURS *THIS CHAPTER'S GEOTOUR EXERCISE (A & B) FEATURES:*

› Transition from Continental Crust to Oceanic Crust › Topography of the Ocean Floor › Continental Drift
› Divergent Boundaries › Convergent Boundaries › Transform Boundaries › Hot Spots

Review Questions

1. What was Wegener's continental drift hypothesis? What was his evidence? Why didn't other geologists accept Wegener's proposal of continental drift, at first?
2. How do apparent polar-wander paths show that the continents have moved?
3. Describe the hypothesis of seafloor spreading.
4. Describe the pattern of marine magnetic anomalies across a mid-ocean ridge. How is this pattern explained?
5. How did drilling into the seafloor contribute further proof of seafloor spreading? How did the seafloor-spreading hypothesis explain variations in ocean floor heat flow?
6. What are the characteristics of a lithosphere plate? Can a single plate include both continental and oceanic lithosphere?
7. How does oceanic lithosphere differ from continental lithosphere in thickness, composition, and density?
8. How do we identify a plate boundary?
9. Describe the three types of plate boundaries.
10. How does crust form along a mid-ocean ridge?
11. Why is the oldest oceanic lithosphere less than 200 Ma?
12. What are the major geologic features of a convergent boundary?
13. Why are transform plate boundaries required on an Earth with spreading and subducting plate boundaries?
14. What is a triple junction?
15. How is a hot-spot track produced, and how can hot-spot tracks be used to track the past motions of a plate?
16. Describe the characteristics of a continental rift and give examples of where this process is occurring today.
17. Describe the process of continental collision and give examples of where this process has occurred.
18. Discuss the major forces that move lithosphere plates.
19. Explain the difference between relative plate velocity and absolute plate velocity.

On Further Thought

20. Why are the marine magnetic anomalies bordering the East Pacific Rise in the Pacific Ocean wider than those bordering the Mid-Atlantic Ridge?
21. The North Atlantic Ocean is 3,600 km wide. Seafloor spreading along the Mid-Atlantic Ridge occurs at 2 cm per year. When did rifting start to open the Atlantic?

Online Resources

Animations
This chapter features simulations of the many forms of plate boundaries and their effects.

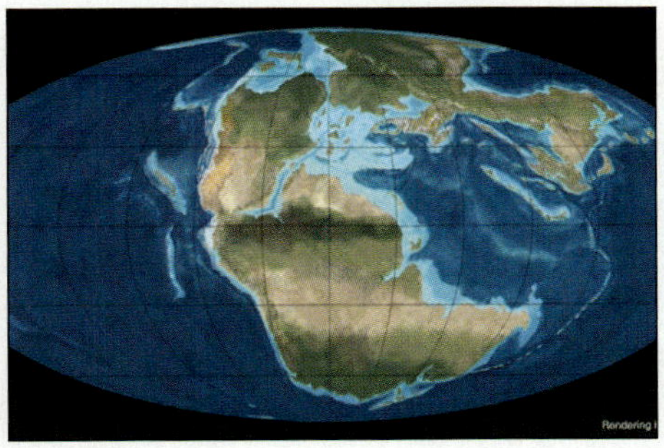

Videos
This chapter features a video explaining the breakup of the supercontinent Pangea.

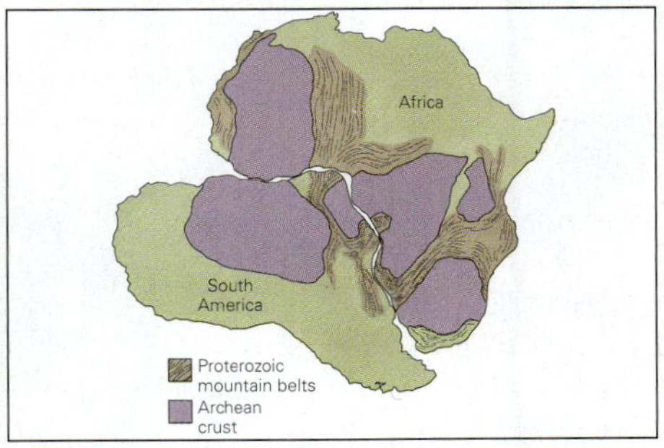

Assessment
This chapter features visual questions on continental drift and collision.

LEARNING OBJECTIVES

By the end of this chapter, you should understand...

1. that the term *mineral* has a very special meaning in geologic contexts.
2. how to organize the thousands of different minerals into just a few classes based on the chemicals the minerals contain.
3. which minerals are the most common ones on Earth, and thus serve as the main building blocks of this planet.
4. how to identify common mineral specimens.
5. why we consider some minerals to be gems, and how their shiny facets are produced.

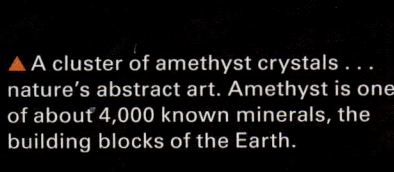

▲ A cluster of amethyst crystals . . . nature's abstract art. Amethyst is one of about 4,000 known minerals, the building blocks of the Earth.

CHAPTER 3

Patterns in Nature: Minerals

3.1 Introduction

In Greek legend, the god Dionysus, in a drunken rage, vowed to kill the next mortal that he saw. Just then, a beautiful young woman named Amethyst walked by, and Dionysus ordered two fearsome tigers to attack her. The goddess Artemis prevented a tragedy by changing Amethyst into a pure white statue made of quartz that was much harder than the tigers' teeth. The statue was so beautiful that Dionysus sobered up and regretted his rashness. In remorse, he spilled his wine onto the statue as an offering, and the wine stained the quartz, turning it into purple amethyst (see chapter-opening photo). The word comes from the Greek *amethustos,* meaning not intoxicated. For centuries afterward, amethyst was thought of as an antidote for drunkenness. It isn't—amethyst has no effect on alcohol and can't prevent inebriation. But legend aside, amethyst is beautiful, one of many minerals that have been used in jewelry making for millennia.

Amethyst, the purple version of a common mineral, quartz, is one of about 4,000 minerals that *mineralogists,* people who specialize in the study of minerals, have identified so far. The vast majority of mineral types are rare. In fact, fewer than 50 are considered common rock-forming minerals of the Earth's crust. Most rocks that you pick up consist almost entirely of only one to six of these common minerals.

> *I died a mineral, and became a plant.*
> *I died as plant and rose to animal, I died as*
> *animal and I was Man. Why should I fear?*
>
> JALAL-UDDIN RUMI (Persian mystic and poet, 1207–1273)

Why study minerals? Without exaggeration, we can say that minerals are the building blocks of our planet. To a geologist, almost any study of Earth materials depends on an understanding of minerals, for minerals make up most of the rocks and sediments comprising the Earth and its landscapes. Minerals are also important from a practical standpoint (see Chapter 12). Industrial minerals serve as the raw materials for manufacturing chemicals, concrete, and wallboard. Ore minerals are the source of valuable metals such as copper and gold (**Fig. 3.1**), and of powerful energy sources, such as uranium. Particularly beautiful forms of minerals—gems—delight the eye in jewelry. Unfortunately, though, some minerals pose environmental or health hazards. No wonder *mineralogy*, the study of minerals, fascinates professionals and amateurs alike.

This chapter, an introduction to mineralogy, begins with the geologic definition of a mineral. We then look at how minerals form and at the main characteristics that enable us to identify minerals. Finally, we describe the basic scheme that mineralogists use to classify minerals. Some of the discussions in this chapter utilize basic concepts from chemistry. If you are rusty on your understanding of these, please study **Box 3.1** before going further.

3.2 What Is a Mineral?

In everyday English, the word *mineral* has many uses. When you play the game 20 Questions, a mineral is anything that's not animal or vegetable, and if you read food ingredients, the word refers to certain nutrients that people need in order to be healthy. Geologists use the term in a very specific way. To a geologist, a **mineral** is a naturally occurring solid, formed by geologic processes, that has a crystalline structure and a definable chemical composition. Almost all minerals are "inorganic." Let's pull apart this mouthful of a definition and examine its meaning in detail.

- *Naturally occurring:* True minerals are formed in nature, not in factories. In recent decades, chemists have learned how to manufacture materials that have characteristics virtually identical to those of real minerals. Such materials can be referred to as "synthetic minerals."
- *Formed by geologic processes:* Traditionally, this phrase implied that minerals were the result only of solidification of molten rock or direct precipitation from a water solution, processes that did not involve living organisms. Increasingly, however, geologists recognize that life is an integral part of the Earth System. So geologists now consider solid, crystalline materials produced by organisms to be minerals, too. To avoid confusion, the term *biogenic mineral* may be used when discussing such materials.
- *Solid:* A solid is a state of matter that can maintain its shape indefinitely, and thus will not conform to the shape of its container. Liquids (such as oil or water) and gases (such as air) are not minerals.
- *Crystalline structure:* The atoms that make up a mineral are fixed in a specific, orderly pattern. A material in which atoms are fixed in an orderly pattern is called a **crystalline solid**. Mineralogists refer to the pattern itself (the imaginary framework representing the arrangement of atoms) as a **crystal lattice**.
- *Definable chemical composition:* This phrase simply means that it is possible to write a chemical formula for a mineral. Some minerals contain only one element, but most are compounds of two or more elements. For example, diamond and graphite both have the formula C because they consist entirely of carbon. Quartz has the formula SiO_2—it contains the elements silicon and oxygen in the proportion of one silicon atom for every two oxygen atoms. Some mineral formulas are more complicated: for example, the formula for biotite is $K(Mg,Fe)_3(AlSi_3O_{10})(OH)_2$. According to this formula, the proportion of magnesium to iron can vary in biotite.
- *Inorganic:* To understand the meaning of *inorganic*, we must first understand what's meant by *organic*. Organic chemicals consist of molecules that include carbon, and either form in living organisms or have structures similar to those that formed in living organisms. Sugar ($C_{12}H_{22}O_{11}$), fat, plastic, propane, and protein, for example, are organic chemicals. Some organic chemicals contain only carbon and hydrogen, while others include other elements, such as oxygen, nitrogen, and/or phosphorus. Almost all minerals are inorganic, in that they are not organic chemicals. But we have to add the qualifier "almost all" because mineralogists now consider a few dozen organic substances formed by "the action of geologic processes on organic materials" to be minerals. Examples include the crystals that grow in ancient deposits of bat guano.

With the geologic definition of a mineral in mind, we can distinguish between a mineral and a **glass**. Both minerals and glasses are solids, meaning that they can retain their shape indefinitely, but a mineral is crystalline, while glass is not. This means that the atoms, ions, or molecules in a mineral are ordered into a crystal lattice, like soldiers standing in formation, but those in a glass are arranged in a semichaotic

FIGURE 3.1 Copper ore is a useful mineral that serves as a source of copper metal.

Malachite grows by precipitation, in a succession of layers.

(a) Malachite is a type of copper ore ($Cu_2[CO_3][OH]_2$); it contains copper plus other chemicals.

(b) The copper for pots is produced by processing ore minerals.

FIGURE 3.2 The contrast between crystalline and noncrystalline solids.

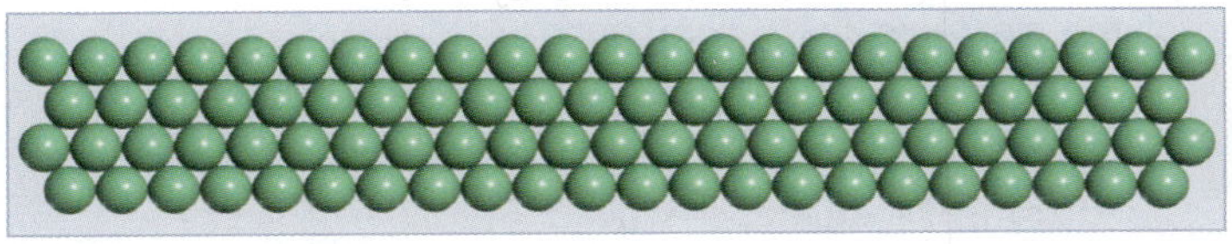

(a) A crystal contains an orderly arrangement of atoms.

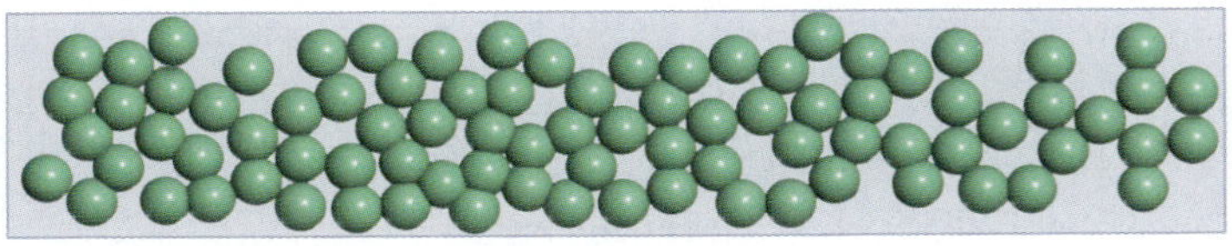

(b) Atoms in noncrystalline solids, such as glass, are not orderly.

way, in small clusters or chains that are neither oriented in the same way nor spaced at regular intervals, like guests at a party (**Fig. 3.2**).

If you ever need to figure out whether a substance is a mineral or not, just check it against the criteria listed above. Is motor oil a mineral? No—it's an organic liquid. Is table salt a mineral? Yes—it's a solid crystalline compound with the formula NaCl. Is the hard material making up the shell of an oyster considered to be a mineral? Microscopic examination of an oyster shell reveals that it consists of calcite, so it can be called a biogenic mineral. Is rock candy a mineral? No. Even though it is solid and crystalline, it's made by people and it consists of sugar (an organic chemical).

Did you ever wonder...
is rock candy a mineral?

TAKE-HOME MESSAGE

Minerals are solids with a crystalline structure (an orderly arrangement of atoms inside) and a definable chemical formula. They form by natural processes in the Earth System.

QUICK QUESTION Is Styrofoam a mineral? Why or why not?

3.3 Beauty in Patterns: Crystals and Their Structure

What Is a Crystal?

The word *crystal* brings to mind sparkling chandeliers, elegant wine goblets, and shiny jewels. But, as is the case with the word *mineral*, geologists have a more precise definition. A **crystal** is a single, continuous (uninterrupted) piece of a crystalline solid, typically bounded by flat surfaces, called **crystal faces**, that grow naturally as the mineral forms. The word comes from the Greek *krystallos*, meaning ice. Many crystals have beautiful shapes that look like they belong in the pages of a geometry book. The angle between two adjacent crystal faces of one specimen is identical to the angle between the corresponding faces of another specimen. For example, a perfectly formed quartz crystal looks like an obelisk, and the angle between the faces of the columnar part of a quartz crystal is always exactly 120° (**Fig. 3.3a, b**). This rule, discovered by one of the first geologists, Nicolas Steno (1638–1686) of Denmark, holds regardless of whether the whole crystal is big or small and regardless of whether all of the faces are the

BOX 3.1 SCIENCE TOOLBOX...

Some Basic Concepts from Chemistry

To describe minerals, we need to use several terms from chemistry. To avoid confusion, terms are listed in an order that permits each successive term to utilize previous terms.

- *Element:* A pure substance that cannot be separated into other materials.
- *Atoms and their components:* The smallest piece of an element retaining the characteristics of the element is an atom. Atoms are so small that over 5 trillion (5,000,000,000,000) could fit on the head of a pin. An atom consists of a nucleus surrounded by a cloud of orbiting electrons. A nucleus is a compact ball of protons and neutrons (except in hydrogen, whose nucleus contains only one proton and no neutrons). The mass of an electron is only about 1/1,836 that of a proton. Electrons have a negative charge, protons have a positive charge, and neutrons have a neutral charge.
- *Neutral atom:* An atom that has the same number of electrons as protons is said to be neutral, in that it does not have an overall electrical charge.
- *Atomic number:* The number of protons in an atom of an element.
- *Atomic mass:* Approximately the number of protons plus neutrons in an atom of an element.
- *Ion:* An atom that is not neutral. An ion with an excess negative charge (because it has more electrons than protons) is an *anion*, whereas an ion with an excess positive charge (because it has more protons than electrons) is a *cation*. We indicate the charge with a superscript. For example, Cl^- has a single excess electron; Fe^{2+} is missing two electrons.
- *Chemical bond:* An attractive force that holds two or more atoms together (**Fig. Bx3.1a–c**). *Covalent bonds* form when atoms share electrons, and *ionic bonds* form when a cation and anion (ions with opposite charges) get close together and attract each other. In materials with *metallic bonds*, some of the electrons can move freely.
- *Molecule:* Two or more atoms bonded together. The atoms may be of the same element or of different elements.
- *Compound:* A pure substance that can be subdivided into two or more elements. The smallest piece of a compound that retains the characteristics of the compound is a molecule.
- *State of matter:* The form of a substance, which reflects the degree to which the atoms or molecules comprising the matter are bonded together. In everyday experience, you see three states of matter—solid, liquid, and gas. There are more bonds in a solid than in a liquid, and more in a liquid than in a gas. Which state exists at a given location depends on pressure and temperature. A fourth state, plasma, exists only at very high temperatures.
- *Evaporation:* When atoms or molecules escape from the surface of a liquid and turn into gaseous form, evaporation has taken place.
- *Freezing:* When a liquid turns into solid form, freezing has taken place.
- *Condensation:* When a gas turns into liquid form, condensation has taken place.
- *Chemical:* A general name used for a pure substance (either an element or a compound).
- *Chemical formula:* A shorthand recipe that itemizes the various elements in a chemical and specifies their relative proportions. For example, the formula for water, H_2O, indicates that water consists of molecules in which two hydrogens bond to one oxygen.
- *Chemical reaction:* A process that involves breaking or forming chemical bonds. Chemical reactions can break molecules apart or yield new molecules and/or isolated atoms. Energy is absorbed or released during a chemical reaction.

same size. Crystals come in a great variety of shapes, including cubes, trapezoids, pyramids, octahedrons, hexagonal columns, blades, needles, columns, and obelisks (**Fig. 3.3c**).

Because crystals have a regular geometric form, people have always considered them to be special, perhaps even a source of magical powers. For example, shamans of some cultures relied on talismans or amulets made of crystals, which supposedly brought power to their wearer or warded off evil spirits. Scientists have concluded, however, that crystals have no effect on health or mood. For millennia, crystals have inspired awe because of the way they sparkle, but such behavior is simply a consequence of how crystal structures interact with light.

Looking Inside a Mineral

What do the insides of a mineral actually look like? This problem was the focus of study for centuries. An answer finally came from the work of a German physicist, Max von Laue (1879–1960), in 1912. He showed that an X-ray beam passing

FIGURE Bx3.1 Types of chemical bonds.

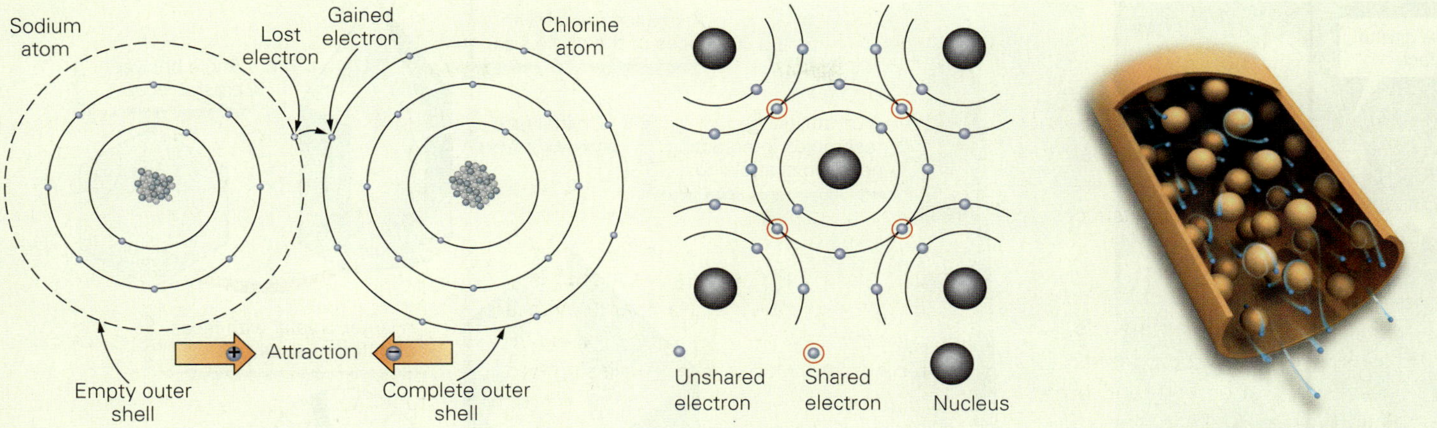

(a) An ionic bond forms between a positive ion of sodium (Na^+) and chloride (Cl^-), a negative ion of chlorine. When sodium gives up one electron to chlorine, so that both have filled shells, halite (NaCl) is produced.

(b) Covalent bonds form when carbon atoms share electrons so that all have filled electron shells.

(c) In metallically bonded material, nuclei and their inner shells of electrons float in a sea of free electrons. The electrons stream through the metal if there is an electrical current.

- *Mixture:* A combination of two or more elements or compounds that can be separated without a chemical reaction. For example, cereal composed of bran flakes and raisins is a mixture—you can separate the raisins from the flakes without destroying either.
- *Solution:* A type of material in which one chemical (the solute) dissolves in another (the solvent). In solutions, a solute may separate into ions during the process. For example, when salt (NaCl) dissolves in water, it separates into sodium (Na^+) and chloride (Cl^-) ions. In a solution, atoms or molecules of the solvent surround atoms, ions, or molecules of the solute.
- *Precipitate:* (noun) A compound that forms when ions in a liquid solution or a gas join together to make a solid that settles out of the solution; (verb) the process of forming solid grains by separation and settling from a solution. For example, when saltwater evaporates, solid salt crystals precipitate out of solution, and form a precipitate of salt.

through a crystal breaks up into many tiny beams to make a pattern of dots on a screen (**Fig. 3.4a**). Physicists refer to this phenomenon as *diffraction*. It occurs when waves interact with regularly spaced objects whose spacing is close to the wavelength of the waves—you can see diffraction of ocean waves when they pass through gaps in a seawall. Von Laue concluded that, for a crystal to cause diffraction, atoms within it must be regularly spaced and the spacing must be comparable to the wavelength of X-rays. Eventually, Von Laue and others learned how to use *X-ray diffraction patterns* as a basis for defining the specific arrangement of atoms in crystals. This arrangement defines the crystal structure of a mineral.

If you've ever examined wallpaper, you've seen an example of a pattern (**Fig. 3.4b**). Crystal structures contain one of nature's most spectacular examples of such a pattern. In crystals, the pattern is defined by the regular spacing of atoms and, if the crystal contains more than one element, by the regular alternation of atoms (**Fig. 3.4c**). (Mineralogists refer to a 3-D geometry of points representing this pattern as a lattice.) The pattern of atoms in a crystal may control the shape

FIGURE 3.3 Some characteristics of crystals.

(a) A quartz crystal can resemble an obelisk. Inside, atoms are arranged in a specific geometric pattern, like the joint points in scaffolding.

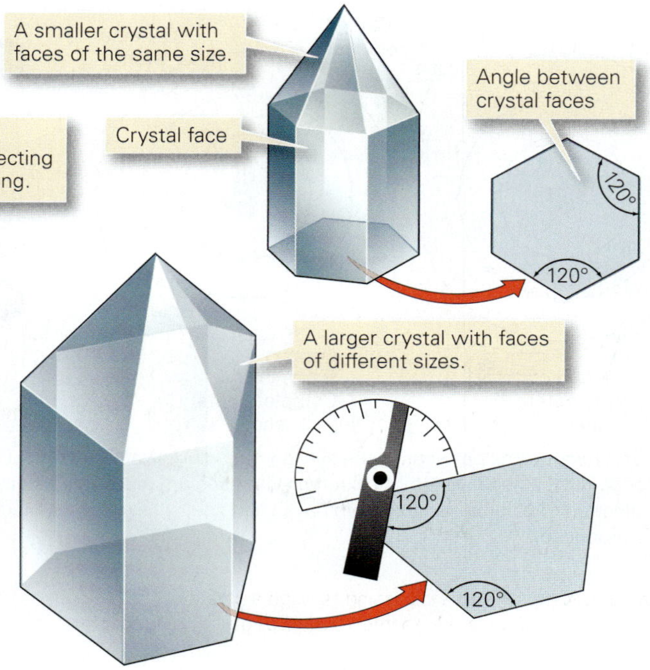

(b) Regardless of specimen size, the angle between two adjacent crystal faces is consistent in a particular mineral.

(c) Crystals come in a variety of shapes, including cubes, prisms, blades, and pyramids. Some terminate at a point and some terminate with flat surfaces.

of a crystal. For example, if atoms in a crystal pack into the shape of a cube, the crystal may have faces that intersect at 90° angles—galena (PbS) and halite (NaCl) have such a cubic shape. Because of the pattern of atoms in a crystal structure, the structure has *symmetry*, meaning that the shape of one part of the structure is the mirror image of the shape of a neighboring part. For example, if you were to cut a halite crystal or a water crystal (snowflake) in half, and place the half against a mirror, it would look whole again (**Fig. 3.4d**).

To illustrate crystal structures, we look at a few examples. Halite (rock salt) consists of oppositely charged ions that stick together because opposite charges attract. In halite, six chloride (Cl^-) ions surround each sodium (Na^+) ion, producing an overall arrangement of atoms that defines the shape of a cube (**Fig. 3.5a, b**). Diamond, by contrast, is a mineral made entirely of carbon. In diamond, each atom bonds to four neighbors arranged in the form of a tetrahedron; some naturally formed diamond crystals have the shape of a double tetrahedron (**Fig. 3.5c**). Graphite, another mineral composed entirely of carbon, behaves very differently from diamond. In contrast to diamond, graphite is so soft that we use it as the "lead" in a pencil; when a pencil moves across paper, tiny flakes of graphite peel off the pencil point and adhere to the paper. This behavior occurs because the carbon atoms in graphite are not arranged in tetrahedra, but rather occur in sheets (**Fig. 3.5d**). The sheets are bonded to each other by weak bonds and thus can separate from each other easily. Of note, two different minerals (such as diamond and graphite) that have the same composition but different crystal structures are **polymorphs**.

The Formation and Destruction of Minerals

New mineral crystals can form in several ways: (1) *Solidification (freezing) of a melt* happens when a liquid cools and turns into a solid. Just as ice crystals form by solidification of water, a great variety of minerals form by solidification of molten rock. (2) *Precipitation from a water solution* takes place when dissolved ions bond together to form crystals that settle out of the water or grow out from the walls of a container. Salt

FIGURE 3.4 Patterns and symmetry in minerals.

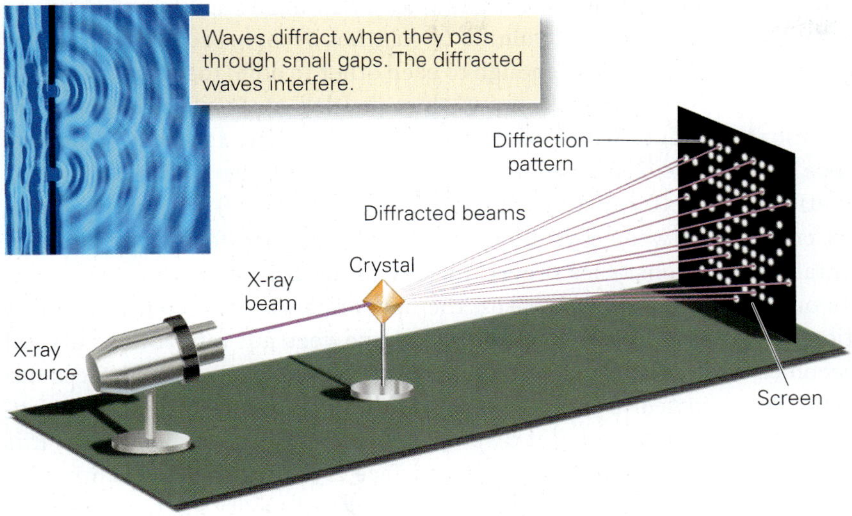

(a) Diffraction of an X-ray beam passing through a crystal produces a pattern of bright spots on a screen. The spots are due to interference of overlapping light waves.

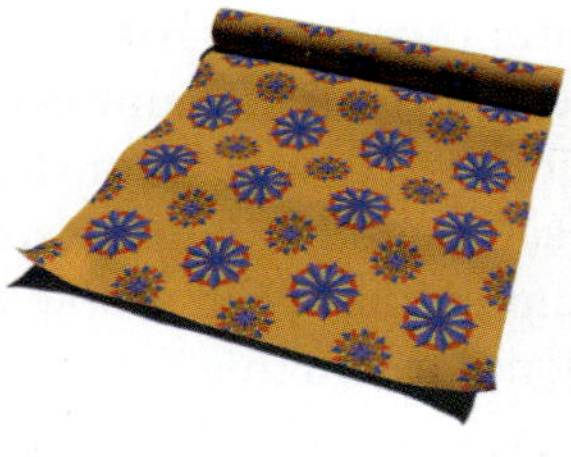

(b) The repetition of a flower motif on wallpaper.

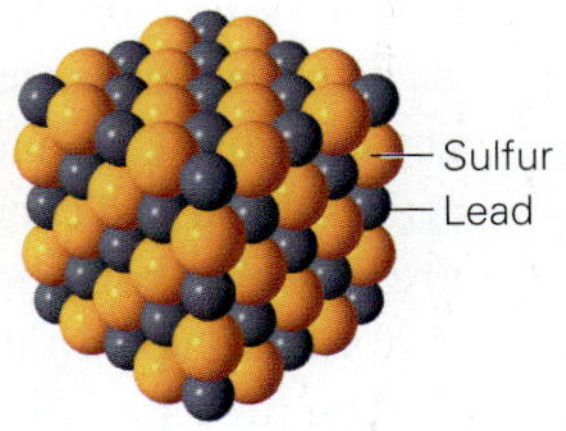

(c) The repetition of alternating sulfur and lead atoms in the mineral galena (PbS).

(d) Minerals display symmetry. One half of a crystal is a mirror image of the other.

FIGURE 3.5 The nature of crystalline structure in minerals.

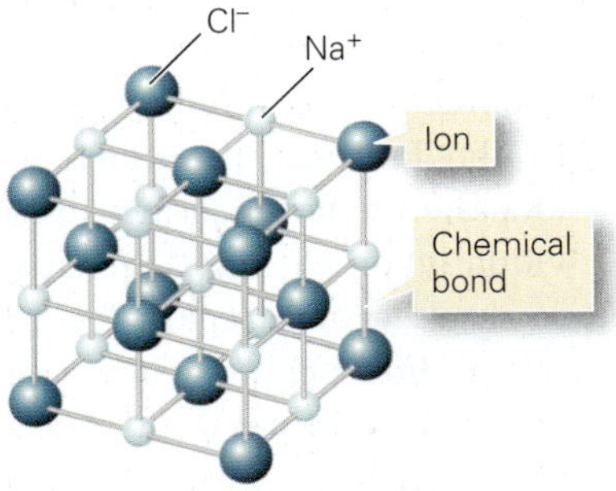

(a) A ball-and-stick model of halite portrays ions as balls, and chemical bonds as sticks.

(b) This packed-ball model gives a sense of how ions fit together in a crystal.

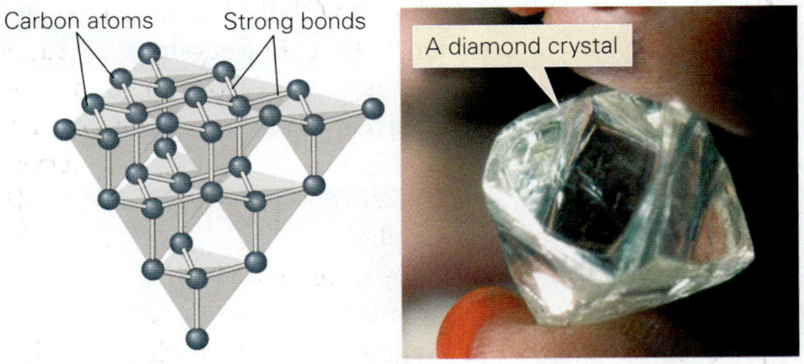

(c) In a diamond, carbon atoms are arranged in tetrahedra. All of the bonds are strong.

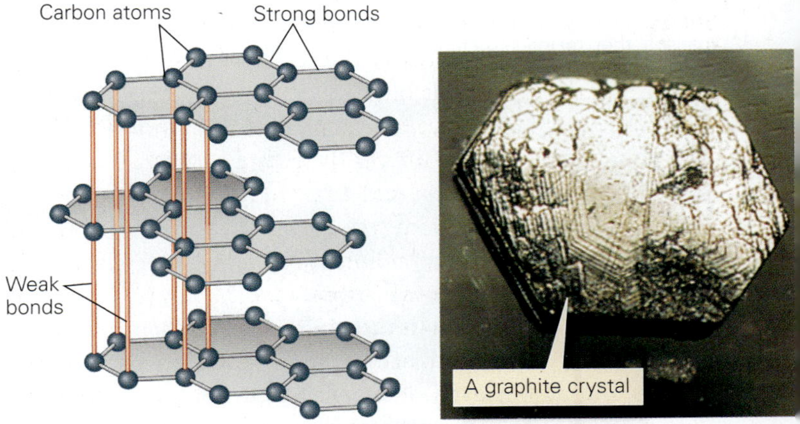

(d) Graphite consists of carbon atoms arranged in hexagonal sheets. The sheets are connected by weak bonds.

crystals, for example, precipitate when salt water evaporates (see Box 3.1). (3) *Solid-state diffusion* results from the movement of atoms or ions through a solid to arrange into a new crystal structure. Garnets, for example, grow by diffusion in solid rock. During this process, they replace preexisting minerals. (4) *Biomineralization* takes place when minerals grow at the interface between the physical and biological components of the Earth System. This process can happen because metabolic processes of some living organisms can cause minerals to precipitate either within their bodies, on their bodies, or immediately adjacent to their bodies. Shells produced by clams and oysters, for example, grow when these organisms extract ions from the water they live in. And (5) *precipitation directly from a gas* can occur around volcanic vents or around geysers, for at such locations volcanic gases or steam enter the atmosphere and cool, so some gas molecules of some elements are able bind together. Some of the bright yellow sulfur deposits found in volcanic regions form in this way.

The first step in forming a crystal is the chance formation of a seed, an extremely small crystal (**Fig. 3.6a**). Once the seed exists, other atoms in the surrounding material attach themselves to the face of the seed. As the crystal grows, crystal faces move outward but maintain the same orientation (**Fig. 3.6b**). The youngest part of the crystal is at its outer edge. In the case of crystals formed by the solidification of a melt, atoms begin to attach to the seed when the melt becomes so cool that thermal vibrations can no longer break apart the attraction between the seed and the atoms in the melt. Crystals formed by precipitation from a solution develop when the solution becomes saturated, meaning the number of dissolved ions per unit volume of solution becomes so great that they can get close enough to each other to bond together.

As crystals grow, they develop their particular crystal shape, based on the geometry of their internal structure. The shape is defined by the relative dimensions of the crystal (for instance, needle-like or sheet-like) and by the angles between crystal faces. Typically, already-formed crystals act as obstacles preventing the unimpeded growth of new minerals. In such cases, the new minerals grow to fill the space that is available, so their shape does not reflect their crystal structures. Minerals without well-formed crystal faces are *anhedral* grains (**Fig. 3.6c**). If a mineral's growth is unimpeded so that it displays well-formed crystal faces, then it is a *euhedral* crystal. The surface crystals of a *geode*, a mineral-lined cavity in rock, may be euhedral (**Fig. 3.6d**).

A mineral can be destroyed by melting, dissolving, or some other chemical reaction. *Melting* involves heating a mineral to a temperature at which thermal vibration of the atoms or ions in the lattice break the chemical bonds holding them to the lattice. The atoms or ions then separate, either individually or in small groups, to move around again freely. *Dissolution* occurs when you immerse a mineral in a solvent, such as water. Atoms or ions then separate from the crystal face and are surrounded by solvent molecules. *Chemical reactions* can destroy a mineral when it comes in contact with reactive materials; for example, iron-bearing minerals react with air and water to form rust. *Microbial metabolism* in the environment can also destroy minerals. In effect, some microbes can "eat" certain minerals; the microbes use the energy stored in the chemical bonds that hold the atoms of the mineral together as their source of energy for metabolism.

FIGURE 3.6 The growth of crystals.

(a) New crystals nucleate and begin to precipitate out of a water solution. As time progresses, they grow into the open space.

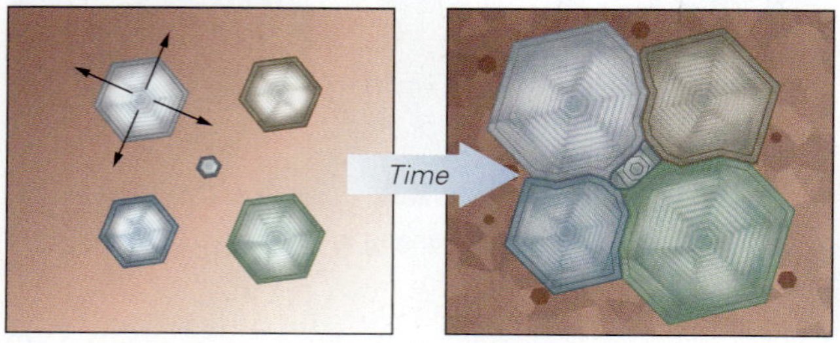

(b) New crystals grow outward from the central seed. As time passes, they maintain their shape until they interfere with each other.

(c) A crystal growing in a confined space will be anhedral.

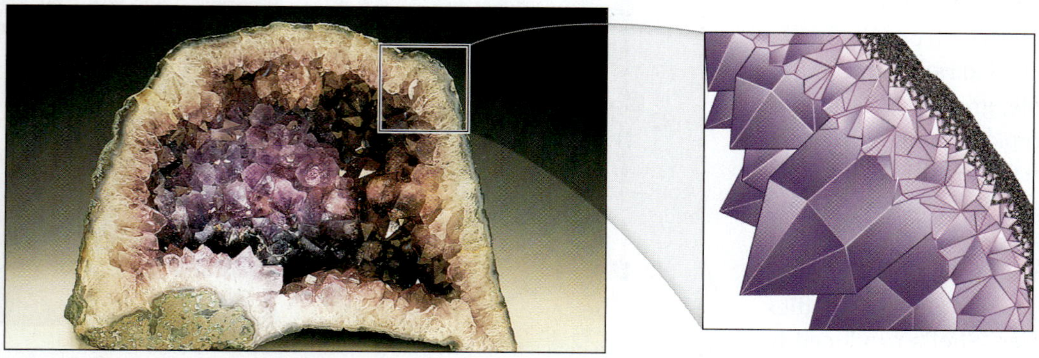

(d) A geode from Brazil consists of purple quartz crystals (amethyst) that grew from the wall into the center. The enlargement sketch indicates that the crystals are euhedral.

> ***TAKE-HOME MESSAGE***
>
> The crystal structure of minerals is defined by a regular geometric arrangement of atoms that has symmetry. Minerals can form by solidification of a melt, by precipitation from a water solution or a gas, or by rearrangement of atoms in a solid.
>
> **QUICK QUESTION** How are minerals destroyed in nature?

3.4 How Can You Tell One Mineral from Another?

Amateur and professional mineralogists alike get a kick out of recognizing minerals. They might hover around a display case in a museum and name specimens without bothering to look at the labels. How do they do it? The trick lies in learning to recognize the basic *physical properties* (visual and material characteristics) that distinguish one mineral from another. Some physical properties, such as shape and color, can be seen from a distance. Others, such as hardness and magnetization, can be determined only by handling the specimen or by performing an identification test on it. Such tests include scratching the mineral by another object, placing it near a magnet, weighing it, tasting it, or placing a drop of acid on it. Let's examine some of the physical properties most commonly used in basic mineral identification.

- *Color:* **Color** results from the way a mineral interacts with light. Sunlight contains the whole spectrum of colors, each with a different wavelength. A mineral absorbs certain wavelengths and reflects others—the color you see when looking at a specimen represents the wavelengths the mineral reflects and does not absorb. Certain minerals always have the same color, but many display a range of colors (**Fig. 3.7a**). Color variations in a mineral are due to the presence of impurities. For example, trace amounts of iron may give quartz a reddish color.
- *Streak:* The **streak** of a mineral refers to the color of a powder produced by pulverizing the mineral. You can obtain a streak by scraping the mineral against an unglazed ceramic plate (**Fig. 3.7b**). The color of a mineral powder tends to be less variable than the color of a whole crystal, and thus provides a fairly reliable clue to a mineral's identity. Calcite, for example, always yields a white streak even though pieces of calcite may be white, pink, or clear.
- *Luster:* When a light beam hits a mineral surface, some of it undergoes scattering, meaning that it divides into smaller beams that bounce off in different directions. **Luster** refers to the way a mineral surface scatters light. Geoscientists describe luster by comparing the appearance of the mineral with the appearance of a familiar substance. For example, minerals that look like metal have a metallic luster, whereas those that do not have a nonmetallic luster—the adjectives are self-explanatory (**Fig. 3.7c, d**). Terms used for types of nonmetallic luster include silky, glassy, satiny, resinous, pearly, or earthy.
- *Hardness:* **Hardness** is a measure of the relative ability of a mineral to resist scratching, and it therefore represents the resistance of bonds in the crystal structure to being broken. The atoms or ions in crystals of a hard mineral are more strongly bonded than those in a soft mineral. Hard minerals can scratch soft minerals, but soft minerals cannot scratch hard ones. Diamond, the hardest mineral known, can scratch most anything, which is why it is used to cut glass. In the early 1800s, a mineralogist named Friedrich Mohs (1773-1839) listed some minerals in sequence of relative hardness; a mineral with a hardness of 5 can scratch all minerals with a hardness of 5 or less. This list, the **Mohs hardness scale**, helps in mineral identification. To make the scale easy to use, we include common items such as your fingernail, a penny, or a glass plate (**Table. 3.1**).

TABLE 3.1 **Mohs Hardness Scale.** Mohs numbers are relative—in reality, diamond is 3.5 times harder than corundum, as the graph shows.

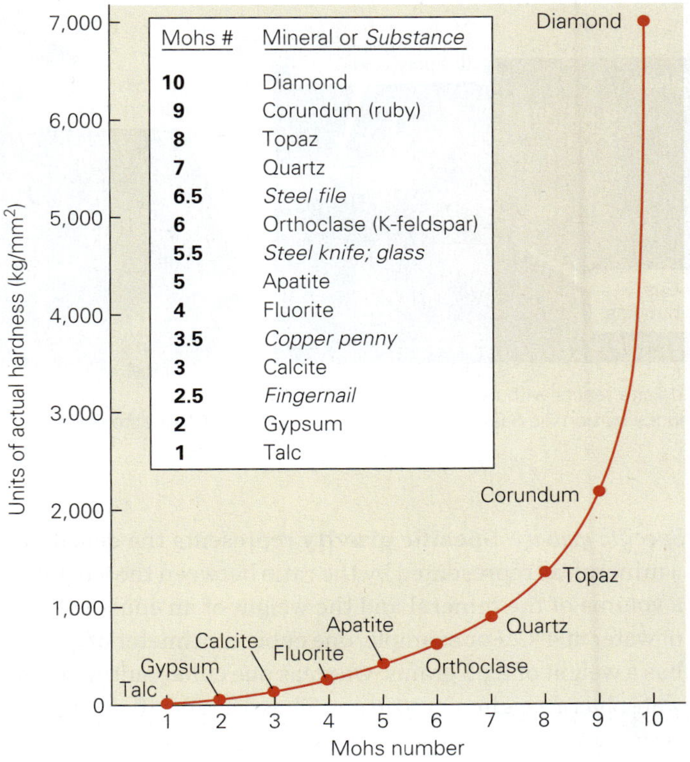

Mohs #	Mineral or *Substance*
10	Diamond
9	Corundum (ruby)
8	Topaz
7	Quartz
6.5	*Steel file*
6	Orthoclase (K-feldspar)
5.5	*Steel knife; glass*
5	Apatite
4	Fluorite
3.5	*Copper penny*
3	Calcite
2.5	*Fingernail*
2	Gypsum
1	Talc

FIGURE 3.7 Physical characteristics of minerals.

(a) Color is diagnostic of some minerals, but not all. For example, quartz can come in many colors.

(b) To obtain the streak of a mineral, rub it against a porcelain plate. The streak consists of mineral powder.

(c) Pyrite has a metallic luster because it gleams like metal.

(d) Feldspar has a nonmetallic luster.

(e) Crystal habit refers to the shape or character of the crystal. The blue kyanite crystals on the right are bladed, and the chrysotile above is fibrous.

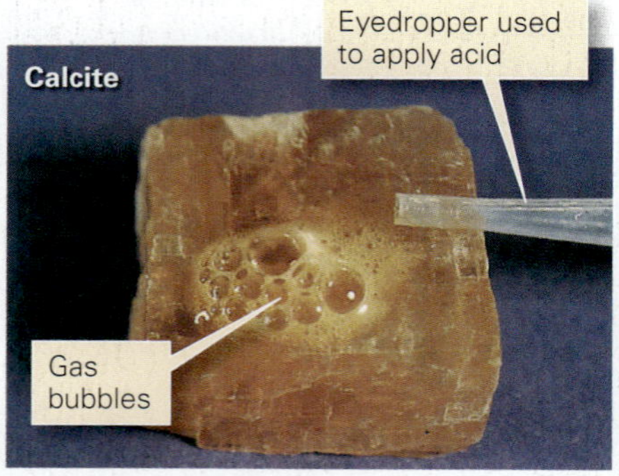

(f) Calcite reacts with hydrochloric acid to produce carbon dioxide gas.

(g) Magnetite is magnetic.

› *Specific gravity:* **Specific gravity** represents the density of a mineral, as represented by the ratio between the weight of a volume of the mineral and the weight of an equal volume of water at 4°C. For example, one cubic centimeter of quartz has a weight of 2.65 grams, whereas one cubic centimeter of water has a weight of 1.00 gram. Thus, the specific gravity of quartz is 2.65. In practice, you can develop a feel for specific gravity by hefting minerals in your hands. A piece of galena (lead ore) feels heavier than a similar-sized piece of quartz.

BOX 3.2 CONSIDER THIS...

Asbestos and Health—When Crystal Habit Matters

There are many types of asbestos minerals, but they all share a key characteristic—all have a fibrous crystal habit (**Fig. Bx3.2a**). Thus, samples of asbestos consist of clusters of needle-like crystals (**Fig. Bx 3.2b**). Asbestos used to be incorporated into floor tiles, roof shingles, brake pads, fireproof clothes, and insulation. Its popularity derived from the fact that, because it's a mineral, it doesn't burn, and because it's fibrous, it can be woven into other materials. Its presence in other materials adds strength because fibers have strong bonds along their length.

When intact, asbestos-bearing materials are not especially hazardous. But during the manufacture of these materials, during the mining of asbestos, or during construction or demolition involving asbestos removal, the fibers can become airborne. If inhaled, the fibers lodge in the lungs where they cause irritation and may trigger cancer. Thus, over the years, a large number of lawsuits involving asbestos have taken place, and its use has been banned since the mid-1980s.

Today, building owners must pay for *asbestos abatement* during construction projects that involve tearing out old asbestos-bearing materials. During abatement, workers handling asbestos must wear protective clothes (**Fig. Bx 3.2c**), and the space in which they're working must be sealed off from its surroundings. Debate continues as to whether all abatement is necessary, because not all asbestos minerals appears to be equally hazardous, and in general, asbestos-bearing materials that have been covered over by paint or other materials, and/or remain undisturbed, may not be problematic.

FIGURE Bx3.2 Asbestos has characteristics that can make it both useful and hazardous.

(a) Asbestos is a fibrous mineral.

(b) At high magnification, asbestos fibers look like tiny needles.

(c) To remove asbestos requires protective gear and for the area to be sealed off.

› *Crystal habit:* The **crystal habit** of a mineral refers to the shape of a single crystal with well-formed crystal faces, or to the character of an aggregate of many well-formed crystals that grew together as a group (**Fig. 3.7e**). The habit depends on the internal arrangement of atoms in the crystal. When describing habit, mineralogists commonly compare the mineral to a common geometric shape, by using adjectives such as cubic, prismatic, bladed, platy, or fibrous (**Box 3.2**). The relative dimensions depend on relative rates of crystal growth in different directions. For example, crystals that grow rapidly in one direction but slowly in the other two directions are needle-like.

FIGURE 3.8 The nature of mineral cleavage and fracture.

(a) Mica has one strong plane of cleavage and splits into sheets.

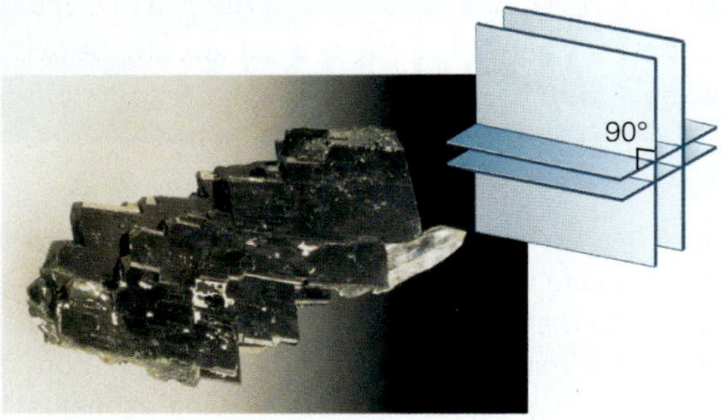

(b) Pyroxene has two planes of cleavage that intersect at 90°.

(c) Amphibole has two planes of cleavage that intersect at 60°.

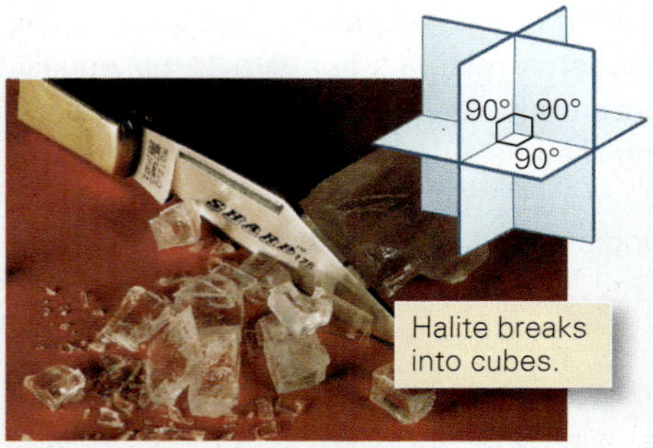

(d) Halite has three mutually perpendicular planes of cleavage.

(e) Calcite has three planes of cleavage, one of which is inclined.

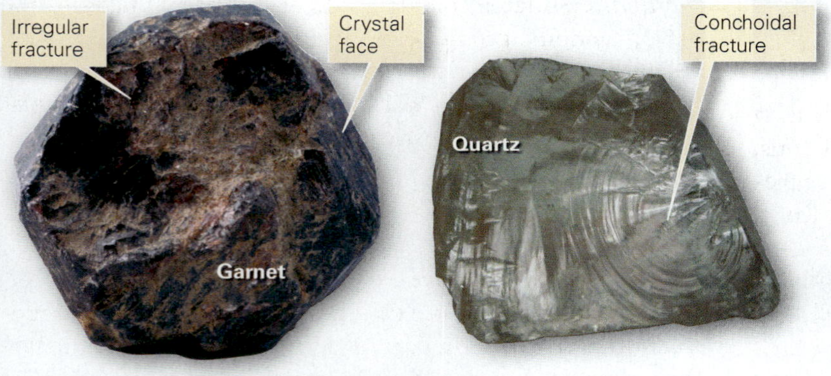

(f) Minerals without cleavage can develop irregular or conchoidal fractures.

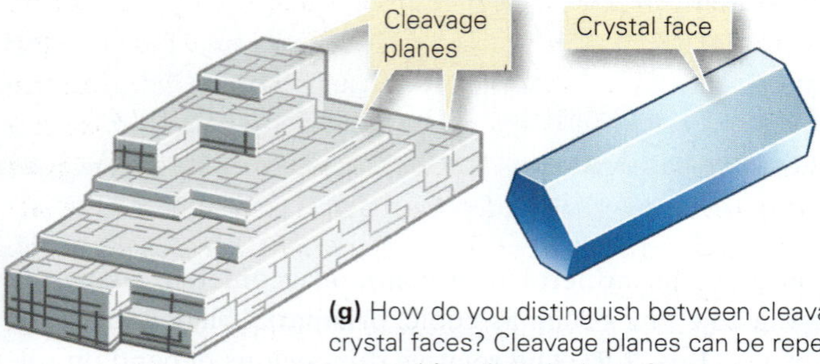

(g) How do you distinguish between cleavage planes and crystal faces? Cleavage planes can be repeated, whereas a crystal face is a single surface.

› *Special properties:* Some minerals have distinctive properties that readily distinguish them from other minerals. For example, calcite ($CaCO_3$) reacts with dilute hydrochloric acid (HCl) to produce carbon dioxide (CO_2) gas (**Fig. 3.7f**). Dolomite ($CaMg[CO_3]_2$) also reacts with acid, but not as strongly. Graphite makes a gray mark on paper, magnetite attracts a magnet (**Fig. 3.7g**), halite tastes salty, and plagioclase has striations (thin parallel corrugations or stripes) on its surface.

› *Fracture and cleavage:* Different minerals fracture (break) in different ways, depending on the internal arrangement of atoms. If a mineral breaks to form distinct planar surfaces that have a specific orientation in relation to the crystal structure, then we say that the mineral has cleavage, and we refer to each surface as a cleavage plane. Cleavage forms in directions where the bonds holding atoms together in the crystal are weaker (**Fig. 3.8a–e**). Some minerals have one direction of cleavage. For example, mica has very weak bonds in one direction but strong bonds in the other two directions. Thus, it easily splits into parallel sheets; the surface of each sheet is a cleavage plane. Other minerals have two or three directions of cleavage that intersect at a specific angle. For example, halite has three sets of cleavage planes that intersect at right angles, so halite crystals break into little cubes. Materials with no cleavage at all (because bonding is equally strong in all directions) break either by forming irregular fractures or by forming conchoidal fractures (**Fig. 3.8f**). *Conchoidal fractures* are smoothly curving, clamshell-shaped surfaces that typically form in glass. Cleavage planes are sometimes hard to distinguish from crystal faces (**Fig. 3.8g**).

TAKE-HOME MESSAGE

The properties of minerals (such as color, streak, luster, crystal habit, hardness, specific gravity, cleavage, magnetism, and reaction with acid) are a manifestation of the crystal structure and chemical composition of minerals and can be used for mineral identification.

QUICK QUESTION Which minerals react with acid to produce CO_2 bubbles?

3.5 Organizing Your Knowledge: Mineral Classification

Just about every object you come across in daily life has been classified in some way, because classification schemes help organize information and streamline discussion. Biologists, for example, classify animals into groups based on how they feed their young and on the architecture of their skeletons, and botanists classify plants according to the way they reproduce and by the shape of their leaves. In the case of minerals, a good means of classification eluded researchers until it became possible to determine the chemical makeup of minerals. A Swedish chemist, Baron Jöns Jacob Berzelius (1779–1848), analyzed minerals and noted chemical similarities among many of them. Berzelius, along with his students, established that most minerals can be classified by specifying the principal anion (negative atom) or anionic group (negative molecule) within the mineral (see Box 3.1). Using this approach, it's possible to divide the 4,000 known minerals into a small number of groups, or **mineral classes**. We now take a look at principal mineral classes, focusing especially on silicates, the class that constitutes most of the rock in the Earth.

The Mineral Classes

Mineralogists distinguish several principal classes of minerals. Here are some of the major ones.

- *Silicates:* The fundamental component of most silicates in the Earth's crust is the SiO_4^{4-} anionic group. A well-known example, quartz (Fig. 3.7a), has the formula SiO_2. We will learn more about silicates in the next section.
- *Sulfides:* Sulfides consist of a metal cation bonded to a sulfide anion (S^{2-}). Examples include galena (PbS) and pyrite (FeS_2; Fig. 3.7c).
- *Oxides:* Oxides consist of metal cations bonded to oxygen anions. Typical oxide minerals include hematite (Fe_2O_3; Fig. 3.7b) and magnetite (Fe_3O_4; Fig. 3.7g).
- *Halides:* The anion in a halide is a halogen ion (such as chloride [Cl^-] or fluoride [F^-]), an element from the second column from the right in the periodic table (a periodic table is provided after Chapter 19). Halite, or rock salt (NaCl; Fig. 3.8d), and fluorite (CaF_2) are common examples.
- *Carbonates:* In carbonate minerals, CO_3^{2-} serves as the anionic group. Elements such as calcium or magnesium bond to this group. The two most common carbonates are calcite ($CaCO_3$; Fig. 3.8e) and dolomite ($CaMg[CO_3]_2$).
- *Native metals:* Native metals consist of pure masses of a single metal. The metal atoms are bonded by metallic bonds (see Box 3.1). Copper and gold, for example, may occur as native metals.
- *Sulfates:* Sulfates consist of a metal cation bonded to the SO_4^{2-} anionic group. Many sulfates form by precipitation out of water at or near the Earth's surface. An example is gypsum ($CaSO_4 \cdot 2H_2O$). Note that in gypsum, water molecules bond to calcium sulfate.

Silicates: The Major Rock-Forming Minerals

Silicate minerals, or **silicates**, make up over 95% of the continental crust and almost 100% of the oceanic crust. Further, nearly all of the Earth's mantle consists of silicates. Thus, silicates are the most common minerals on Earth. As we've noted, silicates in the Earth's crust and upper mantle contain the SiO_4^{4-} anionic group. In this group, four oxygen atoms surround a single silicon atom, thereby defining the corners of a tetrahedron, a pyramid-like shape with four triangular faces (**Fig. 3.9a**). We refer to this anionic group as the **silicon-oxygen tetrahedron** (or, informally, as the silica tetrahedron), and it acts, in effect, as the building block of silicate minerals.

Mineralogists distinguish among several groups of silicate minerals based on the way in which silica tetrahedra are arranged and bonded together (**Fig. 3.9b**). The extent of bonding, in turn, determines the degree to which tetrahedra

FIGURE 3.9 The structure of silicate minerals.

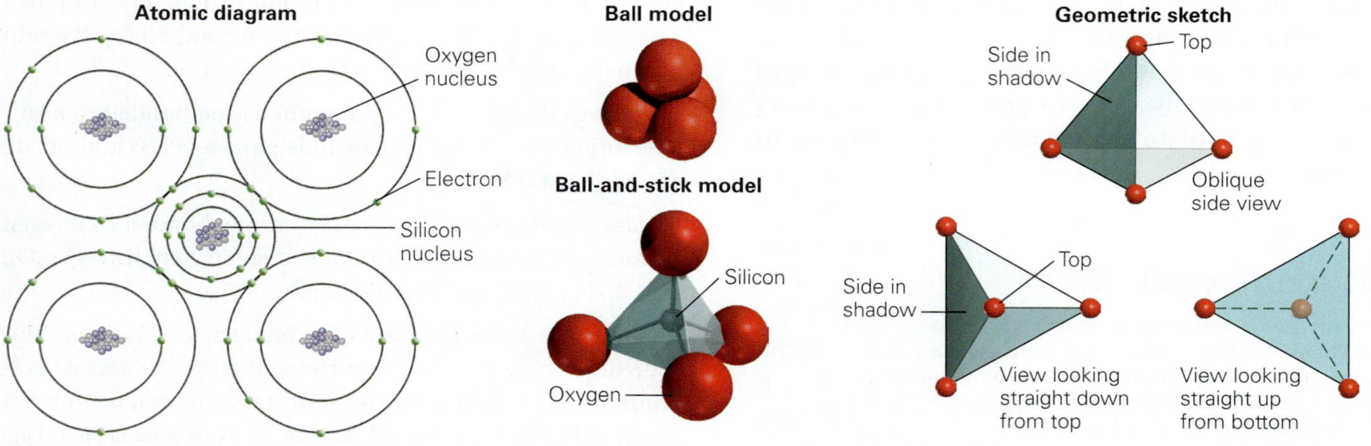

(a) The fundamental building block of a silicate mineral is the silicon-oxygen tetrahedron. In this SiO_4^{4-} group, oxygens occupy the corners of the tetrahedron, and silicon lies at the center. Geologists portray the tetrahedron in a number of different ways.

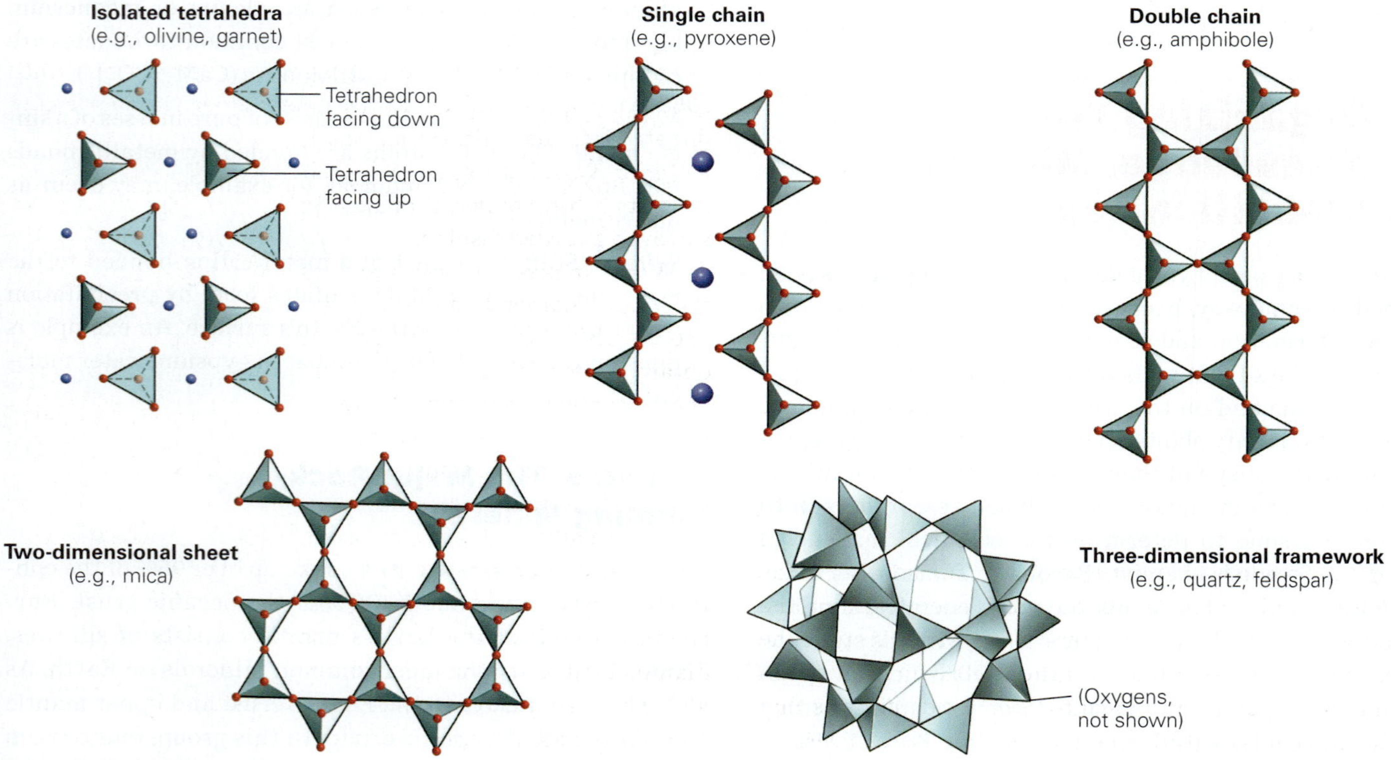

(b) The classes of silicate minerals differ from one another by the way in which the silicon-oxygen tetrahedra are linked. Where the tetrahedra link, they share an oxygen atom. Oxygen atoms are shown in red. Positive ions (not shown) occupy spaces between tetrahedra.

share oxygen atoms. Note that the number of shared oxygens determines the ratio of silicon (Si) to oxygen (O) in the mineral. Here are the groups, in order from fewer shared oxygens to more shared oxygens:

- *Independent tetrahedra:* In this group, the tetrahedra do not share any oxygen atoms. The attraction between tetrahedra and positive ions holds these minerals together. Examples include olivine, a glassy green mineral, and garnet.
- *Single chains:* In a single-chain silicate, the tetrahedra link to form a chain by sharing two oxygen atoms. The most common of the many different types of single-chain silicates are pyroxenes (Fig. 3.8b).

- *Double chains:* In a double-chain silicate, the tetrahedra link by sharing two or three oxygen atoms. Amphiboles are the most common type (Fig. 3.8c).
- *Sheet silicates:* The tetrahedra in this group share three oxygen atoms and therefore link to form two-dimensional sheets. Other ions, and in some cases, water molecules, fit between the sheets in some sheet silicates. Because of their structure, sheet silicates have cleavage in one direction and occur in books of very thin sheets. In this group we find micas (Fig. 3.8a) and clays (which occur only in extremely tiny flakes).
- *Framework silicates:* In a framework silicate, each tetrahedron shares all four oxygen atoms with its neighbors, forming a three-dimensional structure. Examples include feldspar and quartz. The two most common feldspars are plagioclase, which tends to be white, gray, or blue; and orthoclase (also called potassium feldspar, or K-feldspar), which tends to be pink (Fig. 3.7d).

TAKE-HOME MESSAGE

The 4,000 known minerals can be organized into a relatively small number of classes based on chemical makeup. Most minerals are silicates, which contain silicon-oxygen tetrahedra arranged in various ways.

QUICK QUESTION What is the principal anionic group in carbonate minerals?

3.6 Something Precious—Gems!

Mystery and romance follow famous gems. Consider the stone now known as the Hope Diamond, recognized by name the world over (**Fig. 3.10**). No one knows who first dug it out of the ground (**Box 3.3**). Was it mined in the 1600s, or was it stolen off an ancient religious monument? What we do know is that in the 1600s, a French trader named Jean-Baptiste Tavernier obtained a large (112.5 carats, where 1 carat = 200 milligrams), rare blue diamond in India, perhaps from a Hindu statue, and carried it back to France. King Louis XIV bought the diamond and had it fashioned into a jewel of 68 carats. This jewel vanished during a burglary in 1762. Perhaps it was lost forever—perhaps not. In 1830, a 44.5-carat blue diamond mysteriously appeared on the jewel market for sale. Henry Hope, a British banker, purchased the stone, which then became known as the Hope Diamond. It changed hands several times until 1958, when a famous New York jeweler named Harry Winston donated it to the Smithsonian Institution in Washington, DC, where it now sits behind bulletproof glass in a heavily guarded display.

Did you ever wonder... where diamonds come from and how they form?

FIGURE 3.10 The Hope Diamond, now on display at the Smithsonian Institution in Washington, DC.

What makes stones such as the Hope Diamond so special that people risk life and fortune to obtain them? What is the difference between gemstones, gems, and other minerals? A **gemstone** is a mineral that has special value because it is rare and people consider it beautiful. A **gem**, or jewel, is a finished stone ready to be set in jewelry. Jewelers distinguish between precious stones (such as diamond, ruby, sapphire, and emerald), which are particularly rare and expensive, and semiprecious stones (such as topaz, tourmaline, aquamarine, and garnet), which are less rare and less expensive. All precious stones are transparent crystals, though most have some color. The category of semiprecious stones also includes opaque or translucent minerals such as lapis, malachite (see Fig. 3.1a), and opal.

SEE FOR YOURSELF...

KIMBERLEY DIAMOND MINE

Latitude
28°44′17.06″ S

Longitude
24°46′30.77″ E

Zoom to an elevation of 13 km (~8 miles) and look straight down.

The field of view shows the town of Kimberley, South Africa, and its inactive diamond mine. The mine looks like a circular pit. You can also see the tailings pile of excavated rock debris.

In everyday language, pearls and amber may also be considered gemstones. Unlike diamonds and garnets, which form

BOX 3.3 CONSIDER THIS...

Where Do Diamonds Come From?

Diamonds consist of carbon, which typically accumulates only at or near Earth's surface. Experiments demonstrate that the pressures needed to form diamond are so extreme that, in nature, they generally occur only at depths of around 150 km below the Earth's surface. Nowadays, engineers can duplicate these conditions in the laboratory, so corporations manufacture several tons of synthetic diamonds a year.

How does carbon get down to depths of 150 km? Geologists speculate that subduction or collision carries carbon-containing rocks and sediments down to the depth at which it transforms into diamond beneath continents. But if diamonds form at great depth, how do they return to the surface? Some diamonds rise when rifting cracks the continental crust and causes a small part of the underlying mantle to melt. Magma generated during this process rises to the surface, bringing the diamonds with it. Near the surface, the magma solidifies to form an igneous rock called *kimberlite*, named for Kimberley, South Africa. Diamonds brought up with the magma are embedded as crystals in solid kimberlite (**Fig. Bx3.3**). Much of the world's diamond supply comes from mines in this rock. But some sources occur in deposits of sediment formed from the breakdown and erosion of kimberlite that had been exposed at the surface. Rivers and glaciers may transport diamond-bearing sediments far from their original bedrock source.

Not all natural diamonds are valuable; value depends on color and clarity. Diamonds that contain imperfections (cracks, or specks of other material) or are dark gray in color are not used for jewelry. These stones, called *industrial diamonds*, are used as abrasives. Gem-quality diamonds come in a range of sizes. Jewelers measure the size of these gems using carats, where 1 carat equals 200 milligrams (0.2 gram). In English units of measurement, one ounce equals 142 carats. The largest diamond ever found, a stone called the Cullinan Diamond, was discovered in South Africa in 1905, and weighed 3,106 carats (621 grams) before being cut. By comparison, the diamond on a typical engagement ring weighs less than 1 carat. Gem-quality diamonds are actually more common than you might expect—suppliers stockpile the stones in order to avoid flooding the market and lowering the price.

A diamond mine pit.
A diamond embedded in solid kimberlite.

FIGURE Bx3.3 Diamond occurrences.

inorganically in rocks, pearls form in living oysters when the oyster extracts calcium and carbonate ions from water and precipitates them around an impurity, such as a sand grain, embedded in its body. Thus, pearls are a result of biomineralization. Most pearls used in jewelry today are "cultured" pearls, made by artificially introducing round sand grains into oysters in order to stimulate production of round pearls. Amber also forms as a consequence of organic processes—it consists of fossilized tree sap. But because amber consists of organic compounds that are not arranged in a crystal structure, it does not meet the definition of a mineral.

In some cases, gemstones are merely pretty and rare versions of more common minerals. For example, ruby is a special version of the common mineral corundum, and emerald is a special version of the common mineral beryl (**Fig. 3.11a**). As for the beauty of a gemstone, this quality lies basically in its color and, in the case of transparent gems, its "fire"—the way the mineral bends and internally reflects the light passing through it, and disperses the light into a spectrum. Fire makes a diamond sparkle more than a similarly cut piece of glass.

Gemstones form in many ways. Some solidify from a melt, some form by diffusion, some precipitate out of a water

FIGURE 3.11 Cutting gemstones.

(a) Emerald is a green, transparent variety of the mineral beryl.

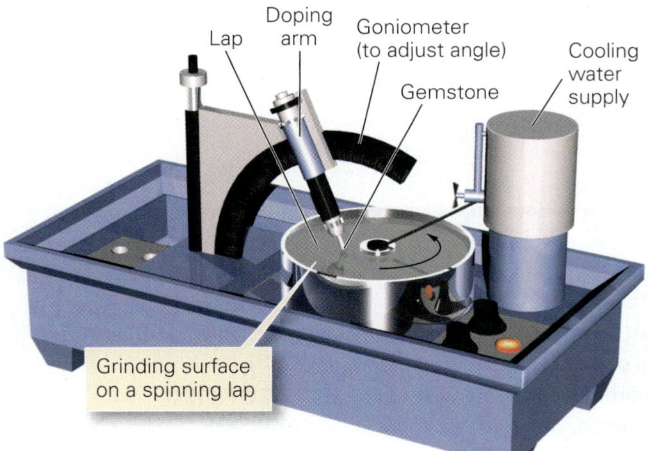

(b) The shiny faces of a gem are made by grinding the stone on a lap.

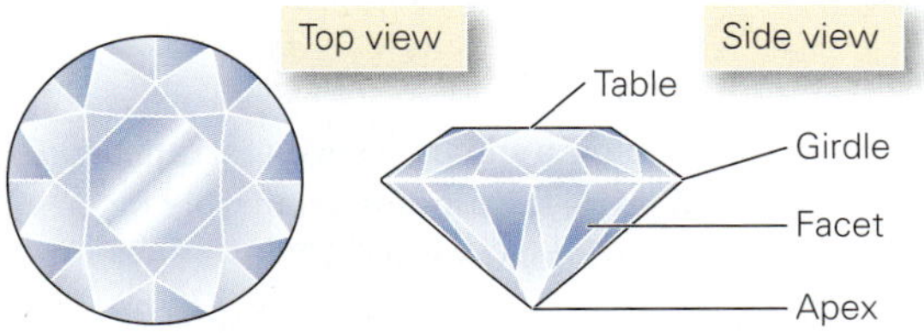

(c) There are many different "cuts" for a gem. Here we see the top and side views of a brilliant-cut diamond.

solution in cracks, and some are a consequence of the chemical interaction of rock with water near the Earth's surface. Several types of gems come from pegmatites, particularly coarse-grained rocks formed by the solidification of steamy melt.

Most gems used in jewelry are "cut" stones, meaning that they are not raw crystals right from the ground, but rather have been faceted. The smooth **facets** on a gem are ground and polished surfaces made with a faceting machine (**Fig. 3.11b**). Facets are not the natural crystal faces of the mineral, nor are they cleavage planes, though gem cutters sometimes make the facets parallel to cleavage directions and will try to break a large gemstone into smaller pieces by splitting it on a cleavage plane. A faceting machine consists of a doping arm, a device that holds a stone in a specific orientation, and a lap, a rotating disk covered with a wet paste of grinding powder and water. The gem cutter fixes a gemstone to the end of the doping arm and positions the arm so that it holds the stone against the moving lap. The movement of the lap grinds a facet. After completing a facet, the gem cutter rotates the arm by a specific angle, lowers the stone, and grinds another facet. The geometry of the facets defines the cut of the stone. Different cuts have names, such as brilliant, French, star, and pear. Grinding facets is a lot of work—a typical engagement-ring diamond with a brilliant cut has 57 facets (**Fig. 3.11c**)!

Did you ever wonder…
how jewelers make the facets on a jewel?

SEE FOR YOURSELF…

EKATI DIAMOND MINE, CANADA

Latitude
64°43′15.44″ N

Longitude
110°36′56.27″ W

Zoom to an elevation of 100 km (~62 miles) and look straight down.

The Ekati Diamond Mine is in a remote, largely uninhabited region of the Northwest Territories of Canada. Prospectors found diamond pipes here in the early 1990s after a 20-year search. The mine opened in 1998 and within 10 years had produced more than 40 million carats (8,000 kg) of diamonds. Zoom down to 10 km (~6 miles) to see details of the mining operation.

TAKE-HOME MESSAGE

Gemstones are particularly rare and beautiful minerals. The gems or jewels found in jewelry have been faceted using a lap—the facets are not natural crystal faces or cleavage surfaces. The fire of a jewel comes from the way it reflects light internally.

QUICK QUESTION What's the difference between a facet on a gem and a crystal face?

Chapter 3 Review

Chapter Summary

- Minerals are naturally occurring, solid substances, formed by geologic processes, with a definable chemical composition and an internal structure characterized by an orderly arrangement of atoms, ions, or molecules in a crystalline lattice. Most minerals are inorganic.
- Biogenic minerals, such as those in clam shells, are produced by organisms through a process called biomineralization.
- In the crystalline lattice of minerals, atoms occur in a specific pattern—one of nature's finest examples of ordering.
- Minerals can form by solidification of a melt, precipitation from a water solution, diffusion through a solid, metabolism of organisms, or precipitation from a gas.
- Mineralogists have identified about 4,000 different types of minerals. Each has a name and distinctive physical properties (such as color, streak, luster, hardness, specific gravity, crystal habit, cleavage, magnetism, and reactivity with acid).
- The unique physical properties of a mineral reflect its chemical composition and crystal structure. By observing physical properties, you can identify minerals.
- The most convenient way to classify minerals is to group them according to their chemical composition. Mineral classes include silicates, oxides, sulfides, sulfates, halides, carbonates, and native metals.
- Silicate minerals are the most common minerals on Earth. The silicon-oxygen tetrahedron, a silicon atom surrounded by four oxygen atoms, serves as the fundamental building block of silicate minerals.
- Groups of silicate minerals are distinguished from each other by the ways in which the silicon-oxygen tetrahedra that constitute them are linked.
- Gemstones are minerals known for their beauty and rarity. The facets on cut gems used in jewelry are made by grinding and polishing the stones with a faceting machine.

Guide Terms

color (p. 91)
crystal (p. 85)
crystal face (p. 85)
crystal habit (p. 93)
crystal lattice (p. 84)
crystalline solid (p. 84)
facet (p. 99)
gem (p. 97)
gemstone (p. 97)
glass (p. 84)
hardness (p. 91)
luster (p. 91)
mineral (p. 84)
mineral classes (p. 95)
Mohs hardness scale (p. 91)
polymorph (p. 88)
silicate (p. 95)
silicon-oxygen tetrahedron (p. 95)
specific gravity (p. 92)
streak (p. 91)

Review Questions

1. What is a mineral, as geologists understand the term? How is this definition different from the everyday usage of the word?
2. Why is glass not a mineral?
3. Salt is a mineral, but the plastic making up an inexpensive pen is not. Why not?
4. Describe several ways that mineral crystals can form.
5. Why do some minerals occur as euhedral crystals, whereas others occur as anhedral grains?
6. List and define the principal physical properties used to identify a mineral. Which minerals react with acid to produce CO_2?
7. How can you determine the hardness of a mineral? What is the Mohs hardness scale?
8. How do you distinguish cleavage surfaces from crystal faces on a mineral? How does each type of surface form?
9. What is the prime characteristic that geologists use to separate minerals into classes?
10. What is a silicon-oxygen tetrahedron? What is the anionic group that occurs in carbonate minerals?
11. On what basis do mineralogists organize silicate minerals into distinct groups?
12. What is the relationship between the way in which silicon-oxygen tetrahedra bond in micas and the characteristic cleavage of micas?
13. Why are some minerals considered gemstones? How do you make the facets on a gem?

GEOTOURS *THIS CHAPTER'S GEOTOUR EXERCISE (C) FEATURES:*

- Diamond Mines
- Mineral Reactions after Coal Mining

On Further Thought

14. Compare the chemical formula of magnetite with that of biotite. Considering that iron is a relatively heavy element, which mineral has the greater specific gravity?

15. Imagine that you find two milky white crystals, each about 2 cm across. One consists of plagioclase and the other of quartz. How can you determine which is which?

16. Could you use crushed calcite to grind facets on a diamond? Why or why not?

Online Resources

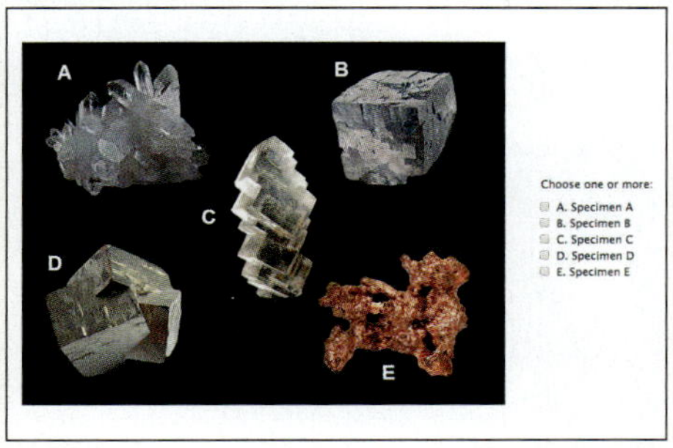

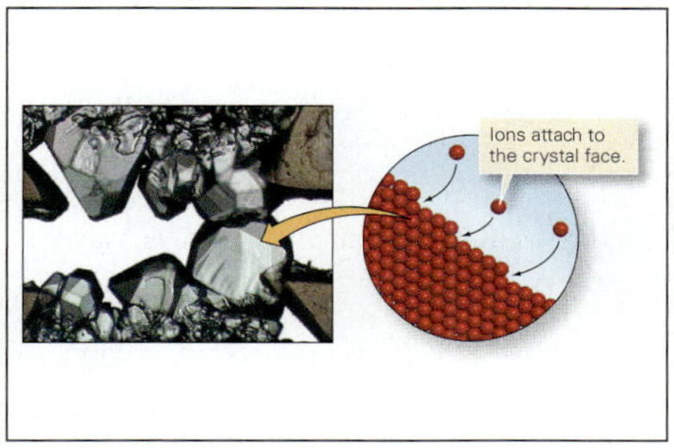

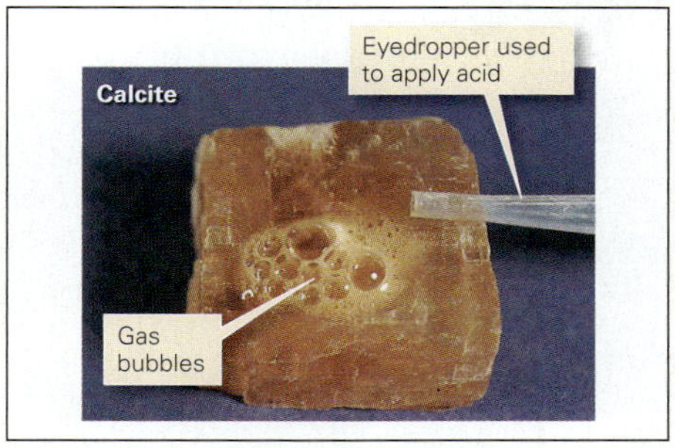

Assessment

This chapter includes visual matching and labeling exercises designed to better understand crystal structure, how crystals grow, and the physical characteristics of minerals.

Another View A museum specimen of wulfenite, a mineral containing lead and molybdenum. It precipitates in cracks to form clusters of small tabular crystals.

INTERLUDE A

Rock Groups

▲ The first railroad through the Sierra Nevada followed the Truckee River. Countless tons of rock had to be blasted and moved by laborers.

LEARNING OBJECTIVES

By the end of this chapter, you should understand...

1. the geologic definition of *rock*.
2. the basis that geologists use to classify rock into three classes.
3. the tools that can be used to study rocks.

A.1 Introduction

During the 1849 gold rush in the Sierra Nevada of California, only a few lucky individuals actually became rich. The rest of the forty-niners either slunk home in debt or took up less glamorous jobs in new towns such as San Francisco. These towns grew rapidly, and soon people from the American west coast were demanding large quantities of manufactured goods from east-coast factories. Making the goods was not a problem, but getting them to California meant either a stormy voyage around the southern tip of South America or a trek with stubborn mule teams through the deserts of Nevada and Utah. The time was ripe to build a railroad linking the east and west coasts of North America, so with much fanfare, a consortium of railroads set to work in 1863. The Union Pacific surveyed a route that crossed the Sierras, and as the Civil War raged, the company transported thousands of Chinese laborers across the Pacific and set them to work chipping ledges around, and blasting tunnels through, the range's towering peaks. Sadly, untold numbers of laborers died of frostbite and exhaustion, or from mistimed blasts, landslides, and avalanches.

Through their efforts, the railroad laborers certainly gained an intimate knowledge of how rock feels and behaves—it's solid, heavy, and hard! They also found that some rocks break easily into layers but others do not, and that some rocks are dark colored while others are light colored. Like anyone who looks closely at rock exposures, they realized that rocks are not just gray, featureless masses, but rather come in a great variety of colors, textures, and configurations.

Why are there so many distinct types of rocks? The answer is simple: rocks can form in many different ways and from many different materials. Because of the relationship between rock types and the process of forming them, rocks provide a historical record of geologic events, and they give insight into interactions among components of the Earth System. The next few chapters are devoted to a discussion of rocks and a description of how rocks form. To provide a general introduction to these chapters, this interlude provides the geological definition of the term *rock*, describes the basic components of rock, and characterizes the three principal classes of rocks. We also describe a few of the methods that geologists use to study rocks.

A.2 What Is Rock?

To geologists, a **rock** is a coherent, naturally occurring solid, consisting of an aggregate of minerals or, less commonly, of glass. Let's take this definition apart to see what its components mean.

- *Coherent:* A rock holds together, and thus must be broken to be separated into smaller pieces. As a result of its coherence, rock can form cliffs or can be carved into sculptures. A pile of unattached mineral grains does not constitute a rock.
- *Naturally occurring:* Geologists consider only naturally occurring materials to be rocks. Manufactured materials, such as concrete and brick, are not rocks.
- *An aggregate of minerals or a mass of glass:* The vast majority of rocks consist of an aggregate (a collection) of many mineral grains, and/or crystals, stuck or grown together. Some rocks contain only one kind of mineral, whereas others contain several different kinds. A few rock types consist of glass.

What holds rock together? Grains in rock stick together to form a coherent mass either because they are bonded by natural **cement**, mineral material that precipitates from water and fills the space between grains (**Fig. A.1a**), or because they interlock with one another like pieces of a jigsaw puzzle (**Fig. A.1b**). Rocks whose grains are stuck together by cement are called **clastic rocks**, whereas rocks whose crystals interlock with one another are called **crystalline rocks**. A glassy rock can be coherent either because it originated as a continuous mass (that is, it does not contain separate grains) or because it formed when separate glass grains welded together while still hot.

At the surface of the Earth, rock occurs either as broken chunks (pebbles, cobbles, or boulders; see Chapter 6) that have moved by falling down a slope or by being transported in

FIGURE A.1 Rocks, aggregates of mineral grains and/or crystals, can be clastic or crystalline.

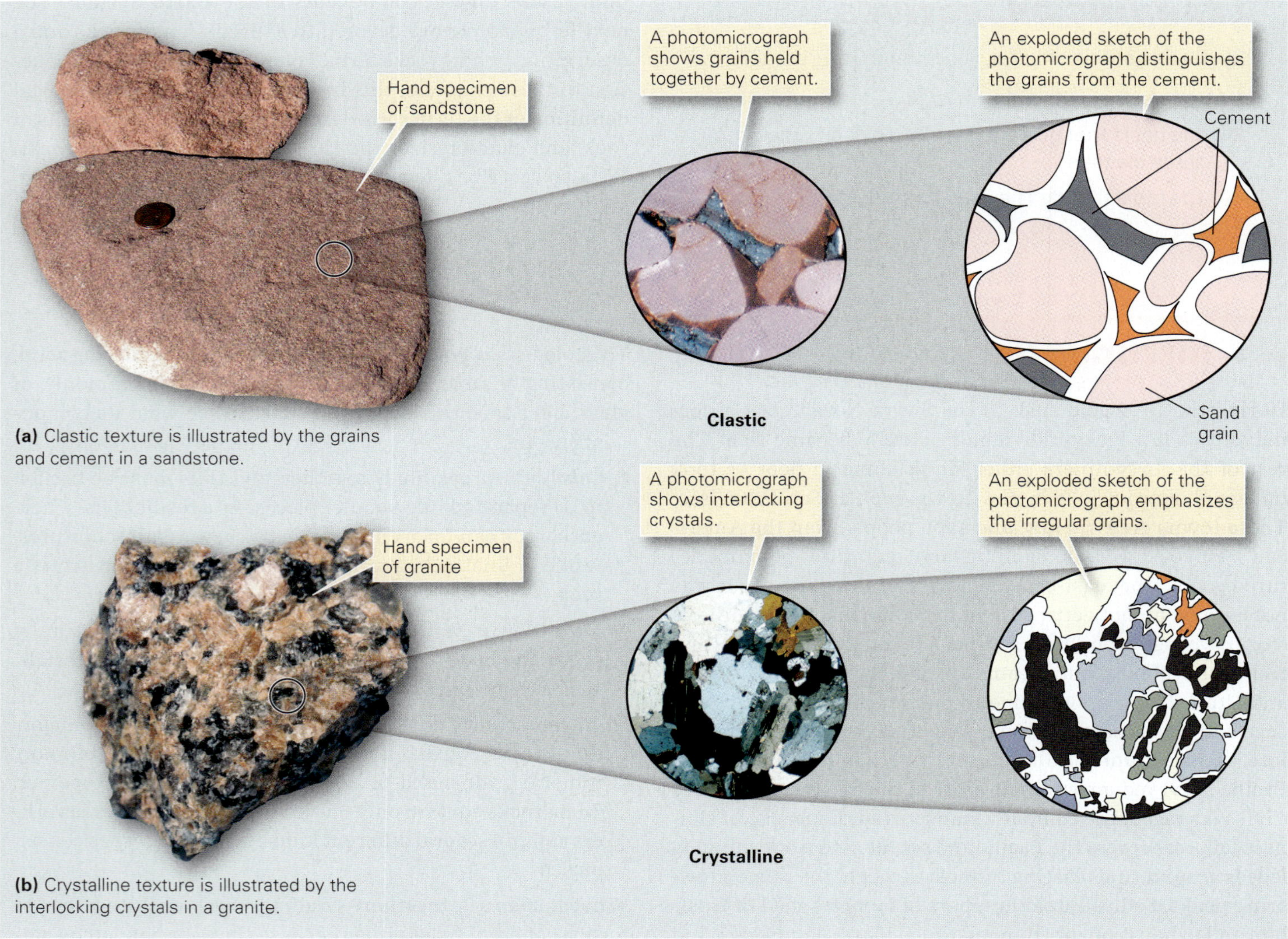
Hand specimen of sandstone
A photomicrograph shows grains held together by cement.
An exploded sketch of the photomicrograph distinguishes the grains from the cement.
Cement
Sand grain
Clastic
Hand specimen of granite
A photomicrograph shows interlocking crystals.
An exploded sketch of the photomicrograph emphasizes the irregular grains.
Crystalline

(a) Clastic texture is illustrated by the grains and cement in a sandstone.

(b) Crystalline texture is illustrated by the interlocking crystals in a granite.

ice, water, or wind, or as **bedrock**, meaning rock, that is still attached to the Earth's crust. Geologists refer to an exposure of bedrock as an **outcrop**. Outcrops may be rounded knobs in a field, cliffs or ridges in the mountains, the carved walls of stream cuts, or faces of roadcuts and other excavations produced by people (**Fig. A.2**). To inhabitants of cities, forests, or farmland, outcrops of bedrock may be unfamiliar, since bedrock may be completely covered by vegetation, loose sand and gravel, water, asphalt, concrete, or buildings. Outcrops are particularly rare in regions such as the midwestern United States, where, during the past million years, melting ice-age glaciers buried bedrock under thick deposits of debris (see Chapter 18).

A.3 The Basis of Rock Classification

Beginning in the 18th century, geologists struggled to develop a sensible way to classify rocks, for like miners, they realized that not all rocks are the same. Classification schemes help us organize information and remember significant details about materials, and they help us recognize similarities and differences among materials. By the end of the 18th century, most geologists had accepted the *genetic scheme* for classifying rocks, which is based on an interpretation of the origin (genesis) of rocks. Using this approach, which we still use

FIGURE A.2 Examples of rock exposures.

(a) This outcrop in arid New Mexico consists of large cliffs.

(b) A small outcrop, mostly hidden by trees, in Illinois.

(c) A stream cut in New York where water stripped away soil and vegetation.

(d) To produce a level grade for a highway in Maryland, engineers excavated this road cut.

today, geologists recognize three basic groups: (1) **igneous rocks**, which form by the freezing (solidification) of molten rock (**Fig. A.3a**); (2) **sedimentary rocks**, which form either by the cementing together of fragments (grains) broken off pre-existing rocks, or by the precipitation of mineral crystals out of water solutions at or near the Earth's surface (**Fig. A.3b**); and (3) **metamorphic rocks**, which form when pre-existing rocks change character in response to a change in temperature, pressure, and/or chemical environment (**Fig. A.3c**). Metamorphic change occurs in the solid state, which means that it does not require melting. In the context of modern plate tectonics theory, different rock types form in different geologic settings, as we will discuss in succeeding chapters (**Fig. A.4**).

Each of the three groups contains many individual rock types, distinguished from one another by physical characteristics, such as the following:

- *Grain size and shape:* The dimensions of individual grains (here used to mean either fragments or crystals) vary greatly. Some grains are so small that they can't be seen without a microscope, whereas others are as big as a car or larger. Grain shape also helps identify rock type; in some rocks grains are **equant** (meaning they have the same dimensions in all directions), whereas in others the grains are **inequant** (meaning the dimensions are not the same in all directions) (**Fig. A.5**).
- *Composition:* A rock is a mass of chemicals. The term **rock composition** refers to the proportions of chemicals that make up the rock. The proportions of chemicals, in turn, affect the proportions of minerals constituting the rock.
- *Texture:* This term refers to the arrangement of grains in a rock, that is, the way grains connect to one another and whether or not inequant grains are aligned parallel to each other. The concept of rock texture will become easier to

FIGURE A.3 Examples of the three major classes of rocks.

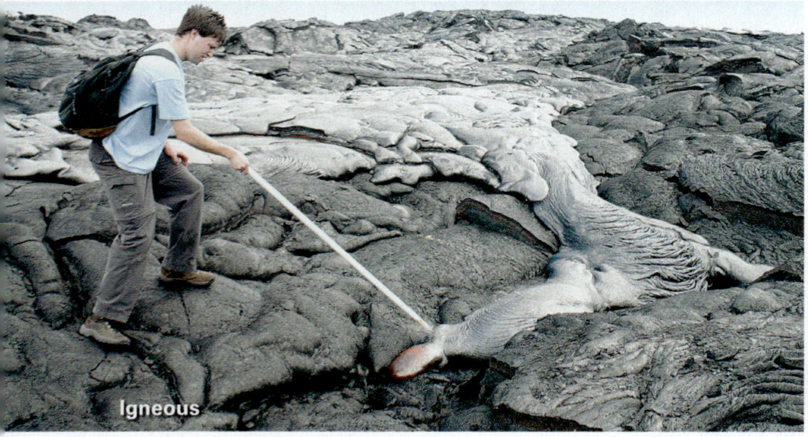

(a) Lava (molten rock) freezes to form igneous rock. Here, the molten tip of a brand-new flow still glows red. Older flows are already solid.

(b) Sand, formed from grains eroded off the rock cliffs, collects on the beach. If buried and turned to rock, it becomes layers of sandstone, such as those making up the cliffs.

(c) Metamorphic rock forms when pre-existing rocks endure changes in temperature and pressure and/or are subjected to shearing, stretching, or shortening. New minerals and textures form.

grasp as we look at different examples of rocks in the following chapters.

- *Layering:* Some rock bodies appear to contain distinct layering, defined either by bands of different compositions or textures, or by the alignment of inequant grains so that they trend parallel to each other. Different types of layering occur in different kinds of rocks. For example, the layering in sedimentary rocks is called **bedding** (**Fig. A.6a**), whereas the layering in metamorphic rocks is called **metamorphic foliation** (**Fig. A.6b**).

Each distinct rock type has a name. Some names reflect the dominant mineral making up the rock, some are derived from the name of the region where the rock was first discovered or

FIGURE A.4 A cross section illustrating various geologic settings in which rocks form.

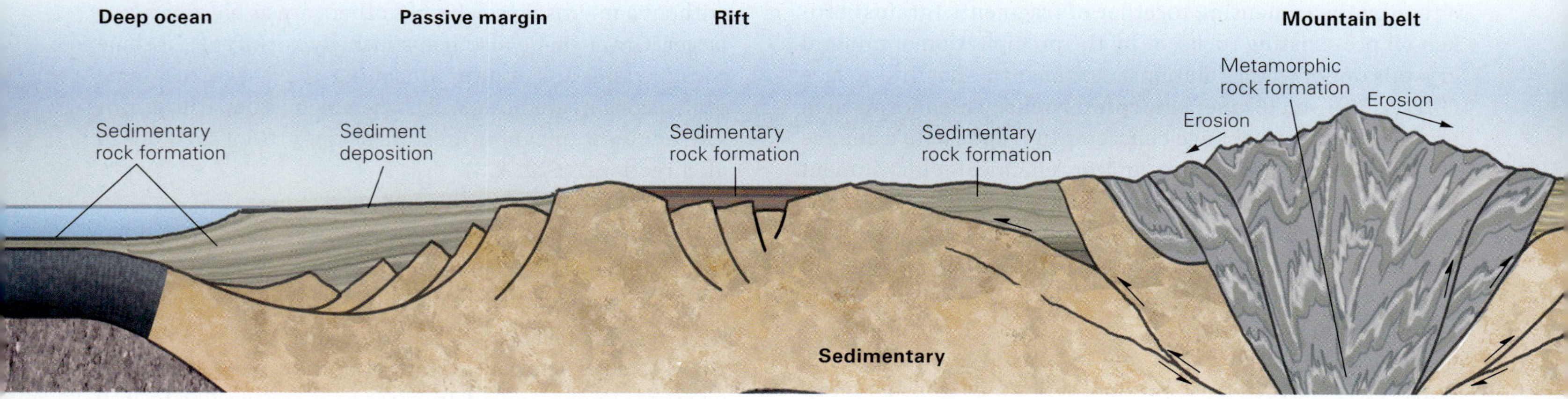

FIGURE A.5 Describing grains in rock.

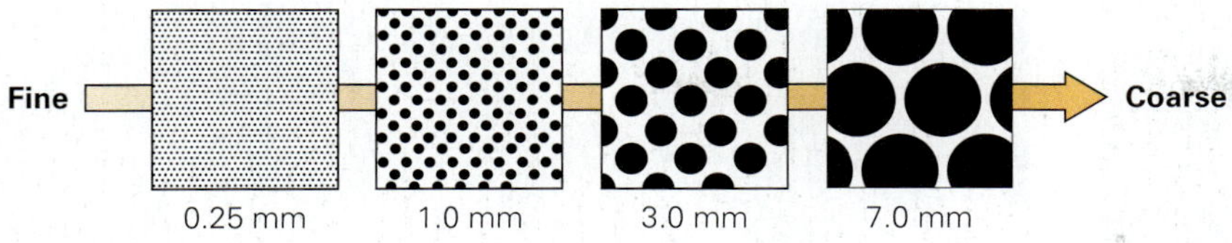

(a) Geologists define grain size by using this comparison chart.

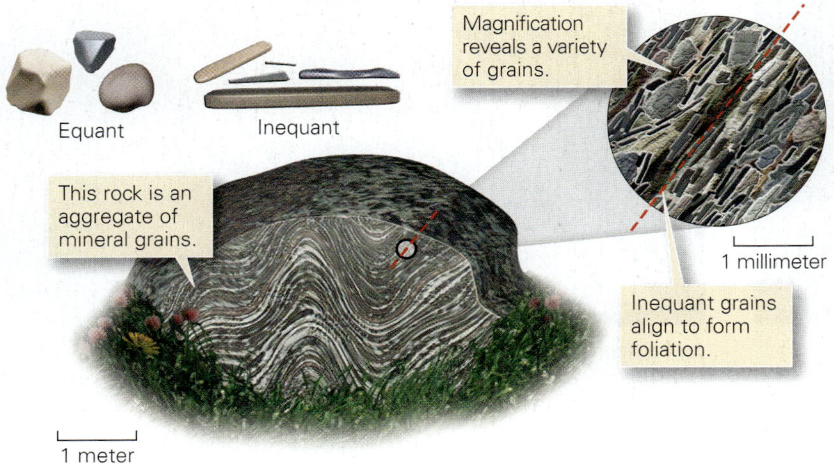

(b) Grains in rock come in a variety of shapes. Some are equant, whereas some are inequant. In this example of metamorphic rock, inequant grains align to define a foliation.

is particularly abundant, some from ancient legends, some from a root word of Latin or Greek origin, and some from a traditional name used by people in an area where the rock is found. Many rock names date back to antiquity, but some were assigned only in recent decades. In this book we will introduce only about 30 out of hundreds of rock names.

A.4 Studying Rock

Outcrop Observations

The study of rocks begins by examining a rock in an outcrop. If the outcrop is big enough, such an examination will reveal relationships between the rock you're interested in and the rocks around it, and will allow you to detect layering. Geologists carefully record observations about an outcrop, and then, using a *rock hammer*, break off a **hand specimen**, a fist-sized piece, that they can examine more closely with a magnifying glass or *hand lens* (**Fig. A.7**). Observation with a hand lens enables geologists to identify sand-sized or larger grains, and may enable them to describe the texture of the rock.

Thin-Section Study

Geologists often must examine rock composition and texture in minute detail in order to identify a rock and develop a hypothesis for how it formed. To do this, they take a specimen back to the lab, make a very thin slice (about 0.03 mm thick, the thickness of a human hair) and mount it on a glass slide to produce a **thin section** (**Fig. A.8a–c**), and study the thin section with a petrographic microscope (*petro* comes from the Greek word for rock). A *petrographic microscope* differs from an ordinary microscope in that it illuminates the thin section with transmitted polarized light. This means that the illuminating light beam first passes through a special polarizing filter

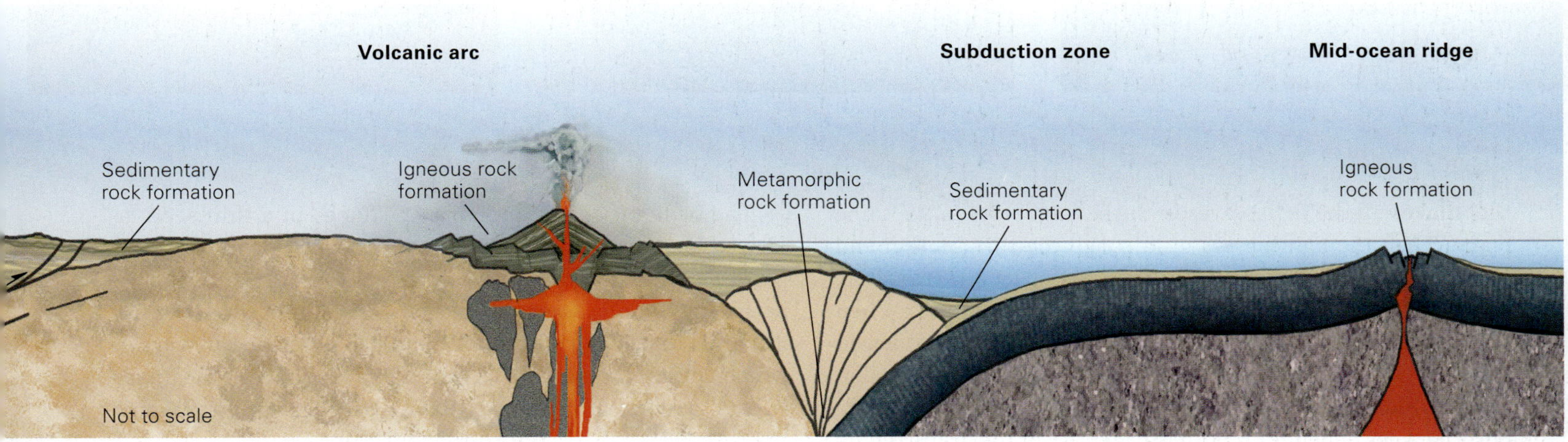

FIGURE A.6 Layering in rock.

(a) Bedding in a sedimentary rock, here defined by alternating layers of coarser and finer grains, as exposed on a cliff along an Oregon beach. Older beds were tilted before younger ones were deposited.

(b) Foliation in this outcrop of metamorphic rock near Mecca, California, is defined by alternating light and dark layers. The color of the layers depends on the minerals comprising the layers.

FIGURE A.7 Two basic tools used for studying rocks in the field.

that makes all the light waves in the beam vibrate in the same plane. Then the light passes up through the thin section and up through another polarizing filter. An observer, therefore, looks through the thin section as if it were a window. When illuminated with transmitted polarized light, and viewed through two polarizing filters, each type of mineral grain displays a unique suite of colors (**Fig. A.8d**). The specific color the observer sees depends on both the identity of the grain and its orientation with respect to the waves of polarized light.

The brilliant colors and strange shapes in a thin section, viewed in polarized light, rival the beauty of an abstract painting or stained glass. By examining a thin section with a petrographic microscope, geologists can identify most of the minerals that make up the rock and can describe the ways in which grains connect to each other. They can record the image with a camera; a photograph taken through a petrographic microscope is called a **photomicrograph**.

FIGURE A.8 Studying rocks in thin section.

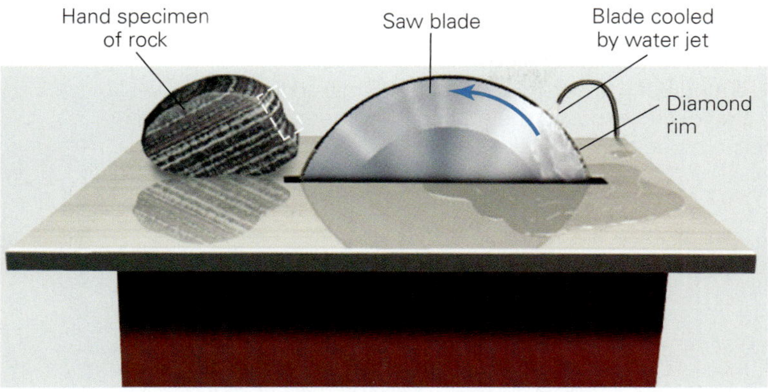

(a) Using a special saw, a geologist cuts a thin chip of a rock specimen.

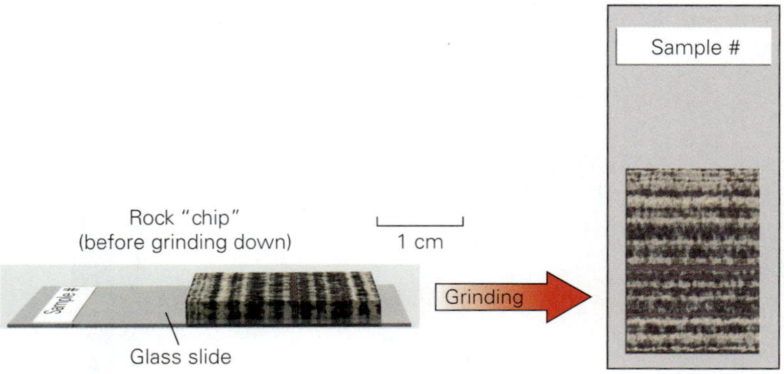

(b) The geologist glues the chip to a glass slide and grinds it down until it is so thin that light can pass through it.

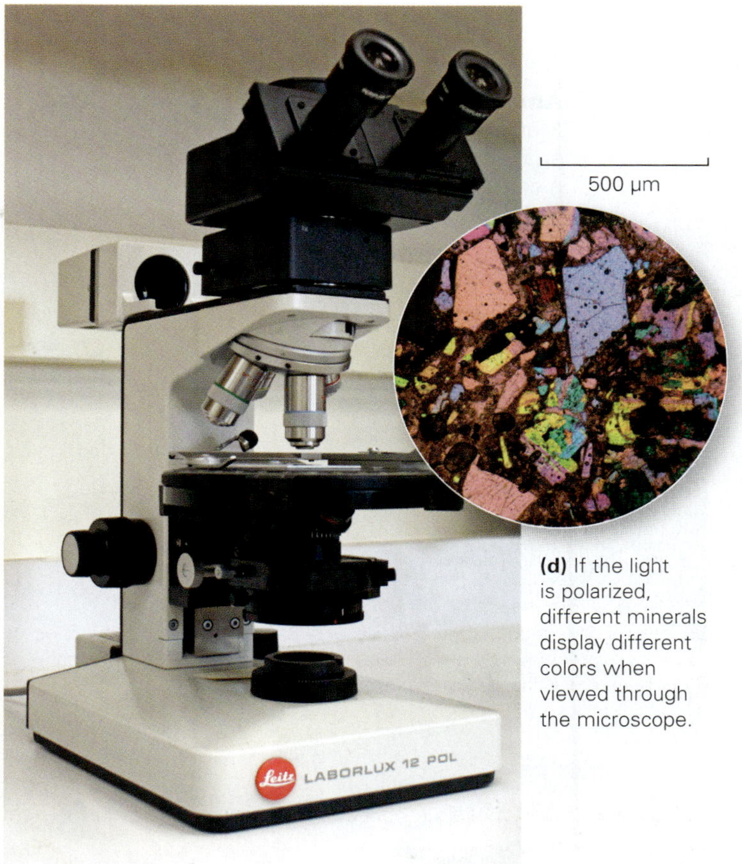

(c) With a petrographic microscope, it's possible to view thin sections with light that shines through the sample from below.

(d) If the light is polarized, different minerals display different colors when viewed through the microscope.

High-Tech Analytical Equipment

Beginning in the 1950s, high-tech electronic instruments became available that enabled geologists to examine rocks on an even finer scale than is possible with a petrographic microscope. Modern research laboratories typically boast such instruments as a *scanning electron microscope* (SEM), which can image the surface of a rock chip at extremely high magnification and can map the distribution of elements in the chip; an *electron microprobe,* which can focus a beam of electrons on a small part of a grain to produce a signal that defines the chemical composition of the mineral (**Fig. A.9**); a *mass spectrometer,* which analyzes the proportions of atoms with different atomic weights contained in a rock; and an *X-ray diffractometer,* which identifies minerals by measuring how X-ray beams interact with crystals. Such instruments, in conjunction with optical examination, can provide geologists with highly detailed characterizations of rocks. This information can in turn help them understand how the rocks formed and where they came from, so the study of rocks serves as a basis for deciphering Earth history.

FIGURE A.9 An electron microprobe uses a beam of electrons to analyze the chemical composition of minerals.

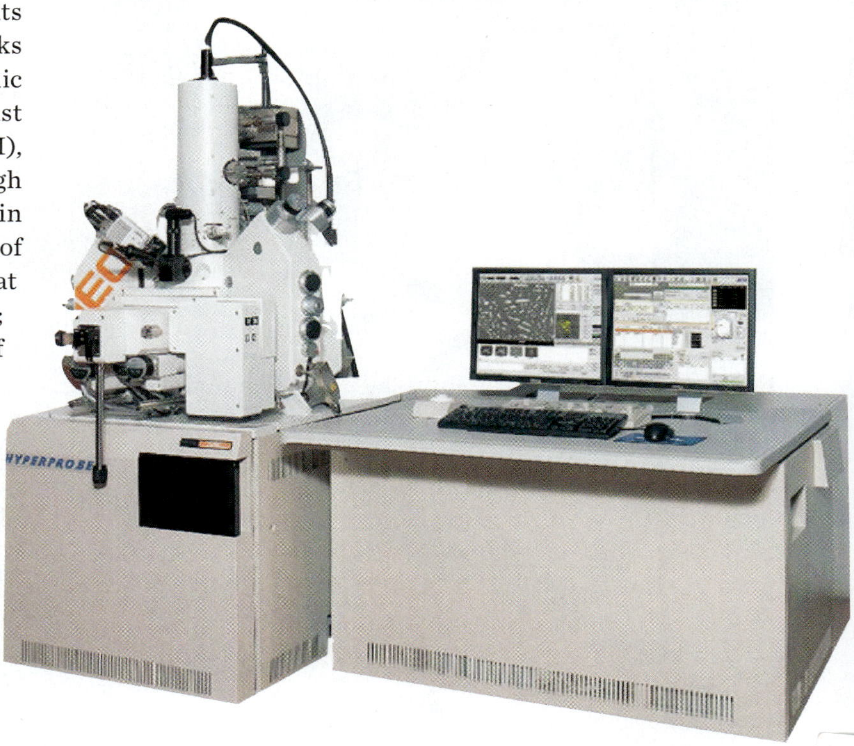

Another View Rocks can be beautiful! Here, we see an outcrop of granite in Joshua Tree National Monument, in the Mojave Desert, California. These rocks formed deep underground, about 85 million years ago. Subsequent erosion removed the rocks that once lay above, and exposed the granite. Weathering of the granite has formed rounded blocks, which resemble shapes in modern sculpture.

Interlude A Review

Interlude Summary

- Rock is a coherent, naturally occurring solid, consisting of an aggregate of minerals or of a mass of glass. Non-glassy rocks can be classified as crystalline or clastic.
- Bedrock consists of in-place rock that is still connected to the underlying crust. Outcrops, exposures of bedrock at the Earth's surface, can be natural or manmade.
- Geologists classify a rock as igneous, sedimentary, or metamorphic based on how the rock formed.
- A variety of characteristics prove helpful in describing rocks. Examples include: grain size, shape, composition, texture, and the nature of layering.
- Geologists make thin sections, very thin slices of rock, to study details of rock texture.
- Hand lenses, microscopes (after making thin sections), and sophisticated electronic equipment help geologists interpret the origin of rocks.

Guide Terms

bedding (p. 106)
bedrock (p. 104)
cement (p. 103)
clastic rock (p. 103)
crystalline rock (p. 103)
equant (p. 105)
hand specimen (p. 107)
igneous rock (p. 105)
inequant (p. 105)
metamorphic foliation (p. 106)
metamorphic rock (p. 105)
outcrop (p. 104)
photomicrograph (p. 108)
rock (p. 103)
rock composition (p. 105)
sedimentary rock (p. 105)
thin section (p. 107)

Review Questions

1. What is the geologist's definition of the term *rock*? Can a brick be considered to be a rock? Explain your answer.
2. Explain the difference between a clastic and crystalline rock?
3. Give examples of different kinds of rock outcrops. Can you find outcrops everywhere? Explain your answer.
4. What is the principal basis that geologists use to classify rocks into three classes? What are these classes?
5. Explain the difference between an equant and an inequant grain.
6. Where are two examples of layering that occur in rock?
7. What are thin sections, how are they examined, and what do they allow you to see?
8. What kinds of high-tech equipment can be used to study rocks?

Online Resources

Hand specimen of sandstone
A photomicrograph shows grains held together by cement.

Assessment

This interlude features visual identification questions on rock classifications and the three main rock groups.

LEARNING OBJECTIVES

By the end of this chapter, you should understand...

1. that the Earth's interior is not entirely molten.
2. why melting only occurs at special places in the Earth.
3. the difference between magma and lava, and the characteristics of each.
4. how and why melt moves to locations where it solidifies.
5. why there are different types of igneous rock, and how to describe and classify them.
6. where igneous activity takes place, in the context of plate tectonics theory.

▲ These peaks in the Wind River Mountains of Wyoming consist of granite, formed by the cooling of molten rock over a long period of time, at depth in the crust. The granite, which formed over 2.5 billion years ago, did not reach the surface of the Earth until kilometers of overlying rock had been eroded away.

CHAPTER 4

Up from the Inferno: Magma and Igneous Rocks

4.1 Introduction

Every now and then, incandescent molten rock (or *melt*) fountains or spills out of the ground on the big island of Hawaii, for this island hosts two active volcanoes—Kilauea and Mauna Loa (**Fig. 4.1a**). Formally defined, a **volcano** is a *vent* or opening from which melt that originates inside the Earth emerges onto the planet's surface, and/or rises into the air, during an episode known as a *volcanic eruption*. The word volcano is also used for the hill or mountain built from the products of an eruption. Geologists refer to melt that has emerged at the surface as **lava**. At a volcano, some lava pools around the vent, while some moves downslope as a syrupy red-yellow stream called a *lava flow* (**Fig. 4.1b**). Near the vent, lava from a Hawaiian volcano has a temperature of about 1,150°C, and it can move swiftly, cascading over escarpments at speeds of up to 60 km per hour. At the base of the volcano, the lava flow slows but continues to advance, engulfing roads, houses, or vegetation in its path (**Fig. 4.1c**). As lava cools, its surface darkens and forms a hardened rind that occasionally cracks open to reveal the hot, sticky mass that continues to ooze within. Finally, the lava flow stops moving entirely, and within days or weeks, cools through and through. As a result, the once red-hot melt becomes a hard, dark-gray rock (**Fig. 4.1d**). We refer to any rock formed by the solidifying (freezing) of a melt

> ***Granite—it seems inevitable to begin with granite, even though so many people have ended with it, lying under those glossy pinkish slabs labeled in gold or black. . . .***
>
> JACQUETTA HAWKES
> (British archaeologist and writer; 1910–1996)

as an **igneous rock**. Considering the fiery heat of the melt from which igneous rocks solidify, their name—from the Latin *ignis*, meaning fire—makes sense. Igneous rocks are very common at the Earth's surface, making up the entire oceanic crust and much of the continental crust.

It may seem strange to speak of "freezing" in regard to forming rock. Most people think of freezing as a change from liquid water to solid ice at a temperature below 0°C (32°F). Nevertheless, the transformation of liquid melt to solid igneous rock represents the same phenomenon, solidification of a liquid. Freezing of molten rock takes place at considerably higher temperatures than does freezing of water.

FIGURE 4.1 Formation and evolution of lava flows.

(a) A view of Hawaii, looking north. Kilauea and Mauna Loa have erupted in the past few decades, and since their recent lava flows are darker than other areas. Mauna Kea last erupted over 4,600 years ago.

Smoke comes from burning vegetation.
STOP

(c) At a distance from the vent, the lava has completely crusted over with new rock, but the interior of the flow remains molten.

Lava fountain
Active lava flow
As the lava cools, it darkens.
Time

(b) Lava erupts as a fountain from a volcanic vent on Hawaii. A fast-moving river of lava then flows downslope.

(d) Eventually, the flow cools completely and becomes a layer of new rock. This flow engulfed a road on Hawaii.

FIGURE 4.2 The intrusive and extrusive realms.

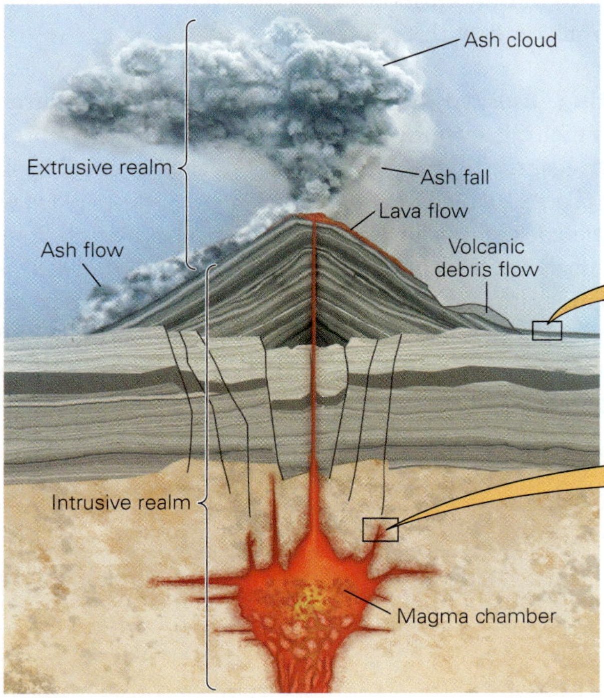

(a) The intrusive realm lies underground and the extrusive realm lies aboveground. Lava flows, as well as various types of ash eruptions, all produce extrusive rocks.

(b) Extrusive rocks include lava flows and pyroclastic layers.

(c) An intrusion of basalt (dark rock) cuts across a mass of granite (light rock).

Igneous rocks freeze at temperatures between 650° and 1,100°C, depending on their chemical makeup. To put such temperatures in perspective, keep in mind that a home oven attains a maximum temperature of only 260°C (500°F).

Geologists distinguish between two main categories of igneous rock, based on where it solidified (**Fig. 4.2a**). Rock that forms by the freezing of lava above ground, in contact with air or water, after it erupts or *extrudes*, is **extrusive igneous rock** (**Fig. 4.2b**). Examples include rock that solidified within a lava flow, as well as rock made from cemented-together fragments of **pyroclastic debris** (from the Greek word *pyro*, meaning fire) that blasted out of a volcano. Pyroclastic debris occurs in many sizes, from very fine particles, called **volcanic ash**, to coarser chunks. The spectacle of a volcano erupting may give the impression that igneous rock forms exclusively at the Earth's surface. In fact, a vastly greater volume of igneous rock forms by solidification of melt underground, after it has pushed its way, or *intruded*, into pre-existing *wall rock*. Geologists refer to underground melt as **magma**, and to the rock formed by solidification of magma as **intrusive igneous rock** (**Fig. 4.2c**).

A great variety of igneous rocks exist on Earth. To understand why, we first discuss the process of forming magma. Then, we examine phenomena that cause magma to rise from depth, factors that determine how easily it flows, and conditions under which magma or lava freezes. Finally, we consider the scheme that geologists use to classify igneous rocks. We will see that, by examining features of an igneous rock, we can characterize the geologic setting in which the rock formed.

4.2 Why Does Magma Form, and What Is It Made Of?

The popular image that the crust, the Earth's outer shell, floats on a sea of molten rock is simply not correct. In fact, geologists knew by the end of the 19th century that the Earth's interior was mostly solid, for if it weren't, the spinning of the Earth on its axis would flatten it more than it does (see Chapter 1). Yet the occurrence in the crust of both extrusive and intrusive igneous rocks of many ages, and of erupting volcanoes today, means that melting took place in the geologic past, and continues to happen today. Let's consider the processes that cause melting, and the nature of magma produced by melting.

Did you ever wonder...
whether Earth's crust floats on a magma sea?

Causes of Melting

What is the source of the heat that can cause magma to form? As we discussed in Chapter 1, much of this heat is a relict of our planet's formation. As the proto-Earth grew, sources of heat included the compression of mass into a smaller volume, the sinking of iron to form the core, the impact of meteorites, and the decay of radioactive elements. The Earth remains hot inside, even after 4.54 billion years, because the decay of radioactive elements in the crust, and heat released by the core, replaces much of the heat lost to space by radiation at the Earth's surface. But even though the Earth's interior continues to be very hot, most of the crust and the mantle remain in solid form. The immense pressure produced by the weight of overlying rock prevents molecules from moving freely relative to one another, which would allow rock to turn into liquid. Magma, therefore, forms only in special places, where conditions trigger melting of pre-existing solid rock. Below, we describe these conditions, and briefly note the geologic settings, in the context of plate tectonics, where melting takes place. We'll wait until the end of this chapter to associate specific rock types with each setting.

Melting Due to Decompression Because pressure prevents melting, even in a very hot rock, a decrease in pressure can trigger melting as long as the rock remains hot (**Fig. 4.3a**). This process, called *decompression melting*, takes place where mantle rock rises slowly, for as the rock moves up, its pressure becomes less (due to the decrease in overburden) while its temperature remains nearly unchanged (for rock acts as an excellent insulator). As we'll see, upward movement causing decompression melting occurs in mantle plumes, beneath rifts, and beneath mid-ocean ridges (**Fig. 4.3b**).

FIGURE 4.3 Decompression melting.

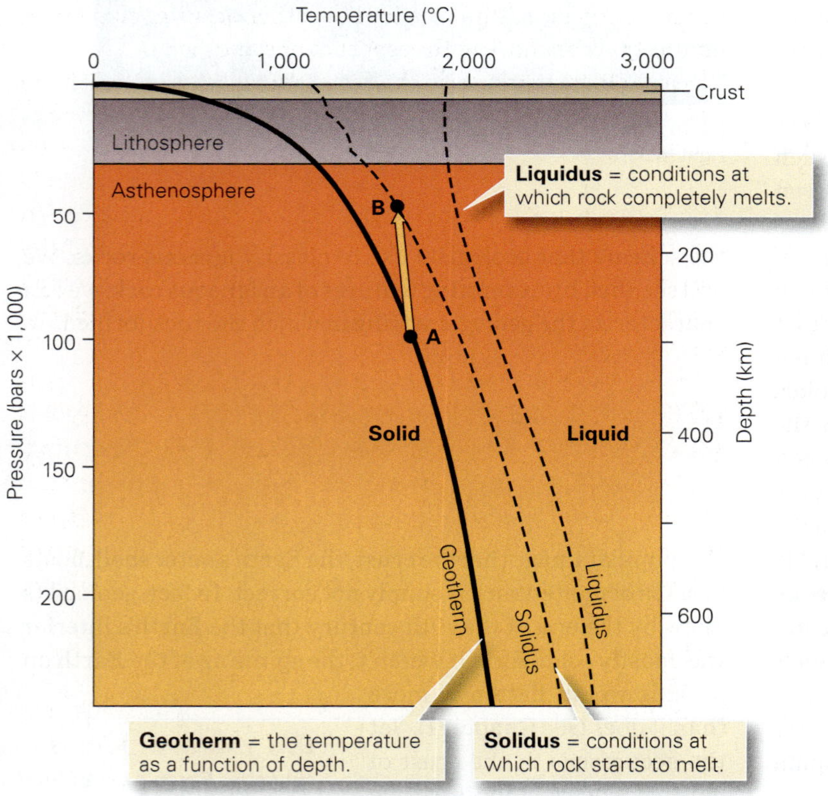

(a) Decompression takes place when the pressure acting on hot rock decreases. As this graph of pressure and temperature conditions in the Earth shows, when rock rises from point A to point B, the pressure decreases a lot, but the rock cools only a little, so the rock begins to melt.

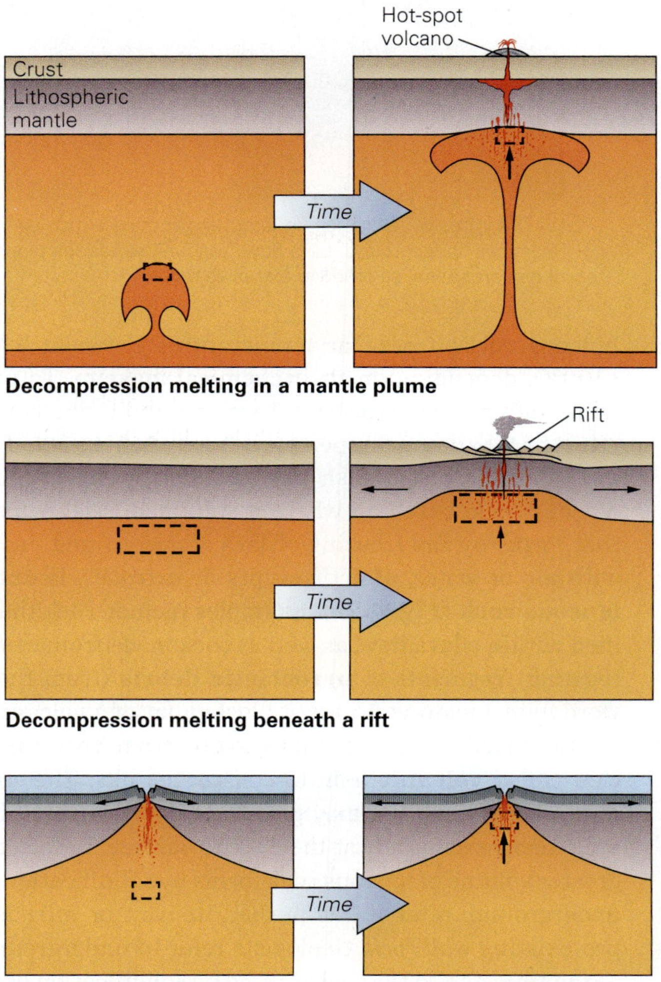

(b) The conditions leading to decompression melting occur in several different geologic environments. In each case, a volume of hot asthenosphere (outlined by dashed lines) rises to a shallower depth, and magma (red dots) forms.

FIGURE 4.4 Flux melting and heat-transfer melting.

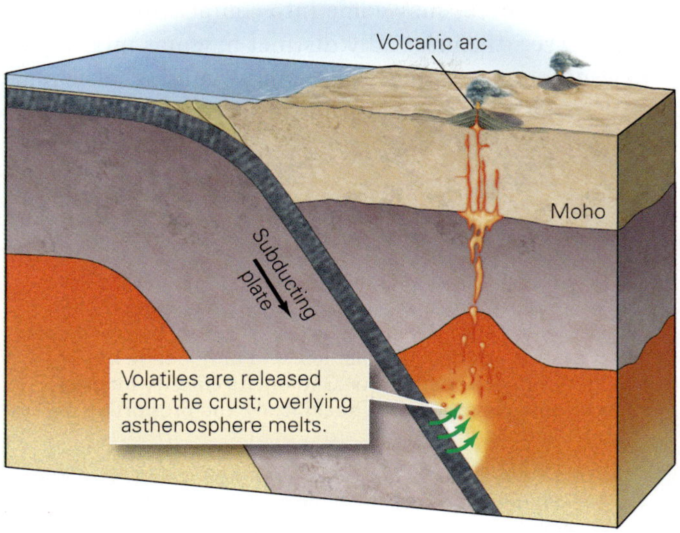

(a) Flux melting occurs where volatiles enter hot mantle; this happens at subduction zones.

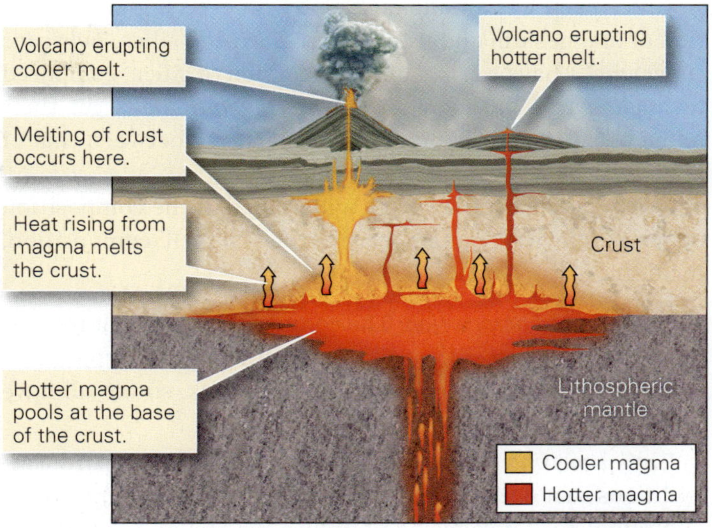

(b) Heat-transfer melting occurs when rising magma brings heat up with it and melts overlying or surrounding rock. (We'll see that the cooler magma is felsic, and the hotter magma is mafic.) (Not to scale.)

Melting Due to Addition of Volatiles Magma also forms at locations where volatiles mix with hot mantle rock. Recall that *volatiles* are substances, such as water (H_2O) and carbon dioxide (CO_2), that evaporate relatively easily. When volatiles mix with hot, dry rock, they cause chemical bonds to break, so that the rock begins to melt, a process sometimes called *flux melting* (**Fig. 4.4a**). In other words, adding volatiles decreases a rock's melting temperature. As we'll see, volatiles seep into hot asthenosphere in the region just above subducting oceanic lithosphere, causing volcanism at convergent plate boundaries.

Melting Due to Heat Transfer When very hot magma from the mantle rises into the crust, the heat it brings raises the temperature of the surrounding crustal rock. In some cases, the rise in temperature may be so significant that it will cause the crustal rock to begin melting, a process called *heat-transfer melting*. To picture the process, imagine injecting hot fudge into a ball of ice cream—the fudge transfers heat to the ice cream, raises its temperature, and causes it to melt (**Fig. 4.4b**).

The Major Types of Melt

All molten rocks (magma or lava) contain *silica*, a compound of silicon and oxygen (SiO_2). They also contain varying proportions of other elements, including aluminum (Al), calcium (Ca), sodium (Na), potassium (K), iron (Fe), and magnesium (Mg). Because molten rock is a liquid, its molecules do not lie in an orderly crystalline lattice but rather occur in clusters or chains that can move with respect to one another.

Geologists distinguish between "dry" melts, which do not contain volatiles, and "wet" melts, which do. Wet melts include up to 15% dissolved volatiles, including water, carbon dioxide, nitrogen (N_2), hydrogen (H_2), and sulfur dioxide (SO_2). These volatiles come out of the Earth at volcanoes in the form of gas. Usually, water constitutes about half of the gas erupting at a volcano. Thus, molten rock contains not only the molecules that constitute solid minerals in rocks, but also the molecules that become water and air.

Molten rocks differ from one another in terms of the proportions of chemicals that they contain (**Table 4.1**). Geologists distinguish four major compositional types depending, overall, on the proportion of silica (SiO_2) relative to the sum of magnesium oxide (MgO) and iron oxide (FeO or Fe_2O_3) in the melt. *Mafic melts* contain a relatively high proportion of magnesium and iron oxide compared to silica—the *ma-* in the

TABLE 4.1 Compositional Categories of Magma

Magma type	Weight percent of silica*
Felsic (or silicic)	66–76%
Intermediate	52–66%
Mafic	45–52%
Ultramafic	38–45%

*Weight percent means the proportion of the magma's weight that consists of silica (SiO_2).

word stands for magnesium, and the *-fic* comes from the Latin word for iron. Ultramafic melts have an even higher proportion of magnesium and iron oxide, relative to silica. *Felsic* (or *silicic*) *melts* have a fairly high proportion of silica, compared to magnesium and iron oxide. *Intermediate melts* get their name because their composition is part way between that of mafic and felsic melts.

Why do we observe such a wide range in the composition of melts? Several factors play a role in melt composition, including:

- *Source rock composition:* The composition of a melt reflects the composition of the solid from which it was derived. Not all melts form from the same source rock, so not all have the same composition.
- *Partial melting:* Under the temperature and pressure conditions that occur in the Earth, only about 2% to 30% of an original rock will melt to produce magma at a given location. The temperatures at sites of magma production simply never get high enough to melt the entire source rock before the magma has had a chance to migrate away from the source. **Partial melting** refers to the process by which only part of an original rock melts to produce magma (**Fig. 4.5a**). Because silica tends to go into the melt preferentially, magmas formed by partial melting are more felsic than the original rock from which they were derived. For example, partial melting of an ultramafic rock produces a mafic magma.
- *Assimilation:* As magma sits underground before solidifying completely, it may incorporate chemicals dissolved from the wall rocks, or from blocks that detached from the wall and sank into the magma (**Fig. 4.5b**). This process is called contamination or **assimilation**.
- *Magma mixing:* Magmas formed in different locations from different sources may come into contact underground. In some cases, the originally distinct magmas mix to yield a new, different magmas. For example, mixing felsic magma with mafic magma could produce intermediate magma.

TAKE-HOME MESSAGE

Though the Earth is hot inside, the crust and mantle are solid, except in special places where pre-existing rock undergoes melting. Melting can be triggered by a decrease in pressure, addition of volatiles, and/or injection of hot magma from deeper below. Geologists classify magma based on its composition, specifically, the proportion of silica that it contains.

QUICK QUESTION Why doesn't a magma formed by melting a particular source rock have the same composition as the source rock?

4.3 Movement and Solidification of Molten Rock

Why Does Magma Rise?

If magma stayed put once it formed, new igneous rocks would not develop in or on the crust. But it doesn't stay—magma tends to move upward, away from the site of melting. In some

FIGURE 4.5 Phenomena that can affect the composition of magma.

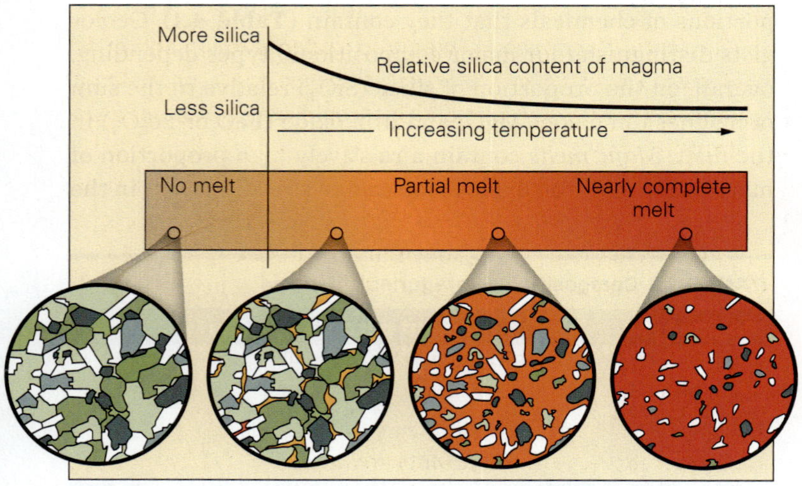

(a) Partial melting: The first-formed melt will be richer in silica than the original rock. As melting continues, magma becomes increasingly mafic.

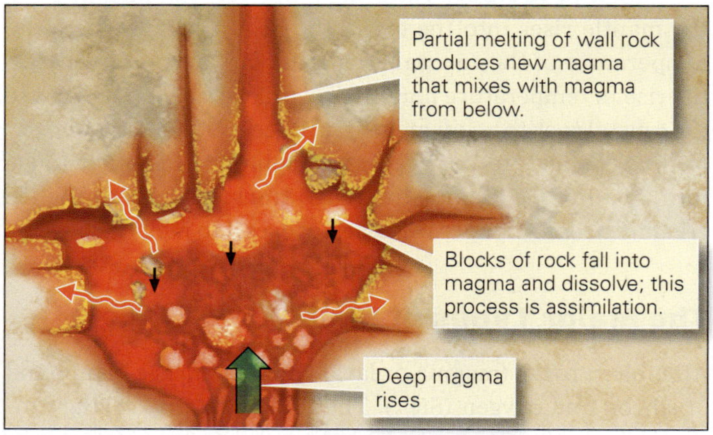

(b) Mixing and assimilation: Heat provided by deep magma partially melts wall rock; the new magma may then mix with deep magma. Also, blocks of wall rock can dissolve (assimilate) in the deep magma, and the wall rock may chemically react with the magma.

cases it reaches the Earth's surface and erupts at a volcano. Movement of magma is an important process in the Earth System, because materials from deeper parts of the Earth flow upward and provide the raw material from which new rocks, the atmosphere, and oceans form.

Magma rises for two reasons. First, buoyancy drives magma upward just as it drives a wooden block up through water, because molten rock is less dense than the surrounding solid rock. Second, magma rises because the weight of overlying rock generates pressure at depth that squeezes magma upward—the same process happens when you step into a puddle barefoot and mud squeezes up between your toes.

What Controls the Speed of Flow?

The resistance to flow, or **viscosity**, of a liquid affects the speed with which the liquid moves. We can say, for example, that molasses is more viscous than water because it flows more slowly than water. All molten rock is much more viscous than molasses, but not all molten rock has the same viscosity. The viscosity of a given melt depends on temperature, volatile content, and silica content. Specifically, hotter melt is less viscous than cooler melt, because thermal energy breaks bonds and allows atoms or molecules to move more easily. A wet melt is less viscous than a dry melt, because volatiles also tend to break apart silicate molecules. Finally, mafic melt is less viscous than felsic melt because relatively more silicon-oxygen tetrahedra occur in felsic melt. These tetrahedra link together to make long molecules that can't move past each other easily. With such relationships in mind, it's not surprising that a very hot mafic lava has relatively low viscosity and can flow to form thin sheets, but a cool felsic lava has relatively high viscosity and clumps up into a bulbous mound (**Fig. 4.6**).

FIGURE 4.6 Viscosity affects lava behavior.

(a) Mafic lava has relatively low viscosity. It can erupt in fountains, move long distances, and form thin lava flows.

(b) Felsic to intermediate lava is very viscous. When it erupts, it may form a mound-like lava dome around the volcano's vent.

Transforming Melt into Rock

When melts rise, they eventually cool, because the temperature of the Earth decreases upward. If magma becomes trapped underground, as an intrusion, it slowly loses heat to surrounding wall rock, drops below its freezing temperature, and solidifies underground. If a magma reaches the Earth's surface and extrudes as lava, it cools because it comes in contact with much cooler air or water.

The time it takes for a magma to cool depends on how fast it can transfer heat into its surroundings. To see why, think about the process of cooling coffee. If you spill coffee on a table, it cools quickly because it loses heat directly to the cold air. For the same reason, lava in an extrusive environment cools relatively quickly. In contrast, if you pour hot coffee into an insulated thermos bottle and seal it, the coffee stays hot for hours, because insulation slows the transfer of heat to the air outside. Like a thermos bottle, wall rock acts as insulation, so magma in an intrusive environment cools slowly. Not all magma trapped in the intrusive realm cools at the same rate, however. Three factors control the cooling time of such magma:

- *The depth of intrusion:* Magma intruded deep in the crust is surrounded by hot wall rock, and thus cools more slowly than does magma intruded into cold wall rock near the ground surface.
- *The shape and size of a magma body:* Heat escapes from magma at an intrusion's surface, so the greater the surface area for a given volume of intrusion, the faster it cools. Thus, an intrusion of magma shaped like a pancake cools faster than one shaped like a melon (**Fig. 4.7a, b**). And since the ratio of surface area to volume increases as size decreases, a body of magma the size of a car cools faster than one the size of a ship.
- *The presence of circulating groundwater:* Water passing through magma absorbs and carries away heat, much like the coolant that flows around an automobile engine.

Changes in Molten Rock Composition during Cooling: Fractional Crystallization

If you cool a tray of water to a temperature of 0°C, crystals of ice start to grow, and if you keep the temperature cold enough for long enough, the water in the tray becomes water ice, a crystalline solid composed entirely of H_2O. The process of freezing magma or lava to form igneous rock is more complex, because unlike the liquid water in the tray, molten rock contains many different compounds. During freezing of molten rock, then, crystals of many different minerals may form, and not all of these grow at the same time. Early-formed minerals tend to be relatively mafic, so their growth preferentially removes iron and magnesium from the magma, and as a consequence, the remaining magma becomes more felsic (**Fig. 4.7c**). This general process, by which different minerals grow in sequence so the melt composition changes progressively, as cooling takes place, is known as **fractional crystallization**. If an originally mafic magma freezes before much fractional crystallization has occurred, a mafic igneous rock forms. Freezing of magma left after much fractional crystallization has occurred, however, yields a felsic igneous rock. **Box 4.1** provides further details about the processes.

FIGURE 4.7 Factors that affect the freezing of molten rock.

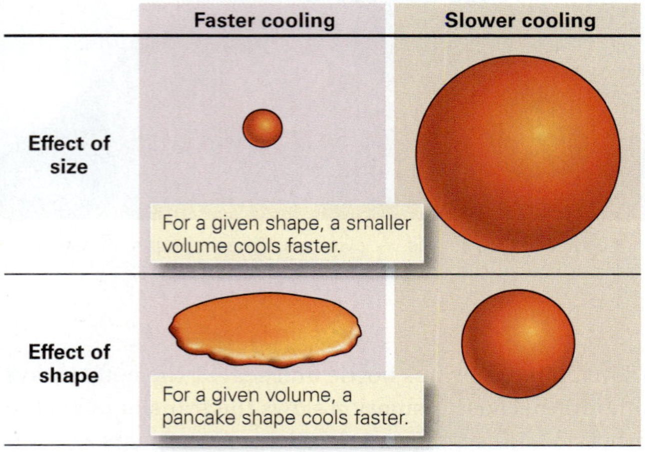
Faster cooling
Slower cooling
Effect of size
For a given shape, a smaller volume cools faster.
Effect of shape
For a given volume, a pancake shape cools faster.

(a) The larger the ratio of its surface area to volume, the faster an intrusion cools. This ratio depends on both size and shape.

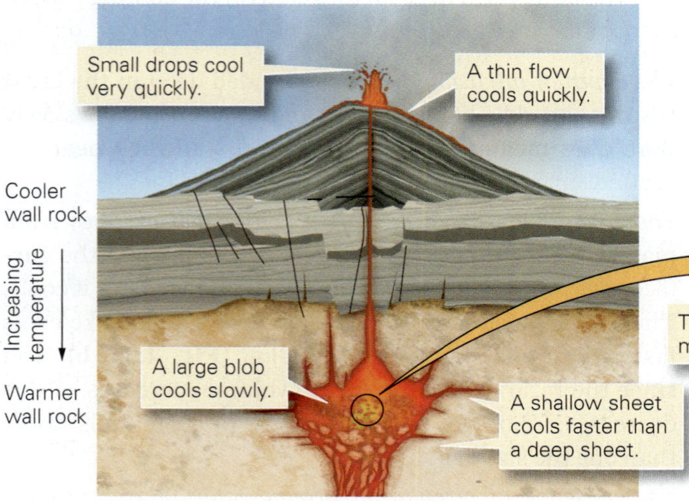
Small drops cool very quickly.
A thin flow cools quickly.
Cooler wall rock
Increasing temperature
Warmer wall rock
A large blob cools slowly.
A shallow sheet cools faster than a deep sheet.

(b) The cooling rate of molten rock depends on the size and shape of the magma or lava body, and on its depth.

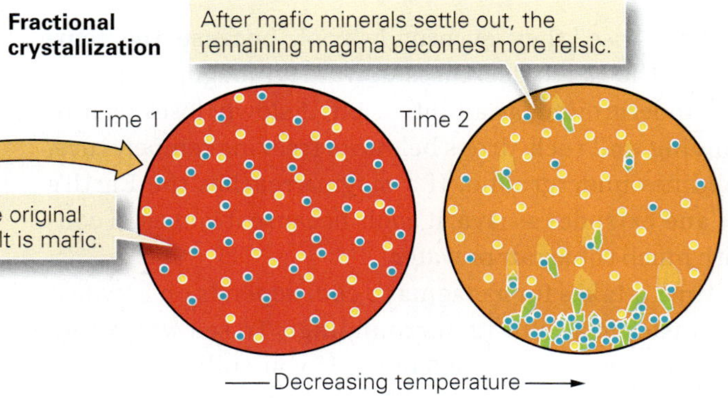
Fractional crystallization
After mafic minerals settle out, the remaining magma becomes more felsic.
Time 1
Time 2
The original melt is mafic.
Decreasing temperature

(c) The process of fractional crystallization results in a progressive change in magma composition during freezing. Blue dots represent mafic components, and yellow dots represent felsic components.

Where Do Melts Cool?

Extrusive Igneous Settings Different volcanoes extrude molten rock in different ways. Some volcanoes erupt streams of low-viscosity lava that flood down the flanks of the volcano and then cover broad swaths of the countryside. When this lava freezes, it forms a relatively thin lava flow. Such flows may cool in days to months. In contrast, some volcanoes erupt viscous masses of lava that pile into rubbly domes. And still others erupt explosively, sending clouds of volcanic ash and debris soaring skyward, and/or avalanches of ash tumbling down the sides of the volcano. Which type of eruption occurs depends largely on a magma's composition and volatile content. Mafic lavas tend to have low viscosity and spread in broad, thin flows (**Fig. 4.8a, b**). Volatile-rich felsic lavas tend to erupt explosively and form thick ash and debris deposits (**Fig. 4.8c, d**). Chapter 5 describes the products of extrusive eruptions, and the causes for their differences, in more detail.

Intrusive Igneous Settings Geologists distinguish among different types of intrusions on the basis of their shape. *Tabular intrusions*, or *sheet intrusions*, are roughly planar and have a fairly uniform thickness. Most are from centimeters to tens of meters thick, and tens of meters to tens of kilometers

FIGURE 4.8 Examples of eruptions and extrusive volcanic materials.

(a) This volcano is producing lava flows and fountains.

(b) A stack of over 50 thin lava flows, capped by debris, visible from inside Mt. Vesuvius, Italy.

(c) This volcanic explosion produced two styles of ash eruption.

(d) Thick layers of ash deposited by explosive eruptions in New Mexico, about 1.14 Ma. Note the highway, for scale.

long. A **dike** is a tabular intrusion that cuts across pre-existing layering (bedding or foliation), whereas a **sill** is a tabular intrusion that injects between layers, and thus is parallel to layering (**Fig. 4.9**). In places where tabular intrusions cut across rock that does not have layering, a nearly vertical, wall-like tabular intrusion is called a dike, and a nearly horizontal, tabletop-shaped tabular intrusion is called a sill. When an intrusion starts to inject between layers but then domes upward, it yields a blister-shaped intrusion known as a **laccolith**. **Plutons** are blob-shaped intrusions that range in size from tens of meters across to tens of kilometers across (**Fig. 4.10a, b**). Intrusion of numerous plutons in a region makes a vast composite body that may be hundreds of kilometers long and over 100 km wide. The resulting immense mass of igneous rock is called a **batholith** (**Fig. 4.10c, d, e**). The boundary between wall rock and any igneous intrusion is an *intrusive contact.*

Where does the space for igneous intrusions come from? Dike intrusion takes place in regions where the crust is stretching horizontally, as happens in a rift. So, as the magma

BOX 4.1 CONSIDER THIS...

Bowen's Reaction Series

In the 1920s, Norman L. Bowen began a series of laboratory experiments designed to determine the sequence in which silicate minerals crystallize from a melt. First, Bowen melted powdered mafic igneous rock by raising its temperature to over 1,200°C. Next, he cooled the melt just enough to cause part of it to solidify. Then he quenched the remaining melt by submerging it quickly in cold mercury. Quenching, which in this context means sudden cooling to form a solid, transformed the remaining liquid into glass, trapping the earlier-formed crystals within it. Bowen made a thin section of the sample and identified the mineral crystals by using a microscope, and he analyzed the chemical composition of the resulting glass. After experiments at different temperatures, Bowen was able to characterize the sequence of mineral-producing reactions that took place during progressive cooling of a mafic magma (**Fig. Bx4.1a**). Geologists now refer to this sequence as **Bowen's reaction series**, in his honor.

Let's examine the sequence more closely. In a cooling mafic melt, olivine and calcium-rich plagioclase form first. The plagioclase reacts with the remaining melt, and new plagioclase, containing more sodium (Na), grows. This new plagioclase may replace or grow around the earlier-formed plagioclase. Meanwhile,

FIGURE 4.9 Igneous sills and dikes, examples of tabular intrusions.

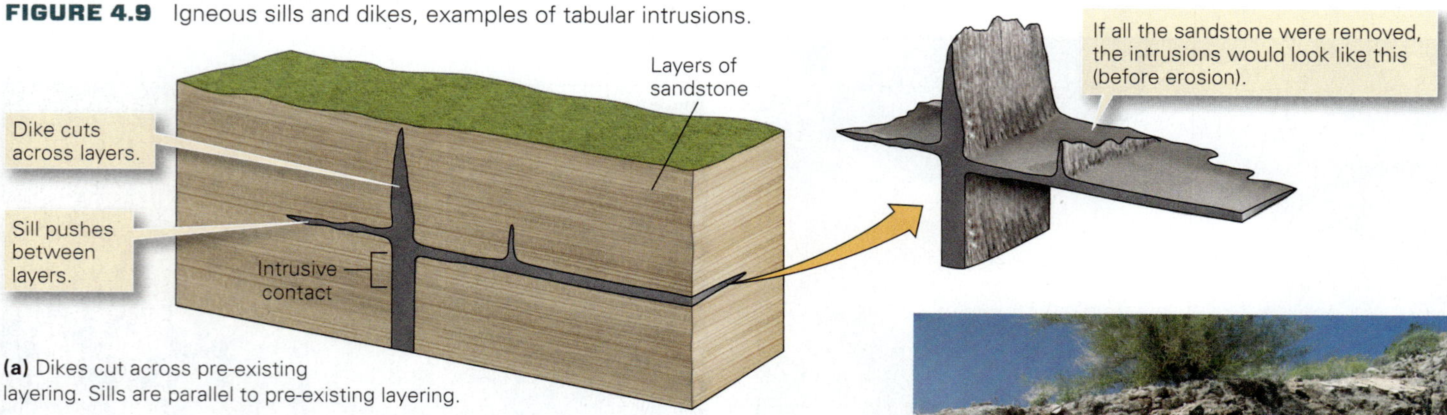

(a) Dikes cut across pre-existing layering. Sills are parallel to pre-existing layering.

(b) A wall-like intrusion cutting into pre-existing igneous rock is also called a dike.

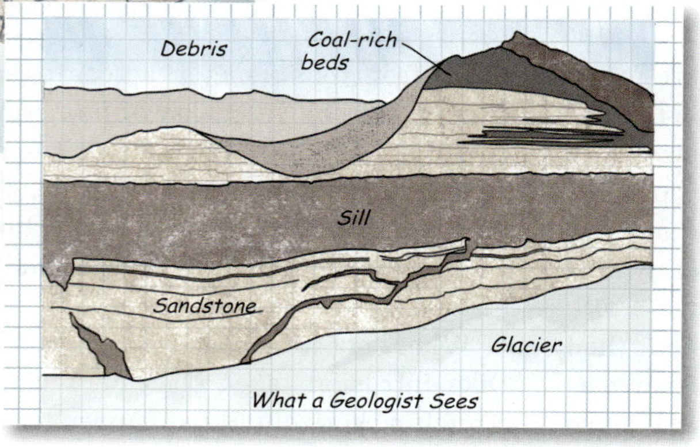

(c) Large sills of basalt intruded sandstone beds in Antarctica, here exposed at Finger Mountain.

FIGURE Bx4.1 Bowen's reaction series indicates the succession of crystallization in cooling magma.

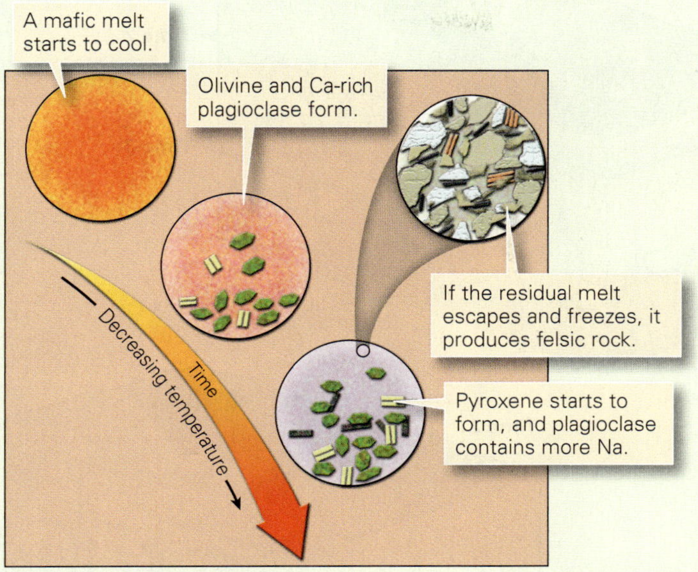

(a) With decreasing temperature, fractional crystallization begins and the composition of the remaining magma becomes more felsic.

some olivine crystals react with the remaining melt to produce pyroxene. Some of the early olivine and pyroxene crystals become isolated from the melt, effectively extracting iron and magnesium, so the remaining melt becomes progressively enriched in silica. As the melt continues to cool, plagioclase continues to form. Later-formed plagioclase has more sodium than earlier-formed plagioclase, pyroxene crystals react with the melt to form amphibole, and then amphibole reacts with the remaining melt to form biotite. All the while, some newly formed crystals become isolated and don't exchange atoms with the melt—some crystals may, in fact, sink and accumulate at the bottom of the melt. As more crystals form, the remaining melt becomes progressively more felsic. Finally, at temperatures between 650° and 850°C, only about 10% melt remains, and this melt has a high silica content, so when the last melt finally freezes, only quartz, potassium feldspar (K-feldspar), and muscovite can form.

Note that the reaction series has two tracks. The *discontinuous reaction series* refers to the sequence of olivine, pyroxene, amphibole, biotite, K-feldspar/muscovite/quartz, in that each step yields a different kind of silicate mineral. The *continuous reaction series* refers to the sequence from calcium-rich to sodium-rich plagioclase, in that each step yields a different version of the same mineral (**Fig. Bx4.1b**).

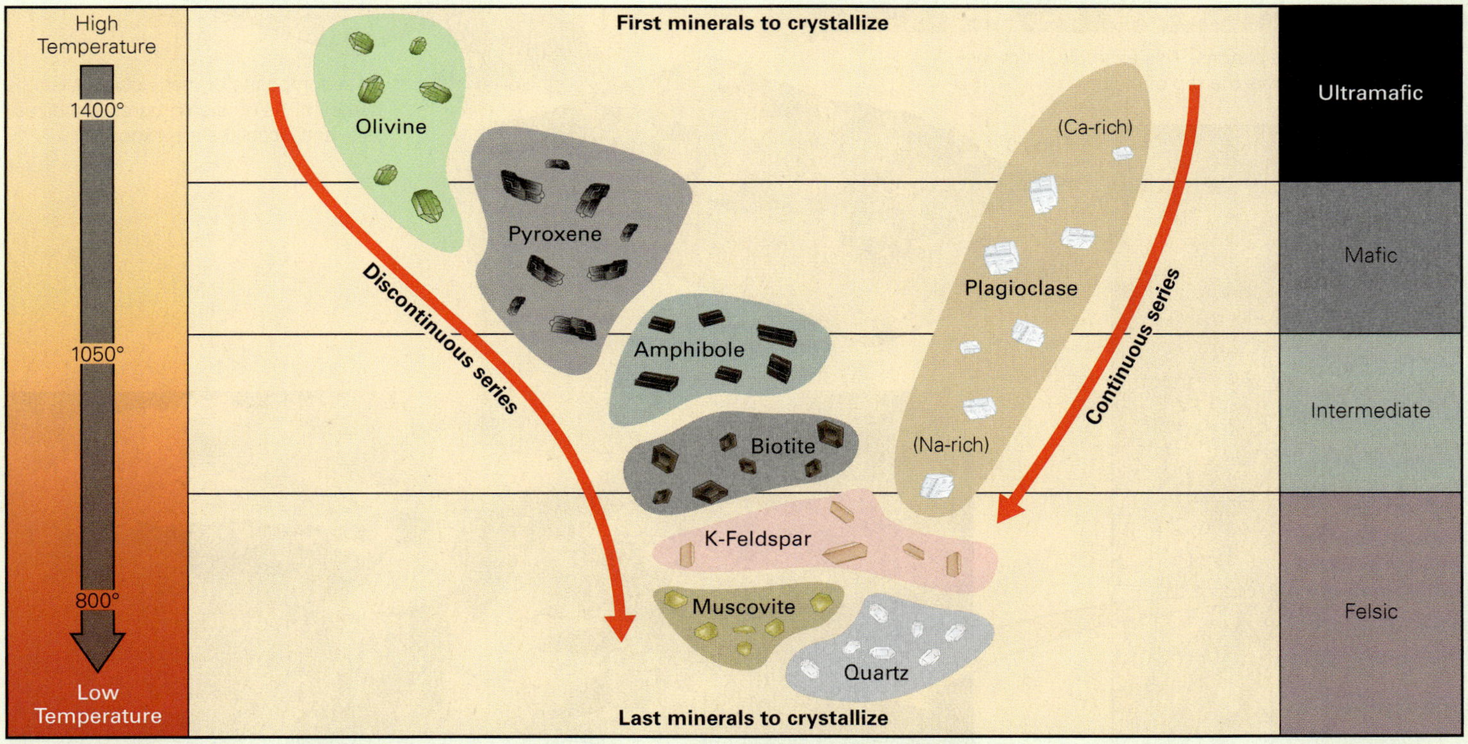

(b) This chart displays the discontinuous and continuous reaction series. Rocks formed from minerals at the top of the series are mafic, whereas rocks formed from the bottom of the series are felsic.

FIGURE 4.10 Igneous plutons, blob-shaped intrusions.

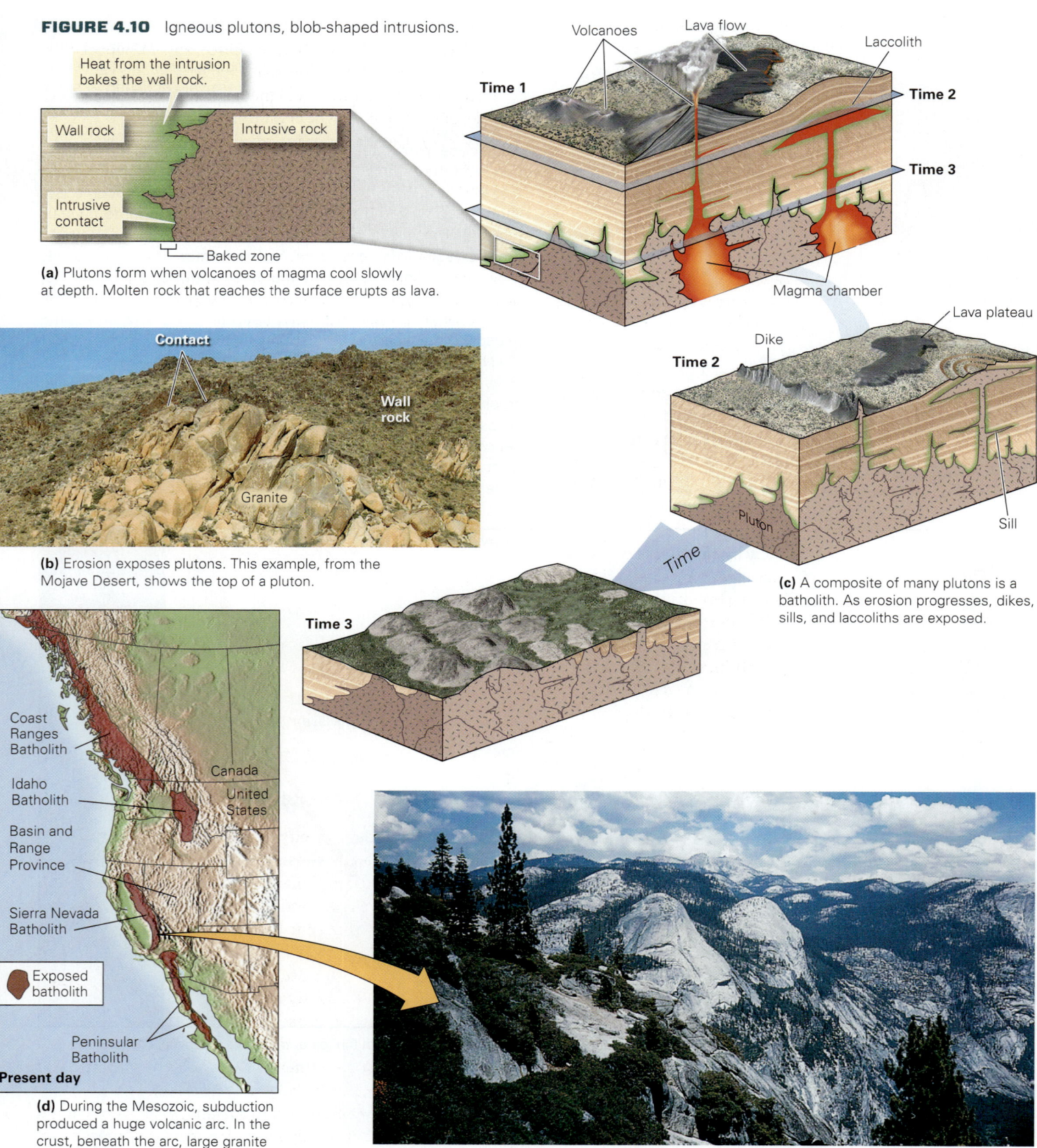

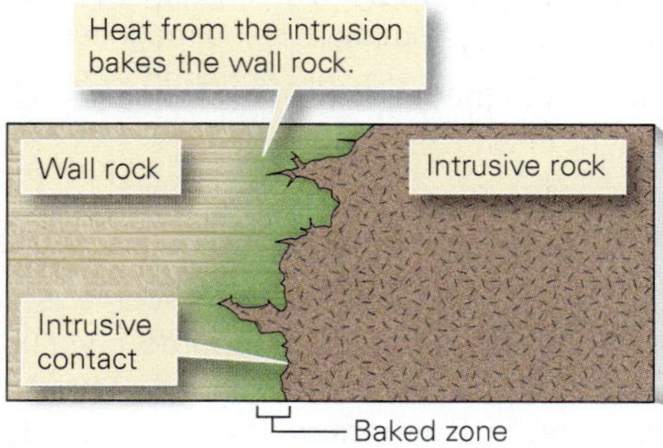

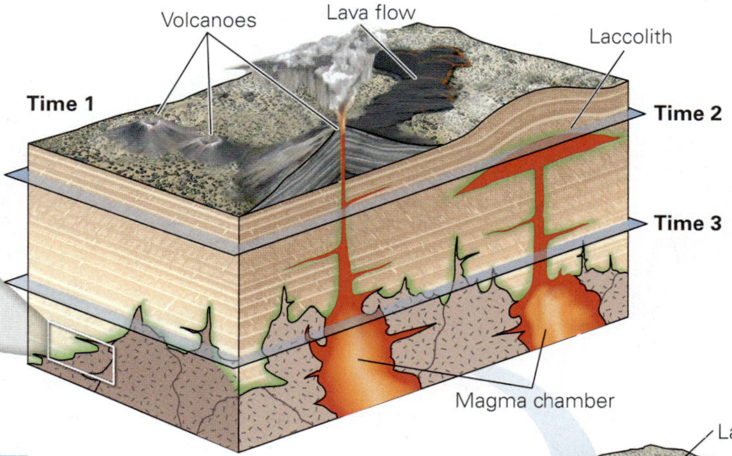

(a) Plutons form when volcanoes of magma cool slowly at depth. Molten rock that reaches the surface erupts as lava.

(b) Erosion exposes plutons. This example, from the Mojave Desert, shows the top of a pluton.

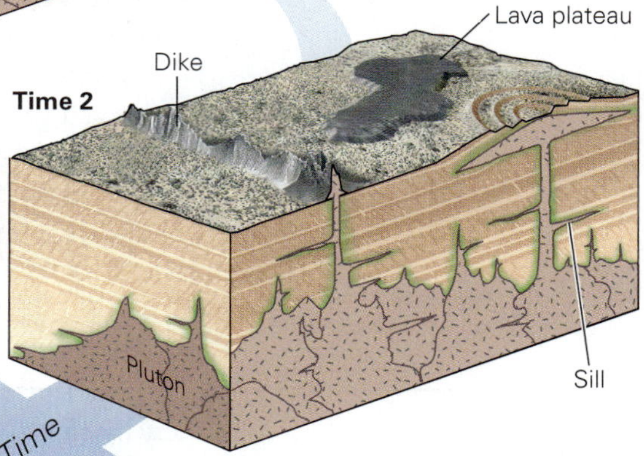

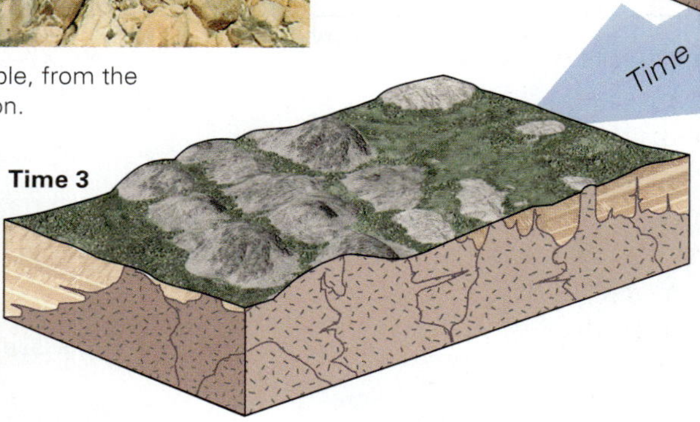

(c) A composite of many plutons is a batholith. As erosion progresses, dikes, sills, and laccoliths are exposed.

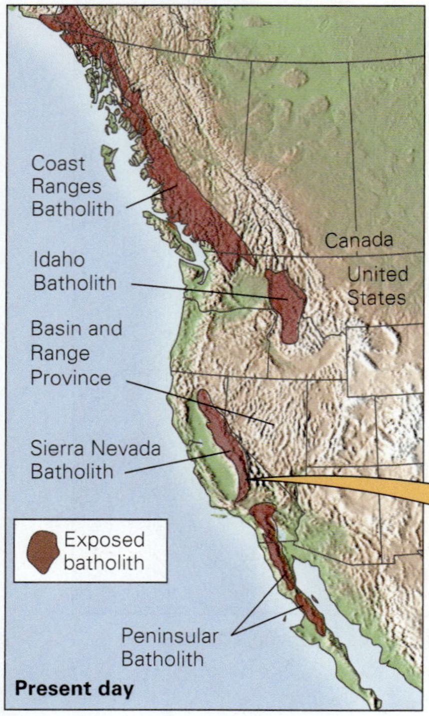

(d) During the Mesozoic, subduction produced a huge volcanic arc. In the crust, beneath the arc, large granite batholiths formed. They are now exposed by erosion.

(e) The Sierra Nevada of California provide exposures of one of these Mesozoic batholiths. The huge granite cliffs, popular with climbers, are part of the batholith.

forces its way into a vertical crack, the walls of the crack can move apart sideways (**Fig. 4.11a**). Sill intrusion occurs near the surface of the Earth, so pressure from magma can push the rock above the sill upward, leading to vertical movement of the Earth's surface (**Fig. 4.11b**). Explaining the processes of pluton intrusion remains controversial. In some cases, a pluton may form as a result of cooling in a large **magma chamber**, an irregularly shaped area inside the crust where magma accumulates. A magma chamber may contain magma only, may contain a mush of crystals and melt, or may be a zone of abundant magma-filled cracks. Some geologists suggest that a magma chamber can become a *diapir*, meaning a light-bulb-shaped blob that flows upward and pushes aside soft wall rock as it rises (**Fig. 4.12a**). Others identify the source of plutons as magma chambers that originate where faulting breaks the crust (**Fig. 4.12b**). Another idea of pluton formation involves *stoping*, a process during which blocks break off from a magma chamber wall and sink into the magma within, where they become partly or entirely assimilated (**Fig. 4.12c**). Recently, some researchers have been studying the possibility that plutons develop by the injection of numerous sills, one underneath the other. These sills may eventually coalesce, and perhaps recrystallize, to become a single, massive body (**Fig. 4.12d**).

In some cases, an intrusive igneous rock contains blocks or chunks of rock that did not crystallize from the same magma (**Fig. 4.12e**). Geologists refer to a body of rock within an intrusion as a **xenolith** (after the Greek word *xeno*, meaning foreign). Some xenoliths are pieces of wall rock that separated from the wall as magma pushed open cracks adjacent to the wall. Others are residual unmelted pieces of rock from the source region of the magma. Still others may be blocks that broke free from the roof of a magma chamber and did not fully assimilate. Typically, the composition of a xenolith differs from that of the rest of the intrusion.

SEE FOR YOURSELF...

GRANITE OF THE SIERRA NEVADA

Latitude
37°44′30.11″ N

Longitude
119°33′59.47″ W

Zoom to 4.7 km (~15,500 feet), tilt the view, and look northwest.

This is Yosemite National Park. Half Dome, an iconic climbing cliff, lies on the right. Bedrock here consists of granite, part of a batholith that intruded during the Mesozoic.

FIGURE 4.11 Making room for igneous dikes and sills.

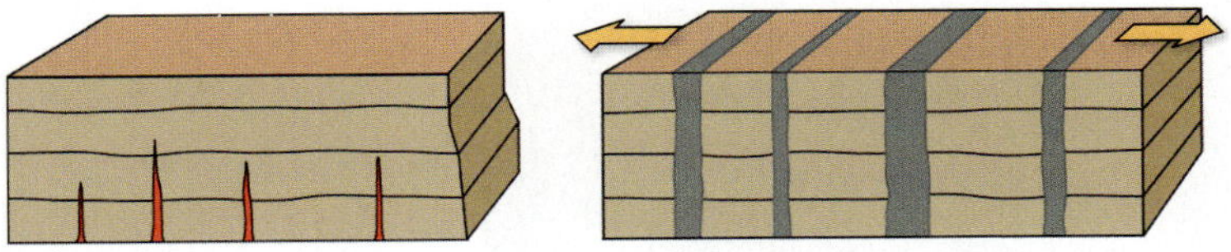

(a) The crust stretches sideways during the intrusion of igneous dikes.

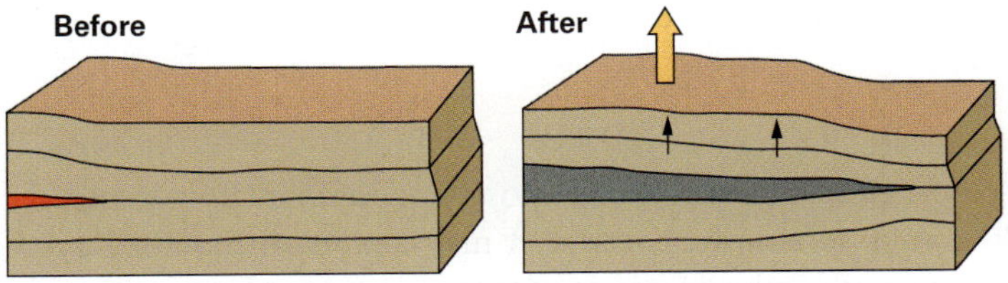

(b) The Earth's surface rises to make room for sills.

TAKE-HOME MESSAGE

Magma rises because it's buoyant and because of pressure due to overlying rocks. The rate of melt movement is affected by viscosity, which depends on composition and temperature. When molten rock enters a cooler environment, it freezes. The rate of cooling depends on the environment and on the shape of the magma body.

QUICK QUESTION Which cools faster—a large blob of magma intruded at depth or a thin flow extruded at the surface? Why?

4.4 How Do You Describe an Igneous Rock?

Characterizing Color and Texture

If you wander around a city you'll find that many building facades and lobbies consist of slices of igneous rock. In recent years polished rock has also become a popular material for home kitchen countertops. Architects like to use such rock, which they refer to in a generic way as "granite" (even though it's not usually a granite in the geologic sense) because it is beautiful and durable. This durability comes from the strong silicate minerals that the rock contains, and from the way in which mineral grains interlock.

Did you ever wonder...
why "granite" used in floors and countertops is so hard?

If you had to describe a rock to a friend, what words might you use? You would probably start by noting the rock's color.

FIGURE 4.12 Making room for igneous plutons.

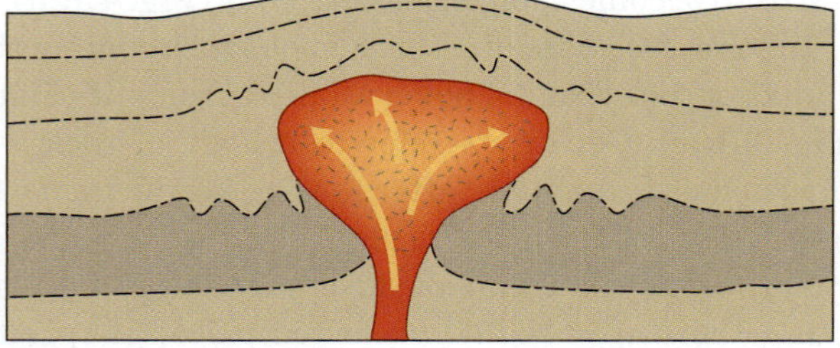

(a) During diapiric rise of a mass of magma, the magma is a blob that forces its way up through wall rock, which plastically deforms to move out of the way.

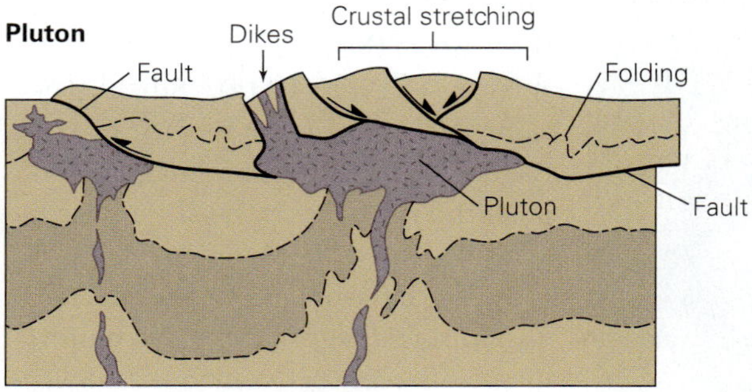

(b) Faulting associated with crustal stretching may accommodate the emplacement of a diapir.

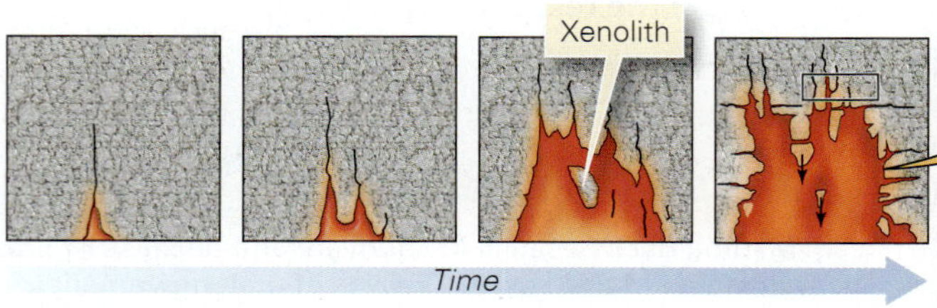

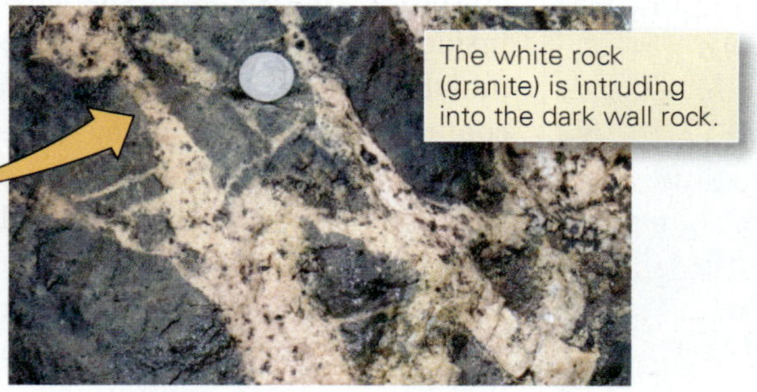

(c) Magma pushes up into cracks, and blocks may be incorporated in the melt by stoping.

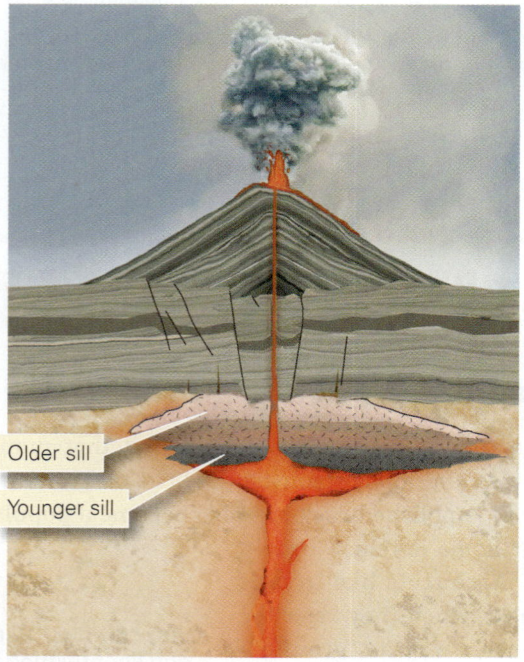

(d) Plutons may intrude as a succession of sill-like sheets.

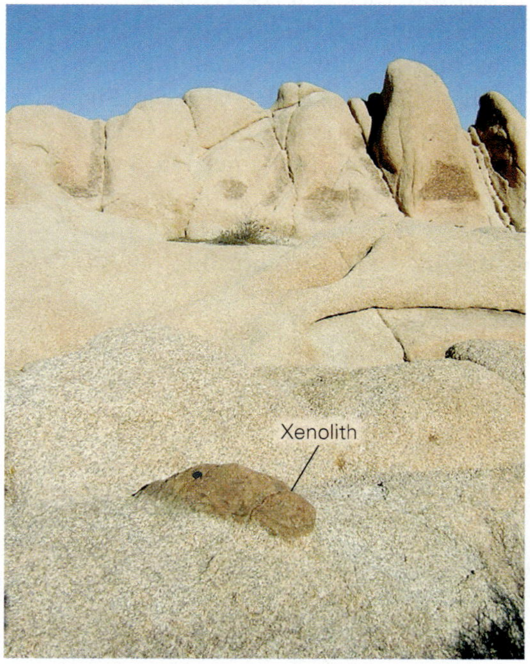

(e) A xenolith of darker rock surrounded by light-colored granite.

Overall, does the rock look dark or light? More specifically, is it gray, pink, white, or black? Describing color may not be easy, because some igneous rocks contain many visible mineral grains, each with a different color; but even so, you'll probably be able to characterize the overall hue of the rock. Generally, the color reflects the rock's composition, but identifying color isn't always so simple, because it may also be influenced by grain size and by the presence of trace amounts of impurities. For example, the presence of a small amount of iron oxide gives rock a reddish tint.

Next, you would probably characterize the rock's texture. A description of *igneous texture* specifies whether the rock consists of interlocking crystals, stuck-together fragments, or solid glass. If the rock consists of crystals or fragments, a description of texture also specifies the grain size. Here are the common terms for defining texture:

- *Crystalline texture*: When a melt solidifies, minerals in some rocks grow and interlock like pieces of a jigsaw puzzle. Geologists refer to such rocks as **crystalline igneous rocks** (**Fig. 4.13a**). The crystals interlock because grains grow around each other where they come in contact, and later-formed grains fill irregular spaces among earlier-formed grains. Subcategories of crystalline igneous rocks are based on the size of the crystals. Coarse-grained (*phaneritic*) rocks have crystals large enough to identify with the naked eye. The crystals of fine-grained (*aphanitic*) rocks are too small to identify with the naked eye. *Porphyritic rocks* have larger crystals (called *phenocrysts*) surrounded by a mass of fine crystals (called *groundmass*).
- *Fragmental texture*: **Fragmental igneous rocks** form from pyroclastic debris and consist of chunks and/or shards that are packed together, welded together, or cemented together after they have solidified (**Fig. 4.13b**).
- *Glassy texture*: Rocks made of a solid mass of glass, or of tiny crystals surrounded by glass, are **glassy igneous rocks** (**Fig. 4.13c**). Glassy rocks typically fracture conchoidally (fractures are curved, like a clam shell).

In glassy or crystalline igneous rocks, grain size tends to reflect the cooling rate. In general, glassy rocks form when a melt cools very rapidly; fine-grained crystalline rocks form when a melt cools at a moderate rate; and coarse-grained igneous rocks form from a slowly cooling melt (**Box 4.2**). Because of the relationship between cooling rate and texture, lava flows, dikes, and sills tend to be composed of fine-grained igneous rock. In contrast, plutons are generally composed of coarse-grained rock. Plutons that intrude into hot wall rock at great depth cool very slowly, and thus they tend to have larger crystals than plutons that intrude into cool country rock at shallow depth where they cool relatively rapidly. Porphyritic rocks form in two stages as a melt cools. First, the melt cools slowly at depth, so that phenocrysts form. Then, the melt rises and intrudes at a shallow depth, or erupts, so that the

FIGURE 4.13 Textures of igneous rocks, as viewed through a microscope. The field of view is about 3 mm. Crystalline rocks have interlocking crystals, fragmental rocks have clasts cemented or welded together, and glassy rocks contain glass (black material) as well as isolated crystals.

(a) Granite is crystalline.

(b) Tuff is fragmental.

(c) Obsidian is glassy.

BOX 4.2 CONSIDER THIS...

The Relationship between Cooling Rate and Grain Size

To understand why grain size is related to cooling rate, let's look again at how a melt becomes a solid. As solidification begins, many crystal *seeds* (tiny specks of solid mineral consisting of a relatively small number of atoms bonded to each other; see Chapter 3) appear. As long as the melt remains hot, atoms throughout the melt have enough *thermal energy*, the energy associated with the movement and vibration of the atoms, to migrate or *diffuse* through the melt. When a diffusing atom bumps into a seed, it can attach, so the seed grows larger. But thermal energy can also break chemical bonds holding atoms to the surface of an existing seed, so the seed can come apart and dissolve back into the melt. Thus, in a hot melt, seeds are constantly growing and dissolving. If, by chance, enough atoms attach to a particular seed, it may grow big enough to become a small crystal. Such small crystals have the potential to survive, but in a hot melt, they may dissolve again.

When a melt cools very quickly, there's not enough time for atoms to diffuse, attach to seeds, and grow into crystals. A melt that solidifies quickly and does not contain an orderly arrangement of crystals, by definition, is a glass. If cooling occurs a bit more slowly, but is still fairly fast, there may be enough time for many seeds to grow and become small crystals. But when the temperature becomes low enough, quickly enough, diffusion slows and there is not enough time for many small crystals either to grow very large, or to dissolve. Then, when solidification is complete, small crystals have replaced all the melt. In a case where cooling happens very slowly, however, not only can seeds dissolve, but small crystals can also dissolve. Only a relatively few crystals may, by chance, become large enough so they don't dissolve completely. And because the melt remains hot for a long time, there's an opportunity for many atoms to diffuse and attach to these crystals. Thus, these lucky crystals become very large, even as many small crystals continue forming and dissolving, and when solidification goes to completion, large crystals have replaced all the melt.

remainder cools quickly and produces groundmass crystals around the phenocrysts.

An exception to the standard cooling rate and grain-size relationship is *pegmatite*, a very coarse-grained igneous rock that contains crystals up to tens of centimeters across. Because it occurs in dikes, which generally cool quickly, its coarse texture may seem surprising. Researchers have shown that pegmatites are coarse because they crystallize from water-rich melts. In water, atoms can migrate (diffuse) rapidly enough so that large crystals grow in a short time.

Classifying Igneous Rocks

Melts have a variety of compositions and can freeze to form igneous rocks in many different environments above and below the surface of the Earth. We can classify igneous rocks according to their texture and composition. Studying a rock's texture tells us about the rate at which it cooled, as we've seen, and therefore about the environment in which it formed (see **Geology at a Glance**, p. 131). The rock's composition tells us about the original source of the magma and the way in which the magma evolved before finally solidifying. Below, we introduce some of the more important igneous rock types.

Types of Crystalline Igneous Rocks The scheme for classifying the principal types of crystalline igneous rocks is fairly simple. We distinguish among the different compositional classes on the basis of silica content—*ultramafic, mafic, intermediate,* or *felsic*—and we distinguish among the different textural classes according to whether the grains are coarse or fine (**Fig. 4.14**). As indicated by the right side of the chart in Figure 4.14, contrasts in chemical composition mean that different rock types contain different groups of minerals. Note that basalt, andesite, and rhyolite could have come from the same magmas as gabbro, diorite, and granite, respectively. But the three fine-grained rock types cooled quickly in lava flows or near-surface dikes and sills, while the three coarse-grained rock types cooled more slowly, in plutons. As a rough guide, the color of an igneous rock indicates its composition—mafic rocks tend to be black or dark gray, intermediate rocks tend to be lighter gray or greenish

FIGURE 4.14 Crystalline igneous rocks are classified based on composition and texture. The right side of the chart shows the percentages of different minerals in the different rock types.

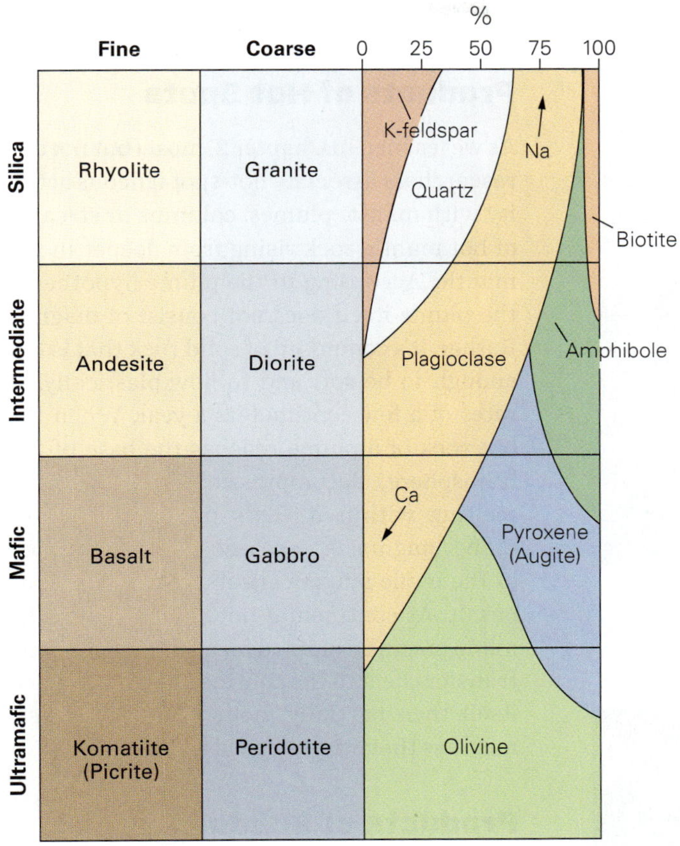

gray, and felsic rocks tend to be light tan to pink or maroon (**Fig. 4.15**). Of the major types of igneous rocks, *peridotite* is most common on Earth, for it makes up the mantle, but it is relatively rare on the Earth's surface.

Types of Glassy Igneous Rocks Glassy texture develops more commonly in felsic igneous rocks because the high concentration of silica inhibits the easy growth of crystals. But basaltic and intermediate lavas can form glass if they cool rapidly enough. In some cases, a rapidly cooling lava freezes while it still contains a high concentration of gas bubbles—these bubbles remain as open holes known as **vesicles**. Geologists distinguish among several different kinds of glassy rocks.

Did you ever wonder...
how the black glass once used for arrowheads formed?

- **Obsidian** is a mass of solid, felsic glass. It tends to be black or brown (Fig. 4.13c). Because it breaks conchoidally, sharp-edged pieces split off its surface when you hit a sample with a hammer. Pre-industrial people worldwide used such pieces for arrowheads, scrapers, and knife blades.
- **Tachylite** is a vesicle-free mass consisting of more than 80% mafic glass. This rock is relatively rare compared to obsidian.
- **Pumice** is a felsic volcanic rock that consists of tiny vesicles, each of which is surrounded by a thin screen of glass. With far more open space than solid glass, it can look like a sponge, and some specimens can actually float on water, like Styrofoam (**Fig. 4.16a**). Pumice forms from quickly cooling, volatile-rich frothy lava that resembles the foam in a glass of beer.
- **Scoria** is a mafic volcanic rock with abundant vesicles (more than about 30%). Generally, the bubbles in scoria are bigger than those in pumice, and the rock looks darker, overall (**Fig. 4.16b**).

Types of Pyroclastic Igneous Rocks When volcanoes erupt explosively, they spew out clots or droplets of lava, as well as glass shards (the broken-up walls of vesicles in pumice), larger fragments of pumice, and other broken-up chunks of recently formed igneous rocks. As we've noted, geologists refer to accumulations of such fragments as pyroclastic debris. When pyroclastic debris becomes consolidated into a solid mass, due either to still-hot clasts welding together during accumulation, or to cementation by minerals precipitating from groundwater long after accumulation, it becomes a **pyroclastic rock**. Pyroclastic rocks have a fragmental texture, and we can distinguish among rock types based on grain size. **Tuff** is one type composed mostly of volcanic ash; larger fragments of pumice may be mixed in with the ash (see Figs. 4.8 and 4.13b).

TAKE-HOME MESSAGE

Geologists divide igneous rocks into three general categories, based on texture—crystalline rocks have interlocking crystals, glassy rocks consist of a mass of glass with no crystalline structure, and pyroclastic rocks have a fragmental texture. Crystalline rocks, in turn, can be classified based on composition and on whether they are fine- or coarse-grained. The color of a rock may provide a clue to its composition.

QUICK QUESTION What kind of rock forms from felsic magma cooled in a large pluton at depth? Could you make a sharp axe out of this rock?

FIGURE 4.15 Examples of igneous rocks, arranged by grain size and composition.

FIGURE 4.16 Some igneous rocks contain an abundance of vesicles, gas bubbles that were frozen into the rock as it cooled.

(a) Pumice is a felsic glassy rock with tiny vesicles.

(b) Scoria is a mafic glassy rock with many vesicles.

4.5 Plate-Tectonic Context of Igneous Activity

Earlier in this chapter, we pointed out that melting occurs only in special locations where conditions lead to decompression, addition of volatiles, and/or heat transfer. The conditions that lead to melting and, therefore, to igneous activity, can develop in four geologic settings (**Fig. 4.17**): (1) at hot spots; (2) along volcanic arcs bordering oceanic trenches; (3) along mid-ocean ridges; and (4) within continental rifts. Let's look more carefully at melting and igneous rock production, in the context of plate tectonics theory, with a focus on the types of igneous rocks that form in each setting. In Chapter 5, we'll add to the story by discussing the types of volcanic eruptions associated with different settings.

Products of Hot Spots

As we learned in Chapter 2, most (but not all) researchers associate hot-spot igneous activity with mantle plumes, columns or streams of hot mantle rock rising from deeper in the mantle. According to the plume hypothesis, the plume itself does not consist of magma. Rather, it's composed of solid rock that is hot enough to be soft and to flow plastically, at rates of a few centimeters a year. When the hot rock of a plume reaches the base of the lithosphere, decompression causes partial melting within it. This process generates mafic magma. At oceanic hot spots, much of the mafic magma erupts at the surface as basalt. At continental hot spots, part of the mafic magma erupts to form basalt, but some transfers heat to the continental crust, which itself then partially melts, producing felsic magmas that erupt to yield rhyolite.

Products of Subduction

A chain of volcanoes, called a **volcanic arc** (or just an *arc*), forms on the overriding plate adjacent to the deep-ocean trenches that mark convergent plate boundaries (see Chapter 2). *Continental arcs* grow along the edge of a continent, where oceanic lithosphere subducts beneath continental lithosphere. *Island arcs* protrude from the ocean at localities where one oceanic plate subducts beneath another oceanic plate. Beneath volcanic arcs, a variety of intrusions—plutons, dikes, and sills—develop, to be exposed only later by erosion of the overlying volcanoes. In some localities, arc-related igneous activity produces huge batholiths.

How does subduction trigger melting? Some minerals in oceanic crust rocks contain volatile compounds (mostly water). At shallow depths, these volatiles are bonded to other elements within mineral crystals. But when subduction carries crust down into the hot asthenosphere, the crust warms up, and at a depth of about 150 km, it becomes so hot that volatiles separate and diffuse up into the overlying hot ultramafic rock (peridotite) of the asthenosphere. Addition of these volatiles causes flux melting of peridotite. As we have seen, only

Formation of Igneous Rocks

Igneous rock forms by the cooling of magma underground or of lava at the surface. Igneous rocks that solidify underground are intrusive, whereas those that solidify at the surface are extrusive. The type of igneous rock that forms depends on the composition of the melt and the environment of cooling.

Stratified volcanic tuff

In the extrusive environment, melt may cool quickly and have a glassy texture. Melt that explodes into the air forms ash and other debris, which may consolidate into rock with fragmental texture.

In the intrusive environment, crystals grow together to form an interlocking texture. Slower cooling makes coarser grains.

Minerals in an igneous rock crystallize in succession as the melt cools.

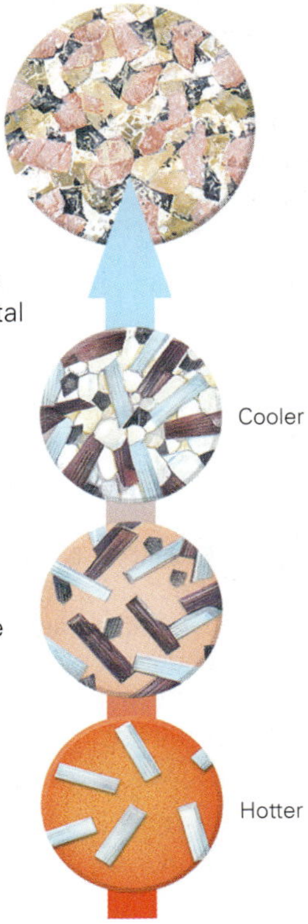

FIGURE 4.17 The tectonic setting of igneous rocks.

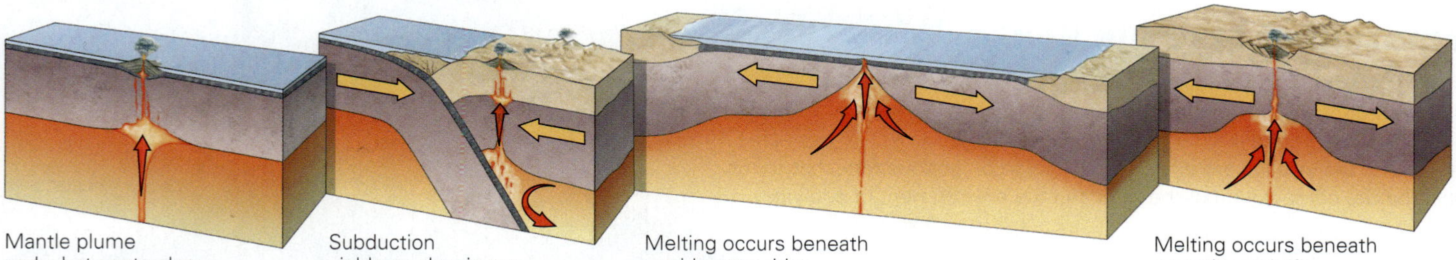

part of the original ultramafic rock actually melts, since silica preferentially goes into the melt, So the process yields mafic magma, some of which rises to form basaltic sills and dikes in the crust, and some makes it all the way to the surface to extrude as basaltic lava.

In continental volcanic arcs, not all mantle-derived basaltic magma rises directly to the surface; some gets trapped at the base of the continental crust, and some resides in magma chambers deep in the crust. Fractional crystallization, as well as assimilation, may cause some of the magma to become progressively more felsic. But in addition, this very hot mantle-derived magma transfers heat into the adjacent continental crust, which causes partial melting of the continental crust. Because much of the continental crust is originally mafic to intermediate in composition, the resulting magma has an intermediate to felsic composition. This resulting magma rises, and it either cools higher in the crust to form plutons, or rises to the surface and erupts as lava. These processes (heat-transfer melting, assimilation, and fractional crystallization) contribute to the abundance of granite, rhyolite, diorite, and andesite at continental arcs.

Forming Igneous Rocks at Mid-Ocean Ridges

The entire oceanic crust, a 7- to 10-km-thick layer of basalt and gabbro that covers 70% of the Earth's surface, forms at mid-ocean ridges (divergent plate boundaries). We can say, therefore, that most igneous rocks present at the Earth's surface are a product of seafloor spreading. Mid-ocean ridge igneous activity happens because, as seafloor spreading takes place, oceanic lithosphere plates drift away from the ridge. Hot asthenosphere rises to keep the resulting space filled. As asthenosphere rises, it undergoes partial melting due to decompression. Partial melting of mantle peridotite yields mafic magma which, as noted in Chapter 2, rises into the crust and collects in a magma chamber. Some cools slowly in the magma chamber to form massive gabbro, while some intrudes upward to fill vertical cracks that appear as newly formed crust splits apart (see Fig. 2.18a). Magma that cools in the cracks forms basalt dikes, and magma that rises as far as the seafloor, and extrudes as lava, forms pillow-basalt flows.

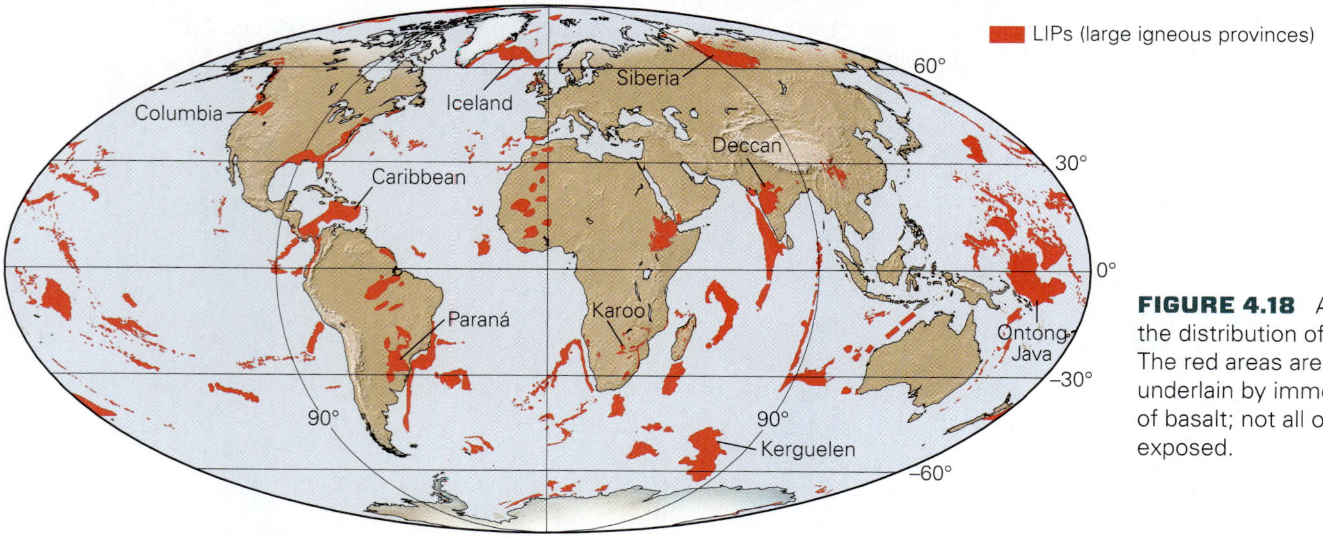

FIGURE 4.18 A map showing the distribution of LIPs on Earth. The red areas are or once were underlain by immense volumes of basalt; not all of this basalt is exposed.

FIGURE 4.19 Flood basalts form when vast quantities of low-viscosity mafic lava "floods" over the landscape and freezes into a thin sheet. Accumulation of successive flows builds a flat-topped plateau.

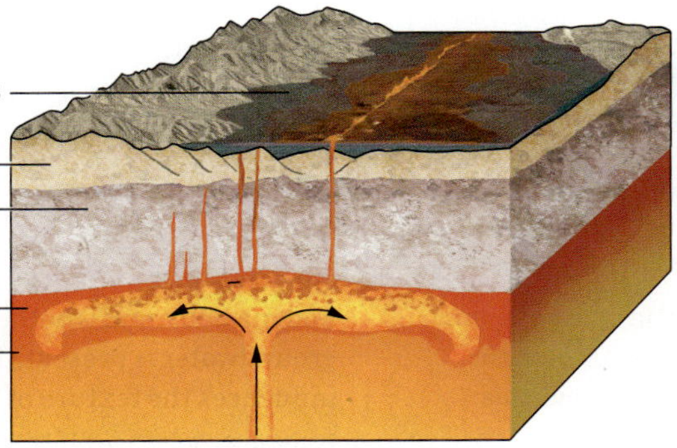

(a) The plume model for forming flood basalts.

(b) Flood basalts form the layers exposed in Palouse Canyon, Washington.

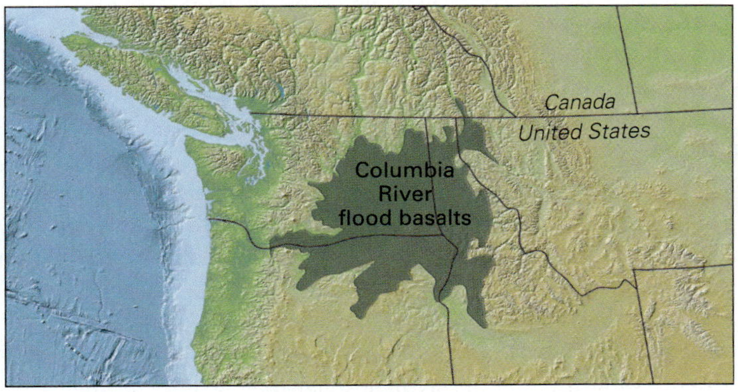

(c) Flood basalts underlie the Columbia River Plateau in Washington and Oregon, the dark area of this map.

Forming Igneous Rocks at Rifts

Rifts are places where horizontal stretching of continental lithosphere takes place. As a result of this stretching, the lithosphere also thins vertically (see Fig. 2.28). As this process takes place, the asthenosphere undergoes decompression. Partial melting due to decompression produces basaltic magma, which rises into the crust. Some magma makes it to the surface and erupts as basalt. However, some becomes trapped in the crust and transfers heat to the crust. The resulting partial melting of the crust yields felsic magmas that erupt explosively as rhyolite. Thus, a sequence of volcanic rocks in a rift generally includes lava flows of basalt and sheets of rhyolitic ash.

Large Igneous Provinces (LIPs)

In many places on Earth, particularly voluminous quantities of mafic magma have erupted or intruded (**Fig. 4.18**). Some of these regions occur along the margins of continents, some in the interior of oceanic plates, and some in the interiors of continents. The largest of these, the Ontong Java Oceanic Plateau of the western Pacific, covers an area of about 5,000,000 km^2 of the seafloor and has a volume of about 50,000,000 km^3. Such provinces also occur on land. It's no surprise that these huge volumes of igneous rock are called **large igneous provinces**, generally abbreviated *LIPs*. The term LIP has also been applied to huge eruptions of felsic ash.

Mafic LIPs may form when the bulbous head of a mantle plume first reaches the base of the lithosphere. More partial melting can occur in a plume head than in normal asthenosphere, because temperatures are higher in a plume head. Thus, a very large quantity of particularly hot mafic magma forms in the plume head, and when this magma reaches the surface, huge quantities of lava spew out of the ground. If the plume head lies beneath a rift, thinning of the lithosphere causes even more decompression, which in turn leads to even more melting (**Fig. 4.19a**).

On land, the hot mafic lava that erupts in LIPs has such low viscosity that it can flow tens to hundreds of kilometers across the landscape, forming a vast sheet of basalt. The process may repeat hundreds of times, yielding thick stacks of thin, broad lava flows known as **flood basalts**. Flood basalts make up the bedrock of the Columbia River Plateau in Oregon and Washington (**Fig. 4.19b, c**), the Paraná Plateau in southeastern Brazil, the Karoo region of southern Africa, and the Deccan region of southwestern India.

TAKE-HOME MESSAGE

The formation of igneous rocks can be understood in the context of plate tectonics. Flux melting happens in the mantle above subducting plates, and decompression melting takes place in the mantle within hot spots, and under both mid-ocean ridges and rifts. Melting of the mantle produces basalt. Other types of igneous rocks most commonly form where rising mantle magma interacts with continental crustal rock and heat-transfer melting takes place.

QUICK QUESTION Which tectonic setting has yielded the most igneous rocks in the past 200 million years?

Chapter 4 Review

Chapter Summary

- Magma is liquid rock (melt) under the Earth's surface. Lava is melt that has erupted from a volcano at the Earth's surface.
- Magma forms when hot rock in the Earth melts. Melting occurs only under certain circumstances—when the pressure decreases, when volatiles are added to hot rock, and when magma rising from the mantle transfers heat into the crust.
- During partial melting, only part of the source rock melts to form magma. Silica preferentially goes into the melt, so magmas tend to contain more silica than the rock from which they were derived.
- Magma occurs in a range of compositions: felsic (silicic), intermediate, mafic, and ultramafic. The composition of magma reflects the original composition of the rock from which the magma formed and the way the magma evolves.
- Magma rises from depth because of its buoyancy and because of pressure caused by the weight of overlying rock.
- Melt viscosity depends on composition. Felsic melt is more viscous than mafic melt, and cooler melt is more viscous than hotter melt.
- Extrusive igneous rocks solidify from lava that erupts out of a volcano. Intrusive igneous rocks develop from magma that freezes inside the Earth.
- Lava may solidify to form flows, or it may explode into the air to form pyroclastic debris, including ash.
- Intrusive igneous rocks form when magma intrudes into pre-existing rock below Earth's surface. Tabular intrusions that cut across layering are dikes, and those that form parallel to layering are sills. Blob-shaped intrusions are called plutons. Huge intrusions, made up of many plutons, are known as batholiths.
- The rate at which intrusive magma cools depends on the depth at which it intrudes, the size and shape of the magma body, and whether circulating groundwater is present. The cooling time influences the texture of an igneous rock.
- Igneous rocks are classified according to texture and composition.
- Plate tectonics theory can explain why igneous rocks form where they do.
- Magma forms at continental or island volcanic arcs along convergent margins, mostly because of the addition of volatiles to the asthenosphere above the subducting slab causes flux melting. Igneous rocks form at hot spots, due to decompression melting of a rising mantle plume.
- Igneous rocks form at rifts as a result of decompression melting of the asthenosphere below the thinning lithosphere or heat transfer from mantle melts into crustal rocks. Igneous rocks form along mid-ocean ridges because of decompression melting of the rising asthenosphere.

Guide Terms

assimilation (p. 118)
batholith (p. 121)
Bowen's reaction series (p. 122)
crystalline igneous rock (p. 127)
dike (p. 121)
extrusive igneous rock (p. 115)
flood basalt (p. 133)
fractional crystallization (p. 120)
fragmental igneous rock (p. 127)
glassy igneous rock (p. 127)
igneous rock (p. 114)
intrusive igneous rock (p. 115)
laccolith (p. 121)
large igneous province (LIP) (p. 133)
lava (p. 113)
magma (p. 115)
magma chamber (p. 125)
obsidian (p. 129)
partial melting (p. 118)
pluton (p. 121)
pumice (p. 129)
pyroclastic debris (p. 115)
pyroclastic rock (p. 129)
scoria (p. 129)
sill (p. 121)
tachylite (p. 129)
tuff (p. 129)
vesicle (p. 129)
viscosity (p. 119)
volcanic arc (p. 130)
volcanic ash (p. 115)
volcano (p. 113)
xenolith (p. 125)

GEOTOURS THIS CHAPTER'S GEOTOUR EXERCISE (C) FEATURES:

- Batholiths and Laccoliths
- Coal Dikes

Review Questions

1. How is the process of freezing magma similar to that of freezing water? How is it different?
2. What is the source of heat in the Earth?
3. Describe the three processes that are responsible for the formation of magmas.
4. Why are there so many different compositions of magmas? Does partial melting produce magma with the same composition as the parent rock from which it was derived?
5. Why do magmas rise from depth to the surface of the Earth?
6. Explain the process of fractional crystallization.
7. What factors control the viscosity of a melt?
8. What factors control the cooling time of a magma within the crust?
9. What is the difference between a sill and a dike, and how do both differ from a pluton?
10. How does grain size reflect the cooling time of a magma?
11. What does the mixture of grain sizes in a porphyritic igneous rock indicate about its cooling history?
12. Describe the way magmas are produced in subduction zones.
13. What process in the mantle may be responsible for causing hot-spot volcanoes to form?
14. Describe how magmas are produced at continental rifts. Why can you find both basalt and rhyolite in such settings?
15. Why does melting take place beneath the axis of a mid-ocean ridge?
16. What is a large igneous province (LIP)?

On Further Thought

16. In Figure 4.14, we list komatiite, a fine-grained ultramafic rock. Komatiite can be found almost exclusively in crust that is > 2.5 Ga. In other words, komatiite doesn't form today, but did form early in Earth history. Why? (*Hint:* Think about how the temperature inside the Earth may have changed over geologic time.)
17. Would you expect felsic pyroclastic flows to erupt from mid-ocean ridge? Explain the reasoning that led you to your answer.
18. What kind of igneous rock would make a dark-colored counter top?

Online Resources

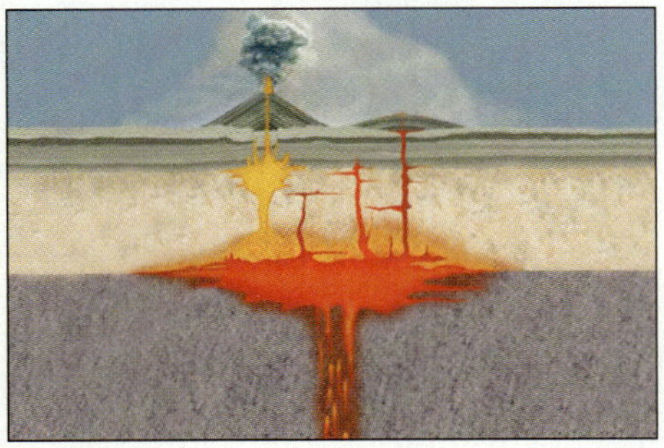

Animations

This chapter features four animations on lava and magma composition and change.

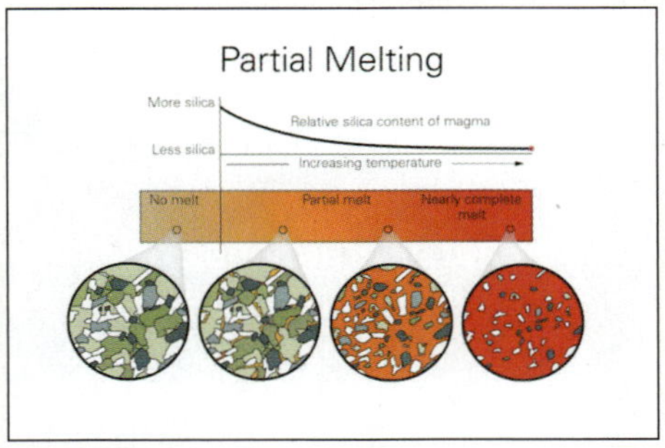

Videos

This chapter features a video on the partial melting of rocks.

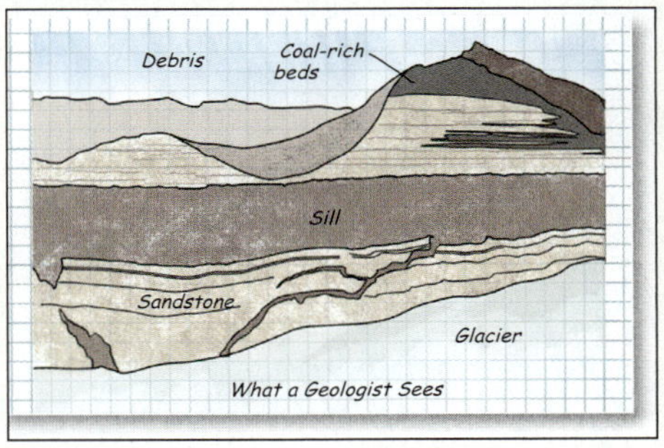

Assessment

Features include questions on recognizing various igneous settings and describing igneous rock.

LEARNING OBJECTIVES

By the end of this chapter, you should understand...

1. that not all eruptions are alike—some yield streams of lava, while others produce catastrophic explosions.
2. how the type of an eruption reflects the character of lava, which in turn depends on the geologic setting.
3. ways in which volcanic eruptions can be dangerous natural hazards.
4. why impending eruptions can be predicted, sometimes.
5. how eruptions may affect climate, evolution, and perhaps the future of civilizations.

▲ An eruption of the Batu Tara Volcano in Indonesia spews glowing bombs of lava skyward. Volcanoes are an important component of the Earth System, serving to transfer rock and gas from our planet's interior to its surface. But, while beautiful to watch from a distance, the force of eruptions can threaten life and property.

CHAPTER 5

The Wrath of Vulcan: Volcanic Eruptions

5.1 Introduction

Every few hundred years, one of the hills on Vulcano, an island in the Mediterranean Sea off the western coast of Italy, rumbles and spews out molten rock, glassy cinders, and dense "smoke" (actually a mixture of gas, ash, and tiny liquid droplets). Ancient Romans thought that such **eruptions**, episodes when volcanoes extrude lava and pyroclastic debris, happened when Vulcan, the god of fire, fueled his forges beneath the island to manufacture weapons for the other gods. Geologic study suggests, instead, that eruptions take place when hot magma, formed by melting inside the Earth, rises through the crust and emerges at the surface. No one believes the Roman myth anymore, but the island's name evolved into the English word **volcano**, which geologists use to designate either an erupting vent through which molten rock reaches the Earth's surface, or a mountain built from the products of eruption.

On the main peninsula of Italy, another volcano, Mt. Vesuvius, towers over the Bay of Naples. Nearly 2,000 years ago, a prosperous Roman resort and trading town named Pompeii sprawled at the foot of Vesuvius. One morning in 79 C.E., earthquakes signaled the mountain's awakening. At 1:00 p.m. on August 24, a dark mottled cloud, streaked by lightning, boiled up above Mt. Vesuvius's summit. The cloud

Glowing waves rise and flow, burning all life on their way, and freeze into black, crusty rock which adds to the height of the mountain and builds the land, thereby adding another day to the geologic past. . . . I became a geologist forever, by seeing with my own eyes: the Earth is alive!

HANS CLOOS (German geologist, 1886–1951), on seeing the eruption of Mt. Vesuvius in Italy

spread over Pompeii, filling the air with fine ash and choking fumes, and turned day into night (**Fig. 5.1a**). Blocks and pellets of rock fell like hail, collecting on streets, plazas, and red tile roofs—eventually, the rubble began to crush buildings. People frantically rushed to escape, but for most it was too late. Suddenly, a turbulent current of scalding ash and pumice fragments surged down the flank of the volcano and swept over Pompeii. By the next day, the town had vanished beneath a 6-m-thick gray-black blanket of debris. This covering protected the ruins of Pompeii so well that when archaeologists excavated the town 1,800 years later, they found artifacts and ruins that gave an amazingly complete record of Roman daily life (**Fig. 5.1b, c**). As they dug, workers discovered open spaces in the debris. When they filled the spaces with plaster and then removed the surrounding ash, casts of Pompeii's unfortunate inhabitants, their bodies twisted in agony or huddled in despair, remained (**Fig. 5.1d**).

Clearly, volcanoes are unpredictable and dangerous. Volcanic activity can build a towering, snow-crested mountain, or it can blast one apart. Eruptions can provide fertile soil that enables a civilization to thrive, or can generate a rain of destruction that can snuff one out. Because the characteristics and consequences of volcanic eruptions are so diverse, this chapter sets out ambitious goals. We first take a close look at the products of volcanic activity and the architecture of volcanoes. Then we discuss the different kinds of volcanic eruptions on Earth and explain why they occur where they do. Finally, we consider the hazards posed by volcanoes, the efforts geoscientists make to predict eruptions and help minimize the damage they cause, and the influence that eruptions may have on climate and civilization.

5.2 The Products of Volcanic Eruptions

The drama of a volcanic eruption transfers materials from inside the Earth to our planet's surface. Products of an eruption come in three forms—lava flows, pyroclastic debris, and gas. Note that we use the term **lava flow** both for a molten, moving layer of lava and for the solid layer of rock that forms when the lava freezes.

Lava Flows

Sometimes lava races down the side of a volcano like a fast-moving, incandescent stream, sometimes it oozes like a sticky but scalding paste, and sometimes it builds into a rubble-covered mound. Clearly, not all lava behaves in the same way, so not all lava flows look the same. Why? The character of a lava reflects its **viscosity** (resistance to flow), and lava viscosity, in turn, depends primarily on chemical composition, for silica in lava tends to bond together into large molecules that can't move easily. Therefore, mafic (basaltic) lava is less viscous than intermediate (andesitic) lava, which in turn is less viscous than felsic (rhyolitic) lava. As a result, basaltic lava can flow farther than andesitic lava, which can flow farther than rhyolitic lava (**Fig. 5.2**). Viscosity also depends on temperature—a hotter magma is less viscous than a cooler lava of the same composition, because thermal energy breaks chemical bonds linking molecules.

Basaltic Lava Flows Basaltic lava has very low viscosity when it first emerges from a volcano because it contains relatively little silica and it is very hot. Thus, on the steep slopes near the summit of a volcano, lava can move at speeds up to 30 km per hour (**Fig. 5.3a**), but it slows down to walking pace after it starts to cool. Most basaltic flows measure less than a few kilometers long, but some extend tens of kilometers from the source (**Fig. 5.3b**). (Flood basalts may flow even farther—up to 600 km from the source.) How can lava travel large distances? Although all the lava of a flow moves when it first emerges, rapid cooling causes the surface to crust over after the flow has moved away from the source. The solid crust serves as insulation, allowing the hot interior of the flow to remain liquid and continue to move. As time progresses, more of the interior solidifies. Eventually, molten lava moves only through a tunnel-like passageway, or **lava tube**, within the flow—the largest of these may be tens of meters in diameter (**Fig. 5.3c**). In some cases, lava tubes drain and eventually become empty tunnels (**Fig. 5.3d**).

The surface texture of a basaltic lava flow depends, in part, on the timing of the lava's freezing relative to its movement. Basalt flows whose surfaces are warm and pasty wrinkle into smooth, glassy, rope-like ridges—geologists use the Hawaiian word **pahoehoe** (pronounced pa-HOY-hoy) for such flows (**Fig. 5.3e**). If a moving lava flow becomes too viscous to contort into pahoehoe ropes, the surface of the lava breaks up into a jumble of sharp, angular fragments, creating a rubbly flow also called by its Hawaiian name, **a'a'** (pronounced AH-ah) (**Fig. 5.3f**). Footpaths made by people living in

FIGURE 5.1 The eruption of Vesuvius buried Pompeii and nearby Herculaneum in 79 C.E.

You can see ruts carved into streets by chariots.
Pre-79 C.E. profile

(a) In this 1817 painting, the British artist J.M.W. Turner depicted the cataclysmic explosion.

(b) Streets and buildings of Pompeii are well preserved.

(c) Excavations exposed ruins of Pompeii with Vesuvius in the distance. The dashed line shows the volcano's profile prior to its eruption.

(d) A plaster cast of a Pompeii resident who was buried in ash.

basaltic volcanic regions follow the smooth surface of pahoehoe rather than the foot-slashing surface of a'a'.

Basaltic flows that erupt underwater look very different from those that erupt on land because lava cools so much more quickly in water. Because of rapid cooling, submarine basaltic lava can travel only a short distance before its surface freezes, producing a glass-encrusted blob, or pillow (**Fig. 5.3g**). The rind of a pillow momentarily stops the flow's advance, but within minutes the pressure of the lava squeezing into the pillow breaks the rind, and a new blob of lava squirts out, freezes, and produces another pillow. A flow made of these blobs is called **pillow lava**. In some cases, successive pillows form worm-like chains.

Even after a flow stops moving and has become solid, it is still warm. As the last of its heat radiates outward, the rock in a lava flows contracts, because like many solids, rock shrinks as it loses heat. Because the lava flow is solid during this final cooling, the shrinkage causes it to crack, forming natural fractures called *joints*. In some cases, the pattern of jointing outlines polygonal columns that have roughly hexagonal cross sections. This type of fracturing is called **columnar jointing** (**Fig. 5.3h**). Of note, such joints can also form in dikes and sills as they cool.

Andesitic and Rhyolitic Lava Flows Because of its greater viscosity, intermediate-composition (andesitic) lava cannot flow as easily as basaltic lava. When erupted, andesitic lava first forms a mound above the vent. This mound advances slowly down the volcano's flank at only about 1 to 5 m a day, in a lumpy flow with a bulbous snout. Typically, andesitic flows are less than a few kilometers long. Because the lava is so viscous, the surface can solidify as it moves, forming angular chunks. The whole flow, called **blocky lava**, looks like a jumble of rubble. If the flow extrudes onto a steep slope, the blocks may tumble down the side of a volcano while still glowing red.

Rhyolitic lava is the most viscous of all lavas because it is the most felsic and the coolest type of lava. Therefore, it tends to accumulate either above the vent in a **lava dome** (**Fig. 5.4**), or in short and bulbous flows rarely more than 1 to 2 km long. Sometimes rhyolitic lava freezes while still in the vent and then pushes upward as a column-like **spire** up to 100 m above the vent. Rhyolitic flows have broken and blocky surfaces.

FIGURE 5.2 The characteristics of a lava flow depends on its viscosity.

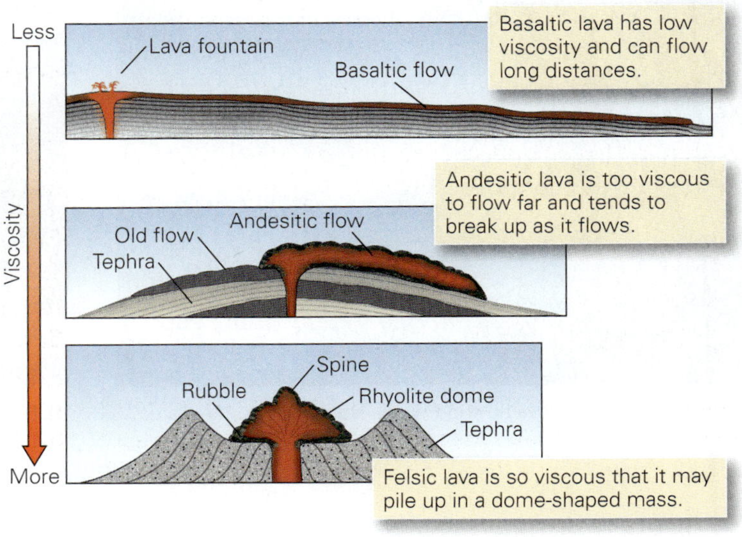
Less
Viscosity
More
Lava fountain
Basaltic flow
Basaltic lava has low viscosity and can flow long distances.
Old flow
Tephra
Andesitic flow
Andesitic lava is too viscous to flow far and tends to break up as it flows.
Spine
Rubble
Rhyolite dome
Tephra
Felsic lava is so viscous that it may pile up in a dome-shaped mass.

FIGURE 5.3 Features of basaltic lava flows. They have low viscosity and thus can flow long distances. Their surface and interior can be complex.

(a) A fast-moving flow coming from Mt. Etna, Sicily.

(b) A basaltic lava flow covers a highway in Hawaii.

(c) In a lava tube, still-molten lava flows beneath a crust of solid basalt.

(d) A drained lava tube exposed in a road cut on Hawaii.

(e) Pahoehoe from a recent lava flow in Hawaii. Note the coin for scale.

(f) The rubbly surface of an a'a' flow, Sunset Crater, Arizona.

FIGURE 5.3 (*continued*)

(g) Pillow basalt develops when lava erupts underwater. Later uplift may expose pillows above sea level, as in this Oregon outcrop.

(h) Columnar jointing develops when the interior of a flow cools and cracks. Such jointing develops in dikes and sills. This example is Devils Postpile in California.

Volcaniclastic Deposits

On a mild day in February 1943, as Dionisio Pulido prepared to sow the fertile soil of his field, 330 km (200 miles) west of Mexico City, an earthquake jolted the ground. Earthquakes had occurred with increasing frequency in recent days, but this time, to Dionisio's amazement, the surface of his field visibly bulged upward by a few meters and then cracked. Ash and sulfurous fumes rose into the air, and the farmer fled. When he returned the following morning, his field lay buried beneath a 40-m-high mound of gray cinders—Dionisio had witnessed the birth of a new volcano, Paricutín. During the next several months, Paricutín erupted continuously, at times blasting clots of lava into the sky like fireworks. By the following year, it had become a steep-sided cone 300 m high. Nine years later, when the volcano ceased erupting, its lava and debris covered 25 square km.

Did you ever wonder... whether anyone has ever seen a brand-new volcano form?

The description of Paricutín's eruption, and that of Vesuvius at the beginning of this chapter, emphasize that volcanoes can erupt fragmental igneous material. Geologists use either the term **tephra** or the term **pyroclastic debris** (from the Greek *pyro*, meaning fire) for unconsolidated fragments resulting from an eruption. Tephra includes the following materials: specks of glass and pieces of rock formed from drops or clots of molten lava that solidified in midair, or on the ground soon after falling; chunks of recently solidified lava or pumice blasted out of the volcano's vent; and/or chunks formed by the forceful fragmentation of pre-existing volcanic rock surrounding the volcano's vent. A broader term, *volcaniclastic debris*, includes not only tephra, but also debris that tumbled down the flank of a volcano in landslides long after an eruption, debris that mixed with water to form a flowing muddy slurry, and debris originally formed at a volcano but later carried and sorted by streams. Let's look at these components in more detail—you'll see that different types form in association with different kinds of eruptions.

FIGURE 5.4 This rhyolite dome formed about 650 years ago, in Panum Crater, California. Tephra (cinders) accumulated around the vent.

FIGURE 5.5 Pyroclastic debris from basaltic eruptions.

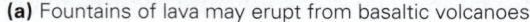

(a) Fountains of lava may erupt from basaltic volcanoes.

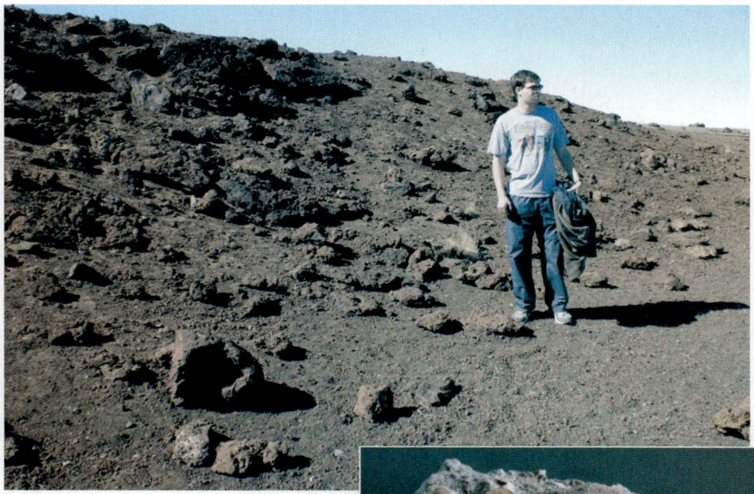

(c) Blocks and lapilli on the flank of a Hawaiian volcano.

(b) A bomb has a smooth, streaked surface.

Pyroclastic Debris from Basaltic Eruptions Basaltic melt rising in a volcano may contain dissolved volatiles (such as H_2O and CO_2). As such melt approaches the surface, the volatiles come out of solution and form gas bubbles. When the bubbles reach the surface of the melt at the vent of the volcano, they burst and eject gas. The gas pressure can drive columns or sprays of basaltic lava upward in dramatic **lava fountains**—clots of lava rain down onto the volcano from the fountains (**Fig. 5.5a**). To picture this process, think of the droplets that spray from a newly opened bottle of soda. The solidified pea-sized to golf-ball-sized fragments of glassy lava and scoria, informally called *cinders*, are a type of **lapilli**, from the Latin word for little stones. Rarely, flying droplets may trail thin strands of lava. The strands freeze into filaments of glass called *Pelé's hair*, after the Hawaiian goddess of volcanoes, and the droplets freeze into tiny streamlined glassy beads known as *Pelé's tears*. Larger blobs of lava that squirt out of a vent and then solidify become streamlined **bombs**, whose surfaces are typically streaked and polished (**Fig. 5.5b**). The erupting basalt also may rip off apple- to refrigerator-sized fragments of already-solid lava from the walls of the vent and eject them as chunky, angular **blocks** (**Fig. 5.5c**). Thus, the pyroclastic debris surrounding a basaltic volcanic vent commonly includes blocks, bombs, and cinder-like lapilli.

Pyroclastic Debris from Andesitic or Rhyolitic Eruptions Andesitic or rhyolitic lavas are more viscous than basaltic lavas, and are generally richer in gas. As we'll see, eruptions of these lavas tend to be explosive and to eject large quantities of pyroclastic debris (**Fig. 5.6a**). Debris ejected during explosive eruptions includes: **ash**, with particles less than 2 mm in diameter, made from glass shards formed either when frothy lava explosively breaks up during an eruption and the glassy walls of bubbles separate into small pieces, or when pre-existing volcanic rock undergoes pulverization (**Fig. 5.6b**); *pumice lapilli*, angular pumice fragments formed when frothy lava breaks into marble-sized chunks (**Fig. 5.6c**); and *accretionary lapilli*, small, snowball-like spheres of ash formed when ash mixes with water in the air and then sticks together, collecting more ash as it falls (**Fig. 5.6d**).

Part of the pyroclastic debris erupted from an exploding volcano shoots upward into the air. Some, however, rushes down the flank of the volcano in an avalanche-like current known as a **pyroclastic flow** (**Fig. 5.6e**), once called a *nuée ardente* (French for glowing cloud) because the debris within may retain heat between 200° and 450°C. Tephra from andesitic or rhyolitic eruptions (ash, or ash mixed with lapilli) becomes **tuff** when buried and transformed into coherent rock. Tuff from material that rains out of an ash

FIGURE 5.6 Pyroclastic debris from andesitic or rhyolitic eruptions.

(a) Pyroclastic debris billowing from the 2008 eruption of Chaitén in Chile.

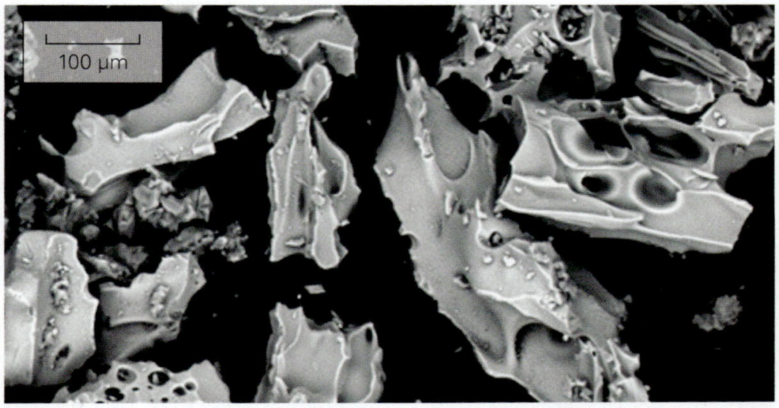

(b) Electron photomicrograph of ash.

(c) Pumice lapilli.

(d) Accretionary lapilli.

(e) A pyroclastic flow rushes down the flank of Mt. Merapi, Indonesia, in 2006.

(f) Water-soaked volcanic debris slid down the side of this volcano in Nicaragua.

(g) A lahar fills a river bed in New Zealand after an eruption in 2007.

FIGURE 5.7 The gas component of volcanic eruptions.

(a) A volcano in Alaska erupting large quantities of steam.

(b) Gas bubbles frozen in lava produce vesicles, as in this block from Sunset Crater, Arizona.

cloud is *air-fall tuff*, whereas a sheet of tuff deposited by a pyroclastic flow is an *ignimbrite*. Tuff generally forms when debris accumulates while still hot enough to weld together, or when its minerals grow in the tephra after deposition, after reacting with groundwater.

Volcanic Debris Flows and Lahars In cases where volcanoes are covered with snow and ice, or are drenched with rain, water can mix with debris to form a **volcanic debris flow** that moves downslope like a sheet of wet concrete (**Fig. 5.6f**). If ash-rich debris becomes very wet, it becomes a soupy, muddy slurry called a **lahar**—these can reach speeds of 50 km per hour and may travel for tens of kilometers (**Fig. 5.6g**). When debris flows and lahars stop moving, they yield a layer consisting of volcanic debris suspended in ashy mud.

Volcanic Gas and Aerosols

Most magma contains dissolved gases such as water, carbon dioxide, and sulfur dioxide (H_2O, CO_2, and SO_2). In fact, up to 9% of a magma may consist of gaseous components—generally, lavas with more silica contain a greater proportion of gas. Volcanic gases come out of solution when the magma approaches the Earth's surface and pressure decreases, just as bubbles come out of solution in a soda when you pop the bottle top off. Commonly, water emitted by a volcano condenses into liquid which dissolves the sulfur-bearing gases to form **aerosols** (tiny droplets) of sulfuric acid.

In low-viscosity magma, gas bubbles can rise faster than the magma moves, and thus most reach the surface of the magma and enter the atmosphere before the lava does. So for a while, some volcanoes may erupt large quantities of gas, without much lava (**Fig. 5.7a**). Lava may still contain gas bubbles as it starts to flow. When the lava freezes, the bubbles are trapped in the rock, and become holes called **vesicles** (**Fig. 5.7b**). In high-viscosity magmas, gas has trouble escaping because bubbles can't push through the sticky lava. As we've noted, some felsic lavas contain so much gas that they are basically froths.

SEE FOR YOURSELF...

MT. ST. HELENS, WASHINGTON

Latitude
46°12′1.24″ N

Longitude
122°11′20.73″ W

Look obliquely from 25 km (~15.5 miles).

Mt. St. Helens, a stratovolcano of the Cascade volcanic arc, erupted explosively in 1980. The explosion blew away the north side of the mountain. Lahars formed during the eruption flooded nearby river valleys.

TAKE-HOME MESSAGE

Volcanoes erupt lava, pyroclastic debris, and gas. The character of a lava flow—whether it has low viscosity and spreads over a large area, or has high viscosity and builds a mound over the vent—depends largely on its composition. Pyroclastic debris includes pumice, lapilli, blocks, and bombs. Some may fall over the countryside like snow, but some surges down the flank of a volcano as a pyroclastic flow.

QUICK QUESTION How can lava travel tens of kilometers or more from a volcanic vent without freezing?

5.3 Structure and Eruptive Style

FIGURE 5.8 Crater eruptions and fissure eruptions come from conduits of different shapes.

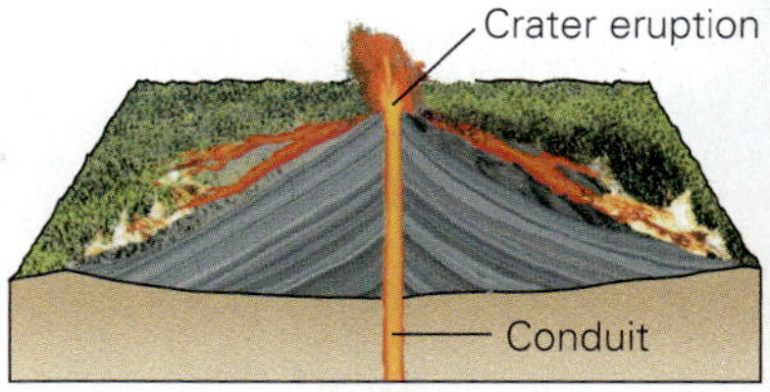

(a) At a crater eruption, lava spouts from a chimney-shaped conduit.

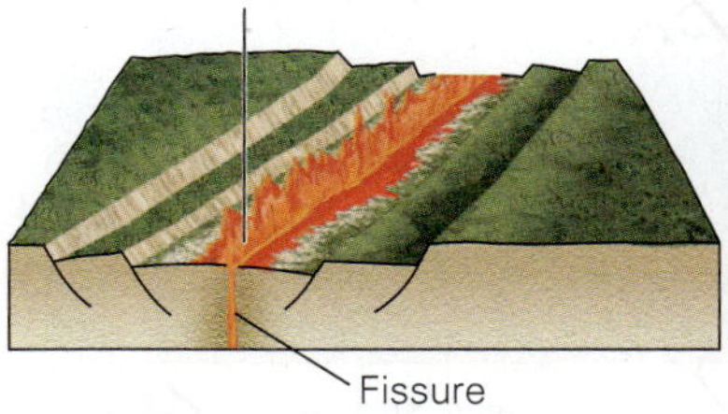

(b) At a fissure eruption, lava comes out in a curtain, along the length of a crack.

Volcanic Architecture

As magma rises from depth, it doesn't always make it all the way to the surface but may accumulate underground in a **magma chamber**, a large, irregularly shaped zone. Inside a magma chamber, melt may mix with crystals to form a crystal mush or may fill pervasive cracks between blocks of solid rock. It may solidify to form a pluton, inject into cracks or between layers to form dikes or sills, or rise along a pathway, called a *conduit,* to the Earth's surface and erupt from a vent. A conduit may resemble a vertical pipe or chimney, or it may take the form of a crack called a **fissure** (**Fig. 5.8**). The products of an eruption, or a succession of eruptions, build into a hill or mountain known formally as a *volcanic edifice* but informally, simply as a volcano. Typically, a circular, bowl-like depression, called a **crater**, occurs at the top of a volcanic edifice. Craters, which may as large as 500 m across and 200 m deep, form either during eruption, as material accumulates around the summit vent, or just after eruption, as the summit collapses into the drained conduit.

During some major eruptions, a large magma chamber beneath a volcano may drain suddenly and produce a **caldera**, a circular depression that is much larger than a crater. Typically, a caldera has steep walls and a fairly flat floor and may be partially filled with ash or solidified lava (**Fig. 5.9**). Examples range from a few kilometers to tens of kilometers across, and they may be several hundred meters deep. Later eruptions can build new volcanoes within a caldera. For example, Wizard Island is a volcano that grew from the floor of the Crater Lake caldera and now protrudes from the lake (see Fig. 5.9d). The caldera collapsed between 6,000 and 8,000 years ago, and Wizard Island last erupted about 4,600 years ago.

Geologists distinguish among several different shapes of subaerial (above sea level) volcanic edifices. **Shield volcanoes** are broad, gentle domes whose shape resembles a soldier's shield lying on the ground (**Fig. 5.10a**). They form when the products of eruption have low viscosity and can't build into a mound at the vent, but rather flow easily and spread out as thin sheets over large areas. **Cinder cones,** also known as *scoria cones,* consist of cone-shaped piles of basaltic lapilli and blocks (**Fig. 5.10b**). The steep slopes of a cinder cone represent the *angle of repose* for lapilli, meaning the maximum slope that the loose fragments can sustain before sliding down. (You can build a similar-shaped pile by pouring dry sand from a bucket onto the ground.) **Stratovolcanoes,** also called *composite volcanoes,* are large volcanoes (up to 4 km high, and 25 km across) that consist of interleaved layers of lava, tephra, and volcaniclastic debris (**Fig. 5.10c**). The prefix *strato-* was coined to emphasize that such volcanoes have internal layering, or stratification. Their shape, exemplified by Japan's Mt. Fuji, serves as the classic image that most people have of a volcano. If you look closely, you'll note that stratovolcanoes generally have steeper slopes near the summit and gentler slopes near the base. This is because the upper part builds up out of successive layers of tephra, deposited at its angle of repose and coated by lava flows, whereas the lower part includes debris flows, lahars, and landslide deposits.

Concept of Eruptive Style: Will It Flow, or Will It Blow?

Kilauea, a volcano on Hawaii, produces rivers of lava that cascade down the volcano's flanks. Mt. St. Helens, a volcano near the Washington–Oregon border, exploded catastrophically in 1980 and blanketed the surrounding countryside with tephra. Clearly, volcanoes erupt in different ways, and as we've noted, successive eruptions from the same stratovolcano may differ markedly in character from one another. Geologists refer to the character of an eruption as **eruptive style**. Below, we describe several distinct eruptive styles and explore why the differences occur.

FIGURE 5.9 The formation of volcanic calderas.

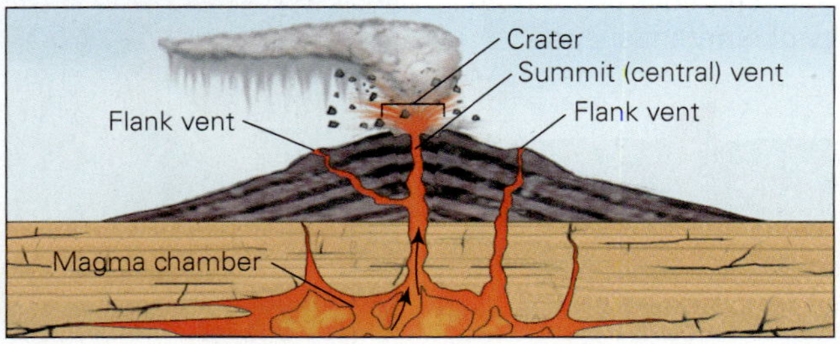

(a) As an eruption begins, the magma chamber inflates with magma. There can be a central vent and one or more flank vents.

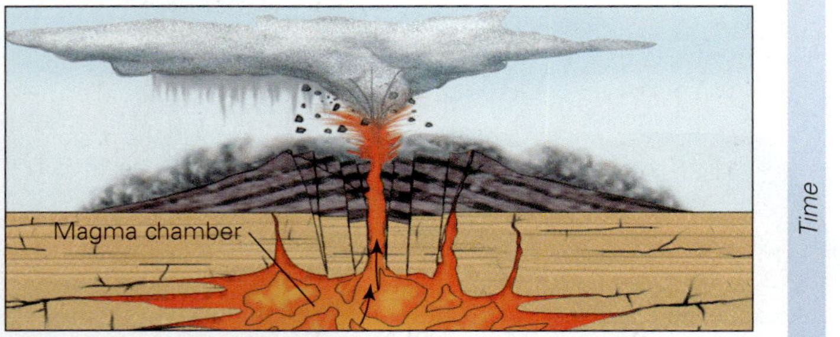

(b) During an eruption, the magma chamber drains, and the central portion of the volcano collapses downward.

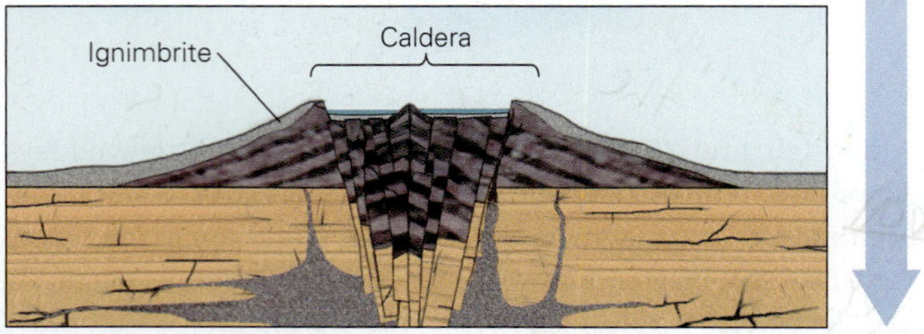

(c) The collapsed area becomes a caldera. Later, a new volcano may begin to grow within the caldera.

Effusive Eruptions The term *effusive* comes from the Latin word for pour out, and indeed that's what happens during an **effusive eruption**—lava pours out from a vent or fissure, filling a lava lake around the crater and/or flowing in molten rivers for great distances (**Fig. 5.11a**). Effusive eruptions occur where the magma feeding the volcano is hot and mafic and therefore has low viscosity.

Explosive Eruptions When pressure builds in a volcano, it can result in an **explosive eruption**. Small explosions take place during basaltic eruptions, when gas builds up and suddenly escapes, spattering lava drops and blobs upward (see Fig. 5.5a). Explosions of basaltic volcanoes erupting in the sea may occur when seawater gains access to the magma chamber via cracks in the edifice, for when water comes in contact with magma, it flashes to steam. The steam instantly expands, which produces enough pressure to blast volcanic debris upward, sometimes even to blast the volcano apart (**Fig. 5.11b**).

Even larger explosions happen in felsic or andesitic volcanoes, where the rising magma tends to be very viscous and rich in gas. We've noted that gas forms bubbles as lava rises. During the final stages of its rise, as magma slowly moves up into the volcano itself, the magma may partly solidify into glass and trap the bubbles. When overlying material erupts, the pressure pushing down on a magma decreases. As external pressure decreases, because bubbles can't expand, their internal gas pressure increases. If cracks occur in the glass,

(d) This caldera in Oregon formed about 7,700 years ago. Afterward, it filled with water to become Crater Lake. Wizard Island, protruding from the lake, is a small volcano that grew on top of the caldera floor.

FIGURE 5.10 Different shapes of volcanoes.

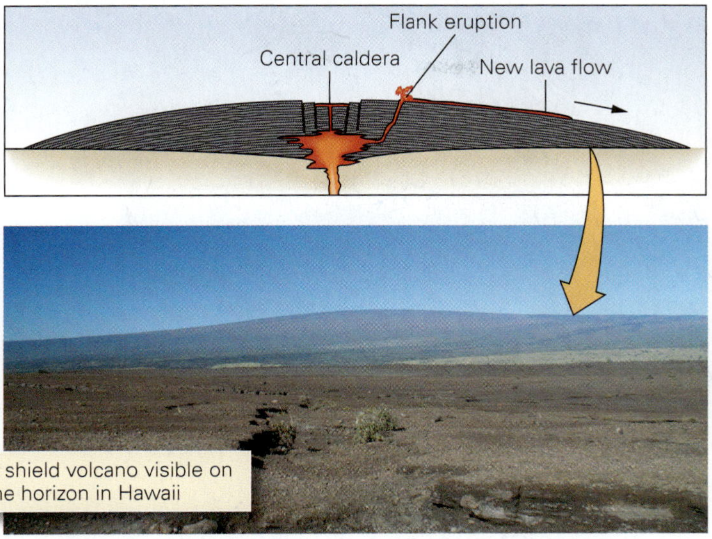

(a) A shield volcano, made from successive flows of low-viscosity basalt, has very gentle slopes.

(b) A cinder cone on the flank of a larger volcano in Arizona. The pile of cinders has assumed the angle of repose. A lava flow covers the land surface in the distance.

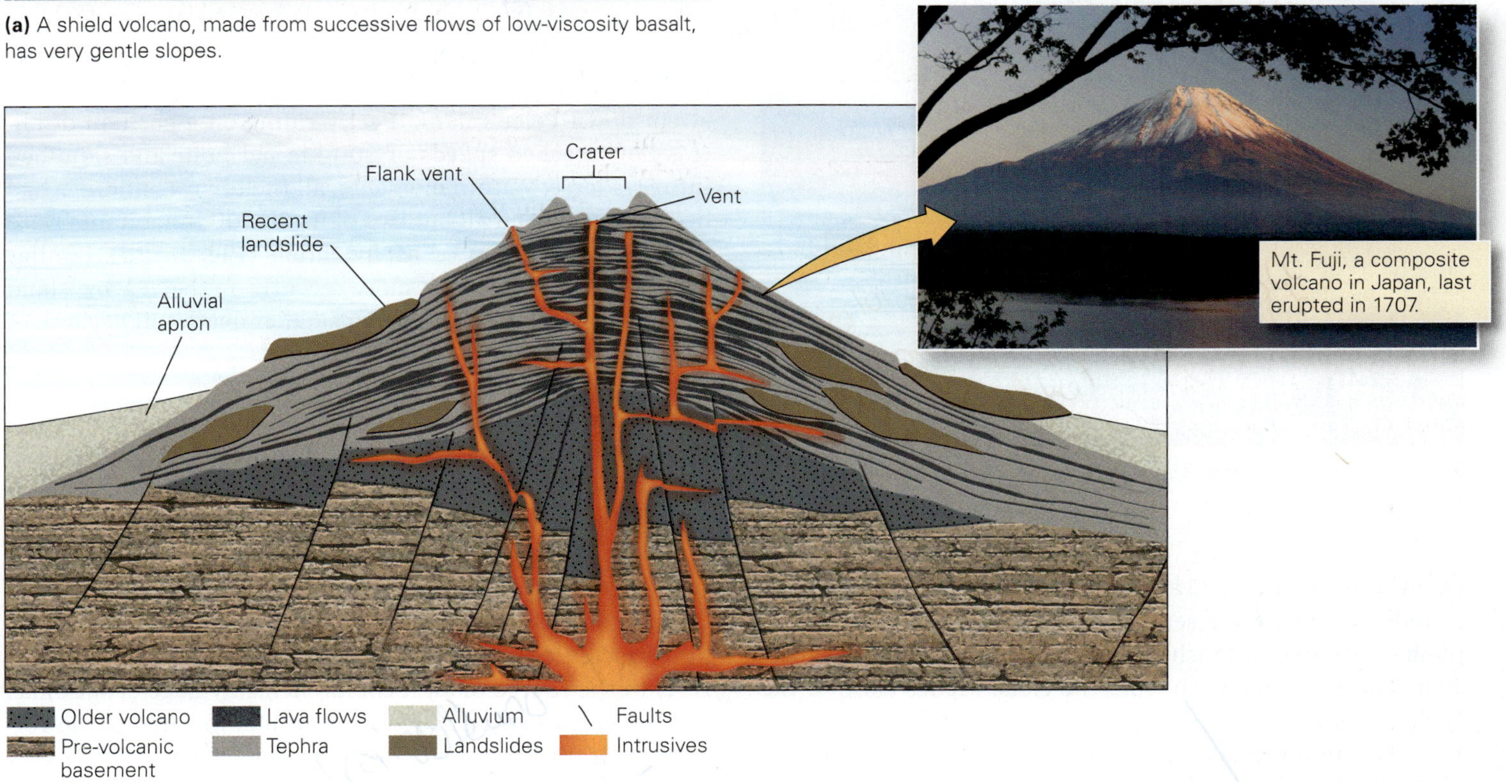

(c) A stratovolcano (composite volcano) consists of layers of tephra and lava. Volcanic debris flows and ash avalanches modify slopes and contribute to the development of a classic cone-like shape.

or screens break between bubbles, the gas escapes and suddenly has room to expand. The sudden expansion blasts out any remaining melt and pulverizes the glass that had surrounded the bubbles, as well as pre-existing rock and debris that had formed the volcanic edifice. The resulting pyroclastic debris shoots out of the volcano like a shotgun blast, or like champagne spouting from a newly opened bottle (**Fig. 5.11c**). Such an explosion, awesome in its power and catastrophic in its consequences, can eject cubic kilometers of debris and may destroy the volcano (**Box 5.1**).

Let's look more closely at the structure of the eruptive cloud. The debris from an explosive eruption (ash, lapilli, and blocks) shoots skyward in a vertical column (**Fig. 5.12a, b**). But eruptive force can push the material only so high. The huge plume of ash that rises to stratospheric heights above a large explosion does so by becoming a billowing *convective cloud* (**Fig. 5.12c**). This means that the warm mixture

FIGURE 5.11 Effusive eruptions compared to explosive eruptions.

(a) A 1986 effusive eruption on Hawaii.

(b) A volcano erupting just beneath the sea surface, near Tonga, sends a mixture of steam and ash skyward.

(c) The eruptive cloud formed during the 1991 eruption of Mt. Pinatubo in the Philippines.

of volcanic ash, gas, and air, which is less dense than the surrounding, cooler air, rises skyward buoyantly. The resulting plume resembles a mushroom cloud above a nuclear explosion. Coarse-grained ash from the cloud settles close to the volcano, whereas fine ash gets carried farther away, where high-elevation winds may transport it around the globe. Not all of the debris ejected by an explosive eruption ends up in the convective cloud; gravity pulls a substantial amount back down around the vent. This phenomenon, known as *column collapse*, produces pyroclastic flows that surge down a volcano's flanks (see Fig. 5.6e).

What is a pyroclastic flow like? Sadly, in 1902 the people of St. Pierre, a town on the Caribbean island of Martinique, found out. St. Pierre was a busy port town, about 7 km south of the peak of Mt. Pelée, a volcano. When the volcano began emitting steam and lapilli, residents of the town became nervous and debated whether to evacuate. Meanwhile, a rhyolite spire rose from the conduit of the volcano. On May 8, the spire suddenly cracked, releasing the immense pressure of gases in the partially solidified magma below. A pyroclastic flow swept down Pelée's flank. Partly riding on a cushion of air, this flow reached speeds of 300 km per hour and slammed into St. Pierre. Within moments, all the town's buildings had been flattened and all but two of its 28,000 inhabitants were dead of incineration or asphyxiation (**Fig. 5.12d**). Similar eruptions have happened more recently on the nearby island of Montserrat, but with a much smaller death toll because of timely evacuation.

Explosive Eruptions The largest explosive volcanic eruption ever observed during human history took place in 1815. This eruption, which blasted the top off Mt. Tambora, a volcano on an island in Indonesia, ejected about 145 cubic km of debris. The geologic record shows that much larger explosions have happened in prehistory (see Fig. Bx5.1a). In fact, during the past few million years, a number of eruptions have ejected over 1000 cubic km of debris. Geologists now refer to the source of an incomprehensibly huge eruption as a **supervolcano**. After such an eruption, the area of the volcano typically collapses to form a large caldera. The largest known supervolcano, which erupted about 73,000 B.C.E. in Indonesia, produced 2,800 cubic km of debris and left behind a caldera now filled with Lake Toba. Supervolcanic eruptions have happened at least twice in the area where Yellowstone Park now lies—in the aftermath of one of these explosions, a 72-km-diameter caldera remained.

Relation of Eruptive Style to Volcanic Type Note that the type of volcano (shield, cinder cone, or composite) depends on its eruptive style. Volcanoes with only effusive eruptions become shield volcanoes, those that generate small

FIGURE 5.12 The components of a large explosive volcanic eruption.

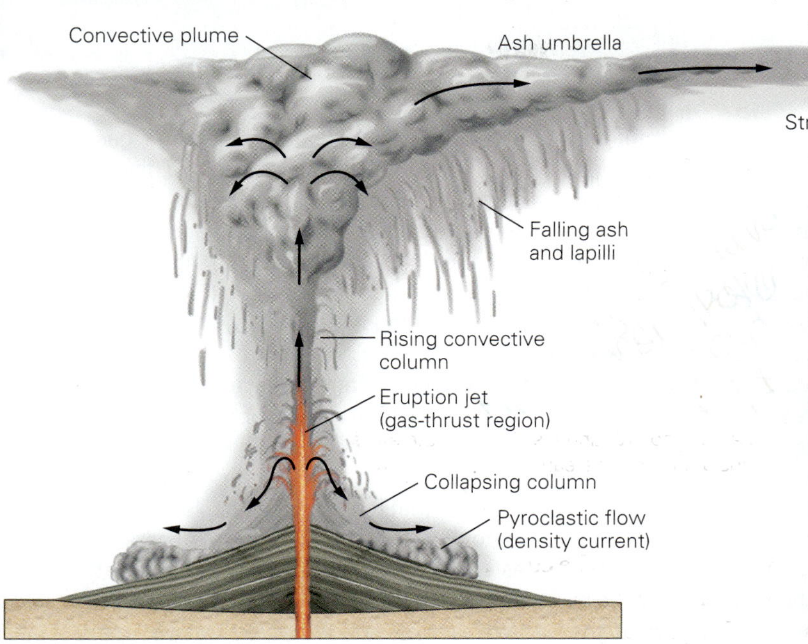

(a) A large explosive eruptive cloud (Plinean-type) contains several components.

(b) The eruptive jet of an eruption on Mt. Etna, Italy.

(c) The 1989 eruption of Redoubt Volcano, Alaska, produced a mushroom cloud that reached the stratosphere.

(d) The aftermath of the 1902 eruption of Mt. Pelée.

pyroclastic eruptions due to fountaining basaltic lava yield cinder cones, and those that alternate between effusive and large pyroclastic eruptions become stratovolcanoes (composite volcanoes). Large explosions yield calderas and blanket the surrounding countryside with ash and/or ignimbrites.

What might explain such contrasts in eruptive style? Eruptive style depends on the viscosity and gas content of the magma in the volcano. These characteristics, in turn, depend on the composition and temperature of the magma and on the environment (subaerial or submarine) where the eruption occurs. Traditionally, geologists classified volcanoes according to their eruptive style, each style named after a well-known example (**Geology at a Glance**, pp. 152–153).

TAKE-HOME MESSAGE

At a volcano, lava rises from a magma chamber and erupts from chimney-like conduits or from crack-like fissures. Low-viscosity basalt lava flows build shield volcanoes. Fountaining basalt spatters tephra to build cones. Successive eruptions of pyroclastic debris and lava build stratovolcanoes, and explosions produce calderas. A large explosive eruption can produce a convecting cloud of ash that rises to stratospheric heights, as well as devastating pyroclastic flows.

QUICK QUESTION Why do cinder cones have steeper slopes than shield volcanoes?

BOX 5.1 Consider This...

Volcanic Explosions to Remember

Explosions of volcanoes generate enduring images of destruction. The historical record shows a vast range in the volume of debris erupted, even though the largest eruption observed by people (Tambora in 1815) was small compared to a superexplosion that took place over 600,000 years ago in what is now Yellowstone National Park, Wyoming (**Fig. Bx5.1a**). Let's look at two notable examples of explosions.

Mt. St. Helens, a snow-crested stratovolcano in the Cascades of the northwestern United States, had not erupted since 1857. However, geologic evidence suggested that the mountain had a violent past, punctuated by many explosive eruptions. On March 20, 1980, an earthquake announced that the volcano was awakening once again. A week later, a crater 80 m in diameter burst open at the summit and began emitting gas and pyroclastic debris. Geologists who set up monitoring stations to observe the volcano noted that its north side was beginning to bulge markedly, suggesting that the volcano was filling with magma that was making it expand like a balloon. Their concern that an eruption was imminent led local authorities to evacuate people in the area.

The climactic eruption came suddenly. At 8:32 a.m. on May 18, geologist David Johnston, monitoring the volcano from a distance of 10 km, shouted over his two-way radio, "Vancouver, Vancouver, this is it!" An earthquake had triggered a huge landslide that caused 3 cubic km of the volcano's weakened north side to slide away. The sudden landslide released pressure on the magma in the volcano, causing a sudden and violent expansion of gases that blasted through the side of the volcano (**Fig. Bx5.1b**). Rock, steam, and ash screamed north at the speed of sound and flattened a forest and everything in it over an area of 600 square km (**Fig. Bx5.1c**). Tragically, Johnston, along with 60 others, vanished forever.

Seconds after the sideways blast, a vertical column carried about 540 million tons of ash (about 1 cubic km) 25 km into the sky, where the jet stream carried it away so that it was able to circle the globe. In towns near the volcano, a blizzard of ash choked roads and buried fields. Water-saturated ash formed viscous slurries, or lahars, that flooded river valleys, carrying away everything in their path. When the eruption was finally over, the once cone-shaped peak of Mt. St. Helens had

FIGURE Bx5.1 Examples of explosive eruptions.

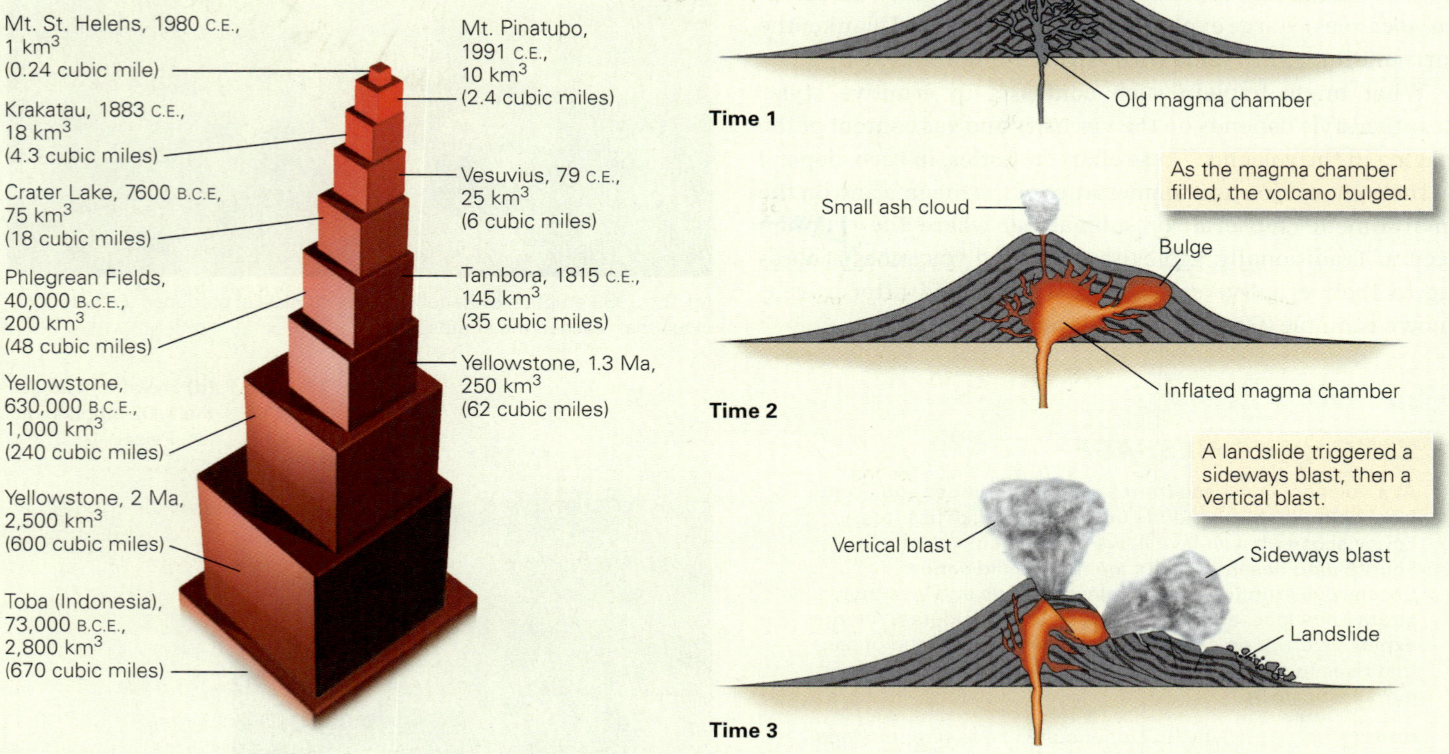

(a) The relative amounts of pyroclastic debris (in cubic km) ejected during major explosive eruptions.

(b) Stages during the eruption of Mt. St. Helens, 1980.

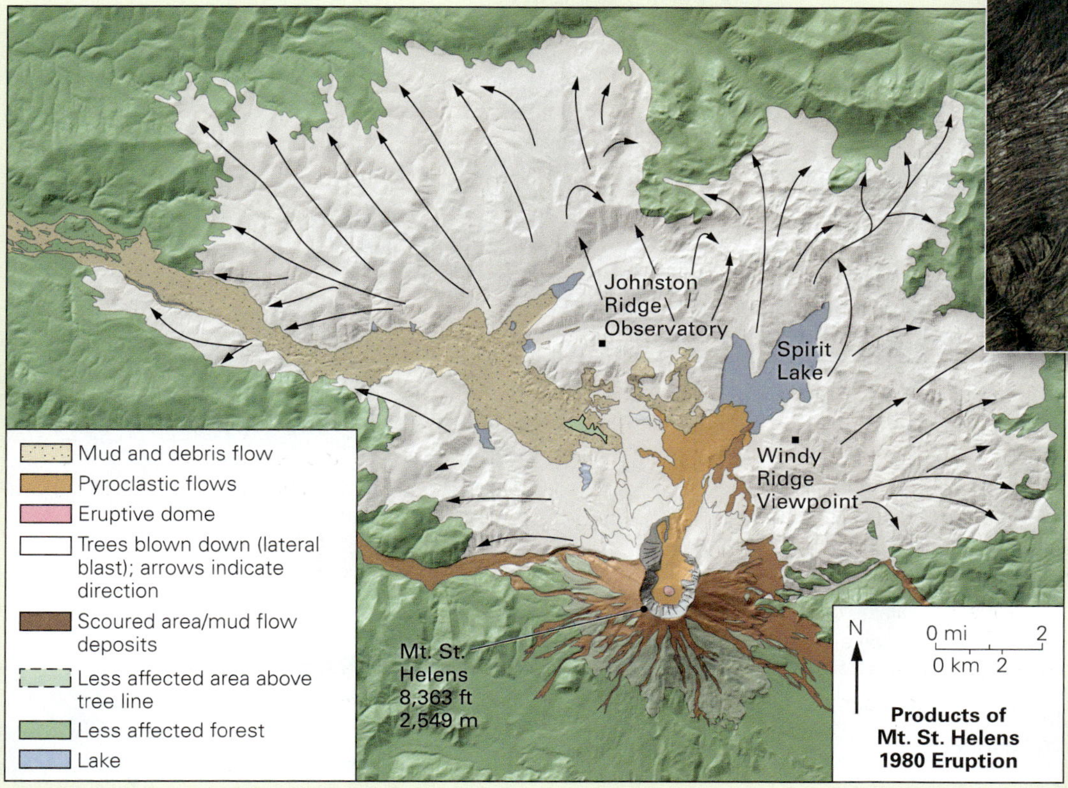

(c) A map shows the dimensions of the region destroyed by the eruption of Mt. St. Helens. The arrows indicate the blast direction. The neighboring forest was flattened by a blast of rock, steam, and ash.

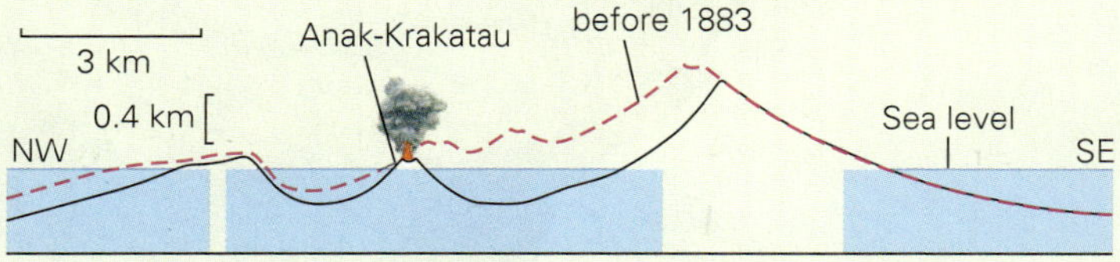

(d) Profile of Krakatau before and after the eruption. Note that a new volcano (Anak-Krakatau) has formed.

disappeared—the summit now lay 440 m lower, and the once snow-covered mountain was a gray mound with a large gouge in one side. The volcano came alive again in 2004, but did not explode.

In 1883, an even greater explosion happened in Krakatau (Krakatoa), a volcano between Indonesia and Sumatra that constituted a 9-km-long island rising 800 m (2,600 ft) above the sea. On May 20, the island began to erupt with a series of large explosions, yielding ash that settled as far as 500 km away. Smaller explosions continued through June and July, and steam and ash rose from the island, forming a huge black cloud that rained ash into the surrounding straits. Ships sailing by couldn't see where they were going, and their crews had to shovel ash off the decks.

Krakatau's demise came at 10 a.m. on August 27, perhaps when the volcano cracked and the magma chamber suddenly flooded with seawater. The resulting blast, 5,000 times greater than the Hiroshima atomic bomb explosion, could be heard as far as 4,800 km away, and subaudible sound waves traveled around the globe seven times. Giant sea waves pushed out by the explosion slammed into coastal towns, killing over 36,000 people. Near the volcano, a layer of ash up to 40 m thick accumulated. When the air finally cleared, Krakatau was gone, replaced by a submarine caldera some 300 m deep (**Fig. Bx5.1d**). All told, the eruption shot 20 cubic km of rock into the sky. Some ash reached elevations of 27 km. Because of this ash, people around the world could view spectacular sunsets during the next several years.

Volcanoes

Beneath a volcano, magma rises to fill a pervasively cracked region of crust and forms a magma chamber. Some of the magma erupts at a surface vent. Once molten rock has erupted at the surface, it is called lava. Some lava spills down the side of the volcano in lava flows. Some fountains out of a vent to form scoria fragments that pile up in a cone around the vent. Eruptions may eject larger chunks as blocks or bombs. The nature of eruptions depends on the viscosity of the lava, which

Cinder cone

Caldera

Vulcanian eruptions occur when a buildup of gas and magma explodes.

Strombolian crater explosions frequently burst through thinly crusted lava.

Hawaiian fountain explosions are caused by escaping gas.

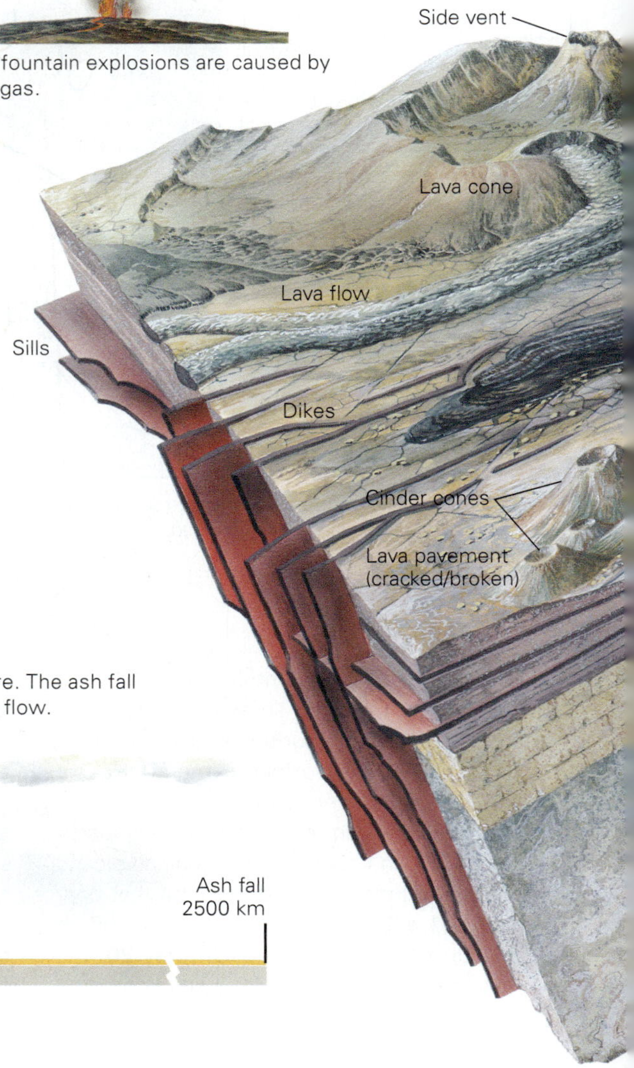

Plinean explosions shoot a huge column of pumice fragments up to 50 km into the atmosphere. The ash fall rains down and the column collapses back around the vent, traveling overland as a pyroclastic flow.

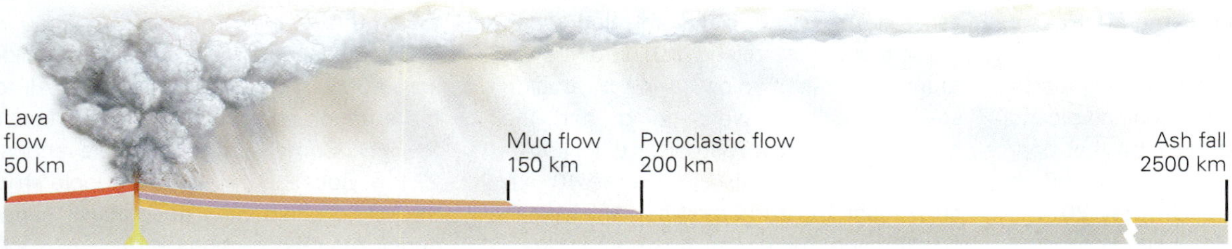

The distance volcanic hazards can travel from an eruption.

in turn depends on lava composition. Volcanic explosions blast up a cloud of ash and lapilli, and may pulverize pre-existing volcanic rock and send it skyward too. Some of the debris from an explosive eruption rises in a turbulent, convective cloud to stratospheric heights. Denser portions of the eruption column collapse and surge down the flanks of the volcano in hot, dangerous pyroclastic flows.

Because of the great variation in the style of eruptions, volcanoes come in many different forms. Cinder cones build from lapilli, shield volcanoes have a gentle dome-like profile built by many layers of basaltic lava, and stratovolcanoes consist of lava flows interlayered with tephra layers. When the ash on the surface of a volcano mixes with water, it forms a slurry that can flow down the sides of the volcano as a lahar. Catastrophic explosions of volcanoes and consequent draining of magma chambers cause huge calderas to form. Traditionally, the style of an eruption has been named based on its similarity to well-known examples.

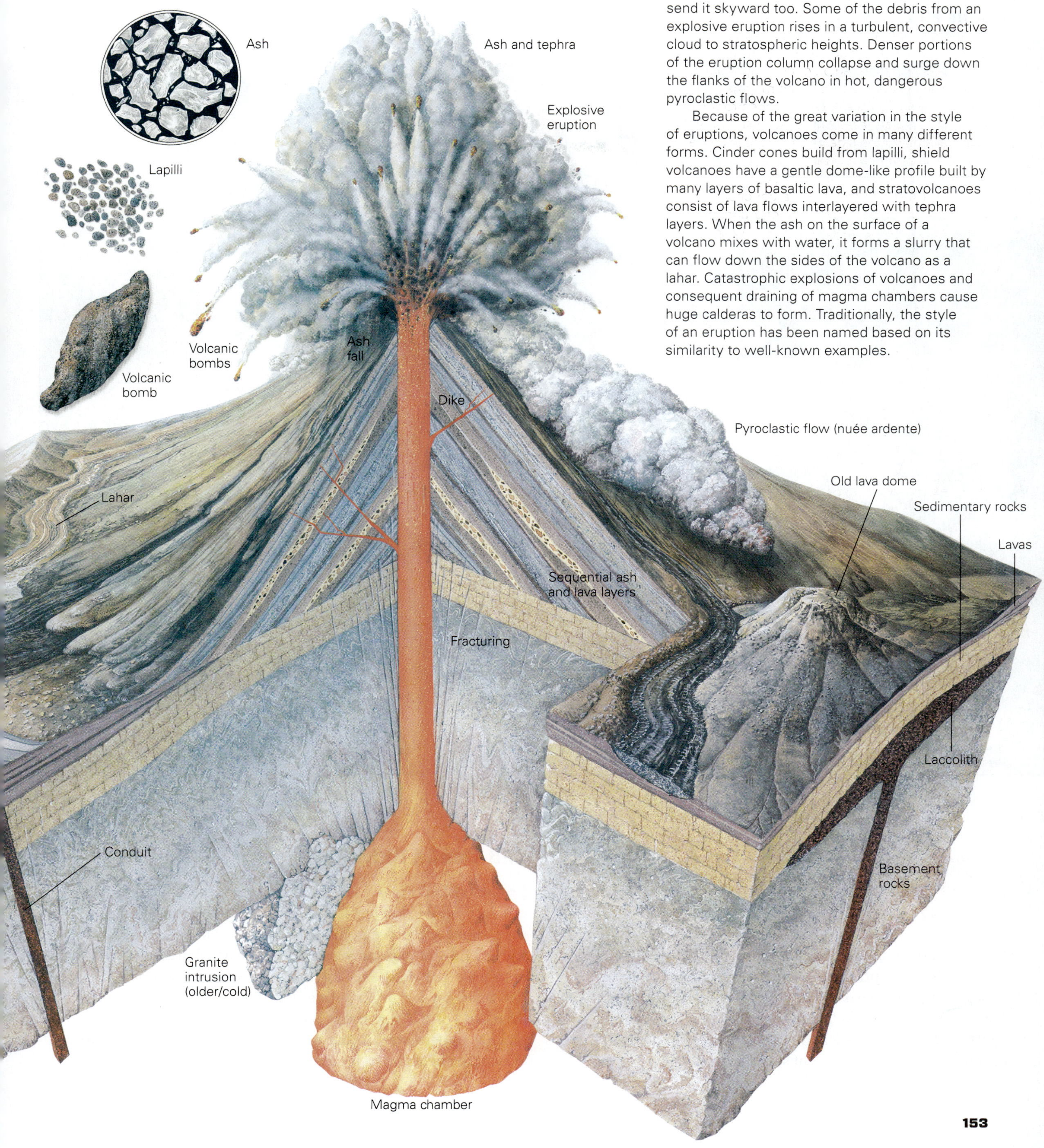

FIGURE 5.13 Volcanoes of the world. The map shows the distribution of volcanoes around the world, and the diagrams illustrate basic geologic settings in which volcanoes form, in the context of plate tectonics theory.

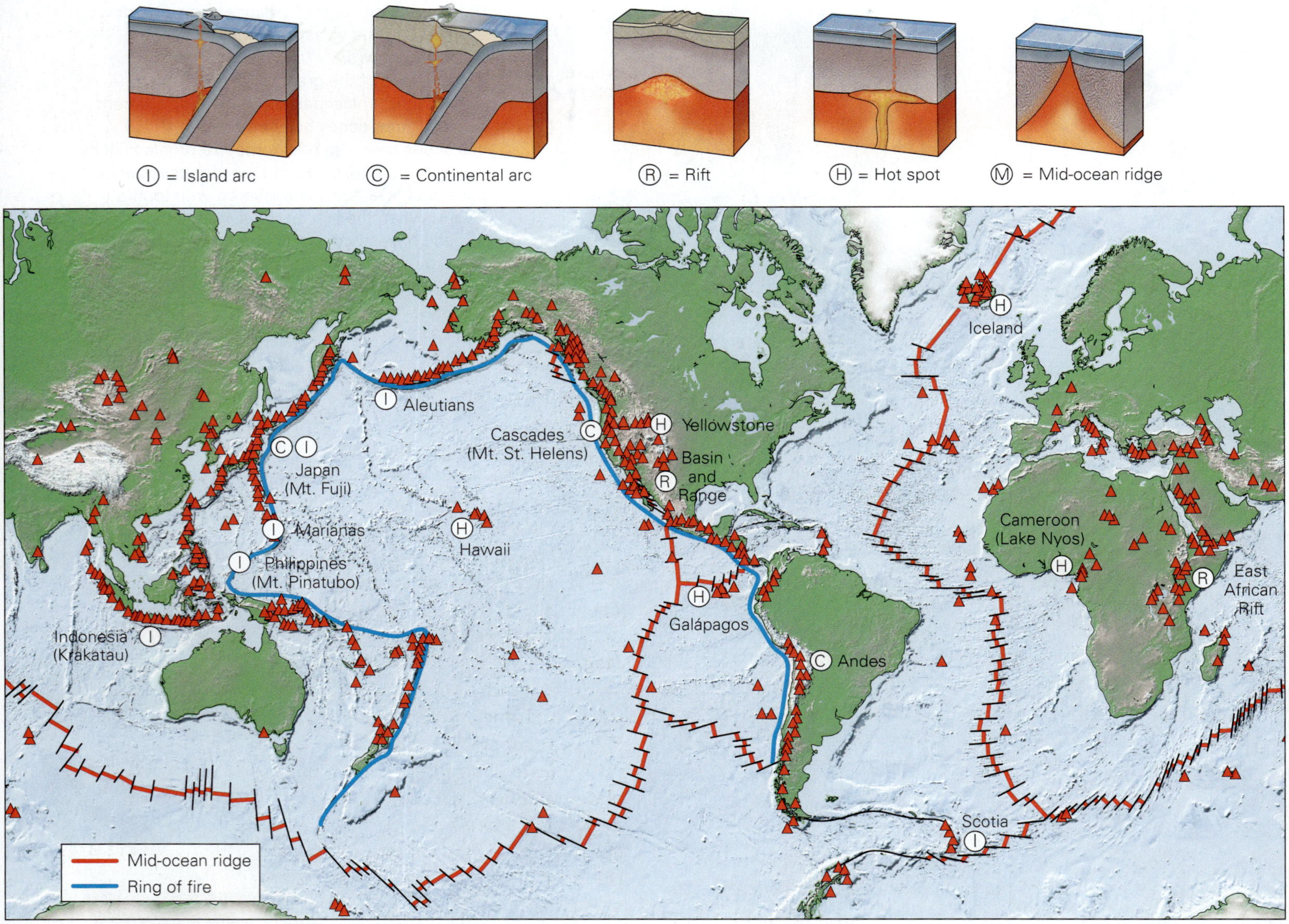

5.4 Geologic Settings of Volcanism

Different styles of volcanism occur at different locations on Earth. Most eruptions occur along plate boundaries, but major eruptions also occur at hot spots and in rifts (**Fig. 5.13**). We'll now look at the settings in which eruptions occur, in the context of plate tectonics theory, and see why different kinds of volcanoes form in different settings.

Mid-Ocean Ridges

Products of mid-ocean ridge volcanism cover 70% of our planet's surface. We don't generally see this volcanic activity, however, because the ocean hides most of it beneath a blanket of water. Mid-ocean ridge volcanoes, which develop along fissures parallel to the ridge axis, are not all continuously active. Each one turns on and off in a time scale measured in tens to hundreds of years. They erupt basalt which, because it cools so quickly underwater, forms pillow-lava mounds (**Fig. 5.14a**). Water that heats up as it circulates through the crust near the

FIGURE 5.14 Igneous activity at plate boundaries.

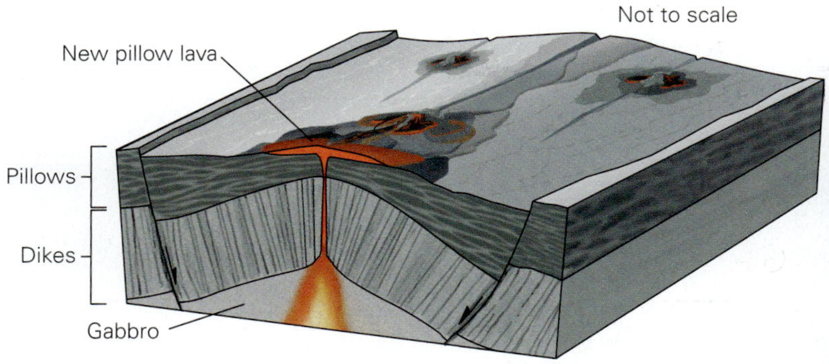

(a) Along a mid-ocean ridge axis at a divergent boundary, small mounds of pillow basalt erupt.

(b) The Aleutian volcanic arc, of Alaska, viewed looking northeast. The white peaks are volcanoes. This chain formed at a convergent plate boundary.

magma chamber bursts out of hydrothermal (hot-water) vents, known as black smokers, along these mounds.

Convergent Boundaries

Most subaerial volcanoes on Earth lie on the edge of an overriding plate along a convergent plate boundary (subduction zone). Convergent boundaries border over 60% of the Pacific Ocean, so a 20,000-km-long chain of volcanoes, known as the *ring of fire*, has developed along the rim of the ocean. Typically, individual volcanoes in volcanic arcs lie about 50 to 100 km apart (**Fig. 5.14b**). Some of these volcanoes grow on oceanic crust and become volcanic island arcs, such as the Marianas of the western Pacific. Others grow on continental crust, building continental volcanic arcs such as the Cascade volcanic chain of Washington and Oregon or the Andes chain of South America. Because of the variety of magma types produced beneath continental volcanic arcs, such arcs can yield large stratovolcanoes.

Continental Rifts

Due to the diversity of magmas that can form beneath rifts, rifts can host both basaltic fissure eruptions, in which curtains of lava fountain up or linear chains of cinder cones develop, and explosive rhyolitic volcanoes. In some places, they even host stratovolcanoes, such as Mt. Kilimanjaro in Africa.

Oceanic Hot-Spot Volcanoes

When a hot-spot volcano forms on oceanic lithosphere, basaltic magma erupts at the surface of the seafloor. First, such submarine eruptions yield an irregular mound of pillow lava, eventually growing above the sea surface. After the volcano has emerged from the sea, effusive eruptions of basalt yield a broad shield shape with gentle slopes (**Fig. 5.15a**). As the volcano grows, portions of it can't resist the pull of gravity and slip seaward, creating large submarine slumps.

Hawaii and other volcanic islands within the Pacific Ocean basin serve as examples of oceanic hot-spot volcanoes. Because the Pacific Plate moves relative to their sources, the active hot-spot volcanoes lie at the ends of chains of now-dead volcanoes and seamounts, defining the hot-spot track (see Figure 2.27). Iceland is one of a few places on Earth where a hot spot lies at the crest of a mid-ocean ridge. Because of the hot spot, vastly more magma erupts at Iceland than at other places along the Mid-Atlantic Ridge, and igneous activity has built a broad submarine plateau (**Fig. 5.15b**). Iceland straddles a divergent plate boundary, which means that it is being stretched apart, and it has been cut up by faulting. Indeed, the central part of the island is a narrow rift, marking the trace of the Mid-Atlantic Ridge axis, where the youngest volcanic rocks of the island were extruded (**Fig. 5.15c**). When they follow cracks parallel to the rift axis, eruptions on Iceland are fissure eruptions, which in some places yield curtains of lava that are many kilometers long and/or linear chains of small cinder cones. Not all volcanic activity on Iceland occurs subaerially—some eruptions take place under glaciers and melt large amounts of ice. When the meltwater bursts through the edge of the glacier it becomes a devastating flood, called a *jokulhlaup* in Icelandic.

SEE FOR YOURSELF...

MAUNA LOA, HAWAII

Latitude
19°27′0.50″ N

Longitude
155°36′2.94″ W

Look straight down from 25 km (~15.5 miles).

You can see the NNE-trending crest of Mauna Loa, a shield volcano associated with the Hawaiian hot spot. A large, elliptical caldera dominates the view. Basaltic lava flows have spilled down its flank.

FIGURE 5.15 Oceanic hot-spot volcanoes.

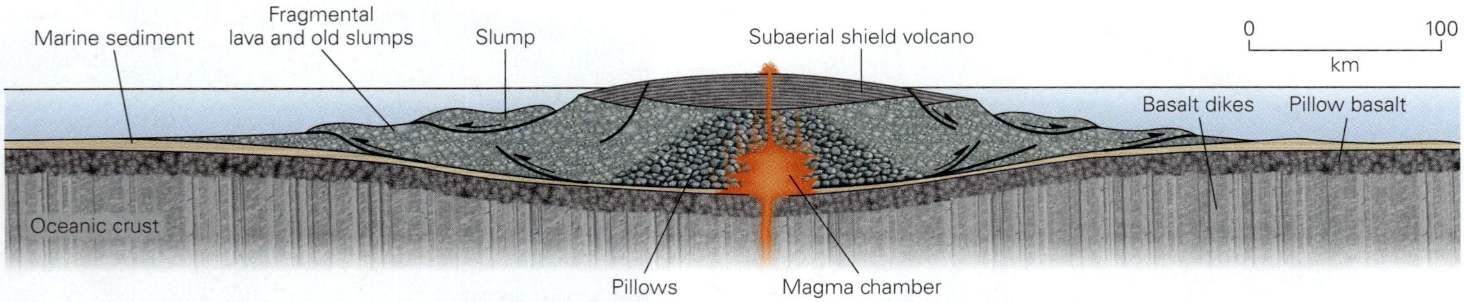

(a) The interior of an oceanic hot-spot volcano is complicated. Initially, eruption produces pillow basalts. When the volcano emerges above sea level, it becomes a shield volcano. The margins of the island frequently undergo slumping, and the weight of the volcano pushes down the surface of the lithosphere. The Hawaiian Islands exemplify this architecture.

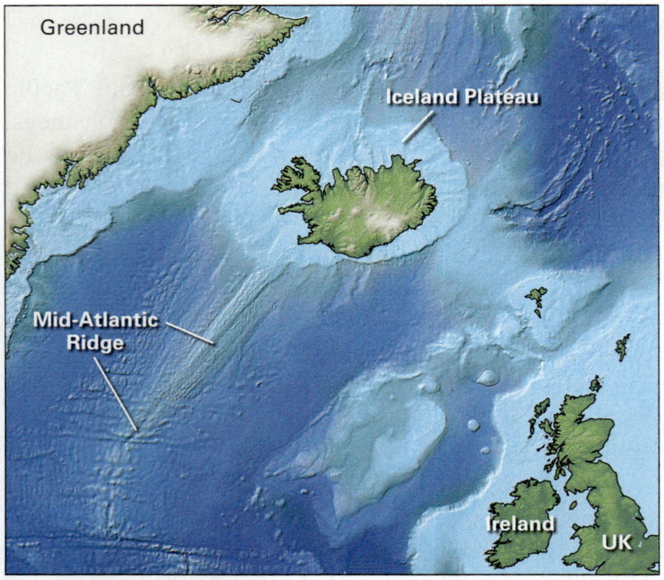

(b) A bathymetric map shows that Iceland sits atop a huge plateau straddling the Mid-Atlantic Ridge. Light blue is shallower water; dark blue is deeper.

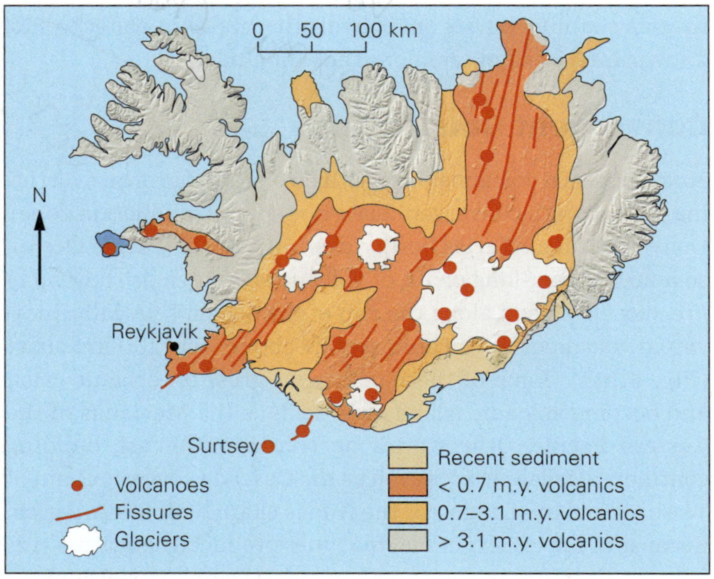

(c) A geologic map of Iceland shows how the youngest volcanoes occur in the central rift, effectively the on-land portion of the Mid-Atlantic Ridge.

Continental Hot-Spot Volcanoes

Yellowstone National Park lies at the northeast end of a string of calderas, the oldest of which erupted 16 million years ago (**Fig. 5.16a,b**). Recent and ongoing activity beneath the Earth's surface has yielded fascinating landforms, volcanic rock deposits, and geysers. Eruptions at the Yellowstone hot spot differ from those in Hawaii in an important way—unlike those of Hawaii, some eruptions at Yellowstone yield basaltic lava, while others, including supervolcanic eruptions, yield rhyolitic pyroclastic debris. The most recent of these, 630,000 years ago, produced an immense caldera that filled with pyroclastic debris and sent up convective clouds of ash. Some ash fell as far east as the Mississippi River (**Fig. 5.16c**). Tuffs from this eruption covered 2,500 square km and within the Yellowstone area, whose name reflects the brilliant yellow-gold color of the debris, reached a thickness of 400 m (**Fig. 5.16d**). Hot magma remains beneath the region today—heat from this magma drives the park's famous geysers, which we will discuss in Chapter 16.

Flood-Basalt Volcanism

In several locations around the world, huge amounts of low-viscosity mafic lava erupted out of fissures and spread out in vast sheets called **flood basalt** (**Fig. 5.17**). Over time, many successive eruptions of flood basalt can build up a broad plateau. The aggregate volume of rock produced by these eruptions can be so great (over 175,000 cubic km) that geologists consider them to be **large igneous provinces (LIPs)**,

FIGURE 5.16 Hot-spot volcanic activity in Yellowstone National Park.

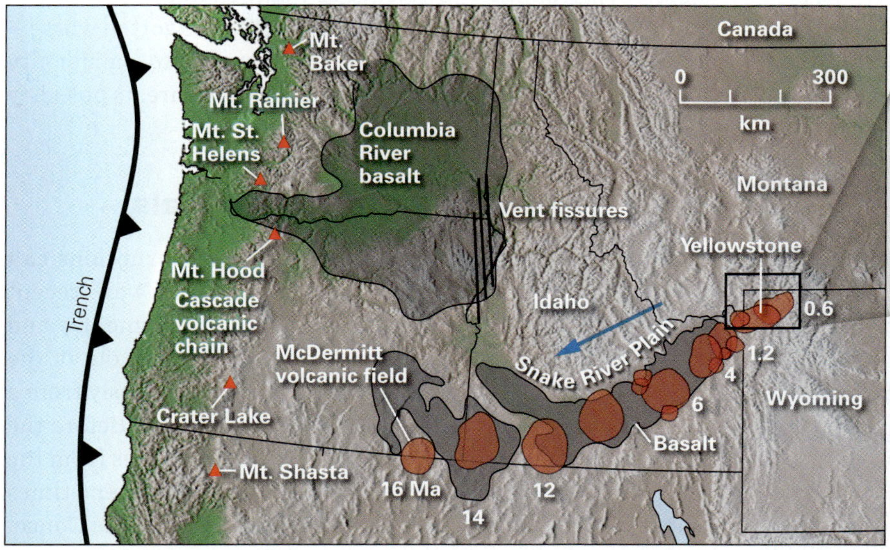

(a) Yellowstone lies at the end of a continental hot-spot track. Progressively older calderas follow the Snake River Plain. The blue arrow indicates plate motion.

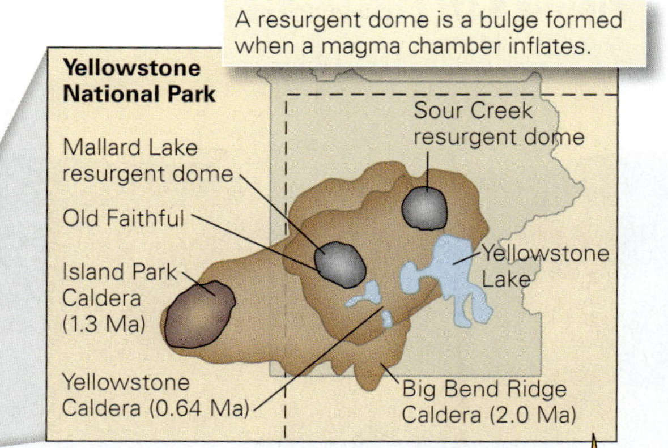

(b) Yellowstone overlies a caldera that exploded twice during the past 2 Ma.

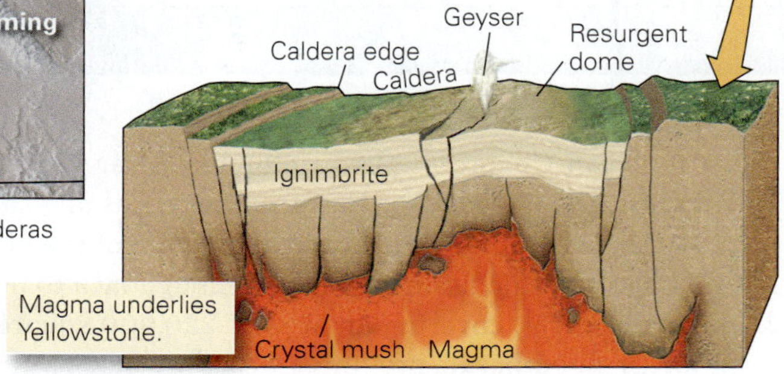

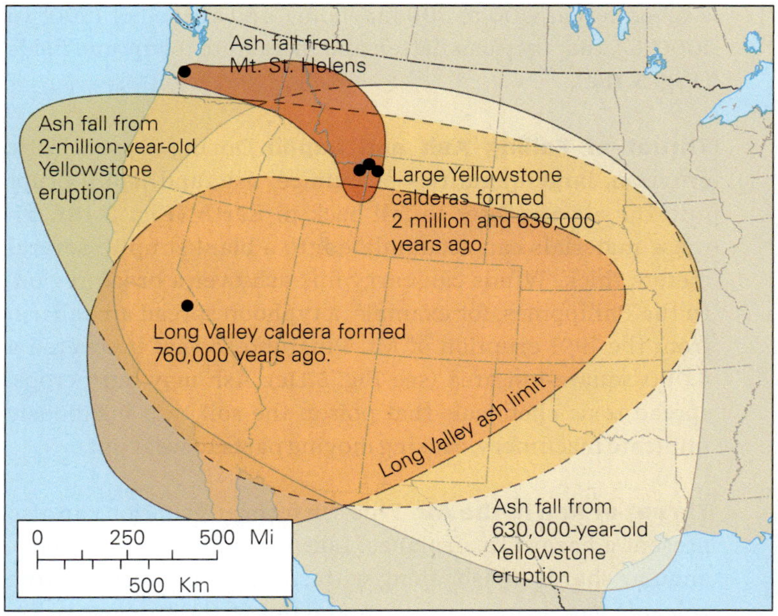

(c) The ash produced by explosions of the Yellowstone calderas covered vast areas—much more than Mt. St. Helens.

(d) Felsic tuffs form the colorful walls of Yellowstone Canyon.

FIGURE 5.17 Flood-basalt layers exposed on the wall of a canyon in Idaho.

as we noted in Chapter 4. The Columbia River Plateau, of Washington and Oregon (see Fig. 4.19), serves as an example of a LIP. About 15 million years ago, the region of the plateau hosted huge volcanic fissures yielding enough lava to build a 3.5 km-thick layer, covering an area of 220,000 square km. Even larger flood-basalt provinces occur in eastern Siberia (an occurrence known as the Siberian Traps), the Deccan Plateau of India, the Paraná Plateau of Brazil, and the Karoo Plateau of southern Africa.

TAKE-HOME MESSAGE

Most volcanic activity takes place on plate boundaries, but some occurs at hot spots. The style of eruption depends on the setting. The sea hides divergent-boundary volcanoes, which erupt pillow basalt. Volcanic arcs rise above sea level and may produce stratovolcanoes. Oceanic hot spots produce shield volcanoes. Continental hot spots and rifts produce both effusive and explosive eruptions.

QUICK QUESTION How does a volcano that formed in a continental island arc differ from one formed at an oceanic hot spot?

5.5 Beware: Volcanoes Are Hazards!

Like earthquakes, volcanoes are natural hazards with the potential to cause great destruction to humanity, in both the short term and the long term. According to one estimate, volcanic eruptions in the last 2,000 years have caused about a quarter of a million deaths—much fewer than those caused by earthquakes, but nevertheless a sizable number. Considering the rapid expansion of cities, far more people live in dangerous proximity to volcanoes today than ever before, so if anything, the hazard posed by volcanoes has gotten worse—imagine if a large explosion were to occur next to a major city today! Let's now look at the different kinds of threats posed by volcanic eruptions.

Hazards Due to Eruptive Materials

Threat of Flows Basaltic lava from effusive eruptions can spread over a broad area (**Fig. 5.18a–d**). In Hawaii, recent lava flows have buried roads, housing developments, and vehicles. Although people usually have time to get out of the way of such flows, they might have to watch helplessly from a distance as an advancing flow engulfs their home. Before the lava even touches it, a building will burst into flames from the intense heat. The most disastrous lava flow in recent times came from the 2002 eruption of Mt. Nyiragongo in the Congo. It destroyed almost half the city of Goma, encasing streets under a 2-m-thick layer of basalt. Pyroclastic flows represent a very dangerous hazard to humans, for they move faster than a speeding car (100 to 300 km/hour), and are so hot (500° to 1000°C), that they can flatten and incinerate towns and fields (**Fig. 5.18e**).

Threat of Falling Ash and Lapilli During a pyroclastic eruption, large quantities of pumice, ash, and lapilli erupt into the air and later to fall back to Earth (**Fig. 5.18f, g**); these materials can accumulate into a blanket up to several meters thick. Winds can carry fine ash over a broad region. In the Philippines, for example, a typhoon spread air-fall ash from the 1991 eruption of Mt. Pinatubo so that it covered a 4,000-square-km area (see Fig. 5.11c). Ash may bury crops, spread toxic chemicals that poison the soil, and insidiously infiltrate machinery, causing moving parts to wear out.

Threat of Ash in the Air Fine ash from an eruption can also present a hazard to airplanes. Like a sandblaster, the sharp, angular shards of ash, along with sulfuric acid formed from volcanic gas, score windows and damage the fuselage. Also, when ingested and heated inside a jet engine, the ash melts, creating a silicate liquid that freezes to glass, coating the interior of the engine. When this happens, temperature sensors indicate that the engine is overheating, so the engine automatically shuts down and the plane loses power.

Encounters between airliners and volcanic plumes have led to terrifying incidents. In 1982, a British Airways 747 flying at 11.5 km (37,000 ft) entered the ash cloud above a volcano in Java. The plane's windshield turned opaque, all four of its engines shut down, and its instruments failed. For 13 minutes,

FIGURE 5.18 Hazards due to lava and ash from volcanic eruptions.

Lava Flows

(a) A lava flow reaches a house in Hawaii and sets it on fire.

(b) Lava from Mt. Etna threatens a town and olive grove in Sicily.

(c) Residents rescue household goods after a lava flow filled the streets of Goma, along the East African Rift.

(d) This empty school bus was engulfed by lava in Hawaii.

Pyroclastic Debris

(e) A pyroclastic flow from the 1991 eruption of Mt. Pinatubo chases a fleeing vehicle.

(f) A blizzard of ash fell from the cloud erupted by Mt. Pinatubo in the Philippines.

(g) Lapilli falls from an eruption in Iceland.

(h) A lahar submerges farmland in Colombia.

the plane silently glided earthward, as the pilots frantically tried to restart the engines, even as they prepared to ditch at sea. Finally, at 3.7 km (12,000 ft), the engines apparently had cooled sufficiently that they roared back to life and the plane headed to Jakarta. By squinting out an open side window, the pilot found the runway and landed safely. A similar event happened to a KLM 747 that encountered the eruptive cloud of Redoubt Volcano in Alaska in 1989 (see Fig. 5.12c).

Because of the lessons learned from such incidents, the 2010 eruption of the Eyjafjallajökull volcano in Iceland had a profound impact on air traffic. All told, the eruption sent only about 0.25 cubic km of pyroclastic material up into the air. But, at the time of the eruption, the jet stream (a high-altitude current of rapidly moving air) was passing over Iceland and dispersed the volcanic ash throughout European air space. Due to concern about possible damage to planes, officials shut down almost all air traffic across Europe for six days. The closure directly cost airlines $200 million/day, disrupted travel plans for countless passengers, and halted air shipment of everything from television sets to flowers, thus impacting economies worldwide.

Other Hazards Related to Eruptions

Threat of the Blast The explosion of a volcano, like the blast of a bomb, flattens everything in its path. After the explosion of Mt. St. Helens, for example, the once-towering trees of the forest around the volcano were stripped of bark and needles and lay scattered over the hillslopes like matchsticks (see Box 5.1). Were such a blast to strike a city, the consequences would be catastrophic.

Threat of Landslides and Lahars Eruptions commonly trigger large landslides along a volcano's flanks. The debris, composed of ash and solidified lava that erupted earlier, can move both fast (as much as 250 km/hour) and far. During the eruption of Mt. St. Helens, 8 billion tons of debris took off down the mountainside, careered over a 360-m-high ridge, and tumbled down a river valley, until the last of it finally came to rest more than 20 km from the volcano.

When volcanic ash and other debris mix with water, the result is a lahar that resembles wet concrete. Because lahars are denser and more viscous than clear water, they pack more force than clear water and can carry away everything in their path. The lahars of Mt. St. Helens, for example, traveled along existing drainages more than 40 km from the volcano, at speeds of 5 to 50 km per hour. When they had passed, they left a gray and barren wake of mud, boulders, broken bridges, and crumpled houses, as if a giant knife had scraped across the landscape. Perhaps the most destructive lahar of recent times accompanied the eruption of the snow-crested Nevado del Ruiz in Colombia on the night of November 13, 1985. The lahar surged down a valley like a 40-m-high wave, hitting the sleeping town of Armero, 60 km from the volcano. Ninety percent of the buildings in the town vanished, replaced by a 5-m-thick layer of mud, which now entombs the bodies of 25,000 people (**Fig. 5.18h**).

Threat of Earthquakes and Tsunamis Earthquakes accompany almost all major volcanic eruptions, for the movement of magma breaks rocks underground. Such earthquakes may trigger landslides on the volcano's flanks and can cause buildings to collapse and dams to rupture, even before the eruption itself begins.

Where explosive eruptions occur in the sea or along a coast, the blast and the underwater collapse of a caldera, or the sudden slip of giant landslides on the flank of the volcano, triggered by the eruption, can generate huge sea waves, or tsunamis, tens of meters high. Most of the 36,000 deaths attributed to the 1883 eruption of Krakatau were not due to ash or lava, but rather to tsunamis that slammed into nearby coastal towns.

Threat of Gas We have already seen that volcanoes erupt not only solid material but also large quantities of gases. The hydrogen sulfide in volcanic gas has a rotten-egg smell, and along with acidic aerosols and other gases, produces a yellowish, choking, caustic cloud that is dangerous to breathe. Occasionally, tourists visiting Japan's Mt. Aso, a volcano known for the clouds of volcanic gas rising from its summit, have inhaled too much of the gas and have fainted; over the years, a few have died.

Carbon dioxide, unlike the sulfurous gases and aerosols that we've just described, although invisible, can be a deadly threat, as illustrated by the catastrophe that happened near Lake Nyos in 1986. Lake Nyos is a small, deep lake that fills the crater of a volcano in Cameroon, western Africa. Though only 1 km across, the lake reaches a depth of over 200 m. Carbon dioxide gas slowly bubbles out of cracks in the floor of the crater and dissolves in the stable layer of cool water at the bottom of the lake. Apparently, by August 21, 1986, the denser, cool bottom-water layer had become supersaturated in carbon dioxide. On that day, perhaps because of a landslide, the layer was disturbed. Cold water was forced upward where it warmed and underwent decompression. As a result, the gas came out of solution, violently bubbled upward, and rose from the lake. Since it's denser than air, the invisible CO_2 flowed down the flank of the volcano and spread out over the countryside for a distance of about 23 km before dispersing. Although it is not toxic, carbon dioxide does not provide the

FIGURE 5.19 The CO_2 disaster at Lake Nyos.

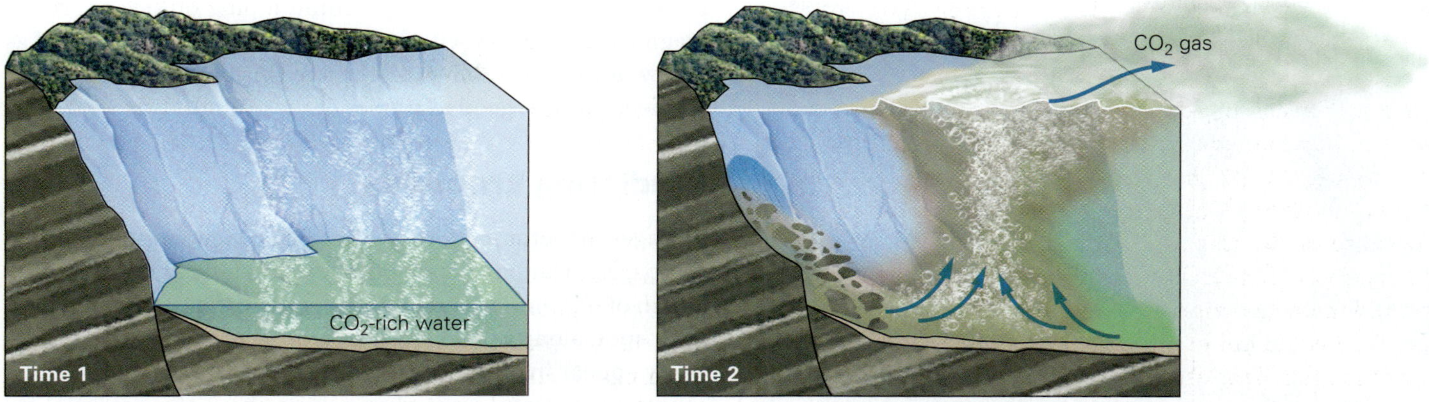

(a) CO_2 dissolved in the colder bottom water. When a landslide or wind disturbed the water, it rose and the CO_2 came out of solution and, in gas form, flowed out of the crater.

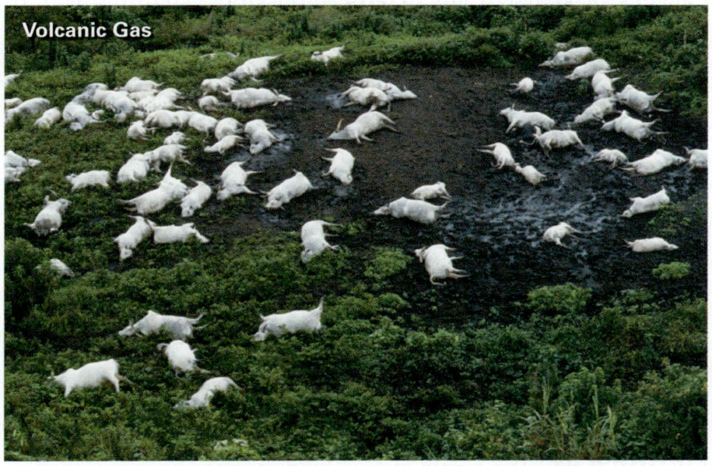

(b) Cattle died by suffocation when the CO_2 cloud engulfed them.

oxygen that is essential for metabolism and oxidation. Sadly, when the gas cloud engulfed the village of Nyos, it quietly put out cooking fires and suffocated 1,742 sleeping people, as well as thousands of cattle (**Fig. 5.19**).

TAKE-HOME MESSAGE

Volcanoes can be dangerous! The lava flows, ash clouds, pyroclastic flows, explosions, mudflows (lahars), landslides, earthquakes, and tsunamis that can be produced during eruptions can destroy towns and farmland. Even the gas emitted by a volcano can lead to catastrophe.

QUICK QUESTION Why can a lahar do so much more damage than a similarly sized flood of clear water?

5.6 Protection From Vulcan's Wrath

Volcanic eruptions are a natural hazard of extreme danger. Can anything be done to protect lives and property from this danger? The answer is yes. Below, we first examine the evidence that geologists can use to determine whether a volcano has the potential to erupt, and we then consider the suite of observations that may allow geologists to predict the timing of an impending eruption.

Active, Dormant, and Extinct Volcanoes

Geologists refer to volcanoes that are erupting, have erupted recently, or are likely to erupt soon, as **active volcanoes**, and consider vents that have not erupted for hundreds to thousands of years but might possibly erupt again in the future to be **dormant volcanoes**. Volcanoes that have shut off entirely and will never erupt again are called **extinct volcanoes**. For example, Hawaii's Kilauea can be considered to be an active volcano, for it is erupting currently and has erupted numerous times during recorded history. In contrast, Mt. Fuji in Japan last erupted in 1708 but has not had activity for over 300 years. Subduction, however, continues along the eastern edge of Japan, so Mt. Fuji could erupt in the future, and geologists consider it to be dormant. The igneous rock mound, known as Devils Tower, in eastern Wyoming, formed when magma intruded beneath the region about 40 million years ago (**Fig. 5.20**). No volcanism can happen near Devils Tower now, for the geologic conditions that caused volcanism there have not existed for almost 40 million years. We can say, therefore, that

FIGURE 5.20 Devils Tower, Wyoming, is not an eroded volcano but probably an intrusion into sedimentary rocks beneath a volcano. It has been exposed by erosion. Volcanism at this location is extinct.

The vertical lines are columnar joints.

any volcanoes that once erupted in the Devils Tower area are now extinct. In fact, they've been extinct so long that they've completely eroded away.

How do we determine if a volcano is active, dormant, or extinct? One way is to examine the historical record. Another is to determine the age of erupted rocks, and to search for evidence that the volcano still lies within a tectonically active area. Finally, we can examine the landscape character of the volcano. Specifically, the shape (shield, stratovolcano, or cinder cone) of an active volcano depends primarily on the eruptive style, since at an erupting volcano, the process of construction happens faster than the process of erosion. Once a volcano stops erupting, erosion attacks. The rate at which erosion destroys a volcano depends on whether it's composed of pyroclastic debris or lava, and on whether the region hosts rivers and/or glaciers that can carve into the landscape. (Cinder cones and ash piles can wash away quickly. In contrast, composite or shield volcanoes, which have been armor-plated by lava flows, can withstand the attack of water and ice for quite some time.) It's possible to distinguish a dormant volcano from an active volcano by the extent to which gullies have been carved into its flanks (**Fig. 5.21**). In some cases, the softer exterior of a volcano completely erodes away, leaving behind a plug of harder frozen magma that once lay within or beneath the volcano, as well as a network of dikes that radiate from this plug. Shiprock, New Mexico, features good examples of these landforms.

Did you ever wonder...
whether a volcano could erupt beneath London, England?

Predicting Eruptions

Changes in tectonic processes at a given location take so long, compared to human history, that we won't likely see them over a span of millennia. Thus, while new volcanoes may appear at a hot spot, along a plate boundary, or in a rift, they won't suddenly appear in other locations. So the question of predicting eruptions is useful only when asking about the potential for an eruption in an active or dormant volcanic region.

Predicting the time of an eruption at a given volcano decades or years in advance is impossible. The only way to constrain such a *long-term prediction* is to determine the *recurrence interval* (the average time between eruptions) for the volcano. If the geologic record indicates that eruptions have happened, on average, every 10 ± 3 years, then a volcano that hasn't erupted for more than 10 years may well erupt within the next several years. If a volcano erupts, on average, every 3,000 ± 600 years, then a volcano that erupted 200 years ago might or might not erupt in the next century. The recurrence interval just gives a sense of the probability, and surprises can happen.

FIGURE 5.21 The shape of a volcano changes as it is eroded.

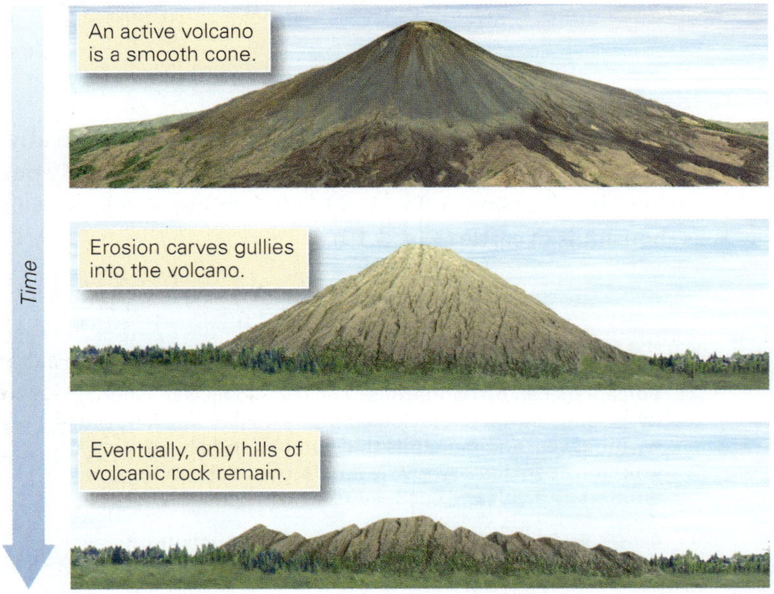
An active volcano is a smooth cone.
Erosion carves gullies into the volcano.
Eventually, only hills of volcanic rock remain.
Time

FIGURE 5.22 To produce this InSAR map of the Three Sisters Wilderness in Oregon, a satellite measures the elevation at two different times. If the land has warped between measurement times, interference of the radar-beam waves, when superimposed, appears as color bands.

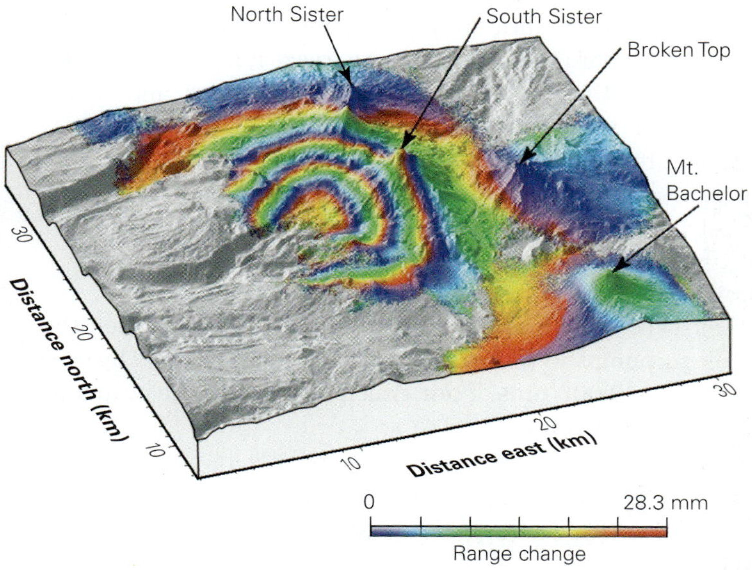

Short-term (days to weeks to months) predictions of impending volcanic activity at an active volcano, unlike short-term predictions of earthquakes, are sometimes feasible. Volcanoes, in fact, may send out distinct warning signals announcing that an eruption may take place very soon, for as magma rises into the ground beneath a volcano, it causes a number of changes that geologists can measure. In many parts of the world, government agencies send monitoring teams to a volcano at the first sign of activity. These teams issue warnings if activity appears to increase. Causes for concern include these:

- *Earthquake activity*: When magma flows into a volcano, wall rocks surrounding the magma chamber crack, and blocks slip with respect to each other. Such cracking and shifting cause small earthquakes. In addition, the movement of magma itself generates vibrations. Thus, in the days or weeks preceding an eruption, the region between 1 and 7 km beneath a volcano becomes seismically active.
- *Changes in heat flow*: The injection of hot magma into the ground beneath a volcano increases the local heat flow, the amount of heat passing up through rock. In some cases, the increase in the heat flow melts snow or ice on the volcano, triggering floods and lahars even before an eruption occurs.
- *Changes in shape*: As magma fills the magma chamber inside a volcano, it pushes outward and can cause the surface of the volcano to bulge. Geologists can use laser sighting, tiltmeters, surveys using the global positioning system (GPS), and a technique called satellite radar interferometry or interferometric synthetic aperature radar (InSAR) to detect modification of a volcano's shape due to the rise of magma (**Fig. 5.22**).
- *Increases in gas and steam emission*: Even though magma remains below the surface, gases bubbling out of the magma, and steam formed when magma heats groundwater, percolate upward through cracks in the Earth and rise from the volcanic vent. So an increase in the volume of gas emission may indicate that magma has entered the ground below.

Mitigating Volcanic Hazards

Danger Assessment Maps and Evacuation Let's say that a particular active volcano has the potential to erupt in the near future. What can we do to prevent the loss of life and property? Since we can't prevent the eruption, the first and most effective precaution is to identify the regions that can be directly affected by the eruption and to compile a *volcanic-hazard assessment map* (**Fig. 5.23**). This type of map shows areas that lie in the path of potential lava flows, lahars, or pyroclastic flows, areas that should be evacuated if a volcano starts to erupt. Unfortunately, because of the uncertainty of prediction, the decision about whether to evacuate is a difficult one. In the case of Mt. St. Helens, hundreds of lives were saved in 1980 by timely evacuation, but in the case of Mt. Pelée in 1902, thousands of lives were lost because warning signs were ignored.

Diverting Flows During a 1669 eruption of Mt. Etna, an active volcano on the Italian island of Sicily, basaltic lava formed a glowing orange river that began to spill down the side of the mountain. When the flow approached the town of Catania, 16 km from the summit, 50 townspeople protected by wet cowhides boldly hacked through the chilled side of the flow to create an opening through which the lava could exit. By doing so, they hoped to cut off the supply of lava feeding the end of the flow near their homes. Their strategy worked, and the flow began to ooze through the new hole in its side. But unfortunately, the diverted flow began to ooze toward the neighboring town of Paterno. Seeing the threat, 500 men from Paterno ran up to the flow and chased away the Catanians who had been keeping the diversion hole open. Eventually, the hole closed, and the flow resumed its path toward Catania, sadly destroying part of the town. More recently, in

FIGURE 5.23 A volcanic-hazard assessment map for Mt. Rainier, Washington, showing the regions that might be affected by flows and lahars.

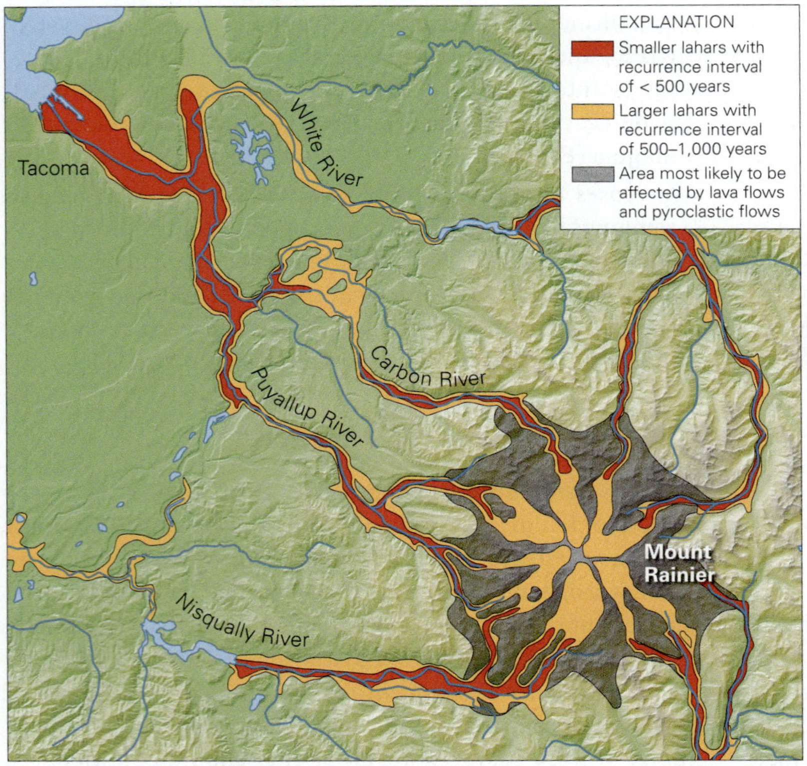

1983 and 1992, people have tried more aggressive techniques to divert Etna's flows—they've used high explosives to blast breaches in the flanks of flows, and used bulldozers to build dams and channels of rubble to divert flows (**Fig. 5.24a**).

Inhabitants of Eldfell, a small volcanic island just south of Iceland, used a particularly creative approach to stop a flow before it completely overran a harbor town. Between February and July, 1973, workers both on land and on ships in the harbor used powerful pumps to spray millions of cubic meters of cold seawater onto the flow, hoping to freeze it in its tracks (**Fig. 5.24b**). The lava boiled away the water, and clouds of steam engulfed the town during the operation, but the pumpers persisted, and probably succeeded in preventing the flows from advancing as far as they might have otherwise. When the pumps were turned off, the lava was covered by over 200,000 tons of salt, precipitated from the evaporating seawater.

Did you ever wonder... whether you can redirect a lava flow?

TAKE-HOME MESSAGE

Volcanoes don't erupt continuously and don't last forever, so we can distinguish among active, dormant, and extinct volcanoes. Once a volcano ceases to erupt, erosion destroys its eruptive shape. Geologists can provide near-term predictions of eruptions, allowing that people to take precautions. In some cases, people have actually been able to divert flows.

QUICK QUESTION How many times do you think Vesuvius has erupted since 79 C.E.? Use the Web to check your estimate.

FIGURE 5.24 Efforts to divert lava flows away from inhabited locations.

(a) Workers spray the lava flow to solidify it and use a bulldozer to build an embankment to divert it on the flanks of Mt. Etna.

(b) Firefighters pumping 6 million cubic meters of water on a lava flow in Iceland, in an effort to freeze it and stop it.

5.7 Effect of Volcanoes on Climate and Civilization

In 1783, Benjamin Franklin was living in Europe, serving as the American ambassador to France. The summer of that year seemed to be unusually cool and hazy. Franklin, who was an accomplished scientist as well as a statesman, couldn't resist seeking an explanation for this phenomenon, and learned that in June of that year, a huge volcanic eruption had taken place in Iceland. He wondered if the "smoke" from the eruption had prevented sunlight from reaching the Earth, thus causing the cooler temperatures. Franklin reported this idea at a meeting, and by doing so, may well have been the first scientist ever to suggest a link between eruptions and climate.

Franklin's idea seemed to be confirmed in 1815, when Mt. Tambora in Indonesia exploded, ejecting over 145 cubic km of ash and pumice into the air. The sky became hazy and stars dimmed by a full magnitude. Temperatures dipped so low in the northern hemisphere that 1816 became known as "the year without a summer." The unusual weather of that year inspired artists and writers. For example, memories of fabulous sunsets and the hazy glow of the sky inspired the luminous atmospheric quality that made the landscape paintings of the English artist J. M. W. Turner so famous (see Fig. 5.1a). Mary Shelley, trapped in her house by bad weather, wrote *Frankenstein*, with its numerous scenes of gloom and doom. Eruption-triggered global temperature drops followed the 1883 eruption of Krakatau and the 1991 eruption of Pinatubo.

How can a volcanic eruption produce these cooling effects? When a large explosive eruption takes place, ash and aerosols enter the stratosphere. In about two weeks, winds distribute the ash and aerosol cloud around the planet. Because they are above the weather and do not get washed away by rainfall, the ash and aerosol may remain suspended for many months to years. The resulting haze keeps the Sun's heat from reaching the Earth—in the case of Pinatubo's eruption, researchers were able to document that the event temporarily diminished the amount of sunlight reaching the Earth by about 10%, therefore affecting global temperatures (**Fig. 5.25a**). Unlike greenhouse gases, such as CO_2, volcanic ash and aerosols do not prevent heat from escaping.

The largest eruption to have happened in the last million years of Earth's history took place at the Toba volcano in Indonesia about 75,000 years ago. The explosion put out huge quantities of ash, covering much of southern Asia like a snowfall. In addition, the eruption injected a huge dose of aerosols into the atmosphere. Not only did the aerosols and ash decrease the solar radiation reaching the Earth, but the white ash on the ground reflected back into space some of the radiation that did reach the planet. As a result, the atmosphere may have cooled for as long as 1,000 years, and this may have caused species of organisms to go extinct. Some anthropologists have speculated that after the cooling event, fewer than 10,000 humans remained on Earth.

In more recent millennia, volcanic eruptions may have led to the demise of civilizations. For example, archaeological research suggests that the Minoan culture, which thrived in the eastern Mediterranean during the Bronze Age, disappeared during the century following explosive eruptions of the Santorini volcano in 1645 B.C.E. (**Fig. 5.25b**). All that remains of Santorini today is a huge caldera whose rim projects above sea level as the Greek island Thera (also known colloquially as Santorini). Ash clouds, tsunamis, and

FIGURE 5.25 Volcanic eruptions may affect climate.

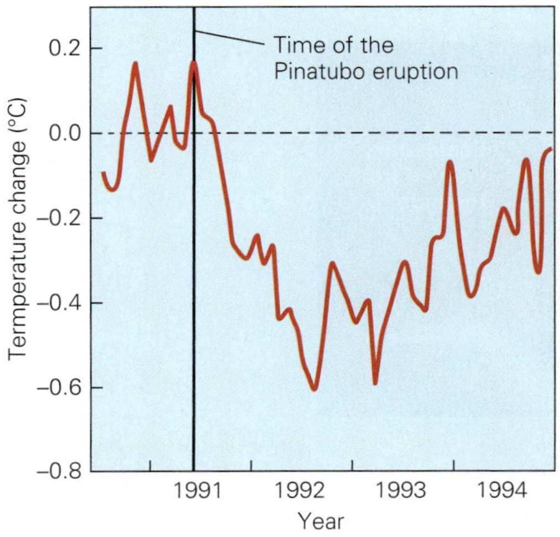

(a) A graph of global average air temperature shows a drop after the Mt. Pinatubo eruption.

(b) The eruption of Santorini may have contributed to the demise of the Minoans, who left behind ruins of elaborate palaces.

earthquakes generated by Santorini may have disrupted their daily life so severely that the Minoan people decided to move elsewhere.

TAKE-HOME MESSAGE

The ash and aerosols produced by explosive eruptions can be blown around the globe. This material can cause significant, but temporary, global cooling by preventing sunlight from reaching the surface. Climatic effects, as well as other consequences of eruptions, may have impacted human evolution and civilization.

QUICK QUESTION Not all eruptions of equivalent size (as defined by the volume of material erupted) trigger equivalent amounts of global cooling. Why? (Hint: Think about eruptive style.)

5.8 Volcanoes Elsewhere in the Solar System

We conclude this chapter by looking beyond the Earth, for ours is not the only place in the Solar System to have hosted volcanic eruptions. We can see the effects of volcanic activity on our nearest neighbor, the Moon, just by looking up on a clear night. The broad darker areas of the Moon, the maria (singular *mare*, after the Latin word for sea), consist of flood basalts that erupted more than 3 billion years ago (**Fig. 5.26a**). Geologists propose that the flood basalts formed when huge meteors collided with the Moon, blasting out giant craters. Crater formation decreased the pressure in the Moon's mantle so that it underwent partial melting, producing basaltic magma that rose to the surface and filled the craters.

FIGURE 5.26 Volcanism on other planets and moons in the Solar System.

(a) Maria of the Moon were seas of basaltic lava.

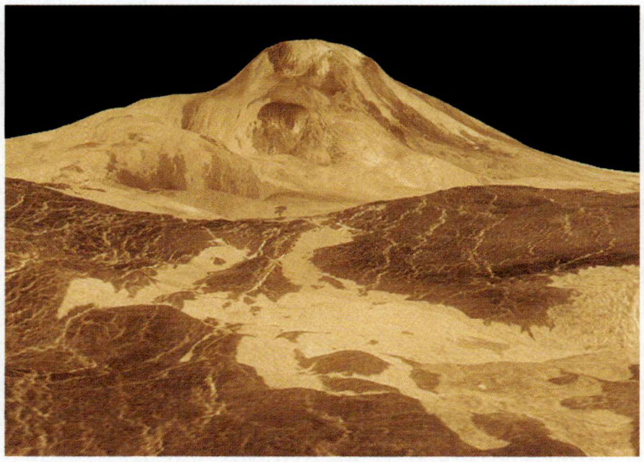

(b) A volcano rises above the plains of Venus.

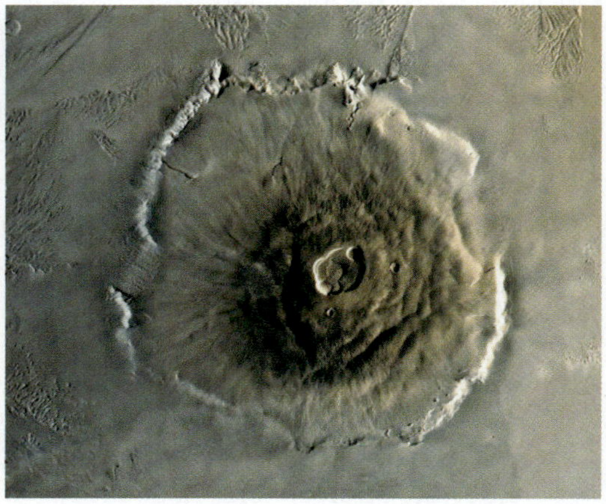

(c) Olympus Mons rises 27 km above the surface of Mars. It's the largest volcano in our Solar System.

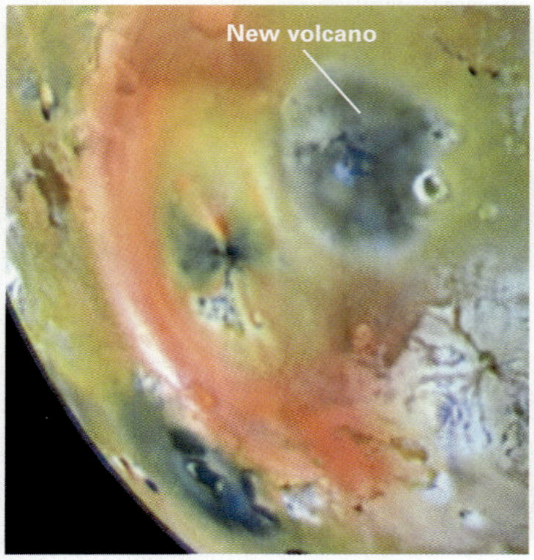

(d) An active volcano erupts sulfur on Io, a moon of Jupiter.

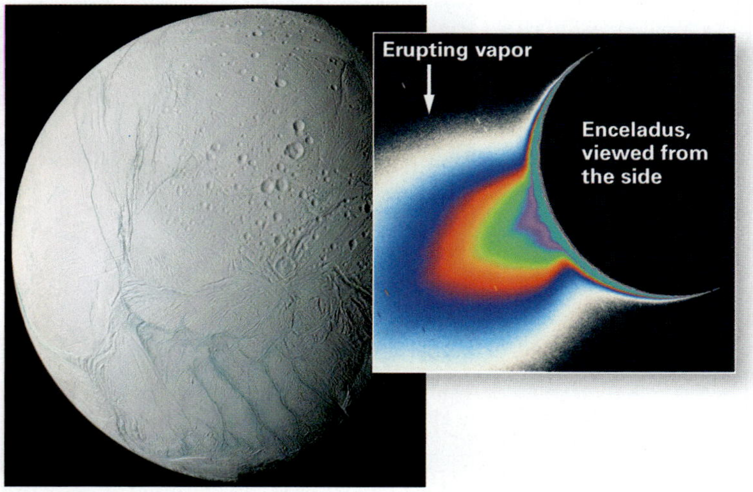

(e) Large cracks at the southern end of Enceladus, a moon of Saturn, erupt water vapor.

The surfaces of both Mars and Venus display distinct volcanic edifices, some of which have calderas at their peaks (**Fig. 5.26b**). In fact, the largest known mountain in the Solar System, Olympus Mons (**Fig. 5.26c**), is an extinct shield volcano on Mars. The base of Olympus Mons is 600 km across, and its peak rises 27 km above the surrounding plains.

Did you ever wonder...
if Earth is the only planet with volcanoes?

Active volcanism currently occurs on Io, one of the many moons of Jupiter. Cameras in the *Galileo* spacecraft have recorded these volcanoes in the act of spraying plumes of sulfur gas into space (**Fig. 5.26d**) and have tracked immense, moving lava flows. Different colors of erupted material make the surface of this moon resemble a pizza. Researchers have proposed that the heat driving this volcanic activity is due to tides, in that the gravitational pull exerted by Jupiter and by other moons alternately stretches and then squeezes Io, generating sufficient friction to keep Io's mantle hot. Geologists have also detected eruptions from moons of Saturn (**Fig. 5.26e**).

TAKE-HOME MESSAGE

Space exploration reveals that volcanism occurs not only on Earth, but has also left its mark on other terrestrial planets and on the moons of giant planets. Spacecraft, such as *Galileo*, have detected active eruptions on the moons of Jupiter and Saturn.

QUICK QUESTION Since most volcanic activity on Earth is a result of plate tectonics, what does the lack of volcanic activity on the Moon tell us about whether plate tectonics happens on the Moon?

Another View The ash cloud from the 2011 eruption of a volcano in the Puyehue-Cordón Caulle chain of Chile. The ash circled the globe within two weeks, disrupting air traffic throughout the southern hemisphere.

Chapter 5 Review

Chapter Summary

- Volcanoes are vents at which molten rock (lava), pyroclastic debris, gas, and aerosols erupt at the Earth's surface.
- The characteristics of a lava flow depend on its viscosity, which in turn depends on its temperature and composition.
- Basaltic lava can flow great distances. Pahoehoe flows have smooth, ropy surfaces, whereas a'a' flows have rough, rubbly surfaces. Andesitic and rhyolitic lava flows tend to pile into mounds at a volcano's vent.
- Pyroclastic debris includes powder-sized ash, marble-sized lapilli, and apple- to refrigerator-sized blocks and bombs. Some debris falls from the air, and some settles from pyroclastic flows.
- The summit of an erupting volcano usually includes a crater. Collapse of a volcano, when a large magma chamber drains, can yield a much larger bowl-shaped depression called a caldera.
- A volcano's shape depends on the style of eruption. Shield volcanoes are broad, gentle domes, and cinder cones are steep-sided, symmetrical hills. Stratovolcanoes (composite volcanoes) can become quite large and consist of alternating layers of pyroclastic debris and lava.
- The type of eruption depends on the lava's viscosity and gas content. Effusive eruptions produce flows, and explosive eruptions produce pyroclastic debris. Typically, the former are basaltic, and the latter are andesitic or rhyolitic.
- Different kinds of volcanoes form in different geologic settings, as explained by plate tectonics theory.
- Volcanic eruptions pose many hazards: lava flows overrun roads and towns, ash falls blanket the landscape, pyroclastic flows incinerate towns and fields, landslides and lahars bury the land surface, earthquakes topple structures and rupture dams, tsunamis wash away coastal towns, and gases suffocate people and animals.
- Eruptions can be predicted by earthquake activity, changes in heat flow, changes in shape of the volcano, and changes in the volume of gas and steam.
- We can minimize the consequences of an eruption by avoiding construction in danger zones.
- Immense flood basalts cover portions of the Moon. The largest known volcano, Olympus Mons, towers over the surface of Mars. Satellites have documented evidence for eruptions on moons of Jupiter and Saturn.

Guide Terms

GEOTOURS *THIS CHAPTER'S GEOTOUR EXERCISE (E) FEATURES:*

- Shield Volcanoes
- Composite Cone Volcanoes
- Cinder Cone Volcanoes

Review Questions

1. Describe the three different kinds of material that can erupt from a volcano.
2. Contrast a pyroclastic flow with a lahar.
3. Describe the differences among shield volcanoes, stratovolcanoes, and cinder cones. How can you explain these differences by the composition of their lavas and other factors?
4. Why do some volcanic eruptions consist mostly of lava flows, while others are explosive and do not produce flows?
5. What processes may lead to hot-spot eruptions? How does an oceanic hot-spot volcano differ from a continental one?
6. Why do some continental-rift eruptions yield flood basalts?
7. What are the characteristics of volcanoes erupting at volcanic arcs? How do they contrast with volcanoes erupting on mid-ocean ridges?
8. Identify some of the major volcanic hazards, and explain how they develop.
9. How do geologists predict volcanic eruptions?
10. Explain how steps can be taken to protect people from the effects of eruptions.

On Further Thought

11. The Long Valley Caldera, near the Sierra Nevada range in California, exploded about 700,000 years ago and produced a huge ignimbrite called the Bishop Tuff. About 30 km to the northwest lies Mono Lake, with an island in the middle and a string of craters extending south from its south shore. Hot springs can be found along the lake. You can see the lake on *Google Earth*™ at latitude 37°59′56.58″ N, longitude 119°2′18.20″ W. Explain the origin of Mono Lake.
12. Mt. Fuji is a 3.6-km-high stratovolcano in Japan formed as a consequence of subduction. See it using *Google Earth*™ at latitude 35°21′46.72″ N, longitude 138°43′49.38″ E. You can see the volcano contains volcanic rocks with a range of compositions, including some andesitic rocks. Why do andesites erupt here? Very little andesite occurs on the Marianas Islands, which are also subduction-related volcanoes. Why?

Online Resources

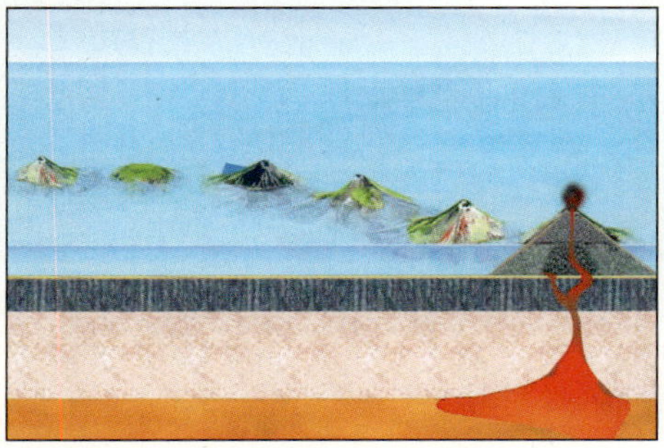

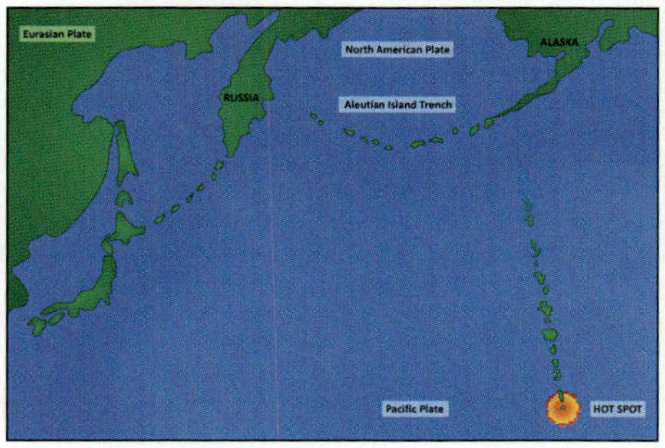

Animations

This chapter features animations about how volcanic hot spots work and the formation of the Hawaiian Islands.

Assessment

This chapter features questions on the different types, shapes, and eruptive styles of volcanoes, and the impact of their hazards.

INTERLUDE **B**

A Surface Veneer: Sediments and Soils

▲ Here at the Earth's surface, in a small canyon in Indiana, a layer of sediment and soil covers bedrock. In stream cuts, bedrock peeks out from under this layer.

LEARNING OBJECTIVES

By the end of this chapter, you should understand...

1. how rocks undergo weathering at or near the Earth's surface to produce sediment.
2. the difference between sediment and soil.
3. factors that affect the character and thickness of soil.

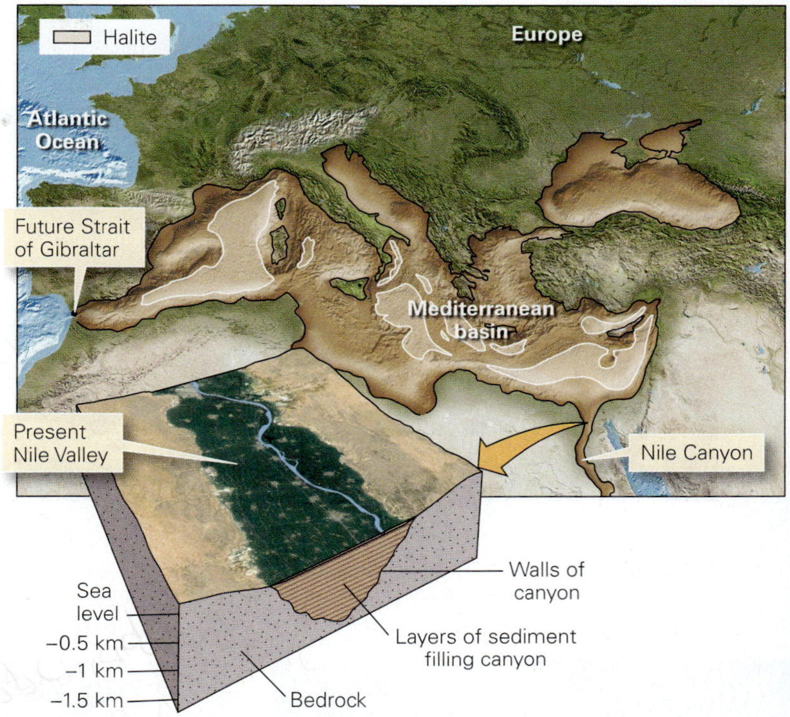

FIGURE B.1 The present Nile River flows on sediment that filled a deep canyon. This canyon was formed when the Mediterranean Basin was dry.

B.1 Introduction

In the 1950s, the government of Egypt decided to build the Aswan High Dam to trap water of the Nile River in a huge reservoir before the water could reach the Mediterranean Sea. In the process of identifying a good site for the dam's foundation, geologists discovered that the present-day Nile River flows on the surface of a 1,500-m-thick layer of gravel, sand, and mud that fills what was once a canyon as large as the Grand Canyon (**Fig. B.1**). How could the river once have carved a canyon this deep, and why did the canyon later fill with debris?

The origin of the pre-Nile canyon remained a mystery until the summer of 1970, when geologists began to study the material that underlies the floor of the Mediterranean Sea. They expected to find layers consisting of the shells of plankton (tiny floating organisms) that had settled out of the water, or of clay that rivers had carried to the sea. To their surprise, however, they also found a 2-km-thick layer of halite and gypsum. Such minerals, types of salts, form when seawater dries up. In order to yield a salt layer that is 2 km thick, the entire Mediterranean must have dried up almost completely several times, with the sea refilling after each drying event. This discovery solved the mystery of the pre-Nile canyon. When the Mediterranean Sea dried up, the Nile River flowed down into a deep lowland, and in the process, carved a canyon. Later, when the sea refilled with water, the river could no longer cut down, and the flooded canyon filled with sand and gravel brought in from upstream.

Why did the Mediterranean Sea dry up? Only 10% of its water comes from rivers, so for the Mediterranean Sea to remain full, water must flow in from the Atlantic Ocean through the Strait of Gibraltar. About 6 million years ago, the northward-drifting African Plate collided with the European Plate, forming a natural dam separating the Mediterranean from the Atlantic. At times when global sea level dropped, water stopped flowing over this dam from the Atlantic, and the Mediterranean evaporated. The salt that had been dissolved in its water precipitated to form a solid deposit of halite and gypsum on the floor of the resulting basin, and the pre-Nile canyon formed. At times when sea level rose, water flooded in from the Atlantic into the Mediterranean, filling the basin again. About 5.5 million years ago, the Mediterranean rose to its present level, and gravel, sand, and mud carried by the Nile River accumulated in the pre-Nile canyon.

Geologists refer to the kinds of deposits just described—sand, mud, gravel, halite and gypsum layers, and shell fragments—as sediment. **Sediment**, broadly defined, consists of: loose fragments of rocks or minerals broken off bedrock; mineral crystals that precipitate directly out of water; and/or shells or shell fragments. Sediments are produced by the *weathering* (physical and chemical breakdown) of pre-existing rock. They form a surface veneer, or cover, over bedrock (**Fig. B.2**). This cover's thickness ranges from nonexistent, in places where bedrock crops out at the Earth's surface, to several kilometers, in rapidly sinking basins. Some sediments transform into soil. In this interlude, we look at how weathering produces sediment, and how soils form and evolve. Note that geologists refer to any loose debris (sediment or soil) as *regolith*.

FIGURE B.2 A layer of unconsolidated sediment (sand, clay, and cobbles), topped by dark soil, overlies bedrock in this outcrop along the coast of western Ireland.

FIGURE B.3 The contrast between fresh and weathered granite in an Arizona road cut. Weathered granite can break apart. Grains fall off and collect as detritus at the base of the outcrop. The inset photos show how weathering visibly changes the rock.

B.2 Weathering: The Process of Forming Sediment

Rock exposed at the Earth's surface sooner or later disintegrates and crumbles away, due to the process of weathering. In a geologic context, **weathering** refers to the combination of processes that corrode and/or break up solid rock, eventually transforming it into loose debris. Weathered rock may look discolored, or rough, compared to unweathered or *fresh rock* (**Fig. B.3**). Just as a plumber can unclog a drain by using physical force (with a plumber's snake) or by causing a chemical reaction (with a dose of liquid drain opener), nature can attack rocks with two types of weathering: physical and chemical.

Physical Weathering

Physical weathering, sometimes called *mechanical weathering*, breaks intact rock into unconnected grains or chunks known as **clasts**, which come in a range of sizes (**Table B.1**). Geologists commonly refer to an accumulation of clasts as *detritus*. Many phenomena contribute to physical weathering, as we now describe.

Jointing A joint is a natural crack in rock—the formation of a joint separates one piece of rock into two separate pieces. Almost all rock outcrops contain joints—for example, large granite plutons typically split into onion-like sheets along *exfoliation joints* that lie parallel to the mountain face, and sedimentary rock layers tend to break into rectangular blocks bounded by joints on the sides and layer boundaries above and below (**Fig. B.4a**). Jointing effectively breaks bedrock into many separate blocks which, when exposed on a cliff, eventually tumble downslope, fragmenting into smaller pieces as they bounce. The resulting chunks may collect in an apron of *talus*, the rock rubble at the base of a slope, or may be carried away by rivers or glaciers at the base of a cliff (**Fig. B.4b**). Joints form for a variety of reasons (see Chapter 9). For example, when rock once buried

TABLE B.1 Clasts Are Classified by Grain Diameter

Boulders	More than 256 millimeters (mm)
Cobbles	Between 64 mm and 256 mm
Pebbles	Between 2 mm and 64 mm
Sand	Between 1/16 mm and 2 mm
Silt	Between 1/256 mm and 1/16 mm
Clay	Less than 1/256 mm

deeply in the crust rises toward the Earth's surface, as **erosion** (the breaking off and removal of rock or sediment) strips away overburden, the rock becomes cooler and the pressure squeezing it decreases. This change causes the rock to change shape only slightly, but enough to cause hard rock to break.

Frost Wedging Freezing water can burst pipes and shatter bottles because water expands when it freezes and pushes the walls of the container apart. The same phenomenon happens in rock. When water trapped in a joint freezes, it forces the joint open and may cause the joint to grow. Such *frost wedging* helps break blocks free from intact bedrock (**Fig. B.5a**).

Salt Wedging In arid climates, dissolved salt in groundwater precipitates in open pore spaces in rocks, forming crystals that push apart the surrounding grains. This process, called *salt wedging*, weakens rock so that when exposed to wind and rain, the rock disintegrates into separate grains. The same phenomenon happens along the seacoast, where salt spray percolates into rock and then dries (**Fig. B.5b**).

Root Wedging Have you ever noticed how the roots of a tree can break up a sidewalk? As roots grow, they apply pressure

Downward pressure
Sedimentary rock layers
Granite pluton
Time
Exfoliation joints are parallel to the ground surface, so rock peels off like layers of an onion.
Joints in granite, California.
Vertical joints intersect bedding to form rectangular blocks.
Vertical joints
Bedding
Intact rock
Talus
100-m-high sandstone cliff, Ireland.

FIGURE B.4 Joints (natural cracks) break bedrock into blocks and sheets, which can tumble down a slope.

(a) When buried deeply, rocks are subjected to a large downward pressure. Later, after erosion removes overburden, rocks expand and crack.

(b) Joint-bounded blocks tumble from a cliff and accumulate in talus aprons near Mt. Snowdon in Wales.

FIGURE B.5 Wedging is one type of physical (mechanical) weathering.

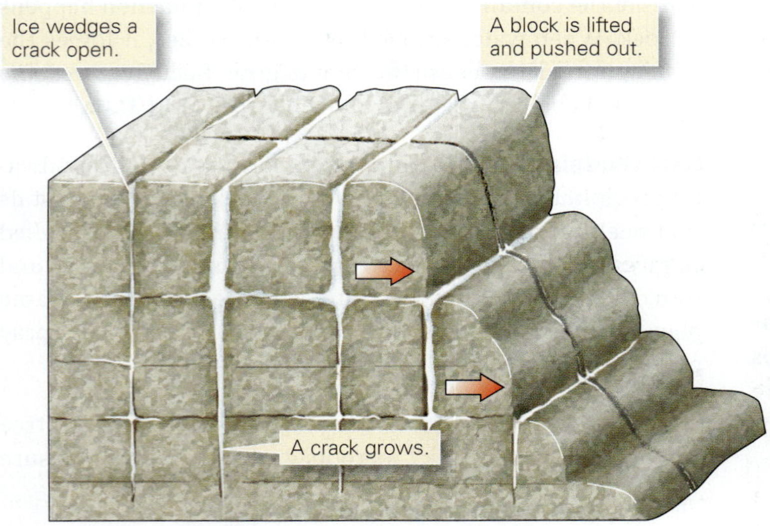
Ice wedges a crack open.
A block is lifted and pushed out.
A crack grows.

(a) When the water that fills cracks freezes, it expands and wedges the cracks open.

(b) Salt wedging led to disintegration of gravestones in Whitby, England.

Tree growing in a joint.
Eventually, the blocks tumble to the base of the cliff.

(c) Root wedging pushes open a joint, slowly separating a block from the cliff.

to their surroundings, and can push joints open in a process known as *root wedging* (**Fig. B.5c**).

Thermal Expansion When the heat of an intense forest fire bakes a rock, the outer layer of the rock expands. On cooling, the layer contracts, which generates force in the rock sufficient to break off the outer part of the rock. Recent research suggests that the heat of the Sun's rays sweeping across dark rocks in a desert may cause cobbles on the ground surface to fracture into thin slices over time.

Animal Attack Animal life also contributes to physical weathering, for burrowing creatures, from earthworms to gophers, can move rock fragments. In the past century, humans have become perhaps the most energetic agent of physical weathering on the planet, for when we excavate quarries, foundations, mines, or roadbeds by digging and blasting, we shatter and displace large volumes of rock that might otherwise have remained intact for millions of years more.

Chemical Weathering

Up to this point, we've taken the plumber's-snake approach to breaking up rock. Now, let's look at the liquid-drain-opener approach. **Chemical weathering** refers to the many chemical reactions that alter or destroy minerals when rock comes in contact with water solutions and/or air. Common reactions involved in chemical weathering include the following:

- *Dissolution:* Water is a good solvent, so when it flows over or through rock, it slowly dissolves minerals. This process of *dissolution* affects primarily salts and carbonate minerals (**Fig. B.6**), but even silicate minerals can dissolve slightly.
- *Hydrolysis:* During hydrolysis, water chemically reacts with minerals and breaks them down (*lysis* means loosen, in Greek) to form other minerals. For example, hydrolysis reactions transform feldspar and many other silicate minerals into clay.
- *Oxidation:* Oxidation reactions in rocks transform iron-bearing minerals, such as biotite and pyrite, into a rusty-brown mixture of various iron-oxide and iron-hydroxide minerals. In effect, oxidation causes iron-bearing rocks to "rust."
- *Hydration:* Hydration, the absorption of water into the crystal structure of minerals, causes some minerals, such as certain types of clay, to swell. Such expansion weakens rock.

Not all minerals undergo chemical weathering at the same rates. Some weather in a matter of months or years, whereas others remain unweathered for millions of years. For example, when a granite undergoes chemical weathering, most of its minerals transform into clay due to hydrolysis. But the quartz it contains does not change, so we can say that quartz is *resistant* to chemical weathering.

Until fairly recently, geoscientists tended to think of chemical weathering as a strictly inorganic chemical reaction, occurring entirely independently of life-forms. But researchers now realize that organisms can play a major role in the chemical-weathering process. For example, the roots of plants, fungi, and lichens secrete organic acids that help dissolve minerals in rocks in order to extract *nutrients* (essential elements needed for growth) from the minerals. Some microbes literally eat minerals by plucking molecules off of mineral surface—they use the energy from the molecules' chemical bonds to supply their own life force.

Physical and Chemical Weathering Working Together

So far we've looked separately at the processes of chemical and physical weathering, but in the real world they happen together, aiding one another in disintegrating rock to form sediment (see **Geology at a Glance**, pp. 176–177). Physical weathering speeds up chemical weathering. To understand why, keep in mind that chemical-weathering reactions take place at a material's surface. Thus, the overall rate at which chemical weathering occurs depends on the ratio of surface area to volume—the greater the surface area, the faster the volume of the whole material can chemically weather. When jointing (physical weathering) breaks a large block of rock into smaller pieces, the surface area increases, so chemical weathering happens faster (**Fig. B.7a**).

FIGURE B.6 Dissolution is one type of chemical weathering.

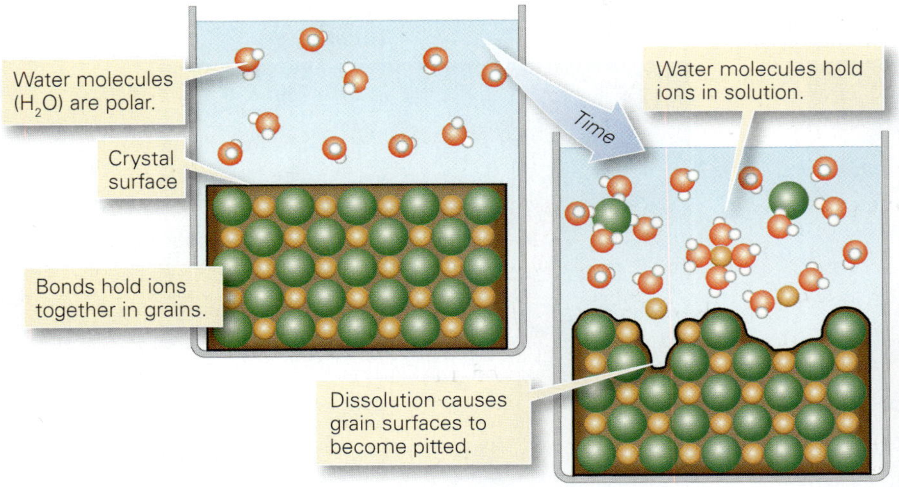
Water molecules (H_2O) are polar.
Crystal surface
Bonds hold ions together in grains.
Time
Water molecules hold ions in solution.
Dissolution causes grain surfaces to become pitted.

(a) Dissolution occurs when water molecules pluck ions off of grain surfaces.

Water seeping into joints in limestone produced troughs.

(b) Dissolution of limestone on an outcrop surface in Ireland.

GEOLOGY AT A GLANCE

Weathering, Sediment, and Soil Production

New boulders tumble from a cliff in New Mexico.

Silt collects along a stream in Indiana.

Tectonic processes uplift the land surface above sea level. Once exposed, rock interacts with air and water and undergoes chemical and physical weathering, ultimately breaking down to produce sediment. Convection in the atmosphere generates wind, rain, and snow. Flowing water, ice, and air erode and transport sediment to sites of deposition. The leaching action of downward-percolating rainwater, along with the addition of organic material and various activities of organisms, produces soil.

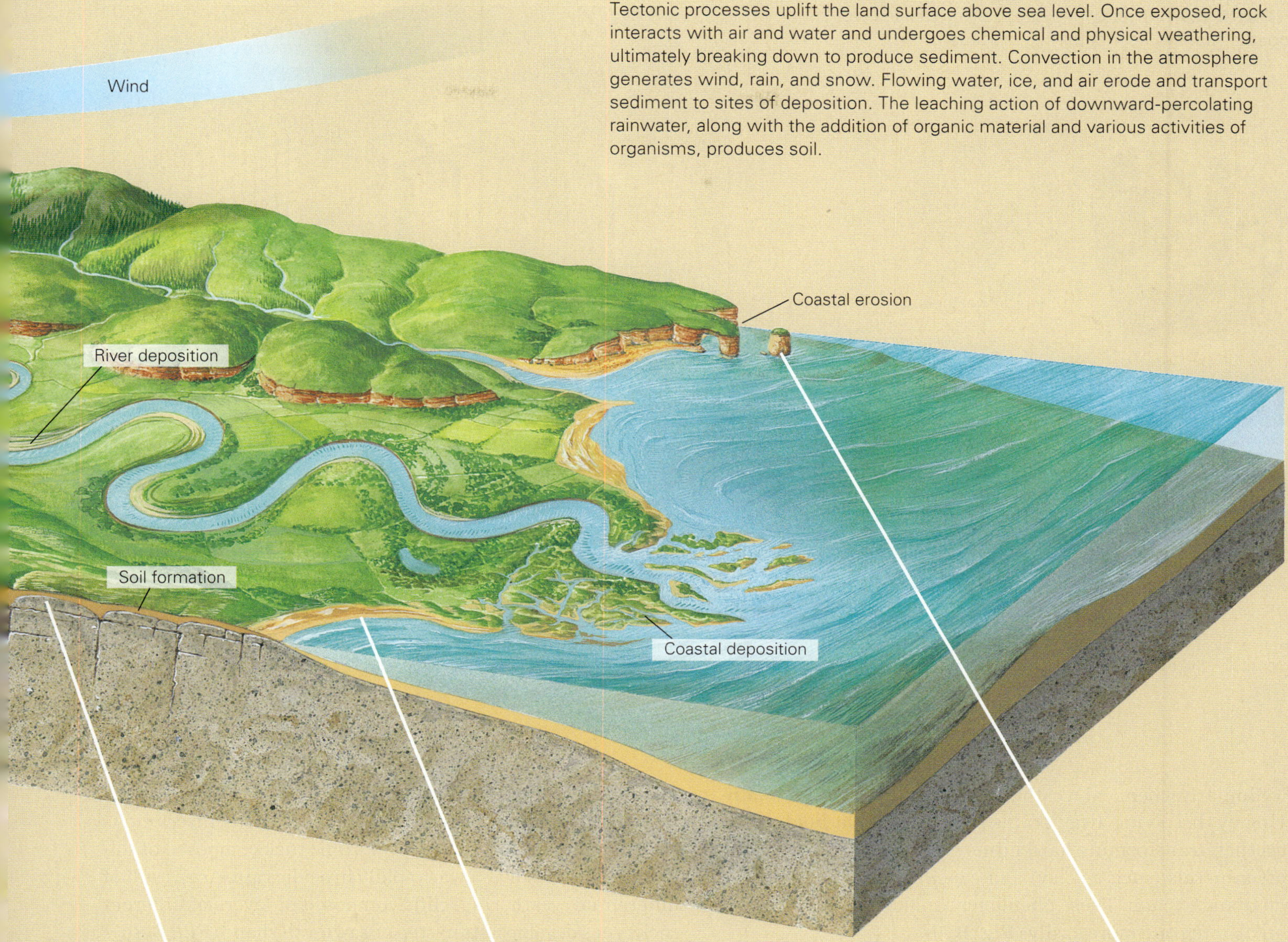

Soil forms on bedrock of chalk in southern England.

Waves move sand on a beach in Brazil.

Erosion carves the coastal cliffs of Ireland.

FIGURE B.7 Physical and chemical processes work together during the weathering process.

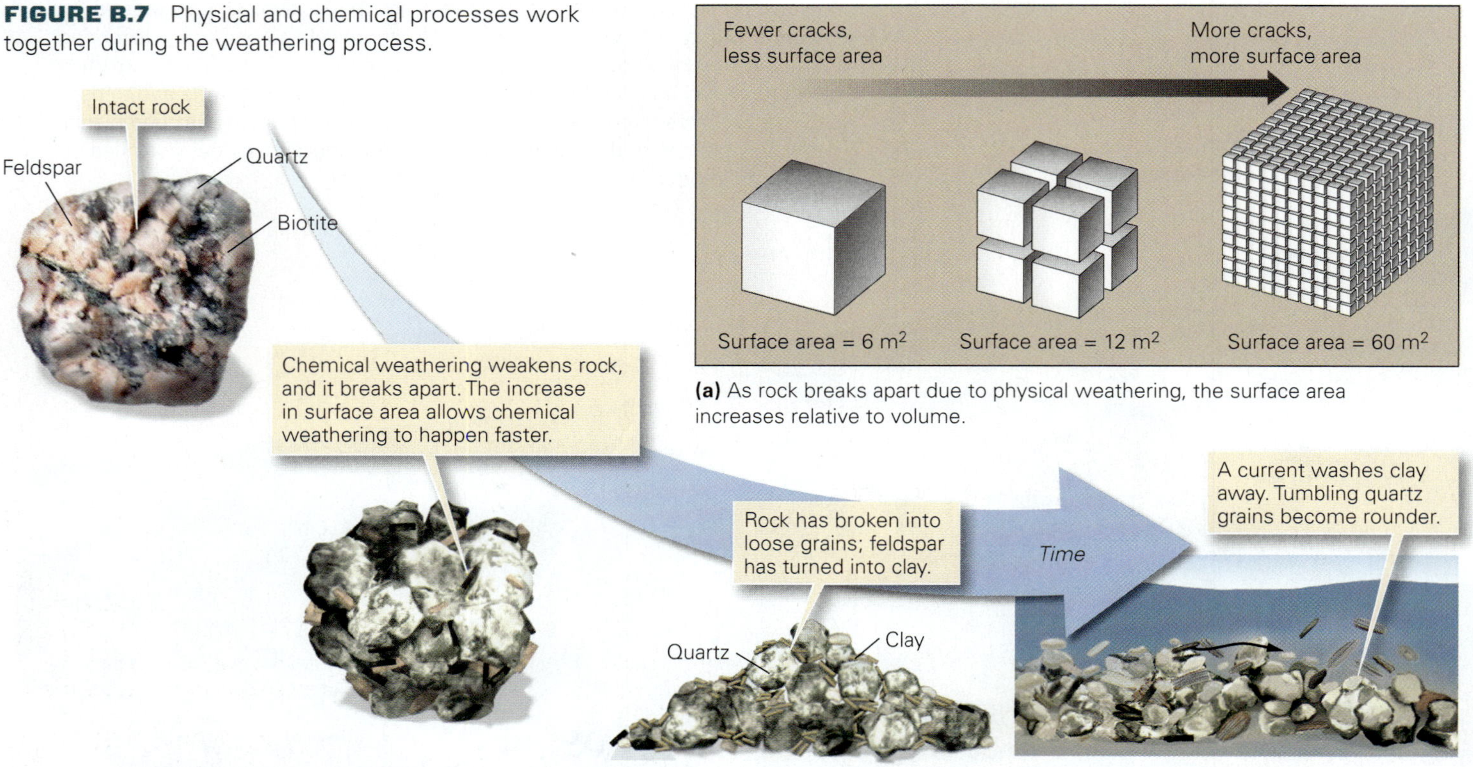

(a) As rock breaks apart due to physical weathering, the surface area increases relative to volume.

(b) Chemical weathering weakens rock, so it breaks apart. As this happens, the surface area increases, so chemical weathering happens still faster. Eventually, the rock completely disaggregates to form sediment. Weathering of granite produces quartz sand and clay.

Similarly, chemical weathering speeds up physical weathering by dissolving away grains or cements that hold a rock together, transforming hard minerals (such as feldspar) into soft minerals (such as clay) and causing minerals to absorb water and expand. These phenomena make rock weaker, so it can disintegrate more easily (**Fig. B.7b**).

Weathering happens faster at edges, and even faster at the corners of broken blocks. This is because weathering attacks a flat face from only one direction, an edge from two directions, and a corner from three directions. Thus, with time, edges of blocks become blunt and corners become rounded (**Fig. B.8a**). In rocks such as granite, which do not contain distinct layers or fabrics that can influence weathering rates, rectangular blocks transform into a spheroidal shape (**Fig. B.8b**).

When rocks in an outcrop undergo weathering at different rates, we say that the outcrop has undergone *differential weathering*. You can easily see the consequences of differential weathering if you walk through a graveyard. The inscriptions on some headstones are sharp and clear, but those on other stones have become blunted or have even disappeared (**Fig. B.8c**). That's because the minerals in different stones have different resistances to weathering. Granite, an igneous rock with a high content of resistant quartz, retains inscriptions the longest. But marble, a metamorphic rock composed of soluble calcite, dissolves away rapidly in acidic rain. As a result of differential weathering, cliffs composed of layers of different rock types develop a stair-step or sawtooth shape (**Fig. B.8d**).

B.3 Soil

If you've ever had the chance to dig in a garden, you've seen firsthand that the material in which flowers grow looks and feels different from beach sand or potter's clay. We call the material in a garden *dirt* or, more technically, soil. **Soil** consists of rock or sediment that has been modified by physical and chemical interaction with organic material, rainwater, and organisms over time. Soil is one of our planet's most valuable resources, for without it there could be no natural forests or grasslands, and no horticulture, agriculture, forestry, or ranching.

FIGURE B.8 Spheroidal weathering and differential weathering.

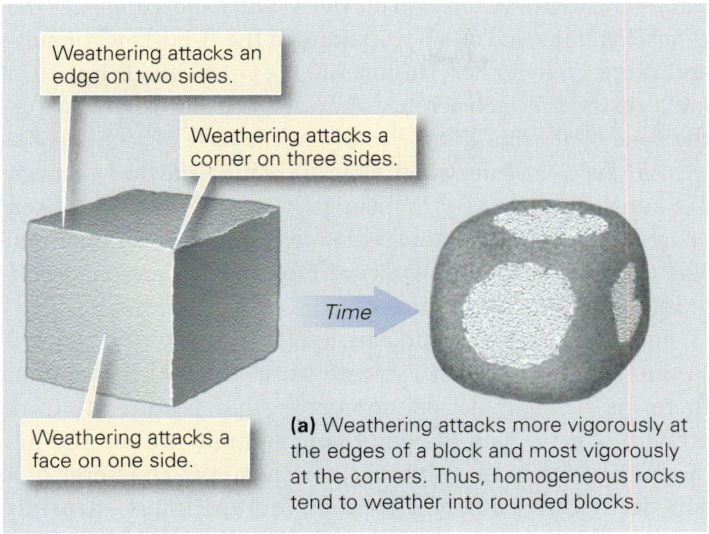
Weathering attacks an edge on two sides.
Weathering attacks a corner on three sides.
Time
Weathering attacks a face on one side.

(a) Weathering attacks more vigorously at the edges of a block and most vigorously at the corners. Thus, homogeneous rocks tend to weather into rounded blocks.

(b) Weathering along cracks in granite of the Mojave Desert led to the formation of rounded blocks. This style is called spheroidal weathering.

AMOS SELLEW
1815 1856
MINERVA K.
HIS WIFE
1820 1877

(c) Inscriptions on a granite headstone (left) last for centuries, but those on a marble headstone (right) may weather away in decades. These gravestones are in the same cemetery and are about the same age.

Weak shale
Strong sandstone

(d) Sawtooth weathering profiles develop in sequences of alternating strong and weak layers on this exposure in New Mexico. Weak layers are indented, whereas strong (resistant) layers protrude.

How Does Soil Form?

Three processes taking place at or just below the surface of the Earth play a role in soil formation. First, chemical and physical weathering produce loose debris and ions in solution. The detritus consists of mineral grains (such as quartz sand or silt), as well as new weathering products (such as clay). Second, downward-percolating water, mostly from rain that has seeped into the ground, redistributes ions and fine clay flakes. Closer to the ground surface, in the **zone of leaching**, the water extracts and absorbs ions and picks up clay, and carries this material further downward. Deeper down, in the **zone of accumulation**, new mineral crystals precipitate out of the percolating water, and because the rate of water movement slows, the water leaves behind its load of fine clay (**Fig. B.9a**). Third, at and just below the ground surface, detritus interacts with organisms. Dead leaves, roots, and animals, along with waste materials from animals, constitute organic material which mixes into the inorganic minerals of the detritus. Microbes, fungi, and insects decompose this organic material, breaking large and complex organic chemicals down into simpler ones. Some of these organisms also absorb nutrients from inorganic minerals, or release acids that dissolve minerals. Finally, plant roots or burrowing animals (insects, worms, and gophers) churn and break up the debris that may have clumped together.

Because different soil-forming processes operate at different depths, soils typically develop distinct zones, known as **soil horizons**, arranged in a vertical sequence called a **soil profile** (**Fig. B.9b, c**). Let's look at an idealized soil profile, from top to bottom, using a soil formed in a temperate forest as our example. The highest horizon is the *O-horizon* (the prefix stands for organic), so called because it consists almost entirely of *humus* (plant debris) with barely any mineral matter. Below the O-horizon, we find the *A-horizon,* in which humus has decayed further and has mixed with clay, silt, and sand. Water percolating down through the A-horizon causes further chemical weathering reactions to occur, yielding ions in solution and new clay minerals. Downward-moving water eventually carries soluble chemicals and fine clay deeper into the subsurface. The O- and A-horizons constitute dark-gray to blackish-brown **topsoil**, the fertile portion of soil that farmers till for planting crops. (In some places, the A-horizon grades downward into the *E-horizon,* a soil level that has undergone substantial leaching but has not yet mixed with organic material.) Beneath the A-horizon (or the A- and E-horizons) lies the *B-horizon.* Ions and clay accumulate in the B-horizon, or **subsoil**. Note from our description that the O-, A-, and E-horizons make up the zone of leaching, whereas the B-horizon makes up the zone of accumulation. Finally, at the base of a soil profile we find the *C-horizon,* which consists of material derived from the substrate that's been chemically weathered and broken apart, but has not yet undergone leaching or accumulation. The C-horizon grades downward into unweathered bedrock, or into unweathered sediment.

Farmers, foresters, and ranchers well know that the soil in one locality can differ greatly from the soil in another, in terms of composition, thickness, and texture. Indeed, crops that grow well in one type of soil may wither and die in another. Such diversity exists because the character of a soil—its texture, nutrient content, and thickness—depends on several soil-forming factors (**Fig. B.10**).

- *Climate:* Climate is the single most important factor in determining the nature of soil development. The aspects of climate that have significant impact on weathering are the amount and distribution of rainfall, and the temperature. Large amounts of rainfall and warm temperatures can accelerate chemical weathering and cause most soluble

FIGURE B.9 Formation of soil horizons.

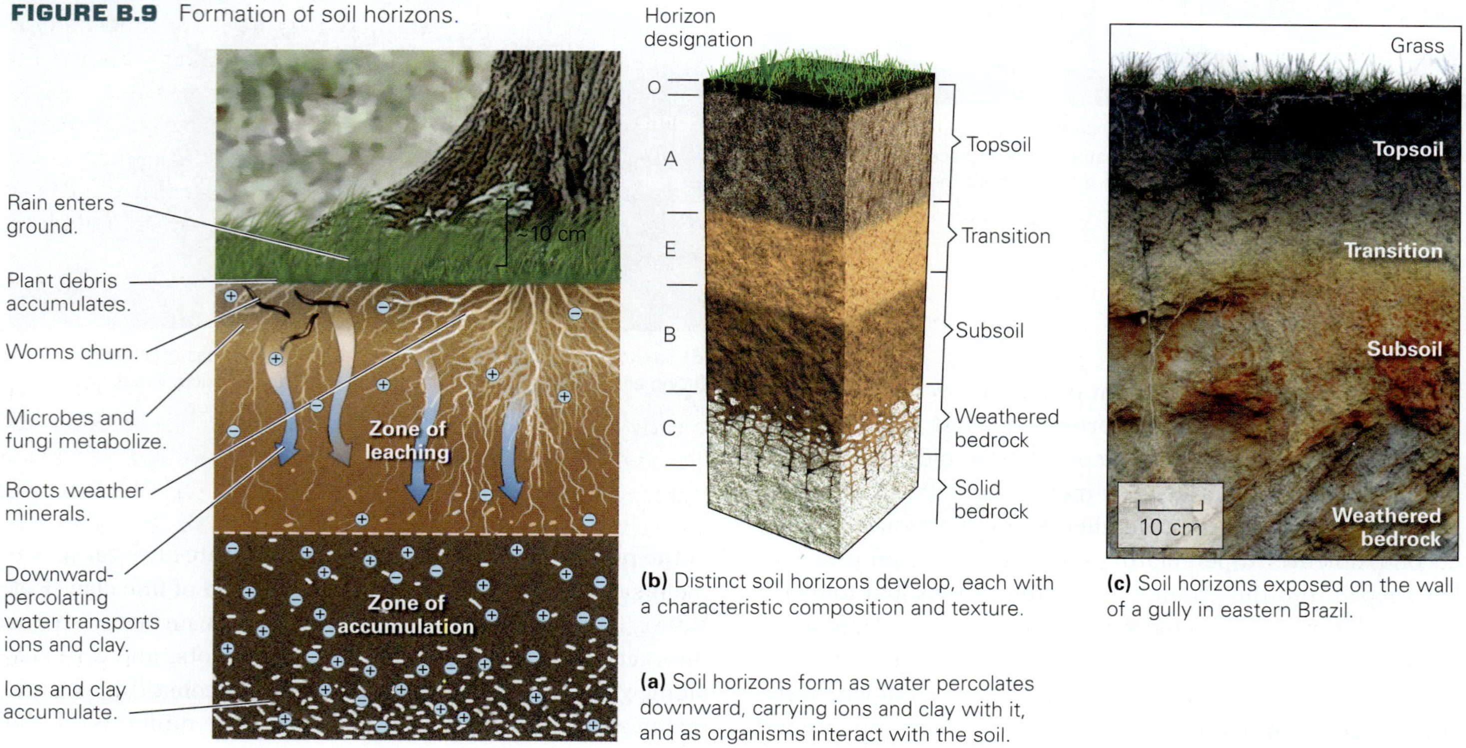

(a) Soil horizons form as water percolates downward, carrying ions and clay with it, and as organisms interact with the soil.

(b) Distinct soil horizons develop, each with a characteristic composition and texture.

(c) Soil horizons exposed on the wall of a gully in eastern Brazil.

FIGURE B.10 Factors that control the character of soil.

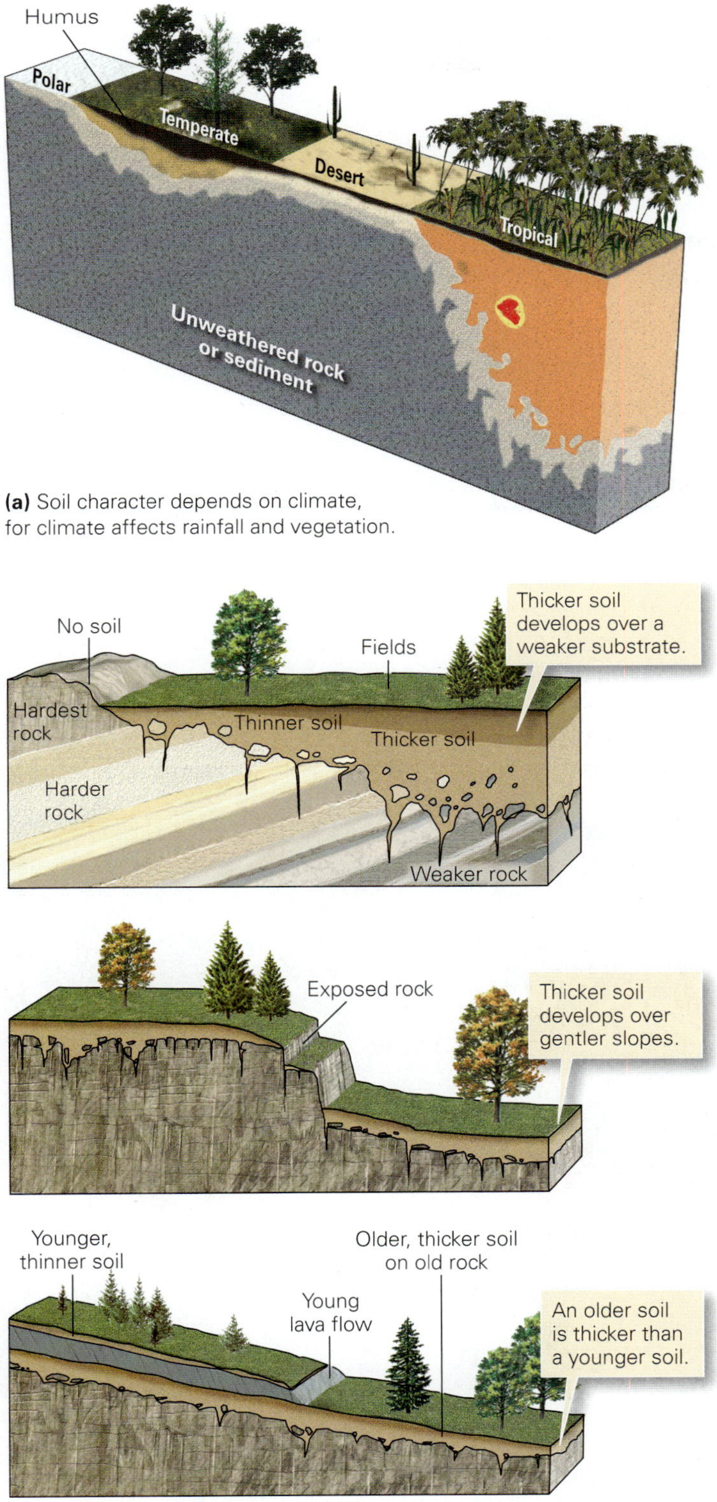

(a) Soil character depends on climate, for climate affects rainfall and vegetation.

(b) Soil character also depends on the strength of the substrate, the steepness of the slope, and the length of time soil has been forming.

elements to be leached. Small amounts of rainfall and cool temperatures result in slower rates of weathering and leaching, so soils take a long time to develop and can retain unweathered minerals and soluble components.

- *Substrate composition:* Some soils form on basalt, some on granite, some on volcanic ash, and some on recently deposited quartz silt. Because these different substrates consist of different materials, the soils formed on each end up with different chemical compositions. Also, substrates have different resistances to erosion—thicker soils develop on less-resistant material.
- *Slope steepness:* A thick soil can accumulate under land that lies flat. But on a steep slope, weathered rock may wash away before it can evolve into a soil. Thus, all other factors being equal, soil thickness increases as slope angle decreases.
- *Wetness:* Depending on the details of local topography and on the depth below the surface at which groundwater occurs, some soil is wetter than other soil in the same region. Wet soils tend to contain more organic material than do dry soils.
- *Time:* Because soil formation is an evolutionary process, a young soil tends to be thinner and less developed (with less distinct horizons) than an old soil. The rate of soil formation varies greatly with environment.
- *Local ecosystem:* The type and quantity of organisms living in a region influence the character of the organic content of soil. For example, where more plants grow, more plant debris is available, and there may be more roots to bind soil and prevent it from washing away. Plants are not the only organisms that matter. Where microbes and maggots thrive, decomposition happens more rapidly to provide organic compounds that mix into soil. If such organisms aren't present, decomposition will not take place, and dead organic matter simply dries out, turns to dust, and blows away.

Soil Classification

Soil scientists worldwide have struggled mightily to develop a rational scheme for soil classification. Not all schemes utilize the same criteria, and even today there's not worldwide agreement on which scheme works best. In the United States, a country that includes many midlatitude climates, soil scientists frequently use the *U.S. Comprehensive Soil Classification System*, which distinguishes among 12 **soil orders** based on the physical characteristics and environment of soil formation (see **Table B.2** and **Fig. B.11**). Canadian soil scientists use a different scheme that focuses on soils that develop in cooler, high-latitude climates.

TABLE B.2 **Soil Orders: U.S. Comprehensive Soil Classification System (see also Fig. B.11)**

Alfisol	Gray/brown, has subsurface clay accumulation and abundant plant nutrients. Forms in humid forests.
Andisol	Forms in volcanic ash.
Aridisol	Low in organic matter, has carbonate horizons. Forms in arid environments.
Entisol	Has no horizons. Formed very recently.
Gelisol	Underlain with permanently frozen ground.
Histosol	Very rich in organic debris. Forms in swamps and marshes.
Inceptisol	Moist, has poorly developed horizons. Formed recently.
Mollisol	Soft, black, and rich in nutrients. Forms in subhumid to subarid grasslands.
Oxisol	Very weathered, rich in aluminum oxide and iron oxide, low in plant nutrients. Forms in tropical regions.
Spodosol	Acidic, low in plant nutrients, ashy, has accumulations of iron and aluminum. Forms in humid forests.
Ultisol	Very mature, strongly weathered soils, low in plant nutrients.
Vertisol	Clay-rich soils capable of swelling when wet and shrinking and cracking when dry.

FIGURE B.11 U.S. Department of Agriculture map of soil types around the world.

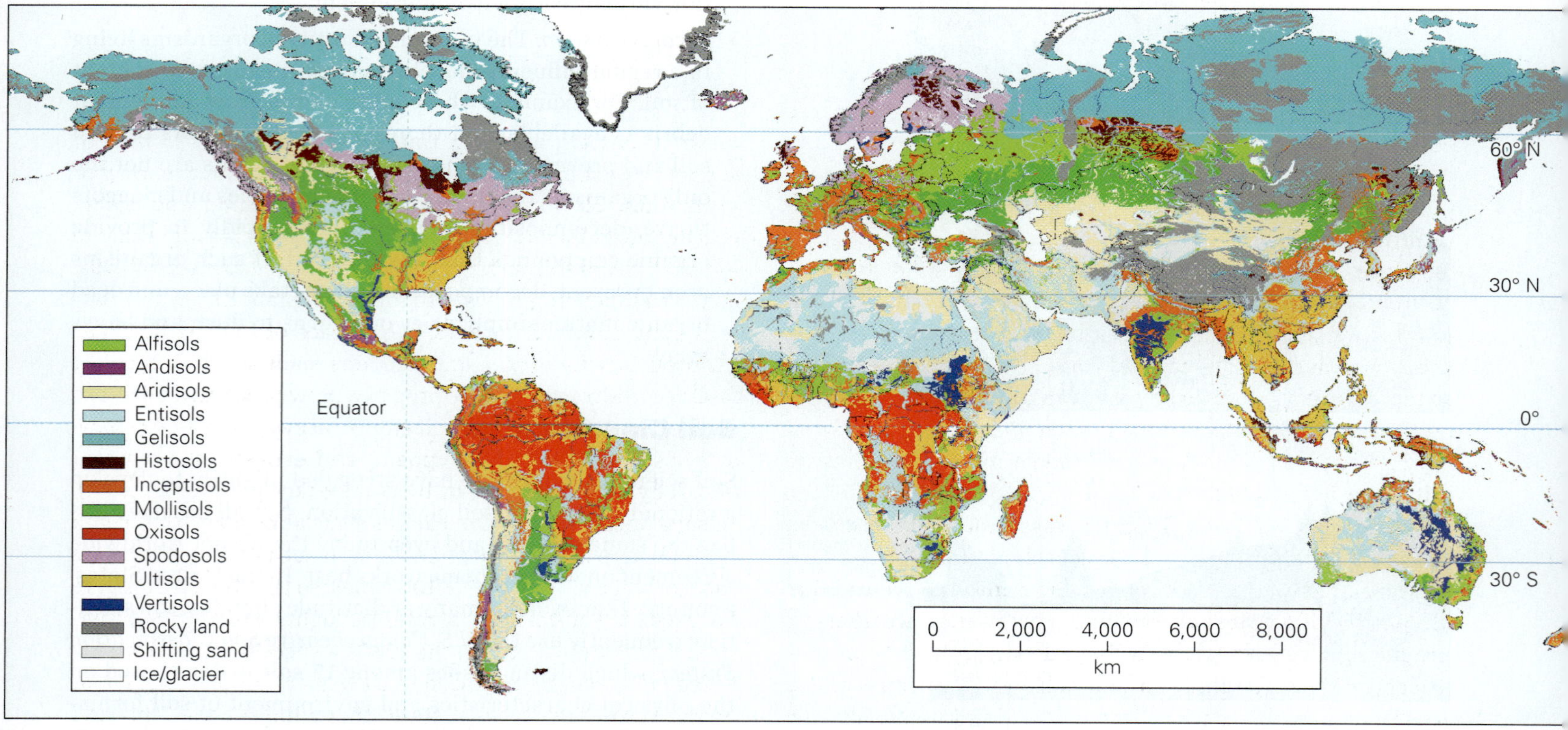

FIGURE B.12 Examples of soil classification.

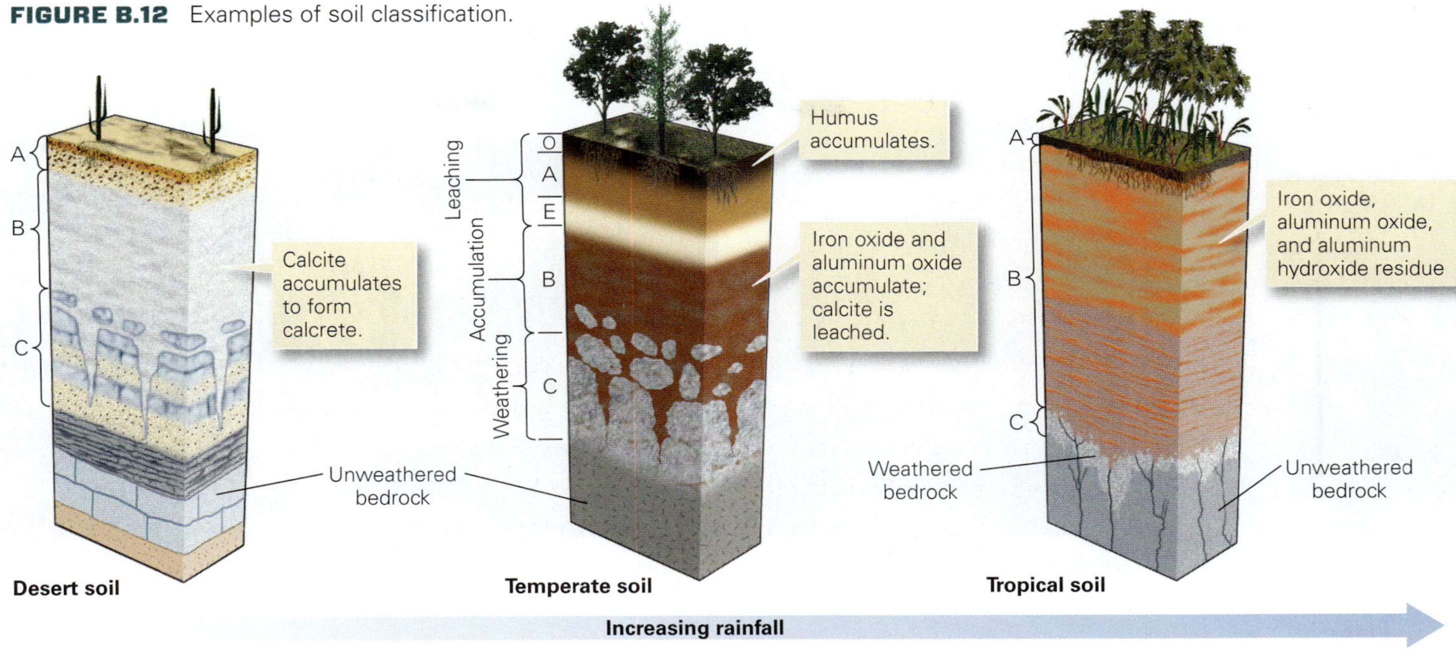

(a) Aridisol forms in deserts. Rainfall is so low that no O-horizon forms, and soluble minerals accumulate in the B-horizon.

(b) Alfisol forms in temperate climates. An O-horizon forms, and less-soluble materials accumulate in the B-horizon.

(c) Oxisol forms in tropical climates where percolating rainwater leaches all soluble minerals, leaving only iron- and aluminum-rich residues.

Let's look at the differences among a few examples of soil types. In deserts, where there is very little rainfall and sparse vegetation, an *aridisol* forms (**Fig. B.12a**). (In older classifications, these were known as pedocal soils.) Aridisols have no O-horizon, because there is so little organic material, and the A-horizon is thin. Soluble minerals, specifically calcite, that would be washed away entirely if there were more rainfall, accumulate instead in the B-horizon. In fact, capillary action may bring calcite up from deeper down as water evaporates at the ground surface. The calcite locally cements clasts together in the B-horizon to form a rock-like mass called *caliche* or *calcrete*. In temperate environments, an *alfisol* forms—this soil has an O-horizon, and because of moderate amounts of rainfall, materials leached from the A-horizon accumulate in the B-horizon (**Fig. B.12b**). (In older classifications, these were known as pedalfer soils.) In a tropical climate, *oxisols* develop. Here, so much rainfall percolates down into the ground that all reactive minerals in the soil undergo chemical weathering, producing ions and clay that flush downward. This process leaves an A-horizon that contains substantial residues of less soluble compounds, such as iron oxide, aluminum oxide, and aluminum hydroxide (**Fig. B.12c**). The resulting soil tends to be brick-red and is traditionally called *laterite*.

Soil Erosion

As we have seen, soils take time to form, and soils capable of supporting crops or forests are a natural resource worthy of protection. However, agriculture, overgrazing, and clear-cutting often lead to soil destruction. Crops rapidly remove nutrients from soil, so if they are not replaced, the soil will not contain sufficient nutrients to maintain plant life. When natural plant cover disappears, the surface of the soil becomes exposed to wind and water. Actions such as the impact of falling raindrops or the rasping of a plow break up soil at the surface, with the result that it can wash away in water or blow away as dust. When this happens, **soil erosion**, the removal of soil by running water or by wind, takes place (**Fig. B.13**). In some localities, erosion carries away almost six tons of soil from an acre of land per year. Human activities can increase rates of soil erosion by 10 to 100 times, so that the erosion rate far exceeds soil-formation rate. Droughts worsen the situation. For example, during the 1930s a succession of droughts killed off so much vegetation in the American plains of Oklahoma, Texas, and adjacent states that wind stripped the land of soil and caused devastating dust storms, transforming the region into the Dust Bowl.

FIGURE B.13 Examples of soil erosion.

(a) When crops aren't growing, the soil of farm fields can be blown away in strong winds.

(b) When vegetation doesn't protect soil, water erosion can carve a badlands of closely spaced gullies.

The consequences of *rainforest destruction* have particularly profound effects on soil. In an established rainforest, lush growth provides sufficient organic debris so that trees can grow. But if the forest is logged, or cleared for agriculture, the humus disappears rapidly, leaving laterite that contains few nutrients. Crop plants consume whatever nutrients remain so rapidly that the soil soon becomes infertile, useless for agriculture and unsuitable for regrowth of rainforest trees.

Another View Bright red laterite, a soil rich in iron and aluminum oxides and hydroxides, exposed in a Brazilian roadcut.

Interlude B Review

Interlude Summary

- Sediment consists of loose fragments, derived from pre-existing rock, precipitated from water, or formed by the breakup of shells.
- Rock at or near the Earth's surface weathers over time due to interaction with air and water.
- Physical weathering breaks larger rock bodies into smaller pieces. Jointing plays an important role.
- Chemical weathering involves a variety of chemical reactions that dissolve and/or alter minerals.
- Soil is regolith that underwent change over time when water percolated down through it. Soil can be modified by interacting with organisms.
- The character of soil depends on the composition of its source material, as well as on climate and time.
- Soil is an essential resource to society, for it provides the basis for agriculture. Various phenomena, some caused by humans, can lead to the loss of soil.

Guide Terms

chemical weathering (p. 174)
clast (p. 172)
erosion (p. 173)
physical weathering (p. 172)
sediment (p. 171)
soil (p. 178)
soil erosion (p. 183)
soil horizon (p. 180)
soil order (p. 181)
soil profile (p. 180)
subsoil (p. 180)
topsoil (p. 180)
weathering (p. 172)
zone of accumulation (p. 179)
zone of leaching (p. 179)

Review Questions

1. Explain the difference between physical and chemical weathering.
2. What processes can cause originally solid rock to break into pieces?
3. What are the various reactions that can contribute to chemical weathering?
4. Why doesn't weathering take place on the Moon?
5. Explain the process of soil formation.
6. Why do soils develop distinct horizons?
7. What factors determine the character (e.g., thickness, texture, types of horizons, etc.) of a soil?
8. How does a soil that forms in a tropical climate differ from one that forms in an arid climate?
9. Explain why soil erosion has been exacerbated by human activity.

Online Resources

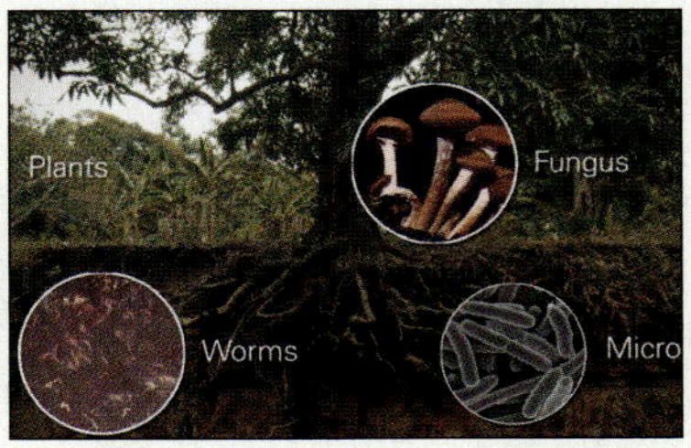

Videos

This interlude features a video on how soil is formed and the role of organisms in this process.

LEARNING OBJECTIVES

By the end of this chapter, you should understand...

1. how sediment produced by weathering and biological activity at the Earth's surface can transform into sedimentary rock.
2. how to interpret past environments and sea-level changes by examining the composition, textures, and layering found in sedimentary rock.
3. clues that geologists use to identify and classify different types of sedimentary rocks.
4. where and why thick accumulations of sedimentary rocks form on Earth.

▲ Look closely at this cliff in southern Nevada. You'll see layer upon layer of sedimentary strata. The ledges expose resistant sandstone. Slopes are underlain by weaker shale. These rocks formed when sediments were buried and lithified long ago.

CHAPTER 6

Pages of Earth's Past: Sedimentary Rocks

6.1 Introduction

In this day when *Google Earth*™ can take you to every nook and cranny of our planet's surface at the touch of a computer mouse, it's hard to imagine a world in which vast regions were blank spaces on a map. But only a century ago, that was the state of affairs that members of the 1910–13 British Antarctic Expedition were facing. Led by Robert Falcon Scott, a team of explorers from the expedition set out to be the first to reach the South Pole. The first part of the journey took them over the Ross Ice Shelf, a broad plain of ice not far above sea level. But to reach the pole, they had to haul their heavy sledges up the Beardmore Glacier, a river of ice that had cut its way down through the rugged Transantarctic Mountains. The cliffs overlooking the glacier, as is the case along much of the length of the Transantarctic Mountains, expose layer upon layer of a light-colored grainy rock. One of the expedition's members, Edward Wilson, served as the team's geologist, and during the journey collected numerous specimens of these rocks. Scott, Wilson, and the others succeeded in reaching the South Pole, on January 17, 1912. But when they arrived, they found to their profound disappointment that a Norwegian explorer, Roald Amundsen, had beaten them there by 34 days. On their return journey, all of the British explorers perished in the blizzards and cold of

> *In every grain of sand there is a story of Earth.*
> RACHEL CARSON
> (American marine biologist and conservationist, 1907–1964)

the southernmost continent. When rescuers eventually came upon Scott's last campsite, they found Wilson's specimens, some of which contained fossils of *Glossopteris* (See Fig. 2.2c), the fossil whose distribution Alfred Wegener would use as evidence of continental drift.

What are the grainy, layered rocks that Wilson collected? They are a type of sedimentary rock called sandstone. Formally defined, **sedimentary rock** is rock that forms at or near the surface of the Earth in one of several ways: by the cementing together of loose clasts (fragments or grains) that had been produced by physical or chemical weathering of pre-existing rock, by the growth of mounds of shells, by the cementing together of shells and shell fragments, by the accumulation and subsequent alteration of organic matter derived from living organisms, or by the precipitation of minerals directly from surface-water solutions. Layers, or beds, of sedimentary rock are like the pages of a book, recording tales of ancient events and environments on the ever-changing face of the Earth. They occur only in the upper part of the crust, and form a *cover* that buries the underlying *basement* of igneous and/or metamorphic rock (**Fig. 6.1**).

In Interlude B, we introduced the concept of weathering and showed how it attacks solid rock to break it down into ions and loose sediment grains. In this chapter, we see how these materials can be buried and transformed into sedimentary rock. We also introduce various specific types of sedimentary rocks and show how geologists use the study of sedimentary rocks to characterize Earth System history. Finally, we discuss special settings, called sedimentary basins, in which particularly thick successions of sedimentary rock accumulate.

6.2 Classes of Sedimentary Rocks

Geologists divide sedimentary rocks into four major classes, based on their mode of origin. (1) **Clastic sedimentary rock** forms from cemented-together *clasts*, solid fragments and grains broken off of pre-existing rocks (the word comes from the Greek *klastos*, meaning broken); (2) **biochemical sedimentary rock** consists of shells; (3) **organic sedimentary rock** consists of carbon-rich relicts of plants or other organisms; and (4) **chemical sedimentary rock** is made up of minerals that precipitated directly from water solutions. Let's now look at each of these major classes in more detail.

Clastic Sedimentary Rocks

Forming Clastic Rocks Until about 800 years ago, a thriving community of Native Americans inhabited the high plateau of Mesa Verde, Colorado. In hollows beneath huge overhanging ledges, they built multistory stone-block buildings that have survived to this day (**Fig. 6.2**). Clearly, the blocks are solid and

FIGURE 6.1 In this close-up of the inner gorge of the Grand Canyon, we see a sedimentary-rock blanket that forms a cover over an older basement of metamorphic and igneous rock.

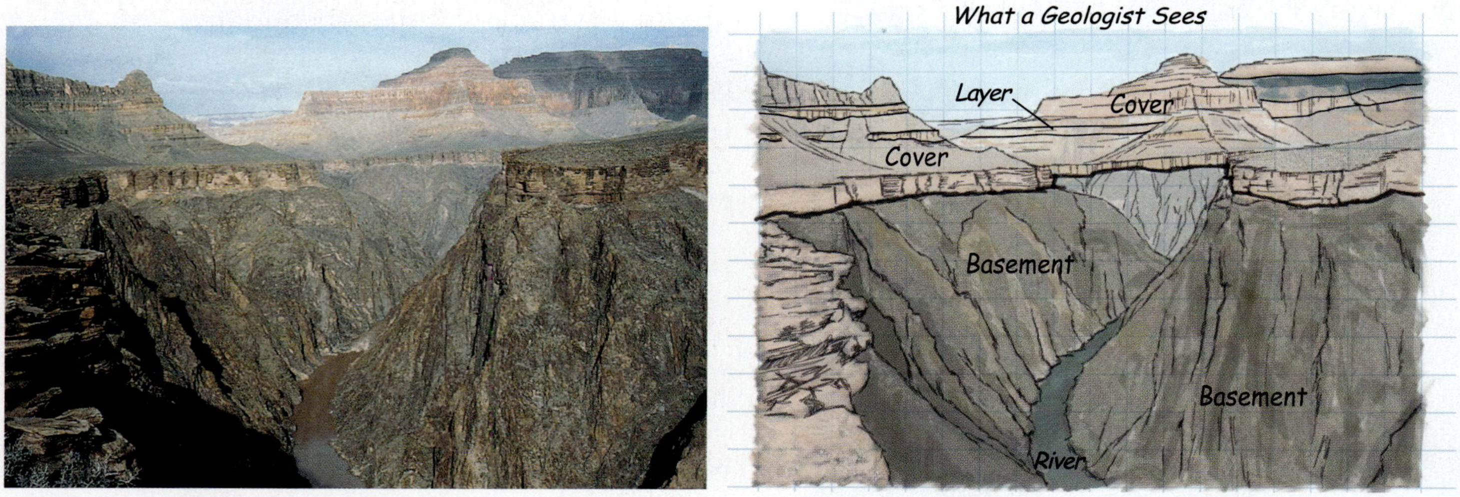

(a) The contact between the sedimentary cover and underlying basement lies at the top of the inner gorge.

(b) A geologist's sketch emphasizes the contact. Here the basement consist of metamorphic and igneous rock.

FIGURE 6.2 The cliff dwellings nestled beneath a ledge at Mesa Verde, Colorado, are made of sandstone blocks. The inset shows the grainy character of the rock.

durable—they are, after all, rock. But if you were to rub your thumb along them, they would feel gritty, and small grains of quartz would break free and roll under your thumb, for the blocks consist of quartz sand grains cemented together. Geologists call such rock a **sandstone**.

Sandstone is an example of clastic sedimentary rock. It consists of loose clasts that have been stuck together to form a solid mass. The clasts, or grains, can consist of individual minerals (such as grains of quartz or flakes of clay) or of chunks of rock (such as pebbles of granite). Forming an accumulation of clasts which then transforms into clastic sedimentary rock takes place via the following five steps (**Fig. 6.3**).

- *Weathering:* Clasts form by disintegration of bedrock into separate grains due to physical and chemical weathering.
- *Erosion:* **Erosion** refers to the combination of processes that separate rock or *regolith* (surface debris) from its substrate. Erosion involves abrasion, falling, plucking, scouring, and dissolution, and is caused by moving air, water, or ice.
- *Transportation:* Gravity, wind, water, or ice can carry sediment. The ability of a medium to carry sediment depends on its viscosity and velocity. Solid ice can transport sediment of any size, regardless of how slowly the ice moves. Very fast-moving, turbulent water can transport coarse fragments (cobbles and boulders); moderately fast-moving water can carry only sand and gravel; and slow-moving water carries only silt and clay. Strong winds can move sand and dust, but gentle breezes carry only dust.
- *Deposition:* **Deposition** is the process by which sediment settles out of the transporting medium. Sediment settles out of wind or moving water when these fluids slow down, because as the velocity decreases, the fluid no longer has the ability to carry sediment. Sediment is deposited by ice when the ice melts.
- *Lithification:* Geologists refer to the transformation of loose clasts into solid rock as **lithification**, a process that takes place in two stages. It begins with **compaction**, when the weight of overburden squeezes air or water out from between grains, so the grains can fit together more tightly. It ends with **cementation**, when minerals (commonly quartz or calcite) precipitate from groundwater and fill the remaining spaces between clasts, to form a "cement" that binds grains together.

Classifying Clastic Sedimentary Rocks Say that you pick up a clastic sedimentary rock and want to describe it in enough detail that, from your words alone, another person can picture the rock. What characteristics should you mention? Geologists find the following characteristics most useful.

- *Clast size:* Size refers to the diameter of fragments or grains making up a rock. The names used for clast size, listed in

FIGURE 6.3 The five steps in clastic sedimentary rock formation.

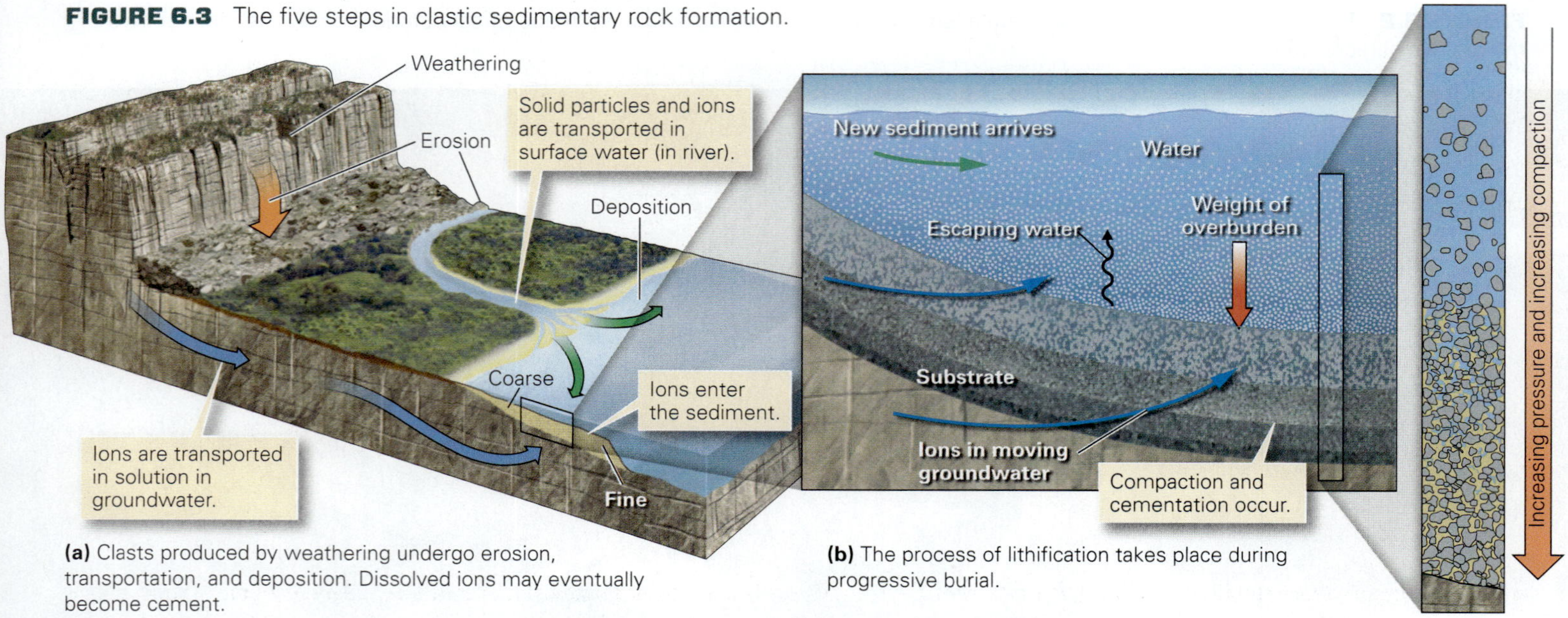

(a) Clasts produced by weathering undergo erosion, transportation, and deposition. Dissolved ions may eventually become cement.

(b) The process of lithification takes place during progressive burial.

order from coarsest to finest, are boulder, cobble, pebble, sand, silt, and clay (see Table B.1 in Interlude B).

- *Clast composition:* Composition refers to the makeup of clasts in sedimentary rock. Clasts may be composed of rock fragments or individual mineral grains.
- *Angularity and sphericity:* Angularity indicates the degree to which clasts have smooth surfaces, or have angular corners and edges (**Fig. 6.4a**). Sphericity, in contrast, refers to how closely the shape of a clast resembles a sphere.
- *Sorting:* **Sorting** of clasts indicates the proportion of clasts in a rock that are the same size. *Well-sorted* sediment consists entirely of clasts that are the same size, whereas *poorly sorted* sediment contains a mixture of clast sizes (**Fig. 6.4b**).
- *Character of cement:* Not all clastic sedimentary rocks contain the same kind of cement. In some, the cement consists predominantly of quartz, whereas in others, it consists predominantly of calcite.

With these characteristics in mind, we can distinguish among several common types of clastic sedimentary rocks, listed in **Table 6.1**. This table provides common rock names, although specialists sometimes use more precise names based on more complex classification schemes.

The size, angularity, sphericity, and sorting of clasts depends on the medium (water, ice, or wind) that transports the clasts and, in the case of water or wind, on both the velocity of the medium and the distance of transport. The composition of the clasts depends on the composition of rock from

FIGURE 6.4 Grain characteristics and their evolution with increasing transport.

Angularity

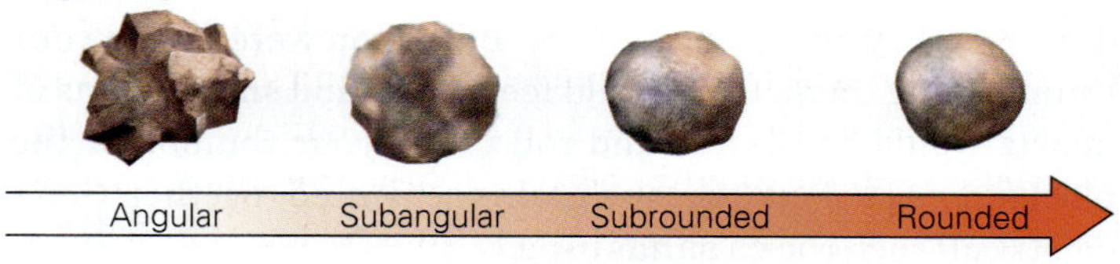

(a) Individual clasts also tend to become more rounded and smoother.

Sorting

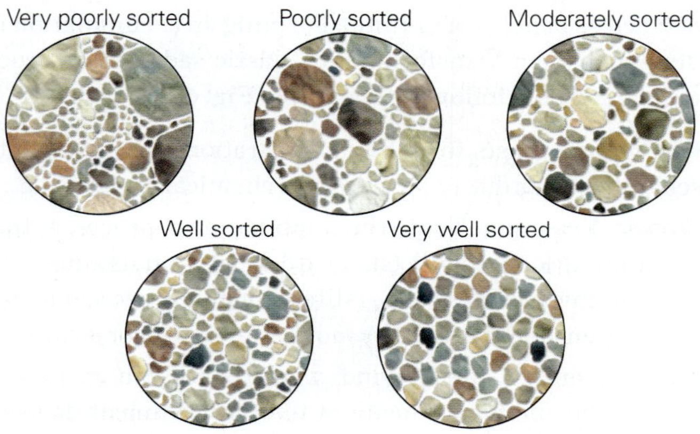

(b) If transport sifts grains, carrying smaller ones farther and leaving coarser ones behind, grains in a sediment tend to be the same size.

TABLE 6.1 Classification of Clastic Sedimentary Rocks

Clast Size*	Clast Character	Rock Name (Alternative Name)
Coarse to very coarse	Rounded pebbles and cobbles Angular clasts Large clasts in muddy matrix	Conglomerate Breccia Diamictite
Medium to coarse	Sand-sized grains › Quartz grains only › Quartz and feldspar sand › Sand-sized rock fragments › Quartz sand and sand-sized rock fragments in a clay-rich matrix	Sandstone › Quartz sandstone (quartz arenite) › Arkose › Lithic sandstone › Wacke (informally called graywacke)
Fine	Silt-sized clasts	Siltstone
Very fine	Clay and/or very fine silt	Shale, if it breaks into platy sheets Mudstone, if it doesn't break into platy sheets

*For precise diameters, see Table B.1 on p. 173.

which the clasts were derived, and on the degree of chemical weathering that the clasts have undergone. Thus, the type of clasts that accumulate in a sedimentary deposit varies with location. To see how, let's follow the fate of rock fragments as they gradually move from a cliff face in the mountains via a river to the seashore. Different kinds of sediment develop along the route. Each kind, if buried and lithified, would yield a different type of sedimentary rock.

To start the discussion of sedimentary rock types, imagine that large blocks of granite tumble off a cliff and slam into other blocks already at the bottom. The impact shatters the blocks, producing a pile of angular clasts. If these fragments were to be cemented together before being transported very far, the resulting rock would be **breccia** (**Fig. 6.5a**). Later, a storm causes the clasts to be carried away by a turbulent river. In the water, clasts bang into each other and into the riverbed, a process that shatters them into still smaller pieces and breaks off their sharp edges. As the clasts are carried downstream, they gradually become rounded pebbles and cobbles. When the river water slows, these clasts stop moving and form a mound or bar of gravel. Burial and lithification of these rounded clasts produces **conglomerate** (**Fig. 6.5b**).

If the gravel stays put for a long time, it undergoes chemical weathering. As a consequence, cobbles and pebbles break apart into individual mineral grains, eventually producing a mixture of quartz, feldspar, and clay. Clay is so fine that flowing water easily picks it up and carries it downstream, leaving sand containing a mixture of quartz and some feldspar grains—this sediment, if buried and lithified, becomes **arkose** (**Fig. 6.5c**). Over time, feldspar grains in sand continue to weather into clay so that gradually, during successive events that wash the sediment downstream, the sand loses feldspar and ends up being composed almost entirely of durable quartz grains. Some of the sand may make it to the sea, where waves carry it to beaches, and some may end up in desert dunes. This sediment, when buried and lithified, becomes quartz sandstone (**Fig. 6.5d**). Meanwhile, some of the silt and clay accumulates in the flat areas bordering streams, regions called *floodplains* (see Chapter 14) that become submerged only during floods. The rest settles in a wedge, called a *delta*, at the mouth of the river, or in lagoons or mudflats along the shore. When lithified, the silt becomes **siltstone**, and the mud becomes **shale** or **mudstone** (**Fig. 6.5e**).

Did you ever wonder...
where beach sand comes from?

Biochemical Sedimentary Rocks

The Earth System involves many interactions between living organisms and the physical planet. Numerous organisms have evolved the ability to extract dissolved ions from seawater to make solid shells. When the organisms die, the solid material in their shells survives. This material, when lithified, comprises biochemical sedimentary rock. Geologists recognize several distinct types of biochemical sedimentary rocks.

Biochemical Limestone A snorkeler gliding above a reef sees an incredibly diverse community of coral and algae, around which creatures such as clams, oysters, snails (gastropods), and lampshells (brachiopods) live, and above which plankton float (**Fig. 6.6a**). Though they look so different from each other, many of these organisms share an important characteristic: they make solid shells of calcium carbonate ($CaCO_3$). The $CaCO_3$ crystallizes either as calcite or aragonite. (These minerals have the same composition, but different crystal structures.) When the organisms die, the shells remain and may accumulate. Rock formed dominantly from this material is *biochemical limestone*. Since the principal compound making up limestone is $CaCO_3$, geologists refer to limestone as a type of *carbonate rock*.

Limestone comes in a variety of textures, because the material that forms it accumulates in a variety of ways. For example, limestone can originate from reef builders (such as corals) that grew in place, from shell debris that was broken up and transported, from carbonate mud, or from plankton shells that settled like snow out of water. Because of this variety, geologists distinguish among *fossiliferous limestone*, consisting of visible fossil shells or shell fragments (**Fig. 6.6b**); *micrite*, consisting of very fine carbonate mud; and *chalk*, consisting of plankton shells. Experts recognize many other types as well.

Typically, limestone is a massive light-gray to dark-bluish-gray rock that breaks into chunky blocks. It doesn't look much like a pile of shell fragments (**Fig. 6.6c**). That's because several processes change the texture of the rock over time. For example, water passing through the rock not only precipitates cement but also dissolves some carbonate grains and causes new ones to grow.

FIGURE 6.5 Different kinds of clasts lithify into different kinds of sedimentary rocks.

(a) Lithification of an accumulation of angular clasts yields breccia.

(b) Layers of river gravel lithify into conglomerate.

(c) Sediment deposited in an alluvial fan, close to its source, can be feldspar rich. Lithification of this sediment yields arkose.

Biochemical Chert If you walk beneath the northern end of the Golden Gate Bridge in California, you will find outcrops of reddish, almost porcelain-like rock occurring in 3- to 15-cm-thick layers (**Fig. 6.7a**). Hit it with a hammer, and the rock cracks to form smooth, spoon-shaped (conchoidal) fractures. This biochemical chert is made from cryptocrystalline quartz (*crypto* is Greek for hidden), meaning quartz grains that are too small to be seen without the extreme magnification of an electron microscope. The chert beneath the Golden Gate Bridge formed from the shells of silica-secreting plankton that accumulated on the seafloor. Gradually, after burial, the shells dissolved, forming a silica-rich gel. Chert then formed when this gel solidified.

(d) Layers of beach or dune sand lithify into sandstone.

(e) Layers of mud, exposed beneath marsh grass, lithify to form shale. Here the thin-bedded shale is interbedded with sandstone.

Organic Sedimentary Rocks

We've seen how the mineral shells of organisms ($CaCO_3$ or SiO_2) can accumulate and lithify to become biochemical sedimentary rocks. What happens to the "guts" of the organisms—the cellulose, fat, carbohydrate, protein, and other organic compounds that make up living matter? Commonly, this organic debris gets eaten by other organisms or decays at the Earth's surface. But in some environments, the organic debris settles along with other sediment and is eventually buried. When lithified, organic-rich sediment becomes organic sedimentary rock. Since the dawn of the industrial revolution in the early 19th century, organic sedimentary rock has provided the fuel of modern industry and transportation, for organic chemicals yield energy when burned. Coal and oil shale are the two most common types of organic sedimentary rocks.

Coal is a black, combustible rock containing between 40% and 90% carbon. The remainder consists of oxygen, nitrogen, hydrogen, sulfur, silica, and minor amounts of other elements. Typically, the carbon in coal occurs in large, complex organic molecules made of many rings. As discussed further in Chapter 12, coal forms when plant remains have been buried deeply enough and long enough for the material to become compacted and to lose significant amounts of volatiles (hydrogen, water, CO_2, and ammonia); as the volatiles seep away, a concentration of carbon remains (**Fig. 6.7b**).

Oil shale is a shale that contains not only clay but also between 15% and 40% organic material in a form called *kerogen*. The kerogen in oil shale comes from the fats and proteins that made up the living part of plankton or algae. If the tiny organisms settle in an environment where they do not immediately rot away or get eaten, they mix with the clay minerals in mud. When the mud gets buried and lithified, to form shale, the organic material transforms into kerogen. The presence of organic material colors oil shale black.

Chemical Sedimentary Rocks

The colorful terraces, or mounds, that grow around the vents of hot-water springs; the immense layers of salt that underlie the floor of the Mediterranean Sea; the smooth, sharp point of an ancient arrowhead—these materials all have something in common. They all consist of rock formed primarily by the precipitation of minerals from water solutions. We call such rocks *chemical sedimentary rocks*. Typically they have a crystalline texture, partly formed during their original precipitation and partly when, at a later time, new crystals grow at the expense of old ones through a process called recrystallization. In some examples, crystals are coarse, whereas in others, they are too small to see. Geologists distinguish among many types of chemical sedimentary rocks, primarily on the basis of composition.

Evaporites—The Products of Saltwater Evaporation In 1965, two daredevil drivers in jet-powered cars battled to be

FIGURE 6.6 The formation of carbonate rocks (limestone).

(a) In this modern coral reef, corals produce shells. If buried and preserved, these become limestone.

(b) Wave energy breaks up shells, forming carbonate clasts that later become cemented together.

(c) This road cut near Kingston, New York, exposes beds of limestone. The vertical stripes are drillholes.

the first to set the land speed record of 600 mph. On November 7, Art Arfons, in the *Green Monster*, peaked at 576.127 mph; but eight days later Craig Breedlove, driving the *Spirit of America*, reached 600.601 mph. Traveling at such speeds, a driver must maintain an absolutely straight line; any turn will catapult the vehicle out of control. Thus, high-speed trials take place on extremely long and flat racecourses. Not many places can provide such conditions—but the Bonneville Salt Flats of Utah do. The salt flats formed when an ancient salt lake evaporated. Under the heat of the Sun, the water turned to vapor and drifted up into the atmosphere, while the salt that had been dissolved in the water stayed behind.

Salt precipitation occurs where salt water becomes supersaturated, meaning that it has exceeded its capacity to contain more dissolved ions. In supersaturated salt water, ions bond to form solid grains that either settle out of the water or grow on the floor of the water body. Supersaturated salt water develops where evaporation removes water from a water body faster than the rate at which new water enters. This process takes place in desert lakes and along the margins of restricted seas (**Fig. 6.8**). For thick deposits of salt to form, large volumes of water must evaporate. Because salt deposits form as a consequence of evaporation, geologists refer to them as **evaporites**. The specific type of salt minerals comprising an evaporite depends on the amount of evaporation. When 80% of the water evaporates, gypsum forms; and when 90% of the water evaporates, halite precipitates.

Travertine (Chemical Limestone) **Travertine** is a rock composed of crystalline calcium carbonate ($CaCO_3$) formed by chemical precipitation from groundwater that has seeped out at the ground surface either in hot- or cold-water springs, or on the walls of caves. What causes this precipitation? It happens, in part, when the groundwater degasses, meaning that some of the carbon dioxide that had been dissolved in the groundwater bubbles out of solution, for removal of carbon dioxide encourages the precipitation of carbonate. Precipitation also occurs when water evaporates, thereby increasing the concentration of carbonate. Various kinds of microbes live in the environments in which travertine accumulates, so biological activity may also contribute to the precipitation

FIGURE 6.7 Examples of biochemical and organic sedimentary rocks.

(a) This bedded chert developed on the deep seafloor by the deposition of plankton that secrete silica shells.

(b) Coal is deposited in layers (beds), just like other kinds of sedimentary rocks.

FIGURE 6.8 The formation of evaporite deposits.

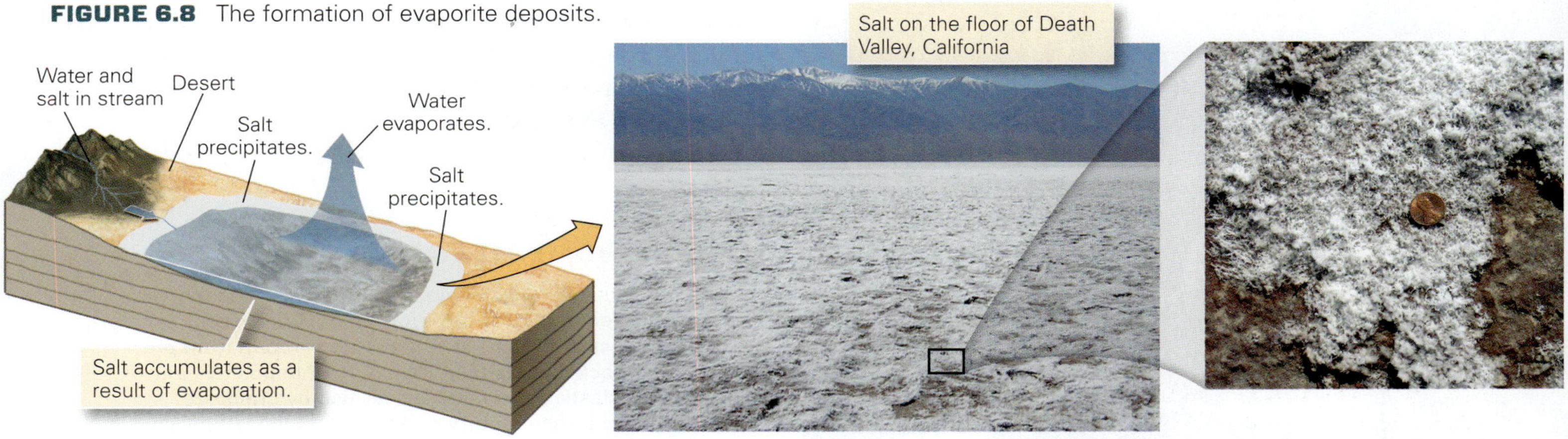

(a) In lakes with no outlet, tiny amounts of salt brought in by streams stay behind as the water evaporates. When the water evaporates entirely, a white crust of salt remains.

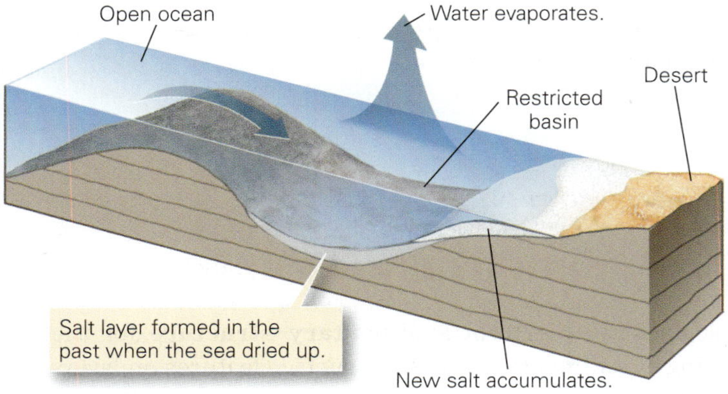

(b) Salt precipitation can also occur along the margins of a restricted marine basin, if salt water evaporates faster than it can be resupplied.

process. Travertine produced at springs forms terraces and mounds that are meters or even hundreds of meters thick, such as those at Mammoth Hot Springs (**Fig. 6.9a**). Travertine also grows on the walls of caves where groundwater seeps out (**Fig. 6.9b**). In cave settings, travertine builds up beautiful and complex growth forms called speleothems (see Chapter 16).

Dolostone Another carbonate rock, **dolostone**, differs from limestone in that it contains the mineral dolomite ($CaMg[CO_3]_2$), which contains equal amounts of calcium and

FIGURE 6.9 Examples of travertine (chemical limestone) deposits.

(a) Travertine accumulates in terraces at Mammoth Hot Springs in Yellowstone Park, Wyoming.

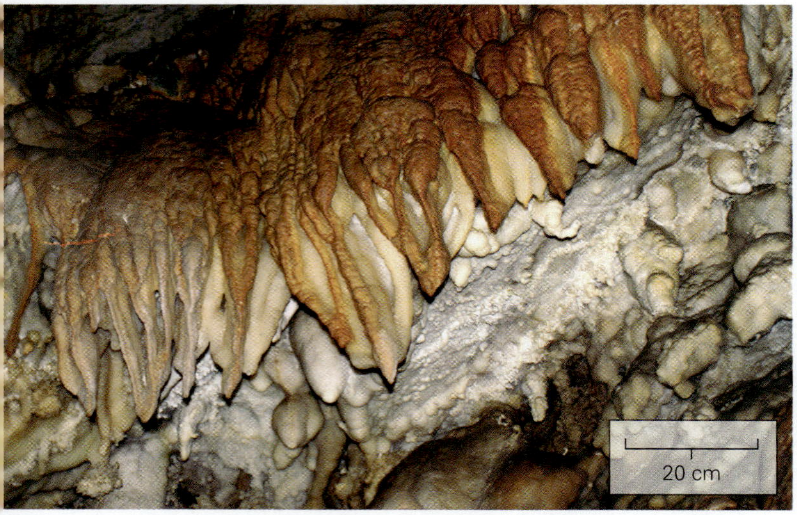

(b) Travertine speleothems form as calcite-rich water drips from the ceiling of Timpanogos Cave in Utah.

magnesium. Where does the magnesium come from? Most dolostone forms by a chemical reaction between solid calcite and magnesium-bearing groundwater. This change may take place beneath lagoons along a shore soon after the limestone formed, or a long time later, after the limestone has been buried deeply.

Chemically Precipitated Chert A tribe of Native Americans, the Onondaga, once lived off the land in eastern New York State. Here, outcrops of limestone contain layers or *nodules* (lenses or lumps) of a black chert (**Fig. 6.10a**). Because of the way this chert breaks, the tribe's artisans could fashion sharp-edged tools (arrowheads and scrapers) from it, so the Onondaga collected it for their own toolmaking industry and for use in trade with other people. Unlike the deep-sea (biochemical) chert described earlier, the chert collected by the Onondaga formed when cryptocrystalline quartz gradually replaced calcite crystals within a body of limestone long after the limestone was deposited; geologists call such material *replacement chert*.

Did you ever wonder... how chert used for arrowheads first formed?

Chert comes in many colors (black, white, red, brown, green, gray), depending on the impurities it contains. Petrified wood is chert that forms when silica-rich sediment, such as ash from a volcanic eruption, buries a forest. The silica dissolves in groundwater and then later precipitates as cryptocrystalline quartz within wood, gradually replacing the wood's cellulose (**Fig. 6.10b**). The chert deposit retains the shape of the wood cells and the growth rings within it. Some chert, known as agate, precipitates in concentric rings inside hollows in a rock and ends up with a striped appearance, caused by variations in the content of impurities incorporated in the chert (**Fig. 6.10c**).

TAKE-HOME MESSAGE

Geologists distinguish among many types of sedimentary rocks based on the mode of formation. Clastic sedimentary rocks consist of grains weathered and eroded from pre-existing rock that are then transported, deposited, and lithified; clastic rocks can be classified by grain size. Biochemical sedimentary rocks, such as limestone, consist of the shells of organisms, and organic sedimentary rocks, such as coal, form from the organic remains of organisms. Chemical sedimentary rocks precipitate from water solutions.

QUICK QUESTION Do all sedimentary rocks have the same composition? Why or why not?

6.3 Sedimentary Structures

Geologists use the term **sedimentary structure** for the layering of sedimentary rocks, for surface features on layers formed during deposition, and for the arrangement of grains within layers.

FIGURE 6.10 Examples of chert that precipitated in place.

(a) Replacement chert forms as layers of nodules between tilted limestone beds in New York.

(b) This 14 cm-diameter log of petrified wood from Wyoming formed about 50 Ma.

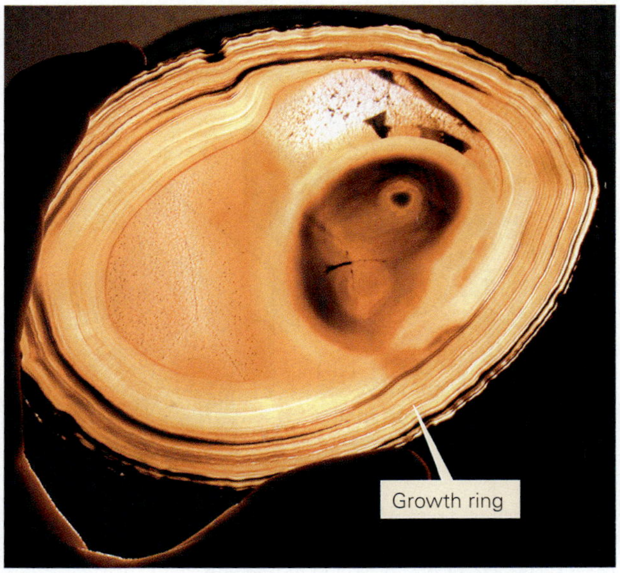

(c) A thin slice of Brazilian agate, lit from the back, shows growth rings.

SEE FOR YOURSELF...

GRAND CANYON, ARIZONA

Latitude
36°8′8.94″ N,

Longitude
112°15′48.56″ W

Zoom to an elevation of 15 km (~9.3 miles) and look straight down.

Here you can see the spectacular color banding caused by the succession of stratigraphic formations exposed on the walls of the Grand Canyon. Some of these units are light gray limestone, some are tan sandstone, some are red shale.

Bedding and Stratification

Let's start by introducing the language geologists use for discussing sedimentary layers. A single layer of sediment or sedimentary rock with a recognizable top and bottom is called a **bed**; the boundary between two beds is a *bedding plane*; several beds together constitute **strata** (singular stratum, from the Latin *stratum*, meaning pavement); and the overall arrangement of sediment into a sequence of beds is *bedding*, or stratification. In some outcrops, stratification can be quite subtle. But commonly, successive beds have different colors, textures, and resistance to erosion, so bedding gives outcrops a striped appearance (**Fig. 6.11a**).

Why does bedding form? To find the answer, we need to think about how sediment accumulates. Changes in the climate, water depth, current velocity, or the sediment source control the type of sediment deposited at a location at a given time. For example, on a normal day a slow-moving river may carry only silt, which collects on the riverbed (**Fig. 6.11b**). During a flood, the river flows faster and carries sand and pebbles, so a layer of sandy gravel forms over the silt layer. Then, when the flooding stops, more silt buries the gravel. If this succession of sediments become lithified and exposed for us to see, they appear as alternating beds of siltstone and sandy conglomerate. During geologic time, the overall character of a depositional environment can change dramatically. For example, if sea level rises, and water submerges an area in which desert sand had been accumulating, layers of carbonate shells may be deposited over layers of sand—when lithified, the sand layers become beds of sandstone, and the shell layers become beds of limestone.

FIGURE 6.11 Bedding in sedimentary rocks.

(a) Beds of sedimentary rock exposed along a road in Utah.

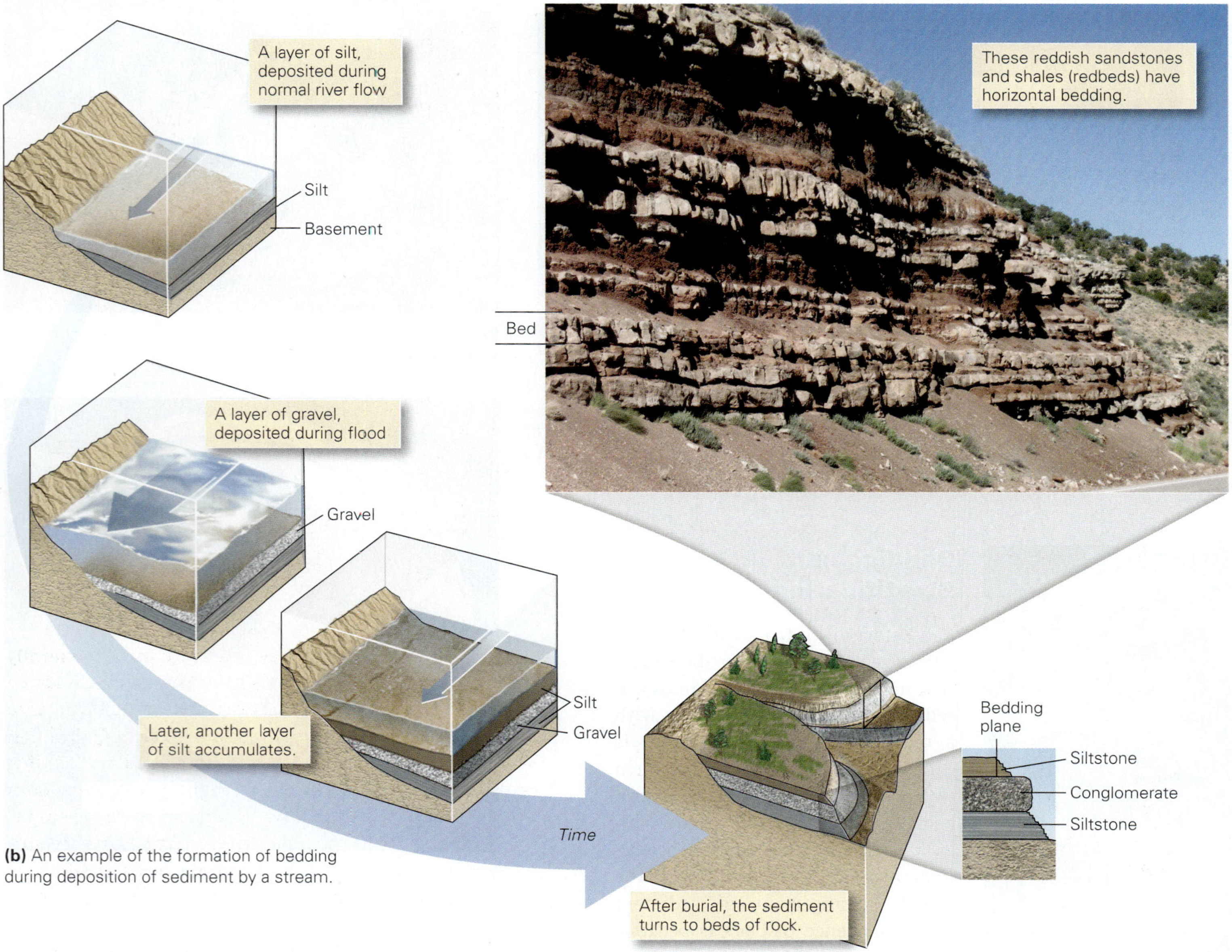
These reddish sandstones and shales (redbeds) have horizontal bedding.
Bed
A layer of silt, deposited during normal river flow
Silt
Basement
A layer of gravel, deposited during flood
Gravel
Later, another layer of silt accumulates.
Silt
Gravel
Time
Bedding plane
Siltstone
Conglomerate
Siltstone
After burial, the sediment turns to beds of rock.

(b) An example of the formation of bedding during deposition of sediment by a stream.

A sequence of strata that is distinctive enough to be traced as a unit across a fairly large region is called a *stratigraphic formation*, or simply a *formation* (**Fig. 6.12a**). For example, a region may contain a succession of alternating sandstone and shale beds deposited by rivers, overlaid by beds of marine limestone deposited later when the region was submerged by the sea. A geologist might identify the sequence of sandstone and shale beds as one formation and the sequence of limestone beds as another. Formations are often named after the locality where they were first found and studied. A *geologic map* portrays the distribution of stratigraphic formations (**Fig. 6.12b**).

Ripple Marks, Dunes, and Cross Bedding: Consequences of Deposition in a Current

Many clastic sediments accumulate in moving fluids (wind, rivers, or waves). Fascinating sedimentary structures develop at the interface between the sediment and the fluid. These structures are called *bedforms*. Bedforms that develop at a given location reflect such factors as the velocity of the flow and the size of the clasts. Though bedforms fall into many types, here we'll focus on only two—ripple marks and dunes. The growth of both produces cross bedding, a special type of lamination within beds.

FIGURE 6.12 The concept of a stratigraphic formation.

(a) A panoramic view of the Grand Canyon reveals many formations; the inset shows a close-up. The names of formations consisting of one rock type may indicate the rock type (e.g., Kaibab Limestone). The name of a formation including more than one rock type includes the word *formation* (Toroweap Formation). Several related formations comprise a group (Supai Group).

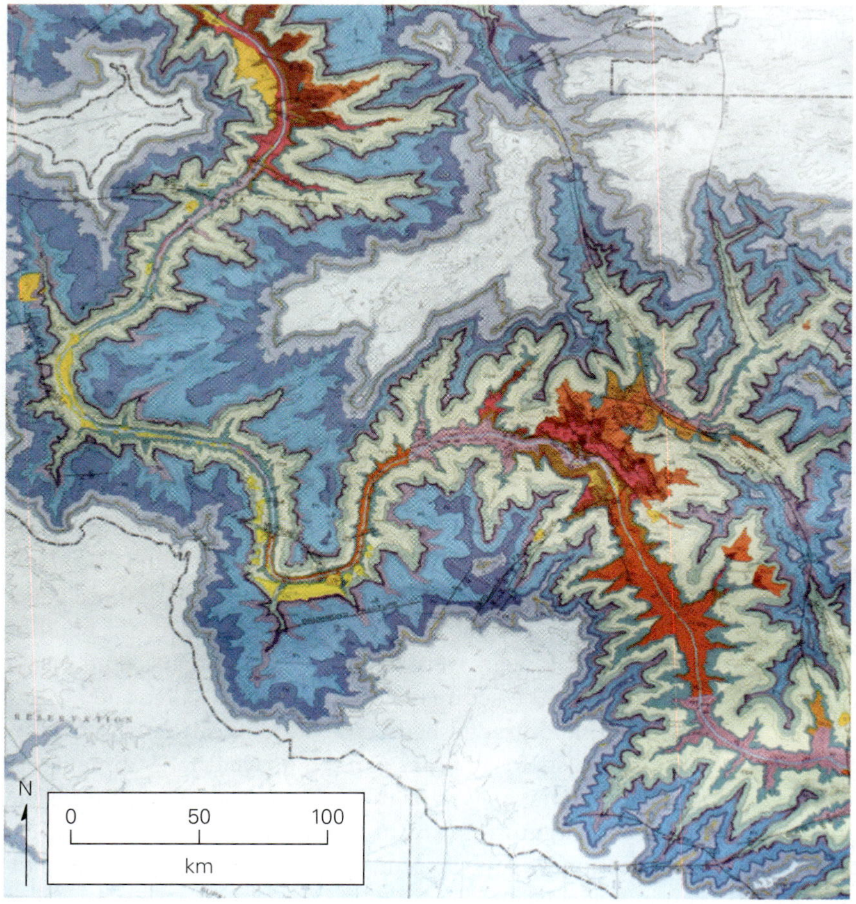

(b) A geologic map portrays the distribution of formations in a portion of the Grand Canyon. Each color band is a specific formation.

Ripple marks are relatively small (generally no more than a few centimeters high), elongated ridges that form on a bed surface at right angles to the direction of current flow. You can find ripples on modern beaches and preserved on bedding planes of ancient rocks (**Fig. 6.13**). A **dune** looks like a ripple, only it's much larger. For example, dunes on the bed of a stream may be tens of centimeters to several meters high, and wind-formed dunes in deserts may be tens of meters to over 100 meters high (**Fig. 6.14a**).

If you examine a vertical slice cut into a ripple or dune, you will find distinct internal laminations that are inclined at an angle to the main sedimentary layer. Such laminations are called **cross beds**. To see how cross beds develop, imagine a current of air or water moving uniformly in one direction (**Fig. 6.14b, c**). The current erodes and picks up clasts from the upstream part of the bedform and deposits them on the downstream or leeward face of the crest. Sediment builds up until gravity causes it to slip down the leeward face. With time, the dune or ripple builds in the downstream direction. The surface of the slip face establishes the shape of the cross beds. Eventually, a new cross-bedded layer builds out over a pre-existing one. The boundary between two

FIGURE 6.13 Ripple marks, a type of sedimentary structure, are visible on the surface of modern and ancient beds.

(a) Modern ripples exposed at low tide along a sandy beach on the shore of Cape Cod, Massachusetts.

(b) These 145-Ma ripples are preserved on a tilted bed of solid sandstone at Dinosaur Ridge, Colorado.

successive layers is called the *main bedding*, and the internal curving surfaces within the layer constitute the *cross bedding* (**Fig. 6.14d**).

Turbidity Currents and Graded Beds

Sediment deposited on a submarine slope tends to be unstable, so an earthquake or storm might disturb this sediment and cause it to slip downslope and mix with water to produce a murky, turbulent cloud. This cloud, which is denser than clear water, flows downslope like an underwater avalanche (**Fig. 6.15**). Geologists refer to such a moving submarine sediment suspension as a **turbidity current**. Downslope, the turbidity current slows, and the sediment that it has carried starts to settle out. Larger grains sink faster through a fluid than do finer grains, so the coarsest sediment settles out first. Progressively finer grains accumulate on top, with the finest sediment (clay) settling out last. This process forms a **graded bed**—that is, a layer of sediment in which grain size varies from coarse at the bottom to fine at the top. Geologists refer to a deposit from a turbidity current as a **turbidite**.

Bed-Surface Markings

A number of features develop on the surface of a bed as a consequence of events that happen during deposition or soon after, while the sediment layer remains soft. Such *bed-surface markings* include the following.

- *Mud cracks:* If a mud layer dries up after deposition, it cracks into roughly hexagonal plates that typically curl up at their edges. We refer to the openings between the plates as *mud cracks* (**Fig. 6.16**).
- *Scour marks:* As currents flow over a sediment surface, they may erode small troughs, called *scour marks*, parallel to the current flow.
- *Fossils:* **Fossils** are relicts of past life. Some fossils are shell imprints or footprints on a bedding surface (see Interlude E).

Burial and lithification of bed-surface markings can preserve the markings.

Why Study Sedimentary Structures?

Sedimentary structures are not just a curiosity, they are important clues that help geologists understand the *depositional environment* in which sediments accumulated. For example, the presence of ripple marks and cross bedding indicates that layers were deposited in a current, mud cracks indicate that the sediment layer was exposed to the air and dried out, and graded beds indicate deposition by turbidity currents. Also, fossil types can tell us whether sediment was deposited along a river or in the deep sea, because different species of organisms live in different environments. In the next section of this chapter, we examine depositional environments in greater detail.

FIGURE 6.14 Cross bedding, a type of sedimentary structure within a bed.

(a) Dunes in Death Valley, California.

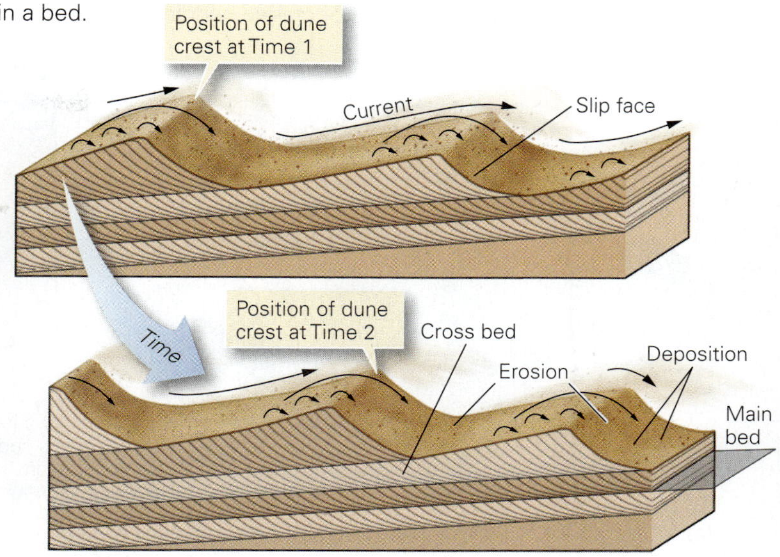

(b) Cross beds form as sand blows up the windward side of a dune or ripple and then accumulates on the slip face. With time, the dune crest moves.

(c) Slip face of a small sand dune, Death Valley. The edges of the slip face are outlined.

(d) A cliff face in Zion National Park, Utah, displays large cross beds formed between 200 and 180 Ma, when the region was a desert with large sand dunes.

TAKE-HOME MESSAGE

Exposures of sedimentary rocks typically contain a variety of sedimentary structures—features or textures formed during deposition. Examples include bedding, ripple marks on bed surfaces, and cross bedding within beds. Features like these provide clues to the environment of deposition. For example, cross beds and dunes form only in a current.

QUICK QUESTION What is the difference between a bed and a stratigraphic formation?

6.4 How Do We Recognize Depositional Environments?

Geologists refer to the conditions in which sediment was deposited as a **depositional environment**. Examples include beach, glacial, and river environments. To identify depositional

FIGURE 6.15 The development of graded bedding from turbidity currents.

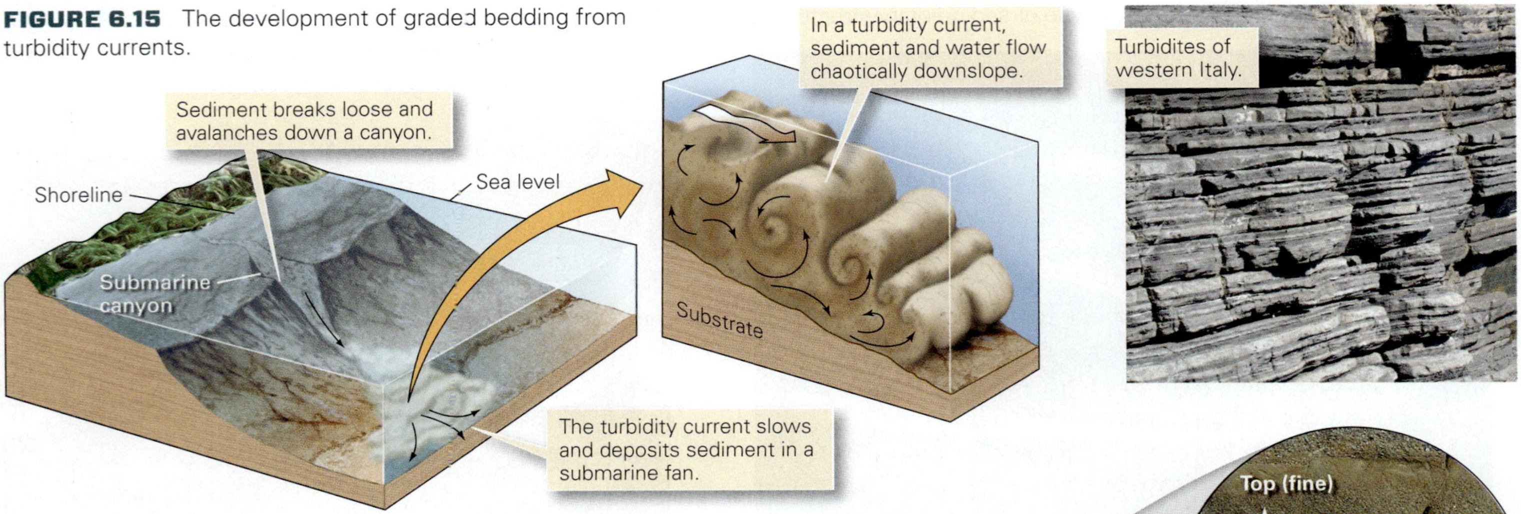

(a) An earthquake or storm triggers an underwater avalanche (turbidity current).

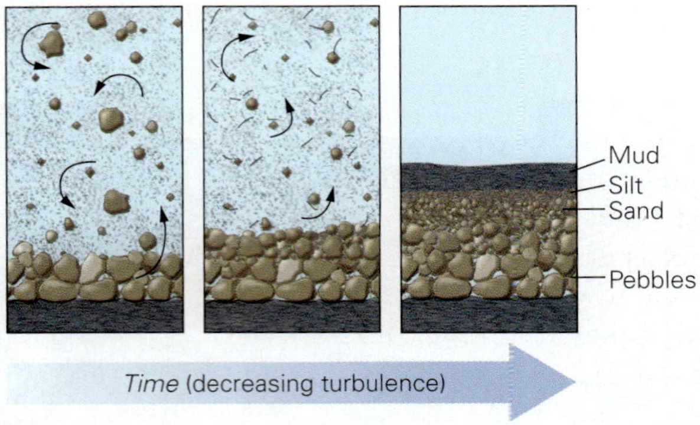

(b) As the turbidity current slows, larger grains settle first, followed by progressively finer grains.

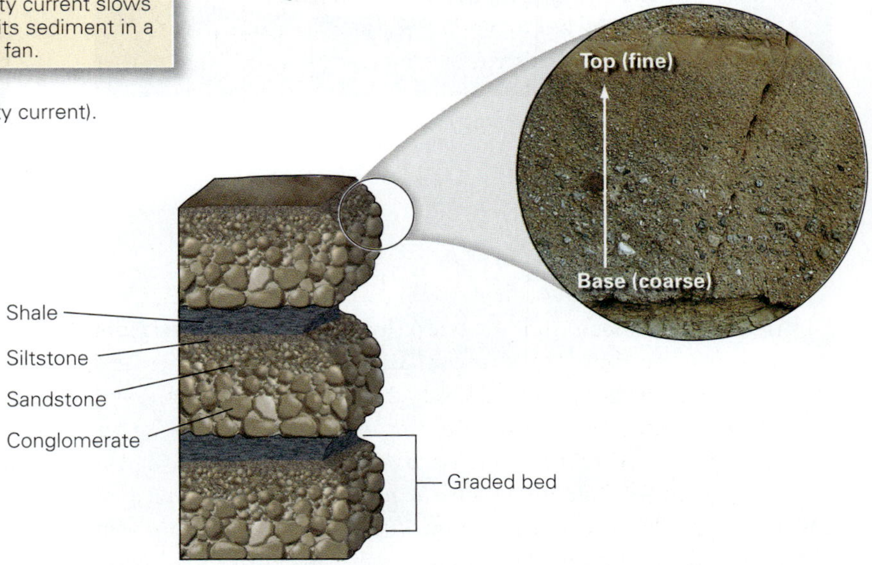

(c) As the process repeats, a succession of graded beds accumulates.

FIGURE 6.16 Mud cracks, a sedimentary structure formed by the drying out of mud.

(a) Mud cracks in red mud at Bryce Canyon, Utah. Note how the edges of the mud plates curl up.

(b) Mud cracks visible on the surface of a 410-Ma bed exposed on a cliff in New York.

environments, geologists, like crime scene investigators, look for clues. Detectives may seek fingerprints and bloodstains to identify a culprit. Geologists examine grain size, composition, sorting, sedimentary structures, and fossils to identify a depositional environment. Geologic clues can tell us if the sediment was deposited by ice, strong currents, waves, or quiet water, and in some cases can provide insight into the climate at the time of deposition. With experience, geologists can examine a succession of beds and determine whether it accumulated on a river floodplain, along a beach, in shallow water just offshore, or on the deep ocean floor.

Let's now explore some examples of depositional environments and the sediments deposited in them, by imagining that we are taking a journey from the mountains to the sea, examining sediments as we go (see **Geology at a Glance** on pp. 206–207). We will see that geologists distinguish among three basic categories of depositional environments: terrestrial, coastal, and marine.

Terrestrial Sedimentary Environments

We begin our exploration with terrestrial depositional environments, those that develop inland, far enough from the shoreline that they are not affected by ocean tides and waves. *Terrestrial sediments,* which are also known as *nonmarine sediments,* accumulate either on dry land or under and adjacent to freshwater. Locally, oxygen in surface water or groundwater reacts with iron in the sediment to produce rust-like iron-oxide minerals, which give the sediment an overall reddish hue. Geologists informally refer to strata with this hue as *redbeds.*

Glacial Environments High in the mountains, where it's so cold that more snow collects in the winter than melts away, glaciers—rivers or sheets of ice—develop and slowly flow. Because ice is a solid, it can move sediment of any size. So as a glacier moves down a valley in the mountains, it carries along all the sediment that falls on its surface from adjacent cliffs, or gets plucked from the ground at its base or sides. When the ice finally melts away, the sediment that had been in or on the ice accumulates as *glacial till* (**Fig. 6.17a**). Till is unsorted and unstratified—it contains clasts ranging from clay size to boulder size all mixed together.

Mountain Stream Environments As we walk down beyond the end of the glacier, we enter a realm where turbulent streams rush downslope in steep-sided valleys. This fast-moving water has the power to carry large clasts; in fact, during floods, boulders and cobbles can tumble down the streambed. Where slopes decrease and water flow slows, the larger clasts settle out to form gravel and boulder beds, while the stream carries finer sediments like sand and mud farther downstream (**Fig. 6.17b**). Sedimentary deposits of a mountain stream would, therefore, include breccia and conglomerate.

Alluvial-Fan Environments Our journey now takes us to the mountain front, where the fast-moving stream empties onto a plain. In arid regions, where there is not enough water for the stream to flow continuously, the stream deposits its load of sediment near the mountain front, producing a wedge-shaped apron of gravel and sand called an **alluvial fan** (**Fig. 6.17c**). Deposition takes place here because when the stream pours from a canyon mouth and spreads out into multiple channels, friction causes the water to slow down, and slow-moving water does not have the power to move coarse sediment. Sediment in alluvial fans may accumulate close to the source, so it will not have undergone much chemical weathering. Thus, sand layers still contains feldspar grains, for these have not yet weathered into clay. Alluvial-fan sediments become conglomerate and arkose.

SEE FOR YOURSELF...

ALLUVIAL FANS AND EVAPORITES, DEATH VALLEY, CALIFORNIA

Latitude
36°7′35.28″N

Longitude
116°45′24.44″W

Fly to 8.75 km, tilt to look obliquely north.

Death Valley is a narrow rift whose floor lies below sea level. You can see alluvial fans and white evaporites.

Desert Environments If the climate is very dry, few plants can grow and the ground surface lies exposed. Strong winds can move dust and sand. The dust gets carried away, and the resulting well-sorted sand can accumulate in dunes. Thus, thick layers of well-sorted sandstone, in which we can find large cross beds, are relicts of desert sand-dune environments (**Fig. 6.17d**).

River Environments In climates where streams flow, we find several distinctive depositional environments. Rivers transport gravel, sand, silt, and mud. The coarser sediments tumble along the bed in the river's channel and collect in cross-bedded, rippled layers while the finer sediments drift along, suspended in the water. This fine sediment settles out along the banks of the river, or on the *floodplain,* the flat land on either side of the river that is covered with water only during floods. On the floodplain, mud layers dry out between floods, so they develop mud cracks. River sediments lithify to form rippled sandstone, siltstone, and shale. Typically, coarser sediments occur in elongate bands, relicts of river channels. Layers of fine-grained floodplain deposits surround

FIGURE 6.17 Examples of nonmarine depositional environments.

(a) Glacial till at the end of a glacier in France.

(b) Boulders and cobbles deposited by a mountain stream in Colorado.

(c) An alluvial fan in Death Valley, California.

(d) Sand dunes in Brazil. Dune buggy for scale.

(e) Deposits of an ancient river channel in Indiana. Note how the floor of the channel cuts across older strata. The geologist's sketch emphasizes the relationship.

(f) Laminated mud from a lake bed.

the relict channels, so in cross section, the channel has a lens-like shape (**Fig. 6.17e**). Geologists commonly refer to river deposits as *fluvial sediments*, from the Latin word *fluvius*, for river.

Lake Environments In temperate climates, where water remains at the surface throughout the year, lakes form. In the offshore portions of a lake, the deeper water is relatively quiet, and clay can settle out to form mud on the lake bed. When lithified, such laucustrine mud turns into shale (**Fig. 6.17f**).

At the mouths of streams that empty into lakes, small deltas may form. A **delta** is a wedge of sediment that accumulates where moving water enters standing water. Deltas were

FIGURE 6.18 A simple "Gilbert-type" delta formed where a stream enters a lake.

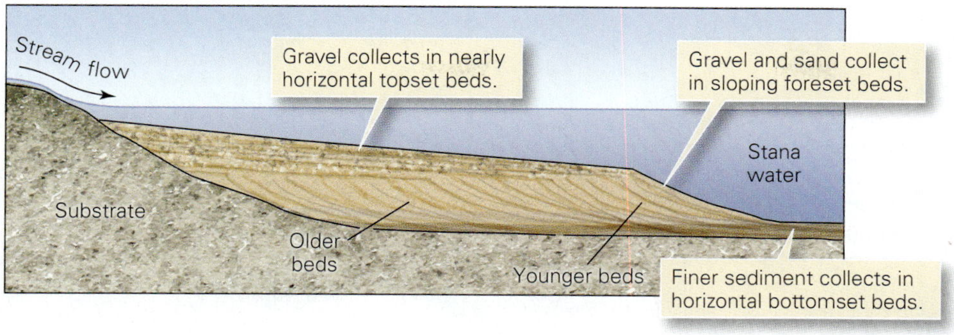
Stream flow
Gravel collects in nearly horizontal topset beds.
Gravel and sand collect in sloping foreset beds.
Stana water
Substrate
Older beds
Younger beds
Finer sediment collects in horizontal bottomset beds.

Thus, deposits of an ocean-margin delta produce a great variety of sedimentary rock types (**Fig. 6.19a**).

Coastal Beach Sands Now we leave the delta and wander along the coast. Oceanic currents transport sand along the coastline. The sand washes back and forth in the surf, so it becomes well sorted (waves winnow out silt and clay) and well rounded, and because of the back-and-forth movement of ocean water over the sand, the sand surface may become rippled (**Fig. 6.19b**). So if you find well-sorted, medium-grained sandstone, perhaps with ripple marks, you may be looking at the remnants of a beach environment.

SEE FOR YOURSELF...

MURRAY RIVER FLOODPLAIN, AUSTRALIA

Latitude
34°7′4.68″ S

Longitude
140°46′9.75″ E

Zoom to 25 km (~15.5 miles) and look down.

The Murray River meanders across the flat-lying landscape of South Australia. In this view, the water is restricted to the channel, exposing the sand and silt deposits that have covered its floodplain.

so named because the map shape of some deltas resembles the Greek letter *delta* (Δ), as we discuss further in Chapter 14. In 1885, an American geologist named G. K. Gilbert showed that small lakeshore deltas contain three components (**Fig. 6.18**): topset beds composed of gravel, foreset beds of gravel and sand, and silty bottomset beds.

Coastal and Marine Sedimentary Environments

Along the seashore, a variety of distinct coastal environments occur; the character of each reflects the nature of the sediment supply and the climate. Marine environments start at the high-tide line and extend offshore, to include the deep ocean floor. The type of sediment deposited at a location depends on the climate, water depth, and whether or not clastic grains are available.

Marine Delta Deposits After following the river downstream for a long distance, we reach its mouth, where it empties into the sea. Here, the river builds a delta of sediment out into the sea. River water stops flowing when it enters the sea, so sediment settles out. Large marine deltas are much more complex than the lake examples that Gilbert studied, for they include many different sedimentary environments such as swamps, channels, floodplains, and submarine slopes. Sea-level changes may cause the positions of the different environments to move with time.

Shallow-Marine Clastic Deposits From the beach, we proceed offshore. In deeper water, where wave energy does not stir the seafloor, finer sediment can accumulate. Because the water here may be only a few meters to tens of meters deep, geologists refer to this depositional setting as a *shallow-marine environment*. Clastic sediments that accumulate in this environment tend to be fine-grained, well-sorted, well-rounded silt, and they are inhabited by a great variety of organisms such as mollusks and worms. Thus, if you see beds of siltstone and mudstone containing marine fossils, you may be looking at shallow-marine clastic deposits.

Shallow-Water Carbonate Environments In shallow-marine settings, where relatively little sand and mud enter the water, warm, clear, nutrient-rich water can host an abundance of organisms with carbonate shells, and these eventually become carbonate sediment (**Fig. 6.20**). Beaches collect sand composed of shell fragments; lagoons are sites where carbonate mud accumulates; and reefs consist of coral and coral debris. Farther offshore from a reef, we can find a sloping apron of reef fragments. Shallow-water carbonate environments transform into various kinds of limestone.

SEE FOR YOURSELF...

SHALLOW MARINE ENVIRONMENTS, RED SEA, EGYPT

Latitude
22°38′17.70″ N

Longitude
36°13′21.27″ E

Zoom to an elevation of 4 km (~2.5 miles) and look straight down.

The desert sands of the Sahara abut the blue waters of the Red Sea. What types of sedimentary rocks would form if the sediments visible in this view were to be buried and preserved?

The Formation of Sedimentary Rocks

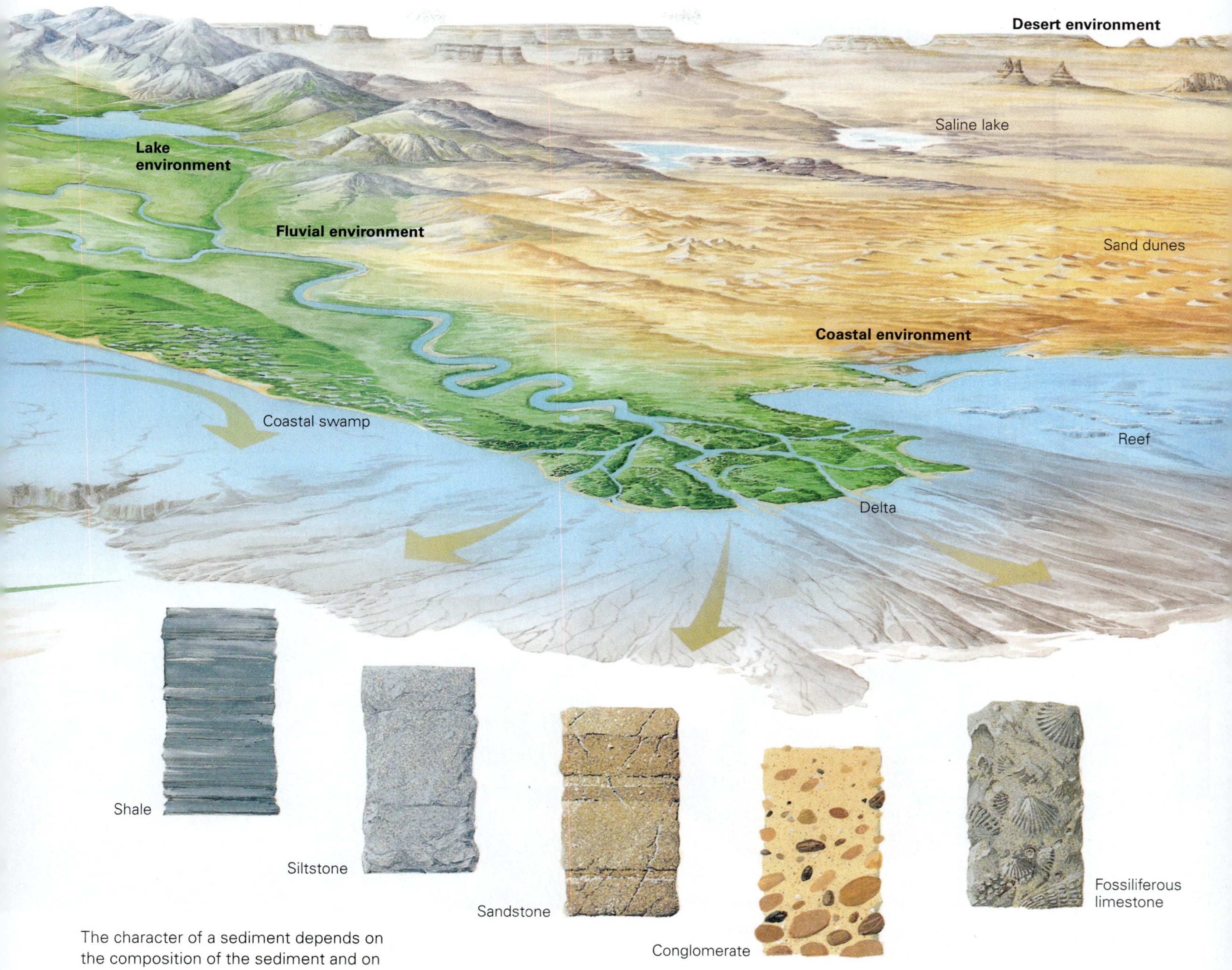

The character of a sediment depends on the composition of the sediment and on the depositional environment in which it accumulated. Glaciers carry sediment of all sizes, so they leave deposits of poorly sorted till. Streams deposit coarser sands and gravels in their channels and finer ones on floodplains. In the quiet water of lakes, fine-grained mud accumulates. In desert environments, sand builds into dunes, evaporates precipitate in saline lakes, and gravel and sand gather in alluvial fans. Wave action along coastal beaches leaves behind well-sorted sand. in swampy areas, large volumes of plant matter accumulate. Where a river flows into the sea, its water slows and deposits a large delta. In warmer coastal marine environments, carbonate sediments, from the shells of organisms, are deposited. Locally, corals and other organisms build carbonate reefs. Offshore, submarine canyons channel avalanches of sediment, or turbidity currents, out to the deep seafloor. Far from shore, pelagic marine sediment, commonly from the shells of plankton, slowly settles out. Burial and lithification of these various types of sediments turn them into examples of the different classes of sedimentary rocks—clastic, chemical, biochemical, and organic.

FIGURE 6.19 Examples of coastal depositional environments.

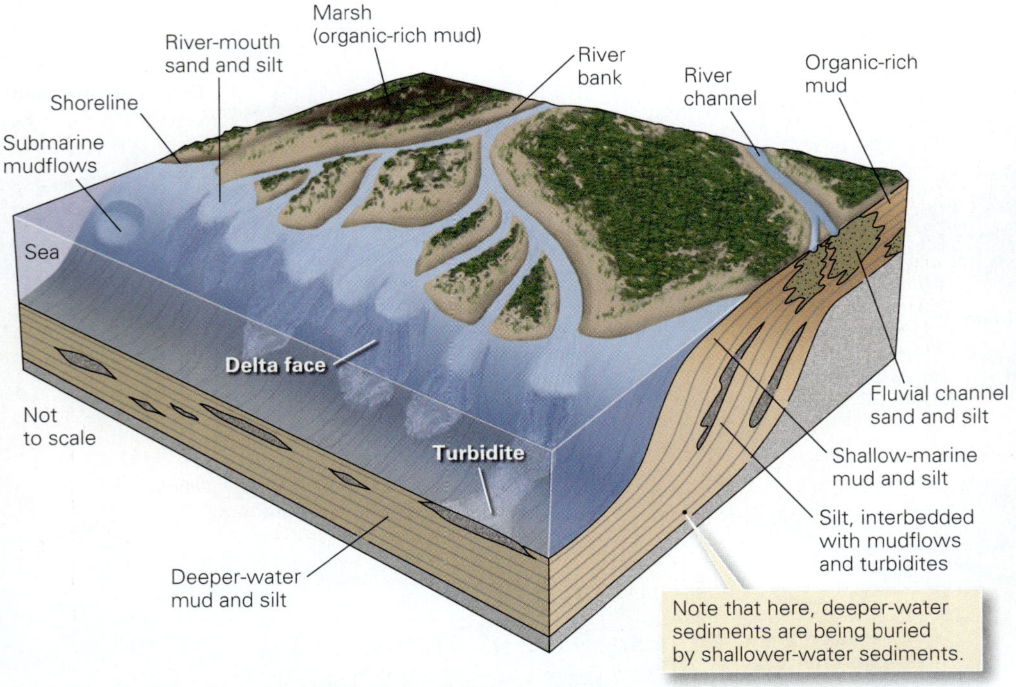

(a) A major river delta along an ocean coast is a complex depositional environment. Sea-level changes affect locations of depositional settings.

(b) Waves on this California beach wash and sort the sand.

FIGURE 6.20 Reef environments for the deposition of carbonate rocks.

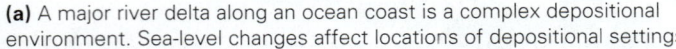

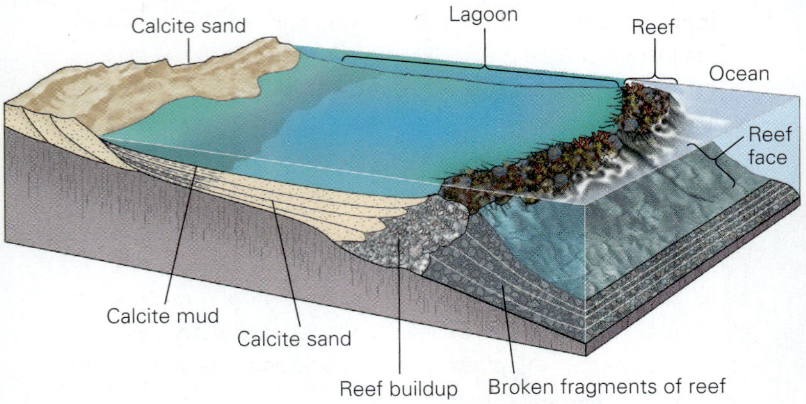

(a) Carbonate reefs form along shorelines in warm-water environments. In detail, reefs include many distinct depositional environments.

(b) A dramatic reef surrounds an island in the tropical Pacific. Deeper water is darker. Note the surf along the edge of the reef.

Deep-Marine Deposits We conclude our journey by sailing far offshore. Along the transition between coastal regions and the deep ocean, turbidity currents deposit graded beds. In the deep-ocean realm, only fine clay and plankton provide a source for sediment (**Fig. 6.21**). The clay eventually settles out onto the deep seafloor, forming deposits of finely laminated mudstones, and plankton shells settle to form chalk (from calcite shells) or chert (from siliceous shells). Thus, deposits of mudstone, chalk, or bedded chert indicate a deep-marine origin.

FIGURE 6.21 Examples of deep-marine sediment.

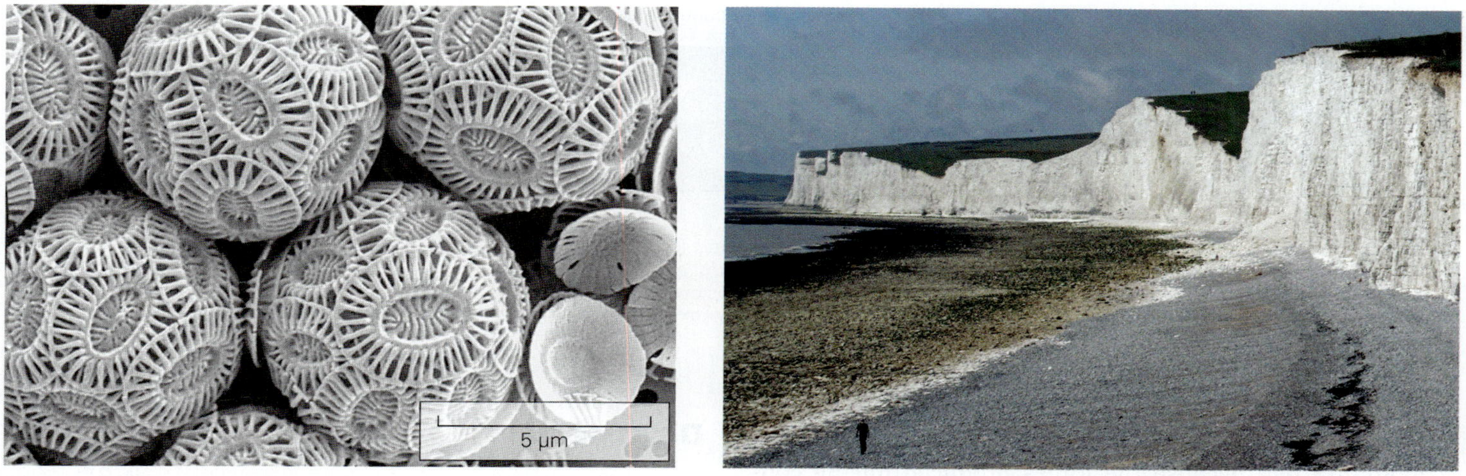

(a) These plankton shells, which make up some kinds of deep-marine sediment, are so small that they could pass through the eye of a needle.

(b) The chalk cliffs of southeastern England consist of plankton shells deposited on the sea floor tens of millions of years ago.

TAKE-HOME MESSAGE

Different types of sedimentary rocks accumulate in different depositional environments. Thus, strata deposited along a river differ from strata deposited by ocean waves, by glaciers, or in the deep sea. By studying sedimentary rocks at a location, geologists can deduce environments that existed at the locality in the past.

QUICK QUESTION How can you distinguish sediment deposited in an alluvial fan from sediment deposited in a shallow-marine environment?

6.5 Sedimentary Basins

The sedimentary veneer on the Earth's surface varies greatly in thickness. If you stand in parts of Siberia or Canada, you will find yourself on igneous and metamorphic basement rocks that are over a billion years old—sedimentary rocks are nowhere in sight. Yet if you stand along the southern coast of Texas, you would have to drill through over 15 km of sedimentary beds before reaching igneous and metamorphic basement. Thick accumulations of sediment form only in special regions where the surface of the Earth's lithosphere sinks, providing space in which sediment collects. Geologists use the term **subsidence** to refer to the process by which the surface of the lithosphere sinks, and the term **sedimentary basin** for the sediment-filled depression. In what geologic settings do sedimentary basins form? Plate tectonics theory can provide a key.

Categories of Basins in the Context of Plate Tectonics Theory

Geologists distinguish among different kinds of sedimentary basins in the context of plate tectonics theory. Let's consider a few examples (**Fig. 6.22**).

- *Rift basins:* These form in continental rifts, regions where the lithosphere is stretching horizontally, and therefore thins vertically. As the rift grows, slip on faults drops blocks of crust down, producing low areas—narrow basins bordered by elongate mountain ridges. These basins fill with terrestrial sediment. In deserts, overlapping alluvial fans line the margins of the basins.
- *Passive-margin basins:* These form along the edges of continents that are not plate boundaries. They are underlain by stretched lithosphere, the remnants of a rift whose evolution successfully led to the formation of a mid-ocean ridge and subsequent growth of a new ocean basin. Passive-margin basins form because subsidence (sinking) of stretched lithosphere continues long after rifting ceases. They fill with sediment carried to the sea by rivers, and with carbonate rocks formed in coastal reefs. Passive margin basins include some of the thickest accumulations of sediment on Earth.
- *Intracontinental basins:* These develop in the interiors of continents. Probably, they initiate because of subsidence over a rift. They continue to subside in pulses even hundreds of millions of years after they formed, for reasons that are not well understood.

FIGURE 6.22 The geologic setting of sedimentary basins.

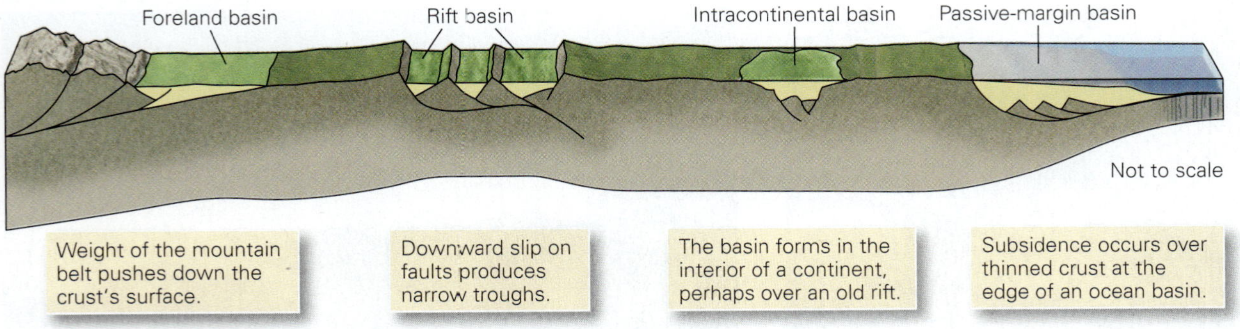
Foreland basin
Rift basin
Intracontinental basin
Passive-margin basin
Not to scale
Weight of the mountain belt pushes down the crust's surface.
Downward slip on faults produces narrow troughs.
The basin forms in the interior of a continent, perhaps over an old rift.
Subsidence occurs over thinned crust at the edge of an ocean basin.

- *Foreland basins:* These form on the continent side of a mountain belt because the forces produced during convergence or collision push large slices of rock up faults and onto the surface of the continent. The weight of these slices pushes down on the surface of the continent, producing a wedge-shaped depression adjacent to the mountain range that fills with sediment eroded from the range. Fluvial and deltaic strata accumulate in foreland basins.

Transgression and Regression

Sea-level changes, relative to the land surface, control the succession of sediments that we see in a sedimentary basin. At times during Earth's history, sea level has risen by as much as a couple of hundred meters, yielding shallow seas that submerge the interiors of continents. At other times, sea level has fallen by hundreds of meters, exposing the continental shelves to air. Global sea-level changes may be due to a number of factors, including climate change, which controls the amount of ice stored in polar ice caps and causes changes in the volume of ocean basins. Sea level at a specific location may also be due to local uplift or sinking of the land surface.

When relative sea level rises, the shoreline migrates inland—we call this process **transgression**. When relative sea level falls, the coast migrates seaward—we call this process **regression**. During transgression and regression, the position of depositional environments migrates, so the depositional environment at a given location changes over time. These processes, acting over time, can lead to the formation of broad blankets of sediment. Note that the age of a given sediment layer deposited during a transgression or regression varies with location (**Fig. 6.23**).

Diagenesis

Earlier in this chapter we discussed lithification, by which sediment hardens into rock. Lithification is an aspect of a broader phenomenon called diagenesis. Geologists use the term **diagenesis** for all the physical, chemical, and biological processes that transform sediment into sedimentary rock and that alter characteristics of sedimentary rock after the rock has formed. This includes changes to a sedimentary rock that happen when it reacts with groundwater underground, even long after lithification. Such reactions can cause existing cements to dissolve and new ones to precipitate, or may cause new minerals to grow in remaining pores.

As temperature and pressure increase still deeper in the subsurface, the changes that take place in rocks become more profound. At sufficiently high temperature and pressure, metamorphism begins, in that a new assemblage of minerals forms and/or mineral grains become aligned parallel to each other. The transition between diagenesis and metamorphism in sedimentary rocks is gradational and occurs between temperatures of 150° and 250°C. In the next chapter, we enter the realm of true metamorphism.

TAKE-HOME MESSAGE

In certain geologic settings, Earth's surface sinks (subsides) to form a depression that fills with sediment. The depression with its thick fill of sediment is a sedimentary basin. As sea level rises and falls, the coast, and therefore depositional environments, can migrate inland or offshore, respectively. Once sediment has been deposited, it undergoes various changes, known as diagenesis, in response to pressure and to interaction with groundwater.

QUICK QUESTION What is the thickest that the fill of a sedimentary basin can become?

FIGURE 6.23 The concept of transgression and regression during deposition of sedimentary sequence.

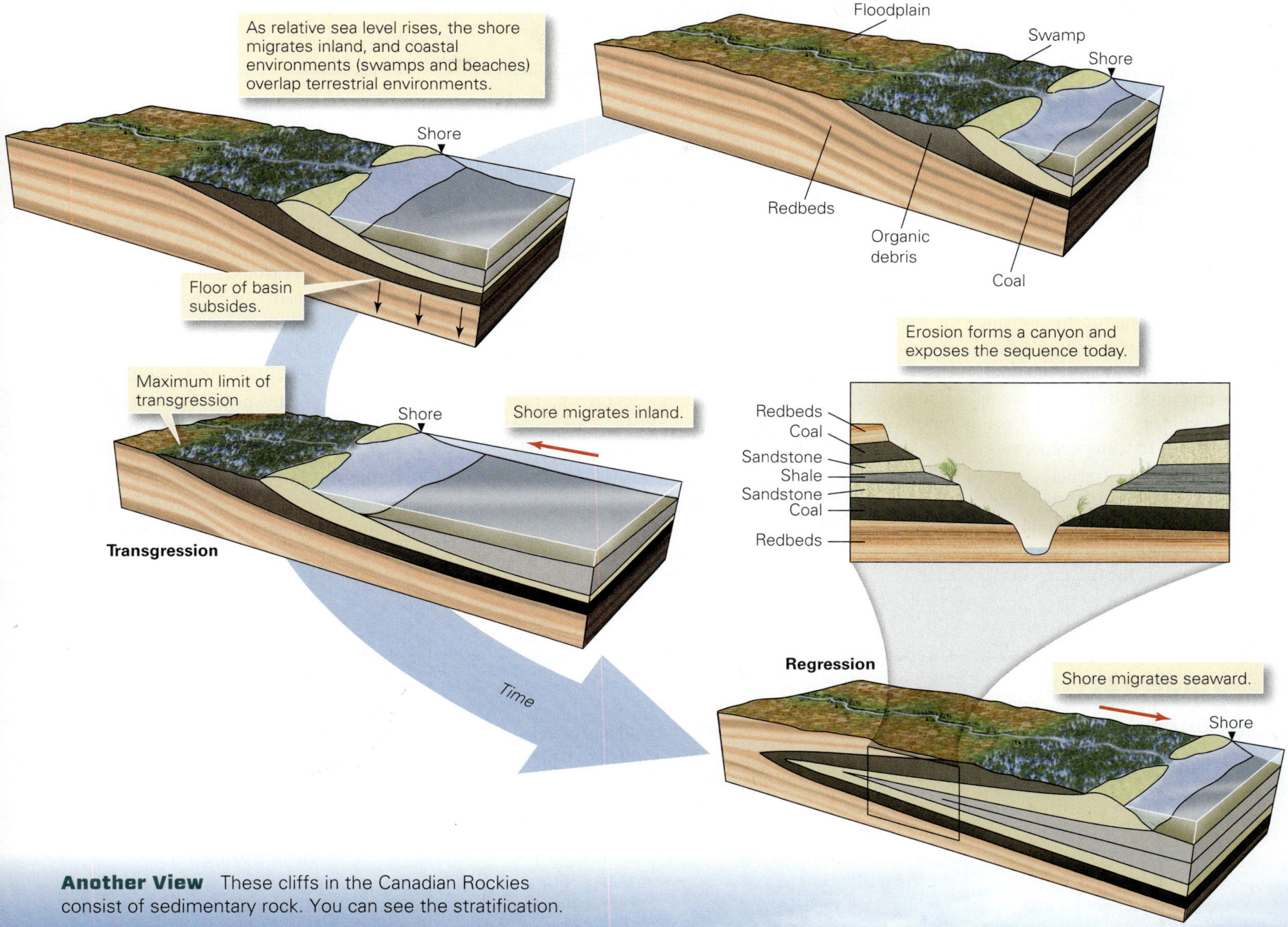

Another View These cliffs in the Canadian Rockies consist of sedimentary rock. You can see the stratification.

Chapter 6 Review

Chapter Summary

- Geologists recognize four major classes of sedimentary rocks. Clastic rocks form from cemented-together grains that were first produced by weathering, then were transported, deposited, and lithified; they are classified based on grain size. Biochemical rocks develop from the shells of organisms. Organic rocks consist of plant debris or of altered plankton remains. Chemical rocks precipitate directly from water, and include evaporates that accumulate when salt water evaporates.
- Sedimentary structures include bedding, cross bedding, graded bedding, ripple marks, dunes, and mud cracks. They serve as clues to depositional settings. Bedding, the fundamental layering in sedimentary strata, can be due to subtle or major changes in depositional conditions at a location.
- The mineralogical and, therefore, the chemical composition is different for different types of sedimentary rocks. For example, limestone consists of calcite, chert of silica, coal of carbon, shale from clay, and sandstone from quartz grains.
- Glaciers, streams, alluvial fans, deserts, rivers, lakes, deltas, beaches, shallow seas, and deep seas each accumulate a different, distinctive assemblage of sedimentary strata. Therefore, study of sedimentary strata provides clues into the environment of a given location at times in the past.
- Thick piles of sedimentary rocks accumulate in sedimentary basins, regions where the surface of the crust sinks or subsides over a long period of time. Such subsidence happens for a variety of reasons. The thickest sedimentary basins form along passive continental margins.
- Transgressions occur when sea level rises and the coastline migrates inland. Regressions occur when sea level falls and the coastline migrates seaward.
- Diagenesis involves processes leading to lithification and processes that alter sedimentary rock at temperatures lower than those that cause metamorphism.

Guide Terms

alluvial fan (p. 203)
arkose (p. 191)
bed (p. 197)
biochemical sedimentary rock (p. 188)
breccia (p. 191)
cementation (p. 189)
chemical sedimentary rock (p. 188)
clastic sedimentary rock (p. 188)
coal (p. 193)
compaction (p. 189)
conglomerate (p. 191)
cross bed (p. 199)
delta (p. 204)
deposition (p. 189)
depositional environment (p. 201)
diagenesis (p. 210)
dolostone (p. 195)
dune (p. 199)
erosion (p. 189)
evaporite (p. 194)
fossil (p. 200)
graded bed (p. 200)
lithification (p. 189)
mudstone (p. 191)
oil shale (p. 193)
organic sedimentary rock (p. 188)
regression (p. 210)
ripple mark (p. 199)
sandstone (p. 189)
sedimentary basin (p. 209)
sedimentary rock (p. 188)
sedimentary structure (p. 196)
shale (p. 191)
siltstone (p. 191)
sorting (p. 190)
strata (p. 197)
subsidence (p. 209)
transgression (p. 210)
travertine (p. 194)
turbidity current (p. 200)
turbidite (p. 200)

GEOTOURS *THIS CHAPTER'S GEOTOUR EXERCISE (F) FEATURES:*

- Sedimentary Rocks Exposed in the Grand Canyon
- Tilted and Folded Sedimentary Rocks
- Arid Depositional Environments
- Fluvial Depositional Environments

Review Questions

1. Describe how a clastic sedimentary rock forms from its unweathered parent rock.
2. Explain how biochemical sedimentary rocks form.
3. How do grain size and shape, sorting, sphericity, and angularity change as sediments move downstream?
4. Describe the two different kinds of chert. How are they similar? How are they different?
5. Do all chemical sedimentary rocks have the same composition? What conditions produce evaporites?
6. How does dolostone differ from limestone, and how does dolostone form?
7. What are cross beds, and how do they form? How can you use cross beds to read the current direction?
8. Describe how a turbidity current forms and moves. How does it produce graded bedding?
9. Compare deposits of an alluvial fan with those of a deep-marine deposit.
10. Why don't sediments accumulate everywhere? What types of tectonic conditions are required to produce sedimentary basins?
11. How is it possible for sandstone derived from sediment deposited in a beach environment to comprise a formation that blankets a broad region?

On Further Thought

12. Recent exploration of Mars by robotic vehicles suggests that layers of sedimentary rock cover portions of the planet's surface. On the basis of examining images of these layers, some researchers claim that the layers contain cross bedding and relicts of gypsum crystals. At face value, what do these features suggest about depositional environments on Mars in the past?
13. The Gulf Coast of the United States is a passive-margin basin that contains a very thick accumulation of sediment. Drilling reveals that the base of the sedimentary succession in this basin consists of redbeds. These are overlaid by a thick layer of evaporite. The evaporite, in turn, is overlaid by deposits composed of sandstone and shale. In some intervals, sandstone occurs in channels and contains ripple marks, and the shale contains mud cracks. In other intervals, the sandstone and shale contain fossils of marine organisms. Be a sedimentary detective, and explain the succession of sediment in the basin.
14. Examine the Bahamas with *Google Earth*™. (You can find a high-resolution image at latitude 23°58′40.98″ N, longitude 77°30′20.37″ W.) Note that broad expanses of very shallow water surround the islands, that white-sand beaches occur along the coast of the islands, and that small reefs occur offshore. What does the sand consist of, and what rock will it become if it eventually becomes buried and lithified?

Online Resources

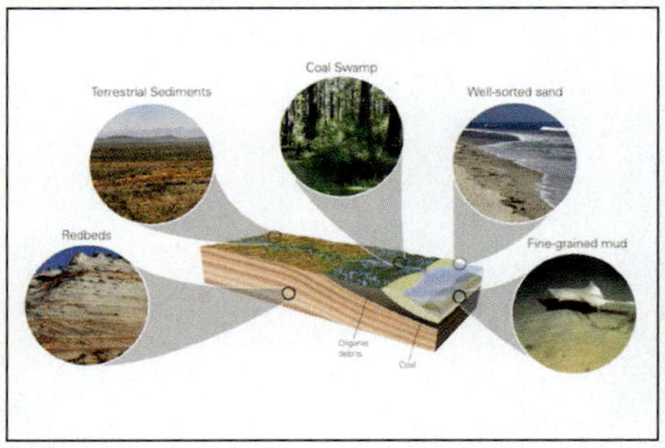

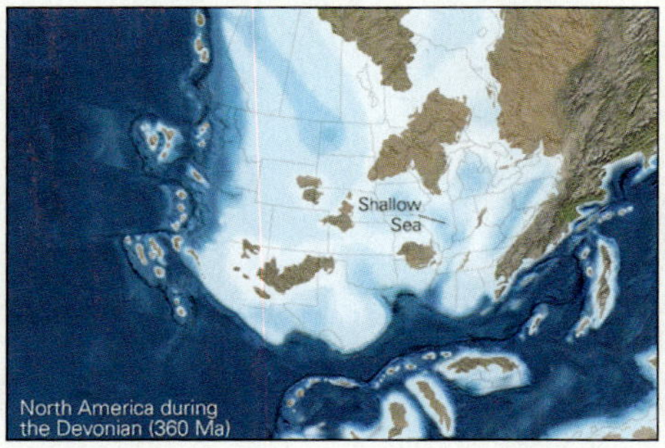

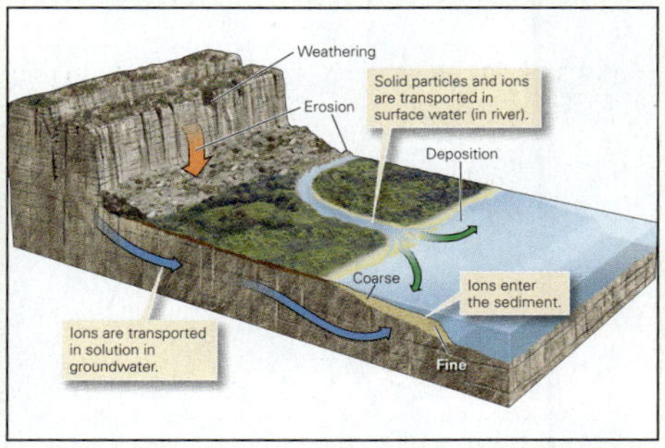

Videos

This chapter features a video on the formation of sedimentary rocks, transgression and regression, and how this appears in the stratigraphic record.

Assessment

This chapter features questions on recognizing types of sedimentary rocks, structures, and environments.

LEARNING OBJECTIVES

By the end of this chapter, you should understand...

1. what metamorphism is, and why it takes place.
2. how changes in temperature, pressure, and stress can cause metamorphism.
3. why metamorphism alters the texture and mineral content of rocks, and how foliation develops.
4. where metamorphism occurs, in the context of plate tectonics theory.
5. how geologists describe and classify metamorphic rocks.

▲ In the stark mountains of northwestern Scotland, outcrops of Lewisian Gneiss poke up from the scruffy grass. The rock formed when pre-existing rocks were subjected to high pressures and temperatures, as well as compression and shear, during Precambrian mountain building.

CHAPTER 7

Metamorphism: A Process of Change

7.1 Introduction

Cool winds sweep across Scotland for much of the year. In this blustery climate, vegetation has a hard time taking hold, so the landscape provides countless outcrops of barren rock. During the mid-18th century, James Hutton, who would come to be known as the "father of geology," examined these outcrops, hoping to learn how the rocks formed. Hutton found that many features in the outcrops resembled the products of present-day sediment deposition or of volcanic eruptions. Thus, he developed a sense for how sedimentary and igneous rock originated. But Hutton also found rocks that contained minerals and textures quite different from those in sedimentary and igneous samples. He described this third, puzzling, kind of rock as "a mass of matter which had evidently formed originally in the ordinary manner . . . but which is now extremely distorted in its structure . . . and variously changed in its composition." These rocks are now known as metamorphic rocks, from the Greek words *meta*, meaning change, and *morphe*, meaning form. In modern terms, a **metamorphic rock** is one that forms when a pre-existing rock, or **protolith**, undergoes a solid-state change in response to the modification of its environment. This process of change is called **metamorphism**.

Let's look at the definition of metamorphic rock more closely, to understand what each component of the definition

> *Nothing in the world lasts, save eternal change.*
>
> HONORAT DE BUEIL (French poet, 1589–1670)

Did you ever wonder…
if, once formed, rocks ever change?

means. By *solid-state,* we mean that a metamorphic rock does *not* form by freezing of magma—rocks that do, by definition, are igneous. By *change,* we mean that a metamorphic rock contains new minerals that did not occur in the protolith, and/or a new texture (arrangement of mineral grains) that differs from that of the protolith. And by *modification of environment,* we mean a change in temperature and/or pressure, an application of stress (a directed push, pull, or shear), and/or exposure to hydrothermal fluids (very hot water solutions). Environmental conditions that drive metamorphism can be called *agents of metamorphism.* In other words, just as caterpillars undergo metamorph*osis* because of hormonal changes in their bodies, rocks undergo metamorph*ism* when they are subjected to *agents of metamorphism.*

From Hutton's day to the present, geologists have undertaken field studies, laboratory experiments, and theoretical calculations to better characterize metamorphism, and to understand the conditions that lead to metamorphism. In this chapter, we introduce the results of this work. We begin by explaining the causes of metamorphism and the basis for classifying metamorphic rocks. We conclude by discussing the geologic settings in which these rocks form. As you will see, the full story of metamorphism couldn't be told until the theory of plate tectonics had been proposed.

7.2 Consequences and Causes of Metamorphism

What Is a Metamorphic Rock?

If someone were to put a rock on a table in front of you, how would you know that it is metamorphic? First, metamorphic rocks may contain **metamorphic minerals**, new minerals that grow within the solid rock only under metamorphic conditions. Commonly, metamorphism yields a group of metamorphic minerals that geologists refer to as a *metamorphic mineral assemblage.* Second, metamorphic rocks can have a *metamorphic texture,* a distinctive arrangement of mineral grains that forms only during metamorphism. In some cases, such textures are manifested by grains that are bigger than the grains that had existed in the protolith, and/or by grains that interlock with each other, even if the grains in the protolith did not. Commonly, metamorphic texture causes rock to have a type of layering called **metamorphic foliation**. In a rock with metamorphic foliation, platy mineral grains lie parallel to one another, elongate mineral grains align with one another, and/or the rock displays alternating light-colored and dark-colored layers. Because of the changes that take

FIGURE 7.1 Metamorphism causes changes in mineral makeup and texture.

(a) A specimen of red shale (left) contains clay, quartz, and iron oxide. When this rock undergoes intense metamorphism, it may change into a gneiss containing different minerals. This gneiss sample (right) contains biotite, quartz, feldspar, and purple garnet.

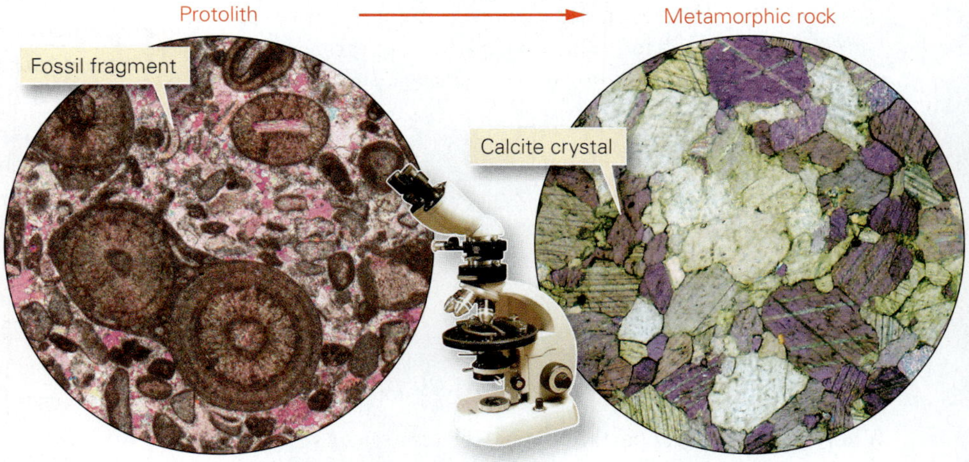

(b) The contrast in texture between a protolith of fossiliferous limestone and a marble formed by metamorphism is evident when viewed through a microscope.

place during metamorphism, a metamorphic rock can differ in appearance from its protolith by as much as a butterfly can differ from a caterpillar. For example, metamorphism of red shale can yield a metamorphic rock consisting of aligned mica flakes, flattened quartz grains, and brilliant garnet crystals (**Fig. 7.1a**), and metamorphism of a fossiliferous limestone can yield a metamorphic rock consisting of large interlocking crystals of calcite (**Fig. 7.1b**).

Processes That Take Place During Metamorphism

The process of forming metamorphic minerals and textures generally takes place very slowly (thousands to millions of years) and involves several processes, which sometimes occur alone and sometimes together. The most common processes are these:

- *Recrystallization*, which changes the shape and size of grains without changing the identity of the mineral making up the grains (**Fig. 7.2a**).
- *Phase change*, which transforms one mineral into another mineral (a polymorph) with the same composition but a different crystal structure. On an atomic scale, phase change involves the rearrangement of atoms.
- *Metamorphic reaction*, or *neocrystallization* (from the Greek *neos*, for new), which results in growth of new minerals that differ from those of the protolith (**Fig. 7.2b**). During neocrystallization, chemical reactions effectively "digest" minerals of the protolith, so the chemicals can go into the production of new minerals.
- *Pressure solution*, which happens where the surface of one mineral grain pushes against the surface of another, under conditions that allow a water film to exist between the grains. The grains preferentially dissolve at the surface of contact and the resulting ions migrate away, through the water film (**Fig. 7.2c**). The ions may precipitate on the sides of grains that aren't being pushed together.
- *Plastic deformation*, which happens when a rock becomes warm enough to behave like soft plastic, so the minerals within it can change shape without breaking (**Fig. 7.2d**).

In the following sections, we'll consider how various agents of metamorphism cause the processes above to take place.

Metamorphism Due to Heating

When you heat cake batter, the batter transforms into a new material—cake. Similarly, when you heat a rock, its ingredients transform into a new material—metamorphic rock.

FIGURE 7.2 Metamorphic processes, as seen through a microscope.

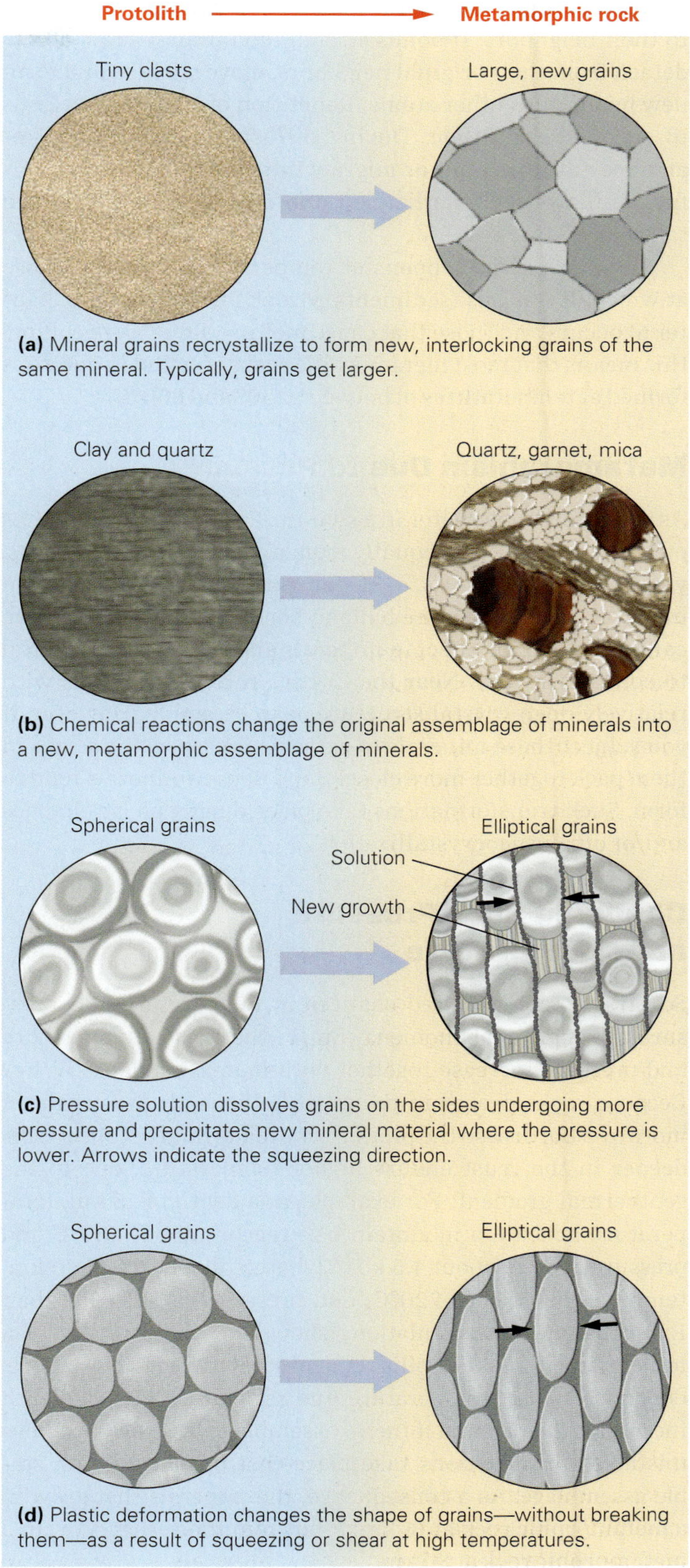

(a) Mineral grains recrystallize to form new, interlocking grains of the same mineral. Typically, grains get larger.

(b) Chemical reactions change the original assemblage of minerals into a new, metamorphic assemblage of minerals.

(c) Pressure solution dissolves grains on the sides undergoing more pressure and precipitates new mineral material where the pressure is lower. Arrows indicate the squeezing direction.

(d) Plastic deformation changes the shape of grains—without breaking them—as a result of squeezing or shear at high temperatures.

Why? Think about what happens to atoms in a mineral grain when the grain warms. Heat causes the atoms to vibrate rapidly, stretching and bending chemical bonds that lock atoms to their neighbors. If bonds stretch too far and break, atoms detach from their original neighbors, move slightly, and form new bonds with other atoms. Repetition of this process leads to *solid-state diffusion*. During diffusion, atoms either rearrange within grains, or migrate into and out of grains. Diffusion allows recrystallization and/or neocrystallization to take place.

Metamorphism happens at temperatures between those at which diagenesis (sedimentary rock formation, see Chapter 6) occurs and those that cause melting. Roughly speaking, this means that most metamorphic rocks you find in outcrops formed at temperatures of between 250° and 850°C.

Metamorphism Due to Pressure

As you swim underwater in a swimming pool, water squeezes or pushes against you equally from all sides—in other words, your body feels *pressure*. If you pull an air-filled balloon underwater in a lake, the balloon becomes smaller. Pressure can have a similar effect in minerals and can cause a material to collapse inward. Near the Earth's surface, minerals with relatively open crystal structures can be stable. However, if you subject these minerals to extreme pressure, the atoms in them pack together more closely, and denser minerals tend to form. Such transformations take place during phase changes and/or during neocrystallization.

Changing Both Pressure and Temperature

So far, we've considered changes in temperature and pressure as separate phenomena. But in the Earth, temperature and pressure increase together with increasing depth. Why? Because rocks deeper in the crust lie beneath the weight of more overburden, so the pressure is greater, and because rocks deeper in the crust endure higher temperature, due to the geothermal gradient. For example, at a depth of 15 km, temperature beneath a mountain belt reaches about 450°C and pressure rises to about 4.5 kbar, whereas at a depth of 30 km, temperature reaches 720°C, and pressure rises to 9.0 kbar. Experiments and calculations show that the stability of a mineral, meaning the ability of a mineral to form and survive, depends on *both* temperature and pressure. Thus, as depth increases, the original mineral assemblage in a rock becomes unstable, and reactions take place that produce a new, stable assemblage. As a consequence, the minerals that grow in a metamorphic rock at 15 km differ from those that grow in a metamorphic rock at 30 km.

Stress, and the Development of Preferred Orientation

In the solid crust of the Earth, the push or pull in one direction is not always the same as it is in other directions. Simplistically, we can say that rocks in the crust may be subjected to *compression* (squeezing) in a given direction, *tension* (pulling) in a given direction, or *shear*, a condition in which one part of a material is pushed sideways past another part. Such compression, tension, and shear are all types of **stress**, formally defined as the force applied per unit area of a material's surface, as we discuss more fully in Chapter 9. To picture compression, imagine that you have just built a house of cards and, being in a destructive mood, you step on it (**Fig. 7.3a**)—the structure collapses because the downward push you apply with your foot exceeds the push provided by air in other directions. Compression can occur both vertically (like the house of cards), or horizontally, like a ball of dough squeezed between two blocks of wood (**Fig. 7.3b**). To picture tension, imagine that you take that same ball of dough, but rather than compressing it, you grab it on each side and pull it until the dough stretches out into a long sausage-like shape. To picture shear, imagine that you place a deck of cards on a table, set your hand on top of the deck, and then move your hand parallel to the table—the cards spread out across the table (**Fig. 7.3c**). (Note that, using the above terminology, we can define *pressure* more precisely as a special condition of stress where compression is the same in all directions.)

At or near the surface of the Earth, application of compression, tension, or shear can break rocks (see Chapter 8). But, when rocks are subjected to such stresses at high temperatures and pressures, they can change shape without breaking, a process called *plastic deformation*. Thus, stress can flatten grains into pancake-like shapes that lie parallel to one another, or stretch them into elongate (sausage-like) shapes that align to point in the same direction. Both platy and elongate grains are *inequant grains*, meaning that their dimensions are not the same in all directions (**Fig. 7.3d**). (In contrast, *equant grains* have roughly the same dimensions in all directions.) When inequant minerals in a rock are parallel or aligned, geologists say that the minerals have a **preferred orientation** (**Fig. 7.3e**). As we've implied, preferred orientation can develop when plastic deformation reshapes pre-existing grains. More commonly, it develops when neocrystallization produces new inequant crystals that become parallel or aligned as they grow.

The Role of Hydrothermal Fluids

Metamorphic reactions commonly take place in the presence of hydrothermal fluids (hot water solutions). Where does the

FIGURE 7.3 Compression and shear change rock grains during metamorphism.

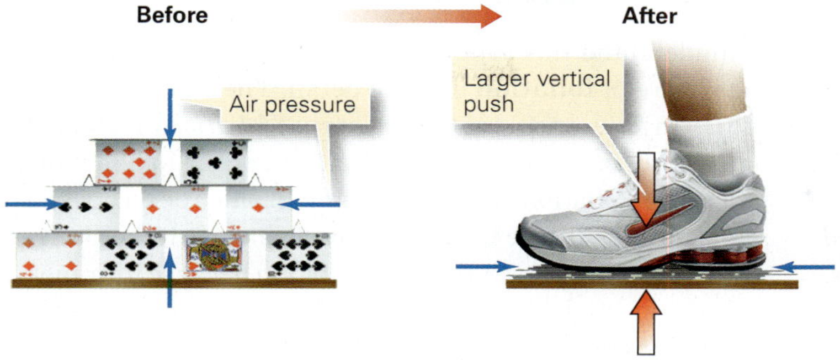

(a) A standing house of cards is subject only to air pressure, which is equal from all sides. When you step on the cards, they are pushed downward because of the vertical compression.

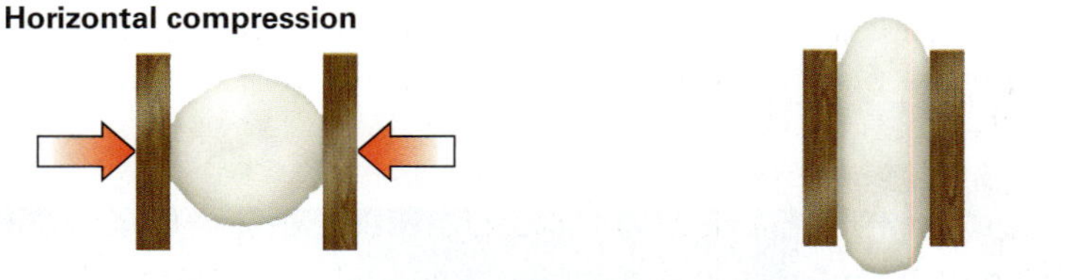

(b) Here horizontal compression flattens a dough ball between wood blocks.

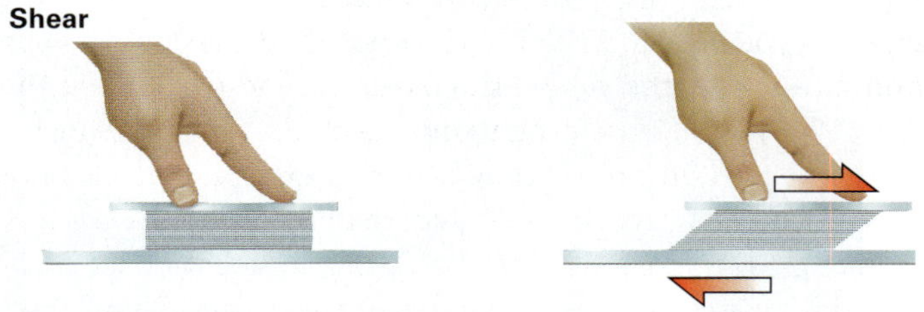

(c) A shear stress acts parallel to a surface. Here shear smears out a deck of cards parallel to a table top.

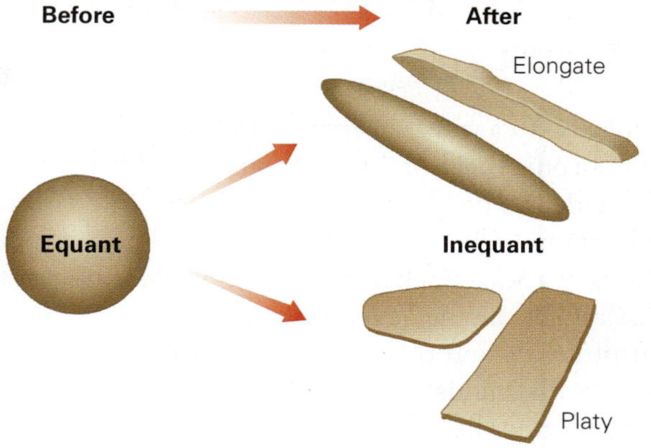

(d) Compression and shear can transform equant grains into inequant grains. Inequant grains can be elongate (cigar-shaped) or platy (pancake-shaped).

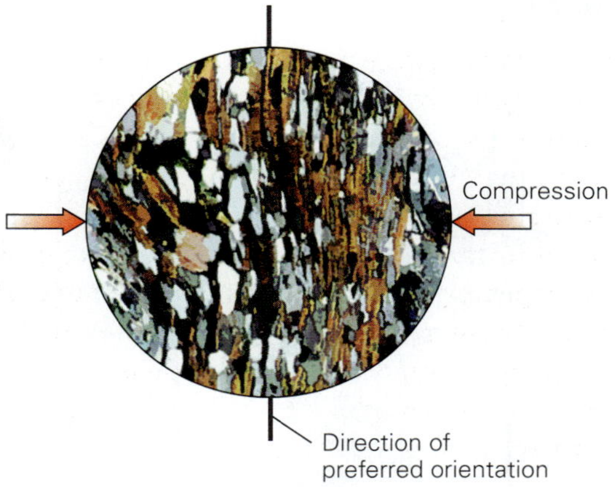

(e) In metamorphic rock, inequant grains may be aligned to form a preferred orientation. As seen through a microscope, the flat planes of grains are perpendicular to the compression direction.

water in hydrothermal fluids come from? Some was originally bonded to minerals in the protolith—heating breaks this water free, so it can diffuse into its surroundings. Some may seep into the protolith from a nearby igneous intrusion, or from groundwater reservoirs. (Notably, under extremely high pressures and temperatures that exist in the metamorphic realm, the water of hydrothermal fluids is in neither a gas nor liquid state, but rather occurs as a *supercritical fluid*, meaning that it has characteristics of both gas and liquid.) Hydrothermal fluids can accelerate metamorphic reactions, because atoms involved in the reactions can migrate faster through a fluid than they can through a solid. Hydrothermal fluids can also provide water that can be absorbed by minerals during metamorphic reactions, and thus can influence which minerals form. And hydrothermal fluids passing through a rock may pick up some dissolved ions and drop off others, much as a bus picks up and drops off passengers, and thus can change the overall chemical composition of a rock during metamorphism. The process of changing a rock's chemical composition by interactions with hydrothermal fluids is called **metasomatism**.

TAKE-HOME MESSAGE

Metamorphism takes place in response to changes in temperature, pressure, application of compression and shear, and/or interaction with hydrothermal fluids. The process involves reactions that take place without melting, and can produce new textures and new minerals.

QUICK QUESTION Can the overall composition of rock change during metamorphism?

7.3 Types of Metamorphic Rocks

Coming up with a way to classify and name the great variety of metamorphic rocks on Earth wasn't easy. Eventually, geologists decided to divide metamorphic rocks into two fundamental groups: *foliated metamorphic rocks* and *nonfoliated metamorphic rocks*. Each group contains several rock types. We distinguish among *foliated* rocks partly by their component minerals and partly by the nature of their foliation, whereas we distinguish among *nonfoliated* rocks only by their component minerals.

Foliated Metamorphic Rocks

To understand this group of rocks, we first need to describe the nature of **foliation** in more detail. The word comes from the Latin *folium*, for leaf. Geologists use *foliation* to refer to the parallel surfaces and/or layers that can occur in a metamorphic rock. Foliation can give a metamorphic rock a striped or streaked appearance in an outcrop, or can give it the ability to split into thin sheets. A rock has foliation because its minerals have a preferred orientation, and/or because it contains alternating dark-colored and light-colored layers.

Foliated metamorphic rocks can be distinguished from one another according to their composition, their grain size,

FIGURE 7.4 Slate is a foliated metamorphic rock that forms at relatively low temperatures and pressures.

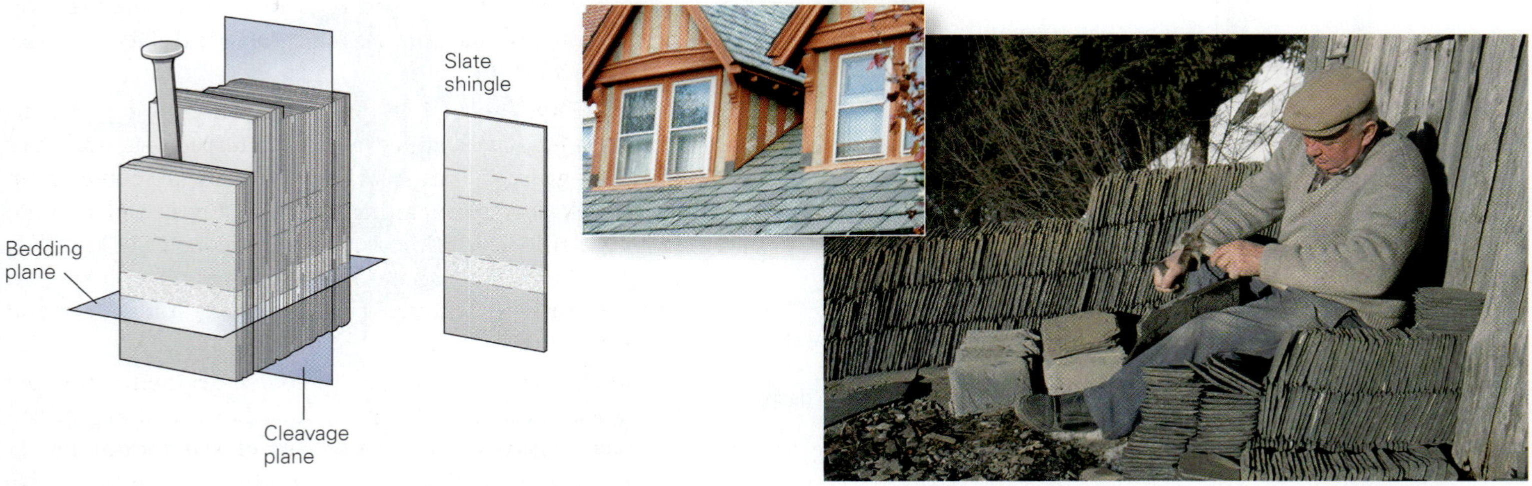

(a) A block of slate splits easily along cleavage planes, which may be at a high angle to the bedding planes. Workers split slate to produce roof shingles that, when overlapped, make a watertight surface (see inset).

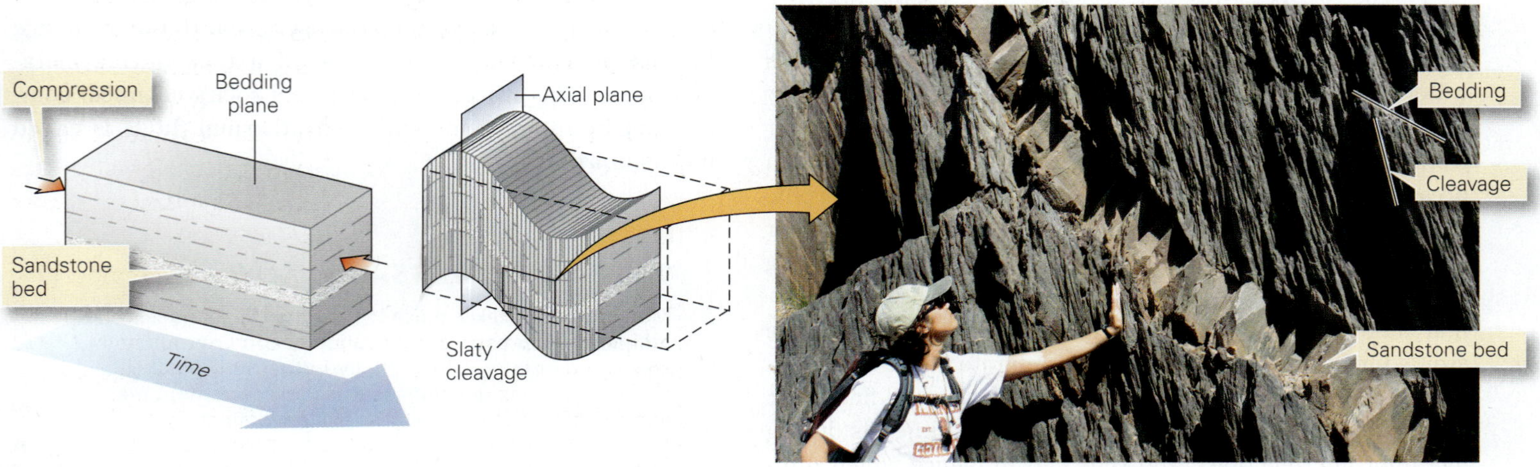

(b) Slaty cleavage forms in response to compressive stress. In this example, layers also bend to form folds, as slaty cleavage develops. The cleavage tends to be oriented parallel to the axial plane, an imaginary surface that, simplistically, divides the fold in half.

FIGURE 7.5 Examples of foliated metamorphic rocks formed at higher temperatures and pressures.

(a) During formation of phyllite, clay recrystallizes to form tiny mica flakes that reflect light, giving the rock a sheen.

(b) A schist contains coarse mica flakes, along with other metamorphic minerals.

(c) In metaconglomerate, pebbles and cobbles flatten into a pancake shape without cracking.

and the nature of their foliation. The most common types include:

- *Slate*: The finest-grained foliated metamorphic rock, **slate**, forms by metamorphism of shale or mudstone (rocks composed dominantly of clay) under relatively low pressures and temperatures. Slate contains a type of foliation called *slaty cleavage*. Slate splits into thin sheets, along slaty cleavage planes, and these sheets make excellent roofing shingles (**Fig. 7.4a**). Slaty cleavage develops when pressure solution removes portions of clay flakes that are not perpendicular to the compression direction, while clay flakes that are perpendicular to the compression direction grow. During the process, some flakes may passively rotate to be parallel with the plane of cleavage, pushed into the new orientation by compression. Horizontal compression of a sequence of shale beds produces vertical slaty cleavage, as well as folds (**Fig. 7.4b**).

Did you ever wonder...
why slate makes such nice roofing shingles?

- *Phyllite*: **Phyllite** is a fine-grained metamorphic rock containing a foliation defined by the preferred orientation of very fine-grained white mica flakes. The word comes from the Greek *phyllon,* meaning leaf. The texture of phyllite gives it a silky sheen known as phyllitic luster (**Fig. 7.5a**). Phyllite forms when slate undergoes metamorphism at a temperature high enough to cause neocrystallization and the growth of white mica.
- *Schist*: **Schist** is a medium- to coarse-grained metamorphic rock containing a type of foliation called *schistosity*. This fabric is defined by the preferred orientation of large mica flakes (typically, muscovite and/or biotite; **Fig. 7.5b**). Schist forms at a higher temperature than does phyllite, for under hotter conditions, mica crystals can grow larger.
- *Metaconglomerate*: Under the metamorphic conditions that produce slate, phyllite, or schist, a conglomerate metamorphoses into **metaconglomerate**. Typically, pressure solution and plastic deformation flatten pebbles and cobbles of such rock into pancake-like shapes, or tension stretches them into cigar-like shapes. The alignment of these inequant clasts defines a foliation (**Fig. 7.5c**).
- *Gneiss*: **Gneiss** is a compositionally layered metamorphic rock, composed of alternating dark-colored layers, typically with a higher concentration of mafic minerals, and light-colored layers, typically with a higher concentration of felsic minerals. The layers range in thickness from millimeters to meters. Such compositional layering, or *gneissic banding*, gives gneiss a striped appearance (**Fig. 7.6a, b**).

FIGURE 7.6 The formation of gneiss, which takes place at very high temperatures and pressures.

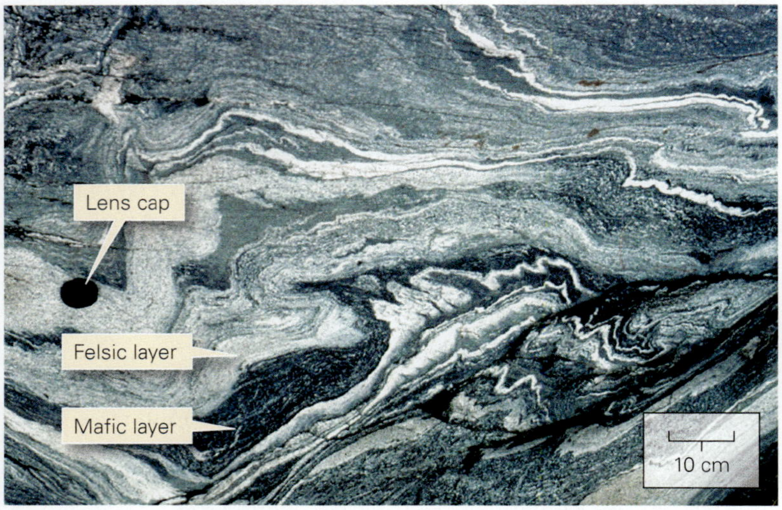

(a) In this outcrop of gneiss in Brazil, some of the layers are only centimeters across.

(b) An outcrop of 2.7-billion-year-old metamorphic rock in Ontario, Canada, shows a distinct foliation, in this case defined by alternating bands of light and dark minerals.

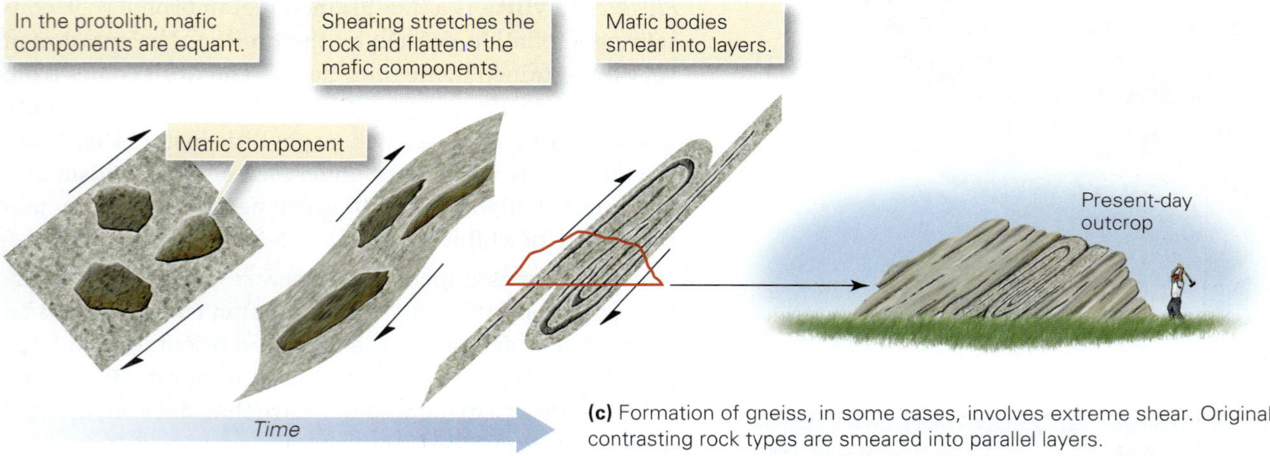

(c) Formation of gneiss, in some cases, involves extreme shear. Original contrasting rock types are smeared into parallel layers.

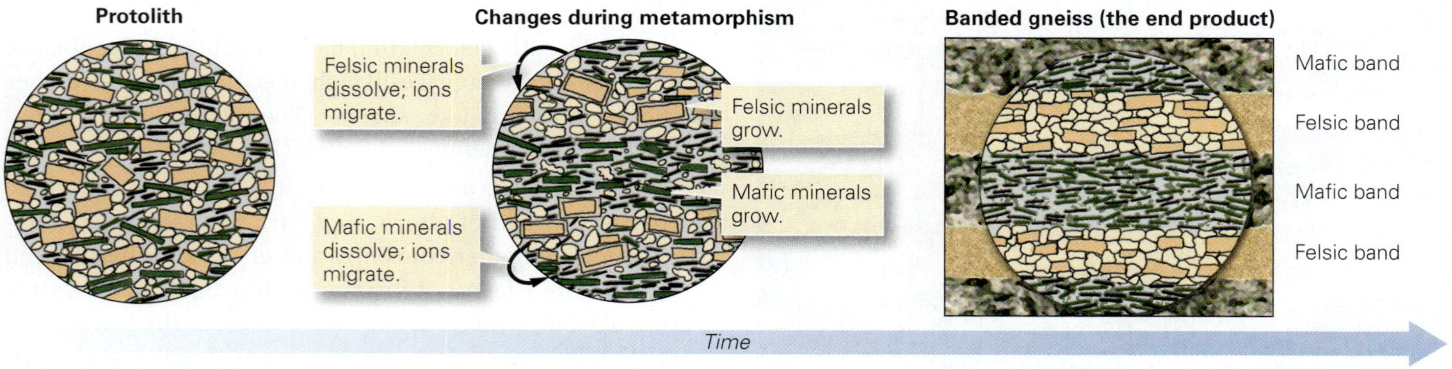

(d) Gneiss may also form by metamorphic differentiation, during which chemical reactions cause felsic and mafic minerals to grow in distinct, separate layers.

How does gneissic banding form? Some banding evolves directly from the original bedding in a rock. For example, metamorphism of a protolith consisting of alternating beds of sandstone and shale could produce a gneiss with alternating bands of quartzite and mica. Most gneissic banding forms when the protolith undergoes an extreme amount of shearing under conditions in which the rock can flow like soft plastic (**Fig. 7.6c**). The shearing stretches and smears out any pre-existing compositional contrasts in the rock, and transforms them into aligned sheets. Formation of banding may also develop by *metamorphic differentiation*, an incompletely understood process during which different metamorphic minerals grow preferentially in different layers (**Fig. 7.6d**).

› *Migmatite*: Under certain conditions (namely, high temperatures in the presence of water), gneiss may *begin* to melt, producing felsic magma and leaving behind still solid, relatively mafic metamorphic rock. If the melt freezes again before flowing out of the source area, a mixture of igneous rock and relict metamorphic rock forms. This mixture is called **migmatite**—in effect, a migmatite is part metamorphic and part igneous. If shear or compression takes place during the formation of migmatite, the felsic and mafic rocks may transform into parallel layers or lenses, defining a gneissic foliation.

Nonfoliated Metamorphic Rocks

Nonfoliated metamorphic rocks contain minerals that recrystallized or grew during metamorphism but do not have a preferred orientation. Therefore, the rock does not possess a foliation. The lack of foliation means either that metamorphism occurred in the absence of compression or shear, or that it resulted in the growth of only equant crystals. Below, we describe some of the rock types that can occur without foliation.

› *Hornfels*: **Hornfels** is a fine-grained nonfoliated rock that contains a variety of metamorphic minerals. The specific mineral assemblage in a hornfels depends on the composition of the protolith and on the temperature and pressure of metamorphism. Inequant mineral grains of a hornfels, in contrast to those of a schist, are randomly oriented.

› *Quartzite*: **Quartzite** forms by the metamorphism of pure quartz sandstone. During metamorphism, pre-existing quartz grains recrystallize, yielding new, larger grains. In the process, the distinction between cement and grains disappears, open pore space disappears, and the grains become interlocking. When quartzite fractures, cracks cut across grain boundaries—in contrast, cracks in a sandstone curve *around* grains. Quartzite looks glassier than sandstone and does not have the grainy, sandpaper-like surface characteristic of sandstone (**Fig. 7.7a**). Quartzite can vary in color—white, gray, maroon, or green— depending on the impurities it contains.

› *Marble*: The metamorphism of limestone yields **marble**. During the formation of marble, calcite of the protolith recrystallizes, so fossil shells, pore space, and the distinction between grains and cement disappear. Thus, marble typically consists of a fairly uniform mass of interlocking calcite crystals. Sculptors love to work with marble because the rock is relatively soft and has a uniform texture that gives it the cohesiveness and homogeneity needed to fashion large, smooth, highly detailed sculptures. Like quartzite, marble comes in a variety of colors—white, pink, green, and black—depending on the impurities it contains. Michelangelo, one of the great Italian Renaissance artists, sought large, unbroken blocks of creamy white marble from quarries in northwestern Italy for his masterpieces (**Fig. 7.7b**). Significantly, not all marble is nonfoliated. If the original protolith contained layers with different impurities, and shear caused the marble to flow plastically, the resulting marble has color banding that makes it a prized decorative stone (**Fig. 7.7c**).

Defining Metamorphic Intensity

The physical conditions under which metamorphism happens vary from place to place. For example, rocks carried to a great depth beneath a mountain range undergo more intense metamorphism than do rocks closer to the surface. Geologists use the term **metamorphic grade** in a somewhat informal way to indicate the intensity of metamorphism, meaning the amount or degree of metamorphic change. (To provide a more complete indication of the intensity of metamorphism, geologists use the concept of *metamorphic facies*; see **Box 7.1**.) The metamorphic grade of a rock depends primarily on the temperature of metamorphism, which plays the dominant role in determining the extent of recrystallization and neocrystallization that take place during metamorphism—temperature determines how easy or difficult it is for chemical bonds to break or form and, therefore, for diffusion to take place. Metamorphic rocks that form at relatively low temperatures (between about 250° and 400°C) are *low-grade rocks*, and metamorphic rocks that form at relatively high temperatures (over about 600°C) are *high-grade rocks*. *Intermediate-grade rocks* form at temperatures between these two extremes (**Fig. 7.8a**). As grade increases, recrystallization and neocrystallization tend to produce coarser grains and new mineral assemblages that are stable at higher temperatures and pressures (**Fig. 7.8b**).

FIGURE 7.7 Examples of quartzite and marble—typically, but not always, these are nonfoliated metamorphic rocks.

Quartz sandstone—the protolith of quartzite.

(a) This unfoliated maroon quartzite from Wisconsin breaks on smooth fractures that cut across grains.

Italian marble quarry sliced into a mountain

(b) Michelangelo used nonfoliated white marble for his spectacular sculptures.

In this unmetamorphosed limestone, fossils and bedding are visible.

(c) This marble floor has color banding inherited from original bedding.

Different conditions of metamorphism yield different metamorphic minerals, so geologists use the identity of minerals as a basis for defining grade. The presence of specific minerals, known as *index minerals*, can delineate the boundary between rocks of a lower grade and rocks of a higher grade. The line on a map along which an index mineral first appears is called an *isograd* (from the Greek *iso*, meaning equal); all points along an isograd have the same metamorphic grade. A **metamorphic zone** is the region between two isograds—zones are named after the index mineral that was not present in the previous, lower-grade zone.

You can see rocks of many different grades by hiking across the Appalachian Mountains in the eastern United States (**Fig. 7.8c**). For example, if you start in central New York State and walk eastward, into central Massachusetts, your path starts in a region where rocks have not undergone metamorphism, and it takes you into a region where rocks underwent intense metamorphism. Thus, during your hike, you cross several metamorphic zones.

TAKE-HOME MESSAGE

Geologists divide metamorphic rocks into classes based on whether the rock contains foliation. Foliated rocks include slate, schist, and gneiss. Nonfoliated rocks include marble and quartzite. The type of metamorphic rock that forms depends on the conditions of metamorphism. Specific grades of metamorphic minerals form under specific temperature and pressure ranges.

QUICK QUESTION Why don't builders use gneiss to make roofing shingles?

FIGURE 7.8 Intensity of metamorphism is indicated by metamorphic grade.

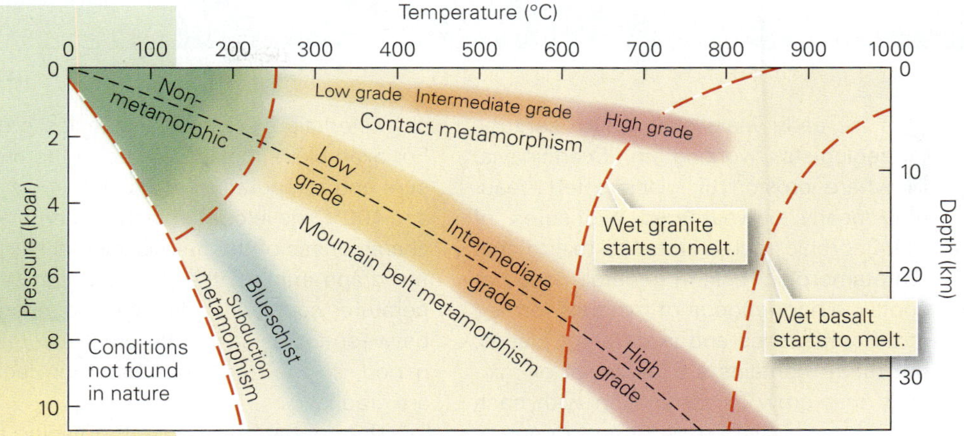

(a) This graph depicts the approximate temperatures and pressures of metamorphic grades. Different conditions occur in different geologic settings.

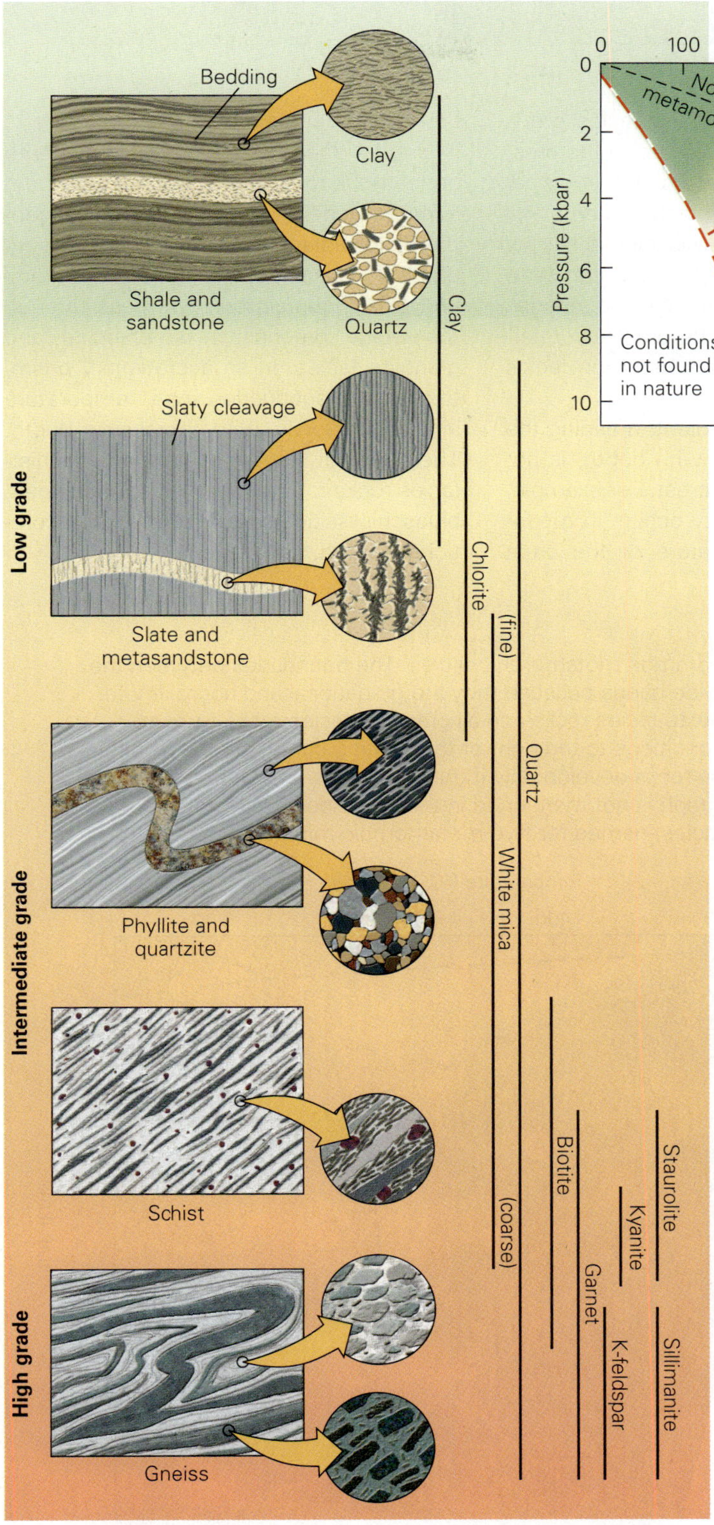

(b) Here we see the consequences of the progressive metamorphism of shale and sandstone from low grade to high grade during mountain building. Lines on the right indicate the range of grades in which key metamorphic minerals form.

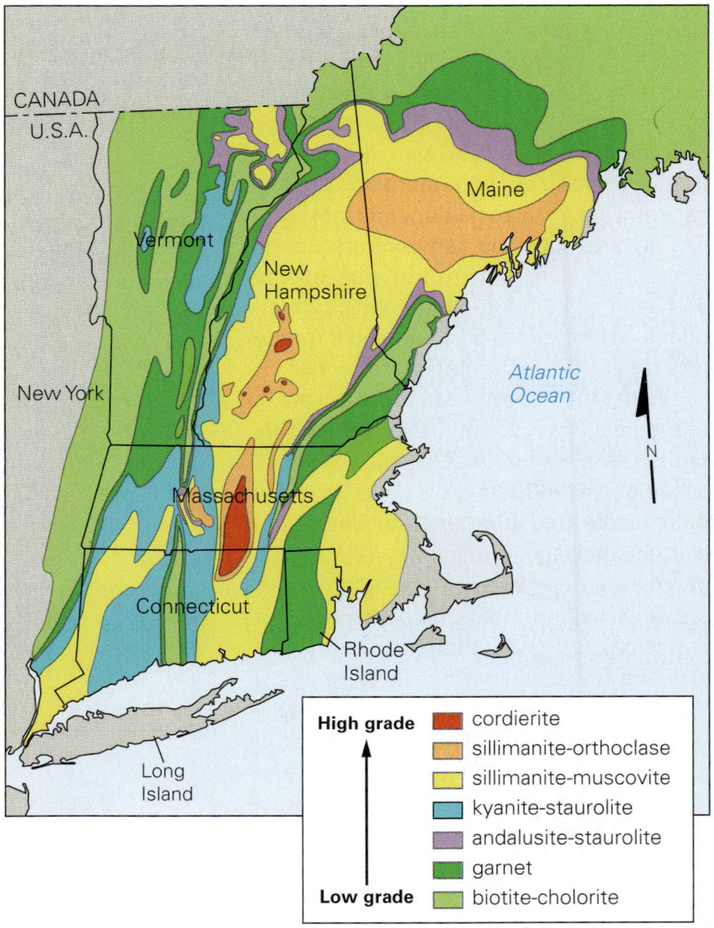

(c) Approximate metamorphic zones and isograds in the northeastern United States. The minerals listed are index minerals.

BOX 7.1 Consider This...

Metamorphic Facies

In the early years of the 20th century, geologists working in Scandinavia, where erosion by glaciers left beautiful, nearly unweathered exposures of metamorphic rocks, came to realize that a metamorphic rock, in general, does not contain a hodgepodge of minerals formed at different times and in different places, but rather contains a distinct set of minerals that grew in association with each other at a certain pressure and temperature. It seemed that such a *mineral assemblage* more or less represents a condition of *chemical equilibrium*, meaning that the chemicals making up the rock had organized into a group of minerals that were—to anthropomorphize a bit—comfortable with each other and their surroundings, and thus did not feel the need to change further. The geologists also determined that a particular metamorphic mineral assemblage in a rock depends not only on the pressure and temperature conditions of metamorphism, but also on the composition of the protolith, which determines what chemicals are available for the growth of new metamorphic minerals.

With the above concept in mind, the geologists defined a **metamorphic facies** as a set of metamorphic mineral assemblages indicative of a certain range of pressure and temperature. Each specific assemblage in a facies reflects the chemical composition of the metamorphic rock in which it occurs. According to this definition, a given metamorphic facies includes several different kinds of rocks that vary in terms of chemical composition, and therefore, mineral content—but all the rocks of a given facies formed under roughly the same temperature and pressure conditions.

Geologists recognize eight metamorphic facies. We can represent the conditions under which different metamorphic facies formed on a pressure-temperature graph (**Fig. Bx7.1**). Each area of the graph, identified with a facies name, represents the approximate range of temperatures and pressures in which mineral assemblages characteristic of the labeled facies grew. For example, a rock subjected to the pressure and temperature at Point A (4.5 kbar and 400°C) develops a mineral assemblage characteristic of the greenschist facies. As the graph implies, the pressure and temperature conditions defining boundaries between facies cannot be precisely determined, so the transitions between facies are gradual.

The geothermal gradient, meaning the change in temperature with depth, is not the same in all types of crust. For example, because igneous activity occurs in mountain belts, the temperature of the crust beneath a mountain belt, at a given depth, is greater than it is at the same depth beneath the middle of a continent, far from a mountain belt. We can plot different geothermal gradients on the graph of Figure Bx7.1 By following the geothermal gradient for a mountain belt, we pass through the zeolite, greenschist, amphibolite, and granulite facies. In an accretionary prism, formed at a subduction zone, temperature increases slowly with increasing depth. Therefore, at depths where greenschist-facies rocks form beneath a mountain belt, blueschist-facies rocks form in an accretionary prism.

FIGURE Bx7.1 The common metamorphic facies. The boundaries between the facies are depicted as wide bands because they are gradational and approximate. Note that some amphibolite-facies rocks and all granulite-facies rocks form at pressure-temperature conditions to the right of the melting curve for wet granite. Thus, such metamorphic rocks develop only if the protolith is dry. One of the facies depicted on the graph is not mentioned in the text: specifically, P-P (prehnite-pumpellyite) facies, named for two metamorphic minerals.

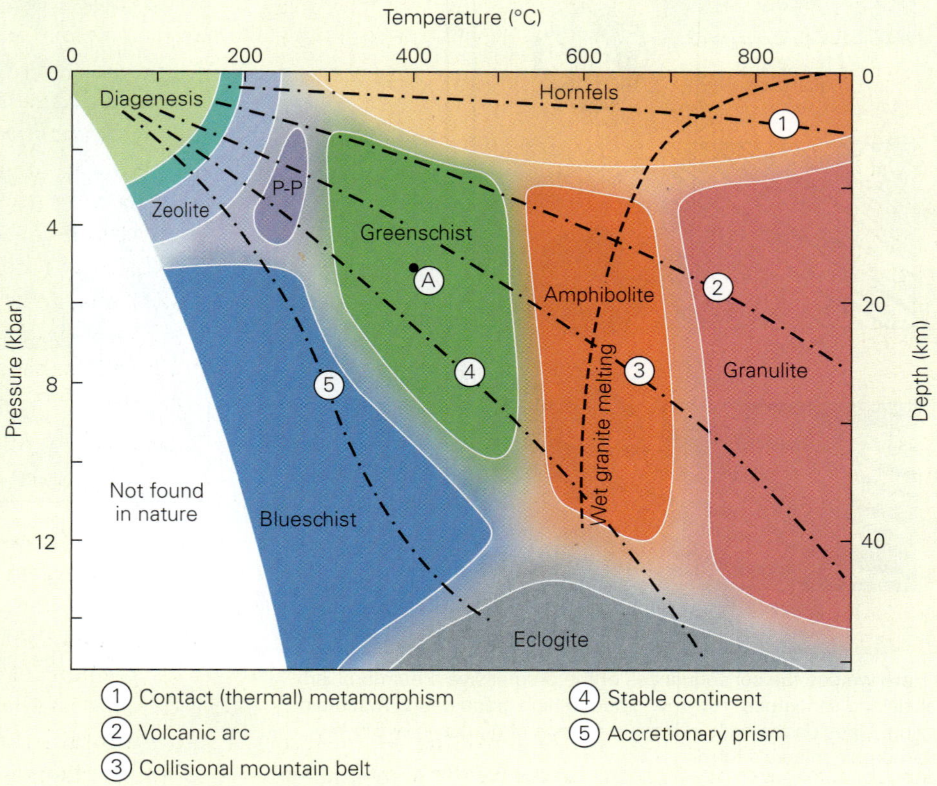

7.4 Where Does Metamorphism Occur?

So far, we've discussed the nature of changes that occur during metamorphism, the agents of metamorphism (heat, pressure, compression, shear, and hydrothermal fluids), the rock types that form as a result of metamorphism, and the concepts of metamorphic grade and metamorphic facies. With this background, let's now examine the geologic settings on Earth where metamorphism takes place, as viewed from the perspective of plate tectonics theory (see **Geology at a Glance**, pp. 228–229). We'll see that the conditions under which metamorphism occurs are not the same in all geologic settings.

Thermal or Contact Metamorphism

Imagine a hot magma that rises from great depth beneath the Earth's surface and intrudes into cooler rock at a shallow depth. Heat flows from the magma into the wall rock, for heat always moves from hotter to colder materials. As a consequence, the magma cools and solidifies while the wall rock heats up. In addition, hydrothermal fluids circulate through both the intrusion and the wall rock. As a consequence of the heat and hydrothermal fluids, the wall rock undergoes metamorphism, with the highest-grade rocks forming immediately adjacent to the pluton, where the temperatures are highest, and progressively lower-grade rocks forming farther away. The distinct belt of metamorphic rock that forms around an igneous intrusion is called a **metamorphic aureole** or *contact aureole* (**Fig. 7.9a**). The width of an aureole depends on the size of the intrusion—larger intrusions produce wider aureoles.

Geologists refer to the local metamorphism caused by igneous intrusion as **thermal metamorphism** (see **Box 7.2**), to emphasize that it develops in response to heat without a change in pressure and without preferential compression or shear. This type of metamorphism is also called **contact metamorphism**, to emphasize that it develops adjacent to the contact between an intrusion with its wall rock. Because such metamorphism takes place without application of compression or shear, aureoles typically contain *hornfels*, a nonfoliated metamorphic rock. Contact metamorphism occurs anywhere that plutons intrude into the crust. In the context of plate tectonics theory, plutons intrude at convergent boundaries, in rifts, and during mountain building, so thermal metamorphism can occur in these settings.

Burial Metamorphism

As sediment undergoes burial in a subsiding sedimentary basin, the pressure increases due to the weight of overburden, and the temperature increases due to the geothermal gradient. At depths greater than about 8 to 15 km, depending on the geothermal gradient, temperatures may be great enough for metamorphic reactions to begin, and low-grade metamorphic rocks form. Metamorphism due only to the consequences of very deep burial is called **burial metamorphism**.

Dynamic Metamorphism

Faults are surfaces on which one piece of crust slides, or shears, past another. Near the Earth's surface (in the upper 10 to 15 km) this movement can fracture rock, breaking it into angular fragments or even crushing it to a powder. But at greater depths, rock is so warm that it behaves like soft plastic when shear takes place along the fault. During this process, the minerals in the rock recrystallize. We call this process **dynamic metamorphism**, because it occurs as a consequence only of shearing under metamorphic conditions, without requiring a change in temperature or pressure. The resulting rock, a *mylonite*, has a foliation that roughly parallels the fault (**Fig. 7.9b**). Mylonites are very fine-grained, due to processes during dynamic metamorphism that replace larger crystals with a mass of very tiny ones. Dynamic metamorphism takes place wherever faulting occurs at depth in the crust. Thus, mylonites can be found at all plate boundaries, in rifts, and in collision zones.

SEE FOR YOURSELF...

PILBARA CRATON, WESTERN AUSTRALIA

Latitude
21°40′39.20″ S

Longitude
119°40′33.69″ E

Zoom to 160 km (99 miles) and look down.

This region of northwestern Australia is a desert where contrasts between different types of bedrock stand out. The north-south stripe consists of meta-sedimentary and meta-volcanic rocks in a narrow belt between two large domes containing granite and gneiss.

Dynamothermal or Regional Metamorphism

During the development of mountain ranges, in response to either convergent-margin tectonics or continental collision, regions of crust undergo compression and large slices of continental crust slip up along faults and are emplaced over other portions of the crust. As a consequence, rock that was once near the Earth's surface ends up at great depth beneath the mountain range (**Fig. 7.9c**). In this environment, the protolith changes in three ways: (1) it heats up because of

Environments of Metamorphism

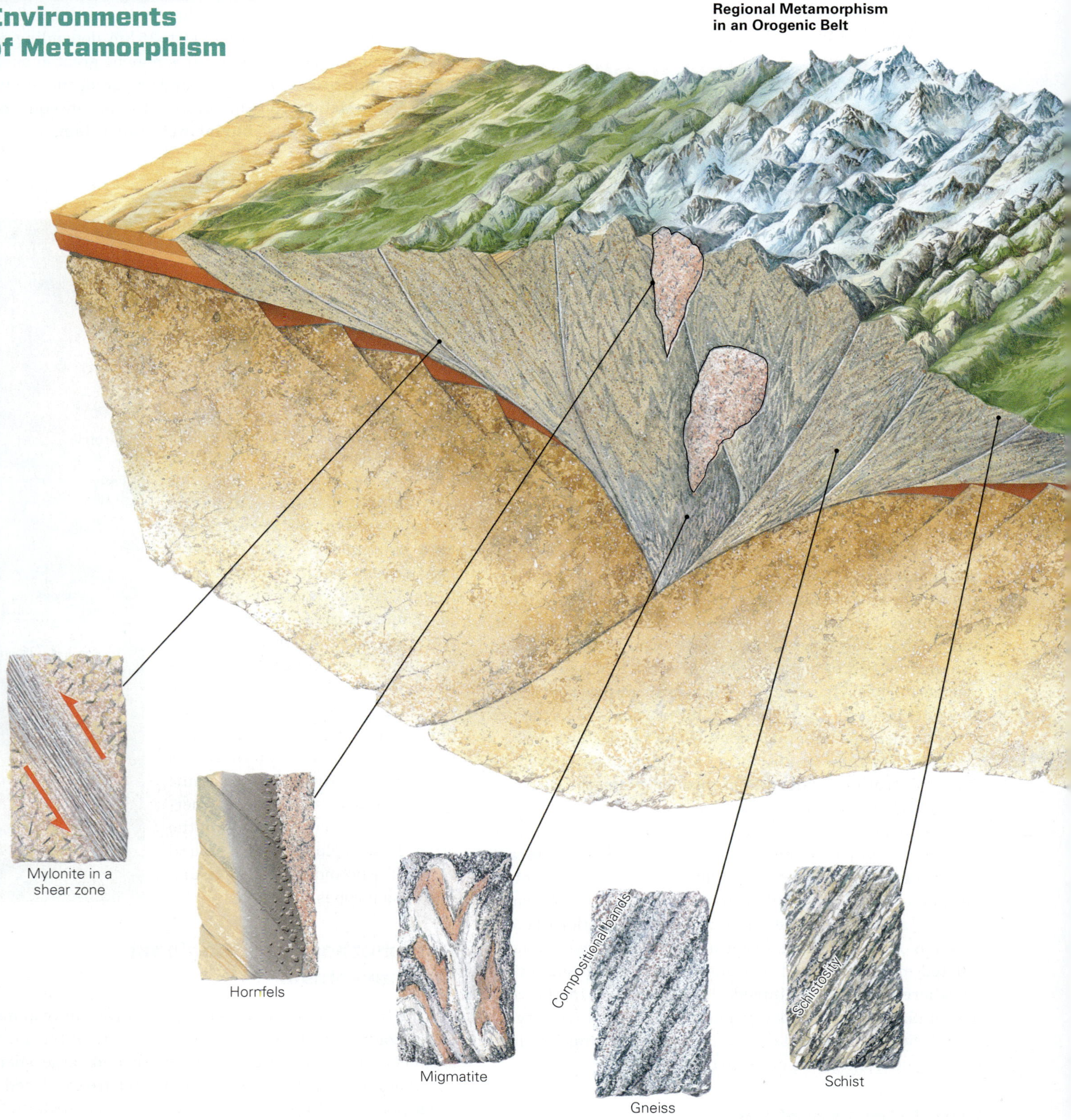

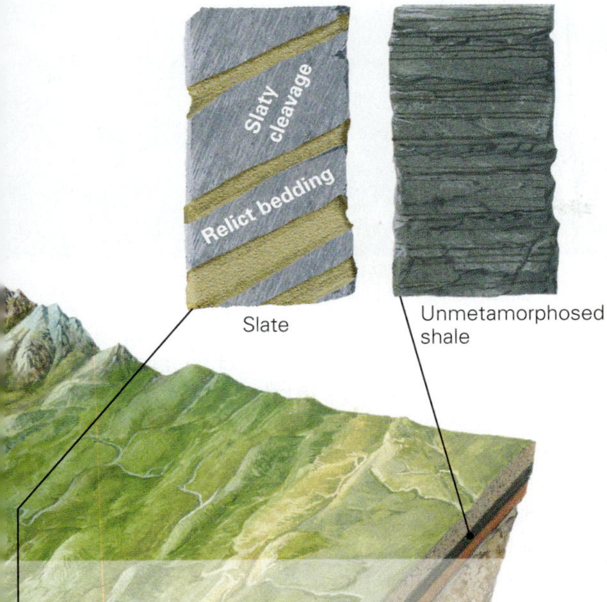

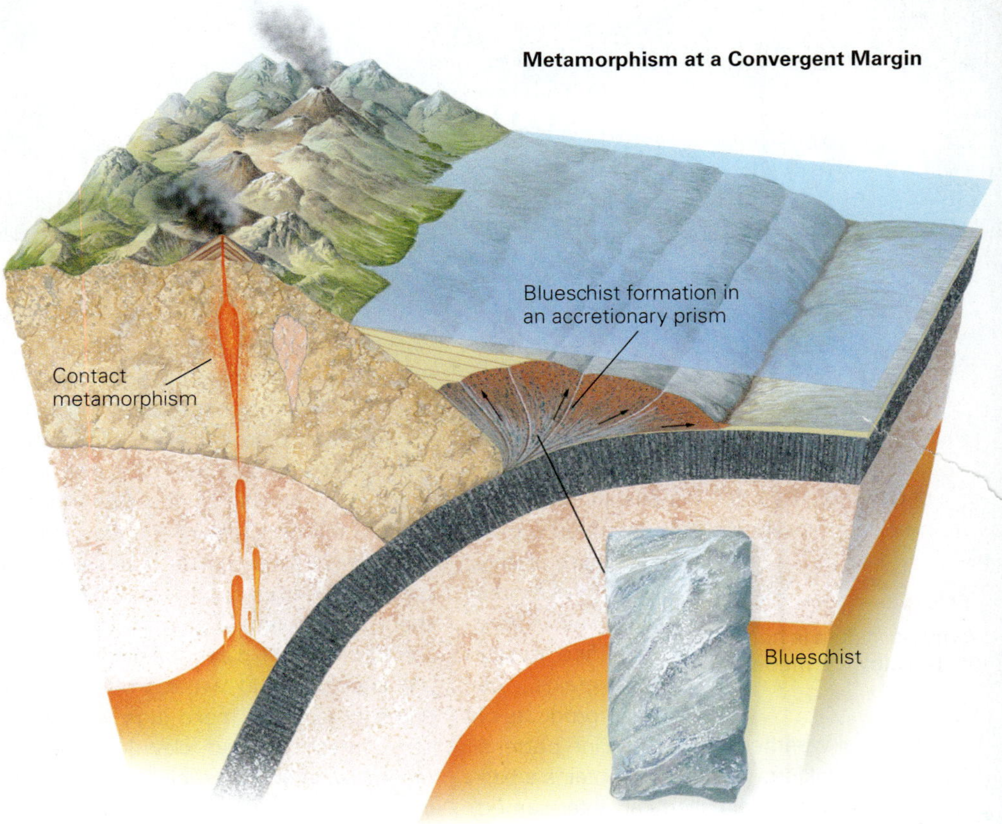

Metamorphic rocks form when a pre-existing rock (a protolith) undergoes changes in texture and/or mineral content in the solid state in response to changes in temperature, pressure, and/or differential stress. Metamorphism may also reflect interaction with hydrothermal fluids. Some metamorphic rocks are nonfoliated (they do not have metamorphic layering), whereas others are foliated (they do have metamorphic layering). Foliation results when rock is compressed or sheared during metamorphism, causing minerals to grow or rotate into parallelism with each other. Dynamothermal (regional) metamorphism occurs during mountain building. Contact metamorphism takes place around an igneous intrusion, or pluton, caused by the heat released by the pluton.

Geologists distinguish among metamorphic rocks according to the type of foliation and the mineral assemblage a rock contains. Hornfels is unfoliated and forms as a result of contact metamorphism. Mylonite develops when shearing produces a foliation but not necessarily a change in types of minerals. Slate, which forms from shale, contains slaty cleavage; clay flakes are typically aligned at an angle to bedding. Schist contains coarse grains of mica (muscovite and/or biotite) aligned parallel to each other. Gneiss has compositional banding. (Migmatite forms when part of the rock melts, and thus it is a mixture of metamorphic and igneous rock.) Quartzite is composed predominantly of quartz (it is metamorphosed sandstone), whereas marble is composed predominantly of calcite or dolomite (it is metamorphosed limestone or dolostone). Quartzite and marble are usually unfoliated.

The types of minerals and foliation in a metamorphic rock indicate the rock's grade. High-grade rocks, such as gneiss, form at higher temperatures and pressures, whereas low-grade rocks, such as schist, form at lower pressures and temperatures. Blueschist is an unusual metamorphic rock that develops under relatively high pressures but relatively low temperatures—the environment of an accretionary prism.

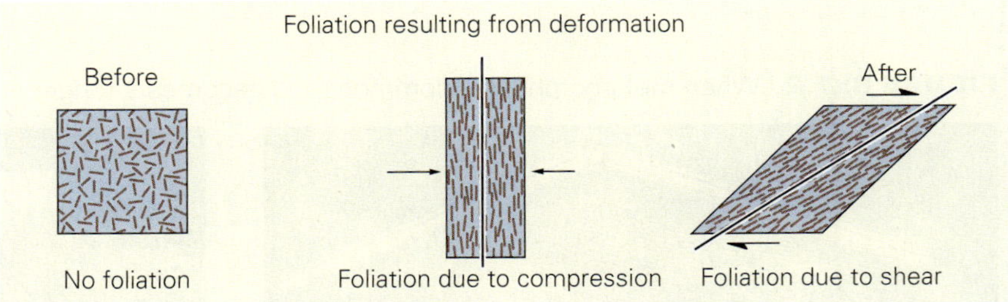

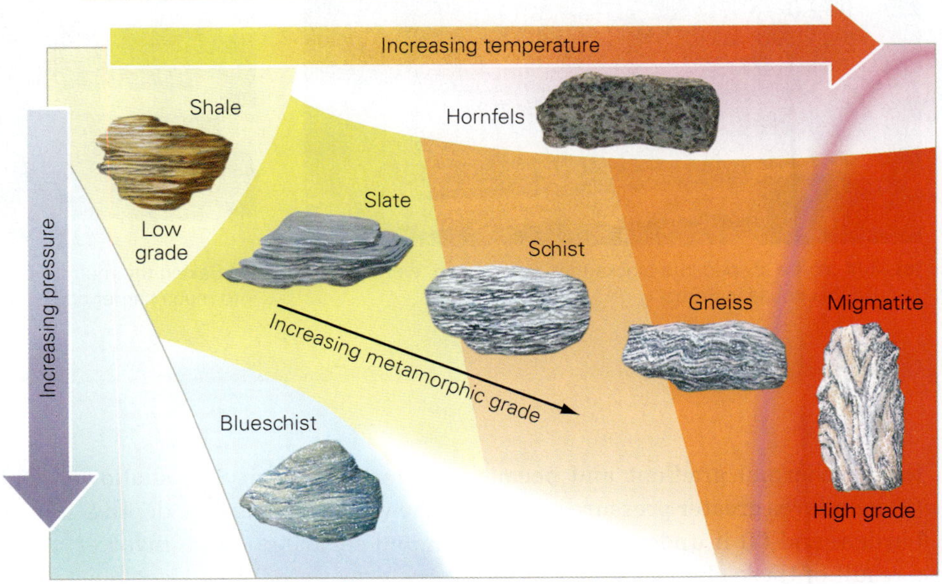

BOX 7.2 *Consider This...*

Pottery Making—An Analog for Thermal Metamorphism

A brick in the wall of an adobe house, an earthenware pot, a stoneware bowl, or a translucent porcelain plate may all be formed from the same lump of soft clay, scooped from the surface of the Earth and shaped by human hands (**Fig. Bx7.2**). This pliable and slimy muck forms when water saturates a mixture of clay minerals and very fine quartz grains, the products of chemical weathering of rock. Fine potter's clay for making white china contains a particular clay mineral called kaolinite, named after Kauling, a locality in China where it was originally discovered.

Did you ever wonder...
why porcelain is harder than the clay it's made from?

People in arid climates make adobe bricks by forming damp clay into blocks, which they then dry in the sun. Such bricks can be used for construction only in arid climates, because if it rains frequently, the bricks rehydrate and turn back into a sticky muck. Drying clay in the Sun does not change the structure of the clay minerals. To make a more durable material, workers place clay blocks in a kiln and bake ("fire") them at high temperatures. Heating transforms blocks of clay into hard bricks that are impervious to water. Potters use the same process to make jugs with impressive durability—fired clay jugs used for storing wine and olive oil have been found intact in sunken Phoenician ships that have rested on the floor of the Mediterranean Sea for thousands of years! Clearly, when clay undergoes intense heating, the process fundamentally and permanently changes it physically (see Fig. 7.9a). In other words, firing causes thermal metamorphism of clay, producing a new mineral assemblage and a new texture. The extent of the transformation depends on the kiln temperature, just as the grade of metamorphic rock depends on the Earth's temperature at a particular time and place. Potters usually fire earthenware at about 1,100°C and stoneware, which is harder than a knife or fork, at about 1,250°C. To produce porcelain—fine china—the clay must be heated to even higher temperatures, up to 1,400°C, for under these conditions the clay begins to melt. If the potter cools the slightly molten clay relatively quickly, glass forms. This glass gives porcelain its translucent, vitreous appearance.

FIGURE Bx7.2 When metamorphosed, common mud becomes stronger and more durable.

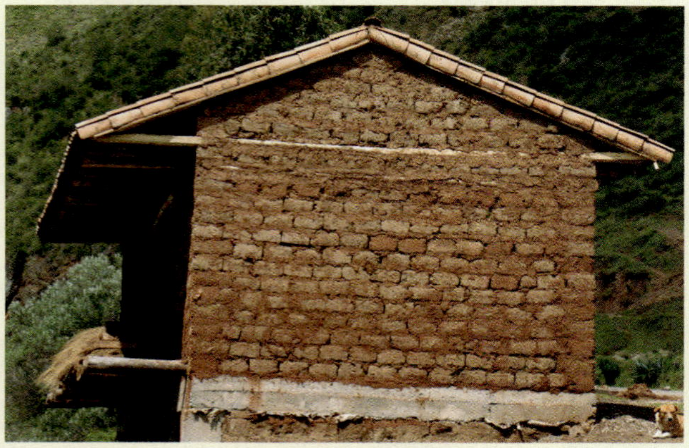

(a) Mud can be shaped into blocks that, when dried, were used to build this house in Peru.

(b) Baking the mud turns it into much harder brick.

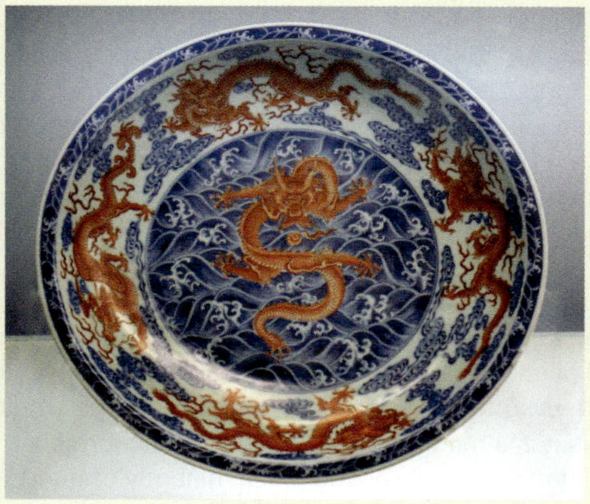

(c) At high temperatures, mud turns into porcelain, as in this plate from China.

the geothermal gradient and because of igneous activity; (2) it endures greater pressure because of the weight of overburden; and (3) it undergoes compression and shearing. As a result of these changes, the protolith transforms into foliated metamorphic rock. The type of foliated rock that forms depends on the grade of metamorphism—slate forms at shallower depths, whereas schist and gneiss form at greater depths. Since the metamorphism we've just described involves not only heat but also compression and shearing, geologists call it **dynamothermal metamorphism**. Since such metamorphism affects a large region, we can also call it **regional metamorphism**.

FIGURE 7.9 Geologic settings of metamorphism.

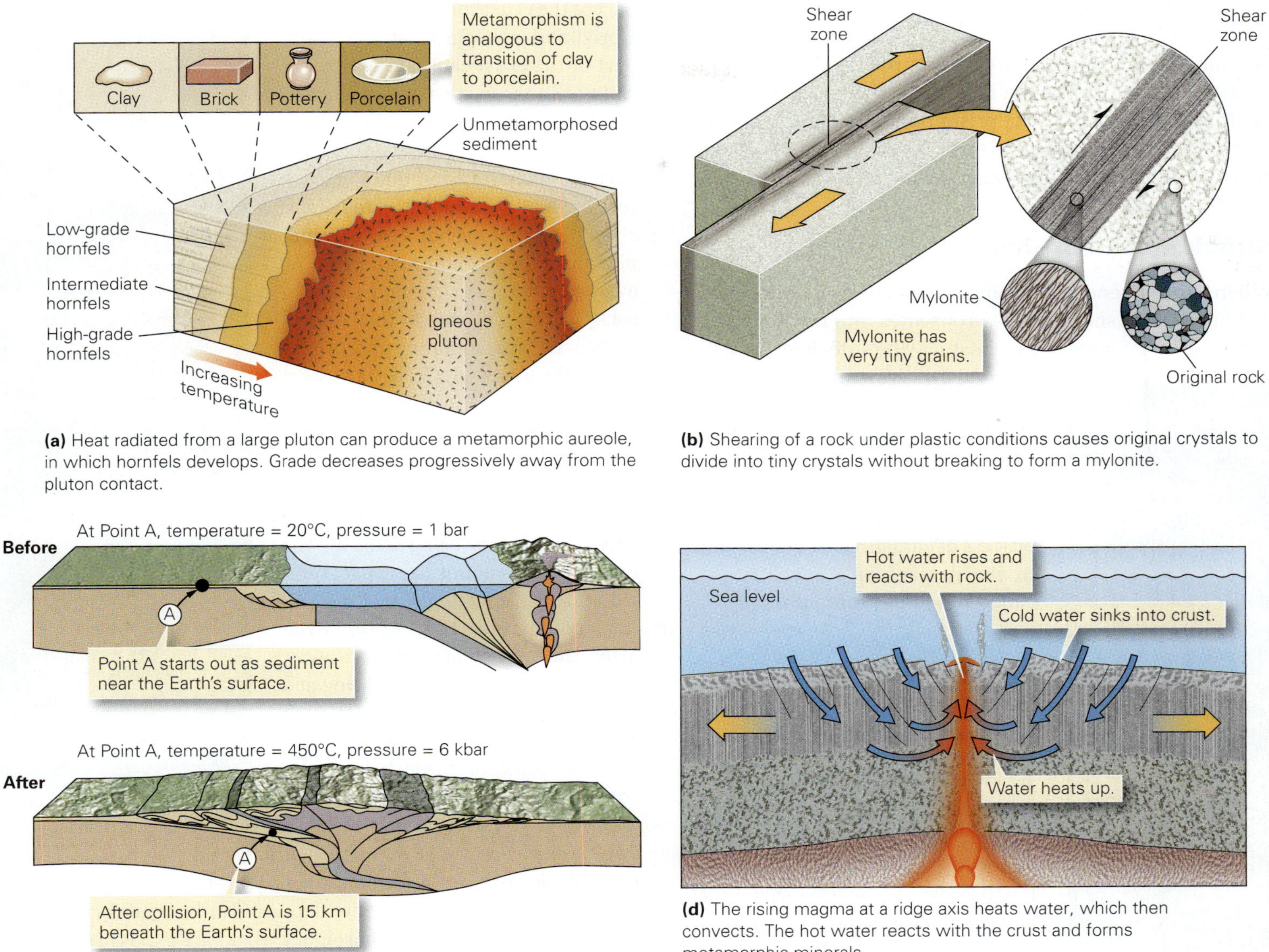

(a) Heat radiated from a large pluton can produce a metamorphic aureole, in which hornfels develops. Grade decreases progressively away from the pluton contact.

(b) Shearing of a rock under plastic conditions causes original crystals to divide into tiny crystals without breaking to form a mylonite.

(c) Dynamothermal metamorphism happens when one part of the crust shoves over another part, so that rocks once near the surface end up at great depth.

(d) The rising magma at a ridge axis heats water, which then convects. The hot water reacts with the crust and forms metamorphic minerals.

Hydrothermal Metamorphism at Mid-Ocean Ridges

Hot magma rises beneath the axis of mid-ocean ridges, so when cold seawater sinks down into the oceanic crust through cracks along ridges, it heats up and transforms into hydrothermal fluid. This fluid then rises through the crust, near the ridge, causing **hydrothermal metamorphism** of ocean-floor basalt (**Fig. 7.9d**). Eventually, the fluid escapes through vents back into the sea at black smokers (see Chapter 2).

Metamorphism in Subduction Zones

Blueschist is a relatively rare rock that contains an unusual blue-colored amphibole. Laboratory experiments indicate that formation of this mineral requires very high pressure but relatively low temperature. Such conditions cannot develop in continental crust, for at the high pressure needed to produce blue amphibole, the temperature in continental crust is also high (see Box 7.1). So to figure out where blueschist forms, we must determine where high pressures can develop at relatively low temperatures.

Plate tectonics theory provides the answer to this puzzle. Researchers determined that blueschist develops only in the accretionary prisms that form at subduction zones (see Geology at a Glance, pp. 228–229). These prisms grow to be over 20 km thick. Rock at the base of the prism endures very high pressure, due to the weight of overburden, but because subducted oceanic lithosphere beneath the prism is cool, temperatures at the base of the prism remain relatively low. The rocks develop a foliation (schistosity), because of the compression and shearing that takes place in an accretionary prism.

Shock Metamorphism

When large meteorites slam into the Earth, a pulse of extreme compression—a shock wave—propagates into the Earth away from the site of impact. The heat may be sufficient to melt or even vaporize rock at the impact site, and the extreme compression of the shock wave causes quartz in rocks below the impact site to undergo a phase change and become a more compact mineral called coesite. The process of changing rock in response to the passage of a shock wave is called **shock metamorphism**.

SEE FOR YOURSELF...

WIND RIVER MOUNTAINS, WYOMING

Latitude
43°6′7.22″ N

Longitude
109°21′36.45″ W

Zoom to 20 km (~12.5 miles) and look obliquely.

Faulting uplifted the Precambrian basement of western Wyoming 40 to 80 million years ago. This rock underwent high-grade metamorphism over 2 billion years ago. You can see that overlying sedimentary strata have warped into a huge fold.

Exhumation and the Exposure of Metamorphic Rocks

When you stand on an outcrop of metamorphic rock, you are standing on material that once lay many kilometers beneath the surface of the Earth. How does metamorphic rock, formed at great depth, return to the Earth's surface? Geologists refer to the overall process by which deeply buried rocks end up back at the surface as **exhumation**. To see how exhumation works, let's look at the specific processes that contribute to bringing metamorphic rocks from below a collisional mountain range back to the surface (**Fig. 7.10**). First, as two continents push together, the rock caught between them squeezes upward, much like dough pressed in a vise—this upward movement takes place by slip on faults and by plastic-like flow of rock. Second, as the mountain range grows, the crust at depth beneath it warms up and becomes softer and weaker, and able to flow like soft plastic. Eventually, the range starts to collapse under its own weight, much like a block of soft cheese placed in the hot Sun. As a result of this collapse, the upper part of the crust spreads out laterally. Horizontal stretching of the upper part of the crust causes it to become thinner in the vertical direction, and as it becomes thinner, the deeper part of the crust ends up closer to the surface. Third, erosion takes place at the surface—weathering, landslides, river flow, and glacial flow together play the role of a giant rasp, stripping away rock at the surface and exposing rock that was once below the surface.

Where are metamorphic rocks presently exposed? You can start your quest to find metamorphic rock outcrops by hiking into a mountain range. As we've seen, the process of mountain building produces and eventually exhumes metamorphic rocks, so the towering cliffs in the interior of a mountain range typically reveal schist, gneiss, quartzite, and marble (**Fig. 7.11a**). Even after the peaks have eroded away, the record of mountain building at a location remains, in the form of a belt of metamorphic rock at the ground surface—some of these belts are a couple of hundred kilometers across and over a thousand kilometers long. Vast expanses of metamorphic rock crop out in continental shields. A **shield** is a broad region of long-lived, stable continental crust where sedimentary "cover" either was not deposited or has been eroded away so that Precambrian "basement" rocks are exposed (**Fig. 7.11b, c**). The basement rocks were metamorphosed during a succession of Precambrian mountain-building events that led to the growth of continents in the first place.

SEE FOR YOURSELF...

CANADIAN SHIELD, EAST OF HUDSON BAY

Latitude
61°09′50.15″ N

Longitude
76°42′43.88″ W

Zoom to 225 km (140 miles) and look down.

The Canadian shield at this locality is covered by sparse vegetation. Differential erosion of the gneiss makes the foliation stand out. The east-west band of foliation delineates a zone of dynamic metamorphism associated with a fault zone.

TAKE-HOME MESSAGE

Thermal (contact) metamorphism develops around igneous intrusions, due to heat from the intrusion. Dynamothermal (regional) metamorphism develops beneath mountain ranges where rock undergoes compression and shear at high temperatures and pressures. Erosion and uplift may eventually expose metamorphic rock in mountain ranges or continental shields.

QUICK QUESTION Could you find a layer of metamorphic rock between the layers of sedimentary rock in a sedimentary basin? Why or why not?

FIGURE 7.10 Processes that bring metamorphic rock back to Earth's surface. Three phenomena contribute to exhumation of rocks at depth. Here the red dot (representing metamorphic rocks formed at the base of a mountain range) gets progressively closer to the surface over time.

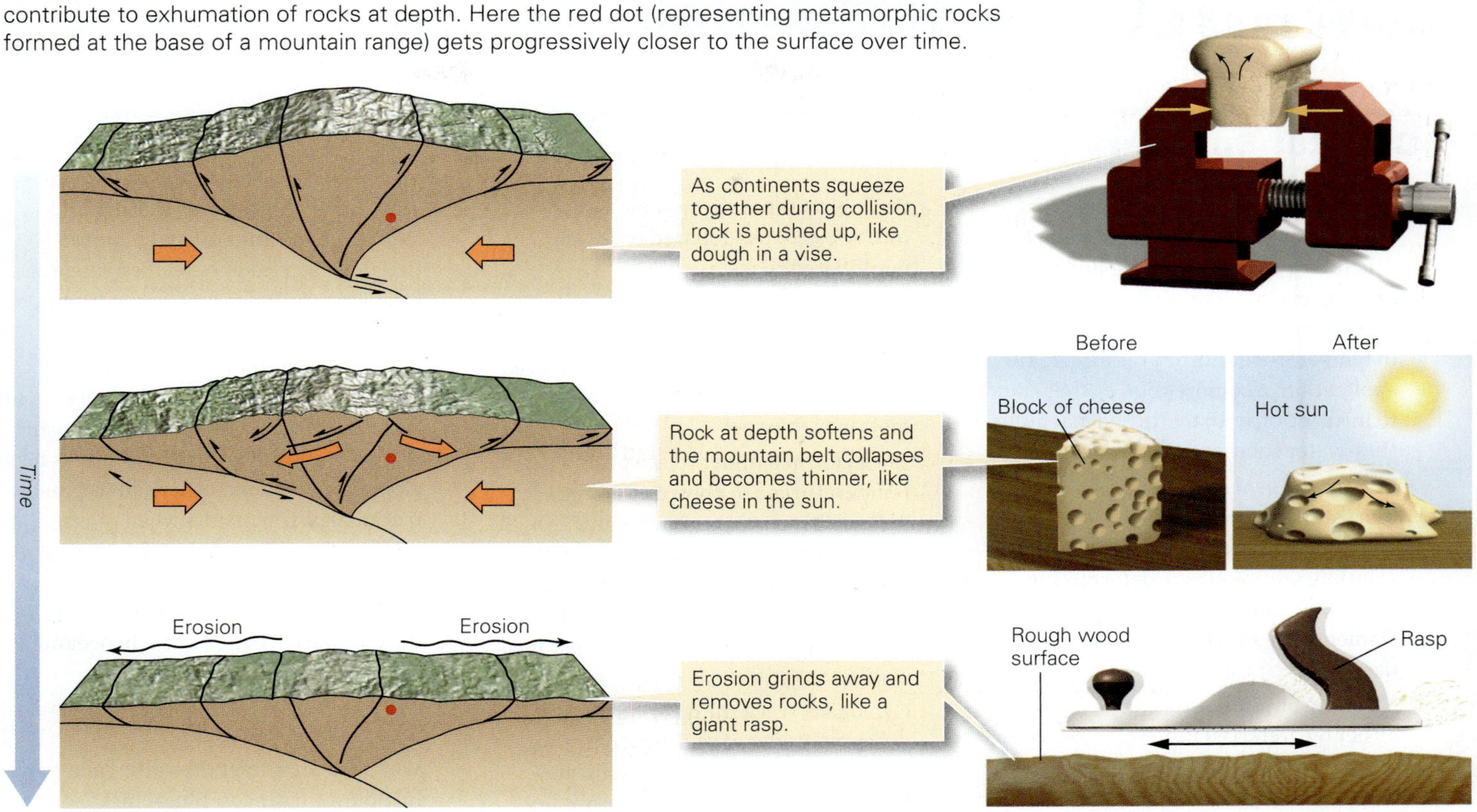

FIGURE 7.11 A map showing the distribution of shields, areas where broad expanses of Precambrian crust, including Precambrian metamorphic rocks, crop out.

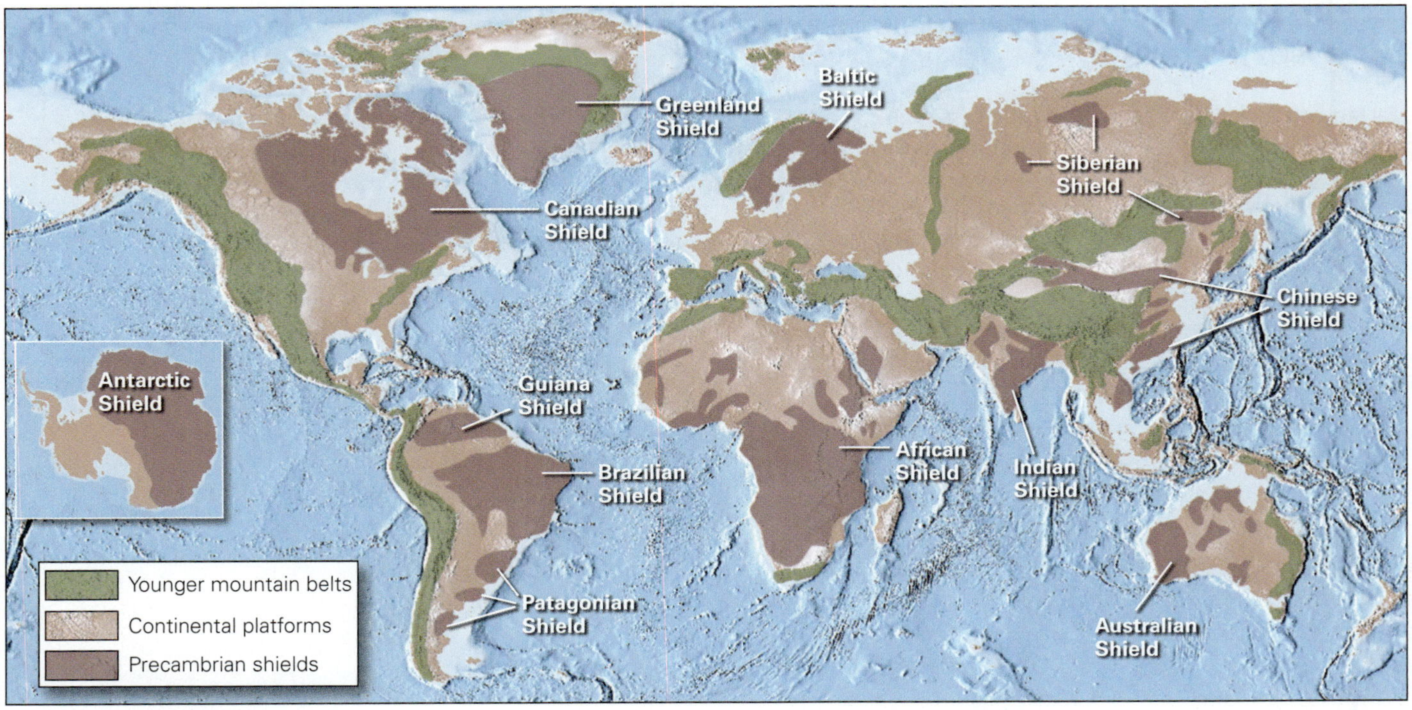

Chapter 7 Review

Chapter Summary

- Metamorphism refers to changes in a rock that result in the formation of a metamorphic mineral assemblage, and/or development of metamorphic texture. It occurs in response to change in temperature and/or pressure, to the application of compression or shear, and to interaction with hydrothermal fluids (hot-water solutions).
- Metamorphism involves recrystallization, phase changes, metamorphic reactions, pressure solution, and/or plastic deformation. If hydrothermal fluids bring in or remove elements, we say that metasomatism has occurred.
- Metamorphic foliation can be defined either by preferred mineral orientation (aligned inequant crystals) or by compositional banding. Preferred mineral orientation develops where compression and shearing take place.
- Geologists separate metamorphic rocks into two groups—foliated rocks and nonfoliated rocks.
- Foliated rocks include slate, phyllite, schist, and gneiss. Nonfoliated rocks include hornfels, quartzite, and marble. Migmatite is a mixture of igneous and metamorphic rock.
- Rocks formed under relatively low temperatures are known as low-grade rocks, whereas those formed under high temperatures are known as high-grade rocks. Intermediate-grade rocks develop between these two extremes. Different metamorphic mineral assemblages form at different grades.
- Geologists track the distribution of different grades of rock by looking for index minerals. An isograd indicates the location at which an index mineral first appears. A metamorphic zone is the region between two isograds.
- A metamorphic facies is a group of metamorphic mineral assemblages that develop under a specified range of temperature and pressure conditions.
- Thermal (contact) metamorphism occurs in an aureole surrounding an igneous intrusion. Burial metamorphism occurs at depth in a sedimentary basin. Dynamically metamorphosed rocks form along faults. Dynamothermal (regional) metamorphism results when rocks undergo heating and shearing during mountain building. Hydrothermal metamorphism can take place due to the circulation of hot water in oceanic crust at mid-ocean ridges. Shock metamorphism happens during the impact of a meteorite.
- We find belts of metamorphic rocks in mountain ranges. Blueschist forms in accretionary prisms. Shields expose broad areas of Precambrian metamorphic rocks.

Guide Terms

burial metamorphism (p. 227)
contact metamorphism (p. 227)
dynamic metamorphism (p. 227)
dynamothermal metamorphism (p. 230)
exhumation (p. 232)
foliation (p. 220)
gneiss (p. 221)
hornfels (p. 223)
hydrothermal metamorphism (p. 231)
marble (p. 223)
metaconglomerate (p. 221)
metamorphic aureole (p. 227)
metamorphic facies (p. 226)
metamorphic foliation (p. 216)
metamorphic grade (p. 223)
metamorphic mineral (p. 216)
metamorphic rock (p. 215)
metamorphic zone (p. 224)
metamorphism (p. 215)
metasomatism (p. 219)
migmatite (p. 223)
phyllite (p. 221)
preferred orientation (p. 218)
protolith (p. 215)
quartzite (p. 223)
regional metamorphism (p. 230)
schist (p. 221)
shield (p. 232)
shock metamorphism (p. 232)
slate (p. 221)
stress (p. 218)
thermal metamorphism (p. 227)

GEOTOURS *THIS CHAPTER'S GEOTOUR EXERCISE (G) FEATURES:*

- Zones of Metamorphic Grade
- Types of Metamorphism
- Metamorphism: Past and present

Review Questions

1. How are metamorphic rocks different from igneous and sedimentary rocks?
2. What two features characterize most metamorphic rocks?
3. What phenomena can cause metamorphism?
4. What is metamorphic foliation, and how does it form?
5. How does slate differ from a phyllite? How does phyllite differ from a schist? How does schist differ from a gneiss?
6. Why is hornfels nonfoliated?
7. What is a metamorphic grade, and how can it be determined? How does grade differ from facies?
8. Describe the geologic settings where thermal, dynamic, and dynamothermal metamorphism take place, respectively.
9. Why does metamorphism happen at the site of meteor impacts and along mid-ocean ridges?
10. How does plate tectonics explain the combination of low-temperature but high-pressure minerals found in a blueschist?
11. Where would you go if you wanted to find exposed metamorphic rocks, and how did such rocks return to the surface of the Earth after being at depth in the crust?

On Further Thought

12. Do you think that you would be likely to find a long belt, hundreds of kilometers across and thousands of kilometers long, in which the outcrop consists of high-grade hornfels? Why or why not?
13. Would we likely find broad regions of gneiss and schist on the Moon? Why or why not?
14. Why don't builders use gneiss to make roof shingles?
15. The geothermal gradient of Mars is 8°C/km, and the crust of Mars is 30 km thick. Do high-grade metamorphic rocks form in this crust?
16. Could you find a layer of metamorphic rock sandwiched between layers of sedimentary rock in a sedimentary basin? Why or why not?

Online Resources

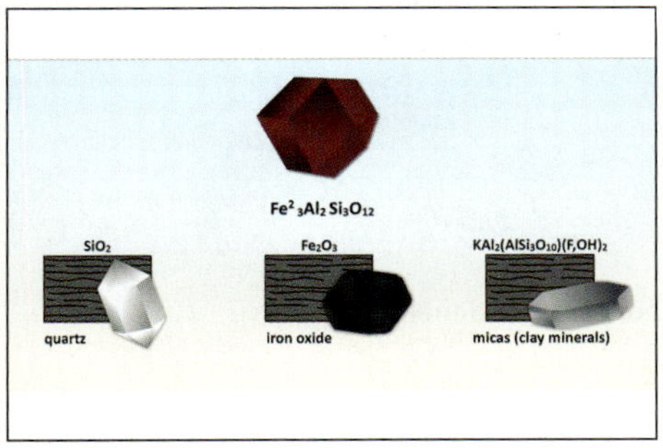

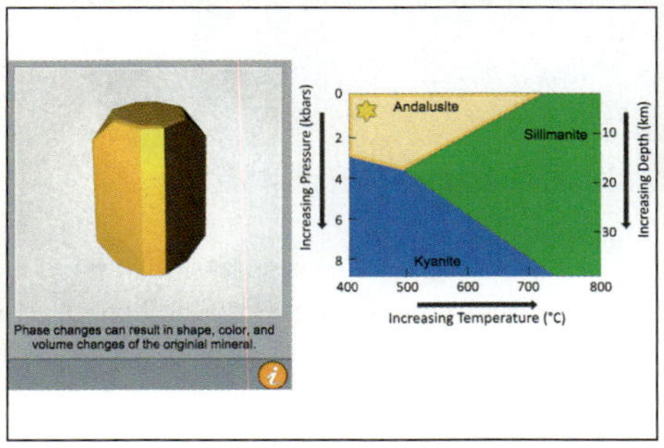

Animations

This chapter features animations on the six major types of metamorphorphic change, focusing on each process close up.

Assessment

This chapter features visual exercises on metamorphism and its effects.

INTERLUDE C

The Rock Cycle

▲ On our dynamic planet, the atoms in a rock of one class may, over time, be incorporated in rock of another class, and then another. Such transformations comprise the rock cycle.

LEARNING OBJECTIVES

By the end of this chapter, you should understand...

1. that rocks don't last forever because the Earth System is dynamic.
2. why phenomena such as weathering, burial, heating, and/or melting can affect rock after it has formed.
3. how components of a given rock may become incorporated in other rocks or even other rock classes, over time, in a progression of change called the rock cycle.
4. why pathways through the rock cycle reflect geologic settings, interpretable in the context of plate tectonics.
5. that steps of the rock cycle can happen at vastly different rates.
6. how energy from inside the Earth, from the Sun, and from gravity drive the rock cycle.

C.1 Introduction

This familiar expression, "stable as a rock," implies that rock, once formed, can last forever. In reality, however, components making up a rock may later be rearranged, and/or moved elsewhere, to form a new rock of the same class, or even one of a different class. In some places, over the course of millions to billions of years, this process has happened many times. Geologists refer to this progressive transformation of Earth materials as the **rock cycle** (**Fig. C.1**). A discussion of the rock cycle helps illustrate relationships among the three different rock types that we described in previous chapters. By following the arrows in Figure C.1, you can see many different paths around or through the rock cycle. Let's look at a few examples.

Imagine the following sequence of events. First, an igneous rock forms by solidification of lava. Then, it undergoes weathering and erosion at the Earth's surface to produce sediment. This sediment then gets deposited and lithified to form sedimentary rock which, in turn, becomes buried to a depth where it transforms into metamorphic rock. Finally, the metamorphic rock becomes so hot that it undergoes partial

FIGURE C.1 The stages of the rock cycle showing various alternative pathways.

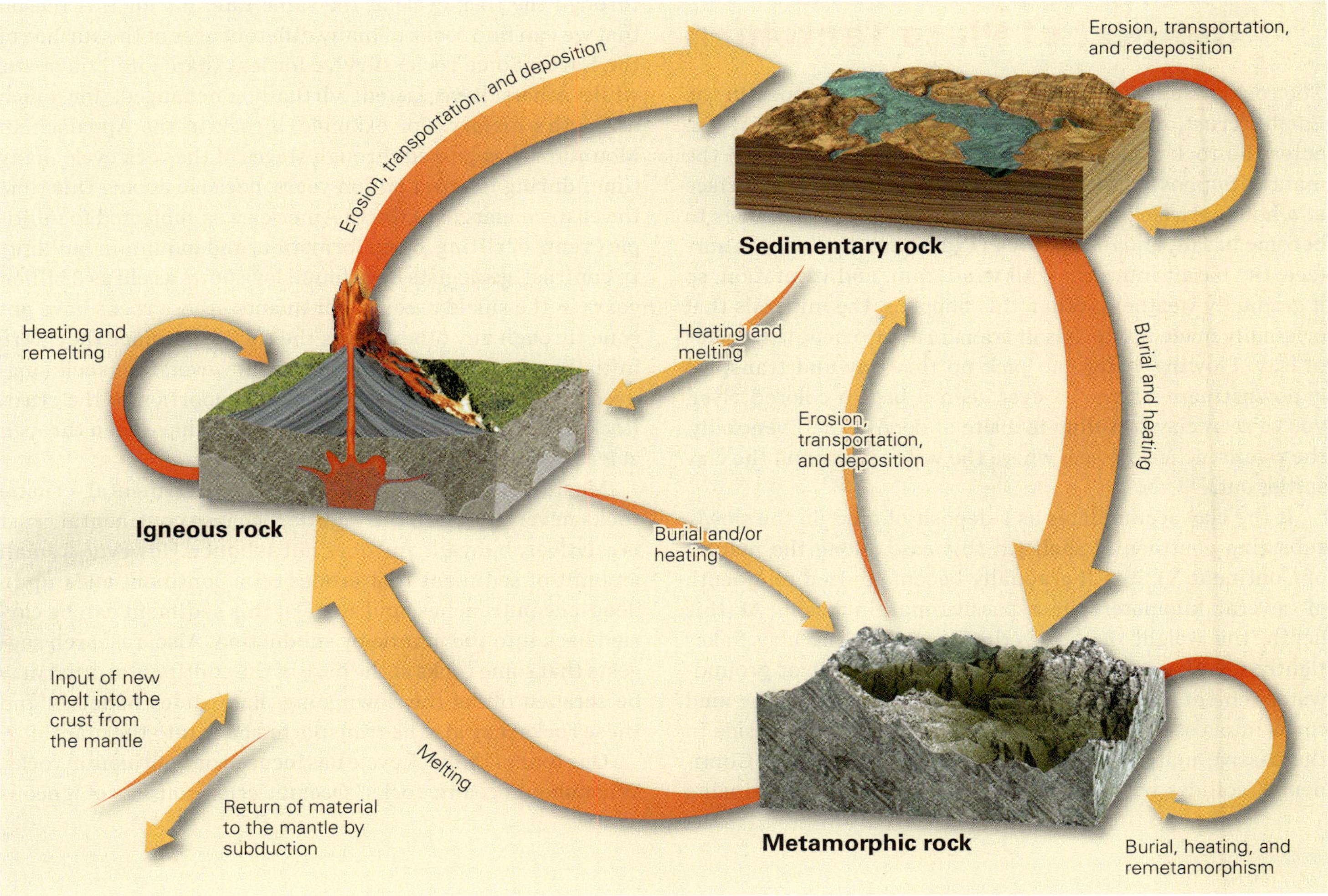

melting to produce a magma that seeps upward and solidifies to form a new igneous rock. We've described a path around the rock cycle that can be symbolized as follows:

Igneous → sedimentary → metamorphic → igneous

Alternatively, imagine that the metamorphic rock mentioned above, instead, undergoes uplift and erosion to form new sediment, which eventually becomes buried and lithified to form a sedimentary rock. We've described a shortcut through the cycle that can be symbolized as

Igneous → sedimentary → metamorphic → sedimentary

If the igneous rock from the above scenario underwent metamorphism directly, as a first step, and the resulting metamorphic rock later underwent uplift and erosion to produce sediment that eventually became sedimentary rock, we'd be seeing a different shortcut through the cycle, symbolized as

Igneous → metamorphic → sedimentary

How do scenarios like the ones we've just described relate to realistic geological events? We can see by interpreting a case study of the rock cycle in the context of plate tectonics theory.

C.2 The Rock Cycle in the Context of Plate Tectonics

The rock cycle mostly involves processes that happen in the Earth's crust, or at the Earth's surface. New material can enter the rock cycle, however, when magma rises from the mantle. Suppose that a magma reaches the Earth's surface at a hot-spot volcano, and extrudes as lava that solidifies to become basalt, an igneous rock (**Fig. C.2a**). Once at the surface, the basalt interacts with wind, rain, and vegetation, so it gradually weathers. When this happens, the minerals that originally made up the basalt transform into new, tiny flakes of clay. Flowing water can pick up this clay and transport it downstream—if you've ever seen a brown-colored river, you've seen clay traveling to a site of deposition. Eventually, the river reaches the sea, where the water slows and the clay settles out.

If the clay accumulates as a deposit of mud on the slowly subsiding continental shelf (in this case, along the margin of Continent X), it will gradually become buried to a depth of several kilometers in a passive-margin basin. At this depth, the weight of the overburden packs the clay flakes tightly together, and minerals precipitating out of groundwater cement the flakes to each other, so eventually the mud turns into a sedimentary rock, shale. The shale can reside in the passive-margin basin for millions of years, until Continent X collides with Continent Y. When the two continents squeeze together, and mountains start to build, slip on faults transports the edge of Continent Y up and over the passive-margin basin of Continent X. The shale may end up at a depth of 20 km below the surface, where temperature, pressure, and stress metamorphose it into schist (**Fig. C.2b**).

The story's not over. Erosion constantly grinds away the mountain range, and when mountain building stops, it may eventually expose the schist at the ground surface. On an outcrop, the schist undergoes erosion to form new sediment, which can be carried off and deposited elsewhere, ultimately to form a new sedimentary rock. Below the ground surface, some of the schist remains intact (**Fig. C.2c**). Imagine now that continental rifting takes place at the site of the former mountain range, and the crust containing the schist begins to split apart. During rifting, injection of hot magma, rising from the mantle below, brings so much heat up into the crust that some of the schist partially melts and a new felsic magma forms. This felsic magma rises to the surface of the crust and freezes into rhyolite, a new igneous rock (**Fig. C.2d**). In terms of the rock cycle, plate interactions have taken Earth materials through a full circle—the atoms that once resided in an igneous rock later resided in a sedimentary rock, then a metamorphic rock, and finally once again in an igneous rock.

It's important to keep in mind that not all atoms pass through the rock cycle at the same rate. It's for this reason that we can find rocks of many different ages at the surface of the Earth. Some rocks survive for less than a million years, while others have lasted, virtually unchanged, for much of Earth's history. For example, a rock in the Appalachian Mountains has passed through stages of the rock cycle many times during the past billion years, because during this time the eastern margin of North America was subjected to multiple events of rifting, basin formation, and mountain building. In contrast, geologists have found lava flows as old as 3 billion years in the shield areas of continents—these rocks have not gone through any other stages the rock cycle since they were initially extruded. Research shows, however, that such long-lived rocks account for a very small proportion of the crust. Rocks of the continental crust, in general, have been through at least a couple of stages in the rock cycle.

Most of the atoms that comprise continental crustal rocks never return to the mantle, because continental crust is relatively buoyant and does not subduct. However, a small amount of sediment that erodes off a continent ends up in deep-ocean trenches, and some of this sediment can be carried back into the mantle by subduction. Also, research suggests that some rocks at the base of the continental crust may be scraped off as the downgoing plate slides beneath, and these rocks may also be transported back into the mantle.

Our tour of the rock cycle has focused on continental rocks. What about oceanic rocks? Oceanic crust consists of igneous

FIGURE C.2 An example of the rock cycle in the context of plate tectonics.

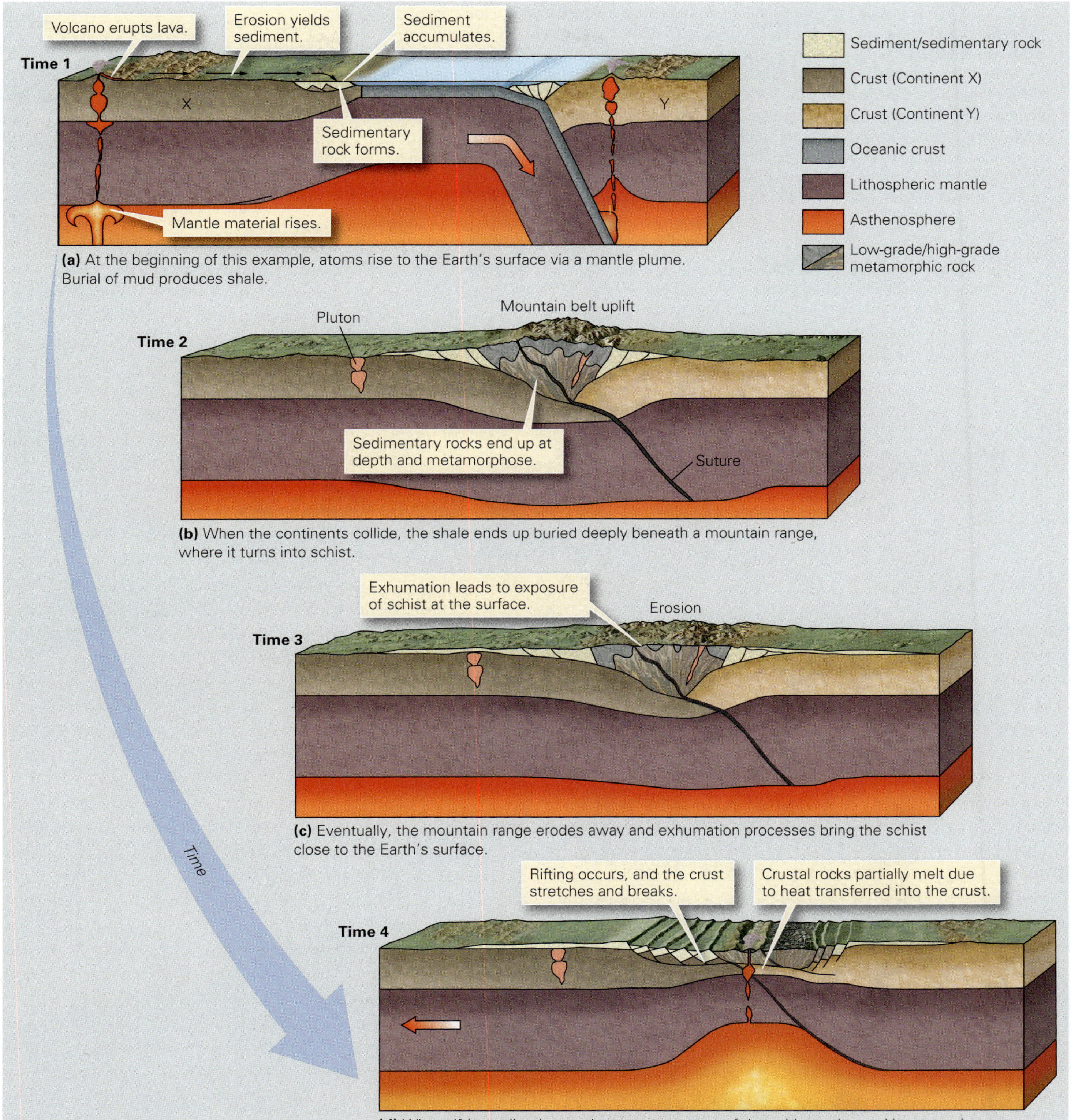
Volcano erupts lava.
Erosion yields sediment.
Sediment accumulates.
Time 1
X
Y
Sedimentary rock forms.
Mantle material rises.
Sediment/sedimentary rock
Crust (Continent X)
Crust (Continent Y)
Oceanic crust
Lithospheric mantle
Asthenosphere
Low-grade/high-grade metamorphic rock
Pluton
Mountain belt uplift
Time 2
Sedimentary rocks end up at depth and metamorphose.
Suture
Exhumation leads to exposure of schist at the surface.
Erosion
Time 3
Time
Rifting occurs, and the crust stretches and breaks.
Crustal rocks partially melt due to heat transferred into the crust.
Time 4

(a) At the beginning of this example, atoms rise to the Earth's surface via a mantle plume. Burial of mud produces shale.

(b) When the continents collide, the shale ends up buried deeply beneath a mountain range, where it turns into schist.

(c) Eventually, the mountain range erodes away and exhumation processes bring the schist close to the Earth's surface.

(d) When rifting splits the continents apart, some of the schist melts, and its atoms become incorporated into magma again.

Rock-Forming Environments and the Rock Cycle

Rocks form in many different environments. Igneous rocks develop where melt rises from depth and cools. Intrusive igneous rocks form where magma cools underground; extrusive igneous rocks form where lava and ash erupt at the surface.

Weathering and erosion break up existing rock and produce sediment. Different kinds of sediments develop in different places, reflecting both the composition of the source and the setting in which the sediment accumulates. When this sediment eventually gets buried and undergoes lithification, new sedimentary rocks form.

Under certain conditions, pre-existing rocks can undergo change in the solid state—metamorphism—which produces metamorphic rocks. Contact metamorphism is due to heat released by an

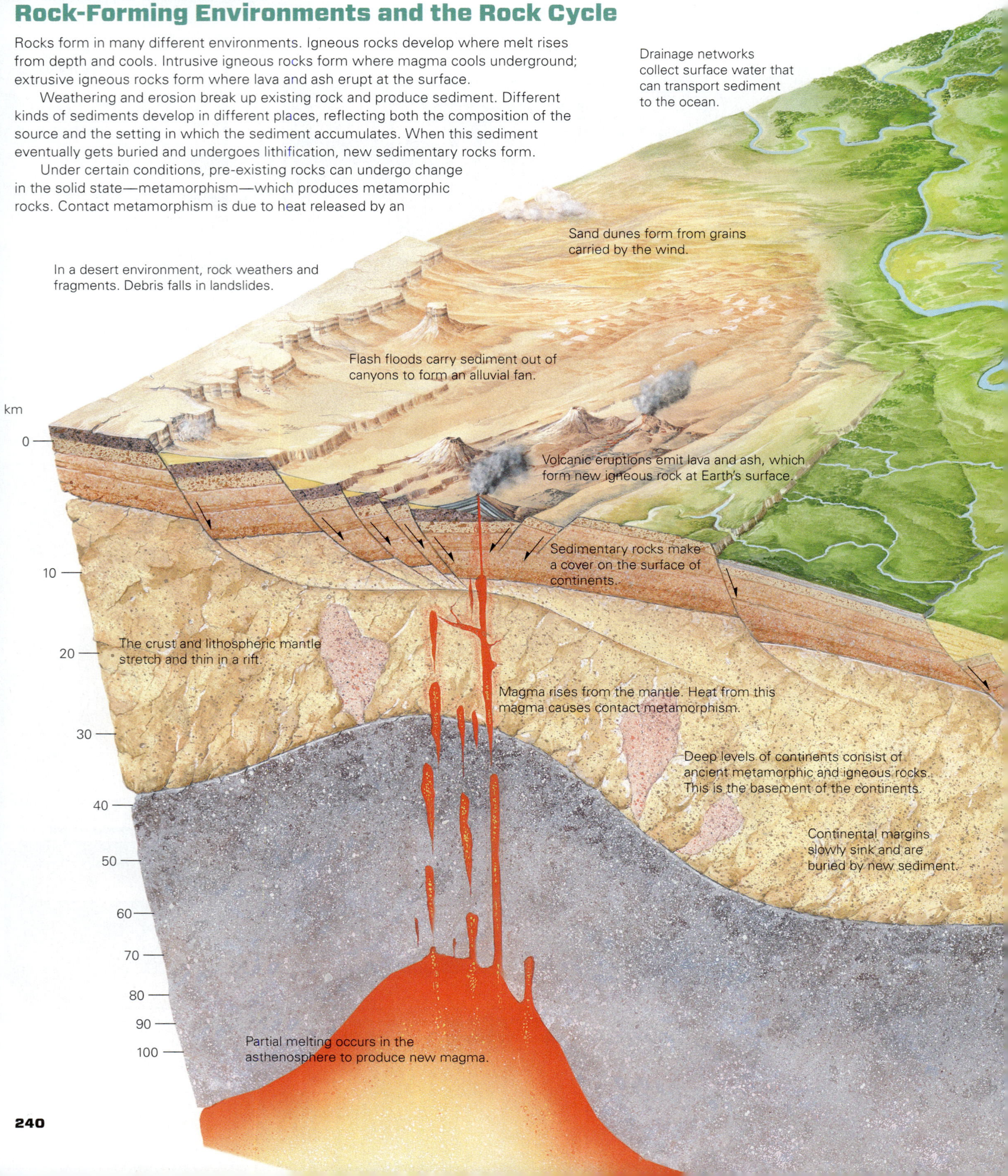

Drainage networks collect surface water that can transport sediment to the ocean.
Sand dunes form from grains carried by the wind.
In a desert environment, rock weathers and fragments. Debris falls in landslides.
Flash floods carry sediment out of canyons to form an alluvial fan.
km
0
10
20
30
40
50
60
70
80
90
100
Volcanic eruptions emit lava and ash, which form new igneous rock at Earth's surface.
Sedimentary rocks make a cover on the surface of continents.
The crust and lithospheric mantle stretch and thin in a rift.
Magma rises from the mantle. Heat from this magma causes contact metamorphism.
Deep levels of continents consist of ancient metamorphic and igneous rocks. This is the basement of the continents.
Continental margins slowly sink and are buried by new sediment.
Partial melting occurs in the asthenosphere to produce new magma.

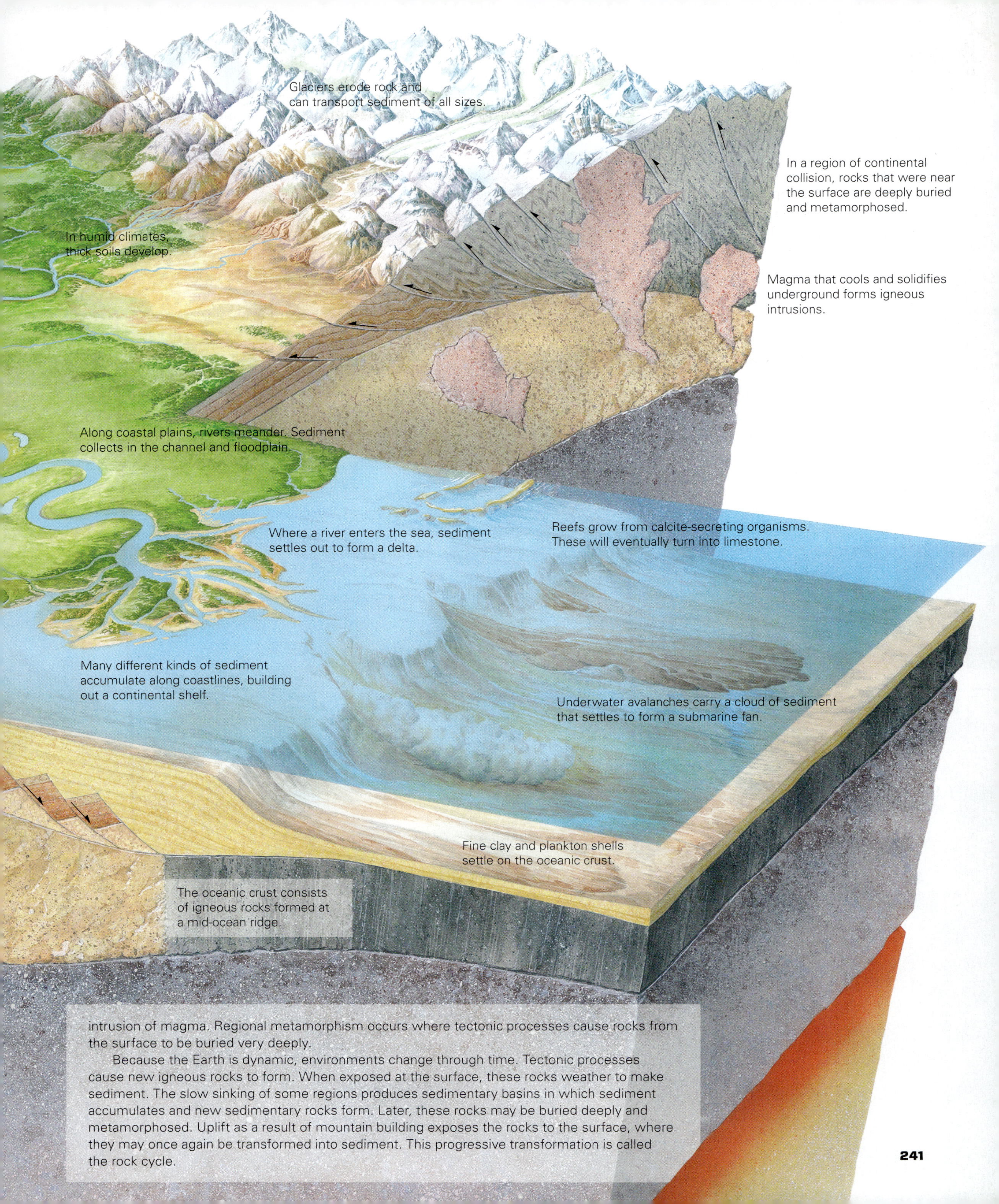

intrusion of magma. Regional metamorphism occurs where tectonic processes cause rocks from the surface to be buried very deeply.

Glaciers erode rock and can transport sediment of all sizes.
In a region of continental collision, rocks that were near the surface are deeply buried and metamorphosed.
In humid climates, thick soils develop.
Magma that cools and solidifies underground forms igneous intrusions.
Along coastal plains, rivers meander. Sediment collects in the channel and floodplain.
Where a river enters the sea, sediment settles out to form a delta.
Reefs grow from calcite-secreting organisms. These will eventually turn into limestone.
Many different kinds of sediment accumulate along coastlines, building out a continental shelf.
Underwater avalanches carry a cloud of sediment that settles to form a submarine fan.
Fine clay and plankton shells settle on the oceanic crust.
The oceanic crust consists of igneous rocks formed at a mid-ocean ridge.

Because the Earth is dynamic, environments change through time. Tectonic processes cause new igneous rocks to form. When exposed at the surface, these rocks weather to make sediment. The slow sinking of some regions produces sedimentary basins in which sediment accumulates and new sedimentary rocks form. Later, these rocks may be buried deeply and metamorphosed. Uplift as a result of mountain building exposes the rocks to the surface, where they may once again be transformed into sediment. This progressive transformation is called the rock cycle.

rock (basalt and gabbro) overlain by sediment. Because a layer of water blankets oceanic crust, oceanic crustal rock does not erode and generally does not follow the path into the sedimentary loop of the rock cycle. But sooner or later, oceanic crust subducts. When this happens, the rock of the crust undergoes metamorphism, for as it sinks, it's subjected to progressively higher temperatures and pressures.

C.3 What Drives the Rock Cycle in the Earth System?

The Earth hosts many rock-forming environments (see **Geology at a Glance**, pp. 240–241). Transfer of material among these environments, during the rock cycle, happens on the Earth because the Earth is a dynamic planet. Our planet's internal heat and gravitational field drive plate movements and plume-associated hot spots. Plate interactions cause igneous activity, as well as the uplift of mountain ranges, and generate settings in which metamorphism occurs and where sedimentary basins develop. At the surface of the Earth, heat from the Sun, together with gravity, drive wind, rain, and ice, which are the agents of weathering and erosion. In the Earth System, life also plays a key role by adding corrosive oxygen to the atmosphere and by directly contributing to weathering. In sum, internal energy (Earth's internal heat), gravity, external energy (solar heat), and life all play a role in driving the rock cycle by keeping the mantle, crust, atmosphere, and oceans in motion. Note that on the Moon, which has no plate tectonics, atmosphere, or oceans, the rock cycle doesn't happen, and the rock and debris of the Moon's surface, once formed, has remained virtually unchanged.

Another View We can see stages of the rock cycle in action along the rocky shore of Brittany, France. Weathering breaks igneous rocks into blocks which eventually break down to sand. If the sand were to be buried and lithified, it would become a new sedimentary rock.

Interlude C Review

Interlude Summary

- A given rock doesn't necessarily last forever. The atoms in a rock may, over time, be incorporated in different rock types. This transfer of atoms progressively from one rock type to another, over time, is the rock cycle.
- Not all atoms follow the same path through the rock cycle. For example, an igneous rock could later become eroded and turned into sediment, which becomes a sedimentary rock that may eventually be metamorphosed. Or the igneous rock could be metamorphosed directly.
- The rock cycle happens because the Earth is dynamic, and there are internal and external sources of energy driving melting, uplift, faulting, weathering, erosion, and burial.

Guide Terms

rock cycle (p. 237)

Review Questions

1. Once formed, does a rock necessarily last for all of Earth's history?
2. Define the rock cycle, and give three examples of pathways through it.
3. Have all rocks on Earth passed through the rock cycle the same number of times? Explain your answer.
4. Is there a rock cycle on the Moon? Why or why not?

Online Resources

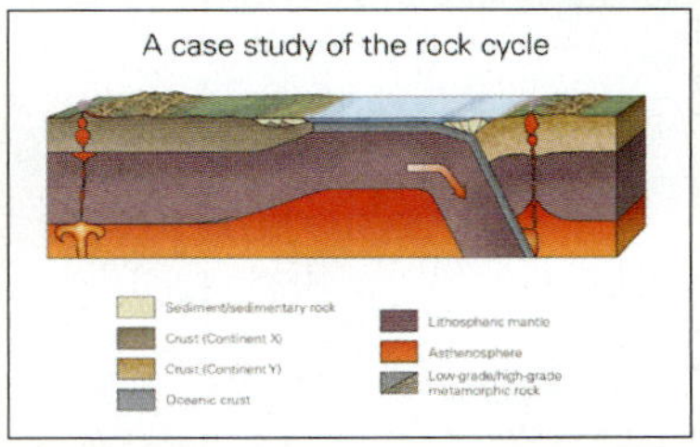

Videos

This chapter features a video that explains the rock cycle, and includes a case study.

LEARNING OBJECTIVES

By the end of this chapter, you should understand...

1. the causes of earthquakes and why they occur where they do.
2. how to specify the size of an earthquake and determine its location.
3. the many ways in which earthquakes trigger damage.
4. the limitations on people's ability to predict earthquakes.
5. steps that people can take to prevent earthquake damage.

▲ A 20-story apartment building tipped over and broke in two during a major earthquake (magnitude 8.8) that struck Chile in 2010. Violent earthquakes, which are inevitable on our dynamic planet, can have catastrophic consequences.

CHAPTER 8

A Violent Pulse: Earthquakes

8.1 Introduction

It was midafternoon of March 11, 2011, and in many seaside towns along the eastern coast of Honshu, the northern island of Japan, fishing fleets unloaded their catch, factories churned out goods, shoppers browsed the stores, and office workers tapped at their computers, all unaware that their surroundings would soon change forever. Honshu lies near a convergent boundary at which the Pacific Plate grinds beneath the edge of Japan and sinks back into the mantle. Averaged over time, the relative movement across this boundary takes place at a rate of about 8 cm per year. But the motion doesn't happen smoothly. Rather, for a while, rocks adjacent to the boundary quietly and subtly bend and warp. Then, suddenly, like a wooden stick that snaps after you've bent it too far, a measureable amount of movement takes place in a matter of seconds to minutes (Fig. 8.1). This movement occurs by slip on a **fault**, a fracture plane on which there has been sliding. On March 11, at 2:46 p.m., the "snap" started on a very large, gently sloping fault at a point located about 130 km (80 miles) east of Japan's coast, and about 24 km (15 miles) below the Earth's surface. Within a few minutes, slip had taken place across a vast area of the fault, and Japan had lurched eastward by as much as several meters, relative to the Pacific Ocean floor. Because of the character of this particular movement, a region of seafloor above the fault rose vertically by as much as 30 cm. The stage had been set for a disaster.

> *We learn geology the morning after the earthquake.*
>
> RALPH WALDO EMERSON (American poet, 1803–1882)

The instant that the slip started on the fault east of Honshu, the bending and warping that had been accumulating in rock adjacent to the fault, over decades to centuries, were released and the rock suddenly straightened out and began to vibrate, just like the two pieces of a bent stick straighten out and vibrate when the stick snaps. Simultaneously, some of the rock bordering the fault fractured and, locally, disintegrated into fragments. Growth of each new fracture yielded more vibrations. Effectively, the slip on the fault, and associated fracturing, released an immense amount of energy that instantly began to propagate through the crust in the form of waves. In solid rock, such waves race along at an average speed of 11,000 km (7,000 miles) per hour, 10 times the speed of sound in air. When the waves reached the surface of the Earth, they caused an episode of ground shaking—an **earthquake**.

So much energy was released by the March 11 event that the resulting earthquake, now known as the Tōhoku earthquake (named for the eastern province of Honshu), had historic consequences. In coastal regions relatively close to the slipped fault, the land lurched back and forth and bounced up and down for over a minute. During the shaking, people became disoriented, panicked, and even seasick—some lost their balance and crouched or fell. Buildings twisted and swayed, bottles and plates flew off shelves and crashed to the floor, bookshelves tipped over, and furniture danced around rooms. In some cases, ceilings and facades fell in a shower of debris, and in a few localities, structures collapsed (**Fig. 8.2a**). Outside, dust rose from the ground to create a fog-like mist, power lines stretched and sparked, and landslides tumbled down hillslopes. In addition, a few natural gas tanks and pipes broke, sending flammable vapors into the air. Some of the gas ignited in billows of flame that set fire to damaged buildings.

During the Tōhoku earthquake, damage due to the shaking itself, though significant, was not devastating because most buildings, due to Japan's stringent building codes, were sturdy enough to resist collapse. Unfortunately, even strong buildings could not resist what happened subsequent to the earthquake shock. The sudden displacement of the seafloor off Japan's coast had displaced the surface of the ocean. This movement produced *tsunamis*, distinctive, very broad, waves that travel rapidly away from a location where the seafloor has been disturbed and water displaced. The first tsunami generated on March 11 reached land 20 to 80 minutes after the earthquake. As it approached, the tsunami locally grew to a height of over 10 m, so it was able to overtop seawalls and wash inland, picking up debris and sediment as it moved. In places, it submerged low-lying land several kilometers in from the shoreline (**Fig. 8.2b, c**). To make matters worse, the tsunamis of March 11 destroyed power sources used to run the pumps that cooled reactors in the Fukushima nuclear power plant, ultimately causing gas explosions that released radioactivity into the environment. In the end, the immensity of the devastation due to the Tōhoku earthquake and its aftermath was almost beyond comprehension.

Earthquakes are a fact of life on planet Earth—almost 1 million detectable earthquakes happen every year. Most are a consequence of plate movement—they punctuate each step in the growth of mountains, the drift of continents, and the opening and closing of ocean basins. Fortunately, most cause no damage or casualties, either because they are too small or because they occur in unpopulated areas. But a few hundred earthquakes per year rattle the ground sufficiently to crack or topple buildings and injure their occupants, and every 5 to 20 years, on average, a great earthquake, such as the Tōhoku event, triggers a horrific calamity. In fact, during the past two millennia, building collapse, tsunamis, landslides, fires, and other phenomena caused by earthquakes have killed over 3.5 million people (**Table 8.1**).

FIGURE 8.1 What happens during an earthquake?

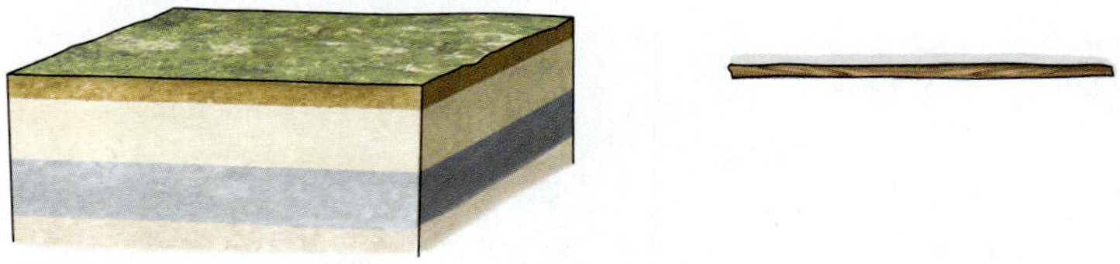

(a) Before deformation, rock layers in this example are not bent.

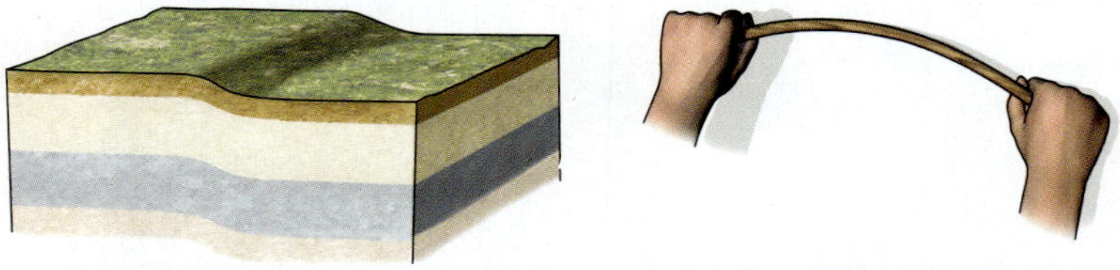

(b) Before an earthquake, rock bends elastically, like a stick that you arch between your hands. The drawing exaggerates the amount of bending.

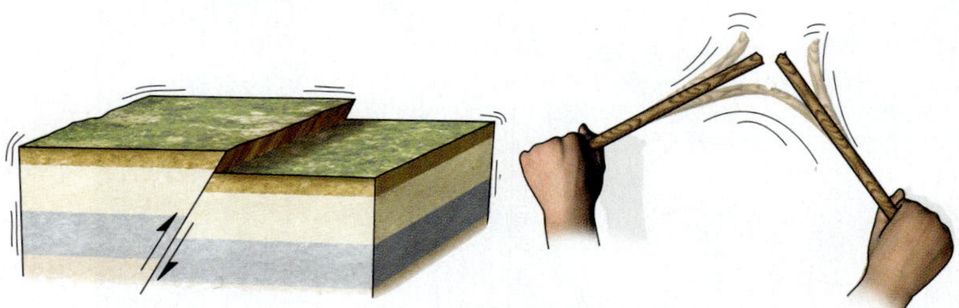

(c) Eventually, the rock breaks, and sliding suddenly occurs on a fault. This break generates vibrations. You feel such vibrations when you break a stick.

FIGURE 8.2 The great Tōhoku earthquake and tsunami in Japan, 2011.

(a) Some buildings collapsed due to ground shaking.

(b) The tsunami filled harbors and spilled over sea walls.

(c) The tsunami washed over low coastal areas. In this photograph, the wave is moving from left to right and is just washing over a canal.

TABLE 8.1 Some Notable Earthquakes

Year	Location	Deaths
2011	Tōhoku, Japan (tsunami)	20,000
2011	Christchurch, New Zealand	180
2010	Haiti	230,000
2010	Concepción, Chile	1,000
2008	Sichuan, China	70,000
2005	Pakistan	80,000
2004	Sumatra (tsunami)	230,000
2003	Bam, Iran	41,000
2001	Bhuj, India	20,000
1999	Calaraca/Armenia, Colombia	2,000
1999	Izmit, Turkey	17,000
1995	Kobe, Japan	5,500
1994	Northridge, California	51
1990	Western Iran	50,000
1989	Loma Prieta, California	65
1988	Spitak, Armenia	24,000
1985	Mexico City	9,500
1983	Turkey	1,300
1978	Iran	15,000
1976	T'ang-shan, China	255,000
1976	Caldiran, Turkey	8,000
1976	Guatemala	23,000
1972	Nicaragua	12,000
1971	San Fernando, California	65
1970	Peru	66,000
1968	Iran	12,000
1964	Anchorage, Alaska	131
1963	Skopje, Yugoslavia	1,000
1962	Iran	12,000
1960	Agadir, Morocco	12,000
1960	Southern Chile	6,000
1948	Turkmenistan, USSR	110,000
1939	Erzincan, Turkey	40,000
1939	Chillán, Chile	30,000
1935	Quetta, Pakistan	60,000
1932	Gansu, China	70,000
1927	Tsinghai, China	200,000
1923	Tokyo, Japan	143,000
1920	Gansu, China	180,000
1915	Avezzano, Italy	30,000
1908	Messina, Italy	160,000
1906	San Francisco	500
1896	Japan	22,000
1886	Charleston, South Carolina	60
1866	Peru and Ecuador	25,000
1811–12	New Madrid, Missouri (3 events)	Few
1783	Calabria, Italy	50,000
1755	Lisbon, Portugal	70,000
1556	Shen-shu, China	830,000

What geologic phenomena trigger earthquakes? Why do earthquakes take place where they do? How do they cause damage? Can we predict when earthquakes will happen or even prevent them from happening? Many of these questions have been addressed by the work of **seismologists** (from the Greek word *seismos*, for shock or earthquake), geoscientists who study earthquakes, during the past century. In this chapter, we present some of the answers that they have obtained.

8.2 Faulting and the Generation of Earthquakes

What causes earthquakes? Ancient cultures offered a variety of explanations for **seismicity** (earthquake activity), most of which involved the restless movements of mythical animals (catfish, elephants, turtles) who supposedly lived underground. Today, we realize that earthquakes happen instead when masses of rock suddenly vibrate, for the energy released by this process travels away the source in the form of waves, known as **seismic waves** or *earthquake waves*, that can reach and shake the surface of the Earth. What phenomena cause the sudden vibration of rock? Although some earthquakes are due to the intrusion and flow of magma beneath a volcano, to underground nuclear tests, to landslides, or even to meteor impacts, most earthquakes, by far, are a consequence of faulting, as was the case for the Tōhoku event. Because fault-generated earthquakes are the most common, we limit our discussion in this chapter to how and why they occur. Below, we first describe faults, and then show how movement on them generates seismic waves.

Faults in the Crust

At first glance, a fault may look simply like a break that cuts across rock or sediment. But on closer examination, you may be able to see evidence of the sliding that occurred on a fault. For example, rock adjacent to the fault may be broken up into angular fragments or may be pulverized into tiny grains, due to the crushing and grinding that can accompany slip, and the surface of a fault may be polished and grooved as if scratched by a rasp. In some localities, a fault cuts through a distinct "marker" (a sedimentary bed, an igneous dike, or a fence)—where this happens, the end of the marker on one side of the fault is offset relative to the end on the other side. The distance between two ends of the offset marker, as measured along the fault surface in the direction of slip, is the fault's **displacement** (**Fig. 8.3**). Most faults are completely underground, so we can't see the displacement across them when it happens. In fact, such faults may become visible only millions of years later, when exposed by the erosion of overlying rock. But some faults do intersect and offset the ground surface, producing a step called a **fault scarp** (**Fig. 8.4a**). The ground surface exposure of a fault, regardless of whether it appears due to

FIGURE 8.3 Examples of fault displacement on the San Andreas fault in California.

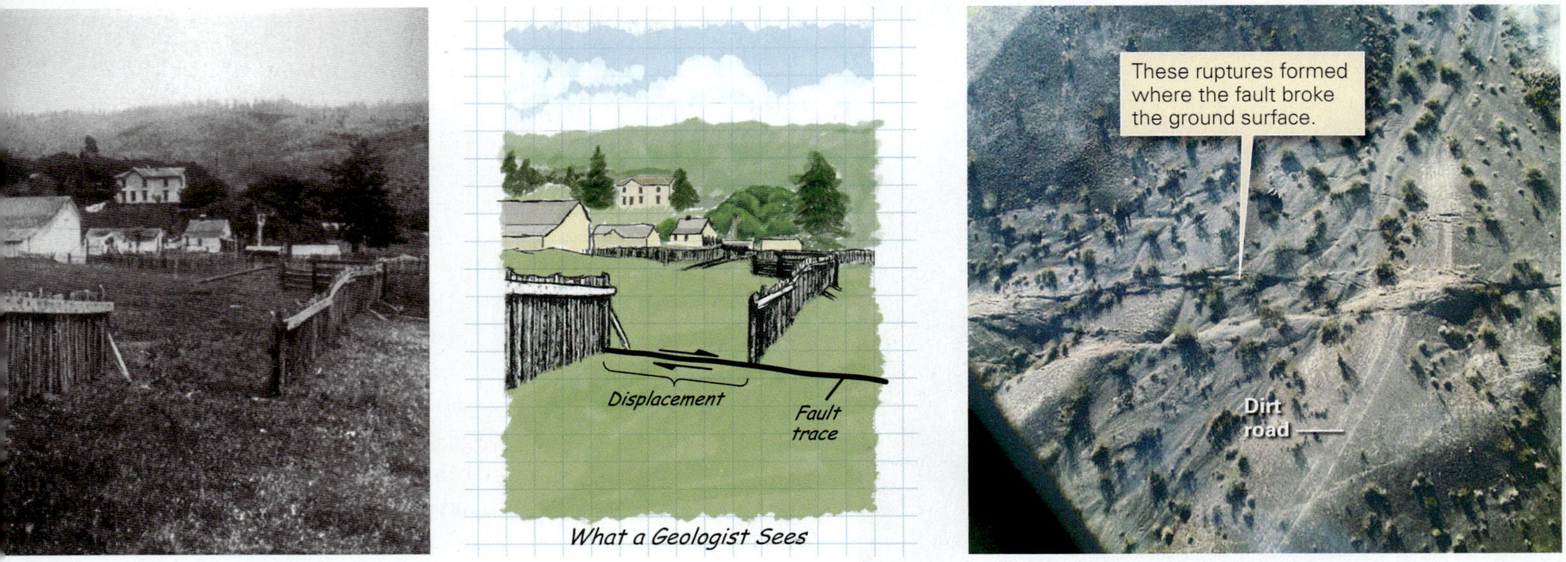

(a) A wooden fence built across the fault was offset during the 1906 San Francisco earthquake. The displacement indicates that the motion was strike slip.

(b) An aerial photograph shows offset of a dirt road by slip on the fault in 1999.

erosion of a once-buried fault or due to displacement of the present-day ground surface by recent movement, is called a *fault line* or fault trace.

In the 19th century, miners who encountered faults in mine tunnels referred to the rock mass above a sloping fault plane as the *hanging wall,* because it hung over their heads, and the rock mass below the fault plane as the *footwall,* because it lay beneath their feet. The miners described the direction in which rock masses slipped on a sloping fault by specifying the *sense of slip,* the direction that the hanging wall moved in relation to the footwall. We still use the terms they developed (see **Geology at a Glance**, pp. 250–251). Specifically, when the hanging wall slips down the slope of the fault, it's a *normal fault,* and when the hanging wall slips up the slope, it's a *reverse fault* or *thrust fault* (**Fig. 8.4a–c**). Note that reverse faults and thrust faults have the same sense of slip—they differ in their slope (dip), in that the former have a steeper dip than the latter. A *strike-slip fault* is a near-vertical fracture on which slip occurs parallel to an imaginary horizontal line, called a strike line, on the fault plane—no up or down motion takes place on a strike-slip fault (**Fig. 8.4d**).

FIGURE 8.4 The basic types of faults. Fault types are distinguished from one another by the direction of slip relative to the fault surface.

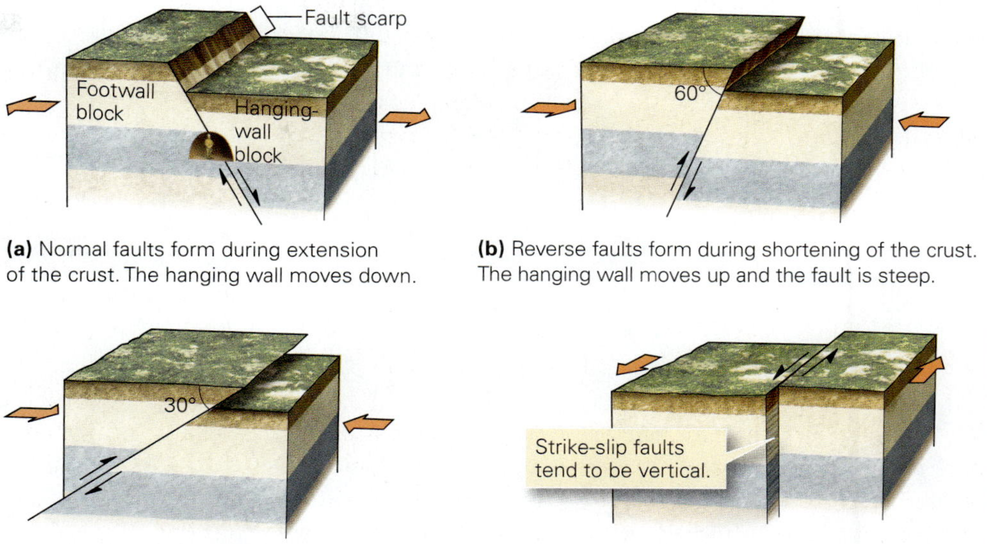

(a) Normal faults form during extension of the crust. The hanging wall moves down.

(b) Reverse faults form during shortening of the crust. The hanging wall moves up and the fault is steep.

(c) Thrust faults also form during shortening. The fault's slope is gentle (less than 30°).

(d) On a strike-slip fault, one block slides laterally past another, so no vertical displacement takes place.

Faults are found in many locations—but don't panic! Not all are likely to be the source of earthquakes. We refer to a fault that has moved relatively recently, or might move in the future, as an *active fault*—if it generates earthquakes, news media sometimes refer to it as an "earthquake fault." A fault that last moved in the distant past and probably won't move again in the near future is an *inactive fault.*

Generating Earthquake Energy

How does faulting generate earthquakes? Seismologists have concluded that the vibrations of an earthquake can be produced in two ways. First, an earthquake can happen when rock breaks and a new fault forms by rupturing previously intact rock, and second, an earthquake can happen when a pre-existing fault suddenly slips again. Let's look more closely at these two causes.

Imagine that you grip each side of a brick-shaped block of rock with a clamp. Now, apply an upward push on one of the clamps and a downward push on the other. By doing so, you have applied a **stress** to the rock—simplistically, we can think of stress as a push, pull, or shear, as we discussed in Chapter 7. At first, the rock bends slightly but doesn't break (**Fig. 8.5a**). In fact, if you were to stop applying stress at this stage, the rock would straighten out and return to its original shape. Geologists refer to such a phenomenon, when a material changes shape when subjected to stress but can return to its original shape when the stress is removed, as elastic deformation—the same phenomenon happens when you stretch a spring or rubber band and then let go. Now repeat the experiment, but bend the rock even more. If you bend the rock far enough, a number of small cracks start to form. Eventually the cracks connect to one another and a fracture that cuts across the entire block of rock develops (**Fig. 8.5b**). The instant this happens, the block breaks in two and the rock on one side slides past the rock on the other side, so the fracture becomes a fault. When the new fault forms and slips, any elastic deformation that had built up prior to breaking gets released, and the rock on either side of the fault straightens out or undergoes *rebound* (**Fig. 8.5c**). Elastic materials don't simply relax smoothly back to their original shape; rather, they vibrate back and forth, before attaining a relaxed shape. (To picture this process, let go of a bent piece of wood that is anchored at one end, and watch it swing back and forth before coming to rest.) Because slip takes place on the fault, rock on one side will have been displaced relative to rock on the other, after rebound ceases. The result of the elastic "twang" that happens when rock fractures and then rebounds is earthquake energy.

When a fault forms and starts to slip, it doesn't continue to slip forever, because **friction**, defined as the force that resists

GEOLOGY AT A GLANCE

Faulting in the Crust

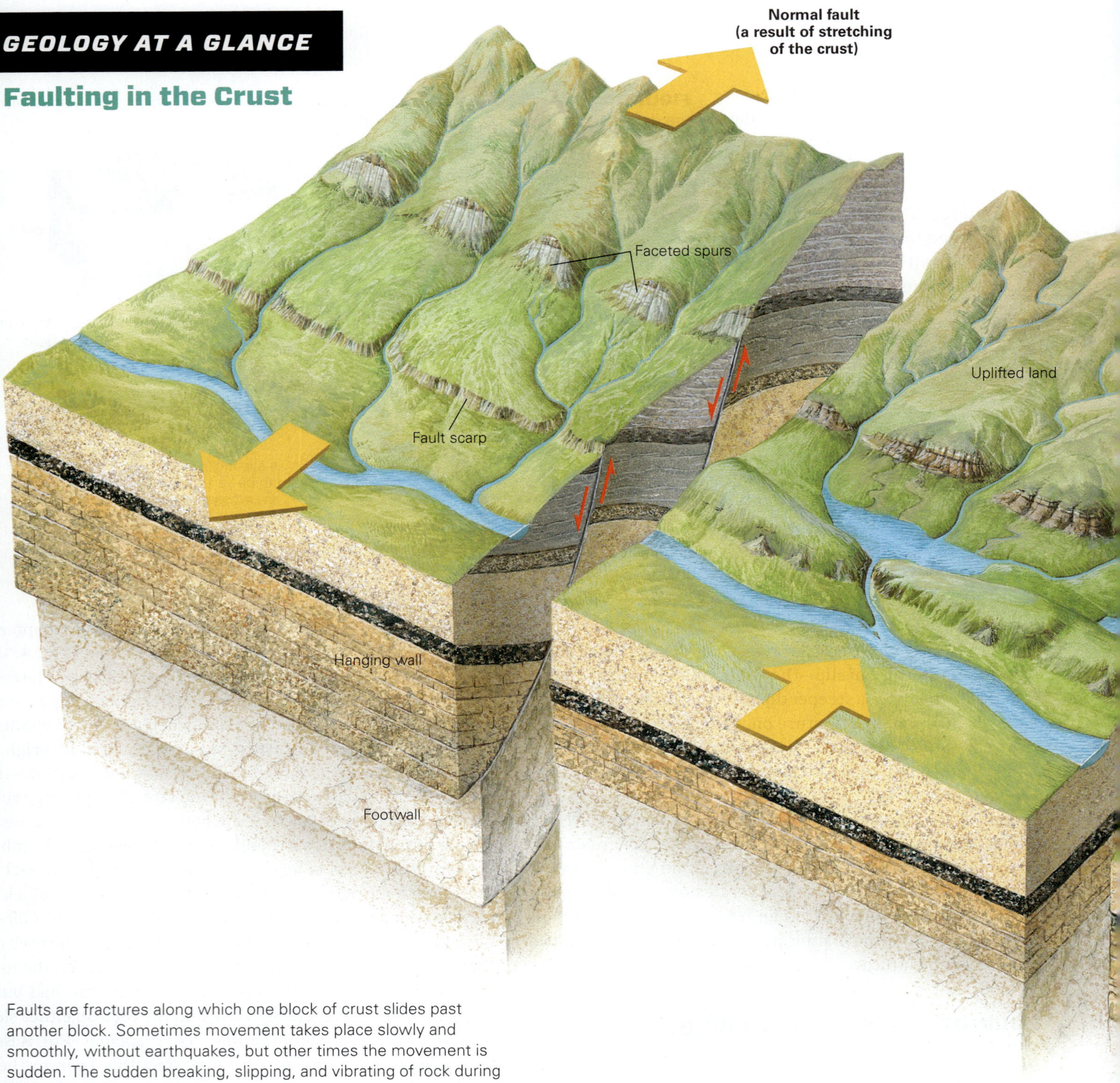

Faults are fractures along which one block of crust slides past another block. Sometimes movement takes place slowly and smoothly, without earthquakes, but other times the movement is sudden. The sudden breaking, slipping, and vibrating of rock during faulting sends shock waves, called seismic waves, through the Earth. Seismic waves that reach the Earth's surface can be felt as earthquakes. The vibrations of an earthquake can be great enough to cause significant damage.

Geologists recognize three types of faults. If the hanging-wall block (the rock above a fault plane) slides down the fault's slope relative to the footwall block (the rock below the fault plane), the fault is a normal fault. Normal faults form where the crust is being stretched apart, as in a continental rift. If the hanging-wall block is being pushed up the slope of the fault relative to the footwall block, then the fault is a reverse fault. Reverse faults develop where the crust is being compressed or squashed, as in a collisional mountain belt. If one block of rock slides past another and there is no up or down motion, the fault is a strike-slip fault. Strike-slip fault planes tend to be nearly vertical.

If displacement on a fault offsets the ground surface, it can yield a ledge called a fault scarp. Where fault scarps cut a system of rivers and valleys, the ridges may be cut off. Strike-slip faults may offset ridges, streams, and orchards sideways. If there is a slight extension along the fault, the land surface sinks, and a sag pond develops.

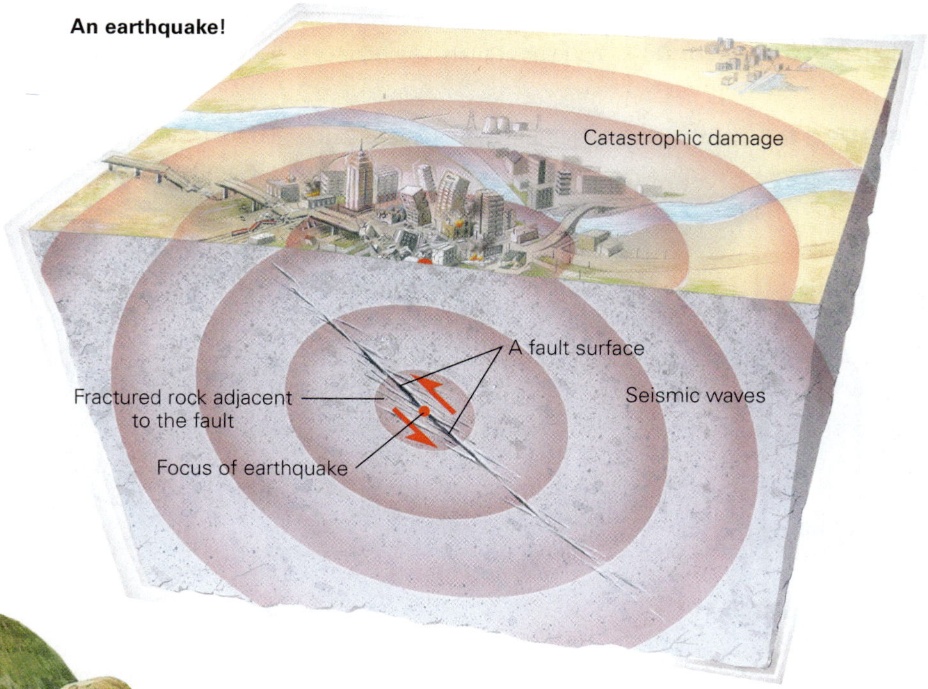

Erosion along the Great Glen fault in Scotland produced a linear valley as shown in this satellite image.

FIGURE 8.5 A model representing the development of a new fault. Rupturing can generate earthquake-like vibrations.

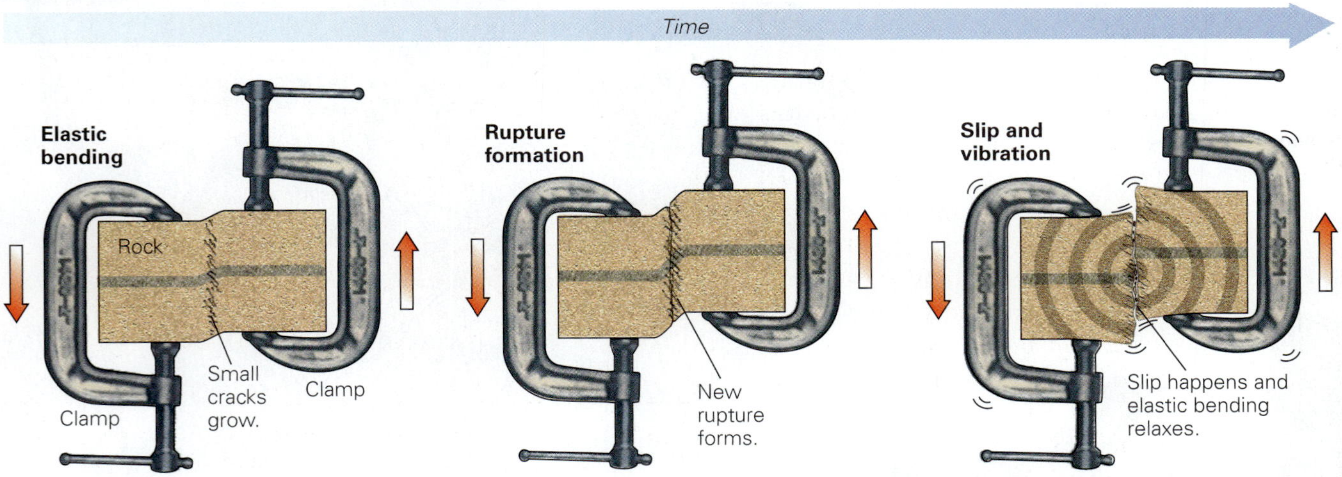

(a) Imagine a block of rock gripped by two clamps. Move one clamp up, and the rock starts to bend. Small cracks develop along the bend.

(b) Eventually, the cracks link. When this happens, a throughgoing rupture forms.

(c) The instant that the rupture forms, the rock breaks into two pieces that slide past each other. The energy that is released generates vibrations (earthquake energy).

sliding on a surface, eventually slows and stops the movement. Why does friction exist? All real surfaces have bumps protruding from them. These bumps act like tiny anchors, snagging and digging into the opposing surface, eventually halting slip.

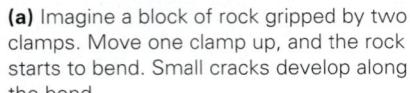

We've just seen how earthquakes can happen when intact rock ruptures. Earthquakes can also result from sudden slip on a pre-existing fault, for once a fault has formed, it remains as a scar in the Earth's crust that can be weaker than surrounding, intact crust. Even though it's weaker, however, friction across a pre-existing fault resists sliding, so slip doesn't occur constantly. As stress builds across a frictionally locked fault, rocks adjacent to the fault bend elastically, just as they did in intact rocks prior to initiation of a new fault. When the stress acting across a pre-existing fault becomes great enough for anchor-like bumps protruding from the walls of the fault either to break off or to plow a furrow in the opposing wall, the fault slips again and the elastic bending in rocks adjacent to the fault rebounds.

The concept that earthquakes occur because stresses build up, causing rock to bend elastically until either a new fault forms or a pre-existing fault slips, at which time the bent rocks suddenly straighten out and vibrate, is called the **elastic-rebound theory**. Because pre-existing faults tend to be weaker than intact rock, slip tends to happen again and again on a pre-existing fault. After each slip event, friction eventually halts displacement and prevents the fault from slipping again until stress builds up enough to overcome friction again. Geologists refer to such alternation between stress buildup and earthquake-generating slip events on a fault as **stick-slip behavior**.

Of note, the major earthquake, or **main shock**, along a fault may be preceded by smaller ones, called **foreshocks**. These possibly result from the development or propagation of smaller cracks in the vicinity of what will be the major fault. In the days to months following a large earthquake, the region affected by the large earthquake will endure a series of **aftershocks**. The largest aftershock tends to be ten times smaller than the main shock, and most are much smaller than that. Aftershocks happen because slip during the main shock does not leave the fault in a perfectly stable configuration. For example, after the main shock, bumps on one side of the fault surface may, in their new position, push so hard into the opposing side that they generate new stresses that are large enough to cause a small portion of the fault around the bump to slip again, or to cause slip on a nearby fault.

The Area and Amount of Fault Slip during an Earthquake

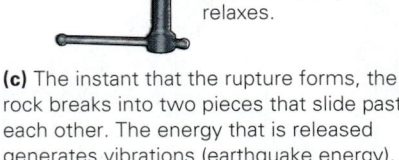

How much of a fault surface slips during an earthquake? The answer depends on the size of the earthquake—generally, the larger the earthquake, the larger the slipped area and the greater the displacement. For example, the major earthquake that hit San Francisco, California, in 1906 rup tured a segment of the San Andreas fault that was 430 km long (measured parallel to the Earth's surface) by 15 km deep (measured perpendicular to the Earth's surface). Thus, the area that slipped was almost 6,500 square km. During the 2011 Tōhoku earthquake, an area 300 km long by 100 km wide (30,000 square km) slipped.

The amount of slip varies along with location on a fault—it tends to be greatest near the place where the slip begins and tends to die out progressively toward the edge of the slipped area. Of course, beyond the end of the slipped area, the displacement is zero. The maximum observed displacement on very large earthquakes can be several meters. For example, a maximum of 30 m of slip happened during the 2011 Tōhoku earthquake, 12 m of slip during the 1964 Good Friday earthquake in southern Alaska, and 7 m during the 1906 San Francisco earthquake. Smaller earthquakes, such as the 1994 event that hit Northridge, California, resulted in only about 0.5 m of slip; even so, this earthquake toppled homes, ruptured pipelines, and killed 51 people. The smallest-felt earthquakes result from displacements measured in millimeters to centimeters.

Although the cumulative movement on a fault during a human life span may not amount to much, over geologic time the movement's impact becomes significant. For example, if the earthquakes taking place on a strike-slip fault cause an average of 1 cm of displacement per year, the fault's movement will yield 10 km of displacement after 1 million years.

SEE FOR YOURSELF...

SAN ANDREAS FAULT, NEAR SAN FRANCISCO, CALIFORNIA

Latitude
37°31′12.66″ N

Longitude
122°21′41.04″ W

Zoom to 2 km (~1.2 miles) and look NE.

Very narrow lakes fill a narrow valley, south of San Francisco. The valley is the trace of the San Andreas fault.

Defining the Focus and Epicenter of an Earthquake

The location where seismic waves first begin to be generated is the **focus,** or *hypocenter,* of an earthquake (**Fig. 8.6a**). For earthquakes associated with faulting, the focus represents the point where the slip on a fault initiates. As we've seen, during a large earthquake, hundreds to thousands of square kilometers of a fault surface will slip within seconds or minutes of the initial movement. Sometimes, the focus lies near the center of the slipped area—but not always. For example, in the case of the huge 2004 earthquake off Sumatra, the focus lay near one end of what would grow to become a slipped area more than 1,500 km (900 miles) long, and the maximum amount of slip on the

FIGURE 8.6 Earthquake hypocenters and epicenters.

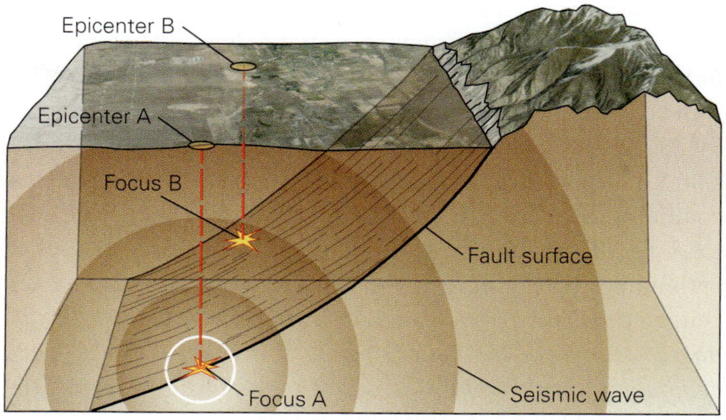

(a) The focus is the point on the fault where slip begins. Seismic energy starts radiating from it. The epicenter is the point on the Earth's surface directly above the focus. Earthquake A just happened; earthquake B happened a while ago.

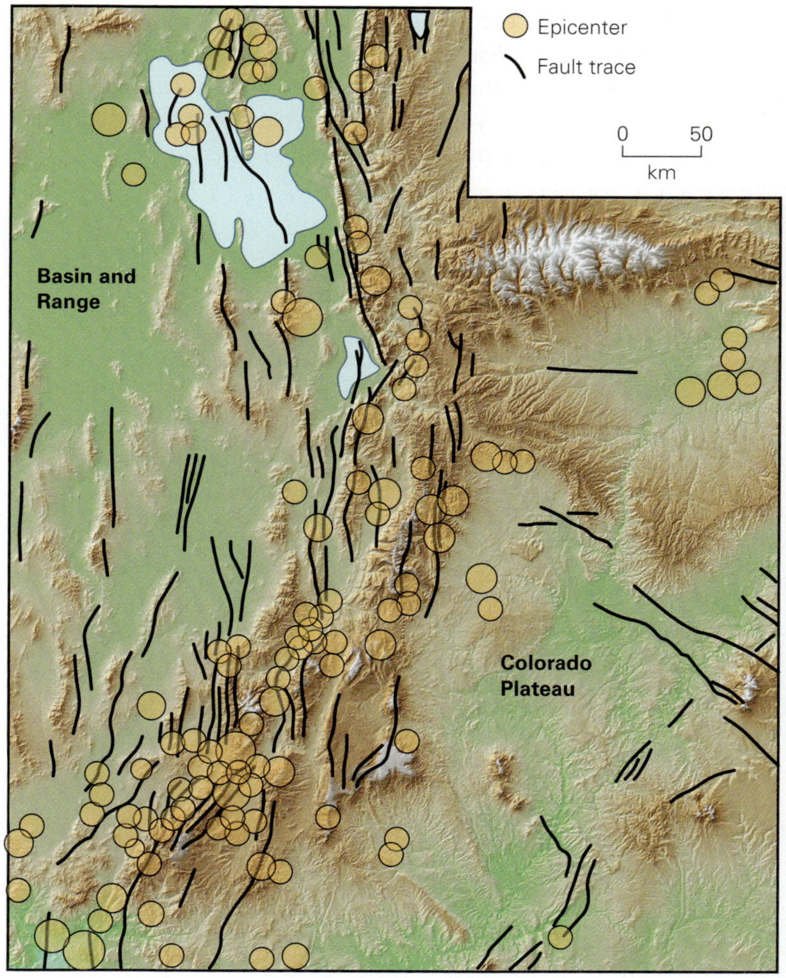

(b) A map of Utah showing the distribution of earthquake epicenters, recorded over several years. The map indicates that seismic activity happens mostly in a distinct belt following the boundary between the Basin and Range rift and the Colorado Plateau. The size of the circle represents the size of the earthquake.

fault plane occurred about 300 km from the focus. Seismologists use a more general term, *seismic source*, or just *source*, to refer to the entire region where energy leading to an earthquake was released.

In continental crust, the focus of most earthquakes generally lies between 5 and 20 km depth, with a few lying as deep as 50 km. As we'll see, the foci of earthquakes in subducting oceanic lithosphere can lie as deep as 660 km below the surface. Overall, seismologists distinguish among three classes of earthquakes based on focal depth: *shallow-focus earthquakes* occur in the top 60 km of the Earth, *intermediate-focus earthquakes* take place between 60 and 300 km, and *deep-focus earthquakes* occur down to a depth of about 660 km. Earthquakes do not happen below 660 km.

Because foci do not lie on the Earth's surface, we can't plot their positions directly on a map. What you are actually looking at, when you see a dot representing the position of an earthquake on a map, is the **epicenter** of the earthquake, the point on the surface of the Earth that lies vertically above the focus (**Fig. 8.6b**). To indicate the depth of a focus, you can either use different colors on the map (with each color indicating a different depth range), or you can plot the focus on a cross section (a representation of a vertical slice through the Earth).

TAKE-HOME MESSAGE

Most earthquakes happen when stress builds to cause either sudden formation of a new fault or slip on a preexisting fault. When the slip takes place, elastically deformed rock on either side of the fault rebounds. This process generates seismic waves that carry energy through rock to the surface, to shake the ground. Once formed, faults display stick-slip behavior in that stress builds until slip takes place and then builds again. The focus is the point within the Earth where slip begins, and the epicenter is the point on the Earth 's surface vertically above the focus. The area of a fault that slips as well as the amount of displacement that takes place are greater for larger earthquakes.

QUICK QUESTION What are foreshocks and aftershocks?

8.3 Seismic Waves and Their Measurement

Types of Seismic Waves

As we've noted, the energy produced by slip on a fault moves through rock and sediment in the form of seismic waves. You can simulate such waves by simply holding one end of a wooden block with your hand while you strike the other end with a hammer—the energy transmitted by the head of the hammer to the block travels through the block as seismic waves, and causes the end of the block in contact with your hand to vibrate. Seismologists distinguish among two categories of seismic waves on the basis of where the waves move. **Body waves** pass through the interior of the Earth, whereas **surface waves** travel along the Earth's surface. To picture the difference, imagine that the focus of an earthquake lies 20 km below the ground surface. The instant that faulting occurs, seismic waves propagate outward in all directions from the fault, like a succession of concentric bubbles. Such waves are body waves, in that they move within the solid Earth. Those that propagate downward go deep into the interior of the Earth and will not reach the surface until they have traversed much of the planet. Those that go upward reach the Earth's surface quickly, and when they do, they cause the surface to move. This movement, in turn, propagates outward as surface waves.

Body waves can cause a material to vibrate in two different ways (**Fig. 8.7a, b**). Those that cause particles of material to move back and forth parallel to the direction in which the wave itself moves are called **compressional waves**. As a compressional wave passes, the material first compresses (squeezes together), then dilates (expands). To see this kind of motion in action, push on the end of a spring and watch as the little pulse of compression moves along the length of the spring. In contrast, body waves cause particles of material to move back and forth perpendicular to the direction in which the wave itself moves. These are called **shear waves**. To see shear-wave motion, jerk the end of a rope up and down and watch how the up-and-down motion travels along the rope. Surface waves also have two different forms—some cause the ground surface to go up and down in rolling undulations, and some cause the surface to go back and forth sideways, like the movement of a snake (**Fig. 8.7c**). Seismologists have assigned names to the different kinds of waves we've just described.

- P-waves (*P* stands for primary) are compressional body waves.
- S-waves (*S* stands for secondary) are shear body waves.
- R-waves (*R* stands for Rayleigh) are surface waves that cause the ground to undulate up and down.
- L-waves (*L* stands for Love) are surface waves that cause the ground to shimmy back and forth.

The different types of seismic waves travel at different velocities. P-waves travel the fastest, which is why they are called primary waves. S-waves are called secondary because they are the next fastest—an S-wave travels at about 60% of the speed of a P-wave. Surface waves are slower than body waves, so both Rayleigh waves and Love waves arrive at a location on the surface of the Earth substantially after the body waves have arrived.

FIGURE 8.7 Different types of earthquake waves.

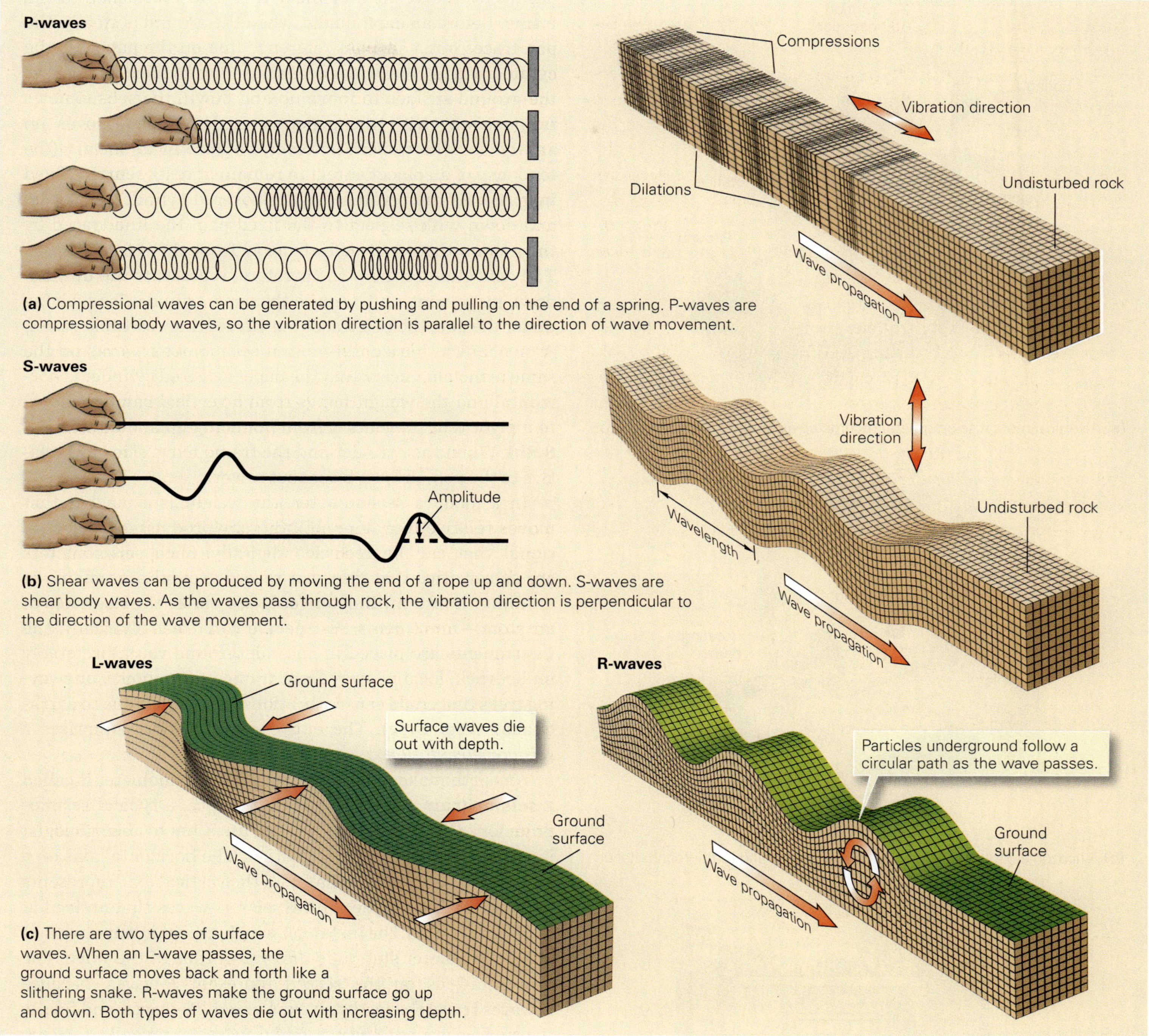

(a) Compressional waves can be generated by pushing and pulling on the end of a spring. P-waves are compressional body waves, so the vibration direction is parallel to the direction of wave movement.

(b) Shear waves can be produced by moving the end of a rope up and down. S-waves are shear body waves. As the waves pass through rock, the vibration direction is perpendicular to the direction of the wave movement.

(c) There are two types of surface waves. When an L-wave passes, the ground surface moves back and forth like a slithering snake. R-waves make the ground surface go up and down. Both types of waves die out with increasing depth.

Seismometers and the Record of an Earthquake

Researchers use a **seismometer** (or seismograph) to measure the ground motion produced by an earthquake. Seismometers can be configured in two ways—a *vertical-motion seismometer* detects up-and-down ground motion, whereas a *horizontal-motion seismometer* detects back-and-forth ground motion. Let's examine how these instruments work.

The heart of a mechanical vertical-motion seismometer consists simply of a heavy weight suspended from a spring. The spring, in turn, hangs from a sturdy frame that has been bolted to the ground (**Fig. 8.8a**) and is kept from swinging sideways by a horizontal bar connected to a pivot. A pen

FIGURE 8.8 The basic operation of a seismometer.

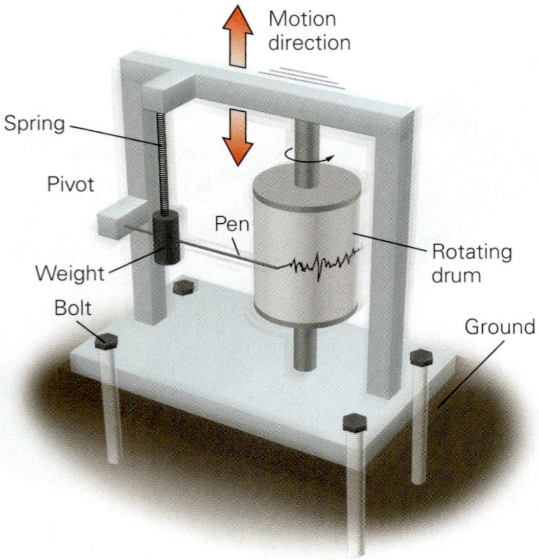

(a) A vertical-motion seismometer records up-and-down ground motion.

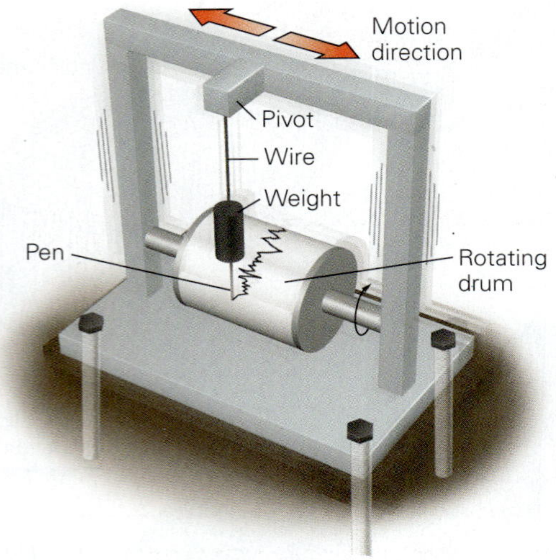

(b) A horizontal-motion seismometer records back-and-forth ground motion.

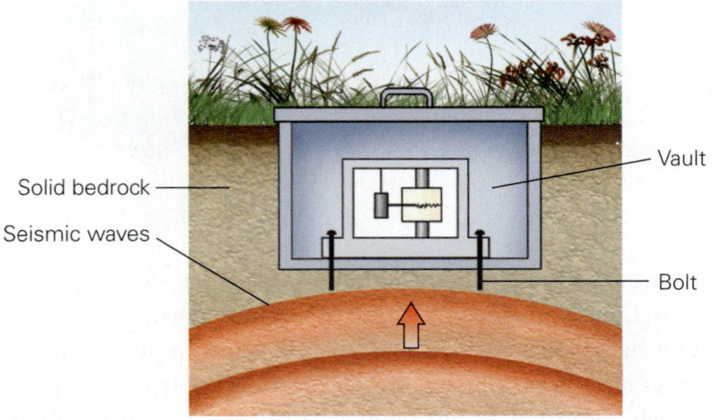

(c) Seismometers are bolted to bedrock in a protected shelter or vault.

extends sideways from the weight and touches a revolving, paper-covered vertical cylinder that is also attached to the frame. Before an earthquake, when the ground is steady, the pen traces out a straight reference line on the paper, as the cylinder turns. But when a seismic wave arrives and causes the ground surface to move up and down, the seismometer frame—along with the paper-covered cylinder—moves up and down. The weight, however, because of its *inertia* (the tendency of an object at rest to remain at rest), remains fixed in space. As the revolving paper-covered cylinder moves up and down with respect to the fixed pen, the line traced by the pen on the paper deflects away from the reference line. The deflection of the pen, therefore, represents the up-and-down movement, and because the paper cylinder is revolving under the pen, the pen traces out curves that resemble waves. A mechanical horizontal-motion seismometer works on the same principle, except that the paper-covered cylinder is horizontal and the weight hangs from a vertical bar, connected to a pivot that prevents it from bouncing up and down (**Fig. 8.8b**). When the cylinder and the frame move sideways relative to the pen, the pen traces out waves.

In a modern seismometer, the weight is a magnet that moves relative to a wire coil, thereby producing an electric signal that can be recorded digitally. Such seismometers are so sensitive that they can record ground movements of a millionth of a millimeter (only ten times the diameter of an atom)—movements that people can't feel. Typically, the instruments are placed in an underground vault, preferably on bedrock, located away from traffic, urban noise, or swaying trees that could cause vibrations that are not due to earthquakes (**Fig. 8.8c**). The entire configuration comprises a *seismometer station*.

An earthquake record produced by a seismometer is called a **seismogram** (**Fig. 8.9**). At first glance, a typical seismogram looks like a messy squiggle of lines, but to a seismologist it contains a wealth of information. The horizontal axis on a seismogram represents time, and the vertical axis represents the *amplitude* (the size) of the seismic waves that arrived at the seismometer. The instant at which a seismic wave appears at a seismometer station is the *arrival time* of the wave. The first squiggles on the record represent P-waves, because P-waves travel the fastest. Next come the S-waves, and finally the surface waves (R-waves and L-waves). Typically, the surface waves have the largest amplitude and arrive over a relatively long interval of time.

Finding the Epicenter

How do we find the location of an earthquake's epicenter? The answer to this question comes from comparing a seismogram recorded closer to the epicenter to one recorded farther

FIGURE 8.9 The nature of seismograms.

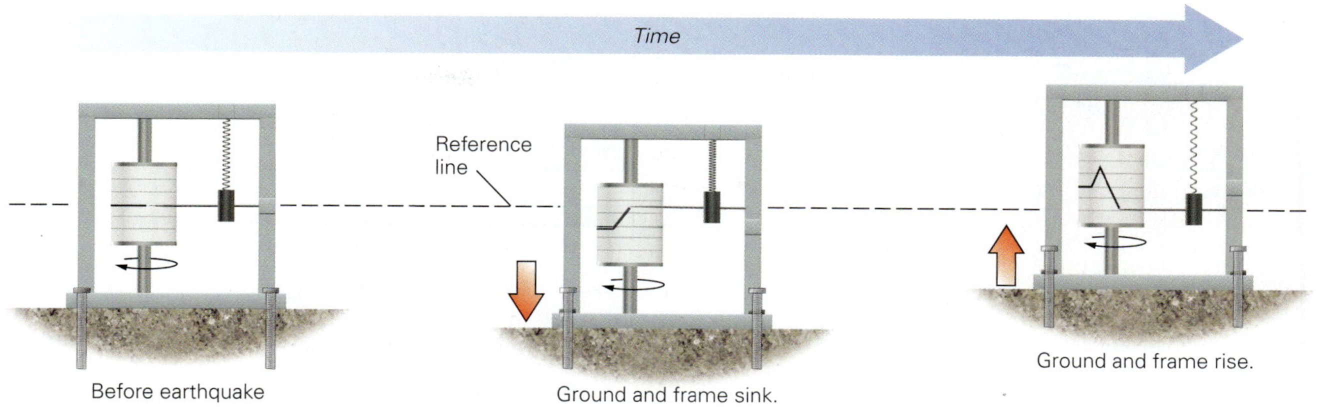

(a) Before an earthquake, the pen traces a straight line. During an earthquake, the paper roll moves up and down while the pen stays in place.

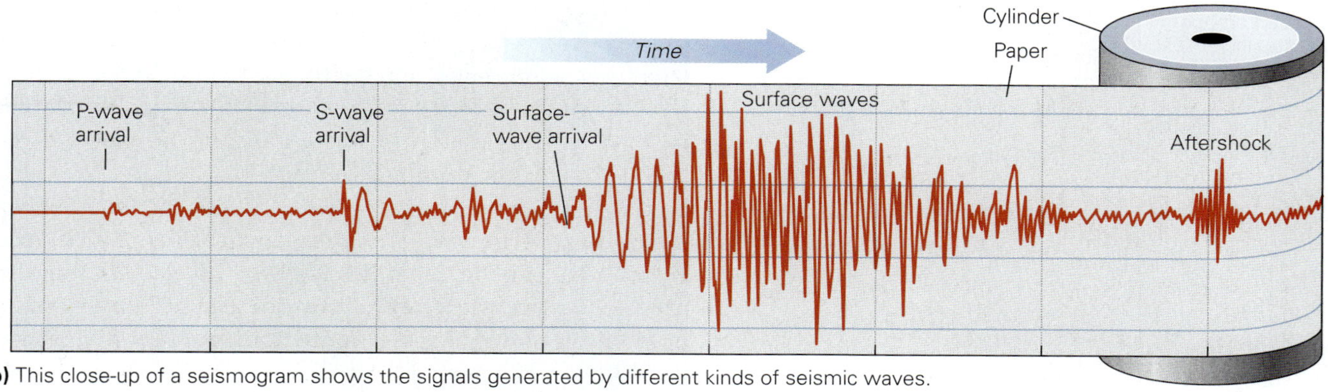

(b) This close-up of a seismogram shows the signals generated by different kinds of seismic waves.

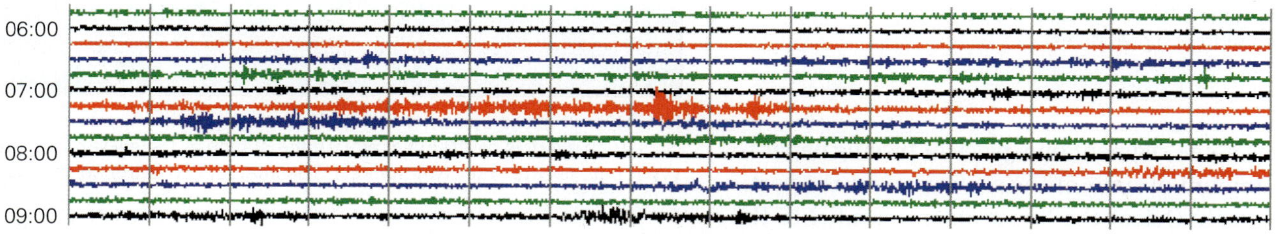

(c) A digital seismic record from a seismometer station in Arkansas. The space between vertical lines represents 1 minute. Colors have no meaning but make the figure more readable. Each color band represents the record of 15 minutes.

from the epicenter. You will see that a given seismic wave takes less time to reach a nearby seismometer than it does to reach a distant one. Also, because different waves travel at different velocities, the difference between the arrival time of a faster wave and the arrival time of a slower wave increases as the distance between the epicenter and the seismometer increases (**Fig. 8.10a**). To picture why, imagine a car race—if one car travels faster than another, the distance between them increases as the race proceeds.

Researchers have found that measuring the difference between the P-wave arrival and the S-wave arrival time provides an easy way to calculate the distance between the epicenter and a seismometer, for these waves are very distinctive. We can represent the time delay between P-waves and S-waves on a graph, called a **travel-time curve**, whose horizontal axis indicates distance from the epicenter and whose vertical axis indicates time. This type of graph shows the increase in time that it takes for an earthquake wave to move from its origin to a seismometer station as the distance between the epicenter and the seismometer station, increases (**Fig. 8.10b**). To use a graph of travel-time curves in order to determine the distance to an epicenter, start by measuring the time difference between the P- and S-waves on a seismogram. This quantity is called the S − P (pronounced "S minus P")

FIGURE 8.10 The method for locating an earthquake epicenter.

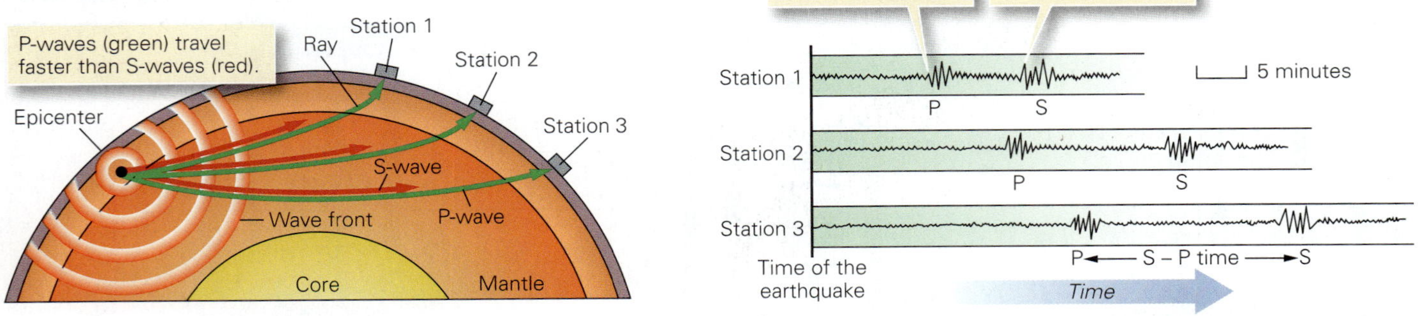

(a) Seismic waves travel at different velocities. The greater the distance between the epicenter and the seismograph station, the longer it takes for earthquake waves to arrive and the greater the delay between the P-wave and S-wave arrival times.

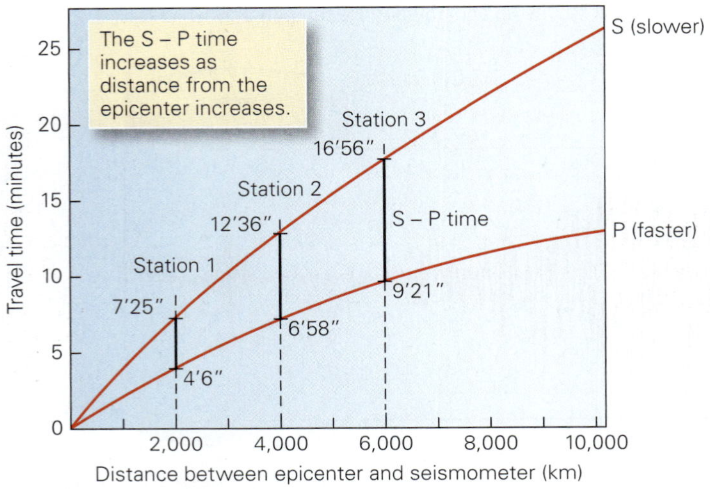

(b) We can represent the different arrival times of P-waves and S-waves on a graph of travel-time curves.

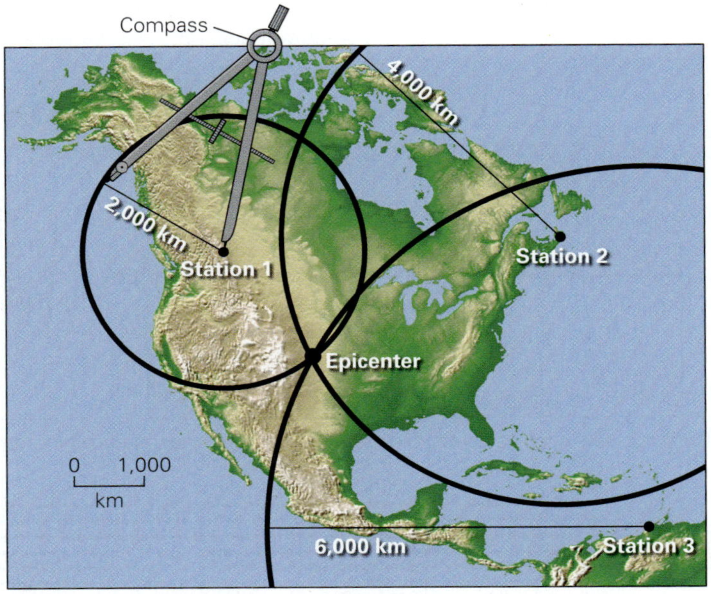

(c) If an earthquake epicenter lies 2,000 km from Station 1, we draw a circle with a radius of 2,000 km around the station at the scale of the map. We repeat for the other two stations. The intersection is the epicenter.

time. Then draw a line segment on a piece of tracing paper to represent this amount of time, at the scale used for the vertical axis of the graph. Orient the line segment parallel to the time axis and move it back and forth until one end lies on the P-wave curve and the other end lies on the S-wave curve (this gives the S – P time). Extend the line down to the horizontal axis, and simply read off the distance to the epicenter from the seismic station.

The analysis of one seismogram tells you only the distance between the epicenter and the seismometer station—it does not tell you in which direction from the station the epicenter lies. To determine the map location of the epicenter, we use a method known as *triangulation*, which requires plotting the distance from the epicenter to three stations. For example, if you know that the epicenter lies 2,000 km from Station 1, 4,000 km from Station 2, and 6,000 km from Station 3, you can draw a circle around each station on a map, such that the radius of the circle is the distance between the station and the epicenter at the scale of the map. The epicenter lies at the intersection of the three circles, for this is the only point at which the epicenter has the appropriate measured distance from all three stations (**Fig. 8.10c**).

TAKE-HOME MESSAGE

Earthquake energy travels as seismic waves. Body waves travel through the interior of the Earth, whereas surface waves travel along the surface. Seismologists distinguish between two kinds of body waves (P-waves and S-waves) and two kinds of surface waves (R-waves and L-waves). Seismometers record ground shaking on a seismogram.

QUICK QUESTION How can you determine the location of an earthquake's epicenter?

8.4 Defining the Size of Earthquakes

Some earthquakes shake the ground violently, whereas others can barely be felt. Seismologists have developed two different scales—the intensity scale and the magnitude scale—to define size in a uniform way and therefore to compare the sizes of earthquakes.

Mercalli Intensity Scale

Earthquake **intensity** at a locality on the Earth's surface refers to the degree of ground shaking at that locality. In 1902, an Italian scientist named Giuseppe Mercalli devised a scale for defining intensity based both on assessing the damage that the earthquake caused and on people's perception of the shaking. A version of this scale, called the **Modified Mercalli Intensity scale**, continues to be used today (**Table 8.2**). On this scale, seismologists represent intensity at a location by a Roman numeral, and an earthquake can have an intensity ranging from I (not destructive) to XII (highly destructive). Note that specifying earthquake intensity depends on a subjective assessment of damage, and how the shaking felt, not on a direct measurement with an instrument.

Significantly, the Mercalli intensity value varies with location for a given earthquake—intensity tends to be greater near the epicenter and to decrease progressively away from the epicenter. Why? First, the energy carried by a seismic wave decreases with increasing distance from the source. Rocks and sediment act as shock absorbers, making a vibration less intense as it passes (much as a car's shock absorbers soften the effect of a bounce when a car goes over a bump). Second, some of the seismic energy passing through the Earth gets reflected back at layer boundaries, and therefore, doesn't reach the seismometer. Seismologists draw contour lines on a map to show zones where an earthquake has a certain intensity. Such maps show how intensity varies over a region for the earthquake (**Fig. 8.11**).

Earthquake Magnitude Scales

When you hear a report of an earthquake disaster in the news, you will likely hear a phrase like, "An earthquake with a magnitude of 7.2 struck the city yesterday at 10:22 in the morning." What does this mean? Earthquake **magnitude** is a number that represents the amount of energy released from the seismic source, based on a measurement of the amplitude of ground shaking as recorded by a seismometer. This *amplitude* means the amount of up-and-down or back-and-forth

TABLE 8.2 Modified Mercalli Intensity Scale

MMI	Destructiveness (Perceptions of the Extent of Shaking and Damage)
I	Detected only by seismic instruments; causes no damage.
II	Felt by a few stationary people, especially in upper floors of buildings; suspended objects, such as lamps, may swing.
III	Felt indoors; standing automobiles sway on their suspensions; it seems as though a heavy truck is passing.
IV	Shaking awakens some sleepers; dishes and windows rattle.
V	Most people awaken; some dishes and windows break, unstable objects tip over; trees and poles sway.
VI	Shaking frightens some people; plaster walls crack, heavy furniture moves slightly, and a few chimneys crack, but overall little damage occurs.
VII	Most people are frightened and run outside; a lot of plaster cracks, windows break, some chimneys topple, and unstable furniture overturns; poorly built buildings sustain considerable damage.
VIII	Many chimneys and factory smokestacks topple; heavy furniture overturns; substantial buildings sustain some damage, and poorly built buildings suffer severe damage.
IX	Frame buildings separate from their foundations; most buildings sustain damage, and some buildings collapse; the ground cracks, underground pipes break, and rails bend; some landslides occur.
X	Most masonry structures and some well-built wooden structures are destroyed; the ground severely cracks in places; many landslides occur along steep slopes; some bridges collapse; some sediment liquifies; concrete dams may crack; facades on many buildings collapse; railways and roads suffer severe damage.
XI	Few masonry buildings remain standing; many bridges collapse; broad fissures form in the ground; most pipelines break; severe liquefaction of sediment occurs; some dams collapse; facades on most buildings collapse or are severely damaged.
XII	Earthquake waves cause visible undulations of the ground surface; objects are thrown up off the ground; there is complete destruction of buildings and bridges of all types.

FIGURE 8.11 This map shows Modified Mercalli Intensity contours for the 1886 Charleston, South Carolina, earthquake. Note that near the epicenter ground shaking reached an intensity of X, and in New York City ground shaking reached an intensity of II to III.

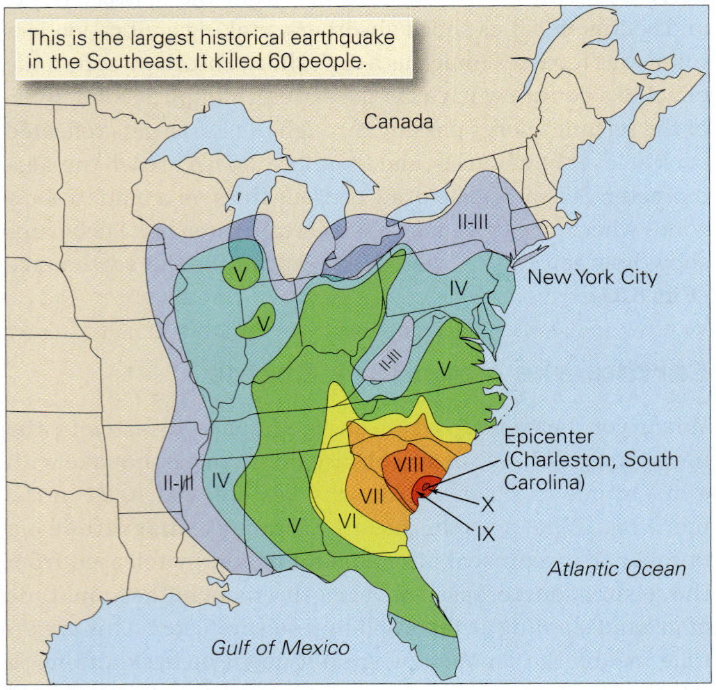

motion of the ground—the larger the amplitude, the greater the deflection of a seismometer pen or needle as it traces out a seismogram. To calculate a magnitude, seismologists first measure the height of the largest deflection on a seismograph. Then, after they have determined the distance between the epicenter and the seismometer, they adjust the measurement to be equivalent to the deflection recorded by a seismometer positioned a certain distance, the reference distance, from the epicenter. Because of this adjustment, ideally seismologists obtain the same magnitude for an earthquake by using a measurement at any seismometer. In other words, a given earthquake should have only one magnitude number; in contrast to earthquake intensity, earthquake magnitude does not depend on distance from the epicenter.

In 1935, an American seismologist, Charles Richter, developed a logarithmic scale for defining earthquake magnitude. His scale, now known as the **Richter scale**, became so widely known that news reports often include wording such as, "The earthquake registered a 7.2 on the Richter scale." Because the Richter scale and all subsequent earthquake magnitude scales are logarithmic, an increase of one unit of magnitude represents a tenfold increase in the maximum amplitude of ground motion. Thus, a magnitude 8 earthquake results in ground motion that is 10 times greater than that of a magnitude 7 earthquake, and 1,000 times greater than that of a magnitude 5 earthquake. Richter used 100 km as the reference distance—and since there's not necessarily a seismometer exactly at 100 km from the epicenter, he developed a simple chart to adjust for distance of the station from the epicenter (**Fig. 8.12**).

These days, seismologists actually use several different scales, not only the Richter scale, to describe earthquake magnitude. The specific way that Richter calculated magnitude actually works well only for shallow earthquakes

FIGURE 8.12 Using the Richter magnitude scale.

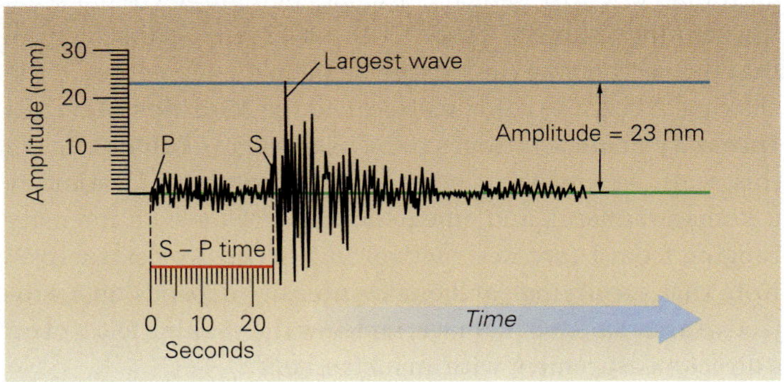

(a) To calculate the Richter magnitude from a seismograph, first measure the S – P (S minus P) time, to determine the distance to the epicenter. Then measure the amplitude of the largest wave.

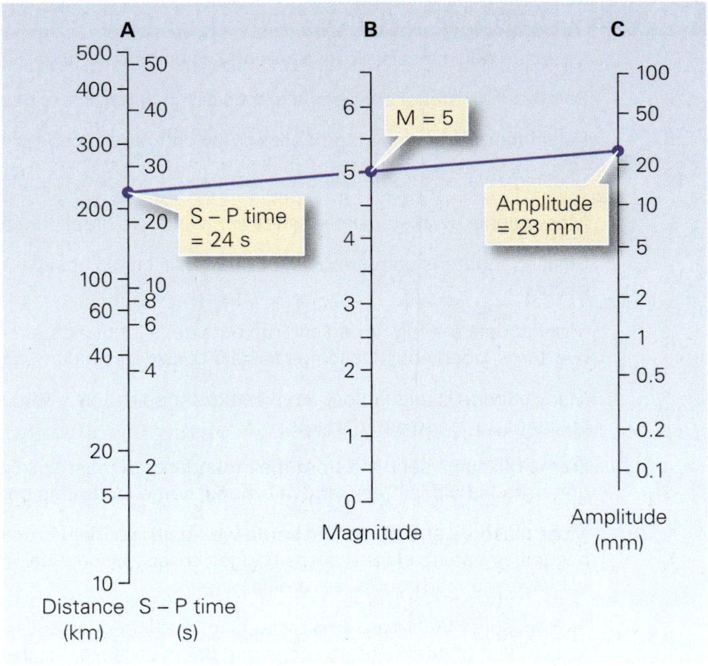

(b) Draw a line from the point on Column A representing the S – P time or the distance to the epicenter, to the point on Column C representing the wave amplitude. Read the Richter magnitude off Column B.

that are relatively close to the seismometer whose data go into the calculation. Because of the distance limitation, a number given by the Richter scale is now called a *local magnitude* (M_L).

To calculate an earthquake's **moment magnitude** (M_W), seismologists measure the amplitude of several different seismic waves, determine the dimensions of the slipped area on the fault, and estimate the displacement that occurred. The moment magnitude scale, like the Richter scale, is logarithmic. Seismologists consider the moment magnitude to be the most accurate representation of an earthquake's size, so M_W serves as the magnitude used in the historic record. When specifying magnitude, seismologists indicate the type of magnitude measurement and the magnitude number, so an earthquake with a moment magnitude of 6.5 can be called "an M_W 6.5 earthquake."

TABLE 8.3 Adjectives for Describing Earthquakes

Adjective	Magnitude	Approximate maximum intensity at epicenter	Effects
Great	> 8.0	X to XII	Major to total destruction
Major	7.0 to 7.9	IX to X	Great damage
Strong	6.0 to 6.9	VII to VIII	Moderate to serious damage
Moderate	5.0 to 5.9	VI to VII	Slight to moderate damage
Light	4.0 to 4.9	IV to V	Felt by most; slight damage
Minor	< 3.9	III or smaller	Felt by some; hardly any damage

To make discussion of earthquakes easier, seismologists commonly use familiar adjectives to describe the size of an earthquake (**Table 8.3**). Earthquakes that most people can feel have a magnitude greater than M_W 4, and earthquakes that can cause moderate damage have a magnitude greater than M_W 5. The largest recorded earthquake in history, the 1960 Chilean quake, registered as an M_W 9.5, and the catastrophic 2011 Tōhoku earthquake registered as an M_W 9.0. The news media sometimes incorrectly state that the magnitude scale goes from 1 to 10. In fact, earthquakes of M_W –1 or M_W –2 can be detected, if they are close to the seismometer. Furthermore, there is no defined upper limit to the magnitude scale—that said, seismologists estimate that an M_W 9.5 is about as big as an earthquake can get, given the known dimensions of faults on Earth.

Energy Release by Earthquakes

To give a sense of the amount of energy released by an earthquake, seismologists compare earthquakes to other energy-releasing events. For example, according to some estimates, an M_W 5.3 earthquake releases about as much energy as the Hiroshima atomic bomb, and an M_W 9.0 earthquake releases significantly more energy than the largest hydrogen bomb ever detonated. Notably, though an increase in magnitude by one integer represents a 10-fold increase in the amplitude of ground shaking, it represents approximately a 32-fold increase in energy release. Thus, an M_W 8 earthquake releases about 1 million times more energy than an M_W 4 earthquake (**Fig. 8.13**). In fact, a single M_W 8.9 earthquake releases as much energy as the entire average global annual release of seismic energy coming from all other earthquakes combined! Fortunately, such large earthquakes occur much less frequently than small earthquakes. There are about 100,000 M_W 3 earthquakes every year, but an M_W 8 earthquake happens only about once or twice a year.

FIGURE 8.13 Energy released by earthquakes increases dramatically with magnitude.

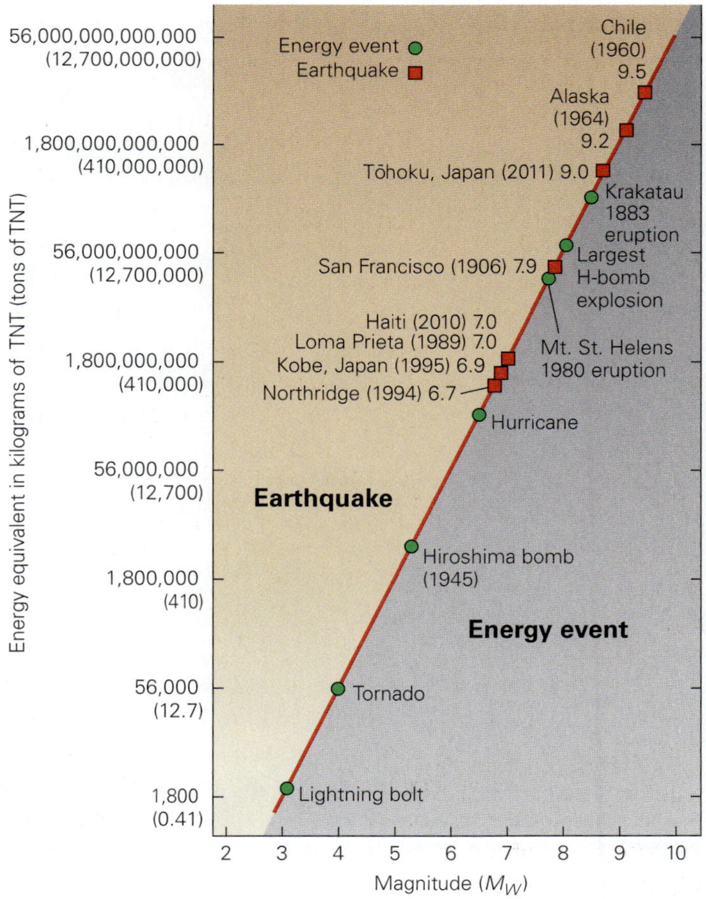

TAKE-HOME MESSAGE

We can specify earthquake size by intensity (based on a perception of damage caused and shaking felt) or by magnitude (a representation of energy released by the source, based on a measurement of ground motion on a seismogram). An increase of one magnitude in number represents a 10 times increase in the amplitude of shaking, at a reference distance, and a 32-fold increase in energy release from the source. A great earthquake releases vastly more energy than a giant hydrogen bomb.

QUICK QUESTION Can you specify the "size" of an earthquake by giving just one intensity number? How about a single magnitude number?

8.5 Where and Why Do Earthquakes Occur?

Earthquakes do not take place everywhere on the globe. By plotting the distribution of earthquake epicenters on a map, seismologists find that most, but not all, earthquakes occur in elongate **seismic belts** (**Fig. 8.14**). (The term *seismic zone* is sometimes used as a synonym, or in cases where the seismic area on a map is more equant in shape.) Most seismic belts correspond to plate boundaries, and earthquakes within these belts are called *plate-boundary earthquakes*. Those earthquakes that occur away from plate boundaries are called *intraplate earthquakes*. Let's look at the characteristics of earthquakes in various geologic settings and learn why earthquakes take place where they do.

Did you ever wonder... if an earthquake might happen near where you live?

Earthquakes at Plate Boundaries

As we've noted, the majority of earthquakes happen at faults along plate boundaries, for the relative motion between plates causes slip on faults. We find different kinds of faulting at different types of plate boundaries.

Divergent-Boundary Seismicity At divergent plate boundaries (mid-ocean ridges), two oceanic plates form and move apart. Divergent boundaries are broken into spreading segments linked by transform faults. Therefore, two kinds of

FIGURE 8.14 A map of epicenters emphasizes that most earthquakes occur in distinct belts along plate boundaries.

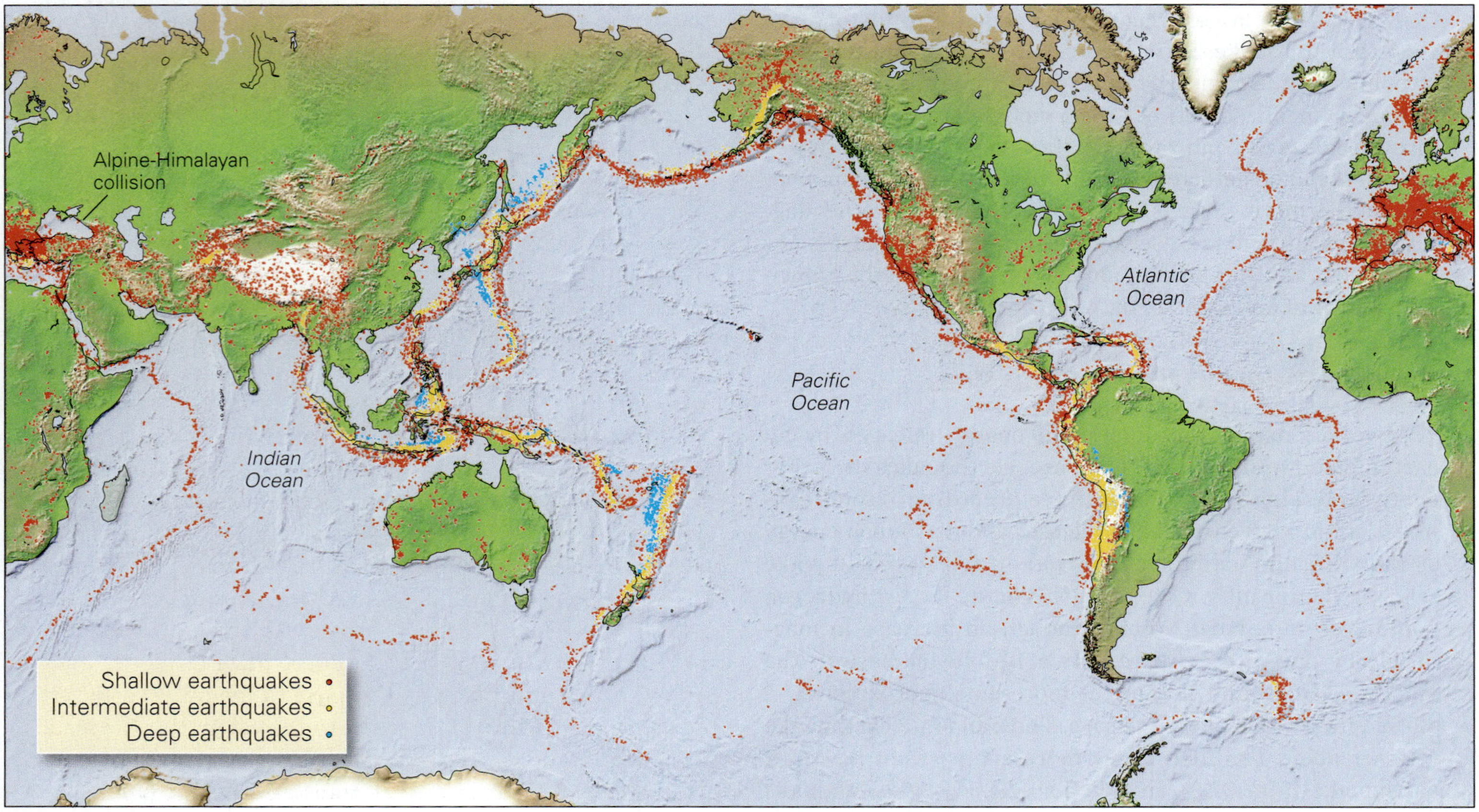

FIGURE 8.15 The distribution of earthquakes at a mid-ocean ridge. Note that normal faults occur along the ridge axis, and strike-slip faults occur along active transform faults. Earthquakes don't take place along inactive fracture zones.

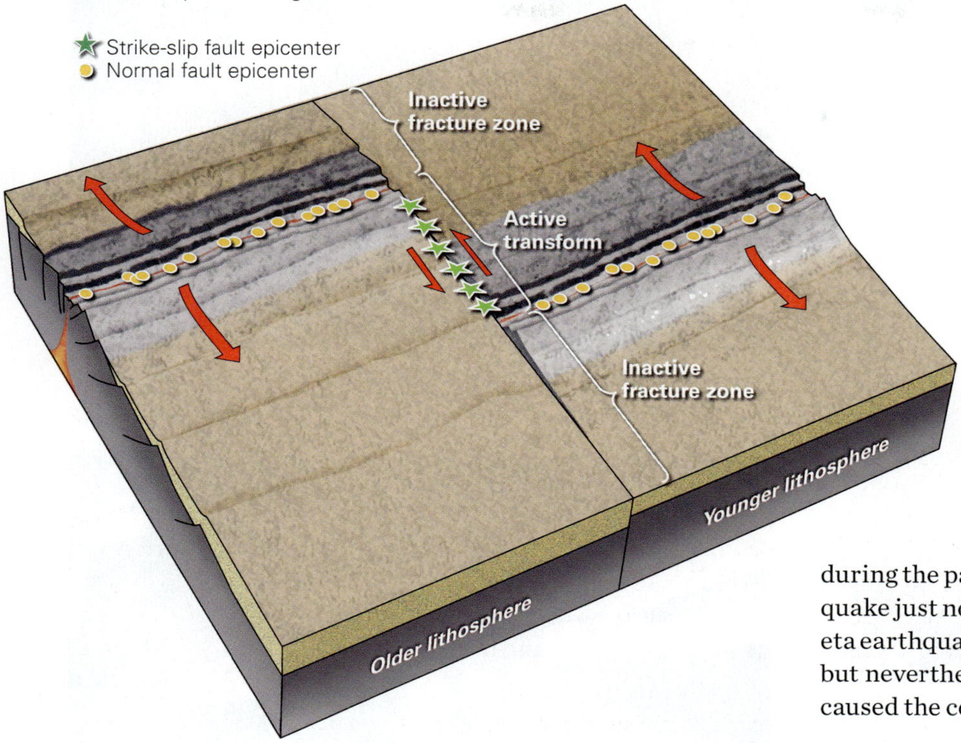

faults develop at divergent boundaries. Along spreading segments, stretching generates normal faults, whereas along transform faults strike-slip displacement occurs (**Fig. 8.15**). Earthquakes along mid-ocean ridges take place at depths of less than 10 km, and thus are shallow-focus earthquakes.

Transform-Boundary Seismicity At transform plate boundaries, where one plate slides past another without producing or consuming oceanic lithosphere, most faulting results in strike-slip motion. The majority of transform faults in the world link segments of oceanic ridges, deep beneath the sea. But a few, such as the San Andreas fault of California, the Alpine fault of New Zealand, and the Anatolian faults in Turkey, cut through continental lithosphere or volcanic arcs. All transform-fault earthquakes have a shallow focus, so the larger ones on land can cause disaster, as was the case when, in 2010, an M_W 7 earthquake struck Haiti, which sits astride the transform fault that delineates part of the plate boundary between the North American and Caribbean plates.

The San Francisco earthquake of 1906 serves as an example of a continental transform-fault earthquake that occurred on the San Andreas fault (**Fig. 8.16a**). In the wake of the gold rush, San Francisco was a booming city with broad streets and numerous large buildings. But it was built on the transform boundary along which the Pacific Plate moves north at an average rate of 6 cm per year, relative to North America. Because of the stick-slip behavior of the fault, this movement happens in sudden jerks, each of which causes an earthquake. At 5:12 a.m. on April 18, the fault near San Francisco slipped by as much as 7 m, and minutes later seismic waves struck the city. Witnesses watched in horror as the streets undulated, buildings swayed and banged together, laundry lines stretched and snapped, and weaker buildings collapsed. Fire followed soon after, consuming huge areas of the city (**Fig. 8.16b**). Seismologists estimate that the earthquake would have registered as an M_W 7.9, had it been recorded by a seismometer.

The San Francisco earthquake has not been the only one to strike along the San Andreas and nearby related faults. Over a dozen major earthquakes have happened on these faults during the past two centuries, including the 1857 M_W 7.7 earthquake just north of Los Angeles, and the 1989 M_W 7.1 Loma Prieta earthquake, which occurred 100 km south of San Francisco but nevertheless shut down a World Series baseball game and caused the collapse of a double-decker freeway (**Fig. 8.16c**).

Convergent-Boundary Seismicity Convergent plate boundaries are complicated regions at which several different kinds of earthquakes take place. The most dangerous type is shallow-focus earthquakes. These occur along the large thrust fault that delineates the boundary between the base of the overriding plate and the top of the subducting plate. Shear on this fault, and on other thrust faults nearby, can produce disastrous earthquakes. Because these earthquakes happen close to the Earth's surface, much of the energy they release will reach the surface. In some cases, the push applied by the downgoing plate compresses the overriding plate and triggers shallow-focus earthquakes within the overriding plate, which can also be dangerous.

In contrast to other types of plate boundaries, convergent boundaries can also host intermediate-focus and deep-focus earthquakes. These occur in the downgoing slab as it sinks into the mantle, defining the sloping belt of seismicity called a **Wadati-Benioff zone**, after the seismologists who first recognized it (**Fig. 8.17a**). Earthquakes of the Wadati-Benioff zone occur partly in response to stresses caused by shear between the sinking lithosphere plate and surrounding asthenosphere, partly in response to the pull exerted by the deeper part of the plate on the shallower part as the plate sinks, partly due to the resistance that the subducting plate encounters as it pushes into the mantle below, and partly due

FIGURE 8.16 The San Andreas fault system, a continental transform.

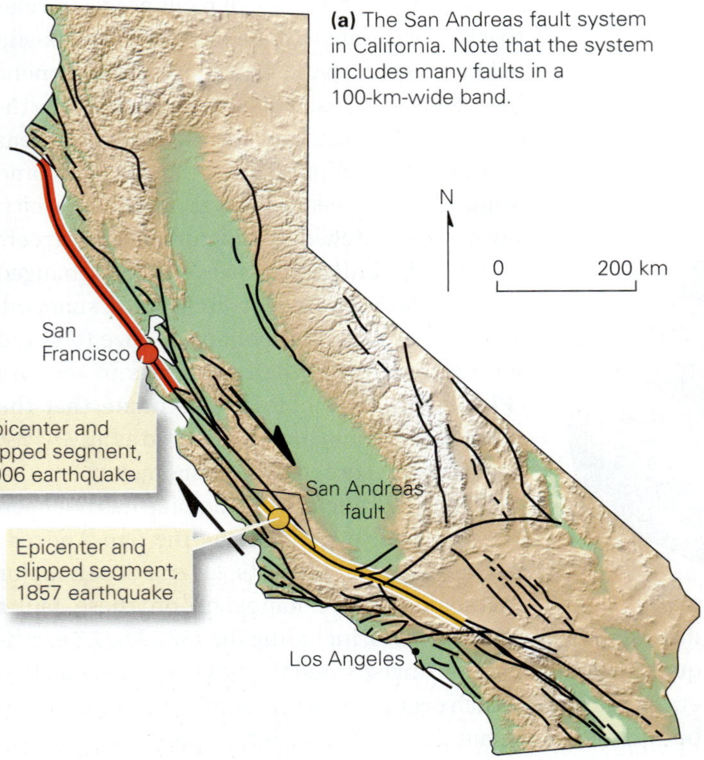

(a) The San Andreas fault system in California. Note that the system includes many faults in a 100-km-wide band.

(b) A street in San Francisco after the 1906 earthquake. Huge fires swept through the city.

(c) The 1989 Loma Prieta earthquake caused a double-decker freeway to collapse, as support columns gave way.

to volume changes that take place when olivine, a mineral in the lithosphere, undergoes a phase change and collapses to form denser minerals, under the extreme pressure that develops in deeply subducted lithosphere.

How are intermediate and deep earthquakes of a Wadati-Benioff zone even possible? Shouldn't the rock of a subducted plate at these depths be too warm and soft to break seismically? Seismologists have determined that rock is such a good insulator that the interior of a subducting plate actually remains cool enough to break seismically, down to the depths of intermediate- and deep-focus earthquakes.

Earthquakes in southern Alaska, eastern Japan, the western coast of South America, the coast of Oregon and Washington, and along island arcs in the western Pacific serve as examples of convergent-boundary earthquakes. Some of these earthquakes are very large, and when they occur near populated areas, they can be devastating. Notable examples include the 1960 M_W 9.5 earthquake in Chile, the 1964 M_W 9.2 Good Friday earthquake in Alaska, the 1995 M_W 6.9 earthquake in Kobe, Japan, which devastated the city (**Fig. 8.17b, c**), the 2004 M_W 9.3 Sumatra earthquake, the 2010 M_W 8.8 Chilean earthquake, and the 2011 M_W 9.0 Tōhoku earthquake.

Earthquakes within Continents

We've seen that earthquakes occur along transform faults that cut across continents. They also occur where continental lithosphere is stretching and pulling apart in rifts, where continents are squeezing together in collisional zones, and where old faults within otherwise stable crust get reactivated, as we now see (**Fig. 8.18**).

Continental Rifts The stretching of crust at continental rifts generates normal faults. Active rifts today include the East African Rift, the Basin and Range Province (mostly in Nevada, Utah, and Arizona), and the Rio Grande Rift (in New Mexico). In all these places, shallow earthquakes occur, similar in nature to the earthquakes at mid-ocean ridges. But in contrast to mid-ocean ridges, these seismic belts can be located under populated areas, and thus can cause major damage.

Collision Zones Two continents collide when the oceanic lithosphere that once separated them has been completely subducted. Such collisions produced great mountain ranges such as the Alps and the Himalayas. The most common cause

FIGURE 8.17 An example of a convergent-plate-boundary earthquake.

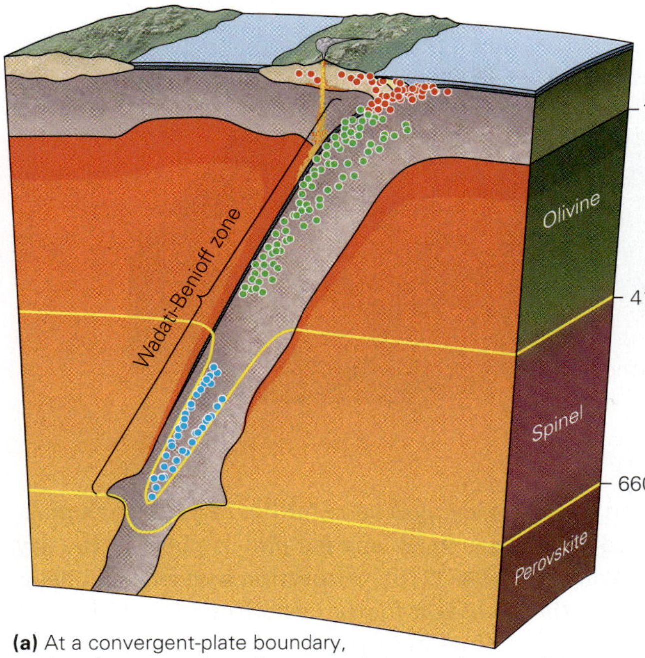

(a) At a convergent-plate boundary, earthquakes occur along the contact between the two plates, as well as in the downgoing plate and overriding plate. Deep earthquakes may be due to the phase change of olivine to spinel. In the asthenosphere, this change happens at 410 km, but in cooler lithosphere, it may happen at greater depths.

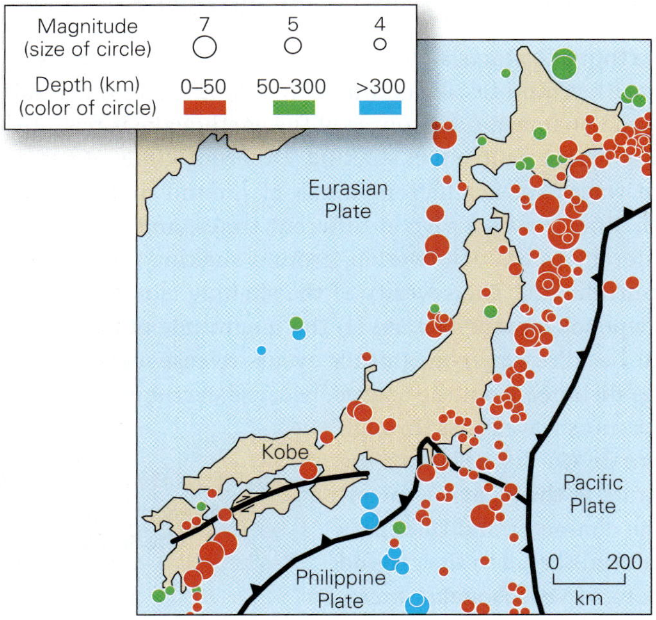

(b) Map of earthquake epicenters and subduction zones in and near Japan.

(c) A collapsed building after the Kobe earthquake in Japan.

of earthquakes in collision zones is movement on thrust faults, as they accommodate crustal shortening. An example of a collision-zone earthquake took place in April 2015, when compression resulting from the northward push of the Indian subcontinent into Asia caused an *Mw* 7.8 earthquake in Nepal, a country that encompasses a portion of the Himalayan Mountains. The event occurred when a 7000 km^2 portion of a large thrust fault, 15 km below the surface, suddenly slipped by up to 3 m (10 ft). Ground shaking resulted in the collapse of whole towns with poorly constructed homes. Tragically, thousands died and many more suffered injury and loss of property. Landslides blocked rivers and ripped out roads, making hard-hit areas difficult to reach. In the capital city of Kathmandu, many monuments recognized as United Nations World Heritage sites crumbled completely. Aftershocks, the largest of which had M_W 6.6, rattled the area for weeks.

Intraplate Earthquakes

Some earthquakes occur in the interiors of plates and are not associated with plate boundaries, active rifts, or collision zones (see Fig. 8.18). These **intraplate earthquakes** account for only about 5% of the earthquake energy released in a year. Almost all have a shallow focus. What causes intraplate earthquakes? Most seismologists favor the idea that stress applied to the boundary of a plate can cause the interior of the plate to break suddenly at weak, pre-existing fault zones, some of which may have initially formed long ago, during the Precambrian.

Intraplate earthquakes happen on all continents but are not uniformly distributed. In North America, most intraplate earthquakes occur near New Madrid, Missouri; Charleston, South Carolina; eastern Tennessee; and Montreal. An M_W 7.3 earthquake occurred near Charleston in 1886, ringing church bells up and down the coast and vibrating buildings as far away as Chicago. In Charleston itself, over 90% of the buildings were damaged, and sixty people died. In 2011, an M_W 5.9 earthquake struck central Virginia, abruptly reminding

FIGURE 8.18 The tectonic settings in which earthquakes occur in continental lithosphere. Subduction-related earthquakes in continental crust are not shown.

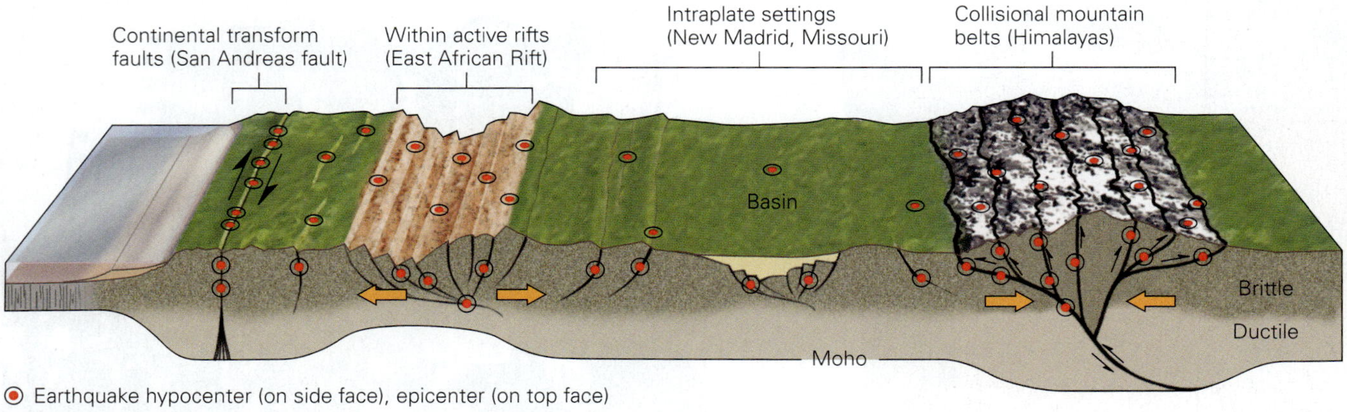

residents of the eastern United States that the region is not immune to seismicity. The tremor was felt from the Carolinas to New England.

The largest intraplate earthquakes to affect the continental United States happened in the early 19th century, near New Madrid, which lies near the Mississippi River in southernmost Missouri. During the winter of 1811–12, three M_W 7.0 to 7.4 earthquakes struck the region. Displacement of the ground surface temporarily reversed the flow of the Mississippi River and toppled cabins (**Fig. 8.19a**). The earthquakes resulted from slip on faults that underlie the Mississippi Valley (**Fig. 8.19b**). Both St. Louis, Missouri, and Memphis, Tennessee, could be damaged if a major earthquake were to take place in the region today.

TAKE-HOME MESSAGE

Most, but not all, earthquakes happen along plate boundaries. Most catastrophic earthquakes occur at convergent boundaries, or along continental transforms. Events also occur in rifts and collision zones, and along weak faults within plate interiors.

QUICK QUESTION On a global basis, why are earthquakes in continental crust more dangerous to society than those along mid-ocean ridges?

8.6 How Do Earthquakes Cause Damage?

On All Saint's Day, November 1, 1755, stresses that had been building due to the northward push of Africa against Europe caused a thrust fault beneath the Atlantic Ocean floor west of Lisbon to slip suddenly. The resulting M_W 9.0 earthquake destroyed Lisbon, Portugal, and led philosophers of the day, such as Voltaire (1694–1778), to question a widely held belief that the way the world is, is for the best. Ground shaking led to the collapse of buildings in Lisbon, but as is the case for many calamitous earthquakes, shaking is only part of the story. Below, we examine the many components of earthquake-related destruction.

Ground Shaking and Displacement

An earthquake starts suddenly and may last from a few seconds to a few minutes, depending on how long the slip on a fault took, and on how far the source of the earthquake is from the locality where people feel shaking. Different kinds of earthquake waves cause different kinds of ground motion (**Fig. 8.20**). Since waves arrive at different times, and the arrival of different waves can overlap, ground shaking overall may be quite chaotic. The severity of the shaking at a given location depends on four factors: (1) the magnitude of the earthquake, because larger-magnitude events release more energy; (2) the distance from the source, because earthquake energy decreases as waves pass through the Earth; (3) the nature of the substrate at the location, meaning the character and thickness of materials beneath the ground surface, for earthquake waves tend to be amplified in weaker substrates; and (4) the wavelength of the earthquake waves, for the wavelength determines the frequency of the waves, meaning the number of waves that reach a point in a specified interval of time.

Did you ever wonder... if all earthquakes feel the same?

If you're out in an open field during an earthquake, ground motion alone won't kill you—you may be knocked off your feet and bounced around a bit, but your body is too flexible to

FIGURE 8.19 New Madrid, Missouri, is an example of intraplate seismic activity.

(a) The earthquakes of 1811–12 destroyed cabins and disrupted the Mississippi River.

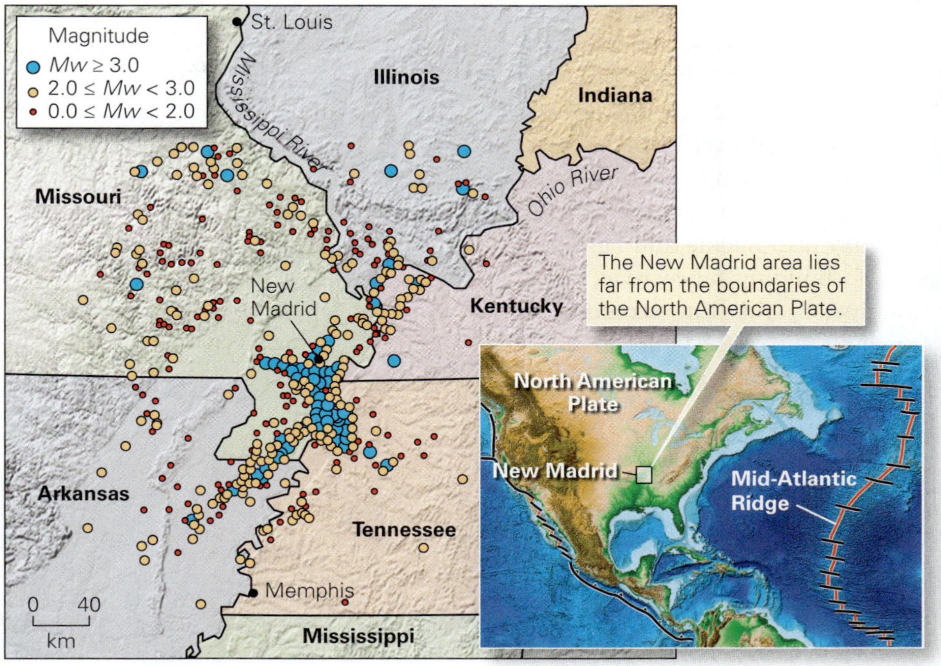

(b) The epicenters of recent small earthquakes in the New Madrid area, as recorded by modern seismic instruments. The region remains active.

FIGURE 8.20 Types of ground motion during earthquakes. The ground can shake in many ways at once, causing surface structures to move.

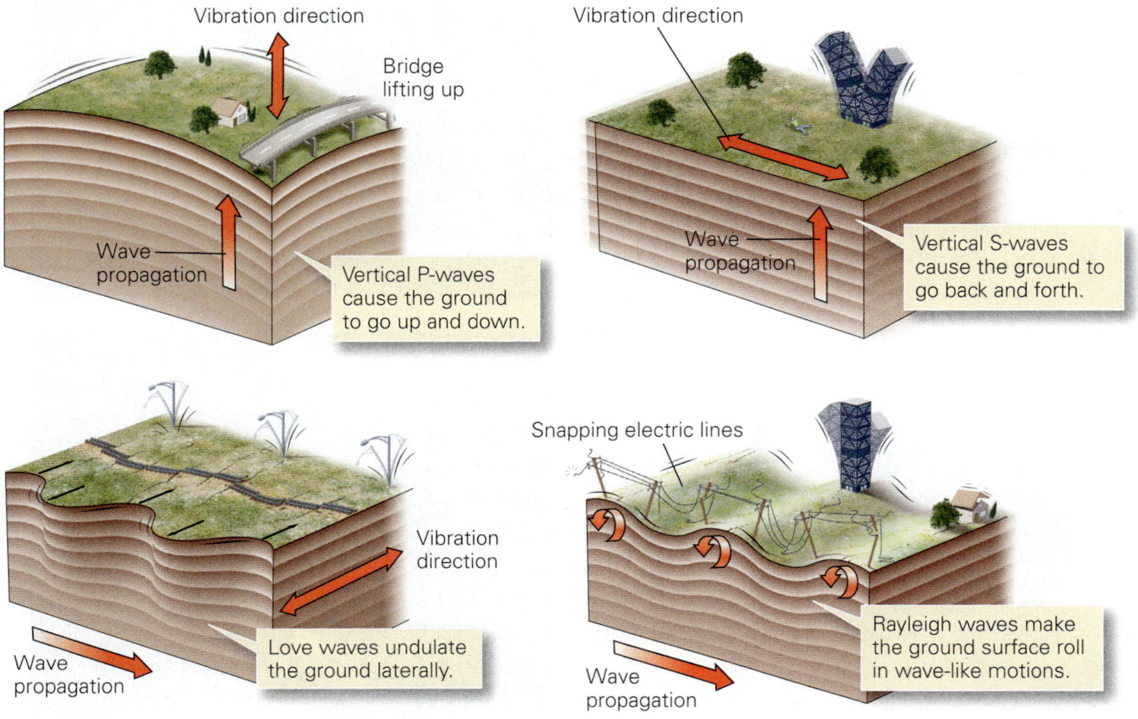

break. Buildings and bridges aren't so lucky (**Fig. 8.21**). When earthquake waves pass, they sway, twist back and forth, or lurch up and down, depending on the type of wave motion. As a result, connectors between the frame and facade of a building may separate, causing the facade to crash to the ground. The flexing of walls shatters windows and makes roofs collapse. Floors or bridge decks may rise up and slam down on the columns that support them, thereby crushing the columns. Some buildings collapse so their floors pile on top of one another like pancakes in a stack, some crumble into fragments, and some simply tip over. The majority of earthquake-related deaths and injuries happen when people are hit by debris or are crushed beneath falling walls or roofs. Aftershocks worsen the problem, because they may topple already weakened buildings, trapping rescuers. During earthquakes, roads, rail lines, and pipelines may also buckle and rupture. If a building, fence, road, pipeline, or rail line straddles a fault, slip on the fault can crack the structure and separate it into two pieces.

FIGURE 8.21 Examples of earthquake damage due to vibration.

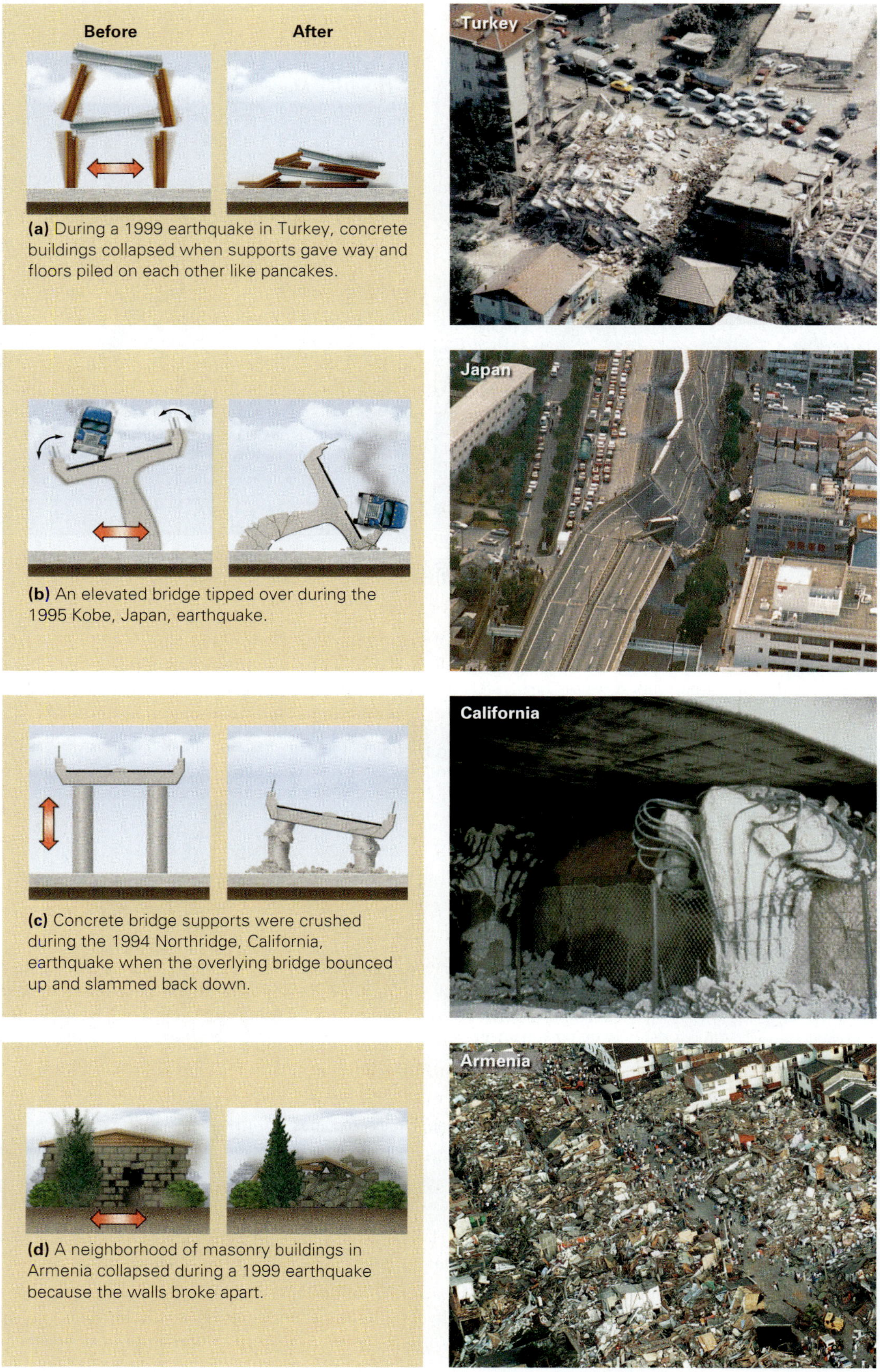

(a) During a 1999 earthquake in Turkey, concrete buildings collapsed when supports gave way and floors piled on each other like pancakes.

(b) An elevated bridge tipped over during the 1995 Kobe, Japan, earthquake.

(c) Concrete bridge supports were crushed during the 1994 Northridge, California, earthquake when the overlying bridge bounced up and slammed back down.

(d) A neighborhood of masonry buildings in Armenia collapsed during a 1999 earthquake because the walls broke apart.

Landslides

The shaking of an earthquake can cause ground on steep slopes or ground underlain by weak sediment to give way. This movement results in a **landslide**, the tumbling and flow of soil and rock downslope (see Chapter 13). Earthquake-triggered landslides occur commonly along the coast of California where expensive homes perch on steep cliffs looking out over the Pacific. When the cliffs collapse, the homes tumble to the beach below (**Fig. 8.22a**). Such events lead to the misperception that California "will someday fall into the sea." Although small portions of the coastline do collapse, the state as a whole remains firmly attached to the continent, despite what Hollywood scriptwriters say. Landslides and slumping are often a major cause of earthquake damage (**Fig. 8.22b**).

Did you ever wonder...
if California will fall into the sea?

FIGURE 8.22 Examples of landslide damage triggered by earthquakes.

(a) Shaking triggers landslides, which caused a steep slope along the coast of California to collapse, carrying part of a home with it.

(b) During the 1964 Alaska earthquake, slumping caused the land to give way beneath parts of Anchorage.

Sediment Liquefaction

In 1964, an M_W 7.5 earthquake struck Niigata, Japan. A portion of the city had been built on land underlain by wet sand. During the ground shaking, foundations of over 15,000 buildings sank into their substrate, causing walls and roofs to crack. Several four-story-high buildings in a newly built apartment complex tipped over (**Fig. 8.23a**). In 2011, an earthquake in Christchurch, New Zealand, caused sand to erupt and produce small, cone-shaped mounds on the ground surface (**Fig. 8.23b**). The transfer of sand from underground onto the surface led to formation of depressions large enough to swallow cars (**Fig. 8.23c**).

These examples are manifestations of a phenomenon called **sediment liquefaction**. Liquefaction in wet sand happens when shaking causes grains to settle together, for this movement causes the pressure in the water filling the pores between grains to increase. The water pushes the grains apart so that they become surrounded by water and no longer rest against each other, making the sand turn into a slurry, incapable of supporting weight. A similar weakening phenomenon happens in a special type of clay called *quick clay*. In a quick clay, the flakes are arranged in an unstable structure, which can be disrupted by ground shaking. When this happens, what had been a stable, gel-like mass transforms into weak, slippery mud. Regardless of the mechanisms of sediment liquefaction, weakening due to liquefaction allows material above the liquefied sediment to settle downward. If cracks open up between the liquefied layer and the ground surface, pressure can squeeze the liquefied sediment upward and out onto the ground surface and build cone-shaped mounds—variously known as *sand volcanoes*, sand boils, or sand blows. The settling of sedimentary layers down into a liquefied layer can also disrupt bedding and can lead to formation of open fissures of the land surface (see Fig. 8.23b).

The 1964 Good Friday earthquake in southern Alaska caused liquefaction beneath the Turnagain Heights neighborhood of Anchorage. The neighborhood was built on a small terrace of uplifted sediment. The edge of the terrace was a 20-m-high escarpment that dropped down to Cook Inlet, a bay of the Pacific Ocean. As the ground shaking began, a layer of quick clay beneath the development turned into mud, and when this happened, the overlying layers of sediment, along with the houses built on top of them, slid seaward. In the process, the layers broke into separate blocks that tilted, turning the landscape into a chaotic jumble, and resulting in the destruction of the neighborhood (**Fig. 8.24**).

Fire

The shaking during an earthquake can make lamps, stoves, or candles with open flames tip over, and it may break wires or topple power lines, generating sparks. As a consequence, areas already turned to rubble and even areas not so badly damaged may be consumed by fire. Ruptured gas pipelines and oil tanks feed the flames, sending columns of fire erupting skyward (**Fig. 8.25a**). Firefighters might not even be able to reach the fires, because the doors to the fire station won't open or rubble blocks the streets. Moreover, firefighters may find themselves without water, for ground shaking and landslides damage water lines.

Once a fire starts to spread, it can become an unstoppable inferno. Lisbon was consumed by fire after the 1755 earthquake, and most of the destruction of the 1906 San Francisco

FIGURE 8.23 Examples of liquefaction triggered by earthquakes.

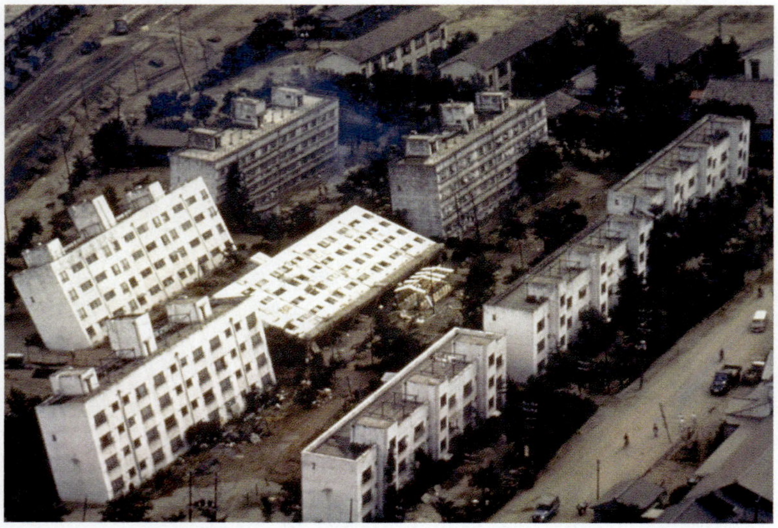

(a) Liquefaction under their foundations caused these apartment buildings in Niigata, Japan, to tip over during a 1964 earthquake.

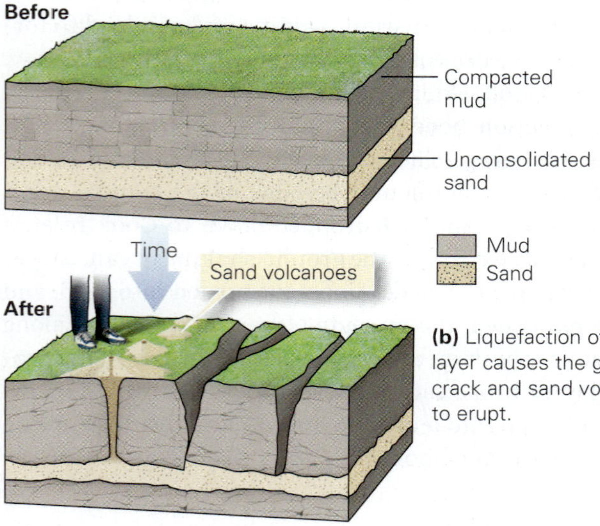

(b) Liquefaction of a sand layer causes the ground to crack and sand volcanoes to erupt.

(c) During the 2011 Christchurch, New Zealand, earthquake, liquefied sand spurted out and spread over the pavement. The process produced open space underground, so the pavement collapsed to form a sinkhole.

FIGURE 8.24 The 1964 Turnagain Heights disaster.

(a) A landslide carried a neighborhood out to sea.

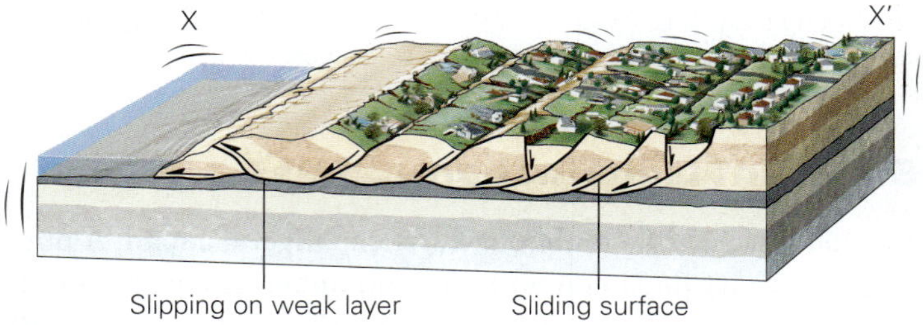

(b) Slip occurred on a weak layer.

earthquake resulted from fire. For three days, the blaze spread through San Francisco until firefighters contained it by blasting a firebreak, but by then, 500 blocks of structures had turned to ash, causing 20 times as much financial loss as the shaking itself. When a large earthquake hit Tokyo in 1923, fires set by cooking stoves spread quickly through the wood-and-paper buildings, creating an inferno—a *firestorm*—that heated the air above the city (**Fig. 8.25b**). As hot air rose, cool air rushed in, creating wind gusts of over 100 mph, which stoked the blaze and incinerated 120,000 people.

Tsunamis

The azure waters and palm-fringed islands of the Indian Ocean's east coast hide one of the most seismically active plate boundaries on Earth—the Sunda Trench. Along this convergent boundary, the Indian Ocean floor subducts at a rate of about 4 cm per year, leading to episodic slip on thrust faults. Just before 8:00 a.m. on December 26, 2004, the crust above a 1,300-km-long by 100-km-wide portion of one of these faults lurched westward by as much as 15 m. The break started at the focus and then propagated north at 2.8 km/s. The rupturing overall took over 9 minutes. This slip triggered a great earthquake (the M_W 9.3 Sumatra earthquake) and pushed the seafloor up by tens of centimeters. The rise of the

FIGURE 8.25 Fire sometimes follows an earthquake.

(a) Broken gas tanks erupt in fountains of flame after the 2011 Tōhoku, Japan earthquake.

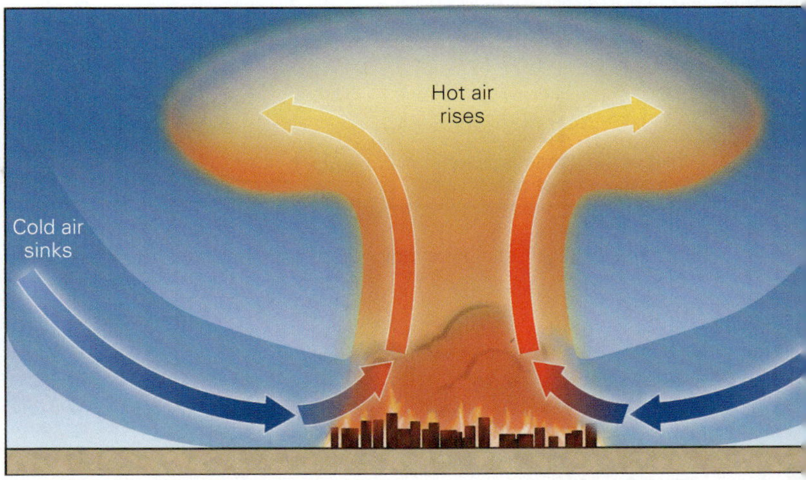

(b) A firestorm develops when cool air rushes in to replace rising hot air above a huge fire. The cool air stokes the blaze, making it larger and hotter.

seafloor, in turn, shoved up the overlying water. Because the area that rose was so broad, the volume of displaced water was immense. As a consequence, a tragedy of an unimaginable extent was about to unfold. Water from above the displaced seafloor began moving outward from above the fault zone, a process that generated a series of broad waves traveling at speeds of about 800 km per hour (500 mph)—almost the speed of a jet plane.

SEE FOR YOURSELF...

BANDA ACEH, BEFORE TSUNAMI

Latitude
5°33′31.82″ N

Longitude
95°17′13.01″ E

Zoom to 1.5 km (1 mile) and look down.

With the historical imagery tool, you can set the clock back to June 22, 2004, and see dense housing built along the shore of the seaside town of Banda Aceh, on the island of Sumatra. Now set the clock to 2006 and compare the views.

Geologists use the term **tsunami** for a wave produced by displacement of the seafloor (**Fig. 8.26a**). The displacement can be due to an earthquake, a volcanic explosion, or as we'll see in Chapter 13, a submarine landslide. *Tsunami* is a Japanese word that translates literally as "harbor wave," an apt name because tsunamis can be particularly damaging to harbor towns. In older literature such waves were called *tidal waves,* because when one arrives, water rises as if a tide were coming in, but in fact the waves have nothing to do with daily tidal cycles.

Regardless of cause, tsunamis are very different from familiar, wind-driven storm waves (**Fig. 8.26b, c**). Large wind-driven waves can reach heights of 10 to 30 meters in the open ocean. But even such monsters have wavelengths of only tens of meters, and thus contain a relatively small volume of water. In contrast, although a wave in deep water may cause a rise in sea level of at most only a few tens of centimeters, which means a ship crossing one wouldn't even notice, tsunamis have wavelengths of 10 to 500 km—so a tsunami involves a *huge volume of water*. In simpler terms, we can think of a tsunami, in map view, as being 100 to 1000 times wider than a wind-driven wave, as measured perpendicular to the wave crest. Because of this difference, a storm wave and a tsunami have very different consequences when they reach the shore.

When any wave approaches the shore, friction between the base of the wave and the seafloor slows the bottom of the wave, so the back of the wave catches up to the front, and the added volume of water builds the wave higher (see Fig. 8.26c). The top of the wave may fall over the front of the wave and cause a breaker (see Chapter 15). In the case of a wind-driven wave, the breaker may be tall when it washes onto the beach, but because the wave doesn't contain much water, it runs out of water and friction slows it to a stop on the beach. Then, gravity causes the water to flow seaward, back down the beach. In the case of a tsunami, the wave is so wide that, as friction slows the wave, it builds into a plateau of water that can be many kilometers wide. If the tsunami is high—and the largest are up to 30 m high—it can cross the beach. If the land is low-lying, the tsunami can keep moving, eventually submerging a huge area. Not all tsunamis are high—one that's only tens of centimeters to a meter high will probably affect only the near-shore area.

Tsunami damage can be catastrophic. For example, when the December 2004 Indian Ocean tsunami (**Fig. 8.27a**) struck Banda Aceh, a city at the north end of the island of Sumatra,

FIGURE 8.26 The formation and propagation of a tsunami.

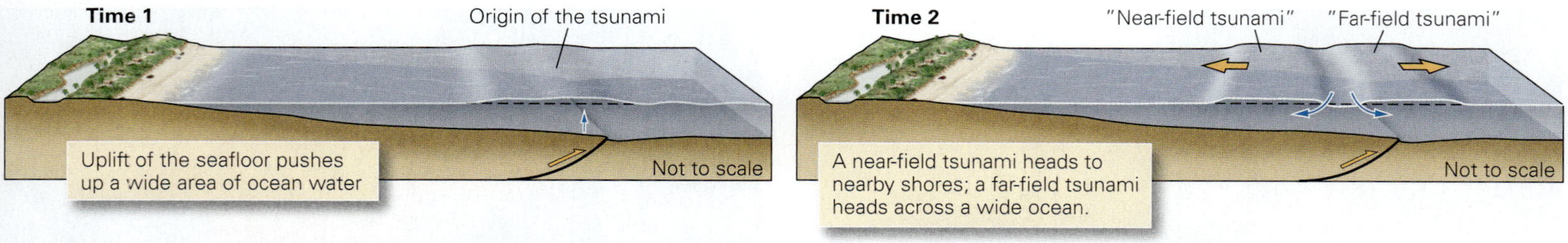

(a) After initial uplift of a mound of water, gravity causes the wave to spread out. The waves travel almost as fast as a jet plane.

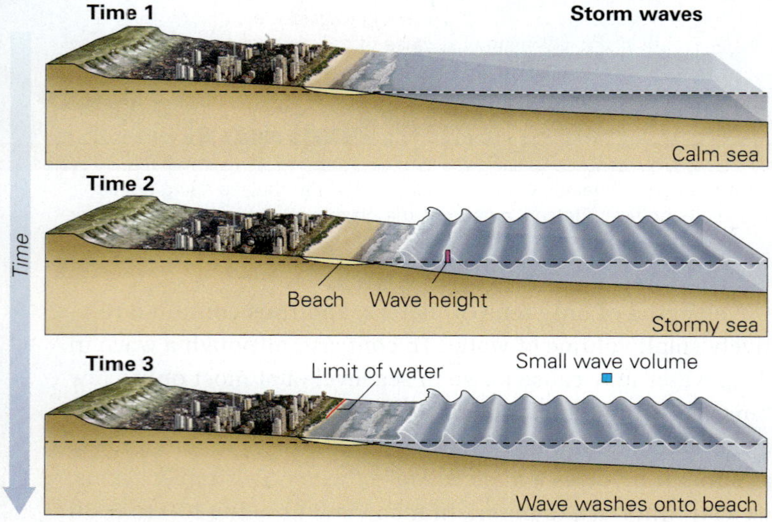

(b) Storm waves can be high, but because they have short wavelengths, they contain relatively little water. Most waves run out of water by the upslope edge of the beach.

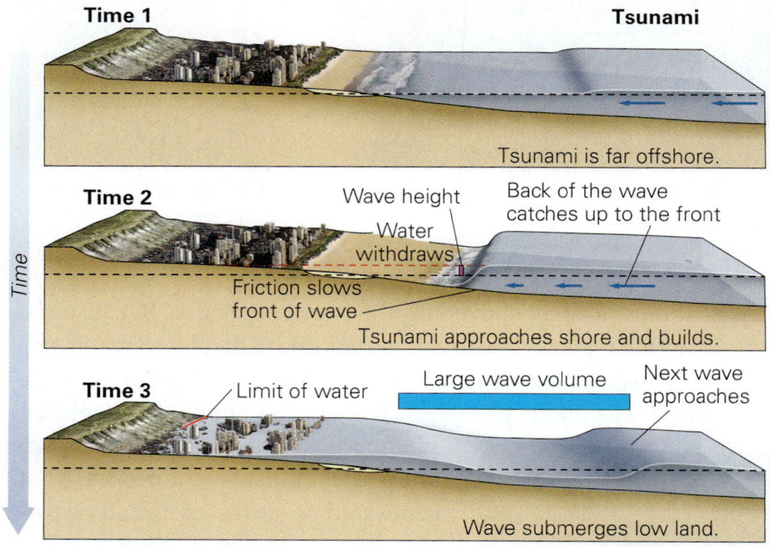

(c) A tsunami is not very high out in the open ocean, but as it approaches the land, friction slows the front of the wave, so the rear catches up and the wave grows. It is so wide that it can cover an extensive area of low-lying land.

much of the town and surrounding fields vanished. The waves arrived on a beautiful, cloudless day. First, the sea receded much farther than anyone had ever seen, exposing large areas of reefs that normally remained submerged even at low tide, and people walked out onto the exposed reefs in wonder. But then, with a rumble that grew to a roar, a wall of frothing water began to build in the distance and approach land (**Fig. 8.27b**). Puzzled bathers first watched, then turned and ran inland in panic when the threat became clear. As the tsunami approached shore, friction with the seafloor had slowed it to less than 30 km an hour, but it still moved faster than people could run (**Fig. 8.27c**). In places, the wave front reached heights of 15 to 30 m (45 to 100 feet) as it slammed into Banda Aceh. The impact of the water ripped boats from their moorings, snapped trees, battered buildings into rubble, and tossed cars and trucks like toys. And the water just kept coming, eventually flooding land up to 7 km inland (**Fig. 8.27d**). In the process, it drenched forests and fields with salt water (deadly to plants) and buried fields and streets with up to a meter of sand and mud. When the water level finally returned to normal, a jumble of flotsam, as well as the bodies of unfortunate victims, were dragged out to sea and drifted away.

Geologists refer to the tsunami that struck Banda Aceh as a *near-field tsunami*, because of its proximity to the earthquake. But sadly, the horror of Banda Aceh was just a preamble to the devastation that would soon visit other stretches of Indian Ocean coast. *Far-field tsunamis* crossed the ocean and struck Sri Lanka 2.5 hours after the earthquake, the coast of India half an hour after that, and the coast of Africa, on the west side of the Indian Ocean, 5.5 hours after the earthquake. In the end, more than 230,000 people died that day.

A tsunami that struck Japan soon after the 2011 Tōhoku earthquake was vividly captured in high-definition video that was seen throughout the world, generating a new level of international awareness. The 10-m-high seawalls that fringe the coast proved to be only a temporary impediment to the advance of the wave, which, in places, was 10 to 30 m high when it reached shore. Racing inland, the wave erased whole

FIGURE 8.27 The great Indian Ocean tsunami of 2004.

(a) A devastating tsunami was triggered by an earthquake off Sumatra. Three hours later, the leading wave struck the coasts of Sri Lanka and India. A computer model shows the wave.

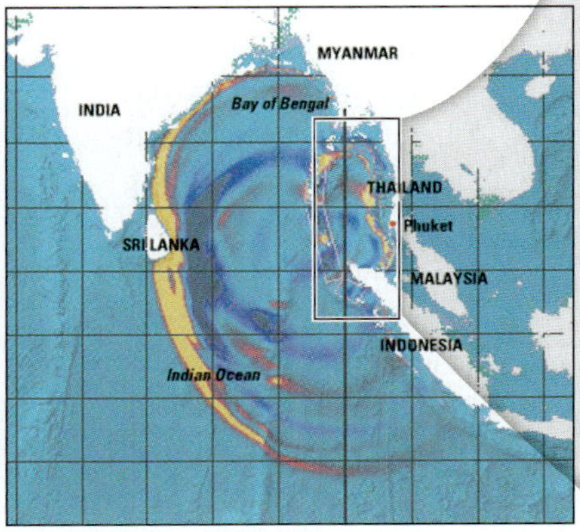

Colors represent wave height—yellow is highest. There were several waves.

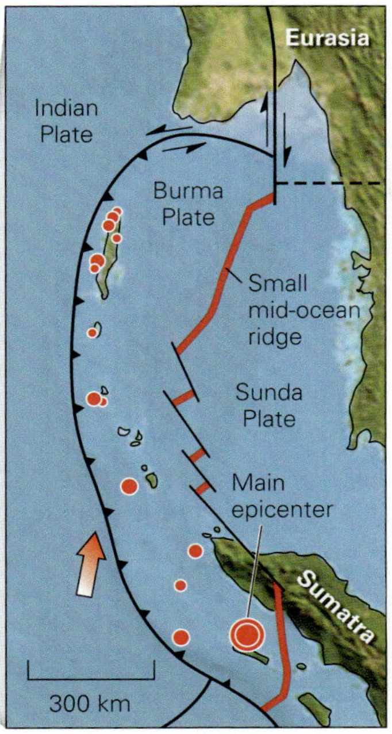

The earthquake was caused by subduction at a trench. Red dots are epicenters; smaller ones are after shocks.

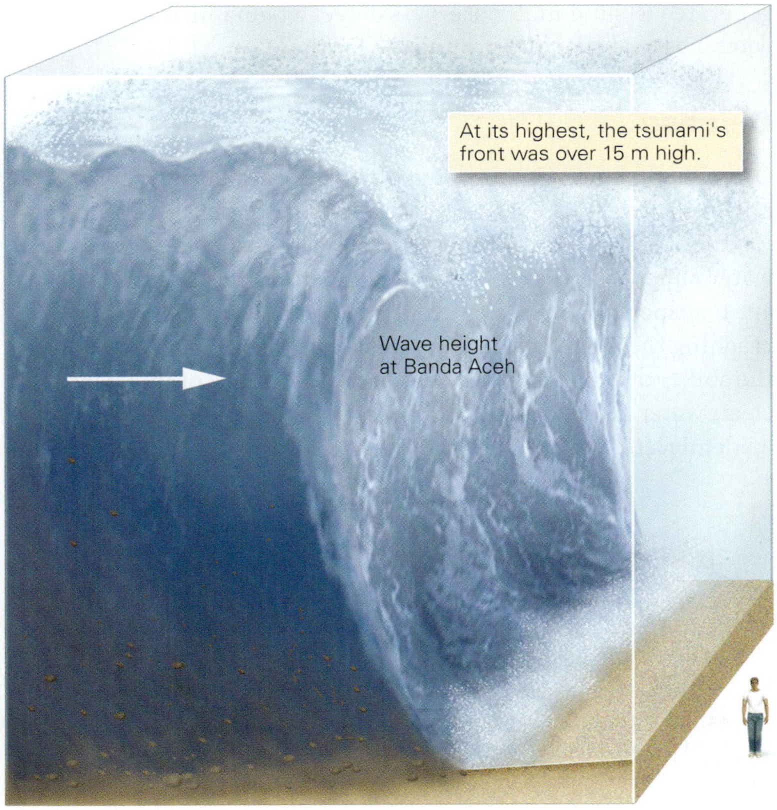

At its highest, the tsunami's front was over 15 m high.

(b) This snapshot shows the wave rushing toward the coast of Sumatra. Recession of water in advance of the wave exposed a reef.

(c) The wave blasts through a grove of palm trees as it strikes the coast of Thailand.

(d) Satellite photos of the Indonesian province of Aceh on Sumatra before and after the tsunami struck. Note that the city of Banda Aceh was washed away and the beach vanished.

FIGURE 8.28 Damage due to the 2011 Tōhoku tsunami.

(a) The complete destruction of a Japanese coastal town by a tsunami, which followed the Tōhoku, Japan, earthquake of March 2011.

(b) Each cubic building houses a reactor of the Fukushima nuclear power plant. The tsunami washed over the seawalls and inundated the plant to a depth of 14 m (46 ft), destroying power to the cooling pumps. Hydrogen explosions destroyed the reactor buildings.

towns, submerging airports and fields (see Fig. 8.2b-c). As the wave picked up dirt and debris, it evolved into a viscous slurry, moving with such force that nothing could withstand its impact. The devastation of coastal towns was so complete that they looked as though they had been struck by nuclear bombs (**Fig. 8.28a**).

But the catastrophe was not over. As we noted earlier, the wave had also struck a nuclear power plant. Though the plant had withstood ground shaking and had automatically shut down, its radioactive core still needed to be cooled by water in order to remain safe. The tsunami not only destroyed power lines, cutting the plant off from the electrical grid, but it also drowned the backup diesel generators, so the cooling pumps stopped functioning. Eventually, water surrounding the heat-producing radioactive core of the reactors, as well as the water cooling spent fuel, boiled away. Some of the superheated water separated into hydrogen and oxygen gas, which then exploded, thereby breaching the integrity of the reactor structure and releasing radioactivity into the environment (**Fig. 8.28b**).

Because tsunamis are so dangerous, predicting their arrival can save thousands of lives. A tsunami warning center in Hawaii keeps track of earthquakes around the Pacific and uses data relayed from tide gauges and seafloor pressure gauges to determine whether a particular earthquake has generated a tsunami. If observers detect a tsunami, they flash warnings to authorities around the Pacific.

Disease

Once the ground shaking and fires have stopped, disease may still threaten lives in an earthquake-damaged region. Earthquakes cut water and sewer lines, destroying clean-water supplies and exposing the public to bacteria, and they cut transportation lines, preventing food and medicine from reaching the area. The severity of such problems depends on the ability of emergency services to cope. The lack of sufficient clean water after the 2010 Haiti earthquake led to a cholera epidemic later that year (**Box 8.1**).

TAKE-HOME MESSAGE

Earthquakes cause devastation in many ways. Ground shaking, landslides, sediment liquefaction, and tsunamis can topple buildings and disrupt the land. Fire and disease may follow.

QUICK QUESTION Is ground shaking the major cause of loss of life in all earthquakes?

BOX 8.1 Consider This...

The 2010 Haiti Catastrophe

On the sunny afternoon of January 12, 2010, at 4:53 p.m., a 70-km-long segment of a large strike-slip fault suddenly slipped by up to 4 m. The motion began at 25 km west-southwest of Port-au-Prince, and 13 km beneath the ground surface, so the shock waves of the resulting M_W 7 earthquake reached the capital city in a matter of seconds, causing the ground to lurch violently over the next 35 seconds.

The ground shaking caused the support columns of poorly constructed buildings to crack and crumble, bringing floors down into gruesome pancake-like stacks (**Fig. Bx8.1a**). Similarly, brick or block walls broke apart and collapsed, roads buckled, and hillslopes slumped (**Fig. Bx8.1b**). Beneath the harbor, sediment liquefied, causing wharfs to sink into the sea. When the shaking, which reached an intensity of IX on the Mercalli scale (**Fig. Bx8.1c**), finally stopped, most of Port-au-Prince had collapsed. As a dense cloud of white dust slowly rose over the rubble, survivors began the frantic scramble to dig out victims, a task made more hazardous by aftershocks, of which there were over 50 between M_W 4.5 and 6.1. The renewed shaking caused still-standing but weakened structures to collapse on rescuers. Some estimates place the death toll at 230,000, about 2.5% of the country's population.

Why did the earthquake occur? Haiti sits astride the transform boundary along which the North American Plate moves westward at about 2 cm per year, relative to the Caribbean Plate. Therefore, earthquakes in Haiti are inevitable. But the last major earthquakes on this plate boundary happened about 240 years ago, so stress has been building for quite some time.

The impact of an earthquake on society depends not only on earthquake size but also on the nature of the substrate, on the steepness of the slopes, on construction practices, and on the quality of emergency services in the affected area. Much of Port-au-Prince was built on a basin of weak sediment, which amplified ground movements, and sadly, the buildings in Haiti were not designed to withstand ground vibration. In addition, many of the city's neighborhoods perch on steep slopes, which slid downhill during the quake. In the days that followed, local emergency services were overwhelmed, and access to the victims was nearly impossible. An air caravan of aid arrived to help out, but even so, in the months that followed, disease spread. On the fifth anniversary of the event, in 2015, recovery from the earthquake remained incomplete.

FIGURE Bx8.1 The disastrous January 2010 earthquake in Haiti, and its geologic setting.

(a) Survivors salvage what they can in Port-au-Prince, the capital of Haiti, after the devastating earthquake of January 2010.

(b) Ground shaking during the 2010 Haiti earthquake caused most of the houses in this residential neighborhood to collapse.

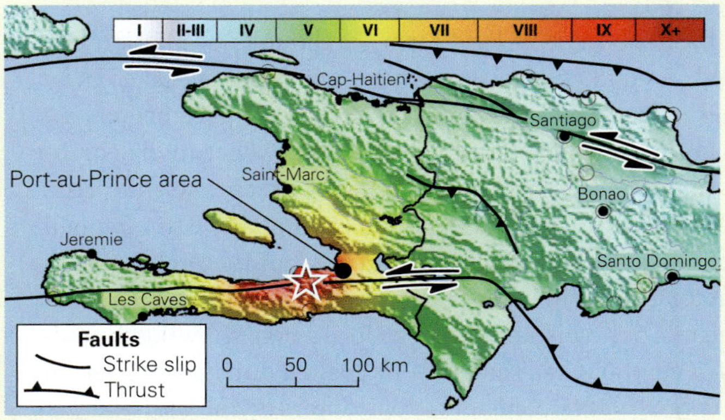

(c) A map showing the Mercalli intensity of shaking in Haiti. The star marks the epicenter of the quake.

8.7 Can We Predict the "Big One"?

Can seismologists predict earthquakes? The answer depends on the time frame of the prediction. With our present understanding of the distribution of seismic belts and the frequency at which earthquakes occur, we can make *long-term predictions* (on the time scale of decades to centuries). For example, with some certainty, we can say that a major earthquake will rattle Istanbul during the next 100 years, and that a major earthquake probably won't strike north-central Canada during the next 10 years. But despite extensive research, seismologists cannot make accurate *short-term predictions* (on the time scale of hours to weeks or even decades). Thus, we cannot say, for example, that an earthquake will happen in Montreal 40 days from now at 3 p.m. In this section, we look at the scientific basis of both long- and short-term predictions and consider the consequences of a prediction. Seismologists refer to studies leading to predictions as *seismic-risk assessment*.

Long-Term Predictions

A long-term prediction estimates the *probability*, or likelihood, that an earthquake will happen during a specified time range. For example, a seismologist may say, "The probability of a major earthquake occurring in the next 50 years in this state is 20%." This sentence implies that there's a 1-in-5 chance that the earthquake will happen before 50 years have passed. Urban planners can use long-term predictions to help create building codes for a region—codes requiring stronger buildings make sense for regions with greater seismic risk. Planners may also use predictions to determine whether it is reasonably safe to build vulnerable structures such as nuclear power plants, hospitals, or dams in a given region. Seismologists base long-term earthquake predictions on two pieces of information: the identification of seismic belts and the recurrence interval.

The basic premise of long-term earthquake prediction can be stated as follows: a region where many earthquakes have occurred in the past will be more likely to experience earthquakes in the future. Seismic belts, regions where many earthquakes happen, are therefore regions of greater seismic risk. This doesn't mean that a disastrous earthquake can't happen far from a seismic belt—they can and do—but the probability is less that an event will happen in a given time window. To identify a seismic belt, seismologists produce a map showing the epicenters of earthquakes that have happened during a set period of time (say, 30 years). Clusters of epicenters define the belt.

The **recurrence interval** is the average time between successive events. Since earthquakes do not happen at predictably spaced times, a recurrence interval is not the exact time between successive events. Because there can be confusion about the meaning of recurrence interval, seismologists sometimes specify the *annual probability* of an earthquake, instead, where annual probability = 1/recurrence interval. Thus, if the recurrence interval is 100 years, the annual probability, meaning the likelihood that an event will happen in a given year, is 1/100, or 1%. Note that, since stress builds up over time on a fault, the elastic-rebound theory hints that the annual probability may increase as time passes.

To determine the recurrence interval for large earthquakes within a given seismic belt, seismologists must determine when large earthquakes happened within the belt in the past. For places where the historical record does not provide information far enough back in time to reveal multiple large events, researchers look for evidence of great earthquakes preserved in the geologic record. For example, recognition of an unweathered fault scarp may indicate that faulting affected an area relatively recently, even if it was before local human habitation. In places where sedimentary strata have accumulated in a basin over a fault, researchers may dig a trench and look for buried layers of sand volcanoes and disrupted bedding in the stratigraphic record. Each layer, whose age can be determined by using radiocarbon dating of plant fragments, records the time of an earthquake (**Fig. 8.29**). By calculating the average of the number of years between successive events, seismologists obtain the recurrence interval. Information on

FIGURE 8.29 Identifying recent fault movement and recurrence interval. The block shows the walls of two trenches cut into the ground at a fault zone.

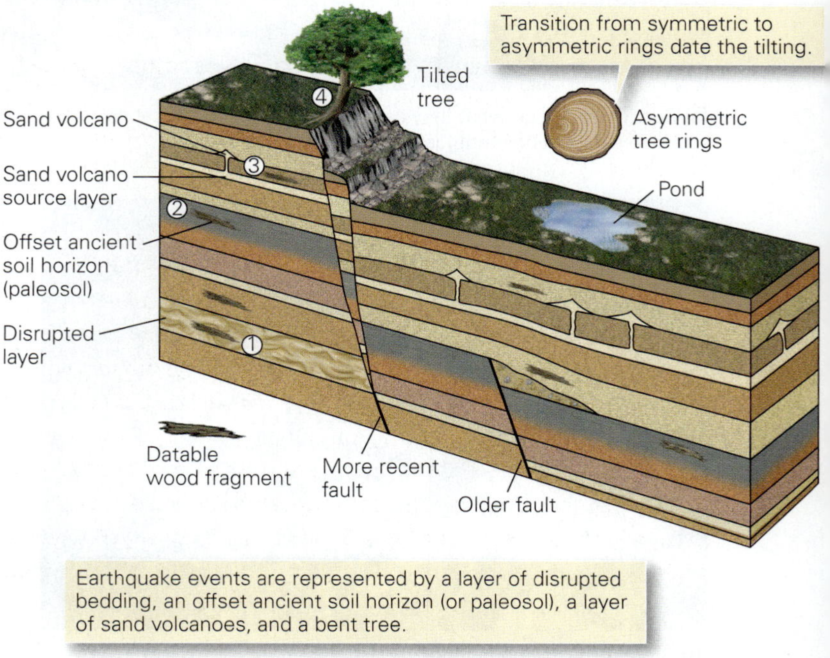

FIGURE 8.30 Examples of seismic-hazard maps.

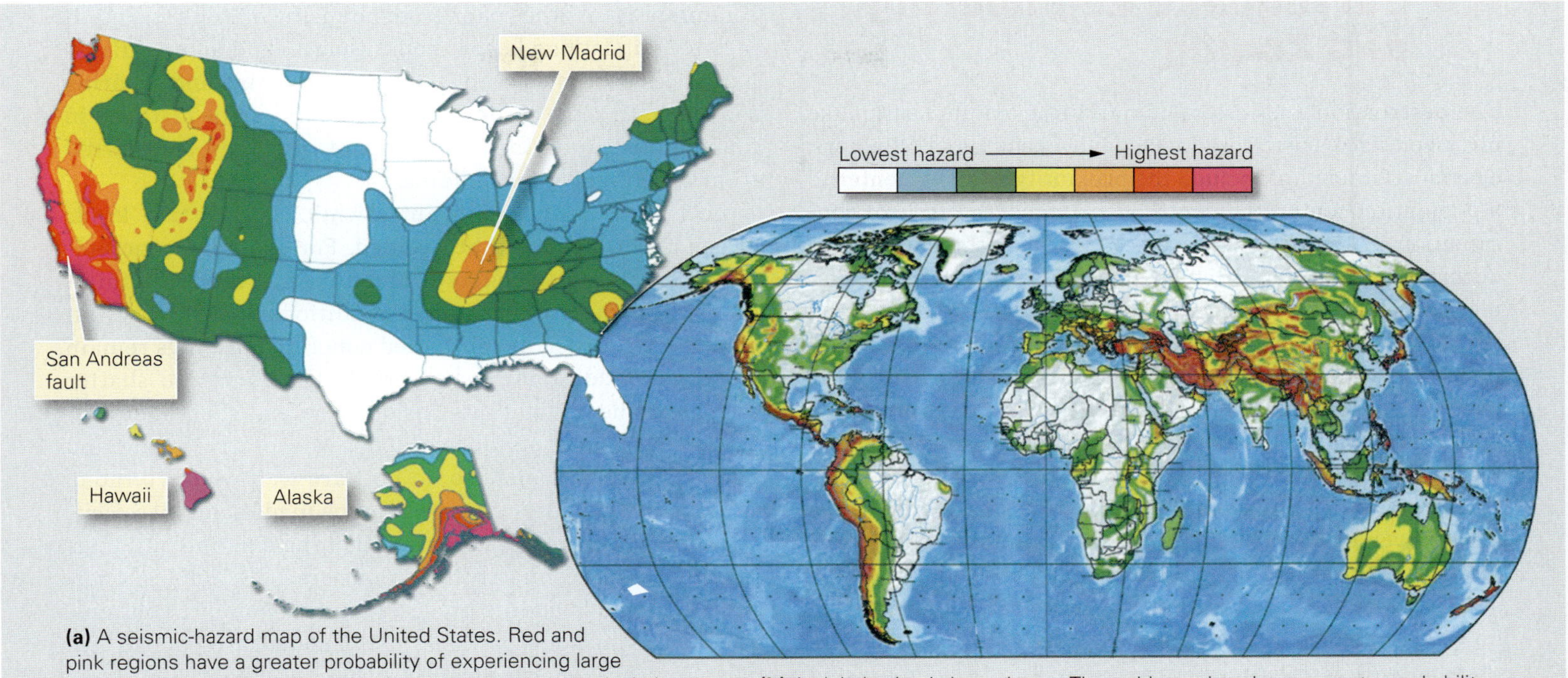

(a) A seismic-hazard map of the United States. Red and pink regions have a greater probability of experiencing large earthquakes. The San Andreas fault plate boundary is particularly hazardous. Risk in the New Madrid area remains subject to debate.

(b) A global seismic-hazard map. The redder regions have a greater probability of experiencing a large earthquake. Probability is high along plate boundaries.

a recurrence interval allows seismologists to refine regional maps illustrating seismic risk (**Fig. 8.30**).

Short-Term Predictions

Short-term predictions, specifying that an earthquake will happen on a given date or within a time window of days to years, are not and may never be reliable. Seismologists have considered, and discounted as unreliable, many supposed bases for short-term prediction. For example, a *swarm* of fore-shocks may indicate that rock is beginning to crack in advance of a main shock. Unfortunately, such swarms can be identified only in hindsight. Precise surveys show that the surface of the ground may warp slightly prior to an earthquake, but no one knows how much warping will take place before an earthquake will happen. Prediction studies focused on measuring changes in water levels in wells, radon gas in spring water, electrical signals emitted by minerals, or agitation of animals have met with similar skepticism.

The concept of a short-term prediction should not be confused with the concept of an **earthquake early warning system**, which is based on a real signal and can potentially save lives. An early warning system works as follows. When an earthquake happens, the seismic waves it produces start traveling through the Earth. Seismometers close to the epicenter may detect an earthquake before the seismic waves have had time to reach populated areas farther from the epicenter. The instant that seismometers detect the earthquake, a computer sends a signal to a control center, which automatically sends out emergency signals to nearby populated areas that might be affected. Potentially, this signal could provide enough warning (several seconds to a minute) for computers to shut down gas pipelines, trains, nuclear reactors, and power lines automatically. The signal could also set off sirens and trigger broadcasts on radio, TV, and cell-phone networks, alerting people to take precautions.

TAKE-HOME MESSAGE

Researchers can determine regions where earthquakes are more likely, but not exactly when and where a given event will occur. Seismic risk is greater where seismicity has happened more frequently in the past and, therefore, where the recurrence interval of large earthquakes is shorter. Early warning systems can provide seconds of warning by sending out signals that travel faster than seismic waves.

QUICK QUESTION What is the relation between recurrence interval and the annual probability of an earthquake?

8.8 Earthquake Engineering and Zoning

The destruction and loss of human life from an earthquake of a given size vary widely and depend on a number of factors. The most important include the proximity of an epicenter to a population center, the depth of the focus, the style of construction in the epicentral region, whether the earthquake occurred in a region of steep slopes, whether faulting displaced the seafloor, whether building foundations are on solid bedrock or on weak substrate, whether the earthquake happened when people were outside or inside, and whether the government was able to provide emergency services promptly.

For example, a 1988 earthquake in Armenia was not much bigger than the 1971 San Fernando earthquake in southern California, but it caused almost 400 times as many deaths (24,000 versus 65). The difference in death toll reflected differences in the style and quality of construction and the characteristics of the substrate. The unreinforced concrete-slab buildings and masonry houses of Armenia collapsed, whereas the structures in California had, by and large, been erected according to building codes that require structures to withstand stresses caused by earthquakes. Most flexed and twisted but did not fall down and crush people. The terrible 1976 earthquake in T'ang-shan, China, killed over a quarter of a million people partly because the ground beneath the epicenter had been weakened by coal mining and collapsed, and because buildings were poorly constructed. During the 2008 M_W 7.9 Sichuan earthquake that rocked China, 70,000 people died and almost 5 million were left homeless because of building failures. Mexico City's 1985 earthquake proved disastrous because the city lies over a sedimentary basin whose composition and bowl-like shape focused seismic energy. During the 1989 Loma Prieta quake in California, portions of Route 880 in Oakland that were built on a weak landfill collapsed, whereas portions built on bedrock remained standing.

Communities can mitigate or diminish the consequences of earthquakes by taking sensible precautions. Earthquake engineering (designing buildings that can withstand shaking) and *earthquake zoning* (determining where land is stable and where it is not) can help save lives and property. In regions prone to large earthquakes, buildings and bridges should be constructed to be somewhat flexible so that ground motions can't crack them, but they should have sufficient bracing so movements don't become too severe (**Fig. 8.31a, b**). Also, supports should be strong enough to maintain loads far in excess of the loads caused by static (nonmoving) weight. Wrapping steel cables around bridge support columns makes them many times stronger. Bolting the bridge spans to the top of a column prevents the spans from bouncing off. Bolting buildings to foundations keeps them in place, and adding diagonal braces to frames keeps them from twisting and shearing too much. In some cases, shock absorbers can be installed in foundations to diminish the amount of energy transferred from the ground to the building.

In regions with significant seismic risk, certain kinds of construction should be avoided. For example, concrete-block, unreinforced concrete, and unreinforced brick buildings crack and tumble under conditions in which wood-frame, steel girder, or reinforced concrete buildings remain standing. Traditional heavy, brittle tile roofs can shatter and bury the inhabitants inside, whereas sheet-metal or asphalt-shingle roofs do not. Loose decorative stone and huge open-span roofs also do not fare well when vibrated. Of note, inadequate structures can be made safer by *seismic retrofitting*, the process of strengthening existing buildings in potentially seismically hazardous areas. Examples of retrofitting include adding braces, jacketing support columns, and coating masonry with resins. In some cases, substrates can be strengthened by draining water.

Also, in regions of significant seismic risk, developers should avoid construction on land underlain by weak mud and clay or wet sand that could liquefy. They should not build on top of, on, or at the base of steep escarpments because the escarpments could fail and produce landslides, and they should avoid locating communities downstream of dams, which could crack and collapse, causing a flood. Critical buildings (schools, hospitals, fire stations, communications centers, power plants) should not be built over active faults, because fault slip could crack and destroy the buildings. Cities in seismic belts should draw up emergency plans to deal with disaster—strategies need to be in place for providing supplies under circumstances where roads may be impassable. And in coastal areas, tsunami warning systems need to be implemented (**Fig. 8.31c**).

Finally, individuals in vulnerable regions should learn to protect themselves during an earthquake. In your home, keep a stock of emergency supplies, bolt bookshelves to walls, strap the water heater in place, install locking latches on cabinets, know how to shut off the gas and electricity, know how to find the exit, have a fire extinguisher handy, and know where to go to find family members. Schools and offices should have earthquake-preparedness drills. When an earthquake strikes, stay away from buildings. If you are trapped inside, crouch beneath a heavy table or solid door frame to be out of the way of falling objects (**Fig. 8.31d**). As long as lithosphere plates continue to move, earthquakes will continue to shake. But we can learn to prepare for them.

FIGURE 8.31 Preventing damage and injury during an earthquake.

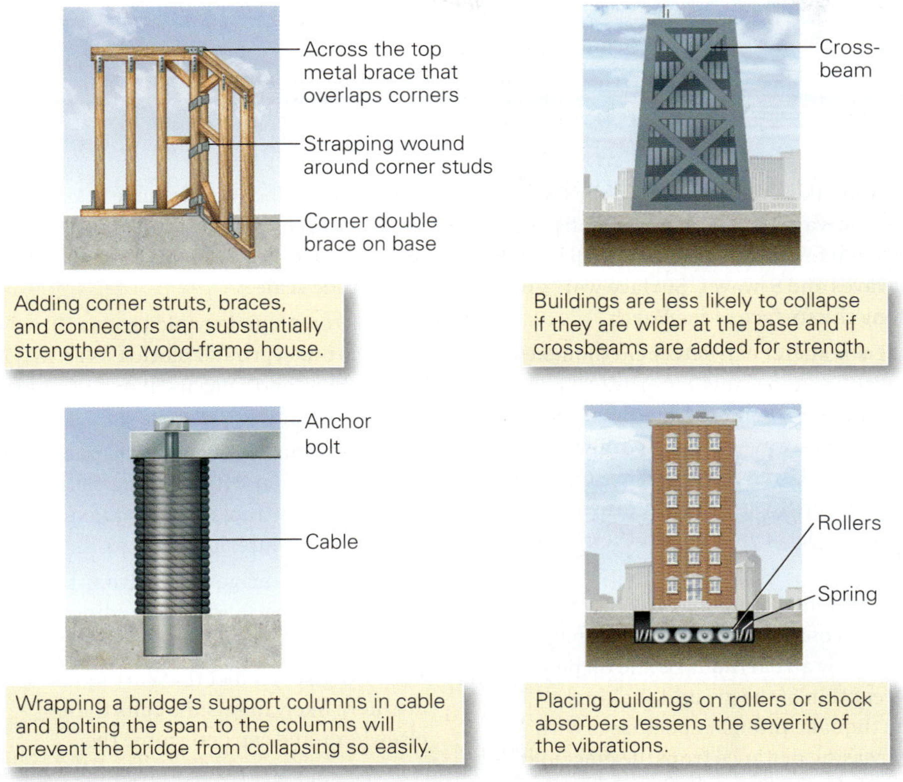
Across the top metal brace that overlaps corners
Strapping wound around corner studs
Corner double brace on base
Adding corner struts, braces, and connectors can substantially strengthen a wood-frame house.
Cross-beam
Buildings are less likely to collapse if they are wider at the base and if crossbeams are added for strength.
Anchor bolt
Cable
Wrapping a bridge's support columns in cable and bolting the span to the columns will prevent the bridge from collapsing so easily.
Rollers
Spring
Placing buildings on rollers or shock absorbers lessens the severity of the vibrations.

(a) Damage can be prevented if buildings are designed to withstand vibration.

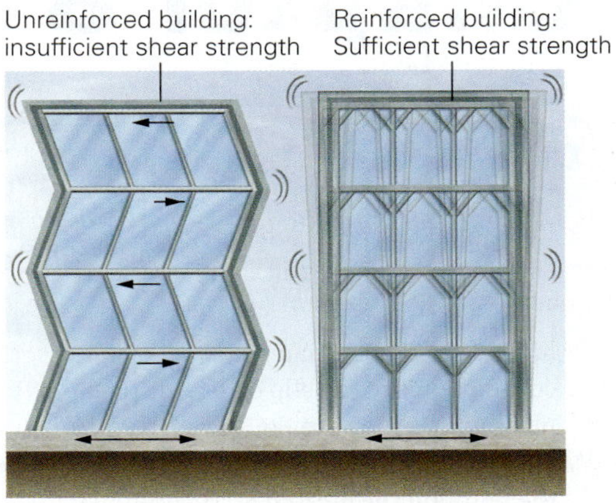
Unreinforced building: insufficient shear strength
Reinforced building: Sufficient shear strength

(b) An unreinforced building will shear side to side in a way that causes floors to shift out of alignment.

TSUNAMI

(c) Buoys can detect a tsunami in the open ocean, so people on land can be warned.

(d) If an earthquake strikes, take cover under a sturdy table near a wall.

TAKE-HOME MESSAGE

Earthquakes are a fact of life on this dynamic planet. People in regions facing high seismic risk should build on stable ground, avoid unstable slopes, and design construction that can survive shaking. Evacuation planning saves lives after an event.

QUICK QUESTION What factors influence the degree of devastation during an earthquake?

Chapter 8 Review

Chapter Summary

- Earthquakes are episodes of ground shaking. Earthquake activity is called seismicity.
- Most earthquakes happen as a consequence of the slip on a fault. The place where rock breaks and earthquake energy is released is called the focus, and the point on the ground directly above the focus is the epicenter.
- Active faults are ones on which movement is likely. Inactive faults ceased being active long ago. Displacement on active faults that intersect the ground surface may yield a fault scarp.
- During fault formation, rock elastically bends, then ruptures. When this happens, the rock vibrates, and this generates an earthquake.
- Most earthquakes happen when stress overcomes friction on a pre-existing fault, and the fault slips again, again causing rocks to vibrate. Faults exhibit stick-slip behavior, in that, once formed, stress builds up until they move in a sudden increment.
- Earthquake energy travels in the form of seismic waves. Body waves, which pass through the interior of the Earth, include P-waves and S-waves. Surface waves pass along the surface of the Earth.
- A seismometer can detect earthquake waves. Seismograms, records of earthquakes, demonstrate that different earthquake waves travel at different velocities. Using the difference between P-wave and S-wave arrival times, seismologists can pinpoint the epicenter location.
- The Modified Mercalli Intensity scale is based on documenting the damage caused by an earthquake and on people's perception of the ground shaking. Earthquake intensity decreases with increasing distance from the epicenter.
- Magnitude scales characterize the amount of energy released at the source, and are based on measuring the amount of ground motion, at a reference distance, as indicated on a seismogram. There is one number for an earthquake magnitude for a given earthquake. The Richter scale is an early version of a magnitude scale. These days, seismologists prefer to use the moment-magnitude scale.
- An M_W 8 earthquake yields about 10 times as much ground motion as an M_W 7 earthquake, and releases about 32 times as much energy.
- Most earthquakes occur in seismic belts, of which the majority lie along plate boundaries. Intraplate earthquakes happen in the interior of plates.
- Earthquake damage results from ground shaking, landslides, sediment liquefaction, fire, and tsunamis.
- Seismologists predict that earthquakes are more likely in seismic belts than elsewhere, and can determine the recurrence interval for great earthquakes. But it may never be possible to pinpoint the exact time and place at which an earthquake will happen.
- Earthquake hazards can be reduced with better construction practices and zoning, and by educating people about what to do during an earthquake.

Guide Terms

aftershock (p. 252)
body wave (p. 254)
compressional wave (p. 254)
displacement (p. 248)
earthquake (p. 246)
earthquake early warning system (p. 277)
elastic-rebound theory (p. 252)
epicenter (p. 254)
fault (p. 245)
fault scarp (p. 248)
focus (p. 253)
foreshock (p. 252)
friction (p. 249)
intensity (p. 259)
intraplate earthquake (p. 265)
landslide (p. 268)
magnitude (p. 259)
main shock (p. 252)
Modified Mercalli Intensity scale (MMI) (p. 259)
moment magnitude (p. 261)
recurrence interval (p. 276)
Richter scale (p. 260)
sediment liquefaction (p. 269)
seismic belt (p. 262)
seismicity (p. 248)
seismic wave (p. 248)
seismogram (p. 256)
seismologist (p. 248)
seismometer (p. 255)
shear wave (p. 254)
stick-slip behavior (p. 252)
stress (p. 249)
surface wave (p. 254)
travel-time curve (p. 257)
tsunami (p. 271)
Wadati-Benioff zone (p. 263)

GEOTOURS *THIS CHAPTER'S GEOTOUR EXERCISE (H) FEATURES:*

- Evidence of Earthquake Activity along a Plate Boundary
- Evidence of Intraplate Earthquake Activity
- Stress Transfer and Earthquake Prediction
- Tsunami Devastation

Review Questions

1. How do normal, reverse, and strike-slip faults differ from each other?
2. Do all earthquakes require that a new fault initiate? Describe elastic-rebound theory, and the concept of stick-slip behavior.
3. What is the difference between a body wave and a surface wave? How do P-waves and S-waves differ from each other? How do Rayleigh waves and Love waves differ from each other?
4. Explain how the vertical and horizontal components of an earthquake are detected by a seismometer.
5. Explain the differences among the scales used to describe the size of an earthquake. Can more than one intensity number be assigned to a given earthquake? Why? Can there be more than one magnitude?
6. How does seismicity on mid-ocean ridges compare with seismicity at convergent or transform boundaries? Do all earthquakes occur at plate boundaries?
7. What is a Wadati-Benioff zone? Why can intermediate-focus and deep-focus earthquakes occur?
8. Describe the types of damage caused by earthquakes. Is all damage due to ground shaking?
9. What is a tsunami, and why do tsunamis form? How do they differ from storm waves?
10. Explain how liquefaction occurs in an earthquake and how it can cause damage.
11. How are long-term and short-term earthquake predictions made? What is the basis for determining a recurrence interval?
12. What types of structures are most prone to collapse in an earthquake? What types are most resistant to collapse? What causes most loss of life during an earthquake?

On Further Thought

13. Is seismic risk greater in a town on the west coast of South America or in one on the east coast? Explain your answer.
14. The northeast-trending Ramapo fault crops out north of New York City. (You can see the fault on *Google Earth*™ by going to latitude 41°10′21.12″ N, longitude 74°5′12.36″ W. The fault trace follows a distinct escarpment.) Where the fault crosses the Hudson River, there is an abrupt bend in the river. A nuclear power plant was built near this bend on sediments deposited by the river. Imagine that you are a geologist with the task of determining the seismic risk of the fault. What evidence of prehistoric seismic activity could you look for?
15. On the seismogram of an earthquake recorded at a seismic station in Paris, France, the S-wave arrives 6 minutes after the P-wave. On the seismogram obtained by a station in Mumbai, India, for the same earthquake, the difference between the P-wave and S-wave arrival times is 4 minutes. Which station is closer to the epicenter? From the information provided, can you pinpoint the location of the epicenter? Explain.
16. Will the duration of shaking recorded at a seismic station be longer or shorter as the distance between the epicenter and the station increases?

Online Resources

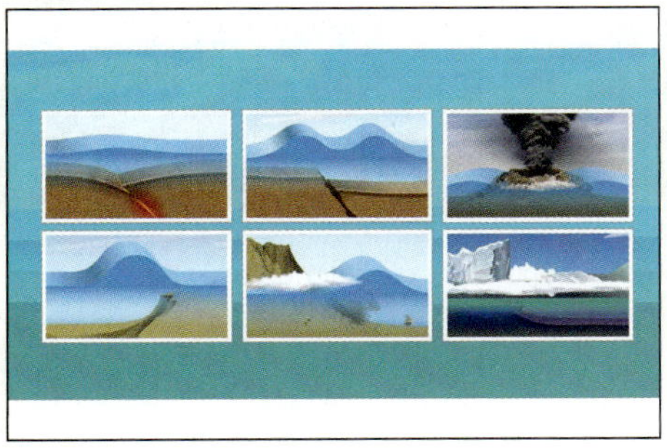

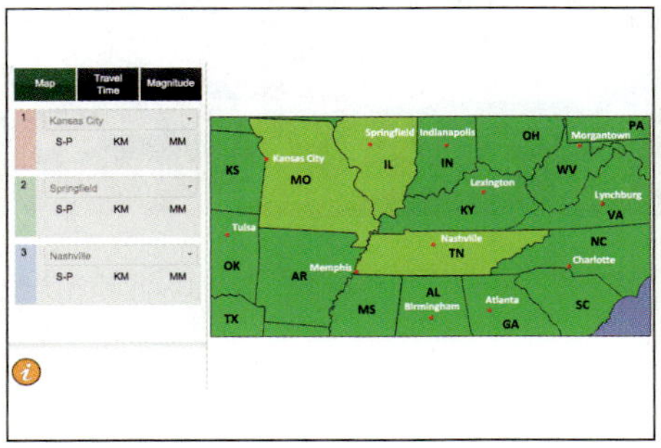

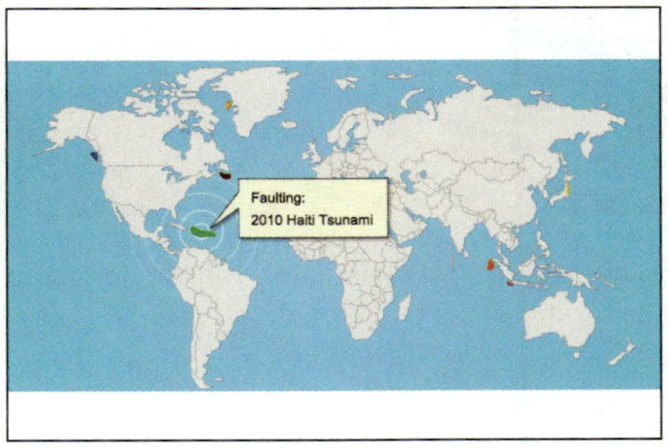

Animations

This chapter features animations that simulate and show the effects of tsunamis and an interactive activity on locating an earthquake's epicenter.

Assessment

This chapter features analysis questions on earthquake magnitude, faults, and tsunamis.

INTERLUDE D

The Earth's Interior Revisited: Insights from Geophysics

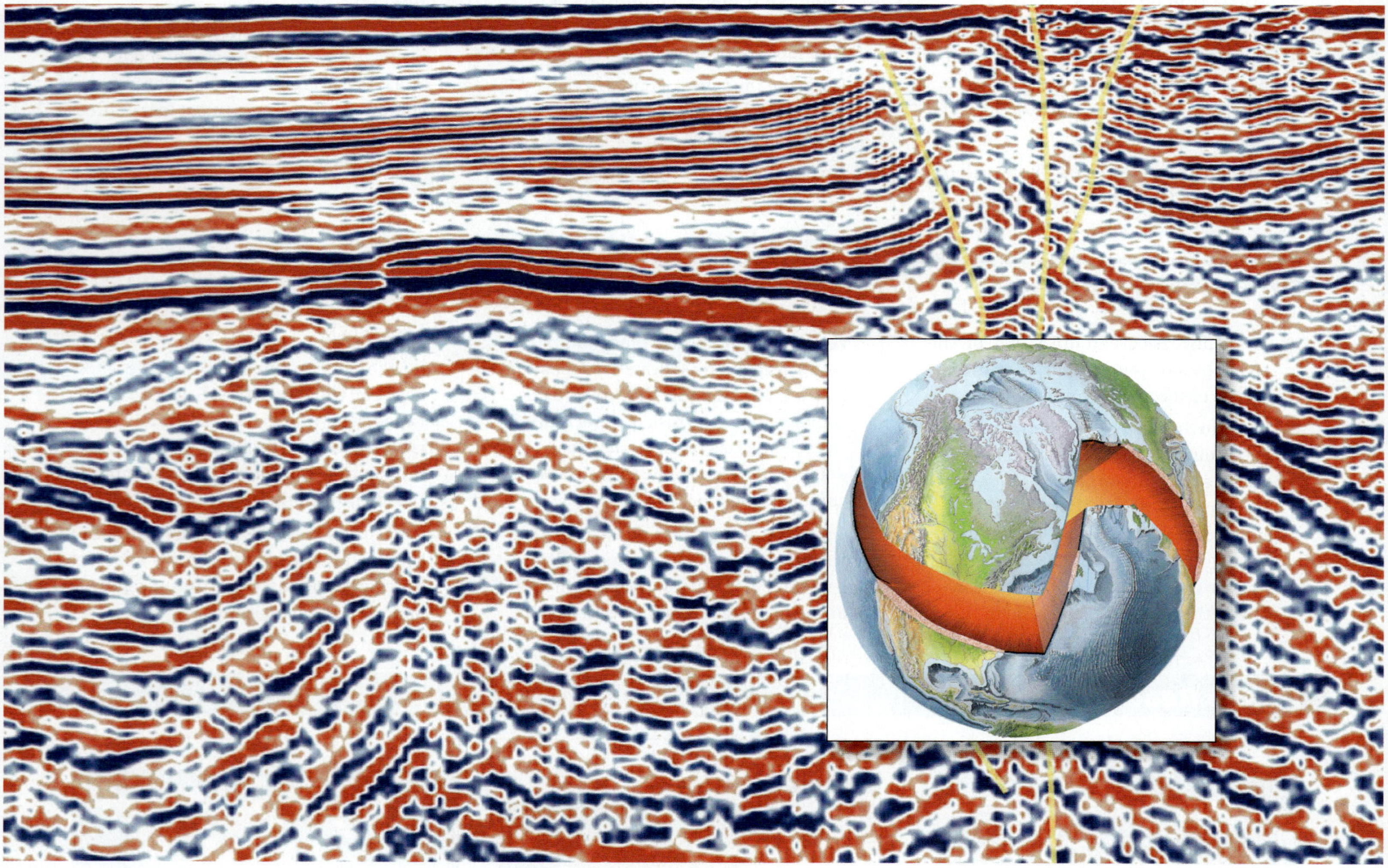

▲ Using geophysical techniques, researchers obtain images of the Earth's interior. Here we see a seismic-reflection profile of sedimentary strata in the upper crust.

LEARNING OBJECTIVES

By the end of this interlude, you should understand...

1. how seismic waves behave as they pass through the Earth's interior.
2. what the study of seismic waves can tell us about layering inside the Earth.
3. that the Earth's gravitational attraction varies with location and what these variations mean.
4. why the Earth's magnetic field exists and why its strength varies with location.

D.1 Introduction

In this interlude, we provide a brief survey of the insight that the study of geophysics can provide to help refine our image of the Earth's interior. **Geophysics** is a subdiscipline of geoscience that focuses on the study of seismic waves, magnetism, gravity, and other physical characteristics of the Earth. Geophysicists use mathematical calculations, measurements with instruments, and computer simulations to carry out their work. (Seismology, the subject of Chapter 8, is one aspect of geophysics.) At the level of this book, we can't delve into the mathematical foundation of geophysics, but we can introduce some of the key concepts that geophysical studies provide.

We begin this interlude by examining how seismic waves interact with layer boundaries, in a general sense, and then discuss how the study of seismic waves can define the specific depths of layer boundaries within the Earth. We next turn our attention to the Earth's gravity field and examine why gravitational pull varies with location. We conclude by revisiting the Earth's magnetic field, this time focusing on the question of why the magnetic field exists. Interlude D ties together several subjects covered in Chapters 1, 2, and 8, so we recommend that you read (or reread) relevant parts of those chapters before proceeding.

D.2 The Basis for Seismic Study of the Interior

Setting the Stage

As we discussed in Chapter 1, the first clues to what's inside our planet came from measurements of the Earth's overall mass and shape. These measurements led researchers to conclude, by the end of the 19th century, that the Earth consists of three concentric layers that differ from each other in terms of relative density (**Fig. D.1**). From the surface down, these are (1) the crust (relatively low density), (2) the mantle (intermediate density), and (3) the core (relatively high density). To go beyond this basic understanding, and to define the specific depths at which layer boundaries occur, researchers searched for a tool that could provide an actual image of the interior. Study of seismic waves provides that tool. By measuring how fast seismic waves travel through the Earth, and how the waves bend or reflect as they travel, it's possible to define the thicknesses of the main layers, and even to recognize sublayers.

Seismic Wave Fronts and Travel Times

The energy released by an earthquake moves through rock in the form of waves, similar to the waves that propagate outward from the impact point of a pebble on the surface of a pond. The boundary between the rock through which a wave has passed and the rock through which it has not yet passed is called a **wave front.** In three dimensions, a wave front initially expands outward from the earthquake focus like a growing bubble. We can represent a succession of wave fronts close to the focus by thinking about a series of concentric spheres or, in cross section, circles (**Fig. D.2a**). The changing position of an imaginary point on a wave front, as the front moves through rock, is called a **seismic ray**. To represent a seismic ray, we draw a line perpendicular to a wave front—each point on a curving wave front follows a slightly different ray. The time it takes for a wave to travel from the earthquake focus to a seismometer along a given ray is the **travel time** along that ray.

The ability of a seismic wave to travel through a material, as well as the velocity at which it travels, depend on several

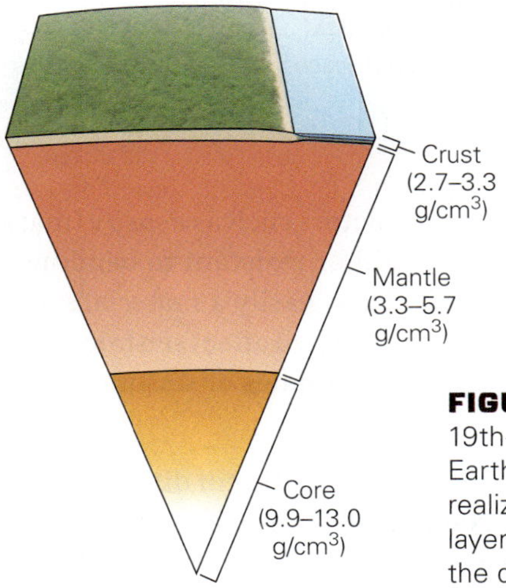

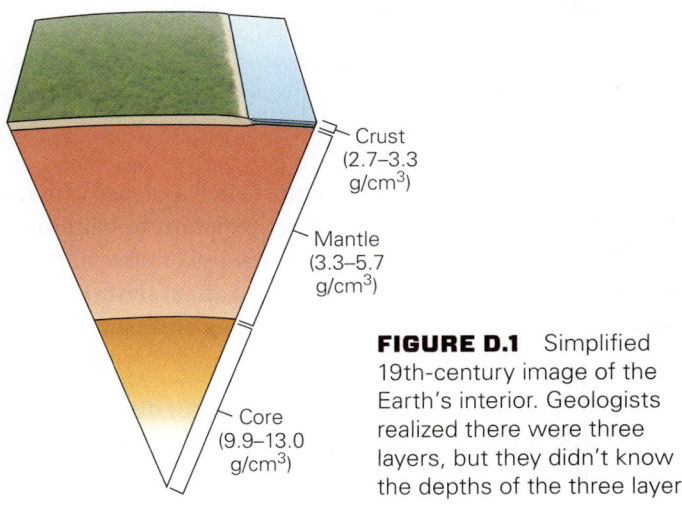

FIGURE D.1 Simplified 19th-century image of the Earth's interior. Geologists realized there were three layers, but they didn't know the depths of the three layers.

FIGURE D.2 The propagation of earthquake waves.

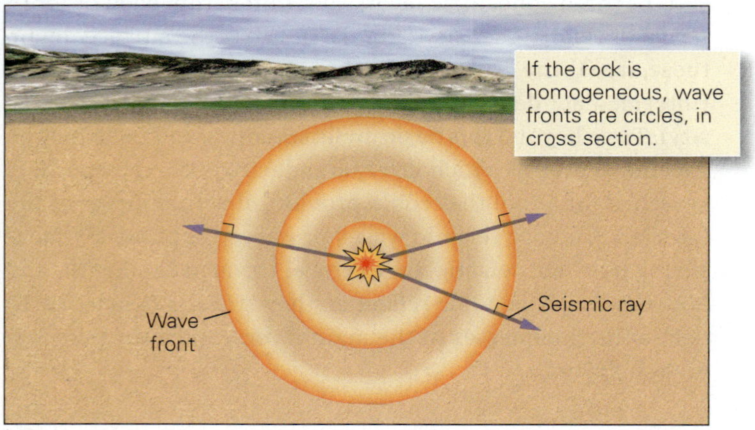

(a) An earthquake sends out waves in all directions. Seismic rays are perpendicular to wave fronts.

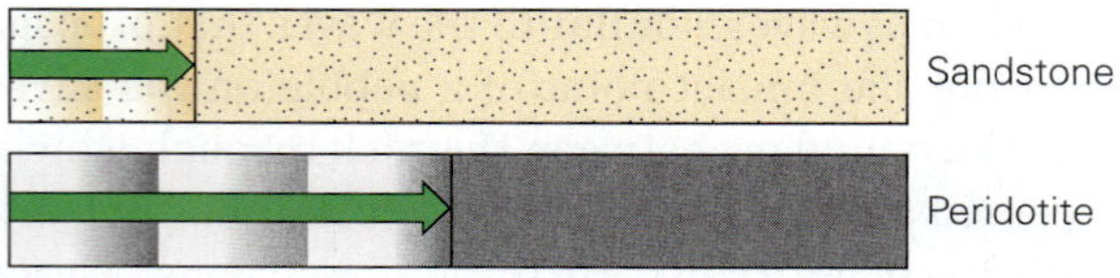

(b) Seismic waves travel at different velocities in different rock types. After a given time, the wave will have traveled farther in peridotite than in sandstone.

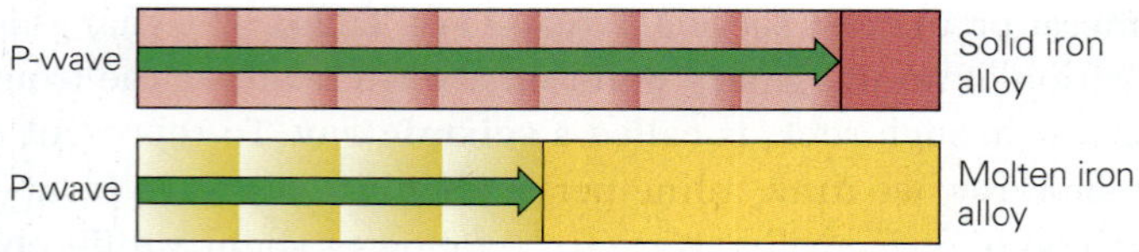

(c) P-waves travel faster in solid iron alloy than in liquid, such as molten iron alloy.

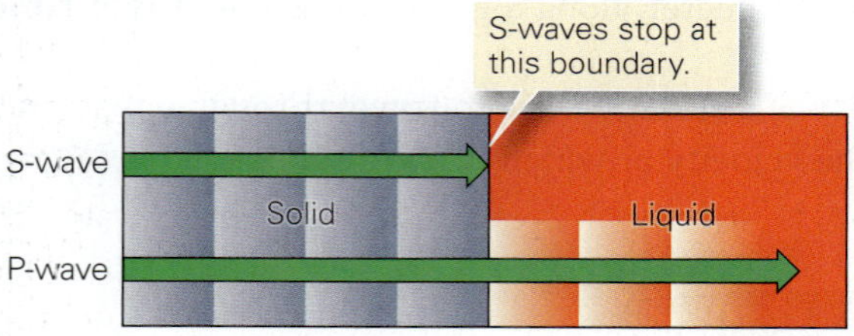

(d) Both P-waves and S-waves can travel through a solid, but only P-waves can travel through a liquid.

characteristics of that material. Factors such as *density* (mass per unit volume), *rigidity* (how stiff or resistant to bending a material is), and *compressibility* (how easily a material's volume changes in response to squashing) all affect seismic-wave velocity and therefore travel times. Studies of seismic waves reveal the following:

- Seismic waves travel at different velocities in different rock types (**Fig. D.2b**). For example, P-waves travel at 8 km per second in peridotite (an ultramafic igneous rock), but at only 3.5 km per second in sandstone (a porous sedimentary rock). Therefore, waves accelerate or slow down if they pass from one rock type into another.
- Seismic waves travel more slowly in a liquid than in a solid of the same composition. Thus, seismic waves travel more slowly in magma than in solid rock, and more slowly in molten iron alloy than in solid iron alloy (**Fig. D.2c**).
- Both P-waves and S-waves can travel through a solid, but only P-waves can travel through a liquid (**Fig. D.2d**). That's because one part of a liquid can easily move sideways relative to another part, so the energy of a shear motion cannot be transmitted from one part of a liquid to another.

Reflection and Refraction of Wave Energy

Shine a light ray into a container of water so that the ray hits the boundary between water and air at an angle. Some of the light bounces off the water surface and heads back up into the air, while some enters the water (**Fig. D.3a**). The light ray that enters the water bends at the air-water boundary, so

FIGURE D.3 Refraction and reflection of waves.

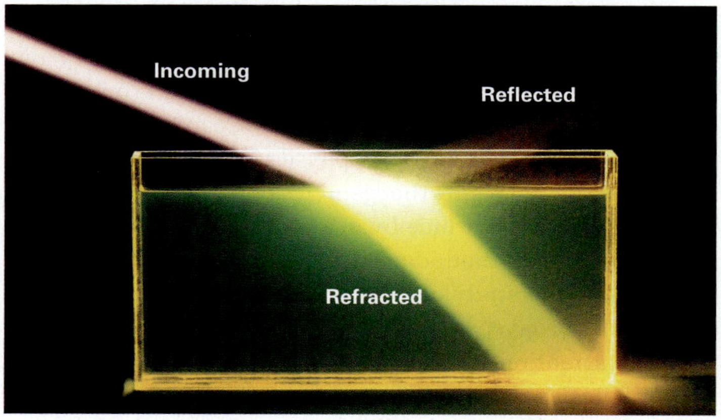

(a) A lab experiment showing how a ray of light reflects and partly refracts when it crosses the boundary between two different materials.

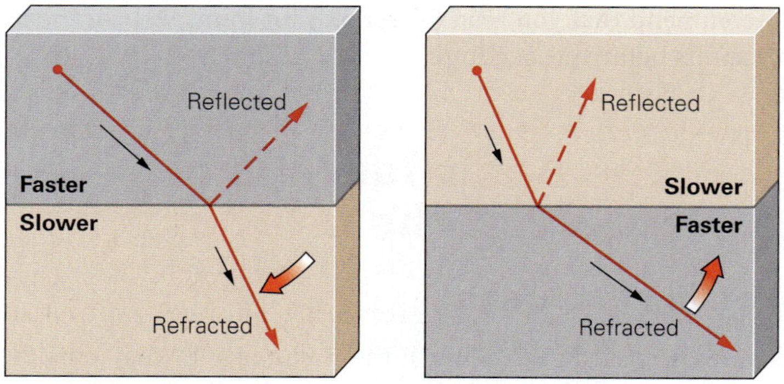

(b) A ray that enters a material through which it travels more slowly bends away from the boundary. A ray that enters a faster medium bends toward the boundary.

that the angle between the ray and the boundary in the air is different from the angle between the ray and the boundary in the water. Physicists refer to the light ray that bounces off the air-water boundary and heads back into the air as the *reflected ray*, and the ray that bends at the boundary as the *refracted ray*. The phenomenon of bouncing is **reflection**, and the phenomenon of bending is **refraction**. Wave reflection and refraction take place at the interface between two materials, if the wave travels at different velocities in the two materials.

The angle at which a reflected ray bounces off a boundary is always the same as the angle at which the incoming, or incident, ray strikes the surface. The angle by which a refracted ray bends at a boundary, however, depends on the contrast in wave velocity between the two materials in contact at the boundary, and on the angle at which a ray hits the interface. As a rule, if waves enter a material through which they will travel more slowly, the ray representing the waves bends down and away from the interface (**Fig. D.3b**). (To see why, picture a car driving from a paved surface diagonally onto a sandy beach—the wheel that rolls onto the sand first slows down relative to the wheel still on the pavement, causing the car to veer toward the sand.) For example, the light ray in Figure D.3b bends down when hitting the air-water boundary because light waves travel more slowly in water. Alternatively, if waves pass from a layer in which they travel slowly into one in which they travel more rapidly, a ray representing the waves would bend up and toward the interface (Fig. D.3b).

D.3 Seismic Study of Earth's Interior

Let's now utilize your knowledge of seismic velocity, refraction, and reflection to see how each of the major layer boundaries inside the Earth was discovered.

Discovering the Crust-Mantle Boundary

In 1909, Andrija Mohorovičić, a Croatian seismologist, noted that P-waves arriving at seismometer stations less than 200 km from the epicenter traveled at an average speed of 6 km per second, whereas P-waves arriving at seismometers more than 200 km from the epicenter traveled at an average speed of 8 km per second. To explain this observation, he suggested that P-waves reaching nearby seismometers followed a shallow path that kept them entirely within the crust, in which they traveled relatively slowly, whereas P-waves reaching distant seismometers refracted at the crust-mantle boundary, so that for part of their route they passed through the mantle, in which they traveled relatively rapidly (**Fig. D.4a, b**). Mohorovičić was able to calculate the depth of the crust-mantle boundary from this observation and proposed that, beneath continents, it occurred at a depth of about 35 to 40 km. Later studies showed that the depth of the crust-mantle boundary beneath continents varies from 25 to 70 km, and beneath oceans varies from about 7 to 10 km (**Fig. D.4c**). As you learned in Chapter 1, the crust-mantle boundary is called the **Moho**, in honor of Mohorovičić.

Defining the Structure of the Mantle

By studying travel times, seismologists have determined that seismic waves travel at different speeds at different depths in the mantle. Between a depth of about 100 and 200 km in the mantle beneath the ocean floor, seismic velocities are

FIGURE D.4 Discovery of the Moho.

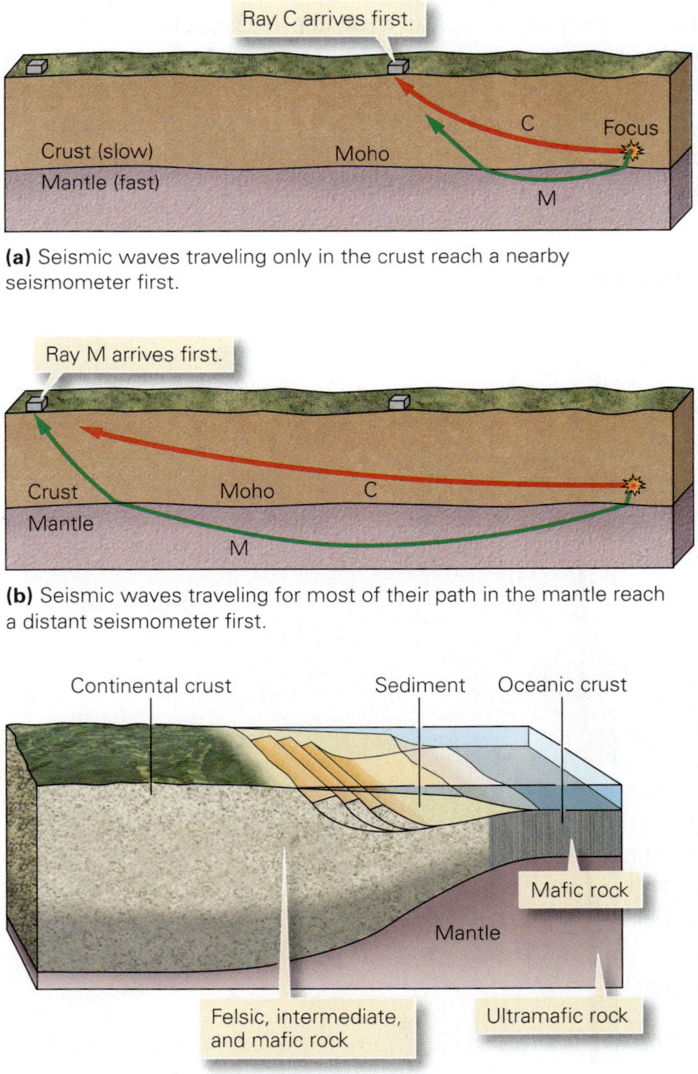

(a) Seismic waves traveling only in the crust reach a nearby seismometer first.

(b) Seismic waves traveling for most of their path in the mantle reach a distant seismometer first.

(c) Oceanic crust is thinner than continental crust and has a different composition. The lower part of the continental crust tends to be more mafic than the upper.

slower than in the overlying mantle (**Fig. D.5a**). This 100- to 200-km-deep layer is called the **low-velocity zone** (LVZ). Researchers suggest that the LVZ corresponds to an interval in which peridotite, the rock comprising the mantle, has undergone partial melting, so the mantle in the LVZ contains about 2% melt. This melt forms films on the surface solid grains. Because seismic waves travel more slowly through liquids than through solids, the coatings of melt slow seismic waves down. In the context of plate tectonics theory, the top of the LVZ delineates the base of the lithosphere beneath oceanic plates—lithospheric mantle lies above this boundary, and asthenospheric mantle lies below. Therefore, the LVZ serves as the weak layer on which oceanic lithosphere plates move. Seismologists do not find a well-developed LVZ beneath continents, meaning that the base of the lithosphere is not delineated by a change in seismic velocity beneath continents.

Below about 200 km, seismic-wave velocities in the mantle, everywhere, increase with depth (**Fig. D.5b**). Because of this change, seismic wave fronts travel faster in the down direction than they do in the up direction, or to the sides. As a result, after a seismic wave front has moved some distance from the focus, it has become elliptical. Seismologists interpret the increase in velocity with depth to mean that mantle peridotite becomes progressively less compressible, more rigid, and denser with depth. This idea makes sense, considering that the weight of overlying rock increases with depth, and as pressure increases, the atoms making up minerals squeeze together more tightly and are less free to move.

The increase in seismic velocity with depth causes seismic waves to refract, so seismic rays curve in the mantle. To understand the shape of a curved ray, let's represent a portion of the mantle by a series of imaginary layers, each of which has a greater seismic-wave velocity than the layer above (**Fig. D.5c**). Every time a seismic ray crosses the boundary between adjacent layers, it refracts a little toward the boundary. After the ray has crossed several layers, it has bent so much that it begins to head back up toward the top of the stack. If we replace the stack of distinct layers with a single layer in which velocity increases with depth at a constant rate, the ray follows a smoothly curving path (**Fig. D.5d**).

At depths between 410 km and 660 km in the mantle, seismic velocity increases in a series of abrupt steps (see Fig. D.5a), so in this interval, the stack of layers depicted in

FIGURE D.5 The velocity of P-waves in the mantle changes because the physical properties of the mantle change with depth.

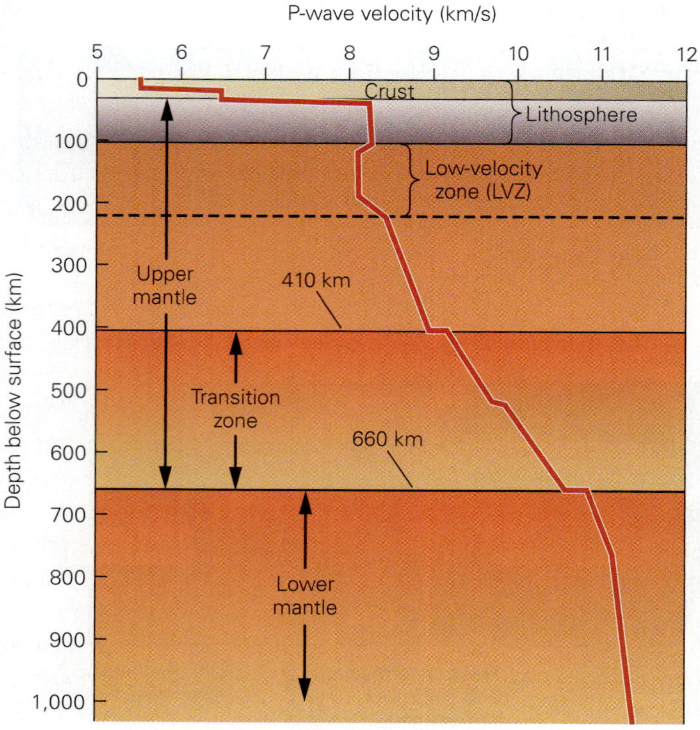

(a) The velocity of P-waves changes with depth in the mantle.

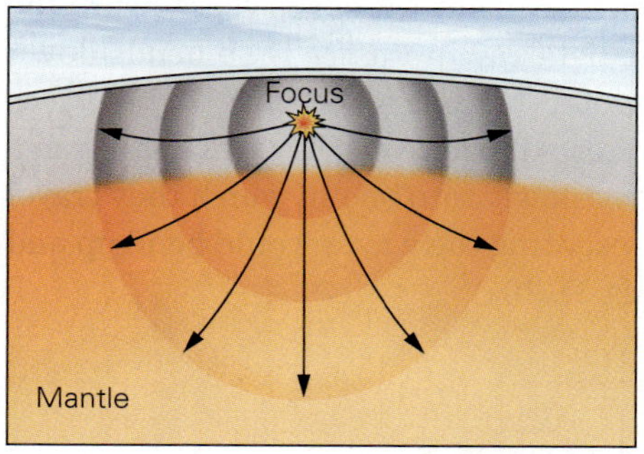

(b) The velocity of seismic waves increases with depth in the mantle, so rays curve and wave fronts are oblong.

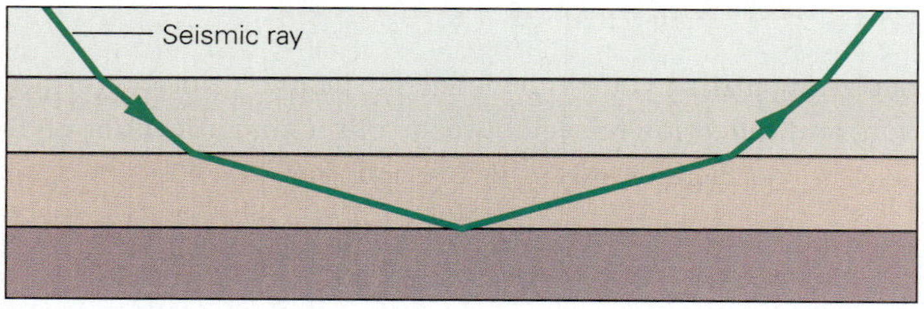

(c) In a stack of discrete layers, rays bend at each boundary. If the velocity is progressively faster in each lower layer, the ray eventually bends back and returns to the surface.

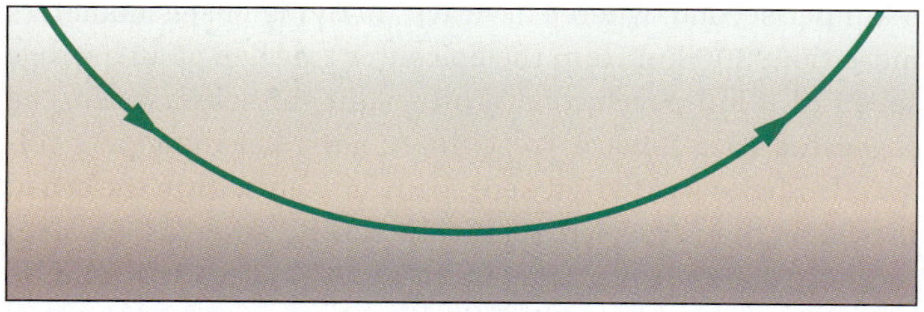

(d) In a material in which the velocity increases gradually with depth, rays curve smoothly and eventually return to the surface.

Figure D.5c is actually a somewhat realistic image. A major step occurs at a depth of 660 km. Experiments suggest that such **seismic-velocity discontinuities** occur at depths where pressure causes atoms in minerals to rearrange and pack together more tightly. This results in a different mineral with a more compact structure—a process called a *phase change* (see Chapter 7). Where such phase changes take place, the overall characteristics of mantle rock change, and seismic waves travel at a different velocity. Researchers are still unsure whether chemical changes also occur at the discontinuities. Because of these seismic-velocity discontinuities, seismologists subdivide the mantle into the **upper mantle** (above 660 km) and the **lower mantle** (below 660 km). The lower portion of the upper mantle (between 410 and 660 km), in which several small seismic discontinuities have been recognized, is called the **transition zone**.

Discovering the Core-Mantle Boundary

In the early 20th century, researchers installed seismometers at many stations around the world, expecting to be able to record waves produced by a large earthquake anywhere on Earth. In 1914, one of these researchers, Beno Gutenberg, discovered that P-waves from a given earthquake did not arrive at seismometers within a band between 103° and 143°, as measured along the surface of the Earth, from the earthquake epicenter. This band is now called the **P-wave shadow zone** (**Fig. D.6a**). If the density of the Earth gradually increased with depth all the way to the center, the shadow zone would not exist, because rays passing into the interior would refract evenly, so rays would reach every point on the Earth's surface.

The presence of a shadow zone means that deep in the Earth a major interface exists where seismic waves abruptly refract down. This implies that the velocity of seismic waves suddenly decreases in this area. We can see this behavior by following two seismic rays, A and B, in Figure D.6a. Ray A curves smoothly in the mantle (we're ignoring seismic-velocity discontinuities in the mantle) and passes just above the core-mantle boundary before returning to the surface at 103° from the epicenter. In contrast, Ray B penetrates the boundary and refracts down into the core. Ray B then curves through the core and refracts abruptly again when it crosses back into the mantle. As a consequence, Ray B intersects the surface at 143° from the epicenter. The dimensions of the shadow zone allowed seismologists to calculate that this interface lies at a depth of 2,900 km, and they interpreted it to be the **core-mantle boundary**. Thus, although researchers back in the 19th century inferred the existence of the core, it took the 20th-century study of seismic waves to determine the specific depth at which its boundary lies.

Discovering the Nature of the Core

Based on the study of meteorites thought to be fragments of a large planetesimal's interior, and on density calculations, researchers have concluded that the core consists of iron alloy. This alloy contains about 85% iron, 5% nickel, and 10% of a lighter element (probably oxygen, silicon, and/or sulfur). The downward bending of seismic waves when they pass from the mantle down into the core indicates that seismic velocities, at least in the outer core, are slower than in the mantle. Thus, even though the core is deeper and denser than the mantle, at least the outer part of the core must be less rigid than the mantle. How can this be?

Seismologists have found that S-waves do not arrive at stations located between 103° and 180° from the epicenter (a band called the **S-wave shadow zone**). This means that S-waves cannot pass through the core at all. If they could, an S-wave headed straight down through the Earth should reach the ground surface on the other side of the planet. S-waves are shear waves, which by their nature can travel only through solids. Thus, the fact that S-waves do not pass through the core means that the core, or at least part of it, consists of liquid (**Fig. D.6b**).

At first, seismologists thought that the entire core might be liquid iron alloy. But in 1936, a Danish seismologist, Inge Lehmann, discovered that P-waves passing through the core reflected off a boundary within the core. She proposed that the core is made up of two parts: an *outer core* consisting of liquid iron alloy, and an *inner core* consisting of solid iron alloy. Lehmann's work defined the existence of the inner core but could not locate the depth at which the inner core–outer core interface occurs. This depth was eventually located by measuring the exact time it took for seismic waves to penetrate the Earth, bounce off the inner core–outer core boundary, and return to the surface (**Fig. D.6c**). The measurements showed that the inner core–outer core boundary occurs at a depth of about 5,155 km.

Why does the core have two layers—a liquid outer layer and a solid inner one? An examination of **Figure D.6d** provides some insight. This graph shows two curves: the *geotherm* indicates how temperature changes with increasing depth, whereas the *melting curve* indicates how the temperature at which a material comprising the Earth's interior begins to melt changes with increasing depth. As the graph shows, the geotherm lies to the left of the melting curve through most of the mantle and in the inner core. This means that the temperature in most of the mantle and in the inner core is not high enough to cause melting, under the very high pressures found in these regions, so these regions are solid. But the geotherm lies to the right of the melting curve in the low-velocity zone of the mantle and in the outer core, so these regions contain molten material.

FIGURE D.6 Shadow zones and the discovery of the Earth's core. (Note that the circumference of a circle is 360°, so it is 180° from a given point to a locality on the other side of the planet.)

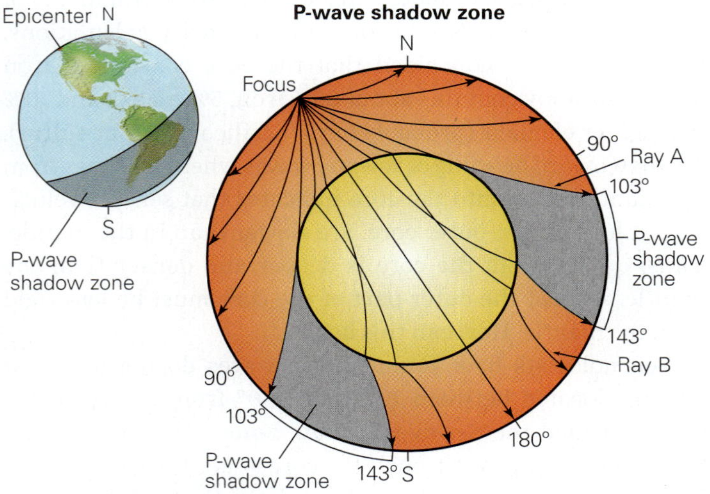

(a) P-waves do not arrive in the P-wave shadow zone because they refract at the core-mantle boundary.

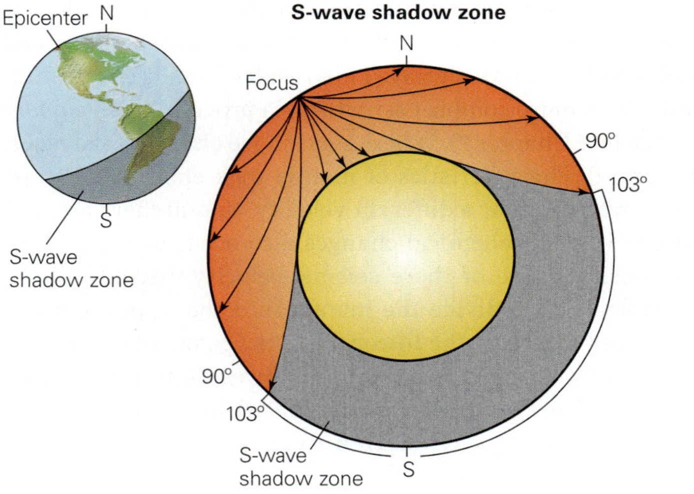

(b) S-waves do not arrive in the S-wave shadow zone because they cannot pass through the liquid outer core.

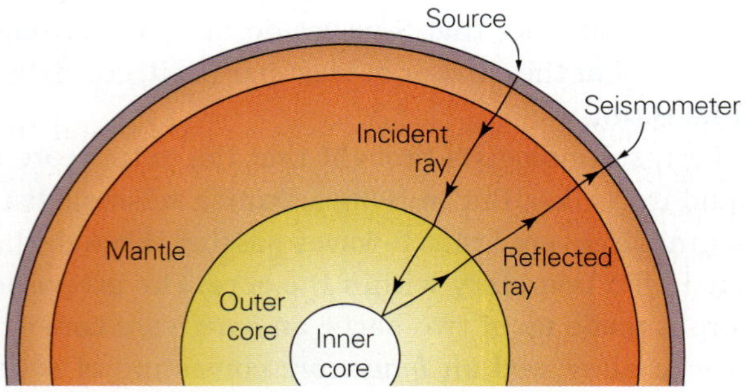

(c) Seismic waves reflect off the inner core–outer core boundary.

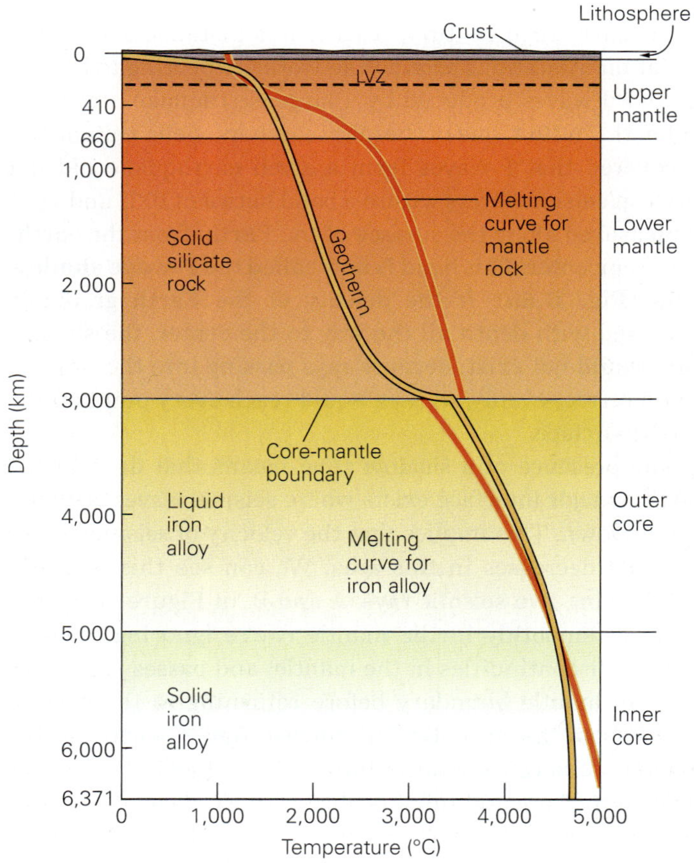

(d) A graph of the geotherm and melting curve for the Earth. Note that the melting temperature is less than the Earth's temperature in the outer core, so the outer core is molten.

A Modern Image of Earth's Layers

Through painstaking effort, seismologists used data on travel times to develop a graph, known as a *velocity-versus-depth curve,* that shows the average depths at which seismic velocity suddenly changes, and the average amount of change. Depths at which major changes take place define the principal layers and sublayers of the Earth, down to its center (**Fig. D.7**). Note that the graph does not show a velocity for S-waves in the outer core, because S-waves cannot travel through molten iron (a liquid).

Seismic Tomography

In recent years, seismologists have developed a technique, called **seismic tomography**, that can produce three-dimensional images of variation in seismic velocities in the Earth's interior. This technique resembles the technique used to produce three-dimensional CAT (or CT) scans of the human body. In seismic tomography studies, seismologists compare the observed travel time of seismic waves following a specific ray path with the predicted travel time that waves following the same path would have if the average velocity-versus-depth model depicted by Figures D.5 and D.7 were completely correct. Seismologists can distinguish locations within the Earth

FIGURE D.7 The velocity-versus-depth profile of the whole Earth.

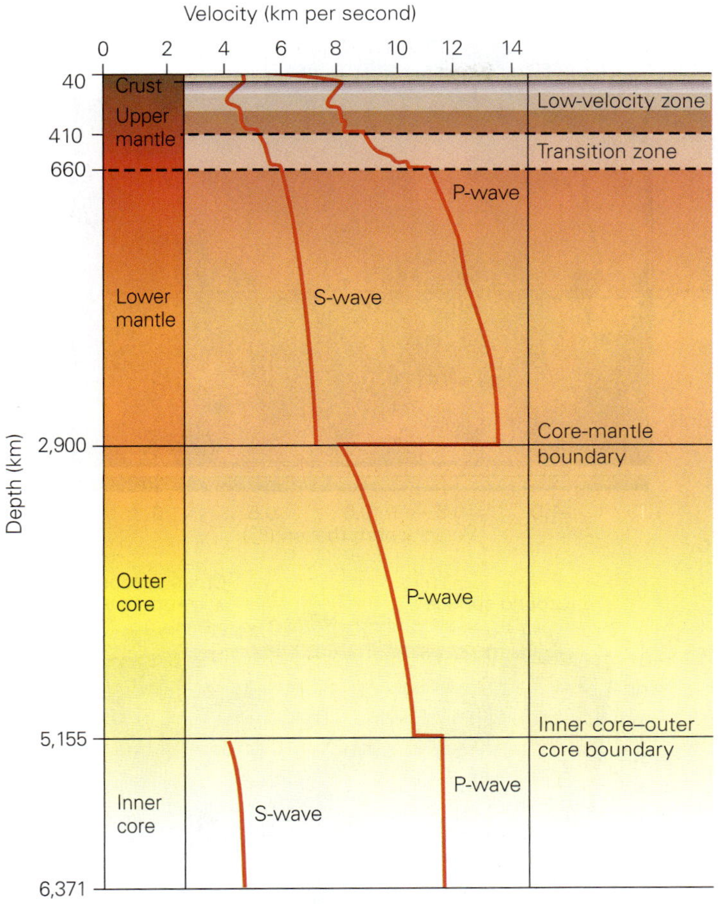

where seismic waves travel faster than expected from regions where the waves travel slower than expected.

Tomographic studies emphasize that the simple onion-like layered image of the Earth that we've shown so far, in which velocities increase with depth at the same rate everywhere, is an oversimplification. In reality, the velocities of seismic waves vary significantly with location at a given depth. Tomographic studies can display these variations in three-dimensional models, cross sections, or maps (**Fig. D.8a, b**). Generally, reds on these images indicate slower regions, whereas blues and purples indicate faster regions. Researchers interpret the slower regions to be warmer regions of the mantle, and faster regions to be cooler regions, for as rock gets hotter and softer, it can't transmit seismic waves as rapidly. The occurrence of warm and cold regions is a consequence of convection in the mantle, so seismic tomography has led researchers to picture the inside of the Earth as a dynamic place (**Fig. D.8c**).

The image of the Earth's interior has become even clearer, in recent years, due to a major research project called **Earth-Scope**. This project involved placing hundreds of seismometers in an array across the United States. Because the seismometers are relatively closely spaced, they record the velocity of seismic waves along a much greater number of ray paths, and therefore can provide a higher resolution (more detailed) image of the interior.

Seismic-Reflection Profiling

During the past half century, geologists have found that by exploding dynamite, by banging large weights against the Earth's surface, or by releasing bursts of compressed air into water, they can produce artificial seismic waves that propagate down into the Earth and reflect off boundaries between layers of rock in the crust. By recording the time at which these reflected waves return to the surface, geologists can determine the depth to these boundaries. With this information, they can produce a cross-sectional view of the crust called a **seismic-reflection profile** (**Fig. D.9**). This image can define subsurface bedding, stratigraphic formation contacts, folds (bends in layers), and faults. Oil companies use seismic-reflection profiles, despite their high cost, to identify likely places where oil and gas reserves have accumulated underground.

In the past decade, computers have become so powerful that geologists can produce three-dimensional seismic-reflection images of the crust. These images provide so much detail that geologists can use them to trace out a ribbon of sand, representing the channel of an ancient stream, now buried several kilometers below the Earth's surface.

D.4 Earth's Gravity

The Earth emanates two field forces, gravity and magnetism. Recall that a **field force** is a push or pull that applies across a distance. A field force can cause an object to accelerate (change speed and/or direction) without ever touching it. In the remainder of this interlude, we see how measurements of variations in the magnitudes of these forces provide important clues to the nature of Earth's interior. In this section, we discuss gravity, and in the next, we turn our attention to magnetism.

The Geoid

Gravity is an attractive field force that one mass exerts on another. As Newton showed, the magnitude of gravitational pull depends on the size of the masses and on the distance between them—larger masses exert stronger pulls than do smaller ones, and closer masses produce stronger pulls than do farther ones. When gravity acts on an object but can't make the object move, then the object stores **gravitational potential energy**. For example, a boulder sitting at rest on a hillslope has gravitational potential energy. A surface on

FIGURE D.8 Tomographic images of the Earth's interior and their interpretation.

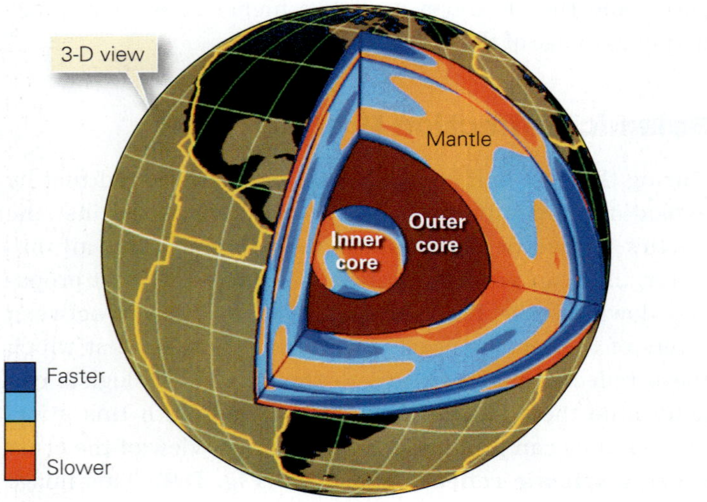

(a) Tomographic image of the Earth. Slower areas may be relatively warmer than their surroundings, and faster areas may be relatively cooler.

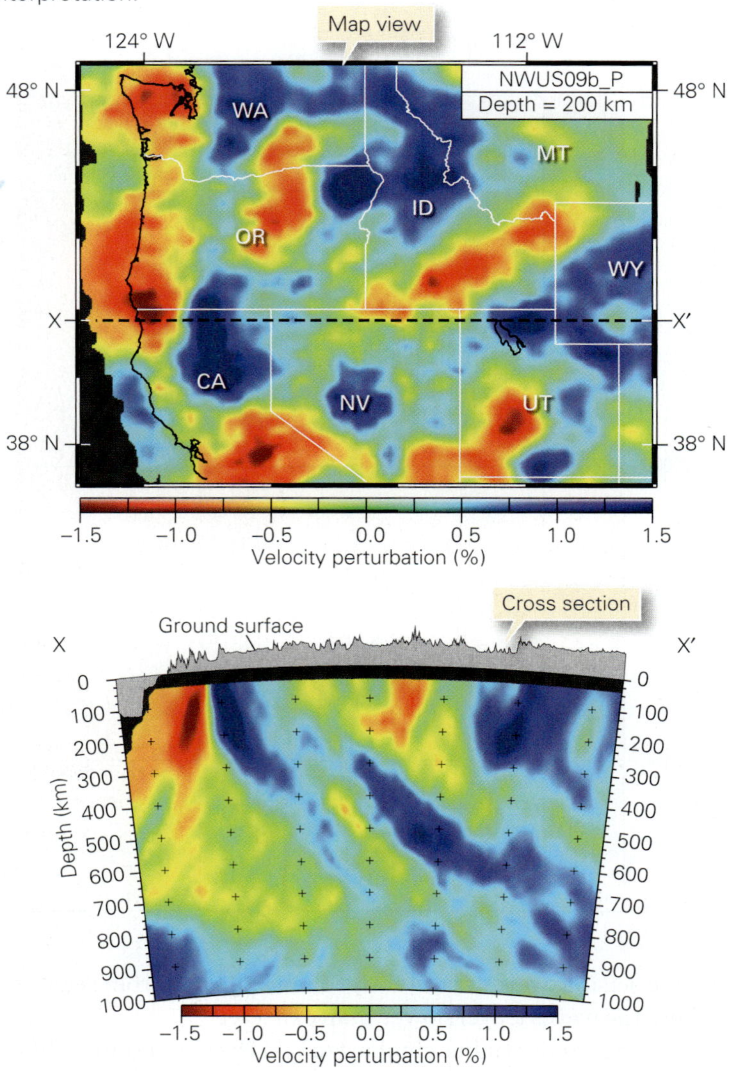

(b) A map (of a surface at a depth of 200 km) and cross section of the Earth beneath the western United States, constructed using tomographic data from the EarthScope array. Red areas have slower-than-expected velocities and probably are warmer, whereas blue areas are cooler.

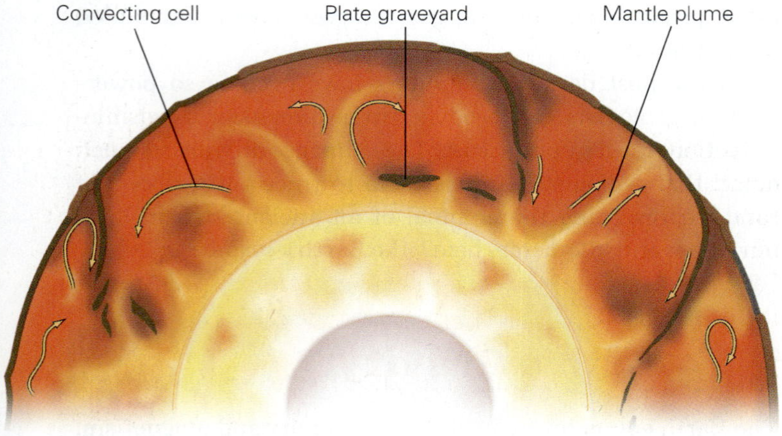

(c) An artist's rendition of a modern view of the complex and dynamic Earth interior. Note the convecting cells, the mantle plumes, and the graveyards of subducted plates.

which all points have the same gravitational potential energy, such as the surface of standing water in a quiet pond, is an **equipotential surface**.

What does the gravitational field produced by the Earth look like? To address this question, let's start by thinking about what the surface of an imaginary ocean that completely surrounded the Earth would look like, if the ocean weren't subjected to winds or current, and the Earth were perfectly spherical and stationary. The surface of this imaginary ocean, representing an equipotential surface, would be a sphere. In reality, however, the Earth is not a perfect sphere, in part because it spins on an axis. This rotation produces centrifugal force, which flattens the Earth. If the Earth were perfectly smooth and homogeneous, this flattening would transform the Earth into a *spheroid*—a slice through a spheroid is an ellipse (**Fig. D.10a**). The imaginary spheroid that serves as the best match to the Earth's overall shape has a radius at the equator of 6,378 km (3,963 miles) and a radius at the pole is 6,357 km (3,950 miles). Geologists refer to this imaginary equipotential surface as the *reference spheroid*.

While the reference spheroid is a starting point for describing the Earth's gravitational field, it doesn't provide a completely accurate picture for many reasons, including (1) the density of material within the Earth's layers is not uniform throughout the layer; (2) the surface of the Earth is not smooth, for the land surface is higher than the seafloor surface, and both the land and the seafloor have mountains and valleys; and (3) rock making up the outermost layer of the

FIGURE D.9 Seismic-reflection profiling. The method allows geologists to see layers underground.

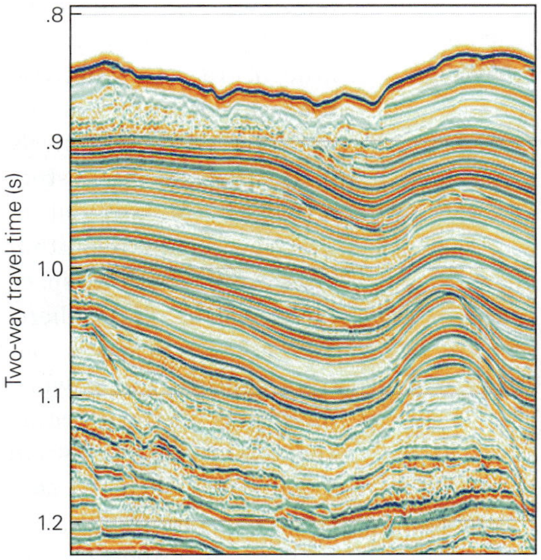
.8
.9
1.0
1.1
1.2
Two-way travel time (s)

(a) After computer analysis, the data yield a seismic-reflection profile. Color bands represent horizons in the stratigraphic sequence.

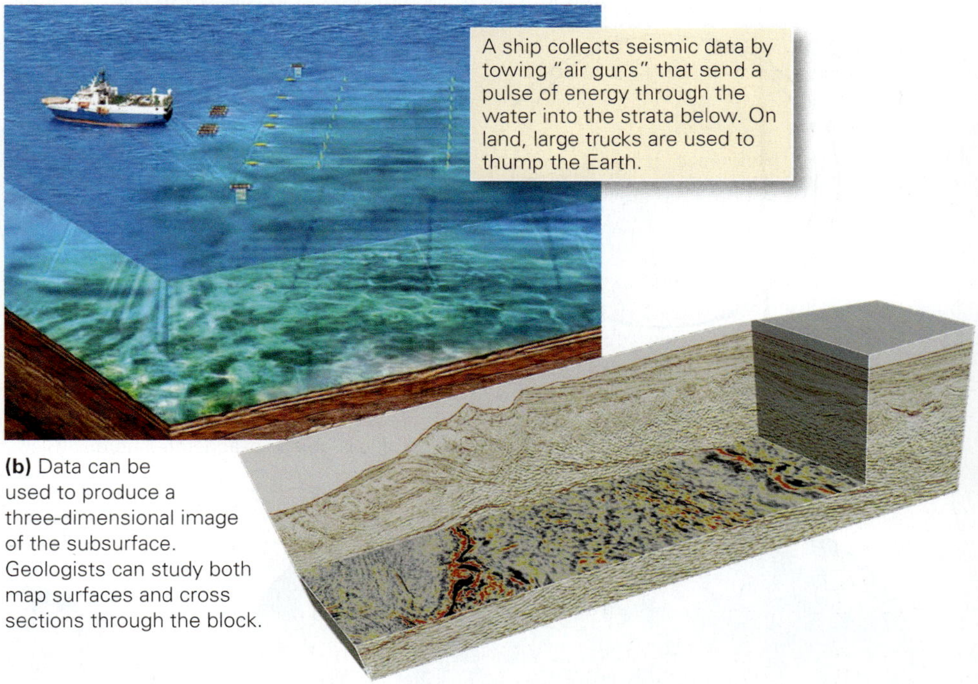
A ship collects seismic data by towing "air guns" that send a pulse of energy through the water into the strata below. On land, large trucks are used to thump the Earth.

(b) Data can be used to produce a three-dimensional image of the subsurface. Geologists can study both map surfaces and cross sections through the block.

Earth, the lithosphere, is strong enough to hold up heavy loads such as volcanoes or glaciers, or to allow depressions such as trenches to persist. Because of these factors, the Earth at a particular location can exert a greater pull or a lesser pull than the reference spheroid predicts. Therefore, the real equipotential surface for the Earth has bumps and dimples. Geologists refer to this irregular surface, which provides the best representation of the Earth's gravity, as the **geoid** (**Fig. D.10b, c**). The surface of the Earth's geoid differs from that of the reference spheroid only slightly—the highest point on the geoid lies 85 m above the reference spheroid, and the lowest point lies 107 m below the reference spheroid. But even these small differences can noticeably influence the orbits of satellites or the accuracy of surveys.

Gravity Anomalies and Isostasy

Geologists refer to the difference between the expected and observed value of gravity at a location as a **gravity anomaly**. Over a *positive gravity anomaly*, the pull of gravity is stronger, whereas over a *negative gravity anomaly*, the pull is weaker. A positive anomaly indicates extra mass below the site, perhaps due to a body of particularly dense rock underground, whereas a negative anomaly means a deficit of mass, perhaps due to the presence of less-dense rock underground (**Fig. D.11**). Some gravity anomalies are due to variations in rock composition within the crust, while others are due to variations in temperature in the mantle, for both composition and temperature can affect density.

To better interpret gravity anomalies, we first need to introduce the concept of isostasy, the basis of which stems from the inspiration of an ancient Greek mathematician, Archimedes (ca. 287–212 B.C.E.). Archimedes had been puzzling over the question of why some objects float and others sink, when one day, according to legend, he noticed the water level rise in a tub as he stepped in to take a bath. He realized that this meant that an object displaces water as it sinks into it. If the density of the object exceeds that of water, it sinks, but if it is less dense, it floats. A floating object sinks into water only until it has displaced an amount of water whose mass equals the mass of the whole object. This relationship is now known as **Archimedes' principle**.

According to Archimedes' principle, a cargo ship anchored in a harbor floats at just the right level so that the mass of the water displaced by the ship equals the mass of the whole ship. Even though the ship consists of heavy steel, the inside of the ship is filled with air. In effect, the ship is a steel-sheathed bubble—it floats because its overall average density is less than that of water. When the ship remains empty, its *freeboard*, the distance that its deck lies above the water, is large (**Fig. D.12**). Addition of heavy cargo causes the ship's keel to sink deeper into the water, the deck to move down, and the freeboard, therefore, to decrease. This movement can take place because the water beneath the ship can flow out of the way as the ship settles downward. When the freeboard of the ship is just right for a given cargo, we say that the ship is in *isostatic equilibrium*, or that a condition of **isostasy** exists. (The word comes from the Greek *iso*, meaning equal.)

FIGURE D.10 Representations of the geoid, the shape of an equipotential surface representing Earth's gravity. Colors represent elevations of the geoid relative to the reference spheroid.

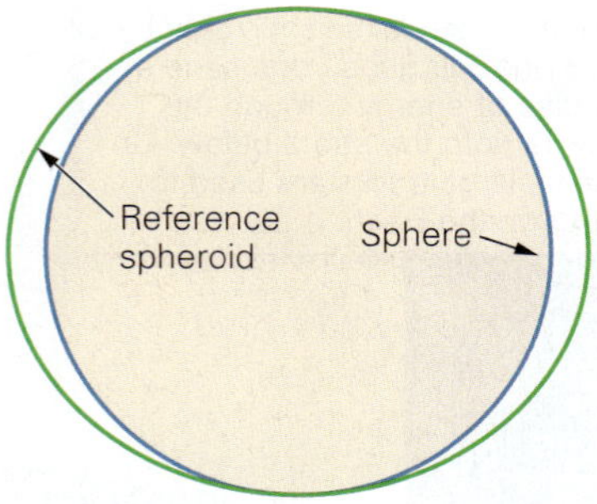

(a) Earth's spin makes it slightly elliptical. The elliptical profile it would have (exaggerated here), if other factors didn't affect gravity, is the reference spheroid.

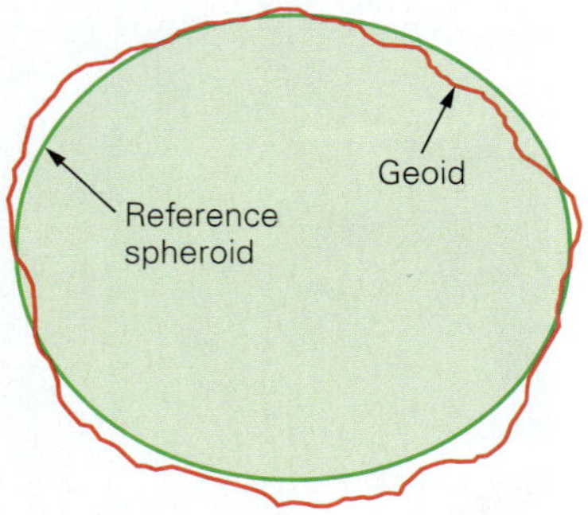

(b) Because the Earth is not homogeneous, there are deviations from the reference spheroid. The actual equipotential surface of the Earth is the geoid; its profile is greatly exaggerated here.

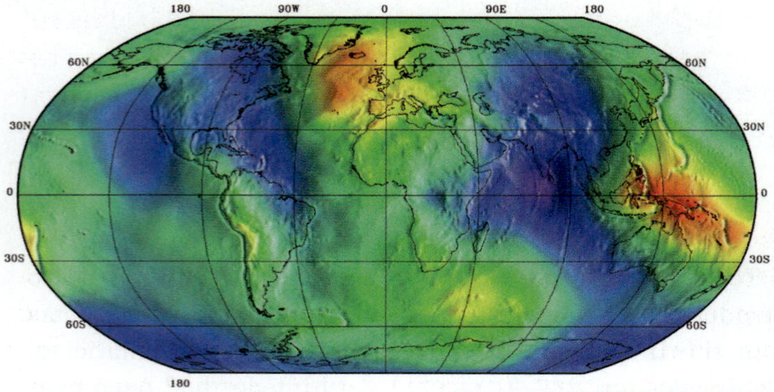

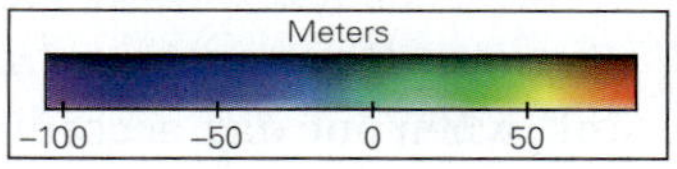

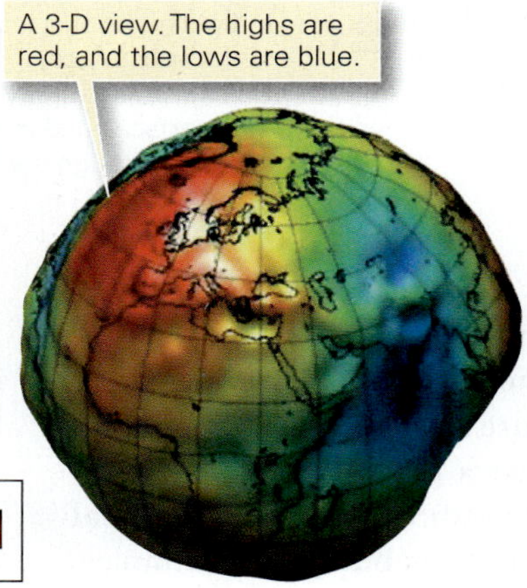

(c) The geoid can also be represented in map view, and in 3-D. Note the geoid low in the Indian Ocean.

As you learned in Chapter 2, we can think of the outer layer of the Earth, the lithosphere, as a relatively rigid shell that "floats" on the underlying, relatively soft asthenosphere. Asthenosphere can flow out of the way of sinking lithosphere or flow in to fill space under rising lithosphere, just as water moves out from beneath a ship when the ship is filled with cargo and sinks down, or flows in to fill space when the ship rises as it's unloaded.

The lithosphere approaches a condition of isostatic equilibrium in many places, and where this happens, the elevation of the lithosphere's surface reflects the density and thickness of the lithosphere below. In such locations, no gravity anomaly exists. In order to maintain isostasy, emplacement of a load, such as growth of an ice sheet or building of a volcano, causes the lithosphere's surface to sink, whereas removal of a load, say by melting of an ice sheet, causes the lithosphere's surface to rise. But, since asthenosphere flows very slowly, isostasy can't be achieved instantly. In addition, because the lithosphere has strength, it can support small, heavy loads in places, or keep relatively low-density materials from rising, without the surface of the lithosphere sinking or rising overall. Places on Earth where isostasy does not exist are the places where we observe gravity anomalies.

FIGURE D.11 A gravity anomaly map of the United States. On this map, colors represent variations in the magnitude of gravitational pull. Red is stronger, blue is weaker.

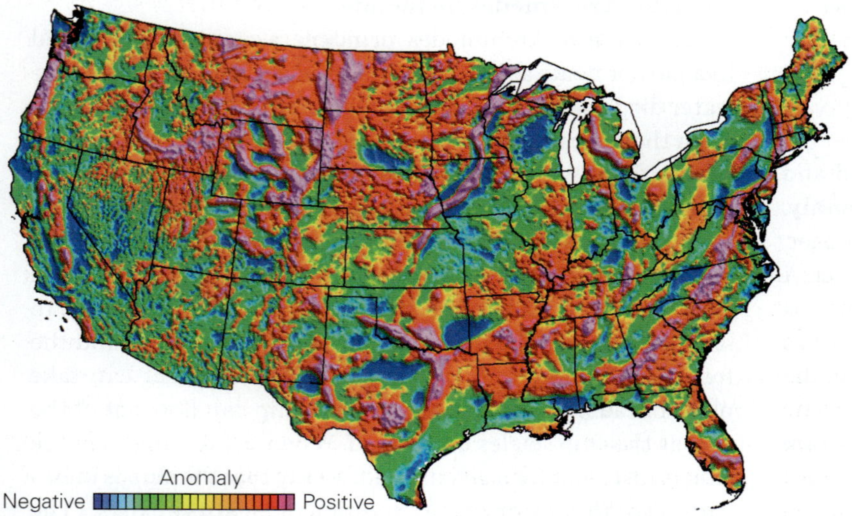

D.5 Earth's Magnetic Field, Revisited

As we discussed in Chapters 1 and 2, the Earth behaves like a weak magnet and thus produces a magnetic field which, like the field produced by a

FIGURE D.12 A ship analog for the concept of isostasy. At isostatic equilibrium, the freeboard reflects Archimedes' principle,so the freeboard decreases as the mass of the ship increases. Water flows out of the way as the ship moves down (right image).

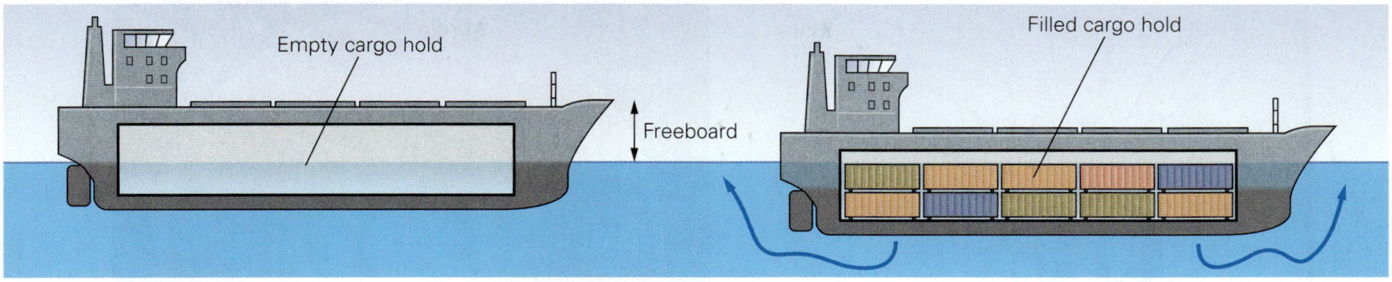

toy bar magnet, can be represented by a dipole with a north end and a south end, and by field lines that curve through space (**Fig. D.13a**). We can represent the dipole by an arrow that points from the north magnetic pole to the south magnetic pole. Here, we discuss why the field exists and why anomalies occur in the strength of the field that we measure at the Earth's surface.

Origin of the Earth's Magnetic Field

Why does the Earth have a magnetic field? The path toward discovering the origin of the Earth's field started in 1926, when researchers realized that our planet's outer core consists of liquid iron alloy. Physicists demonstrated that flow of liquid metal can produce an electric current, so it seemed that the flow of the outer core must be responsible for the magnetic field. But how?

Insight into the generation of the Earth's magnetic field comes from examining how an electric power plant works. In a power plant, the flow of water, wind, or steam spins a coil made of metal (an electrical conductor) around an already magnetic iron bar. The iron bar is a *permanent magnet,* meaning that it always produces a magnetic field, even in the absence of an electric current. The motion of the wire in the bar's magnetic field generates an electric current in the wire—we use that current to power our homes. An apparatus that produces electricity by moving a wire around a magnet is called a **dynamo**. In the case of the Earth, flow in the outer core serves the role of the spinning wire coil, making the outer core behave like an electromagnet. But the inner core can't serve the role of a permanent magnet, because it is way too hot—materials can only be permanent magnets at relatively low temperatures. Researchers suggest, therefore, that the Earth is a *self-exciting dynamo.* Evidently, during the Earth's early history, flow in the outer core took place in the presence of a magnetic field. This flow generated an electric current, which in turn generated a magnetic field. (A physics book provides further detail on the relationship between electricity and magnetism.) Continued flow in the presence of the generated magnetic field produced more electric current, which maintained the magnetic field. Once started, the system perpetuates itself, as long as there's an input of energy to keep the outer core in motion. The input of energy comes from variations in temperature and composition in the outer core that make the outer core convect.

At present, the magnetic poles of the Earth are close to, but not exactly on, the geographic poles. The poles are moving at about 55 km per year, so their position constantly changes. Geophysicists suggest that, when averaged over time, the magnetic poles and geographic poles roughly coincide. What causes this relationship? The key to understanding the orientation of the Earth's magnetic dipole comes from understanding the geometry of the convection cells in the outer core. Researchers suggest that, due to the Earth's rotation, the Coriolis force (see Chapter 15) causes convective cells in the outer core to become spirals that align roughly with the Earth's spin axis (**Fig. D.13b**). The spirals wobble, however, so at any given time, the overall dipole axis they generate is not exactly parallel to the spin axis. Notably, the spirals apply a force to the inner core, causing it to rotate slightly faster than the rest of the Earth, a phenomenon known as *superrotation.*

In Chapter 2, we noted that the polarity of the Earth's magnetic field—the direction that the dipole points—flips (reverses) every now and then. This has happened about two dozen times during the past 5 million years. Today's "normal" polarity (with the dipole pointing to the south) was established 780,000 years ago. Between 780,000 and 900,000 years ago, the field had reversed polarity, with the dipole pointing to the north. Why does the polarity of the Earth's field undergo such reversals every now and then? Computer models suggest that reversals happen because circulation spirals in the outer core are unstable. Over time, they slow and fade away as they interact, causing the magnetic field to weaken or disappear temporarily. As new spirals become established, the field reappears, sometimes with a different polarity. The details of this process remain a subject of active research.

FIGURE D.13 Earth's magnetic field may be due to spiral-like convection cells in the outer core.

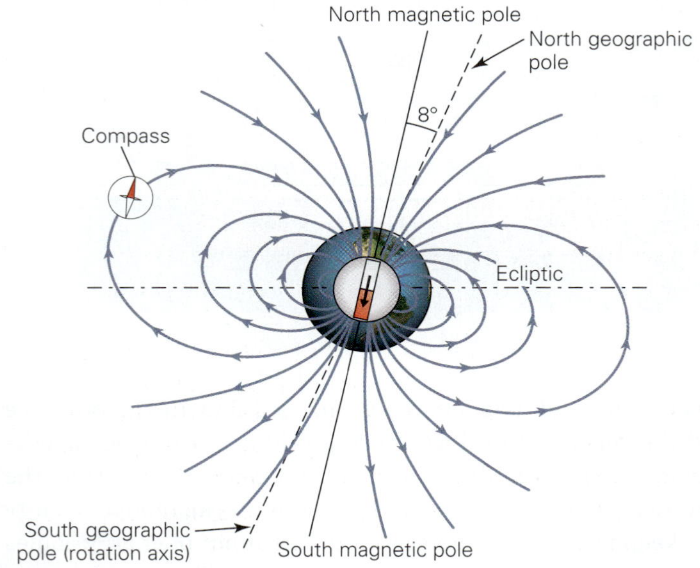

(a) The geometry of Earth's dipole field.

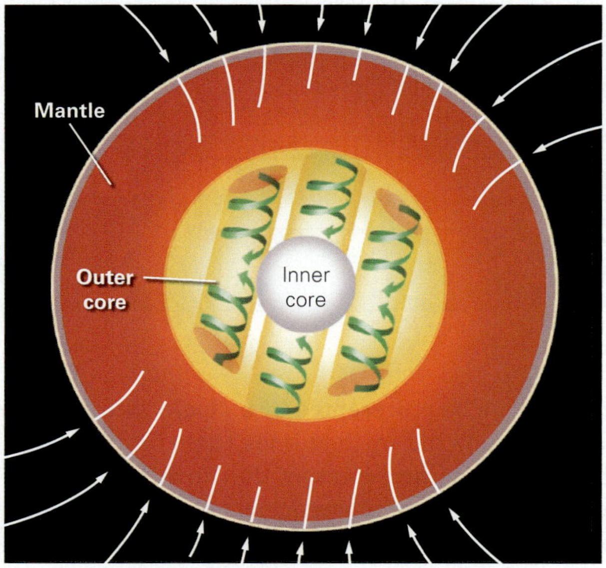

(b) An artist's image of spiral convection cells in the outer core.

Magnetic Anomalies, Revisited

If the Earth's magnetic field were a perfect dipole, geophysicists could predict the field strength at every point on the planet with a high degree of confidence. However, that isn't the case, because magnetization locked into rocks in the crust can contribute to the field strength measured at a given location, causing the field to be either stronger than or weaker than expected at the location. As we discussed in Chapter 2, the difference between the expected field and the measured field at a locality is called a **magnetic anomaly**. A *positive magnetic anomaly* occurs where the field is stronger than expected, and a *negative magnetic anomaly* occurs where the field is weaker than expected.

We've seen in Chapter 2 that magnetic anomalies due to the magnetization of seafloor basalt during seafloor spreading take the form of parallel stripes on the seafloor. On continents, in contrast, the pattern of magnetic anomalies is much more irregular, because the distribution of rock types is more complex (**Fig. D.14**). We see anomalies due to igneous intrusions, lava flows, and concentrations of iron-rich sediments. Rifts, for example, typically stand out as a strong positive magnetic anomaly, because they contain a thick succession of relatively magnetic basalts. Regions with thick accumulations of nonmagnetic sedimentary rocks may appear as regions of negative anomalies.

FIGURE D.14 A magnetic anomaly map of North America, and adjacent seafloor. Note how there is a regular striped pattern on the seafloor, but not within the continent. The white areas are regions where data have not yet been obtained.

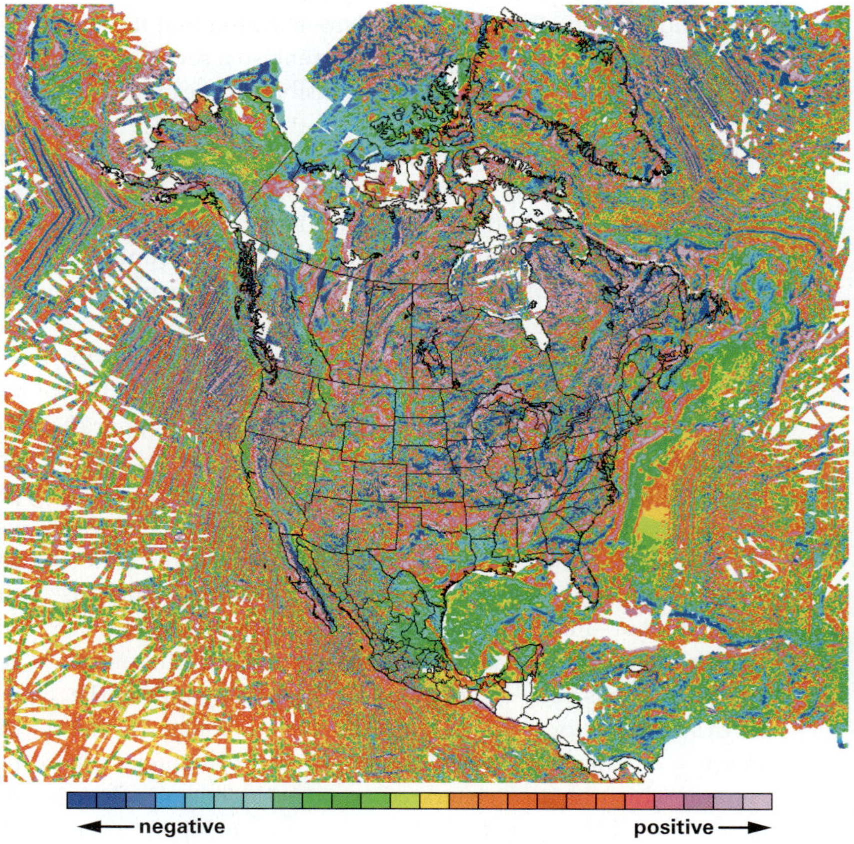

Interlude D Review

Interlude Summary

- Details about the depth to Earth's internal layers, and about divisions within the layers, come from the study of seismic waves passing through the Earth. This is because waves refract and reflect at boundaries.
- The Moho was discovered because seismic waves travel faster through the mantle than through the crust.
- Seismic-velocity discontinuities define boundaries in the mantle. Several discontinuities occur in the transition zone.
- Seismic shadow zones define the depth to the core. S-waves cannot pass through the core, so part of it must be liquid. The reflection of waves off the surface of the solid, inner core defines its depth.
- Seismic tomography reveals that the onion-like model of the Earth's interior is an oversimplification, as the layers are inhomogeneous. Modern EarthScope studies can outline subducted plates.
- Seismic-reflection, using artificially produced seismic waves, allows geologists to identify layering and structures in the sedimentary strata of the upper crust.
- The geoid, the representation of the surface on which the pull of gravity is the same, is not a perfect sphere. Local anomalies may indicate the presence of particularly dense rocks below the surface.
- The elevation of the lithosphere's surface regionally reflects isostatic equilibrium. Local bending or loading, however, yields places that are not in equilibrium.
- The Earth's magnetic dipole exists because of convective circulation in the outer core. Local anomalies may exist where rocks contain magnetic minerals.

Guide Terms

Archimedes' principle (p. 291)
core-mantle boundary (p. 287)
dynamo (p. 293)
EarthScope (p. 289)
equipotential surface (p. 290)
field force (p. 289)
geoid (p. 291)
geophysics (p. 283)
gravitational potential energy (p. 289)
gravity anomaly (p. 291)
isostasy (p. 291)
lower mantle (p. 287)
low-velocity zone (LVZ) (p. 286)
magnetic anomaly (p. 294)
Moho (p. 285)
P-wave shadow zone (p. 287)
reflection (p. 285)
refraction (p. 285)
seismic ray (p. 283)
seismic-reflection profile (p. 289)
seismic tomography (p. 288)
seismic-velocity discontinuity (p. 287)
S-wave shadow zone (p. 287)
transition zone (p. 287)
travel time (p. 283)
upper mantle (p. 287)
wave front (p. 283)

Review Questions

1. What basic layers of the Earth were recognized before the use of seismology?
2. Do seismic waves travel at the same velocities in all rock types?
3. What is the difference between refraction and reflection of waves?
4. What observation led to the discovery of the Moho?
5. How and why do seismic waves bend as they pass through the mantle?
6. What are seismic-velocity discontinuities, and what do they tell us?
7. What are P-wave and S-wave shadow zones, and what do they tell us?
8. Is seismic velocity constant at a given depth? Explain how seismologists learned the answer to this question and what the answer tells us about the mantle.
9. What can we learn from seismic-reflection surveys?
10. What is the principle of isostasy, and why is the lithosphere able to be in isostatic compensation, in general?
11. Explain what the geoid is and why it is so bumpy. What is a gravity anomaly?
12. Why might the Earth's magnetic field exist? What causes magnetic anomalies?

Online Resources

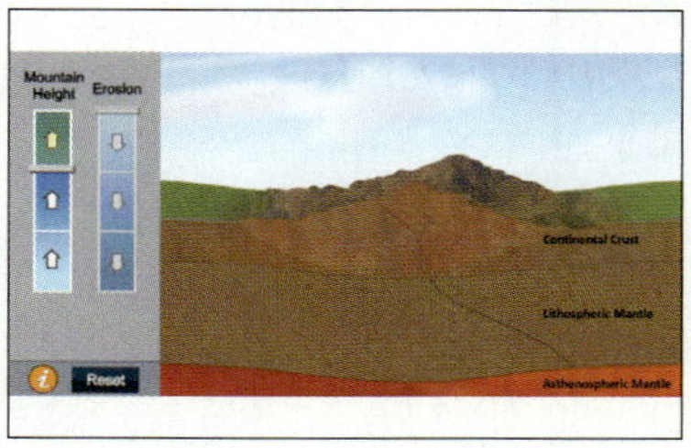

Animations

This chapter features interactive animations on isostasy and the effects of different conditions on isostasy.

LEARNING OBJECTIVES

By the end of this chapter, you should understand...

1. how and why rocks deform (crack, bend, or flow) during mountain building.
2. the characteristics of basic geologic structures (joints, faults, folds, foliations), and how to describe them.
3. where and why mountain belts form, what processes cause their uplift, and how their formation relates to plate tectonics.
4. how nature carves mountain ranges into rugged landscapes.

▲ Look closely at this small cliff face in Arizona. You can see layers that have been contorted into toothpaste-like curves and broken along jagged cracks. We're seeing geologic structures, evidence that these rocks were subjected to deformation during mountain building.

CHAPTER 9

Crags, Cracks, and Crumples: Geologic Structures and Mountain Building

9.1 Introduction

Geographers call the peak of Mt. Everest "the top of the world," for this mountain, which lies in the Himalayas of south-central Asia, rises higher than any other on Earth. The cluster of flags on Mt. Everest's summit flap at 8.85 km (29,029 feet) above sea level, almost the cruising height of modern jets. In 1953, Sir Edmund Hillary, from New Zealand, and Tenzing Norgay, a Nepalese guide, became the first to reach the summit. Since then, thousands of other people have also succeeded, but hundreds have died trying. Mountains appeal to nonclimbers as well, for almost everyone loves a vista of snow-crested peaks. The stark cliffs, clear air, meadows, forests, streams, and glaciers of mountainous landscapes provide a refuge from the mundane.

Geologists have a particular fascination with mountains, for the existence of mountains serves as one of the most obvious indications of Earth's dynamic activity. Think about it . . . to make a single mountain, cubic kilometers of rock must rise skyward against the pull of gravity. Furthermore, with the exception of large volcanoes formed over hot spots, mountains do not occur in isolation but rather as part of

> ***Innumerable peaks, black and sharp, rose grandly into the dark blue sky, their bases set in solid white, their sides streaked and splashed with snow, like ocean rocks with foam. . . . [Mountains] are nature's poems carved on tables of stone.***
>
> JOHN MUIR (American naturalist, 1838–1914)

elongate ranges called **mountain belts** or **orogens** (from the Greek words *oros,* meaning mountain, and *genesis,* meaning formation). Uplift during the growth of an orogen may have moved a million or more cubic kilometers of rock.

A map of the present-day Earth reveals about a dozen major orogens, and numerous smaller ones (**Fig. 9.1**). Many places that are not mountain belts now, were mountainous in the geologic past. So **mountain building**, the process of forming a mountain belt, has happened many times and in many places over Earth's history. Mountain building not only leads to **uplift**, the vertical rise of the land surface and the rock beneath, but also causes rocks to undergo **deformation**, when they bend, break, or flow. Features produced by deformation are known as **geologic structures**. These include *joints* (cracks), *faults* (fractures on which one body of rock slides past another), *folds* (bends, curves, or wrinkles of rock layers), and *tectonic foliation* (a fabric or layering in rock). Mountain building may also cause metamorphism and igneous activity.

A mountain-building event, or **orogeny**, may last for tens of millions of years. As the land rises, erosion starts to grind at it, producing sediment and sculpting spectacular, jagged landscape, eventually bringing the height of a range down near sea level in some cases. But even after the high peaks are gone, a low-lying belt of fractured, contorted, and metamorphosed rock remains. Such crustal scars serve as a permanent monument to what had once been a region of high peaks.

In this chapter, we first learn about the process of deforming rocks, and how to describe and interpret the geologic structures that result from deformation. Then, we turn our attention to the question of why mountain belts form, in the context of plate tectonics theory. Finally, we consider the phenomena that sculpt regions of uplift to form the dramatic terrains that we can see when visiting mountains.

9.2 Rock Deformation in the Earth's Crust

Deformation and Strain

To get a visual sense of what geologists mean by the term *deformation,* let's contrast rock that has not been affected by an orogeny with rock that has been affected. Our undeformed example comes from a roadcut in the Great Plains of North America, and our deformed example comes from a cliff in the Alps of Europe.

FIGURE 9.1 Digital map of world topography, showing the locations of major mountain ranges.

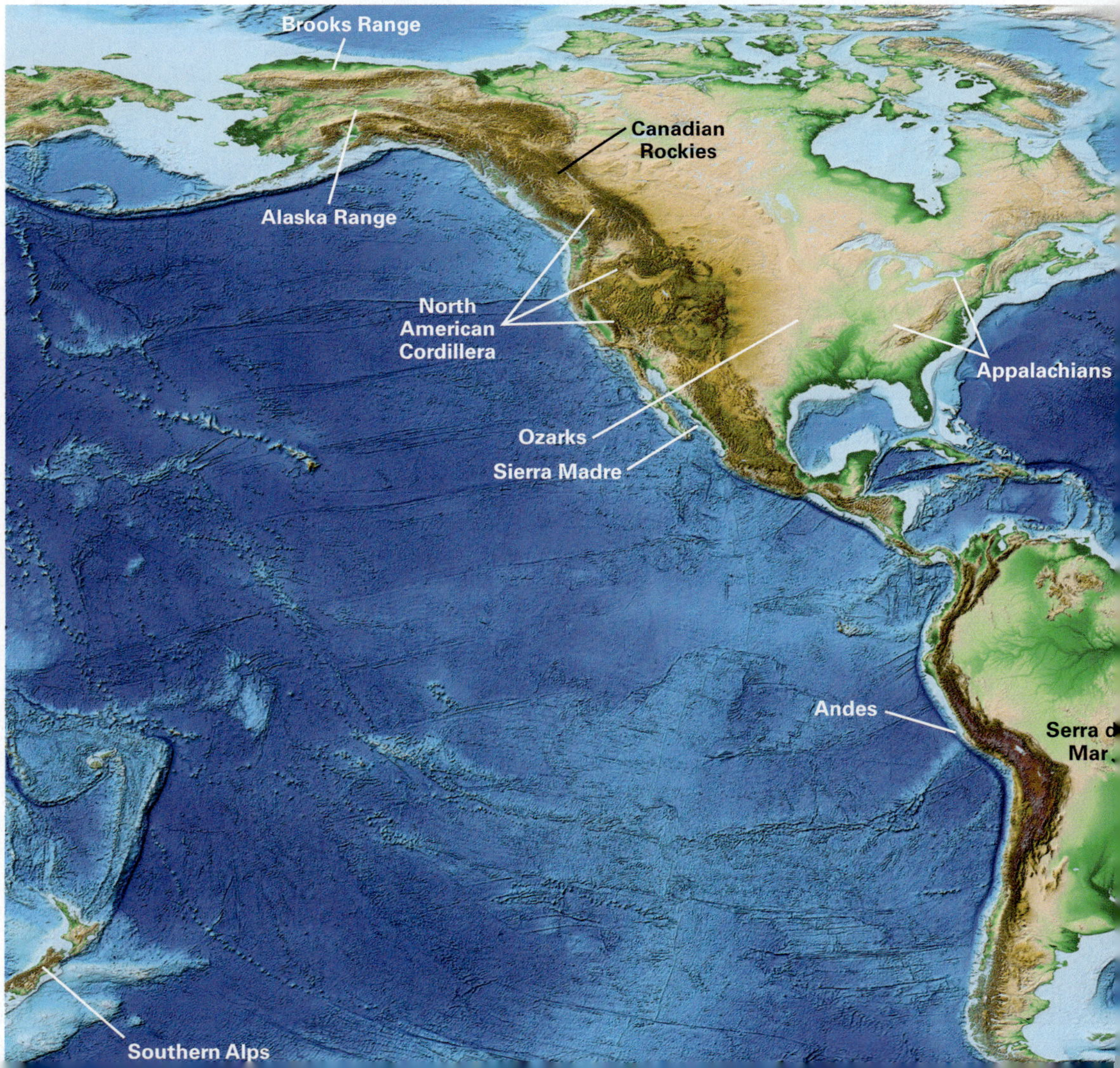

The roadcut exposes nearly horizontal beds of sandstone, shale, and limestone cut by a few joints (**Fig. 9.2a**). These beds have the same orientation that they had when first deposited. Sand grains in the sandstone of this outcrop are nearly spherical (the same shape as when deposited), and clay flakes in the shale lie roughly parallel to the bedding, due to compaction accompanying burial and lithification. When we say that rock of this outcrop is undeformed, we mean that it contains no geologic structures, other than a few joints.

Deformed rocks of the Alpine cliff look very different (**Fig. 9.2b**). Here we find layers of quartzite, slate, and marble (the metamorphic equivalents of sandstone, shale, and limestone). Here too the layers are cut by joints, but they have also been contorted into folds. On microscopic examination, we may find that the grains in the quartzite aren't spherical, but have been flattened into elliptical shapes. And in the slate, clay flakes no longer align with bedding but rather align in a new orientation to define a new foliation; in this example, the foliation is called slaty cleavage (see Chapter 7). Thus, because of changes in grain shapes and alignments, the fabric of this deformed rock has no longer looks like the fabric that the rock had before deformation. Finally, if we try tracing the quartzite and slate layers along the cliff face, we find that they terminate abruptly at a sloping surface marked by broken-up rock. This surface is a fault. In our example, thick layers of marble lie below the fault—the quartzite and slate moved along the fault from where they first formed to get to their present location juxtaposed against the marble.

Study of the types of geologic structures that we see in the Alpine cliff emphasizes that during deformation, rocks can undergo one or more of the following changes (**Fig. 9.3**): (1) a change in location, or *displacement*; (2) a change in orientation, or *rotation*; and (3) a change in shape, or *distortion*. Geologists commonly refer to distortion as **strain** and distinguish among different kinds of strain according to the nature of the shape change observed. If a layer becomes longer, it has undergone *stretching*, whereas if the layer becomes shorter, it has undergone *shortening* (**Fig. 9.4a–c**). And if a change in shape involves the movement of one part of a rock body past another, so that angles between features in the rock change, the result is called *shear strain* (**Fig. 9.4d, e**).

FIGURE 9.2 Deformation changes the character and configuration of rocks.

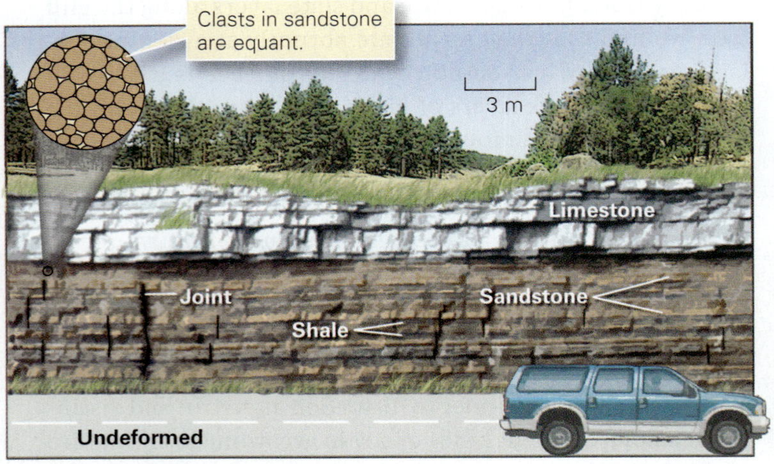
Clasts in sandstone are equant.
3 m
Limestone
Joint
Sandstone
Shale
Undeformed

(a) Flat-lying beds of strata along a highway in the Great Plains of North America are essentially undeformed. A few joints, formed when overlying rock eroded away, are visible.

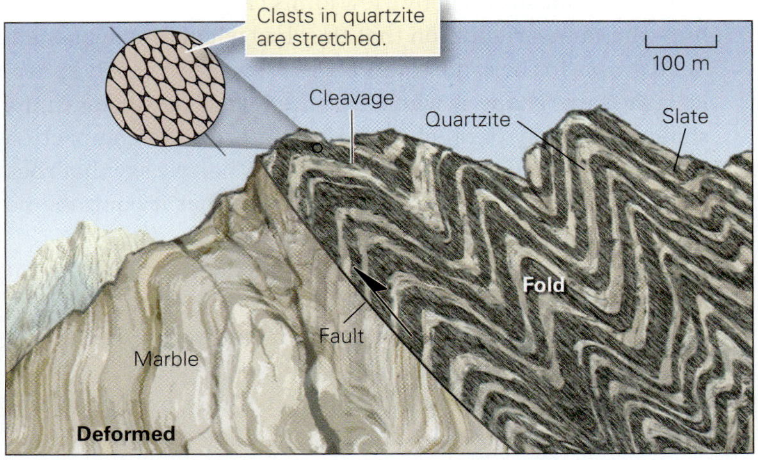
Clasts in quartzite are stretched.
100 m
Cleavage
Quartzite
Slate
Fold
Fault
Marble
Deformed

(b) In a mountain belt, deformation may cause layers to undergo folding and faulting. In addition, foliation (such as slaty cleavage) and stretched clasts may develop.

FIGURE 9.3 The components of deformation include displacement, rotation, and distortion.

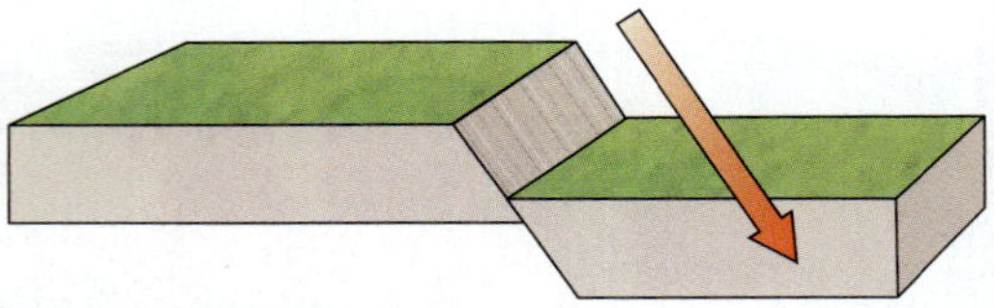

(a) Displacement occurs when a block of rock moves from one place to another.

Slip on a fault transported this rock down from where it was deposited.
~2.2 Ga quartzite
>2.7 Ga greenschist

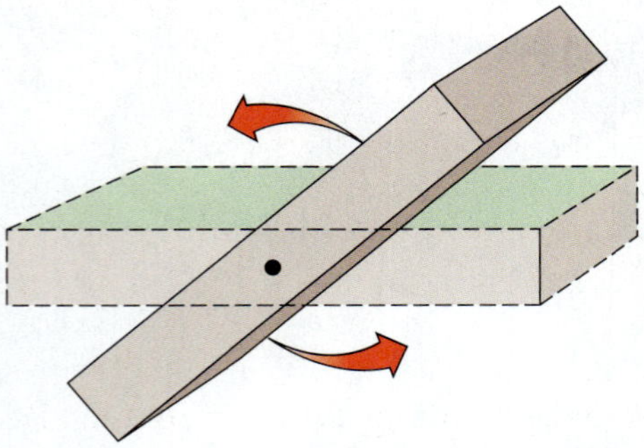

(b) Rotation occurs when a body of rock undergoes tilting.

These beds were horizontal when deposited. They are now tilted.

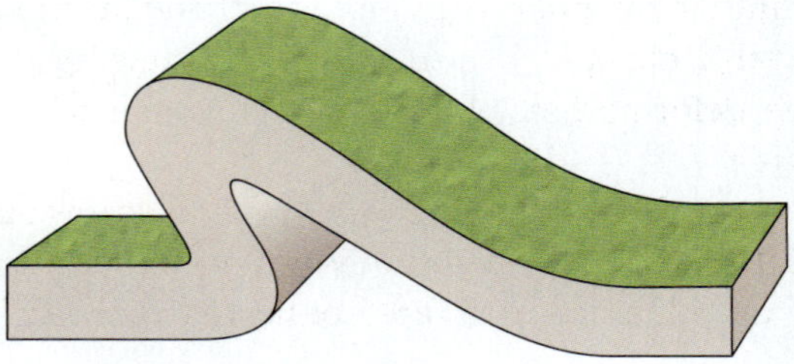

(c) Distortion occurs when rock changes shape. The development of a fold represents one type of distortion.

These beds were once horizontal and of constant thickness. They are now distorted.

FIGURE 9.4 Different kinds of strain in rock. Strain is a measure of the distortion, or change in shape, that takes place in rock during deformation.

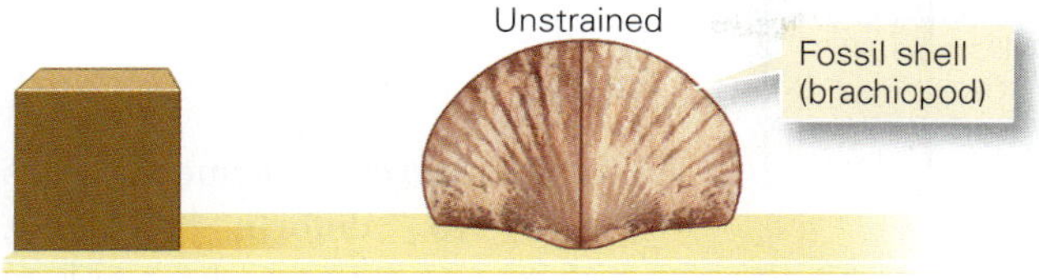

(a) An unstrained cube and an unstrained fossil shell (brachiopod).

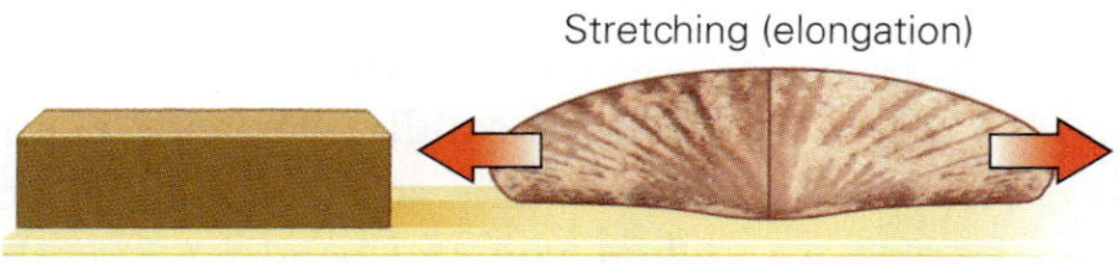

(b) Horizontal stretching changes the cube into a horizontal brick and elongates the shell.

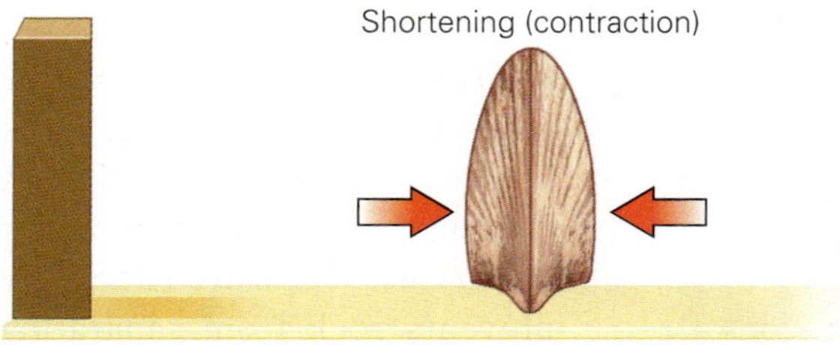

(c) Horizontal shortening changes the cube into a vertical brick and makes the shell narrower.

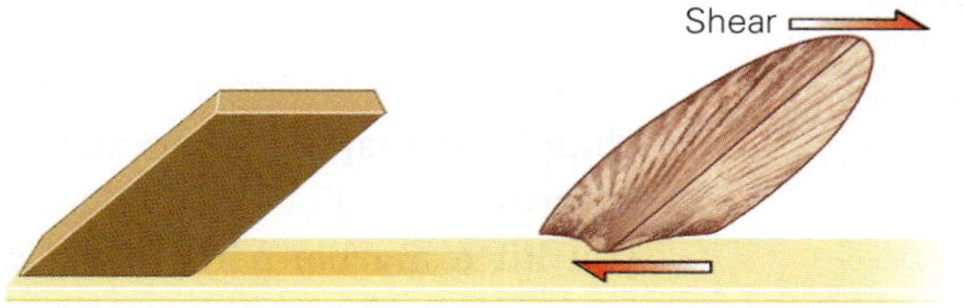

(d) Shear strain tilts the cube and transforms it into a parallelogram and changes angular relationships in the shell.

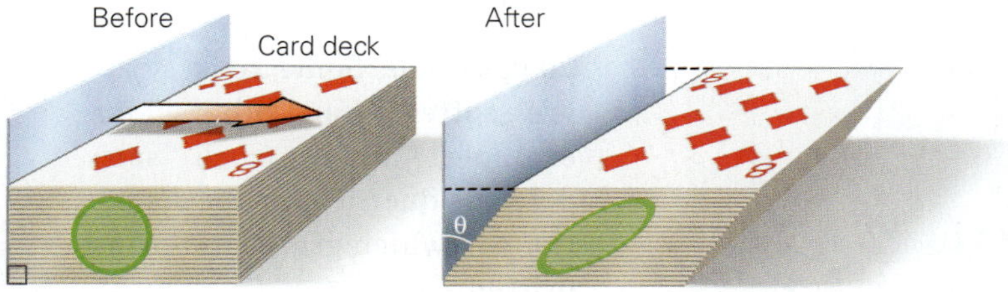

(e) You can produce shear strain by moving a deck of cards so that each card slides a little with respect to the one below.

FIGURE 9.5 Contrast between brittle and ductile deformation.

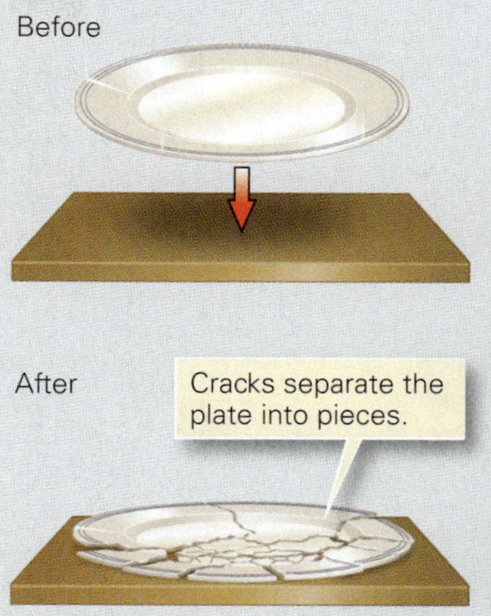

(a) Brittle deformation occurs when you drop a plate and it shatters.

(b) Cracks in these quartzite beds in Utah are due to brittle deformation.

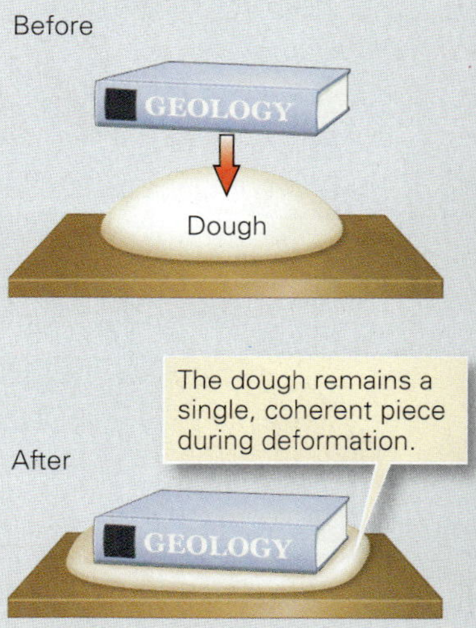

(c) Ductile (plastic) deformation occurs when you press down on a ball of dough.

(d) The quartzite cobbles in the conglomerate were flattened plastically.

Brittle versus Ductile Deformation

Look again at the example of deformed rocks from the Alpine cliff in Figure 9.2b. Forming the joints and the fault involved cracking and breaking rock, a process that geologists refer to as **brittle deformation** (**Fig. 9.5a, b**). You've seen this process in daily life if you've ever dropped a plate on a hard floor, breaking the plate into pieces. Figure 9.2b also shows a fold that formed when the rock changed shape without breaking, a process that geologists refer to as **ductile deformation**, or *plastic deformation* (**Fig. 9.5c, d**). You can produce ductile deformation by squeezing a ball of soft dough between a book and a tabletop, or by bending a stick of chewing gum. During ductile deformation, objects change shape without visibly breaking.

What actually happens within mineral grains during these two different kinds of deformation? Recall that the atoms that make up mineral grains are connected by chemical bonds.

FIGURE 9.6 There are several kinds of stress.

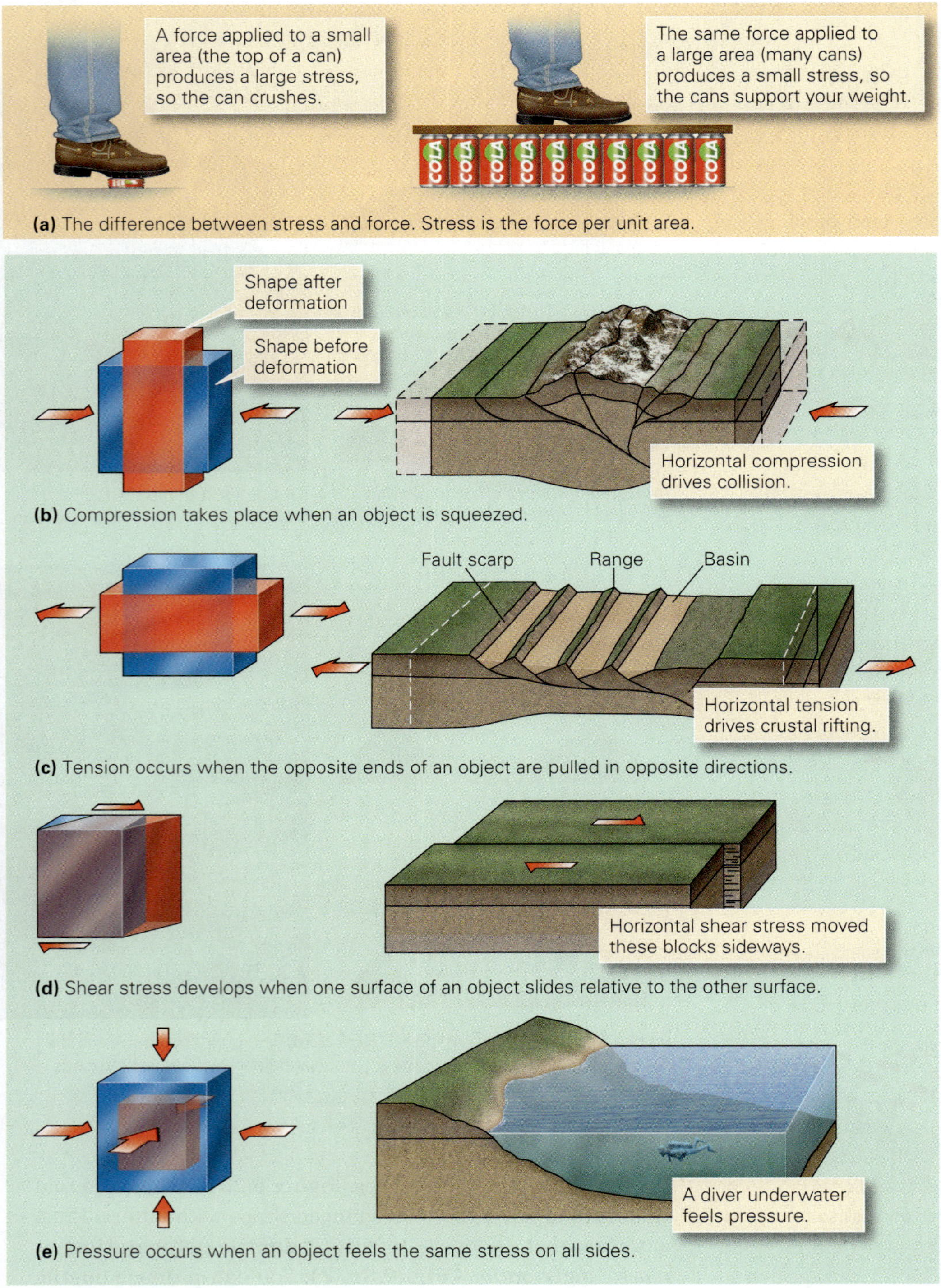
A force applied to a small area (the top of a can) produces a large stress, so the can crushes.
The same force applied to a large area (many cans) produces a small stress, so the cans support your weight.
COLA
Shape after deformation
Shape before deformation
Horizontal compression drives collision.
Fault scarp
Range
Basin
Horizontal tension drives crustal rifting.
Horizontal shear stress moved these blocks sideways.
A diver underwater feels pressure.

(a) The difference between stress and force. Stress is the force per unit area.

(b) Compression takes place when an object is squeezed.

(c) Tension occurs when the opposite ends of an object are pulled in opposite directions.

(d) Shear stress develops when one surface of an object slides relative to the other surface.

(e) Pressure occurs when an object feels the same stress on all sides.

During brittle deformation, many bonds break and stay broken, leading to the formation of a permanent fracture across which material no longer connects. During ductile deformation, in general, some bonds break, but new ones quickly form. In this way, the atoms within grains rearrange, and the grains change shape without permanent cracks forming.

Why do rocks inside the Earth sometimes deform brittlely and sometimes ductilely? The behavior of a rock depends on several factors:

- *Temperature*: Warmer rocks tend to deform ductilely, whereas colder rocks tend to deform brittlely. Heat makes materials softer. To picture the difference, compare the behavior of a candle warmed in an oven to one cooled in a freezer when you try bending it with your hands—the former easily changes shape, but the latter snaps in two.
- *Pressure*: Under great pressures deep in the Earth, rock behaves more ductilely than it does under low pressures near the surface. Pressure effectively prevents rock from separating into fragments as it deforms.
- *Deformation rate*: A sudden change in shape causes brittle deformation, whereas a slow change in shape causes ductile deformation. For example, if you hit a marble bench with a hammer, it shatters, but if you leave the bench alone for a century, it gradually sags without breaking.
- *Composition*: Some rock types are softer than others; for example, halite (rock salt) deforms ductilely under conditions in which granite deforms brittlely.

Considering that pressure and temperature both increase with depth in the Earth, geologists find that, in typical continental crust, rocks generally behave brittlely above a depth of about 10 to 15 km, and ductilely below this depth. The depth at which this change in behavior takes place is called the *brittle-ductile transition*. Earthquakes in continental crust happen mostly above this depth, because earthquakes involve brittle breaking.

Force, Stress, and the Causes of Deformation

Up to this point, we've focused on picturing the consequences of deformation. Describing the causes of deformation in the Earth's crust is a bit more challenging. Captions for museum displays about mountain building typically dispense with the issue by using the phrase, "The mountains were caused by forces deep within the Earth." But what does this mean? Isaac Newton stated that a *force* can cause an object to speed up, slow down, or change direction. In the context of geology, plate interactions and continent-continent collisions apply forces to rock and thus cause rock to change location, orientation, or shape. In other words, the application of forces in the Earth indeed causes deformation.

Geologists, however, commonly use the word *stress* instead of *force* when talking about the cause of deformation. We define the **stress** acting on a plane as the force applied per unit area of the plane. It is important to distinguish between stress and force because the consequences of applying a force depend not only on the total amount of force but also on the region over which the force acts. A pair of simple experiments shows why (**Fig. 9.6a**). Experiment 1: Stand on a single, empty aluminum can. All of your weight—a force—focuses entirely on the can, and the can crushes. Experiment 2: Place a board on top of 100 cans and stand on the board. In this case, your weight distributes across 100 cans, and the cans don't crush. In both experiments, the force caused by the weight of your body was the same, but in Experiment 1 the force was applied over a small area so a large stress developed, whereas in Experiment 2 the same force was applied over a large area so only a small stress developed.

How does the concept of stress apply to geology? During mountain building, the force of one plate interacting with another is distributed across the area of contact between the two plates, so the deformation resulting at any specific location actually depends on the stress developed at that location, not on the total force produced by the plate interaction.

Different kinds of stress can occur in nature (**Fig. 9.6b–e**). **Compression** takes place when a rock is squeezed together, **tension** occurs when a rock is pulled apart, and **shear stress** develops when one part of a rock body moves sideways past another. In general, the stress acting in one direction is not the same as the stress acting in another direction. For example, at a given locality, the horizontal compression in the north-south direction in the crust can be greater than the compression in the east-west or vertical directions. **Pressure** refers to a special stress condition that happens when the same amount of compression acts on all sides of an object.

Note that *stress* and *strain* have very different meanings to geologists, even though we tend to use the words interchangeably in everyday English. Again, stress refers to the amount of force applied per unit area of a rock, whereas strain refers to a change in shape of a rock. Thus, stress causes strain. Specifically, compression causes shortening, tension leads to stretching, and shear stress produces shear strain. When stresses are not the same in all directions, the amount of strain can be different in different directions. For example, if compression is greatest in the north-south direction, shortening will be greatest in the north-south direction. Note that pressure can cause an object to become smaller, but will not cause it to change shape. With our knowledge of stress and strain, we can now look at the nature and origin of various classes of geologic structures.

TAKE-HOME MESSAGE

Stress (compression, tension, shear) can cause rocks to change shape, position, and orientation. During brittle deformation, rocks break, whereas during ductile deformation, rocks bend and distort without breaking. Strain is a measure of shape change.

QUICK QUESTION What is the difference between stress and strain, to a geologist?

9.3 Brittle Structures

Joints and Veins

If you look at the photographs of rock outcrops in this book, you'll sometimes notice thin dark lines that cross the rock faces. These lines are the traces of natural cracks, planes along which the rock broke and separated into two pieces during brittle deformation. (We define a trace as the line of intersection between two planes; the trace of a joint on an outcrop is the line where the joint intersects the face of the outcrop.) Geologists refer to such natural cracks as **joints** (**Fig. 9.7a, b**). Rock bodies break, but do not slide past each other, on joints. Since joints are roughly planar structures, we define their orientation by their *strike* and *dip*, as described in **Box 9.1**.

Joints develop in response to tensile stress in brittle rock: a rock splits open because it has been pulled slightly apart. Joints may form for a variety of geologic reasons. For example, some joints form when a rock cools and shrinks because the process makes one part of a rock pull away from the adjacent part, and others develop when rock formerly at depth undergoes a decrease in pressure as overlying rock erodes away, and thus changes shape slightly. Note that the formation of joints in response to cooling and/or a decrease in pressure can

SEE FOR YOURSELF...

JOINTS IN ARCHES NATIONAL PARK

Latitude
38°47′45.10″ N

Longitude
109°35′34.19″ W

Look straight down from 6 km (~3.75 miles).

The white area of this view is a white sandstone bed; the reddish areas are other stratigraphic units. You can see that the sandstone has been broken by a prominent set of NW–SE-trending joints.

occur even where there has been no mountain building. Some joints form when rock layers bend, and these are associated with mountain building.

If groundwater or hydrothermal fluids seep through cracks in rocks, minerals such as quartz or calcite may precipitate out of the groundwater and fill the cracks. A mineral-filled crack is called a **vein**. Veins typically look like white stripes cutting across a body of rock (**Fig. 9.7c**). Some veins contain small quantities of valuable metals, such as gold.

Geotechnical engineers, people who study the geologic setting of construction sites, pay close attention to jointing when recommending where to put roads, dams, and buildings. Water flows much more easily through joints than it does through solid rock, so it would be a bad investment to situate a water reservoir over rock with lots of joints—the water would leak down into the joints. Also, building a road on a steep cliff composed of jointed rock could be risky, for joint-bounded blocks separate easily from bedrock, and the cliff might collapse.

Faults: Surfaces of Slip

After the San Francisco earthquake of 1906, geologists found a rupture that had torn through the land surface near the city. Where this rupture crossed orchards, it offset rows of trees, and where it crossed a fence, it broke the fence in two—the western side of the fence moved northward by about 2 m (see Fig. 8.3a). The rupture represents the trace of the San Andreas fault. As we have seen, a **fault** is a fracture on which sliding occurs, and slip events, or *faulting*, can generate earthquakes. Faults, like joints, are planar structures, so we can represent their orientation by strike and dip. Geologists refer to the amount of movement that takes place across a fault as the fault's **displacement**.

Faults have formed throughout Earth history. Some are currently *active* in that sliding has been occurring on them in recent geologic time, but most are *inactive*, meaning that sliding on them ceased long ago. Some faults, such as the San Andreas, intersect the ground surface and thus displace the ground when they move. Others accommodate the sliding of rocks in the crust at depth and remain invisible at the surface unless they are later exposed by erosion.

Fault Classification

Not all faults result in the same kind of crustal deformation—some accommodate shortening, some accommodate stretching, and some accommodate horizontal shear. It's important for geologists to distinguish among different kinds of faults in order to interpret their geologic significance.

FIGURE 9.7 Examples of joints and veins.

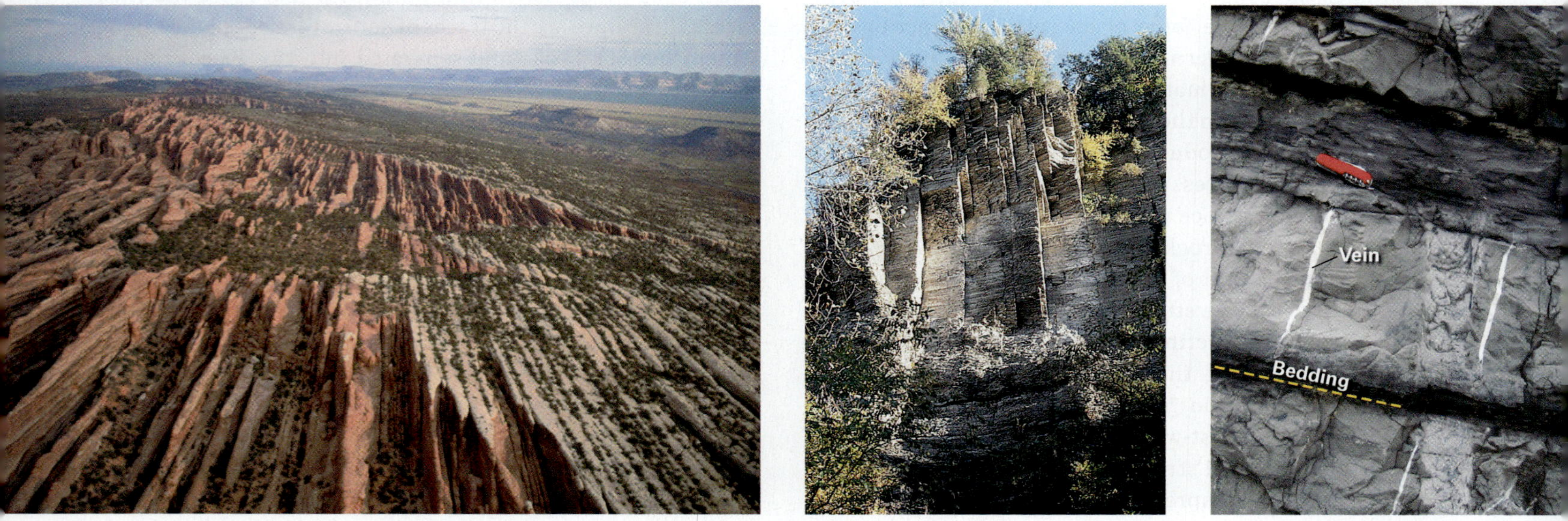

(a) Prominent vertical joints cut red sandstone beds in Arches National Park, Utah, as seen from the air.

(b) Vertical joints on a cliff face in shale near Ithaca, New York.

(c) Milky white quartz veins cut across gray limestone beds.

BOX 9.1 CONSIDER THIS...

Describing the Orientation of Geologic Structures

When discussing geologic structures, it's important to be able to communicate information about their orientation. For example, does a fault exposed in an outcrop at the edge of town continue beneath the nuclear power plant 3 km to the north, or does it go beneath the hospital 2 km to the east? If we knew the fault's orientation, we might be able to answer this question. To describe the orientation of a geologic structure, geologists picture the structure as a simple geometric shape, then specify the angles that the shape makes with respect to a horizontal plane (a flat surface parallel to sea level), a vertical plane (a flat surface perpendicular to sea level), and the north direction (a line of longitude).

Let's start by considering planar structures such as faults, beds, joints, and foliation. We call these structures *planar* because they resemble a geometric plane. A planar structure's orientation can be specified by its strike and dip. The **strike** is the angle between an imaginary horizontal line (the strike line) on the plane and the direction to true north (**Fig. Bx9.1a, b**). The **dip** is the angle of the plane's slope or, more precisely, the angle between a horizontal plane and the dip line (an imaginary line parallel to the steepest slope on the structure), as measured in a vertical plane perpendicular to the strike. A horizontal plane has a dip of 0°, and a vertical plane has a dip of 90°. We can measure the strike with a special type of compass (**Fig. Bx9.1c**), and dip with a clinometer, a type of protractor that uses a bubble to indicate horizontal, and we can represent strike and dip on a geologic map using the symbol shown in Figure Bx9.1b.

A linear structure resembles a geometric line rather than a plane. Examples of linear structures include scratches or grooves on a rock surface, or aligned elongate minerals. Geologists specify the orientation of a line by giving its plunge and bearing (**Fig. Bx9.1d**). The **plunge** is the angle between a line and horizontal in the vertical plane that contains the line. A horizontal line has a plunge of 0°, and a vertical line has a plunge of 90°. The **bearing** is the compass heading of the line, meaning the angle between the projection of the line on the horizontal plane and the direction to true north.

FIGURE Bx9.1 Specifying the orientation of planar and linear structures.

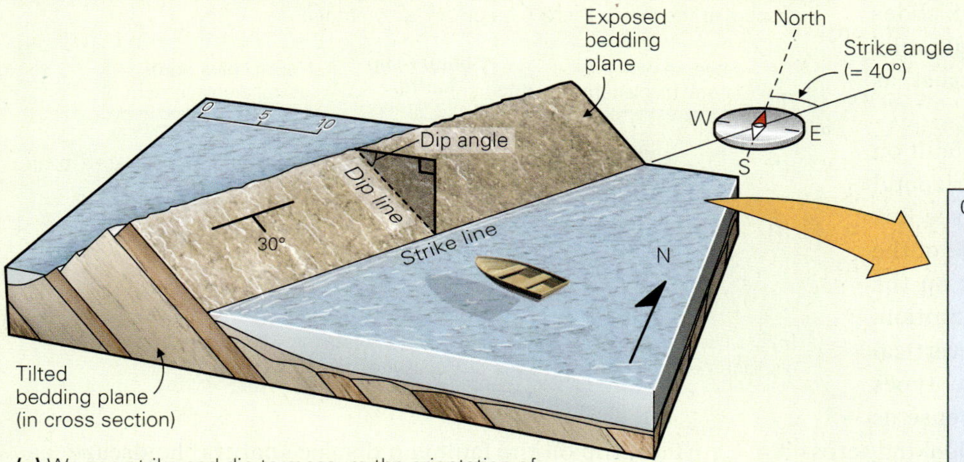

(a) We use strike and dip to measure the orientation of planar structures, such as these tilted beds. The shaded plane is vertical.

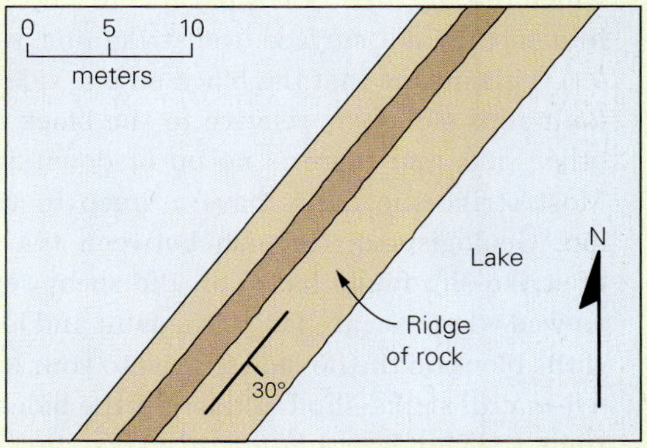

(b) On a map, the line segment represents the strike direction and the tick on the segment represents the dip direction. The number indicates the dip angle as measured in degrees.

(c) Geologists use a Brunton compass to measure strike and dip.

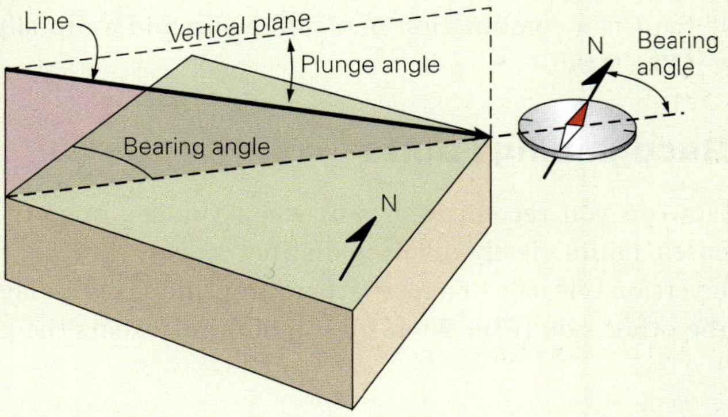

(d) To specify the orientation of a line, we use plunge and bearing.

Fault classification focuses on two characteristics of faults: (1) the *dip* or slope of the fault surface (see Box 9.1)—the dip can be vertical, horizontal, or any angle in between; and (2) the *shear sense* across the fault—by shear sense, we mean the direction that material on one side of the fault moved relative to the material on the other side. With this concept in mind, let's consider the principal kinds of faults, expanding on the discussion presented in Chapter 8.

FIGURE 9.8 The different categories of faults.

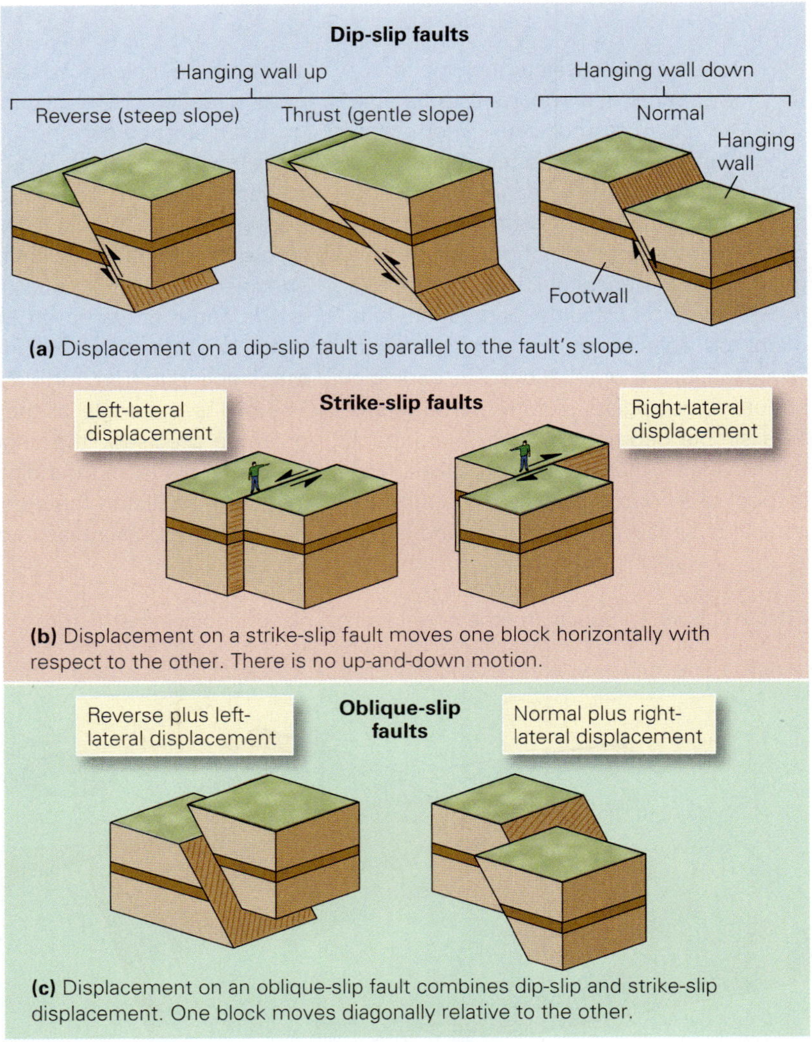

(a) Displacement on a dip-slip fault is parallel to the fault's slope.

(b) Displacement on a strike-slip fault moves one block horizontally with respect to the other. There is no up-and-down motion.

(c) Displacement on an oblique-slip fault combines dip-slip and strike-slip displacement. One block moves diagonally relative to the other.

- *Dip-slip faults:* On a **dip-slip fault**, displacement is parallel to a line going down the slope of the fault surface (the dip line; see Box 9.1), so the *hanging-wall block*, meaning the material above the fault surface, slides up or down relative to the *footwall block*, the material below the fault surface (**Fig. 9.8a**). If the hanging-wall block slides up, the fault is a **reverse fault**; a reverse fault with a gentle dip (less than 30°) is also known as a **thrust fault**. Reverse faults accommodate shortening of the crust, as happens during continental collision. If the hanging-wall block slides down, the fault is a **normal fault**. Normal faults accommodate stretching of the Earth's crust, as happens during rifting.
- *Strike-slip faults*: A **strike-slip fault** is a fault on which the slip direction is parallel to a horizontal line on the fault surface (the strike line; see Box 9.1). This means that the block on one side of the fault slips sideways, relative to the block on the other side, and there is no up-or-down motion. Most strike-slip faults have a steep to vertical dip. Geologists distinguish between two types of strike-slip faults based on the shear sense as viewed when you are facing the fault and looking across it. If the block on the far side slipped to your left, the fault is a *left-lateral* strike-slip fault, and if the block slipped to the right, the fault is a *right-lateral* strike-slip fault (**Fig. 9.8b**). Displacement at a transform plate boundary is accommodated by movement on strike-slip faults.
- *Oblique-slip faults*: On an **oblique-slip fault**, sliding occurs diagonally on the fault plane. In effect, an oblique-slip fault is a combination of a strike-slip and a dip-slip fault (**Fig. 9.8c**).

Recognizing Faults

How do you recognize a fault when you see one? In many cases, faults visibly displace distinct layers in rocks, so that layers on one side of a fault are not continuous with layers on the other side (**Fig. 9.9a, b**). If a fault intersects the ground surface, slip on the fault can displace natural landscape features such as stream valleys or glacial moraines (**Fig. 9.9c**), or human-made features such as highways, fences, or rows of trees in orchards. Displacement on a dip-slip or oblique-slip fault that offsets the ground surface produces a step, known as a **fault scarp**, on the ground surface (**Fig. 9.10a**).

Faulting under brittle conditions may crush or break rock in a band bordering the fault surface. If this shattered rock consists of visible angular fragments, it is called *fault breccia* (**Fig. 9.10b**), but if it consists of a fine powder, it is called *fault gouge*. Some fault surfaces are polished and grooved by movement on the fault. Polished fault surfaces are called **slickensides**, and linear grooves on fault surfaces are *slip lineations* or fault striations (**Fig. 9.10c**). Because faults tend to break up and weaken rock, the *fault trace* (the line of intersection between the fault and the ground surface) may preferentially erode to become a linear valley (see inset photo on p. 251).

FIGURE 9.9 Recognizing fault displacement in the field.

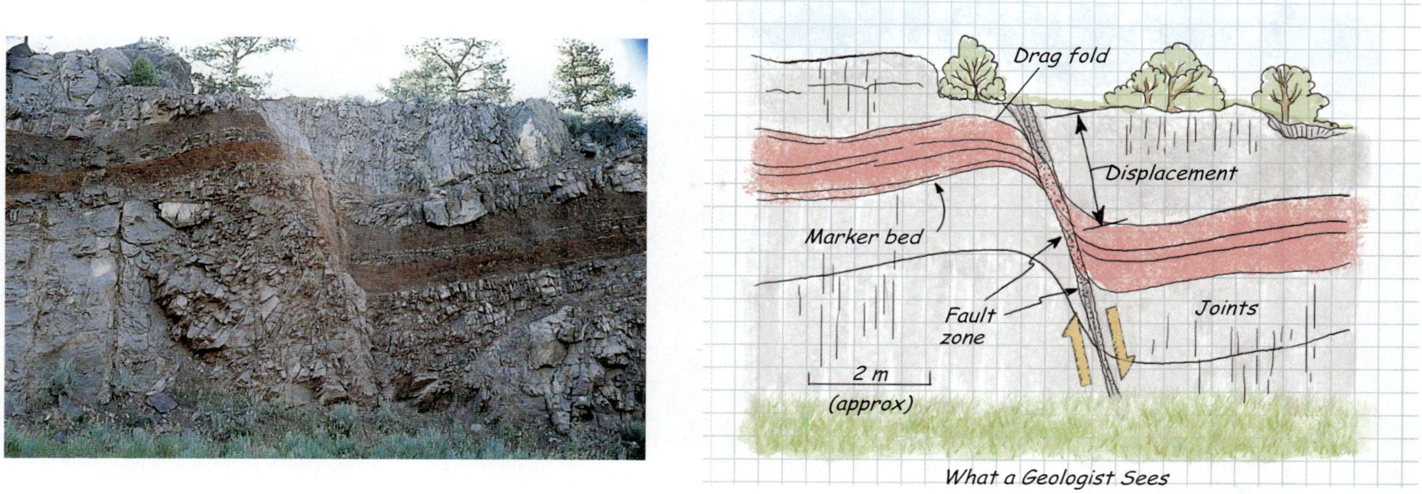

(a) A steep normal fault has displaced a distinctive red bed (a marker layer). Note that the displacement formed a 0.5-m-wide fault zone of broken rock. Drag folds developed adjacent to the fault.

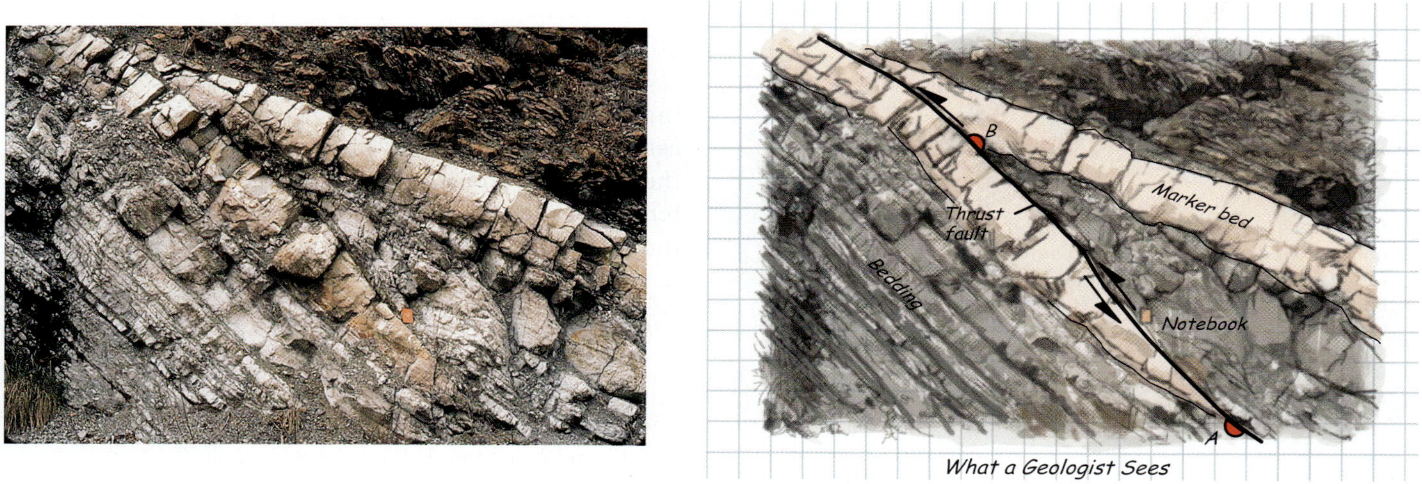

(b) Slip on a thrust fault caused one part of the light-colored marker bed to be shoved over another part, as emphasized in the drawing. Note that the beds are tilted to the right. The distance between the red dots is the displacement.

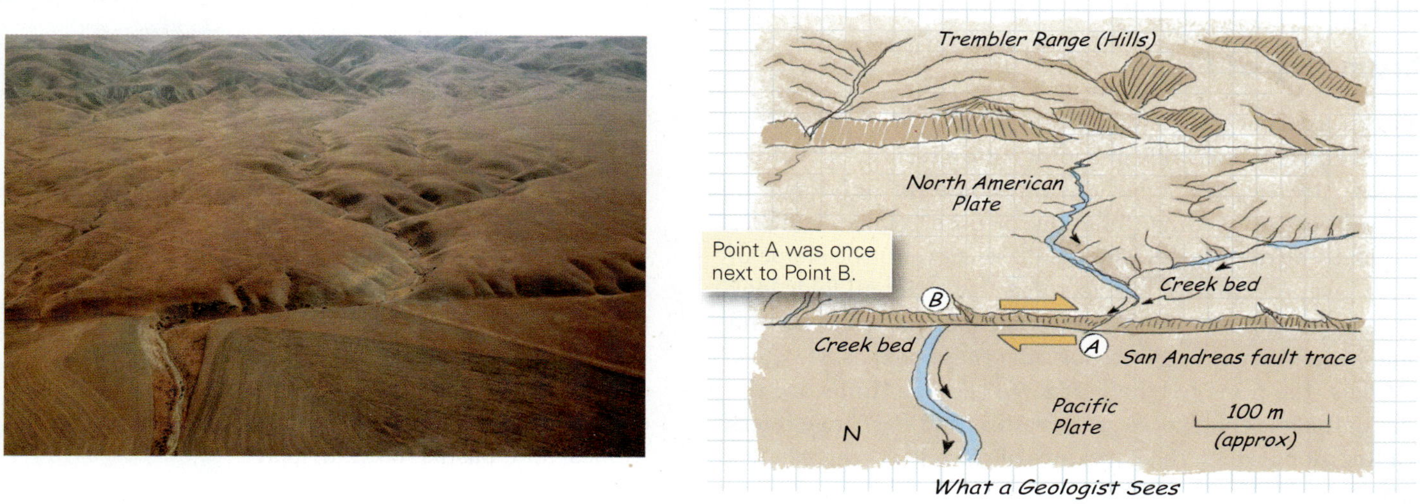

(c) An aerial photograph of a portion of the San Andreas Fault. As emphasized by the sketch, the fault offsets a stream channel in a right-lateral sense.

FIGURE 9.10 Features of exposed fault surfaces.

A scarp due to slip on a normal fault in Nevada.

(a) A fault scarp develops where faulting displaces the land surface.

(b) This fault breccia consists of irregular fragments of light-colored rock.

Striation orientation

(c) Slip lineations or striations on the surface of a strike-slip fault may look like grooves or scratches.

TAKE-HOME MESSAGE

Brittle structures include joints (cracks), veins (mineral-filled cracks), and faults (fractures on which sliding occurs). Geologists distinguish among different kinds of faults based on the relative displacement across the fault. Faulting can break up rock. Fault surfaces themselves may be polished and grooved due to the slip.

QUICK QUESTION How can you distinguish between a joint and a fault in the field?

9.4 Folds and Foliations

Geometry of Folds

Imagine a carpet lying flat on the floor. Push on one end of the carpet, and it will wrinkle or contort into a series of wave-like curves. Stresses developed during mountain building can similarly cause bedding, foliation, or other planar features in rock to warp or bend. The result—a curve in the shape of a rock layer—is called a **fold** (see Fig. 9.2b). Not all folds look the same—some look like arches, some look like troughs, and some have other shapes. To describe these shapes, we must first label the parts of a fold (**Fig. 9.11a**). The **limbs** are the sides of the fold that have less curvature, and the **hinge** refers to a line along which the curvature of the fold is greatest. (As its name suggests, a fold hinge geometrically serves the same role as a familiar metal hinge that allows the movement of a door next to a wall.) The **axial surface** is an imaginary plane that contains the hinge lines of successive layers, and effectively divides the fold into two halves. With these terms in hand, we distinguish among the following:

- *Anticlines, synclines, and monoclines*: Folds that have an arch-like shape in which the limbs dip away from the hinge are called **anticlines** (see Fig. 9.11a), whereas folds with a trough-like shape in which the limbs dip toward the hinge are called **synclines** (**Fig. 9.11b**). A **monocline** has the shape of a carpet draped over a stair step (**Fig. 9.11c**).
- *Nonplunging and plunging folds*: If the hinge is horizontal, the fold is called a nonplunging fold, but if the hinge is tilted, the fold is called a plunging fold (**Fig. 9.11d**).
- *Domes and basins*: A fold with the shape of an overturned bowl is called a **dome**, whereas a fold shaped like an upright bowl is called a **basin** (**Fig. 9.11e, f**). Domes and basins both display circular patterns that look like bull's-eyes on a map, if the land surface cuts across the structure.

SEE FOR YOURSELF...

APPALACHIAN FOLD BELT

Latitude
40°53′24.28″ N

Longitude
77°4′28.11″ W

Zoom to 60 km (37 miles). Tilt the view and look west.

These are relicts of large folds formed when Africa collided with North America, producing the Appalachian Orogen about 280 Ma. Erosion has removed the high mountains. Now, the region is the Valley and Ridge Province; hard sandstone beds resistant to erosion underlie the ridges.

FIGURE 9.11 Geometric characteristics of folds.

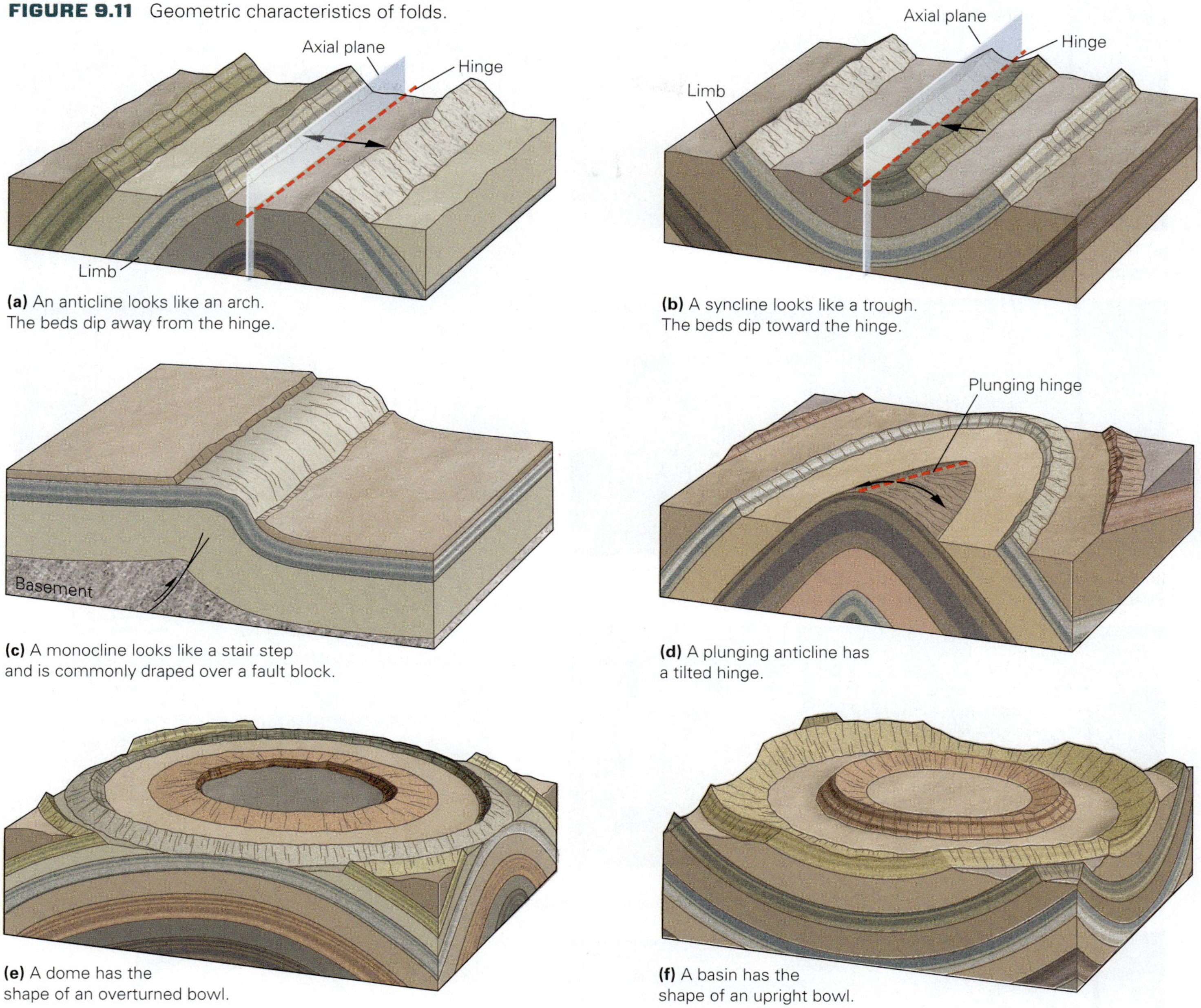

(a) An anticline looks like an arch. The beds dip away from the hinge.

(b) A syncline looks like a trough. The beds dip toward the hinge.

(c) A monocline looks like a stair step and is commonly draped over a fault block.

(d) A plunging anticline has a tilted hinge.

(e) A dome has the shape of an overturned bowl.

(f) A basin has the shape of an upright bowl.

If domes and basins involve sedimentary strata, note that the oldest layer crops out in the center of a dome, whereas the youngest layer crops out in the center of a basin.

Using these terms, see if you can identify the various folds shown in **Figure 9.12**.

Formation of Folds

Folds develop in two principal ways (**Fig. 9.13**). During formation of *flexural-slip folds,* a stack of layers bends, and slip occurs between the layers to accommodate the bending without creating gaps between layers. The same phenomenon happens when you bend a deck of cards—to accommodate the change in shape, the cards slide with respect to each other. *Passive-flow folds* form when the rock, overall, is so soft that it behaves like weak plastic and slowly flows. These folds develop because different parts of the rock body flow at different rates. The layers serve as visual cues to the fold geometry, but they did not control movement during folding.

Why do folds form? Some layers wrinkle up, or buckle, in response to end-on compression (**Fig. 9.14**). In such cases, a "train" of anticlines and synclines may form, with their fold hinges perpendicular to the direction of compression. Others form where shear stress gradually shifts one part of a rock body relative to another—layers caught up in the zone of shear fold because one part of the layer moves past another part.

FIGURE 9.12 Examples of folds on outcrops and in the landscape.

(a) This anticline, exposed in a road cut near Kingston, New York, involves beds of Paleozoic limestone.

(b) This syncline, exposed in a road cut in Maryland, involves beds of Paleozoic sandstone and shale.

(c) This fold, exposed along the coast of Brazil, occurs in Precambrian gneiss.

(d) A train of folds exposed in sea cliffs in eastern Ireland includes anticlines and synclines. The folds affect beds of Paleozoic sandstone and shale.

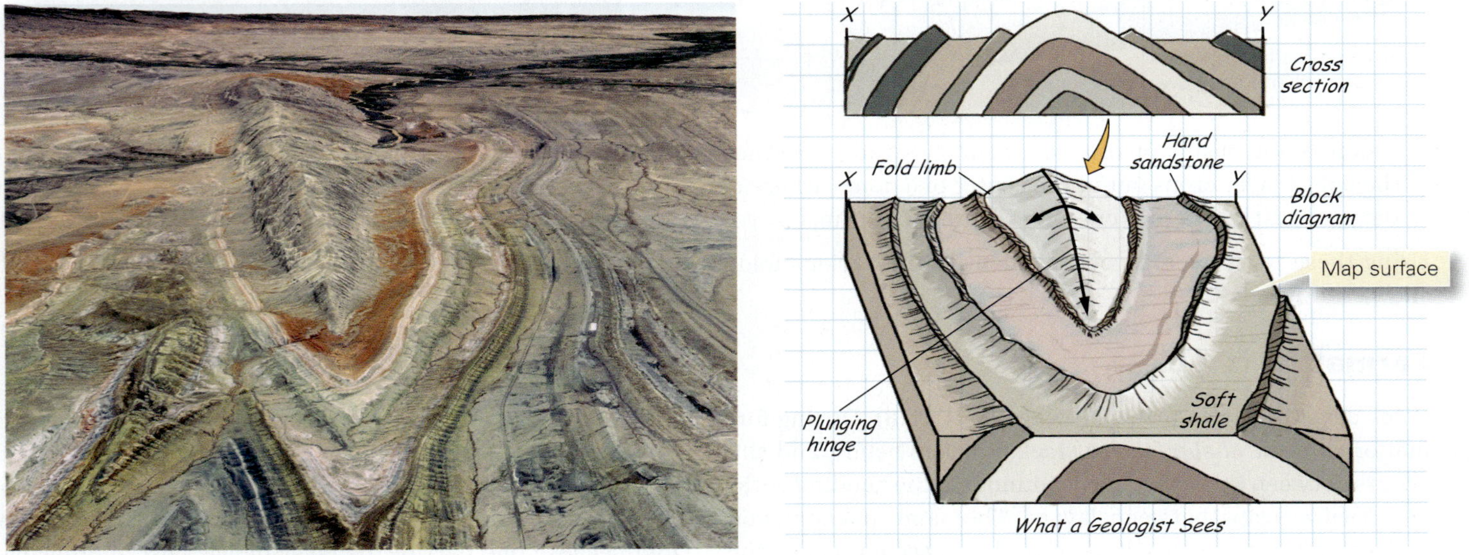

(e) The plunging anticline of Sheep Mountain, Wyoming, is easy to see because of the lack of vegetation. Resistant rock layers (sandstone) stand out as ridges, whereas weak rock layers (shale) erode away. A block diagram shows how surface exposures relate to underground structure.

FIGURE 9.13 Fold development in flexural-slip and passive-flow folding.

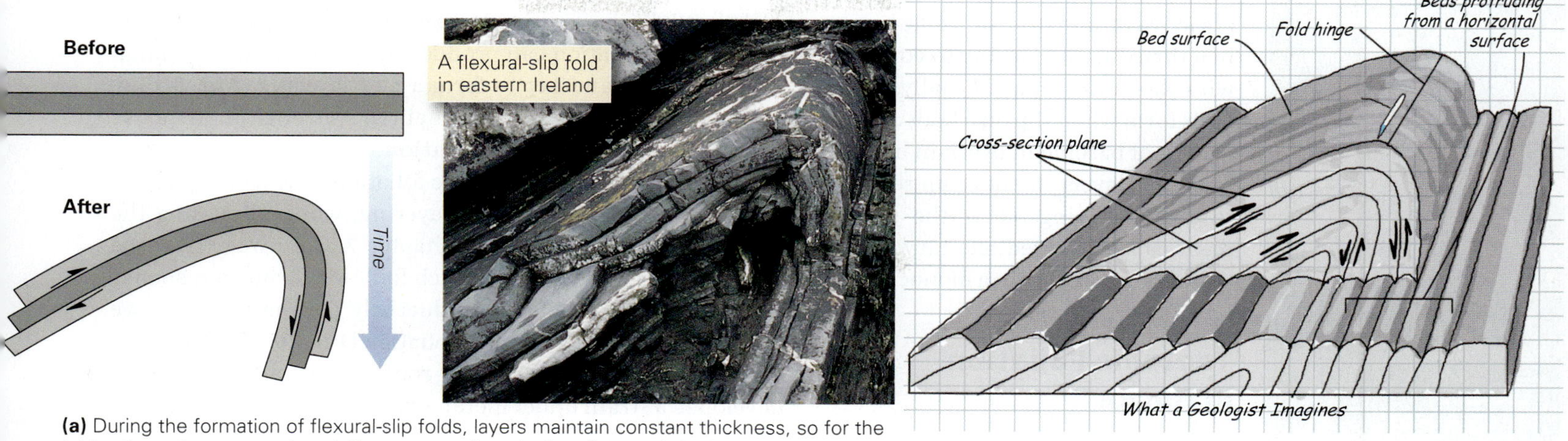

(a) During the formation of flexural-slip folds, layers maintain constant thickness, so for the fold to form, layers must bend. To accommodate the bending, each bed slips relative to its neighbor.

(b) During the formation of passive-flow folds, the rock slowly flows. Different points along a marker line flow at different rates, causing the layer to become folded. Note that the layer's thickness doesn't stay constant during folding.

FIGURE 9.14 Folding is caused by several different processes.

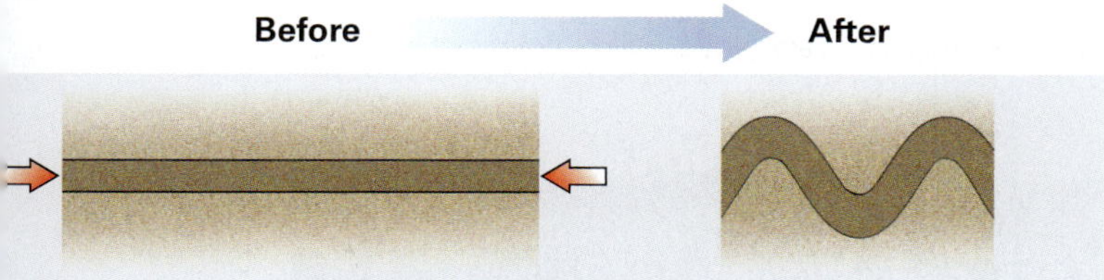

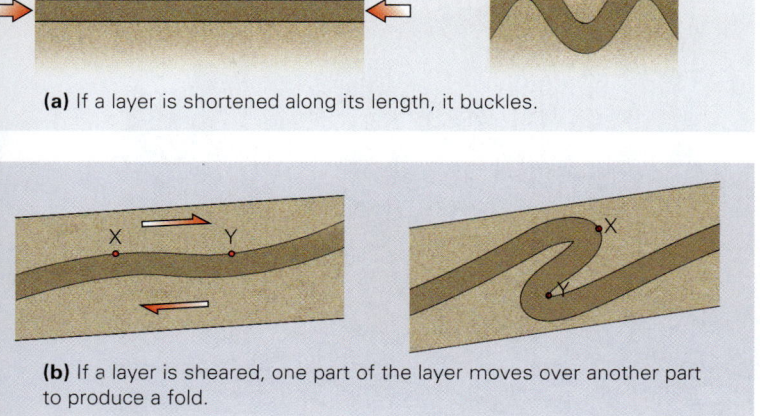

(a) If a layer is shortened along its length, it buckles.

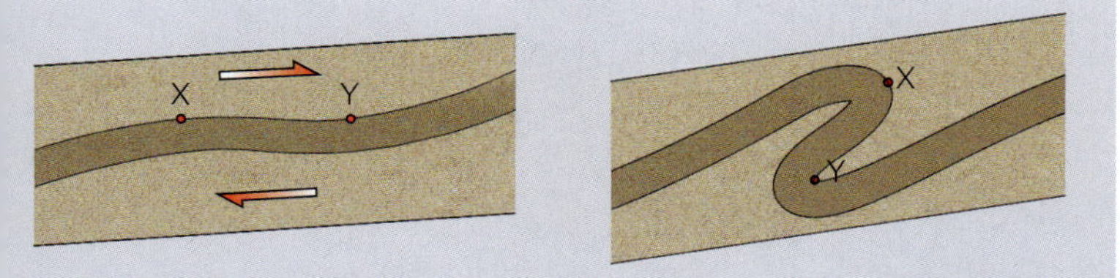

(b) If a layer is sheared, one part of the layer moves over another part to produce a fold.

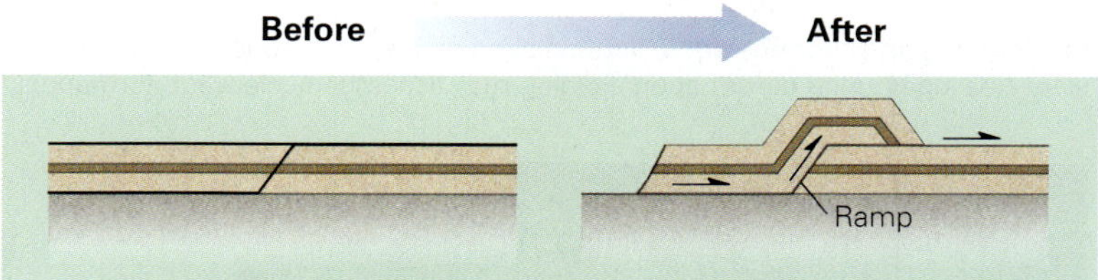

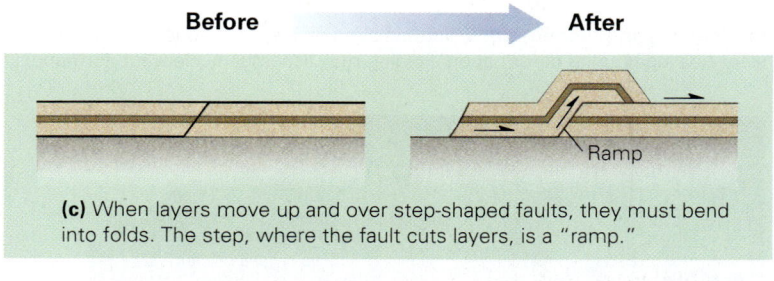

(c) When layers move up and over step-shaped faults, they must bend into folds. The step, where the fault cuts layers, is a "ramp."

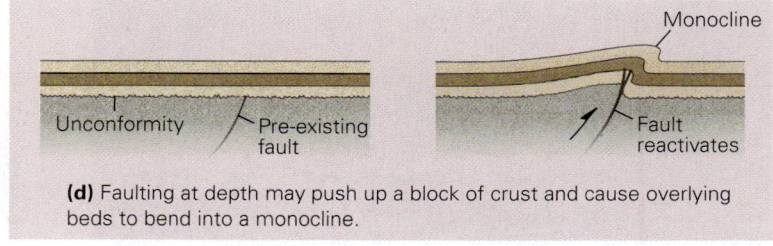

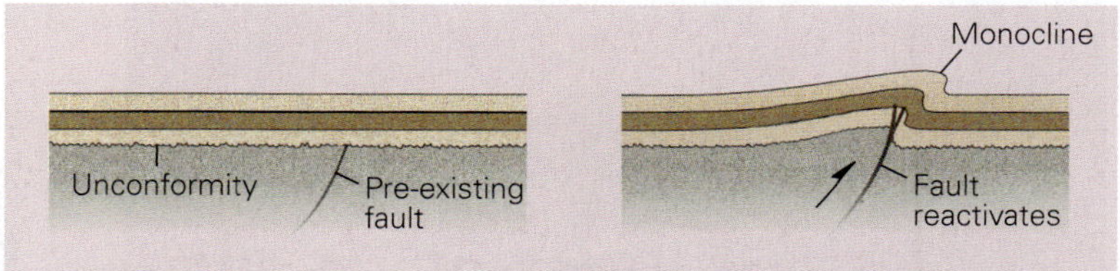

(d) Faulting at depth may push up a block of crust and cause overlying beds to bend into a monocline.

SEE FOR YOURSELF...

FOLDS, CENTRAL AUSTRALIA

Latitude
24°18′44.08″ S

Longitude
132°10′34.02″ E

Zoom to 30 km (~18.5 miles) and look down.

Alternating beds of resistant and nonresistant sedimentary strata have been warped into plunging folds.

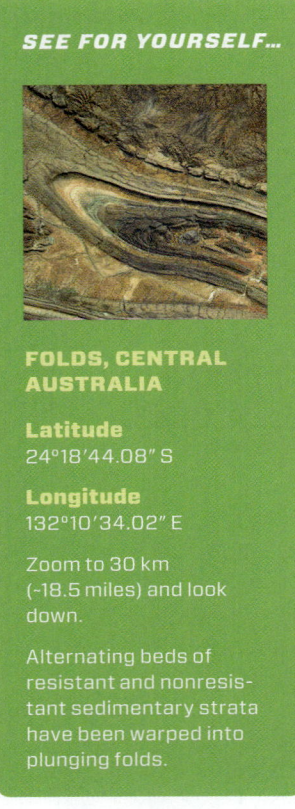

Still others develop where rock layers move up and over step-like bends in a thrust fault, curving to conform with the fault's shape. Finally, some folds form when slip on a fault causes a block of basement, overlain by sedimentary layers, to move up relative to a neighboring block—the movement causes the sedimentary strata to bend, typically into a monocline.

Tectonic Foliation in Rocks

In an undeformed sandstone, the grains of quartz are roughly spherical, and in an undeformed shale, clay flakes press together into the plane of bedding so that shales tend to split parallel to the bedding. During ductile deformation, however, internal changes take place in a rock that gradually modify the original shape and arrangement of grains. For example, quartz grains may transform into tiny pancakes, and clay flakes may recrystallize and undergo reorientation. Overall, deformation can produce inequant grains that align parallel to one another to form a new layering (**Fig. 9.15a**). Layering developed when inequant grains align in response to deformation is **tectonic foliation**.

We introduced metamorphic foliation, such as slaty cleavage, schistosity, and gneissic layering, while discussing the effects of metamorphism in Chapter 7. Here we add to the story by noting that because such foliation forms in response to flattening and shearing in ductilely deforming rocks, we can also refer to it as tectonic foliation (**Fig. 9.15b**). The presence of tectonic foliation in a rock indicates that the rock developed a strain under metamorphic conditions.

TAKE-HOME MESSAGE

Folds are bends or curves defined by the shape of rock layers. Arch-like folds are anticlines, and trough-like folds are synclines. Deformation can change the shape and orientation of grains, aligning them to produce a planar fabric called tectonic foliation.

QUICK QUESTION What mechanisms allow layers to undergo folding?

FIGURE 9.15 The development of tectonic foliation in rock.

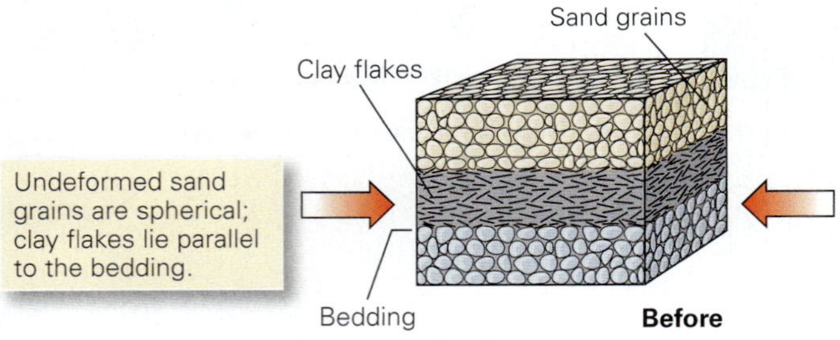

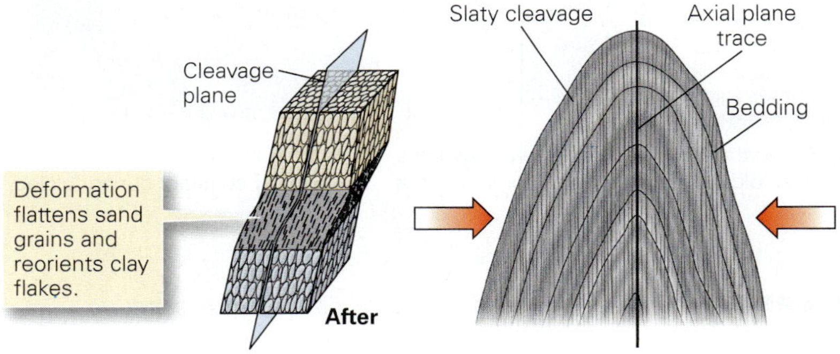

(a) Compression shortens beds, flattens sand grains, and reorients clay flakes. Clay flakes were originally parallel to bedding, but they become parallel to slaty cleavage during deformation. Folding may accompany cleavage formation.

(b) An example of slaty cleavage developed in Paleozoic strata exposed in a stream cut in New York. Relict bedding is still visible, but note that the rock breaks more easily on the cleavage. This cleavage formed in association with folding.

9.5 Causes of Mountain Building

Before plate tectonics theory became established, geologists were just plain confused about how mountains formed. In the context of the new theory, however, the many processes driving mountain building became clear—mountains form primarily in response to convergent-boundary deformation, continental collisions, and rifting. Below, we look at these different settings and the types of mountains and geologic structures that develop in each one.

Did you ever wonder...
why mountains occur in distinct belts?

Mountains Related to Subduction

At convergent boundaries, oceanic lithosphere of the downgoing plate subducts and sinks into the mantle beneath the overriding plate. This process, as we saw in Chapter 2, triggers melting in the mantle and growth of a volcanic arc along the edge of the overriding plate. In many locations where the overriding plate consists of continental crust, the interaction between the downgoing plate and the overriding plate produces compression in the continental crust. This compression, in turn, causes crustal shortening and associated deformation, and yields a *convergent-margin orogen* (**Fig. 9.16a**). Many types of geologic structures form in such orogens. For example, in the warm crust at depth in the interior of the orogen, plastic deformation and associated metamorphism yield passive-flow folds and foliations, as well as reverse faults that gradually carry rocks from deeper crustal levels up toward the Earth's surface. Along the continental side of such a belt, at shallower depths, numerous thrust faults form, each carrying a wide, and relatively thin, sheet of rock called a *thrust slice*. These faults typically merge at depth with a subhorizontal fault called a *detachment* (**Fig. 9.16b**). Each thrust slice moves up and over its neighbor, so the slices overlap like shingles. Notably, rocks within thrust slices undergo flexural-slip folding as the slices move. The overall assemblage of thrust faults and associated folds is called a **fold-thrust belt**. As a result of horizontal shortening, significant uplift takes place in convergent-margin orogens. As the land rises, erosion carves it into rugged topography, as displayed in the Andes (**Fig. 9.16c**).

FIGURE 9.16 Characteristics of convergent-margin orogens.

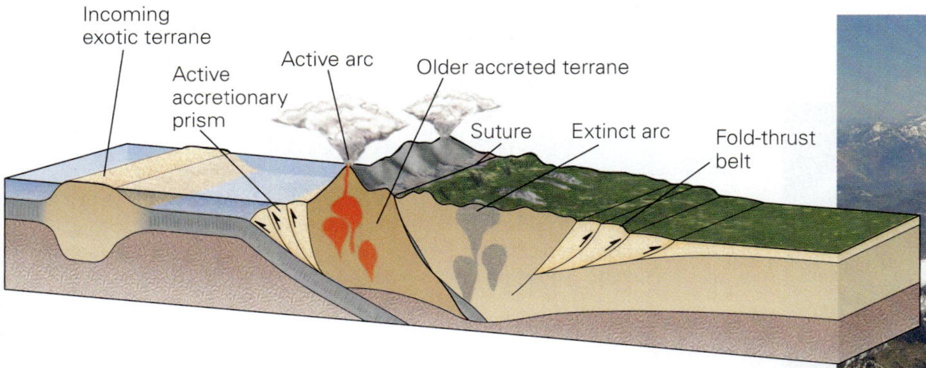

(a) At some convergent margins, compression between the downgoing and overriding plates uplifts a mountain range in which volcanism occurs. Subduction may bring in exotic crustal blocks that collide and become incorporated in the orogen.

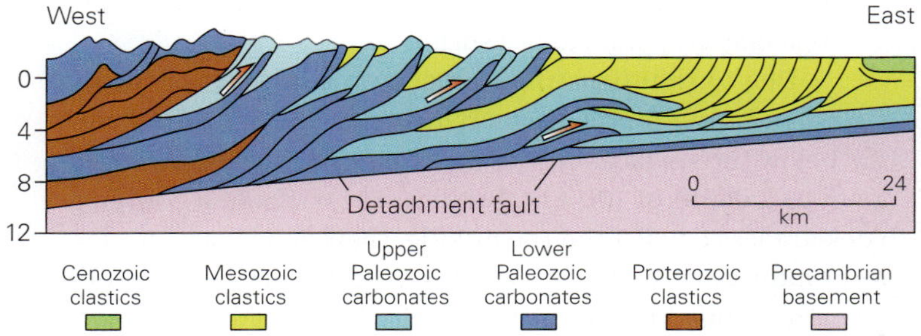

(b) A detailed cross section of a fold-thrust belt, composed of many thrust sheets pushed to the right. The different colors represent different rock ages.

(c) The Andes in Chile formed along a convergent margin.

FIGURE 9.17 Characteristics of collisional orogens.

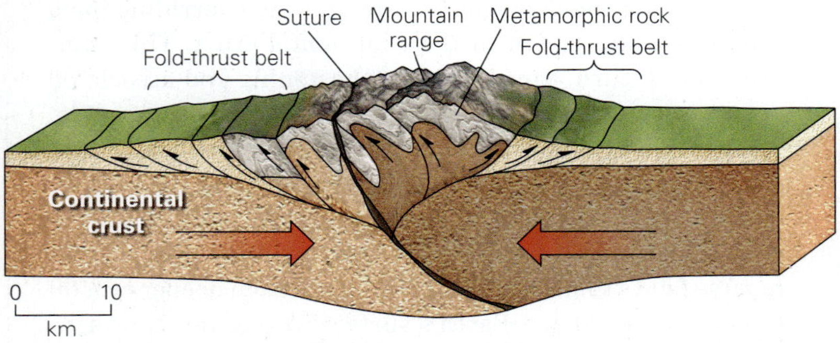

(a) During collision, continents squeeze together and deform. Thrusting brings metamorphic rock up to shallower levels.

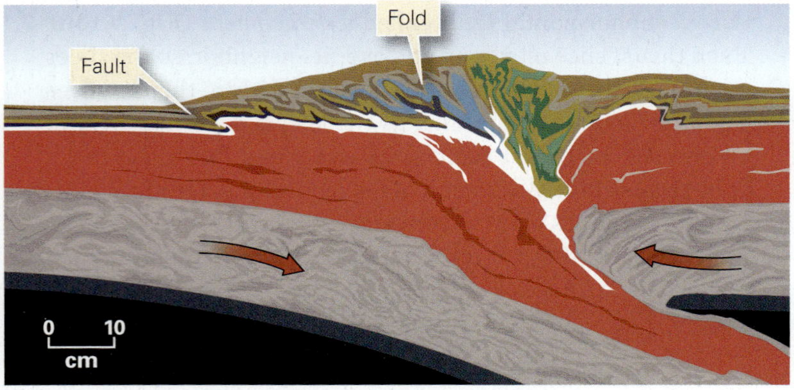

(b) Geologists simulate collision in the laboratory using layers of colored sand. Dragging the left side of the model under the right produces structures and uplift, as shown in this sketch of a model.

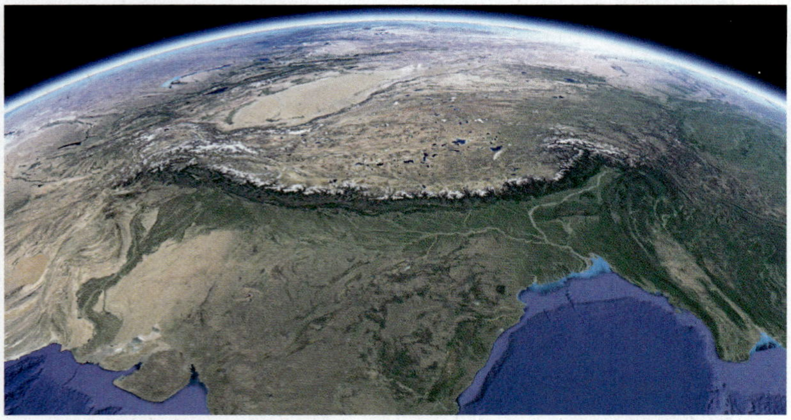

(c) Viewed from space, it's clear that the Himalayas were uplifted when India collided and pushed into Asia.

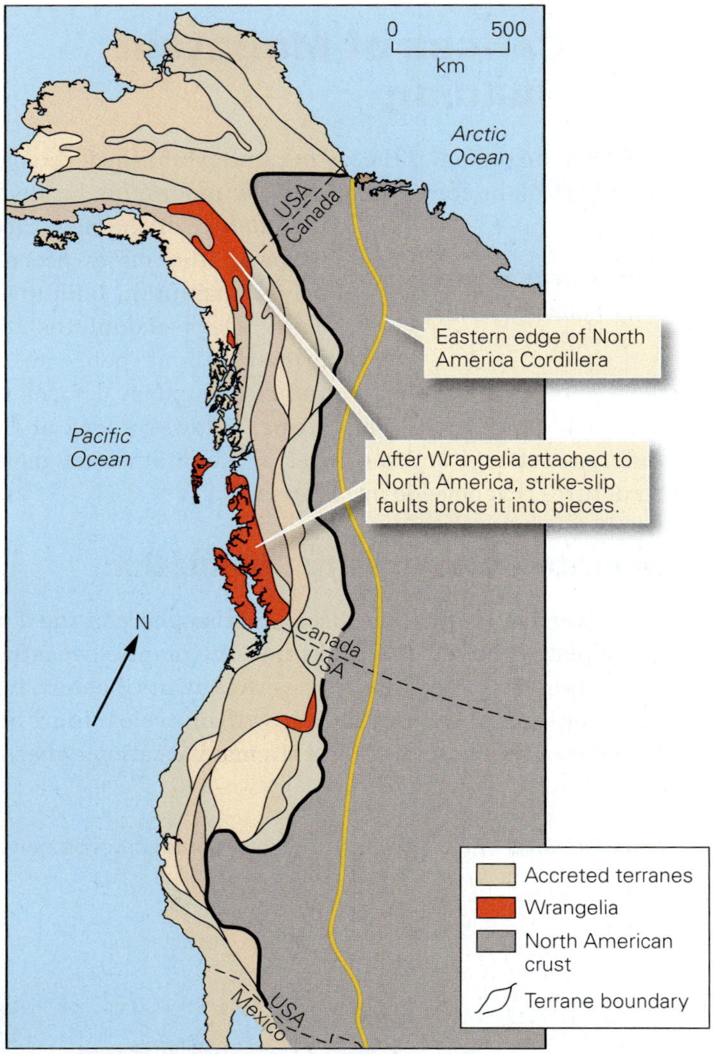

(d) The western portion of the North American Cordillera consists of accreted terranes that attached to the continent during the Mesozoic. During docking, a distinct terrane called Wrangelia (highlighted in red) was sliced into pieces that were displaced by strike-slip faults.

Mountains Related to Collision

Once the oceanic lithosphere between two blocks of relatively buoyant crust subducts completely, the blocks that had once been separated by an ocean collide with each other and produce a **collisional orogen** (**Fig. 9.17a, b**) with large thrust faults in which the edge of one block slips up and over the margin of the other. The boundary between what had been separate blocks is called a **suture**. Slivers of oceanic lithosphere sometimes get caught between colliding blocks, so a narrow belt of sheared mafic and ultramafic rock may crop out along a suture.

Due to the thrusting that occurs at or near a suture, rocks in the footwall block may end up tens of kilometers below the surface—in extreme cases, they may be carried down to a depth of 100 km. But because rocks of a colliding block are less dense than the mantle, their downward motion eventually ceases. Rocks that have been carried to depth beneath a collisional orogen undergo high-grade metamorphism, accompanied by formation of passive-flow folds and tectonic foliation. Collision also results in uplift of the crust's surface by as much as several kilometers. The crust below the orogen thickens; in some cases, total thickness may be 70 km, twice that of normal continental crust. Because folds and reverse faults develop in the interior of the orogen, rocks metamorphosed at great depth eventually squeeze upward and may be exposed at the surface. Broad fold-thrust belts typically develop on the sides of collisional orogens.

The most intense collisions happen when two continents come together. For example, continental collision yielded the Himalayas and Alps during the Cenozoic (**Fig. 9.17c**). Collision produced the Appalachian Mountains during the Paleozoic (see **Geology at a Glance**, pp. 316–317), the final stage in growth resulting when Africa and North America collided. Collisions also occur between island arcs, between small continental blocks, between island arcs and continents, and between oceanic plateaus and continents. Along some convergent boundaries, numerous collisions over geologic time suture blocks to the edge of the overriding plate. Geologists refer to the process of suturing of smaller crustal blocks to a larger one as **accretion** (see Fig. 9.16a). An incoming buoyant crustal block is called an *exotic terrane* when it lies offshore, and an *accreted terrane* once it has sutured to the overriding plate. After a terrane has accreted, the convergent plate boundary may jump to the seaward side of the terrane, so that subduction can continue. The process of accretion can add a substantial width of new crust to the edge of a continent. For example, much of the western half of the North American Cordillera consists of accreted terranes that attached to North America during the Mesozoic (**Fig. 9.17d**).

SEE FOR YOURSELF...

MOUNT EVEREST, NEPAL

Latitude
27°59′17.84″ N

Longitude
86°55′18.01″ E

Zoom to 8.5 km (~5 miles). Tilt the view and look east.

This is the highest mountain on our planet. Note the glaciers carving valleys at lower elevations. From a higher altitude, you can see that the Himalayas lie between the Tibetan Plateau to the north and the Ganges Plain to the south.

Mountains Related to Continental Rifting

A continental rift is a place where a continent is stretching and splitting in two. During rifting, tensional stress causes normal faulting in the upper crust (**Fig. 9.18a**). Movement on normal faults drops down blocks of crust, which typically tilt over as they move. As a result, rifts contain small sets of elongate mountain ranges—the tilted blocks of crust—

FIGURE 9.18 Rift-related mountains.

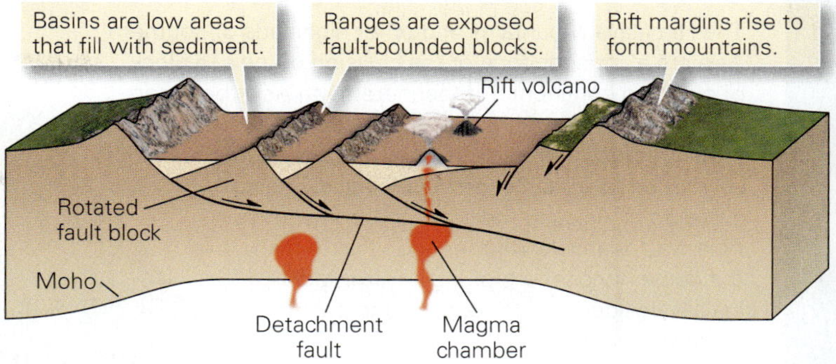

(a) Rifting leads to the development of numerous narrow mountain ranges. Rift-margin mountains may also form.

(b) The Basin and Range Province of the western United States formed during Cenozoic rifting.

separated by deep, sediment-filled basins. (Ranges in a rift are sometimes called *fault-block mountains*.) Stretching during rifting thins the lithosphere, allowing hot asthenosphere to rise and undergo decompression melting (see Chapter 4). This process produces magmas that rise to form volcanoes within the rift. Today, the East African Rift clearly shows the configuration of rift-related mountains and volcanoes. In North America, rifting yielded the broad Basin and Range Province of Utah, Nevada, and Arizona (**Fig. 9.18b**).

Measuring Mountain Building in Progress

Not all mountains are just "old monuments," as John Muir mused. The rumblings of earthquakes and the eruptions of volcanoes in some ranges attest to present-day, continuing movements which geologists can measure through field studies and satellite technology. For example, researchers can determine where coastal areas have been rising relative to

GEOLOGY AT A GLANCE

The Collision of India with Asia

The Himalaya Mountains and other highlands of southern Asia are a consequence of the collision of India, a block of old, cold continental lithosphere, with Asia, between 55 and 40 million years ago. At the time of the collision, the southern margin of Asia consisted of several smaller crustal blocks that had relatively recently accreted to Asia, so the lithosphere of southern Asia was relatively warmer and weaker than the cooler lithosphere of India. After India initially collided, it has continued to push slowly into Asia.

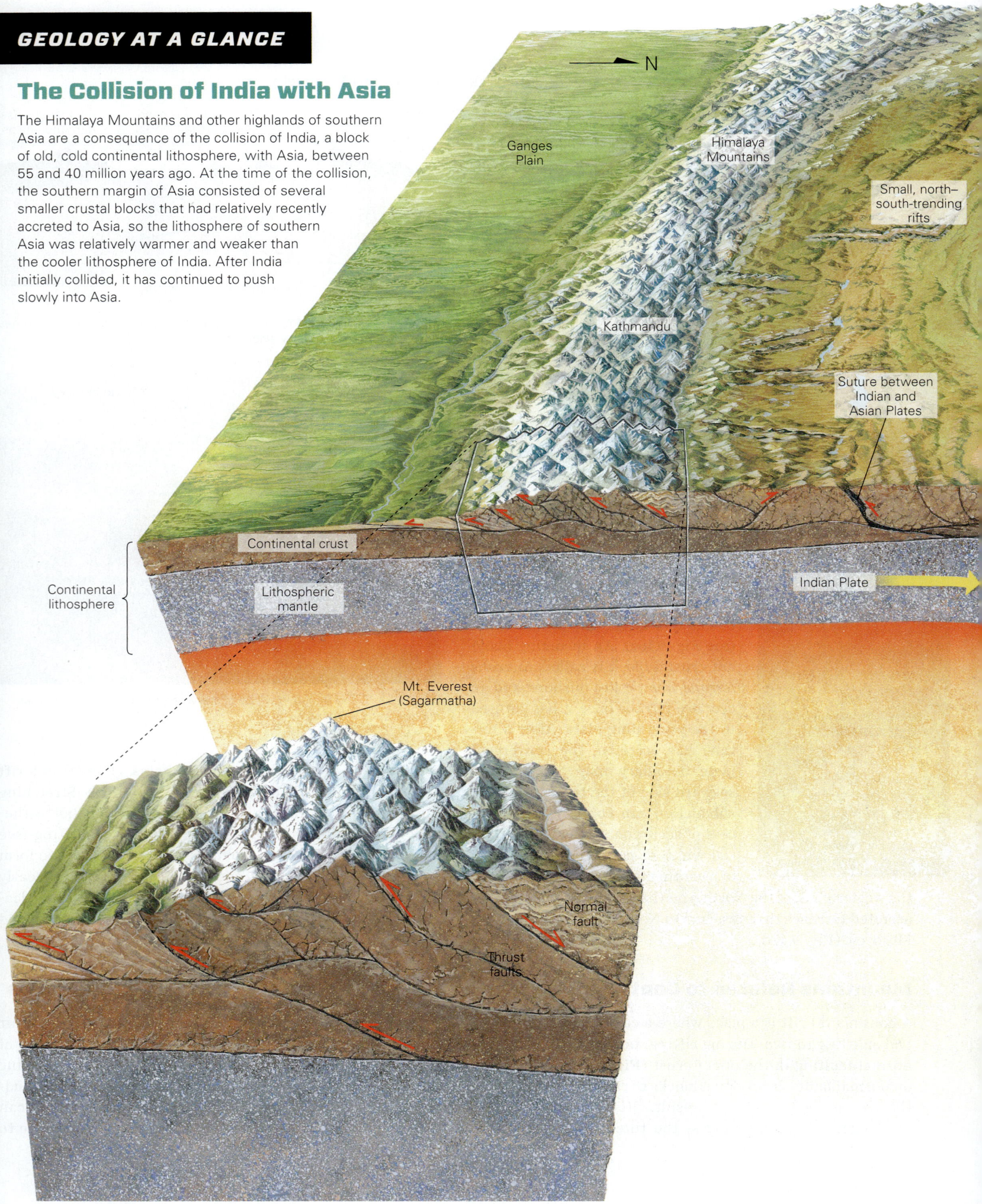

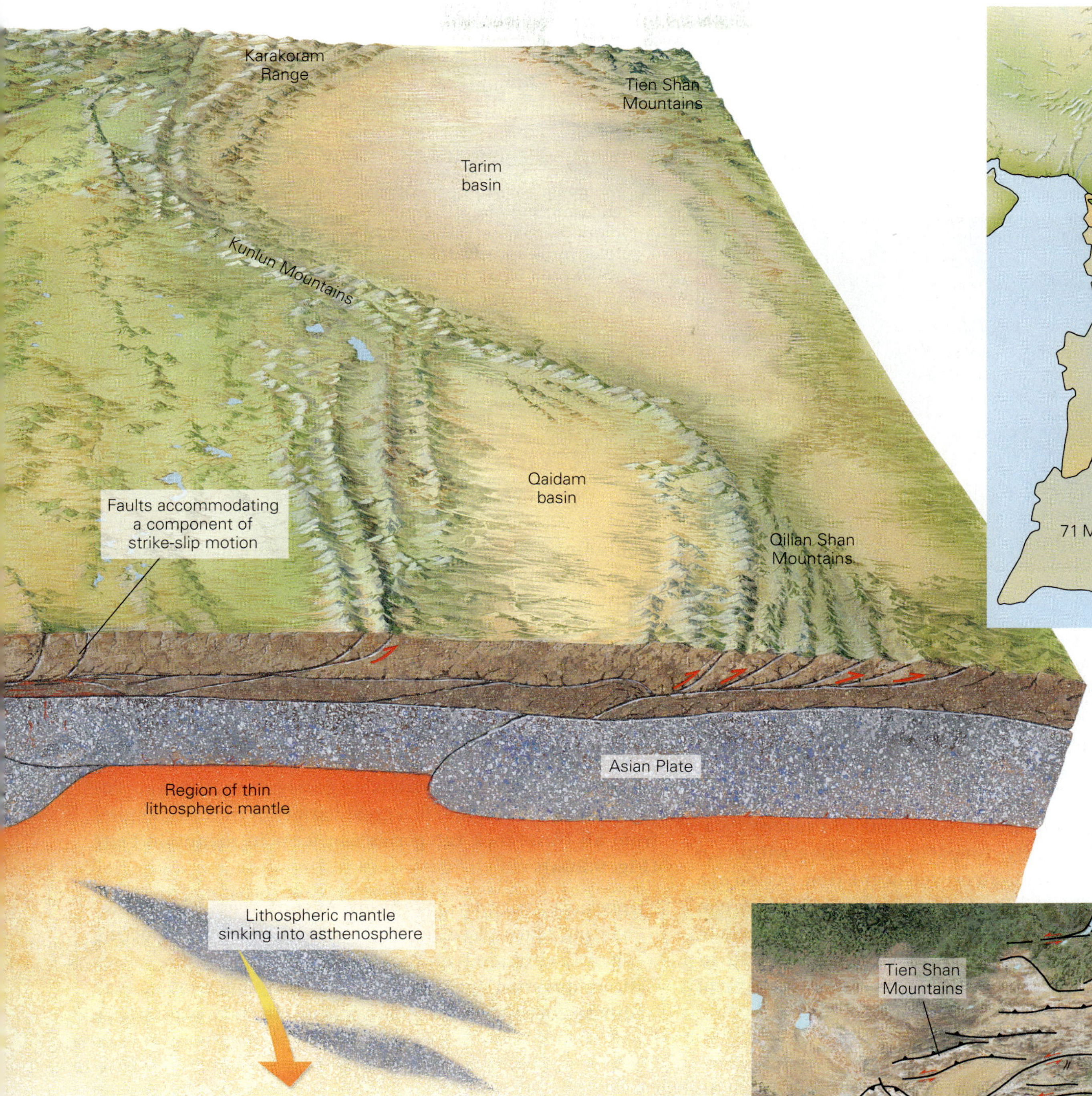

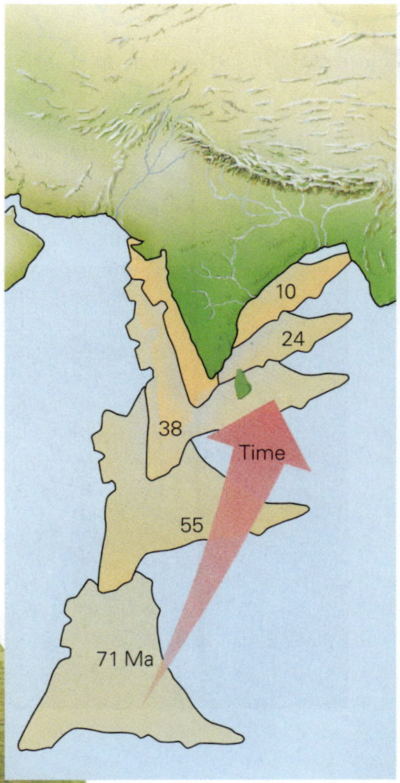

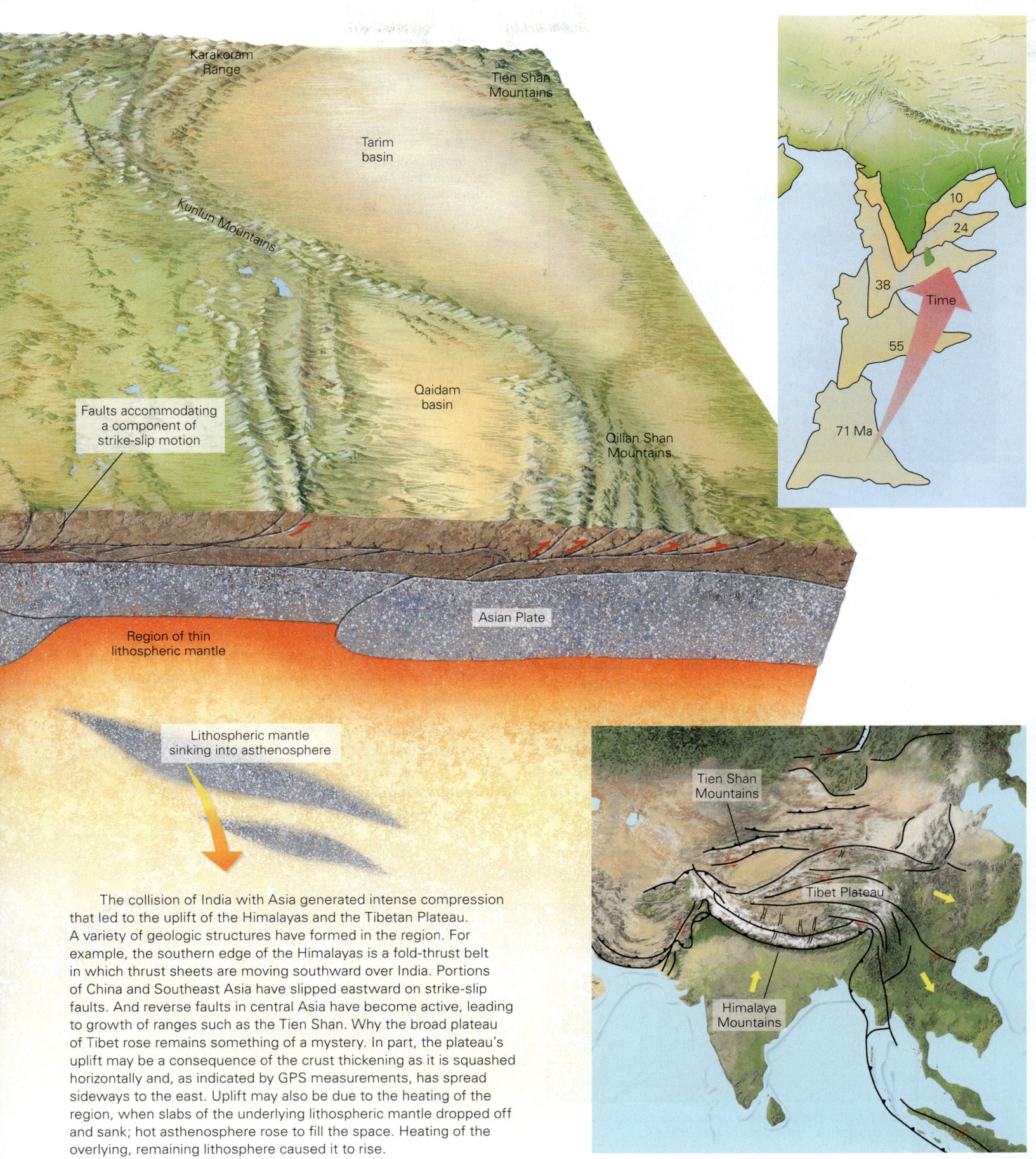

The collision of India with Asia generated intense compression that led to the uplift of the Himalayas and the Tibetan Plateau. A variety of geologic structures have formed in the region. For example, the southern edge of the Himalayas is a fold-thrust belt in which thrust sheets are moving southward over India. Portions of China and Southeast Asia have slipped eastward on strike-slip faults. And reverse faults in central Asia have become active, leading to growth of ranges such as the Tien Shan. Why the broad plateau of Tibet rose remains something of a mystery. In part, the plateau's uplift may be a consequence of the crust thickening as it is squashed horizontally and, as indicated by GPS measurements, has spread sideways to the east. Uplift may also be due to the heating of the region, when slabs of the underlying lithospheric mantle dropped off and sank; hot asthenosphere rose to fill the space. Heating of the overlying, remaining lithosphere caused it to rise.

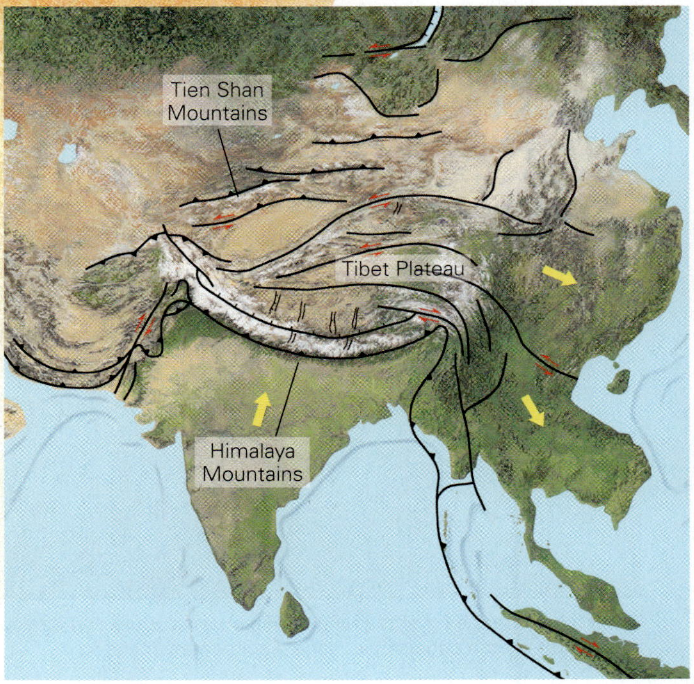

FIGURE 9.19 GPS measurements of shortening in the Andes. The lines indicate the velocity of the red dots relative to the interior of South America. The line at the yellow dot indicates relative plate motion.

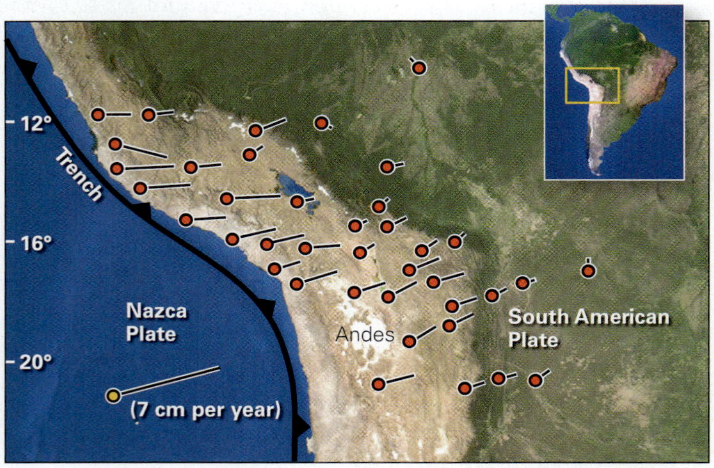

sea level by locating ancient beaches that now lie high above the water. And they can locate where land surface has risen relative to a river by identifying places where a river has recently carved a new valley. In addition, **GPS** (the **global positioning system**) is very useful to measure rates of uplift and horizontal shortening in orogens. While standard hand-held GPS devices provide locations with accuracies of only about ±2 m, research-quality GPS systems can specify locations to within ±2 mm. By comparing a location within an orogen to a location outside an orogen over a time period of a few years, it is possible to detect crustal motion. Thus, we can "see" the Andes shorten horizontally at a rate of a couple of centimeters per year (**Fig. 9.19**), and we can "watch" as mountains along this convergent boundary rise by a couple of millimeters per year.

TAKE-HOME MESSAGE

Mountain belts form in association with convergence, collision, and rifting. During convergent and collisional orogeny, continental crust thickens, large thrust faults and folds form, and regional metamorphism develops. Rifting yields fault-block mountains, separated by narrow basins.

QUICK QUESTION Could fold-thrust belts develop in rifts? Why or why not?

9.6 Other Consequences of Mountain Building

Mountain building not only causes deformation but also leads to conditions that produce new rocks, significant crustal uplift, and distinctive landforms. We've already mentioned these phenomena and have discussed deformation in detail. Now let's look more closely at examples of rock formation, uplift, and topography development.

Forming Rocks in and near Mountains

The process of orogeny establishes geologic conditions appropriate for the formation of a great variety of rocks. Examples from all three rock categories can form in mountain belts (**Fig. 9.20a**):

- *Igneous activity during orogeny*: At convergent boundaries, melting takes place in the mantle above the subducting plate. In rifts, stretching and thinning of lithosphere causes decompression melting of the underlying mantle. And during continental collision, melting may take place where

FIGURE 9.20 An example of the various rocks formed during orogeny.

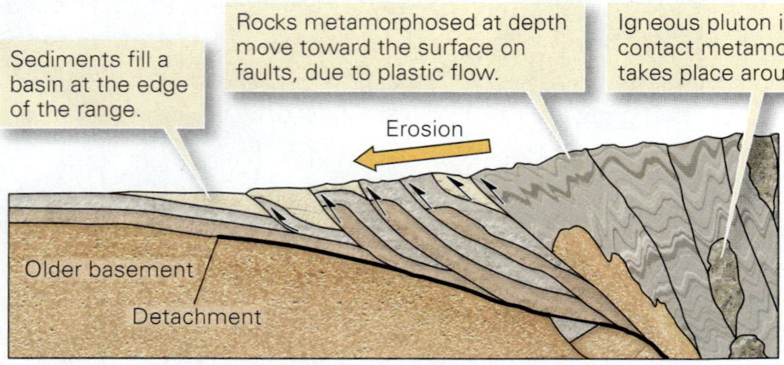

(a) In the internal zone of the range, metamorphic rocks and igneous rocks form. At the edge of the range, sedimentary rocks form.

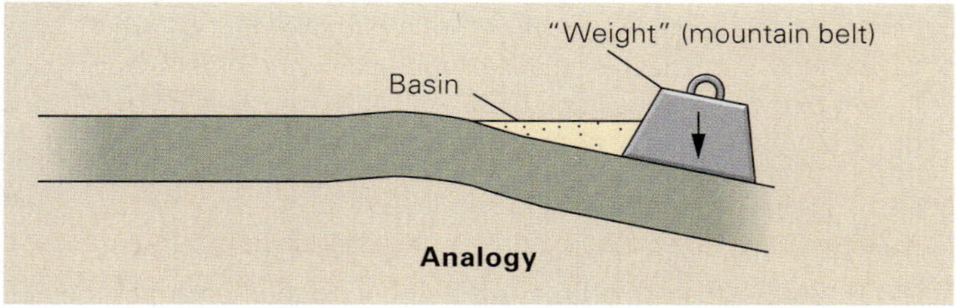

(b) The sedimentary basin develops because the mountain range acts as a weight that pushes down the surface of the lithosphere.

deep portions of the crust undergo heating. All of these melting regimes produce magma, which rises and freezes to form igneous rocks in or on the overlying mountains.

- *Sedimentation during orogeny*: Weathering and erosion in mountain belts generate vast quantities of sediment, which tumbles down slopes. Glaciers or streams transport sediment to low areas where it accumulates in alluvial fans or deltas. In some locations, the weight of the mountain belt itself pushes down the surface of the lithosphere bordering the mountain belt, thereby producing a deep sedimentary basin (**Fig. 9.20b**).
- *Metamorphism during orogeny*: Contact metamorphic aureoles (see Chapter 7) form adjacent to igneous intrusions in orogens. Regional metamorphism occurs where mountain building thrusts one part of the crust over another; when this happens, rock of the footwall ends up at great depth and thus can be subjected to high temperature and pressure. Because deformation accompanies this process, the resulting metamorphic rocks contain tectonic foliation.

Processes Causing Uplift and Producing Mountainous Topography

Leonardo da Vinci, the great Renaissance artist and scientist, enjoyed walking in the mountains, where he sketched ledges and examined the rocks he found there. To his surprise, he discovered marine shells (fossils) in limestone beds cropping out at an elevation of a kilometer above sea level. He puzzled over this observation, and finally concluded that the rock containing the fossils had risen from below sea level up to its present elevation. Modern geologists agree with Leonardo and, as we've noted, refer to vertical movement of the Earth's surface from a lower to a higher elevation as *uplift*. What processes can cause the surface of the Earth to rise? The list is long because, as we have seen, mountain building happens in numerous different geologic settings. To understand how uplift processes work, we must begin by remembering the concept of isostasy introduced in Interlude D.

The lithosphere, which consists of relatively rigid crust and lithospheric mantle, rests on the softer asthenosphere below. In this regard, lithosphere of a region resembles a ship floating in the sea (see Fig. D.12), and the surface of the lithosphere is like the top surface of the ship. If the ship becomes heavier (by adding cargo), or thinner (by removing a deck) the surface becomes lower. In contrast, if the ship becomes lighter (by removing cargo) or taller (by adding a deck) the surface becomes higher. Each change causes the ship's keel to rise or fall until the ship floats at an appropriate level. Geologists refer to the condition that exists when this balance has been achieved as **isostasy**, or *isostatic equilibrium*. Put another way, isostasy exists where the elevation of the Earth's surface reflects the level at which the lithosphere naturally rests.

To picture the relation between isostasy and mountains, let's do an experiment with wood blocks in a bathtub. Start by placing a block of pine (a low-density wood) into a bathtub full of water. Since the block is less dense than water, it floats, with part of the block remaining above the water surface, and most of the block submerged below. Next to it, place a block of oak (a high-density wood), that is the same thickness as the first block. The top of the oak block sits lower than that of the pine block with the same thickness, and the base of the oak block protrudes down further in the water. Next, place a thicker pine block next to the first pine block. The top surface of a thicker block sits higher than the top surface of a thinner block of the same density, and its base protrudes deeper into the water.

From our bathtub experiment, we can deduce that any phenomenon that changes the thickness and/or density of a floating block will affect the elevation of the block's surface above the water surface. Similarly, the elevation of the lithosphere's surface depends on the thickness and density of the lithosphere. So to answer the question of why mountain belts can rise, we must identify geologic processes that can change the thickness and/or density of layers in the lithosphere. Let's consider some examples of why such changes can take place.

Crustal Shortening and Thickening During collisional orogeny or during certain types of convergent-margin orogeny, horizontal compression causes the crust to shorten horizontally and thicken vertically (**Fig. 9.21a**; see also Geology at a Glance, pp. 316–317). To isostatically compensate for the thickening of the crust, the geologic equivalent to adding another block of low-density wood to the top of a floating block, the base of the crust and underlying lithospheric mantle sink, or subside (**Fig. 9.21b**). Indeed, since the Himalayas are about 8 km high, most of the thickened crust extends downward beneath the range just as most of a floating ice cube is underwater. This downward protrusion of crust is called a **crustal root**. We can illustrate this relationship in a bathtub model by lining up a row of floating blocks of different thickness but the same density (**Fig. 9.21c**).

Adding Igneous Rock to the Crust When lava or pyroclastic debris accumulate on the Earth's surface, a volcano grows and may become a mountain. Growth of mountains associated with igneous activity may also occur because intrusions at depth may add material to the crust and, therefore, thicken the crust.

FIGURE 9.21 The concept of isostasy as applied to collisional mountain ranges.

(a) The Himalayas are the world's highest mountain range. The crust beneath them is almost twice the normal thickness.

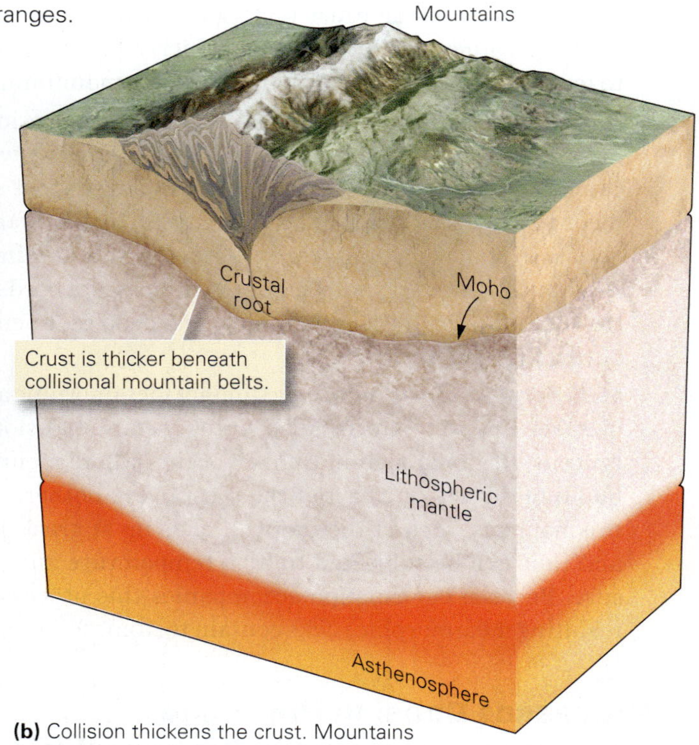

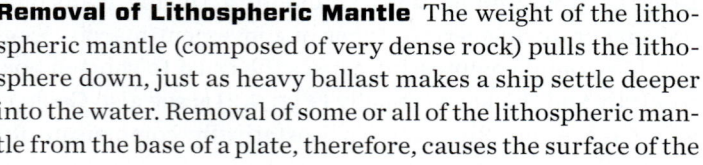

(b) Collision thickens the crust. Mountains form where the low-density crust is thicker.

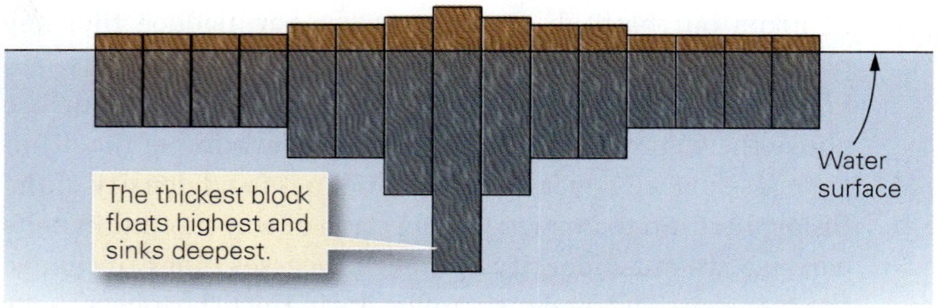

(c) Because of isostasy, blocks of wood floating in water sink to a depth such that the mass of the water displaced is the same as the mass of the block.

Removal of Lithospheric Mantle The weight of the lithospheric mantle (composed of very dense rock) pulls the lithosphere down, just as heavy ballast makes a ship settle deeper into the water. Removal of some or all of the lithospheric mantle from the base of a plate, therefore, causes the surface of the remaining lithosphere to rise to maintain isostasy, even if the thickness of the crustal component remains unchanged (**Fig. 9.22**). Such removal, a process known as *delamination*, resembles removal of ballast from the hold of a ship—as the weight of the ballast disappears, the deck of the ship rises.

FIGURE 9.22 Uplift, due to delamination of the lithosphere root, may happen after collision.

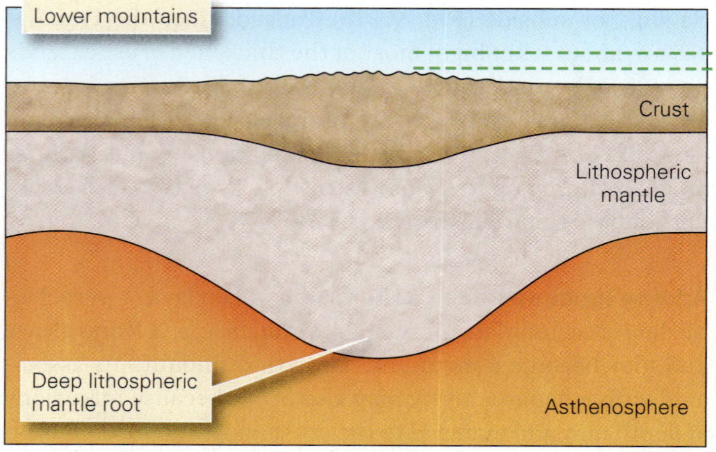

(a) Soon after collision, the crust and lithospheric mantle have thickened. The lithospheric mantle root is like ballast, holding the surface of the crust down.

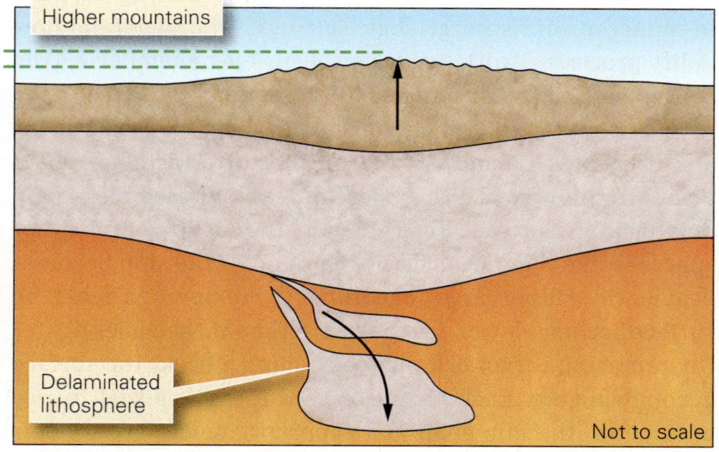

(b) If the lithospheric mantle root detaches and sinks, a process called delamination, the surface of the lithosphere may rise, like a ship offloading cargo.

Thinning and Heating the Lithosphere In rifts, the lithosphere undergoes stretching and thinning. As a result, relatively less-dense asthenosphere rises beneath the rift, and the remaining lithosphere heats up. Replacing denser lithospheric mantle with less-dense asthenosphere, together with heating the remaining thinned lithosphere (which causes the lithosphere to expand and become less dense), causes the overall region of the rift, as well as the margin of the rift, to rise in order to maintain isostasy.

What Goes Up Must Come Down

When the land surface rises significantly, for any reason, gravity and the Sun's heat begin to drive a variety of erosive processes. For example, as slopes steepen, landslides cause rock and debris to tumble from higher to lower elevations (see Chapter 13). When winds blow clouds over the mountains, rain provides water that collects in streams whose flow carries away debris and sculpts valleys and canyons (see Chapter 14). If temperatures remain cold enough during the year, glaciers grow and flow, carving peaks and deepening valleys (see Chapter 18). The net result of all these processes is to grind away elevated areas and produce the jagged landscapes that we associate with mountain terrains (**Fig. 9.23**). It's important to keep in mind that uplift and erosion happen simultaneously in active mountain belts, so for the elevation of a range to increase over time, the rate of uplift must exceed the rate of erosion. If uplift rate becomes less than erosion rate, the elevation of the range decreases.

The highest point on Earth, the peak of Mt. Everest, lies 8.85 km above sea level. Can our planet's mountain ranges get significantly higher? Probably not. Mountains as high as Olympus Mons on Mars, which rises 27 km above the plain at its base, couldn't form on Earth because of the relatively high *geothermal gradient* (the rate of increase in temperature with depth) in Earth's crust. At the high temperatures that occur at depths of 10 to 30 km, quartz-rich crustal rocks become so weak that it is relatively easy for them to flow ductilely. When this flow begins, overlying mountains effectively begin to collapse under their own weight, and spread laterally like soft cheese that has been left out in the summer sun. Geologists call this process **orogenic collapse**. During orogenic collapse, the upper crust breaks and a system of normal faults develops to accommodate the horizontal stretching.

Did you ever wonder... whether mountain belts last forever?

The simultaneous activity of uplift, erosion, and in some cases, organic collapse ultimately brings rock that was metamorphosed at great depth up to the surface of the Earth. This process of revealing deeper rocks by removal of the overlying crust is called unroofing or **exhumation**.

TAKE-HOME MESSAGE

Mountain building is typically accompanied by igneous activity and metamorphism, as well as by deposition of sediment. Mountains exist where major uplift occurs. Beneath some belts, the relatively buoyant crust has thickened, so the lithosphere sits higher. Once uplifted, erosion sculpts rugged topography. Because deep crust is warm and weak, mountain belts may eventually collapse under their own weight.

QUICK QUESTION How do rocks metamorphosed at 20 km below the surface eventually become exposed in a mountain range?

FIGURE 9.23 Erosion sculpts the rugged peaks of mountain ranges. In this example, from the Canadian Rockies, glaciers—which have since melted away—carved the major valleys. Landslides bring debris down the cliff faces. Eventually, rivers carry the debris away.

9.7 Basins and Domes in Cratons

A **craton** consists of crust that has not been affected by orogeny for at least about the last 1 billion years. As a result, cratons have cooled substantially, and therefore have become relatively strong and stable. In North America, the craton forms the interior of the continent—during Phanerozoic time, mountain belts (the North American Cordillera on the west, the Appalachians on the east, and the Ouachitas on the south) formed along the margins, not within the craton. Geologists divide cratons into two provinces: **shields**, in which Precambrian metamorphic and igneous rocks crop out at the ground surface, and **cratonic platforms**, where a relatively thin layer of Phanerozoic sediment covers the Precambrian rocks (**Fig. 9.24**).

In shield areas, we find widespread exposures of intensively deformed metamorphic rocks with abundant examples of flow folds and tectonic foliation. That's because the crust making the cratons was deformed during a succession of orogenies in the Precambrian. These orogens are so old that erosion has worn away the original peaks, in the process exhuming deep crustal rocks.

In cratonic platforms, the pattern of contacts between stratigraphic formations typically defines regional domes and basins—broad areas that gradually subsided or uplifted, respectively, over geologic time (**Fig. 9.25**). For example, in Missouri, strata arch across a broad uplift, the Ozark dome, whose diameter is 300 km. Individual sedimentary layers

FIGURE 9.24 North America's craton consists of a shield, where Precambrian rock is exposed, and a platform, where Phanerozoic sedimentary rock covers the Precambrian.

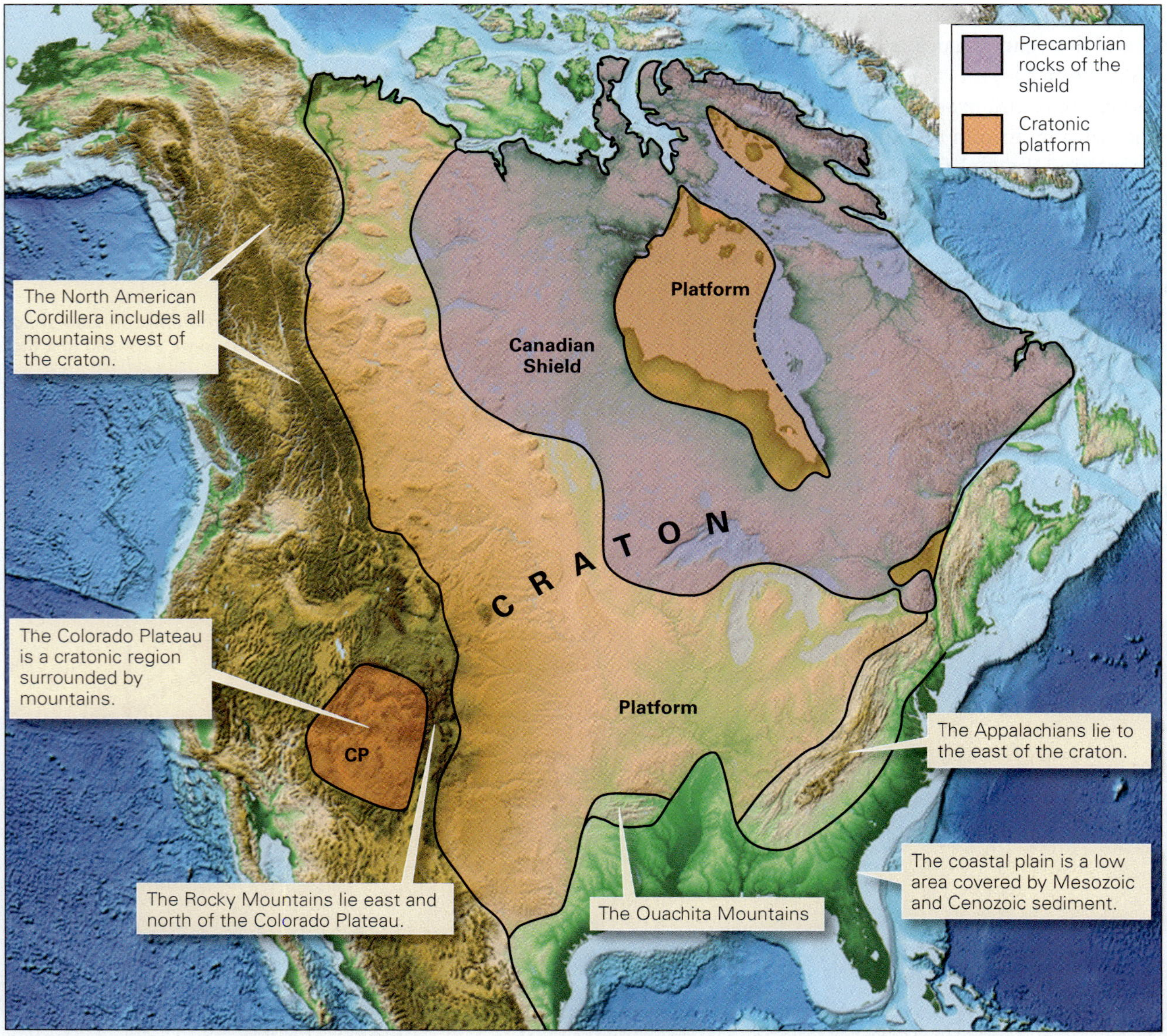
Precambrian rocks of the shield
Cratonic platform
The North American Cordillera includes all mountains west of the craton.
Platform
Canadian Shield
CRATON
The Colorado Plateau is a cratonic region surrounded by mountains.
CP
Platform
The Appalachians lie to the east of the craton.
The coastal plain is a low area covered by Mesozoic and Cenozoic sediment.
The Ouachita Mountains
The Rocky Mountains lie east and north of the Colorado Plateau.

FIGURE 9.25 Domes and basins of the North American cratonic platform.

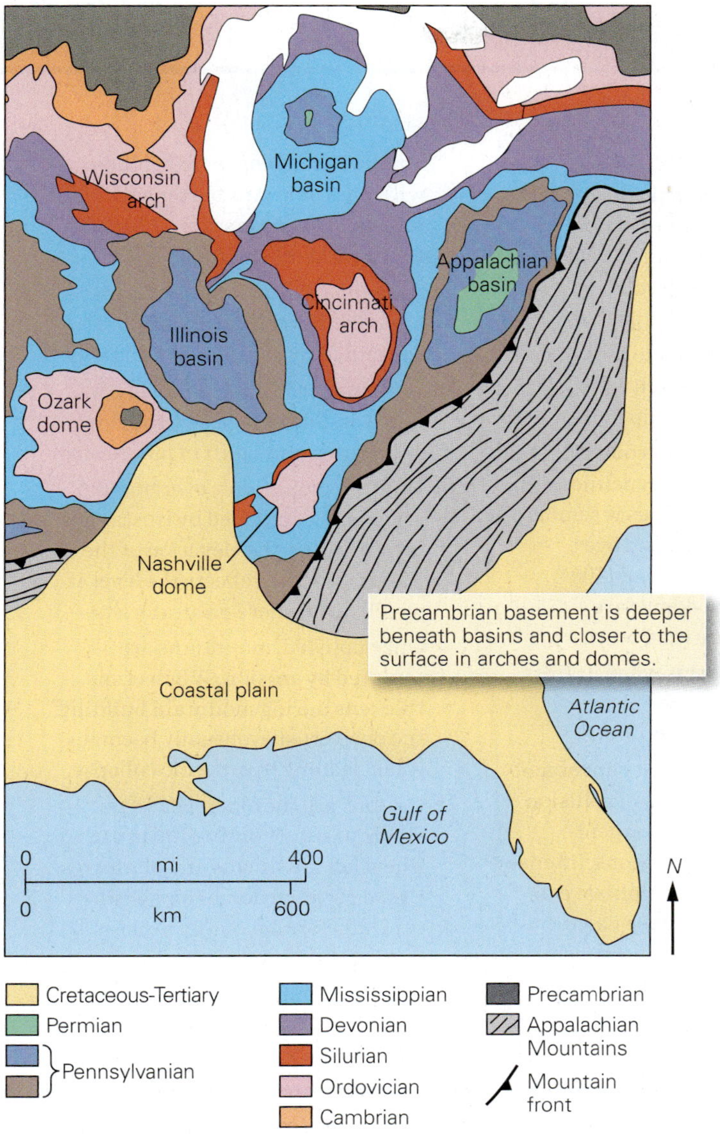

(a) A geologic map of the U.S. mid-continent platform region showing the locations of basins and domes.

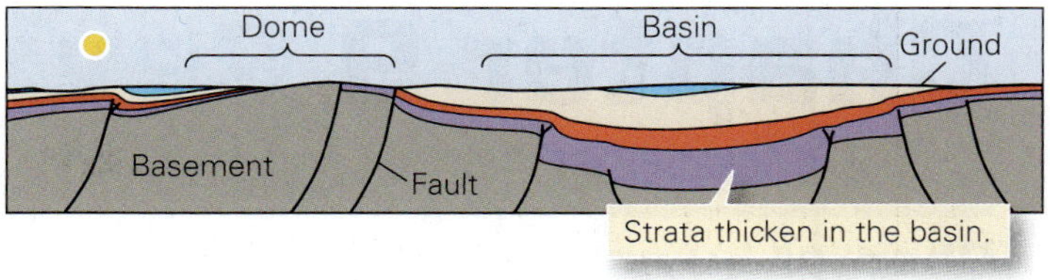

(b) A cross section showing how strata thin toward the crest of a dome and thicken toward the center of a basin. The cross section is vertically exaggerated. This cross section symbolically represents the Ozark dome and Illinois basin.

thin toward the top of the dome, because less sediment accumulated on the dome than in adjacent basins. Erosion during more recent geologic history has produced the characteristic bull's-eye pattern of a dome, with the oldest rocks (Precambrian granite) exposed near the center. In the Illinois basin, strata warp downward into a huge bowl that is also about 300 km across. Strata get thicker toward the basin's center, indicating that the floor of the basin was subsiding at the same time sediment was accumulating—there was more room for sediment to accumulate where the basin subsided the most. The basin also has a bull's-eye shape, but here the youngest strata are exposed in the center. Geologists refer to the broad vertical movements that generate huge, but gentle, midcontinent domes and basins as **epeirogeny** (from the Greek *epeiros*, meaning mainland or continent).

TAKE-HOME MESSAGE

Cratons are portions of continents that consist of old (Precambrian) and relatively stable crust. Parts of cratons may be covered by Paleozoic sedimentary strata. Variations in the dip and thickness of these strata define regional basins, arches, and domes.

QUICK QUESTION Imagine that coal occurs in a particular stratigraphic unit. Will mines to reach the coal be deeper or shallower in the center of a basin?

Another View (left) The Matterhorn of the Alps along the Swiss-Italian border. The strata of the peak lay on the seafloor during the late Mesozoic. Mountain building thrust the strata upward, where they now reach an elevation of 4.5 km; (right) A fold resulting from shearing in South Australia.

Chapter 9 Review

Chapter Summary

- Mountains occur in linear ranges called mountain belts or orogens. An orogen forms during an orogeny, or mountain-building event.
- Mountain building causes rocks to bend, break, shorten, stretch, and shear. Because of such deformation, rocks can change their location, orientation, and shape.
- During brittle deformation, rocks break into pieces. During ductile deformation, rocks change shape without breaking.
- Stress (force applied per unit area) can be compressional, tensional, or shear. Pressure refers to a condition in which a material feels the same amount of compression in all directions.
- Strain refers to the shape change that materials undergo when subjected to a stress. For example, compression can cause shortening, and tension can cause stretching.
- Deformation results in the development of geologic structures.
- Joints are natural cracks in rock, formed in response to tension under brittle conditions. Veins develop when minerals precipitate out of water passing through cracks.
- Faults are fractures on which there has been shear. Geologists distinguish among normal, reverse, strike-slip, and oblique-slip faults.
- Folds are curved layers of rock. Anticlines are arch-like, synclines are trough-like, monoclines resemble the shape of a carpet draped over a stair step, basins are shaped like an upright bowl, and domes are shaped like an overturned bowl.
- Tectonic foliation forms when grains flatten, rotate, or grow so that they align parallel with one another.
- Mountain belts can form at convergent margins along continents, in collision zones, and in rifts. At convergent-margin and collisional orogens, intense deformation and metamorphism can take place. Fold-thrust belts can form along the margins of such orogens. Rifting yields fault-block mountains.
- Small crustal blocks that collide with plate margins become accreted terranes. Accretion can add crust to a continent over time.
- Large collisional mountain ranges are underlain by deep roots and contain folds, faults, and foliations.
- With modern GPS technology, it is now possible to measure the slow shortening and uplift of mountains.
- Uplift in mountains, over broad regions, is controlled by isostasy, meaning that the elevation of the Earth's surface reflects the level at which lithosphere naturally sits.
- Once uplifted, mountains are sculpted by erosion. When crust thickens during mountain building, the deep crust eventually becomes weak, leading to orogenic collapse.
- Cratons are the old, relatively stable parts of continental crust. They include shields and platforms. Broad regional domes and basins typically form in platform areas.

Guide Terms

accretion (p. 315)
anticline (p. 308)
axial surface (p. 308)
basin (p. 308)
bearing (p. 305)
brittle deformation (p. 301)
collisional orogen (p. 314)
compression (p. 303)
craton (p. 322)
cratonic platform (p. 322)
crustal root (p. 319)
deformation (p. 298)
dip (p. 305)
dip-slip fault (p. 306)
displacement (p. 304)
dome (p. 308)
ductile deformation (p. 301)
epeirogeny (p. 323)
exhumation (p. 321)
fault (p. 304)
fault scarp (p. 306)
fold (p. 308)
fold-thrust belt (p. 313)
geologic structure (p. 298)
global positioning system (GPS) (p. 318)
hinge (p. 308)
isostasy (p. 319)
joint (p. 303)
limb (of fold) (p. 308)
monocline (p. 308)
mountain belt (p. 298)
mountain building (p. 298)
normal fault (p. 306)
oblique-slip fault (p. 306)
orogen (p. 298)
orogenic collapse (p. 321)
orogeny (p. 298)
plunge (p. 305)
pressure (p. 303)
reverse fault (p. 306)
shear stress (p. 303)
shield (p. 322)
slickensides (p. 306)
strain (p. 299)
stress (p. 303)
strike (p. 305)
strike-slip fault (p. 306)
suture (p. 314)
syncline (p. 308)
tectonic foliation (p. 312)
tension (p. 303)
thrust fault (p. 306)
uplift (p. 298)
vein (p. 304)

GEOTOURS THIS CHAPTER'S GEOTOUR EXERCISE (I) FEATURES:

- Calculating Strike and Dip from Flatirons
- Brittle Structure—Joints
- Brittle Structures—Faults
- Ductile Structures—Folds

Review Questions

1. What changes do rocks undergo during formation of an orogenic belt such as the Alps?
2. Contrast brittle and ductile deformation.
3. What factors determine whether a rock will behave in brittle or ductile fashion?
4. How are stress and strain different?
5. How is a fault different from a joint?
6. Compare normal, reverse, strike-slip, and oblique-slip faults.
7. How do you recognize faults in the field?
8. Describe the differences among an anticline, a syncline, and a monocline.
9. Discuss the relationship between foliation and deformation.
10. Discuss the processes by which mountain belts form in convergent margins, during collisions, and in continental rifts.
11. Describe the principle of isostasy.
12. How are the structures of a craton different from those of an orogenic belt?

On Further Thought

13. Imagine that a geologist sees two outcrops of resistant sandstone, as depicted in the cross-section sketch. The region between the outcrops is covered by soil. A distinctive bed of cross-bedded sandstone occurs in both outcrops, so the geologist correlated the western outcrop (on the left) with the eastern outcrop. The curving lines in the bed indicate the shape of the cross beds. Keeping in mind how cross beds form (see Chapter 6), sketch how the cross-bedded bed connected from one outcrop to the other, before erosion. What geologic structure have you drawn?

Online Resources

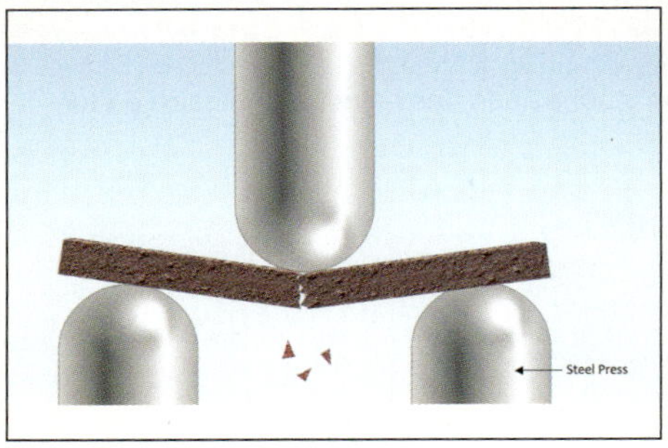

Animations

This chapter features animations on types of rock deformation and faults.

Videos

This chapter features a video on continental collision and the formation of mountains.

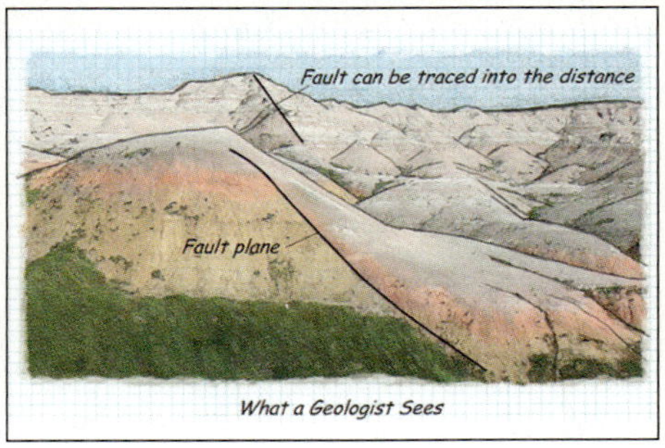

Assessment

Features include visual identification exercises on types of faults, folds, and deformations.

INTERLUDE E

Memories of Past Life: Fossils and Evolution

▲ This fossil of a fish was carefully exposed by removing the overlying layer of rock. Fine details of the fish's bones and fins have been preserved for millions of years, allowing us to see a species that no longer exists.

LEARNING OBJECTIVES

By the end of this interlude, you should understand...

1. what a fossil is, and how fossils form.
2. that it takes special conditions for an organism to be preserved as a fossil, so it's not surprising that the record of known fossils is very incomplete.
3. how fossils are classified, according to the rules of taxonomy, and the names of common fossils.
4. the fundamental principles of the theory of evolution by natural selection.
5. what extinction is, what mass extinctions are, and why mass extinctions may take place.

E.1 The Discovery of Fossils

If you look at bedding surfaces of sedimentary rocks, you may find shapes that resemble shells, bones, leaves, or footprints (**Fig. E.1a**). The origin of these shapes mystified early scientists. Some thought that the shapes had simply grown underground in place. Others considered such **fossils** (from the Latin word *fossilis*, which means dug up) to be remnants or traces of ancient living organisms that were preserved after they had been buried. This second viewpoint, now the standard interpretation of fossils, was first proposed by the Greek historian Herodotus in 450 B.C.E. and revived by Leonardo da Vinci in 1500 C.E. The idea did not, however, become widely accepted until 1669, when a Danish physician named Nicolas Steno (1638–1686) published a book that discussed fossils. Steno argued that after burial, shells, teeth, and bones could retain their distinctive shape after the material containing them had turned into rock. An understanding of fossils increased greatly thanks to the efforts of a British scientist, Robert Hooke (1635–1703), who described the characteristics of many fossil species. (A *species* is distinct group of organisms capable of breeding.) Hooke eventually realized that most fossils represent *extinct species*, those that lived in the past but no longer survive anywhere today. The concept of extinction remained controversial until Georges Cuvier (1769–1832), a French zoologist, carefully demonstrated that the skeletons and teeth of fossil elephant-like mammoths and mastodons differ from those of any modern elephants. These days, we take for granted that species can go extinct, because we've seen a great number vanish during human history. But before Cuvier, most researchers thought instead that even if fossils didn't resemble known species, they must have living relatives hidden somewhere on this planet. Considering that large parts of the Earth remained unexplored, this idea wasn't so far-fetched. But by the end of the 18th century, it became clear that numerous fossil organisms did not have modern-day counterparts anywhere—mastodons and mammoths were too big to hide.

Gradually, **paleontology**, the study of fossils, ripened into a science. Cuvier showed that fossils can be described and classified using an approach similar to the way biologists classify living organisms. **Paleontologists**, scientists who study fossils, described thousands of specimens and established museum collections to preserve and display them (**Fig. E.1b, c**). Work with fossils went beyond description alone when William Smith, a British engineer who surveyed canal construction sites in England during the 1790s, noted that different fossil organisms occur in different layers of strata within a sequence of sedimentary rocks. In fact, Smith realized that strata contain a predictable succession of fossils, and that a given species can be found only in a specific interval of strata. This discovery made it possible for researchers to use fossils as a basis for determining the age of one sedimentary rock layer relative to another, and for documenting the order in which species appeared and went extinct in the geologic record. Thus, fossils became an indispensable tool for studying geologic history, and became the basis for documenting **evolution**, the progressive change over time in characteristics of species that has led to the appearance of new species. In this interlude, we introduce fossils and discuss how their study teaches us about life's evolution. An understanding of fossils serves as essential background for the next two chapters.

E.2 Fossilization

What Kinds of Rocks Contain Fossils?

Most fossils are found in sedimentary rocks. Fossils sometimes occur in tuff, a rock formed from volcanic ash, but they are not found in other igneous rocks and tend to be destroyed by metamorphism. Fossils can form when organisms die and become buried, or when their burrows and footprints become buried. The degree of a fossil's preservation reflects the context of burial. For example, rocks formed from sediment deposited under anoxic (oxygen-poor) conditions in quiet water (such as lakebeds or lagoons) can preserve particularly fine specimens. In contrast, rocks made from sediment deposited in high-energy environments—where strong currents tumble the remains of organisms and break them up—contain at best only small fragments of fossils mixed with other grains.

Forming Fossils

Paleontologists refer to the process of forming a fossil as **fossilization**. To see how a typical fossil develops in sedimentary rock, let's consider an example (**Fig. E.2**). Imagine an

FIGURE E.1 Fossils, and their collection.

(a) Fossil skeleton in 200-Ma sandstone.

(b) A paleontologist collecting specimens.

(c) A drawer of fossil specimens in a museum.

elderly dinosaur searching for food along the muddy shore of a lake on a scalding summer day in the geologic past. The hungry dinosaur succumbs to the heat and collapses dead into the mud. Soon after, scavengers strip the skeleton of meat and may scatter the bones. But before the bones have had time to weather away, the river floods and buries the bones, along with the dinosaur's footprints, under a layer of silt. More silt from succeeding floods buries the bones and prints still deeper, so that the bones cannot be reworked by currents or disrupted by burrowing organisms. Later, sea level rises and a thick sequence of marine sediment buries the fluvial sediment. Eventually, the sediment containing the bones and footprints turns to rock (siltstone and shale). The footprints remain outlined by the contact between the siltstone and shale, while the bones reside within the siltstone. Minerals precipitating from groundwater passing through the siltstone gradually replace some of the chemicals constituting the bones, until the bones themselves have become rock-like. The buried bones and footprints are now fossils.

One hundred million years later, uplift and erosion expose the dinosaur's grave. Part of a fossil bone protrudes from a rock outcrop. A lucky paleontologist observes the fragment and starts excavating, gradually uncovering enough of the bones to permit reconstruction of the beast's skeleton. Further digging uncovers the footprints. The dinosaur rises again, but this time in a museum. In recent years, bidding wars have made some fossil finds extremely valuable. For example, a skeleton of a *Tyrannosaurus rex*, a 67-million-year-old dinosaur, sold at auction in 1997 for $7.6 million. The specimen, named Sue after its discoverer, now stands in the Field Museum of Chicago.

Similar tales can be told for fossil seashells buried by sediment settling in the sea, for insects trapped in *amber* (hardened tree sap), and for mammoths drowned in the muck of a tar pit. In all cases, fossilization involves the burial and preservation of an organism or the trace (a footprint or burrow) of an organism.

The Many Different Kinds of Fossils

Perhaps when you think of a fossil, you picture either a dinosaur bone or the imprint of a seashell in rock. In fact, paleontologists distinguish many different kinds of fossils, according to the specific way in which the organism was fossilized. Let's look at examples of these categories.

- *Frozen or dried body fossils*: In a few environments, whole bodies of organisms may be preserved. Most of these fossils are fairly young, by geologic standards, in that their ages can be measured in thousands, not millions, of years. Examples include woolly mammoths that became incorporated in the permafrost (permanently frozen ground) of Siberia (**Fig. E.3a**). In desert climates, organisms can become desiccated (dried out). Such "mummified" corpses can survive for millennia in caves.
- *Body fossils preserved in amber or tar*: Insects landing on the bark of trees may become trapped in the sticky sap or resin the trees produce. This golden syrup envelops the insects and over time hardens into amber, the semiprecious

FIGURE E.2 The stages in the fossilization of a dinosaur.

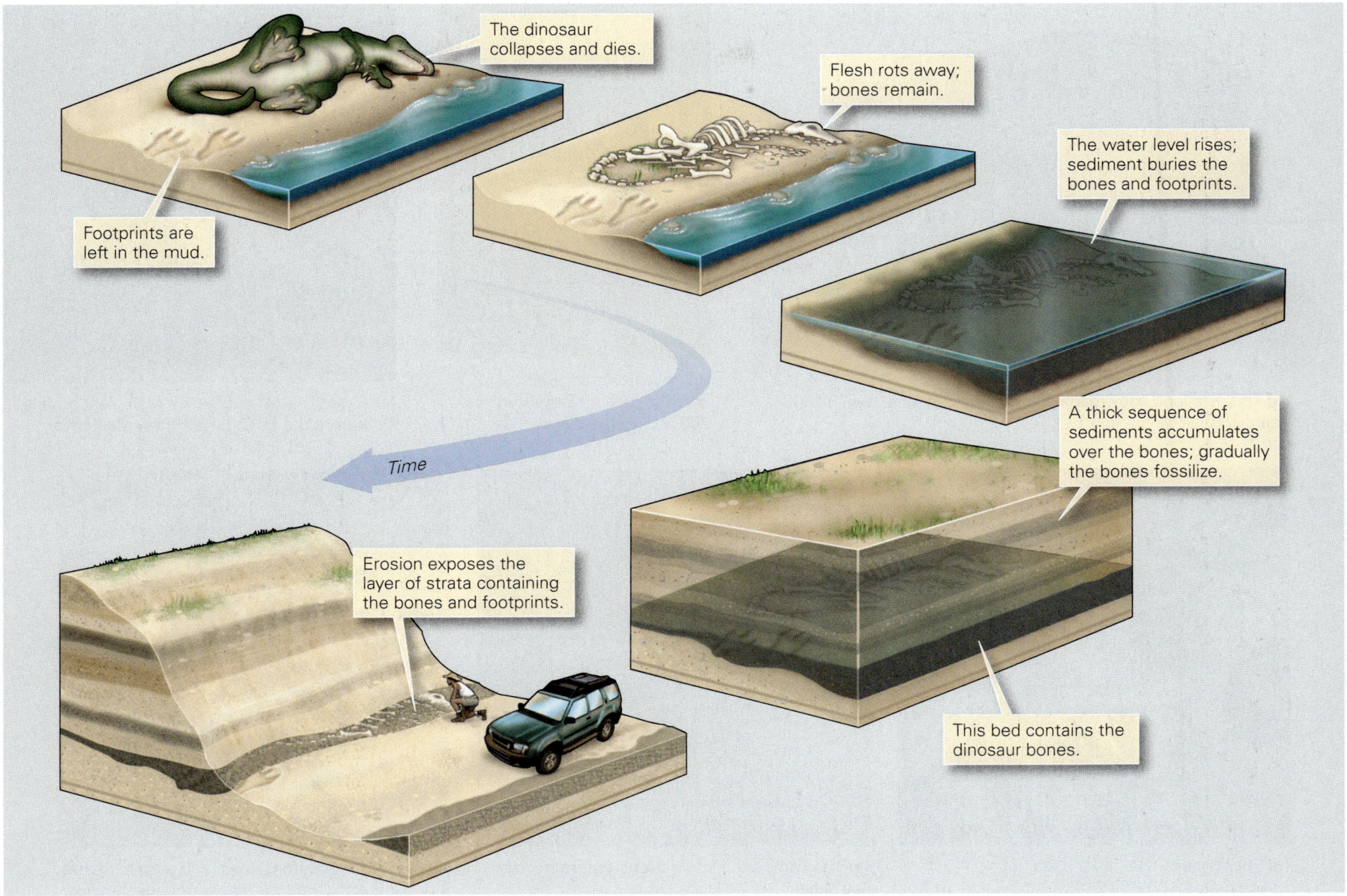
The dinosaur collapses and dies.
Footprints are left in the mud.
Flesh rots away; bones remain.
The water level rises; sediment buries the bones and footprints.
Time
A thick sequence of sediments accumulates over the bones; gradually the bones fossilize.
Erosion exposes the layer of strata containing the bones and footprints.
This bed contains the dinosaur bones.

"stone" used for jewelry. Amber can preserve insects, as well as other delicate organic material such as feathers, for 40 million years or more (**Fig. E.3b**). Tar similarly acts as a preservative. In isolated regions where oil has seeped to the surface, the more volatile components of the oil evaporate away and bacteria degrade what remains, leaving behind a sticky residue of tar. At one such locality, the La Brea Tar Pits in Los Angeles, tar accumulated in a swampy area. While grazing, drinking, or hunting at the swamp, animals became mired in the tar and sank into it. Their bones have been remarkably well preserved for over 40,000 years (**Fig. E.3c**).

- *Preserved or replaced bones, teeth, and shells*: Bones (the internal skeletons of vertebrate animals) and shells (the external skeletons of invertebrate animals) consist of durable minerals, which may survive in rock. Some bone or shell minerals are not stable, and they recrystallize. But even when this happens, the shape of the bone or shell can be preserved.
- *Molds and casts of bodies*: When a shell pushes into soft sediment, a **mold** in the shape of the shell forms. If the inside of the shell then fills with more sediment, that forms a **cast** of the shell (**Fig. E.3d**). Eventually, when the sediment turns into rock, the mold and cast display the shape of the shell and remain in place, even if the shell itself dissolves away. (Similar terms are used by sculptors who fill a mold with molten metal to form a cast of a statue.) Typically, a mold appears as an indentation into a bed of rock, whereas a cast protrudes from the surface of a bed of rock.
- *Carbonized impressions of bodies*: Impressions are simply flattened molds and casts produced when soft or semi-soft organisms (leaves, insects, shell-less invertebrates, sponges, feathers, jellyfish) are pressed between layers of sediment (**Fig. E.3e**). Chemical reactions eventually

FIGURE E.3 Examples of different kinds of fossils.

(a) This 1-m long baby mammoth, found in Siberia, died 37,000 years ago.

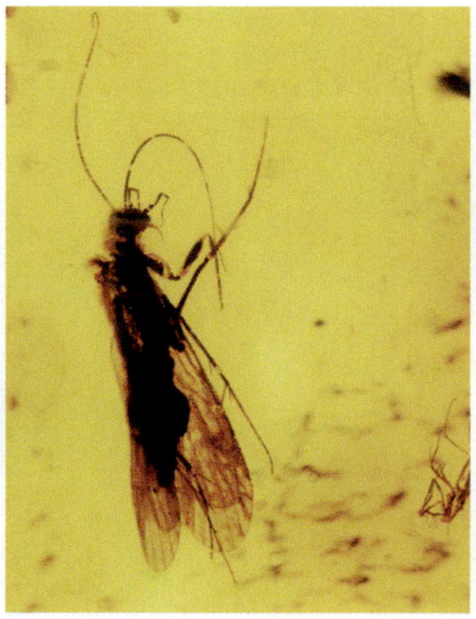

(b) This insect became embedded in amber about 200 million years ago.

(c) A fossil skeleton of a 2-m-high giant ground sloth from the La Brea Tar Pits in California.

(d) Fossil shells, the "hard parts" of invertebrates.

(e) The carbonized impressions of fern fronds in a shale.

(f) Petrified wood from Arizona. It is so hard that it remains after the rock that surrounded it has eroded away.

remove the organic material, leaving only a thin film of carbon on the surface of the impression.

› *Permineralized organisms*: **Permineralization** refers to the process by which minerals precipitate in porous material, such as wood or bone, from groundwater solutions that have seeped into the pores. **Petrified wood** forms when wood is permineralized—in fact, the word *petrified* literally means turned to stone. The formation of petrified wood begins when the wood gets buried by silica-rich sediment or ash. As the buried wood dries out, open spaces form inside the cells because the protoplasm of cells is 80% water. Groundwater carries dissolved silica into these spaces, and chert precipitates from the groundwater to fill the spaces. The organic matter of the cell walls gradually becomes carbonized, and the carbon screens formed by this process outline the fine detail of the wood's cell-wall structure (**Fig. E.3f**).

› *Trace fossils*: These include footprints, feeding traces, burrows, and dung that organisms leave behind in sediment (**Fig. E.4a**). Paleontologists refer to fossilized dung as a coprolite.

› *Chemical fossils*: Living plants or animals consist of complex organic chemicals. When buried with sediment and subjected to diagenesis, some of these chemicals are destroyed, some remain intact, and some break down to form different, but still distinctive, chemicals. A distinctive chemical derived from an organism and preserved in rock is called a *chemical fossil*, a *molecular fossil*, or a **biomarker**.

Paleontologists also find it useful to distinguish among different fossils on the basis of their size. **Macrofossils** are large enough to be seen with the naked eye. But some rocks and sediments also contain abundant **microfossils**, which can be

FIGURE E.4 Examples of trace fossils and microfossils.

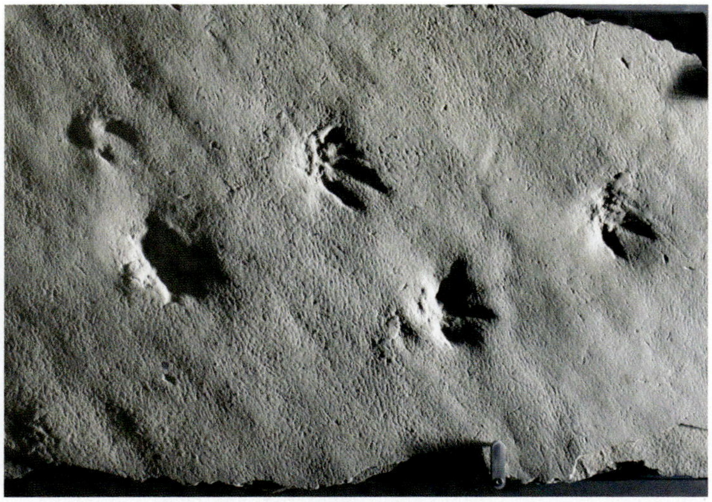

(a) Molds of dinosaur footprints. These push down into the top of a bed.

(b) Examples of fossil plankton shells. Because of their size, geologists refer to these as microfossils.

seen only with a microscope or even an electron microscope (**Fig. E.4b**). Microfossils include remnants of plankton, algae, bacteria, and pollen.

Fossil Preservation

Not all living organisms become fossils when they die—in fact, only a small percentage do. It takes special circumstances to produce a fossil and for a fossil to survive over geologic time. A number of conditions can affect the degree to which a recognizable fossil can be preserved:

- *Death in an anoxic (oxygen-poor) environment*: A dead squirrel by the side of the road will not become a fossil. As time passes, birds, dogs, or other scavengers may come along and eat the carcass. And if that doesn't happen, maggots, bacteria, and fungi infest the carcass and gradually digest it. As the remaining flesh rots in the air, its original organic chemicals undergo oxidation or other reactions, forming new chemicals that wash away in water or escape as gas. Once exposed, the skeleton weathers and turns to dust. Thus, before roadkill can become incorporated in sediment, it has vanished. If, however, a carcass settles into an oxygen-poor environment, oxidation occurs slowly, scavenging organisms are scarce, and bacterial metabolism takes place very slowly. In such environments, the organism will have a chance to be buried and preserved, so the likelihood increases that the organism will become fossilized.
- *Rapid burial*: If an organism dies in a depositional environment where sediment accumulates rapidly, it may be buried before it has time to rot, oxidize, be eaten, or undergo complete weathering. For example, if a storm suddenly buries an oyster bed with a thick layer of sediment, the oysters die and their shells become part of the sedimentary rock derived from the sediment.
- *The presence of hard parts*: Organisms without durable shells, skeletons, or other hard parts usually won't be fossilized, for under most depositional conditions soft flesh decays long before hard parts do. For this reason, paleontologists have learned more about the fossil record of bivalves (a class of organisms, including clams and oysters, with strong shells) than they have about the fossil record of jellyfish (which have no shells) or spiders (which have very fragile shells).

By carefully studying modern organisms, paleontologists have been able to provide rough estimates of the **preservation potential** of organisms, meaning the likelihood that an organism will be buried and eventually transformed into a fossil. For example, in a typical modern-day shallow-marine environment, such as the mud-and-sand seafloor close to a beach, about 30% of the organisms have sturdy shells and thus a high preservation potential, 40% have fragile shells and a lower preservation potential, and the remaining 30% have no hard parts at all and are not likely to be fossilized except in special circumstances. Out of the 30% with sturdy shells, though, only a small number of organisms happen to die in a depositional setting where they become fossilized. Thus, fossilization is the exception rather than the rule.

Extraordinary Fossils: A Special Window to the Past

Although only hard parts survive in most fossilization environments, paleontologists have discovered a few special locations where rock contains fossils of soft parts as well—such fossils are known as **extraordinary fossils** (**Fig. E.5**).

FIGURE E.5 Extraordinary fossils. These fossils are particularly well preserved.

(a) A 50-million-year-old mammal fossil was chiseled from oil shale near Messel, Germany. It still contains the remains of skin.

(b) *Archaeopteryx* from the 150-million-year-old Solnhofen Limestone of Germany. The imprints of feathers are clearly visible.

Examples can include tissue, fur, feathers, or their impressions. We've already discussed how extraordinary fossils of insects and feathers can be preserved in amber. Extraordinary fossils have also been found in deposits of tar pits and in fine-grained limestone or shale lithified from sediment that accumulated on the anoxic floors of quiet-water lakes, lagoons, or the deep ocean.

E.3 Taxonomy and Identification

The study of how to identify and name organisms is called **taxonomy**. Taxonomic classification of fossils follows the same principles used for the classification of living organisms and has a hierarchy of divisions or *taxa*. These principles were first proposed in the 18th century by Carolus Linnaeus, a Swedish biologist.

First, all life is divided into three **domains**: Archaea, Bacteria, and Eukarya. (Note that we capitalize these terms only when talking about domain names.) The domains differ from one another based on fundamental characteristics of their DNA. *Archaea* include a vast array of tiny single-celled microorganisms that occur not only in the mild environments of oceans, soils, and wetlands, but also in the harsh environments of hot springs, black smokers, salt lakes, very acidic streams, and deep subsurface water. Organisms that can survive in harsh environments are known as *extremophiles*—they do not need light, but rather live off the energy stored in the chemical bonds of minerals. *Bacteria* are also tiny single-celled organisms that inhabit almost all livable environments on Earth. Although it may be difficult to distinguish bacteria from archaea visually, on a genetic level they are profoundly different. Both archaea and bacteria are *prokaryotes,* meaning that their cells do not contain a nucleus, a distinct envelope containing the cell's DNA.

Taxonomists divide *Eukarya,* the domain of *eukaryotes,* organisms whose cells do contain a nucleus, into **kingdoms**. Traditionally, these are named as follows: *Protista* (various unicellular and simple multicellular organisms, including algae); *Fungi* (mushrooms and yeast); *Plantae* (trees, grasses, mosses, and ferns); and *Animalia* (examples range widely from sponges to corals to dinosaurs, birds, and humans). (Taxonomists use different names for the kingdoms in modern research literature.) Each kingdom consists of one or more *phyla*. A phylum, in turn, includes several *classes,* a class includes several *orders,* an order includes several *families,* a family includes several *genera,* and a genus includes one or more *species*. Kingdoms, therefore, represent the broadest category of life, and species represent the narrowest. **Table E.1** illustrates how taxonomic classification works for humans.

There's nothing magical about identifying fossils. You can often recognize common fossils in the field simply by examining their **morphology** (form or shape). If the fossil is well preserved and has distinctive features, the process can be straightforward, but if fossils are broken into fragments and parts are missing, or if the fossil shares many characteristics with other fossils, identification can be challenging.

Many fossil organisms resemble modern organisms, so it is relatively easy to figure out how to classify them taxonomically. For example, a fossil clam (class Bivalvia) looks like a clam and doesn't look like, say, a snail (class Gastropoda). At

TABLE E.1 Taxonomic Classification of Humans

Domain	Eukarya	Organisms composed of cells containing nuclei
Kingdom	Animalia	All animals
Phylum	Chordata	Animals with backbones
Class	Mammalia	Chordates with fur, warm blood, and ability to secrete milk
Order	Primates	Great apes, monkeys, and humans
Family	Hominidae	Great apes (gorillas, chimpanzees, orangutans) and humans
Genus	*Homo*	Humans, as well as Neanderthals and other extinct species
Species	*sapiens*	Modern humans (first appearing about 200,000 years ago)

the taxonomic levels below class, identification may involve recognizing such details as the number of ridges on the surface of the shell. Not all fossils, however, resemble living organisms, so figuring out their taxonomic relation to other organisms may be a challenge, and some interpretations remain controversial.

Figure E.6a, which illustrates common invertebrate fossils, can help you identify many of the fossils you might find in a bed of sedimentary rock. We list the names and notable characteristics of these fossils below:

- *Trilobites* have a segmented shell that is divided lengthwise into three parts. They are a type of arthropod.
- *Gastropods* (snails) have a shell that does not contain internal chambers. Some have a spiral shell.
- *Bivalves* (clams and oysters) have a shell that can be divided into two similar halves.
- *Brachiopods* (lampshells) have differently shaped upper and lower shells. They typically have ridges radiating out from the hinge.
- *Bryozoans* are colonial invertebrates whose fossils commonly resemble a screen-like grid. Each opening in the grid is the shell of a single animal.
- *Crinoids* (sea lilies) look like flowers but are actually animals. Their shells have a stalk consisting of numerous circular plates stacked one on top of the other.
- *Graptolites* resemble tiny saw blades, and typically are carbonized. They are remnants of colonial animals that floated in the sea.
- *Cephalopods* are squid-like organisms. They include ammonites, with a spiral shell, and nautiloids, with a straight shell. The shells contain internal chambers and have ridged surfaces.
- *Corals* include colonial species that grow to build distinctive mounds or columns, as well as solitary, cone-shaped species.

E.4 The Fossil Record

A Brief History of Life

Based on laboratory experiments conducted in the 1950s, researchers speculated that reactions in concentrated "soups" of chemicals that formed when seawater evaporated in shallow, coastal pools led to the formation of the earliest protein-like organic chemicals (protolife). More recent studies suggest, instead, that such reactions took place in warm groundwater beneath the Earth's surface or at hydrothermal vents (black smokers) on the seafloor.

While the nature of protolife remains a mystery, the study of fossils in the oldest-known sedimentary rocks has started to provide an image of early life. Specifically, archaea and bacteria fossils appear in rocks as old as about 3.7 billion years. For the first billion years or so of life history, cells of these organisms were the only types of life on Earth. Then, at about 2.5 Ga (Ga = billion years ago), organisms of the protist kingdom first appeared. Early multicellular organisms (fungi, and shell-less invertebrates of the animal kingdom) came into existence at perhaps 1.0 to 1.5 Ga. Green algae appeared by about 750 Ma (Ma = million years ago), and complex multicellular organisms by about 635 Ma. Paleontologists have found the first shelled fauna in rocks as old as 541 Ma. A great variety of shelled invertebrates appeared during next 20 or 30 million years or so, a time interval that researchers refer to as the **Cambrian explosion**, named for the time when it occurred (**Fig. E.6b**; see Chapter 10). The succession of fossils found in the stratigraphic record indicates that fish, land plants, amphibians, reptiles, and finally birds and mammals appeared in sequence over the next several hundred million years.

Researchers have been working hard to understand the **phylogeny** (evolutionary relationships) among organisms, using both the morphology of organisms and, more recently, the study of genetic material. Ideas about which taxa radiated from which ancestors are shown in a chart called the tree of life, or more formally, the **phylogenetic tree** (**Fig. E.7**). Study of the DNA from different organisms has enabled researchers to understand the relationship between molecular processes and evolutionary change.

Is the Fossil Record Complete?

By some estimates, more than 250,000 species of fossils have been collected and identified to date, by thousands of paleontologists working on all continents during the past two

FIGURE E.6 Common types of invertebrate fossils.

(a) Sketches of representative fossils.

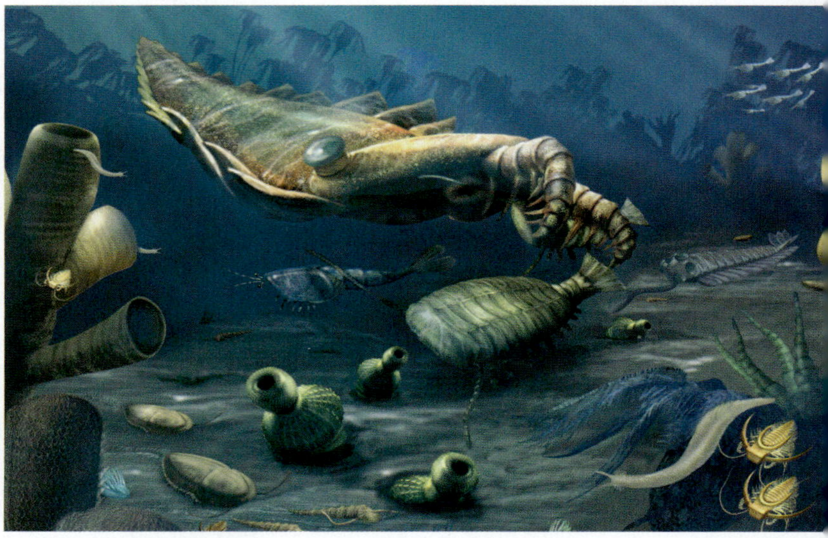

(b) An artist's reconstruction of a Cambrian ecosystem. The painting is based on Burgess Shale fossils.

FIGURE E.7 A phylogenetic tree. The tree of life symbolically illustrates relations among different domains of organisms. Archaea and bacteria are both prokaryotes and don't have nuclei. All other life forms are eukaryotes because they have cells with a nucleus.

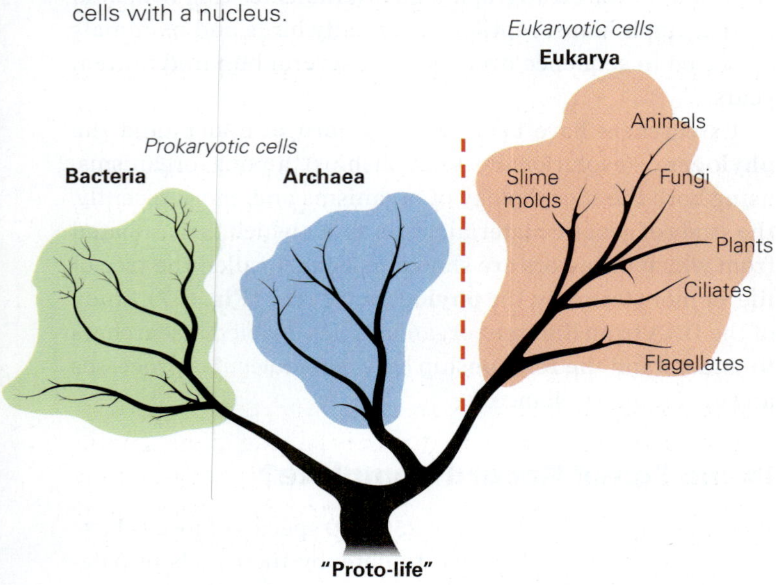

centuries. These fossils define the framework of life's evolution. But the record is not complete—known fossils cannot account for every intermediate step in the evolution of every organism. Over the billions of years that life has existed, 5 billion to 50 billion species of life may have existed. Clearly, known fossils represent at most a tiny percentage of these species. Why is the fossil record so incomplete?

First, despite all the fossil-collecting efforts of the past two centuries, paleontologists have not even come close to sampling every cubic centimeter of sedimentary rock exposed on Earth. Just as biologists have not yet identified every living species of insect, paleontologists have not yet identified every species of fossil. New species and even genera of fossils continue to be discovered every year.

Second, not all organisms are represented in the rock record, because not all organisms have a high preservation potential. As noted earlier, fossilization occurs only under special conditions, and thus only a minuscule fraction of the organisms that have lived on Earth have left a fossil record. We will never be aware of the existence or appearance of the vast numbers of organisms that have not been fossilized.

Finally, as you will learn in Chapter 10, the sequence of sedimentary strata that exists on Earth does not account for every second of time at every location since the formation of our planet. Sediments accumulate only in environments where conditions are appropriate for deposition and not for erosion. They do not accumulate, for example, on dry plains or on steep mountain peaks, but do accumulate in the sea and in the floodplains and deltas of rivers. Because Earth's climate changes through time and because the sea level rises and falls, certain locations on continents are sometimes sites of deposition and sometimes aren't, and on occasion are sites of erosion. Thus, strata accumulate only episodically.

E.5 Evolution and Extinction

Darwin's Grand Idea

As a young man in England in the early 19th century, Charles Darwin had been unable to settle on a career but had developed a strong interest in natural history. Therefore, he jumped at the opportunity to serve as a naturalist aboard HMS *Beagle* on an around-the-world surveying cruise. During the five years of the voyage, from 1831 to 1836, Darwin made detailed observations of plants, animals, and geology in the field and amassed an immense specimen collection from South America, Australia, and Africa. Just before Darwin departed, a friend gave him a copy of Charles Lyell's 1830 textbook *Principles of Geology*, which argued in favor of James Hutton's proposal that the Earth had a long history and that geologic time extended much further into the past than did human civilization.

A visit to the Galápagos Islands, off the coast of Ecuador, led to a turning point in Darwin's thinking during the voyage. The naturalist was most impressed with the variability of Galápagos finches. He marveled not only at the fact that different varieties of the bird occurred on different islands, but at how each variety had adapted to utilize a particular food supply. Keeping in mind Lyell's writings about the long duration of Earth history, Darwin developed a hypothesis that the finches had begun as a single species but had branched into several different species, over a long time, when populations of the birds became isolated on different islands. This can happen because offspring can differ from their parents and new traits can be transferred to succeeding generations. If enough change accumulates over many generations, the living population ends up being so different from its distant ancestors that the population can be classified as a new species. During the course of evolution, old species vanish and new species appear. The accumulation of changes may eventually yield a population that is so different from its ancestors that taxonomists consider it a new genus—greater differences may result in a new class, an order, or even a new phylum. Such change in a population over a succession of generations, due to the transfer of inheritable characteristics, as we noted earlier, is the process of *evolution*.

Darwin and his contemporary, Alfred Russel Wallace, not only proposed that evolution took place, but also came up with an explanation for why it occurs. The crux of the explanation is simply this: populations of organisms cannot increase in numbers forever, because they are limited by competition for scarce resources in the environment. In nature, only organisms capable of survival can pass on their characteristics to the next generation. In each new generation, some individuals have characteristics that make them more fit, whereas some have characteristics that make them less fit. The fitter organisms are more likely to survive and produce offspring. Thus, the survivors' beneficial characteristics get passed on to the next generation. Darwin referred to the survival of the fittest as **natural selection**, because it occurs on its own in nature. According to Darwin, when natural selection takes place over long periods of time—geologic time—it eventually yields a population of new organisms that differ so significantly from their distant ancestors that the new organisms can be considered to constitute a new species. If environmental conditions change, or if competitors enter the environment, species that do not evolve and become better adapted to survive the new conditions eventually die off and become extinct.

Darwin's view of evolution by natural selection has been successfully supported by many observations over time, and has not been definitively disproven by any observation or experiment. Also, it can be used to make testable predictions. Thus, scientists now refer to Darwin's idea as the *theory of evolution by natural selection* or just the *theory of evolution* (see Box P.1 in the Prelude for the definition of a theory). The theory of evolution provides a conceptual framework in which to understand paleontology. By studying fossils in sequences of strata, paleontologists are able to observe progressive changes in a species and in an assemblage of species through time, and can determine when some species die out and other species appear.

Because of the incompleteness of the fossil record, however, many questions remain as to the rates at which evolution takes place during the course of geologic time. Originally, it was assumed that evolution took place at a constant, slow rate—this concept is called **gradualism**. More recently, researchers have suggested that evolution occurs in fits and starts: it goes very slowly for quite a while (the species are in equilibrium), and then, during a relatively short period, it takes place very rapidly. This concept is called **punctuated equilibrium**. Factors that could cause sudden accelerations in the rate of evolution include (1) a sudden *mass extinction event* during which many organisms disappear, leaving ecological niches open for new species to colonize; (2) a relatively rapid change in the Earth's climate that puts stress on organisms—new species that can survive the new climate survive, whereas those that can't become extinct; (3) formation of new environments, as may happen when rifting splits apart a continent and generates a new ocean with new coastlines; and (4) the isolation of a breeding population.

Regardless of the process of evolution, the survivability of different kinds of organisms are not all the same. Some populations are very durable, in that they survive as an identifiable genus or even species for long intervals of geological time (tens of millions to over 100 million years). But others appear and then disappear within a relatively short interval of geologic time (less than a few million years).

Extinction: When Species Vanish

As we've noted, **extinction** occurs when the last members of a species die, so no parents can pass on their genetic traits to offspring. Some species become extinct as a population evolves into new species, whereas other species just vanish, leaving no hereditary offspring. Paleontologists suggest that many different phenomena can contribute to extinction. Some of the geologic factors that cause extinction include the following:

- *Global climate change*: At times, the Earth's mean temperature has been significantly colder than today's, whereas at other times it has been much warmer. Because of a changing climate, an individual species may lose its habitat, and if it cannot adapt to the new habitat or migrate to stay with its old one, the species will disappear.
- *Tectonic activity*: Tectonic activity causes vertical movement of the crust over broad regions, changes in seafloor spreading rates, and changes in the amount of volcanism. These phenomena can modify the distribution and area of habitats. Species that cannot adapt die off.
- *Asteroid or comet impact*: Many geologists have concluded that impacts of large meteorites with the Earth have been catastrophic for life. A large impact would send dust and debris into the atmosphere that could blot out the Sun and plunge the Earth into darkness and cold (see Chapter 19). Such a change, though relatively short lived, could interrupt the food chain.
- *Voluminous volcanic eruption*: Several times during Earth history, incredible quantities of lava have spilled out on the surface and incredible volumes of ash and gas have spewed into the air. These eruptions, perhaps due to the rise of particularly large plumes in the mantle, may have altered the climate globally.
- *The appearance of a new predator or competitor*: Some extinctions may happen simply because a new predator appears on the scene. Researchers suggest that this phenomenon explains the widespread extinction that occurred during the past 20,000 years, when a vast number of large mammal species vanished from North America. The timing of these extinctions appears to coincide with the appearance of the first humans (fierce predators) on the continent. If a more efficient competitor appears, the competitor steals an ecological niche from the weaker species, whose members can't obtain enough food and thus die off.

Some extinctions happen over long time intervals, when the replacement rate of a population simply becomes lower than the mortality rate, but others happen suddenly, when a cataclysmic event leads to the rapid extermination of many organisms. For example, in 1870 the population of passenger pigeons in North America exceeded 3 billion. Due to widespread hunting by people, the population dropped rapidly during the next two decades, and the last member of the species died in 1914.

Of note, paleontologists have found that the total number of genera of fossils, which represents *biodiversity* (the overall variation of life), is different over time and has abruptly decreased at specific times during Earth history. A worldwide abrupt decrease in the number of fossil genera is called a **mass extinction event**. At least five major mass extinction events have happened during the past half-billion years (**Fig. E.8**). These events define the boundaries between some of the major intervals into which geologists divide time. For example, a major extinction event marks the end of the Cretaceous Period, 66 million years ago. During this event, all dinosaur species (with the exception of their modified descendants, the birds) vanished, along with most marine invertebrate species. A huge extinction event also brought the Permian to a close. Significantly, the rate at which species have been disappearing during the past few centuries has been so rapid that some researchers and writers refer to our present time as the *sixth extinction*.

FIGURE E.8 This graph shows how the diversity of life has changed with time. Sudden drops indicate periods when mass extinctions occurred.

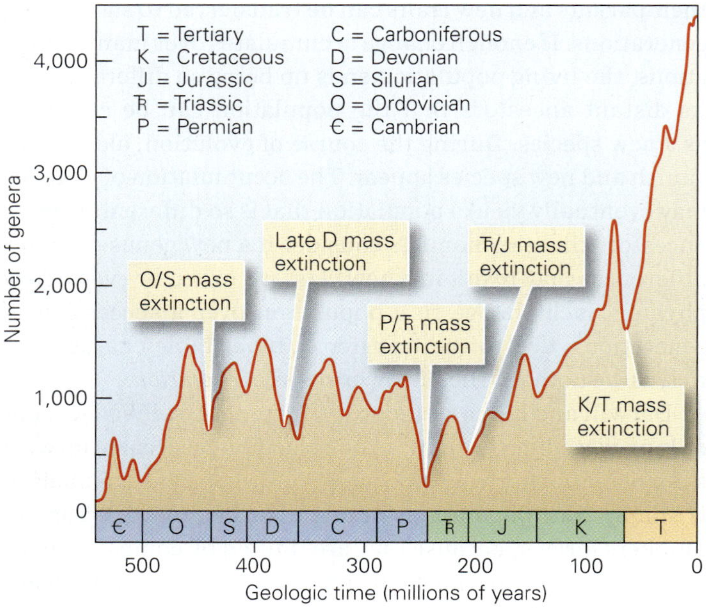

Interlude E Review

Interlude Summary

- Fossils form when organisms, or the traces of organisms (burrows or footprints) are buried and preserved in rock. Typically the hard parts (shells or bones) of fossils are better preserved.
- Examples of fossils include molds or casts of shells, bones, or footprints; permineralized wood or bones; carbonized impressions; preserved shells or bones; and bodies preserved in amber, tar, or permafrost. Certain distinctive molecules can serve as molecular fossils or biomarkers.
- The fossil record is very incomplete, because preserved strata do not record all of geologic time, not all organisms have high preservation potential, and paleontologists have only found a tiny fraction of fossil species.
- Fossils provide a record of life's evolution and a basis for geologists to determine the ages of rock layers relative to one another.
- Fossil organisms can be classified using the same taxonomic concepts used to classify modern organisms.
- Darwin proposed the theory of evolution by natural selection which states that the fittest organisms survive and pass on their traits to their offspring. Over time, organisms become so different from their distant ancestors that they can be considered to be a new species.
- When the last of a species dies out, the species goes extinct. At several times during Earth's history, a high percentage of genera went extinct. These events are called mass extinctions, and may be a consequence of catastrophic events, such as meteorite impacts or unusually abundant volcanism.

Guide Terms

biomarker (p. 330)
Cambrian explosion (p. 333)
cast (p. 329)
domain (p. 332)
evolution (p. 327)
extinction (p. 336)
extraordinary fossil (p. 331)
fossil (p. 327)
fossilization (p. 327)
gradualism (p. 335)
kingdom (p. 332)
macrofossil (p. 330)
mass extinction event (p. 336)
microfossil (p. 330)
mold (p. 329)
morphology (p. 332)
natural selection (p. 335)
paleontologist (p. 327)
paleontology (p. 327)
permineralization (p. 330)
petrified wood (p. 330)
phylogenetic tree (p. 333)
phylogeny (p. 333)
preservation potential (p. 331)
punctuated equilibrium (p. 335)
taxonomy (p. 332)

Review Questions

1. What is a fossil? Give examples of various types.
2. What are the various processes that can yield fossils?
3. According to the fossil record, do strata of all ages contain the same fossils?
4. What conditions can lead to preservation of an organism as a fossil?
5. Is the fossil record complete? Why or why not?
6. Describe the basic principles of the theory of evolution by natural selection.
7. Explain different ideas characterizing the rate of evolution.
8. What phenomena may cause extinction?
9. What is a mass extinction event, and why might such events take place?

Online Resources

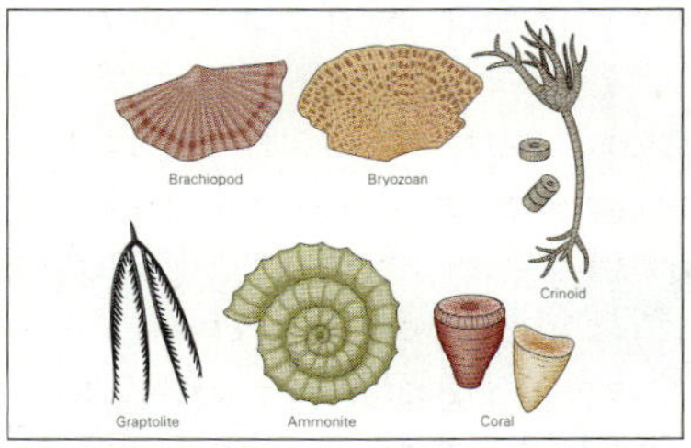

Assessment

This chapter features questions about fossils and life over geologic time.

The view from an airplane window of the Colorado Plateau region in Arizona and Utah reveals layer upon layer of strata deposited long ago and now exposed by erosion. These layers preserve a record of Earth's very long history.

LEARNING OBJECTIVES

By the end of this chapter, you should understand...

1. the meaning of geologic time, and the difference between relative and numerical ages.
2. geologic principles (uniformitarianism, superposition, fossil succession) and their implications.
3. how unconformities form and what they represent.
4. the basis for correlating stratigraphic formations, and how correlation led to development of the geologic column.

5. how geologists determine the numerical age of rocks by using isotopic dating.
6. the basis for determining dates on the geologic time scale, and for determining the age of the Earth.

CHAPTER 10

Deep Time: How Old Is Old?

10.1 Introduction

In May of 1869, a one-armed Civil War veteran named John Wesley Powell set out with a team of nine geologists and scouts to explore the previously unmapped expanse of the Grand Canyon, the greatest gorge on Earth. Though Powell and his companions battled fearsome rapids and the pangs of starvation, most managed to emerge from the mouth of the canyon three months later. During their voyage, seemingly insurmountable walls of rock both imprisoned and amazed the explorers, and led them to pose important questions about the Earth and its history, questions that even casual tourists at the canyon ponder today: Did the Colorado River sculpt this marvel, and if so, how long did it take? When did the rocks making up the walls of the canyon form? Was there a time before the colorful layers accumulated? Such questions pertain to **geologic time**, the span of time since the Earth's formation.

In this chapter, we first learn the geologic principles that allowed geologists to develop the concept of geologic time, and from these principles we develop a reference frame for describing the *relative ages* of rocks, fossils, structures, and

> *If the Eiffel Tower were now representing the world's age, the skin of paint on the pinnacle-knob at its summit would represent man's share of that age; and anybody would perceive that that skin was what the tower was built for. I reckon they would, I dunno.*
>
> —MARK TWAIN (American writer, 1835–1910)

landscapes. This information sets the stage for introducing the *geologic column,* the representation that geologists use to divide geologic time into named intervals. Then, we look at how *isotopic dating* (also known as *radiometric dating*) allows geologists to determine the *numerical age* of rocks, meaning the age of rocks in years, and how numerical ages can be added to the geological column to produce the *geologic time scale.* With the concept of geologic time in hand, a hike down a trail into the Grand Canyon becomes a trip into what many authors have called *deep time.* The geological discovery that our planet's history extends billions of years into the past changed humanity's perception of time and the Universe as profoundly as did the astronomical discovery that the limit of deep space extends billions of light-years beyond the edge of our Solar System.

10.2 The Concept of Geologic Time

Setting the Stage for Studying the Past: Uniformitarianism

Until relatively recently, people in most cultures believed that geologic time began about the same time that human history began, and that our planet has been virtually unchanged since its birth. This view was challenged by James Hutton (1726–1797), a Scottish gentleman farmer and doctor. Hutton lived during the Age of Enlightenment, a time when, sparked by the discovery of physical laws by Sir Isaac Newton, many sought natural, rather than supernatural, explanations for features of the world around them. While wandering in the highlands of Scotland, where rocks are well exposed, Hutton noted that many features (such as ripple marks and cross beds) found in sedimentary rocks resembled features that he could see forming in modern depositional environments. These observations led Hutton to speculate that the formation of rocks and landscapes, in general, were a consequence of processes that he could see happening in his own time.

Hutton's idea came to be known as the principle of **uniformitarianism**. According to this principle, physical processes that operate in the modern world also operated in the past, at roughly the same rates, and these processes were responsible for forming geologic features preserved in outcrops. More concisely, the principle can be stated as "the present is the key to the past." Hutton deduced that the development of individual geologic features took a long time, and that not all features formed at the same time. Therefore, the Earth must have a history that includes a succession of slow geologic events. Since no one in recorded history has seen the entire process of sediment first turning into rock and then later rising into mountains, Hutton also realized that the Earth's processes must have been active for a long time before human history began.

Hutton was not a particularly clear writer, and it took years, and the efforts of subsequent geologists, to interpret and clarify the principle of uniformitarianism and to publicize its significance. Once this had been accomplished, geologists around the world began to apply their growing understanding of geologic processes to define and interpret geologic events of the Earth's past.

Relative versus Numerical Age

Like historians, geologists strive to establish the sequence of events that produced an array of geologic features (such as rocks, structures, and landscapes). When possible, they also try to find the date on which each event took place. We specify the age of one feature with respect to another in a sequence as its **relative age**, and the age of a feature given in years as its **numerical age** (*absolute age,* in older literature). Geologists learned how to determine relative age long before they could determine numerical age, so we look next at the principles leading to relative-age determination.

TAKE-HOME MESSAGE

The principle of uniformitarianism—the present is the key to the past—implies that the Earth must be very old, for geologic processes happen slowly. Geologists distinguish between relative age (older or younger) and numerical age (how many years ago).

QUICK QUESTION What observations led Hutton to propose uniformitarianism?

10.3 Geologic Principles for Defining Relative Age

Building from the work of Steno, Hutton, and others, the British geologist Charles Lyell (1797–1875) laid out a set of formal, usable geologic principles in the first modern textbook of geology. These principles, described below, continue to provide the basic framework within which geologists read the record of Earth history and determine relative ages.

- *The principle of uniformitarianism*: As we've seen, the principle of uniformitarianism means that physical processes we observe operating today also operated in the past, at roughly comparable rates. Put concisely, "the present is the key to the past" (**Fig. 10.1a**).
- *The principle of original horizontality*: Sediments on Earth settle out of fluids in a gravitational field. The surfaces on which sediments accumulate (such as floodplains or the seafloor) are fairly flat. Therefore, layers of sediment, when first deposited, are fairly horizontal (**Fig. 10.1b**). If sediments collect on a steep slope, they typically slide downslope before lithification, so they will not be preserved as sedimentary rocks. With this in mind, we conclude that folding, tilting, and faulting of sedimentary beds must occur after the beds were deposited.
- *The principle of superposition*: In a sequence of sedimentary rock layers, each layer must be younger than the one below, for a layer of sediment cannot accumulate unless there is already a substrate on which it can collect. Thus, the layer at the bottom of a sequence is the oldest, and the layer at the top is the youngest (**Fig. 10.1c**).
- *The principle of lateral continuity*: Sediments generally accumulate in continuous sheets within a given region. When you see a sedimentary layer cut by a canyon, you can assume that the layer once spanned the area, and was later eroded by the river that formed the canyon (**Fig. 10.1d**).
- *The principle of cross-cutting relations*: If one geologic feature cuts across another, the feature that has been cut is older. For example, if an igneous dike cuts across a sequence of sedimentary beds, the beds must be older than the dike (**Fig. 10.1e**). If a fault cuts across and displaces layers of sedimentary rock, then the fault must be younger than the layers. But if a layer of sediment buries a fault, the sediment must be younger than the fault.
- *The principle of baked contacts*: An igneous intrusion "bakes" (metamorphoses) surrounding rocks, so the rock that has been baked must be older than the intrusion (**Fig. 10.1f**).
- *The principle of inclusions*: A rock containing an *inclusion* (fragment of another rock) must be younger than the inclusion. For example, a conglomerate containing pebbles of basalt is younger than the basalt, and a sill containing fragments of sandstone must be younger than the sandstone (**Fig. 10.1g**).

Geologists apply geologic principles to determine the relative ages of rocks, structures, and other geologic features at a given location. They then go further by interpreting the development of each feature to be the consequence of a specific geologic event. Examples of geologic events include deposition of sedimentary beds, erosion of the land surface, intrusion or extrusion of igneous rocks, deformation (folding and/or faulting), and episodes of metamorphism. The succession of events in order of relative age that have produced the rock, structure, and landscape of a region is called the **geologic history** of the region. For example, by applying geologic principles, we can determine relative ages of the features shown in **Figure 10.2a**, and in so doing, we develop a geologic history of the region, defining the relative ages of events that produced the features. In this example, the geologic history includes several events: deposition of the sedimentary sequence in order from beds 1 to 8; intrusion of the sill; folding of the sedimentary beds and the sill; intrusion of the granite pluton; faulting; intrusion of the dike; and erosion to form the present land surface (**Fig. 10.2b**).

Adding Fossils to the Story: Fossil Succession

As Britain entered the industrial revolution in the late 18th and early 19th centuries, new factories demanded coal to fire their steam engines, and they needed an inexpensive means to transport goods. Investors decided to construct a network of canals to transport coal and iron, and they hired an engineer named William Smith (1769–1839) to survey some of the excavations. Canal digging provided fresh exposures of bedrock, which previously had been covered by vegetation. Smith learned to recognize distinctive layers of sedimentary rock and to identify the **fossil assemblage** (the group of fossil species) that they contained. He also realized that a particular assemblage can be found only in a limited interval of strata, and not above or below this position. Thus, once a fossil species disappears at a horizon in a sequence of strata, it never reappears higher in the sequence. Put another way, extinction is forever. Smith's observation has been repeated at millions of locations around the world, and has been codified as the principle of **fossil succession**. It provides the geologic underpinning for the theory of evolution (see Interlude E).

FIGURE 10.1 Examples of the major geologic principles.

(a) Uniformitarianism: The processes that form cracks in the dried-up mud puddle on the left also formed the mudcracks preserved in the ancient, solid rock on the right. We can see the ancient mudcracks because erosion removed the adjacent bed.

(b) Original horizontality: Gravity causes sediments to accumulate in horizontal sheets.

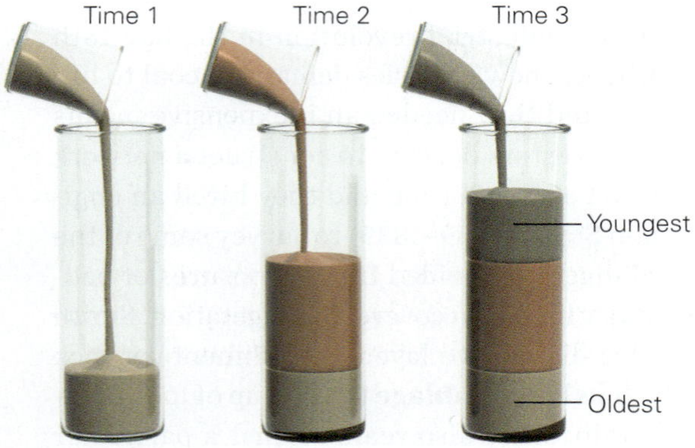

(c) Superposition: In a sequence of strata, the oldest bed is on the bottom, and the youngest on top. Pouring sand into a glass illustrates this point.

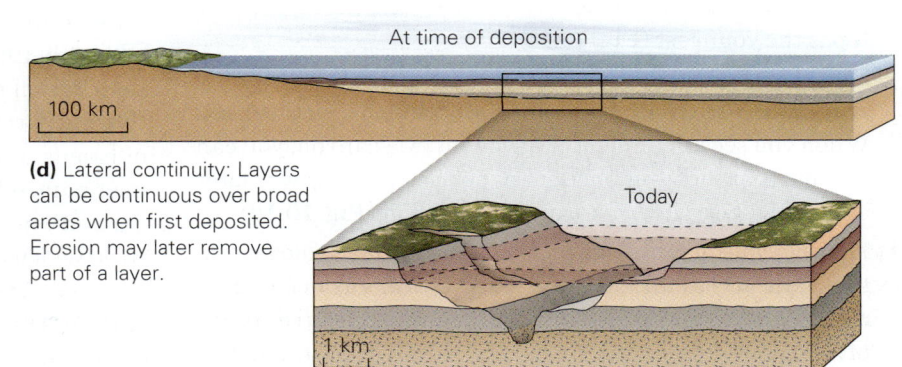

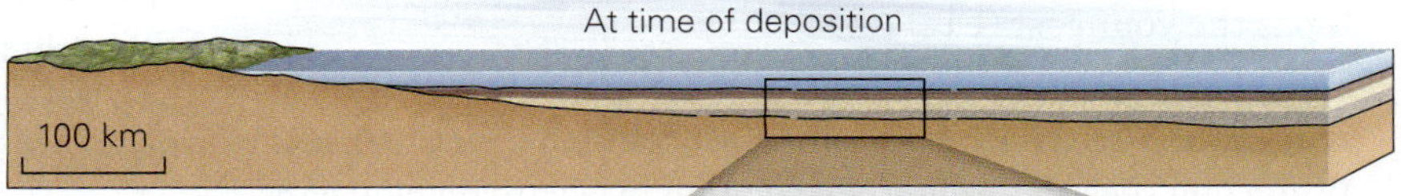

(d) Lateral continuity: Layers can be continuous over broad areas when first deposited. Erosion may later remove part of a layer.

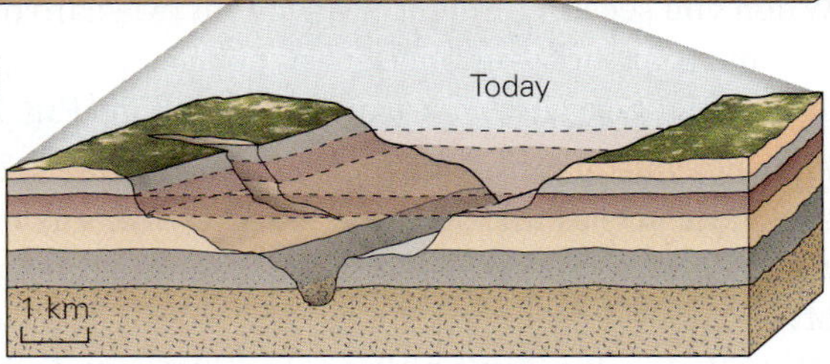

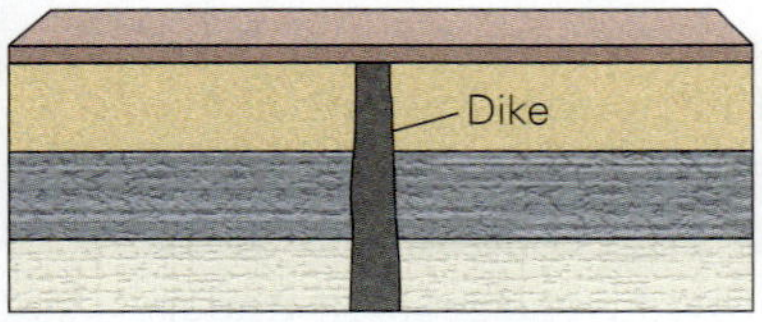

(e) Cross-cutting relations: The dike cuts across the lower three sedimentary beds, so the dike is younger. The top bed buries the dike, so it is younger than the dike.

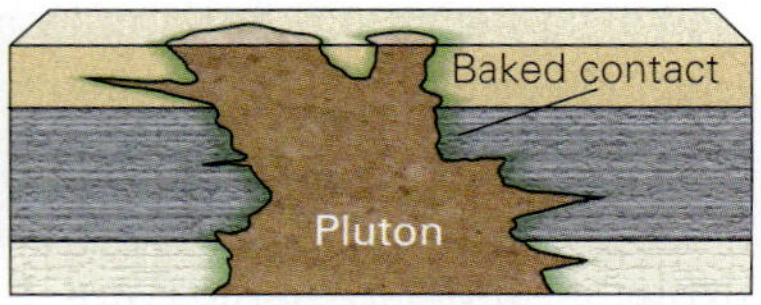

(f) Baked contacts: The pluton baked the adjacent rock, so the adjacent rock is older.

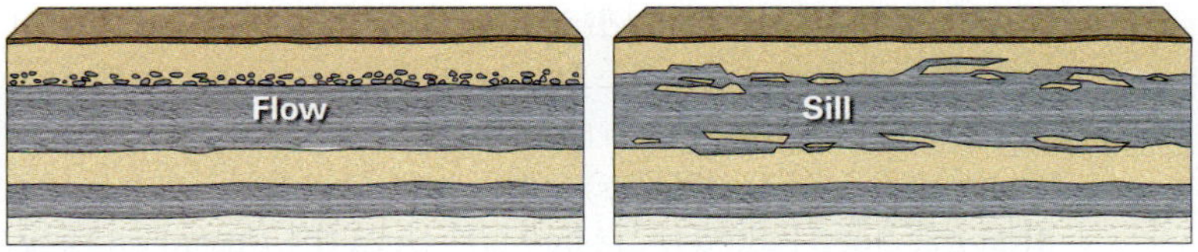

(g) By the principle of inclusions, the pebbles of basalt in a conglomerate must be older than the conglomerate and xenoliths of sandstone must be older than the basalt containing them.

FIGURE 10.2 Interpreting the geologic history of a region, using geologic principles as a guide.

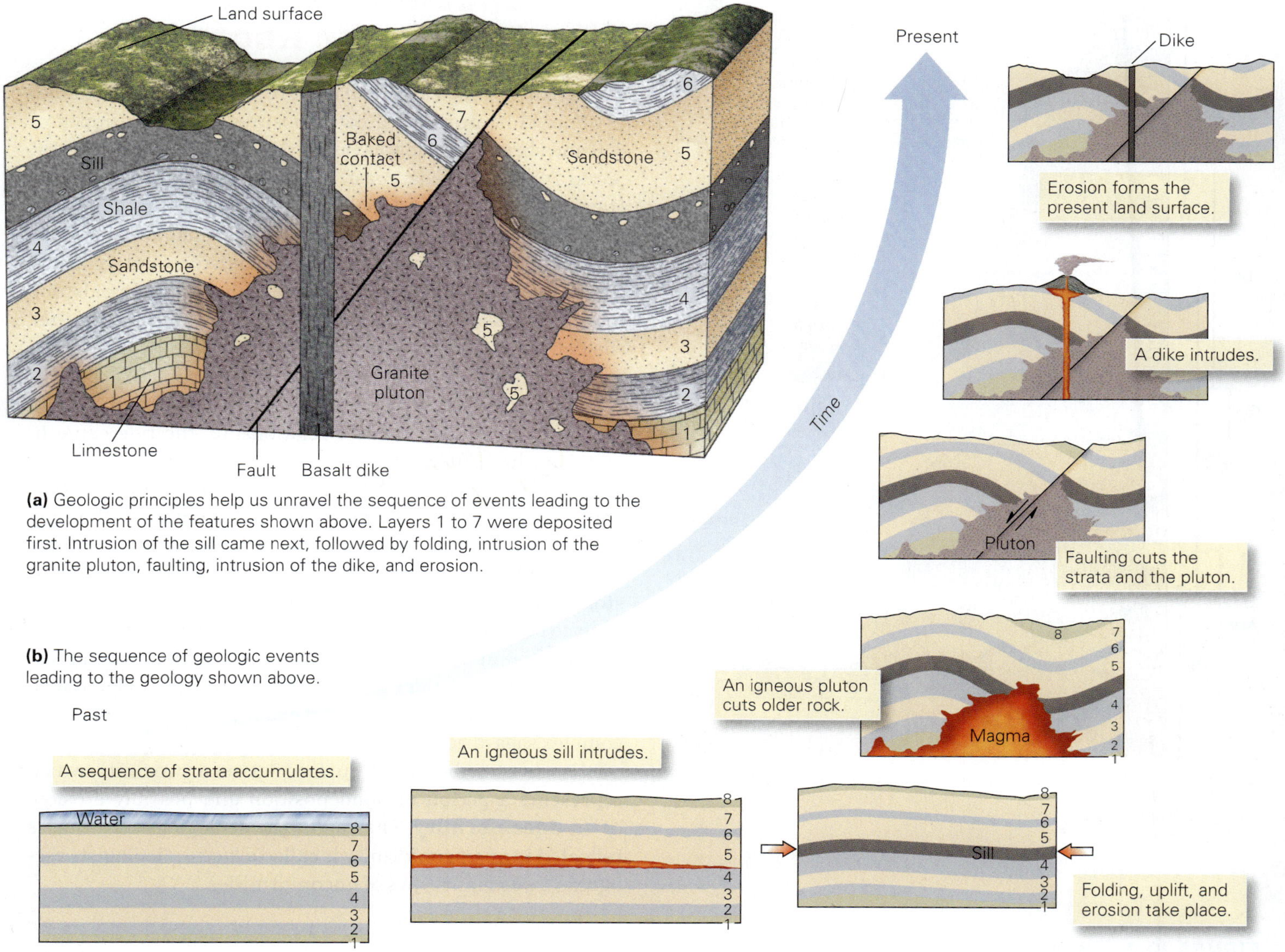

(a) Geologic principles help us unravel the sequence of events leading to the development of the features shown above. Layers 1 to 7 were deposited first. Intrusion of the sill came next, followed by folding, intrusion of the granite pluton, faulting, intrusion of the dike, and erosion.

(b) The sequence of geologic events leading to the geology shown above.

To see how this principle works, examine **Figure 10.3**, which depicts a sequence of strata. Bed 1 at the base contains fossil species A, Bed 2 contains fossil species A and B, Bed 3 contains B and C, Bed 4 contains C, and so on. From these data, we can define the *range* of specific fossils in the sequence, meaning the interval in which the fossils occur. Note that the sequence contains a definable succession of fossils (A, B, C, D, E, F), that the range where a particular species occurs may overlap with the range of other species, and that once a species vanishes, it does not reappear higher in the sequence. Some fossil species are widespread but survived for only a relatively short interval of geologic time. Such species are called **index fossils** (or *guide fossils*), because they can be used by geologists to associate the strata with a specific time interval. Because of the principle of fossil succession, we can define the relative ages of strata by looking at fossils. For example, if we find a bed containing Fossil A, we can say that the bed is older than a bed containing, say, Fossil F. Geologists have now determined the relative ages of over 200,000 fossil species.

TAKE-HOME MESSAGE

Geologic principles (including uniformitarianism, superposition, cross-cutting relations, and fossil succession) provide the basis for determining the relative ages of rocks and other geologic features. By working out relative ages, we can reconstruct the geologic history of a region.

QUICK QUESTION If a fault forms a fault scarp, what are the relative ages of the fault and the landscape surface?

FIGURE 10.3 The principle of fossil succession.

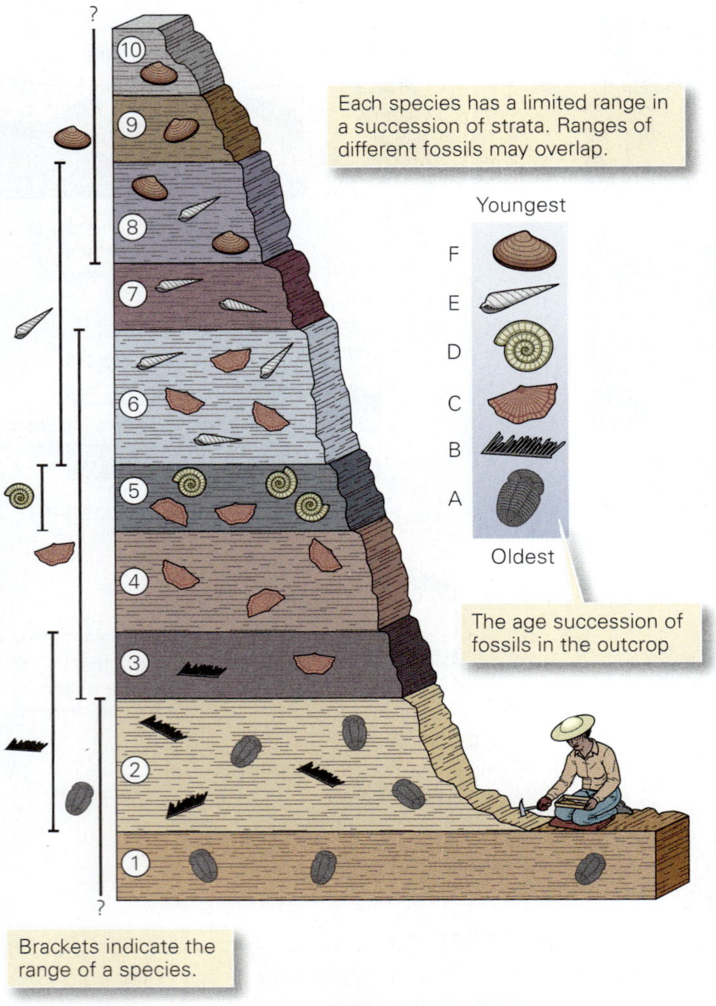

Each species has a limited range in a succession of strata. Ranges of different fossils may overlap.
Youngest
F
E
D
C
B
A
Oldest
The age succession of fossils in the outcrop
Brackets indicate the range of a species.

10.4 Unconformities: Gaps in the Record

James Hutton used a boat to explore the seaside of Scotland, as shore cliffs provided great exposures of rock, stripped of soil and shrubbery. He was puzzled by an outcrop he found at Siccar Point on the east coast. Here he saw a sequence of strata consisting of red sandstone and conglomerate that rested on a distinctly different sequence of gray sandstone and shale (**Fig. 10.4**). Further, the beds of gray sandstone and shale had a nearly vertical dip (slope), whereas the beds of red sandstone and conglomerate had a dip of less than 15°. (See Box 9.1 for the definition of *dip*.) The gently dipping layers seemed to lie across the truncated ends of the vertical layers, like a handkerchief resting on a row of books. Perhaps, as Hutton sat and stared at this odd geometric relationship, the tide came in and deposited a new layer of sand on top of the rocky shore. With the principle of uniformitarianism in mind, Hutton suddenly understood the significance of the outcrop—the gray sandstone–shale sequence had been deposited, turned into rock, tilted, and truncated by erosion, all before the red sandstone–conglomerate beds had even been deposited. Thus, the surface between the gray and red rock sequences represented a time interval when no new strata had been deposited at Siccar Point and the older strata had been partly eroded down. Hutton realized that Siccar Point exposed the record of a long, complex saga of geologic history.

We refer to a boundary surface between two units, which represents a period of nondeposition and possibly erosion, as an **unconformity**. The gap in the geologic record that is reflected in an unconformity is called a *hiatus*. Geologists recognize three main types of unconformities.

FIGURE 10.4 The unconformity at Siccar Point, Scotland.

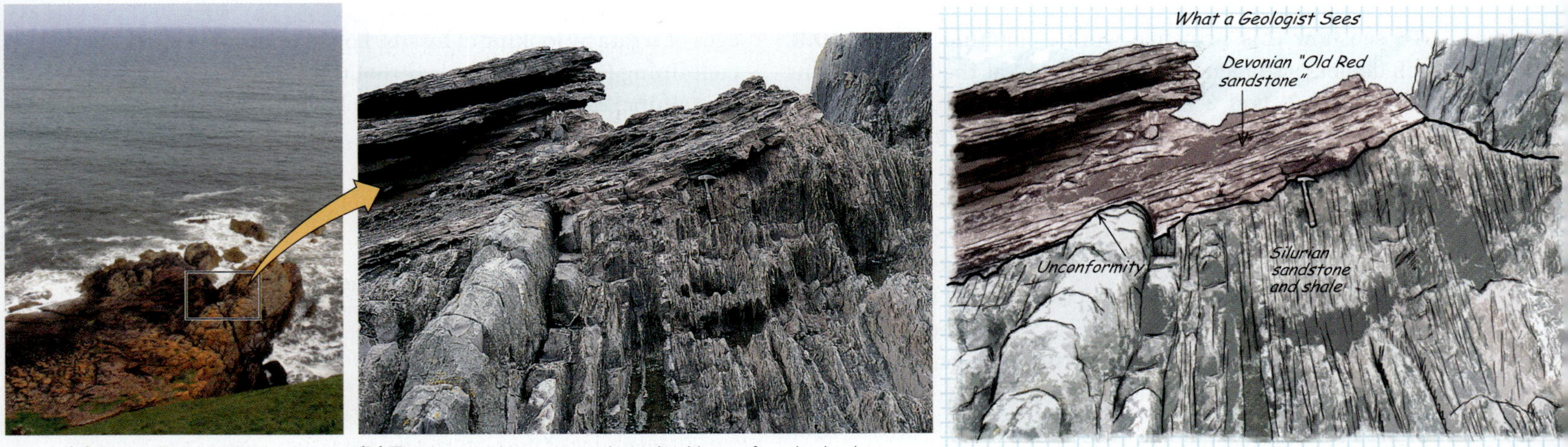

What a Geologist Sees
Devonian "Old Red sandstone"
Unconformity
Silurian sandstone and shale

(a) Siccar Point juts out into the North Sea.

(b) The layers above were deposited long after the beds below had been tilted.

(c) The contact between the two units is an unconformity.

FIGURE 10.5 The three kinds of unconformities and their formation.

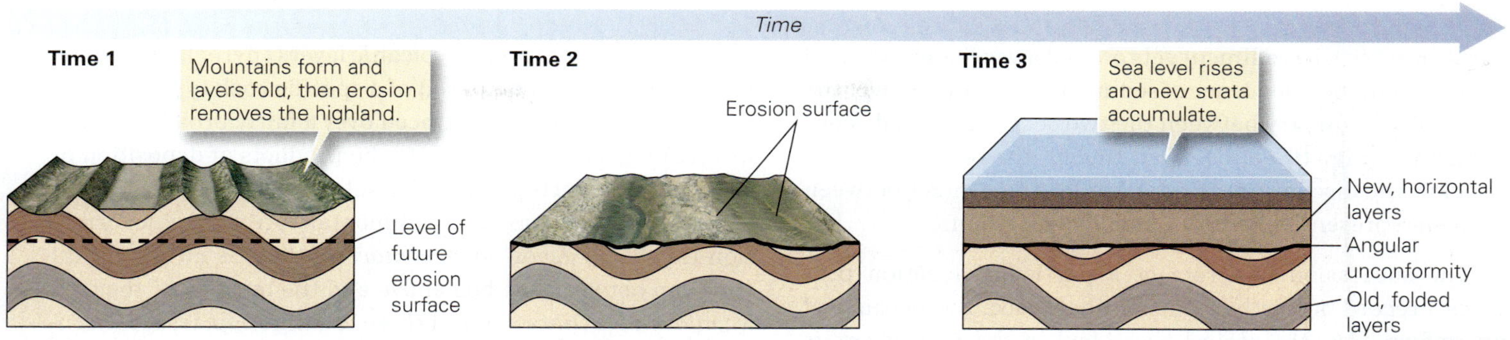

(a) An angular unconformity: (1) layers undergo folding; (2) erosion produces a flat surface; (3) sea level rises and new layers of sediment accumulate.

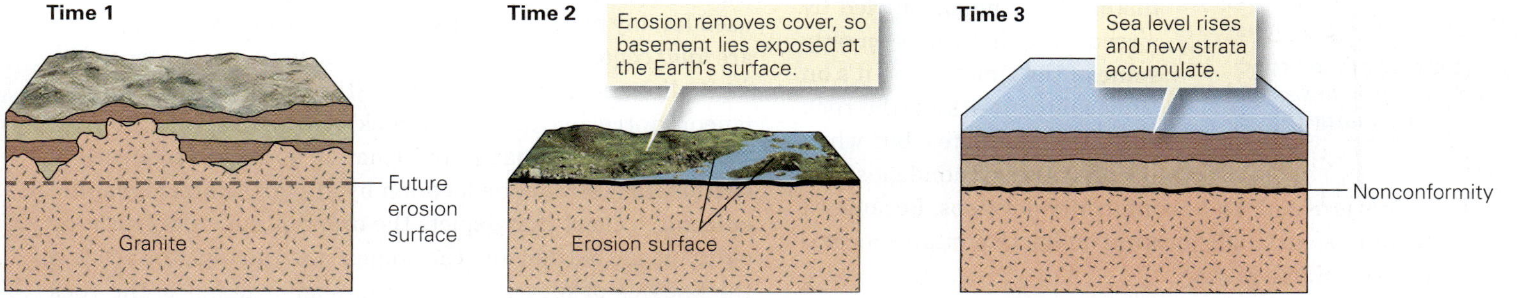

(b) A nonconformity: (1) a pluton intrudes; (2) erosion cuts down into the crystalline rock; (3) new sedimentary layers accumulate above the erosion surface.

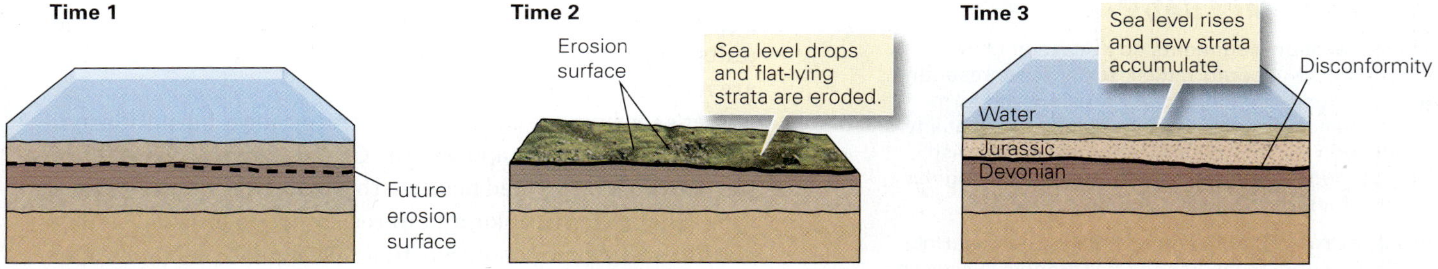

(c) A disconformity: (1) layers of sediment accumulate; (2) sea level drops and an erosion surface forms; (3) sea level rises and new sedimentary layers accumulate.

(d) This road cut in Utah shows a sand-filled channel cut down into floodplain mud. The mud was exposed between floods, and a soil formed on it. When later buried, all the sediment turned into rock; the channel floor is now a disconformity, and the ancient soil is now a paleosol. Note that the channel cut across the paleosol. The paleosol also represents a disconformity, a time during which deposition did not occur.

- *Angular unconformity*: Rocks below an angular unconformity were tilted or folded before the unconformity developed (**Fig. 10.5a**). An angular unconformity cuts across the underlying layers, and the orientation of layers below an unconformity is different from the orientation of the layers above. (The outcrop at Siccar Point exposes an angular unconformity.)
- *Nonconformity*: A nonconformity is a type of unconformity at which sedimentary rocks overlie generally much older intrusive igneous rocks and/or metamorphic rocks (**Fig. 10.5b**). These igneous or metamorphic rocks underwent cooling, uplift, and erosion prior to becoming the substrate, or *basement*, on which new sediments accumulated.
- *Disconformity*: Imagine that a sequence of sedimentary beds has been deposited beneath a shallow sea. Sea level drops, exposing the beds for some time. During this time, no new sediment accumulates, and some of

the pre-existing sediment gets eroded away. Later, sea level rises, and a new sequence of sediment accumulates over the old. The boundary between the two sequences is a disconformity (**Fig. 10.5c, d**). Even though the beds above and below the disconformity are parallel, the contact between them represents an interruption in deposition.

The succession of strata at a particular location provides a record of Earth history in that place. But because of unconformities, the record preserved in the rock layers is incomplete. It's as if geologic history is being chronicled by a recording that turns on only some of the time—when it's on (times of deposition), the rock record accumulates, but when it's off (times of nondeposition and possibly erosion), an unconformity develops. Because of unconformities, no single location on Earth contains a complete record of Earth history.

Did you ever wonder...
if the strata of the Grand Canyon represent all of Earth's history?

TAKE-HOME MESSAGE

At a given location, sediments do not accumulate continuously, so unconformities, surfaces representing intervals of nondeposition and/or erosion, can form. Because of unconformities, the geologic record at any given location is incomplete. Geologists distinguish among nonconformities, disconformities, and angular unconformities.

QUICK QUESTION Is there any one place on the surface of the Earth where the exposed stratigraphic succession represents all of geologic time? Explain.

10.5 Stratigraphic Formations and Their Correlation

The Concept of a Formation

When William Smith began to explore the strata exposed along the newly dug canals of England, he identified distinctive sets of beds, with distinctive assemblages of fossils, at the excavations. He realized that these sets must lie underground over the regions between the excavations, and assigned a name to each set in order to keep track of them. Generations of geologists that have followed now routinely divide successions of sedimentary strata (and/or volcanic deposits) into units, called stratigraphic formations, which other geologists can recognize and identify. Formally defined, a **stratigraphic formation** (*formation*, for short) is an interval of sedimentary strata and/or volcanic layers (pyroclastic deposits and lava flows), composed of a specific rock type or group of rock types, that can be traced over a fairly broad region. A formation generally represents the products of deposition or accumulation during a definable interval of time. The boundary surface at the base of a sedimentary stratigraphic formation is called a *depositional contact*—if there is no significant time gap between the bed above and the bed below, then it's a *conformable* contact, but if there is a time gap, it's an *unconformable* contact or unconformity. A depositional contact is a type of **geologic contact** (*contact*, for short), a broader term used for any boundary between two rock bodies. Other types are *fault contacts* (the boundary between rocks juxtaposed by fault slip) and *intrusive contacts* (the boundary between an igneous intrusion and its wall rock).

Geologists summarize information about the sequence of sedimentary strata at a location by drawing a **stratigraphic column**, a chart that depicts the order of the units (with the oldest at the bottom and the youngest at the top), the relative thicknesses of formations, and in some examples, the rock types within each unit. Geologists may also represent the relative resistance to erosion of beds or formations by making the side of the column irregular to symbolize the way the units might erode on a cliff face—units that stick out further are more resistant.

Let's see how the concepts of a stratigraphic formation and stratigraphic column apply to the Grand Canyon. The walls of the canyon look striped because they expose a variety of rock types that differ in color and in resistance to erosion. Geologists identify major contrasts that distinguish one interval of strata from another and use these as a basis for dividing the strata into formations, each of which may consist of many beds (**Fig. 10.6**). Note that some formations consist of beds of a single rock type, whereas others include interlayered beds of two or more rock types. Note also that not all formations have the same thickness, and that all the beds of a region must be part of a stratigraphic formation. Geologists generally depict conformable contacts in a stratigraphic column by a smooth line, and an unconformity by a wavy line.

Commonly, the name of a formation comes from a locality where it was first identified or first studied. For example, the Schoharie Formation was first defined based on exposures in Schoharie Creek in eastern New York State. If a formation consists of only one rock type, we may incorporate that rock type in the name (such as the Kaibab Limestone), but if a formation contains more than one rock type, we use the word formation in the name (such as the Toroweap Formation). Note that in the formal name of a formation, all words are capitalized. Several adjacent formations in a succession may be lumped together as a **stratigraphic group** (or simply a *group*).

FIGURE 10.6 The stratigraphic column and stratigraphic formation: examples from the Grand Canyon in Arizona.

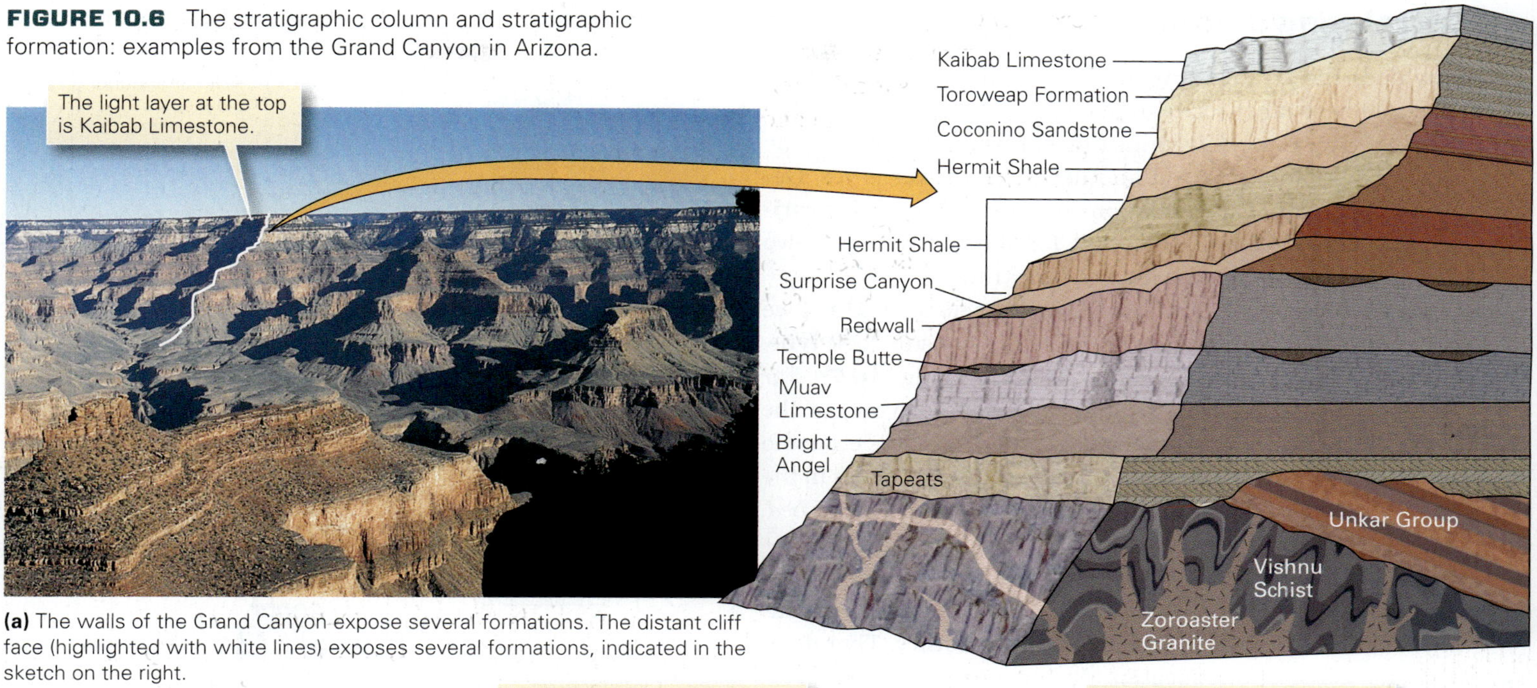

(a) The walls of the Grand Canyon expose several formations. The distant cliff face (highlighted with white lines) exposes several formations, indicated in the sketch on the right.

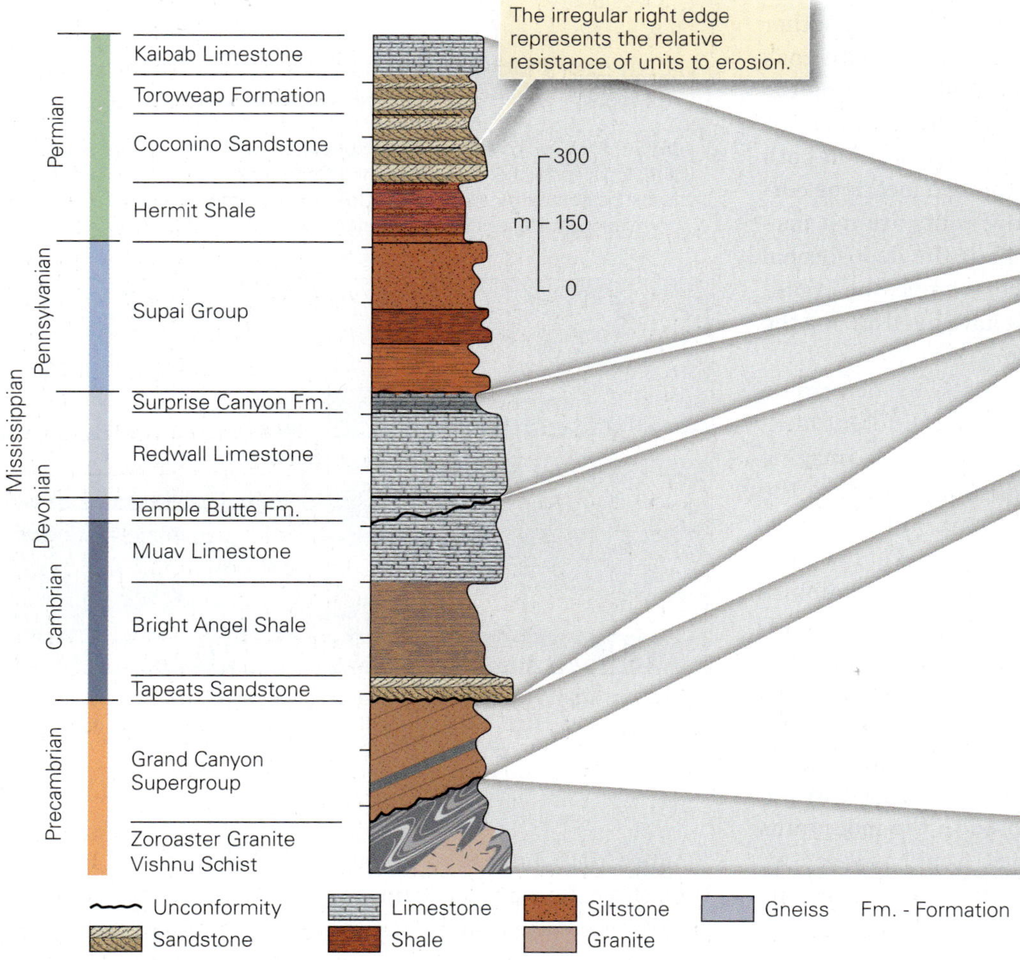

(b) A formalized stratigraphic column for strata of the Grand Canyon. The vertical scale indicates unit thickness. Patterns represent different rock types. The Unkar Group is one of three units that together comprise the Grand Canyon Supergroup.

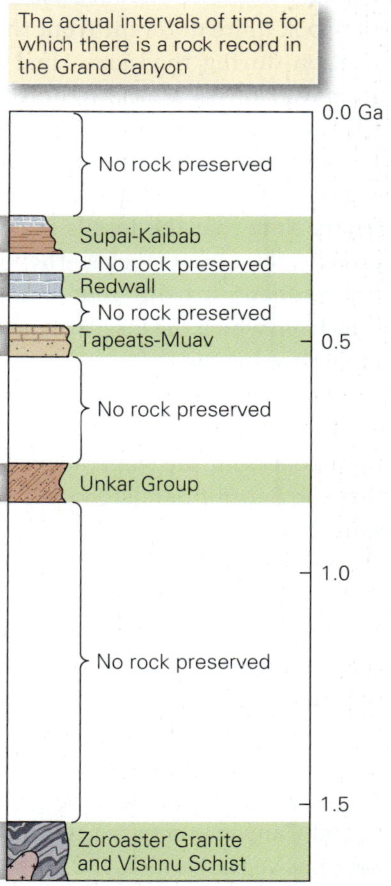

(c) Because of unconformities, the strata represent only a partial record of geologic time. This chart, with a time scale on the vertical axis, shows large gaps in the record.

Correlating Strata

How does the stratigraphy of a sedimentary succession exposed at one locality relate to that exposed at another? Stated another way, can we determine the relative age of strata at one location to that of strata at another? The answer is yes. Geologists determine such relations by a process called **stratigraphic correlation** (*correlation*, for short). There are two approaches for correlating intervals of strata.

When correlating formations among nearby regions, we can simply look for similarities in successions of rock type. We call this method *lithologic correlation* (**Fig. 10.7**). For example, the sequence of strata on the southern rim of the Grand Canyon clearly correlates with the sequence on the northern rim, because they contain the same rock types in the same order. In some cases, a sequence contains a *marker bed*, a unique layer that provides a definitive basis for correlation.

Lithologic correlation doesn't necessarily work over broad areas because the depositional setting and the source of sediments can change from one location to another. Therefore, beds deposited at one location during a given time interval may look quite different from the beds deposited at another location during the same time interval. To correlate rock units over broad areas, we must rely on fossils to define the relative ages of sedimentary units. We call this method *fossil correlation*. If fossils of the same relative age occur at both locations, we can say that the strata at the two locations correlate. Note that the fossils in correlative units are not necessarily the same species—they won't be if the depositional environments are different—but to be correlative, the organisms from which the fossils formed must have lived during the same time interval.

Fossil correlation can allow geologists to determine whether or not beds of the same rock type at different locations represent the same formation. As an example, imagine that the Santuit Sandstone and Oswaldo Sandstone look the same but are of different ages (see Fig. 10.7a). In Location A, the units are separated by the Milo Limestone, but at Location C the units are in direct contact. At first glance, the combination of the two units at Location C may look like a single unit. But a sharp-eyed geologist, looking closely at the strata, will see that the fossils have significantly different ages and will depict the contact between the two units by an unconformity.

Let's now apply correlation principles to the challenge of determining the relative ages of formations exposed in the Grand Canyon compared to those exposed in the mountains near Las Vegas, Nevada, 150 km to the west. Near Las Vegas, we find a sequence of sedimentary rocks that includes a limestone formation called the Monte Cristo Limestone. The Monte Cristo Limestone contains fossils of the same relative age as those of the Redwall Limestone of the Grand Canyon, but the Monte Cristo Limestone is much thicker than the Redwall Limestone. Because the formations contain fossils of the same relative age, we conclude that they were deposited during the same time interval, and thus we say that they correlate with one another. We also find that not only are the units thicker in

FIGURE 10.7 The principles of correlation.

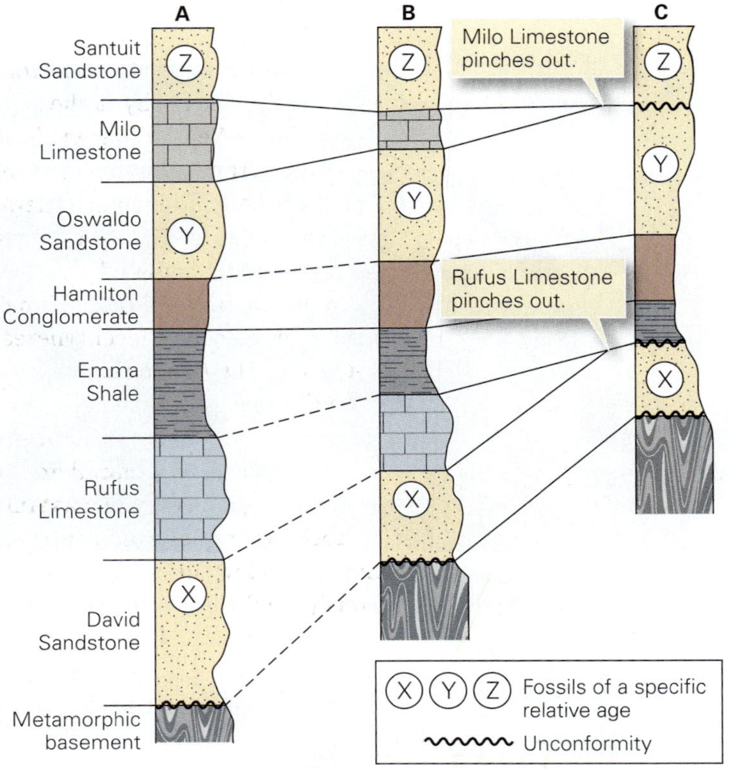

(a) Stratigraphic columns can be correlated by matching rock types (lithologic correlation). The Hamilton Conglomerate is a marker horizon. Because some strata pinch out, Column C contains unconformities. Fossil correlation indicates that the youngest beds in C are Santuit Sandstone.

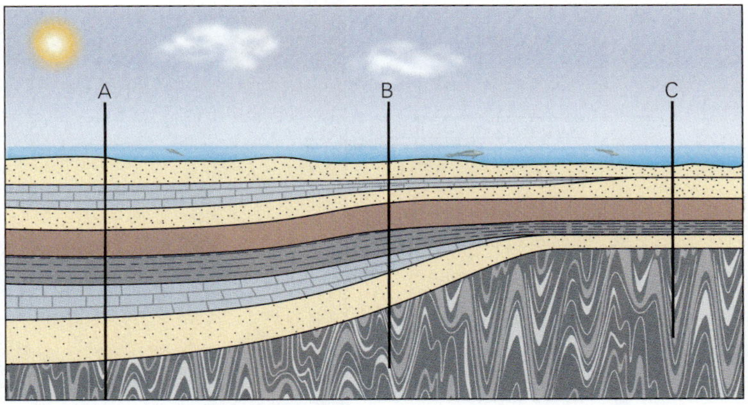

(b) At the time of deposition, locations A, B, and C (which correlate with the columns in part a) were in different parts of a basin. The basin floor was subsiding fastest at A.

SEE FOR YOURSELF...

GRAND CANYON, ARIZONA

Latitude
36°5′53.99″ N

Longitude
112°10′58.58″ W

Zoom to 1.5 km (~1 mile) and look obliquely downstream.

The strata represent an immense amount of geologic time. The dark metamorphic rocks of the inner gorge are ~1.75 Ga. The youngest unit, the layer at the top, is the ~270 Ma Kaibab Limestone.

the Las Vegas area than in the Grand Canyon area, but there are more of them. Thus, the contact beneath the Grand Canyon's Redwall Limestone is an unconformity. Why is the stratigraphy of Las Vegas different from that of the Grand Canyon? During part of the time when thick sediments were accumulating near Las Vegas, no sediments accumulated near the Grand Canyon, because the region of Las Vegas lay below sea level, whereas the region of the Grand Canyon was dry land. Geologists studying this contrast concluded that sediments of the Las Vegas region accumulated in a passive-margin basin that was sinking (subsiding) rapidly and remained submerged below the sea almost continuously. Sediments of the Grand Canyon region, in contrast, accumulated on the crust of a craton that episodically emerged above sea level.

Geologic Maps

Once William Smith succeeded in correlating stratigraphic formations throughout central England, he faced the challenge of communicating his ideas to others. One way would be to create a table that compared stratigraphic columns from different locations. But since Smith was a surveyor and worked with maps, it occurred to him that he could outline and color areas on a map to represent places where strata of a given relative age occurred. He did this using the data he had collected, and in 1815 he produced the first modern geologic map. In general, a **geologic map** portrays the spatial distribution of rock units at the Earth's surface.

The pattern displayed on a geologic map can provide insight into the presence and orientation of geologic structures in the map area (**Fig. 10.8a, b, c**). Formations on a geologic map can be depicted by colors or patterns, contacts on a geologic map by lines, the orientation of layers by strike-and-dip marks (see Box 9.1), and the position of structures by various symbols. With experience, a geologist can use the configuration of contacts and distribution of formations on a map as a basis for identifying folds, faults, unconformities, and other features. For example, repetition of a succession of strata may indicate the presence of a fold: If the oldest strata crop out along the hinge and the strata on opposite limbs of the fold dip away from the hinge, the fold is an anticline. But if the youngest strata crop out along the hinge and the limbs dip toward the hinge, it's a syncline. Using computer technology, it's now possible to plot geologic contacts on *digital elevation maps* (DEMs) that portray the ground surface in three dimensions. *Geologic cross sections* provide an interpretation of geologic relationships underground as they intersect a vertical plane. By combining depictions on a geologic map with depictions on cross sections, a geologist can develop a block diagram of a region, representing contacts and units in 3-D (**Fig. 10.8d**).

TAKE-HOME MESSAGE

A stratigraphic formation is a recognizable sequence of beds that can be mapped across a broad region. Geologists correlate formations regionally on the basis of rock type and fossil content, and portray the configuration of formations in a region on geologic maps, cross sections, and block diagrams.

QUICK QUESTION What will an angular unconformity look like on a geologic map?

10.6 The Geologic Column

As stated earlier, no one locality on Earth provides a complete record of our planet's history, because stratigraphic columns can contain unconformities. But by correlating rocks among localities at millions of places around the world, geologists have pieced together a composite stratigraphic column, called the **geologic column**, that represents the entirety of Earth history (**Fig. 10.9**). The column is divided into segments, each of which represents a specific interval of time. The largest subdivisions break Earth history into the Hadean, Archean, Proterozoic, and Phanerozoic **eons**—the first three together constitute the **Precambrian**. The suffix *-zoic* means life, so Phanerozoic means visible life, and Proterozoic means first life. These names can be a bit misleading, though, because decades after the eons had been named, geologists discovered that the earliest life, cells of bacteria and archaea, actually appeared during the Archean Eon. Eons, in turn, can be subdivided into **eras**. We further divide each era into **periods** and each period into **epochs**.

Where do the names of the periods come from? They refer either to localities where a fairly complete stratigraphic column representing that time interval was first identified (for example, rocks representing the Devonian Period crop out near Devon, England) or to a characteristic of the time (rocks from the Carboniferous Period contain coal). The terminology

FIGURE 10.8 A geologic map depicts the distribution of rock units and structures.

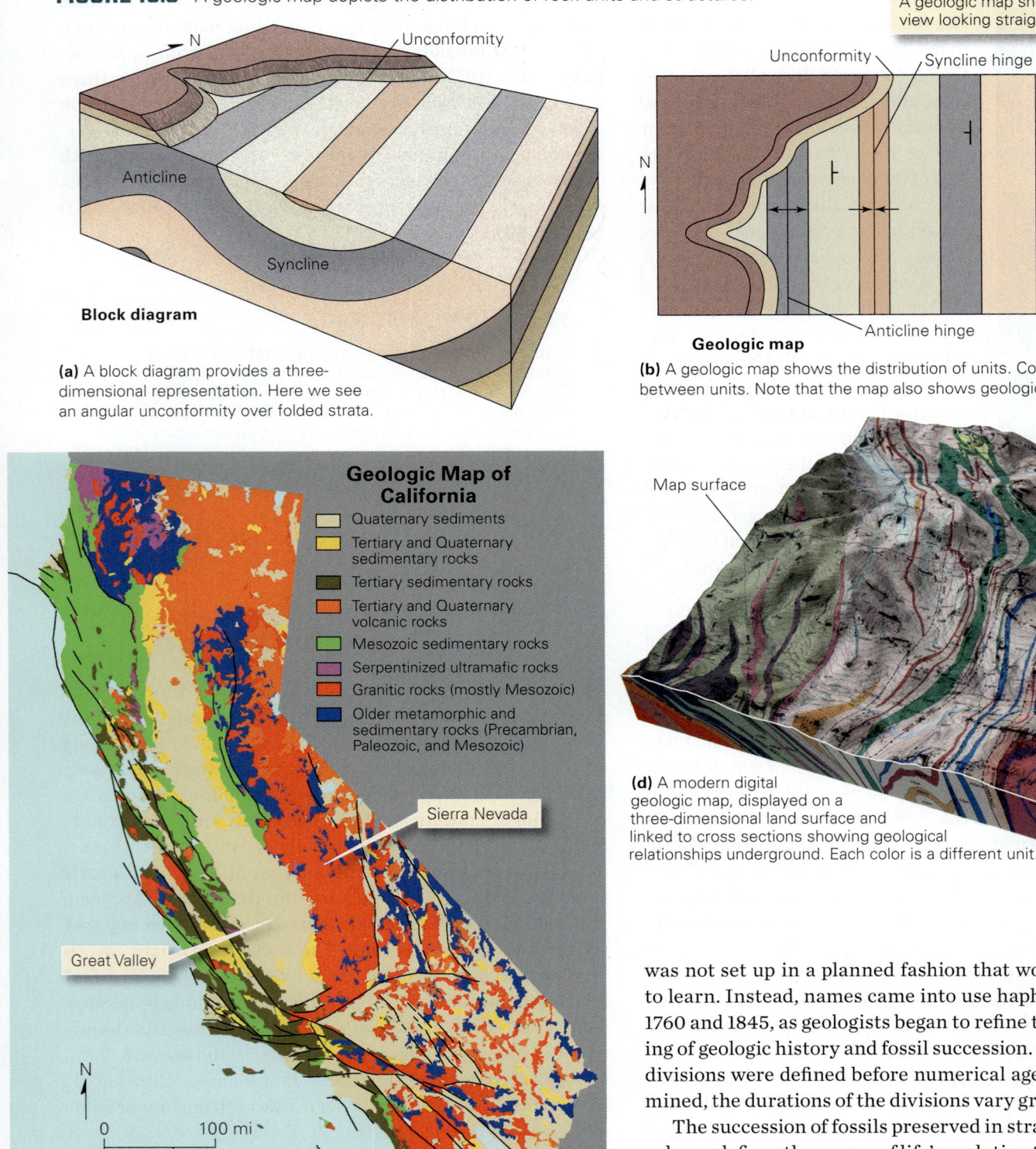

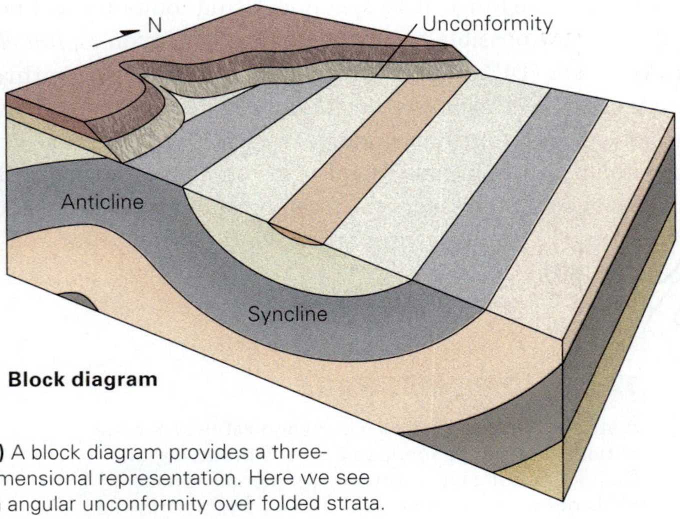

(a) A block diagram provides a three-dimensional representation. Here we see an angular unconformity over folded strata.

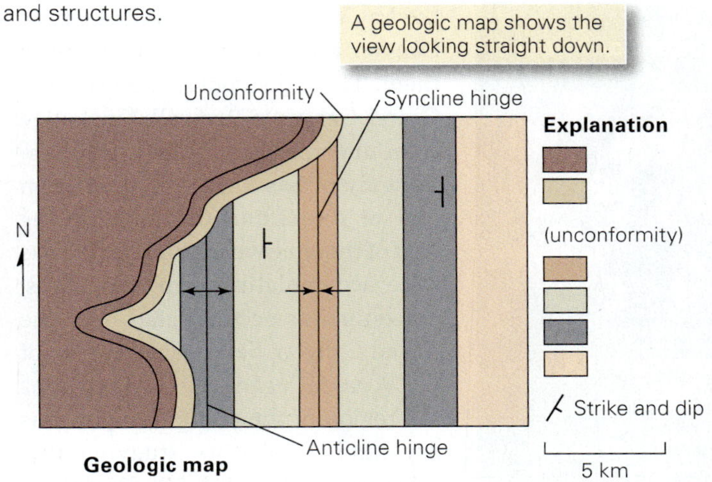

(b) A geologic map shows the distribution of units. Contacts occur between units. Note that the map also shows geologic structures.

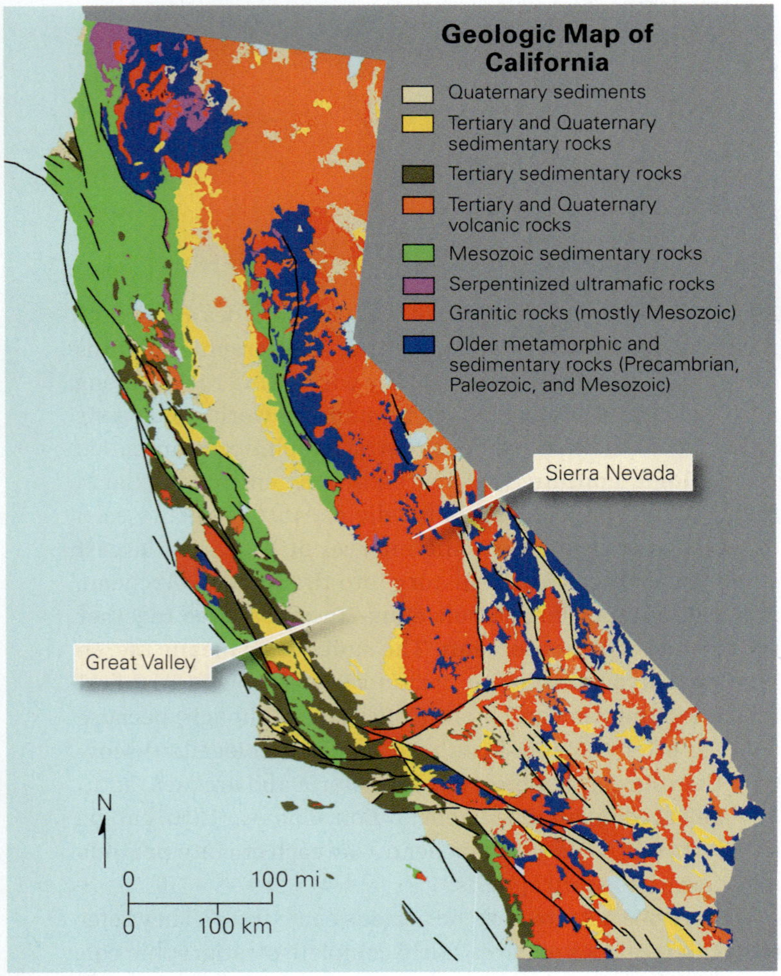

(c) A geologic map of California indicates that the state is underlain by many different rock units. Granite underlies the Sierras, and Quaternary sediments underlie the Great Valley. The black lines are fault traces.

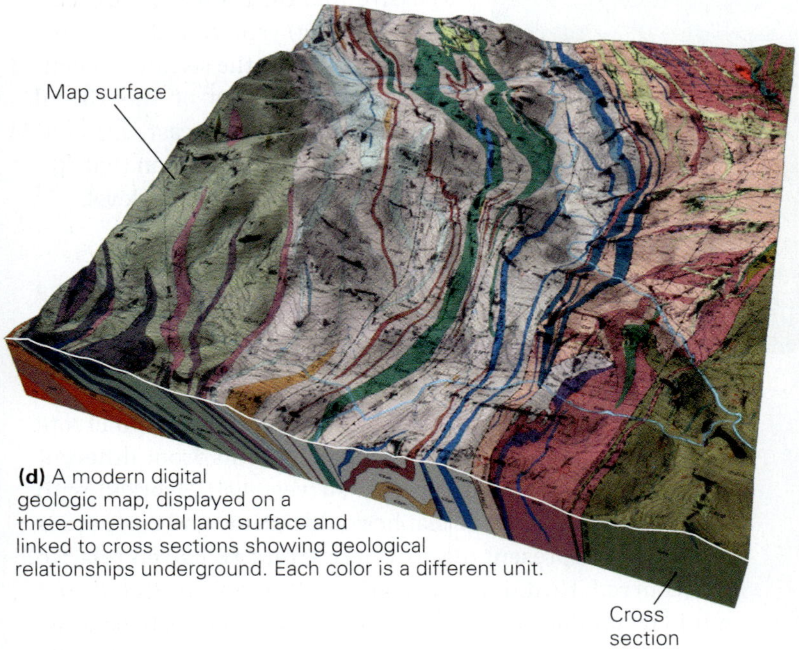

(d) A modern digital geologic map, displayed on a three-dimensional land surface and linked to cross sections showing geological relationships underground. Each color is a different unit.

was not set up in a planned fashion that would make it easy to learn. Instead, names came into use haphazardly between 1760 and 1845, as geologists began to refine their understanding of geologic history and fossil succession. Also, because the divisions were defined before numerical ages could be determined, the durations of the divisions vary greatly.

The succession of fossils preserved in strata of the geologic column defines the course of life's evolution throughout Earth history (**Fig. 10.10**). Simple bacteria and archaea appeared during the Archean Eon, but complex shell-less invertebrates did not evolve until the late Proterozoic. The appearance of invertebrates with shells defines the Precambrian-Cambrian boundary. At this time, there was a sudden diversification

FIGURE 10.9 Global correlation of strata led to the development of the geologic column. (Not to scale.)

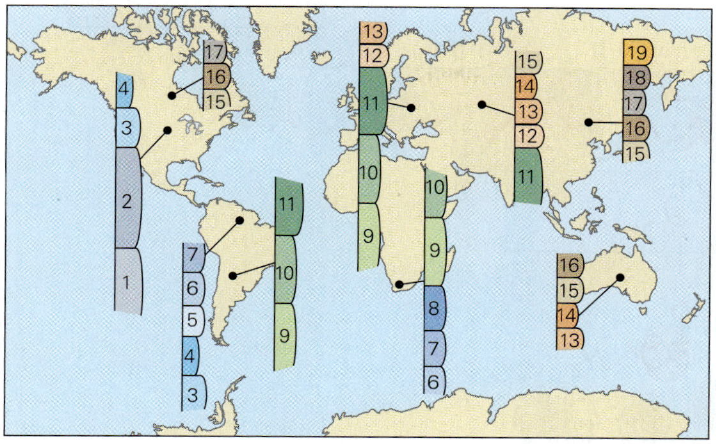

(a) Each of these small columns represents the stratigraphy at a given location. By correlating these columns, geologists determined their relative ages, filled in the gaps in the record, and produced the geologic column.

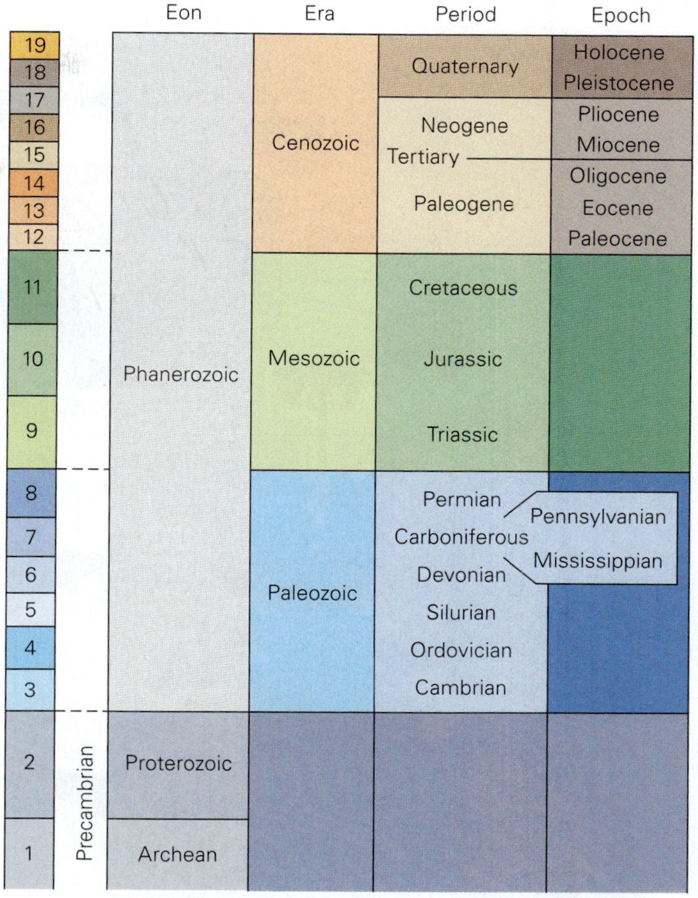

(b) By correlation, the strata from localities around the world were stacked in a chart representing geologic time to create the geologic column. Geologists assigned names to time intervals, but since the column was built without knowledge of numerical ages, it does not depict the duration of these intervals. Subdivisions of eons in the Precambrian are not shown. The Hadean is not shown because rocks do not preserve a record of it.

in life, with many new types of organisms appearing over a relatively short interval—this event is called the **Cambrian explosion**. Progressively more complex organisms populated the Earth during the Paleozoic. For example, the first fish appeared in Ordovician seas, land plants started to spread over the continents during the Silurian (prior to this time the land surface was unvegetated), and amphibians appeared during the Devonian. Though reptiles appeared during the Pennsylvanian Period, the first dinosaurs did not stomp across the land until the Triassic. Dinosaurs continued to inhabit the Earth until their sudden extinction at the end of the Cretaceous Period. For this reason, geologists refer to the Mesozoic Era as the *Age of Dinosaurs*. Small mammals appeared during the Triassic Period, but the diversification (development of many different species) of mammals to fill a wide range of ecological niches did not happen until the beginning of the Cenozoic Era, so geologists call the Cenozoic the *Age of Mammals*. Birds also appeared during the Mesozoic (at the beginning of the Cretaceous Period) but underwent great diversification in the Cenozoic Era.

SEE FOR YOURSELF...

VERMILION CLIFFS, ARIZONA

Latitude
36°49′4.81″ N

Longitude
111°37′56.59″ W

Zoom to an elevation of 2 km (~1.2 miles) and look obliquely north.

Marble Canyon (foreground) is the entry to the Grand Canyon. Outcrops in the distance are the Vermilion Cliffs, exposing reddish-brown sandstone and shale of the Moenkopi Formation. The canyon walls consist of underlying Kaibab Limestone.

Major mass-extinction events mark the boundaries between the Paleozoic and Mesozoic, and between the Mesozoic and Cenozoic. Geologists recognized these boundaries in the 19th century, because of major changes in fossil assemblages, long before they understood the concept and causes of mass extinction.

To conclude our discussion of the geologic column, let's see how it comes into play when correlating strata across a region. We return to the Colorado Plateau of Arizona and Utah, in the southwestern United States (**Fig. 10.11**). Because of the lack of vegetation in this region, you can easily see bedrock exposures on the walls of cliffs and canyons. Some of these exposures are so beautiful that they have become national parks. Using correlation techniques, geologists have determined that the oldest sedimentary rocks of the region crop out near

FIGURE 10.10 Life evolution in the context of the geologic column. The Earth formed at the beginning of the Hadean Eon.

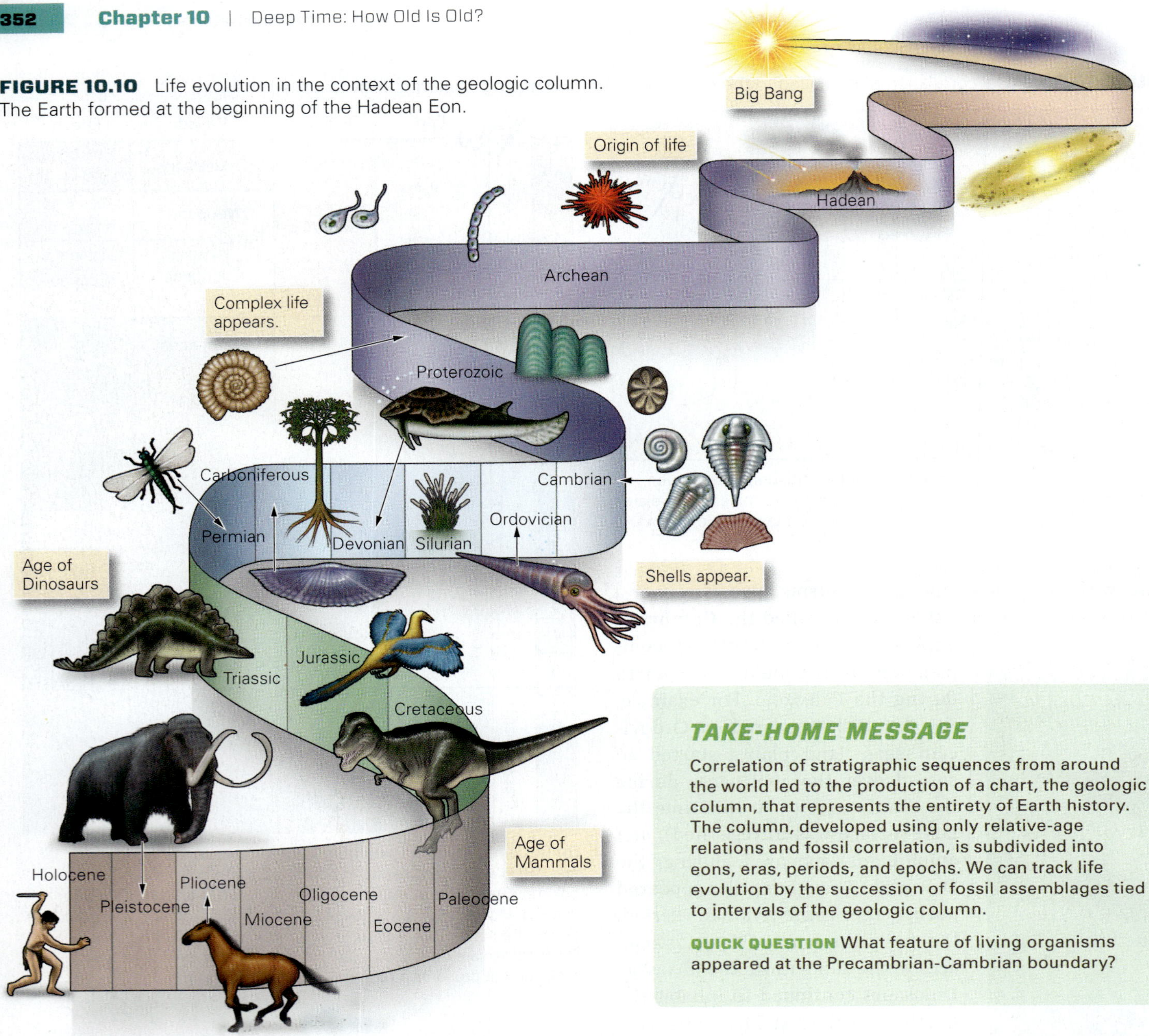

> ***TAKE-HOME MESSAGE***
>
> Correlation of stratigraphic sequences from around the world led to the production of a chart, the geologic column, that represents the entirety of Earth history. The column, developed using only relative-age relations and fossil correlation, is subdivided into eons, eras, periods, and epochs. We can track life evolution by the succession of fossil assemblages tied to intervals of the geologic column.
>
> **QUICK QUESTION** What feature of living organisms appeared at the Precambrian-Cambrian boundary?

the base of the Grand Canyon, whereas the youngest form the cliffs of Cedar Breaks and Bryce Canyon. Walking through these parks is like walking through Earth's history—each rock layer gives an indication of the climate and topography of the region at a time in the past (see **Geology at a Glance**, pp. 354–355). For example, when the Precambrian metamorphic and igneous rocks exposed in the inner gorge of the Grand Canyon first formed, the region was a high mountain range, perhaps as dramatic as the Himalayas today. When the fossiliferous beds of the Kaibab Limestone at the rim of the canyon first developed, the region was a Bahamas-like carbonate reef and platform, bathed in a warm, shallow sea. And when the rocks making up the towering red cliffs of sandstone in Zion Canyon were deposited, the region was a Sahara-like desert, blanketed with huge sand dunes.

10.7 How Do We Determine Numerical Ages?

Geologists since the days of Hutton could determine the relative ages of geologic events, but they had no way to specify numerical ages (called *absolute ages* in older literature)—meaning, the age in years of an event. Thus, they could not determine an accurate timeline for Earth history or the duration of particular events. This situation changed with the discovery of radioactivity. Simply put, a **radioactive element** is one that undergoes radioactive decay reactions that yield a different element. The rate of decay of a population of radioactive atoms is a constant that can be measured in the lab and can be specified in years. In the 1950s, geologists

FIGURE 10.11 Correlation of strata among the national parks of Arizona and Utah.

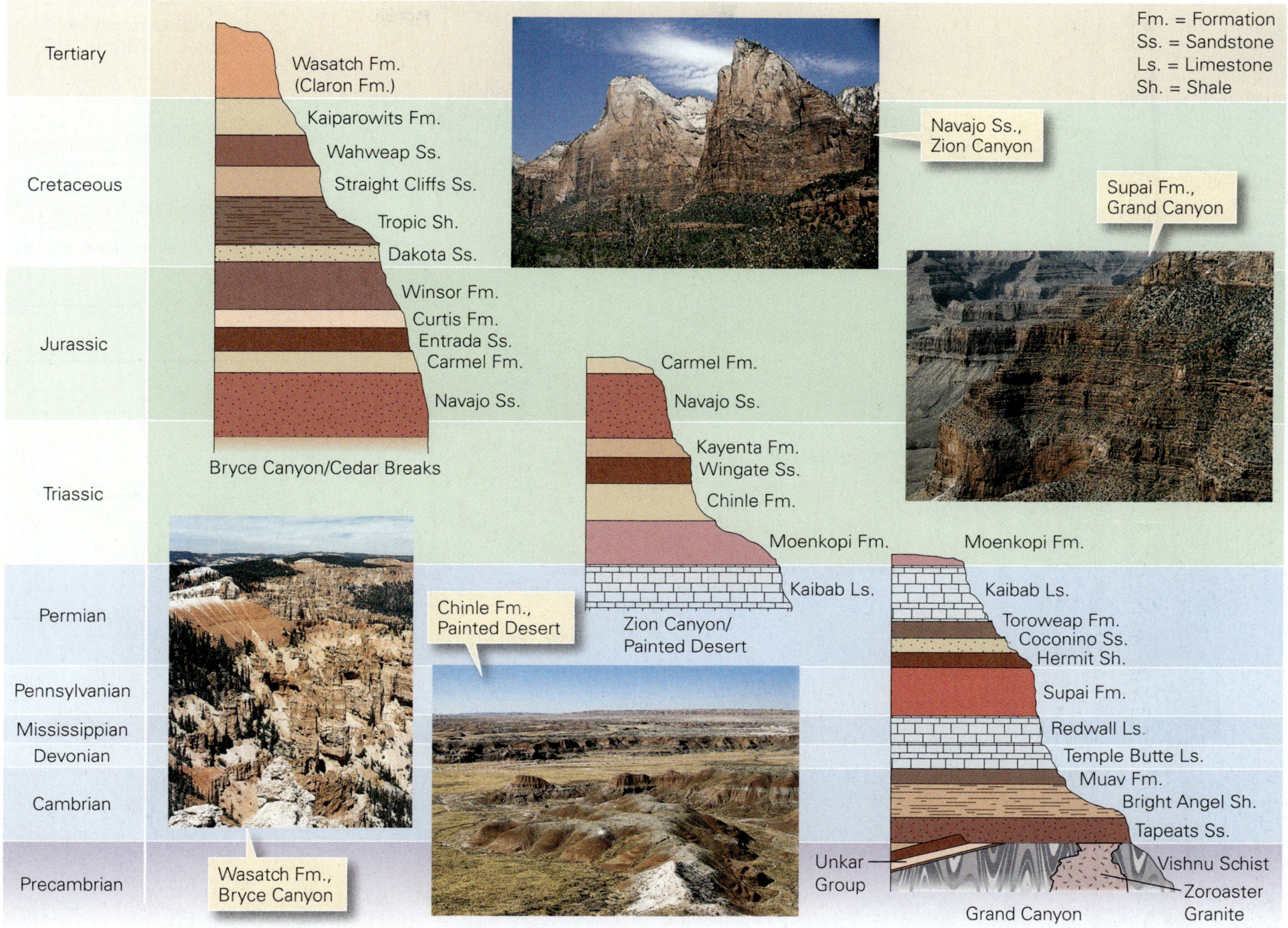

(a) Different intervals of geologic time are represented by the strata of different parks.

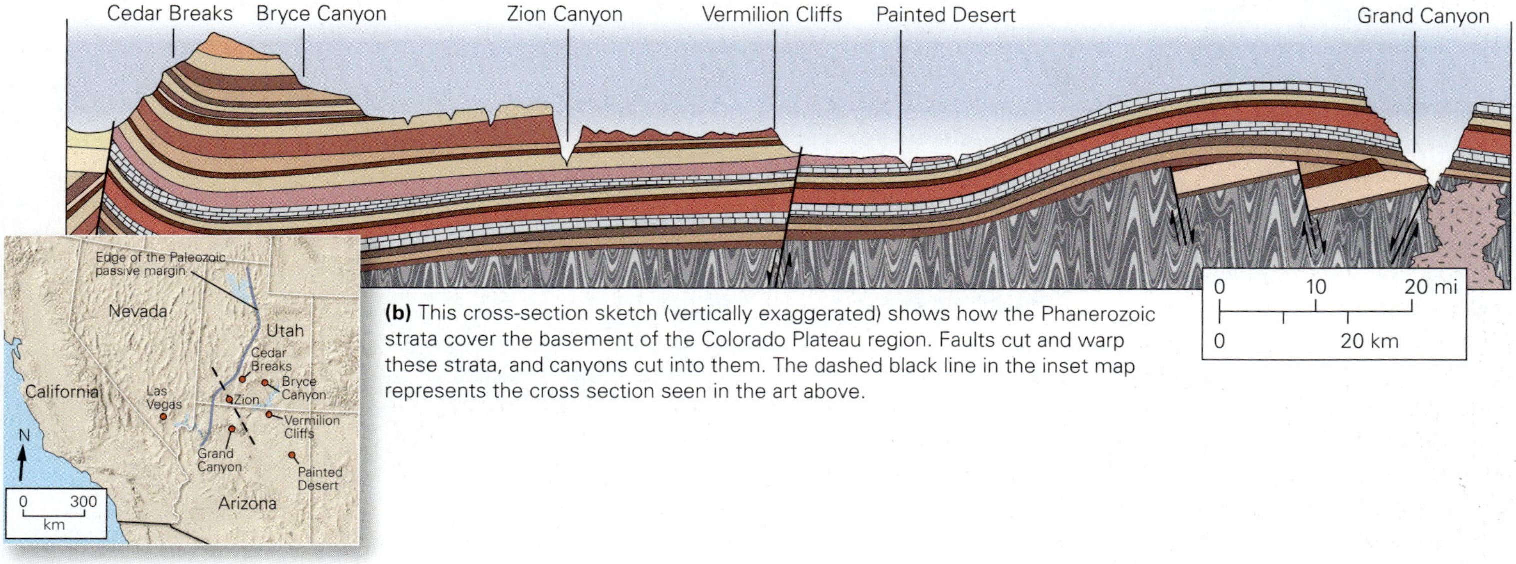

(b) This cross-section sketch (vertically exaggerated) shows how the Phanerozoic strata cover the basement of the Colorado Plateau region. Faults cut and warp these strata, and canyons cut into them. The dashed black line in the inset map represents the cross section seen in the art above.

GEOLOGY AT A GLANCE

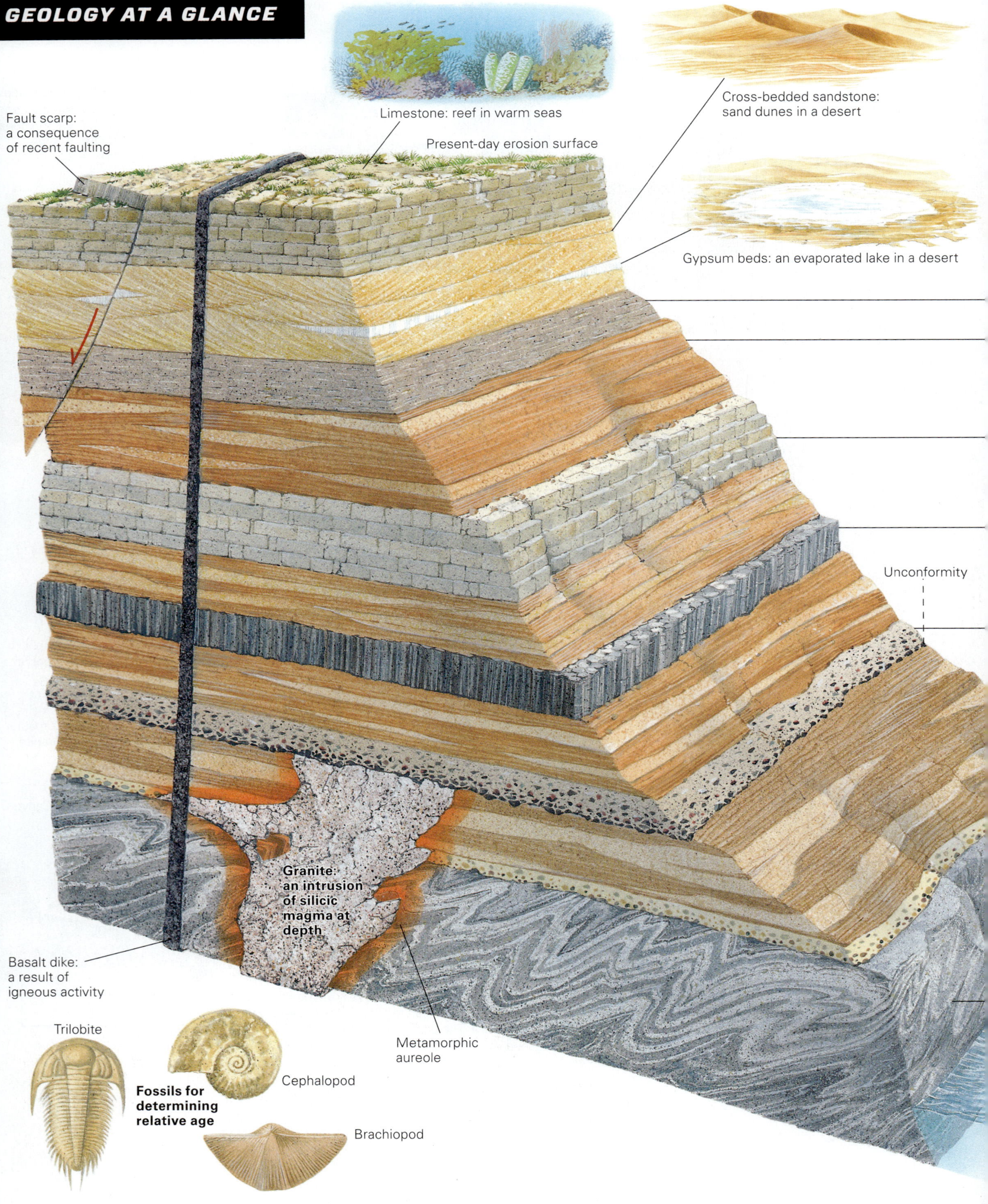

Limestone: reef in warm seas
Cross-bedded sandstone: sand dunes in a desert
Fault scarp: a consequence of recent faulting
Present-day erosion surface
Gypsum beds: an evaporated lake in a desert
Unconformity
Granite: an intrusion of silicic magma at depth
Basalt dike: a result of igneous activity
Metamorphic aureole
Trilobite
Cephalopod
Fossils for determining relative age
Brachiopod

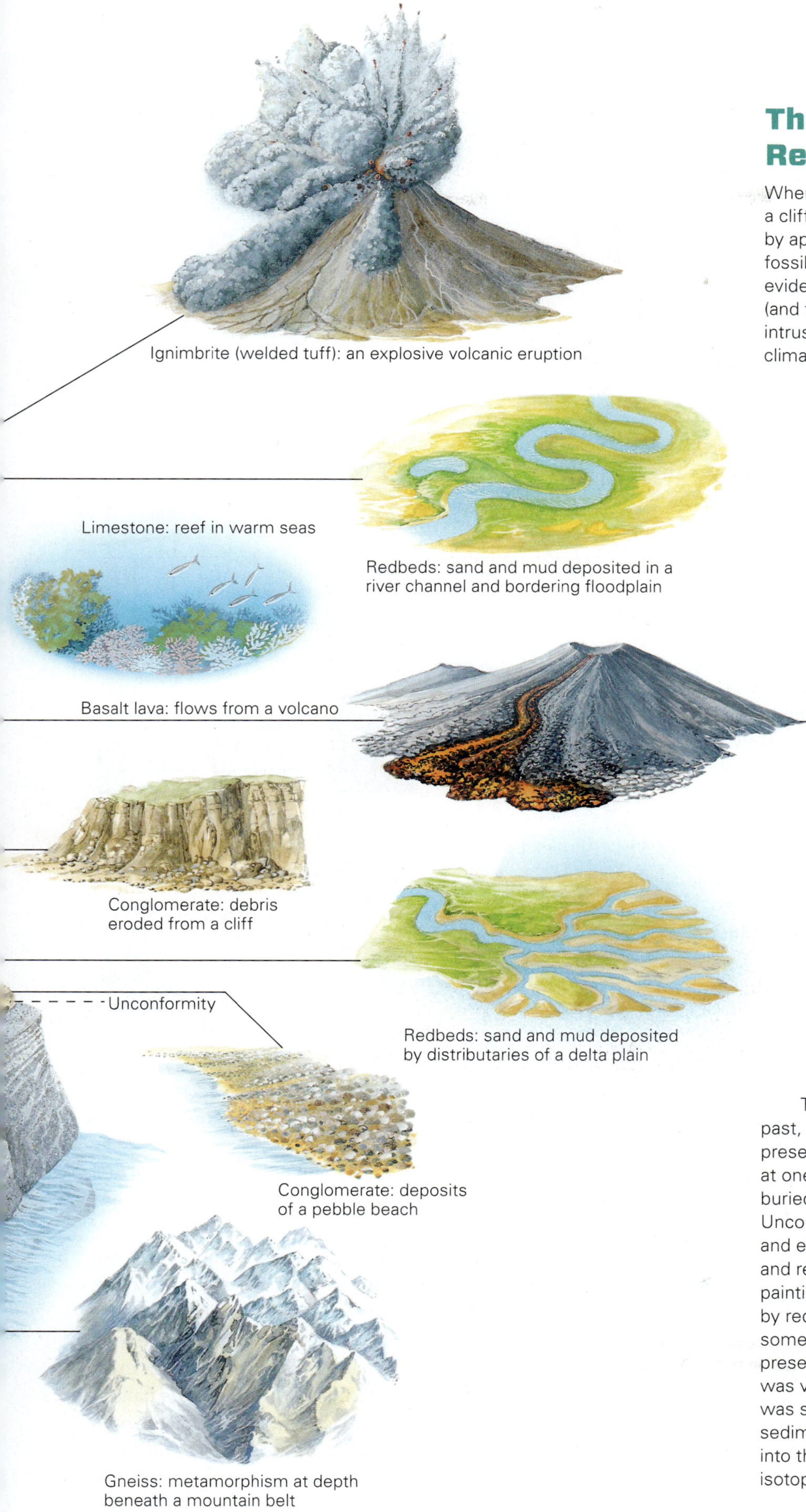

The Record in Rocks: Reconstructing Geologic History

When geologists examine a sequence of rocks exposed on a cliff, they see a record of Earth history that they interpret by applying the basic principles of geology, by searching for fossils, and by using isotopic dating. On this cliff, we see evidence for many geologic events. The layers of sediment (and the sedimentary structures they contain), the igneous intrusions, and the geologic structures tell us about past climates and past tectonic activity.

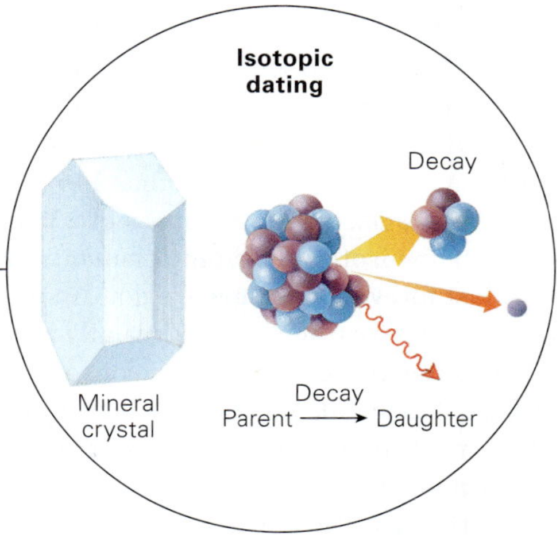

The insets show the way the region looked in the past, based on the record in the rocks. For example, the presence of gneiss at the base of the canyon indicates that at one time the region was a mountain belt where deeply buried crust underwent metamorphism and deformation. Unconformities indicate that the region later underwent uplift and erosion. Sedimentary successions record transgressions and regressions of the sea. The land surface portrayed in this painting was sometimes a river floodplain or a delta (indicated by redbeds), sometimes a shallow sea (limestone), and sometimes a desert dune field (cross-bedded sandstone). The presence of igneous rocks indicates that, at times, the region was volcanically active, and the presence of faults indicates it was seismically active. We can gain insight into the age of the sedimentary rocks by studying the fossils they contain, and into the age of the igneous and metamorphic rocks by using isotopic dating methods.

Did you ever wonder...
how geologists can specify the age of some rocks in years?

developed techniques to calculate the numerical ages of rock by measuring the ratio of radioactive elements to the products of decay, in certain minerals. Originally this process was called **radiometric dating**, but more recently it has also come to be known as **isotopic dating**. Dating techniques have vastly improved over the years, and in some cases can be accurate to within 0.05%. The overall study of numerical ages is the discipline of **geochronology**. We begin our discussion by learning additional details about radioactive decay.

FIGURE 10.12 The concept of a half-life, in the context of radioactive decay.

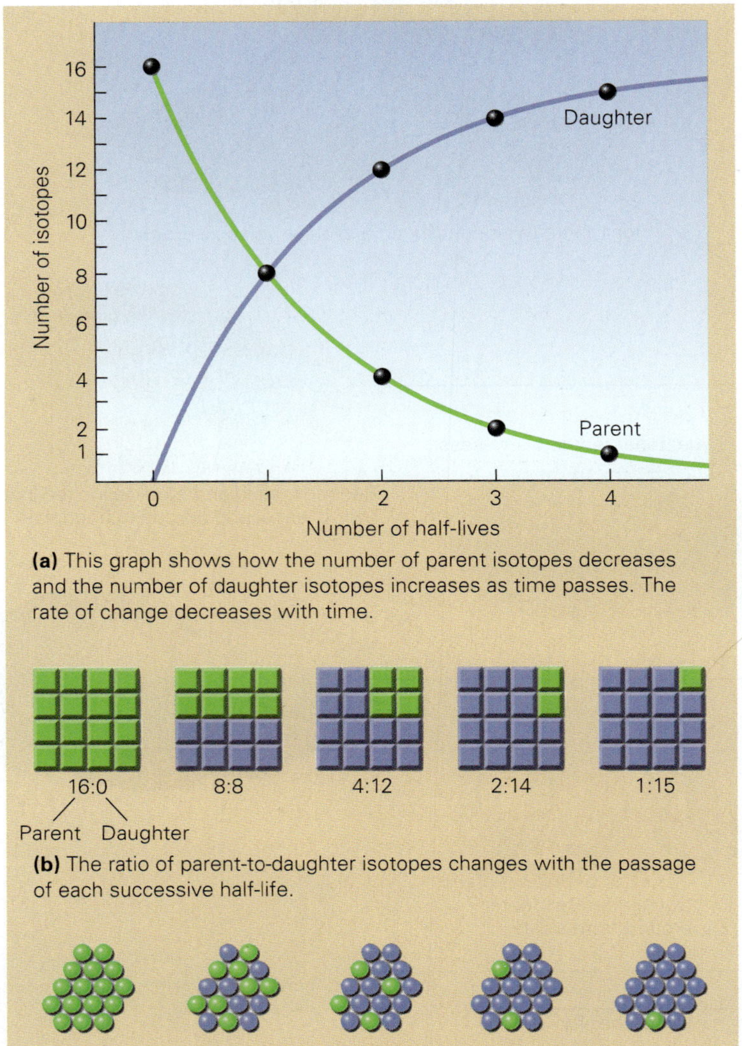

(a) This graph shows how the number of parent isotopes decreases and the number of daughter isotopes increases as time passes. The rate of change decreases with time.

(b) The ratio of parent-to-daughter isotopes changes with the passage of each successive half-life.

(c) In a cluster of isotopes undergoing decay, there is no way to predict which parent will decay next.

Radioactive Decay

Different versions of an element, called **isotopes** of the element, have the same atomic number but different atomic masses (see Box 1.1). To see how this terminology applies, let's consider two atoms of uranium. All uranium atoms have 92 protons, but the uranium-238 isotope (abbreviated ^{238}U) has an atomic mass of 238 and thus has 146 neutrons, whereas the ^{235}U isotope has an atomic mass of 235 and thus has 143 neutrons. Some isotopes of some elements are stable, meaning they last forever, but some are not. An unstable isotope is called a *radioactive isotope*—it eventually undergoes a change called **radioactive decay** which converts it into an atom of a different element. Radioactive decay can take place by a variety of nuclear reactions, but regardless of the details, all these reactions change the atomic number of the nucleus and, therefore, the identity of the element. We refer to an atom that undergoes decay as the *parent isotope* and the decay product as the *daughter isotope* (or daughter product).

Physicists cannot specify how long an individual radioactive atom will survive before it decays, but they can measure how long it takes for half of a group of parent isotopes to decay. This time is called the **half-life**. To visualize the concept of a half-life, imagine a crystal containing 16 radioactive parent atoms (**Fig. 10.12**). (Note that in a real crystal, the number of atoms would be immensely larger.) After one half-life, 8 atoms have decayed, so the crystal now contains 8 parents and 8 daughters. After a second half-life, 4 of the remaining parents have decayed, so the crystal contains 4 parents and 12 daughters. And after a third half-life, 2 more parents have decayed, so the crystal contains 2 parents and 14 daughters. For a given decay reaction, the half-life is a constant, generally measured in years.

Isotopic Dating Techniques

Since radioactive decay of a population of radioactive atoms proceeds at a known rate, like the tick-tock of a clock, it provides a basis for telling time. In other words, because an element's half-life is a constant, we can calculate the age of a mineral by measuring the ratio of parent to daughter isotopes in the mineral.

In practice, how can we obtain an isotopic date? First, we must find the right kind of elements to work with. Although there are many different pairs of parents and daughters among the known radioactive isotopes, only a few have long enough half-lives, and occur in sufficient abundance in minerals, to be useful for isotopic dating. **Table 10.1** lists particularly useful elements. Each radioactive element has its own half-life. Note that *carbon dating* is not used for dating rocks because appropriate carbon isotopes occur only in organisms, not minerals, and radioactive carbon has a very short half-life.

TABLE 10.1 Isotopes Used in the Isotopic Dating of Rock

Parent → Daughter	Half-life (years)	Minerals Containing the Isotopes
$^{147}Sm \rightarrow {}^{143}Nd$	106.0 billion	Garnets, micas
$^{87}Rb \rightarrow {}^{87}Sr$	48.8 billion	Potassium-bearing minerals (mica, feldspar, hornblende)
$^{238}U \rightarrow {}^{206}Pb$	4.5 billion	Uranium-bearing minerals (zircon, uraninite)
$^{40}K \rightarrow {}^{40}Ar$	1.3 billion	Potassium-bearing minerals (mica, feldspar, hornblende)
$^{235}U \rightarrow {}^{207}Pb$	713.0 million	Uranium-bearing minerals (zircon, uraninite)

Sm = samarium, Nd = neodymium, Rb = rubidium, Sr = strontium, U = uranium, Pb = lead, K = potassium, Ar = argon.

Second, we must identify the right kind of minerals to work with. Not all minerals contain radioactive isotopes, but fortunately some fairly common minerals do. Once we have found the right kind of minerals, we can set to work using the following steps.

› *Collecting the rocks*: We need to find unweathered rocks for dating, since the chemical reactions that happen during weathering may lead to the loss of daughter isotopes, and in some cases, parent isotopes.
› *Separating the minerals*: The rocks are crushed, and the appropriate minerals are separated from the debris.
› *Extracting parent and daughter isotopes*: To separate out the parent and daughter isotopes from minerals, we can use several techniques, including dissolving the minerals in acid or evaporating portions of them with a laser.
› *Analyzing the parent-daughter ratio*: Once we have a sample of appropriate isotopes, we pass them through a mass spectrometer, an instrument that uses a strong magnet to separate atoms from one another according to their respective masses (**Fig. 10.13**). The instrument can count the number of atoms of specific isotopes separately.

FIGURE 10.13 In an isotopic dating laboratory, samples are analyzed using a mass spectrometer. This instrument measures the ratio of parent-to-daughter isotopes.

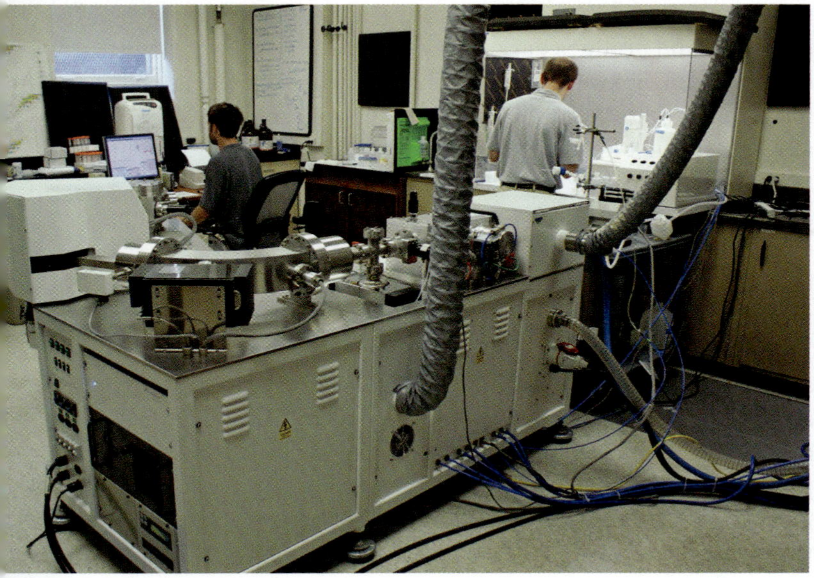

At the end of the laboratory process, we can define the ratio of parent to daughter isotopes in a mineral, and from this ratio calculate the age of the mineral. Needless to say, the description of the procedure here has been simplified—in reality, obtaining an isotopic date requires complex calculations, and it can be time-consuming and expensive.

What Does an Isotopic Date Mean?

At high temperatures, atoms in a crystal lattice vibrate so rapidly that chemical bonds can break and reattach relatively easily. As a consequence, atoms can escape from or move into crystals, so parent-daughter ratios are meaningless. Because isotopic dating is based on the parent-daughter ratio, the "isotopic clock" starts only when crystals become cool enough for atoms to be locked into the lattice. The temperature below which isotopes are no longer free to move is called the **closure temperature** of a mineral. When we specify an isotopic date for a mineral, we are defining the time at which the mineral cooled below its closure temperature.

With the concept of closure temperature in mind, we can interpret the meaning of isotopic dates. In the case of igneous rocks, isotopic dating tells us when a magma or lava cooled to form a solid, cool igneous rock. In the case of metamorphic rocks, an isotopic date tells us when a rock cooled from a metamorphic temperature above the closure temperature to a temperature below. Not all minerals have the same closure temperature, so different minerals in an igneous or metamorphic rock that cools very slowly will yield different dates. Recently, geologists have developed procedures that allow them to date minerals with a very low closure temperature (80° to 100°C). These dates probably represent the time at which the rock was exhumed, meaning the time when uplift and erosion brought a rock that was once deep within an orogen up to within a few kilometers of the Earth's surface.

Can we isotopically date a clastic sedimentary rock directly? No. If we date minerals in a sedimentary rock, we determine only when these minerals first crystallized as part

FIGURE 10.14 Ways to determine numerical ages.

(a) Patterns of tree rings are like bar codes. Each ring in this slice of wood represents the growth of one year.

(c) Dust settling during the summer highlights boundaries between snow layers, now turned to ice in this Oregon glacier.

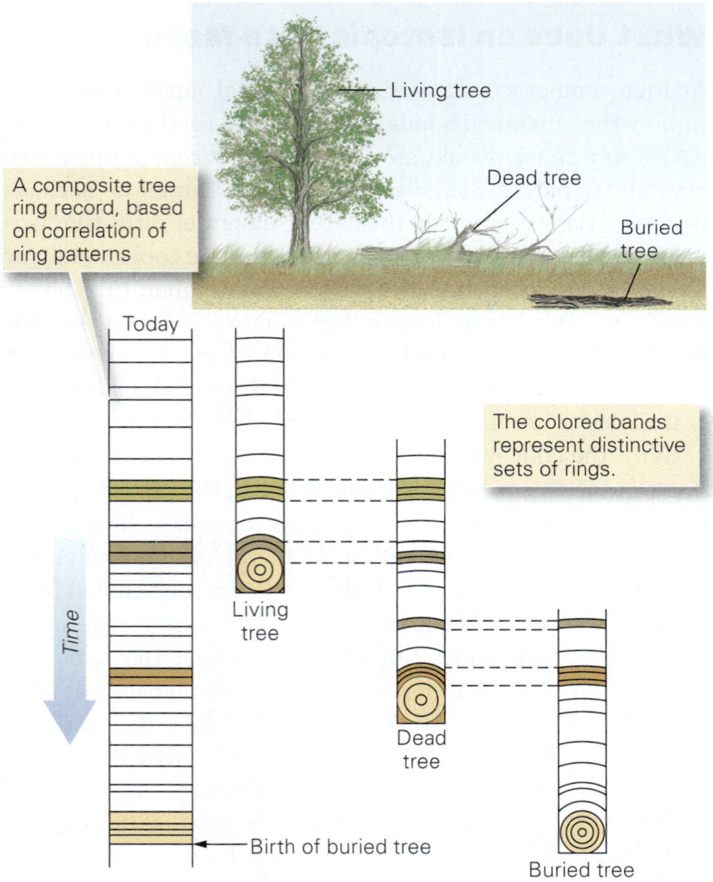

(b) Dendrochronology is based on the correlation of tree rings. Each of the columns in the diagram represents a core drilled out of a tree. Distinctive clusters of closely spaced rings indicate dry seasons. By correlation, researchers extend the climate record back in time before the oldest living tree started to grow.

of an igneous or metamorphic rock, not the time when the minerals were deposited as sediment nor the time when the sediment lithified to form a sedimentary rock. For example, if we date the feldspar grains contained within a granite pebble in a conglomerate, we're dating the time the granite cooled below feldspar's closure temperature, not the time the pebble was deposited by a stream.

Other Methods of Determining Numerical Ages

The rate of tree growth depends on the season. During the spring, trees grow rapidly and produce lighter, less-dense wood, but during the rest of the season trees grow slowly or not at all, and produce darker, denser wood. Thus, wood contains recognizable annual growth rings which provide a basis for determining age. If you've ever wondered about the age of a tree that's just been cut down, look at the stump and count the rings. Notably, by correlating clusters of distinctive rings in the older parts of living trees with comparable clusters of rings in dead logs, scientists can extend the tree-ring record back for many thousands of years, allowing them to track climate changes back into prehistory (**Fig. 10.14a, b**).

Seasonal changes also affect rates of such phenomena as shell growth, snow accumulation, clastic sediment deposition, chemical sediment precipitation, and production of organic material. Geologists have learned to use growth rings in shells, as well as rhythmic layering in sediments and in glacial ice (**Fig. 10.14c**), to date events numerically back through recent Earth history.

TAKE-HOME MESSAGE

Isotopic dating (radiometric dating) specifies numerical ages in years. To obtain an isotopic date, we measure the ratio of radioactive parent isotopes to stable daughter isotopes in a mineral. A date for a mineral gives the time at which the mineral cooled below a closure temperature.

QUICK QUESTION How can we date changes in climate during the past few hundred thousand years?

10.8 Numerical Ages and Geologic Time

Dating Sedimentary Rocks

The mind grows giddy gazing so far back into the abyss of time.

JOHN PLAYFAIR (British geologist, 1747–1819)

We have seen that isotopic dating can be used to date when igneous rocks formed and when metamorphic rocks metamorphosed, but not when sedimentary rocks were deposited. So how do we determine the numerical age of a sedimentary rock? We must answer this question if we want to add numerical ages to the geologic column.

Geologists obtain dates for sedimentary rocks by studying cross-cutting relationships between sedimentary rocks and datable igneous or metamorphic rocks. For example, if we find a sequence of sedimentary strata deposited unconformably on a datable granite, the strata must be younger than the granite (**Fig. 10.15**). If a datable basalt dike cuts the strata, the strata must be older than the dike. And if a datable volcanic ash buried the strata, then the strata must be older than the ash.

The Geologic Time Scale

Geologists have searched the world for localities where they can recognize cross-cutting relations between datable igneous rocks and sedimentary rocks or for layers of datable volcanic rocks interbedded with sedimentary rocks. By isotopically dating the igneous rocks, researchers have been able to provide numerical ages for the boundaries between all geologic periods. For example, information gathered around the world shows that the Cretaceous Period began about 145 million years ago and ended 66 million years ago. So the Cretaceous sandstone bed in Figure 10.15 was deposited during the middle part of the Cretaceous, not at the beginning or end.

The discovery of new data may cause the numbers defining the boundaries of periods to change, which is why the term *numerical age* is preferred to *absolute age*. In fact, around 1995, new dates on rhyolite ash layers above and below the Cambrian-Precambrian boundary showed that this boundary occurred at 542 million years ago, in contrast to previous, less definitive studies that had placed the boundary at 570 million years ago. More recent dating indicates that the boundary lies at 541 Ma. **Figure 10.16** shows the currently favored numerical ages of periods and eras in the geologic column. This dated column is commonly called the **geologic time scale**.

FIGURE 10.15 The Cretaceous sandstone bed was deposited on the granite, so it must be younger than 125 Ma. The dike cuts the bed, so the bed must be older than 80 Ma. Thus, the Cretaceous bed was deposited between 125 and 80 Ma. The Paleocene sandstone was unconformably deposited over the dike and lies beneath a 50-million-year-old layer of ash. Therefore, it must have been deposited between 80 and 50 Ma.

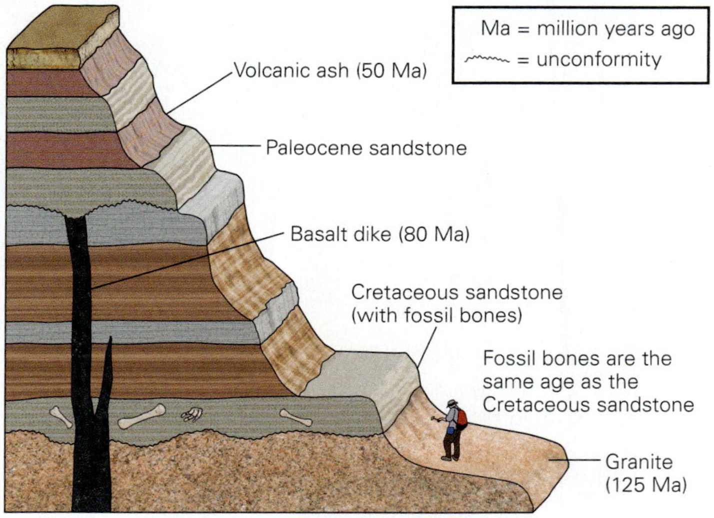

What Is the Age of the Earth?

During the 18th and 19th centuries, before the discovery of isotopic dating, scientists came up with a great variety of clever solutions to the question, "How old is the Earth?" All of these have since been proved wrong. Lord Kelvin (William Thompson, 1824–1907), a 19th-century physicist renowned for his discoveries in thermodynamics, made the most influential scientific estimate of the Earth's age of his time. Kelvin calculated how long it would take for the Earth to cool down from a temperature as hot as the Sun's, and concluded that our planet is about 20 million years old.

Kelvin's estimate contrasted with those being promoted by followers of Hutton, Lyell, and Darwin, who argued that if the concepts of uniformitarianism and evolution were correct, the Earth must be much older. They argued that physical processes that shape the Earth and form its rocks, as well as the

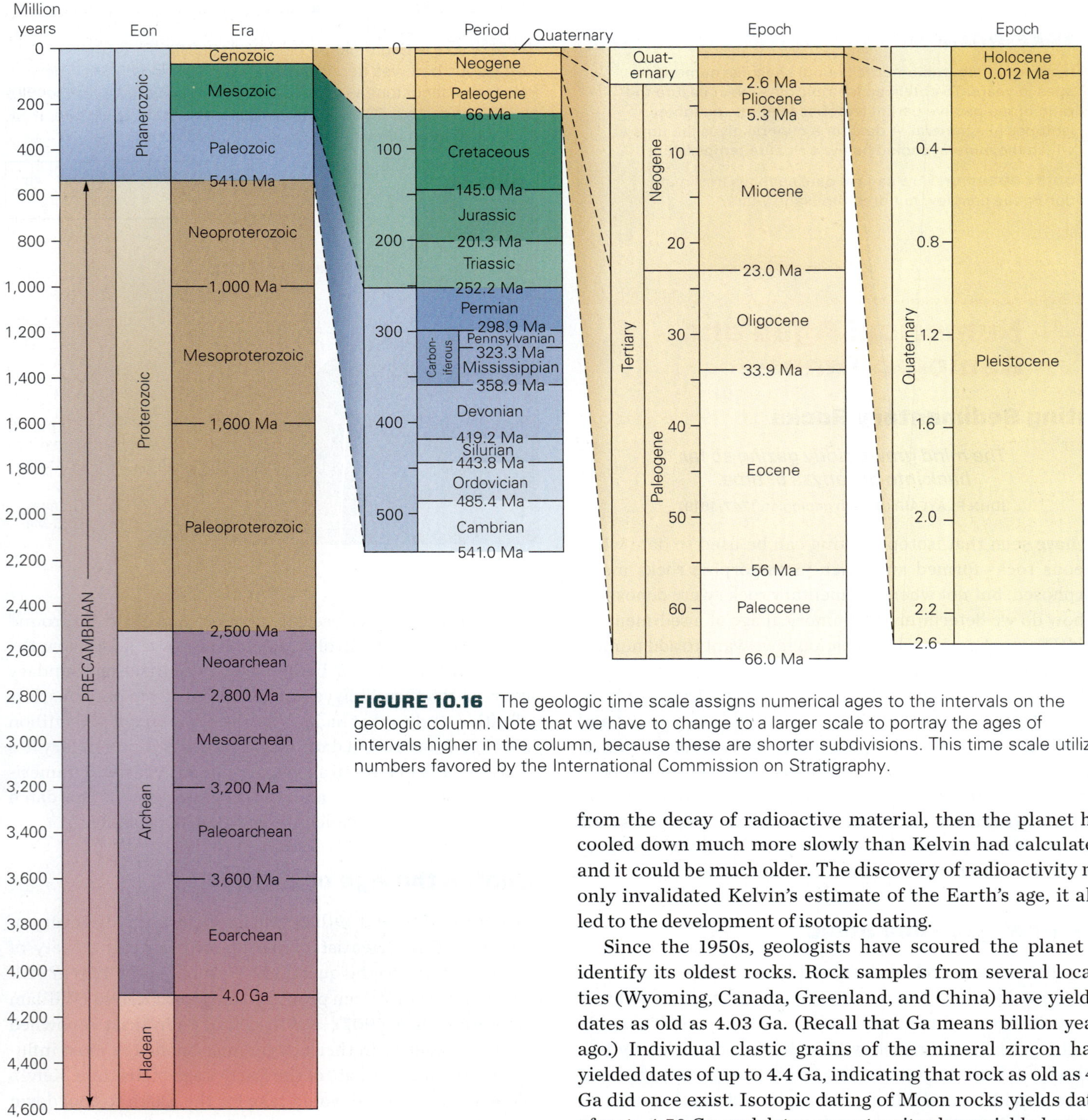

FIGURE 10.16 The geologic time scale assigns numerical ages to the intervals on the geologic column. Note that we have to change to a larger scale to portray the ages of intervals higher in the column, because these are shorter subdivisions. This time scale utilizes numbers favored by the International Commission on Stratigraphy.

process of natural selection that yields the diversity of species, all take a very long time. Geologists and physicists continued to debate the age issue for many years. The route to a solution didn't appear until 1896, when Henri Becquerel (1852–1908) announced the discovery of radioactivity. Geologists immediately realized that if the Earth's interior was producing heat from the decay of radioactive material, then the planet has cooled down much more slowly than Kelvin had calculated, and it could be much older. The discovery of radioactivity not only invalidated Kelvin's estimate of the Earth's age, it also led to the development of isotopic dating.

Since the 1950s, geologists have scoured the planet to identify its oldest rocks. Rock samples from several localities (Wyoming, Canada, Greenland, and China) have yielded dates as old as 4.03 Ga. (Recall that Ga means billion years ago.) Individual clastic grains of the mineral zircon have yielded dates of up to 4.4 Ga, indicating that rock as old as 4.4 Ga did once exist. Isotopic dating of Moon rocks yields dates of up to 4.50 Ga, and dates on meteorites have yielded ages as old as 4.57 Ga. Geologists consider 4.57-Ga meteorites to be fragments of planetesimals. Meteorites from the oldest differentiated planetesimals, those that had separated into a core and mantle, are slightly younger, and these ages are taken to be the same as the age of the Earth itself. Based on these ages, therefore, researchers estimate that the Earth itself differentiated and became a full-fledged planet at 4.54 Ga.

Why don't we find rocks with ages between 4.03 and 4.54 Ga in the Earth's crust? Geologists have come up with several ideas to explain the lack of extremely old rocks. One idea comes from calculations defining how the temperature of our planet has changed over time. These calculations indicate that during the first half-billion years of its existence, the Earth might have been so hot that rocks in the crust remained above the closure temperature for minerals, and isotopic clocks could not start "ticking." More recent studies, looking at isotope ratios in the oldest (4.4 Ga) zircons, suggest alternatively that the Earth had cooled sufficiently to host oceans of water within only a couple of hundred million years of its formation and suggest that intense bombardment of the Earth by meteorites just prior to 4.03 Ga destroyed or remelted any crust that existed and vaporized the earliest oceans. As noted earlier, geologists have named the time interval between the birth of the Earth and the origin of the oldest isotopically dated whole rock as the Hadean Eon, to emphasize that conditions at the surface, at times, resembled literary images of Hades.

Did you ever wonder... how old the Earth's oldest rock is?

Picturing Geologic Time

The number 4.54 billion is so staggeringly large that we can't begin to comprehend it. If you lined up this many pennies in a row, they would make an 87,400-km-long line that would wrap around the Earth's equator more than twice. Notably, at the scale of our penny chain, human history is only about 100 city blocks long.

Another way to grasp the immensity of geologic time is to equate the entire 4.54 billion years to a single calendar year. On this scale, the oldest rocks preserved on Earth date from early February, and the first bacteria appear in the ocean on February 21. The first shelled invertebrates appear on October 25, and the first amphibians crawl out onto land on November 20. On December 7, the continents coalesce into the supercontinent of Pangaea. Birds and the ancestors of mammals appear about December 15, along with the dinosaurs, and the Age of Dinosaurs ends on December 25. The last week of December represents the last 66 million years of Earth history, including the entire Age of Mammals. The first human-like ancestor appears on December 31 at 3 p.m., and our species, *Homo sapiens*, shows up an hour before midnight. The last ice age ends a minute before midnight, and all of recorded human history takes place in the last 30 seconds. To put it another way, human history occupies the last 0.000001% of Earth history. The Earth is so old that there has been more than enough time for the rocks and life forms of Earth to have formed and evolved.

TAKE-HOME MESSAGE

Numerical dates for sedimentary rocks come from isotopic dating of cross-cutting datable rocks. Such work led to the geologic time scale, which assigns dates to periods. The oldest rock of Earth's crust is about 4.0 Ga. Dating of meteorites indicates the Earth became a differentiated planet at 4.54 Ga.

QUICK QUESTION Why don't all periods on the geologic time column have the same duration in years?

Another View The pages of Earth history stand on end in Namibia, southwestern Africa. Here, in a false-color image, the Ugab River cuts across layer upon layer of strata that were tilted to near vertical by a Precambrian mountain-building event. Subsequent erosion exposed the strata, and the desert climate keeps it clear of vegetation. The field of view is 45 km.

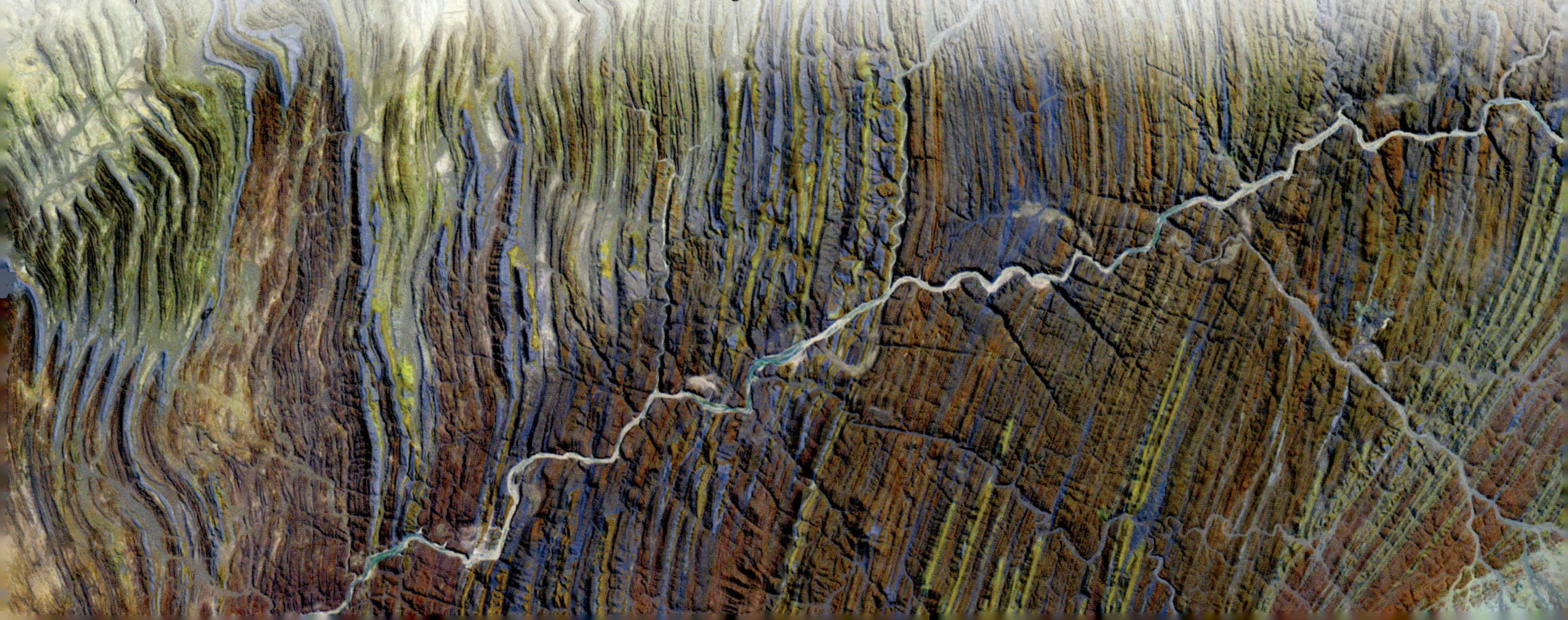

Chapter 10 Review

Chapter Summary

- Geologic time refers to the time span since the Earth's formation.
- Relative age specifies whether one geologic feature is older than or younger than another; numerical age (absolute age) provides the age of a geologic rock or feature in years.
- Using such principles as uniformitarianism, original horizontality, superposition, and cross-cutting relations, we can construct the geologic history of a region.
- The principle of fossil succession states that the assemblage of fossils in strata changes from base to top of a sequence. Once a species becomes extinct, it never reappears higher in the sequence.
- Strata are not necessarily deposited continuously at a location. An interval of nondeposition and/or erosion produces an unconformity. Geologists recognize three kinds: angular unconformity, nonconformity, and disconformity.
- A stratigraphic column shows the succession of strata in a region. A given succession of strata that can be traced over a fairly broad region is called a stratigraphic formation.
- The process of determining the relationship between strata at one location and strata at another is called correlation. A geologic map shows the distribution of formations, as well as of geologic structures.
- A composite chart that represents the entirety of geologic time is called the geologic column. The column's largest subdivisions are eons. Eons are subdivided into eras, eras into periods, and periods into epochs.
- The numerical age of rocks can be determined by isotopic (radiometric) dating. This is because radioactive isotopes decay at a rate characterized by a known half-life.
- The isotopic age of a mineral specifies the time at which the mineral cooled below a closure temperature. We can use isotopic dating to determine when an igneous rock solidified and when a metamorphic rock cooled. To date sedimentary strata, we must examine cross-cutting relations with dated igneous or metamorphic rock.
- Other methods for dating materials include counting growth rings in trees and seasonal layers in glaciers.
- Isotopic dating indicates that the Earth is 4.54 billion years old. This date represents the time when the planet internally differentiated into a core and mantle. A rock record of the first half billion years of Earth history doesn't exist, for the oldest dated whole rock is about 4 Ga.

Guide Terms

Review Questions

1. Explain the concept of uniformitarianism.
2. Compare the numerical age (absolute age) of a rock to the relative age of the rock.
3. Describe the principles that allow us to determine the relative ages of geologic events.
4. How does the principle of fossil succession allow us to determine the relative ages of strata?
5. How does an unconformity develop? Describe the differences among the three kinds of unconformities.
6. Describe two different methods of correlating rock units. How was correlation used to develop the geologic column?
7. What is a stratigraphic formation, and how are they depicted on a geologic map?
8. Is there one place on Earth where we can see the complete geologic column?
9. What does the process of radioactive decay entail?
10. How do geologists obtain an isotopic date? What are some of the pitfalls in obtaining a reliable one?
11. Why can't we date sedimentary rocks directly? Why don't periods on the geologic column have the same duration?
12. What is the age of the oldest rocks on Earth? What is the age of the oldest rocks known? Why is there a difference?

GEOTOURS THIS CHAPTER'S GEOTOUR EXERCISE (J) FEATURES:

- Relative Age Dating & Unconformities
- Stratigraphic Formations in Southern Utah
- Rock Layers and Monoclines, Circle Cliffs, Utah

On Further Thought

13. Imagine an outcrop exposing a sandstone and conglomerate bed. A geologist studying the outcrop notes the following:
    - A layer of volcanic ash overlies the sandstone bed. Isotopic dating indicates that this ash is 300 Ma.
    - A basalt dike, dated at 100 Ma, cuts both the ash and the sandstone-conglomerate sequence.
    - Pebbles of granite in the conglomerate yield radiometric dates of 400 Ma.

    On the basis of these observations, how old is the sandstone and conglomerate? (Specify both the numerical age range and the period of the geologic column.)

14. Examine the photograph and the What a Geologist Sees interpretation of an outcrop in eastern New York State shown below. Write a brief geologic history that explains the relationships displayed in this outcrop. The strata above the unconformity are Late Silurian (Rondout Formation) and Early Devonian (Helderberg Group), whereas the strata below the unconformity are Middle Ordovician (Austin Glen Formation).

What a Geologist Sees

Online Resources

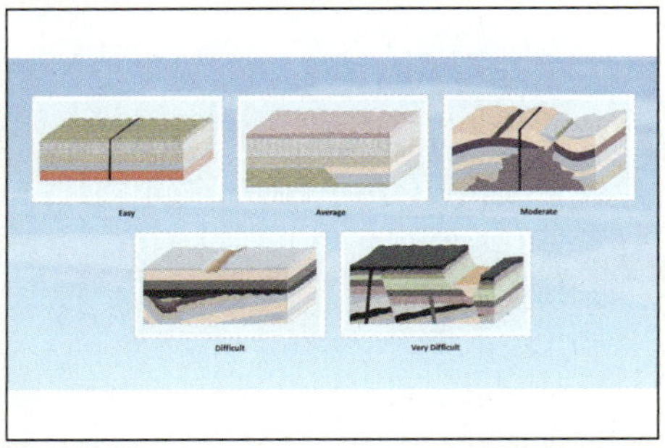

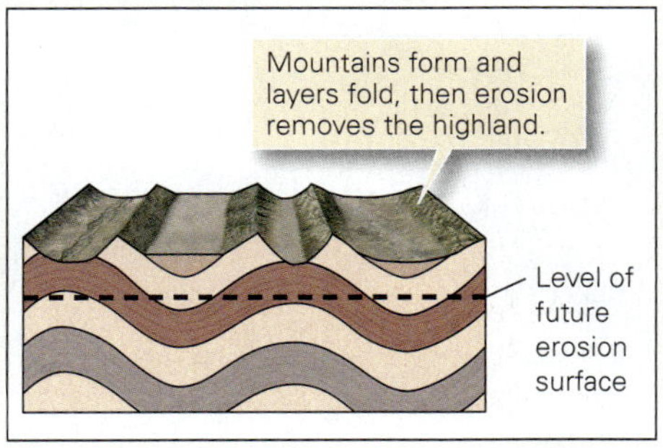

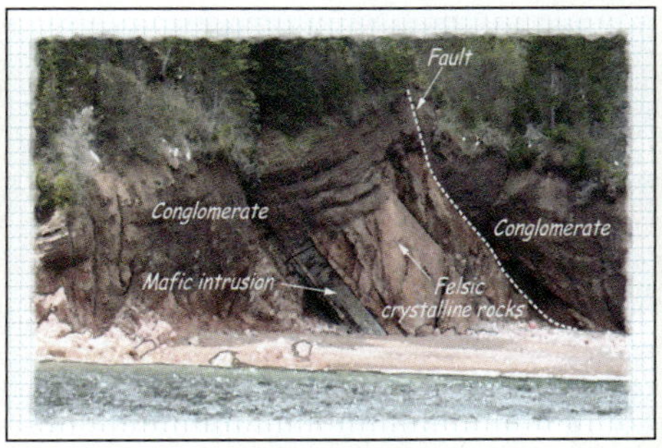

Animations

This chapter features a series of animations demonstrating relative age dating and the different types of geologic unconformities.

Assessment

Questions cover principles for defining relative age, unconformities, and identifying relative ages.

The presence of these fossiliferous limestone beds in Illinois tells us that the middle of North America was once covered by a shallow sea. The dip of the strata tells us that long after the sediment turned to rock, tectonics deformed this part of the continent. Much later, uplift and erosion exposed the rock. Clearly, the Earth has a history!

LEARNING OBJECTIVES

By the end of this chapter, you should understand...

1. geologic clues that provide a basis for interpreting Earth history.
2. why barely any record remains of the earliest half billion years of our planet's existence.
3. that oceans and life have existed since at least 3.8 Ga, and that the atmosphere began to accumulate significant oxygen only since 2.5 Ga.
4. how continental crust first formed, and how it grew in area over time.
5. when supercontinents formed and then rifted apart during Earth history.
6. that early life consisted of archaea and bacteria, but after oxygen levels increased, complex multicellular life became possible, and after shells evolved, organisms diversified.

7. when the land became populated with life, and when the dinosaurs ruled.
8. how, due to sea-level rise and fall, continental interiors sometimes hosted shallow seas.
9. when and why mountain belts formed at various times and places in the past.

CHAPTER 11

A Biography of Earth

11.1 Introduction

In 1868, a well-known British scientist, Thomas Henry Huxley, presented a public lecture on geology to an audience in Norwich, England. Seeking a way to convey his fascination with the subject to people who had no previous geologic knowledge, he focused his audience's attention on the piece of chalk he had been writing with (see the epigraph on the next page). And what a tale the chalk has to tell! Chalk, a type of limestone, consists of microscopic marine algae shells and shrimp feces. The piece of chalk that Huxley held came from beds deposited in the Cretaceous Period—the name *Cretaceous,* in fact, derives from the Latin word for chalk. These beds now form the white cliffs bordering the southeast coast of England. Geologists in Huxley's day knew of similar chalk beds throughout much of Europe, and had discovered that the chalk contains fossils of bizarre swimming reptiles, fish, and invertebrates, species absent in the seas of today. Clearly, when the chalk was deposited, warm seas holding unfamiliar creatures covered some of what is dry land in the cool climate of Europe today. Clues in his humble piece of chalk allowed Huxley to demonstrate to his audience that

> *The man who should know the true history of the bit of chalk which every carpenter carries about in his breeches pocket, though ignorant of all other history, is likely . . . to have a truer and therefore a better conception of this wonderful universe and of man's relation to it than the most learned student who [has] deep-read the records of humanity [but is] ignorant of those of nature.*
>
> THOMAS HENRY HUXLEY (English biologist, 1825–1895), from *On a Piece of Chalk* (1868)

the geography and inhabitants of the Earth in the past differed markedly from those of today, and thus that "the Earth has a history."

In this chapter, we offer a concise geologic biography of our planet, from its birth to the present. During its infancy, the Earth was vastly different from the green and blue globe we know today. We discuss how the oceans and atmosphere first formed, and how continents came into existence and moved about the surface. We also describe mountain-building events, changes in Earth's climate and sea level through time, and the evolution of life. Please remember, we use the following abbreviations: *Ga* (for billion years ago), *Ma* (for million years ago), and *Ka* (for thousand years ago) to streamline discussion.

11.2 Hadean and Archean Time: Earth's Early Days

The Time Before the Rock Record—the Hadean Eon

James Hutton, the 18th-century Scottish geologist who was the first to provide convincing evidence that the Earth was very old, could not measure Earth's age directly, and indeed speculated that there may be "no vestige of a beginning." But, as we discussed in Chapter 10, isotopic dating (radiometric dating) of meteorites has led geologists to consider 4.54 Ga to be the Earth's birth date. At this time, the planet reached nearly its full size and underwent **differentiation** internally, so that metallic iron alloy sank to the center to form a core, leaving behind a mantle of ultramafic rock. Based on dating Moon rocks, researchers have concluded that the Moon formed at about 4.52 Ga, probably when a sizable protoplanet collided with, and may even have merged with, the Earth. The force of the impact blasted material from the Earth's mantle into orbit. Soon after, this debris coalesced into the Moon.

For a time after the Moon's formation, the Earth remained so hot that much of its surface was probably a seething pool of magma, occasionally pummeled by meteorites (**Fig. 11.1a**). But our planet's temperature rapidly decreased, for heat radiated into space from the surface, and the supply of heat from radioactive decay inside diminished as elements with short half-lives had decayed. Calculations suggest that by 4.4 Ga the surface of the Earth had frozen into a skin of solid ultramafic rock. Most likely, this skin underwent rapid recycling, for once it formed, the rock of the skin cooled and became dense enough to sink back into the still very hot and soft mantle. As it sank, it was replaced by new rock at the surface.

During the Earth's early history, rapid *outgassing* also took place, meaning that volatile (gassy) elements or compounds originally incorporated in mantle minerals were released and erupted at the planet's surface, along with lava. These gases accumulated to constitute a toxic atmosphere consisting mostly of water (H_2O), methane (CH_4), ammonia (NH_3), hydrogen (H_2), nitrogen (N_2), carbon dioxide (CO_2), and sulfur dioxide (SO_2). Some researchers speculate that gases from comets colliding with Earth contributed additional gases to the early atmosphere.

Recent studies of oxygen isotopes in the oldest known mineral specimens, 4.4-Ga grains of a durable mineral called zircon found in Western Australia, suggest that some of the H_2O in the young Earth's early atmosphere condensed to form liquid water that rained onto the planet's surface. What might the Earth's surface have looked like at this time? If the hypothesis that liquid water condensed from the early atmosphere is correct, then the surface probably would have been a black, barren, cratered landscape, locally submerged by an acidic sea (**Fig. 11.1b**). Volcanoes spewing lava would rise here and there. Both land and sea would likely have been obscured by murky, dense (H_2O-, CO_2-, and SO_2-rich) air.

Though the oldest dated minerals are 4.4 Ga, the oldest known whole rock, a sample collected in northwestern Canada, has an age of about 4.0 Ga. Geologists use the name **Hadean Eon** to refer to the mysterious time interval between the birth of Earth at 4.54 Ga and the beginning of the rock record by about 4.0 Ga. The name comes from the Greek name Hades, the god of the underworld, for during intervals of this eon, the Earth's surface resembled an inferno. Where did pre-4.0-Ga rocks go? They may have sunk back into the mantle where they were recycled, and/or they may have been pulverized and melted at the end of the Hadean, during a time that astronomers refer to as the *late heavy bombardment*, when the Earth was pummeled intensely by meteorites. This event took place sometime between 4.1 and 3.8 Ga.

FIGURE 11.1 Visualizing the early Earth is a challenge, but by using geologic interpretations, artists have provided useful images.

(a) At a very early stage, during or after differentiation, the surface may have been largely molten. The loss of heat to space would have allowed patches of solid ultramafic solid crust to form. Meteorite impacts would have been fairly frequent.

(b) When Earth's surface fell below the boiling point of water, water from the atmosphere rained onto the surface, submerging it with early oceans.

Birth of the Continents in the Archean Eon

The age of the first whole rocks marks the start of the **Archean Eon** (from the Greek word for beginning). With the advent of the Archean, at 4.0 Ga, crust was locally cool and stable enough, and meteorite impacts rare enough, for rocks to survive and for isotopic clocks to start ticking. Geologists still debate whether plate tectonic activity was taking place in the early Archean Eon. According to some researchers, until sometime between 3.8 Ga and 3.2 Ga, the Earth was so hot that lithospheric mantle could not become cool enough and dense enough to subduct and sink back into the asthenosphere. This means that subduction couldn't happen, and without subduction, plates couldn't move. If this view is correct, then most volcanism of the early Earth may have been related to mantle plumes. Other researchers, in contrast, think that plate tectonic–like processes did began at the end of the Hadean. And still other researchers argue that plate tectonics was operating after 3.2 Ga, and that during the middle of the Archean, there were many rapidly moving small plates, numerous volcanic island arcs, and abundant hot-spot volcanoes.

Significant volumes of new, relatively buoyant rocks formed during the Archean after the Earth's mantle became cool enough that only partial melting could take place. As we discussed in Chapter 4, magmas formed by partial melting tend to be enriched in silica, relative to their source, and thus are buoyant, relative to their source. Partial melting of the ultramafic mantle, in Archean time, produced mafic magma. This melt extruded or intruded in volcanic arcs at convergent plate boundaries, and/or in hot-spot volcanoes

FIGURE 11.2 A model for crust formation during the Archean Eon.

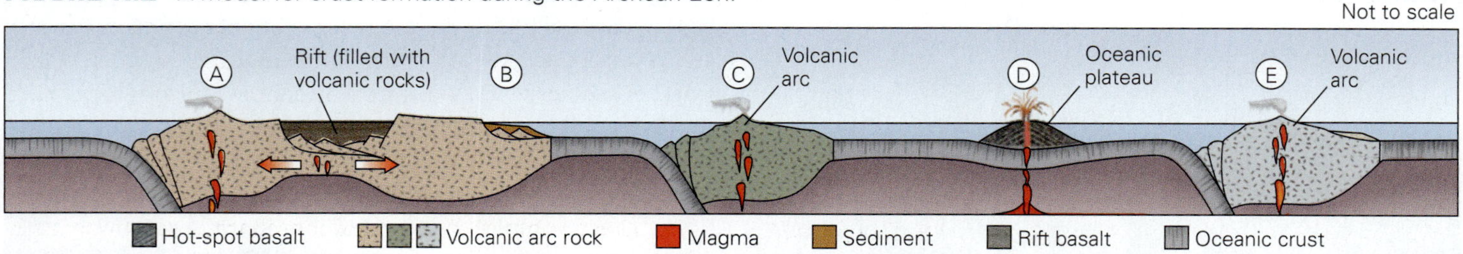

(a) In the Archean, island arcs and hot-spot volcanoes built small blocks of buoyant crust. Rifting of these blocks may have produced flood basalts, and erosion of the blocks produced sediment.

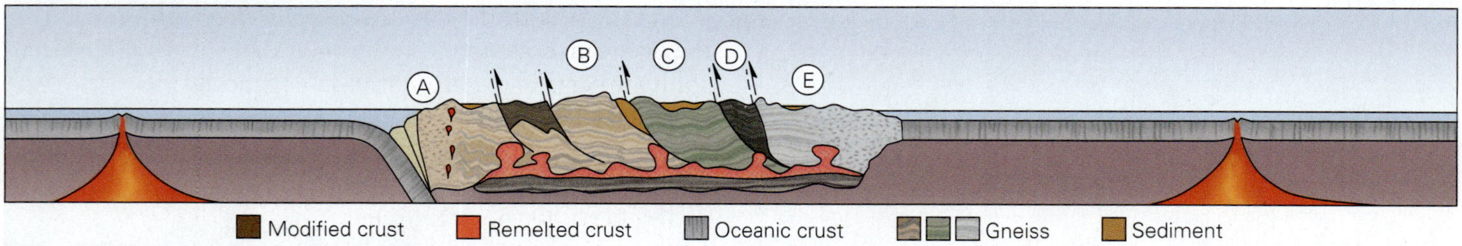

(b) Buoyant blocks collided and sutured together, forming protocontinents. Melting at depth produced granite. Eventually, regions of crust cooled, stabilized, and became cratons.

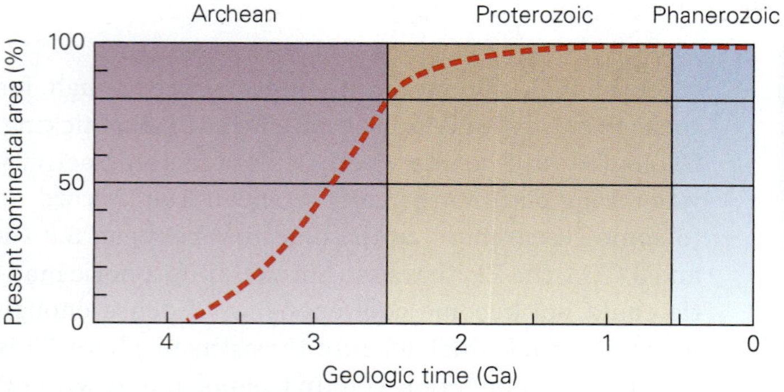

(c) As time progressed, the area of the Earth covered by continental crust increased. Most crust had formed by the beginning of the Proterozoic.

over plumes—plume-related igneous activity may have produced broad basalt plateaus. When these arcs and plateaus collided with one another, they sutured together, forming larger blocks that were buoyant enough to remain at the Earth's surface. When convergent boundaries developed along the margins of these blocks, and rifts and hot spots developed within the blocks, not only did more basaltic magma form, but in addition, felsic and intermediate magma formed, forming even less dense rock. As collisions continued, mountain belts formed broad regions of crustal rock that underwent metamorphism.

Eventually, collisions resulted in the assembly of *protocontinents* (**Fig. 11.2a, b**). Size matters when it comes to the geologic behavior of continents, for the interior of a larger continental block can be isolated from heating by subduction-related igneous activity along its margins. As a result, the crust of the interior region of a protocontinent can slowly cool and strengthen until it becomes a durable and relatively stable crustal block called a **craton**. Between 3.2 and 2.7 Ga, several small Archean cratons came into existence. By the end of the Archean Eon, about 80% of the Earth's continental rock had formed (**Fig. 11.2c**). Notably, much of this rock later passed through stages of the rock cycle one or more times, so only a relative small amount retains Archean isotopic ages. These remnants of Archean crust occur scattered around all of today's continents.

A stratigraphic record of Archean marine sediment deposition has been found in a few localities, indicating that global oceans existed by 3.85 Ga. Oceans have survived ever since, so for most of the Earth's history, surface temperatures have remained between the boiling and freezing point of water. The accumulation of liquid water in oceans in the early Archean changed the Earth's atmosphere dramatically, for prior to ocean formation, H_2O and CO_2 were the dominant gases of the atmosphere. Once the oceans formed, the atmosphere lost most of its H_2O, and once liquid water existed, most atmospheric CO_2 dissolved into the water. Thus, during the Archean, the atmosphere changed from a foggy mixture of H_2O and CO_2 into a transparent blend whose major component was nitrogen (N_2) gas. Since nitrogen is inert (meaning that it doesn't chemically react with or dissolve in other materials), it remained in the form of a gas.

The First Life

The search for the earliest evidence of life on Earth continues to make headlines. Sedimentary rocks as old as 3.8 Ga

FIGURE 11.3 Archean life forms.

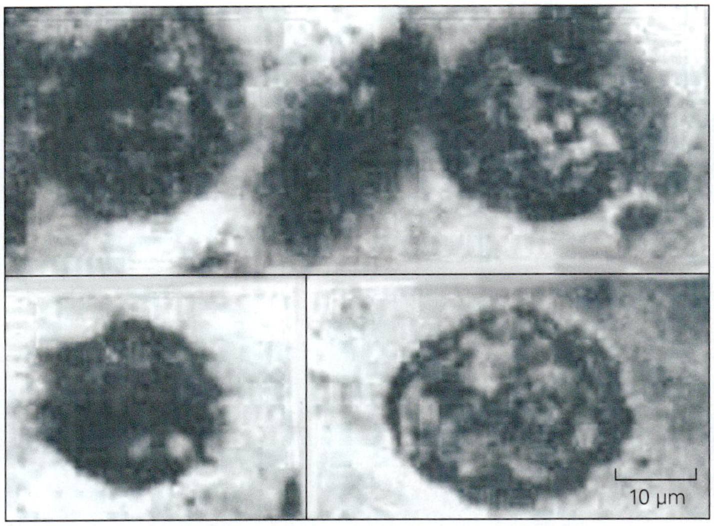

(a) These shapes in 3.2-Ga chert from South America are thought to be fossil bacteria or archaea.

(b) This weathered outcrop of 1.85-Ga dolostone near Marquette, Michigan, reveals the layer-like structure of stromatolites. The delicate ridges represent the fossilized remnants of bacterial mats. Similar stromatolites also occur in exposures of Archean rocks.

contain biomarkers, chemical signatures of organisms, suggesting that life existed in the early Archean. The oldest shapes that look like fossilized cells, however, occur in 3.5-Ga strata, and the oldest undisputed fossil forms of bacteria and archaea occur in 3.2-Ga strata (**Fig. 11.3a**). Sedimentary rocks deposited after about 3.2 Ga locally contain **stromatolites**, distinctive layered mounds of sediment formed when cyanobacteria secrete a mucus-like substance to which sediment settling from water sticks (**Fig. 11.3b**). As the mat gets buried, new cyanobacteria colonize the top of the sediment, building a mound upward. Modern examples of stromatolites can be found in shallow, tropical waters.

Did you ever wonder...
what the oldest relict of life is?

What specific environment on the Archean Earth served as the cradle of life? Laboratory experiments conducted in the 1950s led many researchers to think that life began in warm pools of surface water, beneath a methane- and ammonia-rich atmosphere streaked by bolts of lightning. More recent researchers suggest instead that submarine hot-water vents, so-called *black smokers*, served as the hosts of the first organisms. These vents emit clouds of ion-charged solutions from which sulfide minerals precipitate and build chimneys. The earliest life in the Archean Eon may well have been thermophilic (heat-loving) bacteria or archaea that dined on pyrite at dark depths in the ocean alongside these vents. Later in the Archean, organisms similar to cyanobacteria evolved the ability to carry out photosynthesis, and moved into shallower, well-lit water. Although these organisms produced oxygen, very little accumulated in the atmosphere, for it was either dissolved in the sea or absorbed by weathering reactions with rocks. Oxygen breathers, such as humans, would have suffocated in seconds in the atmosphere of the Archean.

Did you ever wonder...
if the atmosphere has always been breathable?

TAKE-HOME MESSAGE

During the Hadean (4.54–4.0 Ga), the Earth differentiated and the Moon formed. New evidence suggests that liquid water may have appeared as early as 4.4 Ga. But a rock record of the Hadean doesn't exist, because the rock skin of our planet's surface was recycled or was melted and pulverized by meteorite bombardment. During the Archean (4.0–2.5 Ga), the first continental crust formed from colliding volcanic arcs and hot-spot volcanoes. As the oceans filled, the atmosphere lost its H_2O and CO_2, but throughout the Archean, it remained devoid of O_2. Biomarkers indicative of life have been found in 3.8-Ga strata, but the oldest definitive fossils of archaea and bacteria first appear in 3.2-Ga strata.

QUICK QUESTION What did the early atmosphere of the Earth consist of, and where did its gases come from?

11.3 The Proterozoic: The Earth in Transition

Continued Growth of Continents

The **Proterozoic Eon** spans roughly 2 billion years, from about 2.5 Ga to the beginning of the Cambrian Period at

FIGURE 11.4 Major crustal provinces of Earth. The black lines indicate the borders of regions underlain by Precambrian crust. Shields are regions where broad areas of Precambrian rocks are exposed.

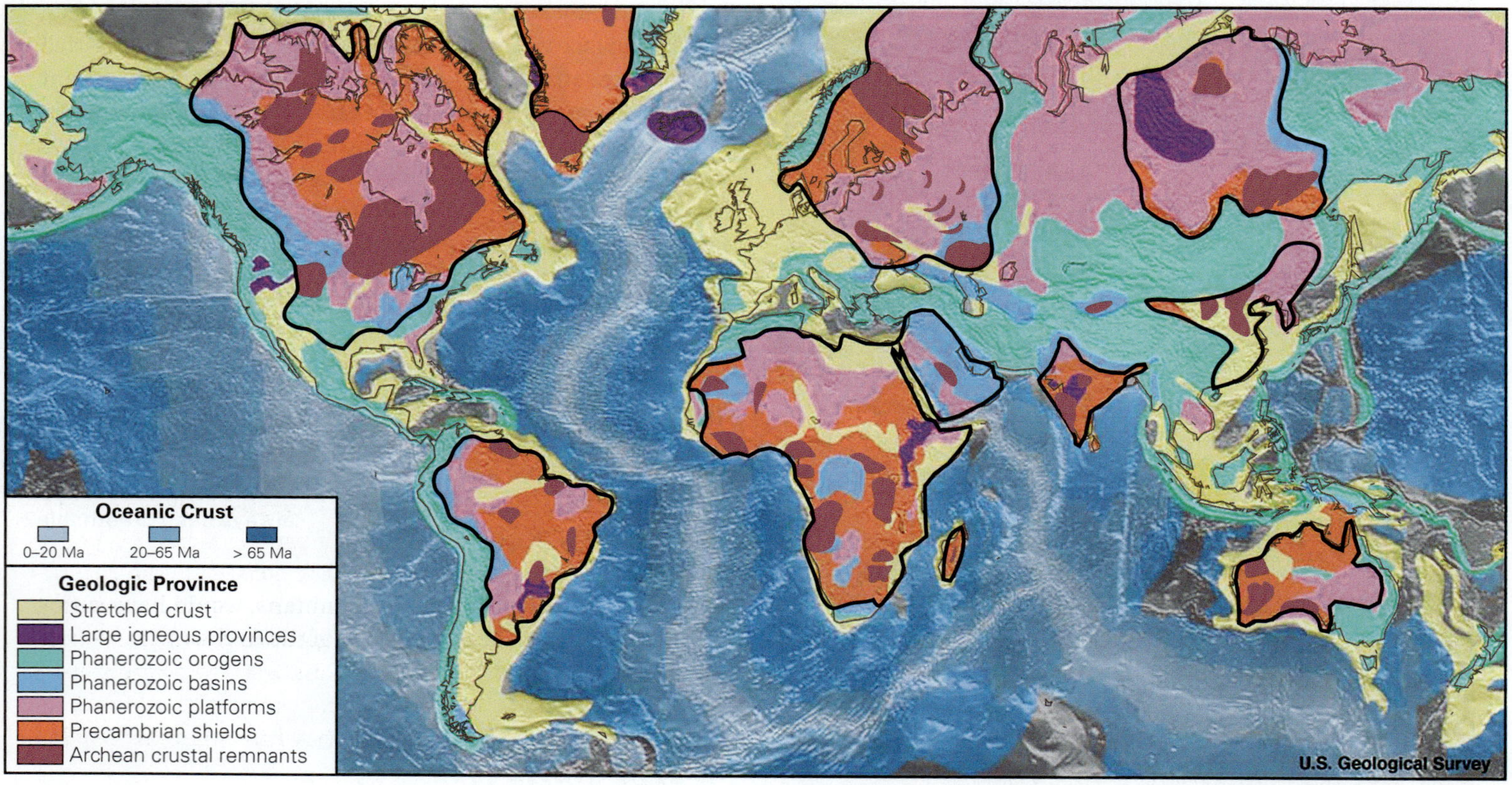

541 Ma—almost half of Earth's history. During Proterozoic time, the Earth's surface environment changed from an unfamiliar world of small, fast-moving plates, small continents, and an oxygen-free atmosphere, to a more familiar configuration of mostly large, slow-moving plates, large continents, and an oxygenated atmosphere.

New continental crust continued to form during the Proterozoic Eon, but at progressively slower rates—by the middle of this eon, over 90% of the Earth's continental crustal area had formed. Collisions between Archean cratons, and between cratons and volcanic island arcs or hot-spot volcanoes, gradually caused larger cratons to grow. All cratons that exist today had formed by about 1 Ga (**Fig. 11.4**), meaning that the crust of cratons ranges from 3.85 Ga to about 1 Ga. Geologists subdivide the basement of North America's craton into distinct, named blocks or provinces, based on the numerical ages of rocks (**Fig. 11.5**). The Canadian Shield, the portion of this craton in which Precambrian rocks are exposed at the Earth's surface, consists of several Archean crust blocks sutured together by Proterozoic orogens. Much of the *cratonic platform* in the United States, a portion of the craton in which a younger sedimentary strata covers Precambrian rock, grew when a series of volcanic island arcs and continental slivers accreted, or attached, to the margin of the Canadian Shield during the time between 1.8 and 1.6 Ga. Granite plutons intruded much of this accreted region, and rhyolite ash flows covered it, between 1.5 and 1.3 Ga.

Successive collisions ultimately brought together most continental crust on Earth into a single supercontinent, named *Rodinia*, by around 1 Ga. The last major collision during the formation of Rodinia was the Grenville orogeny. If you look at a popular (though not universally accepted) reconstruction of Rodinia, you can identify the crustal provinces that would eventually become the familiar continents of today (**Fig. 11.6a**). Several studies suggest that sometime between 800 and 600 Ma, Rodinia "turned inside out," so that Antarctica, India, and Australia broke away from western North America and swung around and collided with the future South America, possibly forming a short-lived supercontinent that some geologists refer to as *Pannotia* (**Fig. 11.6b**).

Life Becomes More Complex

The map of the Earth clearly changed radically during the Proterozoic. But that's not all that changed—fossil evidence

FIGURE 11.5 North America's craton, and the blocks and belts that comprise it.

(a) The craton lies between the Cordillera and the Appalachians. Precambrian rocks crop out in the shield, and are covered by sedimentary strata in the platform.

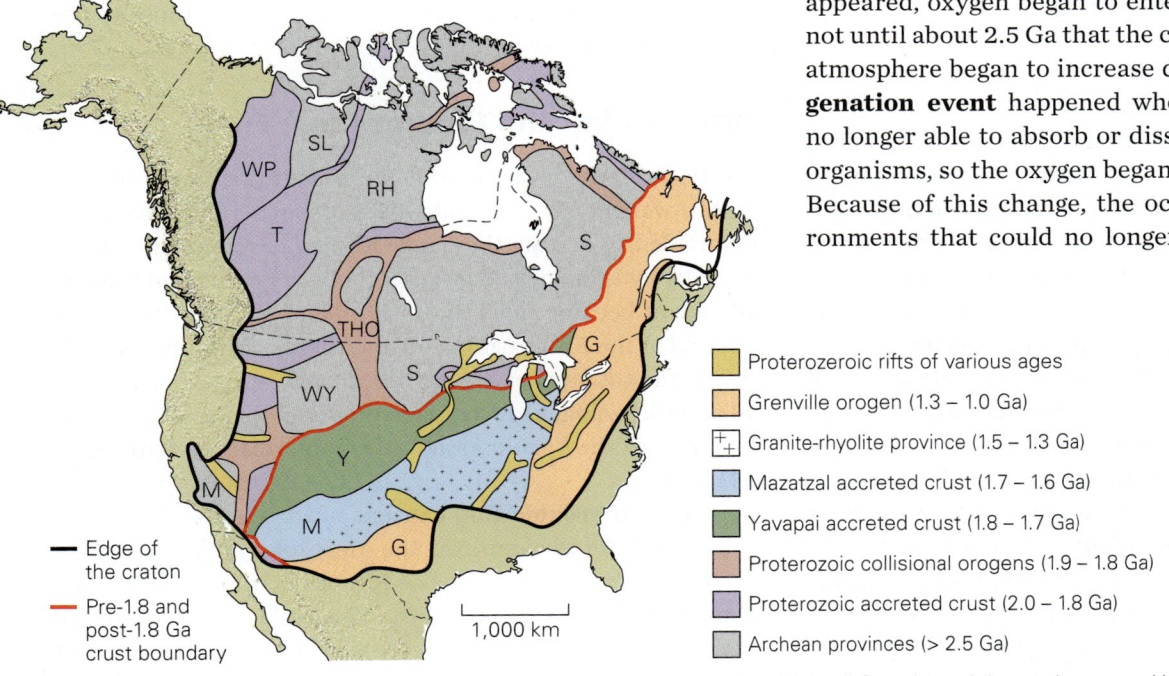

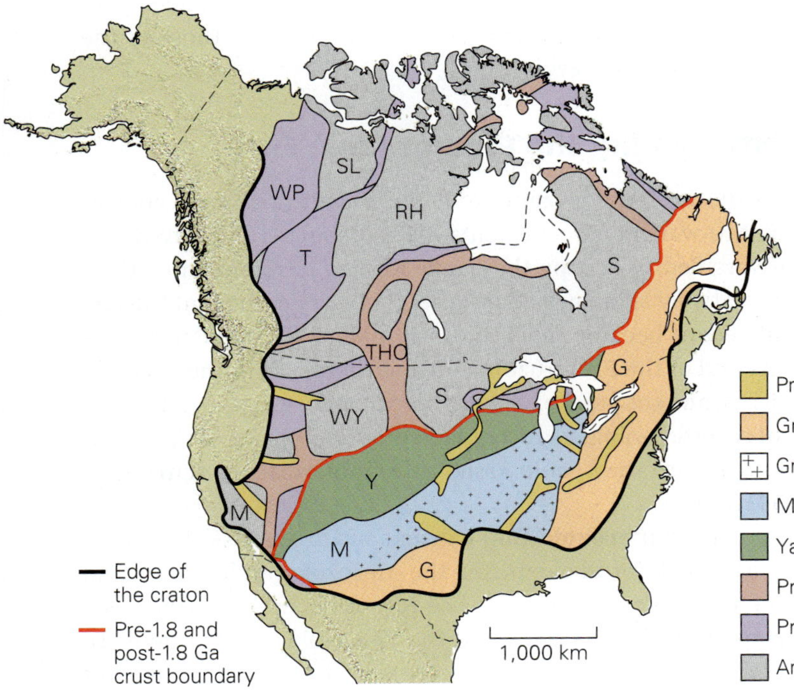

(b) The craton consists of many different blocks, each of which has a name. Most of Canada, and the northwestern United States, had assembled by 1.8 Ga. Belts of younger terranes accreted to southeastern margin nucleus between 1.8 and 1.1 Ga.

suggests that this eon also saw important steps in the evolution of life. When the Proterozoic began, most life was *prokaryotic*, meaning that it consisted of single-celled organisms (archaea and bacteria) without a nucleus. Though studies of chemical fossils hint that *eukaryotic* life, consisting of cells that have nuclei, originated as early as 2.7 Ga, the first possible body fossil of a eukaryotic organism occurs in 2.1-Ga rocks, and abundant body fossils of eukaryotic organisms can be found only in rocks younger than about 1.2 Ga. Thus the proliferation of eukaryotic life, the foundation from which complex organisms eventually evolved, took place during the Proterozoic.

The last half billion years of the Proterozoic Eon saw the remarkable transition from simple to complex organisms. Ciliate protozoans (single-celled organisms coated with fibers that give them mobility) appear at about 750 Ma. A great leap in complexity of organisms occurred during the next 150 million years of the eon, for sediments deposited perhaps as early as 620 Ma and certainly by 565 Ma contain several types of multicellular organisms that together constitute the *Ediacaran fauna*. Ediacaran species, named for a region in southern Australia where fossils of these organisms were first found, survived into the beginning of the Cambrian before becoming extinct. Their fossil forms suggest that some of these invertebrate organisms resembled jellyfish, while others resembled worms (**Fig. 11.7a**).

The evolution of life played a key role in the evolution of Earth's atmosphere. When photosynthetic organisms appeared, oxygen began to enter the atmosphere. But it was not until about 2.5 Ga that the concentration of oxygen in the atmosphere began to increase dramatically. This **great oxygenation event** happened when other environments were no longer able to absorb or dissolve the oxygen produced by organisms, so the oxygen began to accumulate as a gas in air. Because of this change, the oceans became oxidizing environments that could no longer contain large quantities of

FIGURE 11.6 Supercontinents in the late Precambrian.

(a) Rodinia formed around 1 Ga and lasted until about 700 Ma. North America and Greenland together comprise Laurentia.

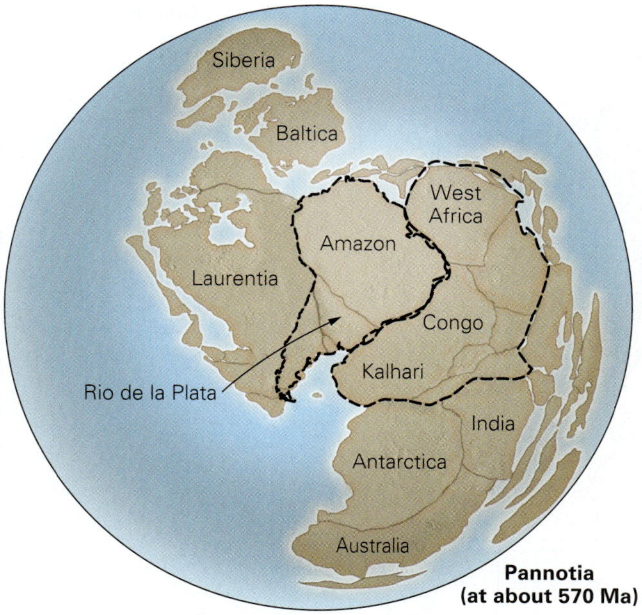

(b) According to one model, by 570 Ma Rodinia had broken apart; continents that once lay to the west of Laurentia ended up to the east of Africa. The resulting supercontinent, Pannotia, broke up soon after it formed.

dissolved iron. Between 2.4 Ga and 1.8 Ga, huge amounts of iron settled out of the ocean to form colorful sedimentary beds known as **banded iron formation (BIF)**. BIF consists of alternating layers of iron-oxide minerals (hematite or magnetite) and jasper (red chert) (**Fig. 11.7b**).

Snowball Earth

Radical climate shifts occurred on Earth at the end of the Proterozoic Eon, and glaciation became widespread, as indicated by the distribution of glacial sediments from this time. Surprisingly, these sediments can be found even in regions that were then located at the equator (**Fig. 11.7c**). This observation implies that the entire planet was cold enough for glaciers to form at the end of the Proterozoic. Geologists speculate that late Proterozoic glaciers not only covered the land, but that the entire ocean surface froze, resulting in **snowball Earth** (**Fig. 11.7d**). Earth might have remained a snowball forever were it not for volcanic CO_2. The icy sheath covering the oceans prevented atmospheric CO_2 from dissolving in seawater, but it did not prevent volcanic activity from adding CO_2 to the atmosphere. CO_2 is a greenhouse gas, meaning that it traps heat in the atmosphere much as glass panes trap heat in a greenhouse (see Chapter 19), so as the CO_2 concentration increased, the Earth's climate warmed up and eventually the glaciers melted. Life may have survived snowball Earth conditions only near submarine black smokers and near hot springs. When the ice vanished, life rapidly expanded into new environments, where new species, such as those of the Ediacaran fauna, evolved.

Did you ever wonder...
if the oceans have ever frozen over entirely?

Introducing the Phanerozoic Eon

As the Proterozoic came to a close, Earth's climate continued to warm and continents drifted apart—life evolved and diversified to occupy the newly formed environments. Over a relatively short period of time, shells appeared and the fossil record became much more complete. This event defines the end of the Proterozoic Eon and therefore, of the Precambrian, and the start of the Phanerozoic Eon. Based on the fossil record, geologists recognized the significance of this event long before they could assign it a numerical age (currently, 541 Ma).

The **Phanerozoic Eon** consists of three eras—the Paleozoic (Greek for ancient life), the Mesozoic (middle life), and the Cenozoic (recent life). The Mesozoic and Cenozoic each consist of three periods and the Paleozoic of six periods. In the sections that follow, we consider changes in the map of our planet's surface (its *paleogeography*), as manifested by the distribution of continents, seas, and mountain belts, as well as life evolution during these three eras.

FIGURE 11.7 Major changes in the Earth System during the Proterozoic Eon.

(a) *Dickinsonia*, a fossil of the Ediacaran fauna. These complex, soft-bodied marine organisms appeared in the late Proterozoic. This specimen is 10 cm long.

(b) An outcrop of BIF in the Iron Ranges of Michigan. The red stripes are jasper (red chert) and the gray stripes are hematite. The rock was folded during a mountain-building event long after deposition.

(c) Layers of Proterozoic glacial till crop out in Africa, indicating that low-latitude landmasses were glaciated during the Proterozoic.

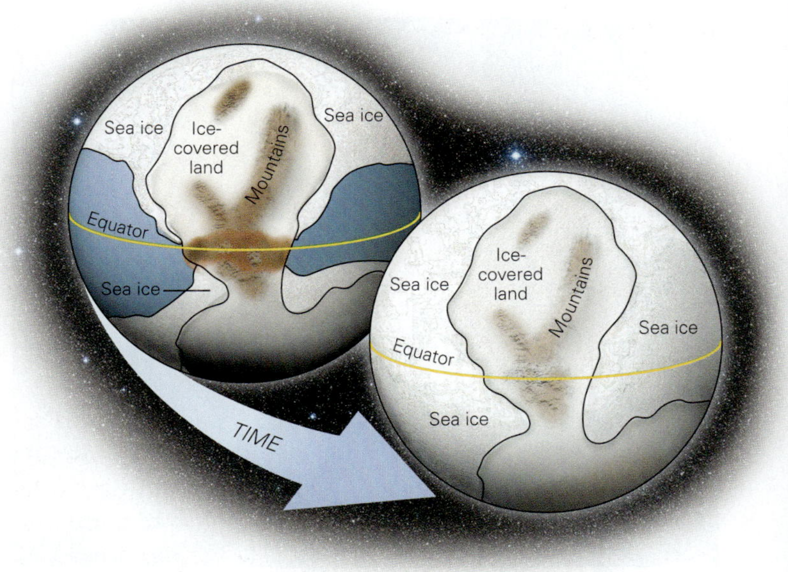

(d) This planet may have frozen over completely to form "snowball Earth." Glaciers first grew on land, and eventually the sea surface froze over.

TAKE-HOME MESSAGE

During the Proterozoic (2.5–0.54 Ga), cratons formed and then sutured together to form continents and, eventually, supercontinents, which later rifted apart. Multicellular organisms appeared, and the atmosphere began to accumulate significant oxygen.

QUICK QUESTION What evidence suggests that oxygen began to accumulate in the air during the Proterozoic?

11.4 The Paleozoic Era: Continents Reassemble, and Life Gets Complex

The Early Paleozoic Era (Cambrian-Ordovician Periods, 541–444 Ma)

Paleogeography At the beginning of the Paleozoic Era, Pannotia broke up, yielding smaller continents including

FIGURE 11.8 Land and sea in the early Paleozoic Era.

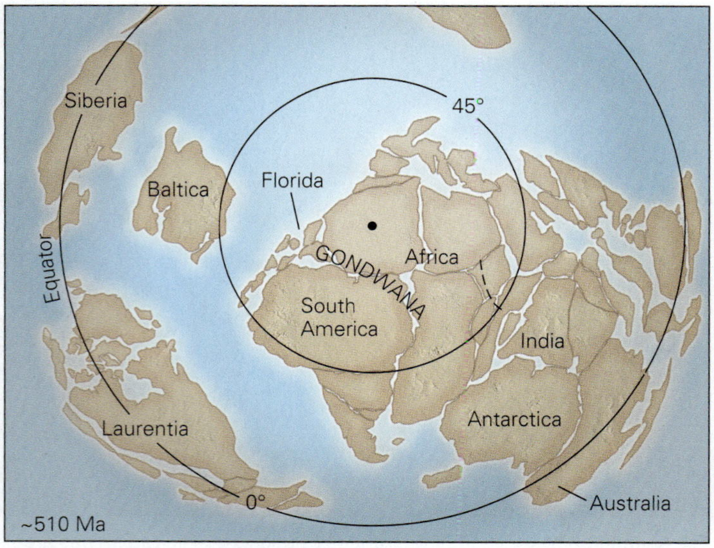

(a) The distribution of continents in the Cambrian Period (510 Ma), as viewed looking down on the South Pole.

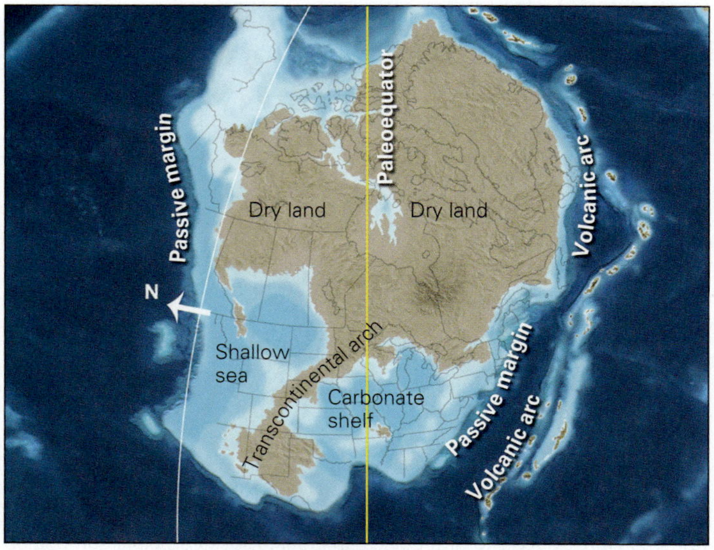

(b) A paleogeographic map of North America shows the regions of dry land and shallow sea in the Late Cambrian Period.

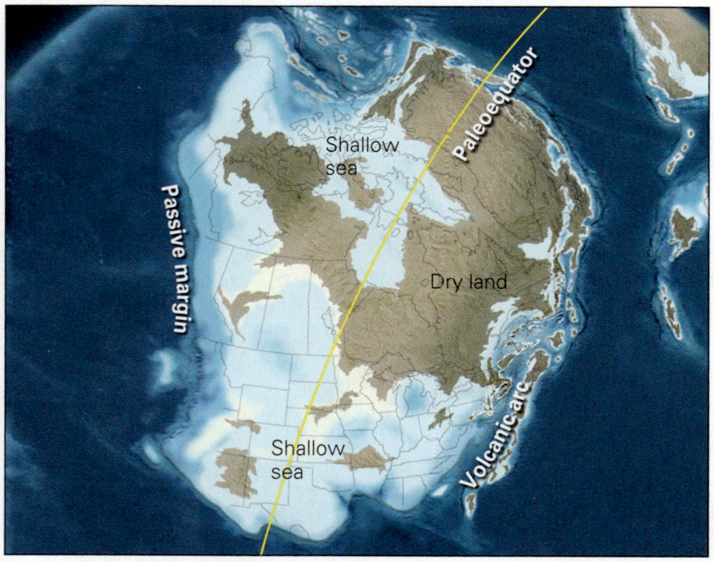

(c) During the Middle Ordovician Period, shallow seas covered much of North America. A volcanic arc formed off the east coast.

Laurentia (North America and Greenland), **Gondwana** (South America, Africa, Antarctica, India, and Australia), Baltica (Europe), and Siberia (**Fig.11.8a**). New passive-margin basins formed along the edges of these new continents. Sea level rose and fell during this time, and at times, vast areas of continental interiors were flooded with shallow, *epicontinental seas* (**Fig. 11.8b**). The sediment deposited from epicontinental seas forms the strata of cratonic platforms. In well-lit shallower waters of epicontinental seas, life abounded, so strata of cratonic platforms are fossiliferous. Sea level, however, did not stay high for the entire early Paleozoic. When regressions took place, land was exposed to weathering and erosion, and unconformities formed.

The geologically peaceful world of the early Paleozoic in Laurentia abruptly came to a close in the Middle Ordovician Period, when its eastern margin rammed into a volcanic island arc and other crustal fragments. The resulting collision, called the *Taconic orogeny*, deformed and metamorphosed strata that had been deposited in the continent's passive-margin basin, and produced a mountain range in what is now the eastern part of the Appalachians (**Fig. 11.8c**).

FIGURE 11.9 A museum diorama illustrates what early Paleozoic marine organisms may have looked like.

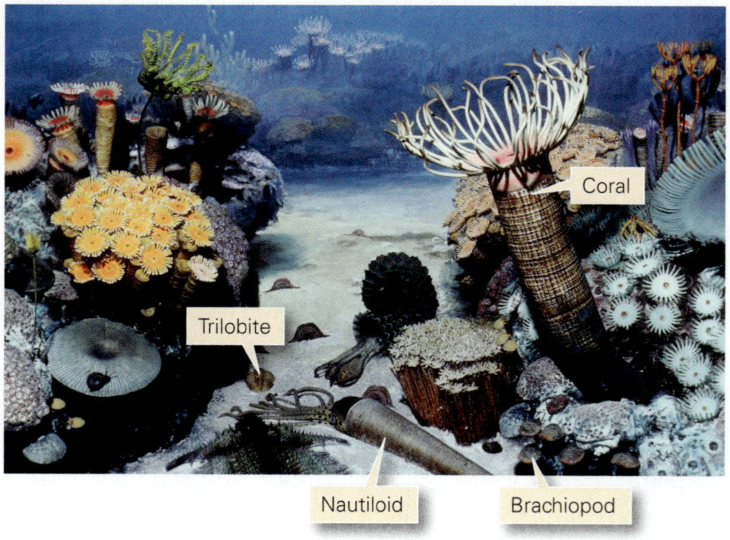

FIGURE 11.10 Paleogeography and fossils of Silurian and Devonian time.

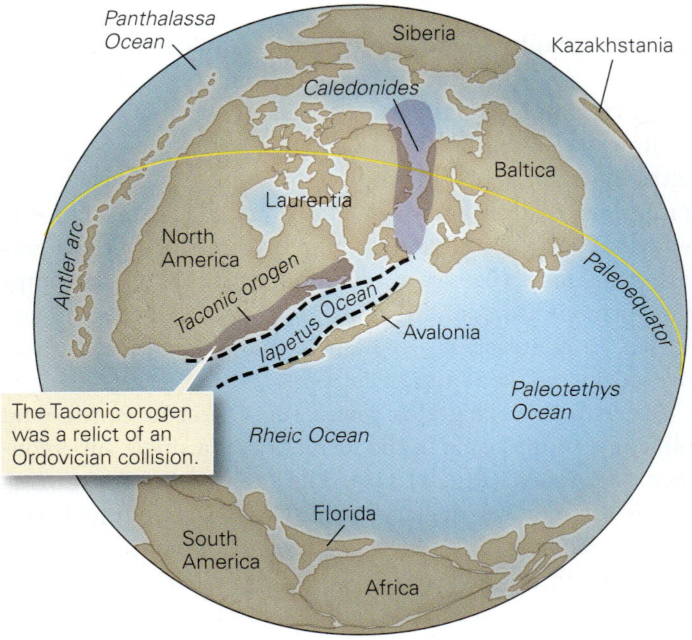

(a) During the Silurian and Devonian Periods, Laurentia collided with Baltica, Avalonia, and South America in succession, as oceans in between were consumed. The Antler arc formed off the west coast.

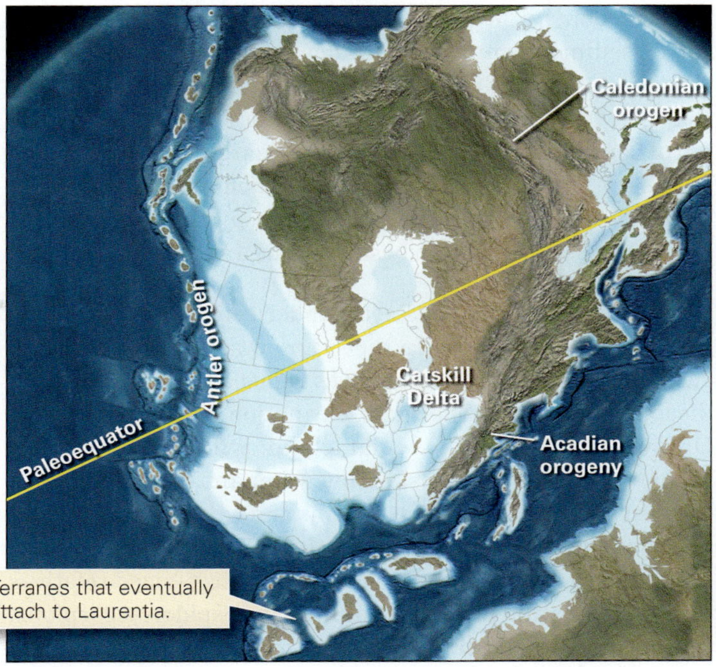

(b) During the Devonian Period, the Acadian orogeny shed sediments into a shallow sea to form the Catskill Delta on the east coast of Laurentia. The Antler orogeny shed sediments in the west.

(c) A Late Devonian fossil skeleton of *Tiktaalik*; this lobe-finned fish was one of the first animals to walk on land.

Life Evolution The fossil record indicates that soon after the Cambrian began, life began to undergo remarkable diversification. This event, which paleontologists refer to as the **Cambrian explosion**, took several million years. What caused this event? No one can say for sure, but considering that it occurred roughly at the time a supercontinent broke up, the explanation may involve the production of new ecological niches and the isolation of populations that resulted when smaller continents formed and drifted apart.

The first new animals to appear in the Cambrian Period had simple tube- or cone-shaped shells, but soon thereafter the shells became more complex. Shells may have evolved as a means of protection against predation by conodonts, small, eel-like organisms with hard parts that resemble teeth. Thus, a complex food chain arose, which included plankton, bottom feeders, and at the top of the chain, predators. By the end of the Cambrian, trilobites were grazing the seafloor, sharing their environment with mollusks, brachiopods, nautiloids, gastropods, graptolites, and echinoderms (**Fig. 11.9**; see Interlude E). Many of the organisms crawled over or swam around reefs composed of mounds of sponges with mineral skeletons. The Ordovician Period saw the first crinoids and the first vertebrate animals, jawless fish. At the end of the Ordovician, a mass-extinction event took place, perhaps because of the brief glaciation and associated sea-level lowering of the time.

The Middle Paleozoic Era (Silurian-Devonian Periods, 444-359 Ma)

Paleogeography As the Earth entered the Middle Paleozoic, global climate warmed, sea level rose, and the continents flooded once again. In parts of Laurentia's interior, huge carbonate reef complexes grew in epicontinental seas. Along the eastern margin of Laurentia convergence and collision produced the *Caledonian orogen* in eastern Greenland, western Scandinavia, and Scotland, as well as the *Acadian orogen* in a region that is now part of the Appalachians (**Fig. 11.10a, b**).

The Caledonian, Acadian, and Antler orogenies all shed deltas of sediment onto the continents—deposits that formed thick successions of redbeds, such as those visible today in the Catskill Mountains of New York State. Along the western margin of Laurentia, in contrast, a passive-margin basin continued to subside until the Late Devonian, when an island arc collided. This event caused the *Antler orogeny*, during which the once quiet passive-margin basin became a site of deformation.

Life Evolution Life on Earth underwent radical changes in the middle Paleozoic. In the sea, new species of trilobites, gastropods, crinoids, and bivalves replaced species that had disappeared during the mass extinction at the end of the Ordovician. On land, vascular plants with woody tissues, seeds, and veins (for transporting water and food) rooted for the first time. The evolution of veins and wood allowed plants to grow much larger, and by the Late Devonian Period the land surface hosted swampy forests with tree-sized relatives of clubmosses and ferns. Also at this time, spiders, scorpions, insects, and crustaceans began to exploit both dry-land and freshwater habitats, and jawed vertebrates such as sharks and bony fish began to cruise the oceans. Finally, at the very end of the Devonian, the first animals crawled out onto land and inhaled air with lungs (**Fig. 11.10c**).

The Late Paleozoic Era (Carboniferous-Permian Periods, 359-252 Ma)

Paleogeography The climate cooled significantly in the late Paleozoic. Seas gradually retreated from the continents, so that during the Carboniferous Period, regions that had hosted the limestone-forming reefs became coastal areas and river deltas in which sand and mud accumulated. During the Carboniferous, Laurentia lay near the equator and rainfall was heavy, so coastal plains and lowlands inshore became swamps harboring lush vegetation. Plant debris of these swamps transformed into coal after burial. Much of Gondwana and Siberia, in contrast, lay at high latitudes, and became ice-covered during the Permian Period.

The late Paleozoic also saw a succession of continental collisions, culminating in the formation of a single supercontinent, **Pangaea** (**Fig. 11.11a**). The largest collision during Pangaea assembly took place during Carboniferous and Permian time. This event, called the *Alleghanian orogeny*, was the final stage in the development of the Appalachians. It happened when eastern North America rammed into northwestern Africa, and what is now the Gulf Coast region of North America squashed against the northern margin of South America (**Fig. 11.11b**). The event generated a broad fold-thrust

FIGURE 11.11 Paleogeography at the end of the Paleozoic Era.

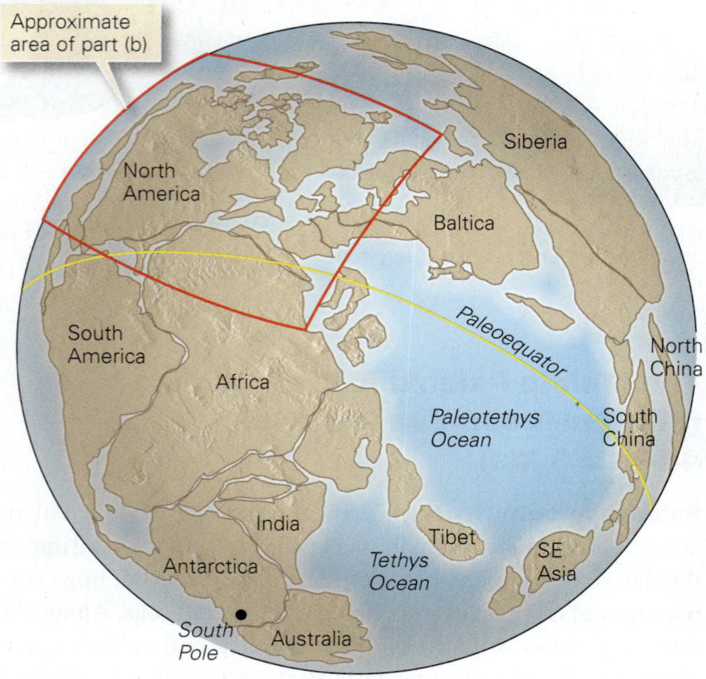

(a) At the end of the Paleozoic, almost all land had combined into a single supercontinent called Pangaea.

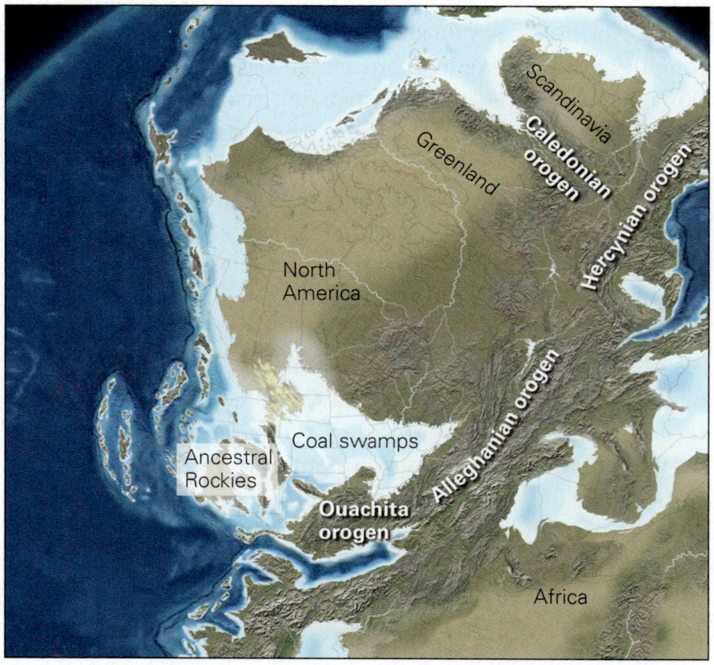

(b) During the Alleghanian and Hercynian orogenies, a huge mountain belt formed. The Caledonian orogeny had formed earlier. Coal swamps bordered interior seas, and the Ancestral Rockies rose.

belt whose eroded remnants underlie the valley-and-ridge provinces of the Appalachian and Ouachita Mountains (**Fig. 11.12**). Stresses generated during the Alleghanian orogeny were so great that pre-existing faults in the continental crust clear across North America became active again. The movement reactivated faulting that produced uplifts, and adjacent basins, in the Midwest and in the region of the present-day Rocky Mountains. Geologists refer to the late Paleozoic uplifts of the Rocky Mountain region as the *Ancestral Rockies*.

The assembly of Pangaea involved other collisions around the world as well. Notably, Africa collided with southern Europe to form the *Hercynian orogen*, and a rift or small ocean in Russia closed, leading to the uplift of the Ural Mountains. In addition, parts of China, along with other fragments of Asia, attached to southern Siberia.

FIGURE 11.12 Features of the Appalachian Mountains in the eastern United States.

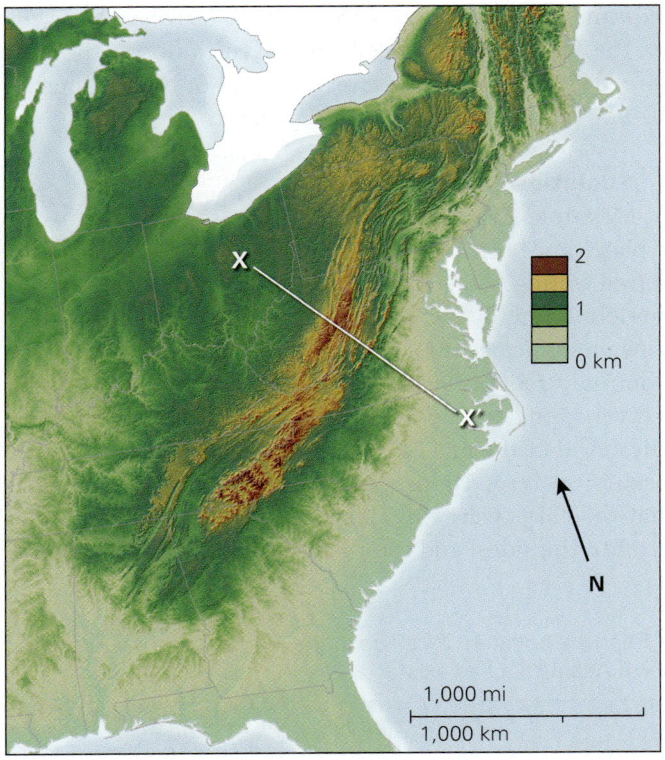

(a) The eroded remnants of the Appalachian orogen stand out in the eastern United States. Line XX′ shows the position of the cross section below.

FIGURE 11.13 A museum diorama of a Carboniferous coal swamp includes a giant dragonfly with a wingspan of about 1 m. The inset gives a sense of its size relative to a human.

Life Evolution The fossil record indicates that during the late Paleozoic, plants and animals continued to evolve toward more familiar forms. In coal swamps, fixed-wing insects including huge dragonflies flew through a tangle of ferns, club-mosses, and scouring rushes, and by the end of the Carboniferous, insects such as the cockroach, with foldable wings, appeared (**Fig. 11.13**). Forests containing gymnosperms ("naked seed" plants, such as conifers) and cycads (trees with a palm-like stalk and fern-like fronds) became widespread in the Permian. Amphibians and, later, reptiles populated the

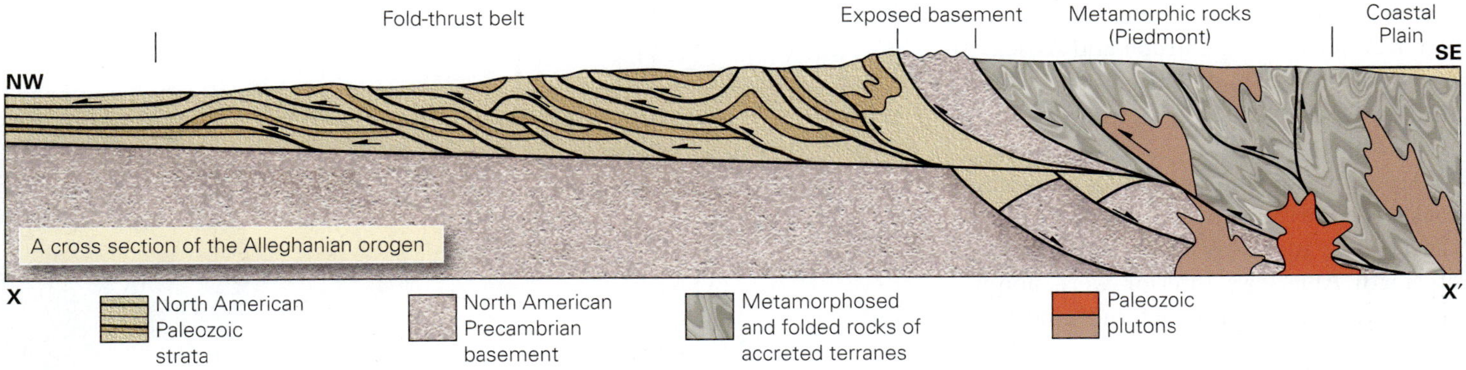

land. The appearance of reptiles marked the evolution of a radically new component in animal reproduction—eggs with a protective shell. Such eggs permitted reptiles to reproduce without returning to the water.

The Permian, and therefore the Late Paleozoic, came to a close with a major mass-extinction event, during which over 95% of marine species disappeared. Why this event occurred remains a subject of debate. According to one hypothesis, the *Permian mass extinction* occurred as a result of an episode of extraordinary voluminous volcanic activity in the region that is now Siberia. The eruptions could have clouded the atmosphere, acidified the oceans, and disrupted the food chain. Another hypothesis relates the mass extinction to a huge meteorite impact.

TAKE-HOME MESSAGE

Breakup of the late Precambrian supercontinent ushered in the Paleozoic. During the Paleozoic, shallow seas covered continents, at times, and life diversified and moved onto land. As the eon ended, collisions led to the assembly of Pangaea.

QUICK QUESTION What event, indicated by the fossil record, marks the end of the Paleozoic?

11.5 The Mesozoic Era: When Dinosaurs Ruled

The Early and Middle Mesozoic Era (Triassic–Jurassic Periods, 252–145 Ma)

Paleogeography Pangaea, the supercontinent formed at the end of the Paleozoic, existed for about 100 million years, until rifting commenced during the Triassic and Early Jurassic Periods and the supercontinent began to break up. During the Early Jurassic, rifting succeeded in splitting North America from Europe and Africa. At this time, the Mid-Atlantic Ridge formed and the North Atlantic Ocean started to grow (**Fig. 11.14**).

According to the record of sedimentary rocks, the Earth had a warm climate during the Triassic and Early Jurassic. Large areas of North America's interior were nonmarine environments in which red sandstones and shales, such as those now exposed in the spectacular cliffs of Zion National Park, were deposited. By the Middle Jurassic, the climate cooled and sea level began to rise. Eventually, a shallow sea submerged much of the Rocky Mountain region of the western United States. Toward the end of the Jurassic, the climate warmed once again.

On the western margin of North America, convergent-margin tectonics became the order of the day. Beginning with Late Permian and continuing through Mesozoic time, subduction generated volcanic island arcs. Over time, as intervening oceanic lithosphere was consumed, several volcanic arcs, microcontinents, and oceanic plateaus collided with western North America. Accretion (attachment) of these crustal fragments added land to the continent. Because these fragments formed elsewhere, not originally on or adjacent to the continent, geologists call them *exotic terranes*. Near the end of the Jurassic, a major continental volcanic arc, the Sierran arc, developed on the western margin of North America itself—you'll learn more about this arc later.

Did you ever wonder... when the dinosaurs lived?

Life Evolution During the early Mesozoic Era, numerous new plant and animal species appeared, filling the ecological niches left vacant by the Late Permian mass extinction. Reptiles swam in the oceans, and new kinds of corals became the predominant reef builders. On land, gymnosperms and reptiles diversified, and the Earth saw its first turtles and flying reptiles. And at the end of the Triassic Period, the first true dinosaurs evolved. Dinosaurs differed from other reptiles in that their legs were positioned under their bodies rather than off to the sides, and they were probably warm-blooded.

FIGURE 11.14 Pangaea began to break up in the Triassic, and by Jurassic time a narrow North Atlantic Ocean existed.

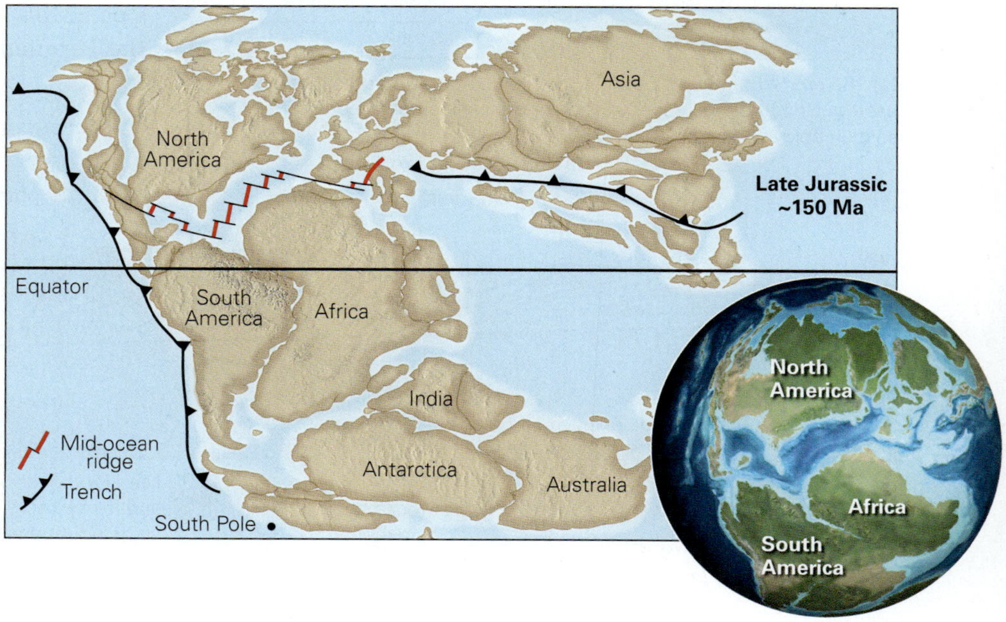

FIGURE 11.15 During the Jurassic, giant dinosaurs roamed the land. This painting shows several species.

FIGURE 11.16 Cretaceous paleogeography.

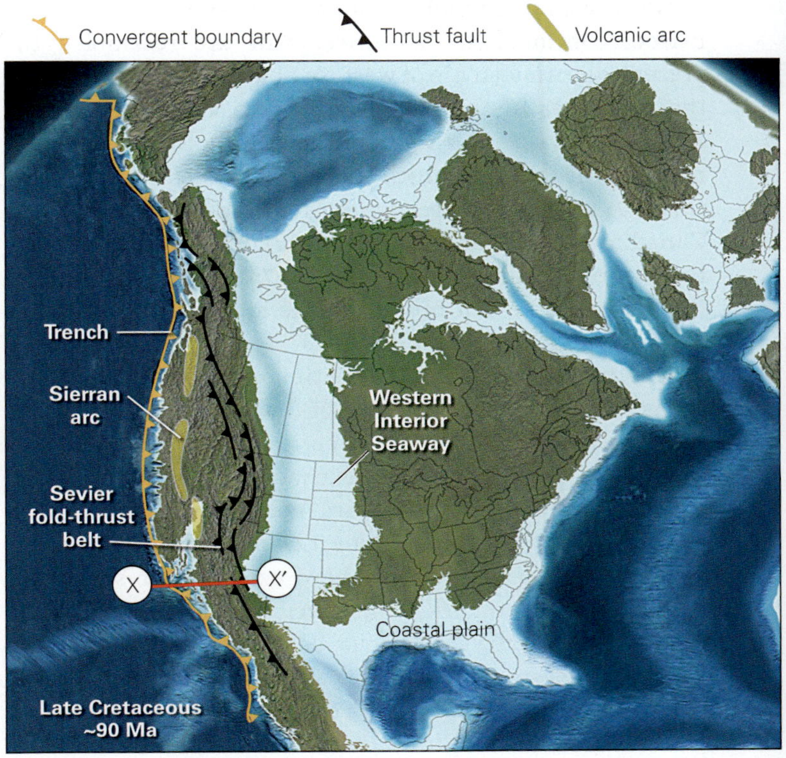

(a) A long seaway flooded the western interior.

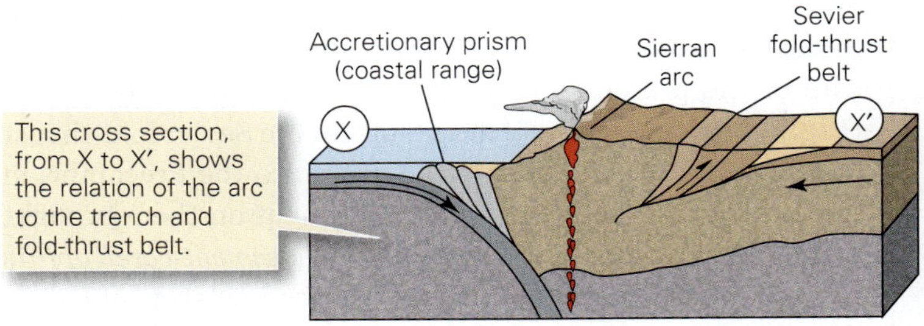

(b) In Late Cretaceous, a continental volcanic arc formed. A fold-thrust belt formed to the east, as did a transcontinental seaway.

SEE FOR YOURSELF...

ROCKY MOUNTAIN FRONT, COLORADO

Latitude
39°46′2.32″ N

Longitude
105°13′45.35″ W

Zoom to 8 km (~5 miles) and look obliquely.

We see the steep face of the Rocky Mountains in Colorado. These mountains were uplifted during the Laramide orogeny. During the event, reactivation of large faults thrust Precambrian rocks up and caused overlying Paleozoic strata to fold.

By the end of the Jurassic Period, gigantic sauropod dinosaurs (weighing up to 100 tons), along with other familiar examples such as stegosaurus, thundered across the landscape, and the first feathered birds, such as archaeopteryx, took to the skies (**Fig. 11.15**). The earliest ancestors of mammals appeared at the end of the Triassic, in the form of small, rat-like creatures.

The Late Mesozoic Era (Cretaceous Period, 145–66 Ma)

Paleogeography During the Cretaceous Period, the Earth's climate remained very warm, and sea level rose significantly, reaching levels that had not been attained for the previous 200 million years. Great shallow seas flooded most of the continents (**Fig. 11.16**). In fact, during the latter part of the Cretaceous, a shark could have swum from the Gulf of Mexico to the Arctic Ocean, or across much of western Europe.

Why was sea level so high and the climate so warm during the Cretaceous? Geologists have determined that seafloor-spreading rates may have been as much as three times faster during the Cretaceous than they are today. As a result, more of the oceanic crust was younger and warmer than it is today, and since young seafloor lies at a shallower depth than does older seafloor (due to isostasy; see Interlude D), Cretaceous mid-ocean ridges occupied more volume than they do today, and thus displaced seawater. Also during the Cretaceous, huge submarine plateaus formed from basalts erupted at hot-spot volcanoes—plateaus also displaced seawater and thus also contributed to sea-level rise. Formation of these plateaus may have been due to mantle upwelling in particularly large plumes, known as **superplumes**. The occurrence of voluminous volcanic activity during the Cretaceous likely released CO_2, a greenhouse gas, whose presence in the atmosphere trapped heat. Rising temperatures caused seawater to expand and polar ice sheets to melt, both of which would make sea level go up even more.

In western North America, the *Sierran arc*, a large continental volcanic arc that initiated at the end of the Jurassic Period, continued to be active. This arc resembled the present-day Andean arc of western South America. Although the volcanoes of the Sierran arc have long since eroded away, we can see their roots in the form of the plutons that now constitute the granitic batholith of the Sierra Nevada range. A fold-thrust belt, whose remnants crop out in the Canadian Rockies and in western Wyoming, formed to the the east of the Sierran arc (**Fig. 11.17a**). Geologists refer to the deformation that produced this fold-thrust belt as the *Sevier orogeny*.

The breakup of Pangaea continued through the Cretaceous Period, with the opening of the South Atlantic Ocean and the separation of South America and Africa from Antarctica and Australia. India broke away from Gondwana and headed rapidly northward toward Asia (**Fig. 11.17b**). Passive-margin basins developed along the margins of the newly formed oceans—these basins have existed ever since, and have filled with immensely thick accumulations of sediments.

At the end of the Cretaceous, deformation in the western United States swept eastward, perhaps due to a decrease in the angle of subduction and/or to the collision of an oceanic plateau with the margin of the continent. Slip occurred on large reverse faults in the region of what is now Wyoming, Colorado, eastern Utah, and northern Arizona. This event, which geologists call the *Laramide orogeny*, generated the structure of the present Rocky Mountains in the United States (**Fig. 11.17c**). In contrast to the faults of the Sevier fold-thrust belt, Laramide faults penetrated deep into the Precambrian basement of the continent, and movement on them brought Precambrian rocks up to the surface in *basement-cored uplifts* (**Fig. 11.17d**). As the basement rose, overlying layers of Paleozoic strata warped into large monoclines, folds whose shape resembles the drape of a carpet over a step.

Life Evolution In the seas of the Cretaceous world, modern fish appeared. In contrast with earlier fish, modern fish had short jaws, rounded scales, symmetrical tails, and specialized fins. Huge swimming reptiles and gigantic turtles (with shells up to 4 m across) preyed on the fish. On land, cycads largely vanished, and angiosperms (flowering plants), including hardwood trees, began to compete successfully with conifers for dominance of the forest. Dinosaurs reached their peak of success at this time, inhabiting almost all environments on Earth. Social herds of grazing dinosaurs roamed the plains, preyed on by the fearsome *Tyrannosaurus rex* (a Cretaceous, not a Jurassic, dinosaur, despite what Hollywood says!). Pterosaurs, with wingspans of up to 11 m, soared overhead, and birds began to diversify. Mammals also diversified and developed larger brains and more specialized teeth, but for the most part, they remained small and rat-like.

Did you ever wonder...
if dinosaurs and humans lived at the same time?

The K-Pg (K-T) Boundary Event Even in the early days of paleontological research, researchers realized that the boundary between Mesozoic and Cenozoic represented a particularly profound change in global fossil assemblages. Until the 1980s, most geologists assumed that the faunal turnover took millions of years. But modern dating techniques indicate that the change happened very rapidly, at 66 Ma. The dinosaurs, which had ruled the planet for over 150 million years, simply vanished, along with 90% of plankton species in the ocean and up to 75% of plant species. This catastrophic mass extinction came to be known as the *K-T boundary event*, or *K-T mass extinction*—K stands for Cretaceous, and T for Tertiary. In recent years, due to changes in terminology used on the geologic time scale, geologists are alternatively referring to it as the **K-Pg boundary event**—Pg stands for Paleogene.

The cause of the K-Pg mass extinction remained a mystery until the late 1970s, when Walter Alvarez, an American geologist, and his colleagues examined a thin shale layer deposited exactly at a horizon representing the transition from Cretaceous to Paleogene time. They found that this shale, exposed on a cliff face near Gubbio, Italy, contained relatively high concentrations of iridium, an element that comes primarily from meteorites. Further study showed that similar shales occurred at the same stratigraphic level worldwide, and that the shale contained other unusual materials, such as tiny glass spheres formed by sudden melting and freezing of rock, and carbon from burned vegetation. The presence of these materials suggested that the shale was deposited immediately following the impact of a huge meteorite with the Earth (**Fig. 11.18a**). The impact caused so much destruction because it not only blasted huge quantities of debris into the sky, but it probably also generated 2-km-high tsunamis that inundated the shores of continents and generated a blast of hot air that set forests on fire. The blast and the blaze together could have ejected masses of debris into the atmosphere that would have caused months of perpetual darkness and winter-like cold. These conditions would cause photosynthesis to all but cease, and thus would break the food chain and trigger extinctions.

Geologists suggest that the meteorite responsible for the K-Pg boundary event landed on the northwestern coast of the Yucatán Peninsula in Mexico. Beneath the reefs and sediments of this tropical realm lies a 100-km-wide by 16-km-deep

FIGURE 11.17 Paleogeography in Late Cretaceous through Eocene time.

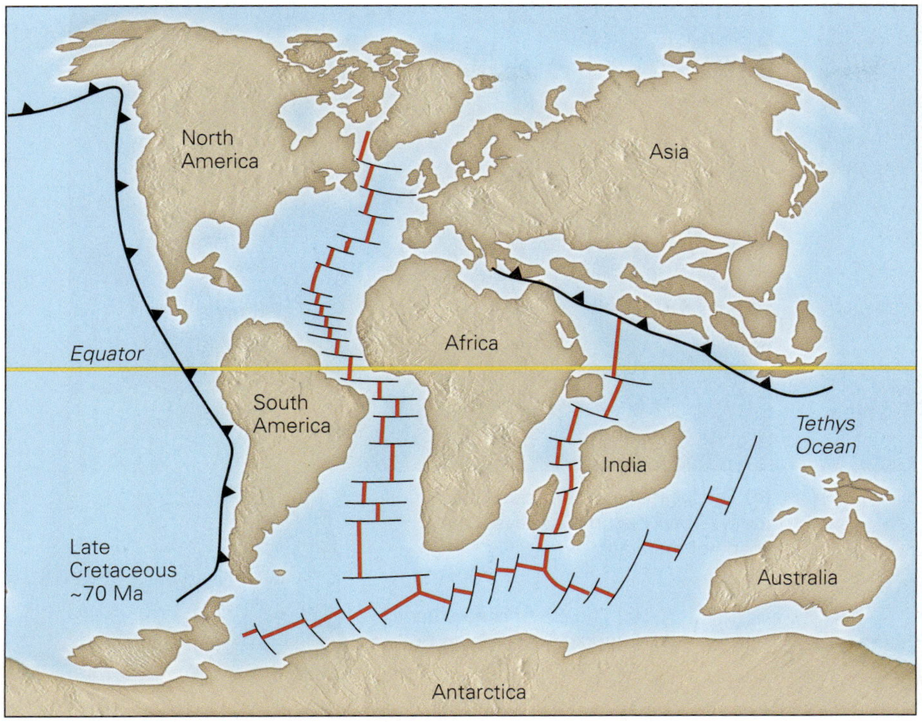

(a) By the Late Cretaceous Period, the Atlantic Ocean had formed, and India was moving rapidly northward to eventually collide with Asia.

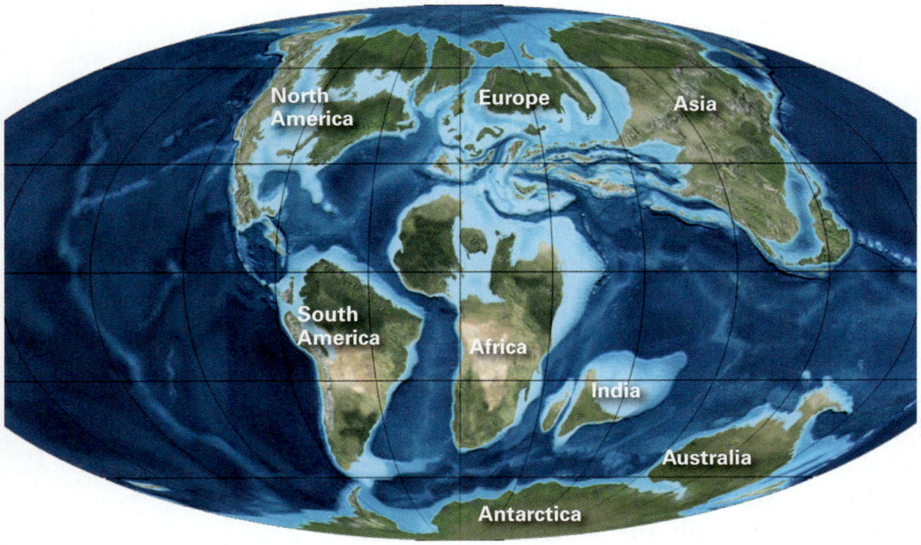

(b) In this Late Cretaceous paleogeographic reconstruction, southern Europe and Asia are beginning to form from a collage of many crustal blocks.

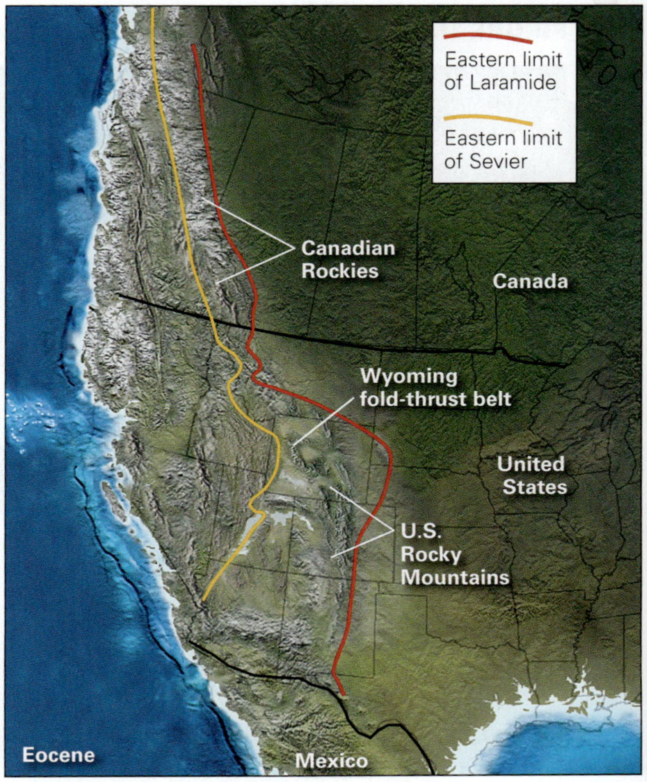

(c) During the Laramide orogeny, deformation shifted eastward in the United States, moving from the Sevier belt to the Rocky Mountains, and the style of deformation changed.

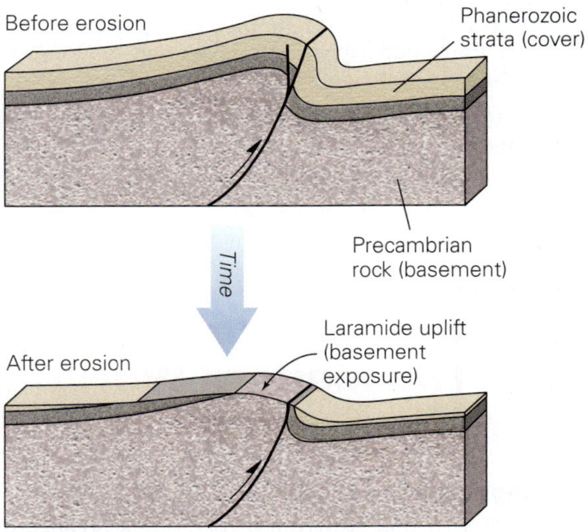

(d) The Laramide orogeny produced "basement-cored uplifts." In these, faults lifted up blocks of basement, causing the overlying strata to bend into a stair-step-like fold.

scar called the Chicxulub crater (**Fig. 11.18b, c**). A layer of glass spherules up to 1 m thick occurs at the K-T boundary in strata near the site, and radiometric dating indicates that igneous melts in the crater formed at 66 Ma. The discovery of this event has led geologists to speculate that other such collisions may have punctuated the path of life evolution throughout Earth history, and has led to modern-day efforts to track asteroids that pass close to the Earth.

FIGURE 11.18 The Cretaceous-Tertiary (K-T) impact. The event caused a mass extinction.

(a) An artist's image of the 13-km-wide object as it hit.

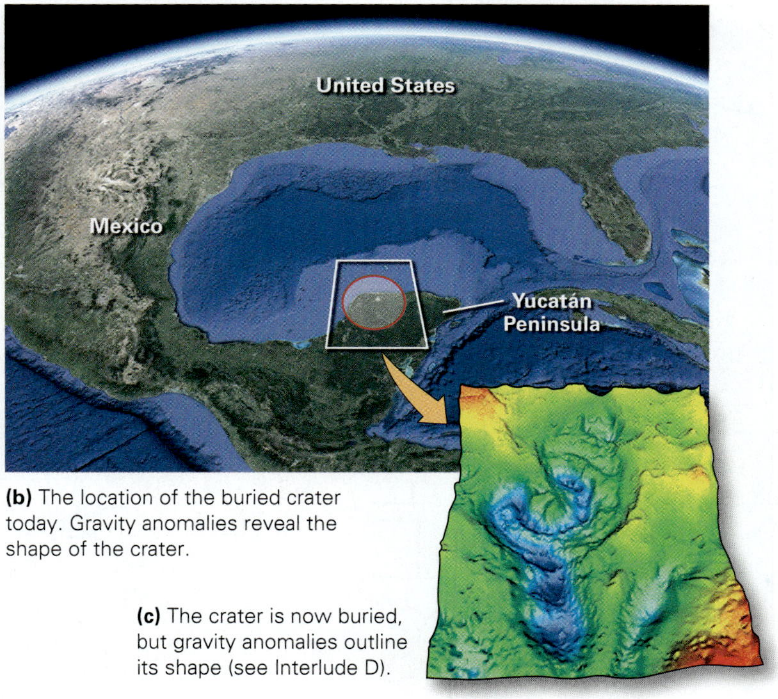

(b) The location of the buried crater today. Gravity anomalies reveal the shape of the crater.

(c) The crater is now buried, but gravity anomalies outline its shape (see Interlude D).

> ***TAKE-HOME MESSAGE***
>
> During the Mesozoic, dinosaurs roamed all continents. In North America, a convergent boundary formed along the west coast. At first a volcanic arc formed, beneath which the granite of the Sierra Nevada formed. Later, deformation moved east, causing uplift of the Rocky Mountains. Mesozoic rifting broke apart Pangaea and produced the Atlantic Ocean. A huge meteorite impact marks the K-Pg (K-T) boundary event, the end of the era. Consequences of this impact may have caused the extinction of the dinosaurs.
>
> **QUICK QUESTION** Did all of today's continents break off of Pangaea at the same time?

11.6 The Cenozoic Era: The Modern World Comes to Be

Paleogeography During the last 66 million years, the map of the Earth has continued to change, gradually producing the configuration of continents and plate boundaries we can see today. The final stages of Pangaea's breakup separated Australia from Antarctica and Greenland from North America. The Atlantic Ocean continued to grow because of seafloor spreading on the Mid-Atlantic Ridge, and thus the Americas have moved progressively away from Europe and Africa. Meanwhile, the continents that once constituted Gondwana drifted northward as the intervening Tethys Ocean was consumed by subduction (see Fig. 11.17). Collisions of the former Gondwana continents with the southern margins of Europe and Asia resulted in the formation of the largest orogenic belt on Earth today, the **Alpine-Himalayan chain** (**Fig. 11.19**). India and a series of intervening volcanic island arcs and microcontinents collided with Asia to form the Himalayas and the Tibetan Plateau to the north, while Africa along with some volcanic island arcs and microcontinents collided with Europe to produce the Alps.

As the Americas moved westward, convergent boundaries evolved along their western margins. In South America, convergent-boundary activity built the Andes, still an active orogen. In North America, convergent-boundary activity continued without interruption until about 40 Ma (the Eocene Epoch). Then, because of the rearrangement of plates off the western shore of North America, a transform boundary gradually replaced the convergent boundary. When this happened, the Laramide orogeny ceased;

> ***SEE FOR YOURSELF...***
>
> 
>
> **BASIN AND RANGE RIFT, UTAH**
>
> **Latitude**
> 39°15′1.83″ N
>
> **Longitude**
> 114°38′32.10″ W
>
> Zoom to 250 km (~155 miles) and look down.
>
> In this region of the Cenozoic Basin and Range rift, darker bands are fault-block mountains, whereas lighter areas are sediment-filled basins. White areas are evaporites, from dried-up lakes.

FIGURE 11.19 The two main active continental orogenic systems on the Earth today. The Alpine-Himalayan system formed when Africa, India, and Australia collided with Asia (inset). The Cordilleran and Andean systems reflect the consequences of convergent-boundary tectonism along the eastern Pacific Ocean.

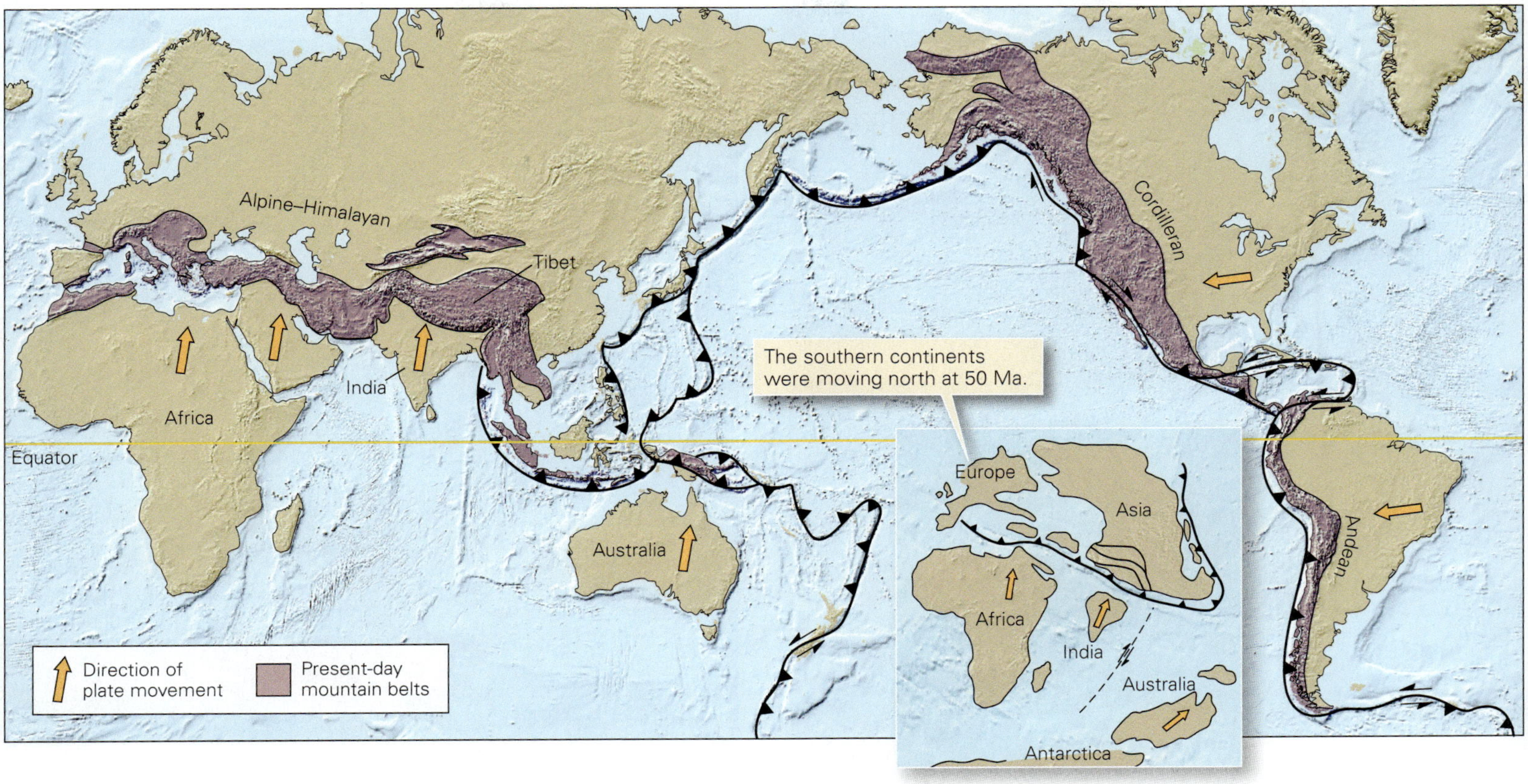

the San Andreas fault system formed along the western coast of the United States and the Queen Charlotte fault system off the coast of Canada. Along these strike-slip faults today, the Pacific Plate moves northward with respect to North America at a rate of about 6 cm per year. In the western United States, convergent-boundary tectonics continues only in Washington, Oregon, and northern California, where subduction of the Juan de Fuca Plate generates the volcanism of the Cascade volcanic chain.

As convergent tectonics ceased in the western United States south of the Cascades, the region began to undergo extension in a roughly east-west direction, and by 25 Ma, the **Basin and Range Province**, a broad continental rift, had been established. It has continued to grow ever since, stretching a portion of the western United States to twice its original width (**Fig. 11.20**). The Basin and Range Province, as its name suggests, encompasses long, narrow mountain ranges separated from each other by flat, sediment-filled basins. This topography formed when the crust of the region was broken up by normal faults (see Chapter 9), and blocks of crust in the hanging wall of these faults slipped down, tilting in the process. Crests of the tilted blocks form the ranges, and the depressions between them, which rapidly filled with sediment eroded from the ranges, became basins. The Basin and Range Province terminates just north of the Snake River Plain. This plain, underlain by products of extrusive volcanism, marks the track of the hot spot that now lies beneath Yellowstone National Park.

Recall that during the Cretaceous Period, the Earth was relatively warm and sea level rose to submerge extensive areas of continents. Starting in the Eocene, in contrast, the global climate rapidly became cooler, and by the early Oligocene Epoch, Antarctic glaciers reappeared for the first time since the Triassic, leading to sea-level fall. In Early Miocene, warming led to thawing of the Antarctic ice cap, but by the Late Miocene Epoch, colder temperatures led to reglaciation of Antarctica. About 2.5 Ma, the Isthmus of Panama formed, separating the Atlantic completely from the Pacific, changing the configuration of oceanic currents, and perhaps leading to freezing over of the Arctic Ocean.

During the overall cold climate of the past 2.6 million years, continental glaciers have expanded and retreated across northern continents at least 20 times. Geologists, therefore refer to this time as the **Pleistocene Ice Age**

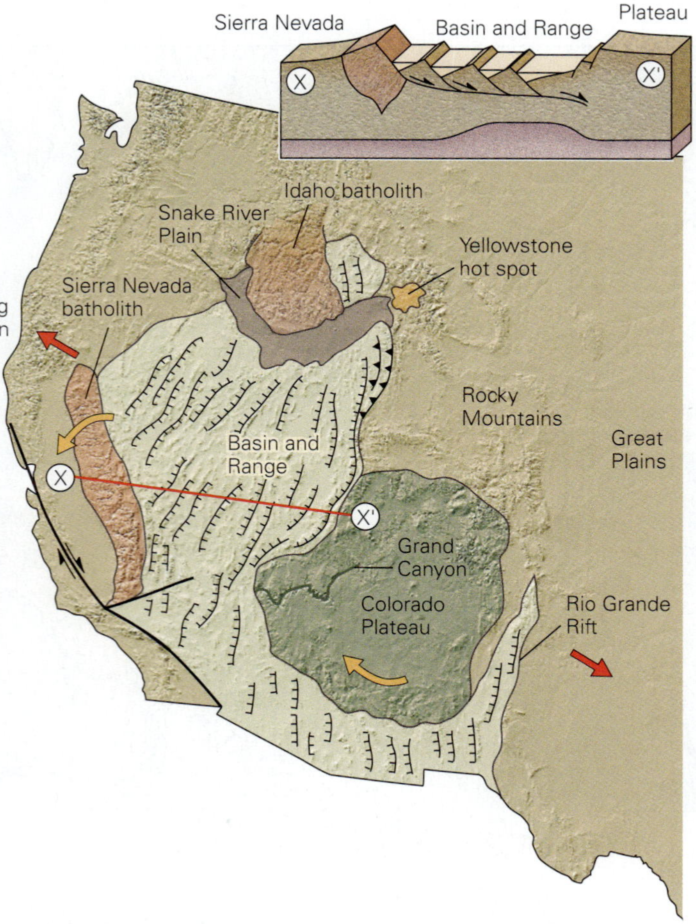

FIGURE 11.20 The Basin and Range Province is a rift. The inset shows a cross section along the red line.

FIGURE 11.21 The maximum advance of the Pleistocene ice sheet in North America.

(**Fig. 11.21**). Each time the glaciers grew, sea level fell so much that the continental shelf became exposed to air, and at times, a *land bridge* formed across the Bering Strait, west of Alaska, providing routes for people from Asia to migrate into North America. A partial land bridge also formed between southeast Asia and Australia, making human migration to Australia easier. Erosion and deposition by the glaciers yielded much of the landscape we see today in northern temperate regions. About 11,000 years ago, the climate warmed, and we entered the *interglacial* time interval we are still experiencing today. The time since the last glaciation is the **Holocene Epoch** (see Chapter 18).

Life Evolution When the skies finally cleared in the wake of the K-Pg boundary catastrophe, plant life recovered, and soon forests of both angiosperms and gymnosperms grew. Grasses, which first appeared in the Cretaceous, spread across the plains in temperate and subtropical climates by the middle of the Cenozoic Era, transforming them into vast grasslands. The dinosaurs, except for their descendants the birds, were gone for good. Mammals rapidly diversified into a variety of forms to take the dinosaurs' place. In fact, most modern groups of mammals originated at the beginning of the Cenozoic Era, giving this time the nickname *Age of Mammals*. During the latter part of the era, huge mammals (such as mammoths, giant beavers, giant bears, and giant sloths) appeared, but these became extinct during the past 10,000 years, probably because of hunting by humans.

It was during the Cenozoic that our own ancestors first appeared. Ape-like primates diversified in the Miocene Epoch (about 20 Ma), and the first human-like primate appeared at about 4 to 5 Ma, followed by the first members of the human genus, *Homo*, at about 2.8 Ma. Fossil evidence, primarily from Africa, indicates that *Homo erectus*, capable of making stone axes, appeared about 1.8 Ma. According to the fossil record, modern people appeared about 200,000 years ago, initially sharing the planet with two other species of the genus *Homo*, the Neanderthals and the Denisovans. When the Neanderthals and Denisovans died out more than 40,000 years ago, *Homo sapiens* remained as the only human species on Earth.

As summarized in **Geology at a Glance** (pp. 386–387), Earth's history reflects the complex consequences of plate interactions, sea-level changes, atmospheric changes, life evolution, and even meteorite impact. In the past few millennia, humans have had a huge effect on the planet, causing changes significant enough to be obvious in the geologic record of the future. In fact, geologists now informally refer to the more recent portion of the Holocene, during which human activities have had a major impact on the Earth System, as the **Anthropocene**. We'll pick up the thread of this story in Chapter 19, where we develop the concept of global change.

TAKE-HOME MESSAGE

During the Cenozoic, the mountain belts of today rose, and modern plate boundaries became established. Mammals diversified. During the Pleistocene, glaciers covered large areas of continents, and humans appeared.

QUICK QUESTION What interval of time does the Anthropocene refer to?

Another View The present-day Bahamas serve as an example of what the interior of the United States might have looked like during intervals of the Paleozoic. Shallow land areas were submerged and became the site of shallow-marine sedimentation.

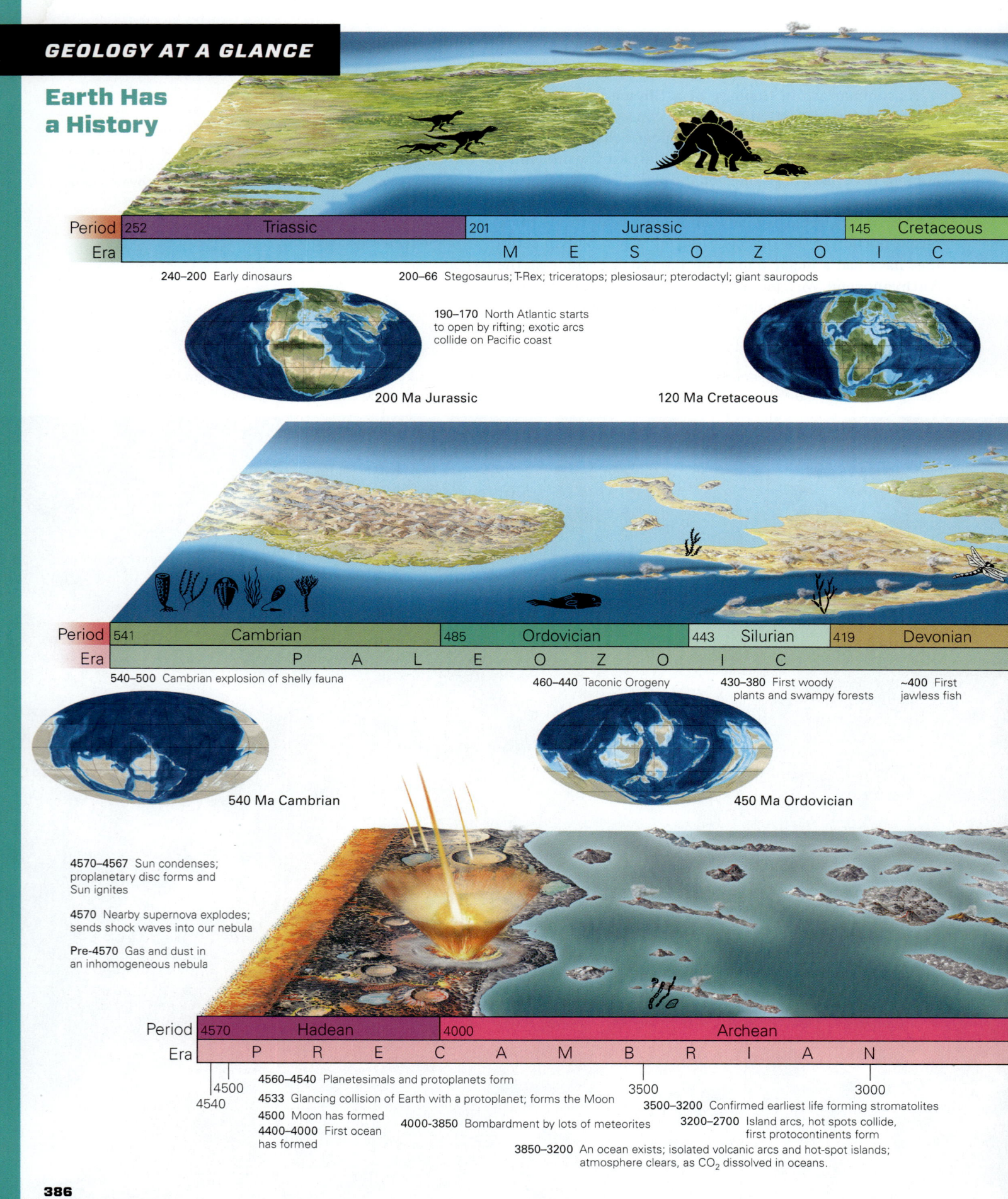

Earth Has a History

Period: 252 Triassic | 201 Jurassic | 145 Cretaceous

Era: M E S O Z O I C

240–200 Early dinosaurs

200–66 Stegosaurus; T-Rex; triceratops; plesiosaur; pterodactyl; giant sauropods

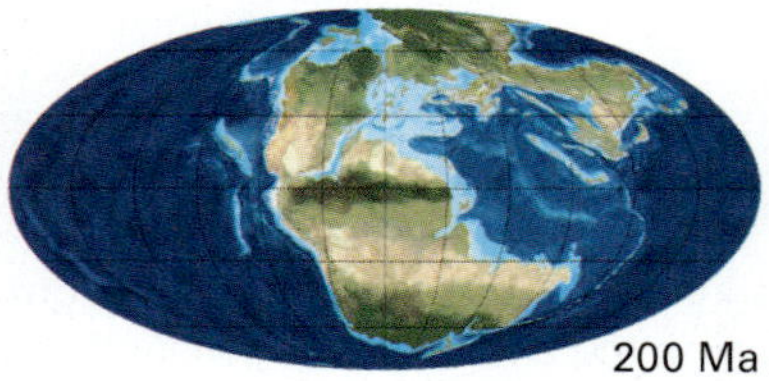

190–170 North Atlantic starts to open by rifting; exotic arcs collide on Pacific coast

200 Ma Jurassic

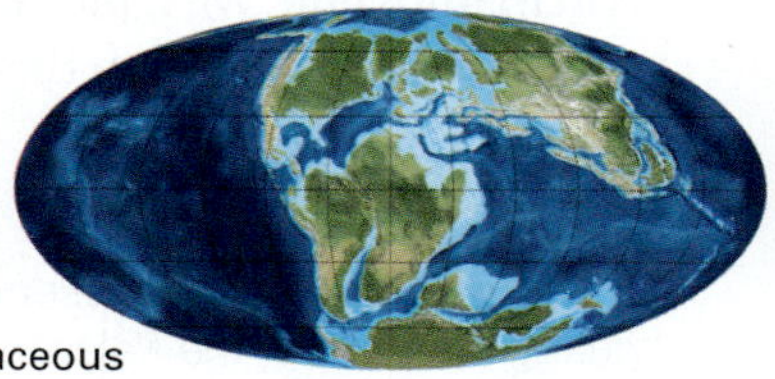

120 Ma Cretaceous

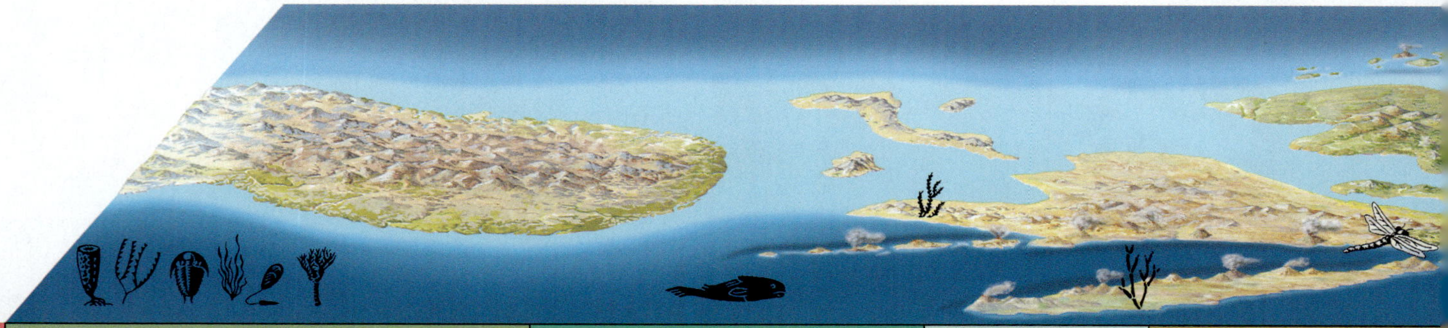

Period: 541 Cambrian | 485 Ordovician | 443 Silurian | 419 Devonian

Era: P A L E O Z O I C

540–500 Cambrian explosion of shelly fauna

460–440 Taconic Orogeny

430–380 First woody plants and swampy forests

~400 First jawless fish

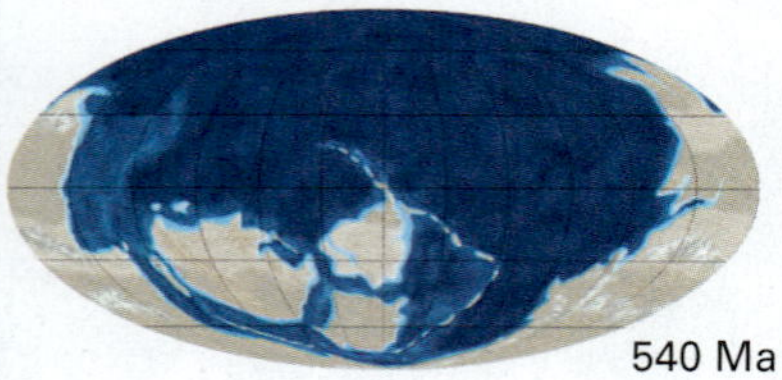

540 Ma Cambrian

450 Ma Ordovician

4570–4567 Sun condenses; proplanetary disc forms and Sun ignites

4570 Nearby supernova explodes; sends shock waves into our nebula

Pre-4570 Gas and dust in an inhomogeneous nebula

Period: 4570 Hadean | 4000 Archean

Era: P R E C A M B R I A N

4540 | 4500 | 3500 | 3000

4560–4540 Planetesimals and protoplanets form

4533 Glancing collision of Earth with a protoplanet; forms the Moon

4500 Moon has formed

4400–4000 First ocean has formed

4000-3850 Bombardment by lots of meteorites

3850–3200 An ocean exists; isolated volcanic arcs and hot-spot islands; atmosphere clears, as CO_2 dissolved in oceans.

3500–3200 Confirmed earliest life forming stromatolites

3200–2700 Island arcs, hot spots collide, first protocontinents form

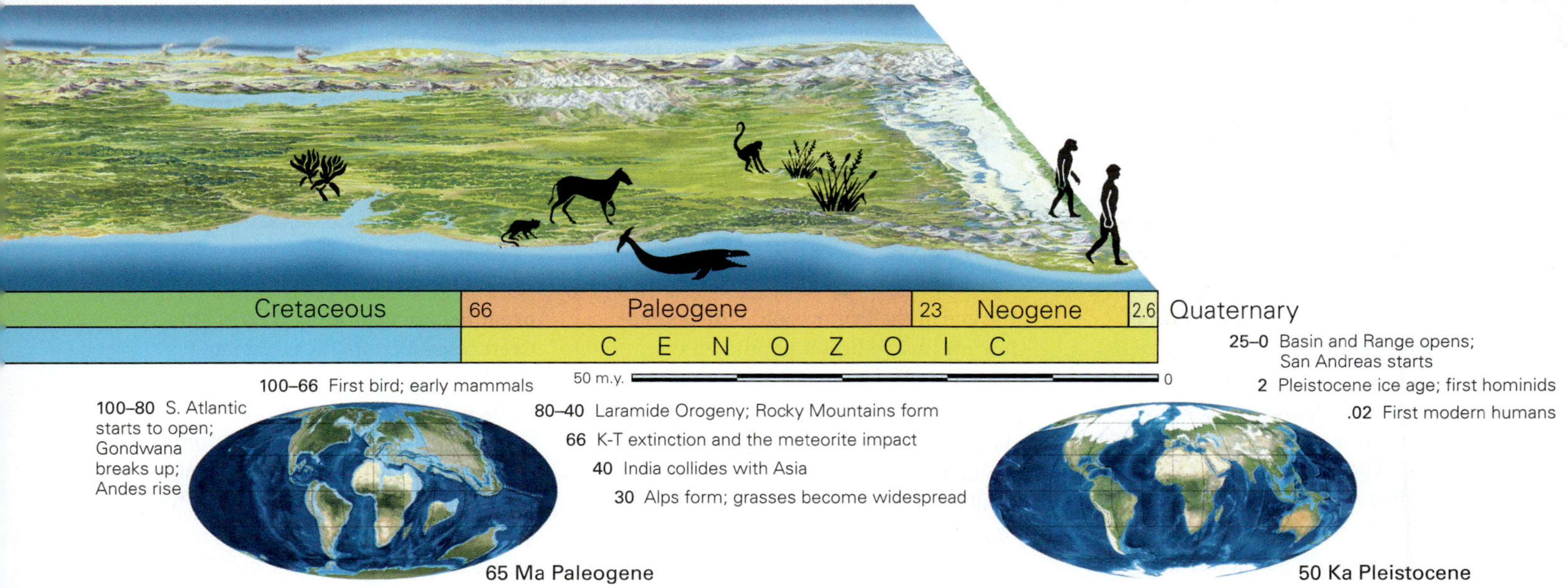
Cretaceous
66 Paleogene
23 Neogene
2.6 Quaternary
C E N O Z O I C
50 m.y.
0
25–0 Basin and Range opens; San Andreas starts
2 Pleistocene ice age; first hominids
.02 First modern humans
100–66 First bird; early mammals
100–80 S. Atlantic starts to open; Gondwana breaks up; Andes rise
80–40 Laramide Orogeny; Rocky Mountains form
66 K-T extinction and the meteorite impact
40 India collides with Asia
30 Alps form; grasses become widespread
65 Ma Paleogene
50 Ka Pleistocene

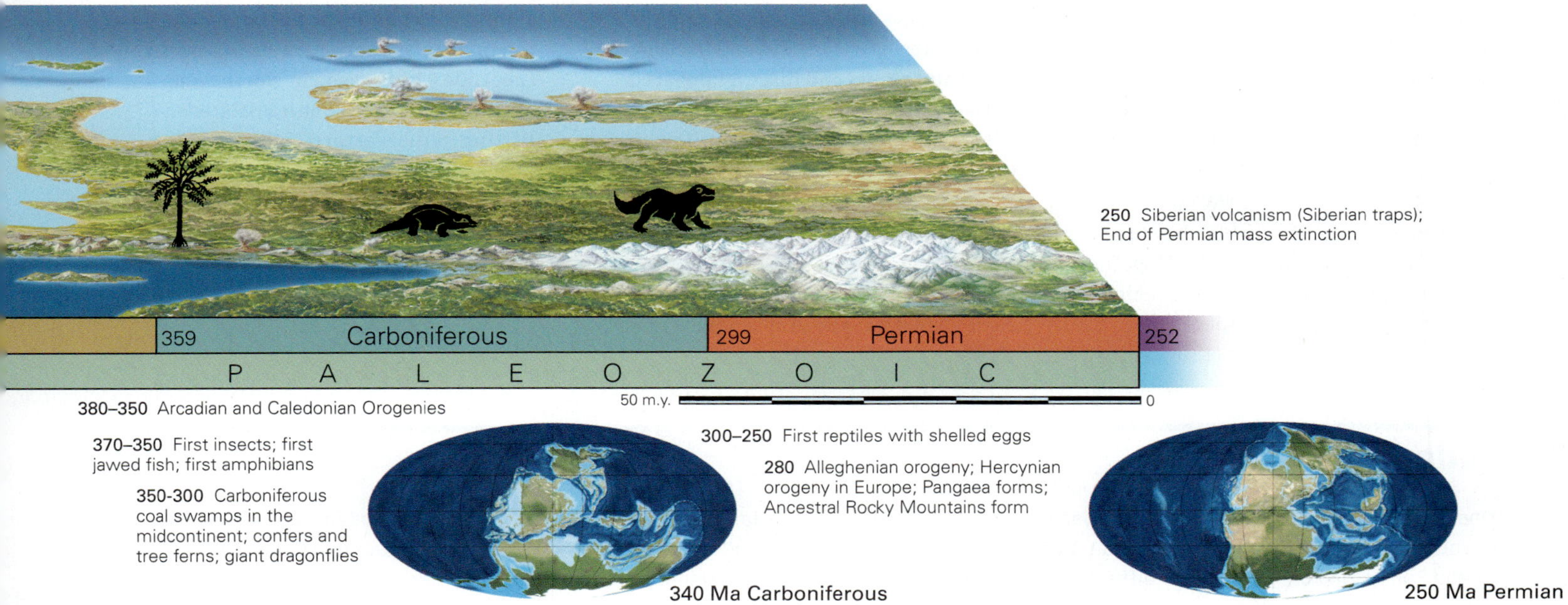
250 Siberian volcanism (Siberian traps); End of Permian mass extinction
359 Carboniferous
299 Permian
252
P A L E O Z O I C
50 m.y.
0
380–350 Arcadian and Caledonian Orogenies
370–350 First insects; first jawed fish; first amphibians
350-300 Carboniferous coal swamps in the midcontinent; confers and tree ferns; giant dragonflies
300–250 First reptiles with shelled eggs
280 Alleghenian orogeny; Hercynian orogeny in Europe; Pangaea forms; Ancestral Rocky Mountains form
340 Ma Carboniferous
250 Ma Permian

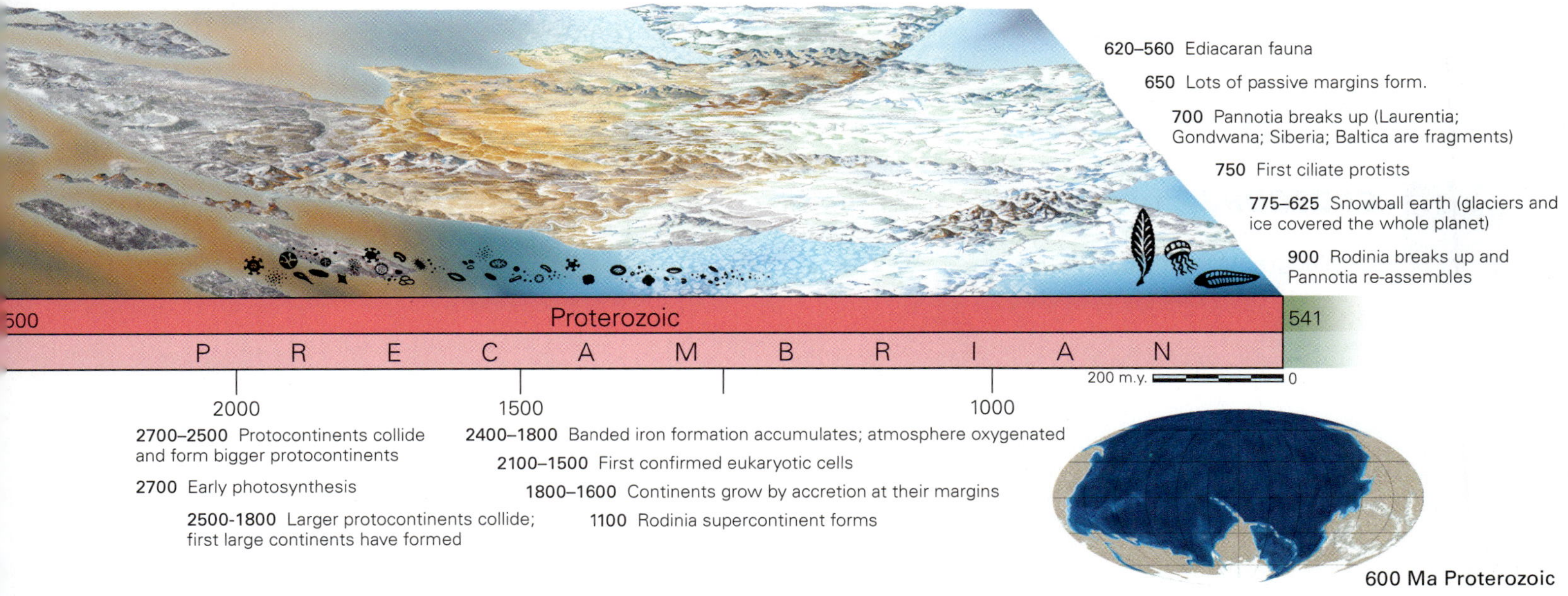
620–560 Ediacaran fauna
650 Lots of passive margins form.
700 Pannotia breaks up (Laurentia; Gondwana; Siberia; Baltica are fragments)
750 First ciliate protists
775–625 Snowball earth (glaciers and ice covered the whole planet)
900 Rodinia breaks up and Pannotia re-assembles
500
Proterozoic
541
P R E C A M B R I A N
200 m.y.
0
2000
1500
1000
2700–2500 Protocontinents collide and form bigger protocontinents
2700 Early photosynthesis
2500-1800 Larger protocontinents collide; first large continents have formed
2400–1800 Banded iron formation accumulates; atmosphere oxygenated
2100–1500 First confirmed eukaryotic cells
1800–1600 Continents grow by accretion at their margins
1100 Rodinia supercontinent forms
600 Ma Proterozoic

Chapter 11 Review

Chapter Summary

- The Earth formed about 4.54 billion years ago. There is no rock record of its first half billion years, the Hadean Eon, even though the planet probably had a solid skin of ultramafic rock. It may even have hosted liquid water for part of this time, perhaps because it was intensely bombarded by meteorites.
- The Archean Eon began about 4.0 Ga, with the formation of the first rocks that still survive. Once plate tectonics started, early continental crust assembled out of volcanic arcs and hot-spot volcanoes that were too buoyant to subduct. The atmosphere contained very little oxygen, and the first life forms—bacteria and archaea—appeared.
- In the Proterozoic Eon, which began at 2.5 Ga, Archean cratons sutured together to form large cratons. Photosynthesis added oxygen to the atmosphere. By the end of the Proterozoic, soft-bodied marine invertebrates populated the planet, and continental crust had accumulated to form a supercontinent.
- As the Paleozoic Era began, rifting yielded several separate continents. Sea level rose and fell, depositing sequences of strata in continental interiors. Continents coalesced again to form another supercontinent, Pangaea. Early Paleozoic evolution produced many invertebrates with shells, and jawless fish. Land plants and insects appeared in the middle Paleozoic, and by the end of the eon, there were land reptiles and gymnosperm trees.
- In the Mesozoic Era, Pangaea broke apart and the Atlantic Ocean formed. Convergent-boundary tectonics dominated along the western margin of North America, and dinosaurs became prominent land animals. During the Cretaceous Period, the continents flooded, and angiosperms appeared along with modern fish. A huge mass-extinction event wiped out the dinosaurs at the end of the Cretaceous, probably due to the impact of a large meteorite.
- In the Cenozoic Era, the collision of Africa and India with Asia and Europe formed the Alpine-Himalayan orogen. Convergent tectonics persisted along the margin of South America, forming the Andes, but ceased in North America as the San Andreas Fault initiated. Rifting in the western United States produced the Basin and Range Province. Mammals filled niches left vacant by the dinosaurs, and the human genus, Homo, appeared and evolved through the Pleistocene Ice Age.

Guide Terms

Alpine-Himalayan chain (p. 382)
Anthropocene (p. 385)
Archean Eon (p. 367)
banded iron formation (BIF) (p. 372)
Basin and Range Province (p. 383)
Cambrian explosion (p. 375)
craton (p. 368)
differentiation (p. 366)
Gondwana (p. 374)
great oxygenation event (p. 371)
Hadean Eon (p. 366)
Holocene Epoch (p. 384)
K-Pg boundary event (p. 380)
Laurentia (p. 374)
Pangaea (p. 376)
Phanerozoic Eon (p. 372)
Pleistocene Ice Age (p. 383)
Proterozoic Eon (p. 369)
snowball Earth (p. 372)
stromatolite (p. 369)
superplume (p. 379)

GEOTOURS *THIS CHAPTER'S GEOTOUR EXERCISE (K) FEATURES:*

- Paleography of the Earth

Review Questions

1. Why are there no whole rocks on Earth that yield isotopic dates older than 4 billion years?
2. Why can we find mineral grains that are older than the oldest whole rocks?
3. Describe the condition of the crust, atmosphere, and oceans during the Hadean Eon.
4. Describe how the first continental crust might have formed, and when plate tectonics started.
5. How did the atmosphere and tectonic conditions change during the Proterozoic Eon?
6. What evidence do we have that the Earth nearly froze over during the Proterozoic Eon?
7. Did supercontinents form in the Proterozoic?
8. How did the Cambrian explosion of life change the nature of the living world?
9. Were there multicellular organisms before the Cambrian?
10. How did the Alleghanian orogeny and Ancestral Rockies event affect North America?
11. What major types of organisms first appeared during the Paleozoic, and in what sequence?
12. What supercontinent formed at the end of the Paleozoic, and what ocean formed when it broke apart?
13. Describe the plate-tectonic conditions that led to the formation of the Sierran arc and the Sevier thrust belt. How did deformation during the Laramide orogeny differ from that of the Sierran/Sevier orogeny?
14. What life forms appeared during the Mesozoic?
15. What may have caused the flooding of the continents during the Cretaceous Period?
16. What could have caused the K-Pg (K-T) mass extinctions?
17. What might have caused the mass extinction at the end of the Paleozoic?
18. What continents formed as a result of the breakup of Pangaea?
19. What caused the Himalayas and the Alps to form?
20. What major tectonic provinces formed in the western United States during the Cenozoic?
21. What major climatic and biologic events happened during the Pleistocene?

On Further Thought

22. Geologists have concluded that 80% to 90% of Earth's continental crust had formed by 2.5 Ga. But if you look at a geological map of the world, you find that only about 10% of the Earth's continental crustal surface is labeled "Precambrian." Why?
23. In the eastern United States, large rift basins formed. Strata in these basins contain dinosaur fossils. When did the basins form, and what tectonic event are they associated with? Are they older or younger than the fold of the Appalachians?

Online Resources

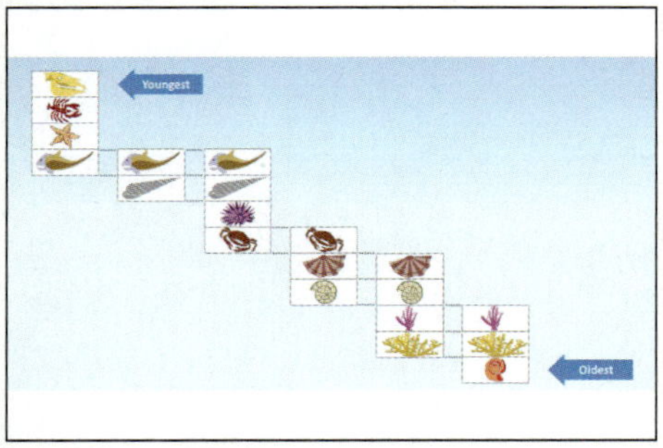

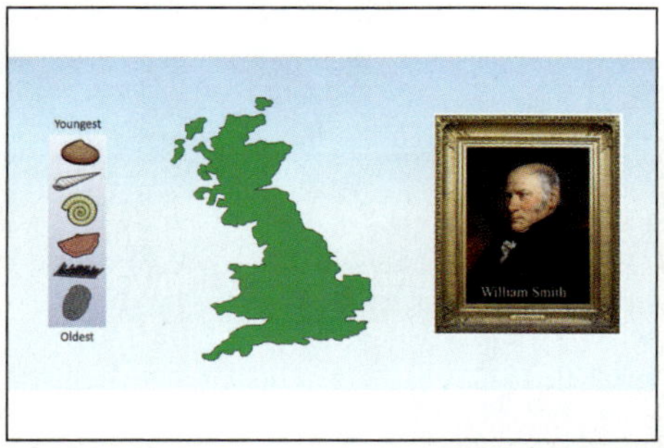

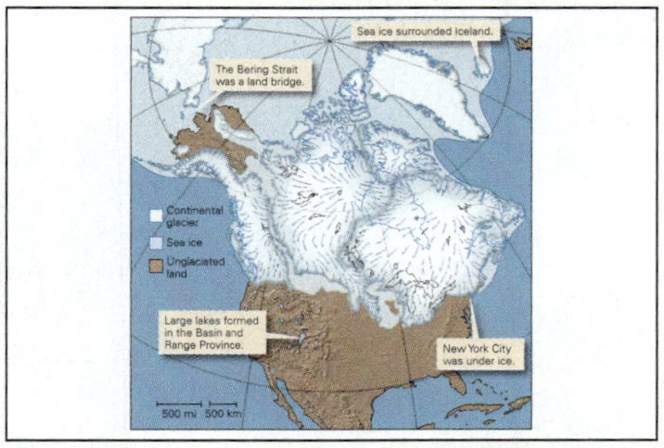

Animations

This chapter features interactive games testing your knowledge of biostratigraphy.

Assessment

This chapter features understanding and ranking exercises about Earth throughout history, including how the Earth and climate have evolved over time.

An exposed fracture surface reveals a weathered coating of malachite, a beautiful green mineral sometimes used for jewelry. Malachite is also a copper ore mineral, meaning that its crystals contain a significant proportion of copper atoms. The copper we use for wires and coins comes from this and other ore minerals. The Earth provides many of the resources essential to society.

LEARNING OBJECTIVES

By the end of this chapter, you should understand...

1. what oil and gas are, how they form, where they come from, and how they are obtained.
2. the difference between conventional and unconventional fossil fuel resources.
3. how coal forms and is classified, obtained, and used.
4. where nuclear fuel comes from, and how nuclear power plants operate.
5. the nature of alternative energy sources, including hydrothermal, geothermal, and wind.

6. what an ore deposit is, where ores form, and how metals can be extracted from ores.
7. where nonmetallic resources (stone, gravel, salts) occur, and how they are used.
8. the challenges that society faces as to the sustainability of energy sources in the future.

CHAPTER 12

Riches in Rock: Energy and Mineral Resources

12.1 Introduction

The earliest humans were hunter-gatherers and needed only food and water to survive. But when people discovered that fire could be used for cooking and heating, weapons could facilitate hunting, shelters could make daily life more comfortable, and farming could make food supplies more reliable, their needs expanded. Specifically, people began to require **energy resources** (sources of heat and/or power) and **mineral resources** (materials from which metals and natural inorganic solids can be derived). Before the 18th century, energy resources were used only to warm homes and to drive pumps or millstones, and mineral resources were used only for constructing buildings or for making weapons and tools. But when industrialization took hold, society's demands increased dramatically, for people began to rely on a huge variety of energy-hungry machines and, eventually, motorized vehicles. It's no surprise that contemporary industrial societies use 100 times more resources per capita than did pre-industrial societies.

Where do people obtain energy and mineral resources? Most come from geologic materials or processes. That's why we discuss resources in a geology textbook. The first half of this chapter focuses on the nature and origin of energy

All the gold which is under or upon the Earth is not enough to give in exchange for virtue.

PLATO (Greek philosopher, ca. 428–347 B.C.E.)

resources. Our discussion begins with a focus on the most widely used energy sources of the past century (oil, gas, coal, and nuclear). Then we address alternative resources that are gradually providing a greater proportion of our needs (hydroelectric, wind, and solar, among others). The second half of the chapter considers both metallic and nonmetallic mineral resources. Citizens of the Earth should understand how resources form, why they occur where they do, how they can be extracted, and how their extraction and use impacts the environment, for this knowledge provides an essential context for facing sustainability challenges of the future. Searching for resources, and dealing with the consequences of their production and use, provides jobs for tens of thousands of geologists worldwide.

12.2 Sources of Energy in the Earth System

Physicists define **energy** as the capacity to do work. In other words, energy causes objects to move or materials to change. Where does the energy in the Earth System originally come from? There are several fundamental sources:

- *Energy directly from the Sun*: Solar energy, resulting from nuclear fusion reactions in the Sun, bathes the Earth's surface. It may be converted directly to electricity or used to heat water.
- *Energy directly from gravity*: The gravitational attraction of the Moon and Sun cause ocean *tides*, the daily up-and-down movement of the sea surface. Flow of water during tidal changes can drive turbines.
- *Energy involving both solar energy and gravity*: Solar radiation heats the air, which becomes buoyant and rises. As this happens, gravity causes cooler air to sink. The resulting air movement—wind—powers sails and windmills. Solar energy also evaporates water, which enters the atmosphere. Wind can blow this moisture over land. When water condenses, forms clouds, and rains on elevated land, it accumulates in streams that flow downhill in response to gravity. This moving water can drive waterwheels that power machines or generators that produce electricity.
- *Energy via photosynthesis*: Algae and green plants absorb solar energy and, through *photosynthesis*, produce sugar. From this sugar, they manufacture complex organic chemicals. Animals eat plants and produce more organic chemicals, as well as waste. For centuries humans have burned **biomass** (plants, wood, manure) to produce energy. When they burn, organic chemicals react with oxygen and break apart to produce carbon dioxide, water, and carbon (soot). These reactions release the energy stored in chemical bonds. More recently, chemists have developed processes to convert biomass into flammable liquids, including ethanol.
- *Energy from fossil fuels*: **Fossil fuels**, such as oil, natural gas, and coal, store solar energy that reached the Earth long ago. This solar energy was converted either directly or indirectly into organic matter. If this organic matter is buried, it can be preserved in rocks over geologic time.
- *Energy from inorganic chemical reactions*: Release of energy stored in chemical bonds does not always require the use of organic chemicals. The heat of a dynamite explosion or the electricity of hydrogen fuel cells, for example, comes from inorganic chemical reactions.
- *Energy from nuclear fission*: Atoms of radioactive elements can split into smaller pieces, a process called nuclear fission. During fission, a tiny amount of mass transforms into a large amount of energy, called *nuclear energy*. This type of energy runs nuclear power plants and nuclear submarines.
- *Energy from Earth's internal energy*: The Earth remains very hot inside—some of this internal energy dates from the birth of the planet, while some comes from later radioactive decay. Internal energy heats underground water, which can warm buildings at the surface or, when transformed to steam, can drive electrical generators.

During the course of civilization, people have relied on various sources of energy (**Fig. 12.1**). Prior to the industrial age, direct burning of wood and other biomass provided most of humanity's energy. But by the second half of the 19th century, deforestation had so depleted this resource, and energy needs had increased so dramatically, that other types of **fuel** (a transportable material that can burn or react to produce energy) came into use. In the next sections, we discuss commonly used energy sources, beginning with fossil fuels, the dominant type in use today.

TAKE-HOME MESSAGE

Energy comes from several sources, such as solar radiation, gravity, chemical reactions, radioactive decay, and Earth's internal heat. The chemicals in living organisms can be preserved in rocks as fossil fuels.

QUICK QUESTION What kinds of energy sources are available on the Moon?

FIGURE 12.1 The proportion of different sources of energy that people use has changed over time, and the amount of energy used almost continuously increases.

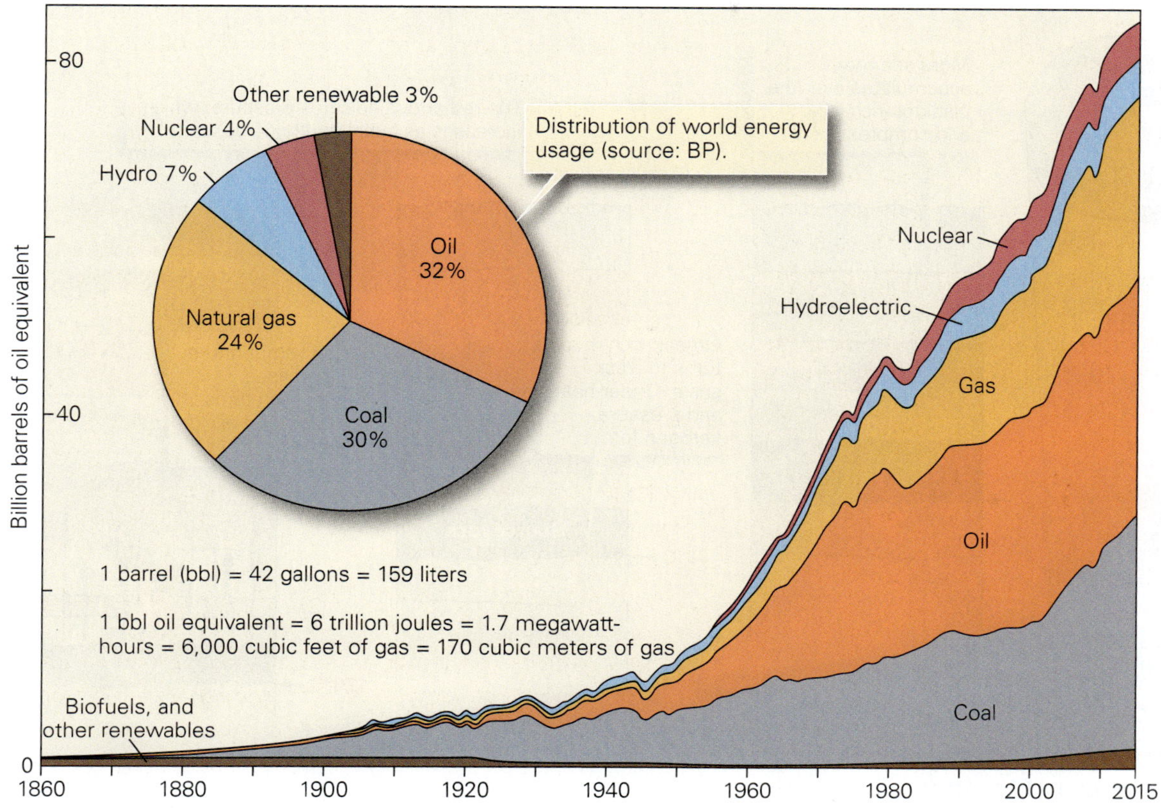

12.3 Introducing Hydrocarbon Resources

What Are Oil and Gas?

Oil and gas are **hydrocarbons**, chain-like or ring-like molecules composed exclusively of carbon and hydrogen atoms. Some hydrocarbons are gaseous and invisible, some resemble a watery liquid, some appear syrupy, and some are solid. The *viscosity* (ability to flow) and the *volatility* (ability to evaporate) of a hydrocarbon product depend on the size of its molecules. Products made of short chains of molecules tend to be less viscous (they can flow more easily) and more volatile (they evaporate more easily) than products composed of long chains, because the long chains tend to tangle up with each other. Thus, at room temperature, short-chain molecules occur in gaseous form (such as cooking gas), moderate-length-chain molecules occur in liquid form (such as gasoline and motor oil), and long-chain molecules occur in solid form (tar).

Industrialized societies today rely heavily on hydrocarbons for their energy needs. Why? Oil and gas have a high **energy density**, meaning that they provide a relatively large amount of energy per unit volume. For example, a liter of jet fuel provides about 67 times as much energy as a lead-acid battery of the same volume. So it's possible to fly an airplane halfway around the world on the jet fuel that it can carry—but the same plane would never get off the ground if it had to be powered by lead-acid batteries.

Hydrocarbon Generation in Source Rocks

News stories commonly imply, incorrectly, that oil and gas are derived from the carcasses of dinosaurs. In fact, the chemicals that make up oil and gas come mostly from *plankton*, a group of very tiny floating organisms including algae, protists, and microscopic animals. When plankton dies, it sinks to the floor of the lake or sea in which it lived. In "quiet" (nonflowing) water, the dead plankton can accumulate on the floor of the lake or sea, and if the water in the depositional environment does not contain much oxygen, the organic chemicals in the plankton can survive and mix with the clay that is also settling from the water. (If the depositional environment contains oxygen, the plankton will be consumed by other organisms, or will be oxidized, before it can accumulate.) The accumulated clay and organic chemicals become an *organic ooze*. Over time, organic ooze may become buried by more sediment, and when buried deeply enough, it undergoes lithification to become a black, organic shale. Organic shale contains the raw materials from which hydrocarbons form, so geologists refer to it as a **source rock**.

Did you ever wonder...
what the "fossils" in fossil fuels are?

At relatively low temperatures, buried organic matter decays, releasing *biogenic natural gas*. As organic shale is buried more deeply, it becomes warmer, for temperature increases with depth in the Earth. Chemical reactions take place in source rocks as they exceed 50°C, slowly

FIGURE 12.2 The formation of oil. The process begins when organic debris settles with sediment. As burial depth increases, heat and pressure transform the sediment into black shale in which organic matter becomes kerogen. At appropriate temperatures, kerogen becomes oil, which then seeps upward.

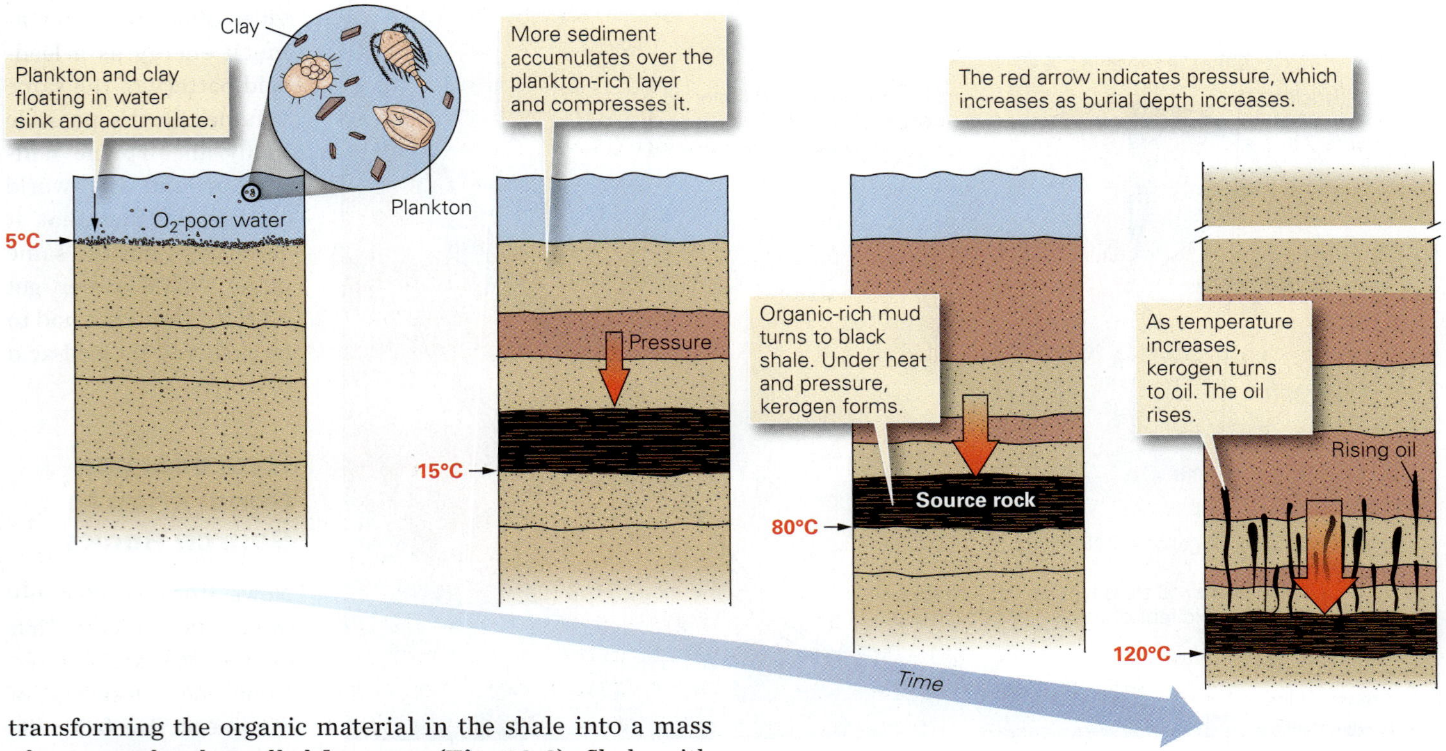

transforming the organic material in the shale into a mass of waxy molecules called **kerogen** (**Fig. 12.2**). Shale with more than 25% to 75% kerogen is called **oil shale**. If oil shale becomes buried even more deeply, warming to temperatures above 90°C, kerogen molecules break into smaller oil and gas molecules, a process known as **hydrocarbon generation**. At temperatures between 150° and 225°C, oil molecules break down into natural gas only, and at temperatures over 225°C, organic matter loses all its hydrogen and transforms into graphite (pure carbon). Thus, oil forms only at temperatures between 90° and 150°C, a range called the **oil window**. *Thermogenic natural gas* can form in the larger *gas window* (90° to 225°C). A typical value for the geothermal gradient, the rate at which temperature increases with depth, is 25°C/km. At this gradient, the oil window occurs at depths of 2 to 6 km, and the gas window at depths of 2 to 9 km.

Conventional versus Unconventional Hydrocarbon Reserves

Oil and gas do not occur in all rocks at all locations. A known supply of oil and gas held underground is a **hydrocarbon reserve**—if the reserve consists dominantly of oil, it's usually called an *oil reserve,* and if it consists predominantly of gas, it's a *gas reserve*. Geologists refer to a region with hydrocarbon reserves underground as an *oil field* or *gas field*, depending on which substance dominates. Until relatively recently, oil companies could economically obtain only those hydrocarbons that could be pumped from the ground easily. We refer to such hydrocarbon reserves as **conventional reserves**. In recent years, however, advances in technology have made it possible to obtain oil and gas from reserves that could not be pumped using traditional methods. Such reserves are known as **unconventional reserves**. The ability to use unconventional reserves has led to major changes in the energy industry. We'll discuss these two types of reserves separately.

TAKE-HOME MESSAGE

Oil and gas are hydrocarbons derived from the remains of plankton buried with clay in an organic ooze. Burial transforms ooze into source rock, a black organic shale, and heating underground transforms organic molecules to kerogen. If a rock attains temperatures within the oil or gas window, kerogen transforms into oil and/or gas. Hydrocarbon reserves are underground accumulations of oil and/or gas.

QUICK QUESTION Could the material that makes up a source rock accumulate in a fast-flowing mountain stream?

12.4 Conventional Hydrocarbon Systems

We've just noted that the hydrocarbons of a conventional reserve can be pumped out of the ground. Developing such a reserve requires a specific association of materials, conditions, and time, collectively known as a *conventional hydrocarbon system*. We have discussed the first two components of the system, namely, the formation of a source rock of organic shale and the existence of thermal conditions that transform the kerogen in the shale into smaller hydrocarbon molecules. Let's now look at the remaining components, the migration of oil into a reservoir rock that lies within a configuration of rocks called an oil trap.

Reservoir Rocks and Hydrocarbon Migration

The clay flakes of an oil shale pack together tightly, so hydrocarbons within the rock cannot flow easily through the rock. Therefore, an energy company can't simply drill a hole into a source rock and pump out oil or gas—the hydrocarbons won't flow into the well fast enough to make the process cost-efficient. Instead, to obtain oil or gas from a conventional oil or gas field, companies drill into **reservoir rocks** with high porosity and high permeability. To understand this statement, let's examine these characteristics.

Porosity refers to the proportion of a rock that consists of open spaces, or **pores**. Primary porosity exists if, during lithification, grains didn't fit together perfectly, and cement didn't fill all spaces available, whereas secondary porosity exists if, after lithification, some of the rock dissolved in groundwater, or the rock underwent fracturing. Not all rocks have the same porosity. For example, shale typically has porosity of less than 10%, whereas some sandstone has porosity of up to 35%. In this example, about a third of a block of sandstone would consist of open space. Note that oil and gas in a reservoir rock resides in the pores, not in underground pools. **Permeability** refers to the degree to which pore spaces connect to one another, either because adjacent pores happen to open into each other, or because fracturing broke apart otherwise solid rock. Permeability allows fluids (water, oil, or gas) to flow through the rock. Thus, when we say that a good reservoir rock has high porosity and permeability, we mean that the rock can contain a lot of oil or gas, and that oil or gas can be pumped out of the rock relatively easily. Reservoir rocks do not contain oil to start with, for the sediments from which reservoir rocks form do not contain organic matter when deposited. So for oil or gas to get into the pores of a reservoir rock, it must first **migrate** (move) upward from the source rock into a reservoir rock, a process that can take thousands to millions of years (**Fig. 12.3**). What causes hydrocarbon migration? Because oil and gas are less dense than water, they try to rise toward the Earth's surface and get above groundwater, just as salad oil rises above vinegar in a bottle of salad dressing. In other words, buoyancy drives oil and gas upward. (Natural gas, being less dense than oil, eventually rises above oil.) Typically, a conventional hydrocarbon system must have a good *migration pathway*, such as a set of permeable fractures, in order for large volumes of hydrocarbons to move from source to reservoir.

Did you ever wonder...
if there are actually pools of oil underground?

Traps and Seals

If oil or gas escapes from the reservoir rock, reaches the Earth's surface, and leaks away at an **oil seep**, none is left underground, available to extract. Thus, for a conventional oil or gas reserve to exist, the hydrocarbons must be held underground in the reservoir rock by means of a geologic configuration called a **trap**. An oil or gas trap has two components. First, a **seal rock**, a relatively *impermeable* (not permeable) rock such as shale, salt, or unfractured limestone, must lie above the reservoir rock and stop the hydrocarbons from rising further. Second, the seal and reservoir rock bodies must be arranged in a geometry that localizes the hydrocarbons in a restricted area. Geologists recognize several types of trap geometries—**Box 12.1** describes four.

Birth of the Conventional Oil Industry

In the United States, during the first half of the 19th century, people collected "rock oil" (later called petroleum, from the Latin words *petra*, meaning rock, and *oleum*, meaning oil) at seeps and used it to grease wagon axles. But such oil was rare and expensive. In 1854, George Bissell, a New York lawyer, realized that oil might have broader uses, particularly as fuel for lamps, to replace increasingly scarce whale oil. Bissell and a group of investors contracted a colorful character named Edwin Drake to find a way to drill for oil in rocks beneath a hill near Titusville, Pennsylvania, where oily films floated on the water of springs. Using the phony title "Colonel" to add respectability, Drake hired drillers and obtained a steam-powered drill. Work was slow and the investors became discouraged, but the very day that a letter arrived ordering Drake to stop drilling, his drillers found that the hole, which had reached a depth of 21.2 m, had filled with oil. They set up

FIGURE 12.3 Initially, oil resides in the source rock. Because it is buoyant relative to groundwater, the oil migrates into the overlying reservoir rock. The oil accumulates beneath a seal rock in a trap.

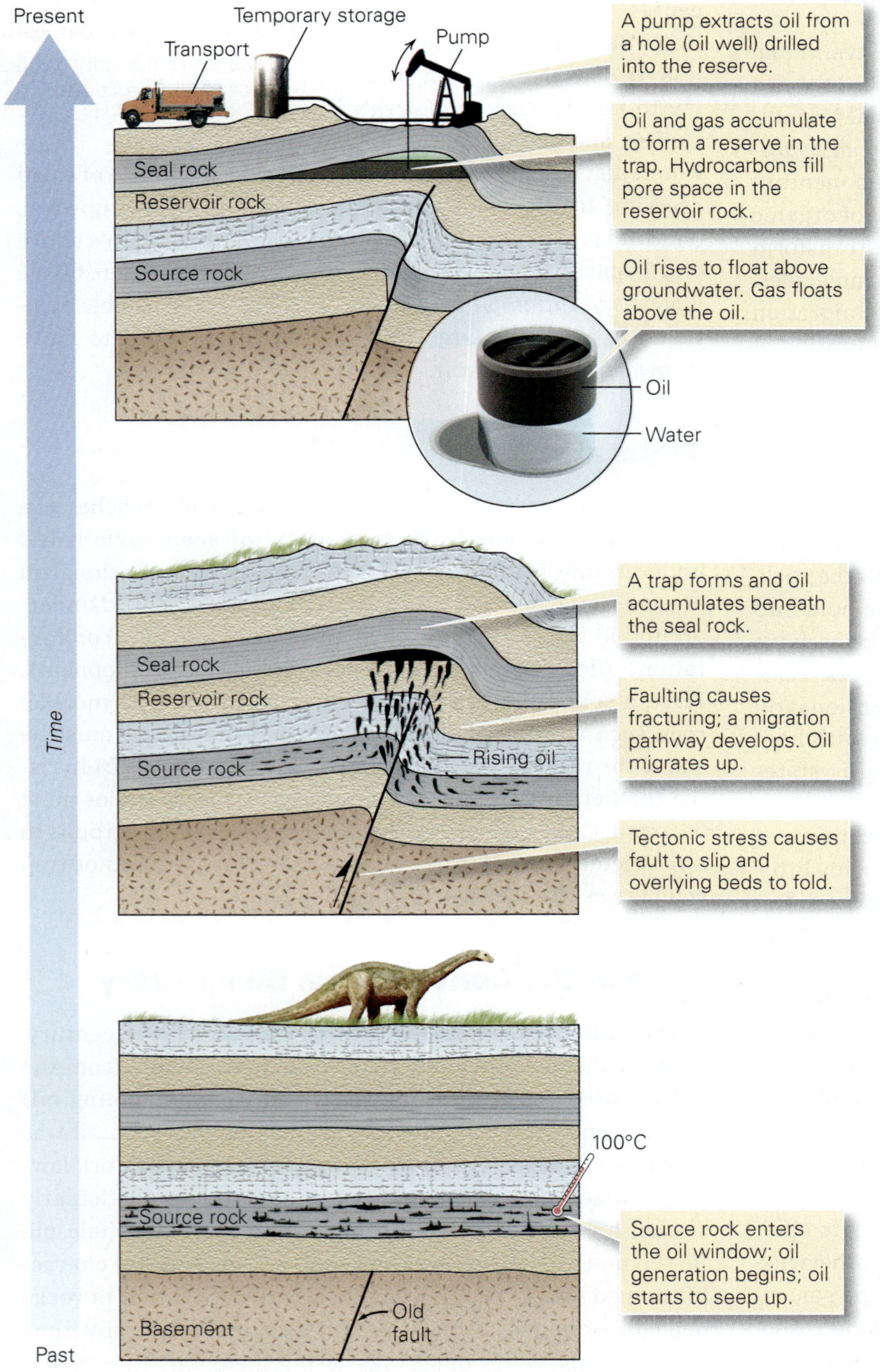
Present
Temporary storage
Transport
Pump
A pump extracts oil from a hole (oil well) drilled into the reserve.
Seal rock
Reservoir rock
Source rock
Oil and gas accumulate to form a reserve in the trap. Hydrocarbons fill pore space in the reservoir rock.
Oil rises to float above groundwater. Gas floats above the oil.
Oil
Water
A trap forms and oil accumulates beneath the seal rock.
Seal rock
Reservoir rock
Source rock
Rising oil
Faulting causes fracturing; a migration pathway develops. Oil migrates up.
Tectonic stress causes fault to slip and overlying beds to fold.
Time
100°C
Source rock
Source rock enters the oil window; oil generation begins; oil starts to seep up.
Basement
Old fault
Past

a pump, and on August 27, 1859, for the first time in history, pumped oil out of the ground. No one had given much thought to the question of how to store the oil, so workers dumped it into empty whisky barrels. A barrel of oil (bbl) has become a standard unit of measurement: 1 bbl = 42 gallons = 159 liters. (Another unit, a metric ton [tonne] of oil, equals 6.5 barrels.) This first oil well yielded 10 to 35 barrels a day.

Within a few years, thousands of oil wells had been drilled in many places in the United States and in other countries, and by the turn of the 20th century, civilization had begun its addiction to oil. Initially, most oil was used to make kerosene for lamps. Later, when electricity became the primary source for illumination, gasoline derived from oil was the fuel of choice for the newly invented automobile. Oil also fueled electric power plants. In its early years, the oil industry was disorganized. When "wildcatters" discovered a new oil field, there would be a short-lived boom during which the price of oil could drop to pennies a barrel. In the midst of this chaos, John D. Rockefeller established the Standard Oil Company, which monopolized the production, transport, and marketing of oil. In 1911, the Supreme Court broke down Standard Oil into several companies, some of which have recombined in recent decades. Oil became a global industry governed by the complex interplay of politics, profits, supply, and demand.

The Modern Search for Oil

Wildcatters discovered the earliest oil fields either by blind luck or by searching for surface seeps. But in the 20th century, when most known seeps had been drilled and blind luck became too risky, oil companies realized that finding new oil fields would require systematic exploration. The modern-day search for oil is a complex, sometimes dangerous, and often exciting procedure with many steps.

Source rocks are always sedimentary, as are most reservoir and seal rocks, so geologists begin exploration by looking for a region with appropriate sedimentary rocks. Then they compile a geologic map of the area, showing the distribution of rock units. From this information, it may be possible to construct a preliminary cross section depicting the geometry of the sedimentary layers underground as they would appear on an imaginary vertical slice through the Earth. To add detail to the cross section, an exploration company makes

BOX 12.1 Consider This...

Types of Oil and Gas Traps

Geologists who work for oil companies spend much of their time trying to identify underground traps. No two traps are exactly alike, but we can classify most into the following four categories.

- *Anticline trap*: In some places, sedimentary beds are not horizontal, as they were when originally deposited, but have been bent by the stresses involved in mountain building. These bends, as we have seen, are called folds. An anticline is a type of fold with an arch-like shape (**Fig. Bx12.1a**; see Chapter 9). If the layers in the anticline include a source rock overlain by a reservoir rock that is overlain in turn by a seal rock, then we have the recipe for an oil reserve. Oil and gas rise from the source rock, enter the reservoir rock, and get trapped at the crest of the anticline.
- *Fault trap*: If the slip on a fault crushes and grinds the adjacent rock to make an impermeable layer along the fault, then oil and gas may migrate upward along bedding in the reservoir rock until they stop at the fault surface. A fault trap may also develop if the slip juxtaposes a seal rock against a reservoir rock (**Fig. Bx12.1b**).
- *Salt-dome trap*: In some sedimentary basins, the sequence of strata contains a thick layer of salt, deposited when the basin was first formed and seawater covering the basin was shallow and very salty. Sandstone, shale, and limestone overlie the salt. The salt layer is not as dense as sandstone or shale, so it is buoyant and tends to rise slowly through the overlying strata. Once the salt starts to rise, the weight of surrounding strata squeezes the salt out of the layer and up into a growing, bulbous **salt dome**. As the dome rises, it bends up the adjacent layers of sedimentary rock. Oil and gas in reservoir rock layers migrate upward until they are trapped against the boundary of the salt dome, for salt is impermeable (**Fig. Bx12.1c**).
- *Stratigraphic trap*: In a stratigraphic trap, a tilted reservoir rock bed "pinches out" (thins and disappears) up-dip between two impermeable layers. Oil and gas migrating upward along the bed accumulate at the pinch-out (**Fig. Bx12.1d**).

FIGURE Bx12.1 Examples of oil traps. A trap is a configuration of a seal rock over a reservoir rock in a geometry that keeps the oil underground.

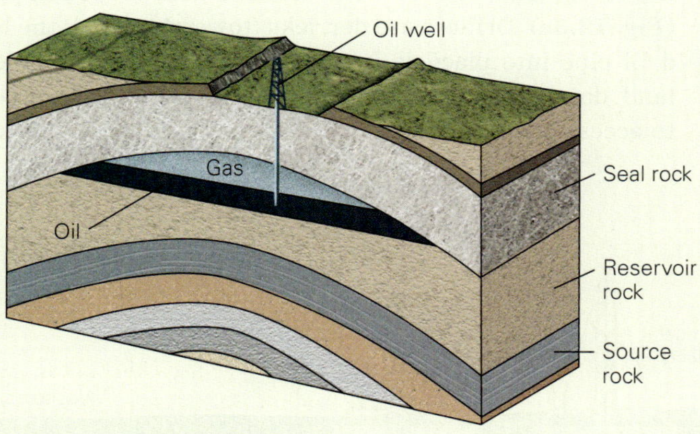

(a) Anticline trap. Oil and gas rise to the crest of the fold.

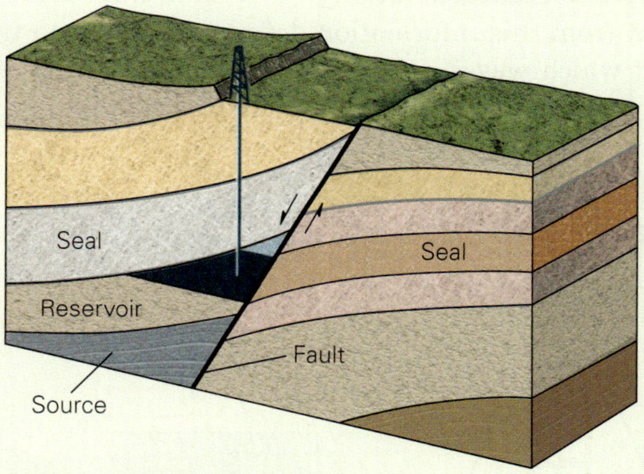

(b) Fault trap. Oil and gas collect in tilted strata adjacent to the fault.

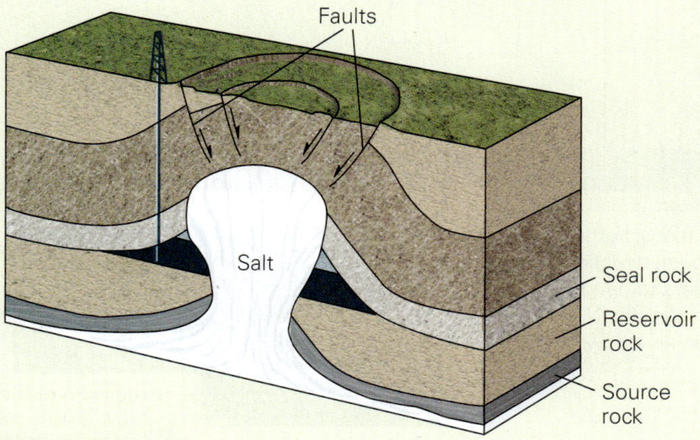

(c) Salt-dome trap. Oil and gas collect in strata on the flanks of the dome, beneath salt.

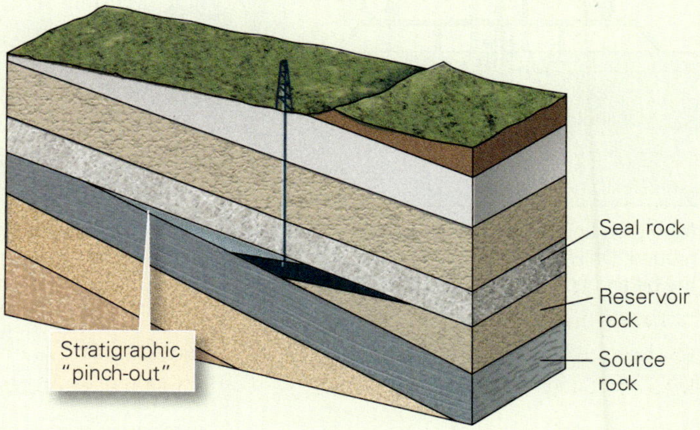

(d) Stratigraphic trap. Oil and gas collect where the reservoir layer pinches out.

FIGURE 12.4 Exploring for oil and gas.

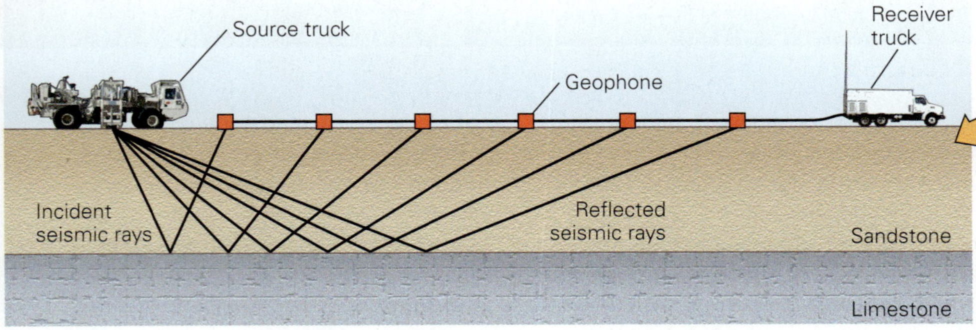

(a) Obtaining a seismic-reflection profile. The source truck sends vibrations downward. These reflect off layer boundaries up to small seismometers (geophones). The travel time indicates the depth to the boundary.

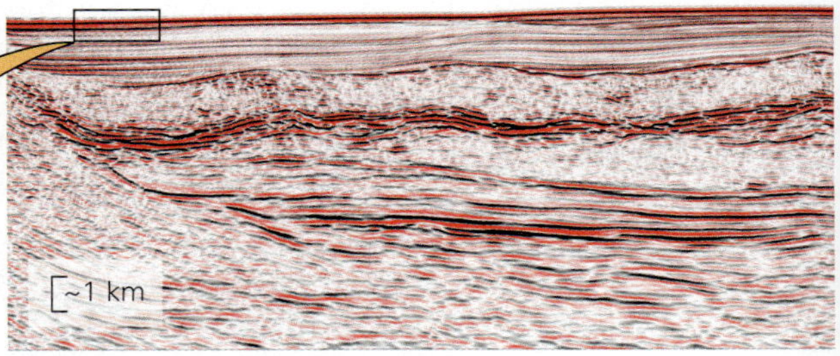

(b) An example of a seismic-reflection profile. The colored bands represent underground sedimentary beds.

a **seismic-reflection profile** of the region (**Fig. 12.4**). This is done by using either special vibrating trucks or dynamite explosions to send artificial seismic waves into the ground (see Interlude D). The seismic waves reflect off contacts between rock layers, just as sonar waves sent out by a submarine reflect off the bottom of the sea. Reflected seismic waves then return to the ground surface, where small seismometers called geophones record their arrival. A computer measures the time between the generation of a seismic wave and its return, and from this information defines the depth to the contacts at which waves reflected. With such information, geologists can construct an image of the underground rock layers and identify traps.

Drilling and Refining

If geologists identify a trap, and if the geologic history of the region indicates the presence of good source rocks, good reservoir rocks, and temperatures within the oil or gas windows, they may recommend drilling. They do not make such recommendations lightly, for drilling a deep well costs millions of dollars. These days, drillers use rotary drills to grind a hole down through rock. A rotary drill consists of a pipe tipped by a rotating bit, a bulb of metal studded with very hard prongs (**Fig. 12.5a**). Drillers use derricks (towers) to hoist the heavy drill pipe into place so it can be lowered into the hole. On land, derricks have concrete or gravel platforms, but in order to access offshore hydrocarbon reserves that occur in strata

FIGURE 12.5 Drilling for oil. The process is challenging.

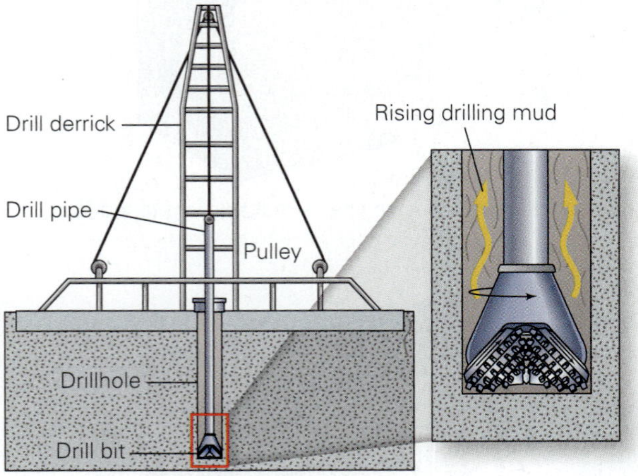

(a) A pipe tipped by a rotating drill bit grinds a hole into the ground. Drilling mud, pumped down through the pipe, cools the bit and flushes up cuttings.

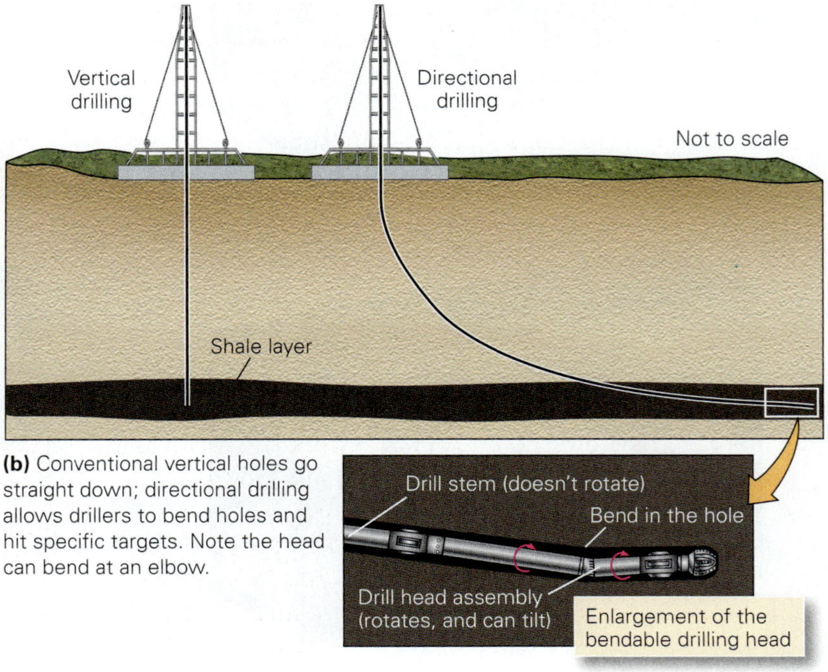

(b) Conventional vertical holes go straight down; directional drilling allows drillers to bend holes and hit specific targets. Note the head can bend at an elbow.

beneath the continental shelf, a derrick must be placed on an *offshore-drilling platform*. Such platforms may be built on huge towers rising from the seafloor, or on giant submerged pontoons. Using *directional drilling* (see below), it's possible to reach multiple targets from the same platform.

During drilling, the bit rotates rapidly. As it does so, it scratches and gouges the rock, turning the rock into powder and chips. Drillers pump **drilling mud**, a slurry of water mixed with clay and other materials, down the center of the pipe. The mud flows down, past the propeller that rotates the drill bit, and then squirts out of holes at the end of the bit. The extruded mud cools the bit head, which otherwise would heat up due to friction as it grinds against rock. Then, the mud flows up the hole on the outside of the drill pipe, carrying *cuttings* (fragments of rock that had been broken up by the drill bit) along with it. Mud also serves another very important purpose—its weight counters the pressure of the oil and gas in underground reservoir rocks, and thus prevents the hydrocarbons from entering the hole. Were it not for the mud, the natural pressure in the reservoir rock would drive oil and/or gas into the hole, and if the pressure were great enough, the hydrocarbons would rush up the hole and spurt out of the ground as a gusher or *blowout*. Blowouts are disastrous, because they spill oil onto the land and, in some cases, ignite into a deadly inferno.

Traditional drilling employed nonadjustable drill bits and thus could yield only a vertical hole. Modern drilling, in contrast, uses adjustable drill bits that can be reoriented during drilling, so a drillhole can now curve to become diagonal or even horizontal (**Fig. 12.5b**). In fact, such **directional drilling** has become so precise that a driller, using a joystick to steer the bit and sensors that specify the exact location of the bit in 3-D space, can hit an underground target that's only 15 cm wide, from a distance of a few kilometers. Once drilling has been finished, workers "complete" the hole. *Hole completion* involves removing the drill pipe, inserting a *casing* (a metal pipe through which rising hydrocarbons ultimately flow), and filling the space between the casing and the walls with concrete. The casing and concrete prevent fluids from beds other than the hydrocarbon-rich target beds from entering the hole. To provide access to specific target beds, drillers puncture holes in the casing at the depth of the target bed.

After completion, workers remove the derrick and either cap the hole or install a pump. When pumping begins, the completed well becomes a *producing oil well* or gas well. Some pumps resemble a bird pecking for grain—their heads move up and down to pull up oil that has seeped out of pores in the reservoir rock into the well (**Fig. 12.6a**). You may be surprised to learn that simple pumping brings only about 20% to 30% of the oil from a reservoir rock out of the ground. Thus, oil companies commonly use *secondary recovery techniques* to coax

FIGURE 12.6 Pumping, transporting, and refiining oil.

The head of the pump goes up and down.
Storage tanks
Hole

(a) When drilling is complete, the derrick is replaced by a pump, which sucks oil out of the ground. This design is called a pumpjack.

(b) The Trans-Alaska Pipeline transports oil from fields on the Arctic coast to a tanker port on the southern coast of Alaska.

(c) An oil tanker capable of carrying 2.6 million barrels (enough to supply the entire United States for 7 hours).

(d) Distilling columns of an oil refinery transform crude oil into gasoline and other hydrocarbon products.

FIGURE 12.7 The distribution of oil reserves around the world.

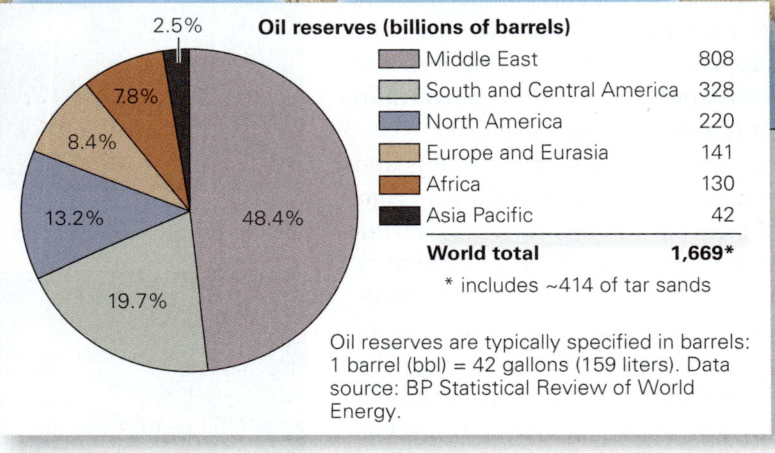

SEE FOR YOURSELF...

OIL FIELD NEAR LAMESA, TEXAS

Latitude
32°33′18.42″ N

Longitude
101°46′55.39″ W

Zoom to 8 km (~5 miles) and look down.

Within a grid of farm fields, 1.6 km or 1 mile wide, small roads lead to patches of dirt. Each patch hosts or hosted a pump for extracting oil. These wells are tapping an oil reserve in Permian sandstone reservoirs.

out as much as 20% more oil. For example, a company may drive oil toward a producing well by forcing steam down nearby wells. The steam heats the oil in the ground, making it less viscous, and pushes it along. In some cases, drillers enhance permeability by opening up existing fractures, or creating artificial fractures in rock adjacent to the hole. They do this by pumping a mixture of water, various chemicals, and sand into a portion of the hole at high-pressure. We'll describe this process, called **hydrofracturing** (also known as hydraulic fracturing, hydrofracking, or fracking) in more detail later, in our discussion of how to pump out unconventional reserves.

Once extracted directly from the ground, *crude oil,* the unrefined oil that comes straight out of the ground, is pumped into storage tanks. Later, pipelines or tankers (**Fig. 12.6b, c**) transport crude oil to a *refinery,* where workers separate crude oil into different-sized molecules by heating it in a vertical pipe called a **distillation column** (**Fig. 12.6d**). Lighter molecules rise to the top of the column, while heavier molecules stay at the bottom. The heat may also "crack" larger molecules to make smaller ones. Chemical factories buy the largest molecules left at the bottom and transform them into plastics.

Natural gas often occurs in association with oil, or in place of oil, in conventional reserves. This gas must be compressed for transport, so transportation requires high-pressure pipelines or special ships. Because compressing and transporting gas can be so expensive, available quantitities must be enough to make the process economical. Where the costs are too high to process natural gas that comes up with oil, rig operators vent the gas from a pipe and burn it as a flare where it enters the air.

Where Do Conventional Reserves of Oil Occur?

Conventional hydrocarbon reserves are not randomly distributed around the Earth (**Fig. 12.7**). They occur in sedimentary basins along passive continental margins, such as the Gulf Coast of the United States and the Atlantic coasts of Africa and Brazil, as well as in intracratonic and foreland basins of continents (see Chapter 6). By far the largest reserves lie beneath countries bordering the Persian Gulf in the Middle East. In fact, this region has almost 50% of the world's reserves.

Why is there so much oil in the Middle East? Reconstructions of past positions of continents indicate that much of this region was situated in tropical areas, between latitude 20° south and 20° north, during the time between the Jurassic (135 Ma) and the Late Cretaceous (66 Ma). Biological productivity was high, so sediments deposited at that time were rich in organic matter. Thick successions of porous sandstone buried the source rocks of the Middle East, and crustal compression due to the collision between Africa and Asia folded the strata to produce excellent traps.

TAKE-HOME MESSAGE

Conventional hydrocarbon reserves are those in which oil has migrated from a source rock into a porous and permeable reservoir rock, situated within an oil trap, so that the oil or gas reserve can be pumped fairly easily. The first oil drilling took place in 1859. Modern methods for finding, drilling, producing, and refining oil are very complex and expensive.

QUICK QUESTION Where do most of the conventional oil reserves in the world occur today?

12.5 Unconventional Hydrocarbon Reserves

In an *unconventional hydrocarbon reserve,* rock contains significant quantities of hydrocarbons. However, because the rock is impermeable or the hydrocarbons are too viscous to flow, hydrocarbons cannot be extracted simply by drilling and pumping. Extracting these resources, therefore, was not possible until engineers developed new technologies and until the products' selling prices reached a level where extraction was profitable when balanced against expensive new technologies. These conditions were met around the turn of this century, and since then, efforts to extract hydrocarbons from unconventional reserves have grown so rapidly that such reserves now provide a significant portion of the global energy supply. Let's examine the geologic context of three types of unconventional reserves: shale oil and shale gas, tar sands, and oil shale.

Source-Bed Oil and Gas

Huge quantities of oil and natural gas remain in source beds of organic shale, even after the strata have been heated to the oil window or gas window. These resources are called **shale oil** or **shale gas**, respectively. Using traditional drilling and producing methods, it was not cost-effective, traditionally, to extract such source-bed hydrocarbons, for two reasons. First, the source beds are so impermeable that the hydrocarbons would not flow to a pumping well. Second, since the hydrocarbons had not migrated into a trap, a vertical well could not access a significant supply. The development of directional drilling and hydrofracturing radically changed this picture. As a result of directional drilling, a single well can follow a source bed horizontally for many kilometers, and several wells can be drilled from the same platform in different directions into the source bed, so a large volume can be accessed (**Fig. 12.8a**). Because of hydrofracturing, a rock's permeability can be increased sufficiently to permit the hydrocarbons to flow into the drillhole (**Box 12.2**).

In 2008, Terry Engelder of Pennsylvania State University, and other geologists, pointed out that previous reports had greatly underestimated the amount of source-bed hydrocarbons available in sedimentary basins that lie near populated areas so that transport costs would be relatively low. An explosion in exploration and drilling began. In the United States, thousands of wells have been drilled into the Marcellus Shale of Pennsylvania, Ohio, and New York, the Bakken Shale of North Dakota and adjacent regions, and the Barnett Shale of Texas (**Fig. 12.8b**).

Energy companies have spent billions of dollars to lease the rights to obtain hydrocarbons from large areas overlying shale deposits, as a result of which some local landowners have become millionaires overnight and thousands have obtained jobs. The boom remains controversial, though, because of lingering questions about economic practicality and environmental concerns. Residents worry that chemicals in hydrofracturing fluids may contaminate water supplies and that gas may enter the groundwater (see Box 12.2). Also, in some cases, extraction brings up large volumes of water, much of which had had resided within rock pores for millions of years. This water tends to be too salty to be used

FIGURE 12.8 Unconventional shale oil and shale gas fields in North America.

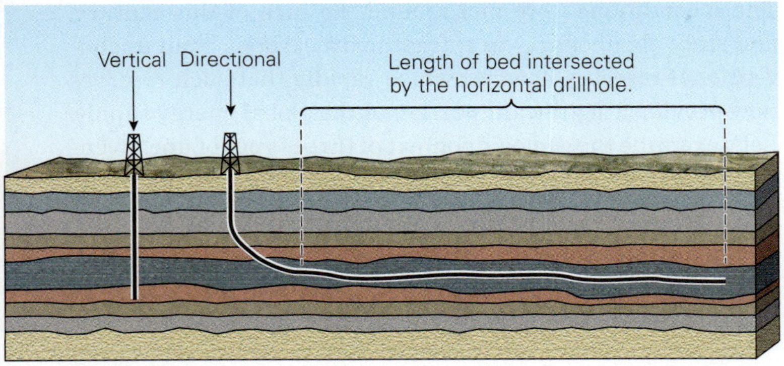

(a) Horizontal directional drilling can allow a very large area of a source bed to be accessed. The length of the hole must be hydrofractured, to provide permeability.

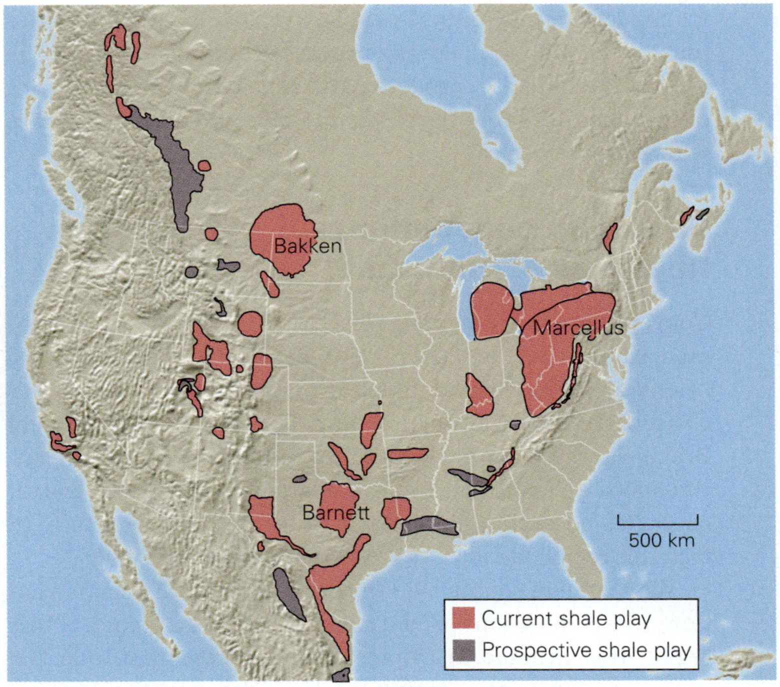

(b) A map of the major unconventional fields in North America. The three largest are labeled. A "play" is a region being drilled.

for irrigation or drinking, so in order to dispose of it, drillers pump it underground into deep injection wells. Researchers have shown that, if injection takes place at very high pressure and at great depth, it may trigger small earthquakes. The water pressure can push apart fault walls slightly, thereby decreasing frictional resistance to sliding.

Tar Sands (Oil Sands)

In several locations around the world, most notably Alberta (in western Canada) and Venezuela, vast reserves of very viscous, tar-like *heavy oil* exist. This heavy oil, known also as *bitumen,* has the consistency of gooey molasses, and thus cannot be pumped directly from the ground. It fills the pore spaces of poorly cemented sandstone, constituting up to 12% of the rock volume. Sandstone containing such high concentrations of bitumen is known as **tar sand** or *oil sand.*

Producing usable oil from tar sand is difficult and expensive, but not impossible. It takes about 2 tons of tar sand to produce one barrel of oil. Oil companies mine near-surface deposits in vast open-pit mines, then heat the tar sand to decrease the viscosity of the oil, so that it can be separated from sand. Producers crack the heavy oil molecules to produce smaller, more usable molecules, and trucks dump the drained sand back into the mine pit. To extract oil from deeper deposits of tar sand, companies drill wells and pump steam or solvents down into the sand to liquefy the oil enough so that it can be pumped out.

Oil Shale

Vast reserves of organic shale have not been subjected to temperatures of the oil window, or if they were, they did not stay within the oil window long enough to complete the transformation to oil. Such rock still contains a high proportion of kerogen. Shale with at least 15% to 30% kerogen is called **oil shale**. (Note that *oil shale,* a rock, is different from *shale oil,* oil extracted from source beds; strata containing the latter have been through the oil window.) Lumps of oil shale can be burned directly and thus have been used as a fuel since ancient times. In general, however, energy companies produce liquid oil from oil shale. The process involves heating the oil shale to a temperature of 500°C, which causes the shale to decompose and the kerogen to transform into liquid hydrocarbons and gas. As is the case with tar sand, producing oil from oil shale is possible but very expensive.

TAKE-HOME MESSAGE

In unconventional reserves, the rock is too impermeable for hydrocarbons to be pumped directly, or the hydrocarbons are too viscous to flow. Directional drilling and hydrofracturing now allow extraction of shale oil and shale gas directly from impermeable source beds. Obtaining hydrocarbons from tar sand or oil shale requires input of heat.

QUICK QUESTION Why does directional drilling allow economical extraction of oil or gas from source beds, even where the source beds do not lie in a trap?

FIGURE 12.9 The formation of coal. Coal forms when plant debris becomes deeply buried.

(a) A museum diorama depicting a Carboniferous coal swamp.

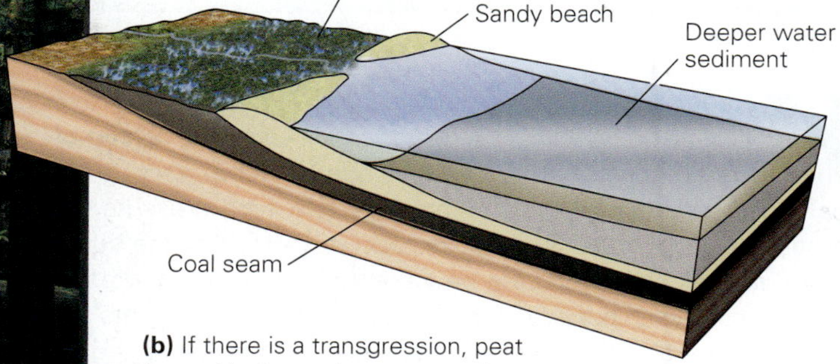

(b) If there is a transgression, peat formed in the coal swamp can be buried and preserved.

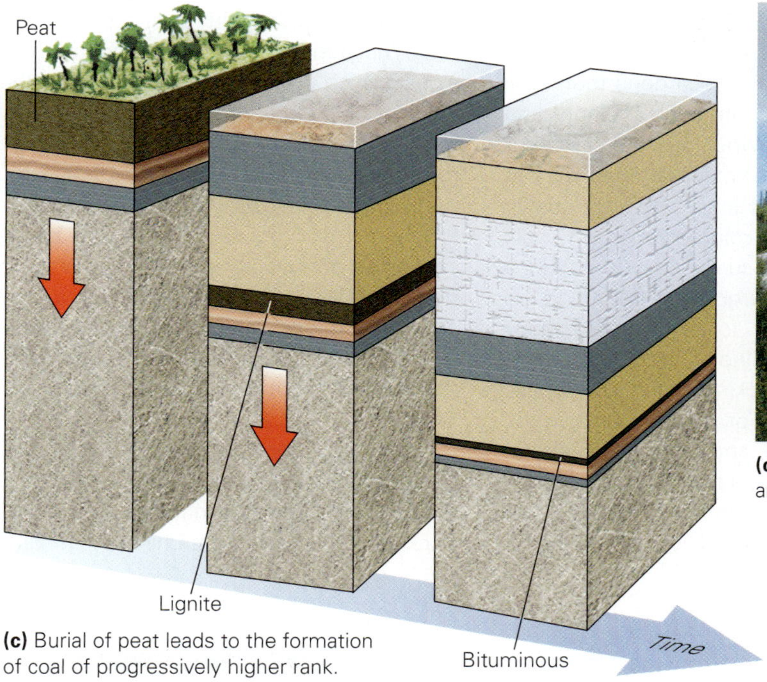

(c) Burial of peat leads to the formation of coal of progressively higher rank.

(d) An example of coal beds interlayered with beds of sandstone and shale.

12.6 Coal: Energy from the Swamps of the Past

Coal, a black, brittle, sedimentary rock that burns, consists of elemental carbon mixed with minor amounts of organic chemicals, quartz, and clay. Like oil and gas, coal is a fossil fuel because it stores solar energy that reached Earth long ago. But coal does not have the same composition or origin as oil or gas—coal is a solid in which carbon atoms have bonded into large, complicated molecules called *coal macerals*, whereas oil and gas consist of hydrocarbons. Also, in contrast to oil and gas, coal forms from plant material (wood, stems, leaves), not plankton. Coal typically occurs in beds, called *coal seams*, that may be centimeters to meters thick and may be traceable over broad regions.

Significant coal deposits could not form until vascular land plants appeared on Earth in the late Silurian Period, about 420 million years ago. The most extensive deposits of coal in the world occur in Carboniferous-age strata (359 to 290 Ma) (**Fig. 12.9a**). In fact, geologists coined the name *Carboniferous* because strata representing this interval of the geologic column contain so much coal. Not all coal reserves, however, are Carboniferous—during the Cretaceous (145 to 66 Ma), large areas of freshwater coal swamps developed in Wyoming and adjacent states.

The Formation of Coal

How do the remains of plants transform into coal? The process begins when vegetation of an ancient swamp dies, falls down, and becomes buried in an oxygen-poor environment,

BOX 12.2 Consider This...

Hydrofracturing (Fracking)

Hydrofracturing (also known as *hydraulic fracturing, hydrofracking,* or *fracking*) is a technique that drillers use to open and propagate existing joints (natural cracks) in rock at depth, as well as to generate new cracks. The open cracks provide a permeability pathway through which hydrocarbons can reach a drillhole and be extracted. Hydrofracturing was invented in the 1950s, but it has become headline news in recent years. This method has led to a substantial increase in the amount of unconventional oil and gas that energy companies are extracting in industrialized countries such as the United States, where conventional reserves have been largely depleted. That's because, while conventional wells can extract hydrocarbons only from reservoir rocks in a trap, hydrofracturing allows extraction of hydrocarbons directly from source beds (black shales).

What is hydrofracturing, how does it work, and what risks are potentially involved in its use? To hydraulically fracture a hole, drillers start by sealing off a length of a well with packers. These inflatable balloons, when filled with high-pressure fluid, press tightly against the wall of the hole (**Fig. Bx12.2a**). Then the drillers insert a pipe through one of the packers into the sealed-off section of the well and pump *fracking fluid* into this section, under high pressure. When the pressure generated by the fluid within the sealed section becomes great enough, it forces open existing joints that intersect the hole and may cause these joints to lengthen at their tip (**Fig. Bx12.2b**). The process may also generate new cracks in the rock adjacent to the hole. The area affected by hydrofracturing can extend tens of meters out from the drillhole. Once hydrofracturing has been completed, drillers pump out the fracking fluid. Hydrocarbons then flow along the fractures into the drillhole and up to the surface, where drillers capture it (**Fig. Bx12.2c**). The volume of fluid used to hydrofracture a section of hole is roughly equivalent to the volume in an Olympic swimming pool. Usually several sections of a hole may undergo hydrofracturing, so the process needs to be repeated many times in a given drillhole.

What's in fracking fluid? A typical example consists of about 90% water, 9.5% quartz sand, and 0.5% other chemicals. The sand props the holes open once the fluid has been removed—without the sand, the cracks opened by hydrofracturing would close up tightly under the pressure applied by surrounding rock, and thus they could not serve as permeability pathways. The 0.5% portion of fracking fluid that is not water or sand contains many chemicals, including: oils that make the fluid slippery so it can inject farther into the rock; acid, which dissolves cement between grains and increases porosity; detergent, which lowers the surface tension of water so it doesn't stick to grains; guar gum, to make the fluid more viscous so that it can carry more sand; antifreeze, to prevent scale buildup or fluid freezing; and biocides, which prevent the growth of bacteria that could clog pores.

Hydrofracturing requires a lot of equipment and materials—at a drilling site, many people will be working, and trucks will come and go carrying water, sand, and other chemicals, as well as portable pumps and giant mixing vats. The site may also have holding tanks or retaining ponds to store fluid that has been removed from the ground, once hydrofracturing has finished (**Fig. Bx12.2d**).

Concerns about hydrofracturing have become the subject of intense public debate. The most common concern is that fracking fluid can contaminate drinking

FIGURE Bx12.2 Hydrofracturing in a horizontal shale bed.

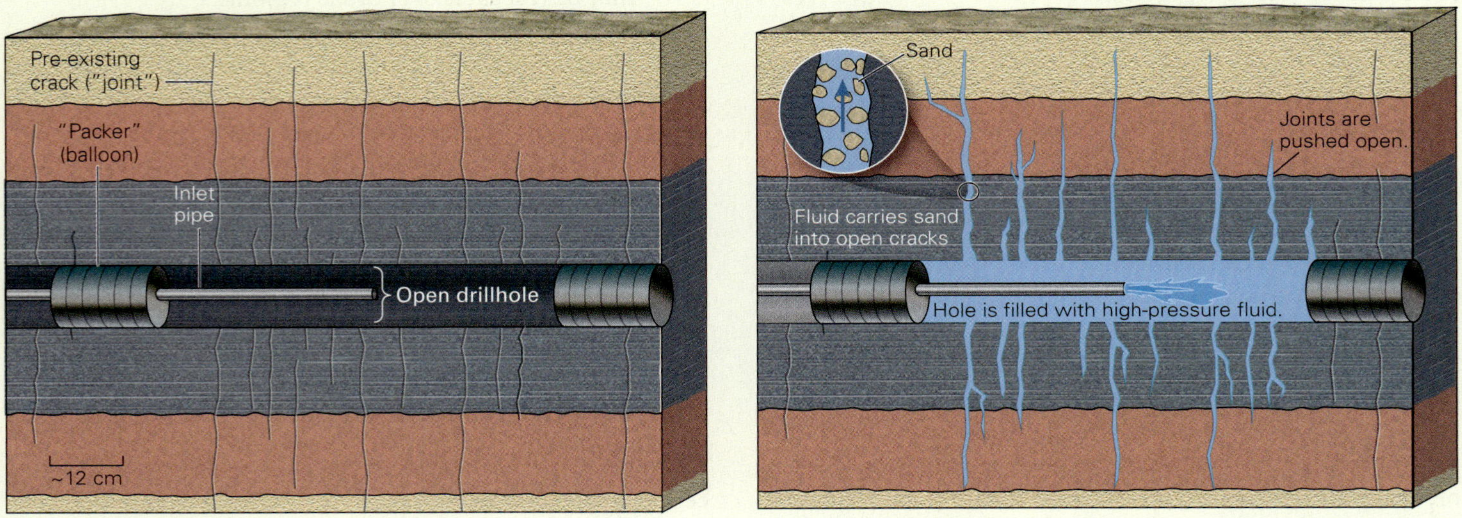

(a) The first step in hydrofracturing. After the hole has been drilled, packers seal off a portion, and a pipe is inserted through one of the packers.

(b) High-pressure fluid is pumped into the segment of hole. The pressure pushes open cracks and forms new ones. Sand injected with the fluid keeps the cracks from closing.

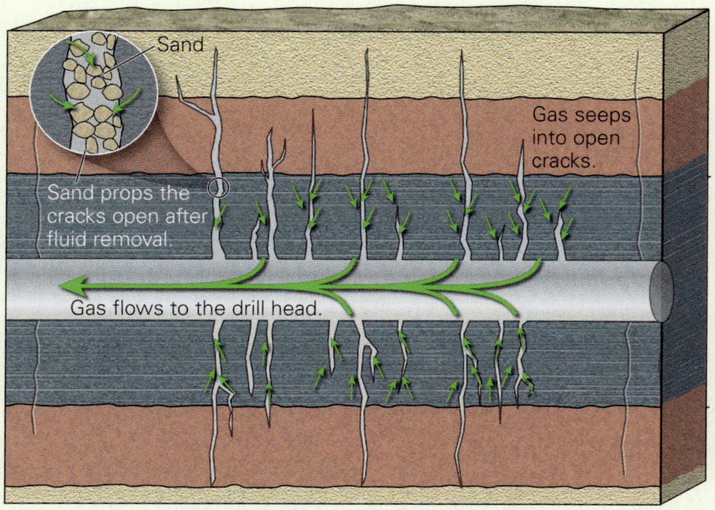

(c) After the packers and the fluid are removed, gas can seep into the pipe and flow to the drill head.

(d) A drilling site. The trucks and the holding pond are used during hydrofracturing. Many holes can be drilled from this site, like spokes of a wheel.

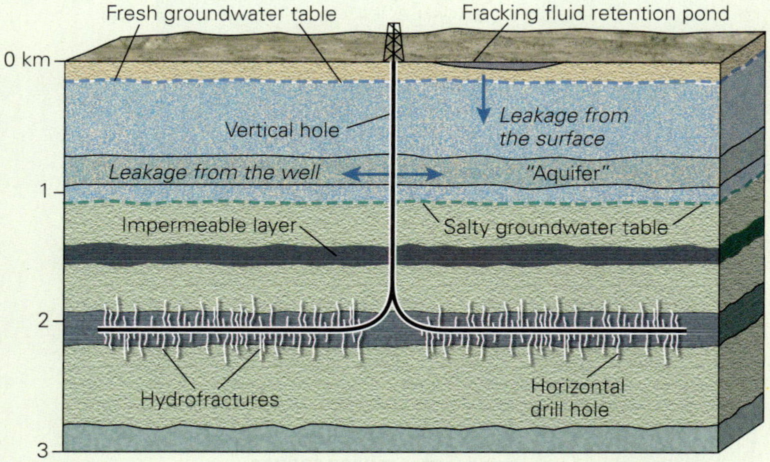

(e) Potential for the contamination of groundwater by hydrofracturing.

water underground. To understand the nature of this risk, it's necessary to understand how groundwater changes with depth. As we'll discuss further in Chapter 16, *groundwater* is water that fills or saturates pores and cracks in rock or sediment underground, beneath a surface called the *water table*—above the water table, pores and cracks contain some air. Typically, groundwater in the upper several hundred to a few thousand meters can be fresh and drinkable, but below that depth, groundwater tends to be saline (**Fig. Bx12.2e**). If the section of the hole subjected to hydrofracturing lies deeper than the saline boundary, fluids from that section probably won't mix with drinkable groundwater, for they are denser than groundwater and thus are not buoyant. Leakage from the portion of the vertical hole above the saline boundary, however, can be problematic, so it is important that this portion of the hole be sealed thoroughly before fracking fluid is pumped in. Leakage at the surface, from tanks, holding ponds, or transporting trucks, is also of concern. To avoid contamination, handling the fluids at the surface must be very carefully monitored.

such as stagnant water. In this way, the organic matter can become incorporated in a sedimentary sequence before it gets eaten or undergoes complete decay. Compaction and partial decay of the vegetation transforms it into **peat**, which consists of up to 50% carbon and can be burned as fuel.

To transform peat into coal, the peat must be buried deeply (4 to 10 km) by overlying sediment. Such deep burial can happen where the surface of the land gradually sinks, creating a sedimentary basin, in which many kilometers of sediment can accumulate over time (**Fig. 12.9b**). Because of transgression and regression (see Chapter 6), many different kinds of sedimentary deposits can accumulate over the coal. At depth in the basin, the weight of overlying sediment compacts the peat and squeezes out any remaining water. Then, because temperature increases downward in the Earth, deeply buried peat gradually heats up. Heat accelerates chemical reactions that gradually destroy plant fiber and release molecules of hydrogen, ammonia, methane, and sulfur dioxide as gases. These gases seep out of the reacting peat layer, leaving behind a residue concentrated with carbon. Once the proportion of carbon in the residue exceeds about 50%, the deposit formally becomes coal. With further burial and higher temperatures, chemical reactions continue, yielding progressively higher concentrations of carbon.

The Classification of Coal

Geologists classify coal based on carbon concentration. Increasing burial transforms peat first into a soft, dark-brown coal called *lignite*. At higher temperatures (about 100°–200°C), lignite, in turn, becomes dull, black *bituminous coal.*

FIGURE 12.10 The distribution of coal reserves.

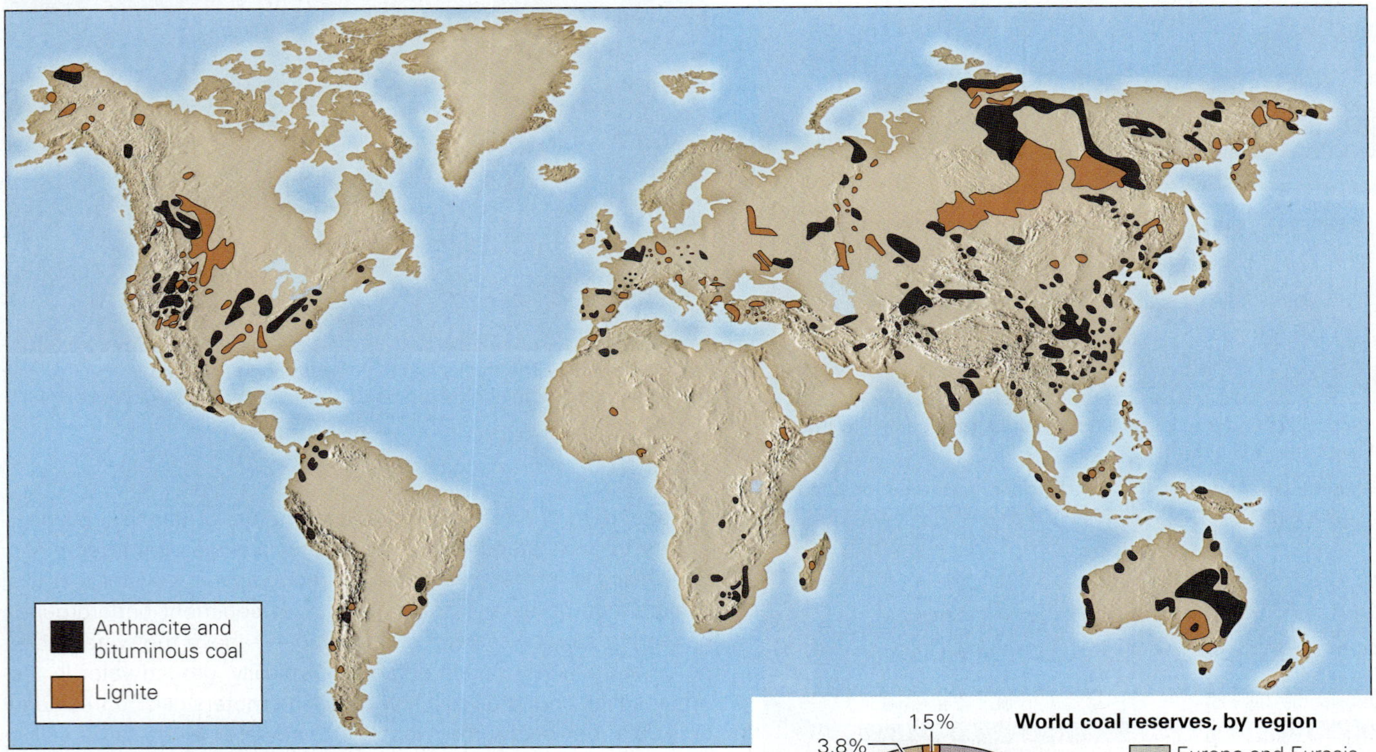

(a) A map showing global distribution of coal reserves. Most coal accumulated in continental interior basins.

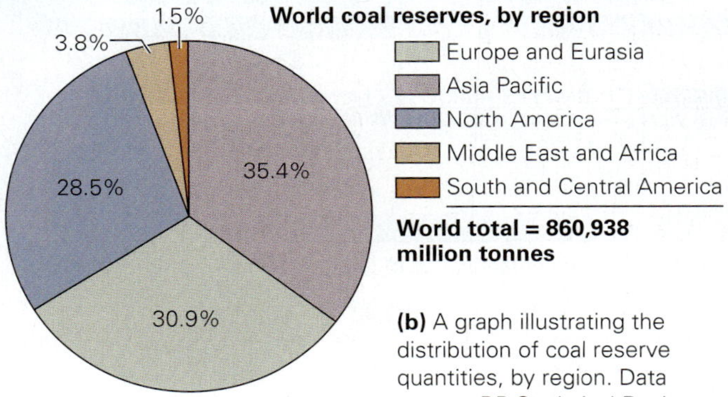

(b) A graph illustrating the distribution of coal reserve quantities, by region. Data source: BP Statistical Review of World Energy.

At still higher temperatures (about 200°–300°C), bituminous coal transforms into shiny, black *anthracite*. The progressive transformation of peat to anthracite coal, as an organic layer undergoes more burial and becomes warmer, reflects the completeness of chemical reactions that remove water and other chemicals, to leave carbon behind (**Fig. 12.9c**). Specifically, lignite contains about 50% carbon, bituminous about 70%, and anthracite about 90%. As the carbon content of coal increases, we say the **coal rank** increases.

Notably, the formation of anthracite coal requires high temperatures that develop only on the borders of mountain belts. In such locations, mountain-building processes push thick sheets of rock up along thrust faults and over the coal-bearing sediment, burying it to depths of 8 to 10 km. In addition, groundwater that passes through the deeply buried coal may flow along a curving path that first takes the water deep into the subsurface, where it heats up. When this groundwater flows back up into the coal, it brings heat with it, and this heat further warms the coal.

Finding and Mining Coal

Because the vegetation that eventually becomes coal was initially deposited in a sequence of sediment, coal seams are interlayered with other sedimentary rocks (**Fig. 12.9d**). To find coal, geologists search for sequences of strata that were deposited in tropical to semitropical, shallow-marine to terrestrial environments—the environments in which a swamp could exist. The sedimentary strata of continents contain huge quantities of discovered coal, or **coal reserves** (**Fig. 12.10**).

The way in which companies mine coal depends on the depth of the coal seam. If the coal seam lies within about 100 m of the ground surface, *strip mining* proves to be the most economical method. In strip mines, miners use a giant shovel called a *dragline* to scrape off soil and layers of sedimentary rock above the coal seam (**Fig. 12.11a, b**). Draglines are so big that the shovel could swallow a two-car garage without a trace. Once the dragline has exposed the seam, miners use smaller power shovels to

FIGURE 12.11 Coal can be mined in strip mines or in underground mines.

(a) A dragline stripping overburden.

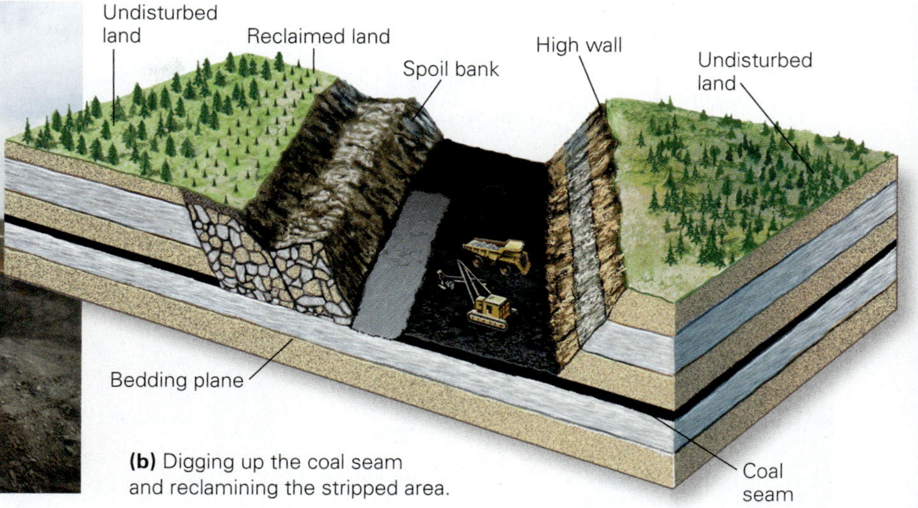

(b) Digging up the coal seam and reclamining the stripped area.

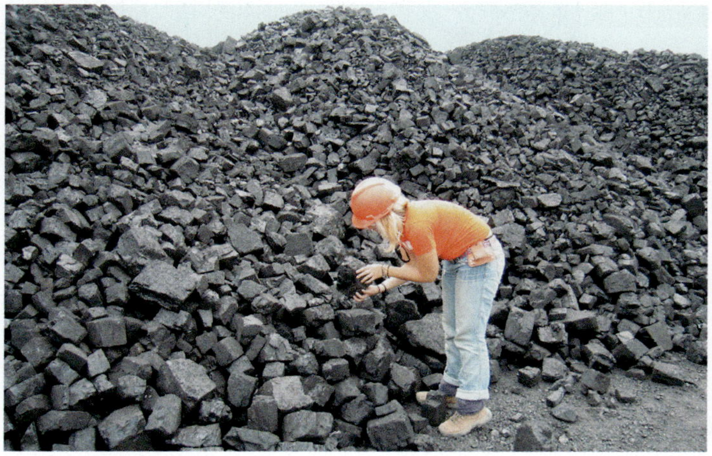

(c) Piles of recently mined coal.

(d) Underground coal mining.

dig out the coal and dump it into trucks or onto a conveyor belt. Before modern environmental awareness took hold, strip mining left huge scars on the landscape. Without topsoil, the rubble and exposed rock of the mining operation remained barren of vegetation. In many contemporary mines, however, the dragline operator separates out the rock that was once above the coal, as well as the soil that once covered bedrock. Then, after the coal has been scraped out, the operator fills the hole with the rock, covers the rock back up with the saved soil, and then plants grass or trees on the soil. The overall process of restoring the landscape of what was once a mine is called *mine reclamation*. In hilly areas, mining companies may use a practice called *mountain top removal,* in which they blast off the top of mountains and dump the debris into adjacent valleys. This practice disrupts the landscape permanently, as the mountain tops cannot be replaced, and the valleys cannot be dug out, so complete reclamation is not possible.

Deep coal can be obtained only by **underground mining**. To develop an underground mine, miners dig a shaft down to the depth of the coal seam and then create a maze of tunnels, using huge grinding machines that chew their way into the coal (**Fig. 12.11c, d**). Underground coal mining can be very dangerous, not only because the sedimentary rocks forming the roof of the mine are weak and can collapse, but because methane gas released by chemical reactions in coal can accumulate in a mine, and a small spark can trigger a deadly mine explosion. Also, mining can produce large amounts of dust, so unless they breathe through filters, underground miners may risk contracting black lung disease from inhaling coal dust.

Coal Gasification

Solid coal can be transformed into various gases, a process called **coal gasification**, which involves the following steps.

FIGURE 12.12 Producing electricity at a nuclear power plant.

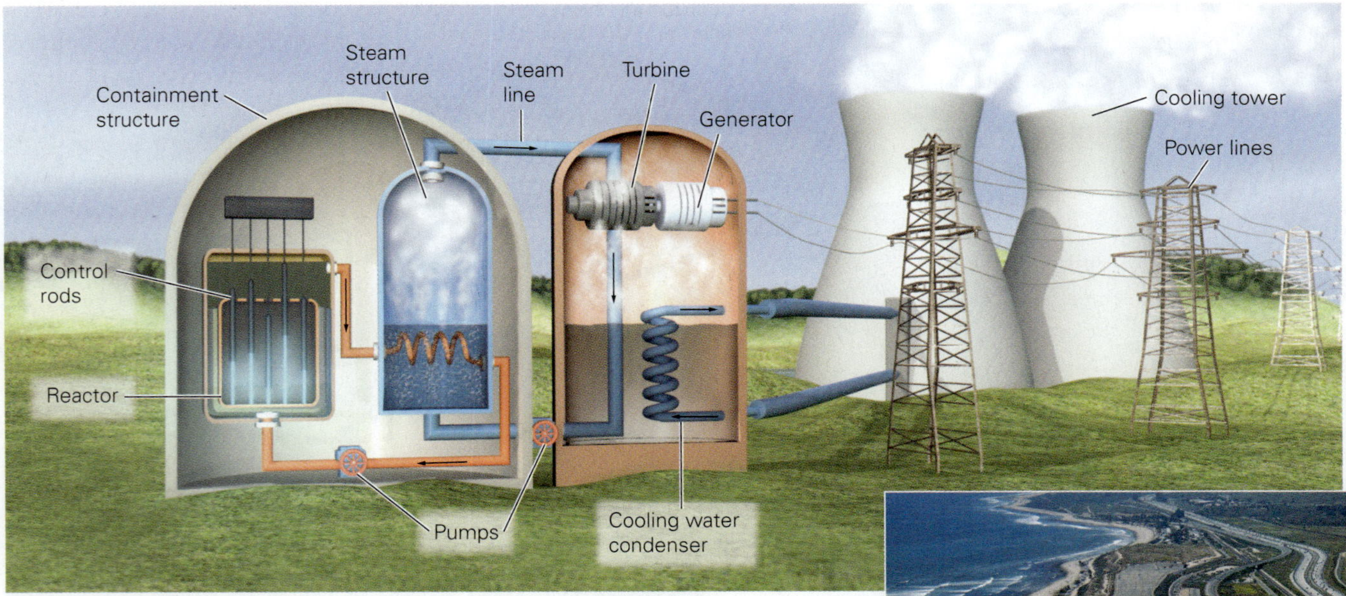

(a) In a nuclear power plant, a reactor heats water, which produces high-pressure steam. The steam drives a turbine that in turn drives a generator to produce electricity. A condenser transforms the steam back into water.

(b) This nuclear power plant in California has two reactors, each in its own containment building.

First, workers place pulverized coal in a large container called a reaction vessel. Then, they send a mixture of steam and oxygen through the coal at high pressure. As a result, the coal heats to a high temperature but does not ignite. Under these conditions, chemical reactions break down and oxidize the molecules in coal to produce carbon monoxide (CO), hydrogen (H_2), H_2O, and CO_2. The CO gas and H_2 gas can then be combined to produce hydrocarbons. During the process of coal gasification, contaminants such as ash, sulfur, and mercury concentrate at the bottom of the reaction vessel, from where they can be removed before the gases produced by coal gasification are burned.

Underground Coal-bed Fires

Coal will burn not only in furnaces but also in surface and subsurface mines, as long as the fire has access to oxygen. Coal mining of the past two centuries has exposed much more coal to the air and has provided many more opportunities for fires to begin. Once started, a coal-bed fire progresses underground, sucking in oxygen from joints that connect to the ground surface. Such fires may be very difficult or impossible to extinguish because there is no easy way to reach them. Some coal-bed fires have made the land above uninhabitable, for they produce toxic fumes that rise through joints and pores to the surface, and may lead to the collapse of the land. For these reasons, a coal-bed fire that began in 1962, near the town of Centralia, Pennsylvania, caused the town to be abandoned. Today, over 100 major coal-bed fires are burning in northern China. Recent estimates suggest that 200 million tons of coal burn underground in China every year, an amount equal to approximately 20% of the annual national production of coal in China.

TAKE-HOME MESSAGE

Coal forms from accumulations of plant material over time in anoxic coal swamps. When buried deeply, this organic material loses water and undergoes chemical reactions that concentrate carbon. Higher-rank coal contains more carbon. Coal occurs as beds, called seams, in sedimentary successions and must be mined underground or in open pits.

QUICK QUESTION How can coal be turned into a gas?

12.7 Nuclear Power

How Does a Nuclear Power Plant Work?

So far, we have looked at fuels (oil, gas, and coal) that release energy when they undergo *chemical burning*. During such burning, a chemical reaction takes place between the fuel and oxygen, releasing the potential energy stored in the *chemical bonds* of the fuel. The energy that drives a nuclear power plant comes from a totally different process—*nuclear fission*—which involves breaking the *nuclear bonds* that hold protons and neutrons together in the nucleus, to split a large atom into smaller pieces. During this process, a small amount of mass transforms into a large amount of thermal and electromagnetic energy, as Einstein recognized. In fact, nuclear fuel has an energy density that is about 48,000 times that of gasoline.

Nuclear power plants were first built to produce electricity during the 1950s. A **nuclear reactor**, the heart of the plant, commonly lies within a *containment building*, a dome-like structure made of reinforced concrete (**Fig. 12.12**). The reactor itself contains nuclear fuel, consisting of pellets of concentrated uranium oxide or a comparable radioactive material, packed into metal tubes called *fuel rods*. Fission occurs when a neutron strikes a radioactive atom, causing it to split. For example, radioactive uranium-235 splits into barium-141 and krypton-92, plus three neutrons. The neutrons released during fission of one atom will then strike three other atoms, thereby triggering more fission in a self-perpetuating process called a **chain reaction** that produces lots of heat. Pipes carry water close to the heat-generating fuel rods, and the heat transforms the water into high-pressure steam. The pipes then carry this steam to a turbine, where it rotates fan blades. The rotation, in turn, drives a dynamo that generates electricity. Eventually, the steam goes into cooling towers, where it condenses back into water that can be reused in the plant or returned to the environment.

The Geology of Uranium

Where does uranium come from? The Earth's radioactive elements, including uranium, probably formed during the explosion of a supernova before the existence of the Solar System. Uranium atoms from this explosion became part of the nebula out of which the Earth grew and thus were incorporated into the planet. They gradually rose into the upper crust, carried along with granitic magma. Although most granite contains uranium, the amount it contains is so tiny that it's not cost-effective to extract. Geologic processes, however, can produce local concentrations of uranium. For example, hot water circulating through a pluton after intrusion may dissolve uranium and carry it to another location where it precipitates in veins as the mineral pitchblende (UO_2). Uranium may be further concentrated once a pluton, with its associated uranium-rich veins, weathers and erodes at the ground surface. When sand derived from a weathered pluton washes down a stream, the current may carry away relatively low-density grains of feldspar and quartz, leaving behind a concentration of high-density uranium-rich grains. Uranium deposits may also form where groundwater percolates through uranium-rich sedimentary rocks—the uranium dissolves in the water and moves with the water to another location, where it precipitates out of solution and fills the pores of the host sedimentary rock.

You can't just mine uranium and put it into a reactor. That's because ^{235}U, the isotope of uranium that serves as the most common fuel for nuclear power plants, accounts for only about 0.7% of naturally occurring uranium—most natural uranium consists of ^{238}U. Thus, to make a fuel suitable for use in a power plant, the ^{235}U concentration in a mass of natural uranium must be increased by a factor of 2 or 3, an expensive process called *uranium enrichment*.

Challenges of Using Nuclear Power

Maintaining safety at conventional nuclear power plants requires work. Operators must constantly cool the nuclear fuel with circulating water, and the rate of nuclear fission must be regulated by the insertion of *control rods*, which absorb neutrons and thus decrease the number of collisions between neutrons and radioactive atoms. If not properly controlled, the fuel could become so hot that it would melt. Such a **meltdown** might cause a steam explosion that could breach the containment building and scatter radioactive debris into the air. In some cases, the water gets so hot that it splits into hydrogen and oxygen gases that can recombine explosively. (Note that a meltdown is *not* the same as an atomic bomb explosion. The fuel of a reactor is not enriched enough for a critical mass to exist, in which fission reactions happen so quickly that the fuel itself explodes; reactor-grade uranium consists of about 3 to 4% ^{235}U, whereas weapons-grade uranium consists of 90% ^{235}U.)

Did you ever wonder...
if a nuclear power plant could explode like an atomic bomb?

Globally, about 435 nuclear power plants currently operate. To date, two major disasters have taken place, in which the reactor buildings were damaged and significant quantities of radiation were released. The most serious nuclear disaster occurred at the power plant in Chernobyl, Ukraine, in April 1986. During a poorly designed test, the fuel pile became extremely hot, triggering a steam or hydrogen explosion that raised the roof of the containment building, dispersed high concentrations of radioactive material around the plant grounds, and spread trace concentrations across broad regions of Europe and Asia. Within 6 weeks, 20 people who had worked at the plant died from radiation sickness. In 2011, a disaster occurred at the multireactor Fukushima power plant along the east coast of northern Japan. This disaster, second only to Chernobyl in terms of the amount of radiation released, occurred when the catastrophic tsunami (giant wave), that followed the *Mw* 9.0 Tōhoku earthquake, knocked out the power providing electricity to pumps that circulated cooling water. As a result, some of the reactors overheated, and they suffered partial meltdowns and hydrogen-gas explosions (see Chapter 8).

The nuclear industry also faces the challenge of managing *nuclear waste*, leftover radioactive material, or substances contaminated by radioactive material. Some of this material decays quickly (in decades to centuries), but some can remain dangerous for thousands of years. Nuclear waste cannot just be stashed in a warehouse or buried in a town landfill, where it could leak into water supplies. In fact, some of the high-level waste is so hot that it needs to be perpetually cooled with water. Ideally, waste should be sealed in containers that will last for thousands of years (the time needed for the short-lived radioactive atoms to undergo decay) and stored in a place where it will not come in contact with the environment. Finding an appropriate storage site is not easy.

TAKE-HOME MESSAGE

Controlled fission in reactors produces nuclear power. The fuel consists of enriched uranium, or other radioactive elements, obtained by mining. Reactors can produce large amounts of energy, but if not properly monitored, they can run the risk of meltdown. They also yield radioactive waste that is challenging to store.

QUICK QUESTION What is the difference between a meltdown in a reactor and the explosion of an atomic bomb?

12.8 Other Energy Sources

Geothermal Energy

As the name suggests, *geothermal energy* comes from heat in the Earth's crust. We can distinguish between high-temperature geothermal energy, which is used for producing heat and electricity at a commercial scale, and low-temperature geothermal energy (also known as *ambient geothermal energy*), which can warm or cool the water used by an individual household.

High-temperature geothermal energy exists because the crust becomes progressively hotter with increasing depth, at a rate defined by the *geothermal gradient*. In active volcanic areas, the increase is so fast that a temperature of 100°C, or more, the boiling point of water, occurs at relatively shallow depths, only several hundred meters from the Earth's surface. Therefore, homeowners in volcanic areas can pump hot groundwater through pipes to heat houses or spas directly, and power companies can use geothermal energy to produce electricity. Geothermal electrical power generation can be accomplished in a few different ways. If the groundwater is hot enough, it turns to steam when pumped up and out of the ground. This happens because the state of water (gas vs. liquid vs. solid) depends both on temperature and pressure—decompression of very hot groundwater, as happens when it is pumped from depth up to the Earth's surface, causes it to change state from liquid to gas. The steam can drive turbines which, in turn, drive electrical dynamos (**Fig. 12.13**). In Iceland and New Zealand, countries which include active volcanic areas, geothermal energy provides a substantial portion of electricity needs, but on a global basis, accessibility to geothermal electricity is limited.

Ambient geothermal energy takes advantage of the fact that below a depth of a few meters, ground temperature remains nearly constant all year, about equal to the average annual temperature of the air above. By running the pipes carrying a home's water supply through the ground, typically at a depth below 5 m, homeowners can precool the water used in their house during the summer, and can preheat the water used in their house in the winter. This process decreases the amount of energy needed to run furnaces or air conditioners, and thus lowers household costs.

SEE FOR YOURSELF...

GEOTHERMAL POWER PLANT, NEW ZEALAND

Latitude
38°37′34.84″ S

Longitude
176°6′18.82″ E

Zoom to 2 km (1.2 miles) and look straight down.

A geothermal power plant sits along the Waikato River in New Zealand. If you fly SSW, you cross Lake Taupo, a caldera marking the site of the world's largest eruption of the last 5,000 years. Further south is Mt. Tongariro, a stratovolcano that has erupted over 70 times in the last two centuries.

Biofuels

In recent years, farmers have begun to produce rapidly growing crops such as corn and sugarcane specifically for the purpose of producing biomass for fuel production. The resulting

FIGURE 12.13 Geothermal power plants utilize hot groundwater. In some cases, the water is so hot that it becomes steam when it rises and undergoes decompression. Most geothermal plants are in volcanic areas.

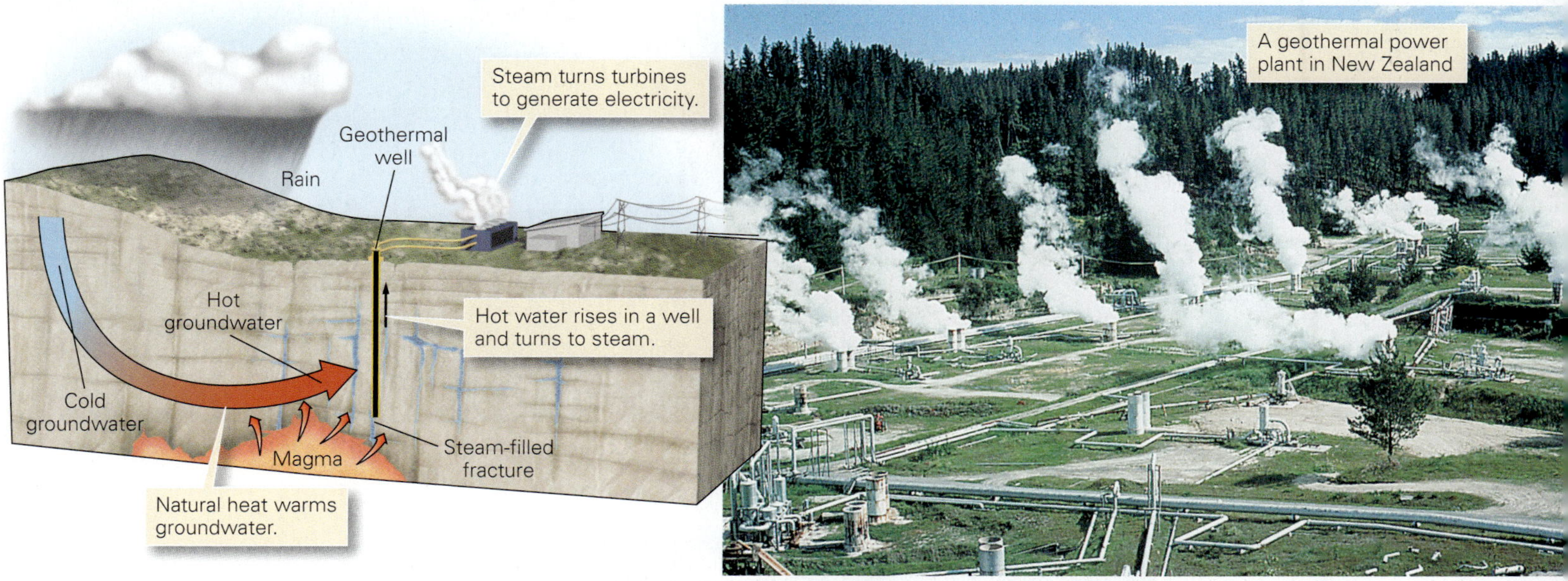
Steam turns turbines to generate electricity.
Geothermal well
Rain
Hot groundwater
Hot water rises in a well and turns to steam.
Cold groundwater
Magma
Steam-filled fracture
Natural heat warms groundwater.
A geothermal power plant in New Zealand

liquids are called **biofuels**. The most commonly used biofuel is ethanol, a type of alcohol, sometimes used as a substitute for gasoline in car engines. The process of producing ethanol from corn includes the following steps. First, producers grind corn into a fine powder, mix it with water, and cook it to produce a mash of starch. They then add an enzyme to the mash, which converts the starch to sugar. The sugar, when mixed with yeast, ferments to produce ethanol, water, and CO_2. Finally, the ethanol is distilled to concentrate the ethanol. Ethanol can be produced directly from sugar extracted from sugarcane, without fermentation, so sugarcane ethanol is less expensive than corn alcohol.

Researchers are working to develop processes that yield ethanol and other hydrocarbons from other sources that are not used as food, so that biofuel production doesn't raise food costs. Examples of alternate biofuel sources include cellulose from perennial grasses, hydrocarbons from algae (which naturally produces fatty organic chemicals), and *biodiesel* (a fuel produced by chemical modification of fats and vegetable oils).

Hydroelectric and Wind Power

For millennia, people have used flowing water and air to produce energy. In fact, many towns were established next to rivers, where streams could rotate the waterwheels that powered mills and factories. And in agricultural areas, farmers used windmills to pump water for irrigation. In the past century, engineers have begun to employ the same basic technology to drive generators that produce electricity.

In a modern hydroelectric power plant, water flows from a higher elevation to a lower elevation. The flowing water turns turbine blades placed in a pipe, and the turbine drives an electrical generator. Most hydroelectric plants rely on water from a reservoir held back by a dam. The largest of these is the Three Gorges Dam on the Yangtze River in China (**Fig. 12.14a**). Dam construction increases the available potential energy of the water because the water level in a filled reservoir is higher than the level of the valley floor that the dam spans. Hydroelectric energy is clean in the sense that its production does not release chemical or radioactive pollutants, and is renewable in that its production does not consume limited resources. Also, reservoirs may have the added benefit of providing flood control, irrigation water, and recreational opportunities. But damming a river does affect the environment. For example, it may flood a spectacular canyon, eliminate exciting rapids, destroy ecosystems, or submerge towns. Further, reservoirs trap sediment and nutrients, thus disrupting the supply of these materials to downstream floodplains or deltas, a process that may adversely affect agriculture.

Not all hydroelectric power generation utilizes flowing river water. Engineers have been developing new means to tap **tidal power**, the energy associated with the daily rises and falls of the tides (see Chapter 15). One approach involves building a dam, called a *tidal barrage*, across the entrance to a bay or estuary (the flooded mouth of a river) in which there are large tides. When the tide rises, water spills into the enclosed area through openings in the dam. When the tide drops, water remains trapped behind the barrage, and then

FIGURE 12.14 Alternative energy sources.

(a) Gravity causes water held back by the Three Gorges Dam to flow through turbines and generate electricity.

(b) A wind farm in southwestern England. The towers are about 50 m high.

(c) A photovoltaic cell produces electricity directly from solar radiation.

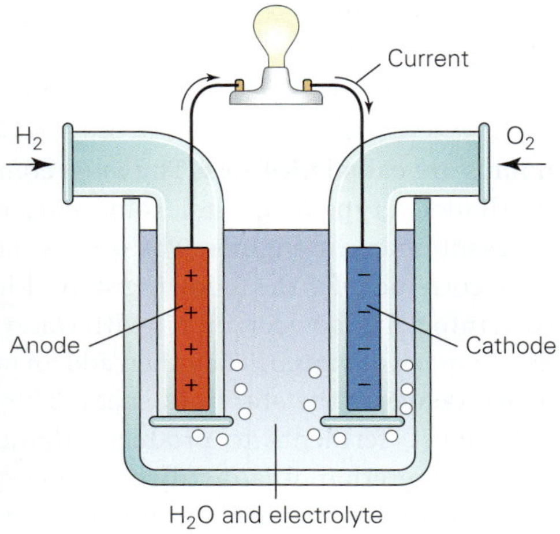

(d) A hydrogen fuel cell.

flows back to the sea via a pipe that carries it through a power-generating turbine.

When you think of wind energy, you may picture a classic Dutch windmill driving a water pump. Modern efforts to harness the wind are on a much larger scale. To produce wind-generated electricity, engineers identify regions, either onshore or just offshore, with steady breezes. These regions host wind farms made of numerous towers, each with a wind turbine, a giant fan blade that turns even in a gentle breeze (**Fig. 12.14b**). Some towers can be 100 m (300 ft) tall, with fan blades over 40 m (120 ft) long, and large wind farms can host hundreds of towers. Wind energy produces no pollutants, but as with any type of energy source, wind power has some drawbacks. Cluttering the horizon with towers may spoil a beautiful view, and the constant loud hum of the towers can disturb nearby residents. The towers may also be a hazard to migrating birds, and offshore towers may interfere with marine life.

Solar Energy

The Sun drenches the Earth with energy in quantities that dwarf the amounts stored in fossil fuels. Were it possible to harness this energy directly, humanity would have a reliable and totally clean solution for powering modern technology. Solar energy used today comes primarily from **solar cells** (also called solar panels or photovoltaic cells), which convert light energy directly to electricity (**Fig. 12.14c**). Solar cells consist of two wafers of silicon pressed together. One wafer contains tiny amounts of arsenic, and the other wafer contains tiny amounts of boron. When light strikes the cell, arsenic atoms release electrons that flow over to the boron atoms. If a wire loop connects the back side of one wafer to the back side of the other, this phenomenon produces an electrical current. Most solar cells have an efficiency of less than 25%, but the cost of solar panels has dropped by more than a factor of 10 over the past 15 years, so in sunny regions, they have become increasingly popular as a means of household-based electricity generation.

Fuel Cells

In a *hydrogen fuel cell*, hydrogen gas flows through a tube, across a platinum anode that has been placed in a water solution containing an electrolyte (a substance that enables the solution to conduct electricity). At the same time, a stream of oxygen gas flows onto a separate platinum cathode that has also been placed into the solution. A wire connects the anode and the cathode to provide an electrical circuit (**Fig. 12.14d**). In this configuration, hydrogen reacts with oxygen to produce water and electricity. Fuel cells are efficient and clean. Their limitation lies in the need for a design that will protect the cells from damage by impact and will enable them to store hydrogen, an explosive gas, in a safe way. Also, it takes a significant amount of energy from other sources to produce and condense the hydrogen used in fuel cells.

TAKE-HOME MESSAGE

A variety of alternative energy resources are now under development. Biomass can be transformed into burnable alcohol and biodiesel. Geothermal energy utilizes groundwater that has been warmed by heat from Earth's interior. Flowing water (in rivers or tides), flowing air, solar panels, and fuel cells can also produce energy.

QUICK QUESTION Is it possible to use ambient geothermal energy in nonvolcanic regions? Why?

12.9 Energy Choices, Energy Challenges

The Age of Conventional Oil, and the Oil Crunch

Energy usage in industrialized countries grew with dizzying speed through the mid-20th century, and during this time people came to rely increasingly on oil. Oil remains the single largest source of energy globally, accounting for about 33% of global energy consumption.

During the latter half of the 20th century, conventional oil supplies within the borders of industrialized countries could no longer match the demand, and these countries began to import more oil than they produced themselves. In 1973, a complex tangle of politics and war led the Organization of Petroleum Exporting Countries (OPEC) to limit its oil exports. In the United States, fear of an oil shortage turned to panic, and motorists lined up at gas stations, in many cases waiting for hours to fill their tanks. The price of oil rose from $3 a barrel to $12 a barrel, and newspaper headlines proclaimed, "Energy Crisis!" Governments in industrialized countries instituted new rules to encourage oil conservation. During the last two decades of the 20th century, the oil market stabilized, with occasional price jumps and short-term shortages. Since 2004, oil prices rose overall, passing the $147/bbl mark in 2008. During the Great Recession of 2008, the price collapsed temporarily, but it then rose again and hovered around $100/bbl until the middle of 2014, when it dropped by 50%, reaching a low of $47/bbl (**Fig. 12.15a**). While this drop caused a significant decrease in the price of gas at the pump, a huge benefit to consumers, it led to a situation where production from some unconventional reserves has become uneconomical, and this has impacted jobs in the industry. The large fluctuations in oil price emphasize that oil is a commodity whose value reflects market rules of supply and demand. When the supply is high, the price drops, and when the supply is low, the price rises. Supply also reflects which products are being produced by refineries.

Will a day come when the supply of conventional oil drops not because of an embargo or limitations on refining capacity, but because there is no more oil to produce? To address such a question, we must keep in mind the distinction between renewable and nonrenewable energy resources. We can call a particular resource *renewable* if nature can replace it within a short time relative to a human life span (in months or, at most, decades), whereas a resource is *nonrenewable* if nature takes a very long time (hundreds to millions of years) to replenish it. Oil is a nonrenewable resource in that the rate at which humans consume it far exceeds the rate at which nature replenishes it, so we will inevitably run out of oil. The question is, when?

Historians in the future may refer to our time as the **Oil Age** because so much of our economy depends on oil. How long will the Oil Age last? A reliable answer to this question is hard to come by, because there is not total agreement on the numbers that go into the calculation, especially as the use of unconventional reserves increases, so estimates vary widely. Geologists estimate that we've already used a substantial proportion of our conventional reserves, but that about 1,650 billion barrels of proven conventional oil reserves remain. Proven means reserves that have been documented and are still in the ground. Optimistically, an additional 2,000 billion barrels of unproven, possible, conventional reserves may exist. Thus, the world conceivably holds between 1,650 and 3,650 billion barrels of conventional oil. Presently, humans

FIGURE 12.15 Price, discovery, and consumption of oil.

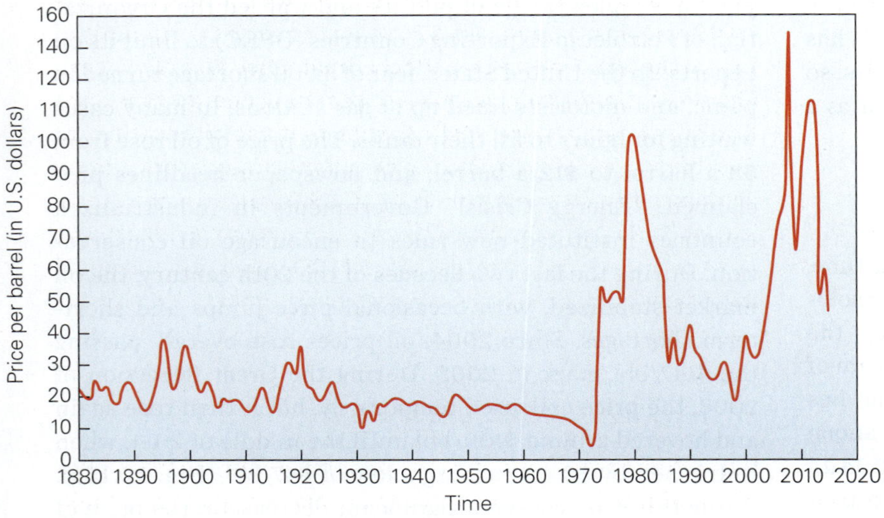

(a) The price of oil held fairly steady for almost 100 years. Starting in 1970, it has risen and fallen dramatically. Data source: BP Statistical Review of World Energy.

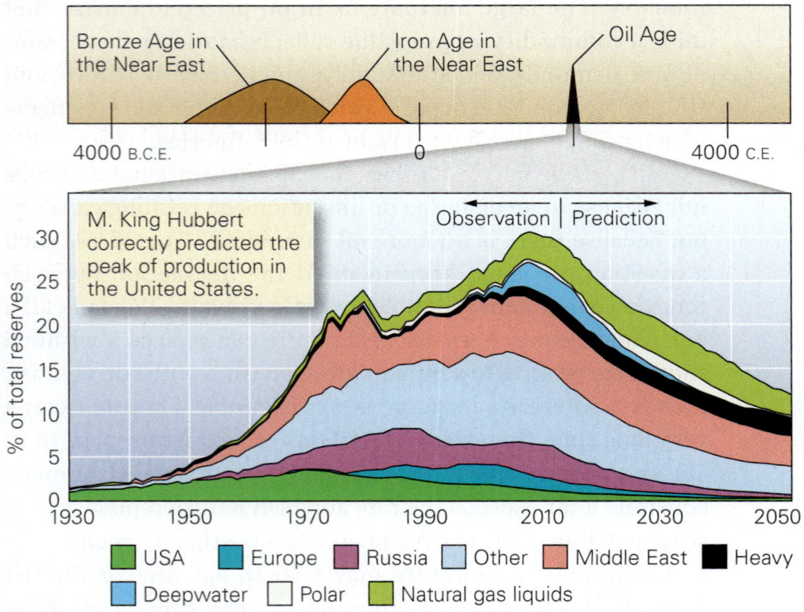

(b) A plot of production vs. time suggests that some time in the early 21st century, the production of oil globally will start to decrease.

> ***Did you ever wonder...***
> how much longer the world's oil supply will last?

consume oil at a rate of about 34 billion barrels per year. At this rate, conventional oil supplies may last until some time between 2070 and 2170.

Some researchers argue that the beginning of the end of the Oil Age has begun, because the rate of consumption now exceeds the rate of discovery, and in many regions the rate of production has already started to decrease. The peak of production for a given reserve is called **Hubbert's Peak**, named for the geologist who first emphasized that the production of nonrenewable resources eventually must decline (**Fig. 12.15b**). Hubbert's Peak for conventional hydrocarbons in the United States appears to have been passed in the 1970s. Some researchers argue that the global peak may have occurred between 2012 and 2015, but this assessment remains uncertain. Conservation measures, such as improving the gas mileage of cars and increasing the amount of insulation in buildings, could stretch out supplies and make them last decades longer.

Of course, the picture of oil supplies changes significantly if unconventional reserves are included in estimates. The huge increase in production of unconventional reserves in the United States has changed the slope of the curve showing production related to time. Current estimates count 420 billion barrels of proven unconventional oil reserves, and more than a trillion additional barrels of unconventional reserves may yet be discovered. However, wide disagreement remains concerning whether it's reasonable to include all of these reserves in estimating hydrocarbon supplies, because a significant proportion would be so difficult and expensive to access that such reserves may never really be an economical energy source.

Even the combination of conventional and accessible unconventional oil reserves means that at current rates of consumption supplies can last for only another 200 years, so the Oil Age could last a total of about 350 years. On a timeline representing the 4,000 years since the construction of the Egyptian pyramids, this looks like a very short blip. We may indeed be living during a unique interval of human history.

Can Other Fossil Fuels Replace Oil?

As true limits to the conventional oil supply approach, societies are looking first at relatively abundant supplies of other fossil fuels, namely, natural gas and coal, as sources of energy. Rough estimates suggest that world natural gas reserves may exceed 180 trillion cubic meters, which would provide approximately the same amount of energy as 1.2 trillion barrels of oil. But tapping into this gas supply requires expensive technologies for extraction and transport. Similarly, worldwide coal reserves are estimated to be about 850 trillion tons, which contains approximately the same amount of energy as 11 trillion barrels of oil. But the stated number for coal reserves does not distinguish clearly between accessible (mineable) coal and inaccessible coal, which is too deep to mine. And, as is the case for oil, there are political and environmental consequences to relying on gas or coal.

Environmental Issues of Fossil Fuel Use

Environmental concerns about energy resources begin right at the source. Oil drilling requires substantial equipment, the use of which can damage the land. And as demonstrated by the 2010 Gulf of Mexico offshore well blowout, oil drilling can lead to tragic loss of life and disastrous marine oil spills (**Box 12.3**). Oil spills from pipelines or trucks sink into the subsurface and contaminate groundwater, and oil spills from ships and tankers create slicks that spread over the sea surface and foul the shoreline (**Fig. 12.16**). Coal and uranium mining also scar the land and can lead to the production of *acid mine runoff,* a dilute solution of sulfuric acid that forms when sulfur-bearing minerals such as pyrite (FeS_2) in mines react with rainwater. The runoff enters streams and kills fish and plants. Collapse of underground coal mines may cause the ground surface to sink.

Numerous air-pollution issues also arise from the burning of fossil fuels, which sends soot, carbon monoxide, sulfur dioxide, nitrous oxide, and unburned hydrocarbons into the air. Coal, for example, commonly contains sulfur, primarily in the form of pyrite, which enters the air as sulfur dioxide (SO_2) when coal is burned. This gas combines with rainwater to form dilute sulfuric acid (H_2SO_4), or **acid rain**. For this reason, many countries now regulate the amount of sulfur that coal can contain when it is burned. Even if pollutants can be decreased, burning fossil fuels still releases carbon dioxide (CO_2) into the atmosphere. As we discuss in Chapter 19, CO_2 is a *greenhouse gas,* so a change in the amount of CO_2 in the atmosphere can affect climate. Because of concern about CO_2 production, people are pursuing energy conservation efforts, improving technologies that use alternatives to fossil fuels (see below), and developing techniques for *carbon capture and sequestration* (*CCS*). CCS involves three steps. First, CO_2 is removed from the effluents sent up smokestacks at powerplants or factories; second, CO_2 gas is transformed into liquid; and third, CO_2 is pumped down wells into reservoir rocks deep underground.

Alternatives to Fossil Fuels

Can other energy sources replace fossil fuels? Vast supplies of uranium, the fuel of traditional nuclear plants, remain untapped. Further, nuclear engineers have designed new generations of power plants that essentially produce their own fuel, and in which the core can be kept cool without water. But concerns about power plants remain because of issues pertaining to radiation, accidents, terrorism, and waste storage, and these concerns have slowed the industry. A substantial increase in hydroelectric power production is not likely, as most major rivers have already been dammed, and industrialized countries have little appetite for taming any more. Similarly, the growth of geothermal-energy output seems limited. Because of these problems, researchers have been increasingly exploring solar power and wind power (see Fig. 12.14b,c). One of the challenges is that these sources can't provide a steady supply that our present-day **energy grid** (the configuration of generators, transformers, and powerlines

FIGURE 12.16 Marine oil spills. These can come from drilling rigs, or from tankers.

(a) An oil tanker leaking oil on the sea surface.

(b) Oil spills can contaminate the shore and can be very difficult to clean.

BOX 12.3 Consider This...

Offshore Drilling and the Deepwater Horizon Disaster

A substantial proportion of the world's oil reserves reside in the sedimentary basins that underlie the continental shelves of passive continental margins. To access such reserves, oil companies must build offshore drilling platforms. In water less than 600 m (2,000 ft) deep, companies position fixed platforms on towers resting on the seafloor. In deeper water, semi-submersible platforms float on huge submerged pontoons. By drilling from these platforms, oil companies can now access fields lying beneath 3 km (10,000 ft) of water. North America's largest offshore fields occur in the passive-margin basin that fringes the coast of the Gulf of Mexico. More than 3,500 platforms operate in the Gulf at present, together yielding up to 1.7 million bbl/day.

During both onshore and offshore exploration, drillers worry about the possibility of a **blowout**. A blowout happens when the pressure within a hydrocarbon reserve penetrated by a well exceeds the pressure that drillers had planned for, causing the hydrocarbons (oil and/or gas) to rush up the well in an uncontrolled manner and burst out of the well at the surface in an oil gusher or gas plume. Blowouts are rare because, although fluids below the ground are under great pressure due to the weight of overlying material, engineers fill the hole with drilling mud with a density greater than that of clear water. The weight of drilling mud can counter the pressure of underground hydrocarbons and hold the fluids underground. But if drillers encounter a bed in which pressures are unexpectedly high, or if they remove the mud before the walls of the well have been sealed with a casing (a pipe, cemented in place by concrete), a blowout may happen.

A catastrophic blowout occurred on April 20, 2010, when drillers on the Deepwater Horizon, a huge semi-submersible platform, were finishing a 5.5-km-long (18,000 ft) hole in 1.5-km-deep (5,000 ft) water south of Louisiana. Due to a series of errors, the casing was not sufficiently strong when workers began to replace the drilling mud with clear water. Thus the high-pressure, gassy oil in the reservoir strata that the well had accessed rushed up the drillhole. A backup safety device called a blowout preventer, which should have clamped the pipe shut, failed, so the gassy oil reached the platform and sprayed 100 m (328 ft) into the sky. Sparks from electronic gear triggered an explosion, and the platform became a fountain of flame and smoke that killed 11 workers. An armada of fireboats could not douse the conflagration (**Fig. Bx12.3a**), and after 36 hours, the still-burning platform tipped over and sank.

Robot submersibles sent to the seafloor to investigate found that the twisted mess of bent and ruptured pipes at the well head was billowing oil and gas. On the order of 50,000 to 62,000 bbl/day of hydrocarbons entered the Gulf's water from the well. Stopping this underwater gusher proved to be an immense challenge, and initial efforts to block the well, or to put a containment dome over the well head, failed. It was not until July 15 that the flow was finally stopped, and not until September 19 that a new relief well intersected the blown well and provided a conduit to pump concrete down to block the original well permanently. All told, about 4.2 million bbl of hydrocarbons contaminated the Gulf from the Deepwater Horizon blowout (**Fig. Bx12.3b**). The spill was devastating to wetlands, wildlife, and the fishing and tourism industries.

FIGURE Bx12.3 The Deepwater Horizon disaster (2010).

(a) Fireboats dousing the burning rig before the rig sank.

(b) A satellite image of the oil slick in the Gulf of Mexico, southeast of the Mississippi delta.

that distributes electricity through a region) requires, since the intensity of sunlight and the strength of the wind varies almost hourly. Clearly, society will be facing difficult choices in the not-so-distant future about where to obtain energy. In the near term, conservation can play an important role by diminishing demand for fossil fuels.

TAKE-HOME MESSAGE

Oil is a nonrenewable resource and conventional supplies may run out in less than a century—we live in the Oil Age, which may last less than 350 years. Unconventional supplies may last longer, but use of fossil fuels has environmental consequences. Thus, research into the use of alternative energy sources has accelerated.

QUICK QUESTION What is Hubbert's Peak, and when was it reached in the United States?

12.10 Mineral Resources

In January of 1848, James Marshall and a crew of workers were putting the finishing touches on a new sawmill in the foothills of the Sierra Nevada range. They had been deepening the channel that would guide flowing water to the wheel that would drive the mill. As Marshall examined the excavation, he noticed a glimmer of metal in the gravel that littered its floor. He picked up a sample and banged it between two rocks, knowing that if the sample crushed into fragments, it was only pyrite (fool's gold), but if it flattened without breaking, it was real gold. Marshall looked up and said to his crew, "I have found it . . . gold!" Word soon spread, and gold fever spread across the country. As a result, 1849 brought 40,000 prospectors to California. These forty-niners had abandoned their friends and relatives on the gamble that they could strike it rich. Marshall himself, sadly, did not, and died penniless.

Gold is but one of many mineral resources, meaning minerals extracted from the Earth's crust for use by people. Without these resources, industrialized societies could not function. Geologists divide mineral resources into two categories: *metallic mineral resources* (materials containing gold, copper, aluminum, iron) and *nonmetallic mineral resources* (building stone, gravel, sand, gypsum, phosphate, salt) used in construction or for chemical production. Below, we look at the nature of mineral resources, the ways people mine them, and the geologic phenomena responsible for their formation. We conclude by considering the sustainability of mineral reserves in coming years.

Metal and Its Discovery

A **metal** is an opaque, shiny, smooth solid that can conduct electricity and is malleable, meaning it can be bent, drawn into wire, or hammered into thin sheets. The way metals look and behave is quite different from wood, plastic, meat, or rock because the atoms that make up metals are held together by *metallic bonds*, so electrons can flow from atom to atom fairly easily and atoms can, in effect, slide past each other without breaking apart.

The first metals that people used—copper, silver, and gold—can occur in rock in the form of *native metal*. Native metal consists only of metal atoms, and thus looks and behaves like metal. Gold nuggets, for example, are chunks of native gold that have eroded free of bedrock (**Fig. 12.17**). Over the ages, people have collected nuggets of native metal from streambeds and pounded them together with stone hammers to make arrowheads, scrapers, and later, coins and jewelry. But if we had to rely solely on native metals as our source of metal, we would have access to only a tiny fraction of our current metal supply. Most of the metal atoms we use today originated as ions bonded to nonmetallic elements in a great variety of minerals that themselves look nothing like metal. To obtain these metals, they have to be extracted from rock by *smelting*. Smelting involves heating ore and reacting it with chemicals—it yields metal, plus gas and a solid, nonmetallic residue called slag.

What Is an Ore?

The minerals from which metals can be extracted are called **ore minerals**, or *economic minerals*. Ore minerals contain metal in high concentrations and in a form that can be easily extracted. Galena (PbS), for example, is about 50% lead, so we consider it to be an ore mineral of lead (**Fig. 12.18a**). We

FIGURE 12.17 Gold occurs as native metal within quartz veins. The quartz breaks up to form sand, leaving nuggets of gold.

Gold bracelets on display in a Kuwaiti marketplace

FIGURE 12.18 Examples of ore minerals.

(a) This lead ore, from Missouri, consists of galena (PbS) crystals that grew in dolostone.

(b) Most copper comes from ore minerals that look nothing like metallic copper. This ore consists of azurite (blue) and malachite (green).

obtain most of our iron from hematite and magnetite. Copper comes from a variety of minerals, none of which look like copper (**Fig. 12.18b**). Geologists have identified a great variety of ore minerals. Many ore minerals are sulfides, in which the metal occurs in combination with sulfur (S); or oxides, in which the metal occurs in combination with oxygen (O). A few are carbonates.

Ore is a rock containing a relatively concentrated accumulation of ore minerals or native metals. Iron constitutes only about 5% to 6% of the continental crust's weight, but it makes up about 30% to 60% of the weight of iron ore. The weight percent of a useful metal in an ore determines the **grade** of the ore—the higher the weight percent, the higher the grade. For particularly valuable metals, the grade of ore can be quite low for the ore to be worth mining. For example, most copper ore used today has a grade of 0.4% to 1.0%.

How Do Ore Deposits Form?

Ore minerals do not occur uniformly through rocks of the crust. If they did, we would not be able to extract them economically. Fortunately for humanity, geologic processes concentrate these minerals into accumulations called **ore deposits**. Geologists distinguish among several different types of ore deposits based on the process by which the deposit formed.

Magmatic Deposits When a magma intrusion cools and starts to solidify, sulfide ore minerals may crystallize preferentially in distinct lenses or bands (**Fig. 12.19a**). Such a concentration is called a *magmatic deposit*, because it forms directly from a magma. Because these concentrations consist almost exclusively of sulfide minerals, they are a type of *massive sulfide deposit*.

Hydrothermal Deposits The heat from a pluton's intrusion can cause groundwater to start convecting through the pluton and the wall rocks surrounding the intrusion. This water heats up as it passes through the intrusion and becomes a hot, hydrothermal fluid that can dissolve metal ions. When the fluid enters a region of lower pressure, lower temperature, different acidity, and/or different availability of oxygen, the metal ions bond to other elements and precipitate as ore minerals in fractures and pores. The resulting concentration of ore is called a *hydrothermal deposit* (**Fig. 12.19b**).

In recent decades, geologists have discovered that hydrothermal activity occurs when seawater enters the crust and convects through the crust in the vicinity of submarine volcanoes along mid-ocean ridges. As it flows through the crust, the fluids dissolve metals, and when the fluids erupt from vents along the ridge axis, they contain high concentrations of dissolved metal and sulfur. The instant the hydrothermal solutions come in contact with cold seawater, the dissolved components precipitate as tiny crystals of metal-sulfide minerals (**Fig. 12.19c**). This phenomenon makes the erupting water look like a black cloud, so the vents are called black smokers (see Chapter 2). Minerals in the cloud sink and form a pile of ore minerals around the vent. Since the ore minerals typically are sulfides, the resulting hydrothermal deposits constitute another type of massive sulfide deposit.

Secondary-Enrichment Deposits Sometimes groundwater passes through ore-bearing rock long after the rock first formed. This groundwater leaches (dissolves) some of the metal in the ore and carries the ions away. When the groundwater eventually flows into a different chemical environment (for instance, one containing a different concentration of oxygen or acid), it precipitates new ore minerals, commonly

FIGURE 12.19 Various processes that form ore deposits.

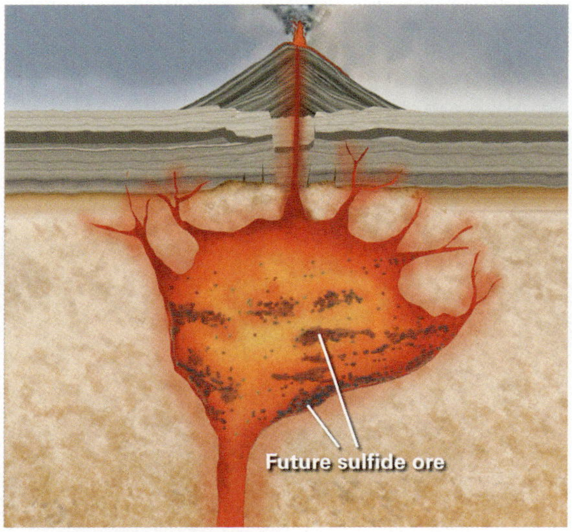

(a) Massive sulfide deposits can form when sulfide ore minerals form concentrations in a magma chamber.

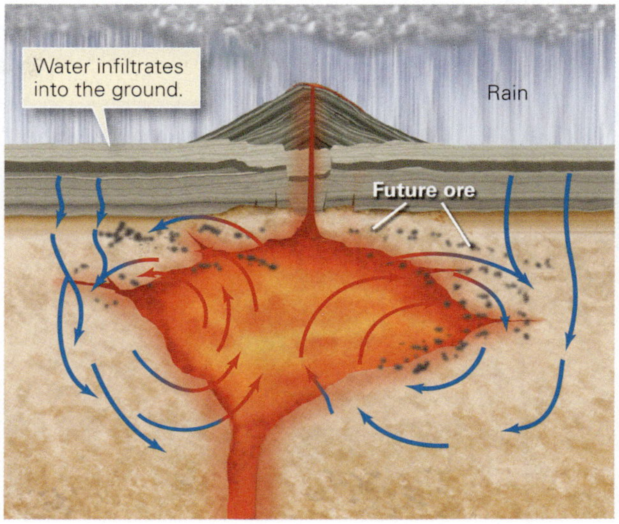

(b) Hydrothermal deposits form when water circulating around and through magma dissolves and redistributes metals (arrows indicate flowing water).

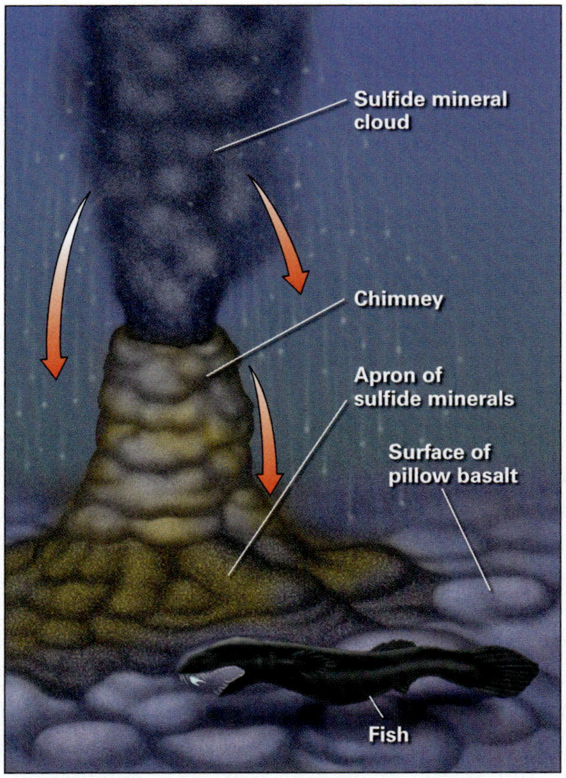

(c) Massive sulfide deposits also form when ore minerals precipitate around hydrothermal vents (black smokers) along a mid-ocean ridge.

in concentrations exceeding that of the original deposit. A new ore deposit formed from metals that were dissolved and carried away from a pre-existing ore deposit is called a *secondary-enrichment deposit*. Some of these deposits contain spectacularly beautiful copper-bearing carbonate minerals, such as azurite and malachite (see Fig. 12.18b).

MVT Ores Rain falling along one margin of a large sedimentary basin may sink into the subsurface and then flow as groundwater along a curving path that takes it first down into rocks in the basin's lower section, several kilometers below the surface, and then eventually back up to the distant margin of the basin. At the base of the basin, temperatures can become high enough that the water dissolves metals. When the water returns to the surface and enters cooler rock, these metals precipitate in ore minerals. Ore deposits formed in this way, typically containing lead- and zinc-bearing minerals, appear in dolomite beds of the Mississippi Valley region of the United States, and thus have come to be known as *Mississippi Valley–type* (*MVT*) ores.

Did you ever wonder...
where the iron in the steel bodies of cars comes from?

Sedimentary Deposits of Metals Some ore minerals accumulate in sedimentary environments under special circumstances. For example, between 2.5 and 1.8 billion years ago, the atmosphere, which previously had contained very little oxygen, started to accumulate oxygen. This change affected the chemistry of seawater so that large quantities of dissolved iron precipitated as iron oxide minerals that settled as sediment on the seafloor. As discussed in Chapter 11, the resulting iron-rich sedimentary deposits are known as *banded iron formation* (*BIF*) (**Fig. 12.20**), because after lithification they consist of

FIGURE 12.20 Precambrian BIF from northern Michigan consists of hematite interbedded with jasper.

Hematite layer
Jasper layer

FIGURE 12.21 Placer deposits form where erosion produces clasts of native metals. Sorting by the stream concentrates the metals.

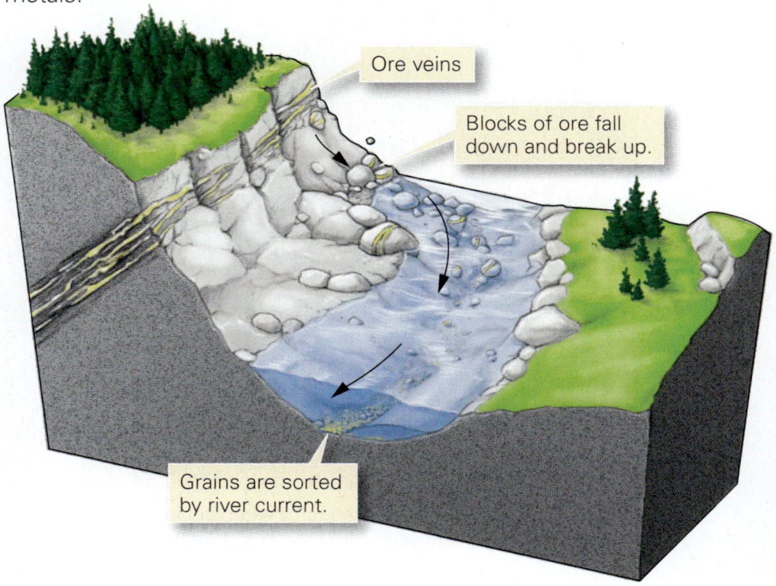
Ore veins
Blocks of ore fall down and break up.
Grains are sorted by river current.

alternating beds of gray iron oxide (magnetite or hematite) and red beds of jasper (iron-rich chert).

Although BIF no longer forms, the chemistry of seawater in some parts of the ocean today can lead to the deposition of manganese-oxide minerals. These minerals grow into lumpy accumulations, known as *manganese nodules*, on the seafloor. Mining companies have begun to explore technologies for obtaining these nodules; researchers estimate that the worldwide supply of nodules contains 720 years' worth of copper and 60,000 years' worth of manganese, at current rates of consumption.

Residual Mineral Deposits Recall from Interlude B that as rainwater sinks into the Earth, it leaches certain elements and leaves behind others, as part of the process of forming soil. In rainy, tropical environments, the residue left behind in soils after leaching includes concentrations of iron or aluminum, which do not dissolve easily in the downward percolating rainwater. Locally, these metals become so concentrated that the soil itself becomes an ore deposit. We refer to such deposits as *residual mineral deposits*. Most of the aluminum ore mined today comes from bauxite, a residual mineral deposit created by the extreme leaching of rocks (such as granite) containing aluminum-bearing minerals.

Placer Deposits Ore deposits may develop when rocks containing native metals erode, producing a mixture of sand grains and metal flakes or nuggets (pebble-sized fragments) that can be concentrated by streams, because the moving water carries away lighter mineral grains (quartz and feldspar) but can't move the heavy metal grains (gold) so easily. Concentrations of metal grains in stream sediments are a type of *placer deposit* (**Fig. 12.21**). Panning further concentrates gold flakes or nuggets, for swirling water in a pan causes the lighter sand grains to wash away, leaving the gold behind.

Where Are Ore Deposits Found?

The Inca empire of 15th-century Peru boasted elaborate cities and temples, decorated with fantastic masks, jewelry, and sculptures made of gold. Then, around 1532, Spanish conquistadors arrived. Within six years, the Inca Empire had vanished, and Spanish ships were transporting Inca treasure back to Spain. Why did the Incas possess so much gold? Or to ask the broader question, what geologic factors control the distribution of ore? Once again, we can find the answer by considering the consequences of plate tectonics.

Several of the ore-deposit types we've mentioned occur in association with igneous rocks. As we learned in Chapter 4, igneous activity does not happen randomly around the Earth, but rather concentrates along convergent plate boundaries (specifically, in the overriding plate of a subduction zone), along divergent plate boundaries (along mid-ocean ridges), along continental rifts, or at hot spots. Thus, magmatic and hydrothermal deposits (and secondary-enrichment deposits derived from these) occur in these geologic settings. Placer deposits are typically found in the sediments eroded from such magmatic or hydrothermal deposits. The Inca gold formed in the Andes along a convergent boundary.

FIGURE 12.22 Mining ore deposits.

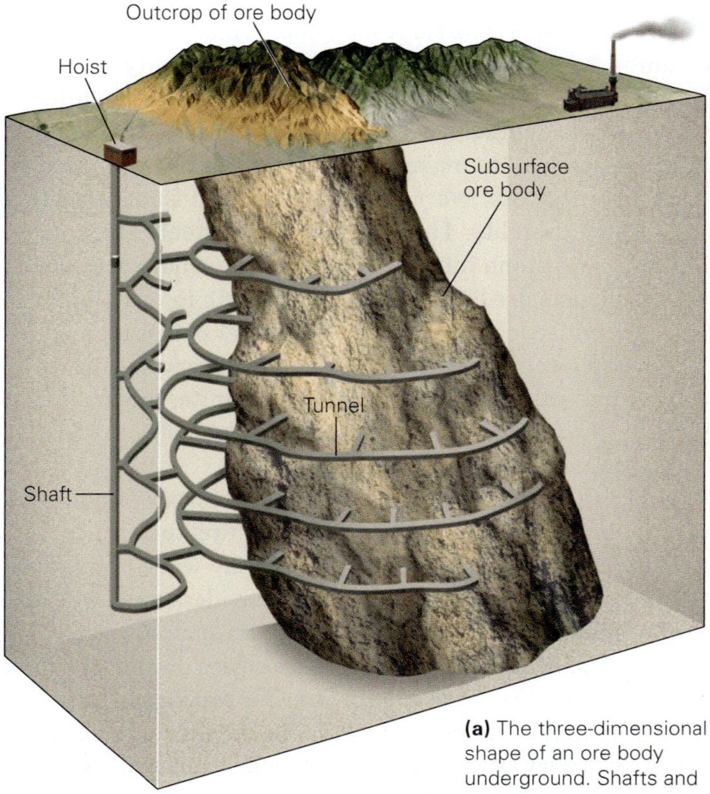

(a) The three-dimensional shape of an ore body underground. Shafts and tunnels access the body.

(b) Open-pit mining utilizes ore that lies fairly close to the ground surface. Terracing the mine walls helps to stabilize them.

Ore-Mineral Exploration and Production

Imagine an old prospector of days past clanking through the desert with a worn-out donkey. To find bedrock ore, such prospectors would eye hillsides for a "show," visible evidence on the ground surface of ore below. What does a show look like? It may be an outcropping of milky-white quartz veins, or brightly colored stains due to oxidation (rusting) of ore minerals. On finding a possible ore, a prospector would take samples back to town for an *assay*, a test to determine how much extractable metal the samples contain. If the assay indicated a significant concentration of metal, the prospector might "stake a claim" by literally marking off an area of ground with stakes.

These days, large mining companies employ professional geologists to survey ore-bearing regions systematically. The geologists focus their studies on rocks that developed in settings appropriate for ore formation. Once such a region has been identified, geologists may measure gravity or magnetic anomalies (see Interlude D), because ore minerals tend to be denser and more magnetic than average rocks, to define the limits of the ore body (**Fig. 12.22a**). They may also sample rocks and soils to test for metal content, and may even analyze plants in the area to detect traces of metals, for plants absorb metals through their roots. Once geologists have identified a possible ore deposit, they drill holes to sample subsurface rock, to help determine the ore deposit's shape, extent, and grade.

If calculations indicate that mining an ore deposit will yield a profit, and if environmental concerns can be addressed properly, a company builds a mine. Mines can be below or above ground, depending on how close the ore deposit is to the surface. To develop an *open-pit mine* (**Fig. 12.22b**), workers first drill a series of holes into the solid bedrock and then fill the holes with high explosives. They must space the holes carefully and must set off the charges in a precise sequence, so that the bedrock shatters into appropriate-sized blocks for handling. When the dust settles, large front-end loaders dump the ore into giant ore trucks, which can carry as much as 200 tons of ore in a single load. (In comparison, a loaded cement mixer at a construction site weighs about 70 tons.) The trucks transport waste rock or *tailings* (rock that doesn't contain ore) to a tailings pile, and the ore to a *jaw crusher*, a large

SEE FOR YOURSELF...

BINGHAM COPPER MINE, UTAH

Latitude
40°31′14.66″ N

Longitude
112°9′1.97″ W

Zoom to 20 km (~12 miles) and look straight down.

The gray patch southwest of Salt Lake City is the largest open-pit mine in the world. Over 17 million tons of copper have been extracted from ore formed when hydrothermal fluids circulated through Cenozoic igneous rock. Zoom in closer to see the pit and tailings pile.

machine that compresses the stone between two metal plates until it shatters into smaller fragments. Workers then separate ore minerals from other minerals in the fragments and send the ore-mineral concentrate to a processing plant, where smelting or treatment with acidic solutions can separate metal atoms from other atoms. Eventually, workers melt the metal and then pour it into molds to make ingots (brick-shaped blocks) for transport to a manufacturing facility.

If the ore deposit lies more than about 100 meters below the Earth's surface, miners must dig an *underground mine.* To do so, they either bore a tunnel into the side of a mountain (the entrance to the tunnel is called an *adit*), or they sink a vertical shaft with an elevator. At the level in the crust where the ore body appears, they build a maze of tunnels into the ore by drilling holes into the rock and then blasting. The rock removed in this process must be carried back to the surface. Rock columns between the tunnels hold up the ceiling of the mine. Miners face danger from mine collapse and rockfalls. To help minimize these risks, operators bolt steel beams and nets against the ceiling, to hold falling rock in place.

TAKE-HOME MESSAGE

Metals are malleable materials in which metallic bonds hold atoms together. While some occur as native metals, most occur in other minerals and have to be extracted by smelting. An ore is a rock that contains a relatively high concentration of a useful metal, and an ore deposit is an accumulation of ore. Ore deposits form for a variety of reasons. For example, some crystalize from melts, some precipitate from hydrothermal fluids, and some collect as sediment. To find ore deposits, geologists search for ore shows, make maps of gravity and rock magnetism, and obtain drill cores. Mining takes place either in open-pit mines or in shafts and tunnels underground.

QUICK QUESTION Why do many ore deposits occur in association with convergent plate boundaries?

12.11 Nonmetallic Mineral Resources

So far, this chapter has focused on resources that contain metal. Society uses many nonmetallic mineral resources, also known as *industrial minerals,* as well. From the ground, we get the stone used to make roadbeds and buildings, the chemicals for fertilizers, the gypsum in drywall, the salt filling salt shakers, and the sand used to make glass—the list is endless. This section looks at a few of these materials and describes their sources.

Dimension Stone

The Parthenon, a colossal stone temple rimmed by 46 carved columns, has stood atop a hill overlooking the city of Athens for almost 2,500 years. No wonder—"stone," an architect's word for rock, outlasts nearly all other construction materials. We use stone to make facades, roofs, curbs, steps, countertops, and floors. We value stone for its visual appeal as well as its durability. The names that architects give to various types of stone may differ from the formal rock names that geologists use. For example, architects refer to any polished carbonate rock as "marble," whether or not it has been metamorphosed. Likewise, they refer to any crystalline rocks containing feldspar and/or quartz as "granite," regardless of whether the rock has an igneous or a metamorphic texture, or a felsic or mafic composition.

To obtain intact slabs and blocks of rock—known as **dimension stone** in the trade—for architectural purposes, workers must carefully cut rock out of the walls of quarries (**Fig. 12.23a, b**). (Note that a *quarry* provides stone, whereas a mine supplies ore.) Traditionally, quarry operators split the blocks off of bedrock by hammering a series of wedges into a crack, causing the crack to grow until it propagates down to an exfoliation joint (a natural, subhorizontal crack). More recently, quarry operators have been able to slice blocks from bedrock by using wireline saws, thermal lances, or water jets. A *wireline saw* consists of a loop of braided wire moving between two pulleys. Operators spill abrasive (sand or garnet grains) and water onto the wire as the wire rubs against the rock surface, so the wire slowly grinds into the rock. A *thermal lance* looks like a long blowtorch—it can apply a flame of burning diesel fuel, stoked by high-pressure air, to a focused spot on the rock surface. The intense heat turns the rock into powder, and thus cuts into the rock like a hot knife into butter. More recently, quarry operators have begun to cut rock using abrasive *water jets that* direct a jet of water and abrasives at very high pressure onto the rock surface. Once blocks have been removed from the ground, they are transported to a workshop where they are sliced into uniform sheets (for countertops or facades) or into smaller blocks for gravestones, monuments, and curbs.

Crushed Stone and Concrete

Crushed stone forms the substrate of highways and railroads and serves as the raw material for manufacturing cement, concrete, mortar, and asphalt. In crushed-stone quarries (**Fig. 12.23c**), operators use explosives to break up bedrock into rubble that then goes by truck to a jaw crusher.

Most of the buildings and highways constructed in the past two centuries consist of bricks attached to each other by mortar, or of walls, floors, columns, and roads made of concrete

FIGURE 12.23 Stone production in quarries.

(a) An active quarrying operation in Missouri that produces large blocks of cut dimension stone.

(b) Sheets of cut dimension stone being measured for cutting to become a kitchen countertop.

(c) A large crushed-stone quarry in Silurian limestone of Illinois. Drillers are working on the shelf in the distance.

Did you ever wonder...
how concrete differs from rock?

that has been spread into a layer or poured into a form. Both mortar and concrete start out as a slurry, but when allowed to set, they harden into a hard, rock-like substance. The slurry from which mortar and concrete form consists of *aggregate* (sand and/or gravel) mixed with water and **cement**. Before mixing, the cement in mortar and concrete is a powder made of lime (CaO), quartz (SiO_2), aluminum oxide (Al_2O_3), and iron oxide (Fe_2O_3). Typically, lime accounts for 66% of cement and silica for 25%. When cement is mixed with water, the chemicals in it dissolve. Mortar and concrete set (solidify) when dissolved chemicals recombine to produce a complex assemblage of new mineral crystals. These minerals bind together the grains of aggregate, much as quartz or calcite bind together grains of quartz in a sandstone.

The ancient Romans were among the first people to use cement—they made it from a mixture of volcanic ash and limestone. In the 18th and early 19th centuries, cement was produced by heating specific types of limestone (which happened to contain calcite, clay, and quartz in the correct proportions) in a kiln up to a temperature of about 1,450°C. The heating releases CO_2 gas and produces "clinker," chunks consisting of lime and other oxide compounds. Manufacturers crushed the clinker into cement powder and packed it in bags for transport. Limestone with the exact composition necessary to make good cement, however, is fairly rare. Most cement used today is **Portland cement**. To create this industrial staple, workers mechanically mix limestone, sandstone, and shale in just the right proportions, then heat it in a kiln to yield a powder with the correct chemical makeup to mix into cement. Isaac Johnson, an English engineer, came up with the recipe for Portland cement in 1844, and chose the name because he thought that concrete made from this material resembled rock exposed in the town of Portland, England.

Nonmetallic Minerals for Homes and Farms

We use an astounding variety of nonmetallic geologic resources in daily life. Consider the materials in a typical house or apartment. The foundation generally consists of concrete, made from limestone mixed with sand or gravel. The bricks in many exterior walls originated as clay, formed from the chemical weathering of silicate rocks and perhaps dug from the floodplain of a stream—to make *bricks*, workers mold wet clay into blocks and then bake it to drive out water and cause metamorphic reactions that recrystallize the clay. The glass used to glaze windows consists largely of silica, formed by melting and then freezing pure quartz sand from a beach deposit or a sandstone formation. Quartz may also produce the silica in solar cells. Gypsum board (drywall), used for many interior walls, is made from a slurry of water and the mineral gypsum, sandwiched between sheets of paper. Gypsum ($CaSO_4 \cdot 2H_2O$) occurs in evaporite strata precipitated from seawater or saline lake water. Evaporites also provide other useful minerals, such as halite, and serve as the source for lithium, a key element in computer batteries.

TAKE-HOME MESSAGE

Society uses a great variety of nonmetallic geologic materials. These include dimension stone, crushed stone, cement (made from roasted limestone), evaporites (including gypsum), and clay (to make bricks).

QUICK QUESTION What is the difference between natural cement and Portland cement?

12.12 Global Mineral Needs

How Long Will Resources Last?

The average citizen of an industrialized country uses 25 kilograms (kg) of aluminum, 10 kg of copper, and 550 kg of iron and steel in a year's time (**Table 12.1**). If you combine these figures with the quantities of energy resources and nonmetallic geologic resources a person uses, the total is about 15,000 kg (15 metric tons) of resources used per capita each year. To create the supply of the materials used in the United States alone, workers must mine, quarry, or pump 18 billion metric tons per year, almost 100 times the weight of sediment transported by the Mississippi River in a year.

Mineral resources, like oil and coal, are nonrenewable resources. Once mined, an ore deposit or a limestone hill disappears forever. Natural geologic processes do not happen fast enough to replace the deposits as quickly as we use them. Geologists have calculated *reserves* (measured and, as yet, unused quantities of a commodity that could be mined economically) for various mineral deposits just as they have for oil. Reserve estimates depend on price, because as price goes up, the grade of ore that can be mined economically goes down. A comparison of reserve estimates to consumption rates suggests that supplies of some metals may run out in only decades to centuries (**Table 12.2**). But these estimates may change as prices rise, if geologists discover new reserves, or if engineers develop new technologies for mining and processing ore. Further, increased efforts at conservation and recycling can cause a dramatic decrease in rates of consumption and thereby stretch the lifetime of existing reserves.

Ore deposits do not occur everywhere because their formation requires special geologic conditions. As a result, some countries possess vast supplies, whereas others have none. In fact, no single country owns all the mineral resources it needs, so nations must trade with each other to maintain supplies, and global politics inevitably affects prices. Many wars and treaties have their roots in competition for mineral reserves. Competition for mineral reserves has increased in recent decades because modern technological innovations have greatly increased the demand for the **strategic minerals** that are essential for the security of a nation. These include certain metals, used for the production of specialized types of steel, and *rare earth elements*, a group of 17 elements including lanthanides, scandium, and yttrium that have become essential in the production of lasers, magnets, X-ray tubes, night-vision goggles, camera lenses, and high-tech lamps.

TABLE 12.1 Yearly per Capita Usage of Geologic Materials (USA)

Material	Weight Used
Stone	4,100 kg
Sand and gravel	3,860 kg
Petroleum	3,050 kg
Coal	2,650 kg
Natural gas	1,900 kg
Iron and steel	550 kg
Cement	360 kg
Clay	220 kg
Salt	200 kg
Phosphate	140 kg
Aluminum	25 kg
Copper	10 kg
Lead	6 kg
Zinc	5 kg

1 kg = 2.205 pounds

TABLE 12.2 Expected Lifetimes of Currently Known Ore Resources (in Years)

Metal	World Resources	U.S. Resources
Iron	120	40
Aluminum	330	–2
Copper	65	40
Lead	50	25
Zinc	30	25
Gold	30	20
Platinum	45	–1
Nickel	75	< 1
Cobalt	50	< 1
Manganese	70	–0
Chromium	75	–0

Mining and the Environment

Mining leaves a big footprint in the Earth System. Some of the gaping holes that open-pit mining creates in the landscape have become so big that astronauts can see them from space. Both open-pit and underground mining yield immense quantities of waste rock, which miners dump in tailings piles. Some of these piles grow into artificial hills as much as 200 m high and many kilometers long. Because they lack soil, tailings piles tend to remain unvegetated for a long time. Mining also exposes ore-bearing rock to the atmosphere, and since many ore minerals are sulfides, they react with rainwater to produce *acid mine runoff*, which can severely damage vegetation downstream (**Fig. 12.24a**). In some cases, mining companies douse tailings with toxic solutions to leach out more metals, and these solutions sometimes escape into the environment. Ore processing also tends to yield pollutants, and if these pollutants enter the atmosphere, they can impact vegetation and animal life downwind (**Fig. 12.24b**).

To help counter some of the problems caused by mining, reclamation efforts have begun in some areas, with the goal of isolating exposed mine surfaces and tailings piles from the environment. And, in recent years, new regulations have been established that require smoke from processing plants to be "scrubbed" before it enters the atmosphere. Finally, engineers are developing new technologies that allow metals to be processed in ways that produce less pollutants in the first place.

FIGURE 12.24 Environmental consequences of producing metallic mineral resources.

Much of the color comes from bacteria and archaea living in the water.

(a) The orange color in this mine runoff is due to iron and sulfide in the water.

(b) Acidic smelter smoke killed off vegetation near Sudbury, Ontario, in the 1970s. A large tailings pile can be seen in the distance.

The "superstack" is 380 m (1,270 ft) high.
Nickel smelter
Tailings pile

TAKE-HOME MESSAGE

People use a vast quantity of mineral resources during the course of a lifetime. Mineral resources are nonrenewable, so minerals have limited reserves, and because reserves are not distributed uniformly around the planet, all supplies are not accessible to all consumers. Also, mineral extraction and utilization have significant environmental consequences that are challenging to address.

QUICK QUESTION If one country restricts the export of a strategic mineral, would that change the minimum concentration of the mineral necessary to make an ore deposit in another country worth mining? Why?

Chapter 12 Review

Chapter Summary

- Energy resources come in a variety of forms: energy directly from the Sun; energy from tides, flowing water, or wind; energy from chemical reactions; energy from nuclear fission; and energy from Earth's internal heat.
- Oil and gas are hydrocarbons formed from the organic remains of plankton, which settle out and become incorporated in black organic shale. Chemical reactions at elevated temperatures convert the organic matter to kerogen and then to oil.
- To form a conventional oil reserve, oil must migrate from a source rock into a reservoir rock. The subsurface configuration of strata that holds oil is called an oil trap.
- Substantial volumes of hydrocarbons also exist in unconventional reserves. These include shale gas and shale oil, tar sand, and oil shale. Extraction of unconventional reserves has increased dramatically in recent years due to the development of directional drilling and hydrofracturing.
- For coal to form, abundant plant debris must be deposited in an oxygen-poor environment. Compaction changes the debris into peat that, when buried deeply and heated, transforms into coal.
- Geologists distinguish among three ranks of coal, based on the amount of carbon the coal contains. Coal occurs in beds, and can be mined by either strip mining or underground mining.
- Nuclear power plants generate energy by using the heat released from the fission of uranium.
- Nuclear reactors must be carefully controlled to avoid overheating or meltdown. The disposal of radioactive nuclear waste can create environmental challenges.
- Geothermal energy uses Earth's internal heat to transform groundwater into steam that drives turbines; hydroelectric power uses the potential energy of water; and solar energy uses solar cells to convert sunlight to electricity.
- We now live in the Oil Age, but oil supplies may last for only another century, and the production and use of energy resources have many environmental consequences.
- Industrial societies use many types of minerals, all of which must be extracted from the upper crust.
- Metals come from ore. An ore is a rock containing native metals or ore minerals in sufficient quantities to be worth mining.
- Magmatic ore deposits form when ore minerals grow during solidification of melt. In hydrothermal deposits, ore minerals precipitate from hot-water solutions. Secondary-enrichment deposits and MVT deposits precipitate from groundwater. Sedimentary deposits settle out of water, residual mineral deposits are the result of soil formation, and placer deposits develop when metal grains accumulate in sediment.
- Nonmetallic resources include dimension stone, crushed stone, clay, sand, and many other materials. A large proportion of materials in your home have a geological ancestry.
- Mineral resources are nonrenewable. Many, including strategic minerals, are now or may soon be in short supply.

Guide Terms

acid rain (p. 415)
biofuel (p. 411)
biomass (p. 392)
blowout (p. 416)
cement (p. 423)
chain reaction (p. 409)
coal (p. 403)
coal gasification (p. 407)
coal rank (p. 406)
coal reserve (p. 406)
conventional reserve (p. 394)
dimension stone (p. 422)
directional drilling (p. 399)
distillation column (p. 400)
drilling mud (p. 399)
energy (p. 392)
energy density (p. 393)
energy grid (p. 415)
energy resource (p. 391)
fossil fuel (p. 392)
fuel (p. 392)
grade (p. 418)
Hubbert's Peak (p. 414)
hydrocarbon (p. 393)
hydrocarbon generation (p. 394)
hydrocarbon reserve (p. 394)
hydrofracturing (p. 400)
kerogen (p. 394)
meltdown (p. 409)
metal (p. 417)
migrate (p. 395)
mineral resource (p. 391)
nuclear reactor (p. 409)
Oil Age (p. 413)
oil seep (p. 395)
oil shale (pp. 394, 402)
oil window (p. 394)
ore (p. 418)
ore deposit (p. 418)
ore mineral (p. 417)
peat (p. 405)
permeability (p. 395)
pore (p. 395)
porosity (p. 395)
Portland cement (p. 423)
reservoir rock (p. 395)
salt dome (p. 397)
seal rock (p. 395)
seismic-reflection profile (p. 398)
shale gas (p. 401)
shale oil (p. 401)
solar cell (p. 412)
source rock (p. 393)
strategic mineral (p. 424)
tar sand (p. 402)
tidal power (p. 411)
trap (p. 395)
unconventional reserve (p. 394)
underground mining (p. 407)

LEARNING OBJECTIVES

By the end of this interlude, you should understand...

1. what a landscape is and what questions geologists ask about landscapes.
2. the difference between uplift and subsidence and the forces that drive them.
3. the contrasts between internal and external energy in the Earth System.
4. how erosional and depositional landscapes differ.
5. the reservoirs and exchange processes of the hydrologic cycle.

Nothing that is can pause or stay;
The moon will wax, the moon will wane,
The mist and cloud will turn to rain,
The rain to mist and cloud again,
To-morrow be to-day.

HENRY WADSWORTH LONGFELLOW
(American poet, 1807–1882)

F.1 Introduction

The Earth's surface is at once a place of endless variety and intricate detail. Observe the height of its mountains, the expanse of its seas, the desolation of its deserts, and you may be inspired, frightened, or calmed. It's no wonder that artists and writers across the ages have sought inspiration from the **landscape**—the character and shape of the land surface in a region—for landscapes engage the full range of human emotion (**Fig. F.1**). Geologists, like artists and writers, savor the impression of a dramatic landscape. But on seeing one, they can't help but ask: How did it come to be, and how will it change in the future?

The subject of landscape development and evolution, and of the **landforms** (individual shapes such as mesas, valleys, cliffs, and dunes) that constitute it, dominate many of the remaining chapters in this book. This interlude, a general introduction to our planet's surface and near-surface realms, characterizes the variety of landscapes, the processes that shape landforms, and the many interrelationships among climate, life, water, rock, and sediment in the Earth System. We will see that water, in both its solid and liquid states, plays a key role in landscape evolution. Thus, we include a focus on the *hydrologic cycle*, the pathway that water molecules follow as they move from ocean to air to land and back to ocean.

F.2 Shaping the Earth's Surface

If the Earth's surface were totally flat, the great diversity of landscapes that embellish our vistas would not exist. But the surface isn't flat, because a variety of geologic processes cause portions of the surface to move up or down relative to adjacent regions. We refer to the upward movement of the Earth's surface as **uplift**, and the sinking or downward movement of the Earth's surface as **subsidence**.

The occurrence of uplift and/or subsidence generates **relief**, the elevation difference between two points separated by a specified horizontal distance on a map (**Box F.1**). Wherever relief develops on land, various components of the Earth System kick into action to modify and shape the land surface. Rock at or near the ground surface weathers, fractures, and weakens. On slopes, this weakened or loose material becomes susceptible to **downslope movement**, gravity-driven tumbling or sliding from higher elevations to lower ones. Moving water, ice, and air cause **erosion** by picking up debris, and by grinding into bedrock, or into previously deposited sediments, at the Earth's surface. Where moving fluids slow down, **deposition**, the accumulation of transported sediment, takes place. Downslope movement, erosion, and deposition redistribute rock and sediment on the Earth, ultimately stripping it from higher areas and collecting it in low areas.

The energy that drives landscape evolution comes from three sources: *internal energy*, the heat within the Earth, which drives the plate motions and mantle plumes that ultimately cause vertical displacements of the surface; *external energy* that comes to the Earth from the Sun, warming the atmosphere and ocean; and *gravitational energy*, which pulls material down slopes at the surface. Gravity, working together with external energy, causes *convection* in the atmosphere and oceans, as manifested by currents and winds. Considering that all these energy sources are always operating, we can think of landscape evolution as a battle between tectonic processes (such as collision, convergence, and rifting) which move land up, and erosional processes, which tear it down. If, in a particular region, the rate of uplift exceeds the rate of erosion, the land surface rises. If the rate of subsidence exceeds the rate of deposition, the land surface sinks. Without uplift,

FIGURE F.1 Examples of the great variety of landscapes on Earth.

(a) The Amazon jungle, Peru.

(b) Glaciated peaks of the Alps, France.

(c) Buttes of sandstone, Monument Valley, Arizona.

(d) Cliffs rise from the forest in the Blue Mountains, Australia.

relief could never develop, and without erosion, high areas would have lasted for the entirety of Earth history.

How rapidly do vertical movements and erosion of the Earth's surface take place? The ground can rise or sink by as much as 3 m during a single major earthquake. But averaged over geologic time, the rates of uplift and subsidence range between 0.01 and 10 mm per year (**Fig. F.2a**). Similarly, a single flood, storm, or landslide can carve out several meters of substrate in an hour or less (**Fig. F.2b**), and deposition during a single event can produce a layer of debris tens of meters thick in a matter of minutes to days. But, averaged over time, erosional and depositional rates also vary between 0.10 and 10 mm per year. Although these average rates seem small, a change in surface elevation of just 0.5 mm (the thickness of a fingernail) per year can yield a net change of 5 km in just 10 million years. Uplift can build a towering mountain range, and erosion can whittle one down to near sea level—it just takes time!

F.3 Factors Controlling Landscape Development

Imagine traveling across a continent. On your journey, you may cross plains, swamps, hills, valleys, mesas, and mountains. Some of these features are **erosional landforms**, in that they result from the breakdown and removal of rock or sediment when moving water, ice, or air carve into the substrate. Of these three **agents of erosion**, water has the greatest effect on a global basis. Other features are **depositional landforms**, in that they result from the deposition of sediment where the medium carrying the sediment evaporates, slows down, or melts. The specific landforms that develop at a given locality, and that together make up the landscape, reflect several factors.

BOX F.1 Consider This...

Topographic Maps and Profiles

We can distinguish one landform from another by its shape—for example, as you will see in succeeding chapters, a river-carved valley does not look like a glacially carved valley. Landform shapes are manifested by variations in elevation within a region. Geologists use the term **topography** to refer to such variations. How can we convey information about topography—a three-dimensional feature—on a two-dimensional sheet of paper? Geologists and cartographers do this by means of a *topographic map*, which uses contour lines to represent elevation (**Fig. BxF.1a, b**). Specifically, a *contour line* represents an imaginary line on the land surface along which all points have the same elevation. You can picture a contour line as the intersection between the land surface and an imaginary horizontal plane. A topographic map that displays many contour lines represents a region with slopes, whereas a map with very few contour lines represents a plain.

The elevation difference between two adjacent contour lines on a topographic map is the *contour interval*. For a given topographic map, the contour interval is constant, so the spacing between contour lines represents the steepness of a slope. Specifically, closely spaced contour lines represent a steep slope, whereas widely spaced contour lines represent a gentle slope. Computer technology can help make slopes on a map stand out by adding shading, to give the effects of shadows cast when the sun lies low in the sky—maps that display slopes this way are called *shaded-relief maps*. If the data used to construct a contour and/or shaded relief map comes in digital form, we refer to the map as a *digital elevation map* (*DEM*).

Geologists represent variations in elevation along a given traverse by means of a **topographic profile**, the trace of the ground surface as it would appear on a vertical plane that sliced into the ground (**Fig. BxF.1c**). If we add a representation of geologic features under the ground surface, then we have a **geologic cross section**. In some cases, researchers can gain insight into subsurface geology simply by looking at the shape of a landform (**Fig. BxF.1d**). For example, a steep cliff in a region of dipping sedimentary strata may indicate the presence of a resistant layer (one that stands up to erosion), whereas low areas may be underlain by a nonresistant layer (one that erodes easily).

FIGURE BxF.1 Topographic maps and profiles.

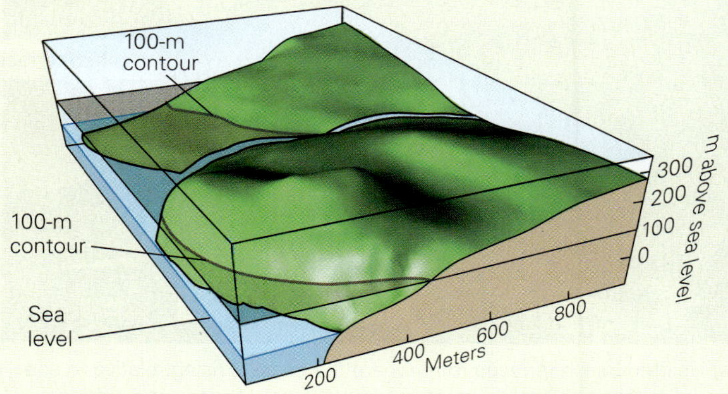

(a) A contour line is the intersection of a horizontal plane with the land surface. This block diagram shows the map area.

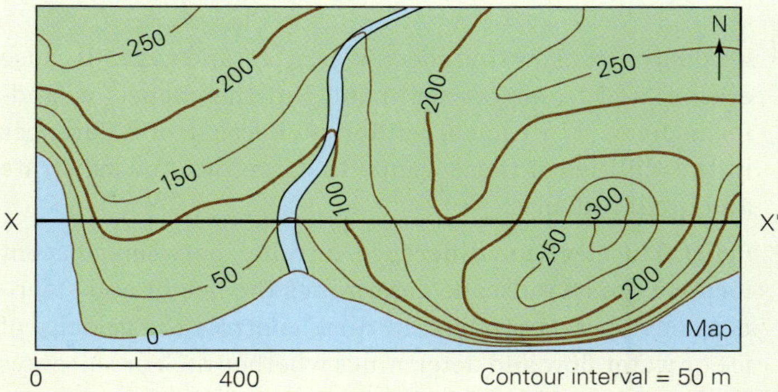

(b) A topographic map depicts the shape of the land surface through the use of contour lines. The difference in elevation between two adjacent lines is the contour interval.

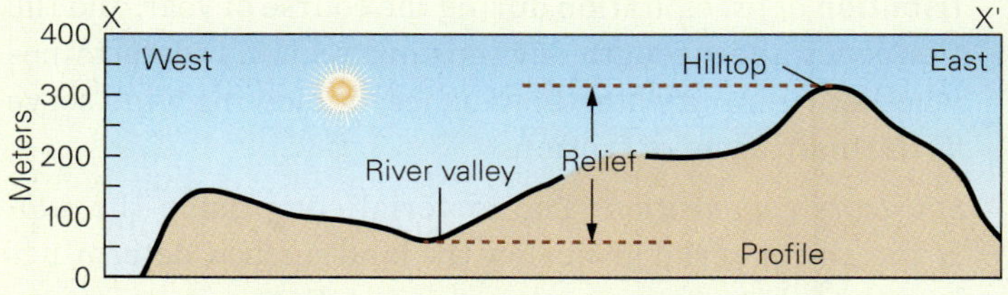

(c) A topographic profile (along section line XX') shows the shape of the land surface as seen in a vertical slice.

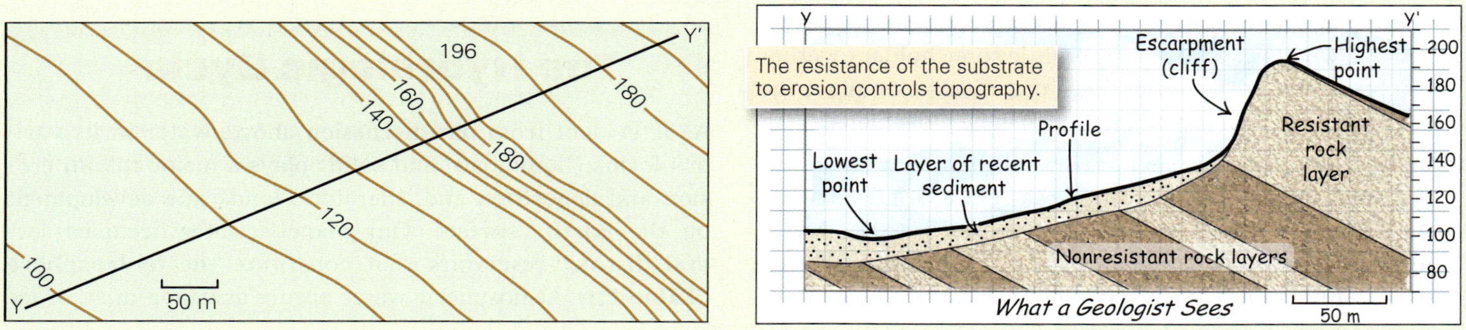

(d) This topographic map shows a distinct cliff. A cross section (right) depicts a geologist's interpretation of the subsurface along YY'. The cliff is the edge of a resistant rock layer.

FIGURE F.2 The processes of uplift, subsidence, erosion, and deposition can be slow or rapid.

(a) Uplifted beach terraces form where the coast is rising relative to sea level. Present-day wave erosion is forming a new terrace and cutting a cliff on the edge of the old one.

(b) So much erosion can take place during a single hurricane that houses built along the beach become undermined. This example happened as a result of "Superstorm Sandy" in 2012.

- *Eroding or transporting agent*: Water, ice, and wind all cause erosion and transport sediment. But the shapes of landforms formed by each are different, because of differences in the abilities of these agents to carve into the substrate and to carry debris.
- *Relief*: The elevation difference, or relief, between adjacent locations in a landscape determines the height and steepness of slopes. Steepness, in turn, controls the velocity of ice or water flow and determines whether rock or soil stays in place or tumbles downslope.
- *Climate*: The climate of a region reflects the average mean temperature, the annual volume of precipitation, the distribution of precipitation during the course of year, and the frequency and strength of winds in a region. It determines whether running water, flowing ice, or blowing wind serve as the main agent of erosion.
- *Substrate composition*: The material comprising the *substrate*, the material just below the land surface, determines how the substrate responds to erosion. For example, strong rocks can stand up to form steep cliffs, while soft sediment collapses and tends to underlie gentle slopes.
- *Life activity*: Some organisms weaken the substrate (by burrowing, wedging, or digesting), while some hold it together (by binding it with roots). Thus, organisms can affect the susceptibility of a substrate to downslope movement.
- *Time*: Landscapes evolve over time, as the substrate weathers, and as erosion or deposition take place. For instance, a gully that has just started to form in response to the flow of a stream does not look the same as a deep canyon that develops after the same stream has been eroding for a long time.

Although water, wind, and ice are responsible for the development of most landscapes, human activities have had an increasingly important impact on the Earth's surface. We have dug pits (mines) where once there were mountains, have built hills (tailings piles and landfills) where once there were valleys, and have made steep slopes gentle and gentle slopes steep (**Fig. F.3**). By constructing concrete walls, we modify the shapes of coastlines, change the courses of rivers, and fill new lakes (reservoirs). In cities, buildings and pavements seal the ground and cause water that might once have seeped down into the soil to spill directly into streams instead, thus increasing the volume of flow. The area of land covered by pavement or buildings in the United States now exceeds the area of the state of Ohio! And in non-urban regions, agriculture, grazing, water usage, and deforestation substantially alter the rates at which natural erosion and deposition take place. For example, agriculture generally increases the rate of erosion, because for much of the year a farm field has no vegetation cover. Clearly, humanity has become a major agent of erosion and deposition.

F.4 The Hydrologic Cycle

As is evident from our discussion above, water in its various forms (liquid, gas, and solid) plays a major role in erosion and deposition and, therefore, landscape development on the Earth's surface. Our planet's water occupies several distinct reservoirs that constitute the **hydrosphere** (**Table F.1**). Atmospheric water occurs as vapor, mist, or ice

FIGURE F.3 Human influence on a geologic scale.

(a) The pyramids of Egypt are human-made hills that rise above the desert sands. They have lasted for thousands of years.

(b) In the process of making highway cuts, deep valleys are cut through high ridges. This example borders a highway near Denver.

(c) This stone dam holds back a reservoir in Colorado. Think about how long it would take a glacier to pile up so much sediment.

TABLE F.1 Major Water Reservoirs of the Earth's Hydrosphere

H_2O Reservoir	Volume (km^3)	% of Total Water	% of Fresh Water
Oceans and seas	1,338,000,000	96.5	—
Glaciers, ice caps, snow	24,064,000	2.05	68.7
Saline groundwater	12,870,000	0.76	—
Fresh groundwater	10,500,000	0.94	30.1
Permafrost	300,000	0.022	0.86
Freshwater lakes	91,000	0.007	0.26
Salt lakes	85,400	0.006	—
Soil moisture	16,500	0.001	0.05
Atmosphere	12,900	0.001	0.04
Swamps	11,470	0.0008	0.03
Rivers and streams	2,120	0.0002	0.006
Living organisms	1,120	0.0001	0.003

Source: Data from P. H. Gleick, *Encyclopedia of Climate and Weather* (Oxford University Press, 1996).

crystals. Surface water collects in oceans, lakes, streams, puddles, and swamps. Frozen water forms snowfields and glaciers. Subsurface water dampens soil and rock near the surface, or sinks deeper to a realm where it fills underground pores and cracks as *groundwater*. A significant amount of water also resides in living organisms—about 60% of your body consists of water. Water constantly flows from reservoir to reservoir—geologists refer to this never-ending passage as the **hydrologic cycle** (see **Geology at a Glance**, pp. 434–435). Perhaps

GEOLOGY AT A GLANCE

The Hydrologic Cycle

Water circulates through a number of reservoirs in the Earth System. The largest reservoir by far is the ocean, which covers 71% of the Earth's surface. Water evaporates from the ocean and enters the atmosphere, where it may remain for quite a while. Thus, the atmosphere serves as another reservoir. Atmospheric water gradually condenses and forms clouds that drop rain or snow onto the oceans or land.

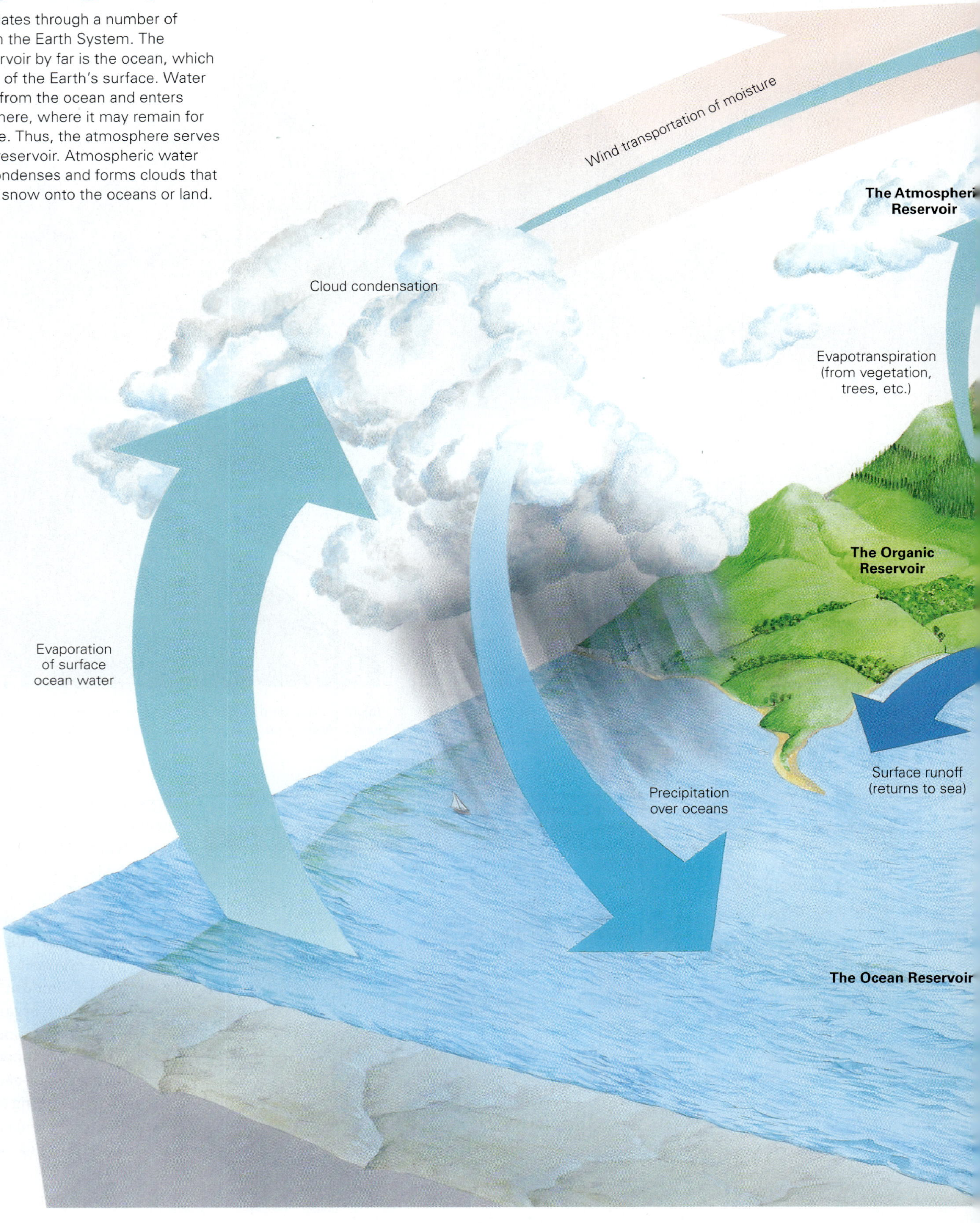

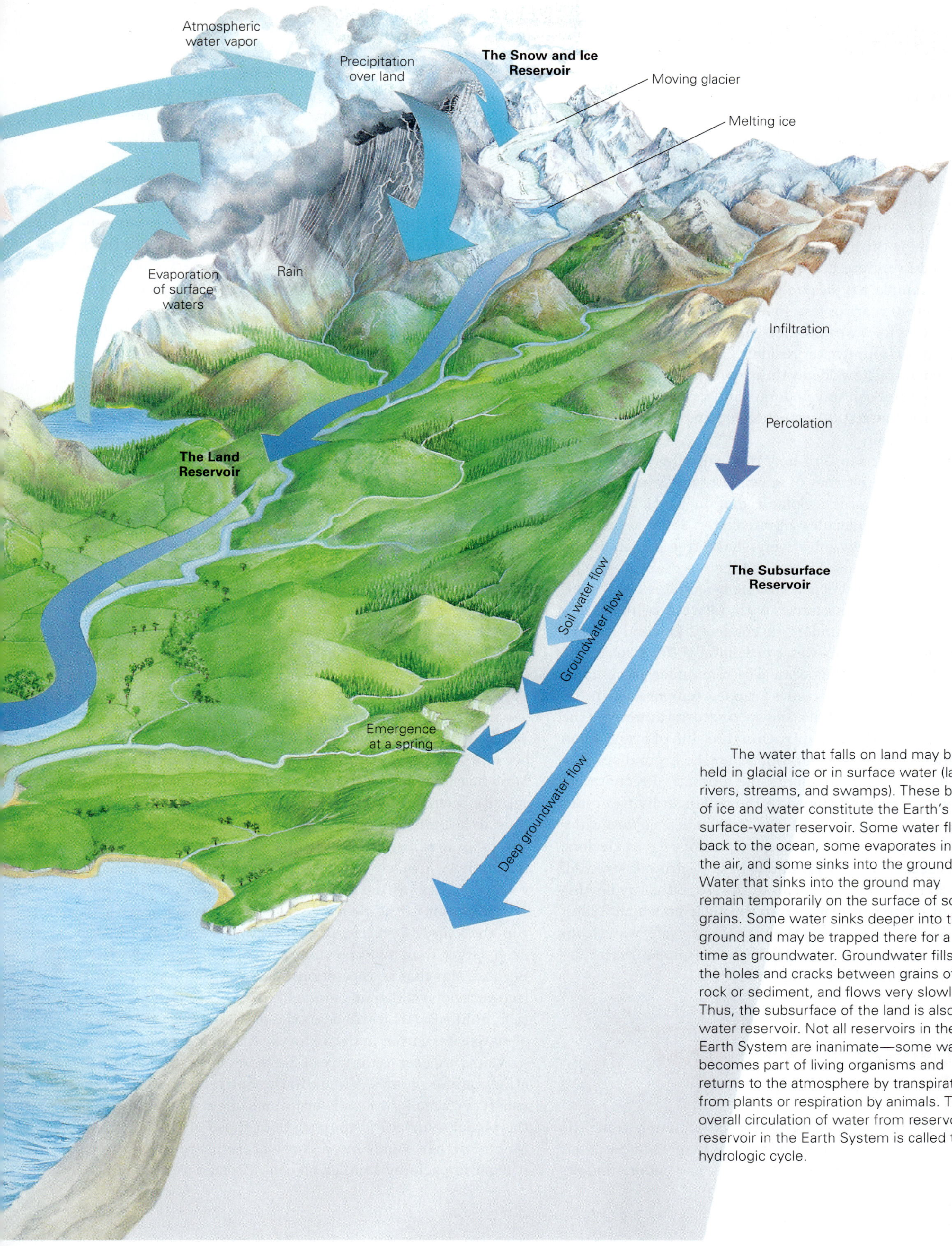

The water that falls on land may be held in glacial ice or in surface water (lakes, rivers, streams, and swamps). These bodies of ice and water constitute the Earth's surface-water reservoir. Some water flows back to the ocean, some evaporates into the air, and some sinks into the ground. Water that sinks into the ground may remain temporarily on the surface of soil grains. Some water sinks deeper into the ground and may be trapped there for a time as groundwater. Groundwater fills the holes and cracks between grains of rock or sediment, and flows very slowly. Thus, the subsurface of the land is also a water reservoir. Not all reservoirs in the Earth System are inanimate—some water becomes part of living organisms and returns to the atmosphere by transpiration from plants or respiration by animals. The overall circulation of water from reservoir to reservoir in the Earth System is called the hydrologic cycle.

without realizing it, Longfellow, an American poet fascinated with reincarnation, provided an accurate if somewhat romantic image of the hydrologic cycle (see quote, p. 429). Without this cycle, the erosive force and transporting activity of running water in rivers and streams, or of flowing ice in glaciers, would not exist.

The average length of time that water stays in a particular reservoir during the hydrologic cycle is called the **residence time**. Water in different reservoirs has different residence times. For example, a typical molecule of water remains in the oceans for 4,000 years or less, in lakes and ponds for 10 years or less, in rivers for 2 weeks or less, and in the atmosphere for 10 days or less. Groundwater residence times are highly variable and depend on how deeply the groundwater flows into the subsurface—water can stay underground for anywhere from 2 weeks to more than 10,000 years before it inevitably moves on to another reservoir.

To get a clearer sense of how the hydrologic cycle operates, let's follow the journey of seawater that has just reached the surface of the ocean. Solar radiation heats the water, making the water molecules vibrate faster. Some of the molecules evaporate—they effectively shake free of the liquid, drift upward in a gaseous state, and mix into the atmosphere. About 30% of the total ocean volume evaporates every year. Atmospheric water vapor moves with the wind to higher elevations, where it cools, undergoes condensation, and rains or snows. About 76% of this water precipitates (falls out of the air) directly back into the ocean. The remainder precipitates onto land, where most becomes trapped temporarily in the soil, or in plants and animals, and soon returns directly to the atmosphere by **evapotranspiration** (the sum of evaporation from bodies of water, evaporation from the ground surface, and release of water from plants and animals). Precipitation that does not become trapped in the soil or in living organisms has several potential routes. It becomes *surface water* (held in lakes, rivers, or swamps), it becomes ice in glaciers, or it sinks deeper into the ground to become groundwater. All of this water eventually ends up in the sea, either by flowing directly, or by first returning into the air from which it later precipitates. In sum, during the hydrologic cycle, water constantly moves among the ocean, the atmosphere, reservoirs on or below the land, and living organisms.

F.5 Landscapes of Other Planets

The dynamic, ever-changing landscape of Earth contrasts markedly with those of other terrestrial planets in our Solar System. Each of the terrestrial planets and moons has its own unique surface landscape features, reflecting the interplay between the object's particular tectonic and erosional processes. Let's look at a few examples: the Moon, Mars, and Venus.

Our Moon has a static, pockmarked landscape generated exclusively by meteorite impacts and volcanic activity. Because no plate tectonics occurs on the Moon, no new mountains form; and because no atmosphere or ocean exists, there is no hydrologic cycle and no erosion from rivers, glaciers, or winds. Therefore, the lunar surface has remained largely unchanged for most of its history (**Fig. F.4a**). Some of the craters that you may see with a telescope formed over 3 billion years ago.

Landscapes on Mars differ from those of the Moon because Mars does have an atmosphere, though one much less dense than that of Earth. Martian winds generate huge dust storms, some of which obscure nearly the entire surface of the planet for months at a time. The two sets of landscapes also differ because Mars once had surface water, and the Moon did not (**Box F.2**; **Fig. F.4b**). Thus, the Martian surface consists of four kinds of materials: volcanic flows and deposits (primarily of basalt), debris from impacts, windblown sediment, and water-laid sediment. There is even evidence that soil-forming processes affected surface materials and hints that small amounts of liquid water exist temporarily on Mars at present. Martian winds not only deposit sediment, they also slowly erode impact craters and polish surface rocks. So landscapes do evolve on Mars, but at a vastly slower rate than on Earth. Mars also has a hydrologic cycle, of sorts, in that it has ice caps that grow and recede on a seasonal basis.

Landscapes on Mars also differ from those on Earth, because Mars does not have plate tectonics. So, unlike Earth, Mars has no mountain belts or volcanic arcs. However, plume activity occurred in the planet's distant past. A plume likely caused the uplift of a 9-km-high bulge (the Tharsis Ridge) that covers an area comparable to that of North America. Thermal activity also led to the eruption of gargantuan hot-spot volcanoes, such as the 22-km-high *Olympus Mons*, the highest mountain in the Solar System. Mars boasts a vast canyon, the *Valles Marineris*, a gash over 3,000 km long and 8 km deep, larger than any known feature on Earth, or anywhere. Because Mars has no vegetation and no longer has rain, its surface does not weather and erode like that of Earth. This means that, unlike Earth, it still bears the scars of impact by swarms of meteorites earlier in the history of the Solar System.

Venus is closer to the size of Earth and may still have operating mantle plumes. Virtually the entire surface of Venus was resurfaced by volcanic eruptions about 3.0 to 1.6 Ga, so the planet's surface is younger than those of the Moon and Mars. Further, Venus has a dense atmosphere that protects it from impacts by smaller objects. Because relatively little

FIGURE F.4 Landscapes of other planets.

(a) Astronomers estimate the crater Clavius, on the Moon, is about 3.8 to 3.9 billion years old, even though its outline remains very clear. It's been around long enough for other craters to have impacted within it.

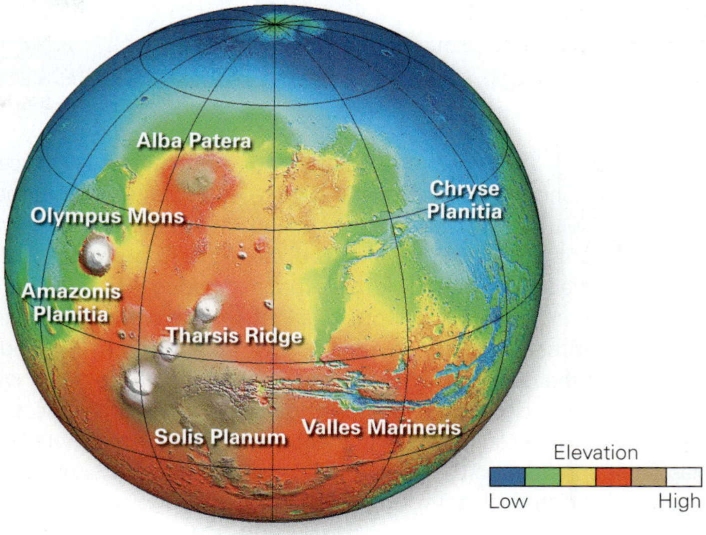

(b) A DEM depicting the surface of Mars. Note the huge bulge of the Tharsis Ridge, the giant Olympus Mons volcano, and the deep Valles Marineris canyon.

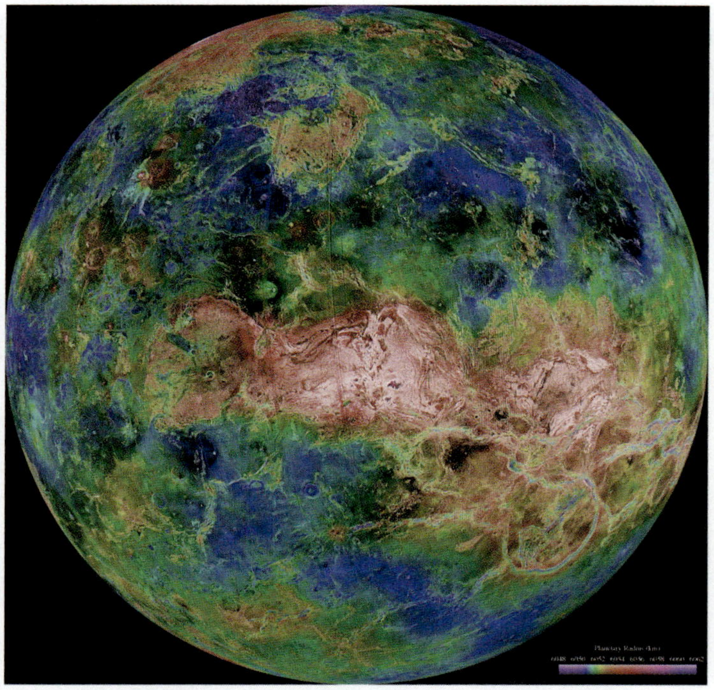

(c) A radar image of Venus. A thick blanket of clouds obscures the planet's surface, so it can't be seen through a telescope. Red areas are higher, blue areas lower.

(d) In 2015, NASA captured the first high-resolution image of Pluto's surface. Pluto is a dwarf planet.

cratering has taken place since the resurfacing event, volcanic and tectonic features dominate the landscape of Venus (**Fig. F.4c**). New images reveal features on the dwarf planet, Pluto, that have yet to be explained (**Fig. F.4d**). Satellites using radar have revealed a variety of volcanic constructions, such as shield volcanoes, lava flows, and calderas. Rifting on Venus produced faults, some of which occur in association with volcanic features. Liquid water cannot survive the scalding temperatures of Venus's surface, so no hydrologic cycle operates there and no life exists. Because of the density of the atmosphere, winds are too slow to cause much erosion or deposition.

BOX F.2 Consider This...

Water on Mars?

In 1877, an Italian astronomer named Giovanni Schiaparelli studied the surface of Mars with a telescope and announced that long, straight *canali* crisscrossed the planet's surface. *Canali* should have been translated into the English word *channel*, but perhaps because of the recent construction of the Suez Canal, newspapers of the day translated the word into the English *canal*, with the implication that the features had been constructed by intelligent beings. An eminent American astronomer began to study the "canals" and suggested that they had been built to carry water from polar ice caps to Martian deserts.

Late-20th-century satellite mapping of Mars showed that the "canals" do not exist—they were simply optical illusions. No lakes, oceans, rainstorms, or flowing rivers exist on the surface of Mars today. The atmosphere of Mars has such low density, and thus exerts so little pressure on the planet's surface, that slight amounts of liquid water that may be locally released at the surface quickly evaporate. Thus, Mars has no hydrologic cycle the way that Earth does. But crucial questions remain: Did Mars ever have significant amounts of running water or standing water in the past? If so, where is the water now? The question of the presence of water lies at the heart of an even more basic question: Given that even the simplest life forms, as far as we know, require water to exist, is there, or was there ever, life on Mars?

Many planetary geologists believe that the case for liquid water on Mars is quite strong. Much of the evidence comes from comparing landforms on the planet's surface with landforms of known origin on Earth. High-resolution images of Mars reveal landforms that look as very much like those formed on Earth in response to flowing water. Examples include networks of river channels resembling those on Earth (**Fig. BxF.2**), scour features, deep gullies, and streamlined deposits of sediment.

Studies by the *Odyssey* satellite in 2003, and by Mars rovers (*Spirit* and *Opportunity*) that landed on the planet in 2004, added intriguing new data to the debate. *Odyssey* detected hints that hydrogen, an element in water, exists beneath the surface of the planet over broad regions, and the Mars rovers have documented the existence of hematite and gypsum, minerals that form in the presence of water. The rovers have also found sedimentary deposits that appear to have been deposited in water. The *Phoenix* lander, in 2008, confirmed the existence of water ice by digging into the surface to expose some. Researchers speculate that Mars was much wetter in its past, perhaps billions of years ago. But since the atmosphere became less dense, almost all water now lies hidden underground or trapped in polar ice caps.

FIGURE BxF.2 Water-carved landscapes on Mars, as photographed by satellites orbiting the planet.

(a) An oblique view showing stream channels on Mars resembling channels cut by rivers on Earth.

(b) Layers of strata, interpreted to be water deposited, in Chasma Canyon, Mars. Note the elongate islands in the channel, resembling islands that have been shaped by rivers on Earth.

Interlude F Review

Interlude Summary

- The character and shape of the land surface in a region is a landscape. Individual shapes are landforms. Topographic maps, shaded-relief maps, and DEMs can portray the shape of landscapes.
- Land can undergo uplift or subsidence, to yield relief. Rock, as well as debris formed by weathering, eventually undergoes downslope movement and collects in lower areas.
- The energy driving landscape evolution comes from three sources: Earth's internal energy, gravitational energy, and energy radiating from the Sun.
- The nature of a landscape depends on climate, time, relief, slope angles, elevation, the activity of organisms, substrate composition, and the rate of tectonic movement.
- Water is the dominant agent of erosion on the Earth. Driven by gravity and by energy from the Sun, water moves among various reservoirs (such as the ocean, the atmosphere, the land surface, the subsurface, and life) during the hydrologic cycle.
- Landscapes on other planets differ markedly from those on Earth. Researchers are working hard to understand how water played a role in forming Martian landscapes, and if water remains on the planet today.

Guide Terms

agent of erosion (p. 430)
deposition (p. 429)
depositional landform (p. 430)
downslope movement (p. 429)
erosion (p. 429)
erosional landform (p. 430)
evapotranspiration (p. 436)
geologic cross section (p. 431)
hydrologic cycle (p. 433)
hydrosphere (p. 432)
landform (p. 429)
landscape (p. 429)
relief (p. 429)
residence time (p. 436)
subsidence (p. 429)
topographic profile (p. 431)
topography (p. 431)
uplift (p. 429)

Review Questions

1. What is the difference between uplift and subsidence?
2. Why do landscapes on the Earth change over geologic time, while they remain static on the Moon?
3. What is topography, and how can we portray it on a sheet of paper?
4. What are the principal agents of erosion on Earth?
5. What factors affect the character of erosional or depositional landforms that develop in a region?
6. Explain the steps in the hydrologic cycle.
7. How do landscapes of other planets differ from those of Earth?

Online Resources

Assessment

This interlude includes questions on features and processes that change the landscape and on the water cycle.

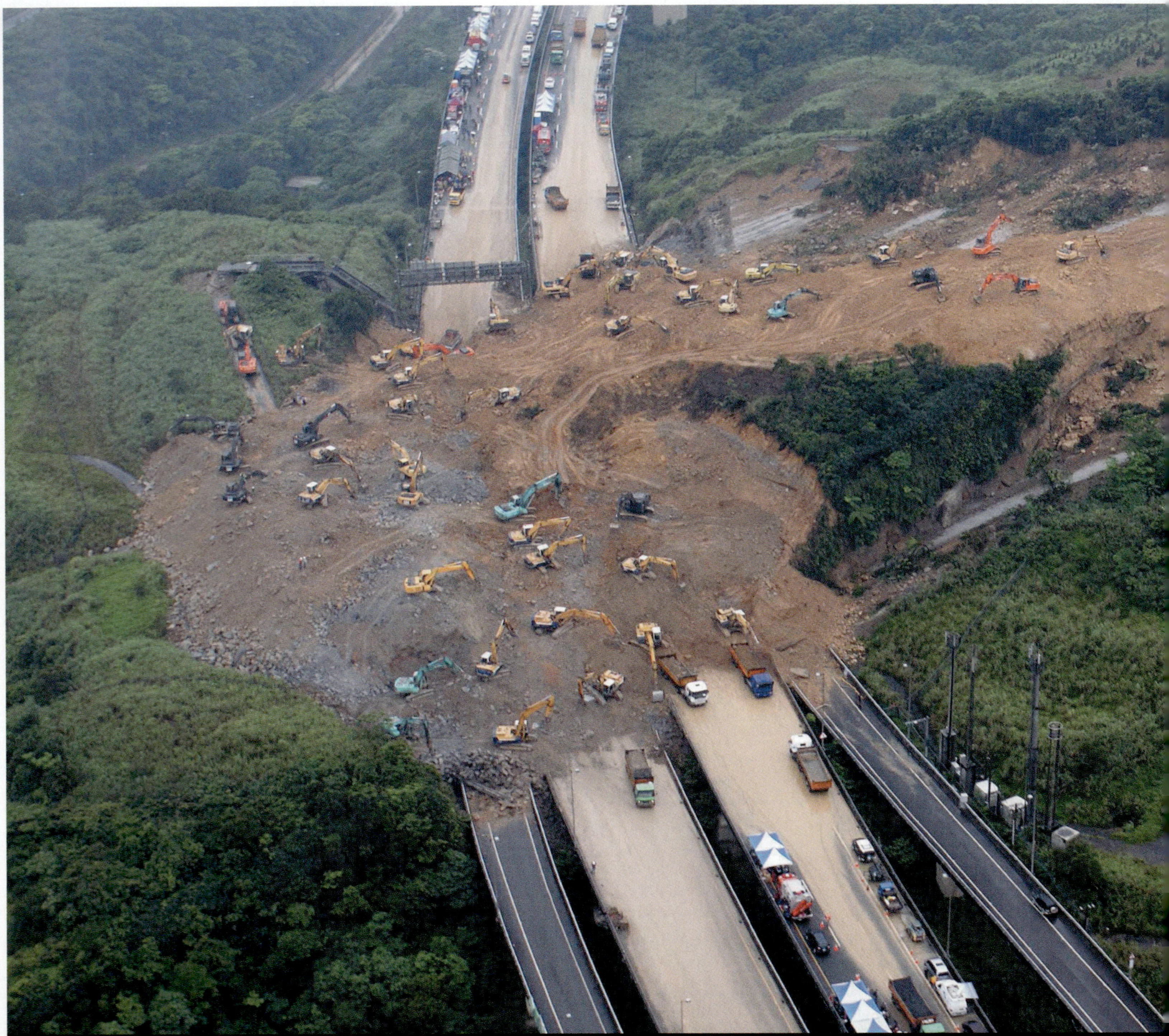

LEARNING OBJECTIVES

By the end of this chapter, you should understand...

1. the characteristics and consequences of different types of mass movements (landslides).
2. factors that determine whether a slope is stable or unstable.
3. events that can trigger a mass-movement event.
4. why some regions are more susceptible to mass movements than are others.
5. how landslide hazards can be identified, evaluated, and in some cases prevented.

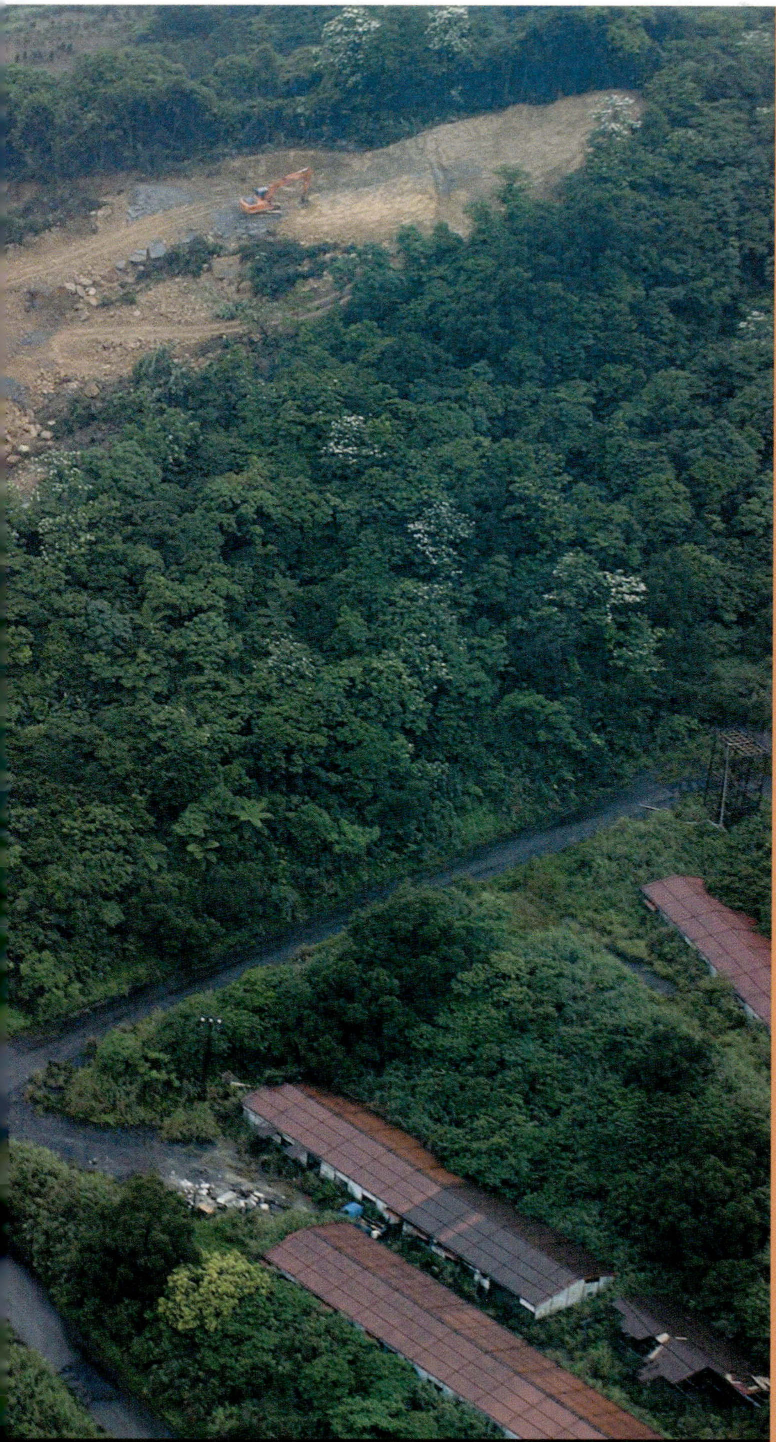

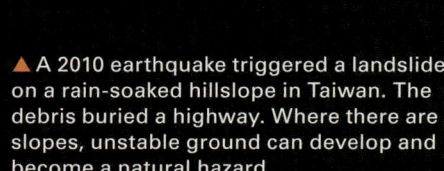

▲ A 2010 earthquake triggered a landslide on a rain-soaked hillslope in Taiwan. The debris buried a highway. Where there are slopes, unstable ground can develop and become a natural hazard.

CHAPTER 13

Unsafe Ground: Landslides and Other Mass Movements

13.1 Introduction

It was Sunday, May 31, 1970, a market day, and thousands of people had crammed into the Andean town of Yungay, Peru, to shop. Suddenly they felt the jolt of an earthquake, strong enough to topple some masonry houses. But worse was yet to come. Vibrations from the earthquake broke an 800-m-wide ice slab off the end of a glacier at the top of Nevado Huascarán, a nearby 6.6-km-high mountain peak. As the ice tumbled down over 3.7 km, it disintegrated into a chaotic avalanche of chunks, parts of which reached speeds of 300 km per hour. Near the base of the mountain, most of the avalanche channeled into a valley and thickened into a churning cloud, as high as a ten-story building, that ripped up rocks and soil along the way. Frictional heating transformed the ice into water, which mixed with rock and dust to produce 50 million cubic meters of a muddy slurry viscous enough to carry boulders larger than houses. This mass, sometimes floating on a compressed air cushion that allowed it to pass without disturbing the grass below, traveled over 14.5 km in less than four minutes.

At the mouth of the valley, most of the debris overran the village of Ranrahirca before coming to rest and creating a dam that blocked the Santa River. But some shot up the sides of the valley and became airborne for several seconds, flying over the ridge bordering Yungay. As the town's inhabitants and visitors stumbled out of earthquake-damaged buildings, they heard a deafening roar and looked up to see a wall of debris burst above the nearby ridge. The debris engulfed the town and buried it. Only the top of the church and a few palm trees remained visible to show where Yungay once stood (**Fig. 13.1**). Today, the site is a grassy meadow with a hummocky (irregular and lumpy) surface, spotted with crosses left by relatives mourning the 18,000 people entombed below.

People often assume that the ground beneath them is terra firma, a solid foundation on which they can build their lives. But the catastrophe at Yungay says otherwise. Some areas of the Earth's surface are unstable and might start moving downslope, if disturbed. Geologists refer to the downslope transport of rock, **regolith** (soil, and loose sediment or debris), snow, and ice as **mass movement**, or *mass wasting*. Like earthquakes, volcanic eruptions, storms, and floods, mass movements are a type of **natural hazard**, meaning a natural feature of the environment that can cause damage to environments and societies. Unfortunately, mass movements have become more of a threat every year, because as the world's population has grown, cities have expanded into areas of unsafe ground. Mass movement also plays a critical role in the rock cycle, as the first step in the production and transportation of sediment and, therefore, in the evolution of landscapes. It's the most rapid geologic process that can change the shape of a hill or mountain.

In this chapter, we look at the types, causes, and consequences of mass movement (both on land and under the sea), and the precautions society can take to protect people and property from its dangers. You might want to consider this information when selecting a site for your home, or when voting on land-use propositions that could affect your community.

FIGURE 13.1 The May 1970 Yungay landslide disaster in Peru.

(a) Before the landslide, the town of Yungay perched on a hill near the ice-covered mountain Nevado Huascarán.

(b) The landslide completely buried the town beneath debris. A landslide scar is visible on the mountain in the distance.

13.2 Types of Mass Movement

Most people refer to any mass movement of rock and/or regolith down a slope as a **landslide**. Geologists and engineers, however, find it useful to distinguish among different kinds of landslides based on four features: (1) the type of material involved (rock or regolith); (2) the velocity of movement (slow, intermediate, or fast); (3) the character of the moving mass (coherent, chaotic, or slurry); and (4) the environment in which the movement takes place (subaerial or submarine). In this section, we first examine mass movements that occur on land, roughly in order from slow to very fast. Then we briefly introduce submarine mass movements.

Creep and Solifluction

Creep refers to the slow, gradual downslope movement of regolith on a slope. Creep happens when regolith alternately expands and contracts in response to freezing and thawing, wetting and drying, or warming and cooling. To see how the process of creep works, let's focus on the consequences of seasonal freezing and thawing. In the winter, when water freezes, the regolith expands, and particles move outward, perpendicular to the slope. During the spring thaw, water becomes liquid again, and gravity makes the particles sink vertically and thus migrate downslope slightly (**Fig. 13.2a, b**). You can't see creep by staring at a hillslope because it occurs too slowly, but over a period of years, creep causes trees, fences, gravestones,

FIGURE 13.2 The process and consequences of slow mass movements (creep and solifluction).

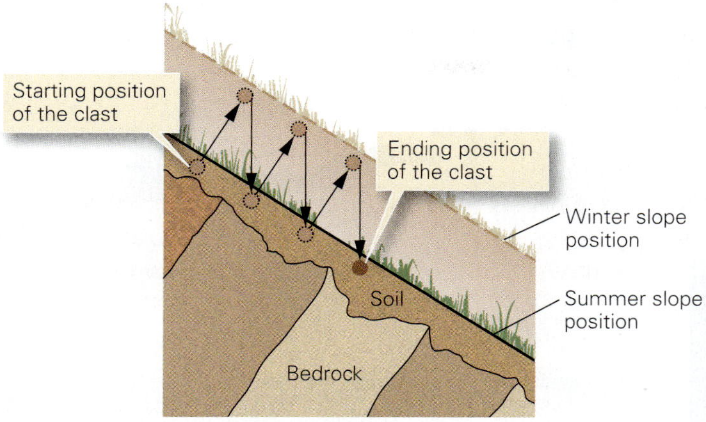

(a) Creep due to freezing and thawing: The clast rises perpendicular to the ground during freezing and sinks vertically during thawing. After 3 years, it migrates to the position shown.

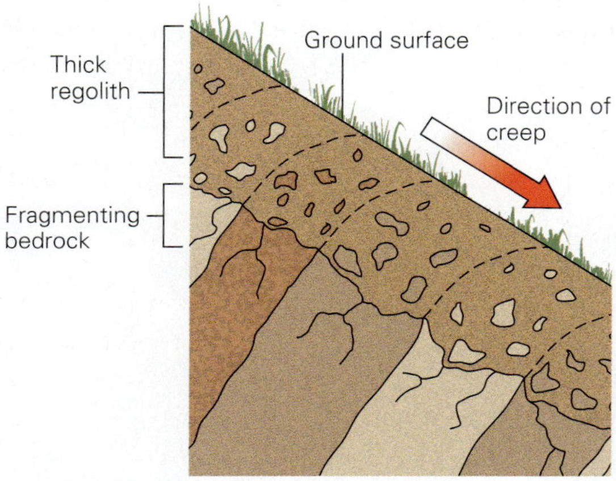

(b) As rock layers weather and break up, the resulting debris creeps downslope.

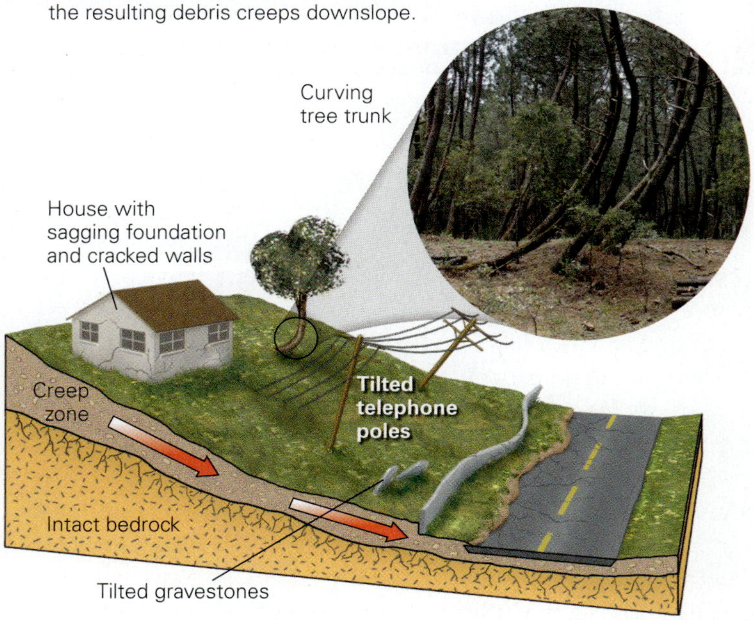

(c) Soil creep causes walls to bend and crack, building foundations to sink, trees to bend, and power poles and gravestones to tilt.

(d) Solifluction on a hillslope in the tundra.

walls, and foundations built on a hillside to tilt downslope (**Fig. 13.2c**). Notably, trees that continue to grow after they have been tilted display a pronounced curvature at their base.

In arctic or high-elevation regions, regolith freezes solid to great depth during the winter, producing permanently frozen ground, or *permafrost*. In the brief summer thaw, only the uppermost 1 to 3 m of the permafrost thaws. Since meltwater cannot sink into permafrost, the melted layer becomes soggy and weak and flows slowly downslope in overlapping sheets. Geologists refer to this kind of creep as **solifluction** (**Fig. 13.2d**).

Slumps

During the summer of 2011, after weeks of drenching rains, a 1.5-km-wide portion of a slope began to move down and out into the floor of Keene Valley, New York. The mass moved at only centimeters to tens of centimeters per day, but even at this slow rate, the accumulated displacement destroyed several expensive homes. The boundary between the moving mass and the unmoving land upslope evolved into a 5-m-high escarpment.

Geologists refer to such relatively slow-moving mass-movement events, during which moving rock or regolith does not disintegrate but rather stays somewhat coherent, as a **slump**, and they refer to the moving mass itself as a *slump block* (**Fig. 13.3**). A slump block slides on a **failure surface**. Some failure surfaces are planar, but commonly they curve

FIGURE 13.3 The process of slumping on a hillslope. Note the scarps that form at the head of the slump.

(a) A head scarp on the hillslope.

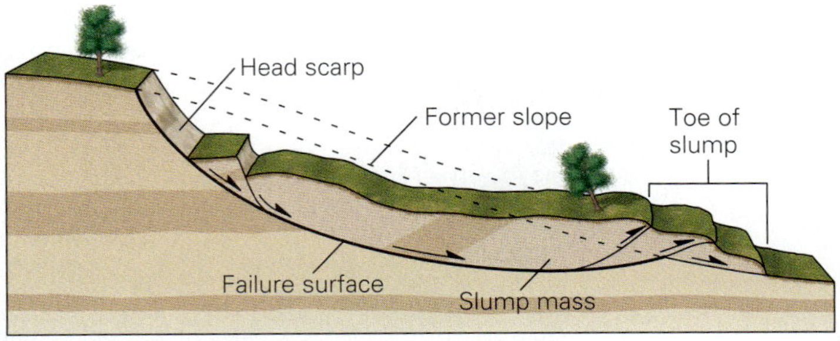

(b) Cross section of a slump.

(c) Slumping dumped sediment into this river in Costa Rica.

(d) A slump beginning to form along a highway in Utah.

and resemble a spoon lying concave side up. The exposed upslope edge of a failure surface forms a *head scarp*, a new cliff face. (Displacement on a head scarp resembles displacement on a normal fault; see Chapter 9.) Geologists refer to the downslope end of a slump block as the block's toe. On a hillslope, the toe can move up and over the pre-existing land surface to form a curving ridge, but along sea coasts or river banks, the toe ends up in the water, where it eventually washes away. In some cases, the upslope and downslope portions of a slump block break into a series of discrete slices, each separated from its neighbor by a sliding surface. Slumps come in all sizes, from only a few meters across to tens of kilometers across. In general, they move at speeds ranging from millimeters per day to meters per minute.

SEE FOR YOURSELF...

SLUMPS NEAR SAN FRANCISCO

Latitude
37°40′47.33″ N

Longitude
122°29′46.14″ W

Zoom to 2.5 km (~8200 feet) and rotate the view so north is on the left (left photo). Then tilt, to look obliquely east (right photo) from 550 m.

Along the coast near San Francisco, waves erode sea cliffs at the base of a terrace. Occasionally, slumps along the cliffs carry debris to the beach. The bowls cut into the terrace are places where slumps occurred in the past.

FIGURE 13.4 Examples of mudflows and lahars.

(a) A 2011 mudslide destroyed a high-rise building at the base of the hill.

(b) Mudslides of 2011 stripped away forests on hillslopes in Brazil.

(c) A recent debris flow in Utah. Note the chaotic mixture of rock chunks and mud.

(d) A lahar that rushed down the side of Mt. St. Helens, Washington.

Mudflows, Debris Flows, and Lahars

Rio de Janeiro, a coastal city in Brazil, originally occupied only the flatlands bordering beautiful crescent beaches that had formed between steep rock hills. But in recent decades, the population has increased so much that, in many places, densely populated communities of makeshift shacks have grown up on steep slopes. These communities, with inadequate storm drains, were built on the thick regolith that resulted from long-term weathering of bedrock in Brazil's tropical climate. Particularly heavy rains can saturate the soil, transforming it into a viscous slurry of mud, resembling wet concrete, that flows downslope. When this happens, whole communities can disappear overnight, replaced by a hummocky muddle of mud and debris. At the base of the cliffs, the flowing mud may knock over and bury buildings of all sizes (**Fig. 13.4a**). In unpopulated areas of Brazil, such mass movements can rip away forests (**Fig. 13.4b**). Geologists refer to a moving slurry of mud as a **mudflow** or *mudslide*, and a slurry consisting of a mixture

SEE FOR YOURSELF...

LA CONCHITA MUDSLIDE, CALIFORNIA

Latitude
34°21′50.29″ N

Longitude
119°26′46.85″ W

Look NE from 250 m (~820 feet).

Tectonic activity has uplifted the coast of California to form a terrace bordered on the ocean side by a steep escarpment. Below, you can see a mudflow that overran houses built at the base of the escarpment. Note the head scarp.

of mud and larger, pebble- to boulder-sized fragments as a **debris flow** or **debris slide** (**Fig. 13.4c**). Some mudflows or debris flows begin as slumps, then evolve when they mix with water, disintegrate, and start moving faster.

The speed at which a mud- or debris flow moves depends on the slope angle and on the water content. Flows move faster if they contain more water and are less viscous, and if they move on steeper slopes. On a gentle slope, drier mudflows move like molasses, but on a steep slope, very wet mud with low viscosity may move at over 100 km per hour. Because mudflows and debris flows have greater viscosity than clear water, they can carry large rock chunks as well as houses and cars. They typically follow channels downslope and at the base of the slope they may spread out into a broad lobe (**Box 13.1**).

Particularly devastating mudflows spill down the river valleys bordering volcanoes. These mudflows, known as **lahars**, consist of a mixture of volcanic ash and water from the snow and ice that melts in a volcano's heat or from heavy rains (**Fig. 13.4d**; see Chapter 5). One of the most destructive lahars occurred on November 13, 1985, in the Andes Mountains in Colombia. That night, an eruption melted a volcano's thick snowcap, producing hot water that mixed with ash. A scalding lahar rushed down river valleys and swept over the nearby town of Armero while most inhabitants were asleep. Of the 25,000 residents, 20,000 perished.

Rockslides and Debris Slides

In the early 1960s, engineers built a huge new dam across a river on the northern side of Monte Toc, in the Italian Alps, to create a reservoir for generating electricity. This dam, the Vaiont Dam, was an engineering marvel, a concrete wall rising 260 m (as high as an 85-story skyscraper) above the valley floor (**Fig. 13.5a**). Unfortunately, the dam's builders did not recognize the hazard posed by nearby Monte Toc. Limestone beds interlayered with weak shale beds underlie the side of Monte Toc facing the reservoir. These beds dip parallel to the surface of the mountain and curve under the reservoir (**Fig. 13.5b**). As the reservoir filled, water seeped into the beds, which weakened them. As a result, the flank of the mountain started to crack, and therefore, to shake. Local residents began to call Monte Toc *la montagna che cammina* (the mountain that walks).

After several days of rain, Monte Toc began to rumble so much that on October 9, 1963, engineers lowered the water

FIGURE 13.5 The Vaiont Dam disaster happened when a landslide suddenly dumped rock debris into a reservoir. The debris pushed the water over the dam.

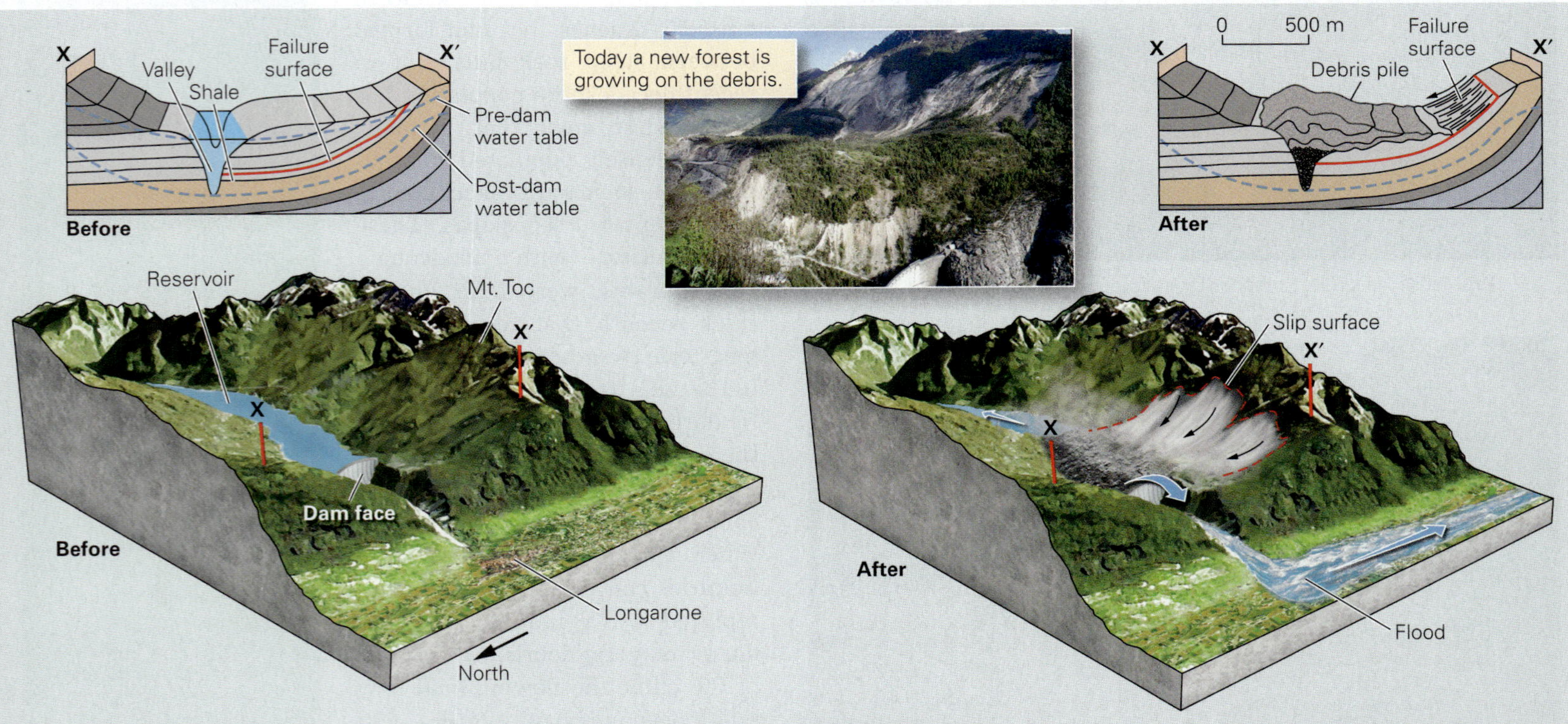

(a) Before the landslide, the north flank of Monte Toc was forested. When the reservoir filled, the slope became unstable. A shale bed a few hundred meters below the ground surface became a failure surface.

(b) Thirty-three million cubic meters of debris slid and displaced water in the reservoir. The water surged over the dam and swept away a village in the valley below.

BOX 13.1 Consider This...

What Goes Up Must Come Down

Along the coast of California, waves slowly erode the land and produce low, flat areas called *wave-cut benches* (see Chapter 15). As this happens, tectonic motions slowly raise the land surface. When uplifted, these benches form small plateaus, or terraces. One such terrace lies at an elevation of 180 m above sea level, about 500 m east of the present-day beach at La Conchita; the west face of this terrace is a cliff-like bluff (**Fig. Bx13.1a**). Repeated slip along the San Andreas plate boundary has broken up the bedrock of the area, and fragments have weathered substantially, so the substrate of the terrace and bluff consists of weak clay and debris.

Relatively little vegetation covers the bluff or the terrace above. Rain that falls on the face of the bluff drains away quickly via a network of small, temporary streams. But the water falling on the terrace infiltrates the ground, sinks down, and saturates clay and debris meters below the surface of the bluff, turning into very weak mud. When this happens, the weight of surface material causes the bluff to give way, and a mass of mud and debris flows downslope at rates of up to 10 m per second.

If the region of La Conchita were uninhabited, such mass wasting would just be part of the natural process of landscape evolution—gravity brings down land that had been raised by tectonic activity. But when downslope movements take place in La Conchita, it makes headlines, because on the modern wave-cut bench between the shore and the base of the bluff, developers built a community housing 350 people. In 1995, a mud- and debris flow overwhelmed 9 houses at the base of the bluff. An even more devastating flow happened in 2005, burying 13 houses and damaging 23, and killing 10 people (**Fig. Bx13.1b**). These events have sent a clear message about the importance of reading the landscape before planning construction.

FIGURE Bx13.1 The 2005 La Conchita mudslide along the coast of California.

(a) A housing development was built in a narrow strip between the beach and steep cliffs.

(b) During heavy rains, the slope gave way and heavy mud flowed down, burying houses and taking several lives.

level in the reservoir. They thought that, at worst, the wet ground might slump a little into the reservoir, with minor consequences. So no one ordered the evacuation of the town of Longarone, a few kilometers down the valley below the dam. Unfortunately, the engineers underestimated the problem. At 10:30 that evening, a huge chunk of Monte Toc—600 million tons of rock—detached from the mountain and slid downslope into the reservoir. Some debris rocketed up the opposite wall of the valley to a height of 260 m above the original reservoir level. The displaced water of the reservoir spilled over the top of the dam and rushed down into the valley below. When the flood had passed, nothing of Longarone and its 1,500 inhabitants remained. Though the dam itself still stands, it holds back only debris and has never provided any electricity.

Geologists refer to such a sudden movement of rock and debris down a nonvertical slope as a **rockslide** if the mass consists only of rock or as a **debris slide** if it includes regolith. Once a slide has taken place, it leaves a scar on the slope and forms a debris pile at the base of the slope. Slides happen when bedrock or regolith detaches from a slope, slips rapidly downhill on a failure surface, and breaks up into a chaotic jumble. Slides may move at speeds of up to 300 km per hour; they are particularly fast when a cushion of air gets trapped beneath, so there is virtually no friction between the slide and its substrate, and the mass moves like a hovercraft. Rockslides and debris slides sometimes have enough momentum to climb the opposite side of the valley into which they fell. Slides, like slumps, come at a variety of scales. In some cases, the material in a rock or debris fall mixes with air and forms a turbulent cloud that races downslope at high velocity. Geologists sometimes refer to such events as *rock avalanches* or *debris avalanches*.

Avalanches

In the winter of 1999, an unusual weather system passed over the Austrian Alps. First it snowed. Then the temperature warmed and the snow began to melt. But then the weather turned cold again, and the melted snow froze into a hard, icy crust. This cold snap ushered in a blizzard that blanketed the ice crust with tens of centimeters (1–2 ft) of new snow. With the frozen snow layer underneath acting as a failure surface, 200,000 tons of new snow began to slide down the mountain. As it accelerated, the mass transformed into a **snow avalanche**, a chaotic jumble of snow surging downslope. At the bottom of the slope, the avalanche overran a ski resort, crushing and carrying away buildings, cars, and trees, and killing over 30 people. It took searchers and their specially trained dogs many days to find buried survivors and victims under the 5- to 20-m-thick pile of snow that the avalanche deposited (**Fig. 13.6a**).

FIGURE 13.6 Examples of avalanches.

(a) Aftermath of a 1999 avalanche in the Austrian Alps. Masses of snow buried several homes.

(b) A dry-snow avalanche in Alaska is a turbulent cloud.

SEE FOR YOURSELF...

ROCKFALLS IN CANYONLANDS, UTAH

Latitude
38°29′50.07″ N

Longitude
110°0′15.63″ W

Zoom to 5 km (3 miles) and look down.

The Green River has carved a deep valley through horizontal layers of strata. Shale forms the slopes, and sandstone forms the top ledge. Erosion of the shale undercuts the sandstone, which breaks away at joints and tumbles downslope to build a talus pile.

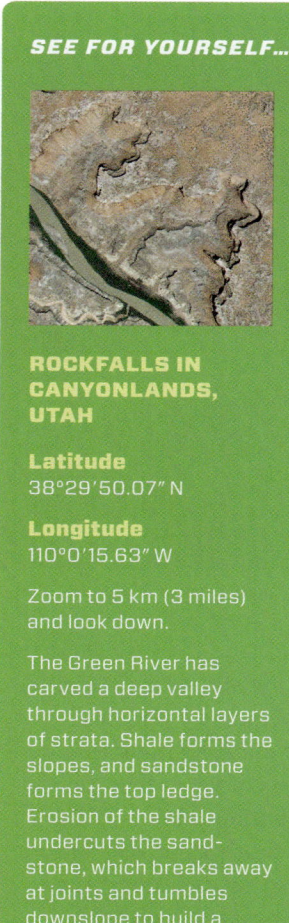

What triggers snow avalanches? Some happen when a cornice, a large drift of snow that builds up on the lee side of a windy mountain summit, suddenly gives way and falls onto slopes below, where it knocks free additional snow. Others happen when a broad slab of snow on a moderate slope detaches from its substrate along an icy failure surface. As we described above, the failure surface is a frozen crust on the surface of older snow. It may also be a layer of rounded grains, consisting of snow that either partly melted or sublimated, and lost its spiky protrusions. Wet avalanches, formed at warmer temperatures, move as a slurry of solid and liquid water, whereas dry avalanches, formed when the snow mass contains no liquid water, tumble as a cloud of powder (**Fig. 13.6b**).

Rockfalls and Debris Falls

Rockfalls and **debris falls**, as their names suggest, occur when a mass free-falls from a cliff for part of its journey (**Fig. 13.7**). We've already described the devastating debris fall that destroyed Yungay, Peru. Rockfalls happen when a body of solid rock separates from a cliff face along a joint. Friction and collision with other rocks may bring some blocks to a halt before they reach the bottom of the slope—these blocks pile up to form a **talus**, a sloping apron of rocks along the base of the cliff. Debris that has fallen a long way can reach speeds of 300 km per hour and may have so much momentum that it keeps moving as an avalanche-like cloud of fragments mixed with air when it reaches the base of a cliff. Large, fast rockfalls push the air in front of them, creating a short blast of hurricane-like wind. For example, the wind in front of a 1996 rockfall in Yosemite National Park flattened over 2,000 trees.

Did you ever wonder...
why highway engineers erect "falling rock" signs?

Small rockfalls happen fairly frequently along steep highway roadcuts, leading to the posting of falling-rock zones. Such rockfalls commonly take place soon after construction because blasting and excavation leave loose rocks on the slope above the road. But rockfalls may continue long after construction, for mechanical and chemical weathering weakens slopes over time (see Interlude B). In temperate climates, rockfalls are particularly common in the spring, after winter ice, which may cause frost wedging, melts.

FIGURE 13.7 Examples of rockfalls.

(a) Successive rockfalls have littered the base of this sandstone cliff with boulders. Note the talus at the base of the cliff.

(b) A rockslide buried the forest bordering a lake in the Uinta Mountains, Utah. Fresh rock exposed by the slide has a lighter color.

Submarine Mass Movements

So far, we've focused on mass movements that occur subaerially, for these are the ones we can see most easily and that affect us the most. But mass movements also happen underwater. Geologists distinguish three types of submarine mass movements, or submarine landslides, according to whether the mass remains coherent or disintegrates as it moves. In **submarine slumps**, semicoherent blocks slip downslope on weak detachments. In some cases, the layers constituting the

FIGURE 13.8 Examples of huge submarine slumps and debris flows.

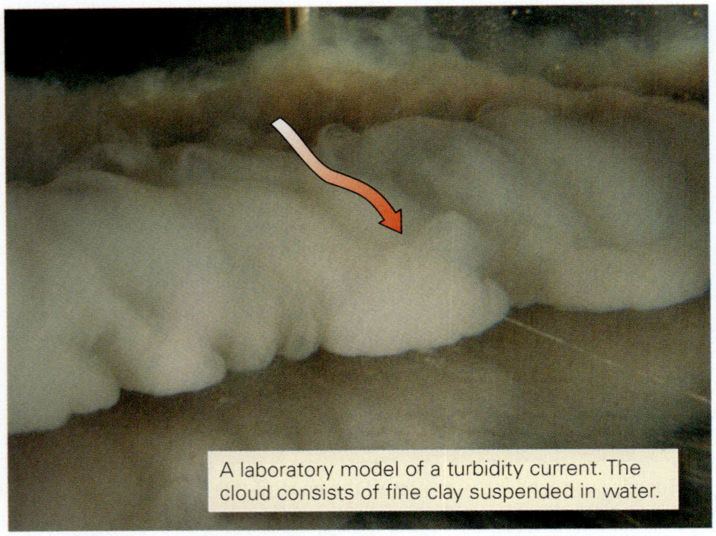

(a) A turbidity current is a cloud of sediment suspended in water; it flows near the seafloor because it is denser than clear water.

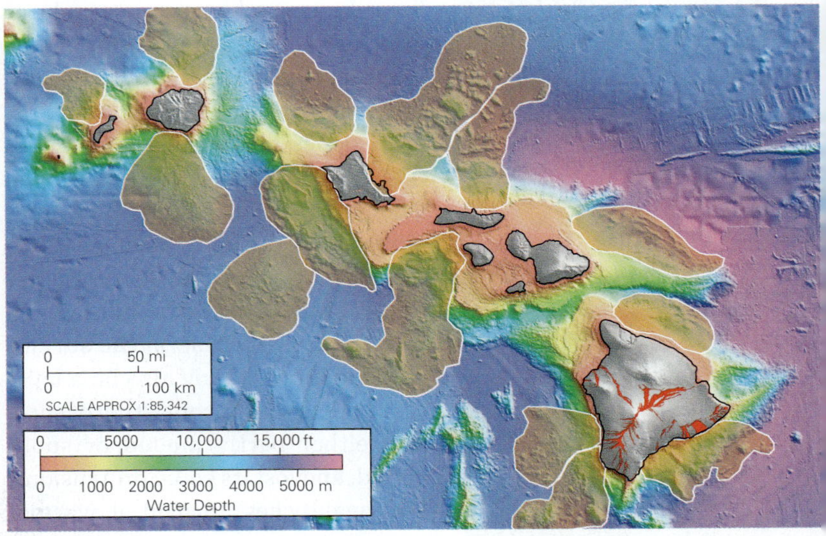

(c) A bathymetric map of the area around Hawaii shows several huge slumps, shaded in tan.

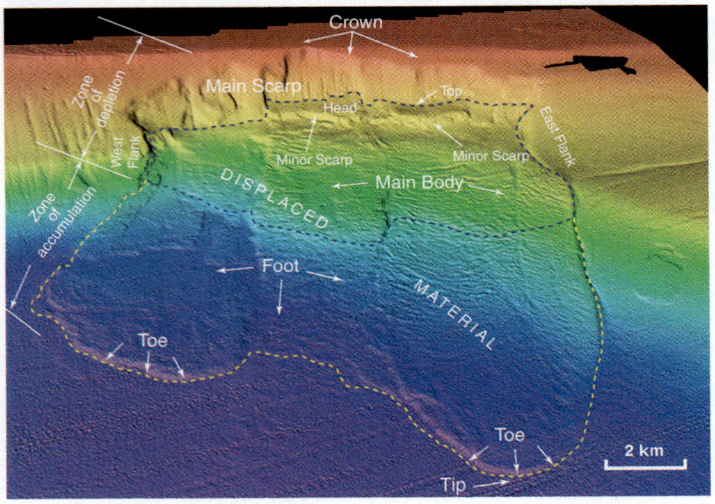

(b) A digital bathymetric map of a slip along the coast of California. The parts of the slump are labeled.

blocks become contorted as they move, like a tablecloth that has slid off a table. In *submarine debris flows*, the moving mass breaks apart to form a slurry containing larger clasts (pebbles to boulders) suspended in a mud matrix. And in **turbidity currents,** sediment disperses in water to create a turbulent cloud of suspended sediment that rushes downslope like an underwater avalanche (**Fig. 13.8a**).

In recent years, geologists have used satellites as well as sonar to map out the extent of submarine landslides. The shapes of slumps and landslide deposits stand out on the resulting new generation of high-resolution seafloor maps (**Fig. 13.8b**). Geologists have found that submarine slopes bordering both hot-spot volcanoes and active plate boundaries are scalloped by many immense slumps and debris flows, because tectonic activity frequently jars these areas with earthquakes that set masses of material in motion. Debris from countless events over millions of years has substantially modified the flanks of the Hawaiian Islands (**Fig. 13.8c**). Some of this debris has moved more than 150 km away from the islands, leaving huge head scarps along the edges of the islands. Significantly, passive-margin coasts are not immune to slumping, and immense slumps have been mapped along the coasts of the Atlantic Ocean.

Since a submarine slump can develop fairly quickly, and since its movement can displace a large area of the seafloor, it can trigger a tsunami. A huge slump called the Storegga Slide occurred west of Norway, along the Atlantic passive margin, thousands of years ago. The area affected by the slump is about 100 km wide, and the slump debris traveled underwater for over 600 km. Tsunamis generated by the Storegga Slide may have wiped out Stone-Age villages all around the coast of the North Sea.

TAKE-HOME MESSAGE

Mass movements differ from one another based on speed and character. Creep, slumping, and solifluction are slow. Mudflows and debris flows move faster, and avalanches and rockfalls move the fastest. Mass movements occur on land and underwater. A large submarine mass movement may generate a tsunami.

QUICK QUESTION In what way is a snow avalanche like a turbidity current?

13.3 Why Do Mass Movements Occur?

We've seen that mass movements travel at a range of different velocities, from slow (creep) to faster (slumps, mudflows and debris flows, and rockslides and debris slides) to fastest (snow avalanches, and rock- and debris falls—see **Geology at a Glance**, pp. 452–453). The velocity depends on the steepness of the slope and the water or air content of the mass. For any movement to take place, the stage must be set by the following phenomena: (1) fracturing and weathering of the substrate, which weakens the substrate so it cannot hold up against the pull of gravity; (2) the development of relief, which provides slopes down which masses move; and (3) an event that sets mass in motion. Let's look at these phenomena more closely.

Weakening the Substrate: Fragmentation and Weathering

If the Earth's surface were covered by completely unfractured rock, mass movements would be of little concern, for intact rock has great strength and could form stalwart mountain faces that would not tumble. But in reality, the rock of the Earth's upper crust has been fractured by jointing and faulting (**Fig. 13.9**), and in many locations the surface has a cover of regolith resulting from the weathering of rock. Regolith and fractured rock are much weaker than intact rock and can indeed collapse in response to gravitational pull. Thus jointing, faulting, and weathering ultimately make mass movements possible.

FIGURE 13.9 Jointing broke up this thick sandstone bed along a cliff in Utah. Blocks of sandstone break free along joints and tumble downslope.

Why are regolith and fractured rocks weaker than intact bedrock? The answer comes from looking at the strength of the attachments holding materials together. A mass of intact bedrock is relatively strong because the chemical bonds within its interlocking grains, or within the cements between grains, can't be broken easily. A mass of loose rocks or of regolith, in contrast, is relatively weak because the grains are held together only by friction, electrostatic attraction, and/or the surface tension of water. All of these forces combined are weaker than chemical bonds holding together the atoms in the minerals of intact rock. To picture this contrast, think about how much easier it is to flatten a sand castle (whose strength comes primarily from the surface tension of water films on the sand grains) than it is to flatten a granite sculpture of a castle.

Slope Stability

Mass movements do not take place on all slopes, and even on slopes where such movements are possible, they occur only occasionally. Geologists distinguish between *stable slopes*, on which sliding is unlikely, and *unstable slopes*, on which sliding will likely happen. When material starts moving on an unstable slope, we say that **slope failure** has occurred. Whether a slope fails depends on the balance between two forces—the *downslope force*, caused by gravity, and the *resistance force*, which inhibits sliding. If the downslope force exceeds the resistance force, the slope fails and mass movement results.

Let's examine slope failure more closely by imagining a block sitting on a slope. We can represent the force of gravitational attraction between this block and the Earth by an arrow, called a vector, that points straight down toward the Earth's center of gravity. This arrow can be separated into two components—the downslope force component is parallel to the slope, whereas the normal force component is perpendicular to the slope. In this representation, we can symbolize the resistance force by an arrow pointing uphill. If the downslope force is larger than the resistance force, then the block moves; otherwise, it stays in place (**Fig. 13.10**). Note that for a given mass, the downslope forces increases as the slope steepens.

What produces a resistance force? As we saw above, chemical bonds in mineral crystals or cement hold intact rock in place, friction holds an unattached block in place, electrical charges and friction hold dry regolith in place, and surface tension holds wet regolith in place. Because of resistance force, granular debris tends to pile up to produce the steepest slope it can without collapsing. The angle of this slope is called the **angle of repose**, and for most dry, unconsolidated materials (such as dry sand) it typically has a value of between 30° and 45°. The angle depends partly on the shape and size of grains, which determine the amount of friction across

GEOLOGY AT A GLANCE

Mass Movement

In Earth's gravity field, what goes up must come down—sometimes with disastrous consequences. Rock and regolith are not infinitely strong, so every now and then slopes or cliffs give way in response to gravity, and materials slide or tumble downslope. This downslope movement is called mass movement, or mass wasting. The kind of mass movement that takes

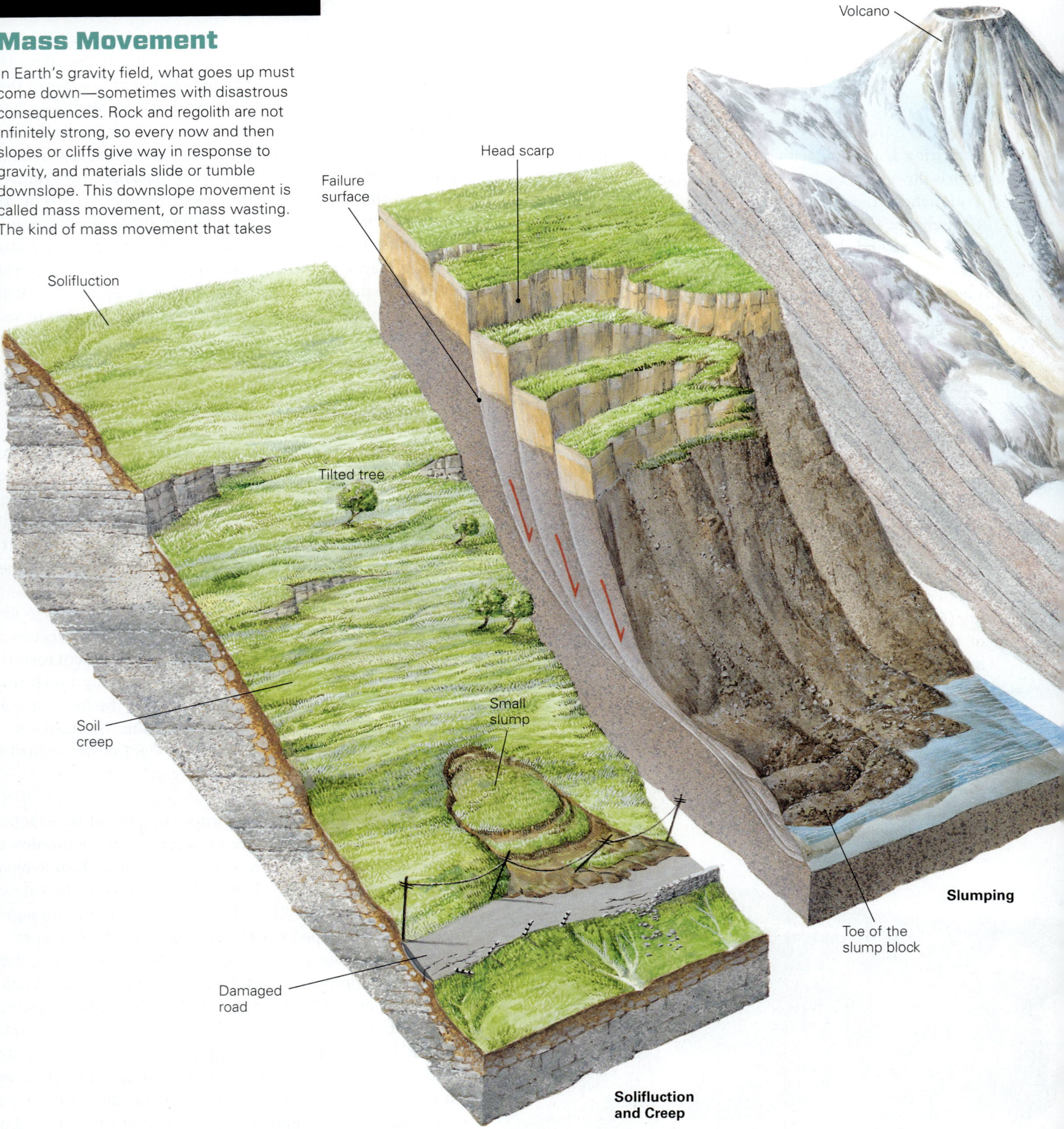

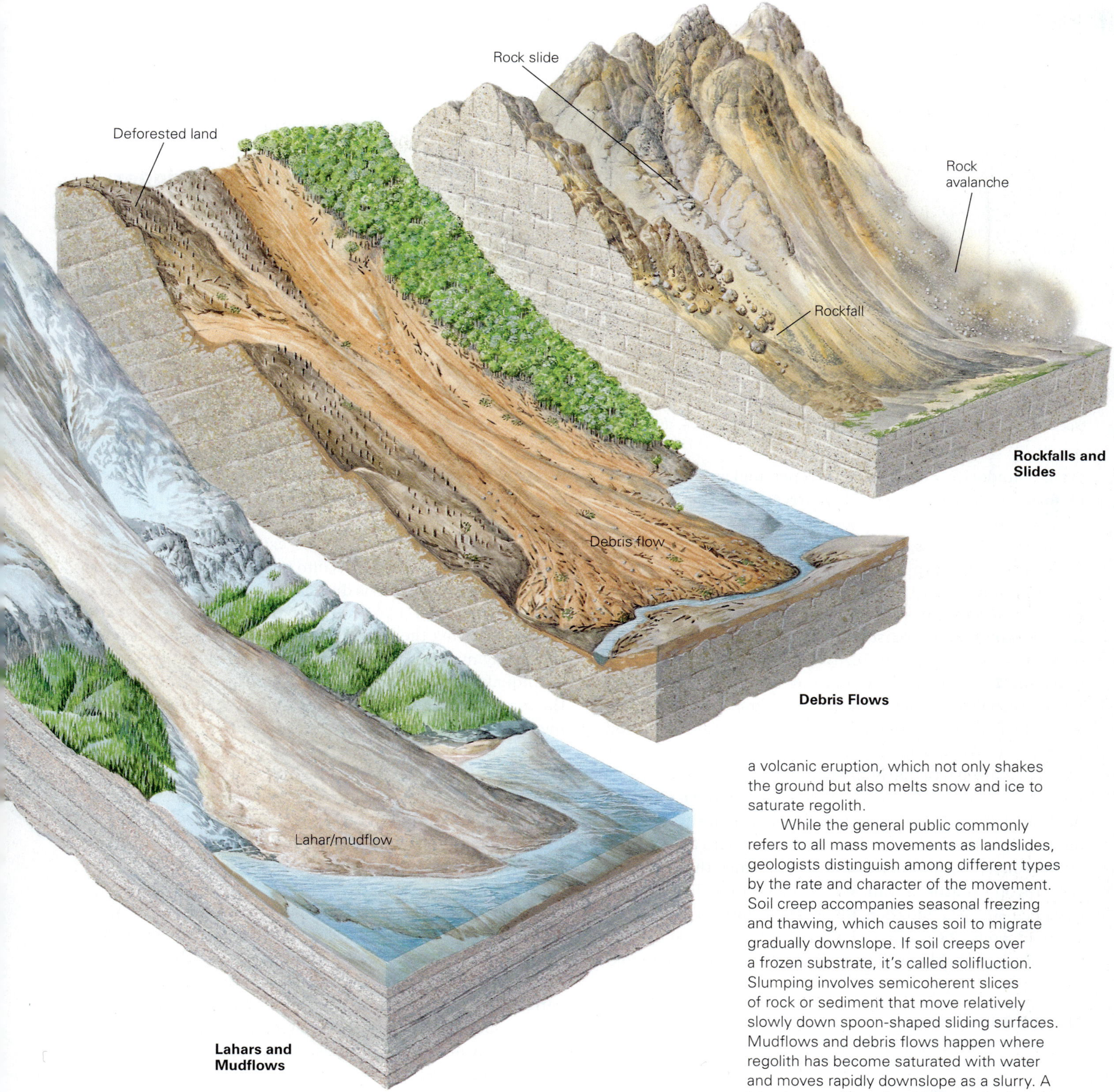

place at a given location reflects the composition of the slope, the steepness of the slope, and the climate. Stronger rocks can hold up steep cliffs, but with time, rock breaks free along joints and tumbles or slides down weak surfaces. Coherent regolith may slowly slide down slopes, whereas water-saturated regolith may flow rapidly. Episodes of mass movement may be triggered by an oversteepened slope, a heavy rainfall that saturates the slope, an earthquake that shakes debris free, or a volcanic eruption, which not only shakes the ground but also melts snow and ice to saturate regolith.

While the general public commonly refers to all mass movements as landslides, geologists distinguish among different types by the rate and character of the movement. Soil creep accompanies seasonal freezing and thawing, which causes soil to migrate gradually downslope. If soil creeps over a frozen substrate, it's called solifluction. Slumping involves semicoherent slices of rock or sediment that move relatively slowly down spoon-shaped sliding surfaces. Mudflows and debris flows happen where regolith has become saturated with water and moves rapidly downslope as a slurry. A lahar is a mudflow that forms on the flank of a volcano when ash mixes with water. Steep, rocky cliffs may suddenly give way in rockfalls. If the rock breaks up into a cloud of debris that rushes downslope at high velocity, it is a rock avalanche. Snow avalanches are similar, but the debris consists only of snow.

FIGURE 13.10 Forces that trigger downslope movement.

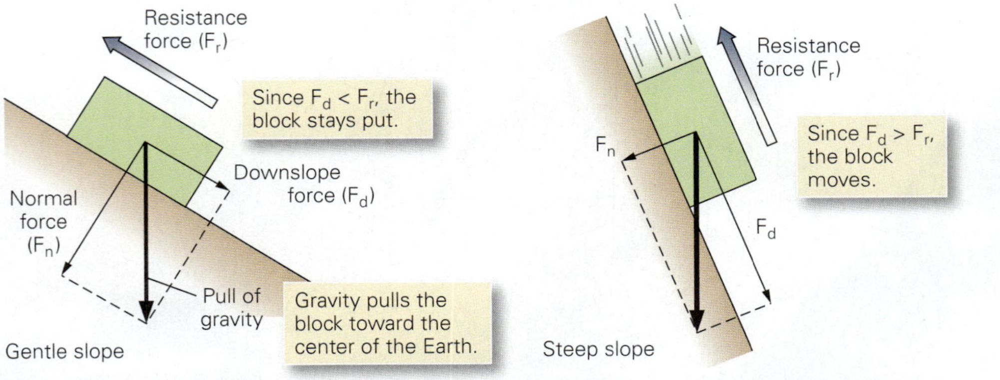

(a) Gravity can be divided into a normal force and a downslope force. If the resistance force, caused by friction, is greater than the downslope force, the block does not move.

(b) If the slope angle increases, the downslope force due to gravity increases. If the downslope force becomes greater than the resistance force, the block starts to move.

grain boundaries. For example, steeper angles of repose tend to form on slopes composed of irregularly shaped grains (**Fig. 13.11**).

In many locations, the resistance force is less than might be expected because a weak surface exists at some depth below ground level. If downslope movement begins on the weak surface, we can say that the weak surface has become a **failure surface**. Geologists recognize several different kinds of weak surfaces that are likely to become failure surfaces (**Fig. 13.12**). These include wet clay layers; wet, unconsolidated sand layers; joints; weak bedding planes (shale beds and evaporite beds are particularly weak); and metamorphic foliation planes. Weak surfaces that dip parallel to the land surface slope are particularly likely to fail. An example of such failure occurred in Madison Canyon, southwestern Montana, on August 17, 1959. That day, vibrations from a strong earthquake jarred the region. Metamorphic rock with a strong foliation that could serve as weak surfaces formed the bedrock of the canyon's southern wall. When the ground vibrated, rock detached along a foliation plane and tumbled downslope. Unfortunately, 28 campers lay sleeping on the valley floor. They were probably awakened by the hurricane-like winds blasting in front of the moving mass, but seconds later they were buried under 45 m of rubble.

FIGURE 13.11 The angle of repose is the steepest slope that a pile of unconsolidated sediment can have and remain stable. The angle depends on the shape and size of grains.

Fingers on the Trigger: What Causes Slope Failure?

What triggers an individual mass-wasting event? In other words, what causes the balance of forces to change so that the downslope force exceeds the resistance force, and a slope suddenly fails? Here, we look at various phenomena—natural and human-made—that trigger slope failure.

Shocks, Vibrations, and Liquefaction Earthquake tremors, storms, the passing of large trucks, or blasting in construction sites may cause a mass that had been on the verge of moving, to start moving. For example, an earthquake-triggered slide dumped debris into Lituya Bay, in southeastern Alaska, in 1958. The debris displaced the water in the bay, creating a 300-m-high (1,200 ft) splash that washed forests off the slopes bordering the bay and carried fishing boats anchored in the bay many kilometers out to sea. The vibrations of an earthquake break bonds that hold a mass in place and/or cause the mass and the slope to separate slightly, thereby decreasing friction. As a consequence, the resistance force decreases, and the downslope force sets the mass in motion. Shaking can also cause *liquefaction* of wet sediment by either increasing water pressure in spaces between grains so that the grains are pushed apart, or by breaking the cohesion between the grains. Liquefaction greatly diminishes the strength of a sediment layer by turning it, effectively, into a fluid.

Changing Slope Loads, Steepness, and Support As we have seen, the stability of a slope at a given time depends on the balance between the downslope force and the resistance force. Factors that change one or the other of these forces can lead to failure. Examples include changes in slope loads, failure-surface strength, slope steepness, and the support provided by material at the base of the slope.

Slope loads change when the weight of the material above a potential failure plane changes. If the load increases, due to construction of buildings on top of a slope or due to saturation of regolith with water, the downslope force increases and may exceed the resistance force. Seepage of water into the ground may also weaken underground failure surfaces, further decreasing resistance force. An example of such failure

FIGURE 13.12 Different kinds of weak surfaces can become failure surfaces.

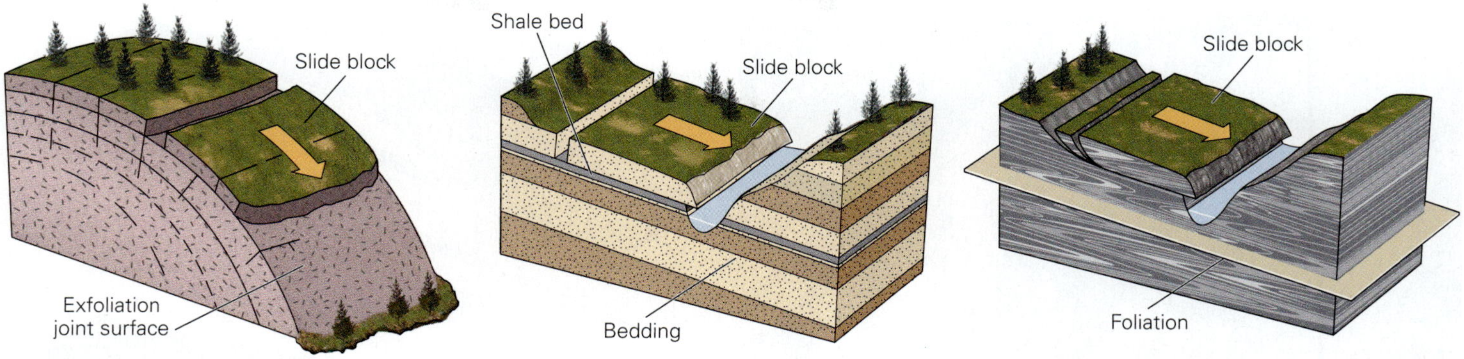

(a) Exfoliation joints form parallel to slope surfaces in granite and become failure surfaces.

(b) In sedimentary rocks, bedding planes (particularly in weak shale) become failure surfaces.

(c) In metamorphic rock, foliation planes (particularly in mica-rich schist) become failure surfaces.

triggered the largest observed landslide in U.S. history, the Gros Ventre Slide, which took place in 1925 on the flank of Sheep Mountain, near Jackson Hole, Wyoming (**Fig. 13.13**). Almost 40 million cubic meters of rock, as well as the overlying soil and forest, detached from the side of the mountain and slid 600 m downslope, yielding a debris flow that filled a valley and formed a 75-m-high natural dam across the Gros Ventre River.

Removing support at the base of a slope due to river or wave erosion or to construction efforts plays a major role in triggering many slope failures. In effect, the material at the base of a slope acts like a dam holding back the material farther up the slope. In some cases, erosion by a river or by waves eats into the base of a cliff and produces an overhang. When such *undercutting* has occurred, rock making up the overhang eventually breaks away from the slope and falls (**Fig. 13.14**).

Changing the Slope Strength The stability of a slope depends on the strength of the material constituting it. If the material weakens with time, the slope becomes weaker and eventually collapses. Three factors influence the strength of slopes:

- *Weathering*: With time, chemical weathering produces weaker minerals, and physical weathering breaks rocks apart. Thus, a formerly intact rock composed of strong minerals is transformed into a weaker rock or into regolith.
- *Vegetation cover*: In the case of slopes underlain by regolith, vegetation tends to strengthen the slope because the roots hold otherwise unconsolidated grains together. Also, plants absorb water from the ground, thus keeping it from turning into slippery mud. The removal of vegetation therefore has the net result of making slopes more susceptible to downslope mass movement. Deforestation in tropical rainforests, similarly, leads to catastrophic mass wasting of the forest's substrate.
- *Water content*: Water affects materials comprising slopes in many ways. Surface tension, due to the film of water on grain surfaces, may help hold regolith together. But if the water content increases, water pressure may push grains apart so that regolith liquefies and can begin to flow. Water infiltration may make weak surfaces underground more slippery, or may push surfaces apart and decrease friction. Some kinds of clays absorb water and expand, causing the ground surface to rise and, as a consequence, break up.

TAKE-HOME MESSAGE

Weathering and fragmentation weaken slope materials and make them more susceptible to mass movement. Failure occurs when the downslope force exceeds resistance force due to shocks, changing slope angles and strength, changing water content, and changing slope support.

QUICK QUESTION If the slope of a sand pile is less than the angle of repose, will the slope of the pile fail?

13.4 How Can We Protect Against Mass-Movement Disasters?

Identifying Regions at Risk

Clearly, landslides, mudflows, and slumps are natural hazards we cannot ignore. Too many people live in regions where mass movements have the potential to kill people and destroy

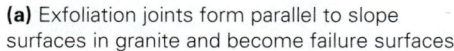

FIGURE 13.13 Stages leading to the 1925 Gros Ventre Slide in Wyoming.

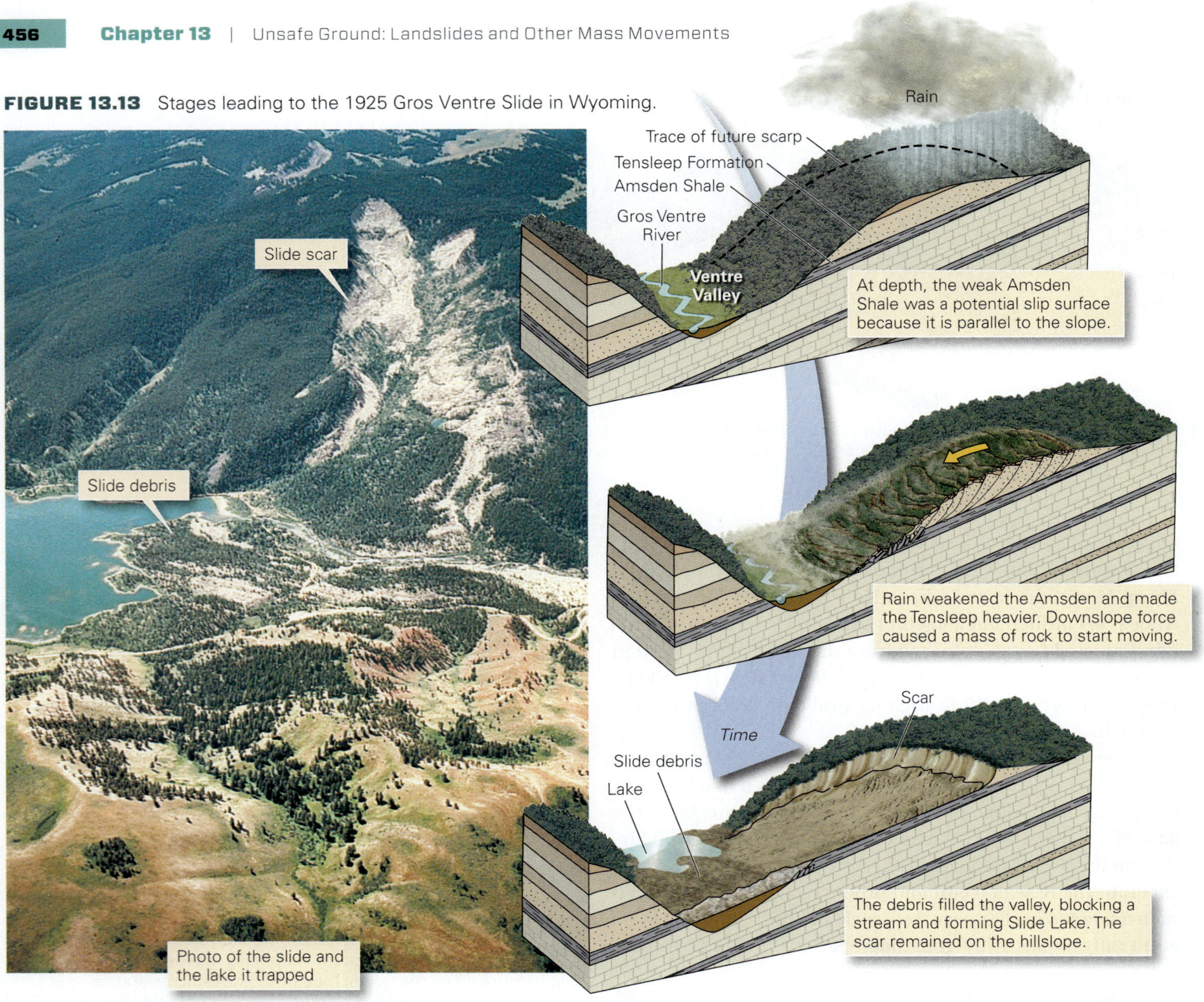
Slide scar
Slide debris
Photo of the slide and the lake it trapped
Rain
Trace of future scarp
Tensleep Formation
Amsden Shale
Gros Ventre River
Ventre Valley
At depth, the weak Amsden Shale was a potential slip surface because it is parallel to the slope.
Rain weakened the Amsden and made the Tensleep heavier. Downslope force caused a mass of rock to start moving.
Time
Scar
Slide debris
Lake
The debris filled the valley, blocking a stream and forming Slide Lake. The scar remained on the hillslope.

FIGURE 13.14 Undercutting and collapse of a sea cliff.

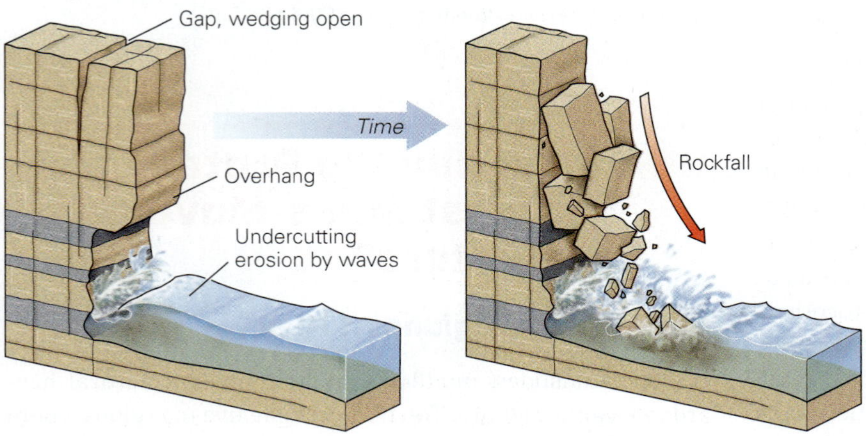
Gap, wedging open
Time
Overhang
Undercutting erosion by waves
Rockfall

(a) Undercutting by waves removes the support beneath an overhang.

(b) Eventually, the overhang breaks off along joints, and a rockfall takes place.

property. In many cases, the best solution is avoidance: don't build, live, or work in an area where mass movement will likely take place. But avoidance is possible only if we know where the hazards are greater.

To pinpoint dangerous regions, geologists look for landforms known to result from mass movements, for where these movements have happened in the past, they might happen again in the future. Features such as slump head scarps, swaths of forest in which trees have been tilted, piles of loose debris at the base of hills, and hummocky land surfaces all indicate recent mass wasting. In some cases, geologists may also be able to detect regions that are beginning to move (**Fig. 13.15**). For example, roads,

FIGURE 13.15 Surface features warn that a large slump is beginning to develop. Cracks that appear at the head scarp may drain water and kill trees. Power-line poles tilt and the lines become tight. Fences, roads, and houses on the slump begin to crack.

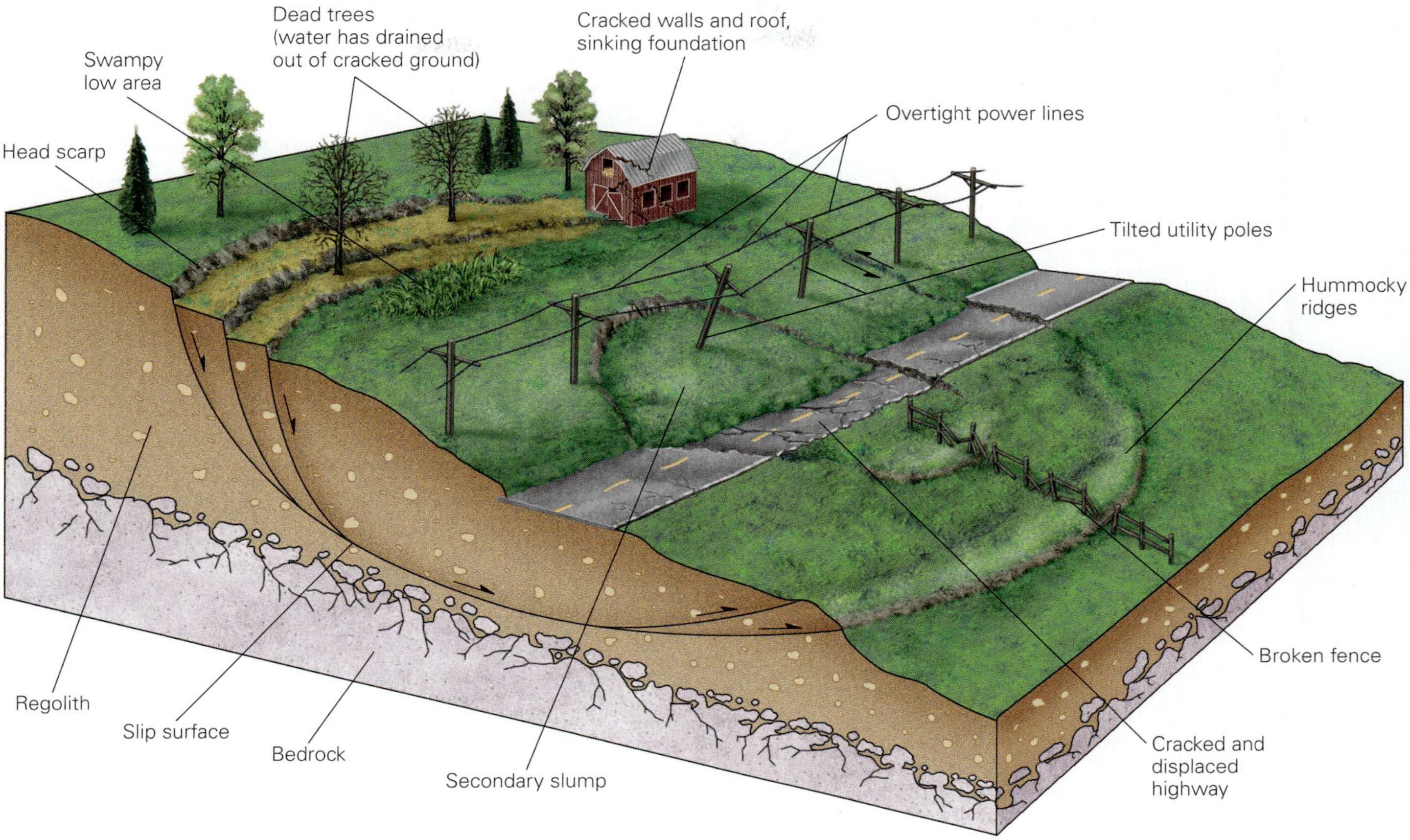

buildings, and pipes begin to crack over unstable ground. Power lines may be too tight or too loose because the poles to which they are attached move together or apart. Visible cracks form on the ground at the potential head of a slump, and the ground may bulge up at the toe of the slump. In some cases, subsurface cracks may drain the water from an area and kill off vegetation, whereas in other areas land may sink and form a swamp. Slow movements cause trees to develop pronounced curves at their base. More recently, new extremely precise surveying technologies have permitted geologists to detect the beginnings of mass movements that may not yet have visibly affected the land surface.

Did you ever wonder...
why parts of the California coast slip into the sea?

Even if there is no evidence of recent movement, a danger may still exist—just because a steep slope hasn't collapsed in the recent past doesn't mean it won't collapse in the future. In recent years, geologists have begun to identify potential hazards by using computer programs to evaluate factors that trigger mass wasting. These factors include the following: slope steepness; strength of substrate; degree of water saturation; orientation of weak surfaces relative to the slope; vegetation cover; potential for heavy rains; susceptibility to undercutting; and likelihood of earthquakes. From such hazard-assessment studies, geologists compile *landslide-potential maps*, which rank regions according to the likelihood that a mass movement will occur (**Fig. 13.16**). In any case, common sense suggests that you should avoid building on or below particularly dangerous landslide-prone slopes.

Preventing Mass Movements

In areas where a hazard exists, people can take certain steps to remedy the problem and stabilize the slope (**Fig. 13.17**).

- *Revegetation*: Stability in deforested areas will be greatly enhanced if landowners replant the region with vegetation that sends down deep roots and binds regolith together.
- *Regrading*: An oversteepened slope can be regraded or terraced so that it does not exceed the angle of repose.
- *Reducing subsurface water*: Because water weakens material beneath a slope and adds weight to the slope, an unstable situation may be remedied either by improving drainage

FIGURE 13.16 A landslide hazard map of the western United States.

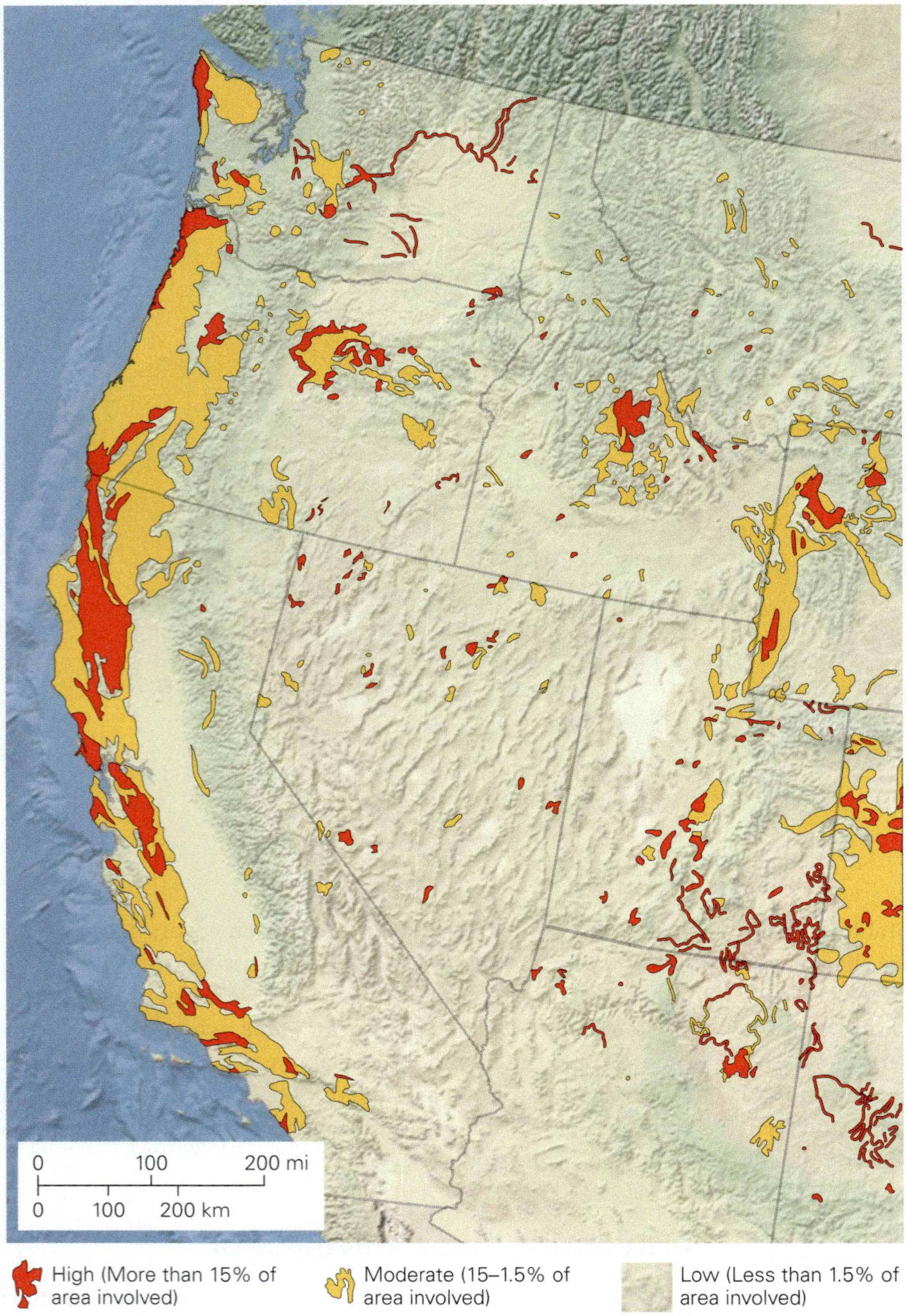

so that water does not enter the subsurface in the first place, or by removing water from the ground.

- *Preventing undercutting*: In places where a river undercuts a cliff face, engineers can divert the river. Similarly, along coastal regions they may build an offshore breakwater or pile **riprap** (loose boulders or concrete) along the beach to absorb wave energy before it strikes the cliff face.
- *Constructing safety structures*: In some cases, the best way to prevent mass wasting is to build a structure that stabilizes a potentially unstable slope or protects a region downslope from debris if a mass movement does occur. For example, civil engineers can build retaining walls or bolt loose slabs of rock to more coherent masses in the substrate in order to stabilize highway embankments. The danger from rockfalls can be decreased by covering a roadcut with chain link fencing or by spraying roadcuts with concrete. Highways at the base of an avalanche-prone slope can be covered by an avalanche shed, whose roof keeps debris off the road.
- *Controlled blasting of unstable slopes*: When it is clear that unstable ground or snow threatens a particular region, the best solution may be to blast the unstable ground or snow loose at a time when its movement can do no harm. Ski resorts routinely blast cornices at the top of slopes to diminish avalanche hazards.

TAKE-HOME MESSAGE

Various features of the landscape may help geologists to identify unstable slopes and estimate risk. Systematic study allows production of landslide-potential maps. Engineers use a variety of techniques to protect localities from mass movements.

QUICK QUESTION Why is mass movement of major concern during production of a large road cut?

FIGURE 13.17 A variety of remedial steps can stabilize unstable ground.

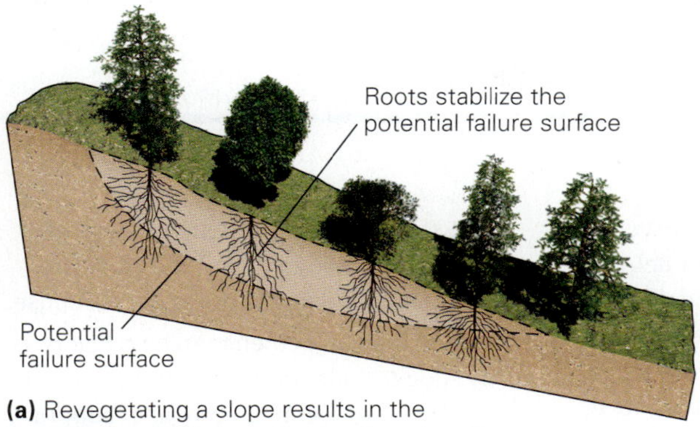

(a) Revegetating a slope results in the growth of roots that can hold a slope together.

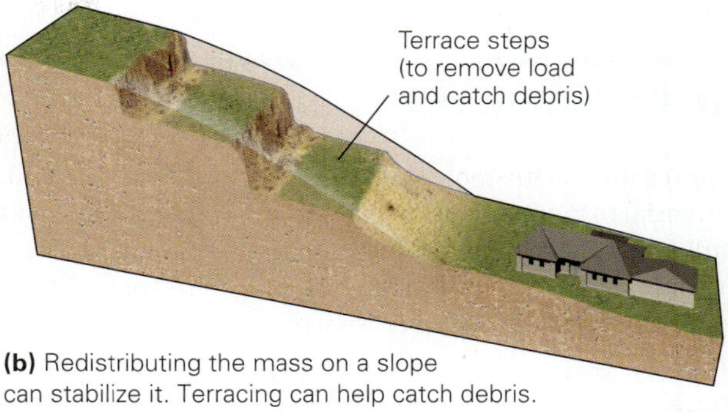

(b) Redistributing the mass on a slope can stabilize it. Terracing can help catch debris.

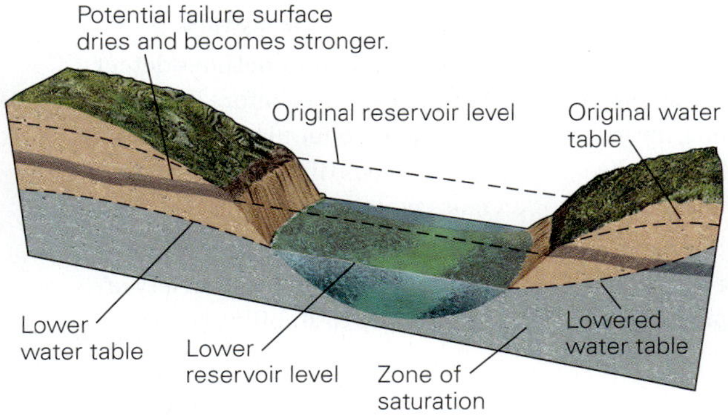

(c) Lowering the level of the water table can strengthen a potential failure surface.

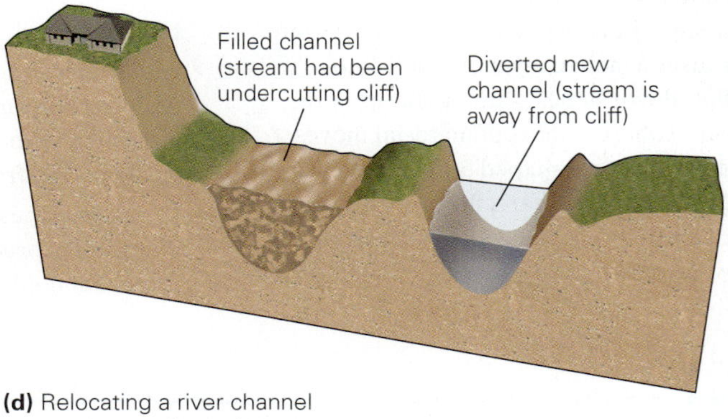

(d) Relocating a river channel can prevent undercutting.

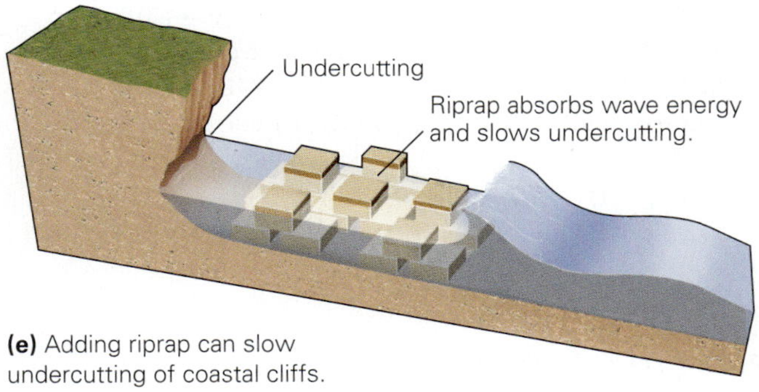

(e) Adding riprap can slow undercutting of coastal cliffs.

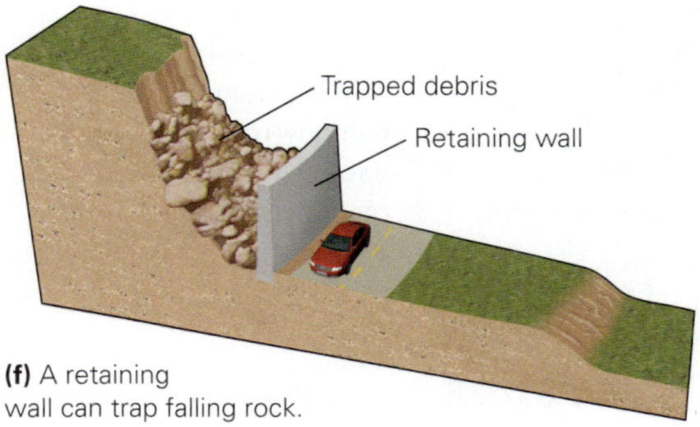

(f) A retaining wall can trap falling rock.

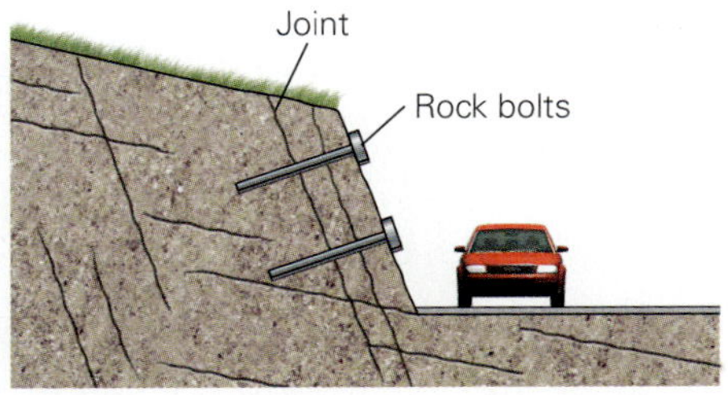

(g) Bolting or screening a cliff face can hold loose rocks in place.

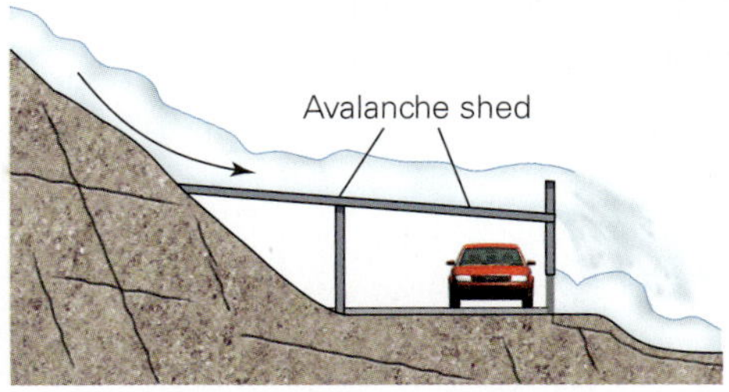

(h) An avalanche shed diverts debris or snow over a roadway.

Chapter 13 Review

Chapter Summary

- Rock or regolith on unstable slopes has the potential to move downslope under the influence of gravity. This process, called mass movement or mass wasting, plays an important role in the erosion of mountains.
- Slow mass movement, caused by the freezing and thawing of regolith, is called creep. In places where slopes are underlain with permafrost, solifluction causes a melted layer of regolith to flow down slopes. During slumping, a semicoherent mass of material moves down a spoon-shaped failure surface. Mudflows and debris flows occur where regolith has become saturated with water and moves downslope as a slurry.
- Rockslides and debris slides move very rapidly down a slope; the rock or debris breaks apart and tumbles. In a debris fall or rockfall, the material free-falls down a vertical cliff. During avalanches, snow or debris mixes with air and moves downslope as a turbulent cloud.
- Large mass movements can take place on underwater slopes. Some may generate tsunamis.
- Intact, fresh rock is usually too strong to undergo mass movement. Thus, for mass movement to be possible, rock must be weakened by fracturing or weathering.
- Unstable slopes start to move when the downslope force exceeds the resistance force that holds material in place. The steepest angle at which a slope of unconsolidated material can remain without collapsing is the angle of repose.
- Downslope movement can be triggered by shocks and vibrations, changes in the steepness of a slope, removal of support from the base of the slope, changes in the strength of a slope, deforestation, weathering, or heavy rain.
- Geologists can sometimes detect unstable ground before it begins to move, and they produce landslide-potential maps to identify areas susceptible to mass movement.
- Engineers can help prevent mass movements by using a variety of techniques to stabilize slopes.

Guide Terms

angle of repose (p. 451)
creep (p. 442)
debris fall (p. 449)
debris flow (p. 446)
debris slide (pp. 446, 448)
failure surface (pp. 443, 454)
lahar (p. 446)
landslide (p. 442)
mass movement (p. 442)
mudflow (p. 445)
natural hazard (p. 442)
regolith (p. 442)
riprap (p. 458)
rockfall (p. 449)
rockslide (p. 448)
slope failure (p. 451)
slump (p. 443)
snow avalanche (p. 448)
solifluction (p. 443)
submarine slump (p. 449)
talus (p. 449)
turbidity current (p. 450)

GEOTOURS *THIS CHAPTER'S GEOTOUR EXERCISE (M) FEATURES:*

- Portuguese Bend Landslide, CA
- La Conchita Mudslide, CA
- Gros Ventre Slide, WY
- Vaiont Dam Slide, Italy

Review Questions

1. What factors do geologists use to distinguish among various types of mass movements?
2. Explain how soil creep operates.
3. Identify the key differences between a slump, a debris flow, a lahar, an avalanche, a rockslide, and a rockfall.
4. Why is intact bedrock stronger than fractured bedrock? Why is it stronger than regolith?
5. Explain the difference between a stable and an unstable slope. What features are likely to serve as failure surfaces?
6. Discuss the variety of phenomena that can cause a stable slope to become so unstable that it fails.
7. Discuss the role of vegetation and water in slope stability. Why can fires and deforestation lead to slope failure?
8. What factors do geologists take into account when producing a landslide-potential map, and how can geologists detect the beginning of mass movement in an area?
9. What steps can people take to avoid landslide disasters?

On Further Thought

10. Imagine that you have been asked by a nongovernmental organization (NGO) to determine whether it makes sense to build a dam in a steep-sided, east-west-trending valley in a small central Asian nation. The local government has lobbied for the dam because the climate of the country has gradually been getting drier, and the farms of the area are running out of water. The NGO is considering making a loan to finance construction of the dam, a process that would employ hundreds of now-jobless people. Initial investigation shows that the rock of the valley wall consists of schist containing a strong foliation that dips toward the valley and is parallel to the slope of the valley wall. Outcrop studies reveal that abundant fractures occur in the schist along the valley floor, and the surface of many fractures are coated with slickensides. Moderate earthquakes have rattled the region. What would you advise the NGO, and why?

Online Resources

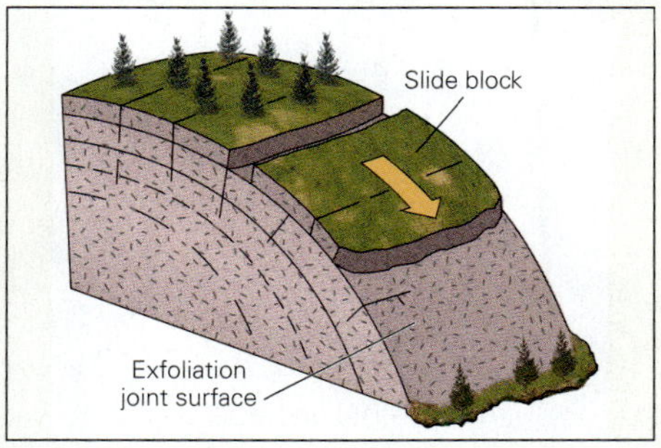

Videos

This chapter features a video on mass movement and slumps.

Assessment

This chapter includes questions on types of mass movement, their causes, and ways that we try to prevent them.

LEARNING OBJECTIVES

By the end of this chapter, you should understand...

1. what streams and drainage networks are, and how they form and evolve.
2. how to describe stream flow and erosion, and how streams transport and deposit sediment.
3. how the character of streams changes from headwaters to mouth.
4. how erosional and depositional landscapes formed by streams evolve over time.
5. the nature and causes of flooding, and the steps that people take to protect against flooding.
6. the environmental issues that pertain to streams.

▲This winter view from an airplane illustrates how running water, flowing down slopes in response to the pull of gravity, cuts channels into the landscape to produce drainage networks.

CHAPTER 14

Streams and Floods: The Geology of Running Water

14.1 Introduction

Wind swooping northward across warm oceans can pick up vast quantities of water vapor. It can then transport that water as moist air into a continent's interior. On a September day in 2013, such a mass of moist air collided with a mass of cold air stalled along the eastern edge of the Rocky Mountains in Colorado. The moist air rose, the moisture in it condensed, and drenching rains began to fall. In Boulder County, 430 mm (17 in.) of rain fell over a period of a few days—normally, the county receives 525 mm (21 in.) over the course of a whole year! Some of the water soaked into the ground, but most ended up in the numerous streams that flow from the mountains toward the plains. One of these streams, the Big Thompson River, carried 30 times more water than normal. The roaring torrent washed away houses and trees, scoured away roads (**Fig. 14.1a**), destroyed rail lines, inundated farmland, broke pipelines, overwhelmed sewage treatment plants, and undermined bridges. During the same year, flooding in Alberta had the same effect in

> *As many fresh streams meet in one salt sea, as many lines close in the dial's center, so may a thousand actions, once afoot, end in one purpose.*
> WILLIAM SHAKESPEARE (1564–1616)

the eastern Canadian Rockies (**Fig. 14.1b**). Where this water spilled out onto the plains, it submerged portions of Calgary (**Fig. 14.1c**).

The people of Colorado and Alberta had, unfortunately, experienced the immense power of *running water,* surface water that flows down sloping land in response to the pull of gravity. Geologists use the term **stream** for any body of running water that flows in a **channel**, an elongate depression or trough. (In everyday English, we tend to refer to large streams as *rivers* and medium-sized ones as *creeks.*) The edges of a channel are the stream's *banks,* the floor of the channel is the *streambed,* and a defined length of the channel is called a *reach.* Water in a stream flows from *upstream* reaches, closer to the source or **headwaters** of the stream, to the *downstream* reaches, closer to the end or **mouth** of the stream. Streams drain water from the landscape and carry it into lakes or to the sea, much as culverts drain water from a parking lot. In the process, streams relentlessly erode the landscape and transport sediment to sites of deposition. Generally, a stream stays within the confines of its channel, but when the supply of water entering a stream exceeds the channel's capacity, water spills out and covers the surrounding land, thereby causing a **flood** like the ones that devastated parts of Colorado and Alberta.

The Earth is the only planet in the Solar System that currently hosts flowing streams of liquid water. Streams are of great importance to human society not only because of how they modify the landscape, especially during floods, but also because they provide avenues for travel and commerce, nutrients and sediment for agriculture, water for irrigation and consumption, and energy to produce electricity and drive factories. In this chapter, we examine how streams operate in the Earth System. First, we learn about the origin of running water and about how the water draining a region organizes into an array of connected streams. Then we look at the process of stream erosion and deposition and at the landscapes that form as a consequence of to these processes. Finally, we focus on the nature and consequences of flooding and see how people can mitigate flooding risk.

FIGURE 14.1 Examples of flooding damage illustrate the power of running water.

(a) A stream roaring down a canyon in Colorado undercut and carried away part of a highway.

(b) Heavy rains in the Canadian Rockies caused extensive flooding in Alberta.

(c) The water inundated parts of Calgary, including the stadium for the city's Stampede (rodeo).

14.2 Draining the Land

Forming Streams and Drainage Networks

Where does the water in a stream come from? To answer this question, we first need to revisit the hydrologic cycle (see Interlude F). Water enters the hydrologic cycle by evaporating from the Earth's surface and rising into the atmosphere. After a relatively short residence time, atmospheric water condenses and falls back to the Earth's surface as rain or snow that accumulates in various reservoirs. Some rain or snow remains on the land as surface water (in puddles, swamps, lakes, snowfields, and glaciers), some flows downslope as a thin film called **sheetwash**, and some sinks into the ground, where it either becomes trapped in soil (as soil moisture) or descends below the water table to become *groundwater*. (As we discuss further in Chapter 16, the *water table* is the level below which groundwater fills all the pores and cracks in subsurface rock or sediment. Above the water table, air partially or entirely fills the pores and cracks.) Streams receive water spilling from the outlets of ponds and lakes, seeping out of the base of melting snowfields or glaciers, and from sheetwash trickling down slopes (**Fig. 14.2**)—you can see this input, because it happens at the Earth's surface. Less visible, but significant, inputs to a stream come from underground. Specifically, the pressure exerted by the weight of new rainfall can squeeze existing soil moisture out of the ground back to the surface and, if the floor of the channel lies below the water table, groundwater can seep out of the channel walls or streambed into the channel.

How does a stream channel form in the first place? The process of channel initiation begins when sheetwash starts flowing downslope due to gravity. Like any flowing fluid, sheetwash erodes its *substrate* (the material it flows over). The efficiency of such erosion depends on the velocity of the flow—faster flows erode more rapidly. In nature, the ground is not perfectly planar, not all substrate has the same resistance to erosion, and the amount of vegetation that covers and protects the ground varies with location. Thus, the velocity of sheetwash also varies with location. Where the flow happens a bit faster, or the substrate is a little weaker, erosion carves a channel. Since the channel is lower than the surrounding ground, sheetwash in adjacent areas starts to head toward it. With time, the extra flow deepens the channel relative to its surroundings, a process called **downcutting**, and a stream forms. Geologists use the term **runoff** for all water flowing on the surface of the Earth—runoff includes sheetwash plus the water in streams.

As flow increases and time passes, a stream channel begins to lengthen at its head (origin). This process, called **headward erosion** (**Fig. 14.3**), occurs for two reasons. First, it happens when the surface flow converging at the entrance to a channel has sufficient erosive power to downcut. Second, it takes place at locations where groundwater seeps out of the ground and enters the head of the stream channel. Such seepage, called *groundwater sapping*, gradually weakens and undermines the soil or rock just upstream of the channel's head until the material collapses into the channel. Each collapse makes the channel longer. The collapsed debris eventually washes away during a flood.

FIGURE 14.2 Excess surface water (runoff) comes from rain, melting ice or snow, and groundwater springs. On flat ground, water accumulates in puddles or swamps, but on slopes it flows downslope in streams.

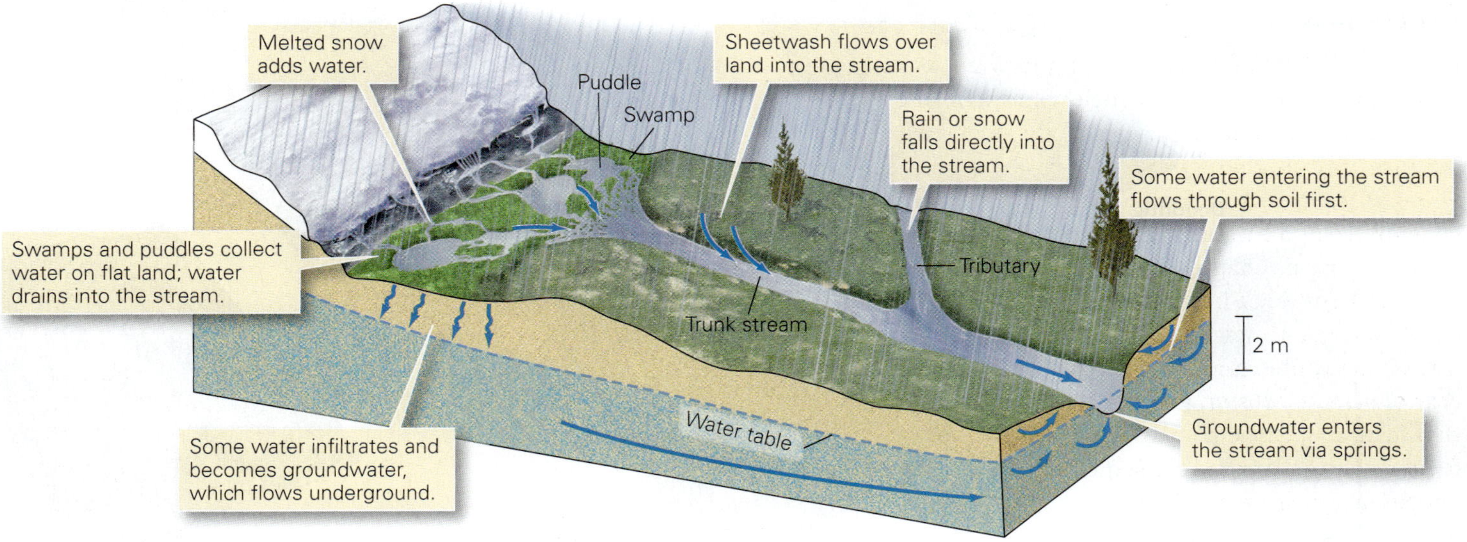
Melted snow adds water.
Puddle
Swamp
Sheetwash flows over land into the stream.
Rain or snow falls directly into the stream.
Some water entering the stream flows through soil first.
Swamps and puddles collect water on flat land; water drains into the stream.
Tributary
Trunk stream
2 m
Water table
Some water infiltrates and becomes groundwater, which flows underground.
Groundwater enters the stream via springs.

FIGURE 14.3 An example of headward erosion. The main stream flows in a deep valley. Side streams are cutting into the bordering plateau.

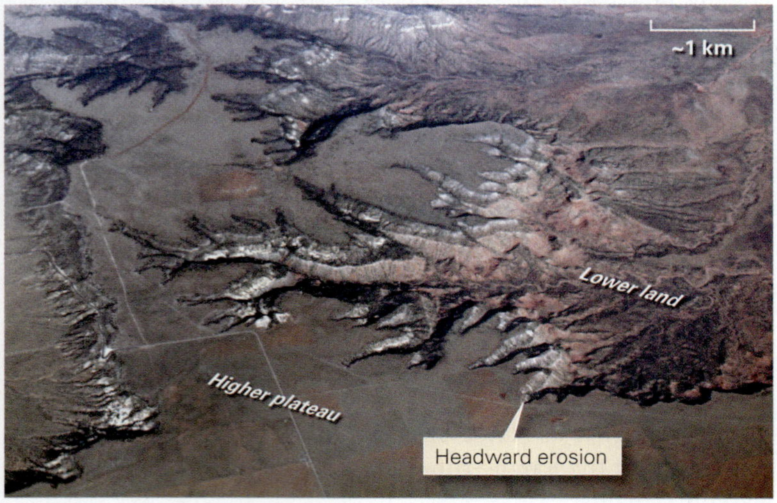

As downcutting deepens the main channel, the surrounding land surfaces start to slope toward the channel. Thus, side channels, or **tributaries**, begin to form, and these flow into the main channel. Eventually, an array of linked streams evolves, with the smaller tributaries flowing into a *trunk stream*. The array of interconnecting streams together constitute the **drainage network**. Like transportation networks of roads, drainage networks of streams reach into all corners of a region, providing conduits for the removal of runoff.

The configuration of tributaries and trunk streams defines the map pattern of a drainage network. This pattern depends on the shape of the landscape and the composition of the substrate. Geologists recognize several types of networks on the basis of the network's map pattern (**Fig. 14.4**).

- *Dendritic*: When rivers flow over a fairly uniform substrate, they develop a dendritic network that looks like the pattern of branches connecting to the trunk of a deciduous tree.
- *Radial*: Drainage networks forming on the surface of a cone-shaped mountain flow outward from the mountain peak, like spokes on a wheel. Such a pattern defines a radial network.
- *Rectangular*: In places where a rectangular grid of fractures (vertical joints) breaks up the ground, channels follow pre-existing fractures and streams join each other at right angles, creating a rectangular network.
- *Trellis*: Where a drainage network develops across a landscape of parallel valleys and ridges, major tributaries flow down a valley and join a trunk stream that cuts across the ridges. The resulting

SEE FOR YOURSELF...

HEADWARD EROSION, UTAH

Latitude
38°17′58.16″ N

Longitude
109°50′18.76″ W

Zoom to 8 km (5 miles) and look down.

Intermittent streams flow down tributary canyons into the Colorado River in Canyonlands National Park, Utah. The tributaries cut horizontal strata. A steep escarpment forms at the head of each tributary canyon. Groundwater sapping at this locality causes collapse of debris into the canyon, resulting in headward erosion.

FIGURE 14.4 Block diagrams illustrating five types of drainage networks.

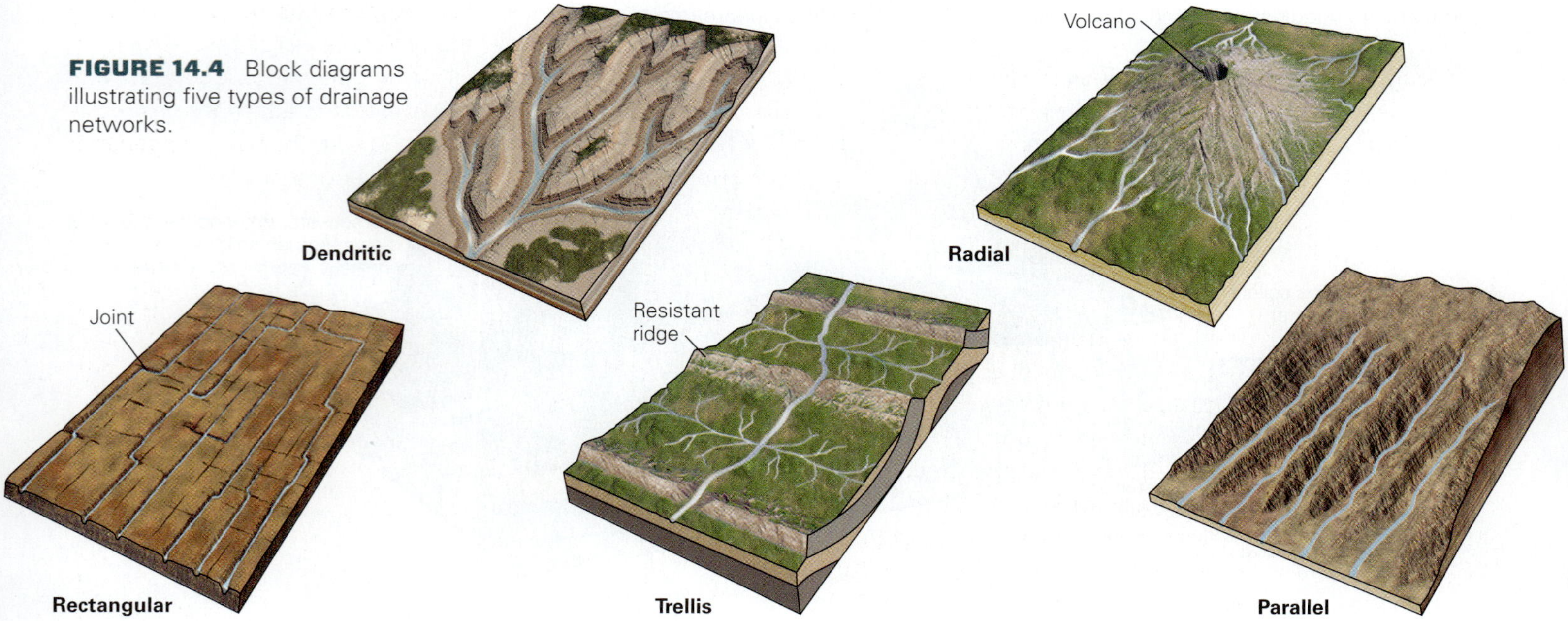

FIGURE 14.5 Drainage divides and basins.

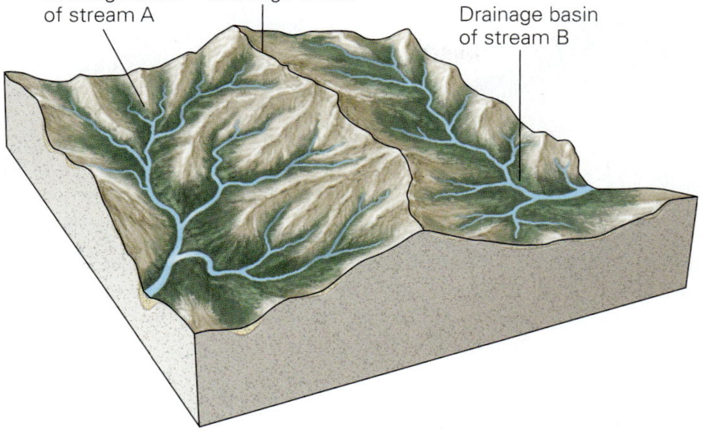

(a) A drainage divide is a relatively high ridge that separates two drainage basins.

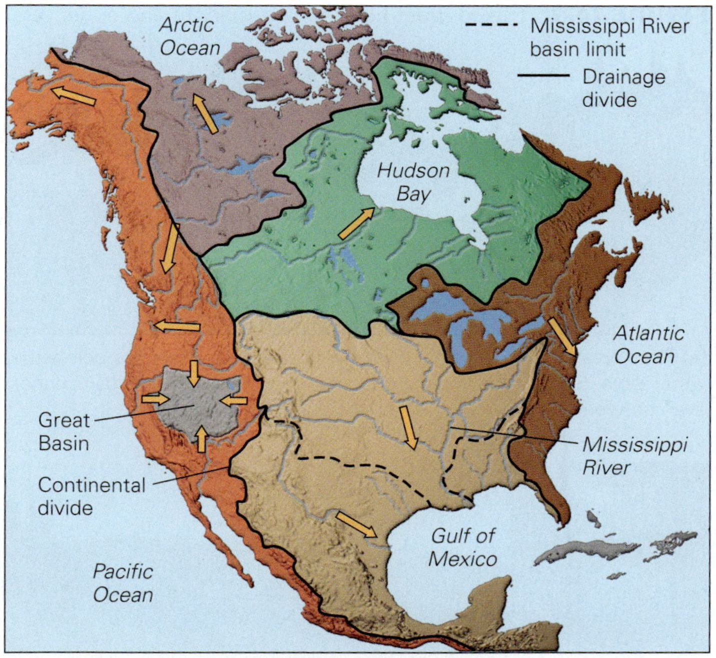

(b) The major drainage basins of North America.

map pattern resembles a garden trellis, so such an arrangement of streams is called a trellis network.

› *Parallel*: On a steep uniform slope, several streams with parallel courses develop simultaneously. The group comprises a parallel network.

Drainage Basins and Divides

A drainage network collects water from a broad region, variously called a *drainage basin, catchment*, or **watershed**, and feeds it into the trunk stream, which carries the water away. The highland, or ridge, that separates one watershed from another is a **drainage divide** (**Fig. 14.5**). A *continental divide* separates drainage that flows into one ocean from drainage that flows into another. For example, if you straddle the continental divide where it runs along the crest of the Rocky Mountains in the western United States, and you pour a cup of water out of each hand, the water in one hand flows to the Atlantic, and the water in the other flows to the Pacific.

Streams That Last, Streams That Don't

Permanent streams contain flowing water all year long (**Fig. 14.6a, b**). Where the bed of a stream lies below the water table, as is typical in temperate or tropical climates, permanent streams can exist because they not only receive a supply of water from upstream and from surface runoff but also fill with groundwater seeping through the stream bed or the channel walls. Rivers such as the Mississippi in the United States and the Amazon in Peru and Brazil serve as examples of such permanent streams. In places where the water table lies below the floor of the channel, a stream can be permanent only if the rate at which water arrives from upstream exceeds the rate at which water seeps down into the ground below. The downstream reach of the Colorado River serves as an example of such a stream. Although this river crosses hundreds of kilometers of the dry Sonoran Desert, it flows all year because enough water enters it from the river's wet headwaters upstream—hardly any water that enters the stream comes from the desert runoff. **Ephemeral streams** contain flowing water for only part of the year. Some survive for the duration of the wet season, a period of many months, but dry up during the dry season. Streams whose watersheds lie entirely within arid regions (deserts) tend to be ephemeral. Such streams may flow for only for tens of minutes to a few hours, following a heavy rain (**Fig. 14.6c, d**). The dry bed of an ephemeral stream is called a *dry wash*, an *arroyo*, or a *wadi*.

TAKE-HOME MESSAGE

Stream channels form by downcutting and lengthen by headward erosion. They carry water from sheetwash, lakes, springs, and melting snow or ice. Drainage networks carry water from a watershed to the sea and have a variety of geometries. Permanent streams flow all year, while ephemeral streams flow only intermittently.

QUICK QUESTION Why do streams develop distinct channels?

FIGURE 14.6 The contrast between permanent and ephemeral streams.

(a) The floor of a permanent stream in a temperate climate lies below the water table. Springs add water from below, so the stream contains water even between rains.

(c) The channel of an ephemeral stream lies above the water table, so the stream flows only when water enters the stream faster than it can infiltrate into the ground.

(b) An example of a permanent stream in the Wind River Mountains, Wyoming.

(d) An example of a dry wash in the Buckskin Mountains, Arizona.

14.3 Describing Flow in Streams: Discharge and Turbulence

Imagine two streams—a larger one in which water flows slowly, and a smaller one in which water flows rapidly. Which stream carries more water? The answer is not obvious. To answer this question completely, geologists must calculate the streams' discharge. Technically speaking, we define **discharge** as the volume of water passing through a cross section of the stream in a given time. We can calculate stream discharge by using a simple formula: $D = A_c \times v_a$. In this formula, A_c is the area of the stream, as measured in an imaginary plane perpendicular to the stream flow, and v_a is the average velocity at which water moves in the downstream direction. For example, if a stream has a cross-sectional area of 100 m^2, and the water in the stream flows at an average velocity of 0.2 m/s, then discharge $D = 100 \text{ m}^2 \times 0.2 \text{ m/s} = 20 \text{ m}^3/\text{s}$. Note that the discharge is specified in units of volume per second. Stream discharge can be determined at a *stream-gauging station,* where instruments measure the velocity and depth of the water at several points across the stream (**Fig. 14.7a**).

A stream's average discharge reflects the size of its drainage basin and the climate. The Amazon River drains a huge rainforest and has the largest average discharge in the world—about 200,000 m^3/s, or 15% of the total amount of runoff on Earth. In contrast, the "mighty" Mississippi River's discharge is only 17,000 m^3/s. The discharge of a given stream varies along its length. For example, discharge in a temperate region increases in the downstream direction, because each tributary that enters the stream adds more water, whereas the discharge in an arid region decreases downstream, for progressively more water seeps into the ground or evaporates. Human activity can affect discharge—for example, if people divert the river's water for irrigation, the river's discharge

FIGURE 14.7 Flow velocity and turbulence in streams.

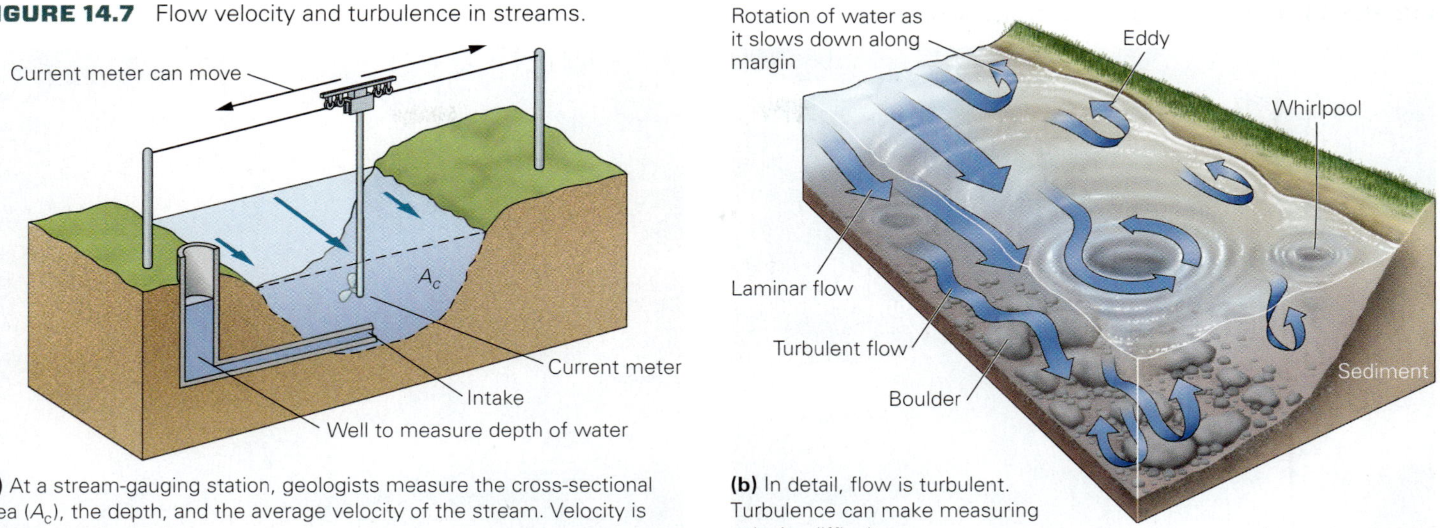

(a) At a stream-gauging station, geologists measure the cross-sectional area (A_c), the depth, and the average velocity of the stream. Velocity is slower near the banks.

(b) In detail, flow is turbulent. Turbulence can make measuring velocity difficult.

decreases. Finally, the discharge at a given location can vary with time. For example, in a temperate climate, a stream's discharge during the spring may be double or triple the amount during a dry summer, and a flood may increase the discharge to more than a hundred times normal.

The average velocity of stream water (v_a) can be difficult to calculate because the water doesn't all travel at the same velocity, for two reasons. First, friction along the sides and floor of the stream slows the flow. Thus, water near the channel walls or the streambed moves more slowly than water in the middle of the flow (see Fig. 14.7a). In fact, the fastest-moving part of the stream lies near the surface in the center of the channel. Second, *turbulence*—the twisting, swirling motion of a fluid—can produce eddies in which water curves and flows upstream or circles in place (**Fig. 14.7b**). Turbulence develops because the shearing motion of one water volume against its neighbor causes the neighbor to spin, and because obstacles such as boulders deflect water flow.

TAKE-HOME MESSAGE

Stream discharge, the amount of water passing through a cross section of the stream in a given time, depends on such factors as watershed area and climate. Water velocity varies across a stream, and tends to be turbulent, so calculating average velocity is complicated.

QUICK QUESTION Relative to a given point along the bank of a stream, where is the velocity of flow in a stream fastest? Why?

14.4 The Work of Running Water

How Do Streams Erode?

The energy that makes running water move comes from gravity. As water flows downslope from a higher to a lower elevation, the gravitational potential energy stored in water transforms into kinetic energy, the energy of motion. About 3% of this kinetic energy goes into the work of eroding the walls and beds of stream channels. Running water causes erosion in four ways:

- *Scouring*: Running water can remove loose fragments of sediment, a process called **scouring**.
- *Breaking and lifting*: In some cases, the push of flowing water can break chunks of solid rock off the channel floor or walls. In addition, the flow of a current over a clast can cause the clast to rise, or lift off the substrate.
- *Abrasion*: Clean water has little erosive effect, but sediment-laden water acts like sandpaper and grinds or rasps away at the channel floor and walls, a process called **abrasion**. In places where turbulence produces long-lived whirlpools, abrasion by sand or gravel carves a bowl-shaped depression, called a **pothole**, into the floor of the stream (**Fig. 14.8a, b**).
- *Dissolution*: Running water dissolves soluble minerals, and it carries the minerals away in solution.

The efficiency of erosion depends on the velocity and volume of water and on its sediment content. A large volume of

FIGURE 14.8 Erosion and transportation in streams.

(a) A pothole in the bed of a stream near Ithaca, New York.

(b) This slot canyon in Arizona formed when many potholes linked together.

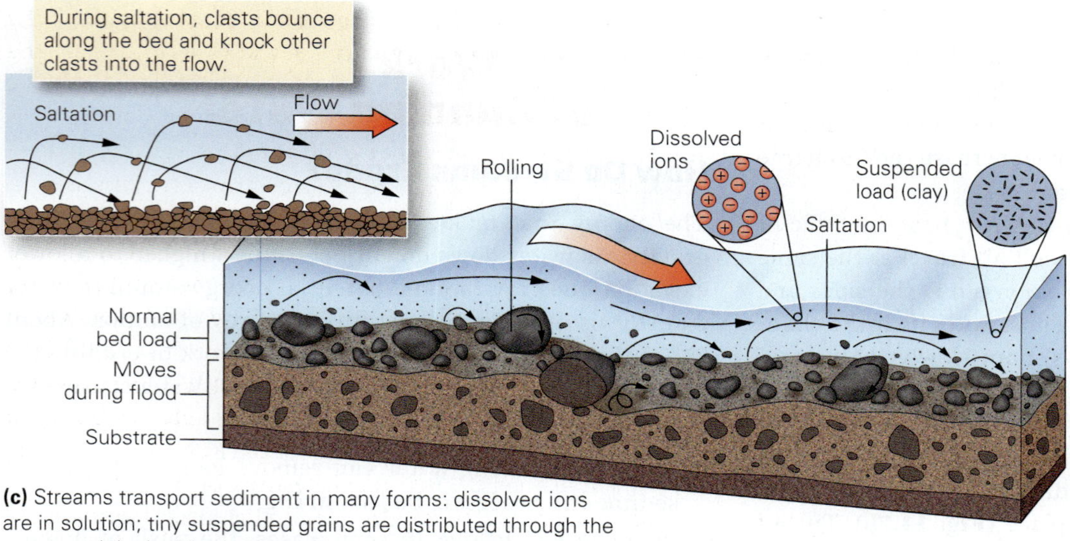

(c) Streams transport sediment in many forms: dissolved ions are in solution; tiny suspended grains are distributed through the water; and the bed load slides, rolls, and/or undergoes saltation.

fast-moving, turbulent, sandy water causes much more erosion than does a trickle of quiet, clear water. Thus, most erosion takes place during floods, when a stream carries a large volume of fast-moving, sediment-laden water.

How Do Streams Transport Sediment?

The Mississippi River received the nickname Big Muddy for a reason—its water can become chocolate brown because of all the clay and silt it carries. Geologists refer to the total volume of sediment carried by a stream as its *sediment load*. The sediment load consists of three components (**Fig. 14.8c**):

› *Dissolved load*: Running water dissolves soluble minerals in the sediment or rock that it flows over, and groundwater seeping into a stream brings dissolved minerals with it. The ions of these dissolved minerals constitute a stream's **dissolved load**.

› *Suspended load*: The **suspended load** of a stream usually consists of tiny solid grains (silt or clay size) that swirl along with the water without settling to the floor of the channel.

› *Bed load*: The **bed load** of a stream consists of large particles (such as sand, pebbles, or cobbles) that bounce or roll along the stream floor. Bed-load movement commonly involves **saltation**. During saltation, a multitude of grains bounce along in the direction of flow, within a zone that extends upward from the surface of the streambed for a distance of several centimeters to several tens of centimeters. Each saltating grain in this zone follows a curved trajectory up through the water and then back down to the bed. When it strikes the bed, it knocks other grains upward, and thus supplies more grains to the saltation zone.

When describing a stream's ability to carry sediment, geologists distinguish between competence and capacity. The **competence** of a stream refers to the maximum particle size it carries—a stream with high competence can carry large particles, whereas one with low competence can carry

FIGURE 14.9 Sediment, carried and deposited by streams. The clast size depends on stream velocity.

(a) Gravel in a streambed of a mountain stream in Denali National Park, Alaska. The large clasts were carried during floods.

(b) Point bars of mud deposited along a gentle, slow-moving stream in Brazil.

(c) An example of a small delta formed where sediment from a stream has spilled into standing water.

only small particles. Competence depends on water velocity, so a fast-moving, turbulent stream has greater competence (it can carry bigger particles) than a slow-moving stream, and a stream in flood has greater competence than a stream with normal flow. In fact, the huge boulders that litter the bed of a mountain creek move only during floods. The **capacity** of a stream refers to the total quantity of sediment it can carry. A stream's capacity depends on both its competence and discharge, so a large river has more capacity than a small creek.

Depositional Processes

A raging torrent of water can carry coarse and fine sediment—finer clasts rush along with the water as suspended load, whereas coarser clasts may bounce and tumble as bed load. If the flow velocity decreases, then the competence of the stream decreases and sediment settles out. The size of the clasts that settle at a particular locality depends on the decrease in flow velocity at the locality. For example, if the stream slows by a small amount, only large clasts settle (**Fig. 14.9a**); if the stream slows by a greater amount, medium-sized clasts settle; and if the stream slows almost to a standstill, the fine grains settle. Because of this process of *sediment sorting*, stream deposits tend to be segregated by size—gravel accumulates in one location and mud and silt in another.

Geologists refer to sediments transported by a stream as *fluvial deposits* (from the Latin *fluvius*, meaning river) or **alluvium**. Alluvium may accumulate along the streambed in elongate mounds, called **bars**. Some stream channels follow broad curves which, as we'll see, are called meanders. Water slows along the inner edge of a meander, so crescent-shaped **point bars** develop along the inner edge (**Fig. 14.9b**). During floods, a stream may overtop the banks of its channel and spread out over its **floodplain**, a broad flat area bordering the stream. Friction slows the water on the floodplain, so a sheet of silt and mud settles out to comprise floodplain deposits. Where a stream empties at its mouth into a standing body of water, the water slows and a wedge of sediment, called a **delta**, accumulates (**Fig. 14.9c**). We will discuss floodplains and deltas in more detail later in this chapter.

TAKE-HOME MESSAGE

Streams erode by scouring, breaking and lifting, abrasion, and dissolution. They carry sediment as dissolved, suspended, or bed loads. Competence, the ability to carry sediment, depends on flow velocity. Where the velocity decreases, sediment settles out.

QUICK QUESTION Why do point bars form on the inner arc of a curve in a stream?

14.5 Streams in the Landscape

How Do Streams Change along Their Length?

In 1803, under President Thomas Jefferson's leadership, the United States bought the Louisiana Territory, a vast tract of land encompassing the western half of the Mississippi drainage basin. At the time, the geography of the territory was a mystery. To fill the blank on the map, Jefferson asked Meriwether Lewis and William Clark to lead a voyage of exploration across the Louisiana Territory to the Pacific. Lewis and Clark, along with about 40 men, began their expedition at the mouth of the Missouri River, where it joins the Mississippi. At this juncture, the Missouri is a wide, languid stream of muddy water. The group found that along the Missouri's downstream reach, where the river's channel is deep and the water smooth, progress was easy. But the farther upstream they went, the more difficult their voyage became, for the **stream gradient** (the slope of the stream channel) became progressively steeper, and the stream's discharge diminished. When Lewis and Clark reached the site of what is now Bismarck, North Dakota, they had to abandon their original boats and haul smaller vessels up *rapids*, where turbulent water plunges over a steep, bouldery bed, and occasionally they had to carry their boats around *waterfalls*, where water free-falls through the air. When they reached what is now southwestern Montana, they abandoned these boats as well and trudged along the stream valley on foot or on horseback, struggling up steep gradients until they reached the continental divide.

If Lewis and Clark had been able to plot a graph showing their elevation above sea level relative to their distance from the river's mouth, they would have found that the **longitudinal profile** of the Missouri, a cross-sectional image showing the variation in the river's elevation along its length, is roughly a concave-up curve (**Fig. 14.10**). This curve emphasizes that, typically, a stream's gradient is steeper near its headwaters (source) than near its mouth. Near its headwaters, an idealized stream flows down deep valleys or canyons, whereas near its mouth, it flows over nearly horizontal plains. Real longitudinal profiles are not perfectly smooth curves, but rather they display little plateaus and steps, representing interruptions by lakes or waterfalls.

The lowest elevation that a stream channel's surface can reach at a locality is the **base level** of the stream. A *local base level* occurs at a location upstream of a stream's mouth, whereas the *ultimate base level*, the lowest possible elevation along the stream's longitudinal profile, lies at sea

FIGURE 14.10 Drainage basins and the change in character of a stream along its longitudinal profile.

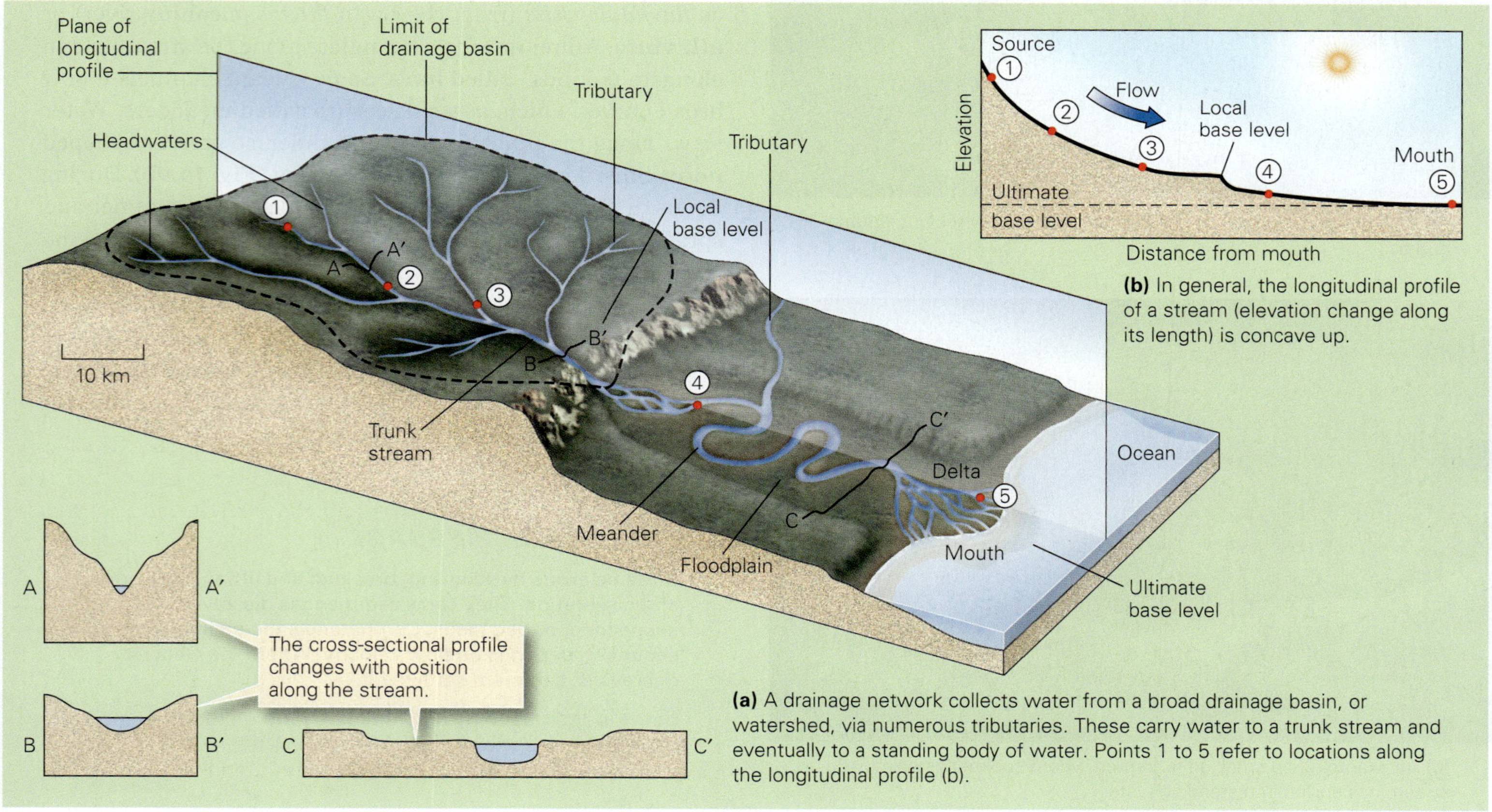

(b) In general, the longitudinal profile of a stream (elevation change along its length) is concave up.

(a) A drainage network collects water from a broad drainage basin, or watershed, via numerous tributaries. These carry water to a trunk stream and eventually to a standing body of water. Points 1 to 5 refer to locations along the longitudinal profile (b).

FIGURE 14.11 The formation of valleys and canyons.

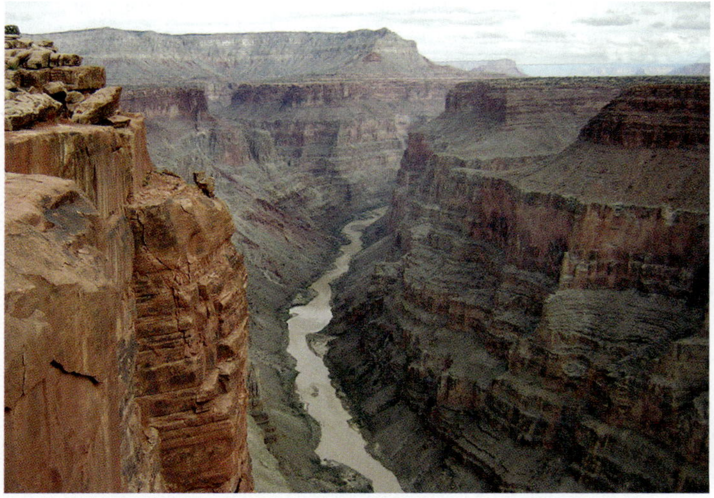

(a) The Colorado River carved the Grand Canyon, here seen near its west end.

(b) V-shaped valleys in the Andes of Peru.

(c) The shape of a canyon or valley depends on the relative resistance of walls to erosion, and the rate of downcutting relative to the rate of mass movement.

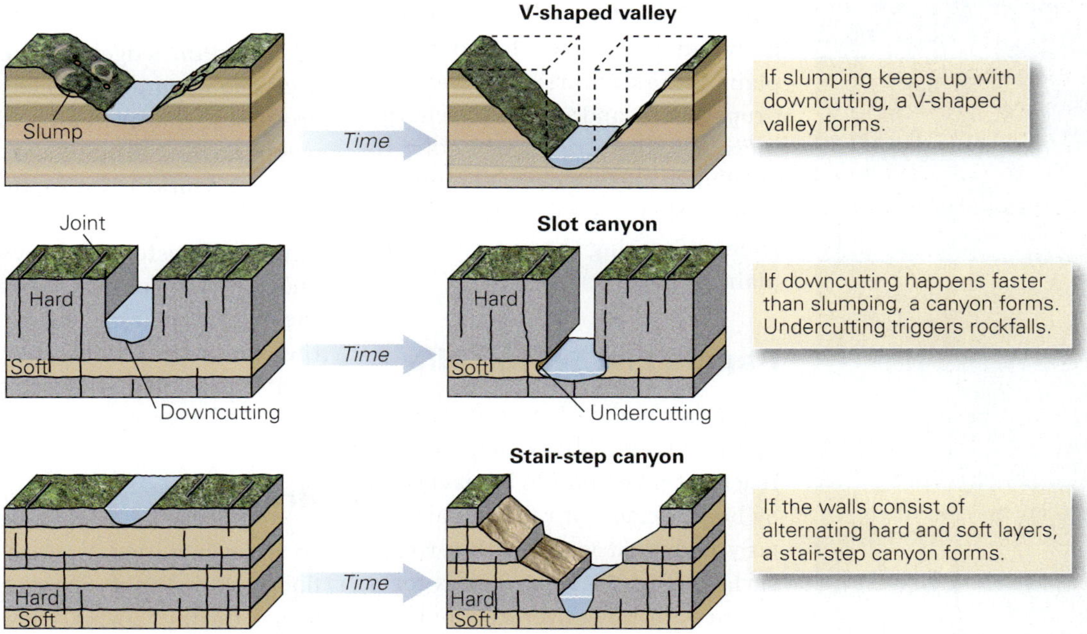

level. Why? The surface of a stream cannot be lower than the body of water that it flows into, for if it were, the stream would have to flow upslope. In a drainage network, the surface of a larger stream acts as the base level for the tributaries that flow into it. Similarly, lakes and reservoirs can act as local base levels along a stream, as can a ledge of resistant rock, for the stream level cannot drop below the ledge until the ledge erodes away.

Valleys and Canyons

Millions of years ago, the region now known as the Colorado Plateau (which covers parts of Arizona, Utah, Colorado, and New Mexico) was a plain whose surface lay close to sea level. The ancestor of the Colorado River flowed across it but could not cause much downcutting, because the stream's surface was already near the ultimate base level. When mountain-building processes caused the region to undergo uplift, the river went to work and began to downcut. Today, the river flows at the base of a steep-walled trough, 1.6 km (1 mile) below the surface of the plateau (**Fig. 14.11a**). The formation of what is now the Grand Canyon illustrates a general phenomenon—if a river flows at an elevation high above its local base level, it can carve a trough that is much deeper than the channel itself. A trough bordered by steep slopes is a **canyon**, whereas one bordered by gentler slopes is a **valley** (**Fig. 14.11b**).

Whether stream erosion produces a valley or a canyon depends on the rate at which downcutting occurs relative to the rate at which mass movement causes the walls on either side of the stream to collapse (**Fig. 14.11c**). In places where a stream downcuts through its substrate faster than the walls of the stream collapse or slump, erosion produces a steep-walled canyon. Such canyons typically form in hard rock, which can hold up cliffs for a long time (see Fig. 14.11a). In places where the walls collapse as fast as the stream downcuts, landslides and slumps gradually cause the slope of the walls to approach the angle of repose. When this happens, the stream channel lies at the floor of a valley whose cross-sectional shape resembles the letter V. Not surprisingly, geologists refer to this landform as a *V-shaped valley*. Where the walls of the stream consist of alternating layers of hard and soft rock, the walls develop a stair-step shape such as that of the Grand Canyon.

In places where active downcutting takes place, the valley floor remains relatively clear of sediment, for the stream—especially when it floods—carries away sediment that has fallen or slumped into the channel from the stream walls. But if the stream's base level rises, its discharge decreases, or its sediment load increases, the valley floor may fill with sediment, creating an *alluvium-filled valley*. The surface of the alluvium becomes a broad floodplain. If the stream's base level later drops again and/or the discharge increases, the stream will start to cut down into its own alluvium, a process that generates **stream terraces** bordering the present floodplain (**Fig. 14.12**).

SEE FOR YOURSELF...

RIVER-CUT GORGE IN THE HIMALAYAS

Latitude
28°9′41.82″ N

Longitude
85°23′1.04″ E

Zoom to 6 km (~4 miles) and look obliquely to the NE.

A deep gorge, about 50 km north of Katmandu, was cut by a river flowing out of the Himalayas. This upper reach of the river has a steep gradient. If you look straight upstream, you'll note the valley's V-shaped profile.

Rapids and Waterfalls

When Lewis and Clark forged a path up the Missouri River, they came to reaches that could not be navigated by boat because of **rapids**, particularly turbulent water with a rough surface (**Fig. 14.13a**). Rapids form where water flows over steps or large clasts in the channel floor, where the channel abruptly narrows, or where its gradient abruptly changes. The turbulence in rapids produces eddies, waves, and whirlpools that roil and churn the water surface and produce *whitewater*, a mixture of bubbles and water. Modern-day whitewater rafters thrill to the unpredictable movement of rapids.

A **waterfall** forms where the gradient of a stream becomes so steep that some or all of the water free-falls above the streambed (**Fig. 14.13b**). The energy of falling water may scour a depression, called a *plunge pool*, at the base of the waterfall. Typically, waterfalls form where a river flows over a resistant ledge of rock. Though a waterfall may appear to be a permanent feature of the landscape, all waterfalls eventually disappear because headward erosion slowly eats back the resistant ledge. We can see a classic example of headward erosion at Niagara Falls, where water flowing from Lake Erie to Lake Ontario drops over a 55-m-high ledge of resistant Silurian dolostone that overlies weak shale. Over time, turbulence in the plunge pool at the base of the waterfall erodes the shale, causing undercutting of the dolostone. Eventually, the overhang of dolostone becomes unstable, breaks along joints, and collapses, with the result that the location of the waterfall migrates upstream (**Fig. 14.14**).

FIGURE 14.12 A rise in base level or a decrease in discharge can cause a valley to fill with alluvium. If base level drops, or discharge increases, the stream will cut down into the alluvium, leaving terraces.

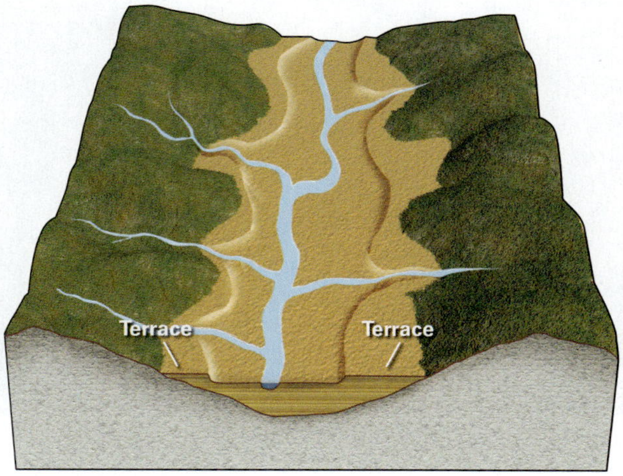

Did you ever wonder...
why a waterfall forms and whether it will always be there?

Braided Streams

In some localities, streams carry abundant sediment during floods but cannot carry this sediment during normal flow. Thus, during normal flow, the sediment settles out and chokes the channel. As a consequence, the stream divides into numerous strands weaving back and forth between elongate bars of gravel and sand. The result is a **braided stream**—the name emphasizes that the streams entwine like strands of hair in a braid (**Fig. 14.15a**).

Alluvial Fans

Where a fast-moving ephemeral stream abruptly emerges from a mountain canyon onto an open plain at the range front, the water that was once confined to a narrow channel can spread out over a broader surface. As a consequence, the water slows and drops its sedimentary load of poorly sorted sand, gravel, and cobbles. This process gradually builds an overall wedge of sediment called an **alluvial fan** (**Fig.

FIGURE 14.13 Examples of rapids and waterfalls.

(a) These rapids in the Grand Canyon formed when a flood from a side canyon dumped debris into the channel of the Colorado River.

(b) Iguaçu Falls, at the Brazil-Argentina border, spills across layers of basalt. The basalt acts as a resistant ledge.

14.15b). Once the fan is built, streams flowing over its surface develop a braided channel. Flow follows a particular set of braided channels until enough sediment accumulates at the mouth of the channels to make the gradient of the channels relatively gentle. When this happens, the flow starts to follow a different set of channels that have a steeper gradient, for water always tries to flow down the steepest slope possible (**Fig. 14.15c**). Over time, the process repeats so that the position of the active channel alternates among different paths out of the canyon mouth. Together, the array of channels on the face of an alluvial fan are called *distributaries*, because they distribute sediment across the face of the fan. If a particularly big flood happens, debris flows may spread over a large portion of the fan, and smooth its surface.

Meandering Streams and Their Floodplains

A riverboat cruising along the lower reaches of the Mississippi River cannot travel in a straight line, for the river channel winds back and forth in a series of snake-like curves, each of which is called a **meander** (**Fig. 14.16a**). In fact, the boat may

FIGURE 14.14 The formation of Niagara Falls, at the border between Ontario, Canada, and New York State. The falls tumble over the Lockport Dolomite, a relatively strong rock layer.

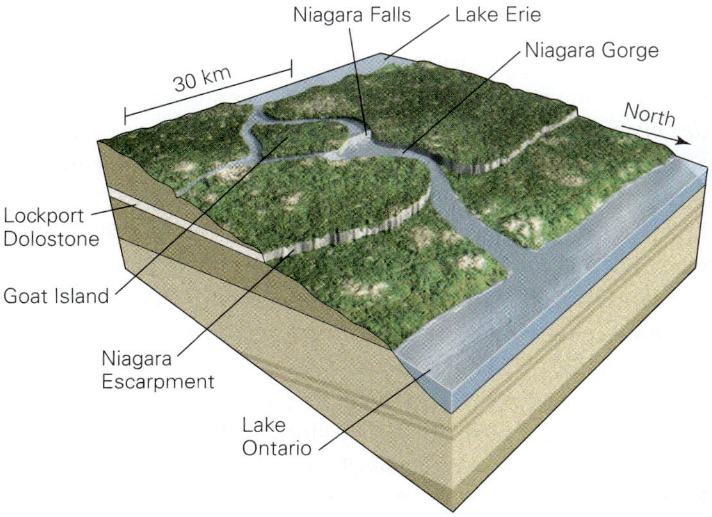

(a) Niagara Falls formed where the outlet of Lake Erie flowed over the Niagara escarpment.

(b) At Horseshoe Falls—part of Niagara Falls—headward erosion, on average, causes the lip of the falls to retreat by about 0.5 m per year.

FIGURE 14.15 Examples of depositional landforms produced from stream sediment.

(a) A braided stream, carrying meltwater from a glacier near Denali, Alaska, deposits elongate bars of gravel.

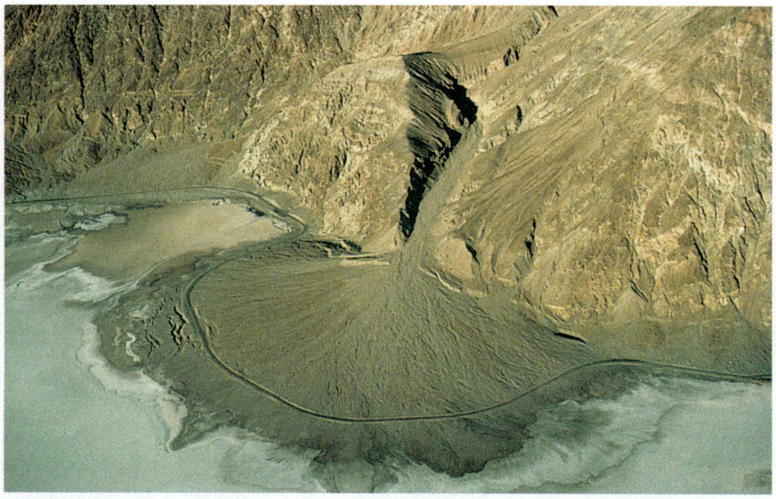

(b) An alluvial fan in Death Valley, California, consists of sand, gravel, and debris flows. The curving black line is a road.

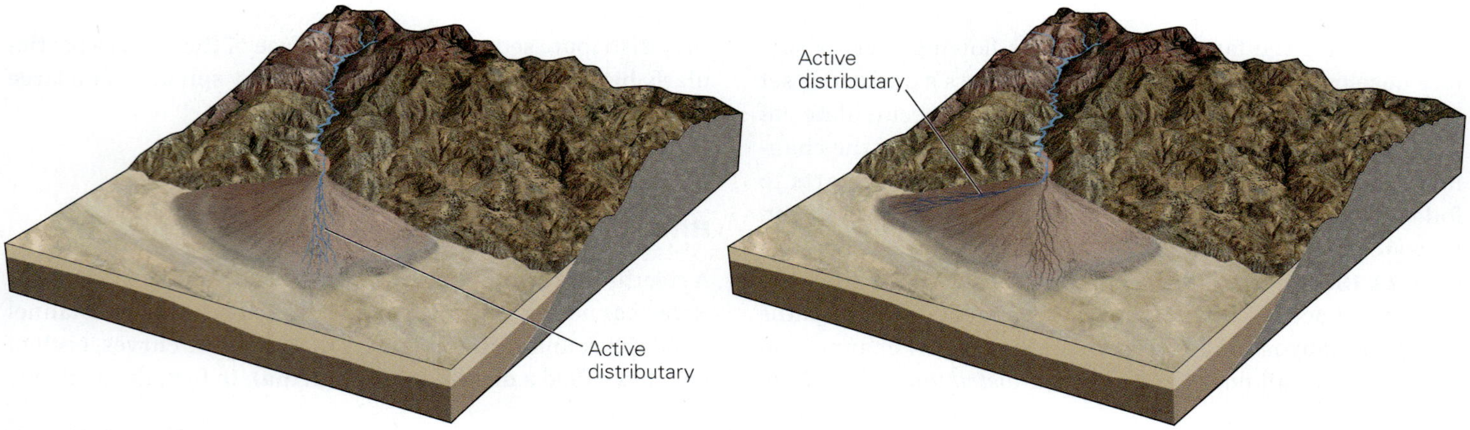

(c) Typically, a given flood, bringing sediment, covers only part of the fan. When the gradient in that location decreases, the flow moves to a different, steeper part of the fan.

have to go 500 km along the river channel to travel 100 km as the crow flies. Geologists refer to a stream with many meanders as a **meandering stream**—typically, meandering streams initiate where the stream's gradient is very gentle.

How do meanders evolve? Even if a stream starts out with a straight channel, natural variations in water depth will cause the fastest-moving part of the current, called the *thalweg*, to swing back and forth. The water erodes the side of the stream more effectively where it flows faster, so it begins to cut away faster on the outer arc of the curve. Thus, each curve begins to migrate sideways and grow more pronounced until it becomes a meander (**Fig. 14.16b**). On the outside edge of a meander, erosion continues to eat away at the channel wall, forming a *cut bank*. On the inside edge, water slows down and the stream's competence decreases, so sediment accumulates, forming a crescent-shaped wedge called a *point bar*, as we noted earlier. With continued erosion, a meander may curve through more than 180°, so that the cut bank at the meander's entrance approaches the cut bank at its end, leaving a *meander neck*, a narrow isthmus of land separating the portions of the meander. When erosion eats through a meander neck, a straight reach called a *cutoff* develops. The meander that has been cut off is called an **oxbow lake** if it remains filled with water, or an *abandoned meander* if it dries out. The course of a meandering stream naturally changes over time, on a time scale of years to centuries, as new meanders grow and old ones are cut off and abandoned.

Most meandering stream channels cover only a relatively small portion of a broad, gently sloping **floodplain** that terminates at its sides along a *bluff*, a small escarpment rising to higher land (**Fig. 14.16c, d**). Floodplains, as we noted earlier, are so named because during a flood, water overtops the edge

FIGURE 14.16 The character and evolution of meandering streams and floodplains.

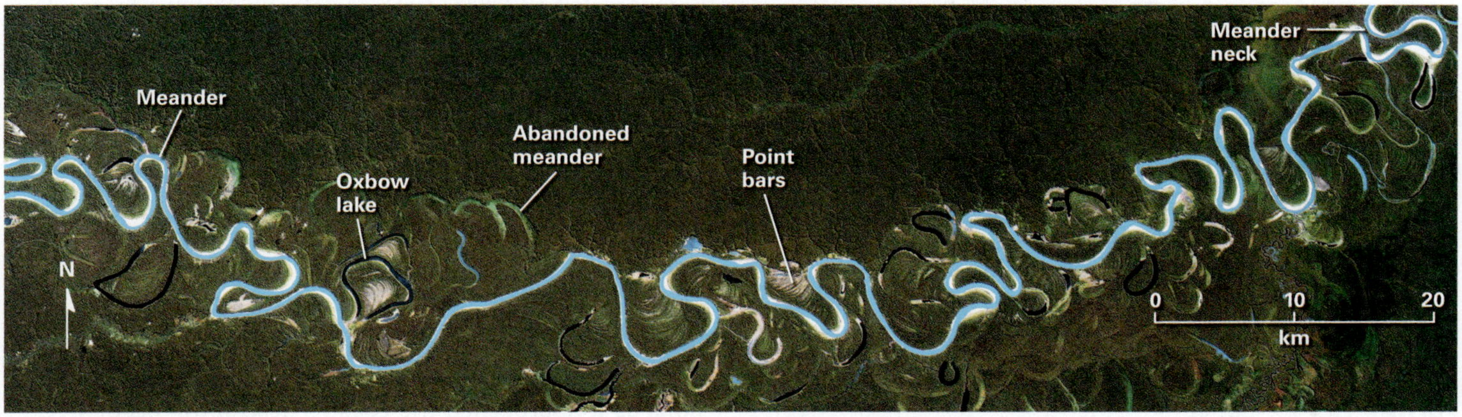

(a) A meandering stream in Brazil, as viewed from space. Note the oxbows, cutoffs, and abandoned meanders.

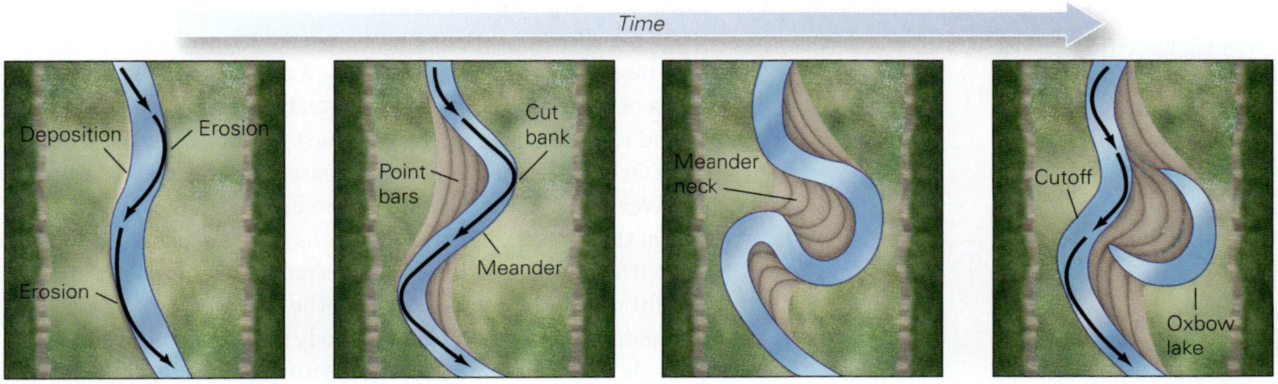

(b) Meanders evolve because erosion occurs faster on the outer bank of a curve, and deposition takes place on the inner curve. Eventually, a cutoff isolates an oxbow lake.

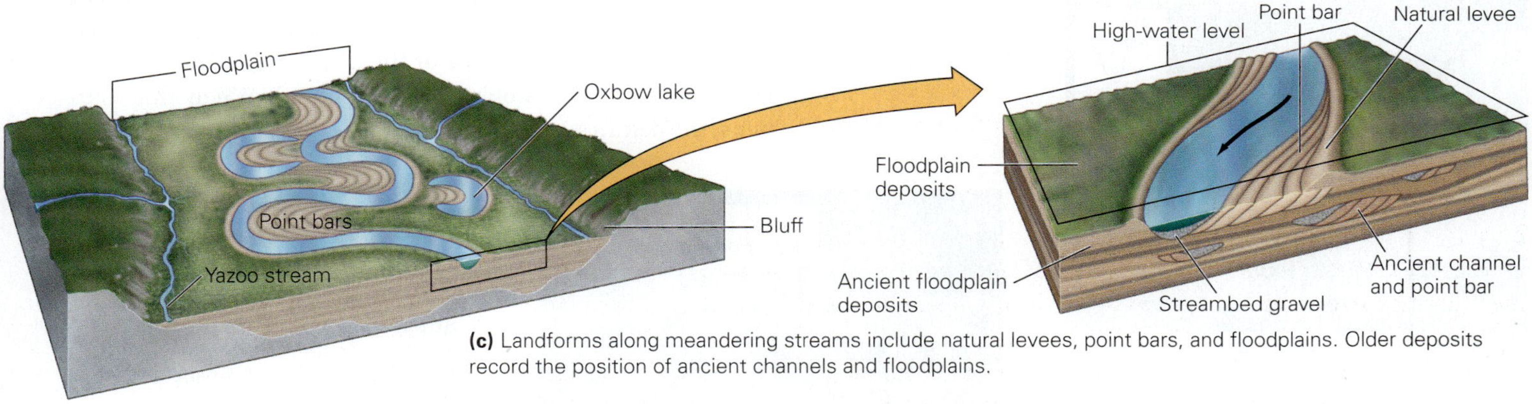

(c) Landforms along meandering streams include natural levees, point bars, and floodplains. Older deposits record the position of ancient channels and floodplains.

(d) The flat land of this floodplain hosts farm fields. Trees delineate the channel.

of the stream channel and spreads out over the floodplain. As the water starts to spread beyond the channel walls, friction slows down the flow, so sediment settles out along the edge of the channel. Over time, the accumulation of this sediment produces a pair of low ridges, called **natural levees**, on either side of the stream. Natural levees may grow so large that the floor of the stream's channel may actually become higher than the surface of the floodplain.

Deltas: Deposition at the Mouth of a Stream

Along most of its length, only a narrow floodplain—covered by irrigated farm fields—borders the Nile River in Egypt. But at its mouth, the trunk stream of the Nile divides into a fan of distributaries, and the area of agricultural lands broadens into a triangular patch. The Greek historian Herodotus noted that this triangular patch resembles the shape of the Greek letter delta (Δ), and so the region became known as the Nile Delta.

Geologists use the name **delta** for a wedge of sediment formed where a stream enters standing water, the stream's flow slows, and therefore, sediment settles out. Deltas come in many shapes. Those with the classic Δ shape form where waves or offshore currents redistribute sediment entering the standing water. Where currents are stronger, less sediment accumulates so the edge of the delta is straight, and where currents are weaker, more sediment accumulates so the edge of the delta arcs outward (**Fig. 14.17a, b**). Deltas that form where the strength of the river current exceeds that of ocean currents are called *bird's-foot deltas,* because they resemble the scrawny toes of a bird—they develop where several distributaries extend far out into relatively calm water (**Fig. 14.17c**). The existence of several overlapping deltas indicates that the main course of the river has shifted on several occasions as it approached the coast. These shifts occur when a toe builds so far out into the sea that the slope of the stream becomes too gentle to allow the river to flow. At this point, the river overflows a natural levee upstream and begins to flow in a new direction, an event called an *avulsion.* The distinct lobes of the Mississippi Delta suggest that avulsions have happened several times during the past 9,000 years (**Fig. 14.18**). New Orleans, built along the Mississippi, may eventually lose its riverfront, for a break in a levee upstream of the city could divert the Mississippi into the Atchafalaya River channel.

With time, the sediment of a large delta compacts, and the weight of the delta pushes down the crust below. As a consequence, the surface of a delta slowly sinks. Distributaries can provide sediment that fills the resulting space so that the delta's surface remains at or just above sea level, forming a broad, flat area called a *delta plain.* If people build up artificial levees to constrain the river to its channel, sediment gets carried directly to the seaward edge of the delta and the delta's interior "starves" (does not receive sediment). When this happens, the delta's surface drops below sea level.

FIGURE 14.17 Delta shape varies depending on current activity, waves, and vegetation.

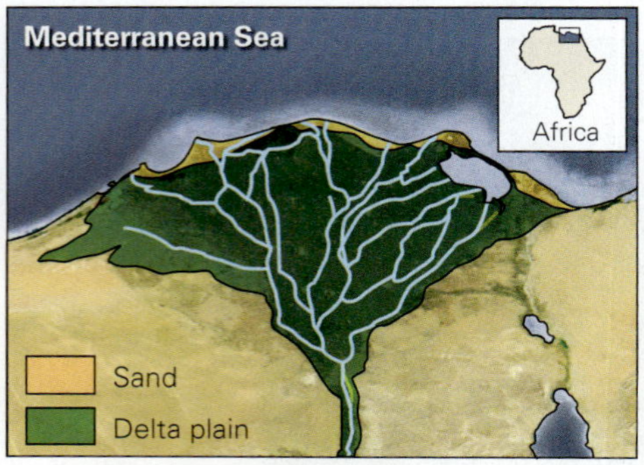

(a) The Nile is a Δ-shaped delta.

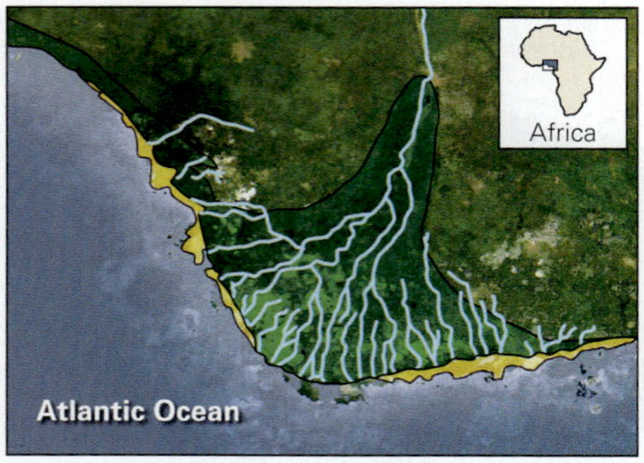

(b) The Niger is an arc-like delta.

(c) The Mississippi is a bird's-foot delta.

TAKE-HOME MESSAGE

Stream erosion and deposition yield distinct landforms. Streams have steeper regional gradients in their headwaters, and carve canyons or valleys. Gradient decreases toward a stream's mouth—the mouth can't be lower than a base level. Streams choked with sediment become braided, whereas those following snake-like paths are meandering. Alluvial fans build out at the mouth of canyons in deserts; deltas build out into standing water.

QUICK QUESTION What factors cause the formation of rapids and waterfalls?

FIGURE 14.18 A map showing ancient lobes of the Mississippi Delta. A major flood could divert water from the Mississippi into the channel of the Atchafalaya.

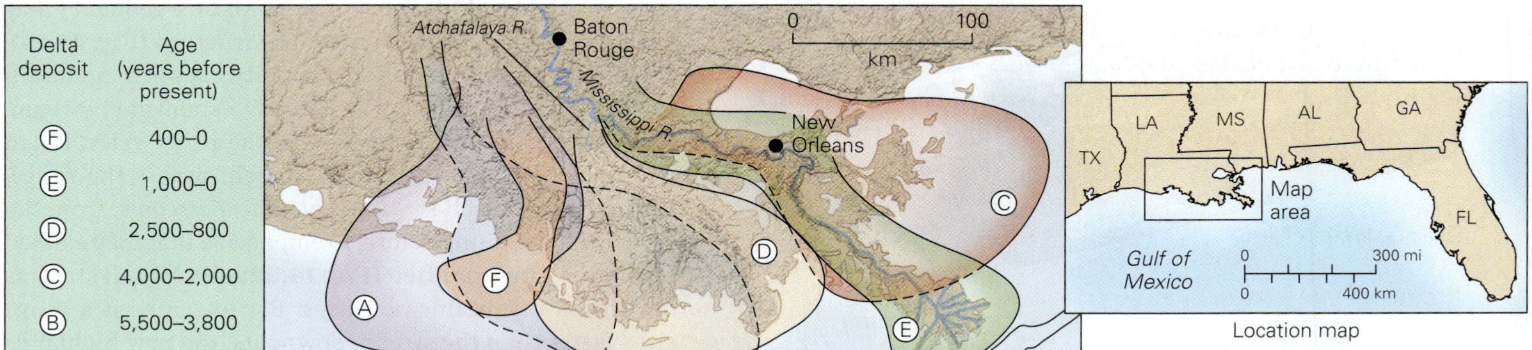

14.6 The Evolution of Drainage

Beveling Topography

Imagine a place where continental collision uplifts a region (**Fig. 14.19a**). At first, in the highlands, rivers have steep gradients, flow over many rapids and waterfalls, and cut deep valleys. But with time, the landscape evolves. Erosion transforms the rugged mountains into low, rounded hills, and the once-deep, narrow valleys broaden into wide floodplains with gentler gradients. As more time passes, erosion bevels even the low hills, and the region becomes nearly planar again, lying at an elevation close to that of a stream's base level. (Some geologists have referred to the resulting landscape as a *peneplain*, from the Latin *paene*, which means almost.)

Although the above model makes intuitive sense, it is an oversimplification. Plate tectonics can uplift the land again, and/or global sea-level rise or fall can change the base level, so in reality peneplains rarely develop before downcutting begins again. **Stream rejuvenation** occurs when a stream starts to downcut into a land surface whose elevation had previously been close to the stream's base level. Triggers for rejuvenation can include factors such as a drop in base level, as happens when sea level falls; an uplift event that causes the land to rise relative to the base level; or an increase in stream discharge that makes the stream more able to erode and transport sediment. As we've seen, rejuvenation can lead to formation of stream terraces in alluvium. In cases where rejuvenation causes a stream to erode deeply into bedrock, a new canyon or valley will develop. If the rejuvenated stream had a meandering course, downcutting produces *incised meanders* (**Fig. 14.19b**).

Stream Piracy and Drainage Reversal

Stream piracy sounds like pretty violent stuff. In reality, it's just a natural process that happens when headward erosion by one stream causes the stream to intersect the course of another stream. When this happens, the pirate stream "captures" the water of the stream that it has intersected, so that the water of the captured stream starts to flow down the pirate stream. Because of piracy, the channel of the captured stream, downstream of the point of capture, dries up (**Fig. 14.20**). In some cases, stream capture changes a *water gap* (a stream-carved notch through a ridge) into a dry *wind gap*. In 1775, Daniel Boone, the legendary pioneer, led settlers through the Cumberland Gap, a wind gap in the Appalachians, to new homesteads in western Kentucky.

The pattern of stream flow in an area can also be altered, over time, on a continental scale. For example, when South America and Africa were adjacent to each other in Pangaea, a highland existed along the boundary between the two continents, and the main drainage network of northern South America flowed westward. Later, when South America rifted away from Africa, a convergent boundary developed along the western margin of the South American Plate, causing the Andes Mountains to rise. The uplift of the Andes caused a **drainage reversal**, in that the overall slope direction of the drainage network became the opposite of what it once had been. As a consequence, westward flow became impossible and the eastward-flowing Amazon drainage network developed.

Superposed and Antecedent Streams

The structure and topography of the landscape do not always appear to control the path, or course, of a stream. For example, envision a stream that carves a deep canyon straight across a resistant ridge of rock. We distinguish two types of streams that cut across resistant topographic highs.

First, imagine a region in which a stream starts to flow over horizontal beds of strata that unconformably overlie folded strata. When the stream eventually erodes down through the unconformity and starts to downcut into the folded strata, it

FIGURE 14.19 Fluvial landscapes evolve over time.

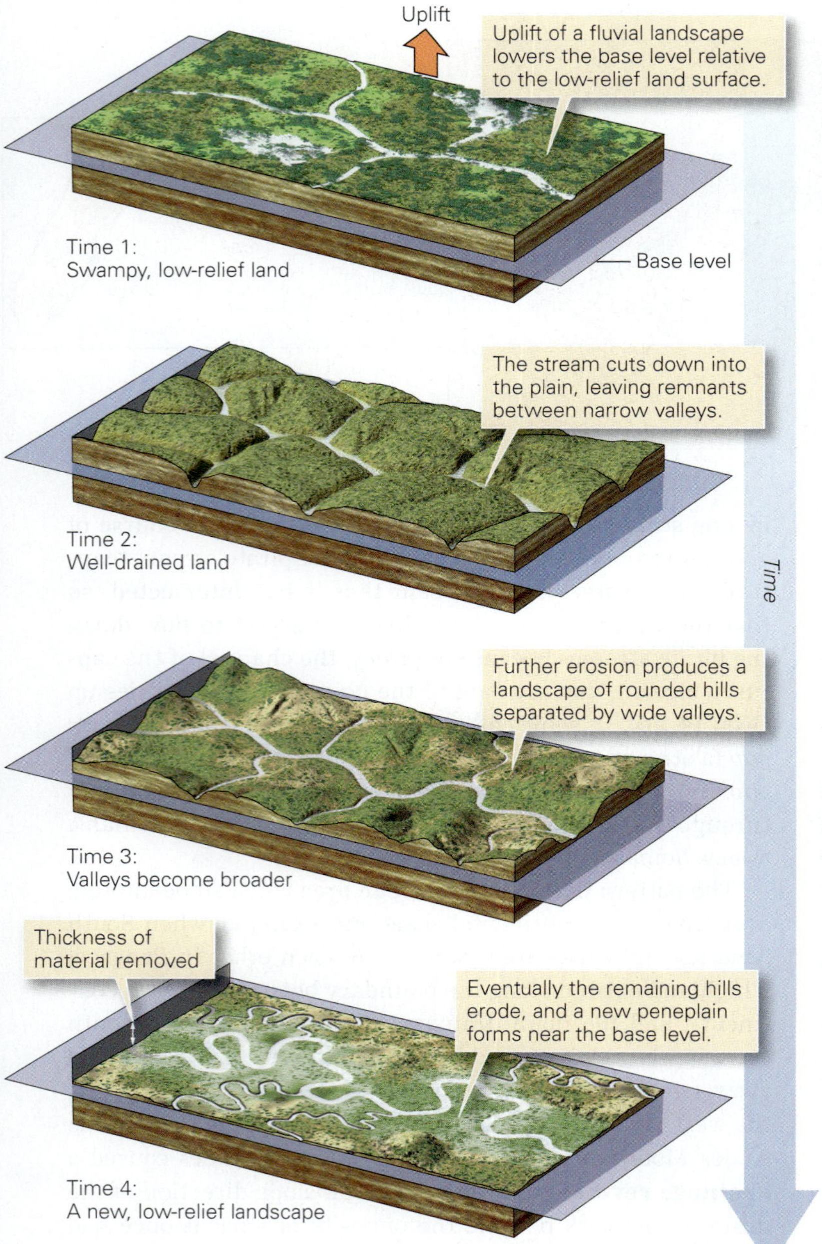
Uplift
Uplift of a fluvial landscape lowers the base level relative to the low-relief land surface.
Time 1:
Swampy, low-relief land
Base level
The stream cuts down into the plain, leaving remnants between narrow valleys.
Time 2:
Well-drained land
Further erosion produces a landscape of rounded hills separated by wide valleys.
Time 3:
Valleys become broader
Thickness of material removed
Eventually the remaining hills erode, and a new peneplain forms near the base level.
Time 4:
A new, low-relief landscape
Time

(a) Evolution of a fluvial landscape when base level drops.

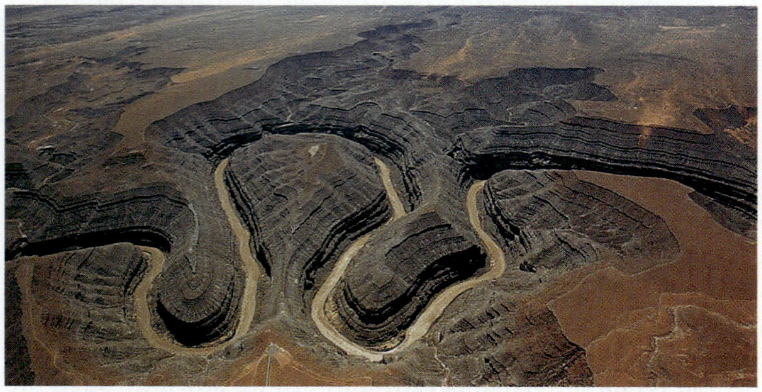

(b) The "goosenecks" of the San Juan River, Utah, are incised meanders.

tends to maintain its earlier course, ignoring the structure of the folded strata. Geologists refer to such a stream as a *superposed stream*, because a pre-existing geometry has been laid down on underlying rock structure (**Fig. 14.21**). Second, imagine that tectonic activity causes a mountain range to rise up beneath an already established stream. If the stream downcuts as fast as the range rises, it can maintain its course and will cut right across the range. Geologists call such streams *antecedent streams*, from the Greek *ante*, meaning before, to emphasize that they existed before the range uplifted (**Fig. 14.22a, b**). Not all streams cut across mountains or ridges. For example, if a range rises faster than the stream downcuts, the new highlands divert (change) the stream's course. The resulting *diverted stream* flows parallel to the range face (**Fig. 14.22c**).

FIGURE 14.20 The concept of stream capture or piracy.

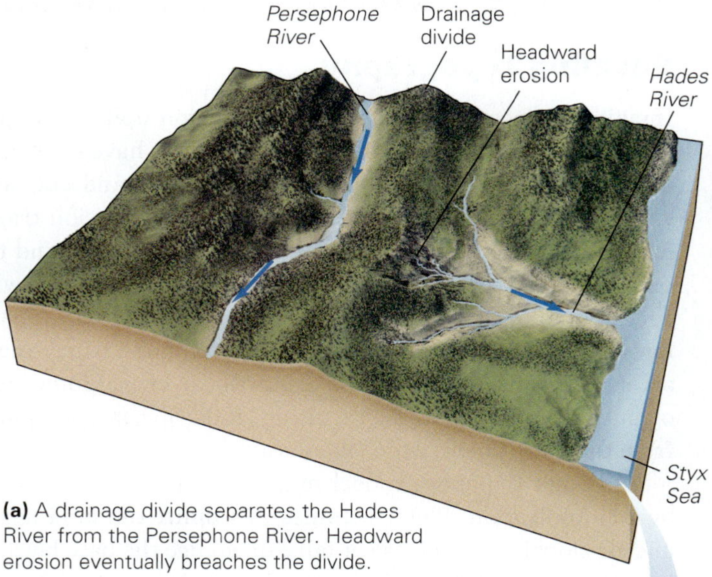
Persephone River
Drainage divide
Headward erosion
Hades River
Styx Sea

(a) A drainage divide separates the Hades River from the Persephone River. Headward erosion eventually breaches the divide.

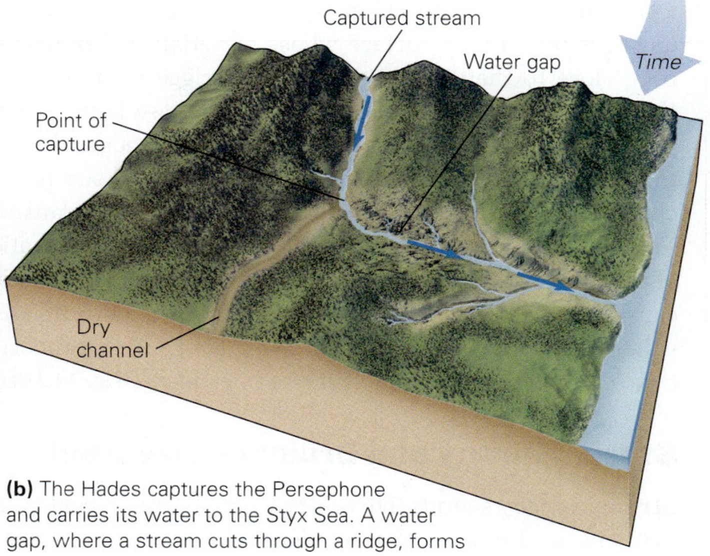
Captured stream
Water gap
Time
Point of capture
Dry channel

(b) The Hades captures the Persephone and carries its water to the Styx Sea. A water gap, where a stream cuts through a ridge, forms and the former Persephone channel becomes a dry canyon.

FIGURE 14.21 Formation of superposed drainage.

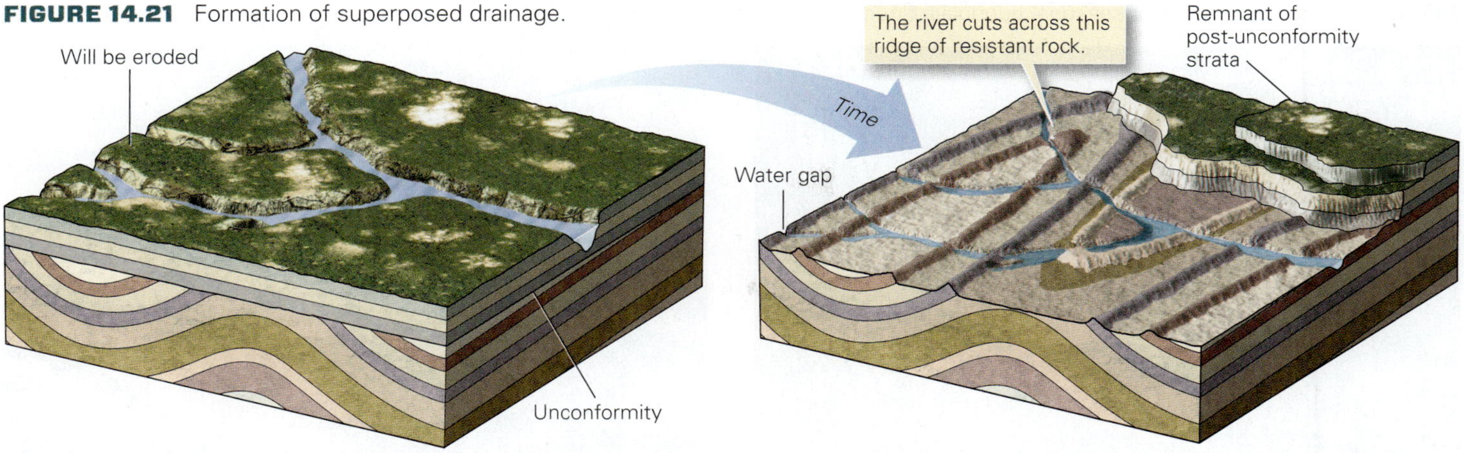

(a) A superposed stream establishes its geometry while flowing over a uniform substrate above an unconformity.

(b) When erosion exposes underlying rock with a different structure, the river is superposed on the structure. As a result, it cuts across resistant ridges instead of flowing around them.

TAKE-HOME MESSAGE

Stream-carved landscapes evolve over time as gradients diminish and ridges between valleys erode away. Superposed streams attain their shape before cutting down into rock structure, whereas antecedent streams cut while the land beneath them uplifts.

QUICK QUESTION What can cause a drainage-reversal event to take place?

14.7 Raging Waters

The Inevitable Catastrophe

Up until now, this chapter has focused on the process of drainage formation and evolution and on the variety of landscape features formed by streams (see **Geology at a Glance**, pp. 486–487). Now we turn our attention to the havoc that

FIGURE 14.22 Development of antecedent and diverted streams.

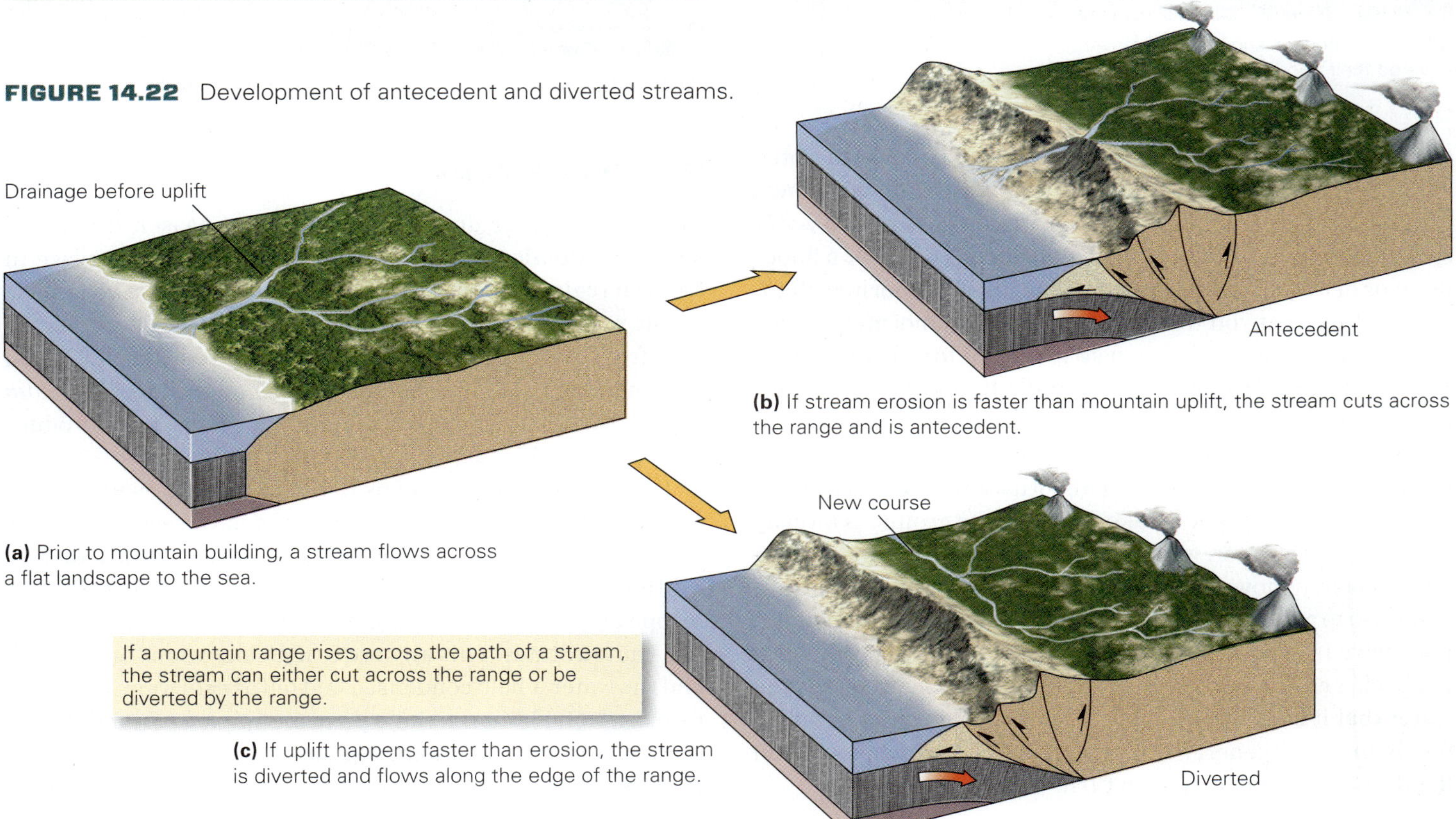

(a) Prior to mountain building, a stream flows across a flat landscape to the sea.

(b) If stream erosion is faster than mountain uplift, the stream cuts across the range and is antecedent.

(c) If uplift happens faster than erosion, the stream is diverted and flows along the edge of the range.

FIGURE 14.23 Examples of seasonal floodplain flooding.

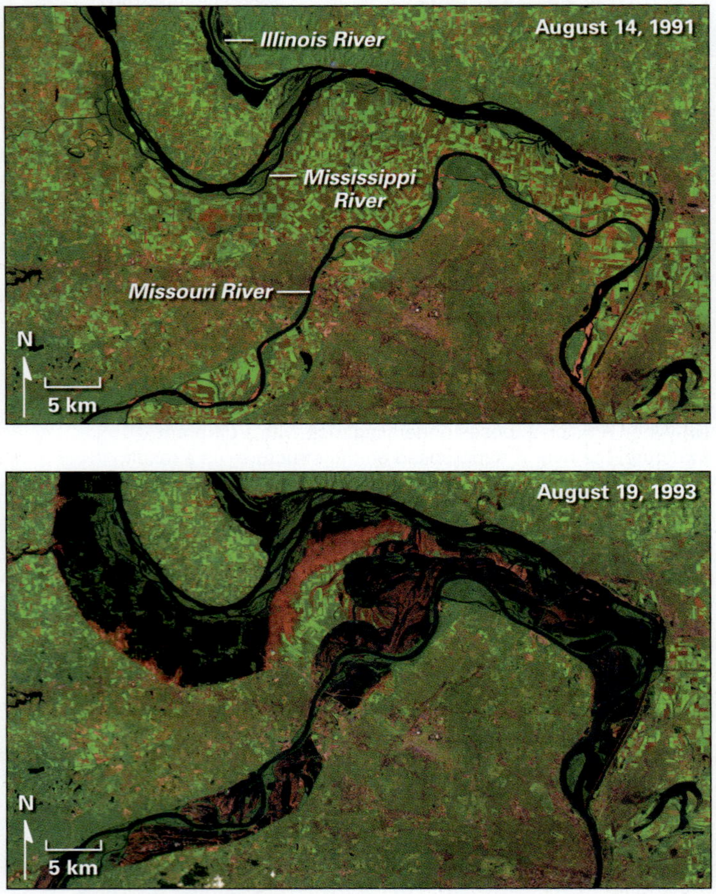

(a) Satellite images show how the rivers in the Midwestern United States covered their floodplains.

(b) Flooding along the Missouri River, in 1993.

(c) Seasonal floods of Bangladesh submerge whole villages.

streams can cause when flooding takes place. As we noted earlier, a flood occurs when the volume of water flowing down a stream exceeds the volume of the stream channel, so water spills out of the normal channel and spreads out over a floodplain or delta plain, or rises above the lip of the normal channel and fills a canyon to a greater depth than normal. Floods can be catastrophic—they can strip land of forests and buildings, they can bury land in clay and silt, and they can submerge cities.

Floods happen if water enters a watershed faster than it can evaporate or be absorbed. This situation can happen when (1) very heavy downpours take place; (2) the ground is already saturated with water, with no room to absorb any more rainfall; (3) heavy snows from the previous winter melt rapidly in response to a sudden hot spell or heavy rain; or (4) a dam holding back a lake or reservoir, or a levee or retaining wall holding back a river or canal, suddenly collapses and releases the water that it held back. Geologists find it convenient to divide floods into two general categories: seasonal floods and flash floods. Let's consider these in turn.

Seasonal Floods

Floods that occur during an annual wet season are called **seasonal floods**. Floods of this type typically take place in tropical regions, during intense monsoonal rains, and in temperate regions, during the spring when downpours drench the land frequently and/or a heavy winter snowpack melts. When seasonal floods submerge floodplains, they produce *floodplain floods*, and when they submerge delta plains they produce *delta-plain floods*.

Seasonal flooding happens in the Mississippi Valley of the central United States, and some years have seen disastrous floods. For example, in 1993, high-altitude winds drifted southward. For weeks they formed an invisible wall that trapped warm, moist air from the Gulf of Mexico over the central United States. This air rose to high elevations and cooled, and the water it held condensed and fell as rain, rain, and more rain. In fact, almost a whole year's supply of rain fell during the spring of 1993, and some regions received 400% more rain than usual. The ground became saturated and could no longer absorb additional water, so the excess entered the region's streams, which carried it into the Missouri and Mississippi Rivers. Eventually, the water in these rivers rose above the height of levees and spread out over the floodplain. By July, parts of nine states were under water (**Fig. 14.23a, b**).

The roiling, muddy flood of 1993 uprooted trees, swept cars away, and even unearthed coffins which floated out of inundated graveyards. All barge traffic came to a halt, bridges and roads were undermined and washed away, and towns along the rivers were submerged in muddy water. In Des Moines, Iowa, 250,000 residents lost their supply of drinking water when floodwaters contaminated the municipal water supply with raw sewage and chemical fertilizers. Rowboats replaced cars as the favored mode of transportation in towns where only the rooftops remained visible. When the water finally subsided, it left behind a thick layer of silt and mud, filling living rooms and kitchens in floodplain towns and burying crops in floodplain fields. For 79 days, the flooding continued. In the end, more than 40,000 square km of the floodplain had been submerged. Similar flooding yielded similar destruction in 2011.

Monsoonal flooding often affects the Ganges Delta of Bangladesh. About 80% of the country occupies a floodplain, and most of this region lies less than 10 m above sea level. During the monsoon season of June through September, wind patterns carry immense amounts of moisture inland from the Indian Ocean. This moisture condenses and rains in deluges over the land. Runoff from this rain, added to melting ice and snow from the Himalayas, fills distributaries of the Ganges River to overflowing as they spread out over the delta. As a result, almost every year about 20% of the country floods, thousands of people die, and millions of homes are destroyed (**Fig. 14.23c**). The flooding often causes drinking water contamination, so many people suffer from disease during and following the flood.

Flash Floods

Events during which the floodwaters rise so fast that it may be impossible to escape from the path of the water are called **flash floods** (**Fig. 14.24a**). They happen during unusually intense rainfall or as a result of a dam collapse or levee failure (**Box 14.1**). During a flash flood, a canyon or valley may fill to a level many meters above normal. Flash floods can be particularly unexpected in arid or semiarid climates, where isolated thundershowers may suddenly fill the channel of an otherwise dry wash. Such a flood may even affect areas downstream that had not received a drop of rain.

The historic Big Thompson River flood of July 31, 1976, illustrates the power of a flash flood. This river, which drains the Front Range of the Rocky Mountains, normally carries clear water in a boulder channel. The channel of the river does not occupy the whole valley floor, so a road follows the course of the river, and in places, vacation cabins and campgrounds dot the land between the river and the road. On July 31, towering thunderheads built up over the Front Ranges, and rain began to pour in quantities that even old-timers couldn't recall. In a little over an hour, 19 cm (7.5 in.) of rain drenched the watershed of the Big Thompson River, so the river's discharge grew to more than four times the maximum recorded during the previous century, and the water level rose by several meters.

Turbulent currents swirled down the canyon at up to 8 m per second and churned up so much sand and mud that the once clear river became a viscous slurry. Landslides of rock and soil tumbled down the steep slopes bordering the river and fed the torrent with even more sediment. The raging water

FIGURE 14.24 Flash floods can occur after torrential rains.

(a) A flash flood in a desert region of Israel has washed over a highway, forcing the evacuation of truckers.

(b) During the 1976 Big Thompson River flash flood, this house was carried off its foundation and dropped on a bridge.

BOX 14.1 Consider This...

The Johnstown Flood of 1889

By the 1880s, Johnstown, built along the Conemaugh River in scenic western Pennsylvania, had become a significant industrial town. Recognizing the attraction of the surrounding hills as a summer retreat, speculators built a mud-and-gravel dam across the river, upstream of Johnstown, to trap a pleasant reservoir of cool water. A group of industrialists and bankers bought the reservoir and established the exclusive South Fork Hunting and Fishing Club, a cluster of lavish 15-room "cottages" on the shore. Unfortunately, the dam had been poorly designed, and debris blocked its spillway (the passageway designed to carry surplus water around the dam), setting the stage for a monumental tragedy. On May 31, 1889, torrential rain drenched Pennsylvania, and the reservoir filled until water began to flow over the dam. Despite frantic attempts to strengthen the dam, the soggy structure abruptly collapsed, and the reservoir emptied into the Conemaugh River Valley. A 20-m-high wall of water roared downstream and slammed into Johnstown, transforming bridges and buildings into twisted wreckage (**Fig. Bx14.1**). When the water subsided, 2,300 people were dead, and Johnstown became the focus of national sympathy. The recently founded Red Cross set to work building dormitories, and citizens nationwide donated everything from clothes to beds. Nevertheless, it took years for the town to recover, and many residents simply picked up and left.

FIGURE Bx14.1 During the 1889 Johnstown flood, raging waters could tumble large houses.

undercut foundations and washed the houses and bridges away (**Fig. 14.24b**). Parts of the road were eroded away, and other parts were buried by debris. Boulders that had stood as landmarks for generations bounced along in the torrent like beach balls, striking and shattering other rocks along the way. Cars drifted downstream until they wrapped like foil around obstacles. When the flood subsided, the canyon had changed forever, and 144 people had lost their lives.

Not all flash floods occur in regions of high relief. For example, in May 2015, intense thunderstorms drenched parts of Texas. Because the region had been suffering from a drought, vegetation had withered and wasn't able to absorb as much water as usual, so the proportion of rain that became runoff was greater than usual. In successive 24-hour periods, about 30 cm (1 foot) of rain fell, much of which entered the drainage network. Some streams rose by up to 12 m, breaking previous records. Low-lying areas, including interstate highways, disappeared beneath the water, and thousands of homes were destroyed.

Living with Floods

Flood Control Mark Twain once wrote of the Mississippi that we "cannot tame that lawless stream, cannot curb it or confine it, cannot say to it, 'go here or go there,' and make it obey." Was Twain right? Since ancient times, people have attempted to control courses of rivers in order to prevent undesired flooding. In the 20th century, flood-control efforts intensified as the population living along rivers increased. For example, since the passage of the 1927 Mississippi River Flood Control Act (after a disastrous flood took place that year), the U.S. Army Corps of Engineers has labored to control the Mississippi. First, engineers built about 300 dams along the river's tributaries so that excess runoff could be stored in the reservoirs and later be released slowly. Second, they built **artificial levees** (elongate mounts of sand and clay) and built concrete *flood walls*, along the banks of the river, to increase the channel's volume. Artificial levees and flood walls isolate a discrete area of the floodplain (**Fig. 14.25**).

Although the Corps' strategy worked for floods up to a certain size, it was insufficient to handle the 1993 and 2011 floods. The river rose until it spilled over the tops of some levees and undermined others. *Levee undermining* occurs when rising water levels increase the water pressure on the river side of the levee, forcing water through sand under the levee. In susceptible areas, water begins to spurt out of the ground on the dry side of the levee, thereby washing away the levee's support. The levee finally becomes so weak that it collapses and water fills in the area behind it. In some cases, the Corps of Engineers intentionally dynamited levees along a relatively unpopulated reach of the river upstream of a vulnerable town, to divert the flow out onto a portion of the floodplain where

FIGURE 14.25 Holding back rivers to prevent floods.

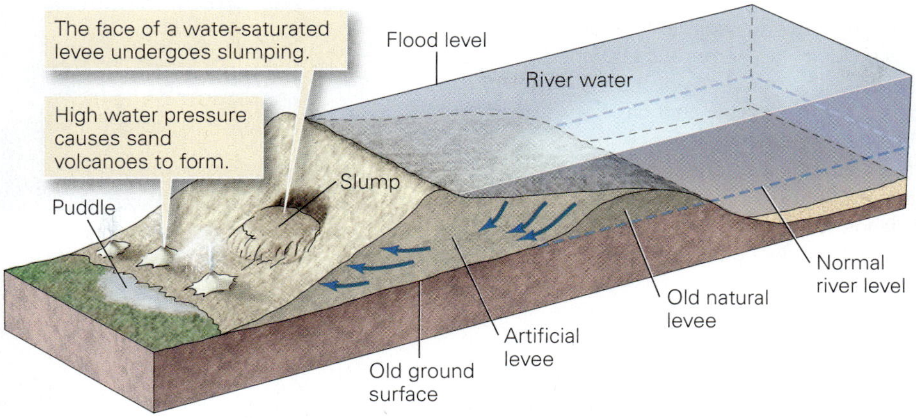

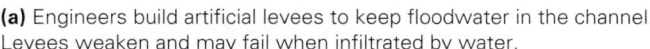

(a) Engineers build artificial levees to keep floodwater in the channel. Levees weaken and may fail when infiltrated by water.

(b) Breaching of a levee lets water spill into the floodplain.

(c) Floodwalls, which can be closed to protect Cape Girardeau, Missouri, from Mississippi River floods.

the water will do less damage, and prevent the floodwaters from overtopping levees close to the town.

Because traditional flood-prevention efforts don't always work, engineers have explored alternatives. For example, restoration of wetland areas along rivers can diminish the volume of water in a channel, for wetlands can absorb significant quantities of floodwater. Also, where appropriate, planners may prohibit construction within designated land areas adjacent to the channel, so that floodwater can fill these areas without causing expensive damage. The existence of such areas, which are known as **floodways**, effectively increases the volume of water that the river can carry and thus helps prevent the water level from rising too high.

Evaluating Flooding Hazard When making decisions about investing in flood-control measures, mortgages, or insurance, planners need a basis for defining the hazard or risk posed by flooding. If floodwaters submerge a locality every year, a bank officer would be ill advised to approve a loan that would promote building there, and insurance, even if available, would be prohibitively expensive. But if floodwaters submerge the locality only very rarely, then the loan or insurance may be worth the risk. Geologists characterize the risk of flooding in two ways. The **annual probability** of flooding indicates the likelihood that a flood of a given size or larger will happen at a specified locality during any given year. For example, if we say that a flood of a given size has an annual probability of 1%, then we mean there is a 1 in 100 chance that a flood of at least this size will happen in any given year. The **recurrence interval** of a flood of a given size is defined as the average number of years between successive floods of at least this size. To understand the concept of a recurrence interval, imagine that a researcher determines that successive floods of a given size happened 96, 32, 200, 14, and 158 years ago. The average of these numbers is 100, so we can refer to a flood of this size as a *100-year flood*. Note that annual probability and recurrence interval are related: Annual probability = 1 ÷ recurrence interval. For example, the annual probability of a 100-year flood is 1/100, which can also be written as 0.01 or 1%. Unfortunately, many people misunderstand the meaning of recurrence interval, and think that they will not face flooding hazard for another century if they buy a home within an area just after a 100-year flood has occurred. Note from our example that such confidence is misplaced. Two 100-year floods can occur in consecutive years, or even in the same year. Alternatively, the interval between such floods could be decades or centuries apart.

Did you ever wonder...
what newscasters mean by a "100-year flood?"

The recurrence interval or annual probability for a flood along a particular river reflects the size of a flood. For example, the discharge of a 100-year flood (annual probability of 1%) is

River Systems

Rivers, or streams, drain the landscape of surface runoff. Typically, an array of connected streams called a drainage network develops, consisting of a trunk stream into which numerous tributaries flow. The land drained is the network's watershed. A stream starts from a source, or headwaters. Some headwaters are in the mountains, collecting water from rainfall or from melting ice and snow. In the mountains, streams carve deep, V-shaped valleys and tend to have steep gradients. Locally, a river may flow over a bouldery bed, forming rapids, or it may drop off an escarpment as a waterfall.

Rivers choked with sediment become braided by dividing into numerous entwined channels separated from one another by gravel bars. Where a stream that has not been choked by sediment has a gentle gradient, it becomes a meandering stream, winding back and forth in snake-like curves called meanders. Because of erosion and deposition, a meandering stream changes

Developing drainage networks

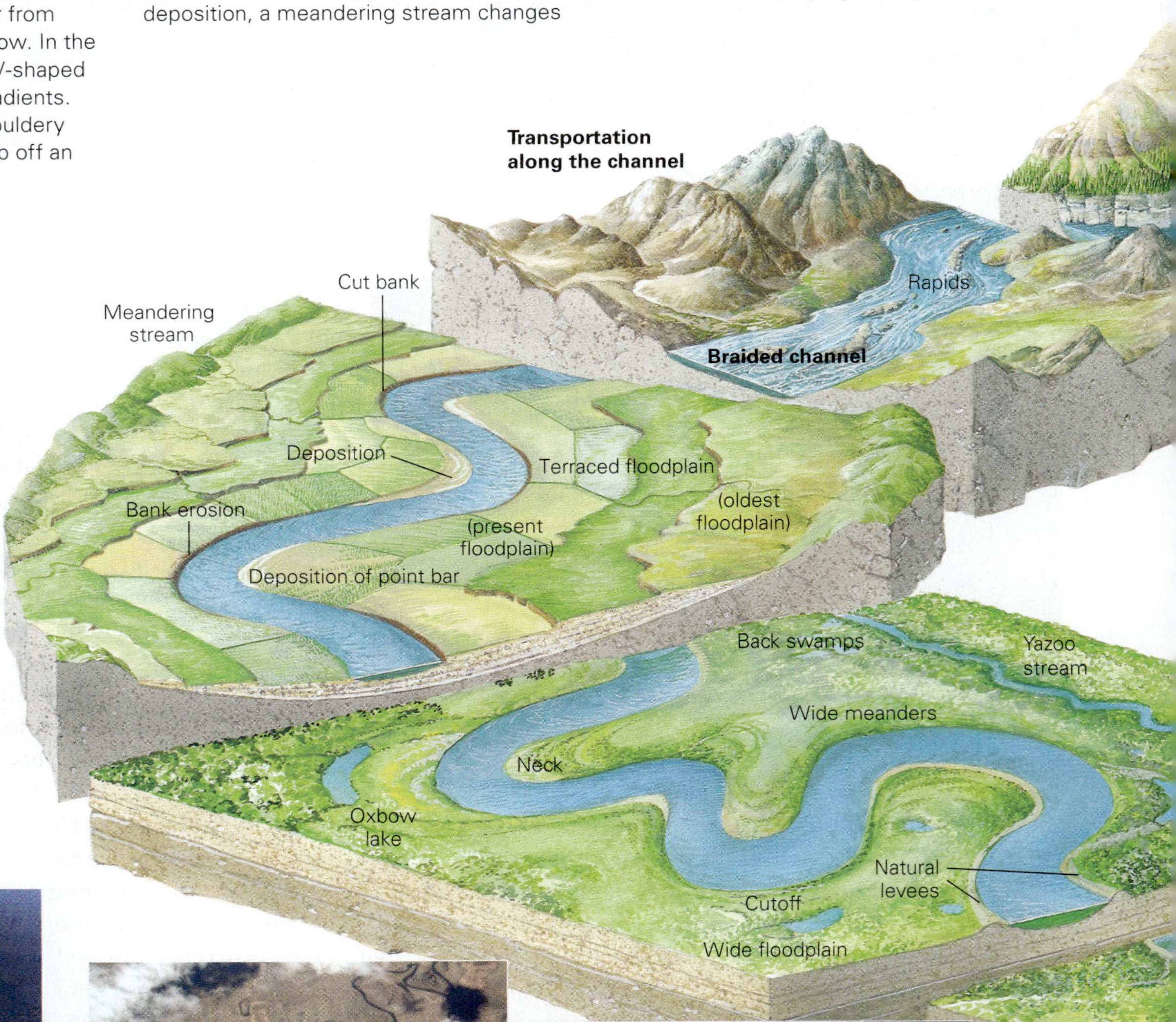

Point bars forming on inner curves

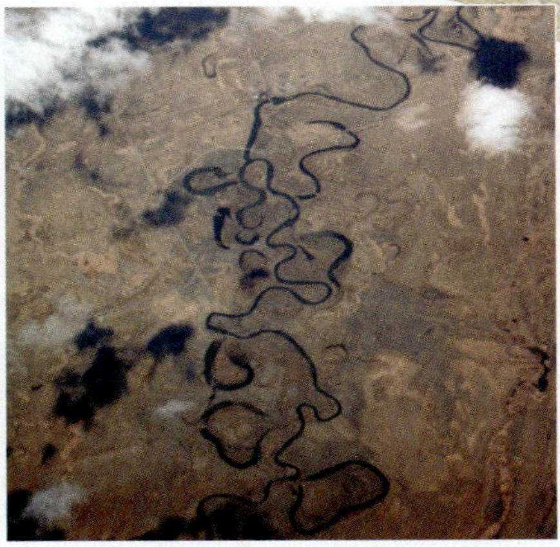

Meanders, abandoned meanders, and cutoffs

A small delta in a mountain lake

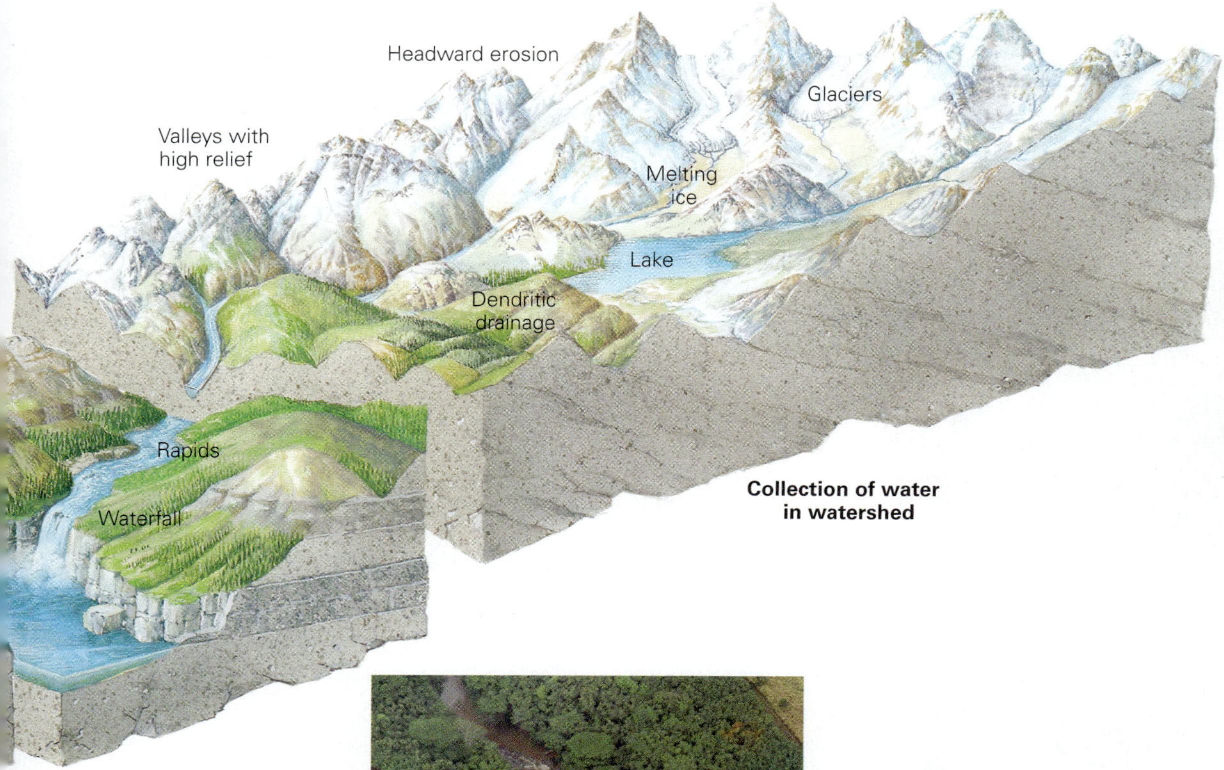

Streams contribute to carving mountains

Waterfall in Hawaii spilling over a basalt ledge

shape over time. Occasionally, a meander may be cut off, leaving a curving lake called an oxbow lake. A broad floodplain, covered with water only during floods, may develop on either side of the stream. Natural levees build up between the channel and the floodplain. They consist of sediment dropped as a flooding river starts to spill out of its channel. Eventually, a river reaches a standing body of water and slows down, and the sediment it carries gets deposited to form a delta.

FIGURE 14.26 The conceptual relationship between flood size and probability.

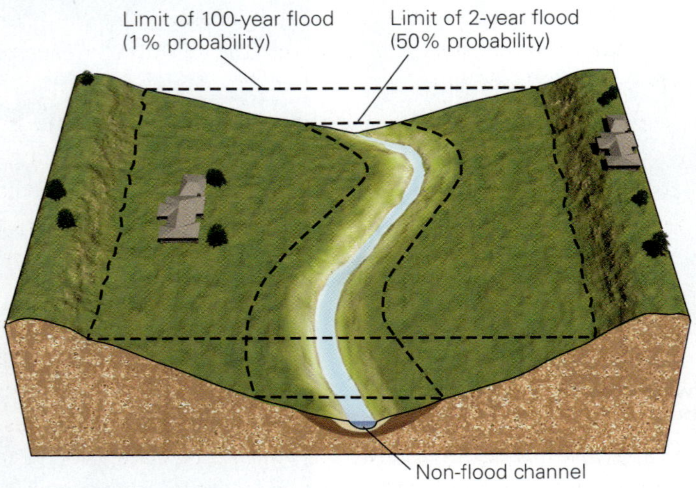

(a) A 100-year flood covers a larger area than a 2-year flood and occurs less frequently.

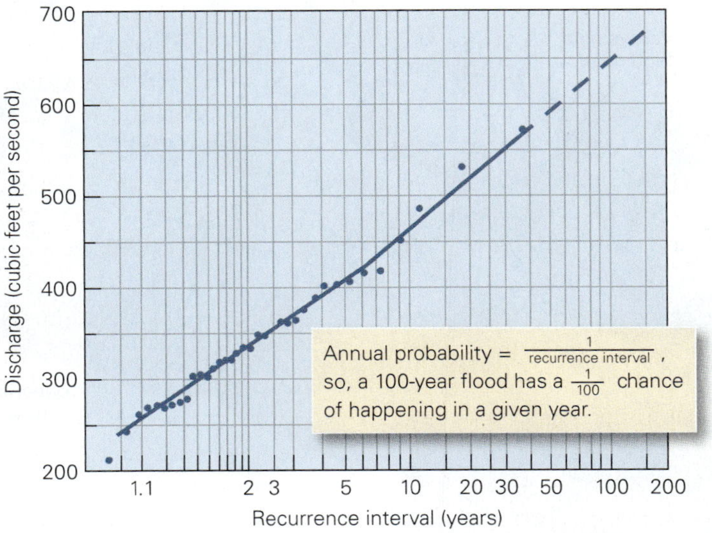

(b) A flood-frequency graph shows the relationship between the recurrence interval and the discharge for an idealized river.

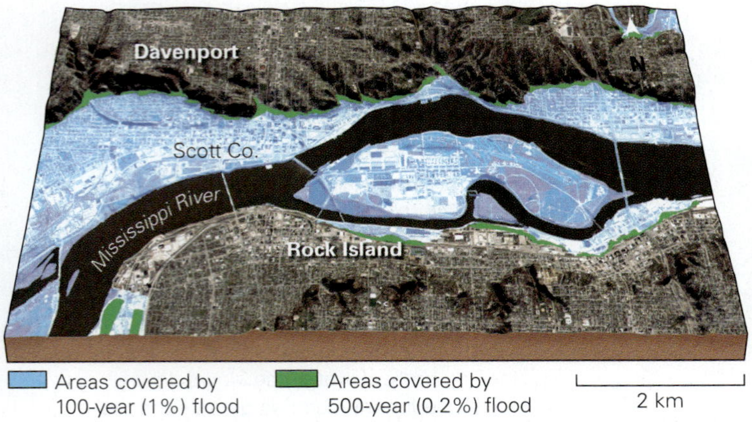

(c) A flood-hazard map shows areas likely to be flooded. Here, near Rock Island, Illinois, even large floods are confined to the floodplain.

larger than that of a 2-year flood (annual probability of 50%), because the former happens less frequently (**Fig. 14.26a**). To define this relationship, geologists construct graphs that plot flood discharge on the vertical axis against recurrence interval on the horizontal axis (**Fig. 14.26b**).

Knowing the discharge during a flood of a specified annual probability, and knowing the shape of the river channel and the elevation of the land bordering the river, hydrologists can predict the extent of land that will be submerged by such a flood. Such data, in turn, permit researchers to produce **flood-hazard maps**. In the United States, the Federal Emergency Management Agency (FEMA) produces maps that show the 1% annual probability (100-year) flood area and the 0.2% annual probability (500-year) flood risk zones (**Fig. 14.26c**). It always important to keep in mind when thinking about flooding hazard that the term floodplain means what it says—it's an area that will eventually flood.

TAKE-HOME MESSAGE

Seasonal floods submerge floodplains and delta plains at certain times of the year. Flash floods are sudden and short-lived. We can specify the probability that a certain size of flood will happen in a given year, but flood-control efforts meet with mixed success. Statements about flooding hazard are, unfortunately, often misinterpreted.

QUICK QUESTION Why might the flood level (height above flood stage) increase for a given discharge when levees are built along the river?

14.8 Human Impact on Rivers

Streams can serve as avenues for transportation, and can be sources of food, irrigation water, drinking water, power, recreation, and (unfortunately) waste disposal. Further, their floodplains provide particularly fertile soil for fields, replenished annually by seasonal floods. Considering the vast resources that rivers provide, it's no coincidence that ancient cultures developed in river valleys and on floodplains. Nevertheless, over time, humans have increasingly tended to abuse or overuse the Earth's rivers. Here we note four pressing environmental issues pertaining to rivers.

- *Pollution*: The capacity of some rivers to carry pollutants has long been exceeded, transforming them into deadly cesspools. Pollutants include raw sewage and storm drainage from urban areas, spilled oil, toxic chemicals from industrial sites, floating garbage, excess fertilizer, and animal waste. Some of these products directly poison aquatic life, some feed algal blooms, and some settle out to be buried along with sediments. At the mouths of major rivers, pollutants spill into the sea, carrying nutrients (particularly nitrogen and phosphorus) that feed microorganisms. These organisms' activity uses up the oxygen in the water, leading to the death of larger organisms, such as fish. The result along the coast can be a *dead zone*, a region of severely reduced biomass and biodiversity. A large dead zone has formed in the Gulf of Mexico, due to nutrients carried in by the Mississippi River.

- *Dam construction*: In 1950, there were about 5,000 large (over 15-m-high) dams worldwide, but today there are over 48,000. Damming rivers has both positive and negative results. Reservoirs provide irrigation water and hydroelectric power, and they trap some floodwaters and create popular recreation areas. But in some locations their construction destroys "wild rivers" (the whitewater streams of hilly and mountainous areas) and alters the ecosystem of a drainage network by forming barriers to migrating fish, by decreasing the nutrient supply to organisms downstream, by removing the source of sediment and nutrients for the floodplain and delta.
- *Overuse of water*: Because of growing populations, our thirst for river water continues to increase, but the supply of water does not. The use of water has grown especially in response to the "green revolution" of the 1960s, during which huge new tracts of land came under irrigation. The flow of rivers may be diverted into canals or pipes that carry water to fields, factories, or towns. In some places, such *river diversion* consumes almost the entire volume of a river's water, so that the channel contains little more than a saline trickle, if that, at its mouth. For example, except during unusually wet years, the Colorado River's channel contains almost no water where it crosses the Mexican border, for huge pipes and canals carry the water instead to Phoenix and Los Angeles (**Fig. 14.27**).
- *Effects of urbanization on streams*: When it rains in a naturally vegetated region or in an agricultural region, much of the water that falls from the sky either soaks into the ground or gets absorbed by plants. Some of the soil moisture or groundwater eventually seeps into a nearby stream, but the remainder flows elsewhere underground. As a result, the amount of water that reaches nearby streams after a storm is less than the total amount of precipitation, and a significant *lag time* (delay) occurs between the time when the water falls and when the stream's discharge increases. Urbanization changes both the volume of water reaching the stream and the duration of the lag time, because when developers transform fields and forests into parking lots, roads, and buildings, a layer of impermeable concrete and asphalt prevents rainfall from infiltrating, and the amount of living biomass available decreases. Storm sewers and streets divert water directly to streams, so not only does the volume of water entering the streams increase, but also the rate at which the volume changes increases.
- *Effects of agriculture*: Although we tend to think of farmland as vegetated land, it actually has less plant cover than does natural grassland or forest. That's because the land surface between the crop rows remains bare during the growing season, and after harvest, entire fields become a broad expanse of exposed soil. Sheetwash flowing across the unprotected land surface erodes and carries with it significant volumes of sediment. Thus, a river's sediment load increases significantly when farms replace forests nearby.

FIGURE 14.27 The Central Arizona Project canal shunts water from the Colorado River to Phoenix.

Rivers are precious resources. People have put immense efforts into improving ways to protect us from them, by preventing floods. Fortunately, researchers are also seeking ways to protect them from us, by preventing pollution, regulating water use, increasing opportunities for stormwater to infiltrate into the ground, and planting off-season crops that keep farm fields being exposed to the erosive effects of runoff.

TAKE-HOME MESSAGE

Society depends on streams for water supplies, irrigation, energy, and transport, but the growth of populations can negatively affect streams. Diversion of flow may decrease stream discharge to a trickle, dam construction can change flow, pollution can foul the water, and urbanization or runoff can change discharge and sediment load.

QUICK QUESTION How can urbanization change the discharge of a stream?

Chapter 14 Review

Chapter Summary

- Streams are bodies of water that flow down channels and drain the land surface. They grow by downcutting and headward erosion. Streams carry water out of a drainage basin; a drainage divide separates two adjacent basins. Drainage networks consist of many tributaries that flow into a trunk stream.
- Permanent streams exist where the water table lies above the bed of the channel or when large amounts of water enter the channel from upstream. Where the water table lies below the channel bed, streams are ephemeral.
- The discharge of a stream is the total volume of water passing a point along the bank in a second. Most streams are turbulent, meaning that their water swirls in complex patterns.
- Streams erode the landscape by scouring, lifting, abrading, and dissolving. The resulting sediment provides dissolved loads, suspended loads, and bed loads. The total quantity of sediment carried by a stream is its capacity. Capacity differs from competence, the maximum particle size a stream can carry. When stream water slows, it deposits alluvium.
- The longitudinal profile of a stream is concave up, meaning that a stream has a steeper gradient at its headwaters than near its mouth. The depth of downcutting is limited by the base level.
- Streams cut valleys or canyons, depending on the rate of downcutting relative to the rate at which the slopes on either side of the stream undergo mass wasting. Where a stream flows down steep gradients and has a bed littered with large rocks, rapids develop, and where a stream plunges off a vertical face, a waterfall forms.
- A meandering stream wanders back and forth across a floodplain. It erodes its outer bank and deposits a point bar on the inner bank. Eventually, a meander may be cut off and turn into an oxbow lake. Natural levees form on either side of the river channel.
- Braided streams form where a stream that carried abundant sediment during floods slow, so the flowing water divides among many entwined channels.
- Where streams or rivers flow into standing water, they deposit deltas. Different deltas have different shapes.
- Fluvial erosion can bevel landscapes to a nearly flat plain. If the base level drops or the land surface rises, stream rejuvenation takes place. The headward erosion of one stream may capture the flow of another.
- If an increase in rainfall or spring melting causes more water to enter a stream than the channel can hold, a flood results. Some floods are seasonal and submerge floodplains or delta plains. Flash floods happen very rapidly. Officials try to prevent floods by building reservoirs and levees.
- Rivers are becoming a vanishing resource because of pollution, damming, and overuse of water.

Guide Terms

abrasion (p. 469)
alluvial fan (p. 474)
alluvium (p. 471)
annual probability (p. 485)
artificial levee (p. 484)
bar (p. 471)
base level (p. 472)
bed load (p. 470)
braided stream (p. 474)
canyon (p. 473)
capacity (p. 471)
channel (p. 464)
competence (p. 470)
delta (pp. 471, 478)
discharge (p. 468)
dissolved load (p. 470)
downcutting (p. 465)
drainage divide (p. 467)
drainage network (p. 466)
drainage reversal (p. 479)
ephemeral stream (p. 467)
flash flood (p. 483)
flood (p. 464)
flood-hazard map (p. 488)
floodplain (pp. 471, 476)
floodway (p. 485)
headward erosion (p. 465)
headwaters (p. 464)
longitudinal profile (p. 472)
meander (p. 475)
meandering stream (p. 476)
mouth (p. 464)
natural levee (p. 477)
oxbow lake (p. 476)
permanent stream (p. 467)
point bar (p. 471)
pothole (p. 469)
rapids (p. 474)
recurrence interval (p. 485)
runoff (p. 465)
saltation (p. 470)
scouring (p. 469)
seasonal flood (p. 482)
sheetwash (p. 465)
stream (p. 464)
stream gradient (p. 472)
stream piracy (p. 479)
stream rejuvenation (p. 479)
stream terrace (p. 474)
suspended load (p. 470)
tributary (p. 466)
valley (p. 473)
waterfall (p. 474)
watershed (p. 467)

GEOTOURS *THIS CHAPTER'S GEOTOUR EXERCISE (N) FEATURES:*

› Headward Erosion
› Stream Patterns
› Meandering Stream Features
› Miscellaneous Stream Features

Review Questions

1. What role do streams serve during the hydrologic cycle?
2. Describe the four different types of drainage networks.
3. Distinguish between permanent and ephemeral streams and explain why the difference exists.
4. Why does the velocity of a stream vary with location, in a given reach of the stream?
5. Describe how streams erode the Earth's surface.
6. What are three components of sediment load in a stream, and how do competence and capacity differ?
7. Describe how the character of a drainage network changes, along its length.
8. What is the difference between a local base level and the ultimate base level of a stream?
9. Why do some streams become braided?
10. Describe how meanders form, develop, are cut off, and then are abandoned. What is a floodplain?
11. Describe how deltas grow and develop. How do they differ from alluvial fans?
12. How does a stream-eroded landscape evolve over time?
13. What is stream piracy? What causes a drainage reversal?
14. How are superposed and antecedent drainages similar, and how do they differ?
15. What is the difference between a seasonal flood and a flash flood?
16. How do people try to protect regions from flood damage?
17. What is the recurrence interval of a flood, and how is it related to the annual probability?
18. How have humans abused and overused the resource of running water?

On Further Thought

19. Records indicate that flood crests for a given amount of discharge along the Mississippi River have been getting higher since 1927, when a system of levees began to block off portions of the floodplain. Why?

Online Resources

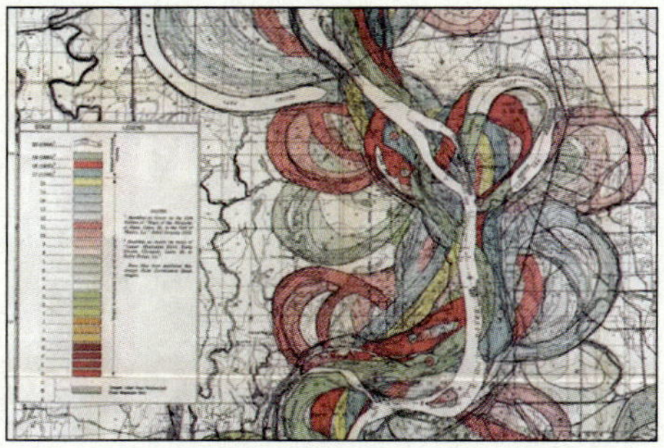

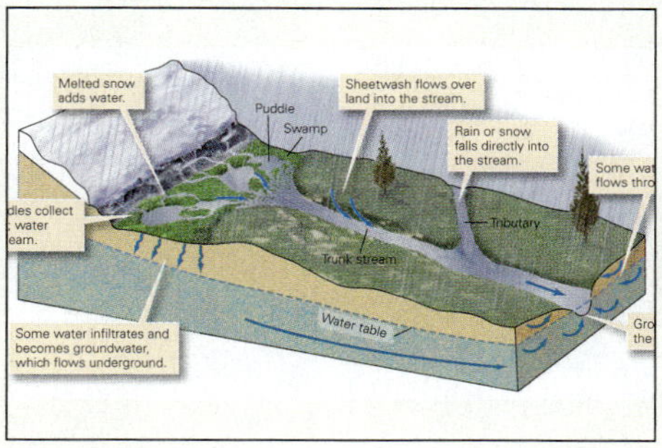
Melted snow adds water.
Puddle
Swamp
Sheetwash flows over land into the stream.
Rain or snow falls directly into the stream.
Tributary
Trunk stream
Water table
Some water infiltrates and becomes groundwater, which flows underground.

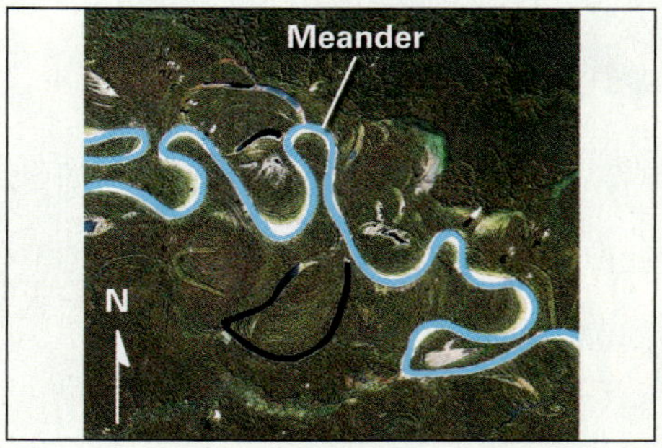
Meander
N

Videos

This chapter features a video about rivers that shows how they change over time and change their surrounding landscapes.

Assessment

This chapter features visual identification exercises on meandering stream systems, erosion, and lake formation.

LEARNING OBJECTIVES

By the end of this chapter, you should understand...

1. why ocean basins exist, and how to explain variations in water depth.
2. the nature and causes of surface and deep currents.
3. the behavior and cause of tides.
4. why waves form and how they behave.
5. how a great variety of different coastal landforms develop and evolve.
6. how rising sea level, wave erosion, and human activities affect coasts.

▲ The rocky coast of Brittany in northwestern France gets pounded by waves, which gradually carve headlands and bays. The ocean is constantly in motion.

CHAPTER 15

Restless Realm: Oceans and Coasts

15.1 Introduction

When seen from space, the Earth glows blue, for the ocean covers most of its surface (**Fig. 15.1**). The ocean incubates life, tempers Earth's climate, and spawns its storms. It also serves as a vast reservoir for water and chemicals that cycle into the atmosphere and crust, and as a resting place for sediment washed off the continents. In this chapter, we begin by addressing the question of why distinct ocean basins exist in the first place. Then, we consider the nature of seawater, and its movements. Finally, we focus on landforms that develop along the *coast,* the region where the land meets the sea, and we learn about some of the natural hazards that challenge people who live in coastal areas.

15.2 Landscapes Beneath the Sea

If the surface of the lithosphere were completely smooth, ocean water would surround the Earth as a uniform, 2.5-km-deep layer. But the surface of the lithosphere is not

> ***The three great elemental sounds in nature are the sound of rain, the sound of wind in a primeval wood, and the sound of the outer ocean on a beach.***
>
> HENRY BESTON (American naturalist, 1888–1968)

smooth—rather, it displays vertical relief of almost 20 km, as measured from the deepest point in the ocean to the highest point on land (**Fig. 15.2a**). Dry land has an average elevation of 0.8 km above sea level, and covers 29.2% of the Earth's surface. In contrast, the seafloor, which accounts for the remaining 70.8% of the Earth's surface, has an average elevation of 3.7 km below sea level. Because of this difference in elevation, we can distinguish between distinct *ocean basins*, the lithosphere surface beneath the sea, and continents, the large land masses between the ocean basins (**Fig. 15.2b**). Due to isostasy (see Interlude D), oceanic lithosphere rests on the asthenosphere so that its surface is lower than the surface of continental lithosphere. On the map of the present-day Earth, geographers recognize seven oceans and numerous smaller seas (see Fig. 15.1). But because all the oceans are interconnected, the boundaries between them are, geologically, somewhat arbitrary.

Have you ever wondered what the ocean floor would look like if all the water evaporated? To gain insight into the **bathymetry** (variation in depth) of the seafloor, researchers, until the mid-20th century, dropped a *plumb line*, a weight at the end of a cable, down to the seafloor. Using this tedious technique, measurements took a long time to obtain. When sonar became available, about the time of World War II, it became possible to make continuous recordings of depth as a ship crossed the sea. Today, satellites survey the ocean and translate very precise measurements of changes in the pull of gravity into measurements of water-depth variation. Taken together, such studies indicate that ocean basins include several *bathymetric provinces*, distinguished from one another by water depth. Let's take a look at these provinces, in order to explain their characteristics in terms of plate tectonics theory.

FIGURE 15.1 The oceans of the world. The Pacific is the largest, covering almost half the planet. The Arctic region is an ocean covered by a thin coating of ice, whereas the Antarctic region is a continent.

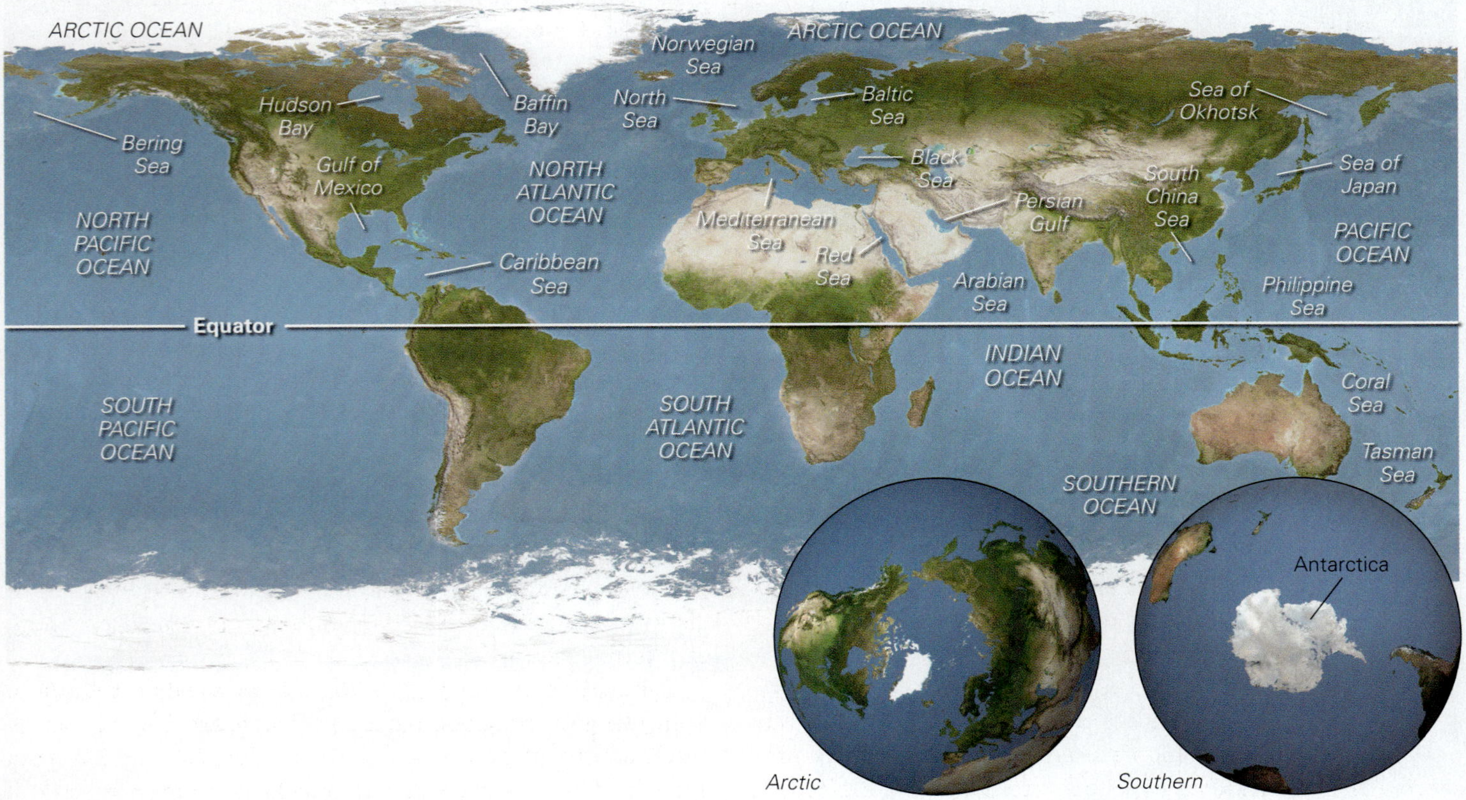

FIGURE 15.2 Contrasts between continental lithosphere and oceanic lithosphere.

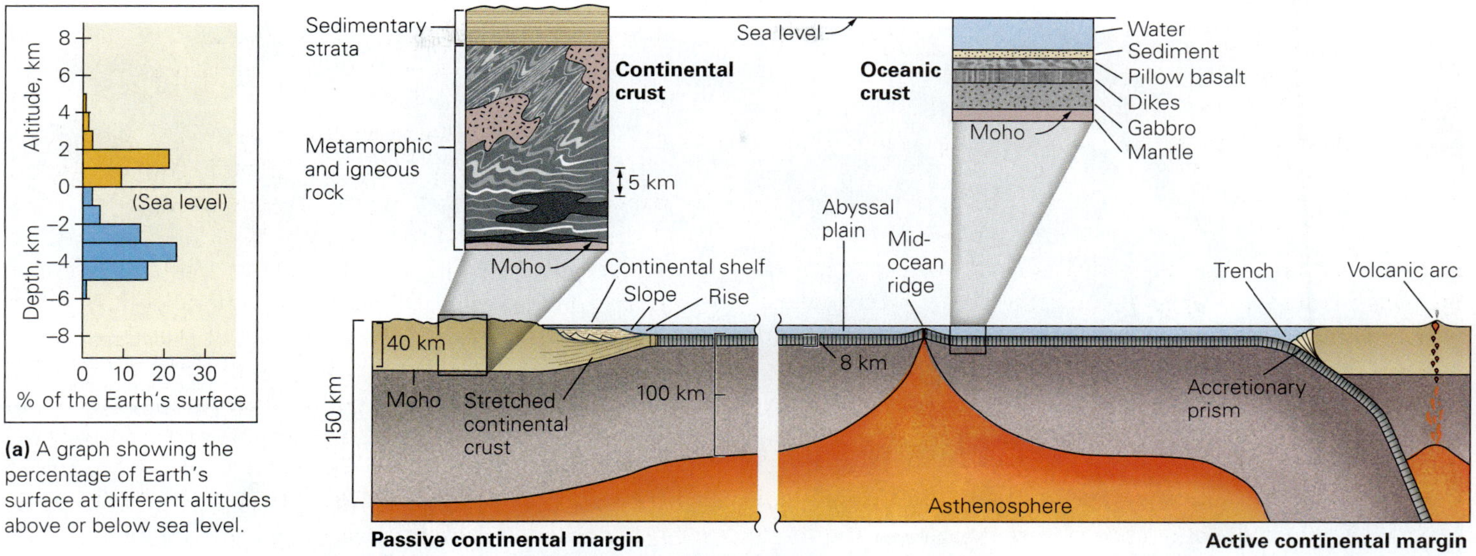

(a) A graph showing the percentage of Earth's surface at different altitudes above or below sea level.

(b) The crustal portion of the continental lithosphere differs markedly from that of oceanic lithosphere.

Continental Shelves, Slopes, and Rises

Imagine you're in a submersible cruising just above the floor of the western half of the North Atlantic. If you start at the shoreline of North America and head east, you will cross the 200- to 500-km-wide **continental shelf**, a relatively shallow portion of the ocean that fringes the continent (**Fig. 15.3a**). Water depth over the continental shelf does not exceed 500 m. At its eastern edge, the continental shelf merges with the *continental slope*, which descends to depths of nearly 4 km. From about 4 km down to about 4.5 km, a province called the *continental rise*, the slope angle decreases until, at 4.5 km deep, you find yourself above a vast, nearly horizontal surface, the **abyssal plain**.

Broad continental shelves, like that of eastern North America, form along *passive continental margins*, margins that are not plate boundaries and thus lack seismicity. A continental shelf is the surface of a very thick accumulation of sediment filling a *passive-margin basin* that originated after rifting broke a continent in two. When rifting stops and seafloor spreading begins, the stretched lithosphere at the boundary between the ocean and continent gradually cools, gets denser, and therefore sinks. Sand and mud that washed off the continent, along with the shells of marine creatures, bury the sinking crust, slowly producing a pile of sediment up to 20 km thick (see Fig. 15.2b).

At many locations, relatively narrow and deep valleys called **submarine canyons** downcut into continental shelves and slopes (**Fig. 15.3b**). Some submarine canyons start offshore of major rivers, for rivers cut into the continental shelf at times when sea level was low and the shelf was exposed. But river erosion cannot explain the great depth of these canyons—some slice almost 1,000 m down into the continental margin, far deeper than the maximum sea-level change. Researchers have determined that much erosion of submarine canyons results from the flow of turbidity currents, submarine avalanches of sediment mixed with water (see Chapter 6).

Bathymetry of Plate Boundaries in the Ocean

Seafloor spreading at a divergent boundary yields a **mid-ocean ridge**, a submarine mountain belt as much as 2 km high. Because crust stretches and breaks as seafloor spreading continues, the ridge axis may be bordered by steep escarpments as a result of normal faulting. *Oceanic transform faults*, strike-slip faults along which one plate shears sideways past another, typically link segments of mid-ocean ridges (see Chapter 2). Transforms are delineated, bathymetrically, by *fracture zones*, narrow belts of steep escarpments and broken-up rock. These fracture zones can be traced into the oceanic plate away from the ridge axis where they are not seismically active but are still distinct because they form the boundary between lithosphere of different ages.

Subduction at convergent boundaries yields a **trench**, a deep, elongate trough bordering a volcanic arc. Some trenches reach depths of over 8 km, with the deepest, the Mariana Trench of the western Pacific, reaching a depth of 11 km.

FIGURE 15.3 Bathymetric features of the seafloor. The maps are produced by computer using measurements from satellites or submersibles.

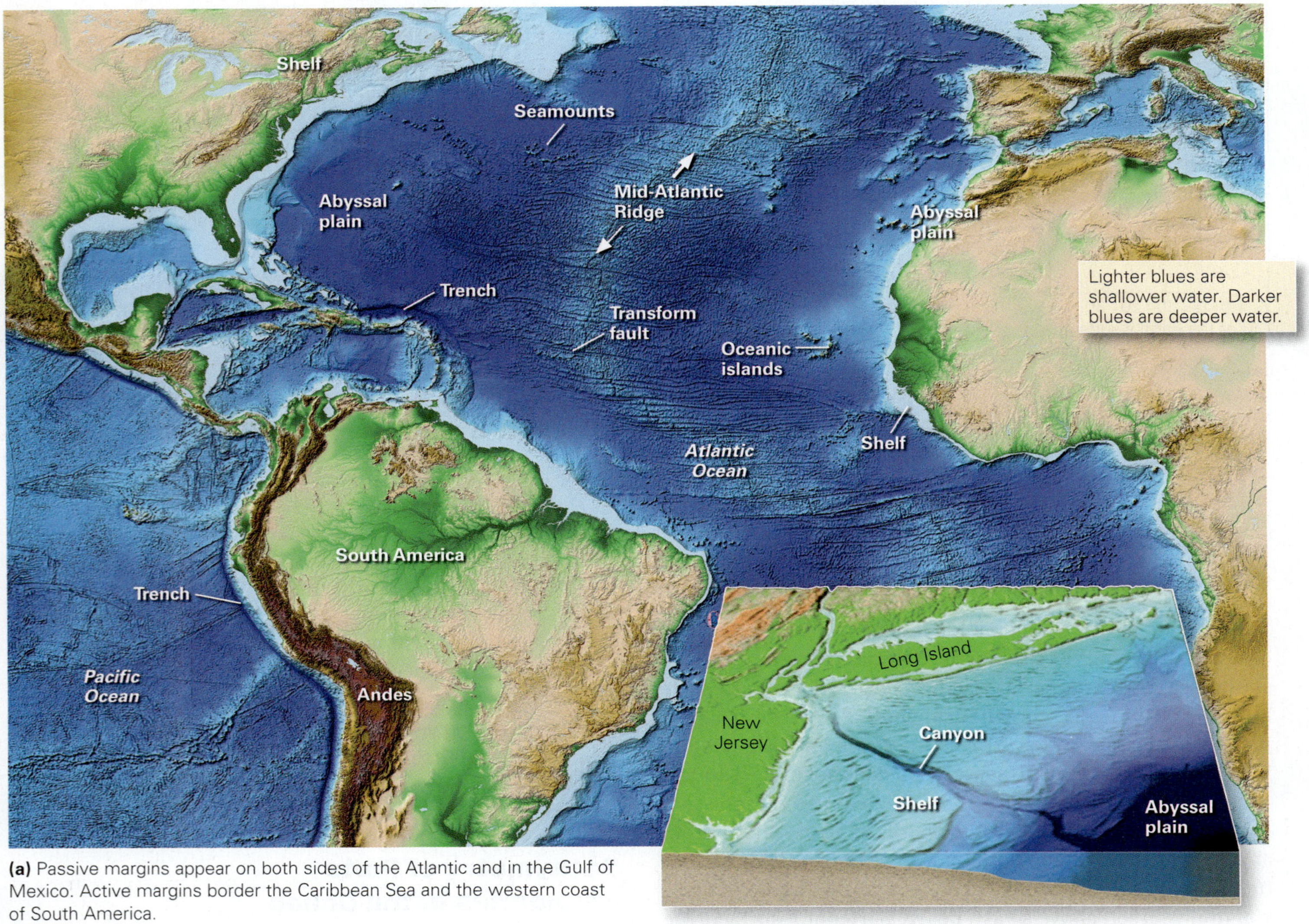

(a) Passive margins appear on both sides of the Atlantic and in the Gulf of Mexico. Active margins border the Caribbean Sea and the western coast of South America.

(b) A submarine canyon along the coast of New Jersey. Turbidity currents in submarine canyons carry sediment to the abyssal plain.

Only three people have descended to the floor of the Mariana Trench—two in 1960 and the third in 2012. Some trenches border continents, whereas others border island arcs, which are curving chains of active volcanic islands that grew on the surface of oceanic lithosphere.

Abyssal Plains and Seamounts

As oceanic crust ages and moves away from the axis of the mid-ocean ridge, two changes take place. First, the lithosphere cools, and as it does so, its surface sinks. Second, a blanket of *pelagic sediment* (deep-marine sediment) gradually accumulates and covers the basalt of the oceanic crust. This blanket consists mostly of microscopic plankton shells and fine flakes of clay, which slowly fall like snow from the ocean water and settle on the seafloor. As the ocean crust moves away from the ridge axis and gets progressively older, sediment thickness increases. Seafloor that has aged more than about 80 million years no longer sinks, and after being buried by a thick layer of pelagic sediment, it tends to be deep, flat, and smooth. Broad areas of old, flat seafloor are called **abyssal plains**.

In dozens of locations around the world, hot-spot eruptions on oceanic lithosphere produced volcanoes. Presently active oceanic hot-spot volcanoes, as well as the remnants of extinct ones, that rise above sea level are *oceanic islands*. Those whose peaks lie below sea level are called **seamounts**. Particularly voluminous eruptions have produced broader buildups of basalt known as *submarine plateaus*. The largest of these are 1,200 km across.

TAKE-HOME MESSAGE

Ocean basins exist because oceanic and continental lithosphere differ in thickness and density. Seafloor bathymetric features (ridges, trenches, and fracture zones) reflect plate-tectonic processes. Abyssal plains overlie old oceanic lithosphere, and continental shelves overlie passive-margin basins.

QUICK QUESTION How do submarine canyons form?

15.3 Ocean Water and Its Movement

Composition and Temperature of Ocean Water

If you've ever had a chance to swim in the ocean, you may have noticed that you float much more easily in ocean water than you do in freshwater. That's because ocean water contains an average of 3.5% dissolved salt, whereas typical freshwater contains less than 0.02% salt. Dissolved ions fit between water molecules without changing the volume of the water, so adding salt to water increases the water's density. According to Archimedes' principle, a buoyant object sinks only until it displaces a mass of water equal to the mass of the object. Therefore, you can float higher in a denser liquid.

The ocean contains so much salt that if all the water suddenly evaporated, a 60-m-thick layer of salt would coat the ocean floor. Though we say "salt," we don't mean that the layer would consist entirely of the mineral halite, common table salt (NaCl). In fact, this layer would consist of only about 75% halite—the remainder would include gypsum ($CaSO_4 \cdot H_2O$), anhydrite ($CaSO_4$), and several other trace salts.

Oceanographers refer to the concentration of salt in water as **salinity**. Although ocean salinity averages 3.5%, a compilation of measurements from around the world reveals that salinity varies with location, ranging from about 1.0% to about 4.1%. Why? Salinity reflects the balance between the addition of freshwater by rivers or rain and the removal of water molecules by evaporation, for when seawater evaporates, salt stays behind. Salinity also depends on water temperature, for warmer water can hold more salt in solution than can cold water. Thus, oceans are saltier along the coasts of hot deserts, and are less salty near the mouths of large, cold rivers.

When the *Titanic* sank after striking an iceberg in the North Atlantic, most of the unlucky passengers and crew who jumped or fell into the sea died within minutes, because the seawater temperature at the site of the tragedy approached freezing, and cold water removes heat from a body very rapidly. Yet swimmers can play for hours in the Caribbean, where sea-surface temperatures reach 28°C (83°F). Though the average global sea-surface temperature hovers around 17°C, it ranges between freezing near the poles to almost 35°C in restricted tropical seas. Average temperatures correlate with latitude because the intensity of solar radiation varies with latitude. Water temperatures in the ocean also vary markedly with depth. Water warmed by the Sun expands and therefore becomes less dense, so it tends to float near the sea's surface. As a result, an abrupt **thermocline**—a boundary below which water temperatures decrease sharply—occurs at a depth of about 300 m in the tropics. Polar seas have no pronounced thermocline, because surface waters there are already so cold. At the floor of the oceans, temperatures approach freezing.

Currents: Rivers in the Sea

Since first setting sail on the open ocean, people have known that ocean water does not stand still, but rather moves in somewhat continuous, recognizable flows, called **currents**, at velocities of up to several kilometers per hour. Unlike rivers on land, currents are not confined by physical channels. Rather, oceanographers define their boundaries by detecting changes in velocity. Currents cause water to circulate around oceans and between oceans at two levels: *surface currents* affect the upper hundred meters of water, whereas *deep currents* keep the remainder of the water column in motion.

Surface currents occur in all the world's oceans (**Fig. 15.4**). They result from interaction between the sea surface and the wind, for as moving air molecules shear across the surface of the water, friction between air and water drags the water along with it. If the Earth did not spin, currents would flow in the same direction as the wind. The Earth's rotation, however, generates the **Coriolis effect** (**Box 15.1**), a phenomenon that causes surface currents in the northern hemisphere to veer toward the right and surface currents in the southern hemisphere to veer toward the left of the average wind direction. Across the width of an ocean, this adjusted motion leads surface currents to make a complete loop, known as a *gyre*. Surface water may become trapped for a long time in the center of the gyre, where currents hardly exist, so these regions tend to accumulate floating plastics, sludge, and seaweed. The Sargasso Sea, named for a kind of floating seaweed, lies at the center of the North Atlantic gyre, and the "Great Pacific Garbage Patch," an accumulation of floating plastic and trash, lies at the center of the North Pacific gyre. Figure 15.4 provides a simplified map of currents. In detail, interactions between currents and coastlines, or between currents and neighboring still water, or between currents flowing in different directions or at

BOX 15.1 CONSIDER THIS...

The Coriolis Effect

Imagine that you're standing at the pole, and you use a huge cannon to fire a projectile from due south, toward the equator (**Fig. Bx15.1a**). If the Earth were standing still, the shot would follow a line of longitude. But the Earth isn't standing still. It rotates counterclockwise around its *axis*, the imaginary line that passes through our planet's center and its geographic poles. To an observer at a fixed location in space, an object sitting on the pole doesn't move at all as the Earth spins, whereas an object at the equator moves at about 1,665 km/h (1,035 mph). Because of this difference, the target on the equator will have moved by the time the projectile reaches it. Therefore, to an observer standing on and moving with the Earth, the projectile follows a curving trajectory. The same phenomenon happens, in reverse, if you place a cannon on the equator and fire an object due north (**Fig. Bx15.1b**). Its path curves because the projectile moves eastward progressively faster than the land beneath it, while it flies northward. This behavior is called the *Coriolis effect*, named for the French engineer who described it in 1835. Because of the Coriolis effect, north-flowing currents in the northern hemisphere deflect to the east, and south-flowing currents deflect to the west. The opposite happens in the southern hemisphere.

FIGURE Bx15.1 The Coriolis effect occurs because the velocity of a point at the equator, in the direction of the Earth's spin, is greater than that of a point near the pole.

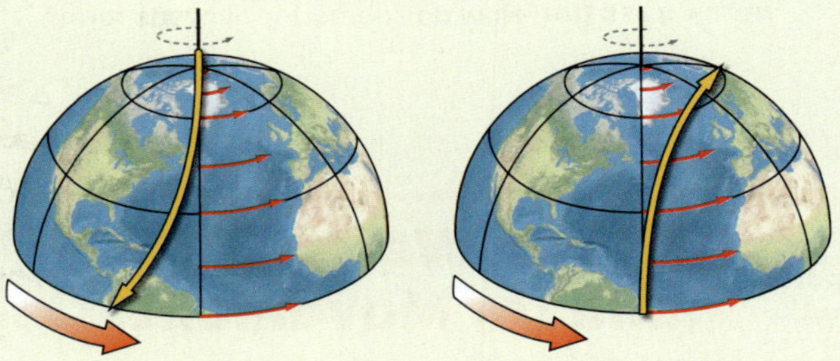

(a) A projectile shot from the pole to the equator deflects to the west.

(b) A projectile shot from the equator to the pole deflects to the east.

different rates, produce *eddies*, in which water circulates in small loops, or local map-view undulations (**Fig. 15.5**).

Surface water exchanges with deeper water in the ocean at a number of locations. Specifically, in *downwelling zones*, surface water sinks, and in *upwelling zones*, deeper water rises. Downwelling and upwelling occur for a number of reasons. For example, in places where winds blow surface water shoreward, an oversupply of water develops along the coast, so surface water must sink to make room. And where winds blow surface water away from the shore, a deficit of water develops along the coast, so deeper water must rise to fill the gap. Upwelling also occurs near the equator, where winds blow steadily from east to west, because the Coriolis effect deflects surface currents to the right in the northern hemisphere and to the left in the southern hemisphere, leading to a water deficit along the equator. The resulting rise of cool, nutrient-rich water fosters an abundance of life in equatorial waters.

Contrasts in water density, caused by differences in temperature and/or salinity, can also drive upwelling and downwelling. Oceanographers refer to this movement as **thermohaline circulation** (**Fig. 15.6a**). During thermohaline circulation, denser (colder and/or saltier) water sinks, while less-dense water (warmer and/or less salty) rises. On a global scale, thermohaline circulation causes warm, equatorial water to move northward at shallow depths. As this water enters polar latitudes it cools, gets denser, and sinks down to the floor of the ocean, where it flows back toward the equator. This process divides ocean water vertically into distinct *water masses* (**Fig. 15.6b**). The combination of surface currents and thermohaline circulation, like a conveyor belt, moves water among the various ocean basins. Because water carries heat, this circulation affects global climate.

Tides—The Daily Rise and Fall of Sea Level

A ship captain hoping to sail out of a port must pay close attention to the **tide**, the generally twice daily rise and fall of sea level. At high tide, the harbor's water may be deep enough that the ship can float over obstacles easily, but at low tide, the water may be so shallow that the ship could run aground. Tides develop in response to the *tide-generating force*, which in simple terms develops in part due to the gravitational attraction of the Sun and Moon, and in part to centrifugal force produced as the Earth-Moon System revolves around its center of mass. The tide-generating force results in two tidal bulges in the oceans on opposite sides of our planet (**Fig. 15.7a**). The larger of these occurs on the side of the Earth closer to the Moon,

FIGURE 15.4 A simplified map showing the major surface currents of the world's oceans. The color key distinguishes cold from warm currents, and the inset shows a detail of the Gulf Stream, which carries warm water to the northeast.

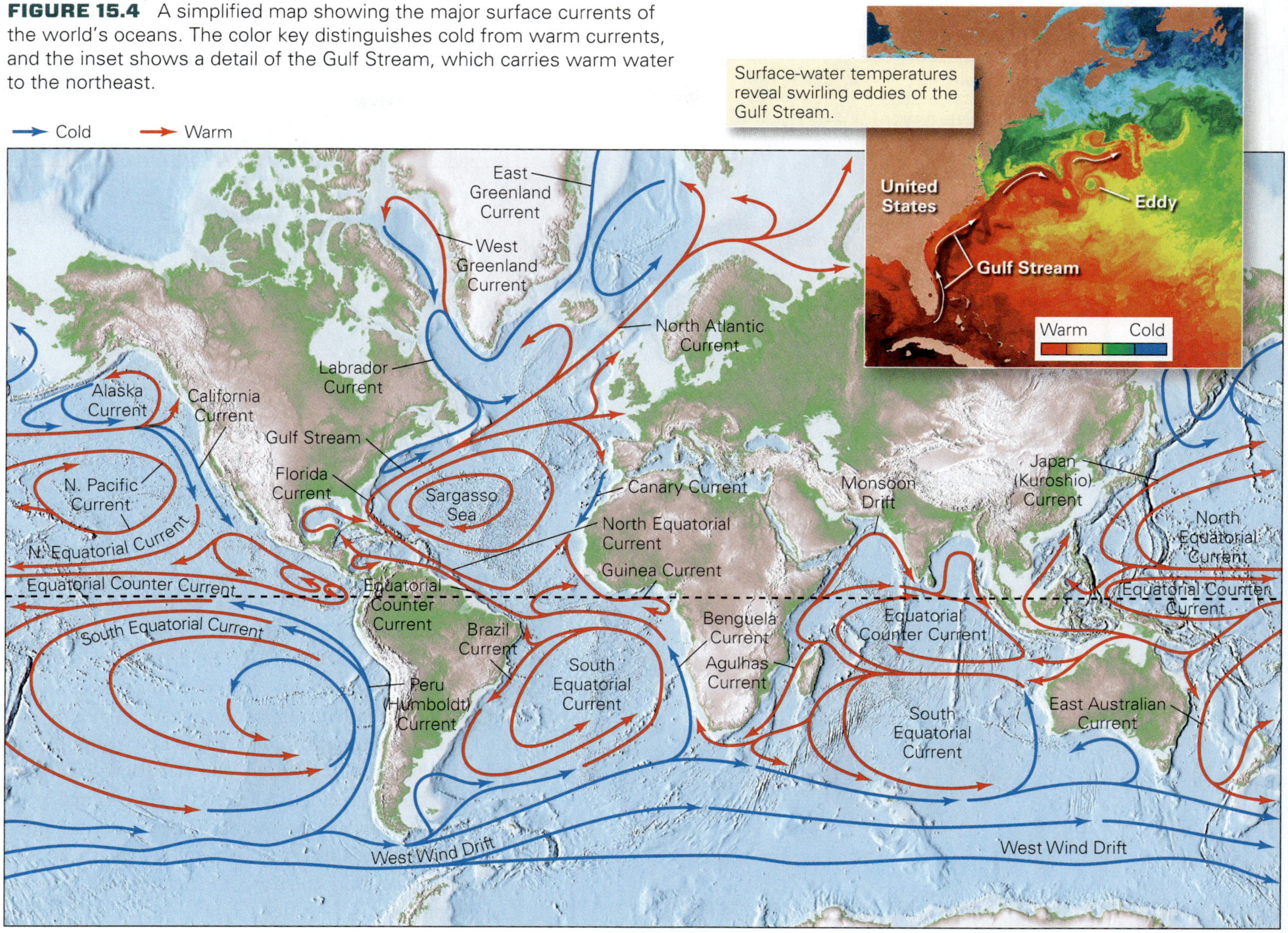

FIGURE 15.5 Animations of ocean currents, prepared by NASA based on data collected over a two-year period, show the complexities that develop where currents shear against the margins of the continents or are squeezed between land masses. The circular flows are eddies.

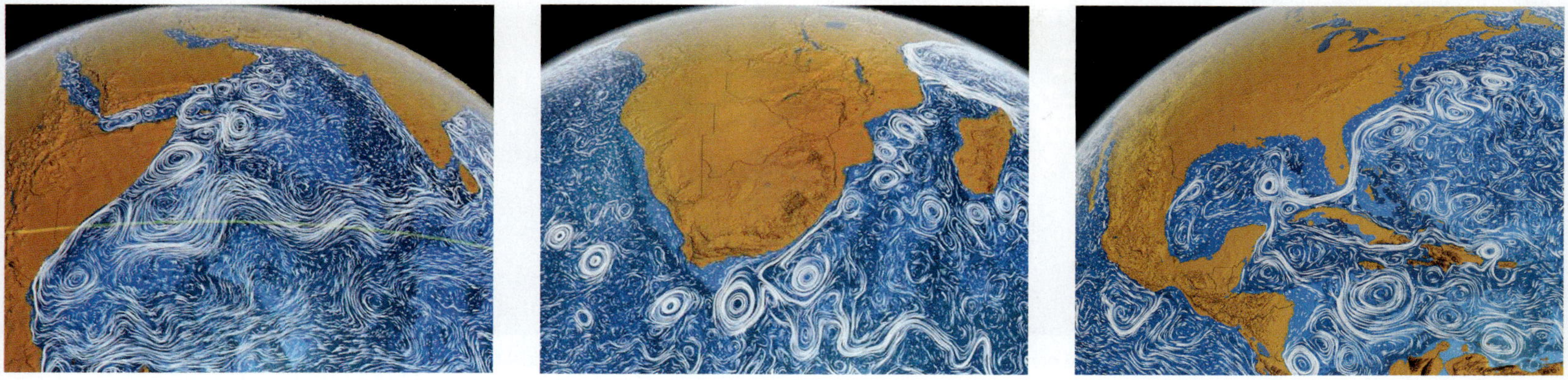

FIGURE 15.6 Global-scale upwelling and downwelling of ocean water.

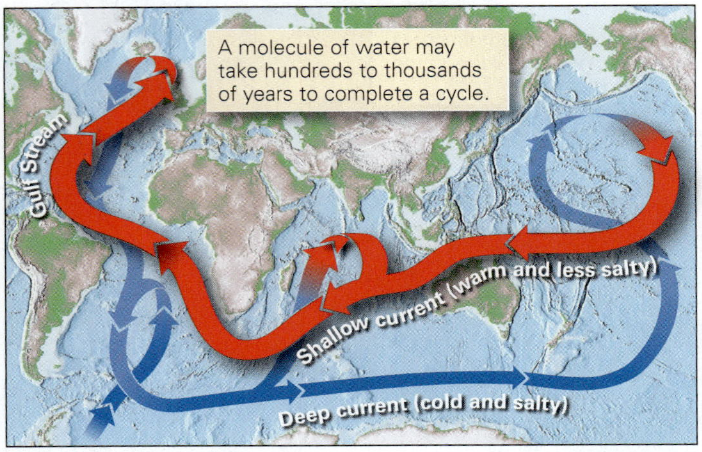

(a) Thermohaline circulation results in a global-scale conveyor belt that circulates water throughout the entire ocean system. The ocean mixes entirely in a 1,500-year period.

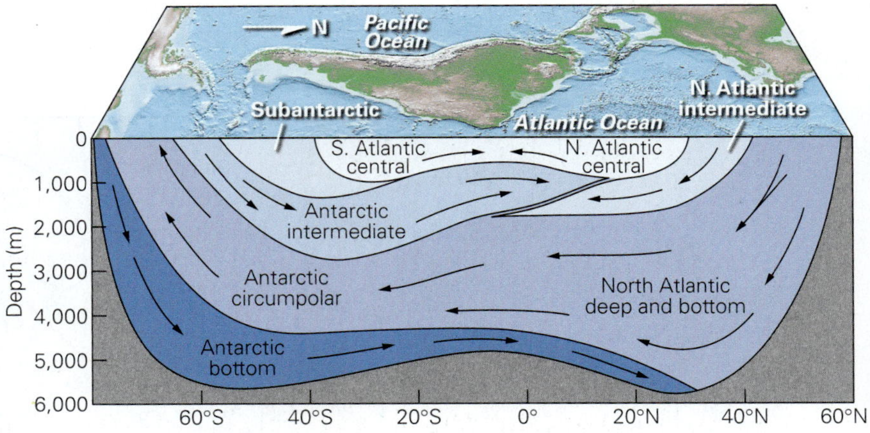

(b) Due to variations in density, the oceans are stratified into distinct water masses.

for the force due to the pull of the Moon's gravity is stronger than that of other contributors to the tide-generating force. When a location lies under a tidal bulge, it experiences a high tide, and when it underlies a depression between two bulges, it experiences a low tide. If the Earth's surface were smooth and completely submerged beneath the ocean, each point on the surface would experience two high tides and two low tides per day, as the Earth spins relative to the tidal bulges. But the complex shape of the seafloor and of the coastline means that the specific timing and magnitude of tides vary from place to place. For example, the **tidal reach**, the elevation difference between sea level at high tide and low tide, can range from less than a meter to several meters at different points along a coast (**Fig. 15.7b**). The largest tidal reach on Earth, 16.8 m (54.6 feet), affects the Bay of Fundy on the Atlantic coast of Canada.

FIGURE 15.7 Ocean tides and their manifestation.

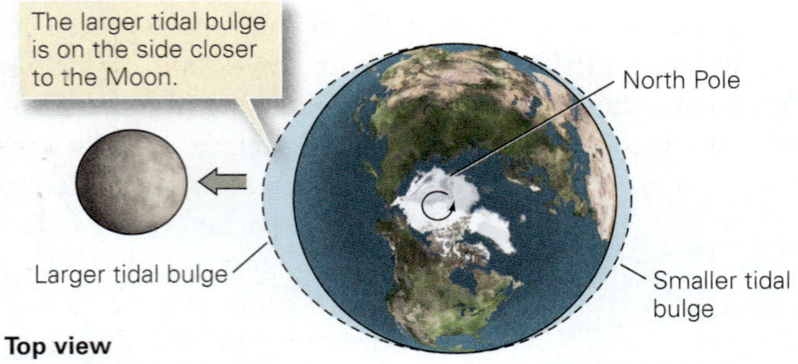

(a) Tides develop as the Earth spins relative to the two tidal bulges.

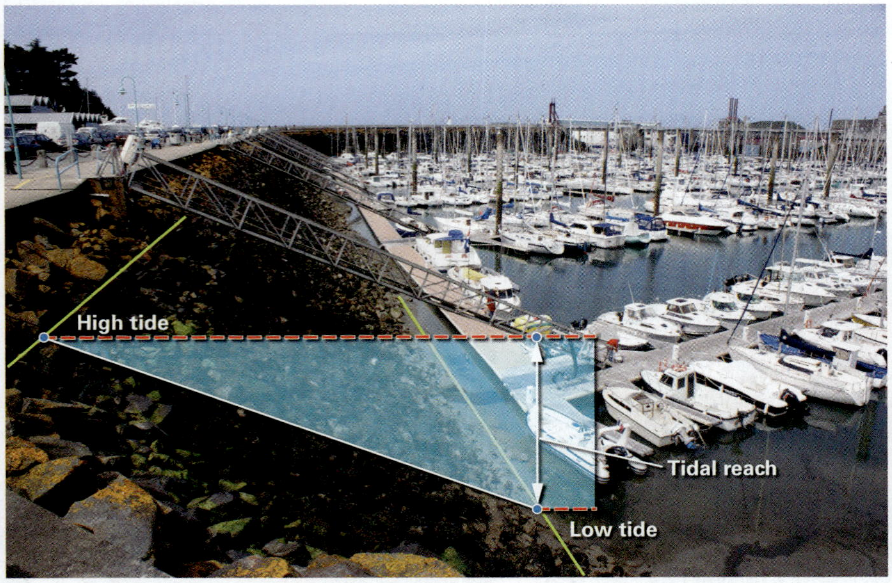

(b) The tidal reach is the vertical distance between low tide and high tide. This photo of a French harbor was taken at low tide.

(c) A broad tidal flat is exposed at low tide around Mont Saint-Michel, on the coast of France.

The manifestation of tides along a shore depends not only on the tidal reach, but on the slope of the shore. Along a vertical cliff, the *intertidal zone* (the region of shore submerged at high tide and exposed at low tide) appears as a ring of stained, barnacle- and seaweed-encrusted rock. Where the coast has a gentle slope, the shoreline moves substantially inland during a rising tide and seaward during a falling tide—in some places, the shoreline at high tide lies a few kilometers inland of the shoreline at low tide. In such locations, vast muddy **tidal flats** may be exposed during low tide (**Fig. 15.7c**).

Wave Action

Wind-driven waves make the ocean surface a restless, ever-changing vista. In technical terms, an ocean wave is a periodic undulation of the water surface—the low part of the wave is its trough, and the high part is its crest. The distance between adjacent crests (or adjacent troughs) is the *wavelength*, and the elevation difference between the trough and crest is the *wave height*. Larger waves move horizontally along the surface of the ocean at speeds of 30 to 50 km per hour (18 to 30 mph). This movement gives the visual impression that water itself is moving horizontally—in fact, it's not. As a wave passes, the molecules of water affected by the wave follow a somewhat circular path. The diameter of the circle is largest for molecules at the surface, where it equals the wave height of the wave. With increasing depth, though, the diameter of the circle decreases until, at a depth equal to about half the wavelength, there is no wave movement at all (**Fig. 15.8a**). Submarines traveling below this **wave base** cruise through smooth water, while ships toss about above. The movement of water in a wave is not a perfect circle, so after about 20 waves have passed, water affected by the wave will have moved a horizontal distance of about 1 wavelength. (This horizontal component of motion leads to the formation of surface currents.)

Ocean waves develop because of shear between air molecules in the wind and water molecules at the surface of the sea. The character of waves in the open ocean depends on the wind's strength (how fast the air moves) and *fetch* (the distance over which it blows). When the wind first begins to blow, it produces ripples in the water surface, small waves with sharp crests. With continued blowing over a long fetch, *swells*, larger waves with heights of 2 to 10 m and wavelengths of 40 to 500 m, form. Hurricane wave heights may reach over 35 m. Swells may travel for thousands of kilometers across the ocean, well beyond the region where they formed. The constructive interference of waves (meaning the additive effect that happens when two wave crests cross each other), the interaction of waves and currents, and the focusing of wave

FIGURE 15.8 Ocean waves build in response to the shear of wind blowing over the water surface.

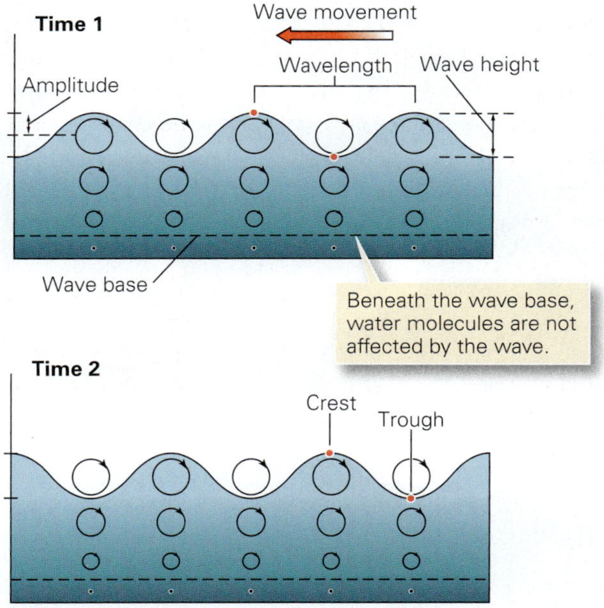

(a) Within a deep-ocean wave, water molecules follow a circular path. The radius of the circle decreases with depth. Note that the wave height is twice the amplitude.

(b) As the wave approaches shore, friction slows its base. Water motion in the wave becomes more elliptical, and the wave becomes a breaker.

energy due to the shape of the seafloor can lead to the growth of **rogue waves**. Such waves are two to five times the size of most of the large waves passing a locality in a given time interval, and can even swamp the decks of ocean liners.

Waves have no effect on the ocean floor, as long as the floor lies below the wave base. However, near the shore, where the wave base just touches the floor, it causes a slight back-and-forth motion of sediment. Closer to shore, as the water gets shallower, friction between the wave and the seafloor slows the deeper part of the wave, and the motion in the wave becomes more elliptical (**Fig. 15.8b**). Eventually, water at the top of the wave curves over the base, and the wave becomes a *breaker*, ready for surfers to ride. Breakers crash onto the shore in the surf zone, sending a surge of water up the beach. This upward surge, or **swash**, continues until friction brings motion to a halt. Then gravity draws the water back down the beach as **backwash**.

Did you ever wonder...
why the breakers that surfers love form near the shore?

Waves may make a large angle with the shore (the edge of the land) as they come in, but they bend as they approach, so that when they reach the shore, wave crests generally make less than a 5° angle with the shore (**Fig. 15.9**). To understand this phenomenon, called **wave refraction**, imagine a wave approaching so that its crest initially makes an angle of 45° with the shore. The end of the wave closer to the shore touches

FIGURE 15.9 Wave refraction and its consequences along the shore.

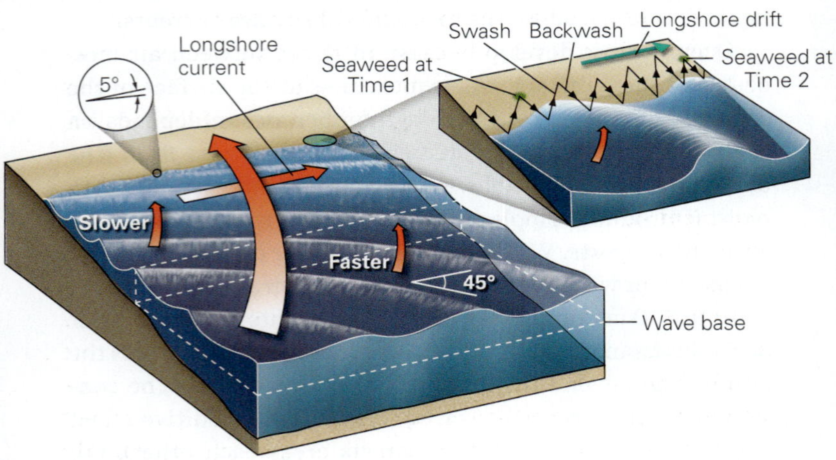

(a) Wave refraction occurs when waves approach the shore at an angle. If the wave reaches the beach at an angle, it causes a longshore current and beach drift of sand.

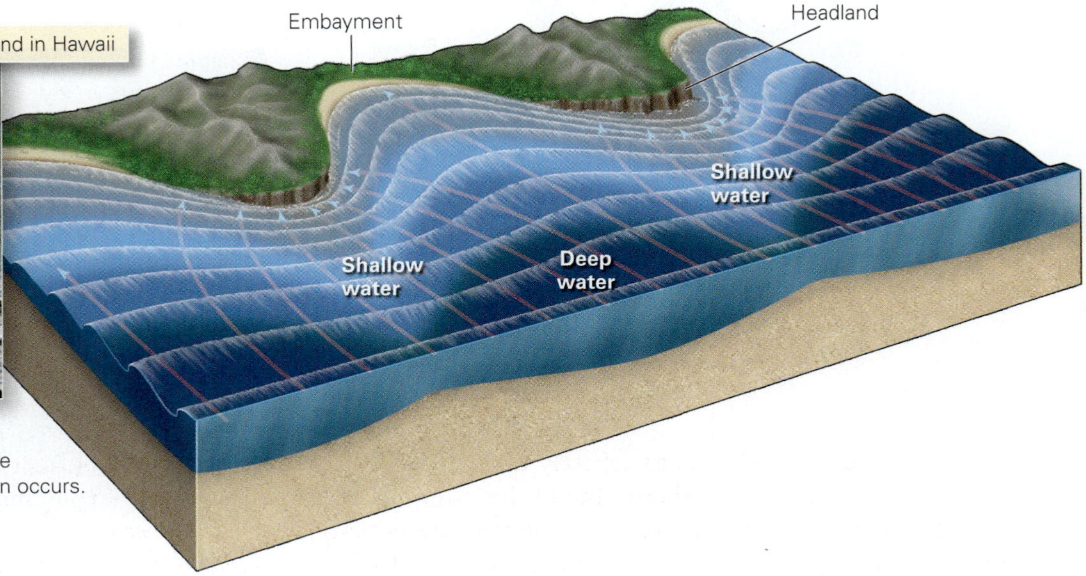

(b) Like a lens, wave refraction focuses wave energy on a headland, so erosion occurs. Wave energy weakens in embayments, so deposition occurs.

bottom first and slows down because of friction, whereas the end farther offshore continues to move at its original velocity, swinging the whole wave around so that it becomes more parallel with the shore. Although wave refraction decreases the angle at which waves move close to shore, it does not necessarily eliminate the angle. Where waves do arrive at the shore obliquely, water in the nearshore region has a component of motion that aligns roughly parallel to the shore. This **longshore current** causes swimmers floating in the water just offshore to drift gradually in a direction parallel to the shore. And where waves roll onto the shore at an angle, sediment in the surf follows a sawtooth pattern of movement that results in a gradual net transport of beach sediment parallel to the shore. Geologists refer to this movement of sediment as **longshore drift**. The sawtooth pattern happens because the swash of a wave moves perpendicular to the wave crest, so an oblique wave carries sediment diagonally onto the shore, whereas the backwash must flow straight down the slope of the shore, due to gravity.

Waves pile water up on the shore incessantly. As excess water moves back to the sea, it may localize into a strong seaward flow, called a *rip current*, perpendicular to the shore. Rip currents cause many people to drown every year, because they can carry unsuspecting swimmers far away from land.

TAKE-HOME MESSAGE

Ocean salinity and temperature vary with depth and location, and ocean water is in constant motion. Wind drives surface currents, upwelling and downwelling zones develop, and thermohaline circulation takes place. The tide-generating force produces two tidal bulges beneath which the Earth spins, so locations generally experience the twice-daily rise and fall of tides, which alternately submerge and expose the intertidal zone. Wind also causes waves to form. Within a wave, water moves approximately in a circle—the amount of motion decreases with depth. Near shore, water piles up into breakers, and the orientation of waves changes.

QUICK QUESTION Why do longshore currents and longshore drift develop?

15.4 Where Land Meets Sea: Coastal Landforms

Tourists along the Amalfi Coast of Italy thrill to the sound of waves crashing on rocky shores. But in the Virgin Islands, sunbathers can find seemingly endless white sand beaches, and along the Mississippi delta, vast swamps border the sea. Large, dome-like mountains rise directly from the sea in Rio de Janeiro, Brazil, but a 100-m-high vertical cliff marks the boundary between the Nullarbor Plain of southern Australia and the Great Southern Ocean. As these examples illustrate, coasts, the belts of land bordering the sea, vary dramatically in terms of topography and associated landforms. Below, we introduce some of the components of coastal landscapes, and explain how they can change over time.

Beaches

For millions of vacationers, the ideal holiday includes a trip to a **beach**, a gently sloping fringe of sediment along the shore. Some beaches consist entirely of sand, others are made up of pebbles or cobbles, and still others have a mixture of sand, pebbles, and cobbles (**Fig. 15.10a, b**). The lack of very fine sediment on beaches is no accident—waves pick up and stir beach sediment, and over time winnow out silt and clay grains. These finer grains get carried to quieter water, where they settle. Cobble beaches exist primarily where new blocks fall from nearby cliffs onto the beach—waves smash the blocks against one another with enough force to round their edges, and may even shatter them into smaller pieces. A beach with a mixture of sand, pebbles, and cobbles typically forms from a sediment supply that had originally been carried to the coast by rivers or glaciers.

Sand, once formed, will not break down into smaller clasts even in a storm wave, for sand grains can't strike each other with enough force to cause fracturing. The composition of sand varies from beach to beach, because different sands come from different sources. Sand derived from the weathering and erosion of felsic to intermediate rocks consist mainly of quartz, for other minerals in these rocks chemically weather to form clay, which washes away. Beaches made from the erosion of limestone, or by the breakup of coral reefs and shells, consist of carbonate sand. And beaches derived from the erosion of basalt boast black sand, made of tiny basalt grains.

Did you ever wonder...
why beautiful sandy beaches don't form along all coasts?

A **beach profile**, a cross section drawn perpendicular to the shore, illustrates the shape of a beach (**Fig. 15.10c**). Starting from the sea and moving landward, a beach consists of several zones: a *foreshore zone* or *intertidal zone*, across which the tide rises and falls; the *beach face*, a steeper, concave-up part of the foreshore zone, where the swash of the waves actively scours the sand; and the *backshore zone*, which extends from a small step cut by high-tide swash to the front of dunes or cliffs that may lie farther inshore. The backshore zone includes one or more *berms*, horizontal to landward-sloping terraces that receive sediment only during a storm.

FIGURE 15.10 Characteristics of beaches, barrier islands, and tidal flats.

(a) A gravel beach along the Olympic Peninsula, Washington. The clasts were derived from erosion of adjacent cliffs.

(b) A sand beach on the western coast of Puerto Rico. Wave action has carried away finer sediment.

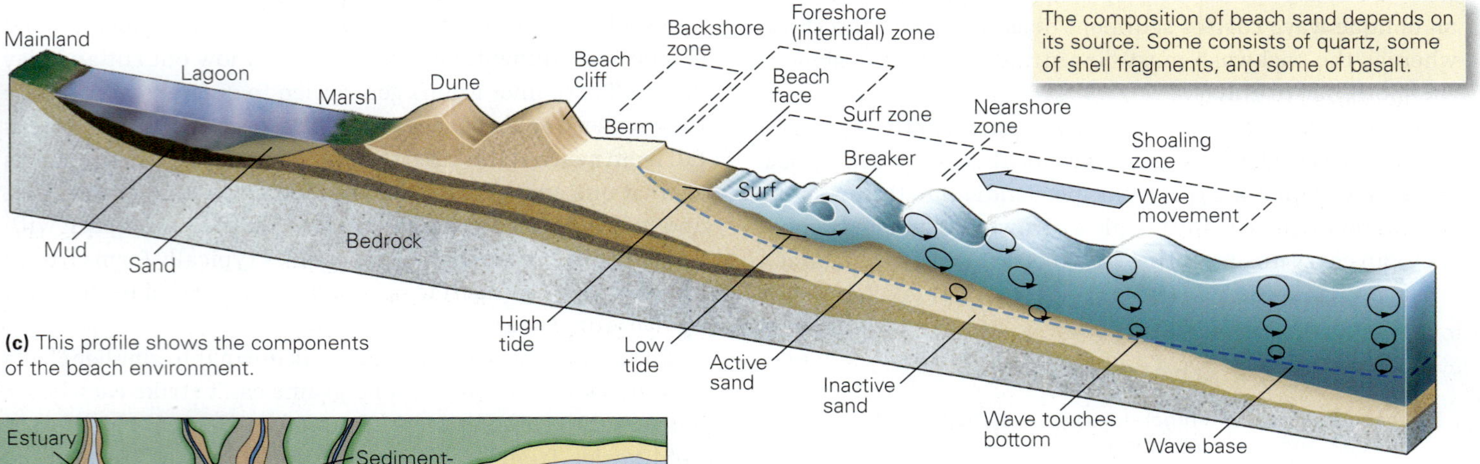

(c) This profile shows the components of the beach environment.

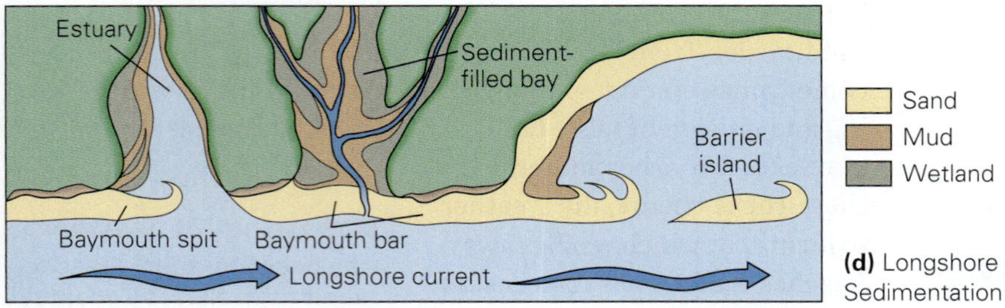

(d) Longshore drift can generate sand spits and baymouth bars. Sedimentation eventually fills in the region behind a baymouth bar.

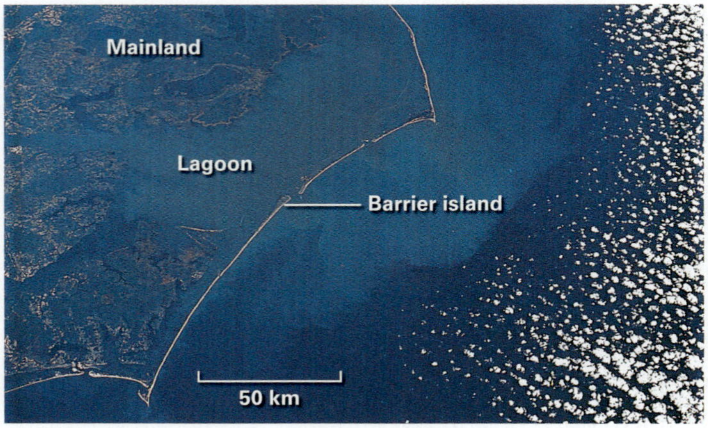

(e) A satellite image showing barrier islands off the coast of North Carolina.

(f) At low tide, boats rest in the mud of a tidal flat on the coast of Brittany.

When waves approach the shore, they stir the sand beginning at the depth where the wave base intersects the seafloor and continuing up the beach face to the limit of the swash. This *active sand layer,* which moves with every incoming wave, overlies inactive sand. In relatively calm weather, the active sand layer may be only a few centimeters deep, but during a large storm, sand down to a depth of a few meters may be set in motion. If the wave crests trend parallel to the shore, the active sand layer only moves back and forth, in a direction perpendicular to the shoreline. But if waves strike the shore at an angle, as we noted earlier, longshore drift transports active sand parallel to the shoreline. Over the course of centuries, sand may move tens to hundreds of kilometers along a coast. For this reason, geologists sometimes refer to beaches as "rivers of sand." Where the coastline indents landward, longshore drift can extend a beach out into open water, and produce a **sand spit**. Some sand spits grow across the opening of a bay, to form a *baymouth bar* (**Fig. 15.10d**).

The component of wave motion perpendicular to the shore sometimes piles sand up in a narrow ridge, called an *offshore bar,* at a distance from the shore. In regions with an abundant sand supply, offshore bars may rise above sea level and become **barrier islands** (**Fig. 15.10e**). Waves lose energy as they wash up onto a barrier island, so the water between a barrier island and the mainland becomes a quiet-water **lagoon**, a body of shallow seawater separated from the open ocean. Though developers have covered some barrier islands with expensive resorts, in the time frame of centuries to millennia, barrier islands are only temporary features that may wash away in a large storm. Tidal flats, as we have seen, are expanses of gooey clay and silt that emerge at low tide but become submerged at high tide; these may develop in lagoons because the lack of wave action (**Fig. 15.10f**).

Rocky Coasts

Many a ship has met its end smashed and splintered in the spray and thunderous surf of a rocky coast, where bedrock cliffs rise directly from the sea. Lacking the protection of a beach, rocky coasts endure the full force of ocean breakers. Water pressure generated during the impact of a breaker can pick up boulders and smash them together until they shatter, and it can squeeze air into cracks, creating enough force to split fragments off of bedrock cliffs. Further, because of its turbulence, the water hitting a cliff face carries suspended sand and thus can abrade the cliff. The combined effects of shattering, wedging, and abrading, together called *wave erosion,* can gradually undercut a cliff face and make a *wave-cut notch* (**Fig. 15.11a**). Undercutting continues until the overhang becomes unstable, breaks away at a joint, and tumbles down. The resulting pile of rubble at the base of the cliff may not last long, for waves immediately attack and break it up. As wave erosion cuts away at a rocky coast, the position of the cliff face gradually migrates inland. Such cliff retreat may leave behind a **wave-cut bench**, that becomes visible at low tide (**Fig. 15.11b**).

Many rocky coasts are irregular, with *headlands* protruding into the sea and *embayments* set back from the sea. Wave refraction focuses wave energy on headlands, which accelerates their erosion, whereas it disperses energy in embayments and allows them to be sites of deposition (**Fig. 15.11c**). In some cases, a headland erodes in stages (**Fig. 15.11d**). As waves curve, they attack the sides of a headland and slowly erode through it to produce a *sea arch* connected to the mainland by a narrow bridge. When the arch eventually collapses, an isolated **sea stack** remains offshore.

Estuaries and Fjords

Along some coastlines, a relative rise in sea level causes the sea to flood a river valley where it merges with the coast and forms an **estuary**, a relatively narrow body of water in which seawater and river water interact. If an estuary fills the mouth of a dendritic drainage network, the estuary will have a complex coastline with many finger-like inlets (**Fig. 15.12**). Oceanic and fluvial waters interact in two ways within an estuary. In quiet estuaries, protected from wave action or river turbulence, the water becomes stratified, with denser oceanic saltwater flowing upstream as a wedge beneath less-dense fluvial freshwater. In turbulent estuaries, oceanic and fluvial water mix to produce nutrient-rich brackish water with a salinity between that of oceans and rivers. Estuaries are complex ecosystems inhabited by unique species of shrimp, clams, oysters, worms, and fish that can tolerate large changes in salinity.

During the last ice age, glaciers carved deep valleys in coastal mountain ranges. When the ice age came to a close, the glaciers melted away, leaving deep, steep-sided valleys. Water stored in the glaciers, along with the water within the vast ice

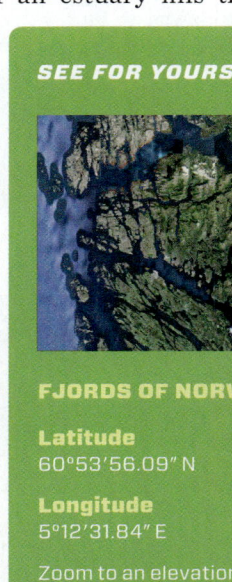

SEE FOR YOURSELF...

FJORDS OF NORWAY

Latitude
60°53′56.09″ N

Longitude
5°12′31.84″ E

Zoom to an elevation of 80 km (50 miles) and look down.

During the last ice age, glaciers carved deep, steep-sided valleys into the mountains of western Norway. When the glaciers melted and sea level rose, the valleys filled with water, forming fjords.

FIGURE 15.11 Erosion landforms of rocky shorelines.

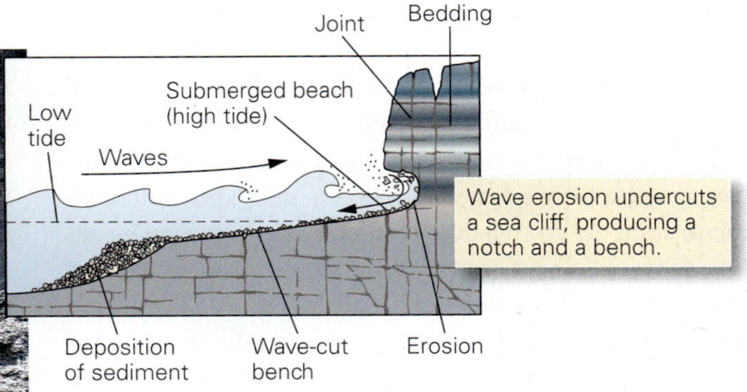

(a) A wave-cut notch.

(b) A wave-cut bench at the foot of the cliffs at Étretat, France.

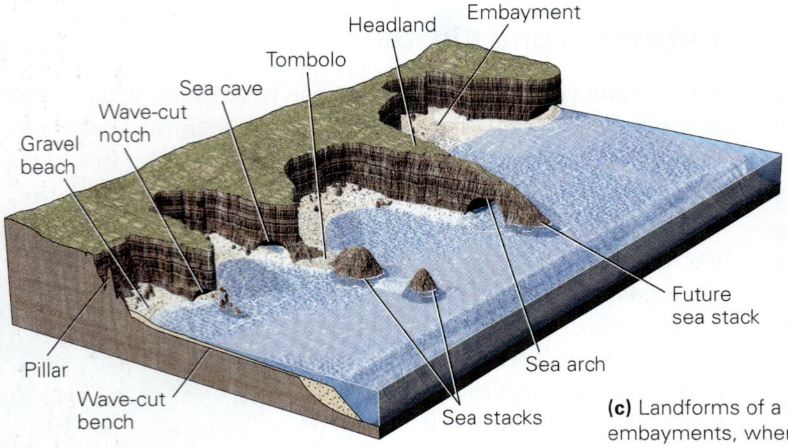

(c) Landforms of a rocky shore. Beaches collect in embayments, whereas erosion concentrates at headlands.

(d) Coastal erosion along Australia's southern coast produced a sea arch (left). Eventually, the bridge will collapse, and only sea stacks will remain (right). These sea stacks are among several that together are known locally as the Twelve Apostles.

sheets that covered continents during the ice age, returned to the sea and caused sea level to rise. The rising sea filled the deep valleys, creating **fjords**, or flooded glacial valleys. Coastal fjords are fingers of the sea surrounded by mountains—because of their deep-blue water and steep walls of polished rock, they boast distinctive beauty (**Fig. 15.13**).

Organic Coasts

Coasts along which living organisms control landforms are called **organic coasts**. The nature of an organic coast depends on the type of organisms that live there, which, in turn, depends on climate and water clarity.

FIGURE 15.12 The Chesapeake Bay estuary formed when the sea flooded river valleys. The region is sinking relative to other coast areas because it overlies a buried meteor crater.

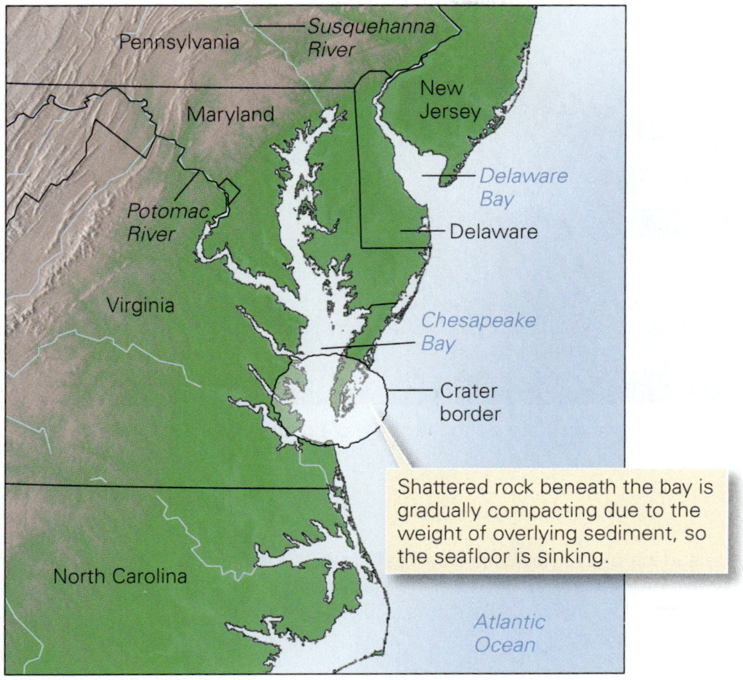

Coastal Wetlands A **coastal wetland** is a vegetated, flat-lying stretch of coast that floods at high tide but generally does not feel the impact of strong waves. In temperate climates, coastal wetlands include *swamps* (wetlands dominated by trees), *marshes* (wetlands dominated by grasses; **Fig. 15.14a**), and *bogs* (wetlands dominated by moss and shrubs). Despite their relatively small area, when compared with the oceans as a whole, wetlands account for 10% to 30% of marine organic productivity.

In tropical or semitropical climates (between 30° north and 30° south of the equator), *mangrove swamps* thrive in wetlands (**Fig. 15.14b**). Mangrove tree roots can filter salt out of water, so the trees have evolved to survive in seawater. Some mangrove species form a broad network of roots above the water surface, making the plant look like an octopus standing on its tentacles, and some send up small protrusions from roots that rise above the water and allow the plant to breathe.

SEE FOR YOURSELF...

ORGANIC COAST, FLORIDA

Latitude
25°9′16.51″ N

Longitude
80°41′33.53″ W

Zoom to 8 km (5 miles) and look down.

Along Florida's southern coast, thickets of mangroves line the shore and separate swamps from the sea. The roots of these trees protect the coast from wave erosion.

Coral Reefs Along the azure coasts of Hawaii, scuba divers can swim through colorful growths of living coral. Some corals look like brains, others like elk antlers, still others like delicate fans (**Fig. 15.15a**). Sea anemones, sponges, and clams grow on and around the coral. Though at first glance coral looks like a plant, it's actually a colony of tiny invertebrates. An individual coral animal, or polyp, has a tube-like body with a head of tentacles. Corals obtain part of their livelihood by filtering nutrients out of seawater—the remainder comes from

FIGURE 15.13 Fjord landscapes form where relative sea-level rise drowns glacially carved valleys.

FIGURE 15.14 Examples of coastal wetlands.

(a) A salt marsh along the coast of Cape Cod.

(b) A mangrove growing along the shore of southern Florida.

FIGURE 15.15 The character and evolution of coral reefs.

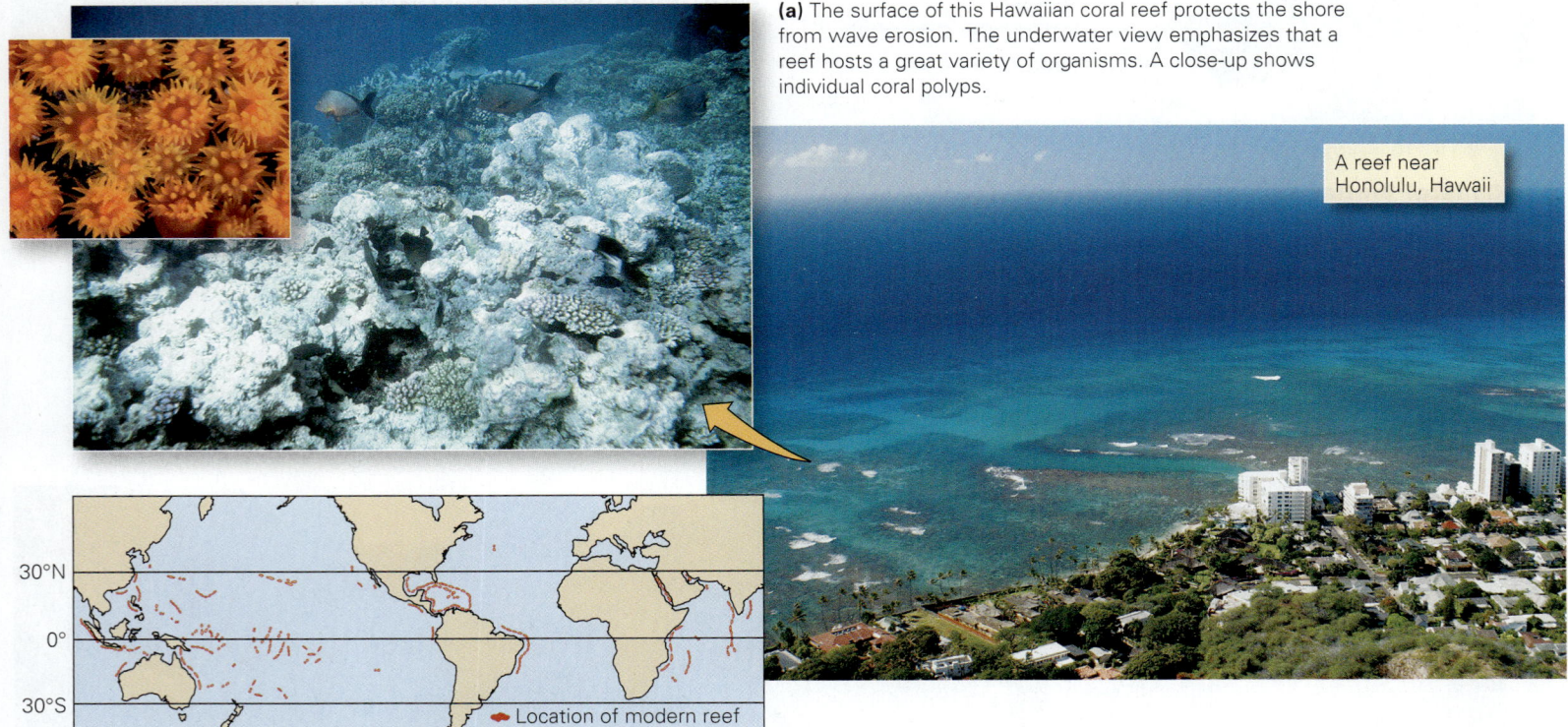

(a) The surface of this Hawaiian coral reef protects the shore from wave erosion. The underwater view emphasizes that a reef hosts a great variety of organisms. A close-up shows individual coral polyps.

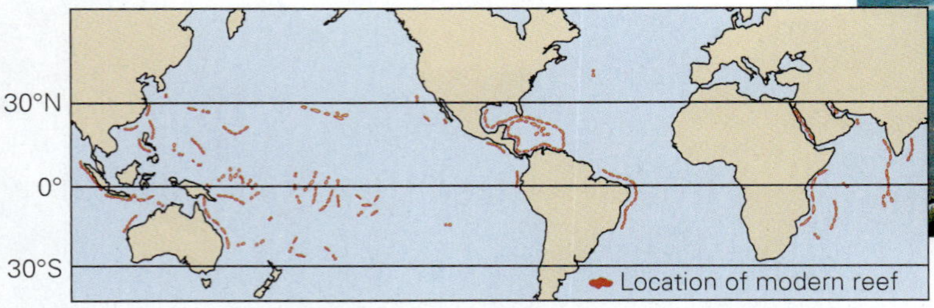

(b) The distribution of warm-water coral reefs on Earth today.

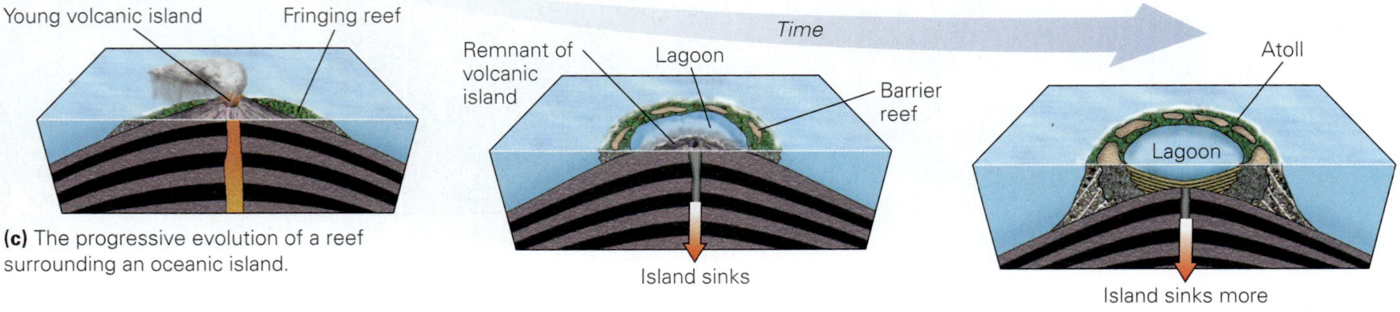

(c) The progressive evolution of a reef surrounding an oceanic island.

SEE FOR YOURSELF...

CORAL REEFS, PACIFIC

Latitude
16°47′26.92″ S

Longitude
150°58′1.27″ W

Zoom to 20 km (~12.5 miles) and look down.

Waves break on the outer edge of the fringing reef of Huahine. Coral buildups lie beneath the surface, and sand partially fills the lagoon. The island itself represents the subsiding remnants of a hot-spot volcano.

algae that live on the corals' tissue. Corals have a symbiotic (mutually beneficial) relationship with the algae in that the algae photosynthesize and provide nutrients and oxygen to the corals while the corals provide carbon dioxide and nutrients to the algae.

Coral polyps secrete calcite shells, which gradually build into a mound of solid limestone whose top surface can extend from just below the low-tide level down to a depth of about 60 m. At any given time, only the surface of the mound is alive—the mound's interior consists of shells from previous generations of now-dead coral. The realm of shallow water underlain by coral mounds, associated organisms, and debris comprises a **coral reef**. Reefs absorb wave energy and thus serve as a living buffer zone that protects coasts from erosion. To thrive, coral reefs need clear, well-lit, warm (18° to 30°C) water with normal oceanic salinity, as can exist along coasts at latitudes of 0° to 30° (**Fig. 15.15b**).

Marine geologists distinguish three different kinds of coral reefs (**Fig. 15.15c**). A *fringing reef* forms directly along a coast, a *barrier reef* develops offshore, and an *atoll* makes a circular ring surrounding a lagoon. As Charles Darwin first recognized back in 1859, coral reefs associated with islands in the Pacific start out as fringing reefs and then later become barrier reefs and finally atolls. This progression reflects the continued growth of the reef as the island around which it formed gradually sinks. Eventually, the reef itself sinks too far below sea level to remain alive and becomes the cap of a flat-topped seamount known as a *guyot*.

TAKE-HOME MESSAGE

A great variety of landscapes can form along coasts. Beaches form from wave-washed sediment, rocky coasts evolve due to wave erosion, fjords are submerged glacial valleys, and estuaries are submerged river valleys. Some coasts host wetlands and reefs.

QUICK QUESTION Why do geologists sometimes refer to beaches as "rivers of sand?"

15.5 Causes of Coastal Variability

Why do we see so much variety in the character of coasts? There's no single cause. Rather, the landscape of a coast at any given locality reflects input from many components of the Earth System. Let's look at the most important factors that control the variability of coastal landscapes.

Relative Sea-Level Changes

Sea level, relative to the land surface, changes during geologic time. Some sea-level changes affect the ocean worldwide. Such *eustatic sea-level changes* may reflect global variations in seafloor spreading rates, for the volume of a ridge depends on spreading rate (see Chapter 2). Changes may also reflect growth or shrinkage of ice-age glaciers, which store water on land and keep it out of the sea (**Fig. 15.16**). Not all sea-level change, however, happens globally. In some cases, local

FIGURE 15.16 Because of sea-level drop during the ice age, there was more dry land exposed along eastern North America.

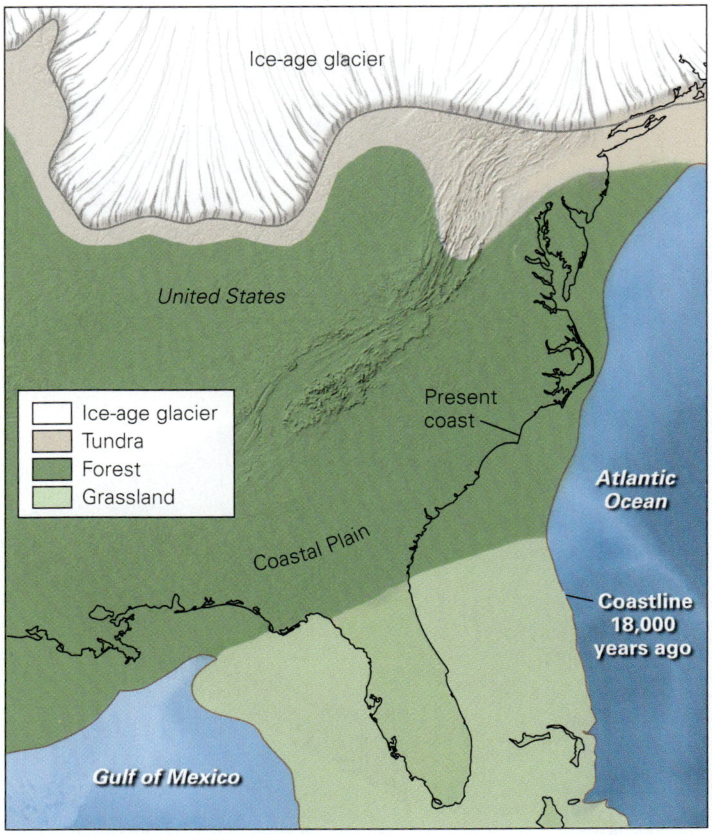

changes reflect the vertical movement of the land—where land rises, relative sea level falls locally, and where land sinks, relative sea level rises locally. Such movement, in turn, may reflect plate-tectonic processes, such as mountain uplift or basin subsidence, or they may be a consequence of the addition or removal of huge weights (glaciers or volcanoes) on the surface of the crust. In the past two centuries, local changes in sea level have also resulted from human activity. For example, where people pump groundwater or oil out of the sediment beneath a delta plain, the sediment packs together more tightly, and thus takes up less volume, so the land surface sinks.

Geologists refer to coasts where the land is currently rising or rose in the past, relative to sea level, as *emergent coasts*. At emergent coasts, steep slopes typically border the shore, and in some locations, a series of step-like terraces form (**Fig. 15.17a**). These terraces reflect episodic changes in relative sea level and/or ground uplift. Coasts where the land has been sinking relative to sea level are *submergent coasts* (**Fig. 15.17b**). At such places, landforms include estuaries and fjords that develop when the rising sea floods coastal valleys.

Plate Tectonic Setting

The tectonic setting of a coast also plays a role in determining whether steep-sided mountain slopes or broad plains border the sea at a given location (see **Geology at a Glance**, pp. 512–513). For example, convergence of plates along the edge of a continent may subject the continent to compression, causing a mountain belt to uplift right along the coast, which may lead

FIGURE 15.17 Features of emergent coastlines (relative sea level is falling) and submergent coastlines (relative sea level is rising).

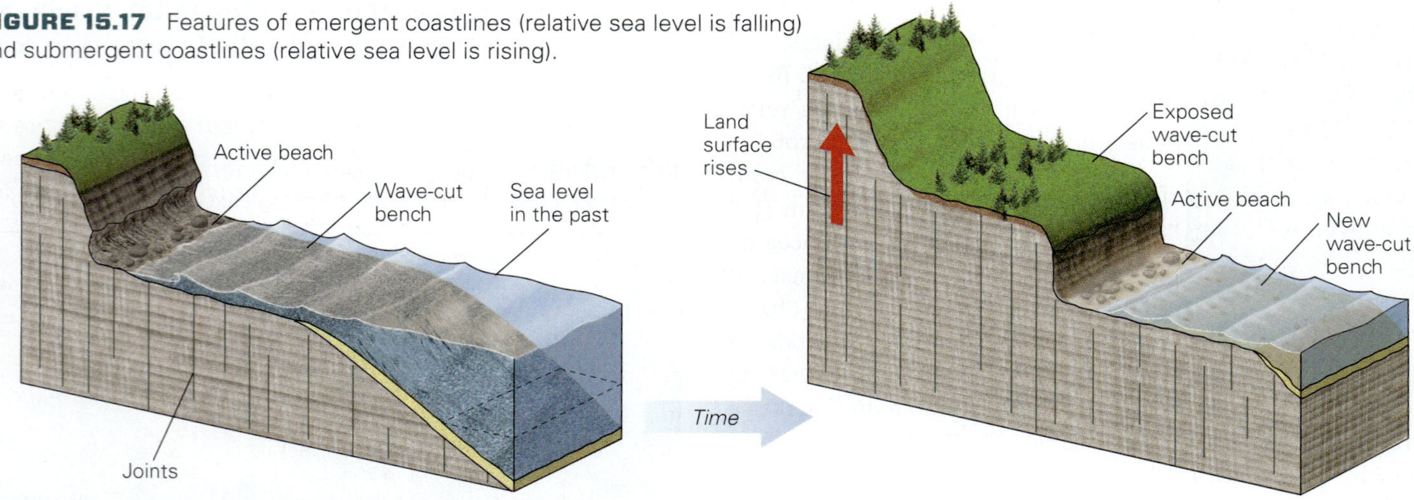

(a) *Emergent coasts*: Wave erosion produces a wave-cut bench along an emergent coast. As the land rises, the bench becomes a terrace, and a new wave-cut bench forms.

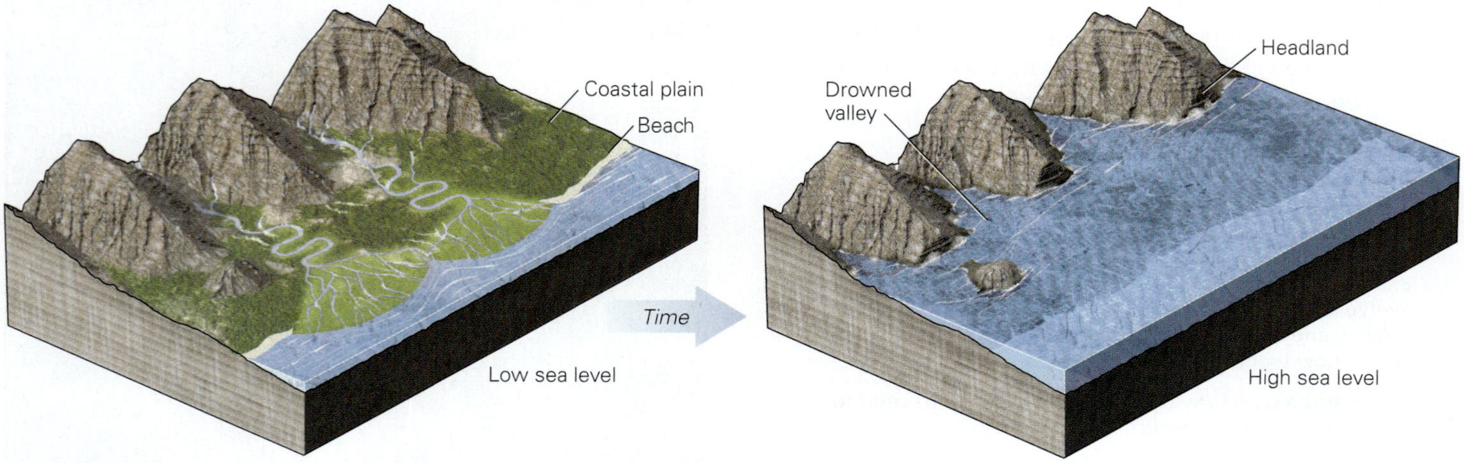

(b) *Submergent coasts*: A coast before sea level rises. Rivers drain valleys and deposit sediment on a coastal plain. As a submergent coast forms, sea level rises and floods the valleys, and waves erode the headlands.

to development of an emergent coast. Where slow subsidence of the lithosphere happens along a coast, a broad *coastal plain*, flat land with very little relief, may develop. Coastal plains merge seamlessly with the continental shelf, and portions will become submergent coasts due to sea-level rise. The Gulf Coast and southeastern Atlantic Coast of the United States host a 200-km-wide coastal plain, formed where the inland edge of the passive-margin basin lies above sea level. Significantly, not all passive margins host coastal plains. For example, highlands that uplifted during or after the rifting that formed the passive margins of the Red Sea, southeastern Brazil, South Africa, and eastern Australia still remain today, with steep escarpments rimming these coastlines.

Sediment Supply and Climate

At *erosional coastlines*, waves wash sediment away faster than it can be supplied, so the coast becomes rocky, whereas at *accretionary coastlines*, more sediment arrives than washes away, so sediment builds outward into a broad beach or even a coastal plain. Whether a local stretch of coastline becomes erosional or accretionary depends on sediment supply and climate. Specifically, for given climate conditions, a coast may become erosional if the sediment supply decreases but may become accretionary if the sediment supply increases. And for a given sediment supply, a coast can become erosional if subjected to frequent storms but accretionary if it enjoys calm weather. Climate also influences rates of erosion by affecting the productivity of organisms. For example, flourishing mangrove swamps and coral reefs of tropical regions can protect coasts from erosion by absorbing wave energy. Lacking such protections, arctic shorelines feel the full brunt of wave action.

TAKE-HOME MESSAGE

Many aspects of the Earth System influence the character of a given coastline. Examples include tectonic activity, sediment supply, and climate. At an emergent coast, land rises relative to sea level, whereas at a submergent coast, it sinks. Relative sea level at a location can rise or fall over time either because of changing mid-ocean ridge volumes, glacial volumes, or tectonic uplift and subsidence. Sediment supply, relative to erosion rates, also impacts the character of a coast by determining whether beaches build outward, or erode away.

QUICK QUESTION What landforms occur at submergent coasts?

15.6 Coastal Problems and Solutions

Contemporary Sea-Level Changes

People tend to view a shoreline as a permanent entity. But, in fact, shorelines are ephemeral geologic features. On a time scale of hundreds to thousands of years, a shoreline moves inland or seaward depending on whether relative sea level rises or falls, or whether sediment supply increases or decreases. In places where sea level is rising today, shoreline towns will eventually be submerged. For example, if present rates of sea-level rise continue along the east coast of the United States, major coastal cities such as Washington, New York, Miami, and Philadelphia may be inundated within the next millennium (**Fig. 15.18**).

Beach Destruction—Beach Protection?

The loss of river sediment due to installation of a dam upstream, or a change in the shape of a shoreline, a change in wave height, or the destruction of coastal vegetation, can all alter the balance between sediment accumulation and sediment removal on a beach. **Beach erosion**, where a beach becomes narrower, happens when sediment removal rates exceed sediment accumulation rates, often due to storms. In only a few hours, a storm can radically alter a landscape that

FIGURE 15.18 Areas that may be submerged if sea level rises by about 1.0 to 1.5 m.

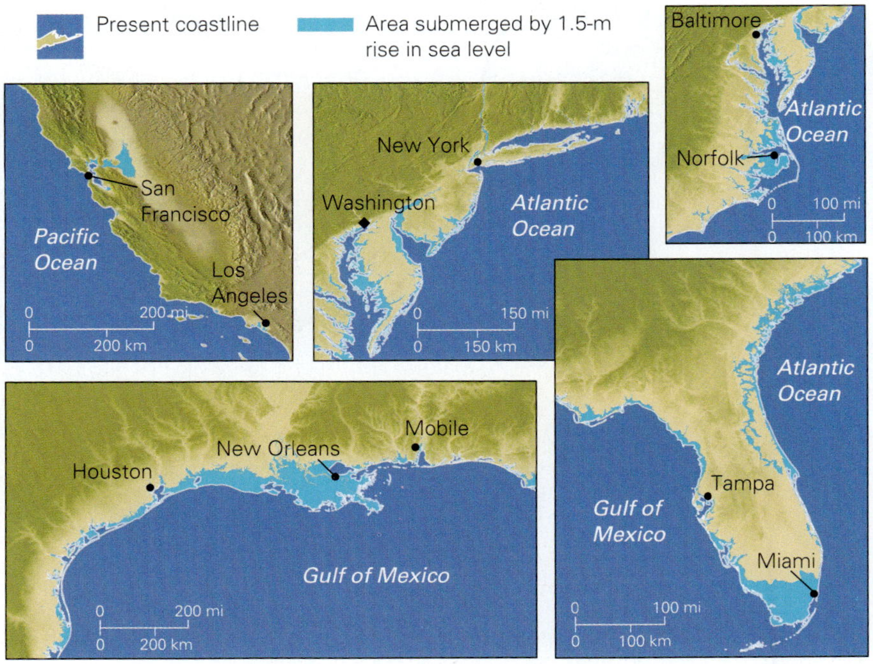

Oceans and Coasts

The oceans of the world host a diverse array of environments and landscapes, illustrating the complexity of the Earth System. Tectonic processes and surface processes, working alone or in tandem, generate unique features beneath the sea and along its coasts.

The major structures of the ocean floor reflect plate-tectonic activity. For example, mid-ocean ridges define divergent boundaries, fracture zones initially form along transform faults, and trenches mark subduction zones. Oceanic islands, seamounts, and plateaus build above hot spots. Along passive continental margins, broad continental shelves develop, locally incised by submarine canyons. Trenches delineate convergent boundaries.

Within the ocean, currents circulate water due to regional variations in both water salinity and temperature (factors that control density) and to the traction between wind and the water surface. Tides cause sea level to rise and fall, and wind builds waves that churn the sea surface, erode shorelines, and transport sediment.

Coastal landscapes reflect variations in sediment supply, relative sea-level rise or fall, and climate. For example, where the supply of sediment is low and the landscape is rising relative to sea level, rocky shores with dramatic cliffs and sea stacks may evolve. Where sediment is abundant, sandy beaches and bars develop. Regions where glaciers carved deep valleys now feature spectacular fjords. Protected coastal areas, especially those in warm climates, host unique coastal ecosystems. Corals may contribute to growth of broad reefs along the shore.

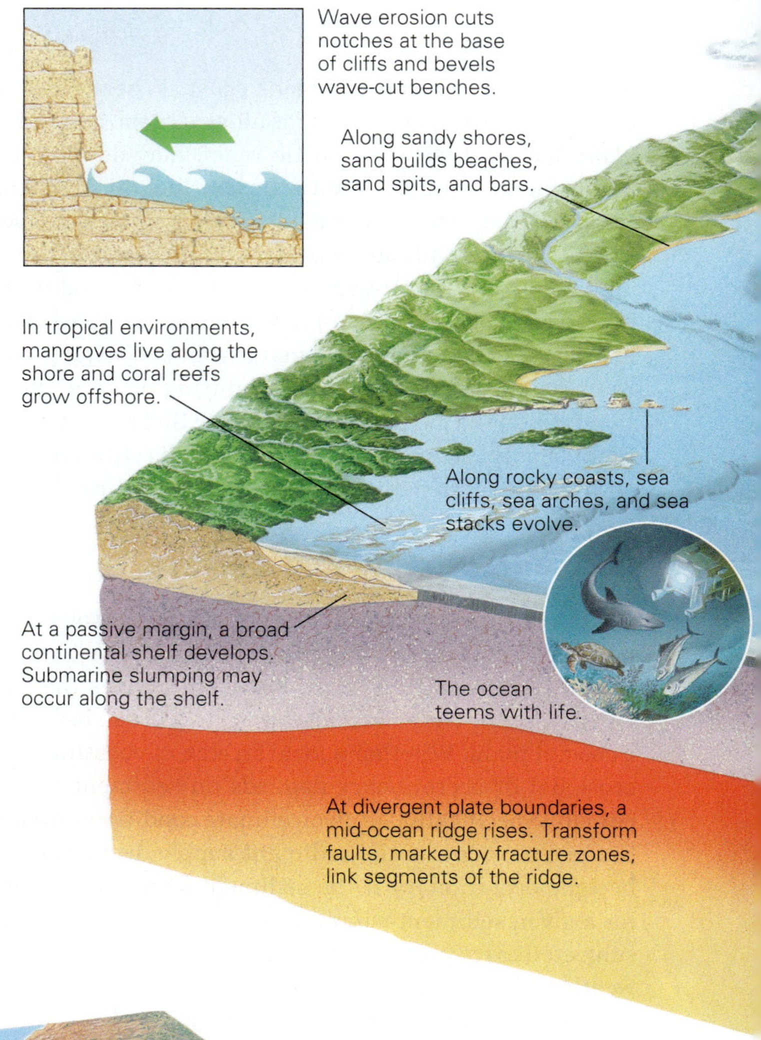

Wave erosion cuts notches at the base of cliffs and bevels wave-cut benches.
Along sandy shores, sand builds beaches, sand spits, and bars.
In tropical environments, mangroves live along the shore and coral reefs grow offshore.
Along rocky coasts, sea cliffs, sea arches, and sea stacks evolve.
At a passive margin, a broad continental shelf develops. Submarine slumping may occur along the shelf.
The ocean teems with life.
At divergent plate boundaries, a mid-ocean ridge rises. Transform faults, marked by fracture zones, link segments of the ridge.

The Global Conveyor

Surface winds drive surface currents in large gyres.

Cold water sinks at polar regions.

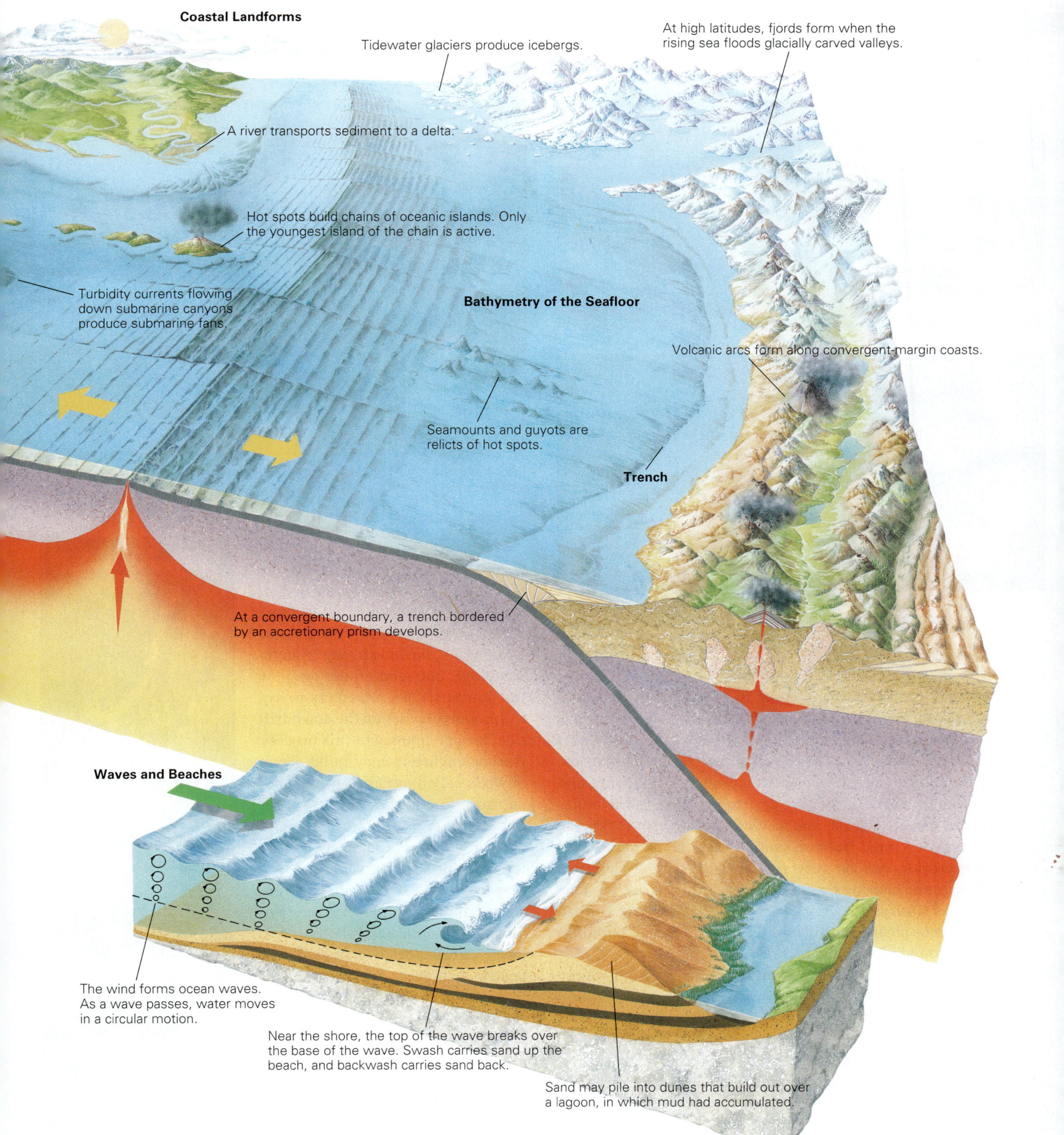
Coastal Landforms
Tidewater glaciers produce icebergs.
At high latitudes, fjords form when the rising sea floods glacially carved valleys.
A river transports sediment to a delta.
Hot spots build chains of oceanic islands. Only the youngest island of the chain is active.
Turbidity currents flowing down submarine canyons produce submarine fans.
Bathymetry of the Seafloor
Volcanic arcs form along convergent-margin coasts.
Seamounts and guyots are relicts of hot spots.
Trench
At a convergent boundary, a trench bordered by an accretionary prism develops.
Waves and Beaches
The wind forms ocean waves. As a wave passes, water moves in a circular motion.
Near the shore, the top of the wave breaks over the base of the wave. Swash carries sand up the beach, and backwash carries sand back.
Sand may pile into dunes that build out over a lagoon, in which mud had accumulated.

FIGURE 15.19 Examples of beach erosion.

(a) When Hurricane Ike hit Galveston, Texas, in September 2008, many beachfront homes were washed away.

(b) Wave erosion has completely removed the beach and has started to erode a beach cliff along the coast of Cape Cod.

had taken centuries or millennia to form. The backwash of large storm waves sweeps vast quantities of sand seaward, leaving the beach a skeleton of its former self. In addition, surf submerges barrier islands and shifts them toward the lagoon. Storms can rip out mangrove swamps and salt marshes and break up coral reefs, thereby destroying the organic buffer that can protect a coast, and leaving it vulnerable to erosion for years to come. Because people build along shorelines, beach erosion can be very costly, undermining shoreline buildings and causing them to collapse into the sea. In some cases, *storm surge*, the very high water levels generated when storm winds push water toward the shore, can float buildings off their foundations (**Fig. 15.19**).

If a resort hotel loses its beach, it probably won't stay in business. The value of a luxury condo on a coast may depend on the accessibility to a scenic beach. And a harbor can't function if its mouth gets blocked by sediment. Thus, property owners often construct artificial barriers to alter the natural movement of sand along the coast. Unfortunately, such constructions can bring undesirable results. For example, beachfront property owners may build *groins*, concrete or stone walls protruding perpendicular to the shore, to prevent longshore drift from removing sand (**Fig. 15.20a**). Sand accumulates on the updrift side of the groin, forming a long triangular wedge, but sand erodes away on the downdrift side. Needless to say, the property owner on the downdrift side doesn't appreciate this process. Harbor engineers may build a pair of walls called *jetties* to protect the entrance to a harbor (**Fig. 15.20b**). But jetties erected at the mouth of a river channel effectively extend the river into deeper water and thus may lead to the deposition of an offshore sandbar. Engineers may also build an offshore wall called a *breakwater*, parallel or at an angle to the beach, to prevent the full force of waves from reaching a harbor. With time, however, sand builds up in the lee of the breakwater and the beach grows seaward, clogging the harbor (**Fig. 15.20c**). To protect expensive shoreline construction, people build *seawalls* out of riprap (large stone or

SEE FOR YOURSELF...

CHICAGO LAKEFRONT PRESERVATION

Latitude
41°54′59.76″ N

Longitude
87°37′36.41″ W

Zoom to 3 km (~9,800 feet) and look down.

Sandy beaches fringe the shores of Lake Michigan at Chicago. Waves strike the shore obliquely. As a result, longshore drift carries sand southward. The city constructed a series of groins in an attempt to prevent beach erosion.

FIGURE 15.20 Techniques used to preserve beaches.

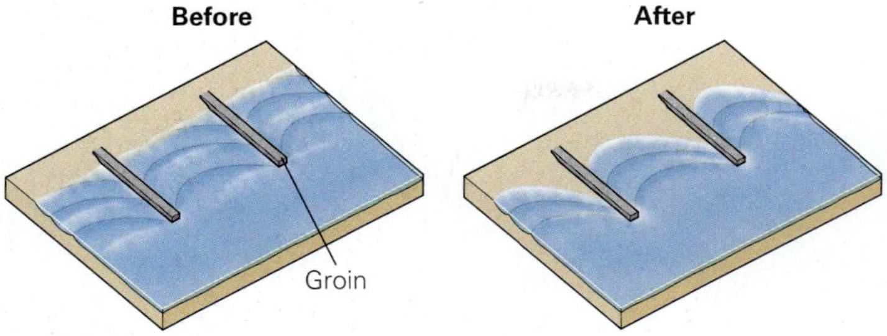

(a) The construction of groins may produce a sawtooth beach.

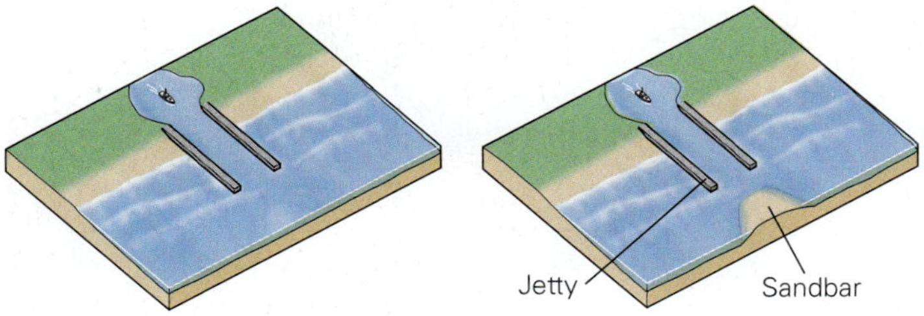

(b) Jetties extend a river farther into the sea but may cause a sandbar to form at the end of the channel.

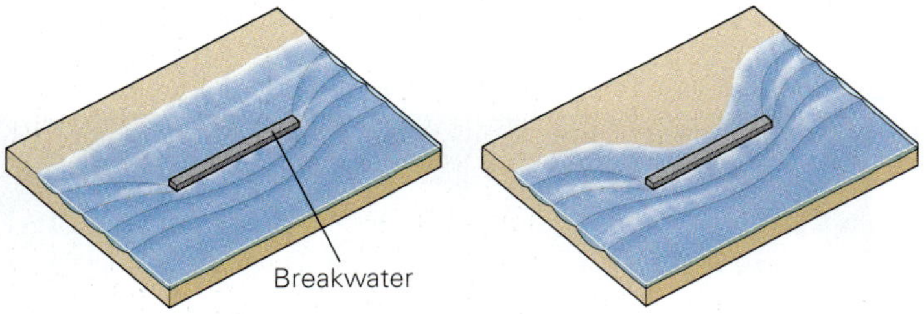

(c) A beach may grow seaward behind a breakwater.

(d) Riprap will slow erosion of a parking lot along a California beach.

concrete blocks) or reinforced concrete on the landward side of the beach (**Fig. 15.20d**), but during a storm, these can be undermined.

In some places, people have given up trying to decrease the rate of beach erosion and instead have worked to increase the rate of sediment supply. To do this, they pump sand from farther offshore, or truck in sand from elsewhere to replenish a beach. This procedure, called **beach nourishment**, can be hugely expensive and at best provides only a temporary fix, for the backwash and beach drift that removed the sand in the first place continue unabated as long as the wind blows and the waves break.

Coastal Pollution, and the Destruction of Wetlands and Reefs

Coastal regions, both onshore and offshore, host fragile ecosystems that can be negatively impacted by the effluents of modern society. For example, local inhabitants may use beaches as dumping grounds for garbage. In some cases, garbage dumped in the sea in an urban area may migrate along the shore, due to longshore drift, and be deposited on a beach tens of kilometers away. Oil spills (from ships that flush their bilges, tankers that foundered in stormy seas, or offshore seeps and leaks) have contaminated shorelines over the years at many locations around the world. The influx of nutrients into coastal waters, from sewage and from agricultural runoff, can produce dead zones along coasts. A *dead zone*, which is a region in which water contains so little oxygen that fish and other organisms within it die, forms when the concentration of nutrients rises enough to stimulate an algal bloom that can deplete oxygen. One of the world's largest dead zones occurs in the northern Gulf of Mexico, where the Mississippi River brings in fertilizer residue from the fields that occupy its watershed.

Coastal wetlands and coral reefs are particularly susceptible to pollution, and many have been destroyed in recent decades. Their loss both increases a coast's vulnerability to erosion and, because they provide spawning grounds for marine organisms, disrupts the global food chain. Destruction of wetlands and reefs happens for many reasons. Wetlands have been filled or drained to be converted to farmland, housing developments, resorts, or garbage dumps. Reefs have been destroyed by boat anchors, dredging, the activities of divers, dynamite explosions intended to kill fish, and quarrying operations in order to obtain construction materials. Chemicals and particulates entering coastal water from urban, industrial, and agricultural areas can cause havoc in wetlands and reefs, for these materials cloud water and/or trigger algal blooms, killing filter-feeding organisms. Toxic chemicals in such runoff can also poison plankton and burrowing organisms and, therefore, other organisms progressively up the food chain.

FIGURE 15.21 The structure and distribution of hurricanes.

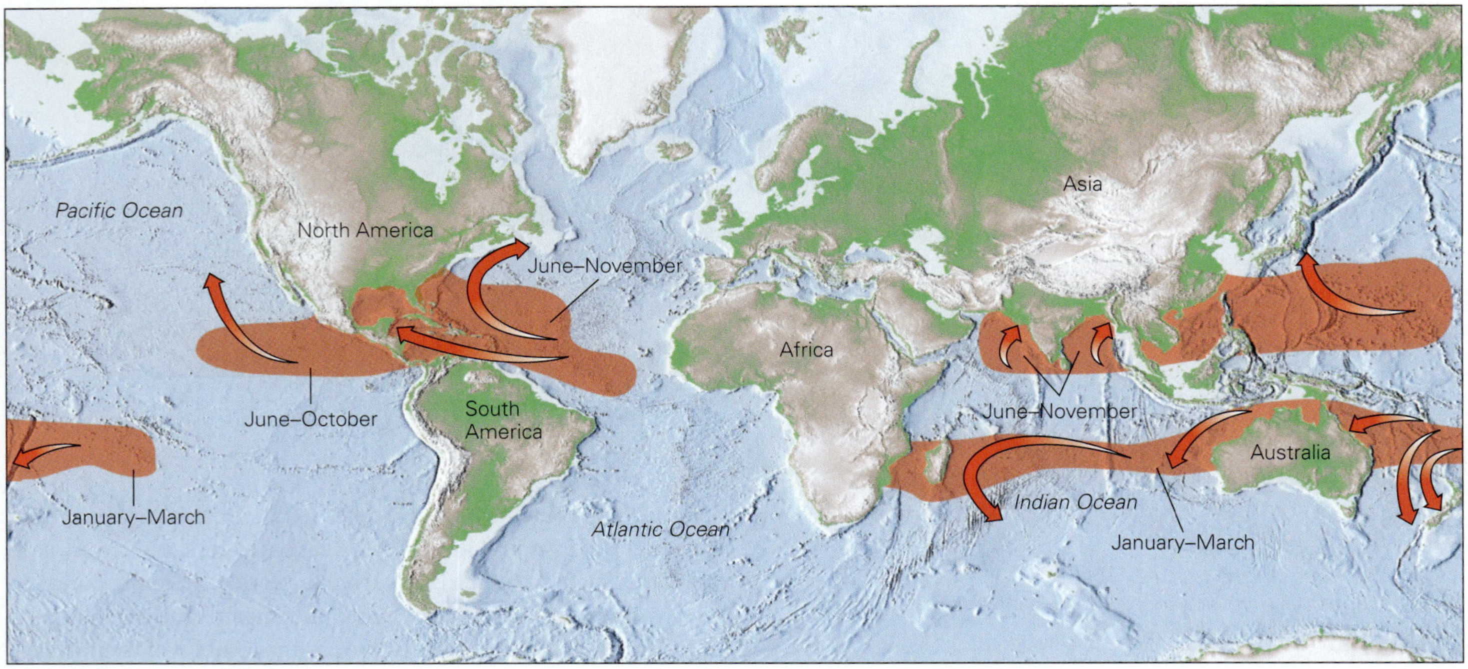

(a) Hurricanes form only in certain regions of the ocean, where water temperatures are high. They generally follow regional tracks and occur during certain times of the year.

(b) Hurricane Sandy approaches the east coast of the United States in 2012, as seen from space.

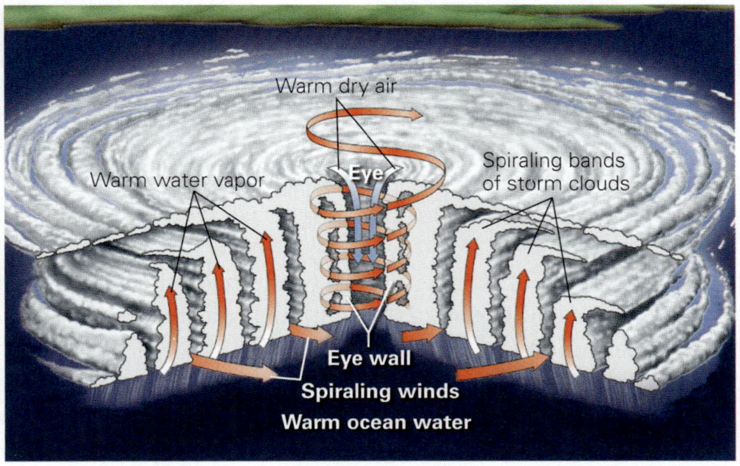

(c) In this cutaway drawing, we can see the spirals of clouds, the eye, and the eye wall of a hurricane. Dry air descends in the eye.

Global climate change also impacts the health of organic coasts (see Chapter 19). For example, transformation of once-vegetated regions into deserts means that the amount of dust carried by winds from the land to the sea has increased. This dust can interfere with coral respiration and can infect them with dangerous viruses. A global increase in seawater temperature may be contributing to *reef bleaching*, the loss of coral color due to the death of the algae that live in coral polyps.

The statistics of wetland and reef destruction worldwide are frightening. Ecologists estimate that between 20% and 70% of wetlands have already been destroyed, and along some coasts, 90% of reefs have died.

Hurricanes—A Coastal Calamity

Global-scale convection of the atmosphere, influenced by the Coriolis effect, causes currents of warm air to flow steadily from east to west in tropical latitudes (**Fig. 15.21a**). As the air flows over the ocean, it absorbs moisture. Because air becomes less dense as it gets warmer, tropical air eventually begins to rise like a balloon. As the air rises, it cools, and the water vapor it contains condenses to form clouds (mists of very tiny water droplets). If the air contains sufficient moisture, the clouds grow into a cluster of large thunderstorms, which consolidate to form a single, very large storm. Because of the Coriolis effect, this large storm starts to rotate, and if it remains over warm ocean water, as can happen in late summer and early fall, it grows, for as it evaporates from the sea water releases energy that feeds the storm. Eventually, the storm evolves into a spiral of intense winds called a *tropical storm*. If a tropical storm becomes powerful enough, it becomes a **tropical cyclone**—a huge rotating storm that forms in tropical latitudes, in which winds exceed 119 km per hour (74 mph). It resembles a giant counterclockwise spiral of clouds, 300 to 1,500 km (930 miles) wide, when viewed from space (**Fig. 15.21b**). Such a storm is called a *hurricane* in the Atlantic and eastern Pacific, a *typhoon* in the western Pacific, and simply a *cyclone* around Australia and in the Indian Ocean.

Atlantic hurricanes generally form in the ocean to the east of the Caribbean Sea, though some form in the Caribbean itself. At first they drift westward at speeds of up to 60 km per hour (37 mph), but they may eventually turn north and head into the North Atlantic or into the interior of North America, where they die when they run out of a supply of warm water. Weather researchers classify the strength of hurricanes using the *Saffir-Simpson scale*, which runs from 1 to 5. On the Saffir-Simpson scale, a Category 5 hurricane has sustained winds of 250 km/h (156 mph) or greater.

A typical hurricane consists of several spiral arms extending inward to a central zone of relative calm known as the *hurricane eye* (**Fig. 15.21c**). A rotating vertical cylinder of clouds, the *eye wall*, surrounds the eye. As winds spiral toward the eye, they accelerate, so they are fastest at the eye wall. Typically, hurricane-force winds occur within a zone that is only 15% to 35% as wide as the whole storm. On the side of the eye where winds blow in the same direction as the whole storm is moving, the ground speed of winds is greatest, because the storm's overall speed adds to the rotational motion.

Hurricanes pose extreme danger in the open ocean, because their winds cause huge waves to build, and thus have led to the foundering of countless ships. They also cause havoc in coastal regions, and even inland. Coastal damage happens for several reasons:

- *Wind*: Winds of weaker hurricanes tear off branches and smash windows. Stronger hurricanes uproot trees, rip off roofs, and collapse walls.
- *Waves*: Winds shearing across the sea surface during a hurricane generate huge waves. In the open ocean, these waves can capsize ships. Near shore, waves batter and erode beaches, rip boats from moorings, and destroy coastal property.
- *Storm surge*: Rising air in a hurricane causes a region of extremely low air pressure beneath. This decrease in pressure causes the surface of the sea to bulge upward over an area with a diameter of 60 to 80 km. Sustained winds blowing in an onshore direction build this bulge even higher. When the hurricane reaches the coast, the bulge of water, or *storm surge*, swamps the land. If the bulge hits the land at high tide, the sea surface will be especially high and will affect a broader area.
- *Rain, stream flooding, and landslides*: Rain drenches the Earth's surface beneath a hurricane. In places, half a meter or more of rain falls in a single day, so streams flood. Saturation of the land with water can trigger landslides.
- *Disruption of social structure*: When the storm passes, the hazard is not over. By disrupting transportation and communication networks, breaking water mains, and washing away sewage-treatment plants, hurricane damage produces severe obstacles to search and rescue, and can lead to the spread of disease, fire, and looting.

Nearly all hurricanes that reach the coast cause death and destruction, but some are truly catastrophic. Storm surge from a 1970 cyclone making landfall on the low-lying delta lands of Bangladesh led to an estimated 500,000 deaths. In 1992, Hurricane Andrew leveled extensive areas of southern Florida, causing over $30 billion in damage and leaving 250,000 people homeless. Hurricane Katrina, in 2005, destroyed much of New Orleans (**Box 15.2**), and Hurricane Sandy in 2012 devastated coastal regions of the eastern United States. Sandy was an immense storm, reaching a diameter of 1,800 km (1,100 miles) at its peak. Storm surge associated with Sandy flattened countless communities and even submerged tunnels and subways in New York City. The most powerful storm on record, Typhoon Haiyan, struck the Philippines in 2013—during its peak, winds gusted to 378 km/h (235 mph).

TAKE-HOME MESSAGE

Coastal landscapes can be impacted by human activities and climate change. Beaches erode, and pollution destroys wetlands. People try to protect beaches by constructing barriers or replenishing sand, but this can lead to other problems. Hurricanes can be particularly destructive hazards along coasts.

QUICK QUESTION What problems can be caused by construction of groins or seawalls?

BOX 15.2 CONSIDER THIS...

Hurricane Katrina

Tropical Storm Katrina came into existence over the Bahamas and headed west. Just before landfall in southeastern Florida, winds strengthened and the storm became Hurricane Katrina. This hurricane sliced across the southern tip of Florida, causing several deaths and millions of dollars in damage. It then entered the Gulf of Mexico and passed directly over an eddy of summer-heated water from the Caribbean that had entered the Gulf of Mexico. This water, which reached a temperatures of 32°C (90°F), stoked the storm, injecting it with a burst of energy sufficient for the storm to morph into a Category 5 monster whose swath of hurricane-force winds reached a width of 325 km (200 miles) (**Fig. Bx15.2a**). When it entered the central Gulf of Mexico, Katrina turned north and began to bear down on the Louisiana-Mississippi coast (**Fig. Bx15.2b, c**). The eye of the storm passed just east of New Orleans, and then across the coast of Mississippi. Storm surges broke records, in places rising 7.5 m (25 feet) above sea level, and they washed coastal communities off the map along a broad swath of the Gulf Coast (**Fig. Bx15.2d, e**).

The winds of Hurricane Katrina ripped off roofs, toppled trees, smashed windows, and triggered the collapse of weaker buildings. Residents of New Orleans gave a sigh of relief when the storm had passed, because wind damage in New Orleans was not catastrophic. But the worst was yet to come.When the winds blew the storm surge into Lake Pontchartrain, the lake's water level rose beyond most expectations, and the weight of elevated water pressed against the network of artificial levees and flood walls that had been built to protect New Orleans, much of which lies below sea level (**Fig. Bx15.2f**). Hours after the hurricane eye had passed, the high water of Lake Pontchartrain found a weakness along the floodwall bordering a drainage canal and pushed out a section. Breaks eventually formed in other locations as well. So, a day after the hurricane was over, New Orleans began to flood. As the water level climbed the walls of houses, brick by brick, residents fled to higher levels of their homes, finally to their roofs. Water spread across the city until the bowl of New Orleans filled to the same level as Lake Pontchartrain, submerging 80% of the city (**Fig. Bx15.2g**). Floodwaters washed some houses away and filled others with debris (**Fig. Bx15.2h**). The disaster took on national significance, as the city's trapped inhabitants sweltered without adequate food, drinking water, or shelter. Lacking communications, medical care, and police protection, the city remained in chaos for days. By the time outside relief reached the city, many people had died and parts of New Orleans, a cultural landmark and major port, had become uninhabitable. It has taken years for the city to rebuild.

FIGURE Bx15.2 Hurricane Katrina and the disaster it caused.

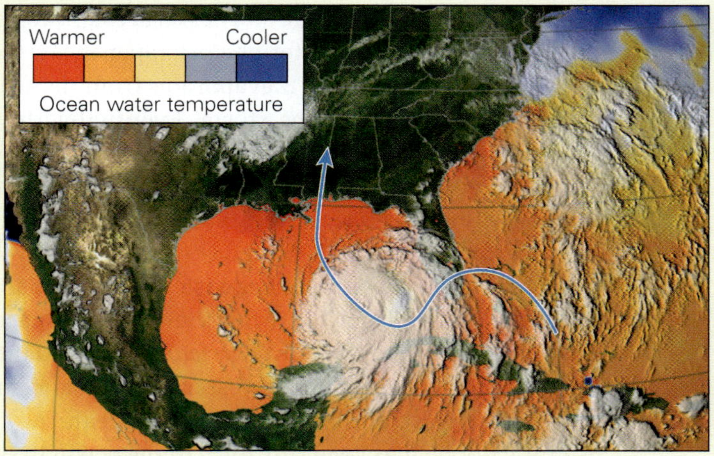

(a) A satellite photo of Hurricane Katrina over the hot water of the Gulf of Mexico. The arrow shows the path of the hurricane.

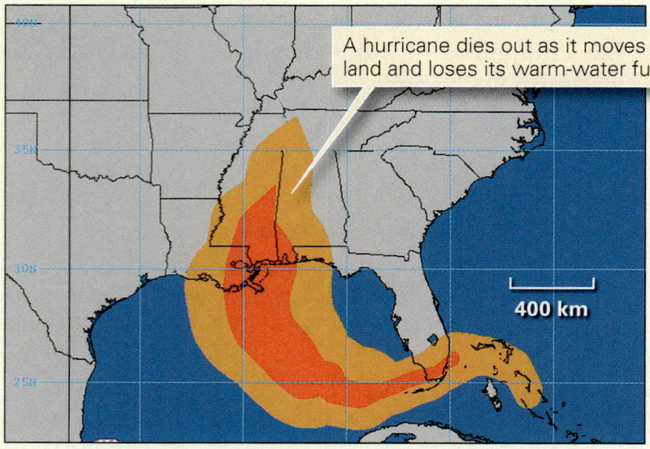

(b) A wind-swath map of Hurricane Katrina. Red areas represent hurricane winds; orange areas represent tropical-storm winds.

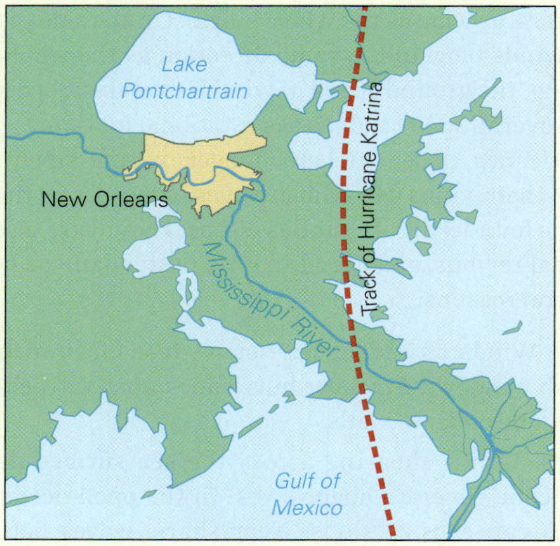

(c) The track of the storm lay just east of New Orleans.

(d) Surge from the storm destroyed homes along the Alabama coast.

(e) Officials survey the storm damage.

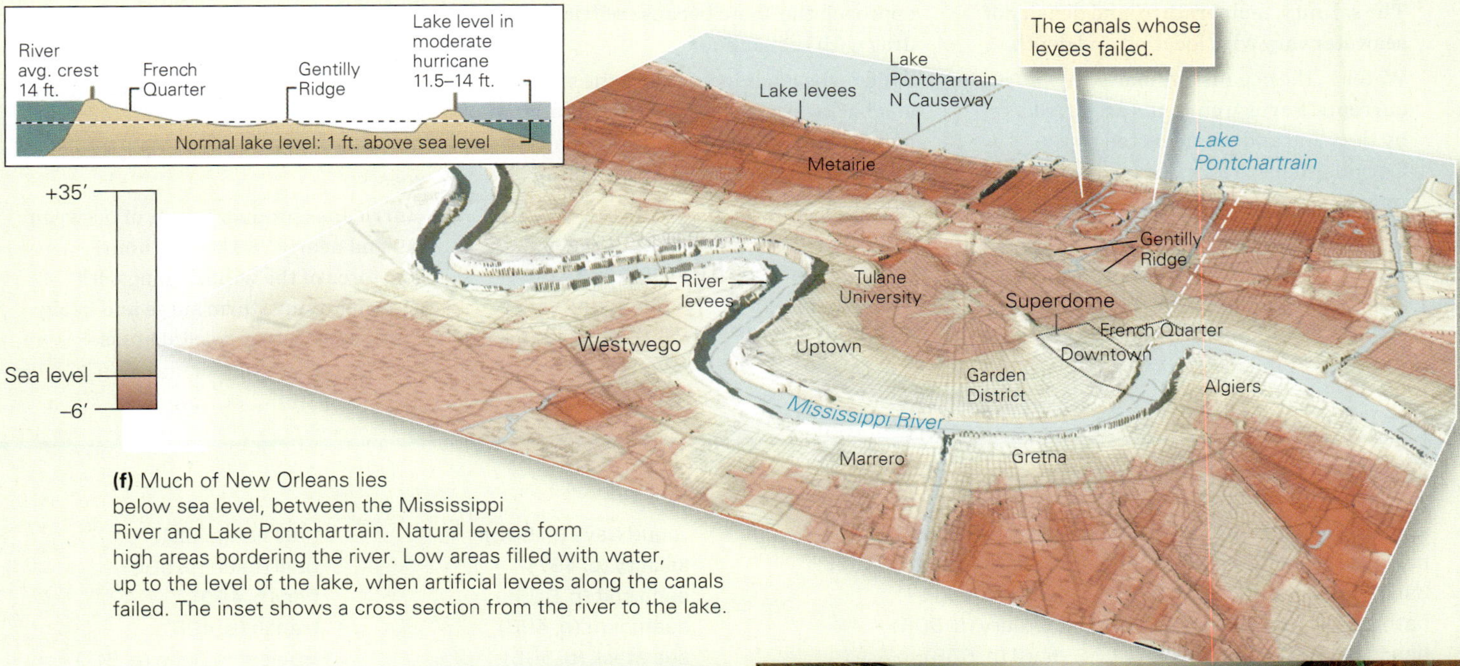

(f) Much of New Orleans lies below sea level, between the Mississippi River and Lake Pontchartrain. Natural levees form high areas bordering the river. Low areas filled with water, up to the level of the lake, when artificial levees along the canals failed. The inset shows a cross section from the river to the lake.

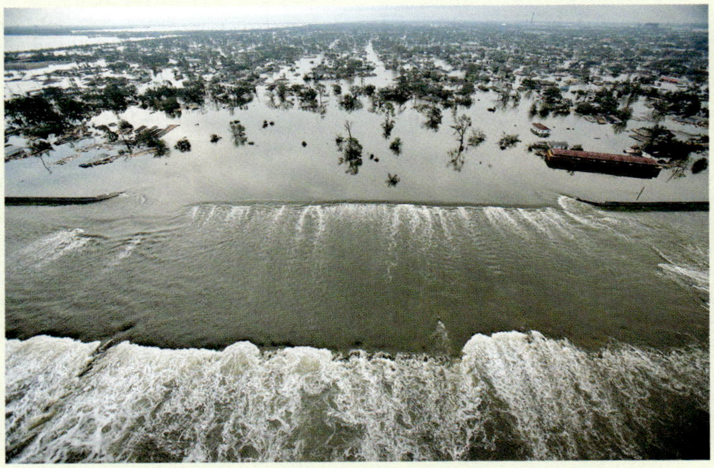

(g) Water flowing across the levees bordering the 17th Street Canal after the hurricane had passed.

(h) Damage due to flooding in a New Orleans home.

Chapter 15 Review

Chapter Summary

- The landscape of the seafloor depends on the character of the underlying crust. Wide continental shelves form over passive-margin basins. Continental shelves may be cut locally by submarine canyons. Abyssal plains develop on old, cool oceanic lithosphere. Seamounts form above hot spots.
- The salinity, temperature, and density of seawater vary with location and depth.
- Water in the oceans circulates in currents. Surface currents are driven by the wind and are deflected by the Coriolis effect to form gyres. The vertical upwelling and downwelling of water produce deep currents. Some of this movement is thermohaline circulation.
- Tides—the daily rise and fall of sea level—are caused by a tide-generating force. The largest contribution to this force comes from the gravitational pull of the Moon.
- Waves are caused by friction where the wind shears across the surface of the ocean. Water particles follow a circular motion in a vertical plane as a wave passes. Waves refract (bend) when they approach the shore because of frictional drag with the seafloor.
- Sand on beaches moves with the swash and backwash of waves. If there is a longshore current, the sand gradually moves along the beach and may build spits.
- At rocky coasts, waves grind away at rocks, yielding such features as wave-cut benches and sea stacks. Some shores are wetlands, where marshes or mangrove swamps grow. Coral reefs grow along coasts in warm, clear water.
- The differences in coasts reflect their tectonic setting, whether sea level is rising or falling, sediment supply, and climate.
- To protect beach property, people build groins, jetties, breakwaters, and seawalls.
- Human activities have led to the pollution of coasts. Reef bleaching has become dangerously widespread, and dead zones have formed along some coasts.
- Hurricanes produce winds of between 119 and nearly 380 km per hour. The force of the winds, along with accompanying storm surge and heavy rains, can destroy coastal areas.

Guide Terms

abyssal plain (pp. 495, 496)
backwash (p. 502)
barrier island (p. 505)
bathymetry (p. 494)
beach (p. 503)
beach erosion (p. 511)
beach nourishment (p. 515)
beach profile (p. 503)
coastal wetland (p. 507)
continental shelf (p. 495)
coral reef (p. 509)
Coriolis effect (p. 497)
current (p. 497)
estuary (p. 505)
fjord (p. 506)
lagoon (p. 505)
longshore current (p. 503)
longshore drift (p. 503)
mid-ocean ridge (p. 495)
organic coast (p. 506)
rogue wave (p. 502)
salinity (p. 497)
sand spit (p. 505)
seamount (p. 496)
sea stack (p. 505)
submarine canyon (p. 495)
swash (p. 502)
thermocline (p. 497)
thermohaline circulation (p. 498)
tidal flat (p. 501)
tidal reach (p. 500)
tide (p. 498)
trench (p. 495)
tropical cyclone (p. 517)
wave base (p. 501)
wave-cut bench (p. 505)
wave refraction (p. 502)

GEOTOURS *THIS CHAPTER'S GEOTOUR EXERCISE (O) FEATURES:*

- Seafloor Bathymetry
- Coral Reefs
- Barrier Islands & Spits
- Coastlines & Sealevel Rise

Review Questions

1. How much of the Earth's surface is covered by oceans?
2. How does the lithosphere beneath a continent differ from that beneath an abyssal plain?
3. How do the shelf and slope of an *active* continental margin differ from those of a *passive* margin? Why do passive-margin basins exist?
4. Where does the salt in the ocean come from? How do the salinity and temperature in the ocean vary?
5. What factors control the direction of surface currents in the ocean? Explain thermohaline circulation.
6. What causes the tides? Why does tidal reach vary with location?
7. Describe the motion of water molecules in a wave. How does wave refraction cause longshore currents?
8. Describe the components of a beach profile. How does beach sand migrate as a result of longshore drift?
9. Describe how rocky coasts evolve.
10. What is an estuary? What is the difference between an estuary and a fjord?
11. Discuss the different types of coastal wetlands. What is a coral reef, and how does the reef surrounding an oceanic island change with time?
12. How do plate tectonics, sea-level changes, sediment supply, and climate change affect the local shape of a coastline? Explain the difference between an emergent and a submergent coast.
13. In what ways do people try to modify or stabilize coasts? How do the actions of people threaten coastal areas?
14. How can hurricanes affect coasts?

On Further Thought

15. In 1789, the crew of the HMS *Bounty* mutinied. Near Tonga, in the Friendly Islands (approximately 20° S and 175° W), the crew forced the ship's commanding officer, Lieutenant Bligh, along with those crewmen who remained loyal to Bligh, into a rowboat and set them adrift in the Pacific Ocean. The castaways, amazingly, survived, and 47 days later, they landed at Timor (near Sumatra), 6,700 km to the west. Why did they end up where they did?

Online Resources

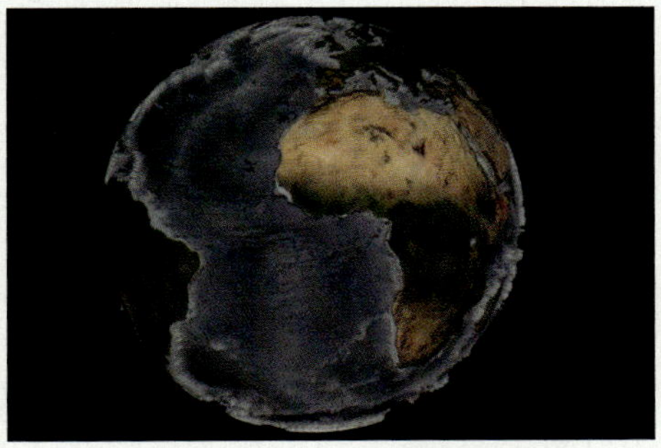

Videos

This chapter features a video on ocean ridges.

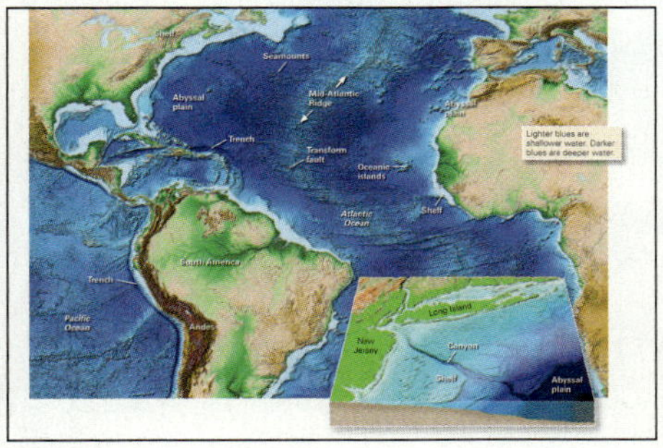

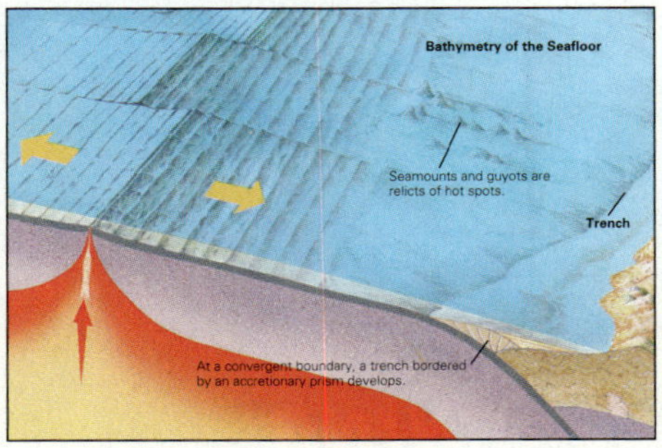

Assessment

This chapter features labeling questions on underwater, oceanic landforms and continental and oceanic lithospheres.

LEARNING OBJECTIVES

By the end of this chapter, you should understand...

1. what groundwater is, where it comes from, and how it flows.
2. the nature of the water table, and factors that determine its depth.
3. how wells and springs provide access to groundwater.
4. why hot springs and geysers exist.
5. how groundwater supplies can be damaged or depleted.
6. why caves and karst landscapes originate and evolve.

▲ Steaming pools and travertine ledges have developed where groundwater bubbles to the surface from the hot springs of Rotorua, New Zealand.

CHAPTER 16

A Hidden Reserve: Groundwater

16.1 Introduction

Imagine Mae Rose Owens's surprise when, on May 8, 1981, she looked out her window and discovered that a large sycamore tree in the backyard of her Winter Park, Florida, home had suddenly disappeared. It wasn't a particularly windy day, so the tree hadn't blown over—it had just vanished! When Owens went outside to investigate, she found that more than the tree had disappeared. Her whole backyard had become a deep, gaping pit. The pit continued to grow for a few days until finally it swallowed Owens's house and six other buildings, as well as the municipal swimming pool, part of a road, and several expensive Porsches in a car dealer's lot (Fig. 16.1a).

What had happened in Winter Park? The bedrock beneath the town consists of limestone. **Groundwater**, the liquid water that resides in sediment or rock under the surface of the Earth, had gradually dissolved the limestone over time, carving open rooms, or *caverns*, underground. On May 8, the roof of a cavern underneath Owens's backyard began to collapse, soil from up above sank into the opening, and a circular depression called a **sinkhole** developed

FIGURE 16.1 Development of sinkholes in central Florida.

(a) The Winter Park sinkhole, as seen from a helicopter.

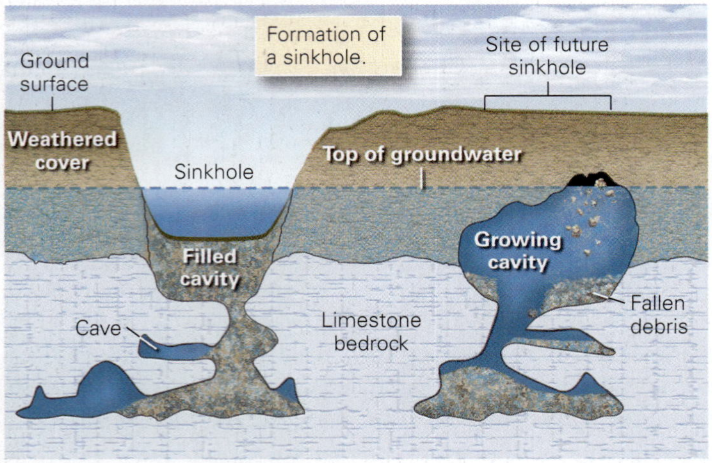

(b) As overburden slowly washes into underlying caves, a cavity forms. When the roof of this cavity collapses, a sinkhole forms (not to scale).

(c) An airplane view of Florida sinkholes that have become lakes.

> *When the rain falls and enters the earth, when a pearl drops into the depth of the sea, you can dive in the sea and find the pearl, you can dig in the earth and find the water.*
>
> MEI YAO-CH'EN (Chinese poet, 1002–1060)

(**Fig. 16.1b**). The sycamore tree and the rest of the neighborhood simply dropped down into the sinkhole. It would have taken too much effort to fill in the sinkhole with rock and soil, so the community allowed it to fill with water, and now it's a circular lake. Similar lakes occur throughout central Florida (**Fig. 16.1c**), and elsewhere worldwide.

The Winter Park sinkhole serves as a dramatic reminder that significant quantities of water reside underground. We can easily see Earth's surface water (in lakes, rivers, streams, marshes, glaciers, and oceans) and atmospheric water (in clouds and rain). Groundwater, however, which accounts for about two-thirds of the Earth's freshwater resources used by homes, agriculture, and industry, lies hidden beneath the surface. In this chapter, we examine where this subsurface water comes from, how it flows, how it interacts with rock and sediment, and how, in special places, it erupts in forceful plumes of steam. We also explore the underground landscapes that can be carved by groundwater.

SEE FOR YOURSELF...

SINKHOLES IN CENTRAL FLORIDA

Latitude
28°37′50.59″ N

Longitude
81°23′13.60″ W

Zoom to 20 km (~12.5 miles) and look down.

You can see several sinkholes, ranging from about 100 to 800 m across, that lie within suburban developments. The sinkholes have filled with water and are now lakes.

16.2 Where Does Groundwater Reside?

The Underground Reservoir

Water moves among various reservoirs during the hydrologic cycle (see Interlude F). One reservoir consists of the open spaces within sediment or rock underground. How does water get into this groundwater reservoir? Some, known as *connate water*, gets buried with sediment grains during deposition, and gets trapped when the rock lithifies. Some bubbles out of magma that has intruded the crust. But most consists of *meteoric water*. This water precipitated from the sky, as rain or snow, and fell on the land. If it doesn't evaporate directly back into the atmosphere, get trapped in glaciers, or run off in a stream, the meteoric water sinks or percolates downward into the ground, a process called *infiltration*. In effect, the upper part of the crust behaves like a giant sponge that can soak up water. Let's look more closely at the interconnected openings that allow infiltration.

Porosity: Open Space in Rock and Regolith

Contrary to popular belief, only a small proportion of groundwater occurs in caves. Most groundwater resides in relatively small open spaces between grains of sediment or between grains of seemingly solid rock, or within cracks of various sizes. Geologists use the term **pore** for any open space within a volume of sediment, or within a body of rock, and the term **porosity** for the total amount of open space within a material. We specify porosity as a percentage. For example, if we say that a block of rock has 20% porosity, we mean that 20% of the block consists of open space.

We can distinguish between two basic kinds of porosity—primary and secondary. *Primary porosity* refers to the open space that remains in a sediment after deposition, or in a rock after its formation (**Fig. 16.2a, b**). Primary porosity exists because grains don't fit together perfectly during deposition, or because grains don't grow to fill all space during crystallization, or because open spaces do not completely fill with cement. *Secondary porosity* refers to new pore space produced in a rock some time after the rock formed. For example, if a solid rock undergoes fracturing, secondary porosity develops because opposing walls of fractures do not fit together perfectly (**Fig. 16.2c**), and if fluids pass through rock after lithification, they may dissolve rock and produce open cavities that also provide secondary porosity.

Permeability: The Ease of Flow

If solid rock completely surrounds a pore, the water in the pore cannot flow to another location. For groundwater to flow, conduits or connections must exist between pores. The ability of a material to allow fluids to pass through an interconnected network of pores is the material's **permeability** (**Fig. 16.2d**). Groundwater flows easily through a high-permeability material, such as loose gravel. So if you pour water into a gravel-filled jar, it quickly trickles down to the bottom of the jar, displaces air, and fills the pores (**Fig. 16.2e**). Groundwater flows slowly through low-permeability material in which it must follow a tortuous path through narrow, crooked conduits. Groundwater doesn't flow at all through an impermeable material, because there are no conduits between pores. We use, in everyday life, materials with different permeability—a household sponge has high permeability, whereas a porcelain bowl has no permeability. A material's permeability depends on several factors:

- *Number of available conduits*: As the number of available conduits increases, permeability increases.
- *Size of the conduits*: Water can travel faster through wider conduits than through narrower ones.
- *Straightness of the conduits*: Water can travel faster through straight conduits than through crooked ones.

The factors that control permeability in Earth materials resemble those that control traffic flow through a city. Traffic can pass quickly in a city with many straight, multilane boulevards, whereas it slows in a city with only a few narrow, crooked streets. Note that porosity and permeability are not the same feature—a material whose pores are isolated from each other can have high porosity but low permeability.

Aquifers and Aquitards

Hydrogeologists, researchers who focus on understanding groundwater, distinguish between **aquifers**, sediment or rock with high permeability and porosity, and **aquitards**, sediment or rock with low permeability regardless of porosity. An aquitard slows, or retards, the motion of water (**Fig. 16.3a**). We can distinguish between an *unconfined aquifer*, one that reaches the Earth's surface, and a *confined aquifer*, one isolated from the Earth's surface by an aquitard.

The Water Table

Infiltrating water moves downward into the subsurface by percolating along cracks and through the conduits that connect pores. Nearer the ground surface, in the *unsaturated zone*, water only partially fills pores, so air remains in some of the open space (**Fig. 16.3b**). In most places, the top of the unsaturated zone consists of soil, so the water in this part of the unsaturated zone can also be called **soil moisture**. Soil moisture may evaporate back into the atmosphere, seep into streams, or get drawn up by plant roots and transpired back into the atmosphere before it even moves below the soil. Deeper down, in the *saturated zone*, water completely fills, or saturates, pore space. In a strict sense, geologists use the term "groundwater" only for subsurface water in the saturated zone. The **water table** is the horizon that separates the unsaturated zone above from the saturated zone below.

FIGURE 16.2 The concepts of porosity and permeability.

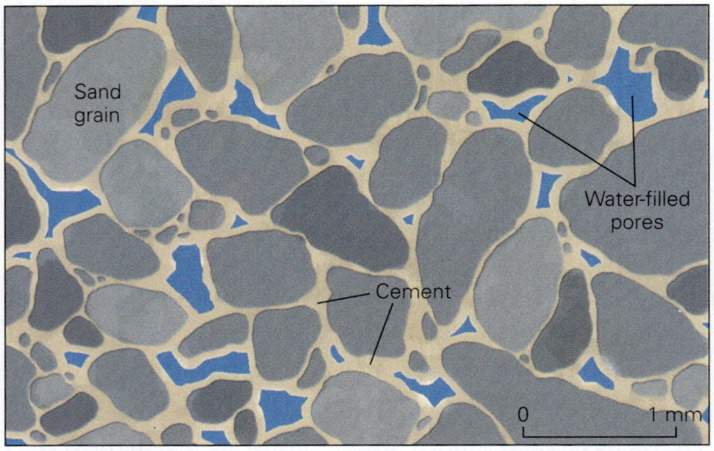

(a) Isolated pores in a sandstone occur in the spaces between grains. Water or air can fill pores.

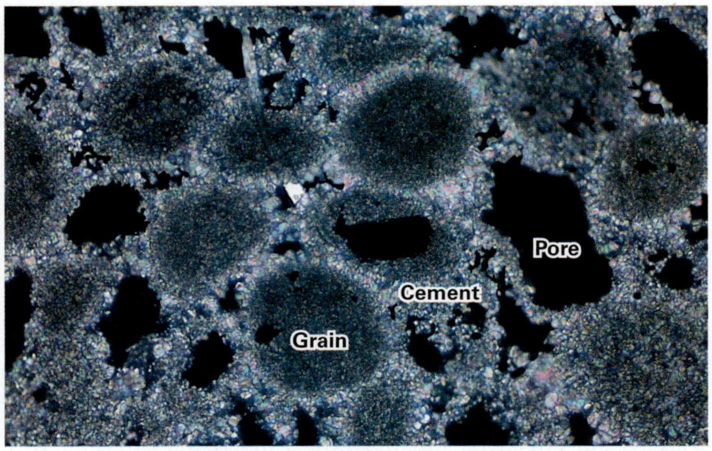

(b) This photo of a sedimentary rock, as seen through a microscope, shows grains, cement, and pores. The field of view is about 3 mm.

(c) Limestone outcrop on the coast of Ireland contains abundant fractures that provide secondary porosity.

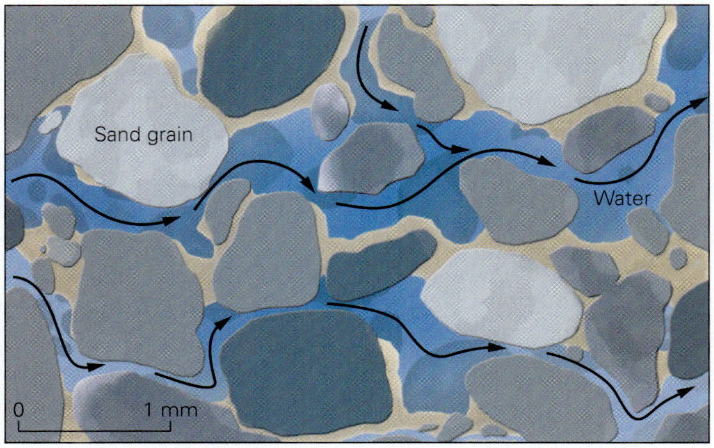

(d) Permemability is the degree to which pores are linked, so water can move from pore to pore and move through rock or sediment.

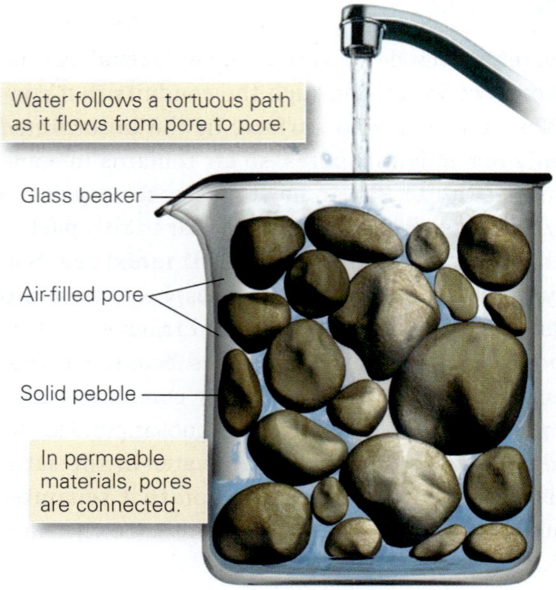

(e) Gravel contains pore space because clasts don't fit together tightly. The connection of pores produces permeability, so water can flow through gravel.

Typically, surface tension, the electrostatic attraction of water molecules to each other and to mineral surfaces, causes some water to seep up from the water table (just as water rises in a thin straw). This water fills pores in a thin layer, known as the **capillary fringe**, just above the water table.

The depth of the water table below the ground surface varies greatly with location. In temperate or tropical regions, where it rains fairly frequently and water often seeps into the ground, the water table lies within a few meters of the Earth's surface. Channels or depressions whose floor lies below the water table will fill with liquid water. Thus, the surface of a permanent stream, lake, or marsh effectively defines the level of the water table in temperate or tropical regions (**Fig. 16.3c**). Rainfall rates affect the water table depth in a given locality—the water table drops during the dry season

FIGURE 16.3 Water underground—aquifers, aquitards, and the water table.

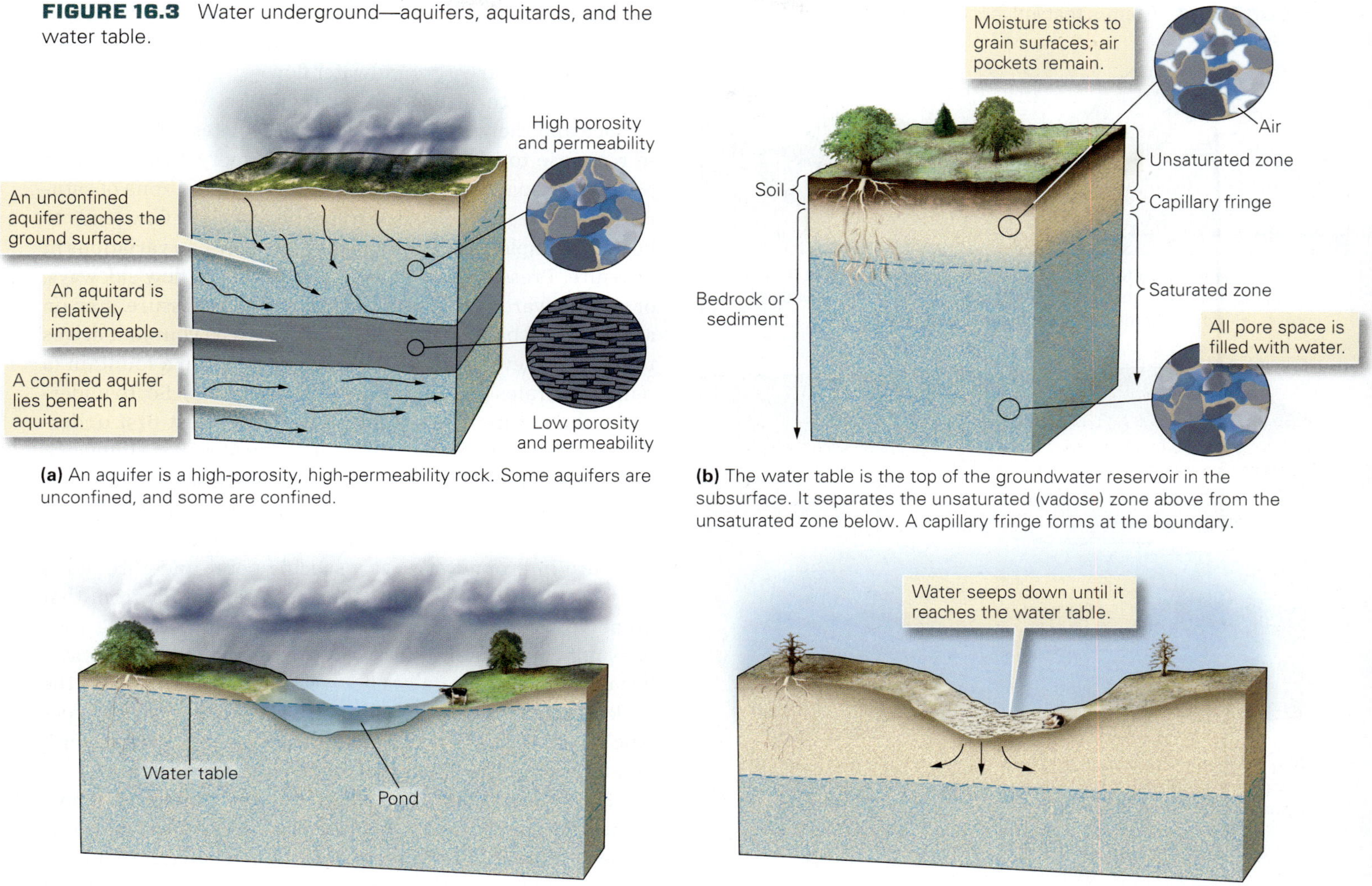
High porosity and permeability
An unconfined aquifer reaches the ground surface.
An aquitard is relatively impermeable.
A confined aquifer lies beneath an aquitard.
Low porosity and permeability
Moisture sticks to grain surfaces; air pockets remain.
Air
Unsaturated zone
Capillary fringe
Soil
Saturated zone
Bedrock or sediment
All pore space is filled with water.
Water table
Pond
Water seeps down until it reaches the water table.

(a) An aquifer is a high-porosity, high-permeability rock. Some aquifers are unconfined, and some are confined.

(b) The water table is the top of the groundwater reservoir in the subsurface. It separates the unsaturated (vadose) zone above from the unsaturated zone below. A capillary fringe forms at the boundary.

(c) Where the water table lies close to the ground surface, ponds remain filled—the water table is the surface of the pond.

(d) In dry regions, the water table sinks deep below the surface. Water that collects temporarily in low areas sinks into the subsurface.

and rises during the wet season. Therefore, streams or ponds that contain water during the wet season may dry up during the dry season, when their water infiltrates into the ground below. In arid regions, where it rarely rains, the water table typically lies tens to hundreds of meters below the ground surface (**Fig. 16.3d**).

Shape of the Water Table

In hilly regions, the water table is not a planar surface. Rather, its shape mimics, in a subdued way, the shape of the overlying land surface (**Fig. 16.4a**). This means that the water table is higher beneath hills than beneath valleys. But the relief of the water table, meaning the vertical distance between the highest and lowest elevations, tends to be less than that of the overlying land, so the surface of the water table tends to be smoother than that of the landscape.

At first thought, it may seem surprising that the shape of the water table reflects ground-surface shape. After all, when you pour a bucket of water into a pond, the surface of the pond immediately adjusts to remain horizontal. Why isn't the water table horizontal? The elevation of the water table varies because groundwater moves so slowly through rock and sediment that it cannot quickly assume a horizontal surface. So when rain falls on a hill and water infiltrates down to the water table, the water table rises a little, and when it doesn't rain, the water table sinks a little, but this movement takes place so slowly that when rain falls again, the water table rises before it has had time to sink very far.

In some locations, lens-shaped layers of impermeable rock (such as shale) may lie within a thick aquifer. A mound of groundwater accumulates above such aquitard lenses, producing a **perched water table**, a local water table that lies above the regional water table (**Fig. 16.4b**).

FIGURE 16.4 Factors that influence the position of the water table.

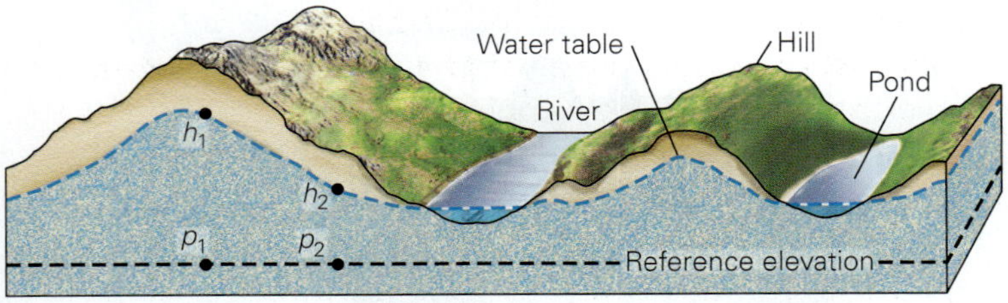

(a) The shape of a water table beneath hilly topography. Point h_1 on the water table is higher than Point h_2, relative to a reference elevation. The pressure at p_1 is, therefore, more than the pressure at p_2.

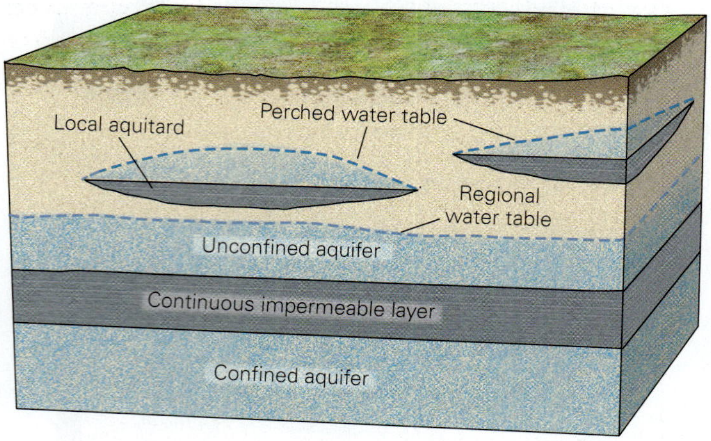

(b) A perched water table occurs where a mound of groundwater becomes trapped above a localized aquitard that lies above the regional water table.

TAKE-HOME MESSAGE

Most underground water fills pores and cracks in rock or sediment. Porosity refers to the total amount of open space within a material, whereas permeability indicates the degree to which pores connect. Aquifers have high porosity and permeability, while aquitards don't. Above the water table, water only partially fills pores, whereas below the water table, water completely saturates pores. The water table tends to be higher beneath hills, and lower beneath valleys.

QUICK QUESTION Why do poorly cemented sandstones make good aquifers?

16.3 Groundwater Flow

What happens to groundwater over time? Does it just sit, unmoving, like the water in a stagnant puddle, or does it flow and eventually find its way back to the surface? Countless measurements confirm that groundwater indeed flows and, in some cases, moves great distances underground. Let's examine factors that drive groundwater flow.

In the unsaturated zone—the region between the ground surface and the water table—water percolates straight down in response to the downward pull of gravity, like the water passing through a drip coffee maker. But in the zone of saturation—the region below the water table—water flow tends to be more complex, for water moves in response to variations in pressure. Pressure can cause groundwater to flow sideways, or even upward. To picture the influence of pressure, step on a water-filled balloon (the pressure caused by your foot drives the water sideways) or watch water spray from a fountain (a pump generates pressure that pushes water up). So, to understand the nature of groundwater flow, we must first understand the origin of pressure in groundwater. For simplicity, we'll consider only the case of an unconfined aquifer.

Let's start by imagining a point in groundwater at a location where the water table over a broad area is horizontal. Pressure in groundwater at a specific point underground is caused only by the weight of the overlying water from that point up to the water table—the weight of overlying rock does not contribute to the pressure exerted on groundwater because the contact points between mineral grains bear the rock's weight. In this case, pressure acting on water at a specific depth below the water table is the same everywhere, with greater pressure at deep points than at shallower points.

Now, imagine a location where the water table is not horizontal, as shown in Figure 16.4a. The pressure at locations along a reference elevation underground changes with location. For example, the pressure at location p_1, which lies below a hill, is greater than the pressure at location p_2, which lies below a valley, even though both p_1 and p_2 lie at the same reference elevation, because there's more water above p_1 than above p_2. The pressure within the water provides energy that can cause groundwater to flow. Hydrogeologists represent the energy available to drive the flow of groundwater, at a given location, by the **hydraulic head**. To measure the hydraulic head at a point in an aquifer, you can drill a vertical hole down to the point and insert a pipe in the hole. The height, above a reference elevation, to which water rises in the pipe represents the hydraulic head—water rises higher in the pipe where the head is higher. As a rule, groundwater flows from regions where it has higher hydraulic head to regions where it has lower hydraulic head. This statement generally implies that groundwater regionally flows from locations where the water table is higher to locations where the water table is lower.

Hydrogeologists have calculated how hydraulic head changes with location underground, taking into account variations in pressure. These rather complicated calculations

FIGURE 16.5 The flow of groundwater.

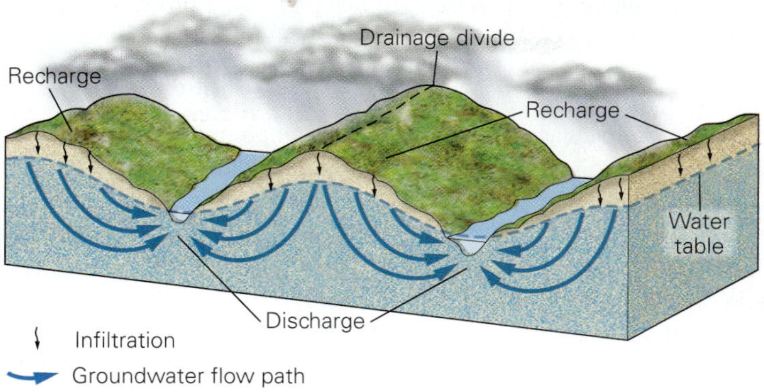

(a) Groundwater flows from recharge areas to discharge areas. Typically, the flow follows curving paths.

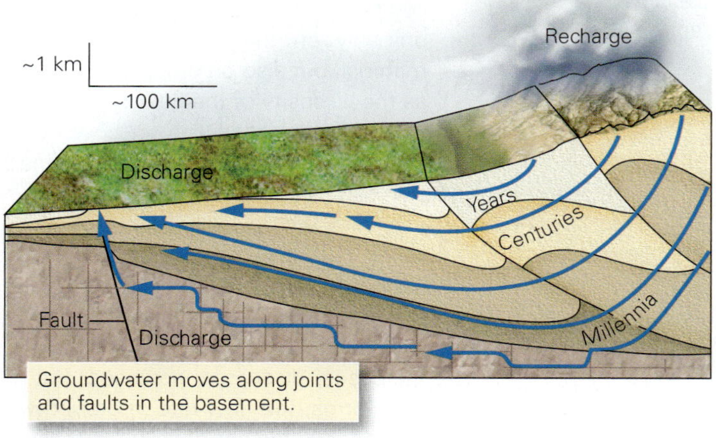

(b) The large hydraulic head resulting from uplift of a mountain belt may drive groundwater hundreds of kilometers across regional sedimentary basins. Deeper flow paths take longer.

indicate that groundwater flows along concave-up curving paths (**Fig. 16.5a**). The curved paths eventually take groundwater from regions where the water table is high (under a hill) to regions where the water table is low (below a valley), but because of flow-path shape, some groundwater first descends along the first part of its path and then flows back up along the final part of its path (**Fig. 16.5b**). A place where water enters the ground and flows down is a **recharge area**, whereas a place where groundwater flows back up to the surface and may seep out at the surface is a **discharge area**.

Water flowing in a steep river channel can reach speeds of up to 30 km per hour, and water flowing in an ocean current moves at up to 3 km per hour. In contrast, groundwater moves at less than a snail's pace, generally from 0.5 to 500 m per year. Groundwater moves much more slowly than surface water, for two reasons. First, groundwater moves by percolating through a complex, crooked network of tiny conduits, so it must travel a much greater distance than it would if it could follow a straight path. Second, friction between groundwater and conduit walls slows down the water flow.

Simplistically, the velocity of groundwater flow depends on the slope of the water table and on the permeability of the material through which the groundwater is flowing—groundwater flows faster through high-permeability materials than it does through low-permeability materials, and it flows faster in regions where the water table has a steep slope than it does in regions where the water table has a gentle slope. For example, groundwater flows relatively slowly through a low-permeability aquifer under the Great Plains, but flows relatively quickly through a high-permeability aquifer under a steep hillslope. The details of calculating groundwater flow rate are a bit more complicated. Hydrogeologists use Darcy's law to calculate the rate for a given location (**Box 16.1**).

TAKE-HOME MESSAGE

Variations in pressure cause groundwater to flow slowly from recharge to discharge areas. Groundwater flow can follow curving flow paths that take it deep into the crust. Simplistically, the rate of flow depends on the water table's slope and on permeability.

QUICK QUESTION Does steepening the water table increase or decrease the rate of groundwater flow?

16.4 Tapping Groundwater Supplies

All living organisms rely on access to water. Of course, while marine life, coastal plants, and certain types of microbes thrive in saline water, most terrestrial organisms, such as our own species, can only use freshwater. Lakes and streams carry freshwater at the Earth's surface, but they don't occur everywhere, and even where they do occur, may not provide a sufficient supply of water for local needs. Groundwater contains the second-largest reservoir of freshwater on the planet, after the glaciers of Antarctica and Greenland. Is there a way to access this groundwater reservoir? Fortunately, yes—groundwater can be obtained from springs and wells.

Springs

Many towns grew up next to **springs**, natural outlets from which groundwater flows or seeps onto the Earth's surface. People can use springs as a source of freshwater for drinking

BOX 16.1 CONSIDER THIS...

Darcy's Law for Groundwater Flow

In 1856, a French engineer named Henry Darcy carried out a series of experiments designed to characterize factors that control the velocity at which groundwater flows between two locations, 1 and 2, each of which has a different hydraulic head, h_1 and h_2, respectively (**Fig. Bx16.1**). Darcy represented the velocity of flow by a quantity called the discharge (Q), meaning the volume of water passing through an imaginary vertical plane perpendicular to the groundwater's flow path in a given time. He found that the discharge depends on several factors:

- *Hydraulic gradient* (i): This quantity represents the change in hydraulic head per unit of distance (j) between two locations, as measured along the flow path. Hydraulic gradient can be represented by the equation: $i = (h_1 - h_2) \div j$. Typically, the slope of the water table is so small that the path length is almost the same as the horizontal distance between two points, so the hydraulic gradient roughly equals the slope of the water table.
- *Area* (A): This quantity represents the cross-sectional area, in square meters, of the imaginary plane through which the groundwater passes.
- *Hydraulic conductivity* (K): This quantity represents the ease with which a fluid can flow through a material. It depends primarily on the permeability of the material, but also on other factors such as the viscosity of the fluid.

The relationship that Darcy discovered is now known as **Darcy's law**. A simplified version can can be written as

$$Q = KAi.$$

In words, this equation means that as the hydraulic gradient increases, discharge increases, and that as hydraulic conductivity increases, discharge increases. Put in simpler terms, the flow rate of groundwater increases as the permeability increases, and as the slope of the water table gets steeper.

FIGURE Bx16.1 The level to which water rises in a drillhole is the hydraulic head (h). The hydraulic gradient (i) is the difference in head divided by the length of the flow path (j).

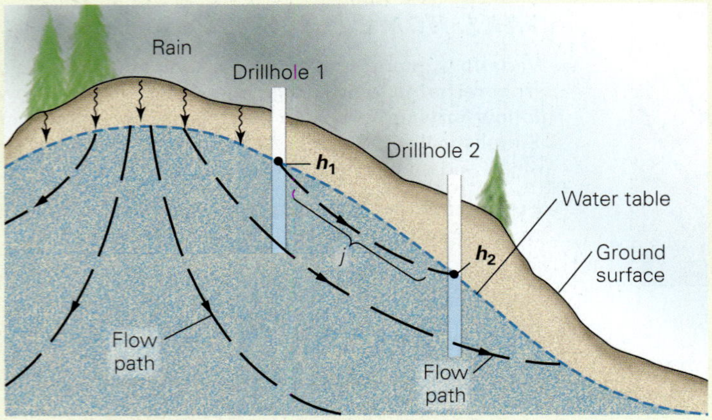

or irrigation, without having to go to the expense of drilling or digging. Some springs spill groundwater onto dry land; others bubble up through the bed of a stream or lake. Springs occur in a variety of locations, including:

- where the ground surface intersects the water table (**Fig. 16.6a**); such springs may add water to lakes or streams.
- where flowing groundwater reaches a steep, impermeable barrier, and pressure pushes the groundwater up to the Earth's surface along the barrier (**Fig. 16.6b**).
- where a perched water table intersects the surface of a hill (**Fig. 16.6c**).
- where downward-percolating water runs into an aquitard and migrates along the top surface of the aquitard to a hillslope (**Fig. 16.6d**).
- where a network of interconnected fractures intersects the water table and provides a conduit for groundwater to reach the ground surface (**Fig. 16.6e**).
- where the ground surface intersects a natural fracture (joint) that taps a confined aquifer in which the pressure is sufficient to drive the water to the surface; such an occurrence is an **artesian spring** (**Fig. 16.6f**).

Did you ever wonder...
where bottled spring water comes from?

Because of the water that springs supply, lush and more diverse vegetation grows around a spring. Where a spring occurs in a desert, an **oasis**—an area where plants flourish in an otherwise bone-dry landscape—stands out as a patch of green against the browns and tans of sand and rock.

FIGURE 16.6 Geologic settings in which springs form.

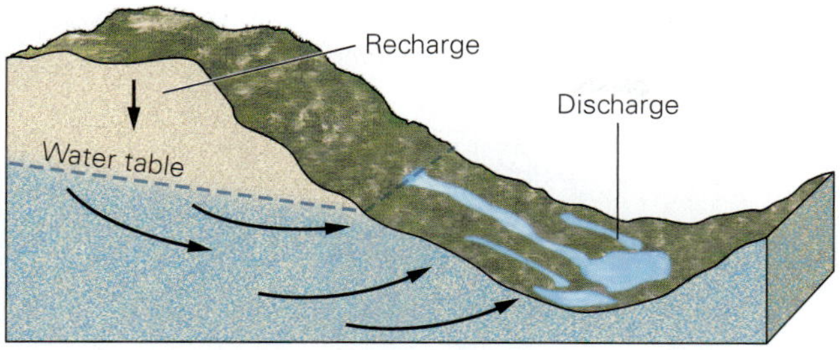

(a) Groundwater reaches the ground surface in a discharge zone.

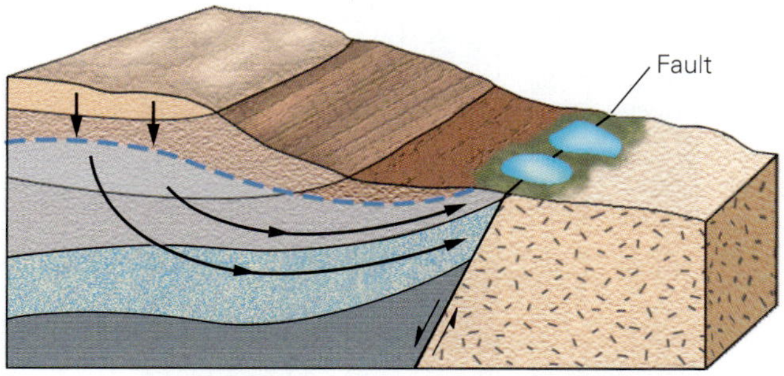

(b) Where groundwater reaches an impermeable barrier, it rises.

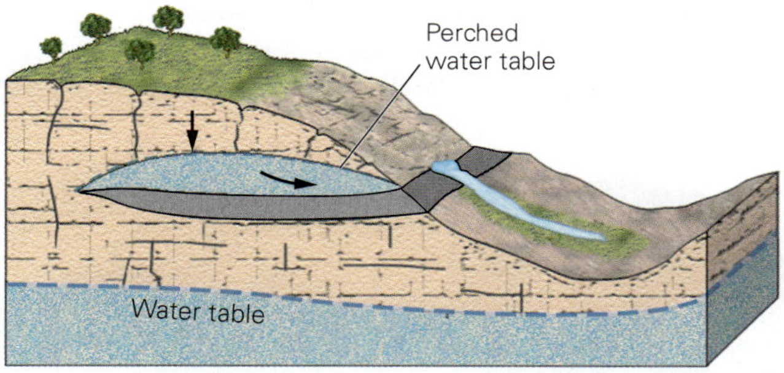

(c) Groundwater seeps where a perched water table intersects a slope.

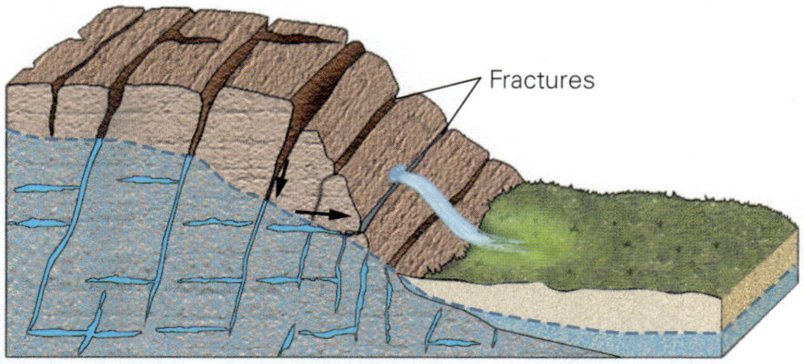

(d) A network of interconnected fractures channels water to the surface of the hill.

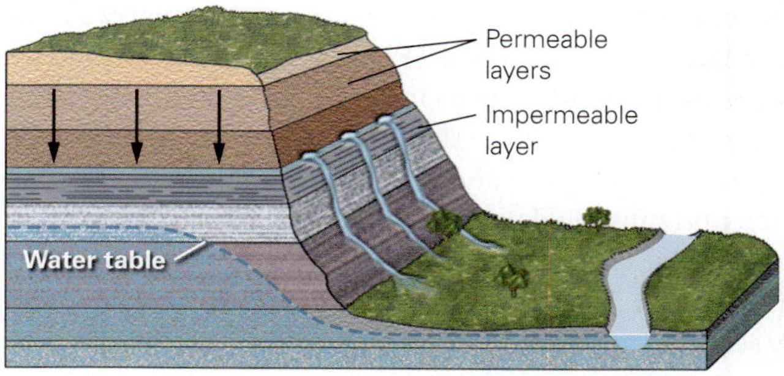

(e) Groundwater seeps out of a cliff face at the top of a relatively impermeable bed.

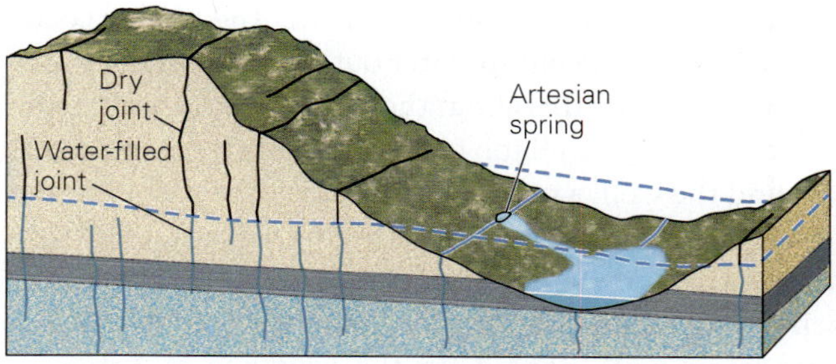

(f) In cases where water under pressure lies below an aquitard, a crack may provide a pathway for an artesian spring to form.

A spring on the wall of the Grand Canyon

Wells

Springs don't occur everywhere, so thousands of years ago, people learned to dig **wells**, holes or pits that provide access to groundwater. Geologists distinguish between ordinary wells and artesian wells, based on whether water in the well rises under its own pressure.

In an **ordinary well**, the base of the well penetrates an aquifer below the water table (**Fig. 16.7a**). Water from the pore space in the aquifer seeps into the well and fills it to the level of the water table. Note that drilling into an aquitard, or into any material that lies above the water table, will not supply water, and thus yields a **dry well**. Some ordinary wells are seasonal in that they function only during the rainy season, when the water table rises above the base of the well. During the dry season, the water table lies below the base of seasonal wells, so the well becomes dry.

To obtain water from an ordinary well, it is necessary either to pull the water up in a bucket, or to pump the water out. As long as the rate at which groundwater fills the well exceeds the rate at which people pump water out, the level of the water table near the well remains about the same. However, if people pump water out of the well too fast, then the water table sinks down around the well, a process called *drawdown*, so that the water table becomes a downward-pointing, cone-shaped surface called a **cone of depression** (**Fig. 16.7b, c**). Drawdown by a deep well may cause nearby shallower wells to run dry.

An **artesian well**, named for the province of Artois in France, penetrates a confined aquifer in which water has enough pressure to rise on its own to a level above the top surface of the aquifer. If this level lies below the ground surface, the well is a *nonflowing artesian well*, whereas if the level lies above the ground surface, the well is a *flowing artesian well*. At a flowing artesian well, water actively spills or fountains out of the ground (**Fig. 16.8a**).

Water rises in an artesian well for the same reason that it comes out of your home faucet. So, to see why artesian wells exist, let's first examine the configuration of a city water supply (**Fig. 16.8b**). The city pumps water into an elevated tank, so that the water has a significant hydraulic head relative to the surrounding areas. A large pipe (the *water main*) runs from the tank into the community network, and a vertical pipe connects the water main to each house. Pressure caused by the height of the water in the tank pushes the water up each vertical pipe and into home plumbing. Now imagine that there were additional pipes, along the water main, that did not end at a faucet in a house. The water in the pipes would rise until it reached an imaginary plane, the *potentiometric surface*, that lies above the ground. At an artesian well (or artesian spring), groundwater is under pressure because it resides in a confined aquifer below the water's potentiometric surface. The confined aquifer serves the role of the water main, with impermeable conduit holding the water down at a depth below the level to which it would rise if not confined. Therefore, where there's a crack or hole in the aquitard above the aquifer, the water rises, and if the potentiometric surface lies

FIGURE 16.7 Pumping groundwater from a normal well can affect the water table

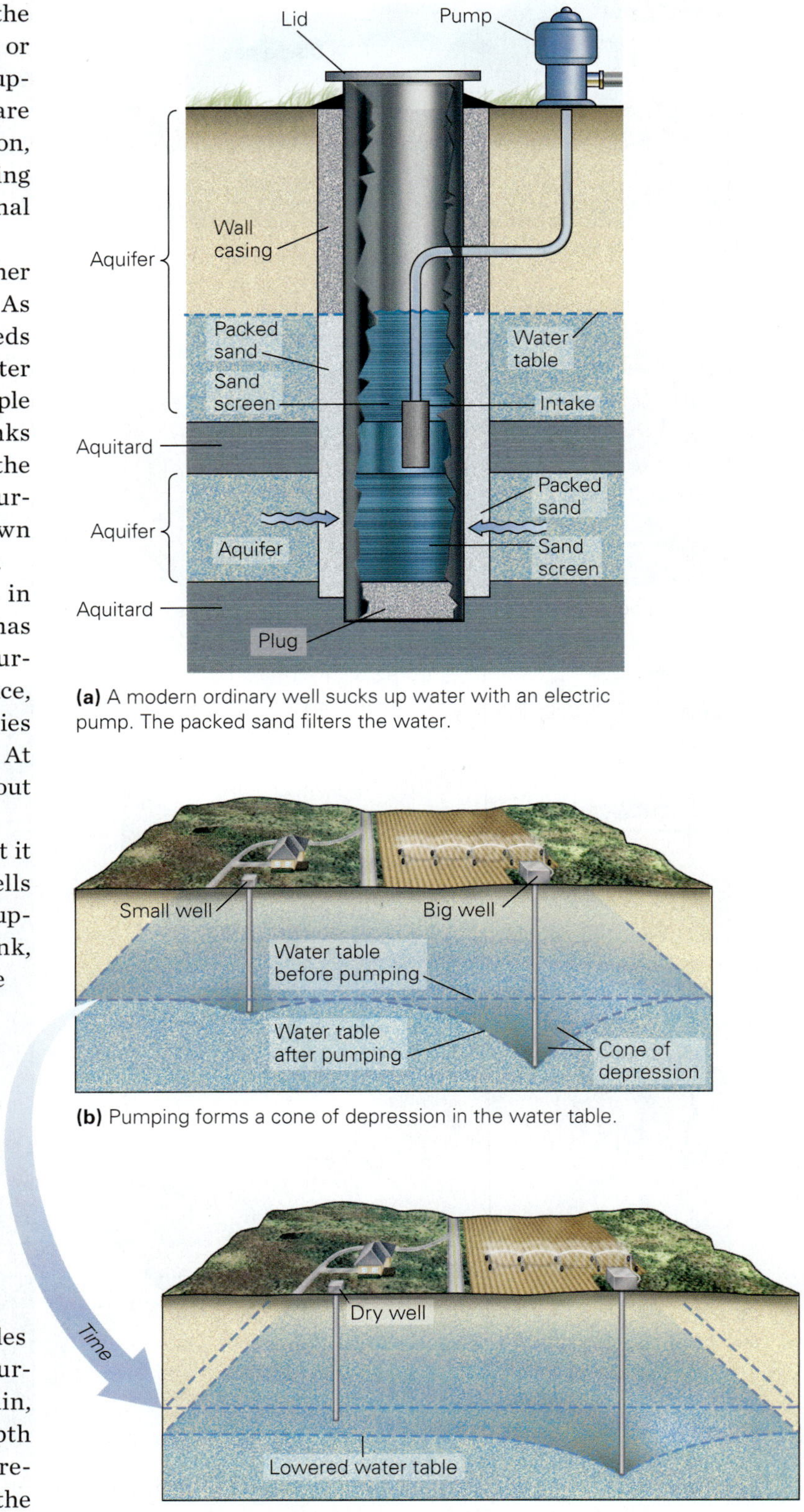

(a) A modern ordinary well sucks up water with an electric pump. The packed sand filters the water.

(b) Pumping forms a cone of depression in the water table.

(c) Pumping by the big well may be enough to make a small well run dry.

above the ground surface, the water flows out of the ground. The geologic situation leading to the occurrence of pressurized water in a confined aquifer happens where the recharge area for the aquifer lies at a high elevation, relative to adjacent lowlands, and is held beneath its potentiometric surface by an aquitard that slopes toward the lowlands (**Fig. 16.8c**).

Natural Groundwater Quality

Much of the world's groundwater is crystal clear, pure enough to drink right out of the ground. Rocks and sediment serve as natural filters capable of removing suspended solids—these solids get trapped in tiny pores or stick to the surfaces of clay flakes. It's no wonder that commercial distribution of bottled groundwater (labeled as "spring water") has become a major business worldwide.

Dissolved chemicals, however, locally may make some natural groundwater unusable. For example, groundwater that has been underground for a long time (in some cases, since the deposition of the strata containing it) or has passed through salt-containing strata may be too saline to be suitable for irrigation or for drinking. Because of its density, this groundwater tends to be deeper, so a boundary between fresh groundwater above and saline groundwater below may exist in the subsurface. Groundwater that has passed through limestone or dolomite contains dissolved calcium (Ca^{2+}) and magnesium (Mg^{2+}) ions. Such water, also known as *hard water*, can be a problem because carbonate minerals precipitate from it to form *scale* that clogs pipes. Also, washing with hard water can be difficult because soap won't develop a lather in hard water. Groundwater that has passed through iron-bearing rocks may contain dissolved iron oxide that precipitates to form rusty stains. Some groundwater contains dissolved hydrogen sulfide (H_2S), a poisonous gas with a rotten-egg smell that forms due to the metabolic processes of certain underground bacteria. This corrosive gas comes out of solution when groundwater rises to the surface and the pressure in it decreases. Groundwater that occurs above hydrocarbon reserves may contain dissolved methane, which also comes out of solution near the Earth's surface, to form flammable bubbles. In recent years, concern has grown about arsenic, a highly toxic chemical that enters groundwater when arsenic-bearing minerals dissolve in groundwater.

TAKE-HOME MESSAGE

Groundwater can be obtained at wells (built by people) and springs (natural outlets). In ordinary wells, water must be lifted to the surface, but in artesian wells and springs, it rises on its own. Pumping groundwater can cause a cone of depression, and may lower the surrounding water table. Groundwater can react with rock and sediment and can become mineralized and, in some cases, toxic.

QUICK QUESTION Why do some springs emerge along the side of a hill?

FIGURE 16.8 Artesian wells, where water rises from the aquifer without pumping.

(a) A flowing artesian well in Wisconsin; the water rises from underground in the corrugated pipe without the need of pumping, and it continuously flows out of the pipe.

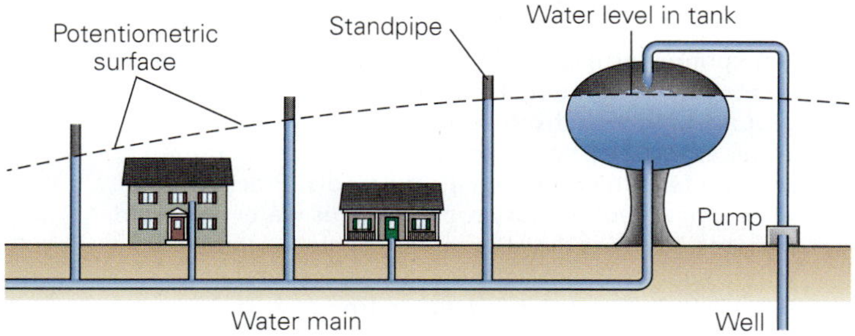

(b) The configuration of a city water supply. Water rises in vertical pipes up to the level of the potentiometric surface.

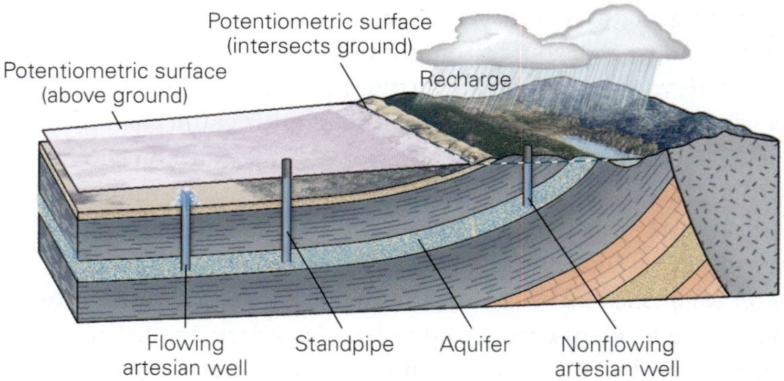

(c) The configuration of a regional artesian system.

16.5 Hot Springs and Geysers

Around 60 C.E., the Romans occupying Britain built a large public bathhouse in what is now the city of Bath, England. Soldiers and officials would spend hours relaxing in elaborate lead-lined pools. They would start in the caldarium (from the Latin *caldus*, meaning hot), and once they had become faint from the heat, they would move on to the tepidarium (from the Latin *tepidus*, meaning warm) to relax, and then would finish their visit by plunging into the frigidarium (from the Latin *frigidus*, meaning cold) to close their pores and cool down. When the Romans left Britain, in the 5th century, the baths were abandoned and eventually they fell to ruin. In the early 19th century, during the reign of Queen Victoria, the baths were resurrected and remain a tourist destination to this day. Why build a spa in Bath? Water at a temperature of 46°C (115°F) bubbles out of a natural spring there. In a general sense, a **hot spring** (or thermal spring) is a conduit from which groundwater ranging in temperature from about 30° to 104°C flows.

Why do hot springs exist? They can be found in two geologic settings. First, hot springs occur in places where faults or joints provide a high-permeability conduit for groundwater to rise from depths of at least a few kilometers within the crust. Rocks deep underground are naturally hot, due to the geothermal gradient, and will warm up any groundwater in contact with them. If this warmed water reaches the surface rapidly without being diluted, it will be still be hot. In some cases, groundwater rising at a hot spring followed a curved flow path that took it from an elevated recharge area down to a level deep within the crust, and then up to a discharge area in the lowlands, tens to hundreds of kilometers away. Second, hot springs develop in **geothermal regions**, places that are currently volcanically active, or where volcanism happened relatively recently. In geothermal regions, magma and/or very hot rock resides fairly close to the Earth's surface, so even shallow groundwater becomes hot (**Fig. 16.9a**).

Hot groundwater dissolves minerals from rock that it passes through. Water becomes a more effective solvent when hot, so hot springs tend to emit mineralized water. Some people believe that immersion in the hot mineral water of hot springs can cure ailments, but although this claim carries no scientific proof, without doubt, hot springs can be relaxing (**Fig. 16.9b**). In nature, the mineral content and temperature of hot springs can incubate life, and as a result, natural pools of geothermal water may be brightly colored—the gaudy greens, blues, and oranges of these pools come from thermophilic (heat-loving) bacteria and archaea that thrive in hot water and eat the sulfur-containing minerals dissolved in groundwater (**Fig. 16.9c**).

Numerous distinctive geologic features form in geothermal regions as a result of the eruption of hot water. In places where the hot water rises into soils rich in volcanic ash and clay, a viscous slurry forms and fills bubbling *mud pots*. Bubbles of steam rising through the slurry cause it to splatter about in goopy drops. Where geothermal waters spill out of natural springs and then cool, dissolved minerals in the water precipitate, forming colorful mounds or terraces of travertine and other chemical sedimentary rocks (**Fig. 16.9d**).

Under special circumstances, geothermal water emerges from the ground in a **geyser**, a fountain of steam and hot water that erupts episodically from a vent (**Fig. 16.9e**). The name comes from an Icelandic example, named Geysir, based on the Icelandic word *geysa*, meaning gush. To understand why a geyser erupts, we first need to picture its underground plumbing. Beneath a geyser, rock contains a network of irregular fractures—one of these serves as a conduit to the vent from which a geyser erupts. After a geyser eruption, the fractures are temporarily empty. They immediately begin to fill with groundwater, and soon, the geyser's conduit contains water again. Water at the top of the conduit cools down, because it's in contact with the air, while the water at depth remains *superheated*, meaning it remains liquid though its temperature exceeds 100°C, the boiling point of water at the Earth's surface. This deeper water can be superheated because the boiling point of water increases with increasing pressure. Cooler water near the top of the conduit cannot mix with the superheated groundwater below because the conduit is too narrow for convection to take place—the weight of the cooler water keeps the superheated water deeper down under pressure. Eventually, however, the top layer of superheated groundwater gets hot enough or shallow enough to start boiling. Expansion due to the formation of rising vapor bubbles pushes some of the cooler water at the top of the conduit out of the vent. This spillage suddenly decreases the weight of the water column above the superheated groundwater, and the sudden drop in pressure causes the superheated water to flash into vapor. This vapor quickly rises, forcefully ejecting all the water and steam above it out of the vent, as a geyser eruption. Once the conduit has emptied, the cycle starts again.

TAKE-HOME MESSAGE

Hot springs form either in geothermal regions, heated by magma below, or in localities where discharged groundwater rises from several kilometers down. The water emitted by hot springs tends to be mineralized, and can feed microbes in colorful pools. Precipitation of chemical sedimentary rock from this water can build terraces. Geysers develop under special circumstances where sudden boiling of superheated groundwater forcefully ejects water and steam from a vent.

QUICK QUESTION Why do geysers generally erupt episodically instead of continuously?

FIGURE 16.9 Geothermal waters and examples of their manifestation in the landscape.

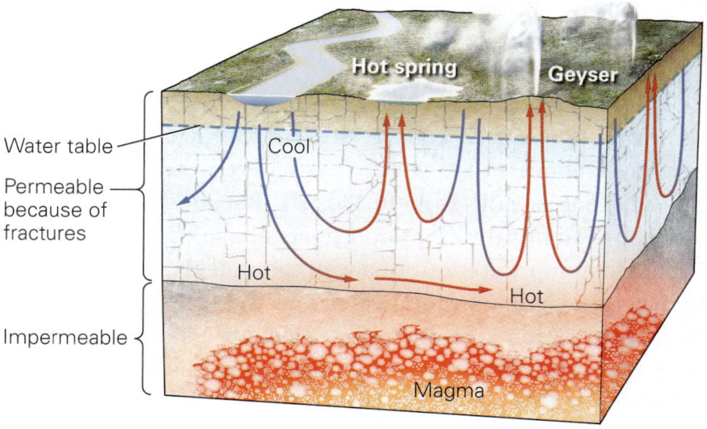

(a) Geysers and hot springs occur where groundwater, heated at depth, rises to the surface.

(b) Hot springs in Iceland, warmed by magma below, attract tourists from around the world.

(c) Colorful bacteria- and archaea-laden pools, Yellowstone National Park, Wyoming.

(d) Terraces of minerals precipitated at Mammoth Hot Springs, Yellowstone.

(e) The Old Faithful geyser in Yellowstone erupts predictably.

16.6 Groundwater Problems

Since prehistoric times, groundwater has been an important resource that people have used for drinking, irrigation, and industry. Dependence on groundwater has increased dramatically in the past century, because of population growth, because of the human preference to live in warmer climates, and because of the need to increase production of crops that require lots of water. (It takes about 20 liters, or 5.4 gallons, of water to grow just one head of broccoli.) Though groundwater accounts for about 95% of the liquid freshwater on the planet, accessible groundwater cannot always be replenished quickly enough, and this leads to shortages. Pollution of groundwater has also become a major concern. Such pollution may be invisible to us but may ruin a water supply for generations to come. In this section, we'll take a look at problems associated with the use of groundwater supplies.

Depletion of Groundwater Supplies

Is groundwater a renewable resource? In a time frame of 10,000 years, the answer is yes, for if we were to stop using groundwater, the natural hydrologic cycle would eventually resupply depleted reserves. But in a time frame of 10 to 1,000 years—the span of a human lifetime or a civilization—groundwater in many regions may behave like a *nonrenewable* resource. By pumping water out of the ground at a rate faster than nature replaces it, people effectively "mine" the groundwater supply and cause **groundwater depletion**, a decrease in the volume of groundwater. In fact, in portions of the desert southwest region of the United States, supplies of young groundwater have already been exhausted, and deep wells are now extracting 20,000-year-old groundwater. Some ancient water has been in rock so long that it has become too mineralized to be usable. A number of problems accompany the depletion of groundwater.

Lowering the Water Table When overpumping takes place, meaning that people extract groundwater from wells at a rate faster than it can be resupplied by nature, the water table drops. First, a cone of depression forms locally around the well, and then the water table gradually sinks over a broad region. As a consequence, existing wells, springs, rivers, and swamps dry up (**Fig. 16.10a, b**), and to continue tapping into the water supply requires drilling progressively deeper. In northern India, water demands of a rapidly growing population have caused the water table to drop at astounding rates, in some places as much as 6 m (20 ft) per year. Due to an intense drought in California, starting about 2011 and continuing through 2015, the water table within the agricultural district of the state's central valley has locally dropped by tens of meters (**Fig. 16.10c**). Without enough rain or snow in the recharge area, water is not available to infiltrate to the water table. Needless to say, this water-table drop is headline news, and a major political issue.

Notably, the water table can also drop when people divert surface water from the recharge area. Such a problem has developed in the Everglades of southern Florida. In this huge swamp, before the expansion of Miami and the development of agriculture, the water table lay at the ground surface. Diversion of water from the Everglades' recharge area into canals has significantly lowered the water table, causing parts of the Everglades to dry up (**Fig. 16.10d, e**).

Reversing the Flow Direction of Groundwater The cone of depression that develops around a well produces a local slope to the water table. The resulting hydraulic gradient may locally reverse the flow direction (**Fig. 16.11a, b**). Such reversals can allow contamination seeping out of a septic tank to head toward the well.

Saline Intrusion In coastal areas, fresh groundwater lies in a layer above saline water that entered the aquifer from the adjacent ocean (**Fig. 16.11c, d**). Because freshwater is less dense than saline water, it floats above the saline water. If people pump water out of a well too quickly, the boundary between the saline groundwater and the fresh groundwater rises. And if this boundary rises above the base of the well, then the well will start to yield useless saline water. Geologists refer to this phenomenon as *saline intrusion*.

Pore Collapse and Land Subsidence When groundwater fills the pore space of a rock or sediment, it holds the grains apart, for water cannot be compressed. The extraction of water from a pore eliminates the support holding the grains apart, because the air that replaces the water is a gas, and gases can be compressed. As a result, the grains pack more closely together. Such *pore collapse* decreases a rock's porosity and permeability (**Fig. 16.11e, f**). Some pores may refill if the water table rises again, but some collapses may be permanent.

Pore collapse also decreases the volume of materials underground, with the result that the ground above sinks. Such land subsidence may cause fissures at the surface to develop and the ground to tilt. Buildings constructed over regions undergoing land subsidence may themselves tilt, or their foundations may crack. In the San Joaquin Valley of California, the land surface subsided by 9 m between 1925 and 1975, because water was removed to irrigate farm fields. Land subsidence associated with groundwater removal in the Phoenix, Arizona, area has led to the development of open fissures at the land surface.

Human-Caused Groundwater Contamination

As we've noted, some contaminants (such as salts, methane, and arsenic) occur in groundwater naturally. Unfortunately, in recent decades, people have been responsible for introducing vast quantities of new contaminants into aquifers (**Fig. 16.12a**). These contaminants include agricultural waste (pesticides, fertilizers, and animal sewage), industrial waste (organic and inorganic chemicals), effluent from landfills and septic tanks (including bacteria and viruses), petroleum products and other chemicals that do not dissolve in water, radioactive waste (from weapons manufacture, power plants, and hospitals), and acids leached from sulfide minerals in coal and metal mines. The cloud of contaminated groundwater

FIGURE 16.10 Effects of human modification of the water table.

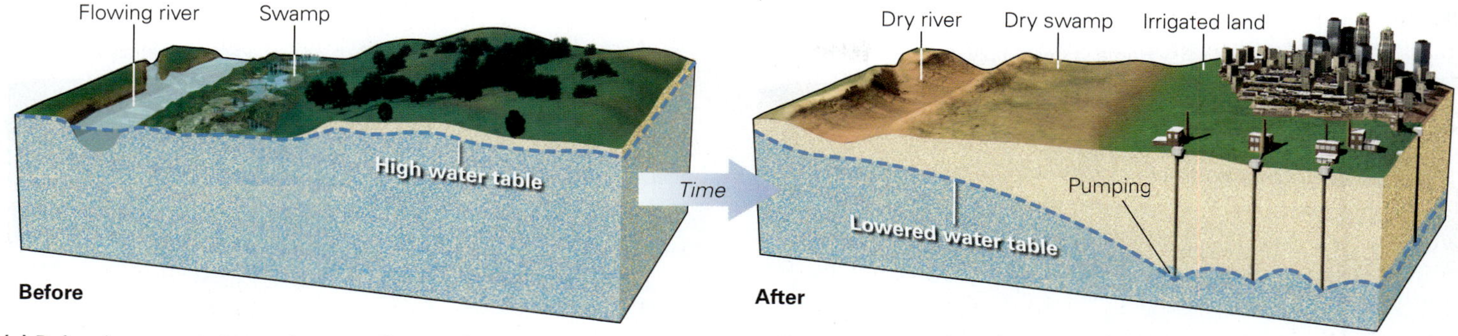

(a) Before humans start pumping groundwater, the water table is high. A swamp and permanent stream exist.

(b) Pumping for consumers in a nearby city causes the water table to fall, so the swamp dries up.

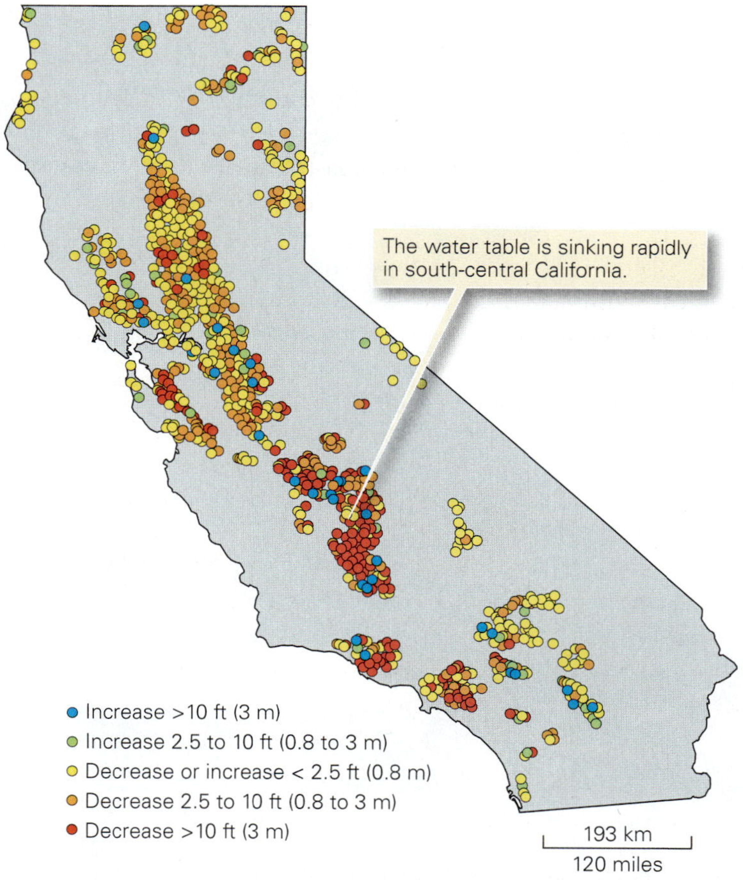

(c) A map showing variations in the drop of the water table in California, from the spring of 2013 to the spring of 2014, as measured in wells.

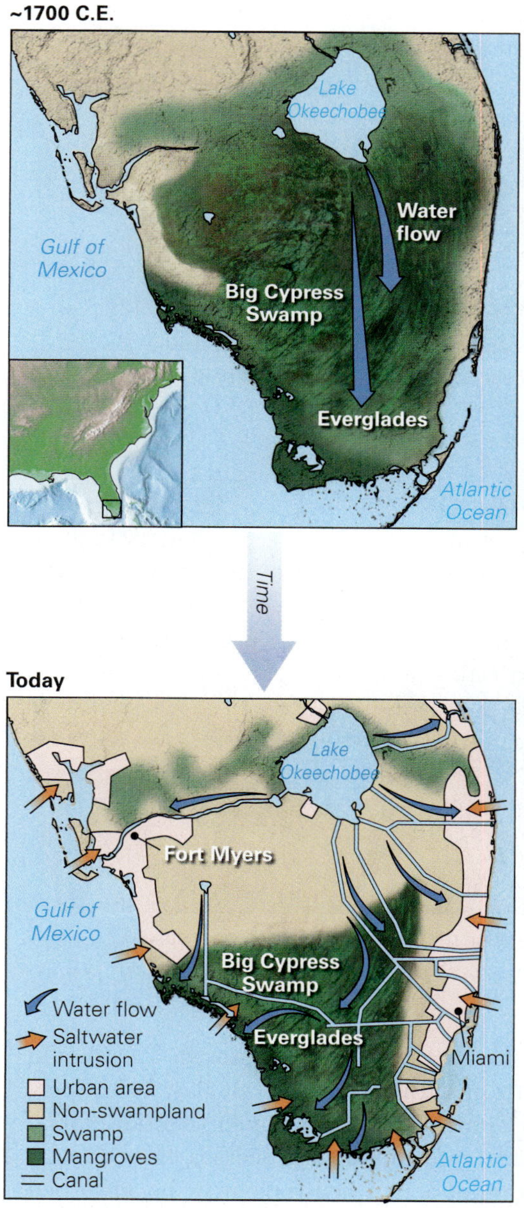

(d) The Florida Everglades before the advent of urban growth and intensive agriculture.

(e) Channelization and urbanization have removed water from recharge areas, disrupting flow paths.

that moves away from the source of contamination is called a **contaminant plume** (**Fig. 16.12b**).

The best way to avoid such **groundwater contamination**, or groundwater pollution, is to prevent waste products from entering groundwater in the first place. This can be done by placing contaminants in sealed containers or above impermeable bedrock so that they are isolated from

FIGURE 16.11 Some causes of groundwater problems.

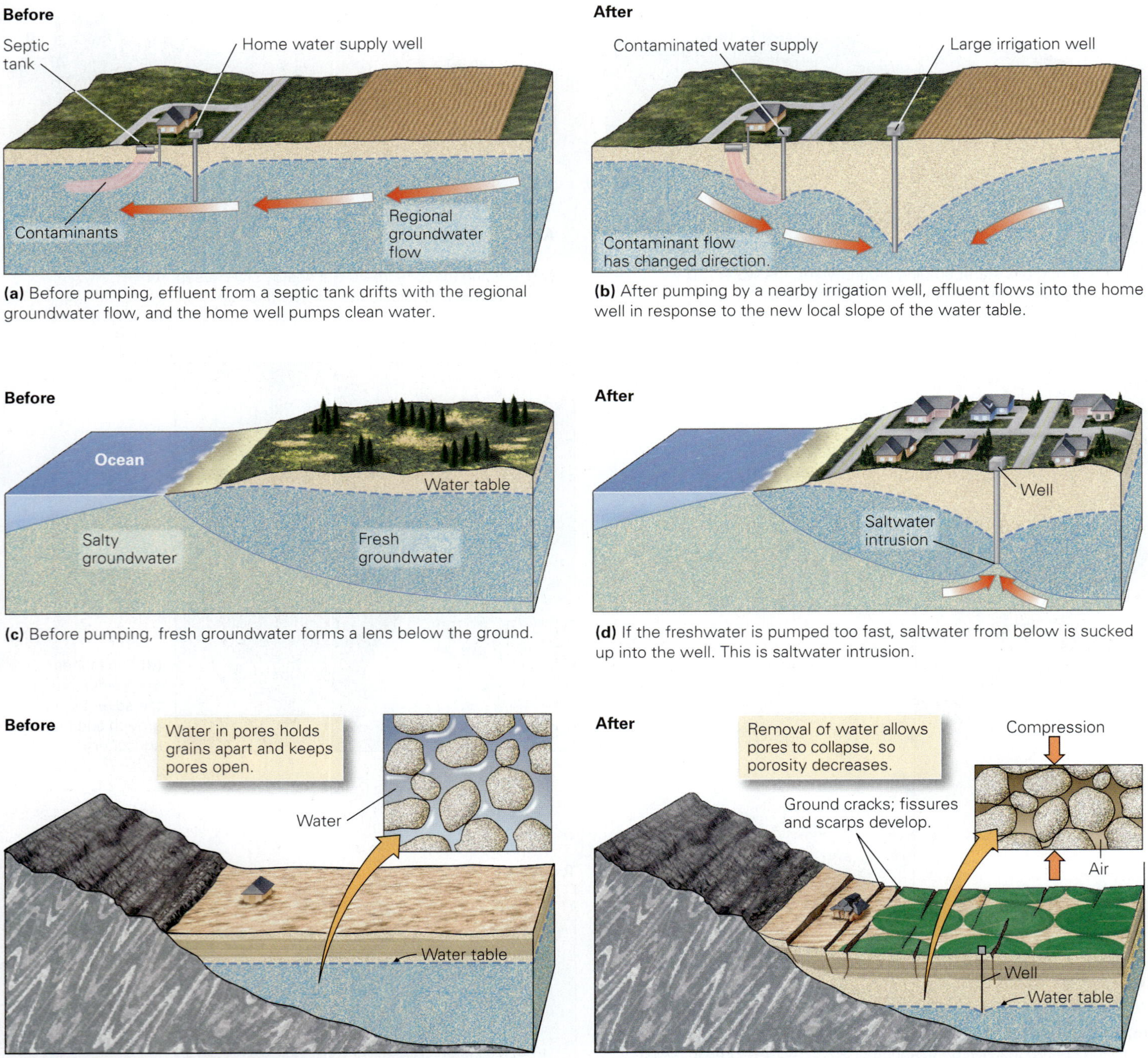

(a) Before pumping, effluent from a septic tank drifts with the regional groundwater flow, and the home well pumps clean water.

(b) After pumping by a nearby irrigation well, effluent flows into the home well in response to the new local slope of the water table.

(c) Before pumping, fresh groundwater forms a lens below the ground.

(d) If the freshwater is pumped too fast, saltwater from below is sucked up into the well. This is saltwater intrusion.

(e) When intensive irrigation removes groundwater, pore space in an aquifer collapses.

(f) As a result, the land surface sinks, leading to the formation of ground fissures and causing houses to crack.

aquifers. If such a site is not available, the storage area should be lined with plastic or with a thick layer of clay, for the clay not only acts as an aquitard, but it can bond to contaminants. Fortunately, in some cases, natural processes can clean up groundwater contamination. Chemicals may be absorbed by clay, oxygen in the water may oxidize the chemicals, and bacteria in the water may metabolize the chemicals, thereby turning them into harmless substances.

At locations where contaminants have entered an aquifer, environmental engineers can drill test wells to determine

FIGURE 16.12 Contamination plumes in groundwater.

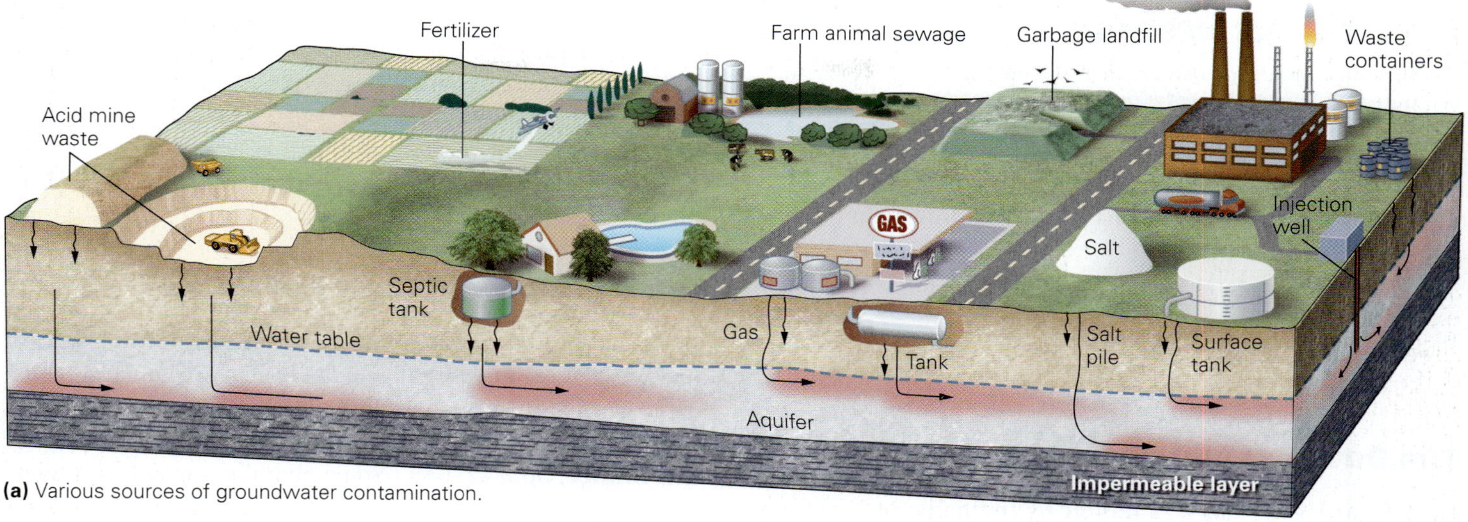

(a) Various sources of groundwater contamination.

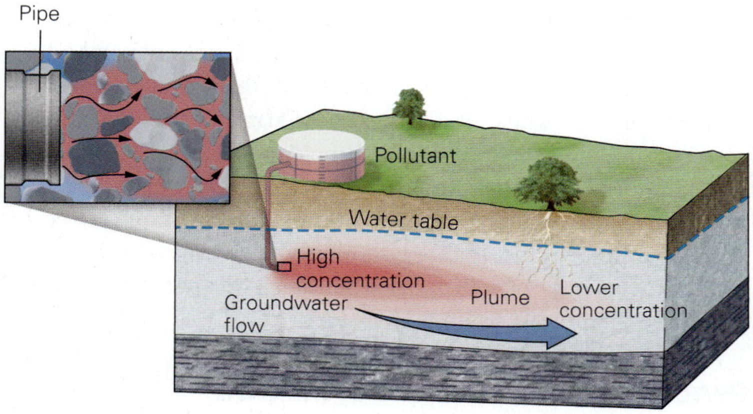

(b) A contaminant plume as seen in cross section. The darker the color, the greater the concentration of contaminant.

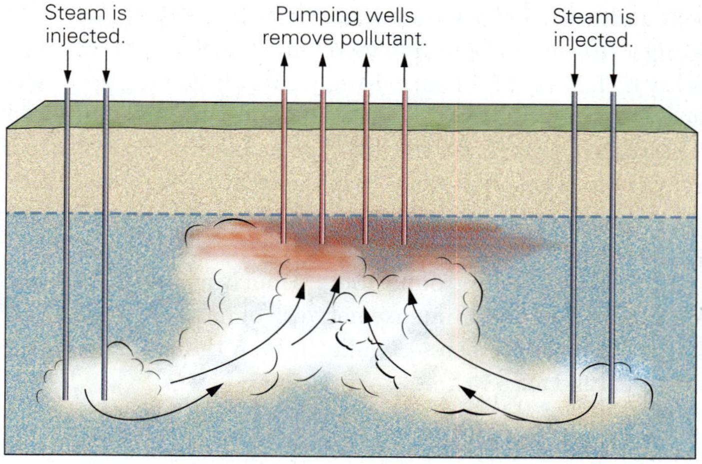

(c) Steam injected beneath the contamination drives the contaminated water upward in the aquifer, where pumping wells remove it.

which way and how fast the contaminant plume is flowing. Once they know the flow path, they can close wells in the path to prevent consumption of contaminated water. Engineers may also attempt to clean the groundwater by drilling a series of extraction wells to pump it out of the ground. If the contaminated water does not rise fast enough, steam or water can be pumped down injection wells into the ground beneath the contaminant plume. Such injection can push the contaminants up into extraction wells, where they can be pumped out (**Fig. 16.12c**).

More recently, environmental engineers have begun exploring techniques to stop plumes from migrating by burying or injecting certain chemicals in the path of the contaminated groundwater—the contaminants react with these chemicals to form solids that can't flow. And in some locations, engineers have tried **bioremediation**, a technique that involves injecting oxygen and nutrients into a contaminated aquifer to foster growth of bacteria that can break down molecules of contaminants.

Unwanted Effects of Rising Water Tables

We've seen the negative consequences of sinking water tables, but what happens when the water table rises? Is that necessarily good? Sometimes, but not always. If the water table rises above the level of a house's basement, water seeps through the foundation and floods the basement floor. Catastrophic damage occurs when a rising water table weakens the base of a hillslope or a failure surface underground triggers landslides and slumps.

TAKE-HOME MESSAGE

Groundwater usage can cause problems. Too much pumping lowers the water table, causing land subsidence and/or saltwater intrusion. Due to population growth and drought, loss of groundwater can have severe impacts on society. Contamination can ruin a groundwater supply.

QUICK QUESTION Can contaminants be removed from groundwater?

16.7 Caves and Karst

The Development of Caves

In 1799, as legend has it, a hunter by the name of Houchins was tracking a bear through the woods of Kentucky when the bear suddenly disappeared on a hillslope. Baffled, Houchins plunged through the brambles trying to sight his prey. Suddenly he felt a draft of surprisingly cool air flowing down the slope from uphill. Now curious, he climbed up the hill and found a dark portal into the hillslope beneath a ledge of rocks. Bear tracks were all around—was the creature inside? Houchins returned later with a lantern and cautiously stepped into the passageway. After walking a short distance, he found himself in a large, underground room. Houchins had discovered Mammoth Cave, an immense network of natural tunnels and subterranean chambers—a walk through the entire network would extend for 630 km!

Most large cave networks develop in limestone bedrock because limestone dissolves relatively easily in corrosive groundwater. Generally, the corrosive component in groundwater is dilute carbonic acid (H_2CO_3), which forms when water absorbs carbon dioxide (CO_2) from materials, such as soil, that it has passed through as it percolates down. When carbonic acid comes in contact with calcite ($CaCO_3$) in limestone, it reacts to produce HCO_3^{-1} and Ca^{2+} ions, which then dissolve. In recent years, geologists have discovered that about 5% of limestone caves around the world form due to reactions with sulfuric-acid-bearing water—Carlsbad Caverns in New Mexico serves as an example. Such caves form where limestone overlies strata containing oil, because microbes can convert the sulfur in the oil to hydrogen sulfide gas, which rises and reacts with oxygen to produce natural sulfuric acid. When this acid comes in contact with limestone, it reacts to produce gypsum and CO_2 gas.

Geologists debate about the depth at which limestone cave networks form. Some limestone dissolves above the water table. However, it appears that most cave formation takes place in limestone that lies just below the water table, for in this interval the acidity of the groundwater remains high, the mixture of groundwater and newly added rainwater has not yet saturated with dissolved ions, and groundwater flow is fastest. The association between cave formation and the water table helps explain why openings in a cave network align at the same level.

Did you ever wonder... why huge underground caverns form?

The Character of Cave Networks

As we have noted, caves in limestone usually occur as part of a network. Cave networks include rooms, or *chambers*, which are large, open spaces, sometimes with cathedral-like ceilings, and also tunnel-shaped or slot-shaped *passages*. (See **Geology at a Glance**, pp. 542–543.) Some chambers may host underground lakes, and some passages may serve as conduits for underground streams. The shape of the cave network reflects variations in permeability and in the composition of the rock from which the caves formed. Larger open spaces developed where the limestone was most soluble and where groundwater flow was fastest. Thus, in a sequence of strata, caves develop preferentially in the more soluble limestone beds. Passages in cave networks typically follow preexisting joints, for the joints provide openings along which groundwater can flow faster (**Fig. 16.13a**). Because joints commonly occur in orthogonal systems (two sets of joints oriented at right angles to each other; see Chapter 9), passages may form a grid.

Precipitation and the Formation of Speleothems

When the water table drops below the level of a cave, the cave becomes an open space filled with air. In some places downward-percolating groundwater containing dissolved calcite emerges from the ceiling or walls of the cave, which gradually changes the surface of the cave. As this water emerges into the air, it evaporates slightly and releases some dissolved carbon dioxide. As a result, calcite precipitates out of the water, producing *travertine* and various intricately shaped formations called **speleothems**.

Cave explorers (spelunkers) and geologists have developed a detailed nomenclature for different kinds of speleothems (**Fig. 16.13b**). Where water drips from the ceiling of the cave, the precipitated limestone builds *dripstone*. Initially, calcite precipitates around the outside of the drip, forming a delicate, hollow tube called a *soda straw*. But eventually, the soda straw fills up, and water migrates down the margin of the cone to form a more massive, solid icicle-like cone called a **stalactite**. Where the drips hit the floor, the resulting precipitate builds an upward-pointing cone called a **stalagmite**. If the process of dripstone formation in a cave continues long enough,

FIGURE 16.13 Development of karst, dripstone, and flowstone.

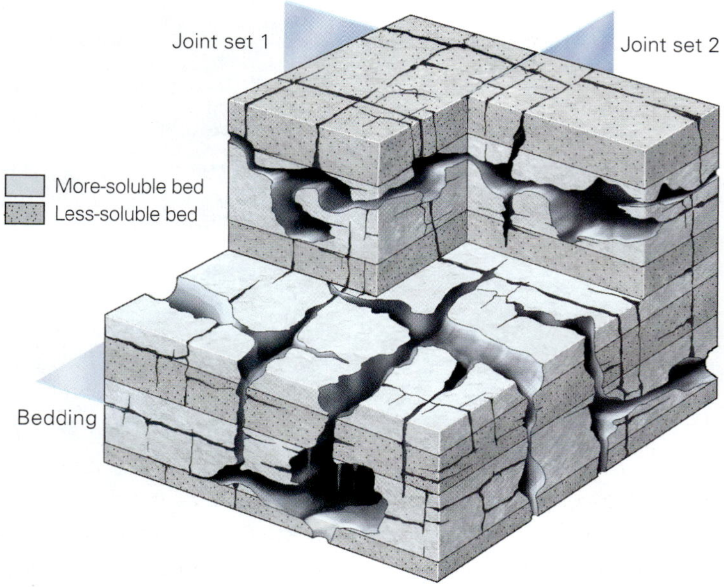

(a) Joints act as conduits for water in cave networks. Caves and passageways follow joints and preferentially form in more-soluble beds.

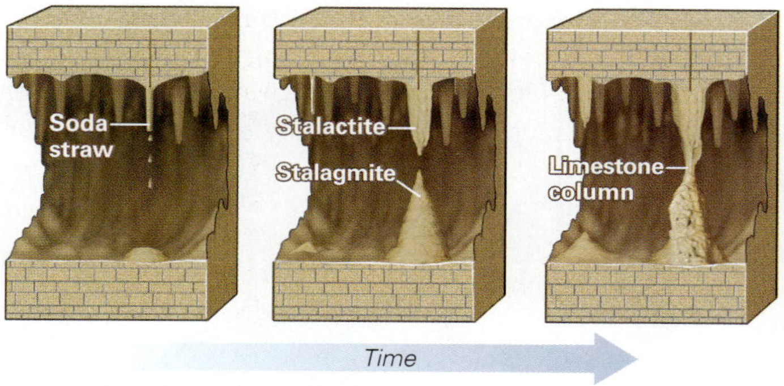

(b) The evolution of a soda straw stalactite into a limestone column.

(c) Flowstone on the wall of a cave in Vietnam.

stalagmites merge with overlying stalactites to produce travertine *columns*. In some cases, groundwater flows along the surface of a wall and precipitates to produce drape-like sheets of travertine called *flowstone* (**Fig. 16.13c**).

The Formation of Karst Landscapes

Limestone bedrock underlies most of the Kras Plateau in Slovenia, along the east coast of the Adriatic Sea. The name *kras*, meaning rocky ground, is appropriate for a region with abundant rock exposures (**Fig. 16.14a**). Geologists refer to such regions, where surface landforms develop when limestone bedrock dissolves both at the surface and in underlying cave networks, as **karst** landscapes or karst terrains—from the Germanized version of kras.

FIGURE 16.14 Features of karst landscapes.

(a) Karst terrains typically have a rough, rocky surface.

(b) A small disappearing stream in the Hudson Valley region of New York; the water is dropping into a subsurface cave.

GEOLOGY AT A GLANCE

Limestone pavement, Ireland

Sinkholes

Underground pool, Mexico

Natural Bridge, Virginia

Spelunker crawling in a cave

Caves and Karst Landscapes

Limestone is soluble in acidic water. Underground openings that develop by dissolution are called caves or caverns. Some of these may be large, open rooms, whereas others are long, narrow passages. Underground lakes and streams may cover the floor. A cave's location depends on the orientation of bedding and joints, for these features localize the flow of groundwater.

Caves originally form at or near the water table. As the water table drops, caves empty of most water and become filled with air. In many locations, groundwater drips from the ceiling of a cave or flows along its walls. As the water evaporates and loses its acidity, new calcite precipitates as speleothems (stalactites, stalagmites, columns, and flowstone).

Distinctive landscapes, called karst landscapes, develop at the Earth's surface over limestone bedrock. In such regions, the ground may be rough where rock has dissolved along joints. Where the roofs of caves collapse, sinkholes develop. If a surface stream flows into a cave network, we say that the stream is disappearing. In some places, the collapse of subsurface openings leaves behind natural bridges.

FIGURE 16.15 Tower karst forms a spectacular landscape in southern China.

The landscape is treeless today, a consequence of industrialization policies in the 1950s.

Chinese artists painted scrolls depicting forested towers of karst.

FIGURE 16.16 The progressive formation of caves and a karst landscape.

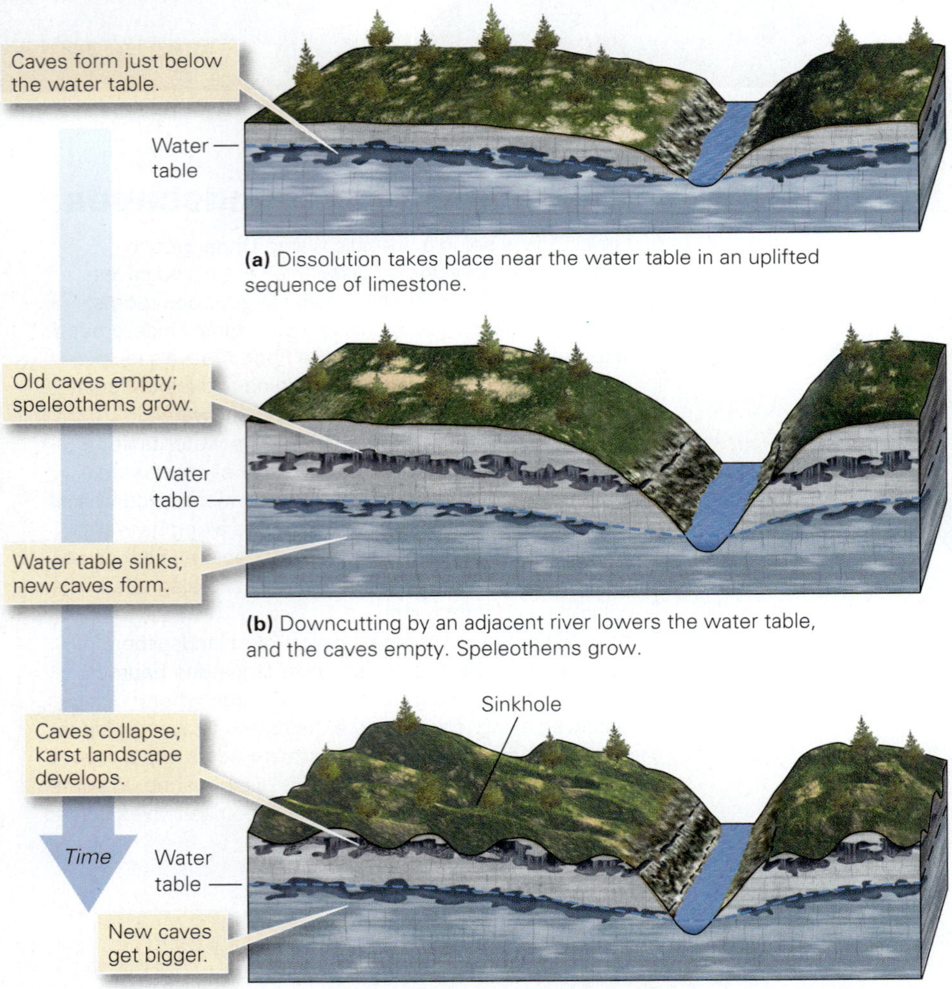
Caves form just below the water table.
Water table
Old caves empty; speleothems grow.
Water table
Water table sinks; new caves form.
Sinkhole
Caves collapse; karst landscape develops.
Time
Water table
New caves get bigger.

(a) Dissolution takes place near the water table in an uplifted sequence of limestone.

(b) Downcutting by an adjacent river lowers the water table, and the caves empty. Speleothems grow.

(c) After roof collapse, the landscape becomes pockmarked with sinkholes.

SEE FOR YOURSELF...

KARST LANDSCAPE IN PUERTO RICO

Latitude
18°23′53.04″ N

Longitude
66°25′49.43″ W

Zoom to 7 km (~4.3 miles) and look down.

We see a karst terrain in central Puerto Rico. Each of the rounded depressions is a sinkhole. Ridges of limestone separate adjacent sinkholes.

Karst landscapes typically display a number of distinct features. Where surface streams intersect cracks (joints) or holes that link to caverns or passageways below, the water cascades downward into the subsurface and disappears (**Fig. 16.14b**). Such **disappearing streams** may flow through passageways underground and later emerge from a cave entrance downstream. As we noted earlier, where the ground collapses into an underground cave below, a sinkhole develops (see Fig. 16.1). In cases where the ground collapses over a long, joint-controlled passage, sinkholes may be elongate and canyon-like. Over time, most of the cave eventually collapses, leaving the walls that once existed between caves as steep-sided ridges, and remnants of cave roofs as natural bridges. Over time, even the walls erode, leaving only jagged, isolated spires—a karst landscape dominated by such spires is called *tower karst*. The surreal collection of pinnacles constituting the tower karst landscape in the Guilin region of China has inspired generations of artists, who portray them in scroll paintings (**Fig. 16.15**). Karst terrains evolve over time (**Fig. 16.16**). In fact, while karst is forming by the collapse of a cave network closer to the surface, a new cave network may be developing at depth.

Life in Caves

Despite their lack of light, caves are not sterile, lifeless environments. Caves that are open to the air provide a refuge for bats as well as for various insects and spiders. Similarly, fish and crustaceans enter caves where streams flow in or out. Species living in caves have evolved some unusual characteristics. For example, cave fish lose their pigment and in some cases their eyes. Recently, explorers discovered caves in Mexico in which warm, mineral-rich groundwater currently flows. Colonies of bacteria metabolize sulfur-containing minerals in this water and produce thick mats of living ooze in the complete darkness of the cave. Long gobs of this bacteria slowly drip from the ceiling. Because of the mucus-like texture of these drips, they have come to be known as *snottites*.

TAKE-HOME MESSAGE

Reaction with natural acids dissolves limestone underground to form caverns. Most dissolution takes place near the water table. If the water table sinks, dripping of water in caves can produce speleothems. Collapse of a cavern network produces karst terrain.

QUICK QUESTION Can organisms live in the pitch black of caves? If so, how?

Another View Swamps are swamps because the land surface lies just below the water table. The bald cypress trees of this swamp in southern Illinois have adapted to life in soggy ground. During dry seasons, the water table sinks below the ground surface and the land dries out.

Chapter 16 Review

Chapter Summary

- During the hydrologic cycle, water infiltrates the ground and fills the pores and cracks in rock and sediment. This subsurface water is called groundwater. The amount of open space in rock or sediment is its porosity, and the ease with which liquid can flow through is its permeability.
- Aquifers are relatively permeable and aquitards are relatively impermeable.
- The water table is the surface in the ground above which pores contain mostly air, and below which pores are filled with water. The shape of a water table is a subdued mimic of the shape of the overlying land surface.
- Groundwater flows wherever the water table has a hydraulic gradient, and moves from recharge areas to discharge areas. The velocity of flow depends on permeability and the hydraulic gradient.
- Groundwater contains dissolved ions. Hard water contains a relatively high concentration of ions; scale in pipes forms when these ions precipitate.
- At a spring, groundwater exits the ground on its own. Springs form for many reasons.
- Groundwater can be extracted in wells. An ordinary well simply penetrates below the water table, so the water level in the well is the water table. In an artesian well, water rises under its own pressure.
- Hot springs and geysers release hot water to the Earth's surface. This water may have been heated by residing very deep in the crust, or by the proximity of a magma chamber.
- Groundwater is a precious resource, used for municipal water supplies, industry, and agriculture. In recent years, some regions have lost their groundwater supply because of overuse or contamination. Pumping water out of a well too fast causes drawdown, yielding a cone of depression.
- When limestone dissolves just below the water table, underground caves develop. Soluble beds and joints determine the location and orientation of caves. If the water table drops, caves empty out. Limestone then precipitates out of water dripping from cave roofs, and produces speleothems.
- Regions where abundant caves have collapsed to form sinkholes are called karst landscapes. These terrains can contain sinkholes, natural bridges, and disappearing streams.

Guide Terms

aquifer (p. 525)
aquitard (p. 525)
artesian spring (p. 530)
artesian well (p. 532)
bioremediation (p. 539)
capillary fringe (p. 526)
cone of depression (p. 532)
contaminant plume (p. 537)
Darcy's law (p. 530)
disappearing stream (p. 544)
discharge area (p. 529)
dry well (p. 532)
geothermal region (p. 534)
geyser (p. 534)
groundwater (p. 523)
groundwater contamination (p. 537)
groundwater depletion (p. 536)
hot spring (p. 534)
hydraulic head (p. 528)
karst (p. 541)
oasis (p. 530)
ordinary well (p. 532)
perched water table (p. 527)
permeability (p. 525)
pore (p. 525)
porosity (p. 525)
recharge area (p. 529)
sinkhole (p. 523)
soil moisture (p. 525)
speleothem (p. 540)
spring (p. 529)
stalactite (p. 540)
stalagmite (p. 540)
water table (p. 525)
well (p. 531)

GEOTOURS *THIS CHAPTER'S GEOTOUR EXERCISE (P) FEATURES:*

- Irrigation in the Saudi Desert
- Surface and Groundwater Flow
- Hot Springs
- Karst Features

Review Questions

1. How do porosity and permeability differ? Give examples of substances with high porosity but low permeability.
2. What is a water table, and what factors affect the level of the water table? What factors affect the flow direction of groundwater below the water table?
3. How does the rate of groundwater flow compare with that of moving ocean water or river currents? What factors control the rate of groundwater flow?
4. How does the chemical composition of groundwater change with time? What is hard water?
5. How does excessive pumping affect the local water table?
6. How is an artesian well different from an ordinary well?
7. Why do natural springs form? Explain why hot springs form and why geysers erupt.
8. Is groundwater a renewable or nonrenewable resource? What are the consequence of drought or overpumping?
9. Describe some of the ways in which human activities can adversely affect the water table.
10. What are some sources of groundwater contamination? How can contamination be prevented?
11. Describe the process leading to the formation of caves and the speleothems within caves.
12. Describe the various features of a karst landscape, and explain how they evolve.

On Further Thought

13. The population of Desert Paradise (a fictitious town in the southwestern United States) has been doubling every seven years. The town has no permanent streams or lakes anywhere nearby. In fact, the only standing water in the town occurs in the ponds of golf courses. The water in these ponds needs to be replenished almost constantly, for without supplementing it, the water seeps into the ground quickly and the ponds dry up. Desert Paradise has been growing on a flat, gravel-filled basin between two small mountain ranges. Where does the town water supply come from? What do you predict will happen to the water table of the area in coming years, and how might the land surface change as a consequence? Is there a policy that you might suggest to the residents of Desert Paradise that could slow the process of change?

Online Resources

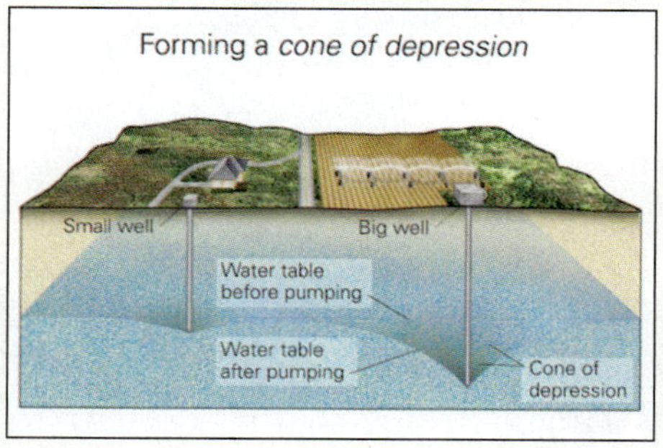

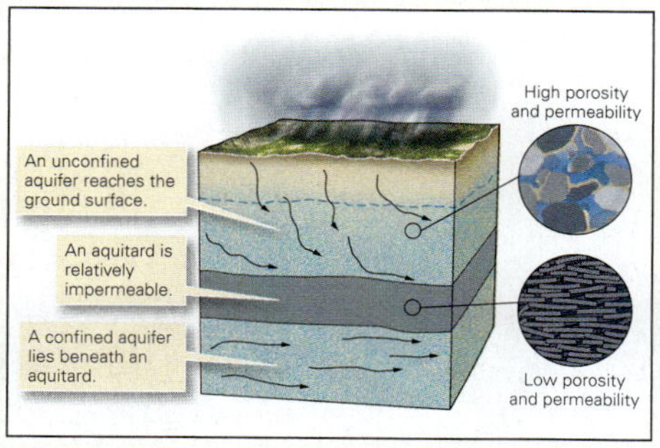

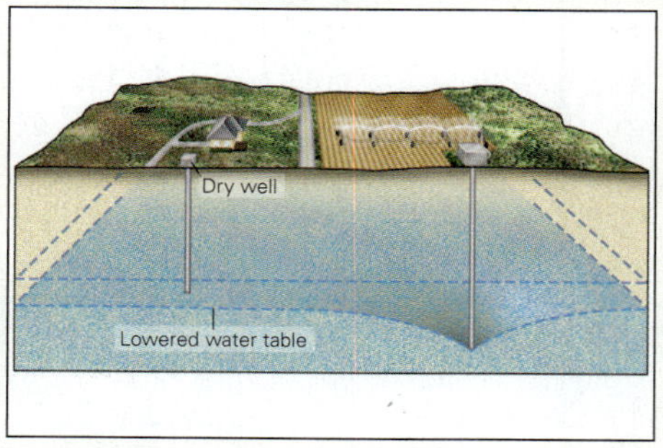

Videos

This chapter features a video on groundwater and the consequences of removing it too quickly.

Assessment

Questions in this chapter include visual labeling exercises on water tables, wells, and porosity and permeability.

LEARNING OBJECTIVES

By the end of this chapter, you should understand...

1. why certain regions are classified as deserts, and what factors cause these regions to be arid.
2. how weathering and erosional processes in deserts differ from those in temperate lands.
3. the nature and origin of landforms and landscapes in deserts.
4. why human activity may transform vegetated regions into deserts.

▲ Only particularly hardy shrubs can survive in this field of sand dunes, at the base of a barren rock ridge in the desert of Death Valley, California.

CHAPTER 17

Dry Regions: The Geology of Deserts

17.1 Introduction

For generations, nomadic traders have saddled camels to traverse the Sahara Desert in northern Africa (**Fig. 17.1a**). The Sahara, the world's largest desert, receives so little rainfall that it has hardly any surface water or vegetation. So camels must be able to walk for up to three weeks without drinking or eating. They can survive these journeys because they sweat relatively little, they have the ability to metabolize their own body fat to produce new water, and they can withstand severe dehydration.

The survival challenges faced by a camel emphasize that deserts are lands of extremes—extreme dryness, heat, cold, and, in some places, beauty. Desert vistas include everything from sand seas to sagebrush plains, cactus-covered hills to endless stony pavements. Although less populated than other regions on Earth, deserts cover about 25% of the land surface, and thus make up an important component of the Earth System (**Fig. 17.1b**). In this chapter, we take a look at deserts and their landscapes. We learn why deserts occur where they do, and how erosion and deposition shape their surface. We conclude by exploring life in the desert and by examining the problem of desertification, the gradual transformation of temperate lands into desert.

FIGURE 17.1 Deserts and their hardy inhabitants.

(a) Camels can survive the harsh conditions of a desert.

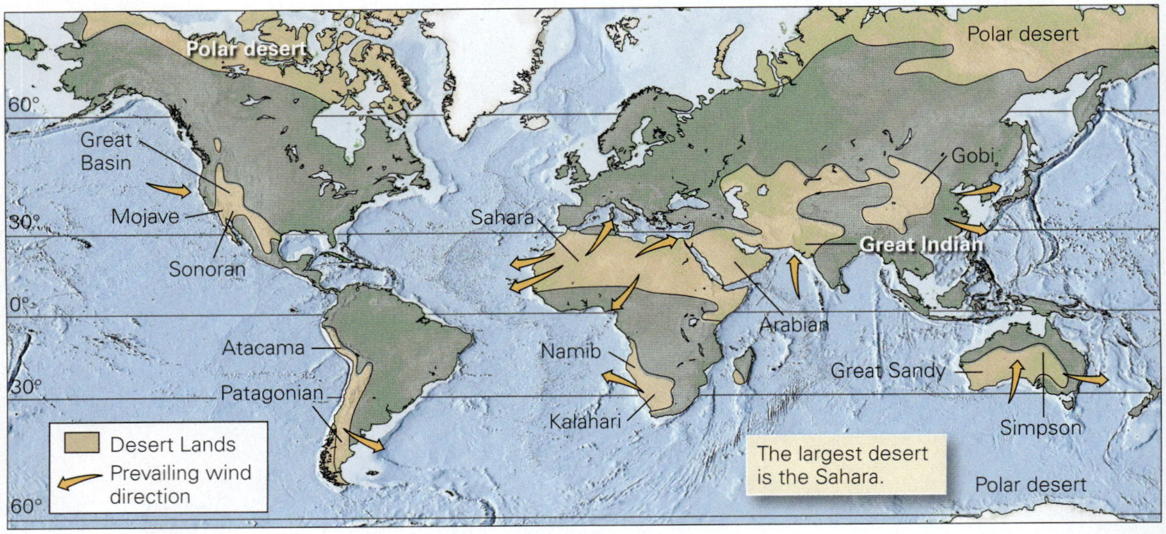

(b) The global distribution of deserts. Arid regions cover 25% of Earth's land surface.

> *The bare hills are cut out with sharp gorges, and over their stone skeletons scanty earth clings. . . . A white light beat down, dispelling the last trace of shadow, and above hung the burnished shield of hard, pitiless sky.*
>
> CLARENCE KING (1842–1901), first director of the U.S. Geological Survey, describing a desert

17.2 The Nature and Locations of Deserts

What Is a Desert?

Formally defined, a *desert* is a region that is so arid (dry) that it supports vegetation on no more than 15% of its surface. In general, desert conditions exist where less than 25 cm of rain falls per year, on average. Because of the lack of water, deserts contain no permanent streams, except for those that bring water from temperate regions elsewhere.

Note that the definition of a desert depends on a region's aridity, not on its temperature. Geologists distinguish between *cold deserts*, where temperatures generally stay below about 20°C all year, and *hot deserts*, where daytime temperatures exceed 35°C in the summer. Cold deserts develop at high latitudes where the Sun's rays strike the Earth obliquely and thus don't provide much energy, at high elevations where thin air can't hold much heat, or in lands adjacent to cold ocean currents where seawater can absorb heat from the air above. Hot deserts develop at low latitudes where the Sun's rays strike the land at a high angle, at low elevations where dense air can hold a lot of heat, and in regions distant from the cooling effect of cold ocean currents. The hottest recorded temperatures on Earth—58°C (136°F) in Libya, and 57°C (135°F) in Death Valley, California—occur in low-latitude, low-elevation deserts.

Did you ever wonder... how hot it can become in a desert?

Types of Deserts

Each desert on Earth has unique landscapes and vegetation that distinguish it from others. Geologists group deserts into five different classes, based on the environment in which the desert forms (**Fig. 17.2**).

› *Deserts formed in the subtropics*: The world's largest deserts (such as the Sahara, Arabian, Kalahari, and Australian) form because of the global pattern of air circulation in the atmosphere. At the equator, near-surface air becomes warm and humid, for sunlight is intense and water rapidly evaporates from the ocean. The hot, moisture-laden air rises to great heights above the equator. As this air rises, it expands and cools, and can no longer hold so much moisture, so the water it contains condenses and falls in downpours that feed the lushness of the equatorial rainforest.

FIGURE 17.2 Subtropical deserts form because the air that convectively flows downward in the subtropics warms and absorbs water as it sinks.

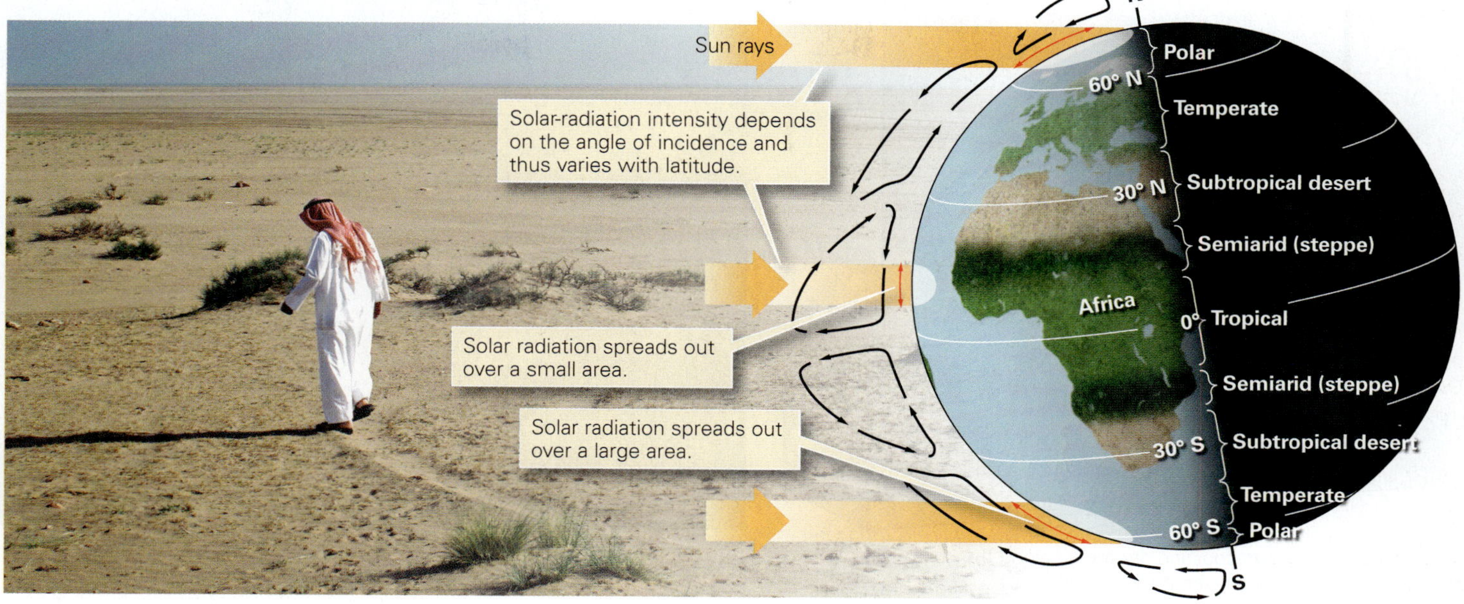

The now-dry air high in the troposphere spreads laterally north or south. When this air reaches latitudes of 20° to 30°, a region called the subtropics, it has become cold and dense enough to sink. Because the air is dry, no clouds form, and intense solar radiation strikes the Earth's surface all day. The sinking, dry air becomes denser and warmer, soaking up any moisture present. In a *subtropical desert* swept by this hot air as it heads back to the equator, evaporation rates greatly exceed rainfall rates, so the land becomes parched.

› *Deserts formed in rain shadows*: As moist air flows from the sea toward a coastal mountain range, the air must rise (**Fig. 17.3**). The rising air expands and cools, so the water it contains condenses and falls as rain on the seaward flank of the mountains, where it drenches a coastal rainforest. When the flowing air finally reaches the inland side of the mountains, it has lost all its moisture and can no longer provide rain. As a consequence, a *rain shadow* forms, and the land beneath the rain shadow becomes a desert. An example of such a *rain-shadow desert* lies east of the Cascade Mountains in the state of Washington.

› *Deserts formed near cold ocean currents*: Cold ocean water cools the overlying air by absorbing heat, thereby decreasing the capacity of the air to hold moisture. For example, the cold Humboldt Current, which carries water northward from Antarctica to the western coast of South America, cools the air that blows east, over the coast. The air becomes so dry, by the time it reaches the coast, that it can't provide any water to coastal areas of Chile and Peru, so this region hosts *coastal deserts*, including the Atacama Desert (**Fig. 17.4**). Astonishingly, portions of the Atacama received no rain at all between 1570 and 1971.

› *Deserts formed in the interiors of continents*: As air masses move across a continent, they progressively lose moisture by dropping rain, even in the absence of a coastal mountain range. Thus, by the time an air mass reaches the interior of a broad continent, it has become so dry that the land beneath becomes arid. The largest present-day example of such a *continental-interior desert* is the Gobi, in central Asia.

FIGURE 17.3 The formation of a rain-shadow desert. Moist air rises and drops rain on the coastal side of the range. By the time the air has crossed the mountains, it is dry.

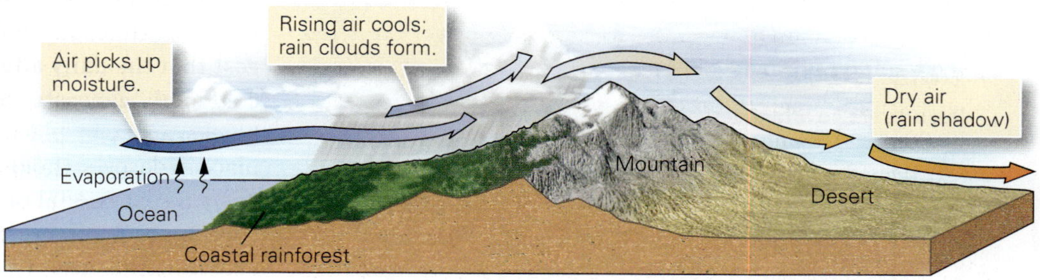

FIGURE 17.4 The formation of a coastal desert.

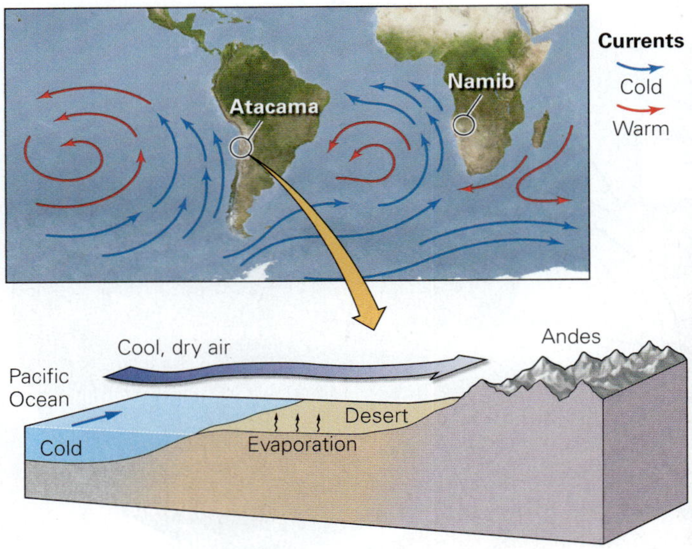

(a) Cold ocean currents cool and dry the air along the coast. This air absorbs moisture from adjacent coastal land.

(b) The Atacama Desert is the driest place on Earth.

› *Deserts of the polar regions*: So little precipitation falls in Earth's polar regions (north of the Arctic Circle and south of the Antarctic Circle) that these areas are, in fact, arid. Such *polar deserts* are dry, in part, for the same reason that the subtropics are dry (the global pattern of air circulation means that the air flowing over these regions is dry), and in part, for the same reason that coastal areas along cold currents are dry (cold air holds little moisture).

Different regions of the land surface have become deserts at different times in the Earth's history, because plate movements change the latitude of landmasses, the position of landmasses relative to the coast, and the proximity of landmasses to a mountain range. Because of plate tectonics, some regions that were deserts in the past are temperate or tropical regions now, and vice versa.

TAKE-HOME MESSAGE

Deserts are so arid that they host only very sparse vegetation. They occur in several settings: subtropical dry climates, rain shadows, coasts bordered by cold currents, continental interiors, and polar regions.

QUICK QUESTION Why does the world's largest desert, the Sahara, exist?

17.3 Producing Desert Landscapes

Weathering and Soil Formation in Deserts

If you stand in the Mojave Desert of California and look around, you'll see barren cliffs exposing fractured rock, and slopes or plains littered with gravel (**Fig. 17.5a**). Clearly, physical weathering happens in deserts—over time, blocks break off from escarpments at joints, and tumble downslope, perhaps shattering into smaller pieces as they collide with other blocks on the way down. In some cases, **talus**, a steeply sloping apron of debris, forms at the base of a cliff (**Fig. 17.5b**). A few tumbling blocks may bounce and roll out beyond the talus onto gentler slopes, then sit as isolated blocks resting on finer sediment. Blocks of rock in a desert can remain unchanged for a long time, but they don't last forever. Recent studies suggest that the daily alternation from midday heat to midnight cold might generate sufficient stress to gradually break small blocks apart in place. And chemical weathering does take place in deserts, though it happens very slowly. Moisture from dew and occasional rain allows oxidation, hydrolysis, and dissolution reactions to destroy cements and to transform silicate minerals into clay, thereby causing rocks to disaggregate into pebble- or sand-sized debris.

FIGURE 17.5 Rocky surfaces in deserts.

(a) The dark gravel on this slope, beneath a cliff in the Mojave Desert, washed down from the cliffs above.

(b) This talus apron along the base of a desert cliff formed from rocks that broke off and tumbled down the cliff.

FIGURE 17.6 Soil formation and chemical weathering in deserts.

(a) The red hues of the Painted Desert in Arizona are due to the oxidation of iron in the rock.

(b) By chipping away desert varnish to reveal the lighter rock beneath, Native Americans created art and symbols.

Does soil develop in a desert? Yes. In places where regolith stays put long enough, unique desert soils develop slowly. After a rare downpour, infiltration of water leaches soluble ions, such as Ca^{+} and CO_3^{2-}, and transports them downward (see Interlude B). But since such infiltration events happen infrequently, the leached ions don't flush away entirely. Rather they precipitate underground to form new calcite cement that can bind regolith into a solid, rock-like material known as *caliche* or *calcrete*.

Bedrock color controls land-surface color in deserts. Variations in the amounts of iron, or in the extent of iron oxidation, can result in spectacular color banding. Specifically, well-oxidized strata are dark red or maroon, strata with moderate iron are light red or orange, strata in which iron has been chemically reduced are whitish or greenish, and strata without much iron are tan or gray. The Painted Desert of northern Arizona earned its name from the brilliant and varied hues of oxidized iron in the region's shale bedrock (**Fig. 17.6a**). Where desert soils do form, soil color tends to resemble that of the bedrock from which the soil was derived, for without organic content mixed in, soils don't display the black or brown colors typical of temperate soils.

Desert Varnish

Shiny **desert varnish**, a dark, rusty brown coating of iron oxide, manganese oxide, and clay, locally coats the surface of rocks in deserts. Desert varnish may form when wind-borne dust sticks to the surface of rock. Microbes in the dust extract metal ions and convert them into oxides. Desert varnish doesn't form in humid climates because frequent rain washes dust away before it has time to be converted into varnish.

Desert varnish takes a long time to form. By measuring the thickness of a desert varnish layer, geologists can provide an estimate of how long a rock has been exposed at the ground surface. Over the centuries, people have used desert-varnished rock as a medium for art. Chipping away the varnish to reveal underlying lighter-colored rock is a way to create figures or symbols known as **petroglyphs** (**Fig. 17.6b**).

Water Erosion

Although rain rarely falls in deserts, when it does come, it can radically alter a landscape in a matter of minutes. In the absence of plant cover, rainfall, sheetwash, and stream flow are all extremely effective agents of erosion. It may seem surprising, but more erosion in deserts is due to running water than to wind (**Fig. 17.7a**).

Water erosion begins with the impacts of raindrops, which send bits of sediment from the ground into the air. On a hill, the ejected sediment lands downslope. The ground quickly becomes saturated with water during a heavy rain, so water starts flowing across the surface, carrying the loose sediment with it. Within minutes after a heavy downpour begins, dry stream channels fill with a turbulent mixture of water and sediment, which rushes downstream as a flash flood. Because it is fast and relatively viscous (owing to a load of suspended sediment), water in flash floods can cause intense erosion. Streams in deserts tend to be ephemeral, meaning that when the supply of rainwater stops, the water sinks into the gravel of the stream bed and disappears (see Chapter 14). The dry channels of ephemeral streams in desert regions are called *dry washes* or *arroyos* in southwestern North America, and *wadis* in the Middle East and North Africa (**Fig. 17.7b**).

FIGURE 17.7 Evidence of erosion by running water in deserts.

(a) Even though these hills in the desert near Las Vegas, Nevada, are bone dry, their shape indicates erosion by water. Note the numerous stream channels.

(b) A flash flood left behind gravel and sand on the floor of a dry wash after a flash flood in Death Valley. Erosion by the water is cutting a channel.

Wind Erosion

In temperate and humid regions, plant cover protects the ground surface from the wind, but in deserts, the wind has direct access to the ground and can cause substantial **wind erosion**, both by picking up and transporting sediment, and by abrading and scouring the surface. The sediment moved by wind includes both suspended load and surface load. **Suspended load**, fine-grained sediment such as dust and silt, held aloft due to air flow, can be carried so high into the atmosphere (up to several kilometers above the Earth's surface) and so far downwind (tens to thousands of kilometers) that it may move completely out of its source region. A particularly strong wind can generate a dramatic *dust storm*, known as a *haboob* in the Middle East, that can be 100 km long and 1.5 km high. A dust storm, as it approaches, looks like a roiling, opaque wave or cloud. Dust storms can be very hazardous, because they decrease visibility and foul machinery. When a particularly large dust storm engulfed Phoenix, Arizona, in 2011, the airport had to shut down (**Fig. 17.8a**). Strong winds can also drive a **surface load**, sediment that undergoes **saltation**, the process of rolling and/or bouncing along the ground (**Fig. 17.8b**). Saltation begins when turbulence caused by wind shearing along the ground surface lifts sand grains. The grains move downwind, following an asymmetric, arch-like trajectory. Eventually, they return to the ground, where they strike other sand grains, causing the new grains to bounce up and drift or roll downwind. The collisions cause sand grains to become rounded and frosted. Saltating grains generally rise no more than 0.5 m, but where they bounce on bedrock, grains may rise as much as 2 m.

Since most sand consists of quartz, a hard mineral, saltating sand not only can strip the paint off a car, but it can also abrade rock surfaces in the desert and, over long periods, can carve smooth faces, or *facets*, on pebbles, cobbles, and boulders. If a rock rolls or tips relative to the prevailing wind direction after it has been faceted on one side, or if the wind shifts direction, a new facet with a different orientation

FIGURE 17.8 Sediment transport by wind in deserts.

(a) A huge dust storm approaching Phoenix, Arizona.

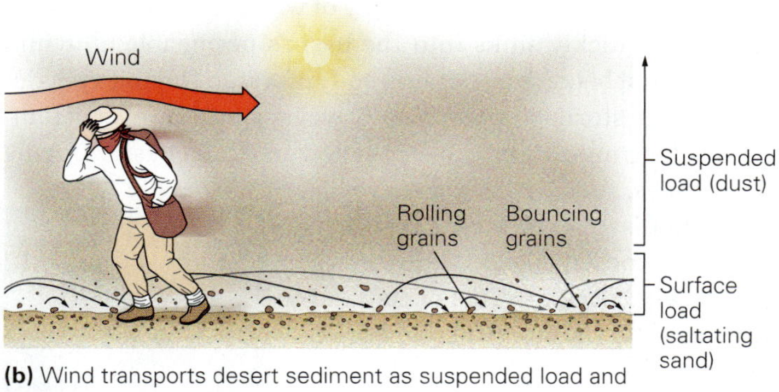

(b) Wind transports desert sediment as suspended load and surface load.

FIGURE 17.9 The progressive development of a ventifact.

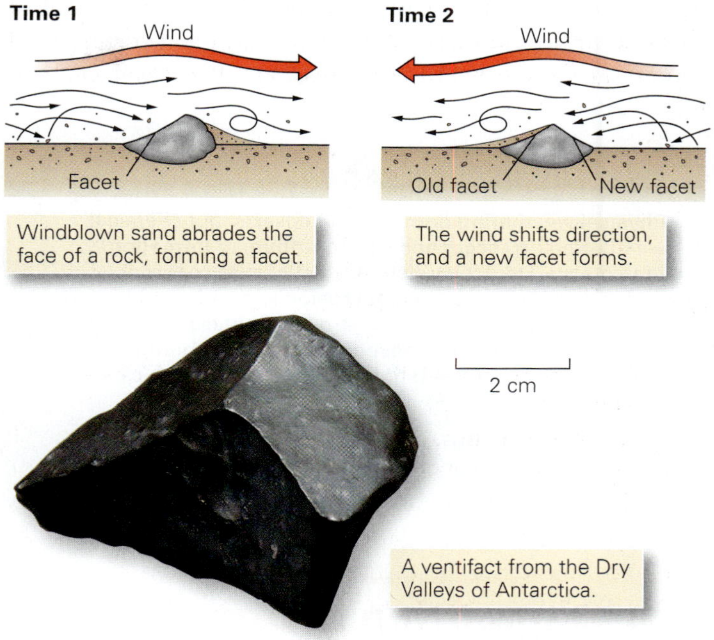

forms, and the two facets join at a sharp edge. Rocks whose surface has been faceted by the wind are called **ventifacts** (**Fig. 17.9**). Such abrasion also gradually polishes the surfaces of desert-varnished outcrops into a reflective sheen. Where wind blowing constantly in the same direction carves strata of alternating resistant and nonresistant layers, it may produce *yardangs*, streamlined caps of resistant rock resting on pillars of nonresistant rock.

The size of clasts that wind can carry depends on the wind velocity, so wind does an effective job of sorting sediment. Dust-sized particles loft skyward and sand-sized particles bounce along the ground, while pebbles and larger grains remain behind. Over time, in regions where the substrate consists of soft sediment, wind picks up and removes so much sediment that the land surface becomes lower, a process known as **deflation**. Shrubs can stabilize a small patch of sediment with their roots, so after deflation a forlorn shrub with its residual pedestal of soil stands isolated above a lowered ground surface (**Fig. 17.10a**). In some places, wind carries away so much fine sediment that pebbles and cobbles become concentrated at the ground surface (**Fig. 17.10b**). Geologists refer to

FIGURE 17.10 Erosion of sediment by desert winds.

(a) Wind erosion in Death Valley has left bushes perched on mounds; roots keep soil from blowing away.

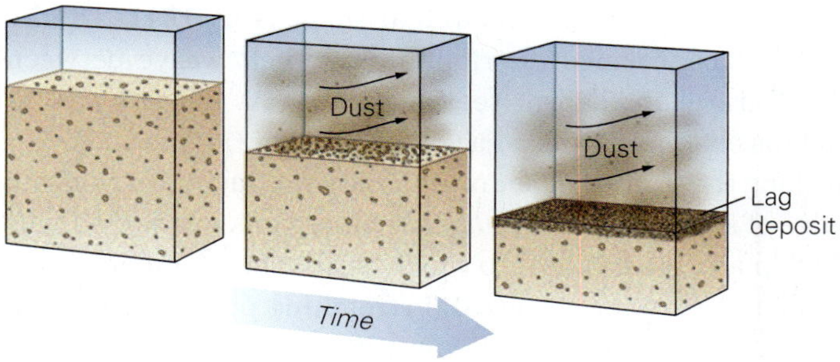

(b) A lag deposit develops when wind blows away finer sediment, leaving behind a layer of coarser grains.

an accumulation of coarser sediment left behind when fine-grained sediment blows away as a **lag deposit**. If the shape of the land surface twists the wind into a turbulent vortex, wind erosion can scour a bowl-like depression called a *blowout*.

TAKE-HOME MESSAGE

Physical weathering dominates in deserts, but chemical weathering can occur and can contribute to formation of thin soils. Desert varnish may develop to coat rock surfaces. Water causes most erosion and sediment transport but rarely flows, so stream channels are usually dry washes. Wind can transport sediment, generating dust storms and causing saltation. Saltating grains can carve and polish rock.

QUICK QUESTION How is weathering in a desert similar to or different from that of a temperate region?

17.4 Deposition in Deserts

We've seen that erosion relentlessly eats away at bedrock and sediment in deserts. Where does the debris go? Below, we examine some desert settings where sediment accumulates.

Alluvial Fans

Flash floods can carry large amounts of sediment downstream in a steep-walled canyon. When the turbulent floodwater—sometimes viscous enough to be called a debris flow—emerges onto a plain at the mouth of a canyon, it spreads out over a broader surface and slows so that, as a consequence, sediment in the water settles out. The single channel that emerges from the mountains tends to subdivide into subchannels or distributaries that spread outward in different directions. Which distributary becomes active during a given flood varies over time, because as sediment gets deposited at the mouth of a distributary, the channel's gradient decreases until, eventually, a flood will find a different, steeper distributary to follow. Sediment (alluvium) deposited by the network of distributaries yields a broad **alluvial fan**, a wedge- or apron-shaped pile whose downslope edge has a semicircular shape (**Fig. 17.11**). On the fan's surface, braided distributary channels weave between bars of gravel and debris (see Chapter 14). As they grow, alluvial fans emerging from adjacent valleys may merge and overlap along the front of a mountain range, producing an elongate wedge of sediment called a *bajada*. In regions where rifting is taking place, the basins into which bajadas build undergo slow subsidence (sinking) as crustal stretching proceeds. In such regions, alluvium may accumulate into a layer that is several kilometers thick (see **Geology at a Glance**, p. 558–559).

FIGURE 17.11 This alluvial fan accumulated at the mouth of a small canyon in Death Valley, California.

Playas and Salt Lakes

Water from larger flash floods may flow to the center of an alluvium-filled basin. If the supply of water is relatively small, the water quickly sinks into the basin's permeable alluvium and does not become a standing body of water. During a particularly large storm or an unusually wet spring, however, a temporary lake may develop over the low part of a basin. During drier times, such desert lakes evaporate and disappear, leaving behind a dry, flat, exposed lake bed known as a **playa** (**Fig. 17.12**). A crust of clay and various salts (halite, gypsum, borax, and other minerals) accumulates on the surface of a playa.

Where sufficient water flows into a desert basin, a permanent lake forms. If the basin has no outlet, the lake becomes very salty because, although the lake's water escapes by evaporation in the desert sun, its salt stays in place. The Great Salt Lake, in Utah, serves as an example. Even though the streams feeding the lake are fresh enough to drink, their water contains trace amounts of dissolved ions. Because the lake has no outlet, these ions have become concentrated in the lake over time, making it even saltier than the ocean.

SEE FOR YOURSELF...

DEATH VALLEY, CALIFORNIA

Latitude
36°12′42.34″ N

Longitude
116°48′25.43″ W

Zoom to 35 km (~20 miles) and look down.

Death Valley, a narrow basin whose floor lies below sea level, hosts many desert landforms. The white patch is a playa. A bajada forms the slope between the playa and the mountains to the west. Small alluvial fans spill into the basin on the east.

Deposition from the Wind

As mentioned earlier, wind carries two kinds of sediment loads—a suspended load of dust-sized particles

FIGURE 17.12 Playas form where a shallow, salty lake dries up.

(a) This playa in California formed at the base of a bajada.

(b) White salt crystals encrust the floor of a playa in Death Valley.

and a surface load of sand. Much of the dust and silt lofts out of the desert and accumulates elsewhere. Sand, however, cannot travel far, and accumulates within the desert in elongate mounds called **dunes**, ranging in size from less than a meter to over 300 m high. In certain locations, dunes accumulate to form vast "sand seas." We'll look at dunes in more detail later in this chapter.

TAKE-HOME MESSAGE

Sediment carried by water and wind in deserts accumulates in a variety of landforms. Alluvial fans form at the outlets of canyons, playas form where water temporarily collects in basins, and dunes form where large amounts of sand are available.

QUICK QUESTION How do bajadas develop?

17.5 Desert Landscapes and Desert Life

In popular media, writers often depict deserts as endless vistas of sand, punctuated by the occasional palm-studded oasis. In reality, sand doesn't cover all desert landscapes. Some deserts host rocky plains, others sport a stubble of cacti and other hardy plants, and still others display intricate rock formations that look like medieval castles. Early explorers of the Sahara recognized these differences, and distinguished among hamada (barren, rocky highlands), reg (vast, stony plains), and erg (sand seas in which large dunes form). In this section, we'll see how the erosional and depositional processes described above can produce such contrasting landscapes.

Did you ever wonder... if all deserts are completely covered by sand?

Rocky Cliffs and Mesas

In desert regions with significant relief, the lack of soil exposes rocky ridges and cliffs. Where bedrock consists of unfoliated rock, such as granite, joint-bounded blocks tend to become rounded, for weathering happens faster at corners and edges than it does at flat faces. Where bedrock consists of horizontal strata, cliff faces tend to form along long, vertical joints. These are only temporary landscape features, for when rockfalls happen at a cliff face, the position of the cliff effectively steps back. Since rock breaks away from the cliff along a joint that is parallel to the cliff face, the cliff overall retains roughly the same shape as its position migrates. Note that this process, called **cliff retreat**, occurs in fits and starts—a cliff may remain unchanged for decades or centuries until a slab of rock breaks off crumble into rubble downslope. Cliffs exposing alternating layers of strata with contrasting strength develop a step-like shape—thick, strong layers of sandstone or limestone become vertical cliffs, and weak layers of thin-bedded shale become rubble-covered slopes.

GEOLOGY AT A GLANCE

The Desert Realm

The desert of the Basin and Range Province in Utah, Nevada, and Arizona consists of alternating basins (grabens or half-grabens) separated by narrow ranges (tilted fault blocks), depicted schematically here. The Sierra Nevada, underlain largely by granite, borders the western edge of the province, while the Colorado Plateau, underlain by flat-lying sedimentary strata, borders the eastern edge. The overall climate of the region is dry.

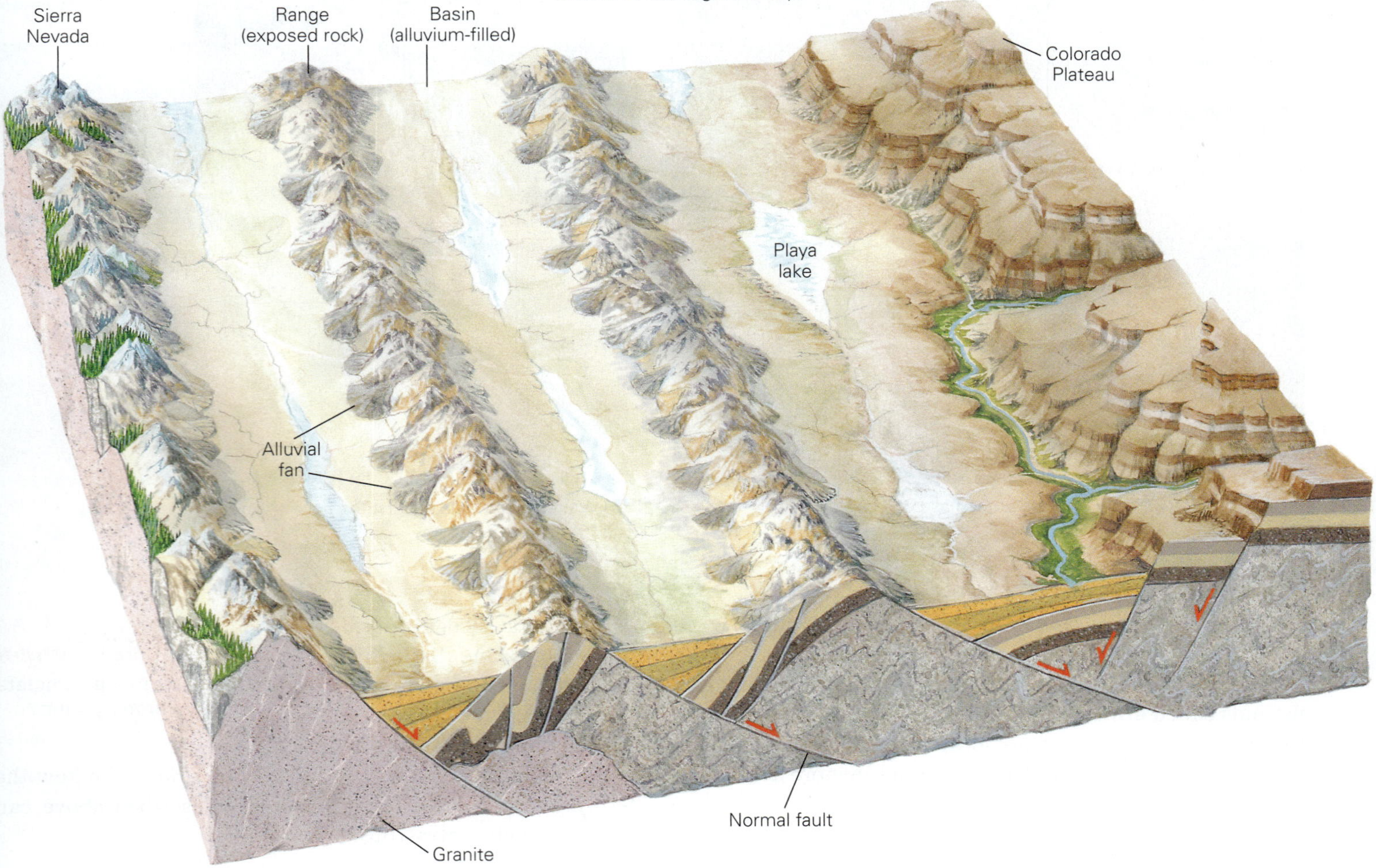

A great variety of distinct landforms occur in deserts, including sand dunes (with cross beds inside), mesas, alluvial fans, arches, and chimneys. Wind, carrying sand and dust, can be an effective agent of erosion in the desert, and can carve yardangs. In places, desert pavements develop. Water may temporarily fill playa lakes; when this water evaporates, it leaves salt.

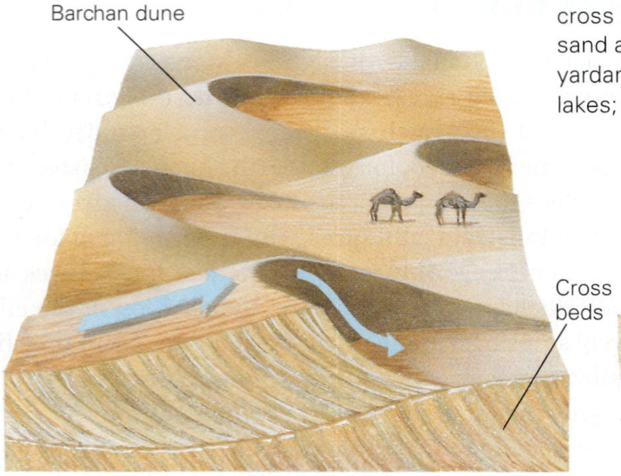

Most stream channels in deserts fill with water only after heavy rains. At other times, they are dry washes (also known as arroyos or wadis). When rain is heavy, runoff enters dry washes and fills them very quickly, yielding a flash flood.

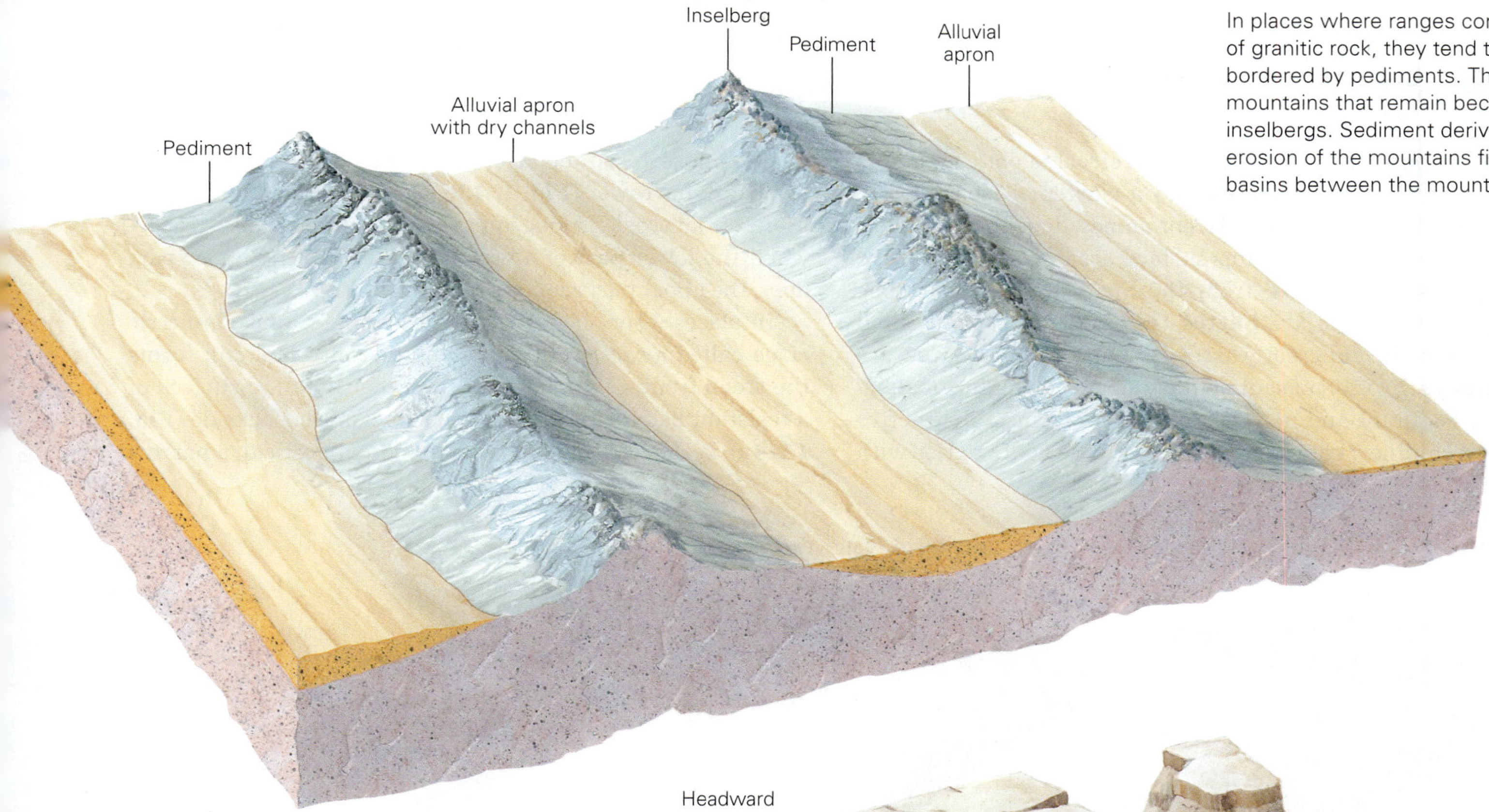

In places where ranges consist of granitic rock, they tend to be bordered by pediments. The isolated mountains that remain become inselbergs. Sediment derived by the erosion of the mountains fills the basins between the mountains.

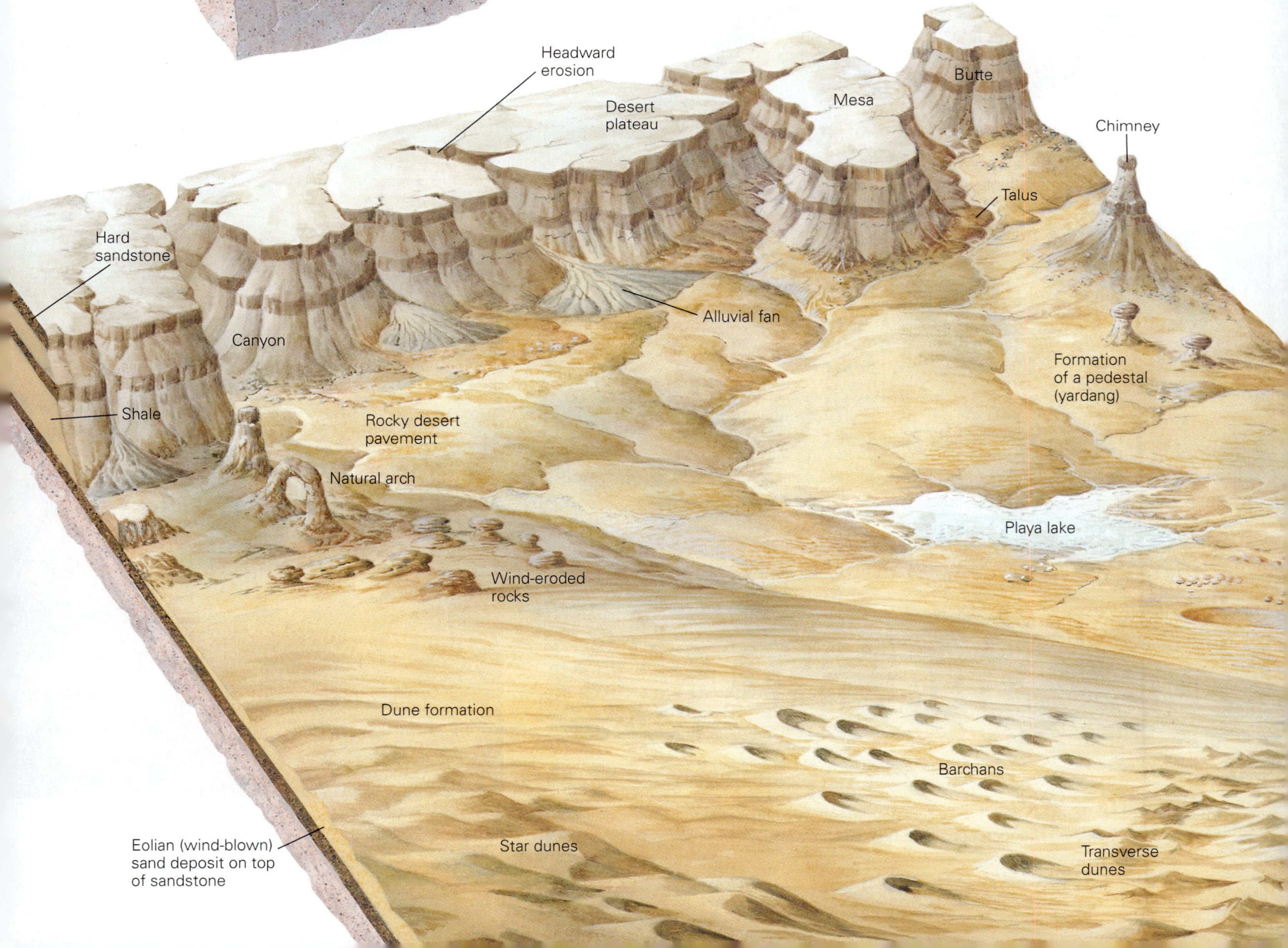

With continued erosion and cliff retreat, a plateau of horizontal strata slowly evolves into a cluster of isolated flat-topped hills. These go by different names, depending on their size—large hills (with a top surface area of several square kilometers) are **mesas**, from the Spanish word for table; medium-sized hills are **buttes** (**Fig. 17.13a, b**); and small ones, whose height greatly exceeds their top surface area, are **chimneys**. Erosion of sedimentary strata consisting of alternating resistant and nonresistant layers forms a distinctive type of chimney, known as a *hoodoo*, whose width varies with height—more resistant layers are wider, and less resistant layers are narrower (**Fig. 17.13c**). *Natural arches* form when preferential erosion along joints leaves narrow walls of rock—when the lower part of a wall erodes before the upper part does, the upper part remains as an arch (**Fig. 17.13d**).

In places where bedding dips at an angle, an asymmetric ridge called a *cuesta* develops. A joint-controlled cliff forms the steep front side of a cuesta, and the tilted top surface of a resistant bed forms the gradual slope on the backside (**Fig. 17.14a**). If the bedding dip is steep to near vertical, a narrow

FIGURE 17.13 Mesas and buttes form in deserts as cliffs retreat over time.

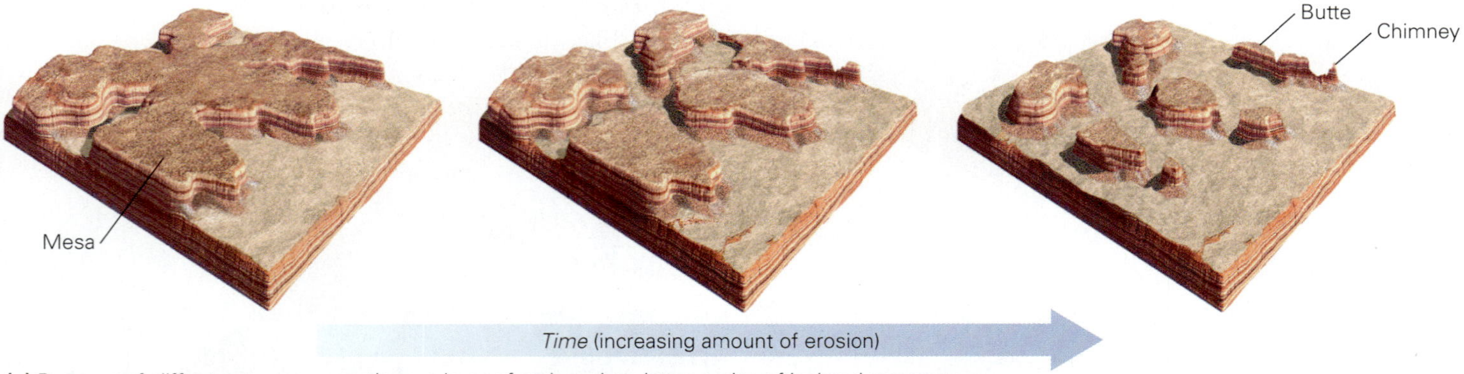

(a) Because of cliff retreat, a once-continuous layer of rock evolves into a series of isolated remnants. If the bedding is horizontal, the resulting landforms have flat tops.

(b) Buttes and mesas tower above the floor of Monument Valley, Arizona.

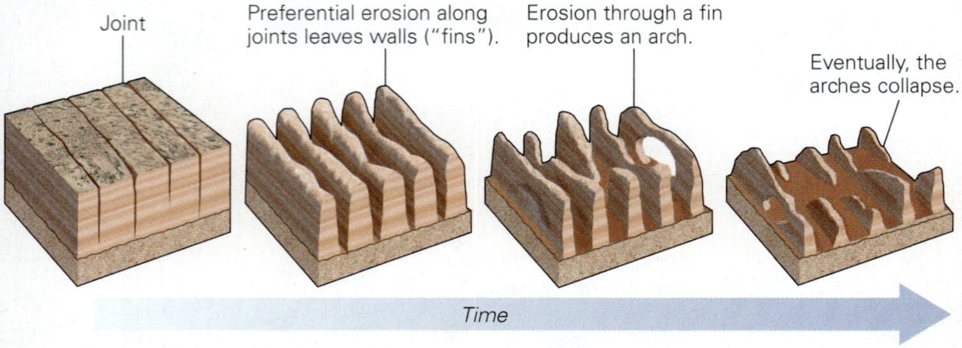

(d) Natural arches form when erosion occurs preferentially along joints, to produce wall-like fins of rock. When the lower part of a fin erodes, an arch remains.

(c) Erosion produced hoodoos, chimney-like columns of rock, in Bryce Canyon, Utah.

FIGURE 17.14 Examples of erosional landscapes in deserts.

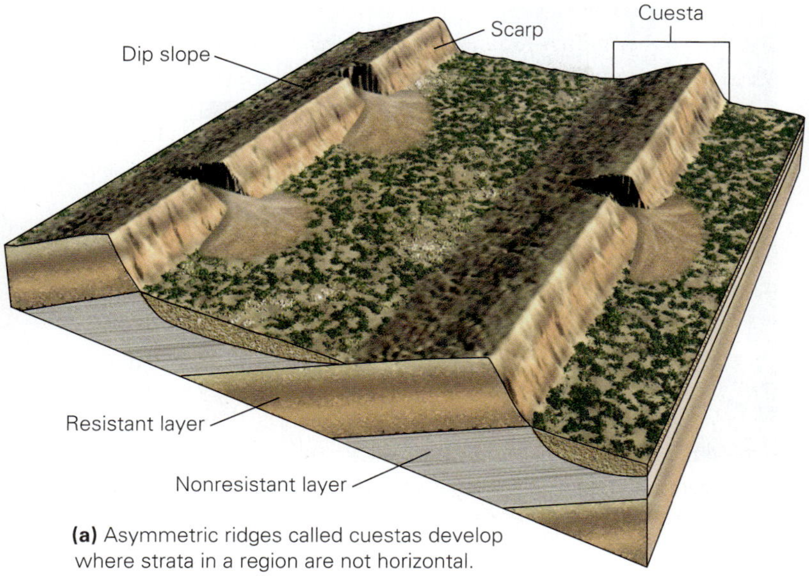

(a) Asymmetric ridges called cuestas develop where strata in a region are not horizontal.

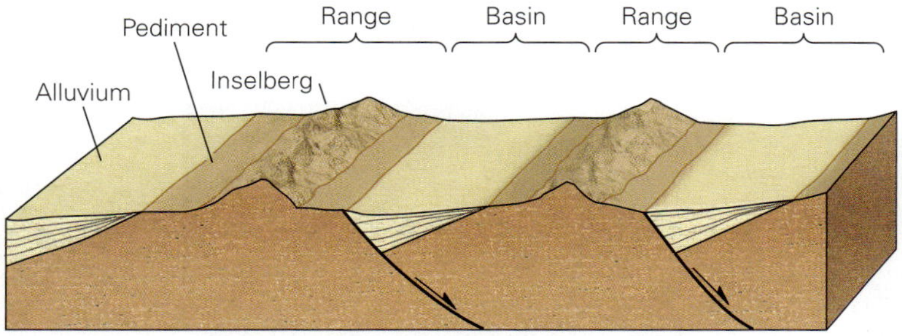

(b) In the Basin and Range Province of the southwestern United States, tilted fault-block ranges evolve into inselbergs, bordered by sediment-filled basins.

symmetrical ridge, called a *hogback*, forms.

After a long period of erosion, all that may remain of a once-broad region of uplifted land is an isolated ridge or mound of exposed bedrock, surrounded by alluvium-filled basins (**Fig. 17.14b**). Geologists refer to such "islands of rock" by the German word **inselberg** (island mountain). Depending on the rock type or the orientation of stratification in the rock, and on rates of erosion, inselbergs may be sharp-crested, plateau-like, or loaf-shaped (meaning that they have steep sides and a rounded crest). Inselbergs with a loaf geometry, as exemplified by Uluru (Ayers Rock) in central Australia (**Fig. 17.15**), are also known as *bornhardts*.

SEE FOR YOURSELF...

ULURU, AUSTRALIA

Latitude
25°20′47.64″ S

Longitude
131°2′28.96″ E

Zoom to 5 km (~3 miles) and look down.

The red sandstone of Uluru (Ayers Rock), a bornhardt, rises above the stony plains of the central Australian desert. The NW-trending diagonal lines are traces of vertical bedding in the sandstone.

Stony Plains, Desert Pavement, and Pediments

The coarse sediment eroded from desert mountains and ridges eventually washes into adjacent lowlands, covering the toes of bajadas and even regions beyond the end of bajadas with pebbles, cobbles, and boulders. Such areas constitute *stony plains*. Portions of stony plains may evolve into **desert pavements**, places where the desert surface, made up of

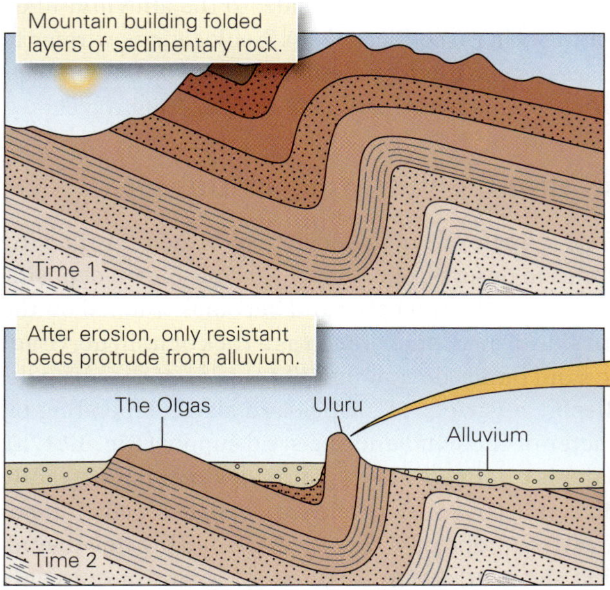

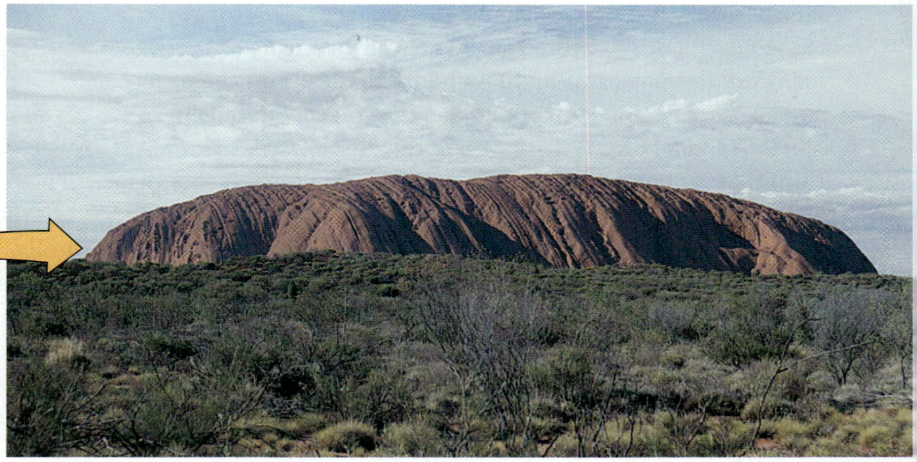

FIGURE 17.15 Uluru (Ayers Rock) in central Australia is an inselberg. The red sandstone beds comprising it are the eroded remnant of the vertical limb of a large syncline. Alluvium buried the surrounding bedrock.

FIGURE 17.16 Desert pavement and a hypothesis for how it forms by building up a soil from below.

(a) A well-developed desert pavement in the Sonoran Desert, Arizona. The inset shows a close-up of the pavement.

(b) Students standing at the edge of a trench cut into desert pavement. Note the soil between the pavement and the underlying alluvium.

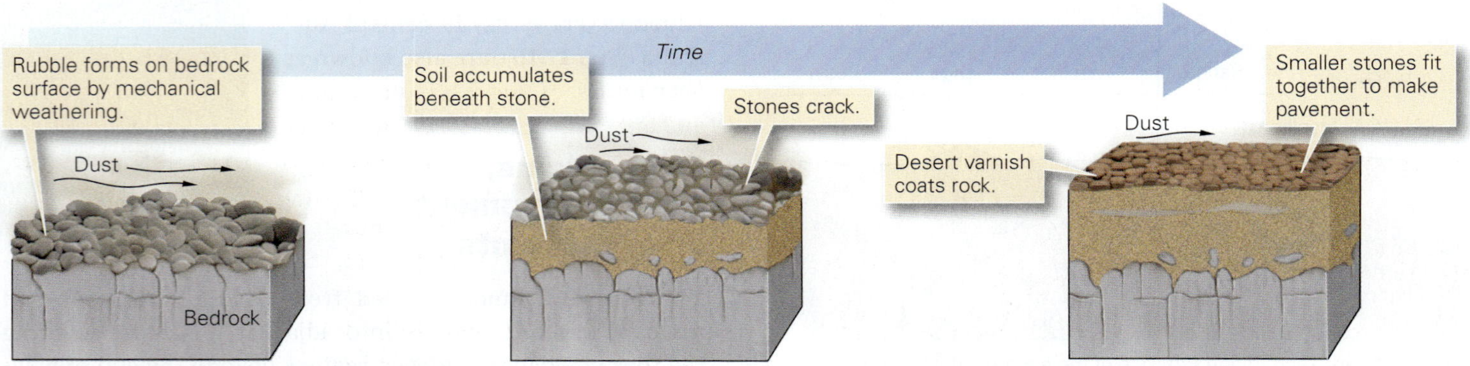

(c) Desert pavement forms in stages. First, loose pebbles and cobbles collect at the surface. Dust settles among the stones and builds up a soil layer below. The stones eventually crack into smaller pieces and settle to form a mosaic-like pavement.

small stones that fit together tightly, resembles a tile mosaic and forms a fairly smooth surface layer (**Fig. 17.16a**). Typically, desert varnish coats the top surfaces of the stones that form desert pavement, so the overall surface has a reflective sheen. Recently, researchers have suggested that desert pavements form when the rocks at the surface of a stony plain crack, perhaps due to differential heating by the desert sun, to form smaller pieces. Windblown dust that falls on the desert surface slowly sinks down between the pieces, to form a soil beneath them (**Fig. 17.16b**). When the dust gets wet, during a rainfall, it expands, and when it subsequently dries, it shrinks. This movement causes the clasts to settle slowly so that their top surfaces align. Sheetwash carries away any fine sediment left at the surface, or in spaces between the rocks, until the rocks fit into a stable, jigsaw-like arrangement (**Fig. 17.16c**).

When travelers began trudging through the desert of the southwestern United States during the 19th century, they found that in many locations the wheels of their wagons were rolling over flat or gently sloping bedrock surfaces. These bedrock surfaces extended outward like ramps from the steep cliffs of a mountain range on one side, to the alluvium-filled valleys on the other (see Fig. 17.13b). Geologists now refer to such surfaces as **pediments**. Pediments develop when sheetwash carries sediment away as the mountain front retreats.

Seas of Sand: The Nature of Dunes

A *sand dune* is a pile of sand deposited by wind or by flowing water. Dunes in deserts, of course, are wind-blown deposits. They may start to form where sand becomes trapped on the windward side of an obstacle, such as a rock or a shrub. Gradually, the sand builds downwind into the lee of the obstacle. Dunes display a variety of shapes and sizes, depending on the character of the wind and the sand supply (**Fig. 17.17a**). Where sand is relatively scarce and the wind blows steadily in one direction, beautiful crescents called *barchan dunes* develop, with the tips of the crescents pointing downwind. If the wind shifts direction frequently, a group of crescents

FIGURE 17.17 The types of sand dunes and the cross beds within them.

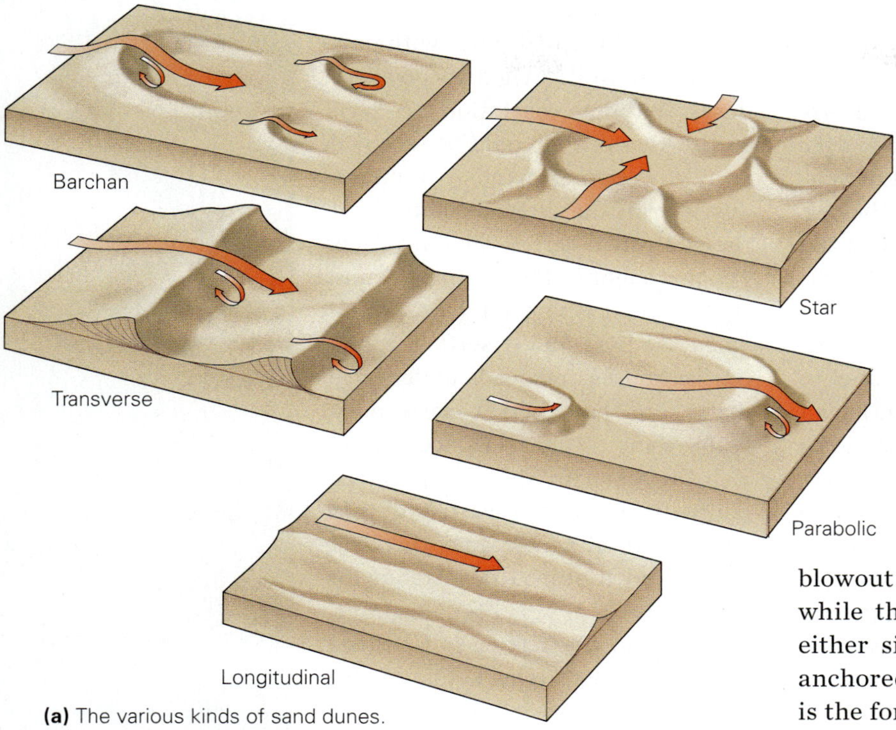

(a) The various kinds of sand dunes.

(b) A sand dune with surface ripples.

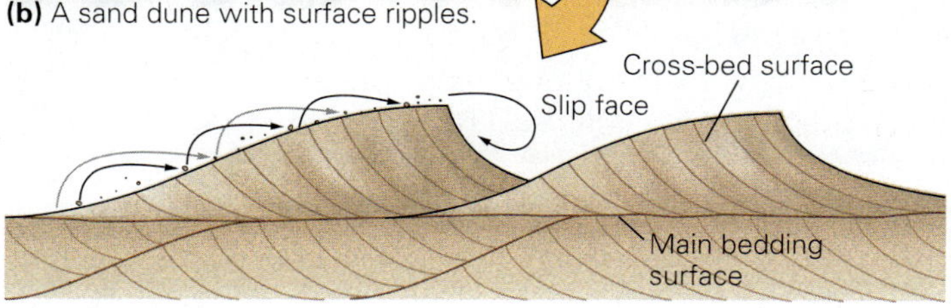

(c) Cross bedding inside a dune.

pointing in different directions overlap one another, yielding a *star dune*. Where enough sand accumulates to bury the ground surface completely, and moderate winds blow, sand piles into simple, wave-like shapes called *transverse dunes*; the crests of transverse dunes lie perpendicular to the wind direction. When strong winds cause a blowout in the middle of a vegetated transverse dune, the sand from the blowout gets deposited downwind, while the portions of the dune on either side of the blowout remain anchored by vegetation. The result is the formation of a *parabolic dune*, whose ends point in the upwind direction. Finally, with abundant sand and a strong, steady wind, the sand streams into *longitudinal dunes*, whose axes lie parallel to the wind direction.

In a sand dune, sand saltates up the windward side of the dune, blows over the crest of the dune, and then settles on the steeper, lee face of the dune. The slope of this face attains the angle of repose, the slope angle of a freestanding pile of sand. As sand collects on this surface, it eventually becomes unstable and slides down the slope, so geologists refer to the lee side of a dune as the *slip face*. As more and more sand accumulates on the slip face, the crest of the dune migrates downwind, and former slip faces become preserved inside the dune. In cross section, these slip faces appear as cross beds (**Fig. 17.17b, c**).

SEE FOR YOURSELF...

NAMIB DESERT, NAMIBIA

Latitude
24°44′27.80″ S

Longitude
15°27′58.92″ E

From 20 km (~12.5 miles), look down and zoom in.

Huge dunes of orange sand border a dry wash in western Africa. The slip faces of dunes are recognizable. A road in the wash provides scale.

TAKE-HOME MESSAGE

Over time, landscapes evolve in deserts. As rocks collapse along cliffs, the position of a cliff face can migrate, and if bedrock consists of horizontal layers, buttes and mesas form. Erosion of tilted layers yields cuestas. Desert landscapes are variable. Some include vast stony plains, some of which evolve into desert pavements. In sandy areas, many types of dunes will grow.

QUICK QUESTION What factors control the shape, dimensions, and orientation of sand dunes?

17.6 Desert Problems

Natural droughts, overpopulation, overgrazing, careless agricultural practices, and diversion of water supplies can transform a semiarid grassland into a desert landscape in a matter of years to decades. The consequences of such **desertification**, for example, have devastated parts of the Sahel, the belt of semiarid grassland that fringes the southern margin of the Sahara (**Fig. 17.18a**). In years past, the Sahel provided sufficient vegetation to support a small population of nomadic people and animals. But during the second half of the 20th century, large numbers of people migrated into the Sahel to escape overcrowding in central Africa. The immigrants began farming and maintained large herds of cattle and goats. As a consequence, plowing and overgrazing removed soil-preserving grass and caused the soil to dry out. In addition, trampling animal hooves compacted the ground so it could no longer soak up water. A series of natural droughts has hit the region, bringing catastrophe (**Fig. 17.18b**). Wind erosion stripped off the remaining topsoil. Without vegetation, the air grew drier, the semi-arid grassland of the Sahel became desert, and its inhabitants endured a terrible famine.

Desertification can also happen in industrialized nations. People in the western Great Plains of the United States and Canada suffered from desertification of an agricultural area beginning in 1933, the fourth year of the Great Depression. Banks had failed, workers had no jobs, the stock market had crashed, and hardship burdened all. No one needed yet another disaster—but that year, even nature turned hostile. All through the fall, so little rain fell in the plains, from Texas and Oklahoma up to south-central Canada, that the region's grasslands and croplands browned and withered, and the topsoil turned to powdery dust. Without vegetation to protect it, the topsoil could be stripped away by strong winds. Immense dust storms that literally blotted out the sun rolled across the landscape. People caught in these storms found themselves gasping for breath, and when the dust finally settled, it had buried houses and roads under huge drifts, and had dirtied every nook and cranny. What had once been a rich farmland in the southwestern plains turned into a wasteland that soon acquired a nickname, the Dust Bowl. And it stayed devastated for years.

Why did the fertile soils of the southern Great Plains suddenly dry up? The causes were complex—some natural, and some human-induced. Typically, the region has a semi-arid climate in which only thin soil develops. But the plains were settled in the 1880s and 1890s, which were unusually wet years. Not realizing its true character, far more people moved into the region than it could

SEE FOR YOURSELF...

SAHARA DESERT, EGYPT

Latitude
25°41′49.34″ N

Longitude
32°40′38.76″ E

Zoom to x km (~y miles) and look down.

The green stripe is the Nile Valley near Luxor. Except for the land that can be irrigated with river water, the desert hosts no vegetation. Flash floods carve the intricate canyons in nearby hills but do not provide enough water to sustain life. The Nile brings in water from wetter lands to the south.

FIGURE 17.18 Desertification has been happening in the Sahel, a region of Africa.

(a) The Sahel is the semi-arid land along the southern edge of the Sahara. Large parts have undergone desertification.

(b) In the Sahel, forests and grasslands have dried up.

sustain, and the land was farmed too intensively. Plowing destroyed the fragile grassland root systems that held the thin soil in place. So when the drought of the 1930s came, it brought catastrophe.

Desertification has an additional dangerous side effect. The dust in dust storms doesn't consist of only clay and silt, but can include salts, toxic chemicals, microbes, and fungi, and these materials can be carried great distances. For example, recent studies suggest that dust blown off the Sahara traverses the Atlantic and settles onto the Caribbean Sea (**Fig. 17.19**). Geologists are concerned that this dust may infect corals with disease or in some other way inhibit their life processes. So windblown dust may contribute to the destruction of coral reefs.

TAKE-HOME MESSAGE

In lands bordering deserts, droughts and population pressures have led to desertification, transforming vegetated land into desert. As more regions turn into deserts, more wind-transported dust enters the atmosphere, and this dust can include not only fine grains of minerals, but also toxins and pathogens. These damaging materials can be carried across oceans.

QUICK QUESTION What factors transformed Oklahoma and Texas into the Dust Bowl during the 1930s?

FIGURE 17.19 In this satellite image, a huge dust cloud that originated in the Sahara blows across the Atlantic.

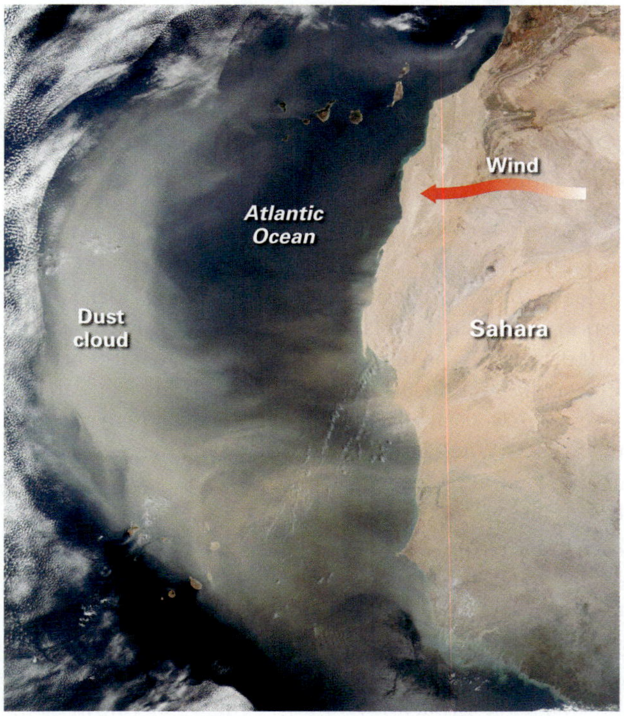

Another View The machinery of an abandoned farm in South Dakota lies buried in dust, as illustrated by this 1936 photograph.

Chapter 17 Review

Chapter Summary

- Deserts generally receive less than 25 cm of rain per year. Vegetation covers less than 15% of their surface.
- Subtropical deserts form between latitudes of 20° and 30°, rain-shadow deserts are found on the inland side of mountain ranges, coastal deserts are located on the land adjacent to cold ocean currents, continental-interior deserts exist in landlocked regions far from the ocean, and polar deserts form at high latitudes.
- Physical weathering in deserts produces rocky debris. Chemical weathering happens slowly; it produces ions that precipitate as new minerals just below the surface.
- Water causes significant erosion in deserts, mostly during heavy downpours. Flash floods carry large quantities of sediments down ephemeral streams. When rain stops, streams dry up, leaving steep-sided dry washes.
- Wind picks up dust and silt as suspended load, and it causes sand to saltate. Where wind blows away finer sediment, a lag deposit remains. Windblown sediment abrades the ground, creating a variety of features such as ventifacts.
- Desert pavements are mosaics of varnished stones armoring the surface of the ground.
- Talus aprons form when rock fragments accumulate at the base of a slope. Alluvial fans form at a mountain front where water in ephemeral streams deposits sediment on a plain. When temporary desert lakes dry up, they leave playas.
- In some desert landscapes, erosion causes cliff retreat, eventually resulting in the formation of mesas, buttes, cuestas, and inselbergs. Pediments of nearly flat or gently sloping bedrock surround some inselbergs.
- Where sand is abundant, the wind builds it into dunes. Common types include barchan, star, transverse, parabolic, and longitudinal dunes.
- Changing climates and land abuse may cause desertification, the transformation of semiarid land into deserts. Windblown dust, sometimes carrying microbes and toxins, may waft from deserts across oceans.

Guide Terms

alluvial fan (p. 556)
butte (p. 560)
chimney (p. 560)
cliff retreat (p. 557)
deflation (p. 555)
desertification (p. 564)
desert pavement (p. 561)
desert varnish (p. 553)
dune (p. 557)
inselberg (p. 561)
lag deposit (p. 556)
mesa (p. 560)
pediment (p. 562)
petroglyph (p. 554)
playa (p. 556)
saltation (p. 554)
surface load (p. 554)
suspended load (p. 554)
talus (p. 552)
ventifact (p. 555)
wind erosion (p. 554)

GEOTOURS *THIS CHAPTER'S GEOTOUR EXERCISE (Q) FEATURES:*

- Deserts and Desert Features
- Water Use in Arid Regions

Review Questions

1. What factors determine whether a region can be classified as a desert?
2. Explain the various conditions that can lead to the formation of deserts.
3. Have today's deserts always been deserts? (*Hint*: Keep in mind the consequences of plate tectonics.)
4. How do weathering processes in deserts differ from those in temperate or humid climates?
5. Describe how water modifies the landscape of a desert. Be sure to discuss both erosional and depositional landforms.
6. Explain the ways in which desert winds transport sediment.
7. What phenomena may lead to formation of desert varnish, desert pavement, and ventifacts?
8. Describe the process of formation of alluvial fans, bajadas, and playas.
9. Describe the process of cliff (scarp) retreat and the landforms that result from it.
10. What are the various types of sand dunes, and what factors determine which type of dune develops at a particular location?
11. What is the process of desertification, and what causes it? How can desertification in Africa affect the Caribbean?

On Further Thought

12. You are working for an international nongovernmental organization (NGO) and have been charged with the task of providing recommendations to an African nation that wishes to slow or halt the process of desertification within its borders. What are your recommendations?
13. The Namib Desert and Kalahari Desert both formed in southern Africa. Look at Google Earth to find these deserts and examine some of the features in them. Did both of these deserts form for the same reason? Explain your answer.

Online Resources

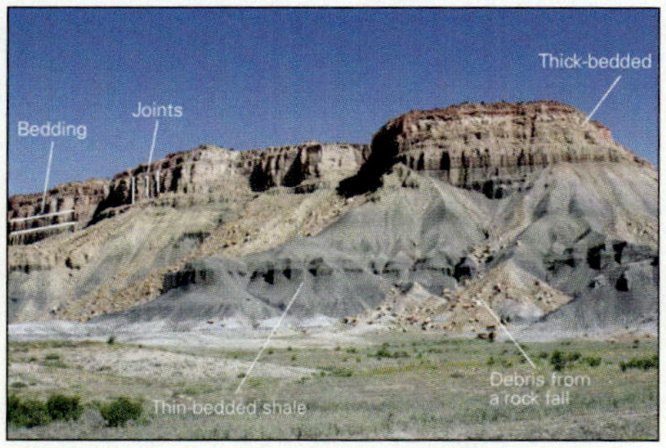

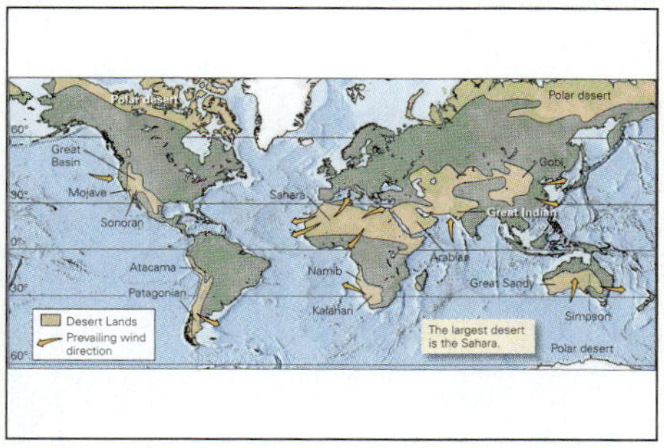

Videos

This chapter features a video on the geologic features of deserts.

Assessment

Questions in this chapter include labeling exercises about desert landscapes and the different types of deserts.

LEARNING OBJECTIVES

By the end of this chapter, you should understand...

1. how glacial ice forms and flows, and how to distinguish among various kinds of glaciers.
2. how glaciers advance and retreat, and how their flow erodes the landscape.
3. how to recognize sedimentary deposits and associated landforms left by glaciers.
4. that glaciers cover large areas of continents and that sea level drops during ice ages.
5. why ice ages happen, and why glaciations during an ice age take place periodically.

▲ A small glacier, all that remains of what was once a large ice cap, clings to the side of a mountain in the Canadian Rockies. The flow of glaciers contributed to carving this rugged landscape.

CHAPTER 18

Amazing Ice: Glaciers and Ice Ages

18.1 Introduction

There's nothing like a good mystery, and one of the most puzzling in the annals of geology came to light in northern Europe early in the 19th century. When farmers of the region prepared their land for spring planting, they occasionally broke their plows by running them into hidden boulders buried randomly through otherwise fine-grained sediment. Many of these boulders did not consist of local bedrock, but rather came from outcrops hundreds of kilometers away. Because the boulders had apparently traveled so far, they came to be known as **erratics** (from the Latin *errare*, to wander).

The mystery of the wandering boulders became a subject of great interest to early-19th-century geologists, for they realized that such deposits of extremely *unsorted sediment* (containing a variety of clast sizes) could not be examples of typical stream alluvium, for running water sorts clasts by size. Most attributed the deposits to a vast flood that they imagined had been powerful enough to spread a slurry of boulders, sand, and mud across the continent. In 1837, however, a young Swiss geologist named Louis Agassiz proposed a radically different interpretation. Agassiz often hiked among

I seemed to vow to myself that some day I would go to the region of ice and snow and go on and on till I came to one of the poles of the earth.

ERNEST SHACKLETON (British polar explorer, 1874–1922)

glaciers in the Alps near his home. He observed that these slowly flowing masses of ice could carry enormous boulders, as well as sand and mud, because ice is a solid that's strong enough to support the weight of rock. Agassiz realized that, because solid ice does not sort sediment as it flows, glaciers leave behind unsorted sediment when they melt. On the basis of these observations, he proposed that the mysterious erratics of Europe were left by vast glaciers that had once covered much of the continent (**Fig. 18.1**). In Agassiz's mind, Europe had once been in the grip of an *ice age,* a time when glaciers spread over areas that presently have temperate climates.

Agassiz's radical proposal initially faced intense criticism, but by the late 1850s geologists everywhere agreed that, indeed, Europe once had Arctic-like climates. Later in life, Agassiz traveled to the United States and identified many glacial features in North America's landscape, proving that the last ice age did not affect just Europe. Glaciers cover only about 10% of the land on Earth today, but during the last ice age, which ended less than 12,000 years ago, as much as 30% of continental land surface was covered by an *ice sheet,* a layer of ice hundreds to thousands of kilometers across and as much as a few kilometers thick.

The work of Louis Agassiz brought the subject of glaciers and ice ages into the realm of geologic study, and ultimately led people to recognize that major climate changes do happen during Earth history. In this chapter, after considering the nature of ice, we see how glaciers form, why they move, and how they modify landscapes by erosion and deposition. We also discuss the evidence for ice ages, with a particular focus on the most recent one, known as the *Pleistocene Ice Age,* whose impact on the landscape can still be seen today. The chapter concludes by considering hypotheses that geologists have put forth to explain why ice ages happen.

FIGURE 18.1 Agassiz envisioned that extensive areas of the northern hemisphere were once covered by vast ice sheets comparable to the one covering Antarctica today.

18.2 Ice and the Nature of Glaciers

What Is Ice?

Ice consists of solid water, formed when liquid water cools below its freezing point. In effect, ice crystals are minerals—they are naturally occurring with a definite chemical composition (H_2O) and a regular crystal structure. Ice crystals have a hexagonal shape, so snowflakes grow into six-pointed stars (**Fig. 18.2a**). We can consider natural occurrences of ice to be comparable to various types of rock. Specifically, a layer of fresh snow, like a layer of silt, forms when snow (clasts of ice) settle out of the air, and a layer of snow that has been compacted, so that the grains stick together, resembles a layer of sedimentary rock (**Fig. 18.2b**). Ice that appears on the surface of a pond, like an igneous rock, forms when molten ice (liquid water) solidifies, and glacial ice, like a metamorphic rock, develops when a mass of pre-existing ice *recrystallizes* in the solid state so that it develops a new texture (**Fig. 18.2c**).

How a Glacier Forms

In order for a glacier to form, the local climate must be cold enough that winter snow does not melt entirely away. There must be enough snowfall that a large amount of snow can accumulate even during the summer, the surface on which snow accumulates must have a gentle slope so that snow falling on it does not slide away in avalanches, and accumulated snow must be protected from the wind so that it doesn't blow away. These conditions can be met in polar regions, and in mountains. Glaciers develop in polar regions, even though these areas are relatively arid, because temperatures remain so cold that any snow that falls can remain. Glaciers develop in mountains, even at low latitudes, because temperature decreases with elevation, so at high elevations, the mean temperature remains cold enough for ice and snow to last all year. Since the temperature of a region depends on latitude, the elevation at which mountain glaciers grow, and the elevation down to which they can flow, depends on latitude. In the Earth's present-day climate, glaciers at the equator do

FIGURE 18.2 The nature of ice and the formation of glaciers. Snow falls like sediment and metamorphoses to ice when buried.

(a) The hexagonal shape of snowflakes. No two are alike.

(b) Layers of snow accumulate. They recrystallize to become ice.

(c) As revealed by a microscope, glacial ice has coarse grains and contains air bubbles.

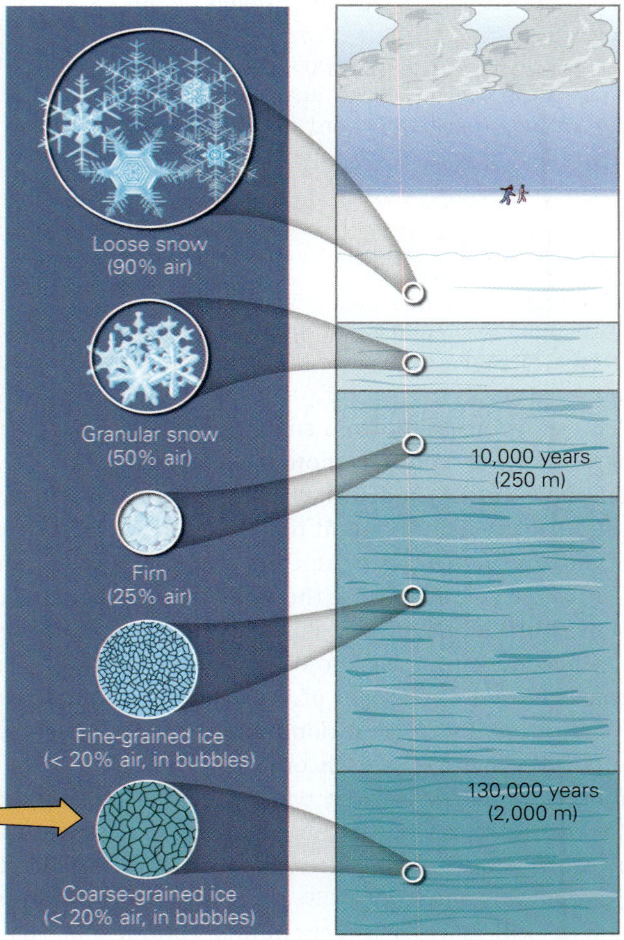

(d) Snow compacts and melts to form firn, which recrystallizes into ice. Crystal size increases with depth.

BOX 18.1 CONSIDER THIS...

Polar Ice Caps on Mars

Mars has white polar ice caps whose sizes change with the season, suggesting that they partially melt and then refreeze (**Fig. Bx18.1**). The question of what the ice caps consist of remained a puzzle until fairly recently. It now appears that the Martian ice caps consist mostly of H_2O ice mixed with a small amount of dust, and they attain a maximum thickness of 3 km. During the winter, atmospheric carbon dioxide freezes and covers the north polar cap with a 1-m-thick layer of frozen CO_2 (dry ice). During the summer, this layer melts away. The south polar cap has a dry-ice blanket that is 8 m thick and doesn't melt away entirely in the summer. The difference between the north and south polar caps may reflect elevation—the south pole is 6 km higher and therefore remains colder.

FIGURE Bx18.1 The ice caps of Mars.

(a) During the winter, the ice caps expand to lower latitudes.

(b) A close-up of the northern polar cap in summer.

High-resolution photographs reveal that distinctive canyons, up to 10 km wide and 1 km deep, spiral outward from the center of the north polar ice cap. Why did this pattern form? Recent calculations suggest that if the ice sublimates on the sunny side of a crack and refreezes on the shady side, the crack will migrate sideways over time. If the cracks migrate more slowly closer to the pole, where it's colder, than they do farther away, they will naturally evolve into spirals.

not descend lower than an elevation of 5 km, whereas glaciers formed at latitudes between 60° and 90° can flow down to sea level.

The transformation from snow to glacial ice takes place progressively, as younger snow buries older snow. Freshly fallen snow consists of delicate hexagonal crystals with sharp points. The crystals do not fit together tightly, so fresh snow contains up to 90% air. With time, the points of the snowflakes become blunt because they either melt (turn into a liquid) or **sublimate** (evaporate directly into vapor). When the flakes become blunt, the snow packs more tightly. As snow becomes buried, the weight of the overlying snow increases the pressure. Under conditions of increased pressure, ice grains slide relative to each other to fill pore spaces, and grains melt slightly at their contact points, forming liquid water that refreezes in pore space. As a result, deeply buried snow transforms into a packed granular material called **firn**, which contains only about 25% air (**Fig. 18.2d**). As pressure increases still more, firn recrystallizes into a solid mass of new, coarser interlocking ice crystals. Such glacial ice, which may still contain up to 20% air trapped in bubbles, tends to absorb red light and thus has a bluish color. The transformation of fresh snow to blue glacial ice can take as little as tens of years, in regions with abundant snowfall, or as long as thousands of years, in regions with little snowfall.

Categories of Glaciers

Formally defined, a **glacier** is a stream or sheet of recrystallized ice that stays mostly frozen all year long and flows under the influence of gravity. Today, glaciers highlight coastal and mountain scenery in Alaska, the Cordillera of western North America, the Alps of Europe, the Southern Alps of New Zealand, the Himalayas of Asia, and the Andes of South America, and they cover most of Greenland and Antarctica. (Of note, Earth is not alone in hosting ice sheets at its poles—Mars has them too; **Box 18.1**). Geologists distinguish between two main categories of glaciers on the Earth, mountain glaciers and continental glaciers, based on their relation to topography and on their area.

Mountain glaciers (also known as *alpine glaciers*) exist in or adjacent to mountainous regions, and overall, glaciers flow from higher elevations to lower elevations (**Fig. 18.3a**). Geologists distinguish among many types of mountain glaciers, based on their relation to the landscape. Examples include: *cirque glaciers*, which fill bowl-shaped depressions, or **cirques**, on the flank of a mountain; *valley glaciers*, rivers of ice that flow down valleys; *ice caps*, mounds of ice that submerge peaks and ridges at the crest of a mountain range; and *piedmont glaciers*, fans or lobes of ice that form where a valley glacier emerges from a valley and spreads out into the adjacent plain (**Fig. 18.3b–d**). Mountain glaciers range in size from a few hundred meters long to a few hundred kilometers long.

Continental glaciers are vast ice sheets that spread over thousands of square kilometers of continental crust. Today, they exist only on Antarctica and Greenland. Antarctica is a continent, so its ice sheet rests mostly on solid ground (**Fig. 18.4**). Locally, lakes of liquid water exist at the base of the Antarctic ice sheet. In 2012, Russian geologists drilled into one of these lakes, Lake Vostok, a lens of freshwater that has an area of 12,500 square km and a depth of about 430 m. Continental glaciers flow outward from their thickest point, where ice may be up to 4 km thick, and thin toward their margins, where ice may be only a few hundred meters thick.

How Do Glaciers Move?

Glacial ice is solid, yet glaciers flow. To understand this paradox, keep in mind that under appropriate conditions, glacial ice, like the metamorphic rocks that we described in Chapter 7, can undergo **plastic deformation**, a process during which existing grains slowly change shape without breaking, and/or new grains grow while old ones disappear. Simplistically, plastic deformation happens when, in response to stress, chemical bonds break and then immediately reattach in a somewhat different configuration. In silicate rocks, plastic deformation can only occur at depths in the crust where temperatures exceed a few hundred degrees

Did you ever wonder...
how a glacier moves?

FIGURE 18.3 A great variety of glaciers form in mountainous areas.

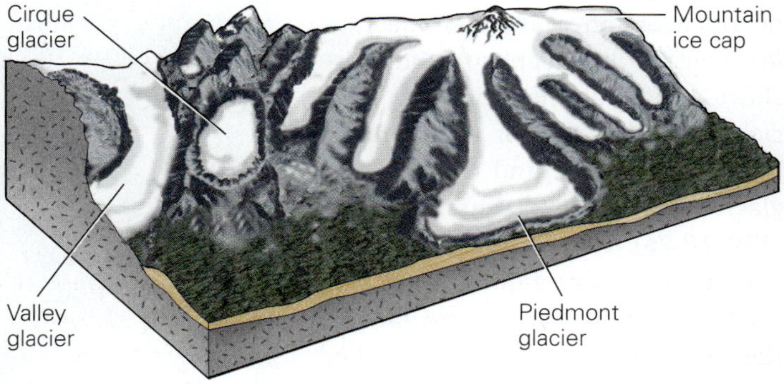

(a) Mountain glaciers are classified based on shape and position.

(b) A valley glacier and cirque glaciers in Switzerland.

(c) Valley glaciers draining a mountain ice cap in Alaska.

(d) A piedmont glacier near the coast of Greenland.

FIGURE 18.4 Antarctica is an ice-covered continent.

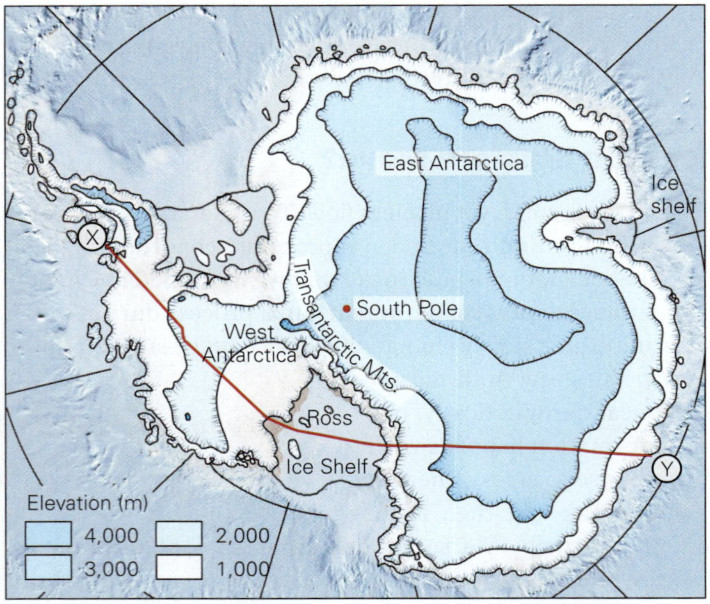

(a) A contour map of the Antarctic ice sheet. Valley glaciers carry ice from the ice sheet of East Antarctica down to the Ross Ice Shelf.

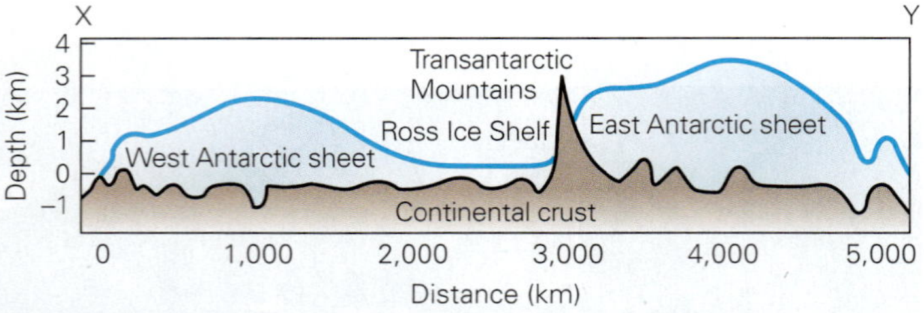

(b) A cross section X to Y of the Antarctic ice sheet. The Transantarctic Mountains separate East Antarctica from West Antarctica.

and pressures become huge. But the bonds among water molecules in ice are fairly weak, so plastic deformation in a glacier can happen at depths of only about 60 m (200 ft), a depth known as the *brittle–plastic transition*. Since glaciers can be hundreds of meters to several kilometers thick, most of the ice in a glacier can flow plastically. In *polar glaciers*, those in which ice remains below its melting temperature throughout the year, plastic deformation takes place completely in the solid state. In *temperate glaciers*, where ice tends to be at or near its melting temperature during at least part of the year, plastic deformation also involves grains sliding past their neighbors on thin water films.

Above the brittle–plastic transition, glacial ice tends to be too brittle (breakable) to flow easily. So, as the deeper parts of a glacier flow, the shallower parts may crack. A crack that opens into a downward-tapering gash is called a **crevasse** (**Fig. 18.5**). In large glaciers, crevasses can be hundreds of meters long, and at the surface of the glacier, they may be up to 15 m across. People traversing glaciers must take safety precautions to avoid falling into crevasses, which in some cases may be hidden by snow drifts that could collapse under the weight of a snowmobile, or even of a single person.

Why do glaciers move? Simply, the force of gravity can overcome the strength of ice (**Fig. 18.6**). A glacier flows in the direction in which its top surface slopes. Therefore, mountain glaciers flow down slopes or valleys, and continental ice sheets spread outward from their thickest point. To picture the movement of a continental ice sheet, imagine pouring honey on a tabletop. The honey spreads out until the puddle reaches an even thickness. In the case of a continental ice sheet, a thick pile of ice builds up, and gravity causes the top of the pile to push down on the ice at the base. Eventually, the basal ice can no longer support the weight of the overlying ice and begins to deform plastically. When this happens, the basal ice starts squeezing out to the side, carrying the overlying ice with it. The greater the volume of ice that builds up, the wider the ice sheet can become. Note that because the slope of a glacier's surface determines the direction that it flows, glaciers can move up and over ridges or hills in the substrate at their base.

Overall, glaciers flow at rates of 10 to 300 m per year. But not all parts of a glacier move at the same rate, for friction between rock and ice slows a glacier. Therefore, the interior of a valley glacier moves faster than its margins, and the top of a glacier moves faster than its base (**Fig. 18.7a**). When glaciers flow into conditions where liquid water can accumulate and mix with sediment at the base of a glacier, a process called *basal sliding* takes place. During basal sliding, the glacier glides along on a wet sediment slurry without coming into frictional contact with bedrock and, as a consequence, may undergo a *surge*, meaning that it moves much faster than normal. During a surge, ice flow has been clocked at speeds of 10 to 110 m per day!

Glacial Advance and Retreat

Glaciers resemble bank accounts: snowfall accumulates and adds to the account, while **ablation**—the removal of ice by melting (turning to

SEE FOR YOURSELF...

GLACIALLY-CARVED PEAKS, MONTANA

Latitude
48°56′33.66″ N

Longitude
113°49′54.59″ W

Zoom to 8 km (~5 miles) and look down.

Three cirques bound a horn in the Rocky Mountains, north of Glacier National Park. Note the knife-edge arêtes between the cirques. The glaciers that carved the cirques have melted away. The thin stripes on the mountain face are traces of bedrock sedimentary beds.

FIGURE 18.5 Crevasses form in the upper layer of a glacier, in which the ice is brittle. Commonly, cracking takes place where the glacier bends while flowing over steps or ridges in its substrate.

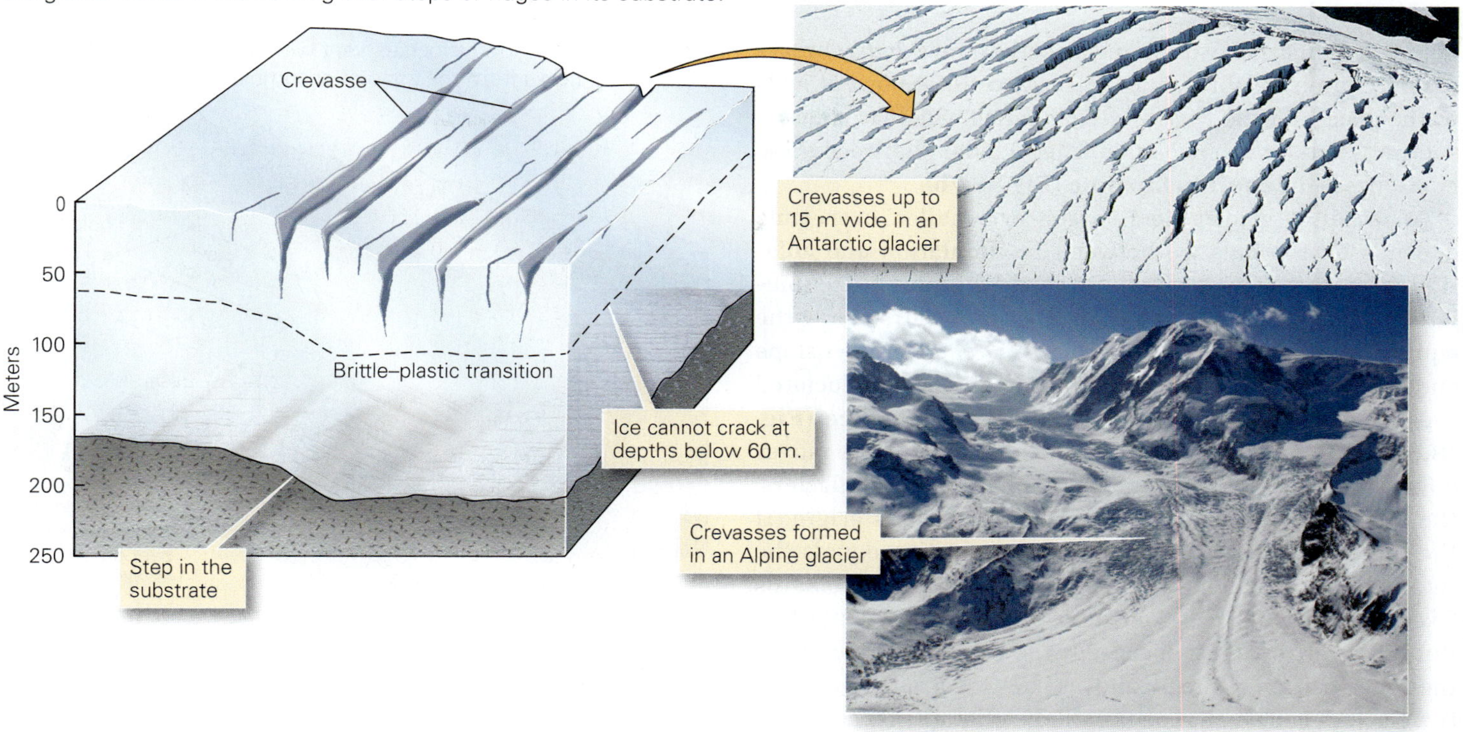

Crevasse
0
50
100
150
200
250
Meters
Brittle–plastic transition
Ice cannot crack at depths below 60 m.
Step in the substrate
Crevasses up to 15 m wide in an Antarctic glacier
Crevasses formed in an Alpine glacier

FIGURE 18.6 Forces that drive the movement of glaciers.

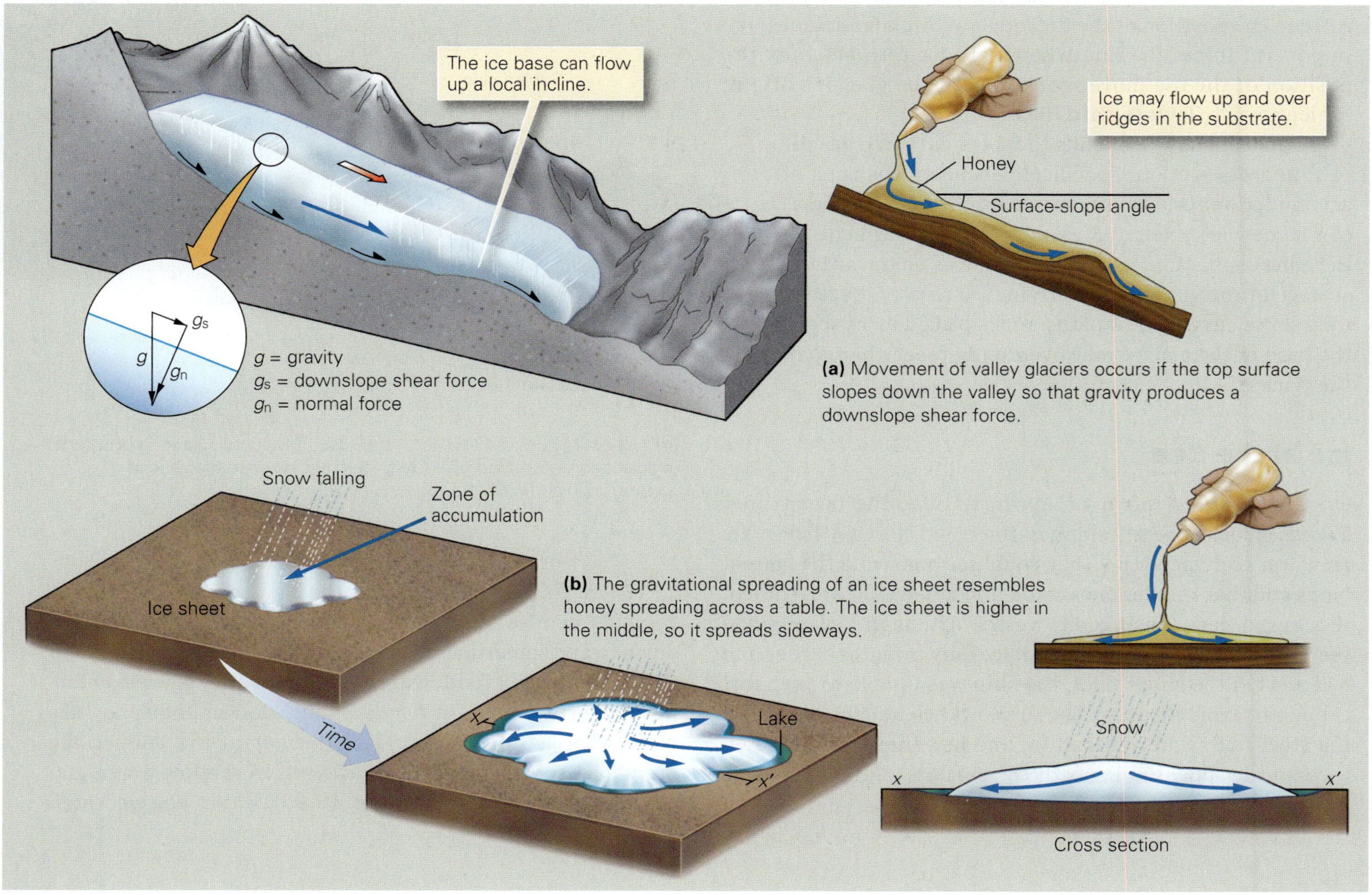

The ice base can flow up a local incline.
g = gravity
gs = downslope shear force
gn = normal force
Ice may flow up and over ridges in the substrate.
Honey
Surface-slope angle
(a) Movement of valley glaciers occurs if the top surface slopes down the valley so that gravity produces a downslope shear force.
Snow falling
Zone of accumulation
Ice sheet
(b) The gravitational spreading of an ice sheet resembles honey spreading across a table. The ice sheet is higher in the middle, so it spreads sideways.
Time
Lake
X
X′
Snow
Cross section

liquid), sublimation (vaporizing directly), and *calving* (breaking off chunks)—subtracts from the account. Snowfall adds to the glacier in the *zone of accumulation*, whereas ablation subtracts in the *zone of ablation*. The boundary between these two zones is the **equilibrium line** (**Fig. 18.7b**).

Whether a glacier gets larger or smaller over time depends on the balance between accumulation and ablation. If the rate at which ice builds up in the zone of accumulation exceeds the rate at which ablation occurs below the equilibrium line, then the *terminus* or *toe* (the downslope end of a glacier) moves forward into previously unglaciated regions—such a change is called a **glacial advance** (**Fig. 18.8a, b**). In mountain glaciers, the position of a toe moves downslope during an advance, and in continental glaciers, the toe moves outward, away from the glacier's origin. If the rate of ablation below the equilibrium line equals the rate of accumulation, then the position of the toe remains fixed, even though ice continues to flow toward the toe. But if the rate of ablation exceeds the rate of accumulation, then the position of the toe moves back toward the origin of the glacier—such a change is called a **glacial retreat** (**Fig. 18.8c**). During a mountain glacier's retreat, the position of the toe moves upslope, and during a continental glacier's retreat, the position of the toe moves to higher latitudes. It's important to realize that when a glacier retreats, only the position of the toe moves back toward the glacier's origin, for ice always flows toward the toe.

As glaciers flow, ice tends to follow curved trajectories, in cross section. Specifically, beneath the zone of accumulation, ice flows down toward the base of the glacier as new ice accumulates above it, whereas beneath the zone of ablation, ice gradually moves up toward the surface of the glacier, as overlying ice ablates. Because ice follows curving flow paths, rocks picked up by ice at the base of the glacier may eventually reach the top of the glacier.

FIGURE 18.7 Flow velocities vary with location in a glacier. Overall, ice flows from the zone of accumulation to the toe.

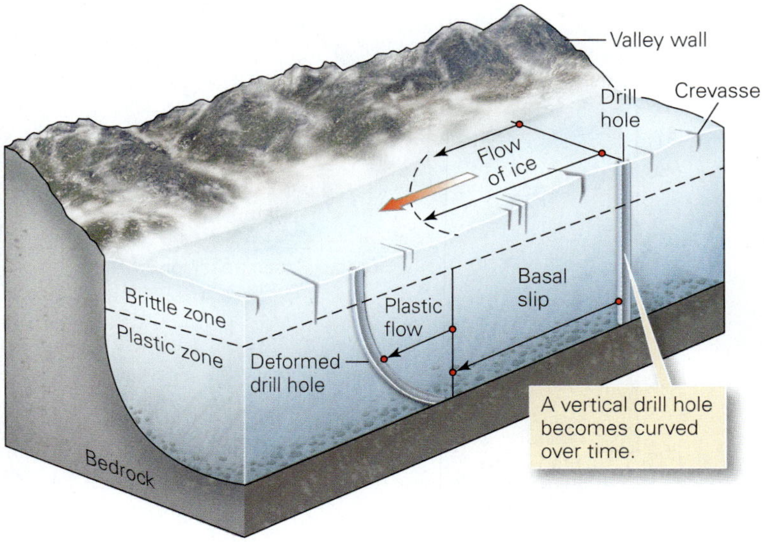

(a) Different parts of a glacier flow at different velocities due to friction with the substrate. The top and center regions flow fastest.

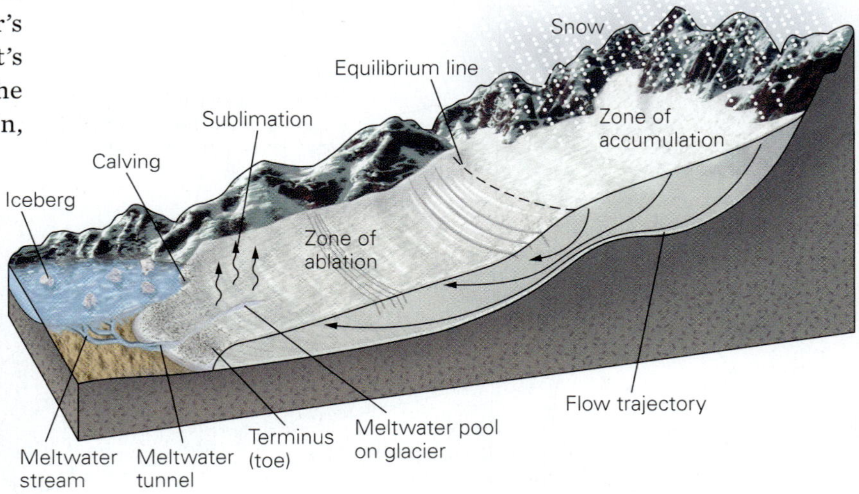

(b) The equilibrium line separates the zone of accumulation from the zone of ablation. As indicated by arrows, ice flows down in the zone of accumulation and up in the zone of ablation.

Ice in the Sea

On the moonless night of April 14, 1912, the ocean liner *Titanic* struck a large iceberg in the frigid North Atlantic. Its crew was convinced that they could see and avoid the biggest bergs and that smaller ones could not damage the steel hull of this brand new "unsinkable" vessel. Although lookouts had seen the ghostly, floating mass of ice only minutes earlier and had alerted the ship's pilot, the ship was unable to turn fast enough to avoid disaster. The force of the glancing blow split the steel hull, water gushed in, and less than 3 hours later, the ship disappeared beneath the surface and 1,500 people perished.

Where do icebergs, such as the one responsible for the *Titanic*'s demise, originate? In high latitudes, glaciers flow down to and sometimes into the sea. Glaciers whose toe lies in the water are called **tidewater glaciers.** Valley glaciers entering the sea become *ice tongues,* much longer than they are wide, whereas continental glaciers entering the sea become broad, flat sheets known as *ice shelves.* In shallow water, glacial ice remains grounded (**Fig. 18.9a**). But in deeper water,

FIGURE 18.8 Glacial advance and retreat.

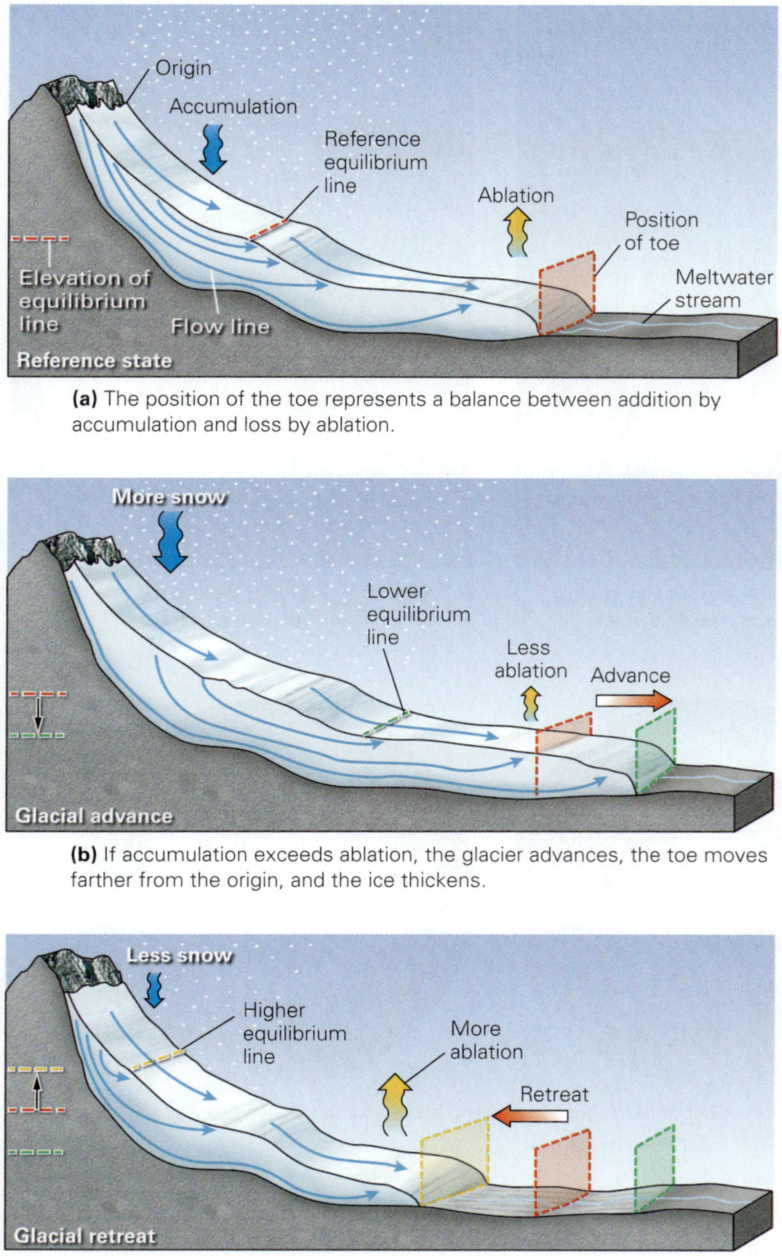

(a) The position of the toe represents a balance between addition by accumulation and loss by ablation.

(b) If accumulation exceeds ablation, the glacier advances, the toe moves farther from the origin, and the ice thickens.

(c) If ablation exceeds accumulation, the glacier retreats and thins. The toe moves back, even though ice continues to flow toward the toe.

the ice floats such that about 88% ($^7/_8$) of its mass lies below the water's surface. At the toe of a tidewater glacier, blocks of ice calve off and tumble into the water with an impressive splash. A free-floating chunk that rises 6 m above the water, and is at least 15 m long, is formally called an *iceberg*. Since most of its mass lies below the surface of the sea, the base of a large iceberg may actually extend down for a few hundred meters below the water's surface (**Fig. 18.9b, c**).

Not all ice floating in the sea originates in glaciers on land. In polar climates, the surface of the sea itself freezes, forming **sea ice** (**Fig. 18.9d**). The Earth's north polar ice cap consists of sea ice, formed on the surface of the Arctic Ocean. Some sea ice floats freely, but some protrudes outward from the shore.

TAKE-HOME MESSAGE

Glaciers form when buried snow lasts all year and turns to ice. Mountain glaciers form at high elevation and flow downslope. Ice sheets form in high latitudes and spread over continents. Flow of ice can be accomplished by plastic deformation, or by basal gliding. The balance of accumulation and ablation determines whether a glacier advances or retreats. In polar regions, ice shelves, ice tongues, and sea ice cover large areas.

QUICK QUESTION Does ice actually flow uphill during a retreat of a valley glacier?

18.3 Carving and Carrying by Ice

The Process of Glacial Erosion

During the last ice age, valley glaciers carved deep, steep-sided valleys into the Sierra Nevada range of California. In the process, some granite domes were cut in half, leaving a rounded surface on one side and a steep cliff on the other. Half Dome, in Yosemite National Park, formed in this way (**Fig. 18.10a**)—its steep cliff has challenged many rock climbers. Such glacial erosion also produces the knife-edge ridges and pointed spires of high mountains (**Fig. 18.10b**) and broad expanses where rock outcrops have been stripped of overlying sediment and polished smooth (**Fig. 18.10c**). In many localities, the rock surface visible today is the same surface that was once in contact with ice, but elsewhere, subsequent rockfalls and river erosion have substantially modified the surface.

Glaciers pick up fragments of their substrate in several ways. During *glacial incorporation*, ice surrounds loose rock and carries it away, whereas during *glacial plucking*, a glacier breaks off fragments of bedrock. Plucking occurs when ice flows into joints that intersect the bedrock's surface. Freezing and thawing, along with the push of moving ice, causes joints to grow until a joint-bounded block of rock finally breaks free of its substrate and starts to move with the ice. As glaciers flow, sand and silt embedded in the ice act like the teeth of a giant rasp and grind away the substrate. This process, *glacial abrasion*, pulverizes rock into a fine powder known as *rock*

FIGURE 18.9 Ice shelves, tidewater glaciers, and sea ice.

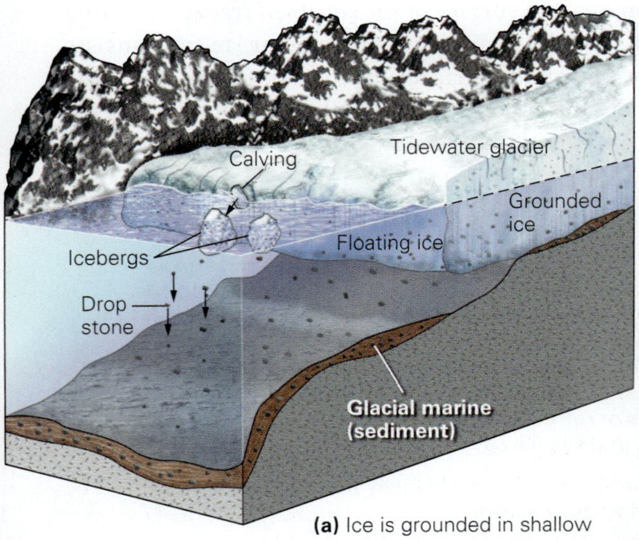

(a) Ice is grounded in shallow water but floats in deep water.

(b) This artist's rendition of an iceberg emphasizes that most of the ice is underwater.

(c) In summer, some of the sea ice of Antarctica breaks up to form tabular icebergs.

(d) Sea ice covers most of the Arctic Ocean (left) and surrounds Antarctica (right).

flour, and yields shiny *glacially polished surfaces* on bedrock. Clasts moving with the ice may also carve grooves or scratches, known as **glacial striations**, into bedrock (**Fig. 18.10d**). Striations range from 1 cm to 1 m across and may be tens of centimeters to tens of meters long. As you might expect, striations run parallel to the flow direction of the ice.

Let's now look more closely at erosional landscape features associated with a mountain glacier (**Fig. 18.11a**). Freezing and thawing during the fall and spring help fracture the rock bordering the *head* of the glacier (the ice edge high in the mountains). This rock falls on the ice or gets picked up at the base of the ice, and moves downslope with the glacier. As a consequence, a bowl-shaped depression, or **cirque**, develops on the side of the mountain. If the ice later melts, a lake called a *tarn* may form at the base of the cirque. The shape of a cirque may be maintained or even amplified by rockfalls after the glacier has melted away. An **arête**, a residual knife-edge ridge of rock, separates two adjacent cirques. A pointed mountain peak surrounded by at least three cirques is called a **horn**. The Matterhorn, a peak in Switzerland, is a particularly beautiful example of a horn; each of its faces is a cirque (**Fig. 18.11b**).

Glacial erosion drastically modifies the shape of a valley. To see how, compare a river-eroded valley with a glacially eroded valley. A topographic profile across a river in unglaciated mountains shows that a river typically flows down a **V-shaped valley**, with the river channel itself forming the point of the V. The V develops because river erosion occurs only in the channel, and mass wasting causes the valley slopes to approach the angle of repose. In contrast, a profile across a glacially eroded valley resembles a U, with steep walls and a curved floor. A **U-shaped valley** forms because the combined processes of glacial abrasion and plucking not only lower the floor of the valley but also bevel its sides (**Fig. 18.11c**).

Glacial erosion in mountains also modifies the intersections between tributaries and the trunk valley. In a river system, the trunk stream serves as the local base level for tributaries (see Chapter 14), so the mouths of the tributary valleys lie at the same elevation as the trunk valley. The ridges (spurs) between valleys taper to a point when they join the trunk valley floor. During glaciation, tributary glaciers

FIGURE 18.10 Products of glacial erosion. Ice is a very aggressive agent of erosion.

(a) Half Dome in Yosemite National Park, California.

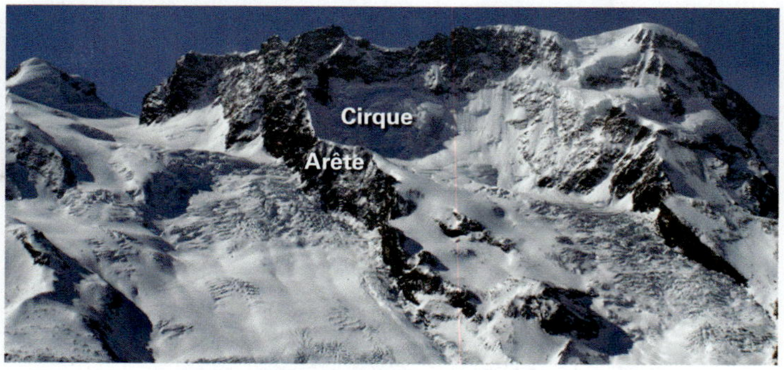

(b) Examples of a cirque and an arête in the Swiss Alps.

(c) Glacially polished outcrop in Central Park, New York City.

(d) Glacial striations in Victoria, British Columbia.

flow down side valleys into a trunk glacier. But the trunk glacier cuts the floor of its valley down to a depth that far exceeds the depth cut by the tributary glaciers. As a consequence, when the glaciers melt away, the mouths of the tributary valleys perch at a higher elevation than the floor of the trunk valley. Geologists refer to such side valleys as **hanging valleys**. The water in a post-glacial stream that flows down a hanging valley cascades over a spectacular waterfall to reach the post-glacial trunk stream (**Fig. 18.11d**). As they erode, trunk glaciers also chop off the ends of ridges between valleys, to produce truncated spurs.

Now let's look at the erosional features produced by continental ice sheets. To a large extent, these depend on the nature of the pre-glacial landscape. Where an ice sheet spreads over a region of low relief, such as the Canadian Shield, glacial erosion creates a vast region of polished, flat, striated surfaces. Where an ice sheet spreads over a hilly area, it deepens valleys and smooths hills. Glacially eroded hills may become elongate in the direction of flow and may be asymmetric, for glacial rasping bevels the upstream part of the hill, to produce a gentle slope, whereas glacial plucking eats away at the downstream part, yielding a steep slope. Ultimately, the hill's profile resembles that of a sheep lying in a meadow—such a hill is called a **roche moutonnée**, from the French for sheep rock (**Fig. 18.12**).

Fjords: Submerged Glacial Valleys

As noted earlier, where a valley glacier meets the sea, the glacier's base remains in contact with the ground until the water depth exceeds about seven-eighths of the glacier's thickness, at which point the glacier floats. Thus, glaciers can carve U-shaped valleys even below sea level. During an ice age, water extracted from the sea becomes locked in the ice sheets on land, so sea level drops significantly. Because coastal glaciers can carve valleys below sea level, and because

SEE FOR YOURSELF...

BAFFIN ISLAND, CANADA

Latitude
67°8′27.56″ N

Longitude
64°49′49.31″ W

Zoom to 40 km (~25 miles) and look down.

Two valley glaciers draining the Baffin Island ice cap merge into a trunk glacier that flows NE and then into a fjord, partly filling a U-shaped valley. Note the lateral and medial moraines. In some nearby fjords, meltwater has deposited a delta of outwash at the end of the glacier.

FIGURE 18.11 Features formed by the glacial erosion of a mountainous landscape.

Tributary valley
V-shaped valley
Before glaciation, valleys are V-shaped, and tributary mouths are the same elevation as the trunk stream.
Trunk valley
Trunk valley
Tributary valley
Cirque
Tarn
Mt. Snowdon, Wales
During glaciation, the valleys fill with ice.
Time
After glaciation, the region contains U-shaped valleys, hanging valleys, truncated spurs, and horns.
Horn
Cirque
Arête
U-shaped valley
Hanging valley
Truncated spur

(a) Stages in the development of a glacially carved mountainous landscape.

(b) The Matterhorn in Switzerland. The first ascent was in 1865.

(c) A U-shaped glacial valley in the Tongass National Forest, Alaska.

(d) A waterfall spilling out of a U-shaped hanging valley in the Sierra Nevada.

FIGURE 18.12 A roche moutonnée is an asymmetric bedrock hill shaped by the flow of glacial ice.

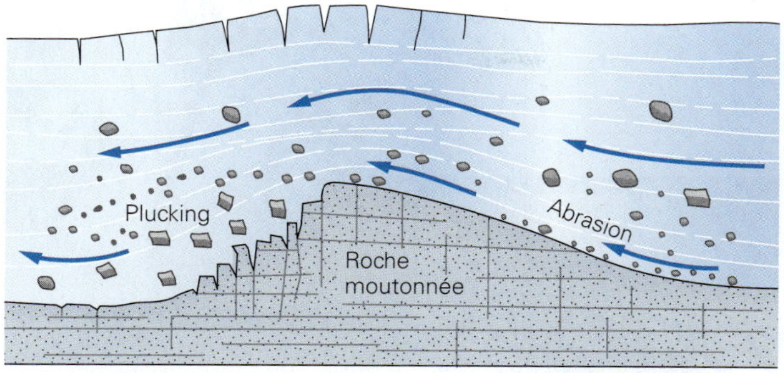

(a) Abrasion rasps the upstream side, and plucking carries away fracture-bounded blocks on the downstream side.

(b) An example of a roche moutonnée. The glacier flowed from right to left.

sea level is so low during an ice age, the floors of valleys cut by coastal glaciers during the Pleistocene Ice Age were cut hundreds of meters deeper than present sea level. Today, the sea has flooded these deep valleys, producing **fjords** (see Chapter 15). In the spectacular fjord-land regions along the coasts of Norway, Scotland, New Zealand, Chile, Maine, and Alaska, the walls of submerged U-shaped valleys rise straight from the sea as vertical cliffs up to 1,000 m high (**Fig. 18.13**). Fjords also develop where an inland glacial valley fills to become a lake.

FIGURE 18.13 One of the many spectacular fjords of Norway. The water is an arm of the sea that fills a glacially carved valley. Tourists are standing on Pulpit Rock (Prekestolen).

TAKE-HOME MESSAGE

A glacier scrapes up and plucks rock from its substrate, and carries debris that falls on its surface. Glacial erosion polishes and scratches rock, and carves distinctive landforms, such as U-shaped valleys, cirques, and horns. U-shaped valleys that fill with water become fjords.

QUICK QUESTION Why do we find hanging valleys spilling waterfalls into trunk valleys, in regions that have been eroded by mountain glaciers?

18.4 Deposition Associated with Glaciation

The Glacial Conveyor

Glaciers, in general, do not consist of pure ice. All glaciers incorporate clasts at their base and sides, where ice comes in contact with its substrate. And in the case of valley glaciers, the ice also carries clasts that tumble down from bordering mountain slopes onto the surface of glaciers. All this debris, regardless of grain size, moves with the ice, so glaciers—like giant conveyor belts—transport large quantities of unsorted sediment in the direction of flow (**Fig. 18.14a, b**). Geologists refer to a pile of sediment either carried on or left behind by a glacier as a **moraine**. Sediment dropped on the side margins of a glacier becomes a *lateral moraine*, a stripe of debris adjacent to the valley wall on the top of the glacier (**Fig. 18.14c**). When a glacier melts, lateral moraines become stranded along the side of the glacially carved valley, like bathtub rings. Where two valley glaciers merge, the lateral moraines along

FIGURE 18.14 The glacial conveyor and the formation of lateral and medial moraines on glaciers.

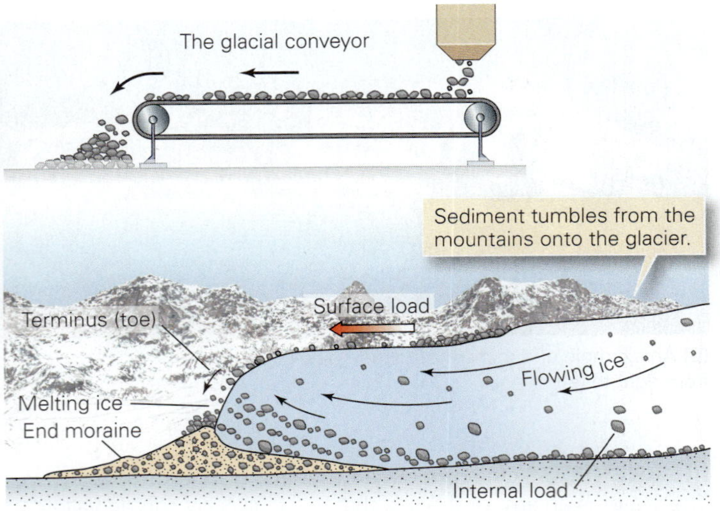

(a) Sediment falls on a glacier from bordering mountains and gets plucked up from below. Glaciers are like conveyor belts, moving sediment toward the toe of the glacier.

(b) This glacier in the French Alps carries lots of sediment.

(c) A medial moraine forms where lateral moraines of two valley glaciers merge.

the sides of the two merging glaciers join to become a *medial moraine*, a stripe down the interior of the composite glacier. Trunk glaciers formed by the merging of many tributary glaciers contain several medial moraines. Sediment left at the base of a glacier when the glacier melts away becomes *ground moraine*, and sediment transported to a glacier's toe accumulates to form an *end moraine*.

Types of Glacial Sedimentary Deposits

Several different types of sediment can be deposited in glacial environments—all of these types together constitute **glacial drift**. The term dates from pre-Agassiz studies of glacial deposits, when geologists thought that the sediment had "drifted" into place during an immense flood. Specifically, glacial drift includes the following:

- *Glacial till* consists of sediment transported by ice and deposited beneath, at the side, or at the toe of a glacier in moraines. Till is unsorted, because the solid ice of glaciers can carry clasts of all sizes (**Fig. 18.15a**).
- *Glacial erratics* are relatively large cobbles and boulders that have been dropped by a glacier. Some lie within or on moraines, whereas others rest directly on glacially polished surfaces (**Fig. 18.15b**).
- *Glacial marine* consists of clasts carried out to sea by icebergs. When the icebergs melt, the clasts sink to the seafloor and mix with marine sediment.

FIGURE 18.15 Sedimentation processes and products associated with glaciation. Glacial sediment is distinctive.

(a) This glacial till in Ireland is unsorted, because ice can carry sediment of all sizes.

(b) Glacial erratics resting on a glacially polished surface in Wyoming.

(c) Braided streams choked with glacial outwash in Alaska. The streams carry away finer sediment and leave the gravel behind.

(d) Thick loess deposits underlie parts of the prairie in Illinois.

(e) In the quiet water of an Alaskan glacial lake, fine-grained sediments accumulate. Alternating layers in the sediment (varves), now exposed in an outcrop near Puget Sound, Washington, reflect seasonal changes.

FIGURE 18.16 The formation of depositional landforms associated with continental glaciation.

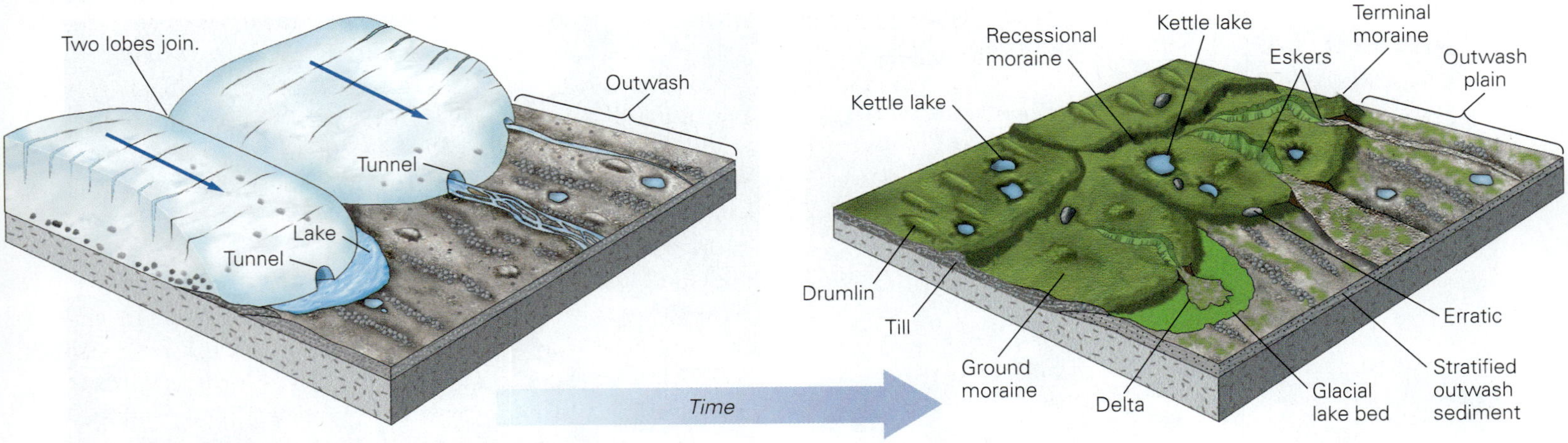

(a) The ice in continental glaciers flows toward the toe; sediment accumulates at the base and at the toe of the ice sheet.

(b) Several distinct depositional landforms form during glaciation; some developed under the ice and some at the toe.

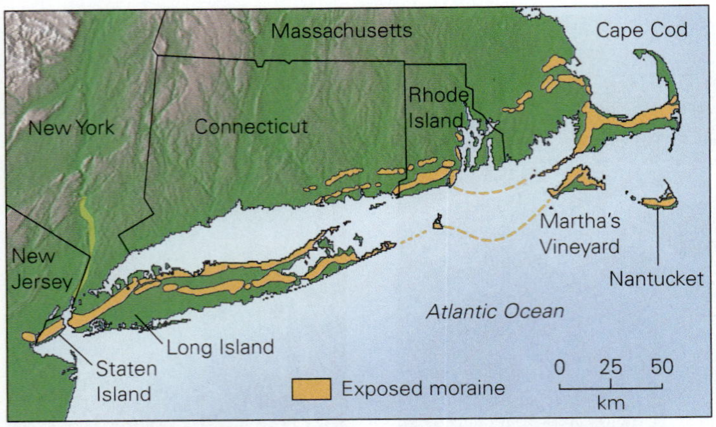

(c) Cape Cod, Long Island, and other landforms in the northeastern United States formed at the end of the continental ice sheet.

- *Glacial outwash* forms when meltwater streams carry and sort till. The clasts are deposited by a braided stream network to form a broad area of gravel and sandbars (**Fig. 18.15c**).
- *Loess* forms from clay and silt that had been transported by strong *katabatic winds* that develop when warmer air above ice-free land beyond the toe of a glacier rises and the cold, denser air from above the glacier rushes in to take its place. Deposits of loess can be held together by cement composed of iron oxide and calcite, so erosion of the deposits yields steep escarpments (**Fig. 18.15d**).
- *Glacial lake-bed sediment* consists of very fine-grained sediment, including rock flour, that accumulates on the floor of a meltwater lake. Such deposits commonly contain varves. A **varve** is a pair of thin layers deposited during a single year—one layer consists of silt brought in during spring floods and the other of clay deposited in winter when the lake's surface freezes over and the water is still (**Fig. 18.15e**).

Depositional Landforms of Glacial Environments

Picture a group of hunters wearing reindeer skins, gazing southward from the crest of an ice cliff at the toe of a continental glacier in what is now southern Canada. It's about 12,000 years ago, and the glacier has been receding for at least a millennium. The hunters would have been able to see a variety of landscape features, some formed by glacial erosion and some by deposition, due to moving ice and meltwater (see **Geology at a Glance,** pp. 586–587). We've already described erosional features, so now let's focus on the depositional features of the landscape (**Fig. 18.16a, b**).

From their vantage point, the hunters would probably see a few curving ridges of sediment in the region between the glacier's toe and the horizon. Each of these ridges is an end moraine, formed from till deposited when the position of the glacier's toe remained in the same location for a while. As we've seen, ice keeps flowing to the toe, so the longer the toe stays in the same place, the bigger the end moraine becomes. Geologists refer to the end moraine at the farthest limit of glaciation as the **terminal moraine**. In the northeastern United States, a large terminal moraine built up during the most recent Ice Age—this ridge of sediment now underlies Long Island, New York, and Cape Cod, Massachusetts (**Fig. 18.16c**). When a glacier starts receding, the position of its toe may pause several times—an end moraine that forms when a glacier stalls while receding is a *recessional moraine.* Till that accumulates in the region between end moraines constitutes *ground moraine.* Since this till was deposited by moving ice, clasts within it may be aligned and scratched.

Landscapes composed of glacial deposits tend to be hummocky (bumpy), in that they consist of small hills interspersed with depressions. In some cases, the random ups and downs of a moraine surface simply reflect local variations in the amount of sediment supplied by the ice during deposition, or removed subsequently by meltwater streams. But in some places, each depression on a moraine has a distinct circular shape, in map view. Such depressions, known as **kettle holes**, form when a block of ice that calved off the toe of a glacier became buried by till, for when the block melts, an empty space remains (**Fig. 18.17a, b**). Geologists refer to a land surface spotted with many kettle holes separated by rounded hills or ridges of sediment as *knob-and-kettle topography* (**Fig. 18.17c**). In places where the water table lies close to the ground surface, kettles fill with water and become lakes.

Locally, flowing glacial ice reshapes the underlying till into elongate hills known as **drumlins** (from the Gaelic word for small hill or ridge). Drumlins commonly occur in swarms, and tend to be about 50 m high. Their long axis trends parallel to the flow direction of the glacier. Notably, drumlins taper in the direction of flow—a drumlin's upstream end is steeper than its downstream end (**Fig. 18.17d, e**).

As we've noted, not all of the sediment associated with glacial landscapes was deposited directly by ice. Meltwater also carries and deposits sediment. Water-transported sediment, in contrast to till, tends to be sorted and stratified. Sediment deposited in meltwater tunnels beneath a glacier may remain as a sinuous ridge, known as an **esker**, when the glacier melts away (**Fig. 18.18**). Braided meltwater streams that flow

FIGURE 18.17 Drumlins and knob-and-kettle topography characterize some areas that were once glaciated.

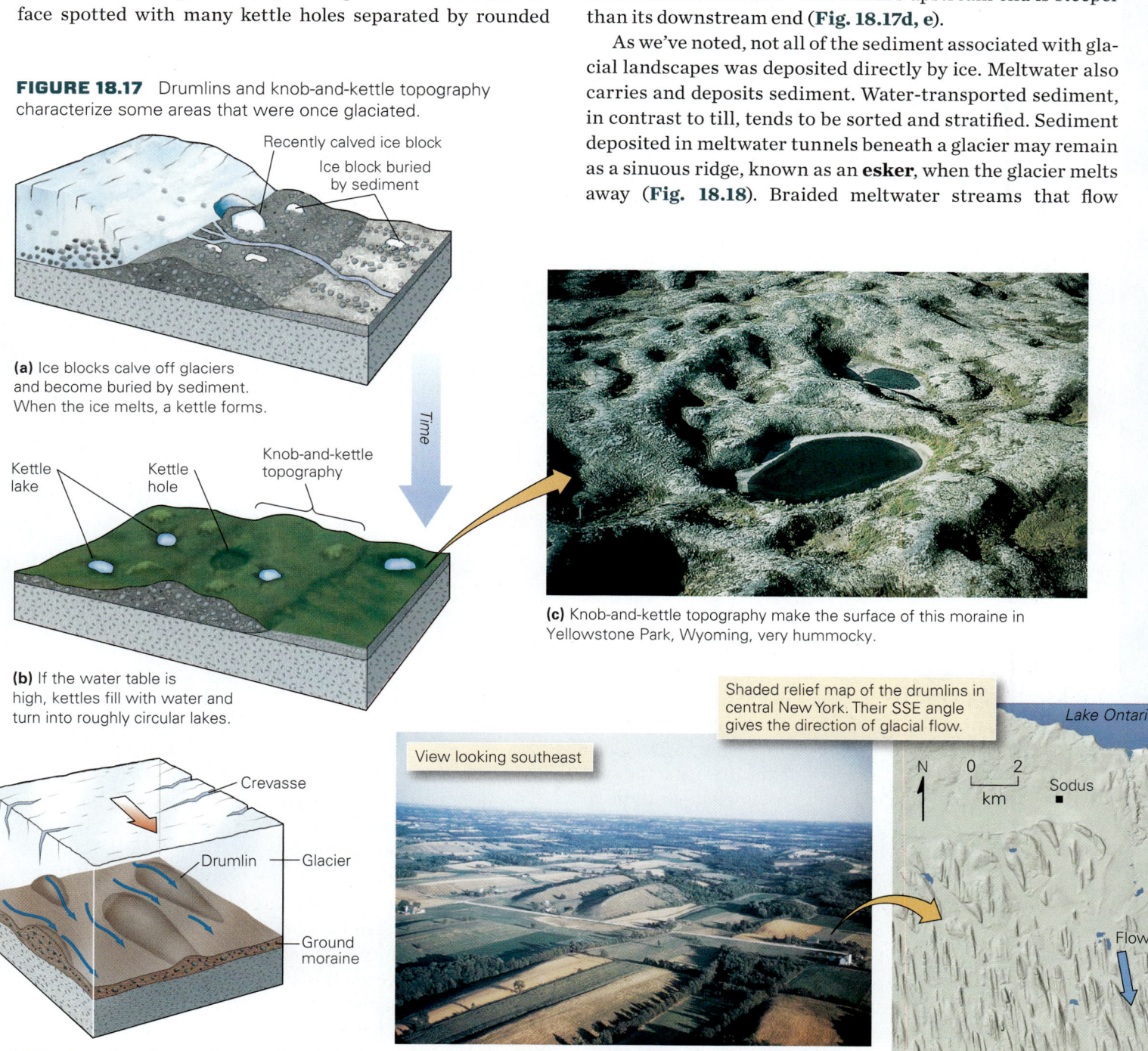

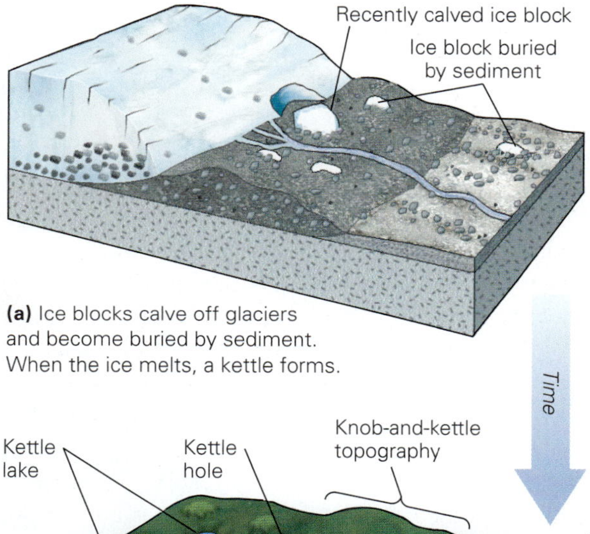

(a) Ice blocks calve off glaciers and become buried by sediment. When the ice melts, a kettle forms.

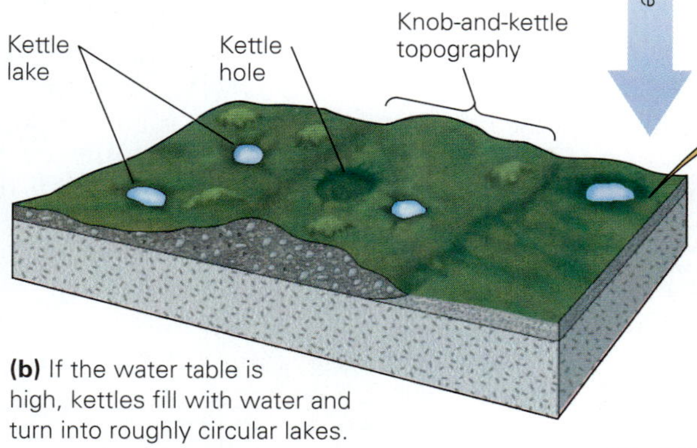

(b) If the water table is high, kettles fill with water and turn into roughly circular lakes.

(c) Knob-and-kettle topography make the surface of this moraine in Yellowstone Park, Wyoming, very hummocky.

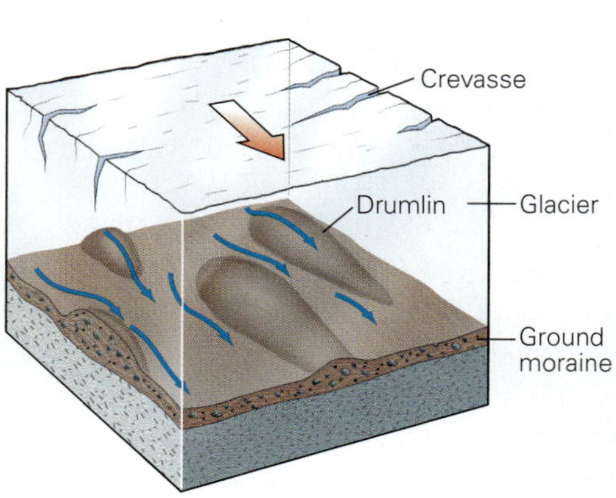

(d) The formation of a drumlin beneath a glacier.

(e) Drumlins dominate this landscape near Rochester, New York.

GEOLOGY AT A GLANCE

Glaciers and Glacial Landforms

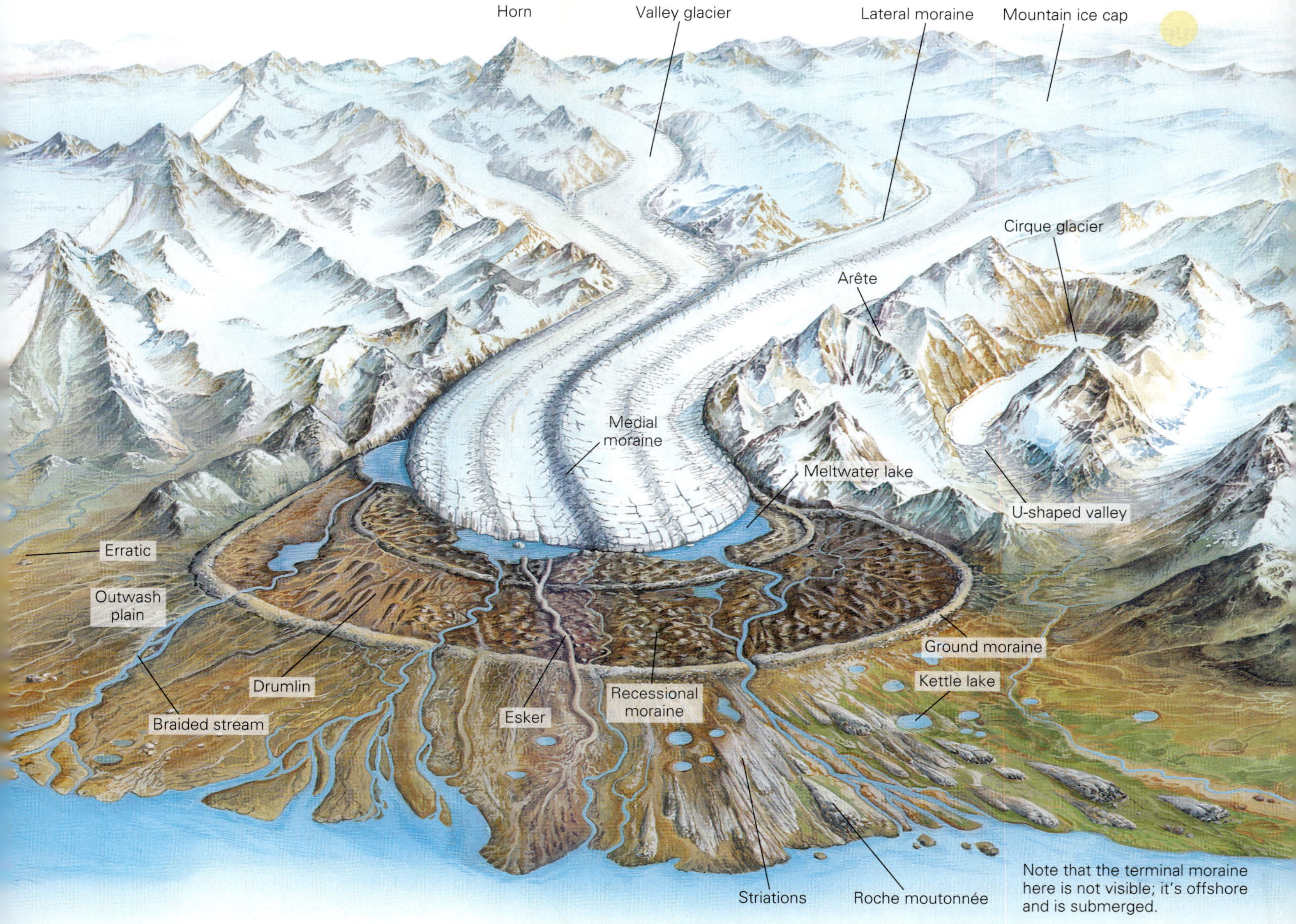

Continental glaciers, vast sheets of ice up to a few kilometers thick, covered extensive areas of land during times when Earth had a colder climate. They form from snow that accumulates at high latitudes. When buried deeply enough, the snow packs together and recrystallizes into glacial ice. Although it is solid, ice is weak, so ice sheets spread over the landscape like syrup over a pancake, though much more slowly. When a continental glacier reaches the sea, it becomes an ice shelf. At the edge of the shelf, icebergs calve off and float away, and sediment carried by the ice falls to the seafloor. A second category of glaciers, called mountain or alpine glaciers, grow in mountainous areas because snow can last all year at high elevations. Valley glaciers are confined to valleys, and ice caps cover the peaks of mountains. These glaciers carve a landscape containing U-shaped valleys, cirques, horns, and hanging valleys. Lateral moraines form along their sides.

During an ice age, mountain glaciers advance and may eventually flow out onto the land surface beyond the mountain front. When climate warms, the glacier starts to recede. The landscape in front of the glacier retains erosional features, such as striations and roches moutonée—formed when the area was ice covered by flowing ice.

When the glacier pauses, till (unsorted glacial sediment) accumulates to form an end moraine. Meltwater lakes gather at the toe, and streams pick up till, sort it, and redistribute it as glacial outwash. Sediment that accumulates in ice tunnels, exposed when the glacier melts, make up sinuous ridges called eskers. Flowing ice can mold underlying sediment into drumlins. Ice blocks buried in till melt away and leave behind kettle holes.

FIGURE 18.18 Eskers are snake-like ridges of sand and gravel that form when sediment fills meltwater tunnels at the base of a glacier.

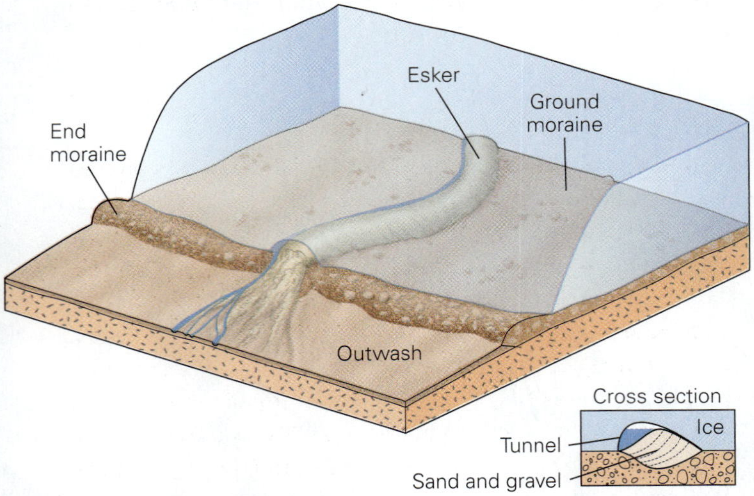

(a) At the time of formation, an esker develops beneath an ice sheet. In cross section (inset), wedges of sand accumulate in the tunnel.

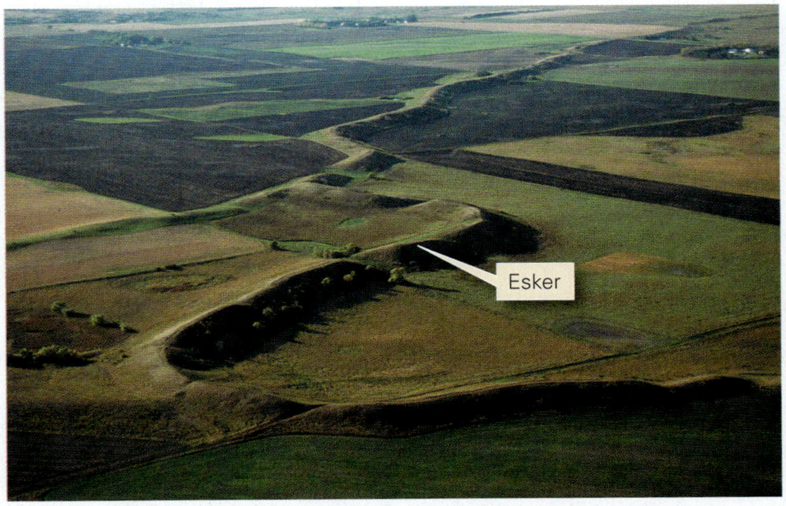

(b) An example of an esker in an area once glaciated but now farmed.

beyond the end of a glacier deposit layers of sand and gravel that underlie *glacial outwash plains*. Meltwater may collect in a lake adjacent to the glacier's toe, to form an *ice-margin lake*, and additional meltwater lakes and swamps may form in low areas on the ground moraine. Sediments deposited in eskers and glacial outwash plains serve as important sources of sand and gravel for construction, and the fine sediment of former glacial lake beds evolves into fertile soil for agriculture.

TAKE-HOME MESSAGE

Glaciers, like giant conveyor belts, transport vast amounts of sediment. When the ice melts, the sediment accumulates as unsorted till. Meltwater streams and wind transport and sort the sediment to form outwash-plain gravels and loess deposits, respectively. Deposition by glaciers produces distinctive landforms, such as moraines, eskers, and kettle holes.

QUICK QUESTION How does knob-and-kettle topography develop?

18.5 Other Consequences of Continental Glaciation

Ice Loading and Glacial Rebound

When a large ice sheet (more than 50 km in diameter) grows on a continent, its weight causes the surface of the lithosphere to sink. In other words, ice loading causes **glacial subsidence**. Lithosphere, the relatively rigid outer shell of the Earth, can sink because the underlying asthenosphere is soft enough to flow slowly out of the way (**Fig. 18.19**). For example, because of ice loading, the bedrock surface of East and West Antarctica now lies as far as several hundred meters below sea level, even though the top surface of the ice sheets covering the continent lies up to 3 km above sea level (see Fig. 18.4). If the overlying ice instantly melted away, much of the continent would be flooded by a shallow sea.

What happens when continental ice sheets melt? Gradually, the surface of the underlying continental lithosphere rises back up, by a process called **post-glacial rebound**. As this happens, the asthenosphere flows back underneath the lithosphere to fill the space. Because asthenosphere flows so slowly—at rates of a few millimeters per year—it may take

FIGURE 18.19 The concept of subsidence and rebound due to continental glaciation and deglaciation.

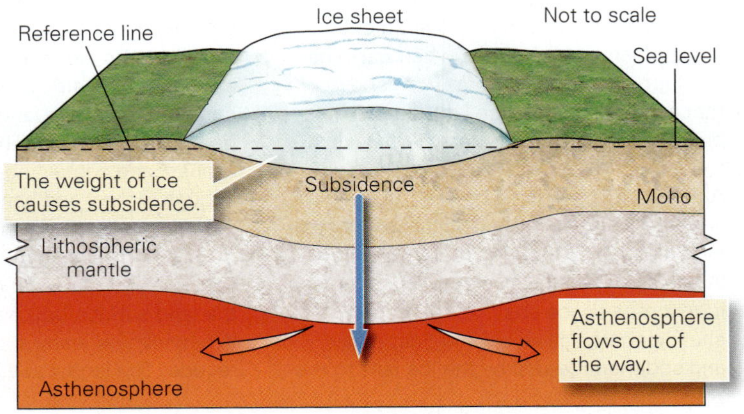

thousands of years for ice-depressed continents to rebound completely. In fact, glacial rebound is still happening in north-central Canada, a region that became ice-free about 6,500 years ago.

Sea-Level Changes: Filling and Emptying the Glacial Reservoir

More of the Earth's surface and near-surface freshwater resides in glacial ice than in any other reservoir. When, during the Pleistocene Ice Age, glaciers covered almost three times as much land area as they do today, they trapped more than three times as much water than they do now. In the context of the hydrologic cycle, this means that large quantities of water transferred from the ocean reservoir to the glacial reservoir. As a consequence, sea level during the Ice Age dropped by as much as 100 m. As the coastline migrated seaward, and extensive areas of continental shelves became exposed as dry land (**Fig. 18.20a**). Exposure of shallow seafloor resulted in the appearance of a land bridge across the Bering Strait, providing routes over which prehistoric humans walked as they migrated into the Americas (**Fig. 18.20b**). When the last glacial advance ended, about 17,000 years ago, sea level rose rapidly until the continental glaciers of North America and Eurasia had vanished, about 6,000 B.C.E (**Fig. 18.20c**). Sea level remained fairly constant (with a few ups and downs) until about 150 years ago, when it began to rise again. If the present-day continental glaciers of Antarctica and Greenland were to completely melt, ocean water would submerge North America's coastal plain (see Fig. 18.20a).

Changing Drainage, and Lake Formation

As we noted earlier, the weight of continental glaciers pushes the land surface down. This subsidence results in a low area, or depression, along the toe of the glacier. In some cases, the depressions fill with meltwater and become ice-margin lakes. The largest known ice-margin lake during the Pleistocene Ice Age submerged portions of Manitoba and Ontario, in south-central Canada, and North Dakota and Minnesota in the United States (**Fig. 18.21a**). This body of water, called Glacial Lake Agassiz, existed between 11,700 and 9,000 years ago, and had an area of 250,000 square km (100,000 square miles), more than that of all present-day Great Lakes combined. Growth of continental glaciers can also disrupt the configuration of drainage networks, for when an ice sheet overruns a drainage network, rivers seek different courses, abandon their old valleys, and carve new ones.

In the toe region of an ice sheet or a mountain glacier, a lobe of the glacier can serve as an *ice dam*, which blocks the outlet of an unglaciated watershed to form a deep lake. Inevitably, when the glacier retreats, the ice dams break, and in a

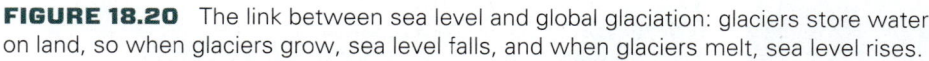

FIGURE 18.20 The link between sea level and global glaciation: glaciers store water on land, so when glaciers grow, sea level falls, and when glaciers melt, sea level rises.

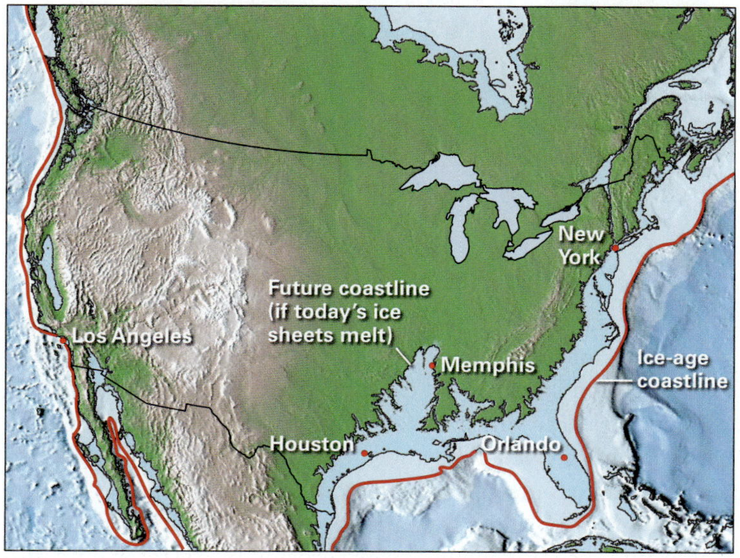

(a) The red line shows the coastline during the last ice age; much of the continental shelf was dry. If present-day ice sheets melt, coastal lands will flood.

(b) Prehistoric people migrated across the Bering Strait land bridge.

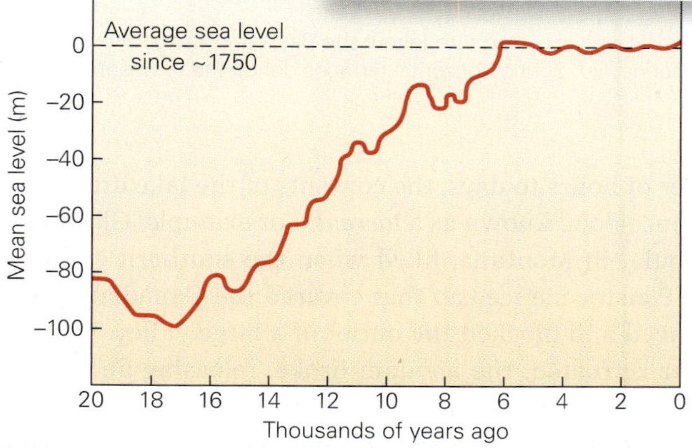

(c) Sea-level rise between 17,000 and 7,000 B.C.E. was due to the melting of ice-age glaciers.

FIGURE 18.21 Ice-age lakes in North America.

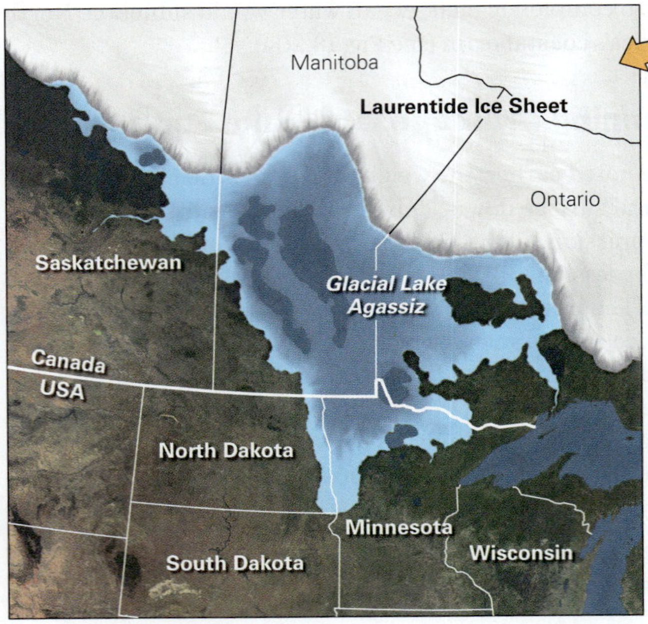

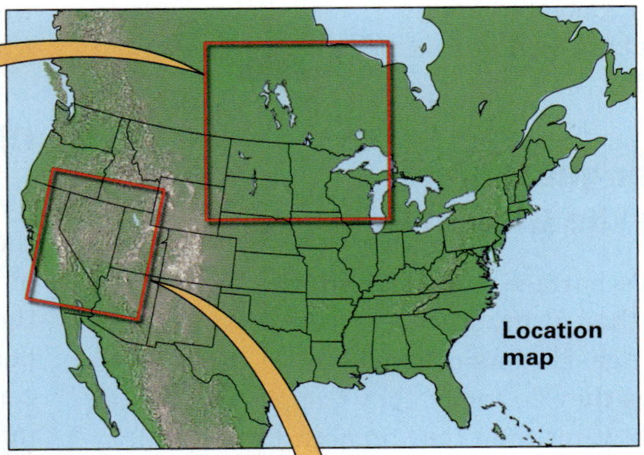

(a) Glacial Lake Agassiz was an ice-margin lake that formed near the end of the last ice age.

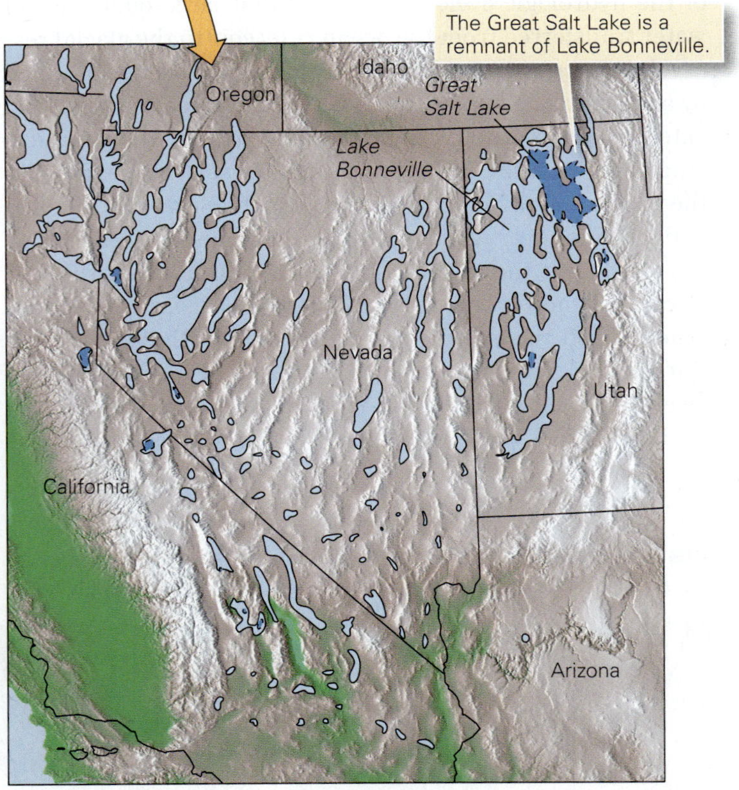

(b) Pluvial lakes occurred throughout the Basin and Range Province during the last ice age due to the wetter climate. The largest of these was Lake Bonneville. Subtle horizontal terraces define the remnants of beaches, now over 100 m above the present level of the Great Salt Lake.

matter of hours to days, the contents of the lake drains in an immense flood known as a *torrent*. For example, Glacial Lake Missoula, in Montana, filled when the southern end of the huge Pleistocene ice cap that covered the Canadian Rockies advanced and blocked the outlet of a large valley. When the glacier retreated, the ice dam broke, releasing an immense torrent—the Great Missoula Flood—that scoured eastern Washington, creating a barren, soil-free landscape called the *channeled scablands*. Because, as we will see, glaciers advance and retreat many times during an ice age, two dozen such torrents may have occurred in eastern Washington. When the ice dam blocking Glacial Lake Agassiz broke, a huge torrent carried so much water down the St. Lawrence Seaway that global sea level may have risen by 1 to 3 m during a single year. Torrents may also take place when lakes held back by end moraines suddenly drain due to the breaching of the end moraine, either when headward erosion by a stream cuts into the moraine, or when the lake overtops the moraine.

Pluvial Features

During the Pleistocene Ice Age, the climate in regions to the south of continental glaciers was wetter than it is today. Fed by enhanced rainfall, lakes accumulated in low-lying land at a great distance from the ice front. Many such **pluvial lakes** (from the Latin word *pluvia*, meaning rain) flooded interior basins of the Basin and Range Province in Utah and Nevada, a region that is now a desert (**Fig. 18.21b**). The largest pluvial lake, Lake Bonneville, covered almost a third of western Utah. When this lake suddenly drained after a natural dam broke, the former shoreline remained like a bathtub ring rimming the mountains near Salt Lake City. Today's Great Salt Lake is but a small remnant of Lake Bonneville.

Periglacial Environments

In polar latitudes today, and in regions adjacent to the fronts of continental glaciers during the last ice age, the mean annual temperature stays low enough (below −5°C) that soil moisture and groundwater freeze and, except in the upper few meters, stay solid all year. Regions underlain by such permanently frozen ground, or **permafrost**, are called *periglacial environments* (from the Greek word *peri*, meaning encircling), because periglacial environments develop around the edges of glacial environments (**Fig. 18.22a**).

The upper few meters of permafrost may melt during the summer months, only to refreeze again when winter comes. As a consequence of the freeze-thaw process, the ground of some permafrost areas splits into pentagonal or hexagonal shapes, resulting in a landscape called **patterned ground** (**Fig. 18.22b**). Melting of permafrost presents a unique challenge to people who live or work in polar regions. For example, heat from a building may melt underlying permafrost, producing a mire into which the building settles. For this reason, buildings in permafrost regions must be placed on stilts, so that cold air can circulate beneath them to keep the ground underneath frozen.

TAKE-HOME MESSAGE

Growth of glaciers can affect many aspects of the Earth System. The weight of a growing continental ice sheet causes underlying lithosphere to subside, then melting allows it to rebound. Continental ice sheets store so much water that their growth or melting affects global sea level. Sudden drainage of lakes related to glaciation leads to torrents that can scour the landscape.

QUICK QUESTION How did the landscape south of the Pleistocene ice sheet differ from the landscape of today?

18.6 The Pleistocene Ice Age

The Pleistocene Glaciers

Today, most of the land surface in New York City lies hidden beneath concrete and steel, but in Central Park it's still possible to see land in a seminatural state. If you stroll through the park, you'll find that the top surfaces of many outcrops are smooth and polished, and in places they have been grooved and scratched (Fig. 18.10c). Here and there, glacial erratics rest on the bedrock. You are seeing evidence that an ice sheet once

FIGURE 18.22 Periglacial regions are not ice covered but do include substantial areas of permafrost.

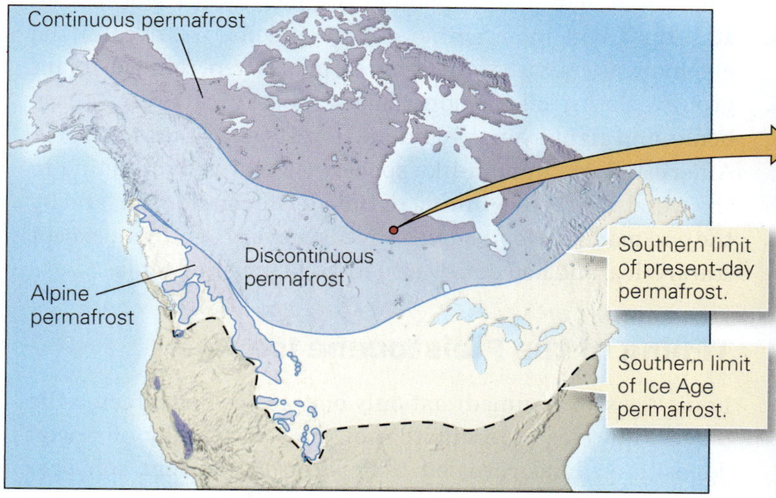

(a) The distribution of periglacial environments in North America.

(b) An example of patterned ground near a pond in Manitoba, Canada.

scraped along this now-urban ground. Geologists estimate that the ice sheet that overrode the New York City area may have been 250 m thick, tall enough to bury a 75-story building.

The fact that glaciated landscapes still decorate the surface of the Earth means that the last ice age occurred fairly recently during Earth's history. Otherwise, these landscape features would have been either eroded away or buried by now. The ice age responsible for the glaciated landscapes of North America, Europe, and Asia happened during the Pleistocene Epoch, which began about 2.6 million years ago (Ma), so as we've noted earlier (see Chapter 11), this event is commonly known as the **Pleistocene Ice Age**.

Based on studying patterns of glacial striations and of the sources of erratics, geologists have developed an approximate idea of where the Pleistocene ice sheets originated, and the directions in which the ice sheets flowed. In North America, major ice sheets began to grow in at least three locations (**Fig. 18.23**)—the Labrador ice sheet over northeastern Canada, the Keewatin ice sheet in northwestern Canada, and the Baffin ice sheet over Baffin Bay. These sheets, together with one or more smaller ones, merged to form the giant *Laurentide ice sheet* that covered all of Canada east of the Rocky Mountains, and spread southward over the northern portion of the United States. The *Cordilleran ice sheet*, which originated in the mountains of western Canada, spread westward to the Pacific coast and eastward until it merged with the Laurentide ice sheet. Other Pleistocene ice sheets formed in Greenland, Scandinavia, Britain, central Asia, and Siberia.

In addition to continental ice sheets, sea ice in the northern hemisphere expanded during the Pleistocene to cover all of the Arctic Ocean and parts of the North Atlantic. In fact, sea ice surrounded Iceland and approached Scotland and also fringed most of western Canada and southeastern Alaska.

FIGURE 18.23 Pleistocene ice sheets of the northern hemisphere.

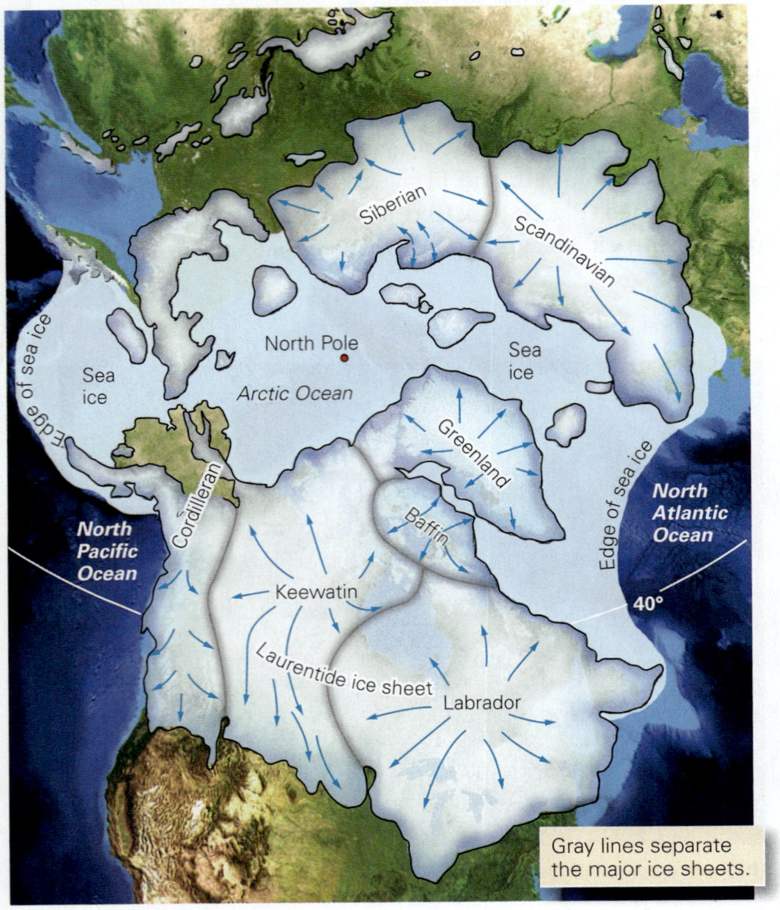

Life and Climate in the Pleistocene World

During the Pleistocene Ice Age, all climatic belts shifted southward (**Fig. 18.24a, b**). Geologists can document this shift by examining fossil pollen, which can survive for thousands of years if preserved in the sediment of bogs. Presently, the southern boundary of North America's *tundra*, a treeless region supporting only low shrubs, moss, and lichen capable of living on permafrost, lies at a latitude of 68° N—during the Pleistocene Ice Age, it moved down to 48° N. Much of the interior of the United States, which now has temperate, deciduous forest, instead harbored cold-weather spruce and pine forest. Ice-age climates also changed the distribution of rainfall on the planet. As we noted earlier, increased rainfall in North America, south of the glaciers, led to the filling of pluvial lakes in Utah and Nevada. In contrast, decreased rainfall in equatorial regions led to shrinkage of the tropical rainforest. And because glaciers trapped so much water, as we have seen, sea level dropped.

Fossils also tell us that numerous species of now-extinct large mammals inhabited the Pleistocene world (**Fig. 18.24c**). Giant mammoths and mastodons, relatives of the elephant, along with woolly rhinos, musk oxen, reindeer, giant ground sloths, bison, lions, saber-toothed cats, giant cave bears, and giant hyenas wandered forests and tundra in North America. Early human-like species were already foraging in the woods by the beginning of the Pleistocene Epoch, and by the end modern *Homo sapiens* lived on every continent except Antarctica, and had discovered fire and invented tools.

Timing of the Pleistocene Ice Age

Louis Agassiz assumed that only one ice age had affected the planet. But close examination of the stratigraphy of glacial deposits on land revealed that *paleosols* (ancient soil preserved in the stratigraphic record), as well as beds containing fossils of warmer-weather animals and plants, lay between

FIGURE 18.24 Climate belts during the Pleistocene.

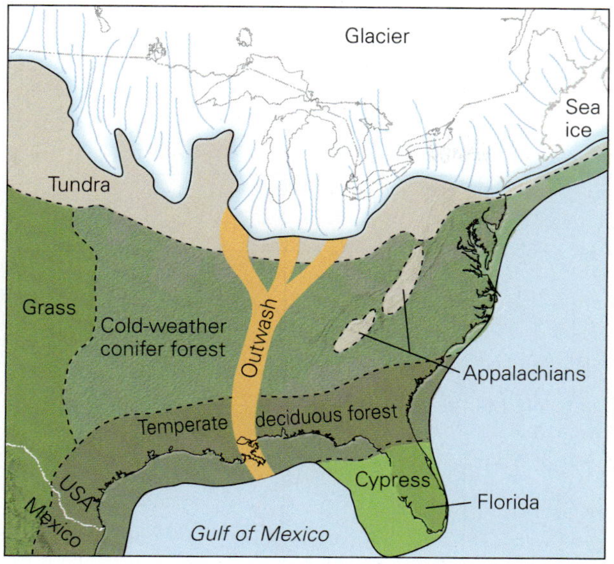

(a) Tundra covered parts of the United States, and southern states had forests like those in New England today.

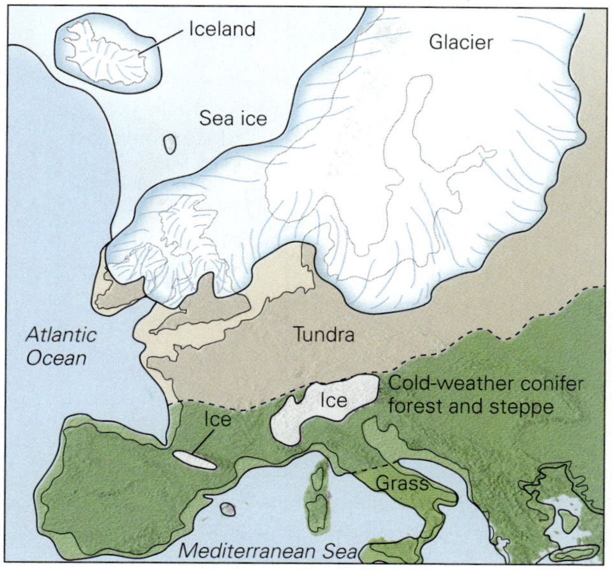

(b) Regions of Europe that support large populations today would have been barren tundra during the Pleistocene.

(c) Cold-adapted large mammals, now extinct, roamed regions that are now temperate.

distinct layers of glacial sediment. This observation suggested that between episodes where glaciers covered large regions, the glaciers disappeared and temperate climates prevailed. In the second half of the 20th century, when modern methods for dating geological materials became available, the difference in ages between the different layers of glacial sediment could be confirmed. Clearly, glaciers advanced and then retreated more than once during the Pleistocene. Times during which the glaciers grew and covered substantial areas of the continents are called glacial periods, or *glaciations*, and times between glacial periods are called interglacial periods, or *interglacials*.

Using the on-land sedimentary record, geologists traditionally recognized five Pleistocene glaciations in Europe and four in the Midwestern United States (Wisconsinan, Illinoian, Kansan, and Nebraskan, named after the southernmost states in which their till was deposited; **Fig. 18.25**). Of note, geologists no longer distinguish Nebraskan and Kansan, but instead refer to glacial sediments deposited before the Illinoian simply as "pre-Illinoian." In the 1960s, geologists realized that this traditional chronology of glaciations was an oversimplification. They discovered evidence of glaciations preserved in the stratigraphic record of Pleistocene marine sediment. By searching for layers containing glacially transported grains, or for layers containing distinctive cold-water plankton, they determined that 20 to 30 glaciations took place during the Pleistocene Epoch. The stratigraphic record of the ice age that had been preserved on land turned out to be very incomplete.

FIGURE 18.25 Pleistocene glacial deposits in the north-central United States. Curving moraines reflect the shape of glacial lobes.

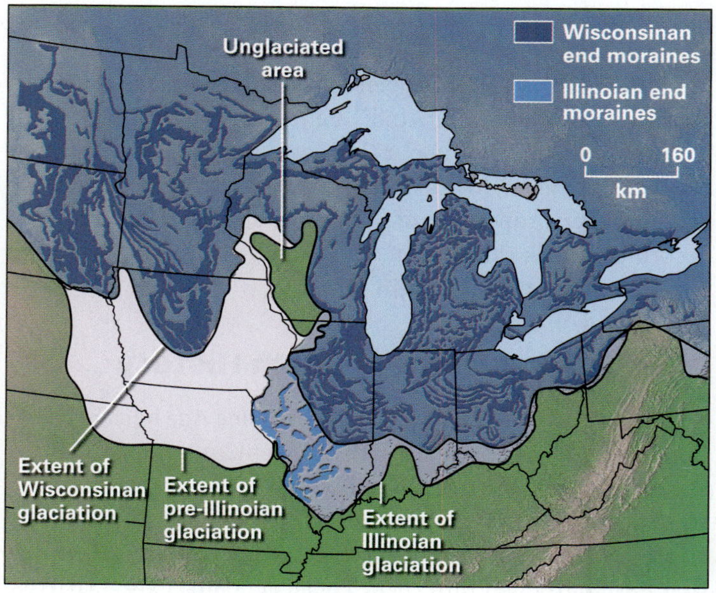

FIGURE 18.26 The timing of glaciations. Ice ages have occurred at several times in the geologic past.

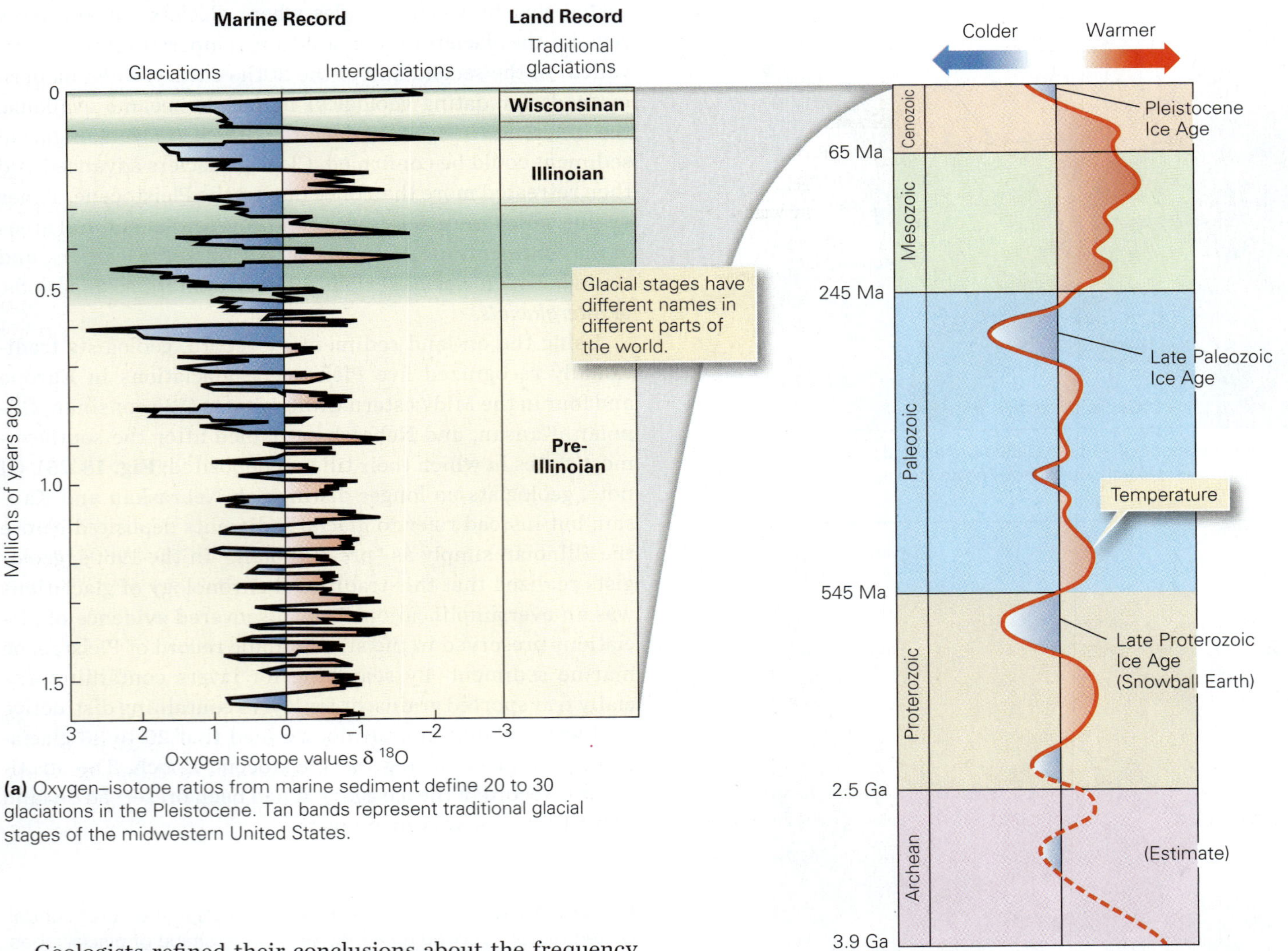

(a) Oxygen–isotope ratios from marine sediment define 20 to 30 glaciations in the Pleistocene. Tan bands represent traditional glacial stages of the midwestern United States.

(b) The Pleistocene is not the only ice-age time in Earth history. Glacial events happened in colder intervals of earlier eras, too.

Geologists refined their conclusions about the frequency of Pleistocene glaciations by examining the isotopic composition of fossil shells. Shells of many plankton species consist of calcite ($CaCO_3$). The oxygen in the shells includes two isotopes, a heavier one (^{18}O) and a lighter one (^{16}O). The ratio of these isotopes tells us about the water temperature in which the plankton grew, for as water gets colder, plankton incorporate a higher proportion of ^{18}O into their shells. The isotope record confirms that lots of cold events occurred during the last 2.6 million years (**Fig. 18.26a**).

Older Ice Ages during Earth History

So far, we've focused on the Pleistocene Ice Age because of its importance in developing Earth's present landscape. Was this the only ice age during Earth history, or do ice ages happen frequently? To answer such questions, geologists study the stratigraphic record and search for ancient glacial deposits that have hardened into rock. These deposits, called **tillites**, consist of larger clasts distributed throughout a matrix of sandstone and mudstone. In many cases, tillites are deposited on glacially polished surfaces.

By using the stratigraphic principles described in Chapter 10, geologists have determined that tillites associated with continental glaciers were deposited during the Late Paleozoic (these are the deposits Alfred Wegener studied when he argued in favor of continental drift; **Fig. 18.26b**), at the end of the Proterozoic Eon (about 600 to 700 Ma), near the beginning of the Proterozoic (about

Did you ever wonder...
how many ice ages have happened during Earth history?

2.2 Ga), and perhaps in the Archean Eon (about 2.7 Ga). Strata deposited at other times in Earth history do not contain widespread tillites. Thus, it appears that glacial advances and retreats have not occurred steadily throughout Earth history, but rather are restricted to four or five specific time intervals, or ice ages. Of particular note, some tillites of the late Proterozoic event were deposited at equatorial latitudes, suggesting that, for at least a short time, the continents worldwide were largely glaciated, and the sea may have been covered worldwide by ice. Geologists refer to the ice-encrusted planet as **snowball Earth**.

TAKE-HOME MESSAGE

During the Pleistocene (2.6 Ma to 11 Ka), ice sheets advanced and retreated up to 30 times. The record on land is less complete than that in marine strata, for evidence of only four or five major glaciations is preserved on land. Ice ages also happened earlier in Earth history. During the late Proterozoic, ice may have covered all of snowball Earth.

QUICK QUESTION What is the evidence for multiple Pleistocene glaciations?

18.7 The Causes of Ice Ages

Ice ages occur only during restricted intervals of Earth history, hundreds of millions of years apart. During an ice age, glaciers advance and retreat periodically, with a frequency measured in tens of thousands to hundreds of thousands of years. Thus, there must be both long-term and short-term controls on glaciation. What are they?

Long-Term Causes

As we noted earlier, the most recent ice age, the Pleistocene Ice Age, happened during the last 2.6 million years. The previous ice age took place near the end of the Paleozoic, with the most widespread glaciations occurring between about 305 and 325 Ma, the time when Pangaea existed. The ice age prior to that was between 630 and 780 Ma, before complex multicellular animals had appeared, and the ice age before that was 2100 to 2400 Ma, when the Earth was just beginning to have oxygen in its atmosphere. Clearly, ice ages happen only rarely during Earth history, suggesting that their appearance may in some way link to major tectonic and/or evolutionary events during Earth history. Geologists have suggested many possible links, though all for now remain in the realm of speculation. Hypotheses under debate suggest that ice ages may be triggered when the following occur:

- Plate movements place large areas of continents at high latitudes, so the land surface receives less solar energy and can become cold;
- High-latitude continents move over mantle upwelling zones, so land elevation overall rises to higher, cooler elevations;
- The global volume of mid-ocean ridges decreases so the capacity of ocean basins increases—when this happens, shallow seas, whose presence could moderate temperatures, disappear from the land;
- The growth of island arcs blocks ocean currents from carrying warm water to high-latitude regions, so the regions become cold enough to host ice-sheet formation;
- Organisms that extract CO_2 (a greenhouse gas) from the atmosphere flourish—the resulting lower CO_2 concentration would decrease the ability of the atmosphere to retain heat;
- The overall amount of chemical weathering on land increases, due to the uplift of mountain ranges, for weathering absorbs CO_2 and would decrease the ability of the atmosphere to retain heat;
- The amount of volcanic activity increases, for volcanic activity can put aerosols into the atmosphere that prevent solar energy from reaching the Earth;
- Uplift of mountain belts changes the pattern of atmospheric circulation, preventing warm air from reaching high latitudes.

Short-Term Causes

Once an ice age has started, for whatever reason, continental glaciers advance and retreat many times, and there is a periodicity to these changes. Why? In 1920, Milutin Milanković, a Serbian astronomer and geophysicist, published work that provides the basis for an explanation. Milanković studied how the Earth's orbit changes shape and how its axis changes orientation through time, and he calculated the frequency of these changes. In particular, he evaluated three aspects of Earth's movement around the Sun, and found the following: (1) The *eccentricity of the Earth's orbit* gradually changes from being more circular (low eccentricity) to being more elliptical in shape (high eccentricity) over a time period of around 100,000 years (**Fig. 18.27a**). When the Earth's orbit has higher eccentricity, it spends part of the year farther from the Sun. (2) The *tilt of the Earth's axis* varies between 22.5° and 24.5°, over a time period of 41,000 years (**Fig. 18.27b**)—we experience seasons because the Earth's axis is not perpendicular to the plane of its orbit, so when the tilt is greater, seasonal differences in high latitudes are greater. And (3) the *precession of the Earth's axis* changes over the course of about

FIGURE 18.27 Milankovitch cycles influence the amount of insolation received at high latitudes.

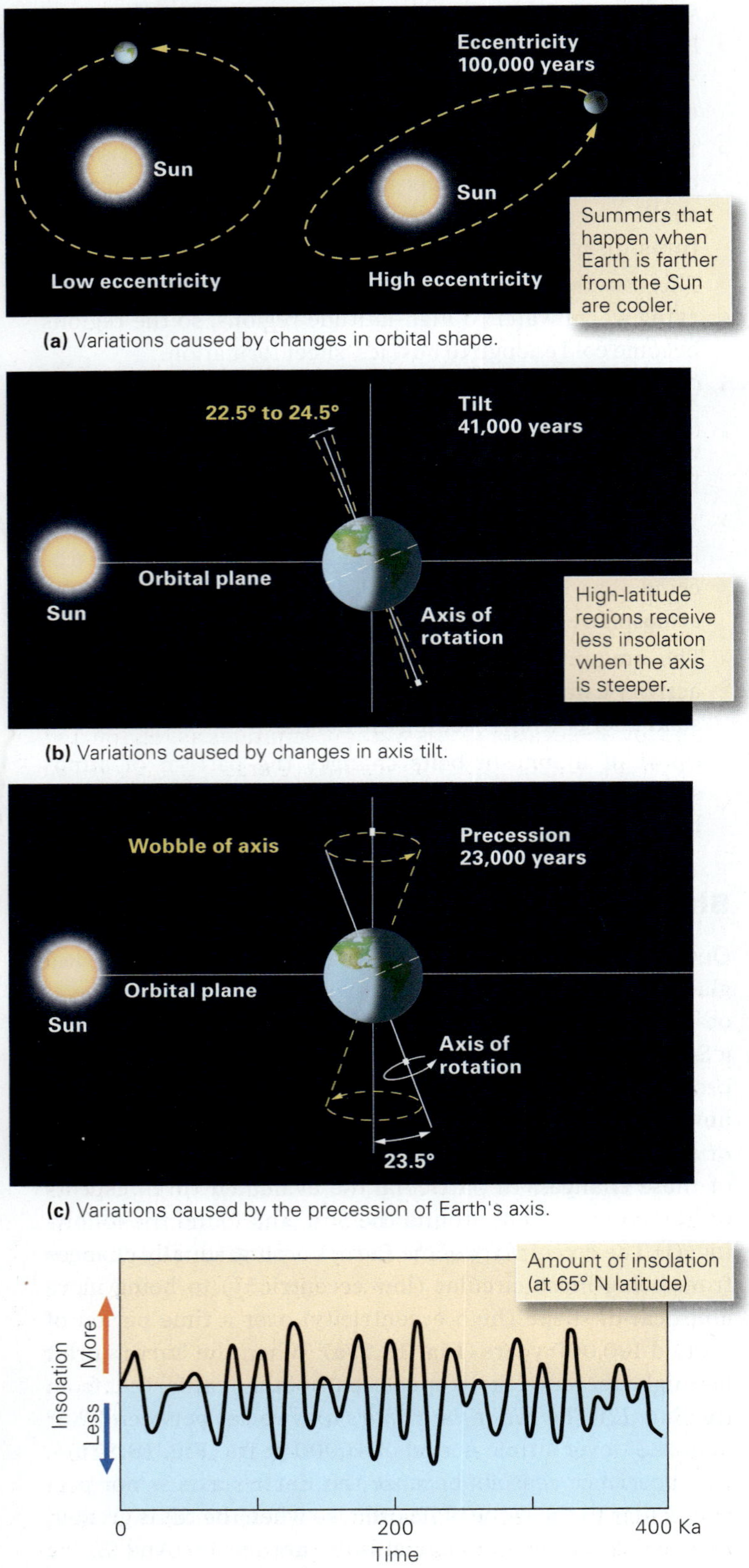

(a) Variations caused by changes in orbital shape.

(b) Variations caused by changes in axis tilt.

(c) Variations caused by the precession of Earth's axis.

(d) Combining the effects of eccentricity, tilt, and precession produces distinct periods of more or less insolation.

23,000 years, because the Earth wobbles, or *precesses*, like a spinning top (**Fig. 18.27c**)—this affects the relationship between the timing of the seasons and the position of Earth along its orbit around the Sun.

Taken together, precession, along with variations in orbital eccentricity and tilt, combine to affect both the total annual amount of *insolation* (exposure to the Sun's rays) and the seasonal distribution of insolation that the Earth receives at mid- to high-latitude regions by as much as 25%. According to Milanković, these changes are enough to cause periodic cooler summers at mid- to high latitudes on a time frame measured in tens of thousands of years (**Fig. 18.27d**). When geologists began to study the climate record, they found climate cycles with the frequency predicted by Milanković. These climate cycles are now called **Milankovitch cycles**.

Milankovitch cycles, however, cannot be the whole story. Geologists suggest that several other factors may come into play in order to trigger or amplify a glacial advance, including these:

- *Changing albedo*: When snow remains on land throughout the year, or clouds form in the sky, the *albedo* (reflectivity) of the Earth increases, so Earth's surface reflects incoming sunlight and thus becomes even cooler.
- *Interrupting the global heat conveyor*: As climate cools, evaporation rates from the sea decrease, so seawater becomes less salty. Decreasing salinity might stop the system of thermohaline currents that brings warm water to high latitudes (see Chapter 15), so high-latitude regions become even colder than otherwise.
- *Biological processes that change CO_2 concentration*: As we have noted, biological processes may amplify climate changes by altering the concentration of carbon dioxide in the atmosphere.
- *Solar variability*: Changes in the radiation output of the Sun could affect the amount of energy the Earth receives.

A Model for Pleistocene Ice-Age History

Long-Term Cooling in the Cenozoic Era Taking all of the above causes into account, geologists have proposed a scenario that may explain how the Pleistocene Ice Age began, and how a cycle of glacial advance and retreat takes place. The story begins in the Eocene Epoch, about 55 Ma (**Fig. 18.28**). At that time, the Earth's climate was relatively balmy, even above the Arctic Circle. Near the end of the Middle Eocene (about 40 Ma), however, climate began to cool. This long-term climate change may have been caused, in part, by a change in the pattern of atmospheric circulation that happened when uplift of the Himalayas and Tibet diverted winds in a way that cooled the climate, and exposed more rock to CO_2-absorbing chemical weathering reactions. In addition, plate motions opened the southern ocean and isolated Antarctica from warm currents, allowing the cold circum-Antarctic current to develop, which in turn may have cooled the global ocean.

So far, we've examined hypotheses that explain long-term cooling since about 40 Ma, but what caused the sudden birth of the Laurentide ice sheet growth at about 2.6 Ma? This event may coincide with other plate-tectonic events. For example, the closing of the gap between North and South America by the growth of the Isthmus of Panama separated the waters of the Caribbean from those of the tropical Pacific for the first time, and when this happened, warm currents that previously flowed out of the Caribbean into the Pacific were blocked and diverted northward to merge with the Gulf Stream. The Gulf Stream transfers warm water from the Caribbean, up the Atlantic Coast of North America, and ultimately to the British Isles. As the warm water moves up the Atlantic Coast, it provides moisture to the air above, and this moisture in turn serves as a source for the snow that falls over New England, eastern Canada, and Greenland. In other words, perhaps the Arctic has long been cold enough for ice caps, but it wasn't until the Gulf Stream was diverted northward, by the growth of Panama, that enough snow fell to trigger glacial growth.

Short-Term Advances and Retreats in the Pleistocene Epoch Once the Earth's climate had cooled overall, short-term climate changes due to Milankovitch cycles led to periodic advances and retreats of the glaciers. To understand how, let's look at a possible case history of a single advance and retreat of the Laurentide ice sheet. (Note that such models remain the subject of vigorous debate.)

FIGURE 18.28 Until recently, Earth's atmosphere has been gradually cooling, overall, since the Cretaceous.

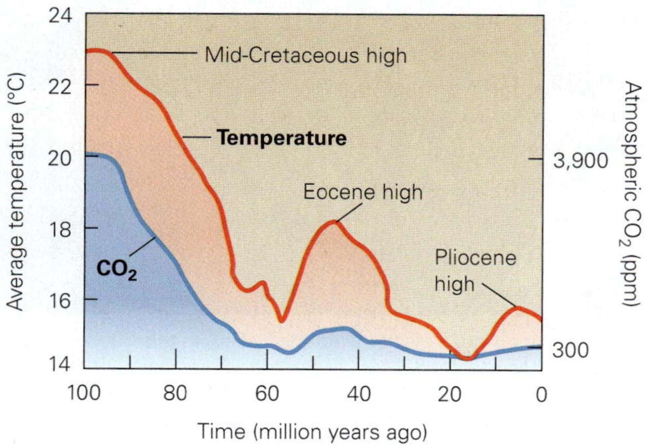

- *Stage 1*: When the Earth reaches a point in the Milankovitch cycle when the average mean temperature in temperate latitudes drops, not all winter snow melts away during the summer in northern Canada. The snow reflects sunlight, so the region grows still colder and even more snow can accumulate. When the snow gets deep enough, the base of the pile turns to ice which begins to spread outward as a new continental glacier.
- *Stage 2*: The ice sheet continues to grow as more snow piles up in the zone of accumulation, and the atmosphere continues to cool because of the albedo effect. Note that increasing albedo is a positive feedback, meaning that its development enhances the effect that caused it. Eventually, however, the weight of the ice loads the continent and makes it sink, so the elevation of the glacier's surface decreases, and it becomes warmer. Also, sea ice covers the ocean offshore, cutting off the source of moisture for snow, so the amount of snowfall diminishes. In effect, the glacial advance chokes on its own success—the rate of ablation exceeds the rate of accumulation, and the glacier begins to retreat.
- *Stage 3*: As the glacier retreats, albedo decreases, temperatures gradually increase, and the sea ice begins to melt. The supply of water to the atmosphere from evaporation increases once again, but with the warmer temperatures and lower elevations, this water precipitates as rain during the summer. The rain drastically accelerates the rate of ice melting, so the retreat progresses rapidly.

Will There Be Another Glacial Advance?

What does the future hold? Considering the periodicity of glacial advances and retreats during the Pleistocene Epoch, we may be living in an interglacial period of an ice age that could continue for millions of years yet to come. Pleistocene interglacials lasted about 10,000 years, and since the present interglacial began about 12,000 years ago, the time may be ripe for a new glaciation to begin. If a glacier on the scale of the Laurentide ice sheet were to develop, major cities and agricultural belts would be overrun by ice, and their inhabitants would have to migrate southward. Of course, long before the ice front arrived, the climate in high-latitude regions would become so hostile that some cities might already have been abandoned.

The Earth actually had a brush with ice-age-like conditions between the 1300s and the mid-1800s, when average annual temperatures in the northern hemisphere fell sufficiently for mountain glaciers to advance significantly. During this period, now known as the *Little Ice Age,* sea ice surrounded Iceland and canals froze in the Netherlands—it was during this time that ice skating became a popular pastime in the country

FIGURE 18.29 The Little Ice Age and its demise. Glaciers that advanced between 1550 and 1850 have since retreated.

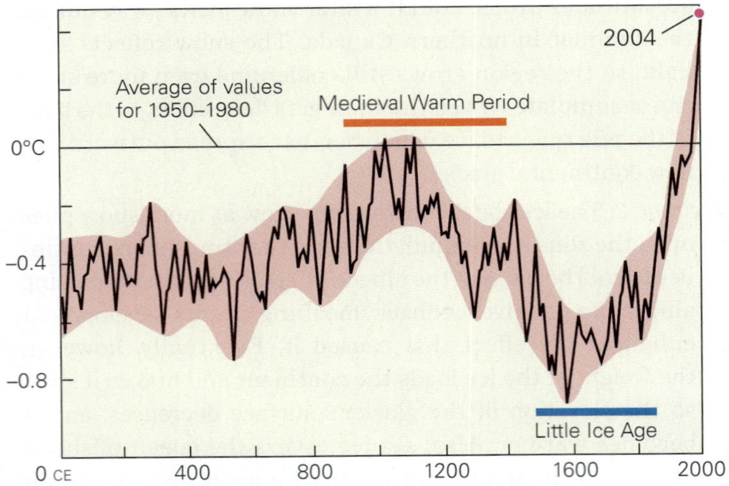

(a) A model of global temperature for the past 2,000 years. Overall trends display the Medieval Warm Period followed by the Little Ice Age. Since 1850, temperatures have warmed.

(b) Skaters (ca. 1600) on the frozen canals of the Netherlands during the Little Ice Age.

(c) During the Little Ice Age, a glacier filled this valley. In this 2003 photo, most of the glacier has vanished. Most of the retreat has happened in the last century.

(**Fig. 18.29a, b**). During the past 150 years, temperatures have warmed, and most mountain glaciers have retreated significantly (**Fig. 18.29c**). Large slabs have been calving off Antarctic ice shelves. In fact, the Larsen B Ice Shelf of Antarctica, an area larger than Rhode Island, disintegrated in 2002 over a period of only one month. Greenland's glaciers, in particular, are showing signs of accelerating retreat (**Fig. 18.30**). Large meltwater ponds are forming on the surface of the ice sheet, and some of these drain abruptly through cracks to the base of the glacier a kilometer below. The vast majority of researchers suggest that this global-warming trend is due to the increased CO_2 in the atmosphere from the burning of fossil fuels. Such global warming could conceivably cause a "super-interglacial" period, one that lasts much longer than the Milankovitch cycle predicts. The next chapter addresses the causes and consequences of such change.

TAKE-HOME MESSAGE

Ice ages may be triggered when the distribution of continents, the ocean currents, and the concentration of atmospheric CO_2 are appropriate. Advances and retreats during a given ice age are controlled by Milankovitch cycles, determined by periodic variations in Earth's orbital shape and rotation-axis orientation. We may be living in a super-interglacial period.

QUICK QUESTION How does positive feedback contribute to a glaciation during an ice age, and why does a glaciation eventually cease?

FIGURE 18.30 Greenland's melting glaciers. Melting has accelerated in the last few decades.

(a) Lakes of meltwater accumulate on the surface of the ice sheet during the summer.

(b) Lakes suddenly drain through cracks that carry the water to the base of the glacier, 1 km down. Addition of liquid water to the base allows the glacier to move faster, causing a "surge."

(c) Where glaciers meet the sea, huge masses calve off and crash into the water. This is happening so fast that the ice front is retreating.

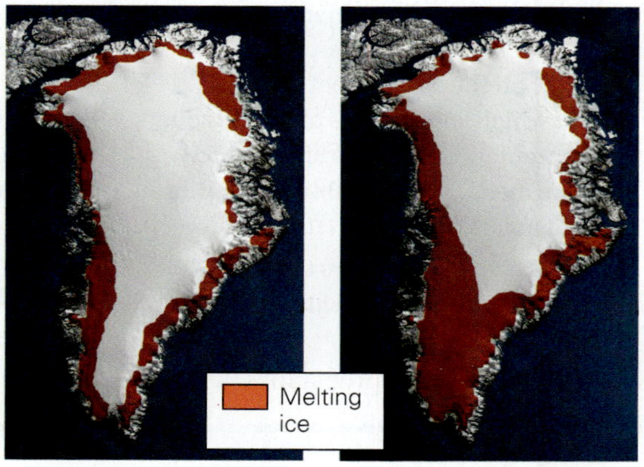

(d) The melting area has increased dramatically in recent years.

Another View This mountainous landscape of southern Alaska, as viewed from an airplane, was carved by glaciers. We can see several cirques and arêtes. The view includes several valley glaciers, each cut by crevices, which look like dark gashes cut into the ice.

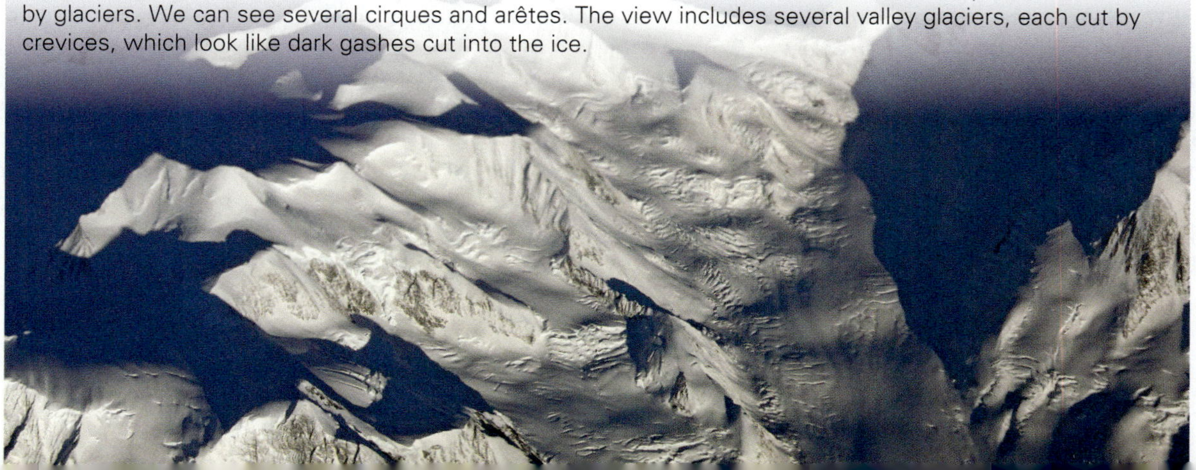

Chapter 18 Review

Chapter Summary

- Glaciers are streams or sheets of recrystallized ice that survive for the entire year and flow in response to gravity. Mountain glaciers exist in high regions and fill cirques and valleys. Continental glaciers (ice sheets) spread over substantial areas of the continents.
- Glaciers form when snow accumulates over a long period of time. With progressive burial, the snow first turns to firn and then to ice.
- Temperate glaciers melt during part of the year. Polar glaciers do not. Glaciers move by basal sliding over water or wet sediment, and/or by plastic deformation of ice grains. In general, glaciers move tens of meters per year.
- Gravitational pull causes glaciers to flow in the direction of their overall surface slope.
- Whether the toe of a glacier stays fixed in position, advances, or retreats depends on the balance between the rate at which snow builds up in the zone of accumulation and the rate at which glaciers melt, calve, or sublimate in the zone of ablation.
- Icebergs break off glaciers that flow into the sea. Continental glaciers that flow into the sea along a coast make ice shelves. Sea ice forms where the ocean's surface freezes.
- As glacial ice flows over sediment, it incorporates clasts. The clasts embedded in glacial ice abrade the substrate, and can polish bedrock and cut striations into it.
- Mountain glaciers carve numerous landforms, including cirques, arêtes, horns, U-shaped valleys, hanging valleys, and truncated spurs. Fjords are glacially carved valleys that filled with water when sea level rose after an ice age.
- Glaciers can transport sediment of all sizes. Till consists of unsorted sediment dropped by melting ice. Glacial marine collects underwater, by melting of icebergs. Streams sort till and deposit it as glacial outwash. Some of the finer sediment accumulates as lake-bed mud, or as windblown loess.
- Glacial depositional landforms include moraines, knob-and-kettle topography, drumlins, eskers, meltwater lakes, and outwash plains. Lateral moraines accumulate along the sides of valley glaciers, medial moraines down the middle, and end moraines at a glacier's toe.
- Continental crust subsides as a result of ice loading. When the glacier melts away, the underlying lithosphere rebounds. When continental glaciers store substantial amounts of water, global sea level drops. When these glaciers melt, sea level rises.
- During past ice ages, the climate in regions south of the continental glaciers was wetter, and pluvial lakes formed. Permafrost exists in periglacial environments.
- During the Pleistocene Ice Age, large continental glaciers covered much of North America, Europe, and Asia.
- The stratigraphy of Pleistocene glacial deposits indicates that glaciers advanced and retreated many times during the ice age. The record of glaciations is more complete in oceanic sediment.
- Over the long term, plate tectonics and changes in the concentration of atmospheric CO_2 may set the stage for ice ages. Short-term advances and retreats may be caused by Milankovitch cycles.

Guide Terms

ablation (p. 574)
arête (p. 578)
cirque (pp. 573, 578)
continental glacier (ice sheet) (p. 573)
crevasse (p. 574)
drumlin (p. 585)
equilibrium line (p. 576)
erratic (p. 569)
esker (p. 585)
firn (p. 572)
fjord (p. 581)
glacial advance (p. 576)
glacial drift (p. 582)
glacial retreat (p. 576)
glacial striation (p. 578)
glacial subsidence (p. 588)
glacier (p. 572)
hanging valley (p. 579)
horn (p. 578)
kettle hole (p. 585)
Milankovitch cycle (p. 596)
moraine (p. 581)
mountain (alpine) glacier (p. 573)
patterned ground (p. 591)
permafrost (p. 591)
plastic deformation (p. 573)
Pleistocene Ice Age (p. 592)
pluvial lake (p. 591)
post-glacial rebound (p. 588)
roche moutonnée (p. 579)
sea ice (p. 577)
snowball Earth (p. 595)
sublimate (p. 572)
terminal moraine (p. 584)
tidewater glacier (p. 576)
tillite (p. 594)
U-shaped valley (p. 578)
V-shaped valley (p. 578)
varve (p. 584)

GEOTOURS *THIS CHAPTER'S GEOTOUR EXERCISE (R) FEATURES:*

› Continental Glacier Features › Alpine/Valley Glacial Features › Piedmont Glacial Features

Review Questions

1. What evidence did Louis Agassiz offer to support the idea that an ice age happened?
2. How do mountain glaciers and continental glaciers differ?
3. Describe the transformation from snow to glacial ice.
4. Explain how arêtes, cirques, and horns form, and how they differ in shape.
5. Describe the mechanisms that enable glaciers to move, and explain why glaciers move.
6. How fast do glaciers normally move? How fast can they move during a surge?
7. Explain how the balance between ablation and accumulation controls advances and retreats.
8. How can a glacier continue to flow toward its toe even though its toe is retreating?
9. How does a glacier transform a V-shaped valley into a U-shaped valley? Discuss how hanging valleys form.
10. Describe the various kinds of glacial deposits. Be sure to note the materials from which the deposits are made and the landforms that result from deposition.
11. How does the lithosphere respond to the weight of glacial ice? How does sea level change during an ice age?
12. How was the on-land chronology of glaciations developed? Why was it so incomplete? How was it modified with the study of marine sediment?
13. Were there ice ages before the Pleistocene? If so, when?
14. What are some of the long-term causes that lead to ice ages? What are the short-term causes that trigger glaciations and interglacials?

On Further Thought

15. If you fly over the barren cornfields of central Illinois during the early spring, you will see slight differences in soil color due to variations in moisture content—wetter soil is darker. These variations outline the shapes of polygons that are tens of meters across. What do these patterns represent, and how might they have formed?
16. An unusual late Precambrian rock unit crops out in the Flinders Range, a small mountain belt in South Australia. This unit consists of clasts of granite and gneiss, in a wide range of sizes, suspended through a matrix of slate. The rock unit lies unconformably above a basement of granite and gneiss, and if you dig out the unconformity surface, you will find that it is polished and striated. What is the unusual rock?

Online Resources

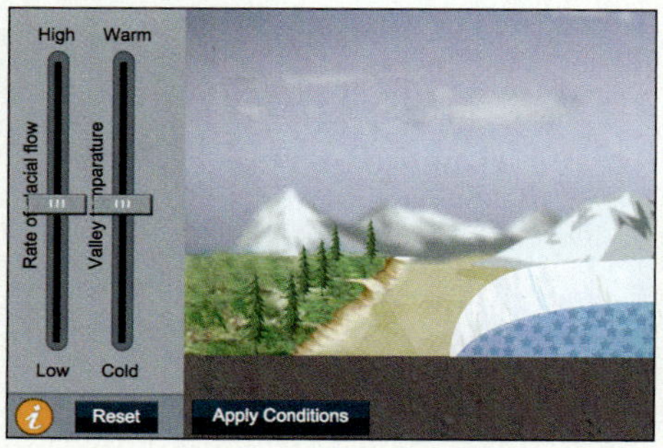

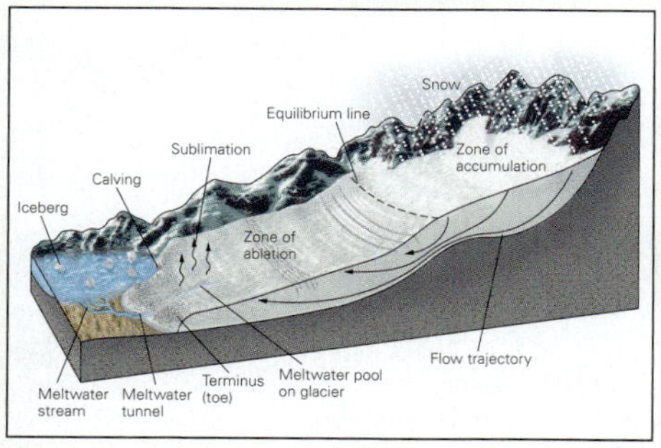

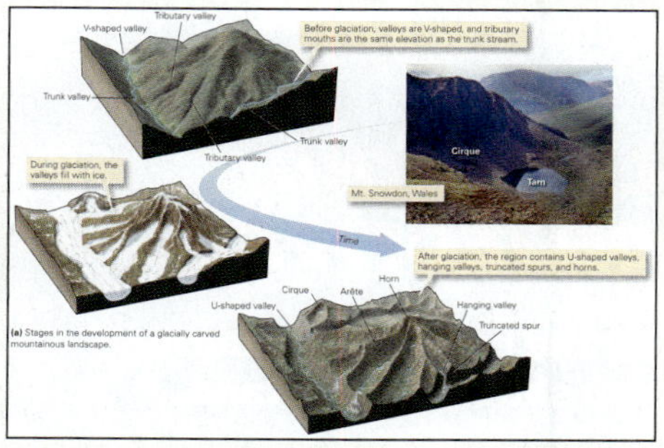

Animations

This chapter features an interactive animation that simulates glacial bodies under different conditions.

Assessment

Question topics include visual exercises on glacial systems and their effects on the landscape.

LEARNING OBJECTIVES

By the end of this chapter, you should understand...

1. that the Earth is a dynamic planet.
2. that some changes are unidirectional, while others are cyclic, and that changes happen at different rates.
3. how elements or compounds flow among various reservoirs.
4. how the Earth's climate changes, and the role greenhouse gases serve in regulating climate.
5. evidence for contemporary climate and sea-level change.
6. diverse ways in which humans change the Earth.
7. how the Earth may change in the future.

▲ Rice paddies and villages cover the countryside near Shanghai, China. The landscape here would have looked vastly different before the arrival of humanity. Land-use change affects many aspects of the Earth System.

CHAPTER 19

Global Change in the Earth System

19.1 Introduction

Did the Earth's surface look the same in the Jurassic Period as it does today? Definitely not! Two hundred million years ago, the North Atlantic Ocean was a narrow sea and the South Atlantic Ocean didn't exist at all, so most dry land lay within a single vast supercontinent (**Fig. 19.1**). Today, both parts of the Atlantic are wide oceans, and the Earth has seven separate continents. Moreover, during the Jurassic, the call of the wild rumbled from the throats of dinosaurs, whereas today, the largest land animals are mammals. In essence, what we see of the Earth today is just a snapshot, an instant in the life story of a constantly changing planet. This idea arguably stands as geology's greatest philosophical contribution to humanity's understanding of our Universe.

Why has the Earth changed so much over geologic time, and why does it continue to change? Ultimately, change happens both because the Earth's internal heat makes the asthenosphere weak enough to flow, and because the Sun's radiation can keep most of the Earth's surface at temperatures above the freezing point of water. Flow in the asthenosphere permits plate tectonics, which in turn leads to continental drift, volcanism, and mountain building. Solar

> *All we in one long caravan*
> *are journeying since the world began,*
> *we know not whither, but we know . . . all must go.*
> BHARTRIHARI (Indian poet, ca. 500 CE)

radiation keeps streams, glaciers, waves, and wind in motion, thereby causing erosion and deposition, and it also fuels photosynthesis. If the Earth did not have just the right mix of tectonic activity and solar heat, it would be a frozen dust bowl like Mars, a crater-pocked wasteland like the Moon, or a cloud-choked oven like Venus, and could not host life as we know it. Many of the changes that take place on Earth reflect complex interactions among geological and biological phenomena. For example, photosynthetic organisms affect the composition of the atmosphere by providing oxygen, and atmospheric composition, in turn, determines the nature of chemical weathering reactions that can take place in rocks.

We've referred to the global interconnecting web of physical and biological phenomena on Earth as the *Earth System* (see **Geology at a Glance,** pp. 608–609). In this context, we can now define **global change**, in a general sense, as the modifications of physical and biological components in the Earth System over time. Geologists distinguish among different types of global change, on the basis of the rate or way in which change progresses: *gradual change* takes place slowly, over long periods of geologic time (millions to billions of years); *catastrophic change* takes place relatively rapidly (seconds to millennia); *unidirectional change* involves transformations that never repeat; *cyclic change* repeats the same steps over and over, though not necessarily with the same results or at the same rate; and *periodic change* repeats steps with a definable frequency.

In this chapter, we begin by reviewing examples of global change involving phenomena discussed earlier in this book. Then we introduce the concept of a *biogeochemical cycle,* the exchange of chemicals among various living and nonliving reservoirs, for some kinds of global change reflect modifications in the proportions of chemicals held in different reservoirs. Finally, we focus on climate change, the transformations of Earth's climate over time. We conclude this chapter, and this book, by considering hypotheses that describe the ultimate global change—the end of the Earth in the very distant future.

FIGURE 19.1 The map of Earth's surface changes over time because of plate motions.

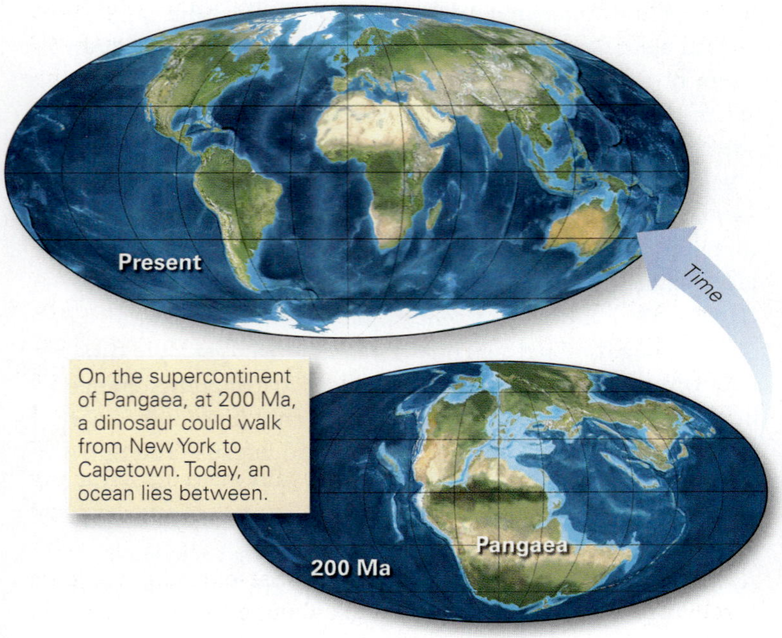

19.2 Unidirectional Changes

The Evolution of the Solid Earth

Recall from Chapter 1 that Earth began as a fairly homogeneous mass, formed by the coalescence of planetesimals. The homogeneous proto-Earth did not last long, for within about 100 million years of its birth, when it had attained a diameter approaching 1000 km, the planet began to melt, yielding liquid iron alloy that sank to the center to form the core. This process, the *differentiation of the Earth's interior,* represents a major unidirectional change in our planet—it produced a layered planet with an iron alloy core surrounded by a rocky mantle.

According to a widely held model, another protoplanet collided with the newborn Earth soon after differentiation. This collision caused a catastrophic change, in that a significant portion of the Earth fragmented or vaporized, yielding a ring of debris that quickly coalesced to form the Moon (**Fig. 19.2**). For a time after this collision, the Earth's mantle remained largely molten, and the planet's surface was a sea of magma. New evidence suggests that our planet's surface changed rapidly, solidifying to form a scum-like crust of ultramafic rock by about 4.4 billion years ago (Ga). Liquid water may have collected on the surface soon after. The early surface of the Earth didn't last forever, though. Intense bombardment by asteroids and comets, between about 4.0 and 3.9 Ga, destroyed any crust and any sea that had existed prior to 3.9 Ga, by pulverizing and melting it. Eventually, however, bombardment did cease, and our planet gradually

FIGURE 19.2 A popular model attributes the formation of the Moon to a collision with a planetesimal.

Time
A planetesimal collides with Earth.
Debris sprays into space.
The debris coalesces to form the Moon.

cooled, permitting a new crust to stabilize at its surface, the oceans formed, and an early form of plate tectonics to begin operating. Over time, igneous activity produced lasting continental crust, blocks of which collided and sutured together, and by about 2.7 Ga, most of the continental crust that exists today had been distilled from the mantle. Differentiation, Moon formation, crust formation, and ocean formation all represent major unidirectional changes that took place early in Earth's history.

The Evolution of the Atmosphere and Oceans

Like its surface, the Earth's atmosphere has also changed unidirectionally over time. Early on, it formed from gases released by volcanic activity. More gases may have arrived when comets collided with our new planet. Eventually, Earth accumulated an early atmosphere composed dominantly of carbon dioxide (CO_2) and water (H_2O). Other gases, such as nitrogen (N_2), composed only a minor proportion of the early atmosphere. When the Earth's surface cooled, however, water condensed and fell as rain, collecting in low areas to form oceans. Permanent oceans have existed since about 3.8 Ga. Gradually, CO_2 dissolved in the oceans and also was absorbed by chemical-weathering reactions on land, so its concentration in the atmosphere decreased. Nitrogen, which doesn't react with other chemicals, was left behind. Thus, the atmosphere's composition changed to become dominated by nitrogen. Photosynthetic organisms appeared early in the Archean. It probably wasn't until between 2.5 and 2.0 Ga, the early Proterozoic, that oxygen became a significant proportion of the atmosphere, even though it didn't reach breathable concentrations for another billion years. This last major unidirectional change is known as the *great oxygenation event*.

Did you ever wonder...
if the Earth's atmosphere has always been breathable?

The Evolution of Life

The fossil record indicates that life had appeared, in the form of single-celled archaea and bacteria, at least by 3.7 billion years ago, and has undergone unidirectional change (*evolution*) ever since (see Interlude E). Complex multicellular organisms appeared in the late Proterozoic, and life diversified substantially at the beginning of the Cambrian, during an event called the *Cambrian explosion*.

Initially, multicellular life inhabited only the sea, but in the Middle Ordovician, plants succeed in invading the land, and animals followed in the Silurian. Early land animals included millipedes and scorpions. Reptiles appeared in the Carboniferous, and placental mammals in the Cretaceous. When dinosaurs became extinct, at the end of the Mesozoic, mammals became the largest land animals (**Fig. 19.3**). While primates became a distinct order of mammals as early as 55 million years ago (Ma), our own species appeared only about 150,000 years ago. Once a species has gone extinct, it never reappears, and new species appear over time. Therefore, we can consider life evolution to be unidirectional change. As a result of evolution, life now inhabits regions from a few kilometers below the surface to a few kilometers above, yielding a diverse and complex biosphere.

The rate at which life evolution takes place remains the subject of debate. As we noted in Interlude E, paleontologists have concluded that pulses of rapid evolution punctuate times of relative equilibrium. Notably, the stratigraphic record shows that Earth history includes several **mass-extinction events**, during which large numbers of species abruptly vanish—these events define boundaries between geologic periods. During a mass-extinction event, *biodiversity* (the number of different species that exist) decreases abruptly. In the succeeding few million years, biodiversity increases again, with a great many new species appearing, for extinction leaves many ecological niches open for repopulation.

Geologists speculate that some mass-extinction events reflect catastrophic change brought about either by unusually

FIGURE 19.3 Because of evolution, organisms inhabiting the Earth today are not the same as those that inhabited the planet in the past.

The largest Mesozoic land animal was a dinosaur. The inset shows an elephant at the same scale.

The largest land animal today is the elephant.

voluminous volcanic eruptions, or by the impact of a comet or an asteroid with the Earth. Either of these events could eject enough debris into the atmosphere to block sunlight. Without the warmth of the Sun, winter-like or night-like conditions would last for weeks to years, long enough to disrupt the food chain. In addition, either type of event could eject aerosols that would turn into global acid rain, scatter hot debris that would ignite forest fires, or give off chemicals that would make the ocean either toxic or overly nutritious and thus disrupt its ecosystems. At present, many geologists favor the hypothesis that the Permian-Triassic extinction was due to the eruption of 3 million cubic km of basalt in Siberia, and that the Cretaceous-Paleogene extinction was due to a giant meteorite impact.

TAKE-HOME MESSAGE

Since it first formed, our planet has undergone major changes that are unidirectional, in that they modified the Earth System forever and will never repeat. Examples include internal differentiation (core formation), Moon formation, atmospheric evolution, ocean formation, and life evolution. Catastrophic changes have taken place on occasion.

QUICK QUESTION How are life evolution and atmospheric evolution linked?

19.3 Cyclic Changes

We're used to seeing daily or monthly cyclic changes—the Sun rises and sets every day, seasons come and go, the tides rise and fall, and the Moon waxes and wanes. During such cyclic changes, a sequence of stages may be repeated over time in the same order with a definable frequency, so we can consider them to be periodic. But not all familiar cyclic changes are periodic. For example, while flooding cycles may occur in a drainage basin, floods of a given size don't happen on a predictable schedule—some may be months apart, and some may be centuries apart (see Chapter 14).

The Earth System undergoes cyclic changes on a much longer time frame as well. Below, we look at several examples of geologic-time-scale cyclic change—you'll see that some involve movements of physical components of the Earth, whereas others involve transfer of chemicals among both living and nonliving reservoirs. You'll also see that most are not periodic.

The Supercontinent Cycle

Over the course of Earth history, the map of our planet's surface has changed constantly. During some intervals of geologic time, almost all continental crust lay within one or two supercontinents, and during other intervals, the planet's surface has hosted many smaller continents. The process of change during which a supercontinent forms by continental collision and later breaks apart by rifting is the *supercontinent cycle*. Geologists have found evidence that supercontinents existed at least three or four times during the past 3 billion years of Earth history—no two were alike. The most recent supercontinent, Pangaea, formed 300 Ma, at the end of the Paleozoic Era, and began to break apart by about 200 Ma, in the Mesozoic. The supercontinent cycle is not periodic, and it can take hundreds of millions of years to pass through a single complete cycle.

The Sea-Level Change Cycle

Global sea level has gone up and down by as much as 300 m during the Phanerozoic, and likely did the same in the

FIGURE 19.4 This chart provides one interpretation of sea-level change during the past half billion years, based on the stratigraphic record.

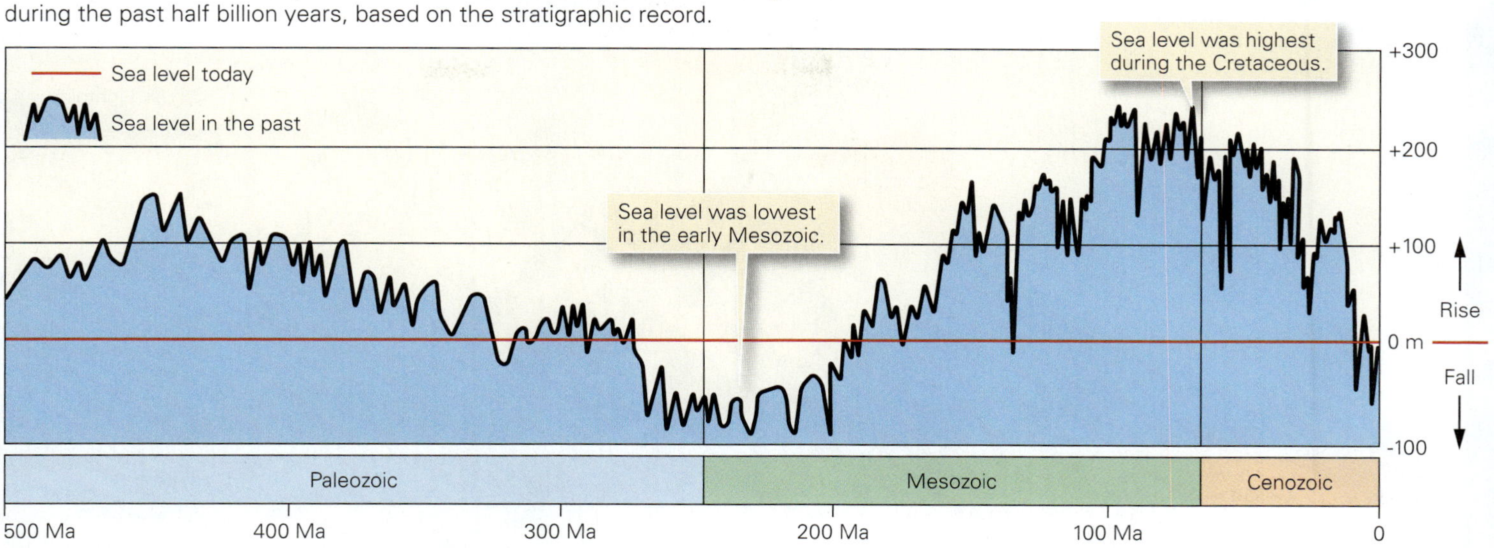

Precambrian. When sea level rises, the shoreline migrates inland, resulting in a transgression during which low-lying plains become submerged. At times of particularly high sea level, more than half of Earth's continental area has been covered by shallow seas. When this happens, new layers of sediment bury broad portions of continents. When sea level falls, a regression takes place, the continents become dry again. The lack of deposition and/or the occurrence of widespread erosion after a regression produces regional-scale unconformities.

After studying sedimentary sequences around the world, geologists have pieced together a chart defining the succession of global transgressions and regressions during the Phanerozoic Eon. The global sedimentary cycle chart may largely reflect the cycles of *eustatic sea-level change* (**Fig. 19.4**). Eustatic, in this context, means worldwide—such sea-level changes may be due to a variety of factors, including advances and retreats of continental glaciers, changes in the volume of mid-ocean ridge systems, and changes in overall continental elevation and area. The chart of Figure 19.4 probably does not give us an exact image of sea-level change over time, because the sedimentary record reflects other factors as well, such as changes in sediment supply.

The Rock Cycle

We learned early in this book that the crust of the Earth consists of three rock types: igneous, sedimentary, and metamorphic. Atoms making up the minerals of one rock type may later—after erosion and deposition, after metamorphism, or after melting—become part of another version of the same rock type, or of a different rock type. In effect, rocks serve as reservoirs of atoms, and the atoms can move from reservoir to reservoir over time, a process called the *rock cycle*. Each stage in the rock cycle changes the Earth by redistributing and modifying material.

Biogeochemical Cycles

A **biogeochemical cycle** involves the passage of chemicals among nonliving and living reservoirs in the Earth System, mostly on or near the planet's surface. Nonliving reservoirs include the atmosphere, rock, magma, soil, and the ocean, whereas living reservoirs include plants, animals, and microbes.

Some stages in a biogeochemical cycle may take only hours, some may take thousands of years, and others may take millions of years. The transfer of a chemical from reservoir to reservoir during these cycles doesn't really seem like a change in the Earth in the way that the movement of continents or the metamorphism of rock seems like a change. In fact, for intervals of time, biogeochemical cycles attain a **steady-state condition**, meaning that the proportions of a chemical in different reservoirs remain fairly constant, even though there is a constant flux (flow) of the chemical among reservoirs. When we speak of global change in a biogeochemical cycle, we mean a change in the relative proportions of a chemical held in different reservoirs at a given time—in other words, a change in the steady-state condition. A great variety of chemicals (water, carbon, oxygen, sulfur, ammonia, phosphorus, and nitrogen) participate in biogeochemical cycles. Here, we look at two: water (H_2O) and carbon (C).

The Hydrologic Cycle As we learned in Interlude F, the hydrologic cycle involves the movement of water from reservoir to reservoir on or near the surface of the Earth. As a

GEOLOGY AT A GLANCE

The Earth System

At the Earth's surface, we can see interactions among the many components of the Earth system—rocks, sediments, and soils of the solid earth (the lithosphere); the liquid water in oceans, lakes, streams, and groundwater (the hydrosphere); the solid water of glaciers and permafrost (the cryosphere); and the planet's gaseous envelope (the atmosphere). Countless species of life (the biosphere), ranging from nearly invisible bacteria to giant whales and trees, populate complex ecosystems, mostly on the surface or in the sea, but also including the upper few kilometers of the crust and the lower few kilometers of the atmosphere. Two key sources of energy fuel the dynamic exchanges that take place among components of the Earth System. External energy comes from solar radiation, and internal energy comes from the Earth's interior.

Features of the Earth System undergo change. Various materials cycle among living and nonliving components of the Earth System during the hydrologic cycle, the rock cycle, and various biogeochemical cycles (such as the carbon cycle). The proportions of these materials held in different reservoirs of the Earth System can change over time. In addition, climate change and sea-level change have markedly affected the character of the planet's surface. Plate interactions constantly, though slowly, alter the map of the planet, and over geologic history, life and the atmosphere have evolved. Despite its immensity, the Earth System is fragile—human activity has caused major changes in the Earth System in just a few hundred years.

consequence, a chemical compound, H_2O, passes through both nonliving and living entities—the oceans, the atmosphere, surface water, groundwater, glaciers, soil, and living organisms. Global change in the hydrologic cycle occurs primarily when a change in global climate alters the ratio between the amount of water held in the ocean and the amount held on land in continental ice sheets. For example, during an ice age, water that had been stored in oceans moves into glacial reservoirs. When the continents become covered with ice, sea level drops. When the climate warms, water returns to the oceans, and sea level rises.

The Carbon Cycle Most carbon in the near-surface realm of Earth originally bubbled out of the mantle in the form of CO_2 gas released by volcanoes (**Fig. 19.5**). Once it enters the atmosphere, carbon moves through various reservoirs of the Earth system through the **carbon cycle**. Some dissolves in seawater to form bicarbonate (HCO_3^-) ions, and some of this carbon eventually becomes incorporated in the shells of invertebrates or precipitates directly from water. Atmospheric CO_2 can also be absorbed by photosynthetic organisms that convert it to sugar and other organic chemicals—this carbon enters the food chain and ultimately makes up the flesh, fat, and sinew of animals.

Some of the carbon incorporated by living organisms returns directly to the atmosphere through the respiration of animals (again as CO_2), by the flatulence of animals as methane (CH_4), or by the decay of dead organisms (which can produce both CH_4 and CO_2). But some of this carbon can be stored underground for long periods of time in fossil fuels (oil and coal), in organic shale, and in methane hydrates. Similarly, carbon incorporated into solid minerals, such as calcite, may be stored for long periods in limestone or marble. The carbon in these reservoirs may eventually return to the atmosphere in the form of CO_2, due to the burning of fossil fuels, to the metamorphism of rocks containing carbonate, and to the production of concrete. Or it can return to the sea, after undergoing dissolution in river water or groundwater.

TAKE-HOME MESSAGE

Some changes in the Earth System are cyclic, though not necessarily periodic. Examples include the supercontinent cycle (accumulation of continents by collision, and then dispersal by rifting), the sea-level cycle (rise and fall of the sea), and the rock cycle. During biogeochemical cycles, carbon, water, and other elements cycle through living and nonliving reservoirs. Some carbon may be trapped underground for geologic time.

QUICK QUESTION What evidence indicates that sea level rises and falls over time?

FIGURE 19.5 In the carbon cycle, carbon transfers among various reservoirs at or near the Earth's surface. Red arrows indicate release to the air, and green arrows indicate absorption from air.

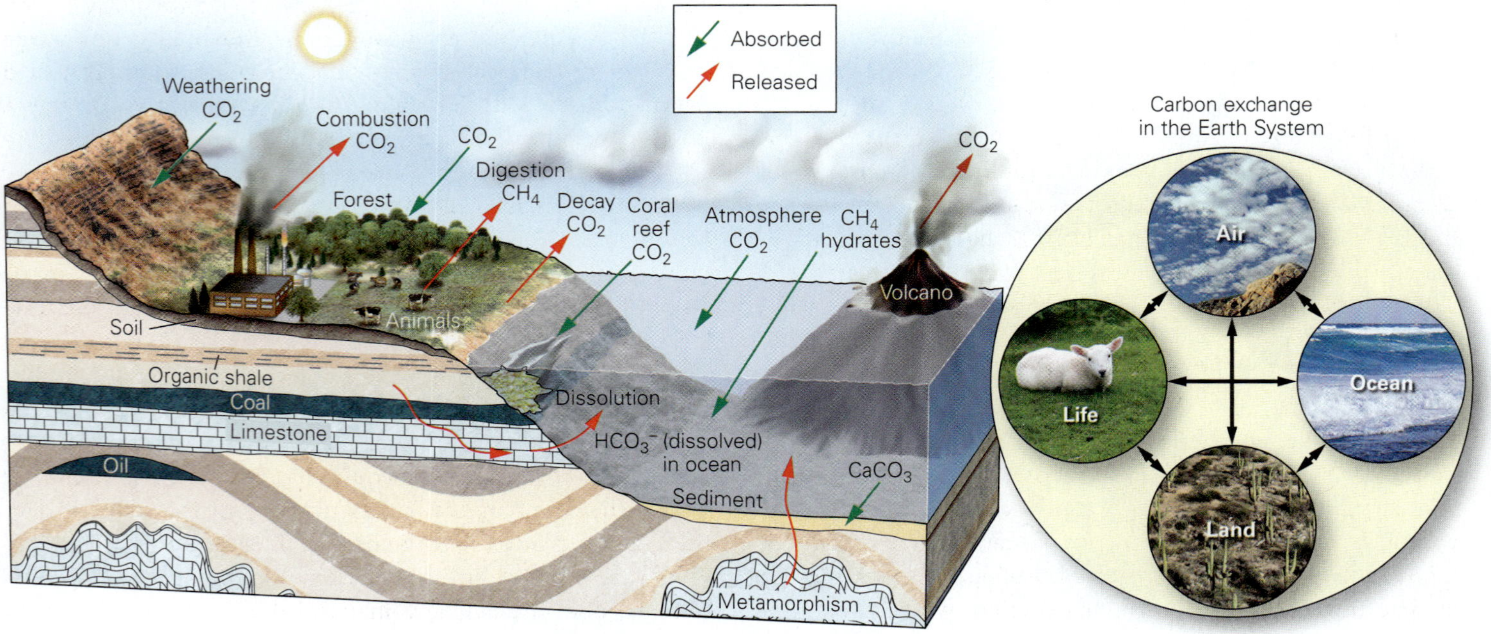

19.4 Global Climate Change

What Is Climate Change?

How often have you seen a news report proclaim "Record High Temperatures!" when thermometers register temperatures several degrees above "normal" for days on end. Do such headlines mean that the climate is changing? To start addressing this question, we first need to distinguish between two common terms: *weather* and *climate*. Atmospheric conditions during a specific time interval at a given locality define the region's **weather**. Weather may be "windy, rainy, and cool" in the morning, and "calm, sunny, and dry" in the afternoon. The term **climate**, in contrast, refers to an overall range of weather conditions, as well as the typical daily to seasonal variability of weather conditions, as observed over a period of decades for a region. A place with a *tropical climate* tends to have hot, humid days all year, whereas a place with a *temperate climate* tends to have contrasting seasons and, therefore, a wide range of temperature and humidity over the course of a year. So a newspaper's headline about a hot spell or cold snap does not mean that the climate is changing. But if a new set of conditions—say, an overall increase in average temperature, a rising snow line, a longer growing season, or a change in storm frequency or intensity—becomes the new norm for a region, then **climate change** has occurred. And if such changes happen worldwide, then *global climate change* has occurred. An increase in global average atmospheric and sea-surface temperatures represents **global warming**, and a decrease represents **global cooling**.

To gain insight into how global climate changed throughout Earth's long history, researchers study **paleoclimate** (past climate), by searching for climate indicators—such as depositional environments, fossil species, or geochemical signatures—that can be dated accurately. *Long-term global climate change*, which happens in a time frame of tens to hundreds of millions of years, has been recognized since geologists first learned how to interpret the stratigraphic record. The record shows that depositional settings at a locality change significantly over time, so such change is not news. But global climate change has entered general discussion in recent years, because research during the past few decades has led to the conclusion that this type of change is now happening, not at the slow pace typical of most geologic phenomena, but quickly enough to have significant impacts on human society in as little as the next few decades. In other words, many readers of this book may see the effects of global climate change in their lifetimes. In this section, we first examine the geologic manifestations and possible causes of long-term climate change. Then, we turn our attention to *short-term global climate change*, meaning changes that take place over time frames of less than 1 million years. Of note, the resolution (detail) retained in the geologic record generally permits us to recognize short-term changes that have taken place only during the past few million years. This discussion sets the stage for the next section of the book, in which we focus on *contemporary global warming*, meaning changes that have taken place the past few centuries and may impact lifestyles during the next century.

Long-Term Climate Change

The stratigraphic record can provide geologists with a basic history of global climate, represented by mean global temperature, over the span of geologic time. That's because the depositional setting in which sediment accumulates, and the assemblages of organisms that live in, on, or above the sediment, depend on the climate. For example, layers of organic debris typically develop only in warm climates, while layers of till accumulate only in cold climates. Therefore, when we examine a succession of strata that includes coal beds overlain by tillite, we're observing a change from warmer to cooler climate.

The stratigraphic record shows that the Earth's surface temperature has stayed between the freezing point of water and the boiling point of water for almost all time since the beginning of the Archean. At some times in the past, the Earth's atmosphere has been significantly warmer than it is today, whereas at other times, it has been significantly cooler. The warmer periods have come to be known as **greenhouse periods** (or *hothouse periods*), and the colder ones are called **icehouse periods**. (The more familiar term, *ice age*, refers to the specific times during an icehouse period when the Earth was cold enough for ice sheets to advance and cover substantial areas of the continents.) As the chart in **Figure 19.6** shows, at least five major icehouse periods have occurred during geologic time. During greenhouse periods, even lands at polar latitudes were largely free of ice. During icehouse periods, glaciers covered land at midlatitudes. Of note, during the late Proterozoic icehouse period, land at equatorial latitudes became ice covered, and this planet may have frozen over entirely to become "snowball Earth."

What causes long-term global climate change? The answer may lie in the complex relationships among the various solar, geologic, and biogeochemical components of the Earth System, as described earlier. For example, the positions of continents, due to plate movements, influences the climate by

FIGURE 19.6 The Earth has experienced icehouse and greenhouse periods at various times during Earth history.

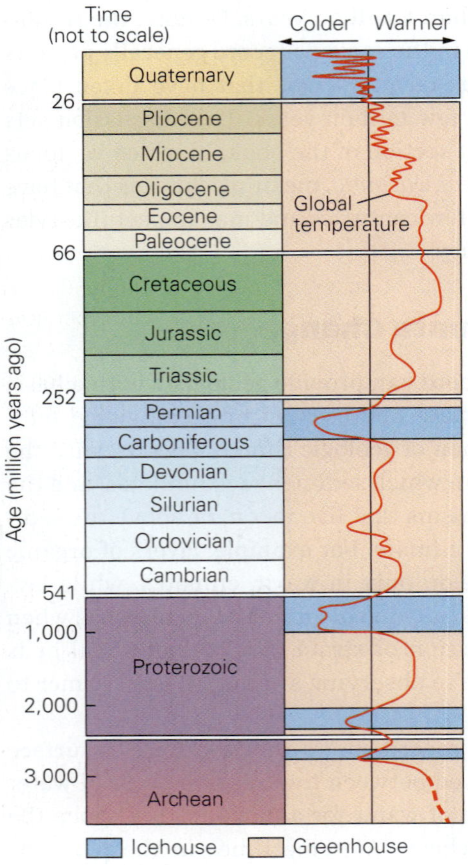

controlling the pattern of oceanic currents, which redistribute heat around the planet's surface. Continental movement also determines whether the land lies at high or low latitudes (and thus how much solar radiation strikes it), and whether or not large continental interior regions exist where extremely cold winter temperatures can develop. Changes in the concentration of *greenhouse gases* also play an important role. Greenhouse gases, such as CO_2 and CH_4, constitute only a small proportion of the atmosphere, but they play a major role in regulating the temperature of the atmosphere (**Box 19.1**). Over geologic time, the concentration of these gases can change as a consequence of many factors, including: life evolution (such as the appearance of photosynthetic organisms, which remove CO_2); the degree to which carbon-containing materials (coal, oil, and limestone) form and become buried and, therefore, able to trap the carbon in underground reservoirs; and whether plate interactions have caused the uplift of mountain ranges, for uplift exposes rock to weathering reactions that involve greenhouse gases and can affect atmospheric circulation.

Short-Term Climate Change

The detail of paleoclimate information retained in the stratigraphic record decreases as rocks get older, partly because methods to provide age constraints have a range of uncertainty, and partly because geologic phenomena such as compaction, diagenesis, and metamorphism can destroy the record. Thus, geologists can obtain a very detailed record of paleoclimate changes for only the past few thousand to few million years. The following data sources provide this record:

- *Microfossils*: In Pleistocene and Holocene deposits, a record of climate change can come from the study of fossil plankton and pollen. Studies of plankton shells preserved in marine deposits at a given location, for example, can define variations in sea-water temperature over time at the location, for plankton shells have distinctive shapes that allow species to be identified accurately, and the species that live at a locality reflect the water temperature. Similarly, studies of delicate pollen grains preserved in lake deposits have allowed researchers to determine how the type of forest at a given latitude changed in association with the advance and retreat of continental glaciers (**Fig. 19.7**).
- *Oxygen-isotope ratios*: Geologists have found that the ratio of ^{18}O to ^{16}O in glacial ice indicates the atmospheric temperature at which the snow that made up the ice formed—simplistically, the ratio is larger in snow that forms in warmer air, but smaller in snow that forms in colder air. Because of this relationship, the isotope ratio of the oxygen in H_2O measured in a succession of ice layers in a glacier indicates temperature change over time. Researchers have now obtained ice cores down to a depth of almost 3.5 km in Antarctica, providing a record that spans up to 800,000 years (**Fig. 19.8a**). (A record back to 1.5 Ma may be possible.) For similar reasons, the $^{18}O/^{16}O$ ratio in the $CaCO_3$ making up plankton shells also gives geologists an indication of past temperatures. Measurement of oxygen-isotope ratios in drill cores of marine sediment, therefore, extends the record of temperature change back over millions of years (**Fig. 19.8b, c**).
- *Growth rings*: If you've ever looked at a tree stump, you will have noticed the concentric rings visible in the wood. In general, each ring represents one year of growth, and the thickness of the ring indicates the rate of growth in a given year. Trees grow faster during warmer, wetter years and more slowly during cold, dry years. Thus, the succession of ring widths provides a record of climate during the lifetime of the tree. Growth rings in corals and shells may provide similar information.

As we discussed in Chapter 18, paleoclimate data led to the realization that continental glaciers advanced and retreated

BOX 19.1 Consider This...

The Role of Greenhouse Gases and Feedback Mechanisms

The Sun constantly bathes the Earth in electromagnetic radiation. Some of this energy reflects off of the atmosphere, or off the Earth's surface, so that our planet shines when viewed from space, and some gets absorbed by gases in the atmosphere. The remaining radiation (mostly visible light and ultraviolet light) reaches the surface and gets absorbed by rocks, sediments, soils, or water. These materials heat up and then release the energy back into the atmosphere in the form of thermal energy (infrared radiation) that heads back upward. If the Earth had no atmosphere, all of this thermal energy would escape back into space. But our planet does have an atmosphere, and certain gases in the air absorb thermal radiation and re-radiate it. Some of the re-radiated energy heads up into space, but some heads downward and warms the lower atmosphere (**Fig. Bx19.1**). In effect, these gases trap infrared radiation and keep the lower atmosphere warm, somewhat like glass traps heat in a greenhouse. The overall trapping process is called the **greenhouse effect**, and gases (such as H_2O, CO_2, CH_4, and N_2O) that cause it are called **greenhouse gases**. An increase in the concentration of greenhouse gases in the atmosphere will cause the atmosphere to become warmer, whereas a decrease in the concentration of these gases will allow the atmosphere to become cooler.

Atmospheric warming or cooling due to changes in the concentration of greenhouse gases can be amplified by positive or negative **feedback mechanisms**. *Negative feedback* slows a process down or even reverses it, whereas *positive feedback* enhances a process and amplifies its consequences. Let's consider an example of a feedback mechanism involving the greenhouse effect. Imagine that global average atmospheric temperature increases due to an increase in CO_2. This increase will, in turn, cause the oceans to warm and evaporate more, so more water transfers into the atmospheric reservoir globally, causing still more warming—this is a positive feedback. Warming may also cause permafrost in arctic regions and methane hydrates on the seafloor to melt and, therefore, release CH_4 into the atmosphere to cause even more warming, another example of positive feedback.

FIGURE Bx19.1 The greenhouse effect shows how thermal energy can be trapped in the atmosphere.

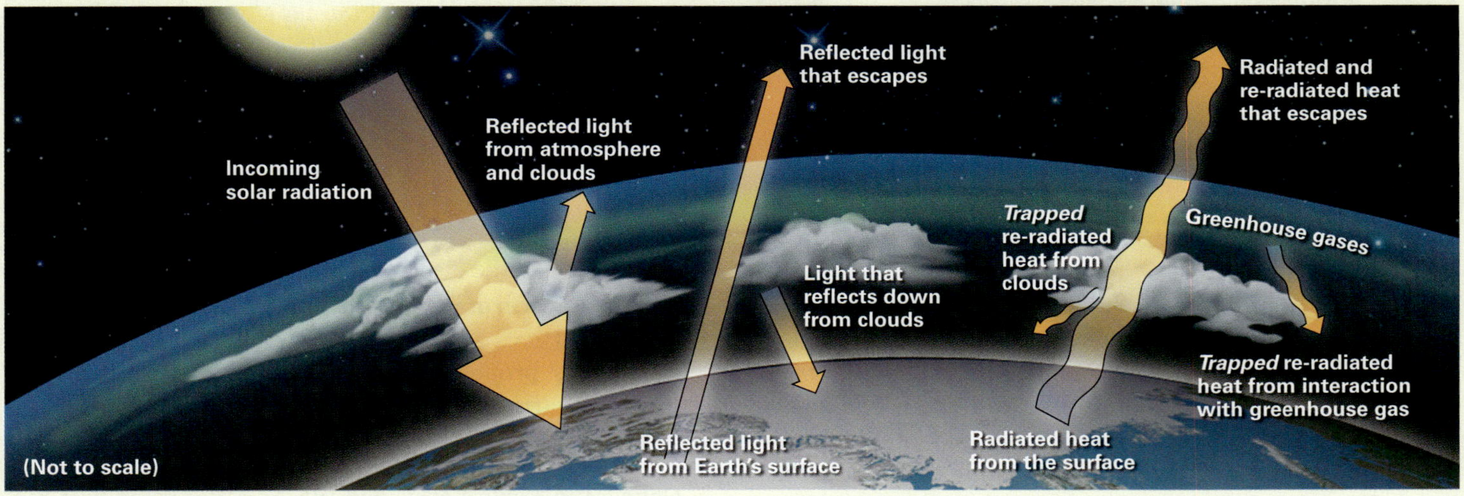

up to 30 times in the northern hemisphere during the Pleistocene Ice Age. Each advance represents an interval of global cooling, and each retreat represents an interval of global warming. If we focus on the last 15,000 years, we see trends of cooling or warming that last thousands of years, within which there are events whose duration lasts centuries or less (**Fig. 19.9a**). Specifically, the time between about 15,000 and 10,500 B.C.E. was a warming period during which the last ice-age glaciers retreated. This warming trend was followed by the *Younger Dryas*, an interval of cooler temperatures

FIGURE 19.7 Changes in the assemblage of pollen in sediment indicate a shift in climate belts.

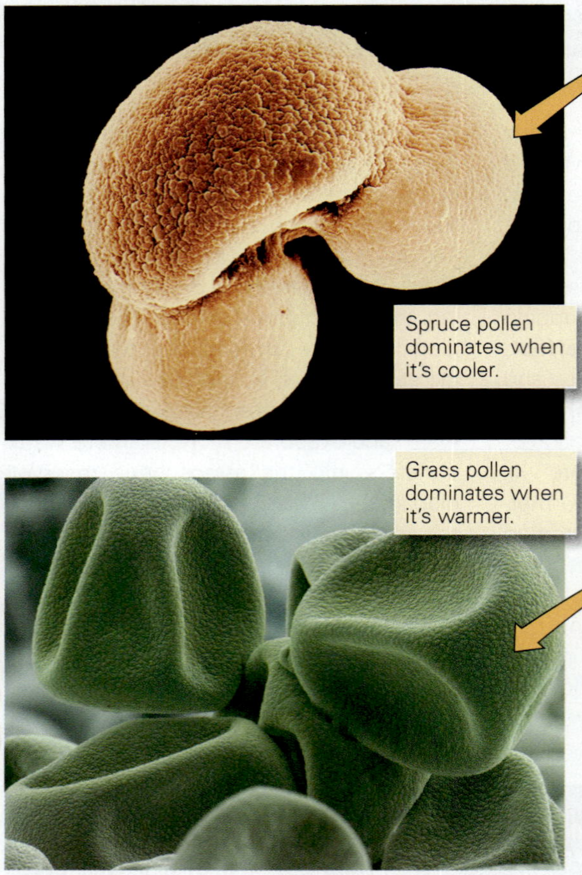

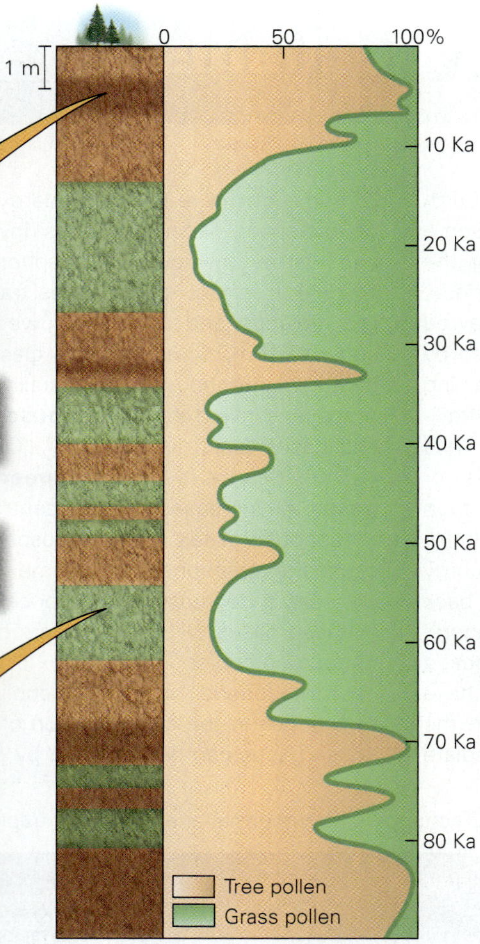

(a) The proportion of tree pollen relative to grass pollen can change in a sedimentary sequence through time. Researchers plot changes in the proportion of pollen types over time by examining samples from a column of sediment.

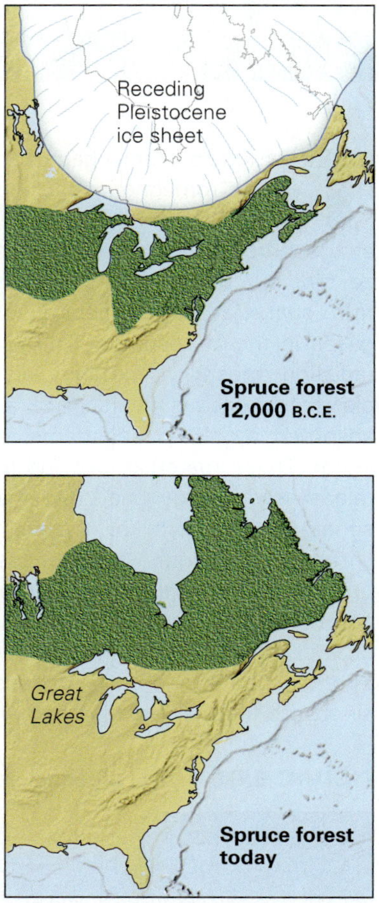

(b) Spruce forests (green) grew farther south 10,000 years ago than they do today.

named for an Arctic flower that became widespread during the interval. Then climate warmed again, reaching a peak at 5,000 to 7,000 years ago, a period called the *Holocene maximum,* when average temperatures were about 2°C above temperatures of today. This warming led to increased evaporation and therefore precipitation, making the Middle East unusually wet and fertile—conditions that may have contributed to the rise of civilization in this part of the world. The temperature dipped to a low about 3,000 years ago, before returning to a high during the Middle Ages, a time called the *Medieval Warm Period*. During this time, Vikings established self-supporting agricultural settlements along the coast of Greenland. The temperature dropped again from 1500 C.E. to about 1800 C.E., a period known as the *Little Ice Age,* when Alpine glaciers advanced and the canals of the Netherlands froze over in winter (**Fig. 19.9b;** see Fig. 18.29). Overall, the climate has warmed since the end of the Little Ice Age, and today it is as warm or warmer than it was during the Medieval Warm Period.

Geologists propose that several factors contributed to short-term warming and cooling events apparent in the near-term paleoclimate record:

- *Changes in Earth's orbit and tilt*: As Milanković first recognized in 1920, the change in the tilt of Earth's axis, the Earth's precession cycle, and changes in the eccentricity of its orbit together cause variation in the amount of summer heat at high latitudes. They also cause variation in the overall amount of heat reaching the Earth (see Chapter 18). These changes correlate with observed ups and downs in atmospheric and oceanic temperature.
- *Changes in ocean currents*: Recent studies suggest that the configuration of currents can change quite quickly, and that this configuration affects the climate.

FIGURE 19.8 The proportion of isotopes transferred between reservoirs during evaporation or precipitation depends on temperature. The $^{18}O/^{16}O$ ratio can be studied in glacial ice (H_2O) and fossil shells ($CaCO_3$).

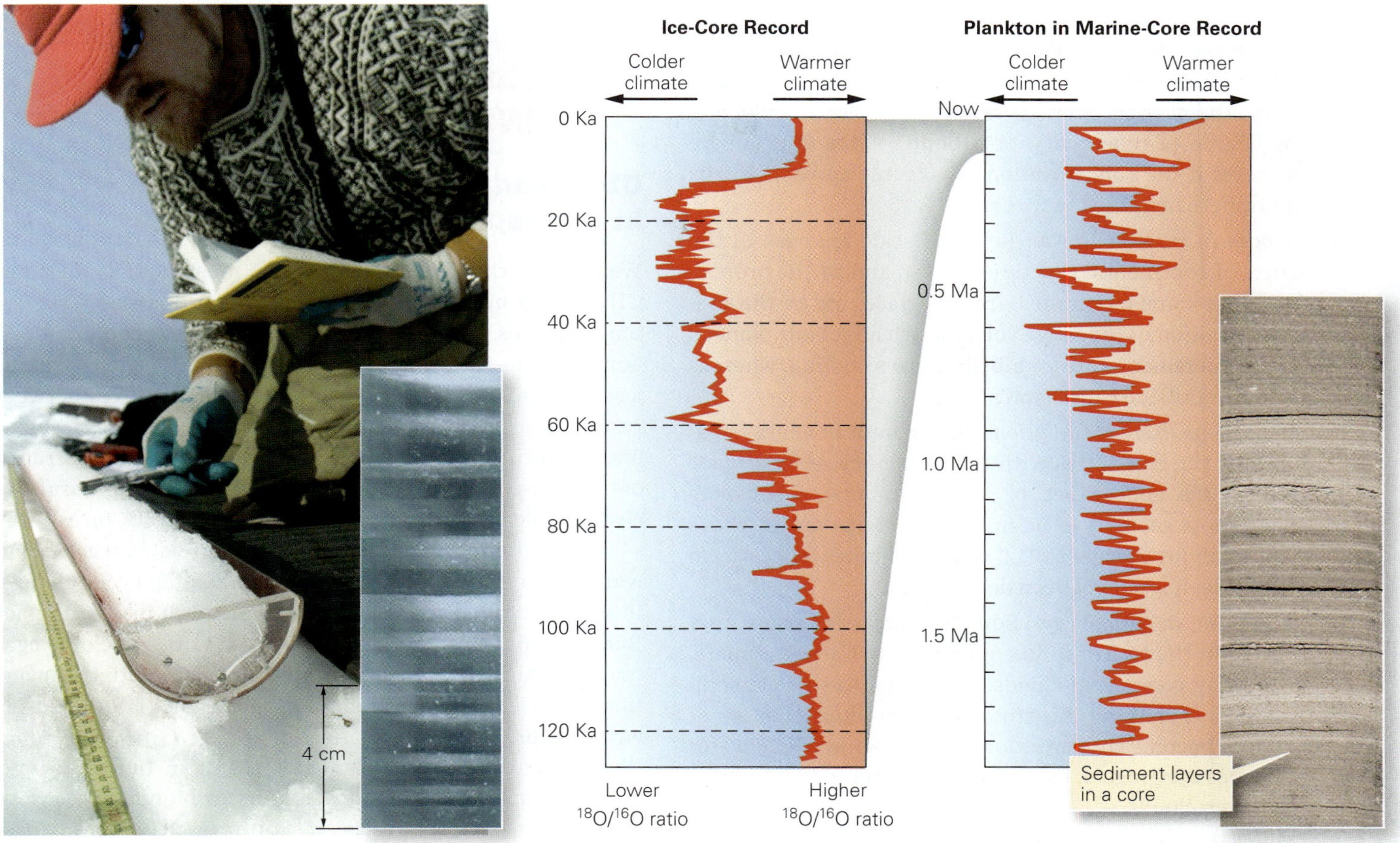

(a) A researcher examines an ice core in the field. Lab photos reveal annual layers.

(b) The $^{18}O/^{16}O$ ratios in an ice core represent temperature changes.

(c) Studies of $^{18}O/^{16}O$ ratios in deep-sea cores also provide a climate record.

FIGURE 19.9 Climate during the Holocene. Measurements suggest that temperature has varied significantly.

(a) There were several temperature highs and lows during the Holocene.

(b) Glaciers advanced in Europe during the Little Ice Age, but have since been retreating due to warmer temperatures.

- *Large eruptions of volcanic aerosols*: Not all of the sunlight that reaches the Earth penetrates its atmosphere and warms the ground. Some is reflected by the atmosphere. The degree of reflectivity, or **albedo**, of the atmosphere increases if the concentration of volcanic aerosols in the atmosphere increases.
- *Changes in surface albedo*: Regional-scale changes in the nature of continental vegetation cover, and/or the proportion of snow and ice on the Earth's surface, and/or the sudden deposition of reflective volcanic ash affect our planet's albedo. Increasing surface albedo causes cooling, whereas decreasing albedo causes warming.
- *Fluctuations in solar radiation*: The amount of energy produced by the Sun varies with the *sunspot cycle*. This cycle involves the appearance of large numbers of sunspots (black spots thought to be magnetic storms on the Sun's surface) about every 9 to 11.5 years. Whether longer-term cycles exist has not yet been determined.
- Fluctuation in *cosmic rays*: Some researchers suggest that changes in the rate of influx of cosmic rays affect climate, perhaps by generating clouds. According to this concept, cosmic rays striking the atmosphere produce clusters of ions that become condensation nuclei around which water molecules congregate, leading to the formation of droplets making up clouds.
- *Abrupt changes in concentrations of greenhouse gases*: A sudden change in greenhouse gas concentration in the atmosphere could affect climate. One such change might happen if sea temperature warmed or sea level dropped. This could cause sudden melting of a large amount of the methane hydrate that crystallized in sediment on the seafloor, releasing CH_4 to the atmosphere. Sudden melting of permafrost could have a similar effect, for the resulting rot of organic material preserved in the permafrost produces CH_4. Algal blooms and reforestation absorb CO_2 and could cause cooling.

TAKE-HOME MESSAGE

Geologic study shows that, over the long term, climate on the Earth alternates between greenhouse and icehouse conditions. Factors including life evolution, plate movements, and plate interactions likely cause this long-term change. Study of the near-term record shows that short-term changes in climate can also take place. These changes may relate to the Milankovitch cycle, to sudden release of methane, or to volcanic activity.

QUICK QUESTION How does continental drift contribute to climate change?

19.5 Contemporary Global Warming

Observed Changes in Atmospheric CO_2 and CH_4

We've seen that greenhouse gases, most notably CO_2 and CH_4, play a major role in the regulation of Earth's surface temperatures. Without these gases, the Earth could not be a home for life. Both gases cycle through various biogeochemical reservoirs of the Earth System, and the fluxes between reservoirs determine the amount in any given reservoir at any given time. For most of geologic time, the concentration of these gases in the atmosphere was governed by natural processes—volcanic eruptions, life-evolution events, forest fires, sea-level rise or fall, orogenic uplift and related weathering, warming and cooling due to the Milankovitch cycle, changes in solar activity or cosmic-ray flux, and even meteor impact. But beginning around 8,000 years ago, human society began to modify the environment significantly, first with the invention of agriculture and then, during the past two centuries, with the spread of industrialization.

Both agriculture and industry result in the transfer of carbon from underground reservoirs or biomass reservoirs into the atmospheric reservoir. For example, burning fossil fuels oxidizes vast quantities of carbon that had previously been held in oil, gas, or coal underground to produce CO_2 that mixes into the atmosphere. Heating calcite ($CaCO_3$) to produce the lime (CaO) of concrete takes carbon that had been locked in limestone and produces CO_2 that also mixes into the air. Clear-cutting of forests to make way for grazing land or fields replaces high-biomass vegetation (trees) with low-biomass vegetation (grasses or crops), thereby leaving CO_2 in the atmosphere. Finally, decay of organic material in soggy rice paddies, as well as the flatulence of cattle herds, produces significant quantities of CH_4 that would otherwise have remained locked in biomass.

It may seem strange that human society can have such a significant effect on Earth's greenhouse gases. But a comparison of the quantity of greenhouse gases produced by humans to those produced by volcanoes demonstrates that society's production of these gases are indeed significant. Specifically, researchers estimate that all volcanic eruptions together in a given year—including both submarine and subaerial eruptions—emit about 0.15 to 0.26 billion tons of CO_2. By comparison, activities of people emit about 35 billion tons of CO_2 every year (about 135 times as much). This means that human activities now produce more CO_2 in three days than all of the volcanoes on Earth in a typical year. A major volcanic explosion emits about as much CO_2 as human society produces in one day, and even a "supervolcanic" explosion (see Chapter 5), which happens on the order of once every 100,000 to 300,000 years, produces about as much CO_2 as society produces in one year.

Of the anthropogenic (human-produced) CO_2 that society sends into the atmosphere, 85% comes from burning fossil fuels and producing cement, while the remaining 15% is a consequence of deforestation. About 40% to 50% of this gas dissolves in the ocean, reacts with rocks during chemical weathering, or gets incorporated into plants during photosynthesis. The remainder stays in the atmosphere and mixes thoroughly with other gases. Evidence that atmospheric CO_2 concentration has been increasing became clear starting in the early 1960s, when a chemist named Charles Keeling began to analyze air samples that he collected every month at the summit of Mauna Kea, an inactive volcano in Hawaii that lies far from local sources of atmospheric pollution. After completing monthly measurements over many years, Keeling showed not only that distinct seasonal variations take place (CO_2 concentration goes down in the warm summer when rates of photosynthesis increase, and it goes up in the winter when organic matter dies and decays), but also that the average annual concentration of CO_2 was steadily rising (**Fig. 19.10a**)! Specifically, the average yearly CO_2 concentration was 320 ppm in 1965, and it reached 360 ppm in 1995. Charles Keeling died in 2005, but measurements at Mauna Kea have continued, and in 2015, the average yearly CO_2 concentration measured there was 399 ppm, with the midyear peak just above 401 ppm.

Using measured CO_2 concentrations in air bubbles trapped in ancient ice obtained by drilling into glaciers, researchers have extended the record of atmospheric CO_2 concentration further back in time (**Fig. 19.10b**). They found that in 1750, CO_2 concentration was only 280 ppm, and thus, atmospheric CO_2 concentration has increased by about 40% since the

FIGURE 19.10 Changes in carbon dioxide (CO_2) and methane (CH_4) concentrations over time.

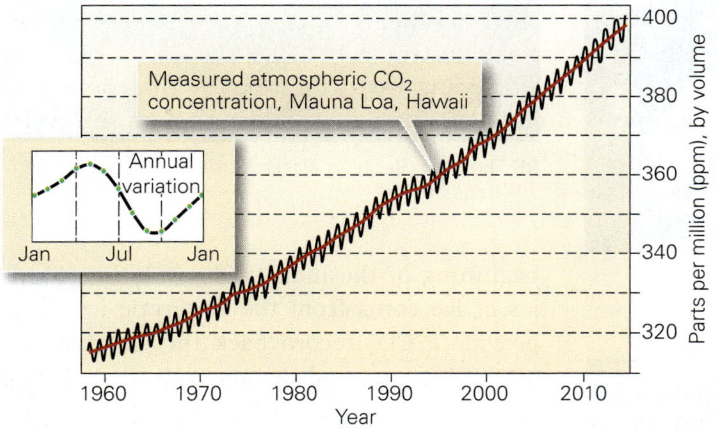

(a) Monthly measurements begin in 1960. There is an annual cycle, related to seasons, but the overall increase is clear.

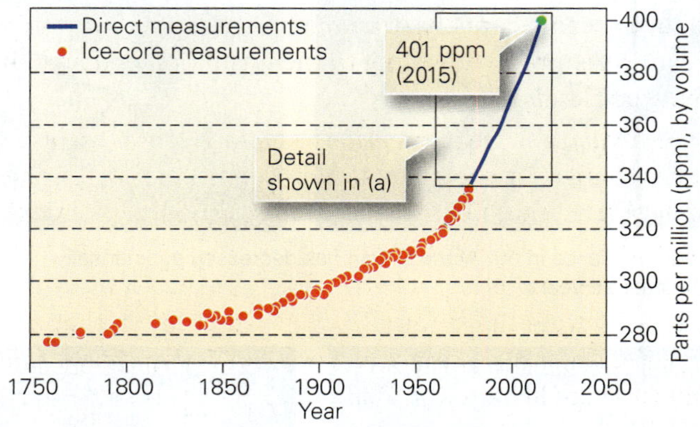

(b) Since the industrial revolution, CO_2 concentration has steadily increased.

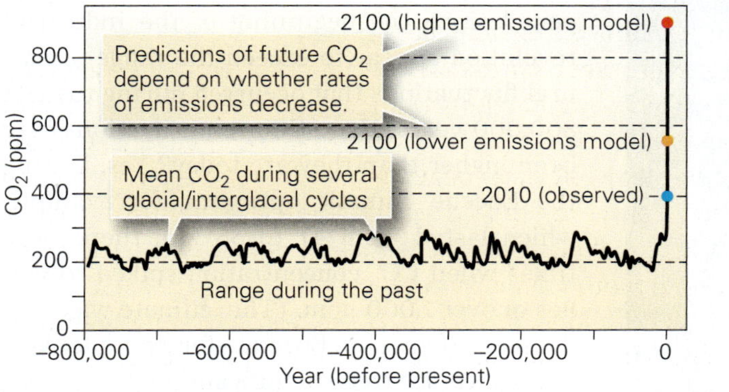

(c) Based on ice-core studies in glaciers, researchers find that CO_2 concentration varied between 180 and 300 ppm throughout glacial advances and retreats. The current value, 401 ppm, is above this range.

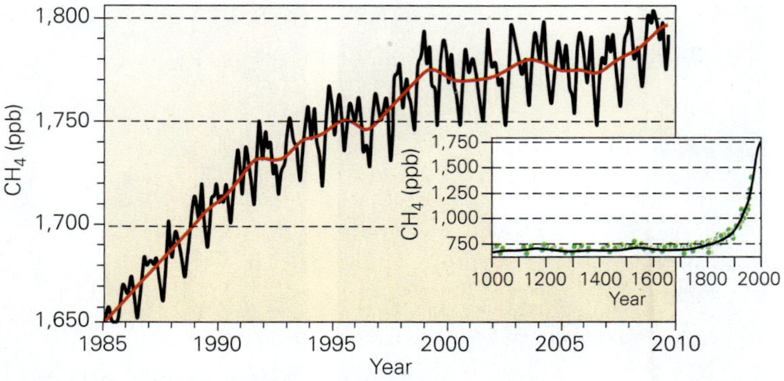

(d) Methane, another greenhouse gas, has been increasing, too. The inset shows change over the past millennium.

FIGURE 19.11 Visible examples of changes in Earth's ice cover.

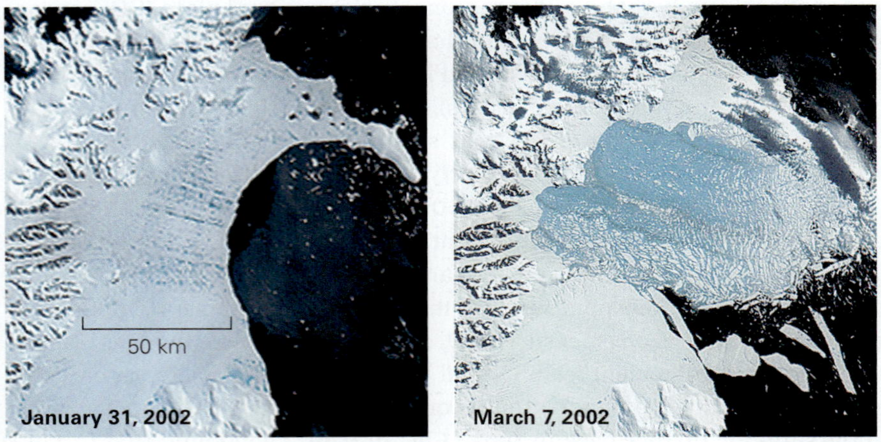

(a) In 2002, the Larson B Ice Shelf of Antarctica disintegrated over the course of a month.

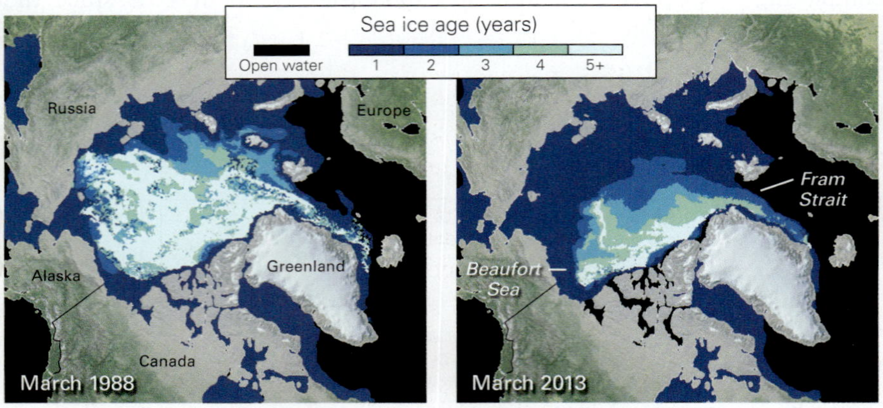

(b) The age of sea ice in the Arctic Ocean has decreased substantially during the past 25 years.

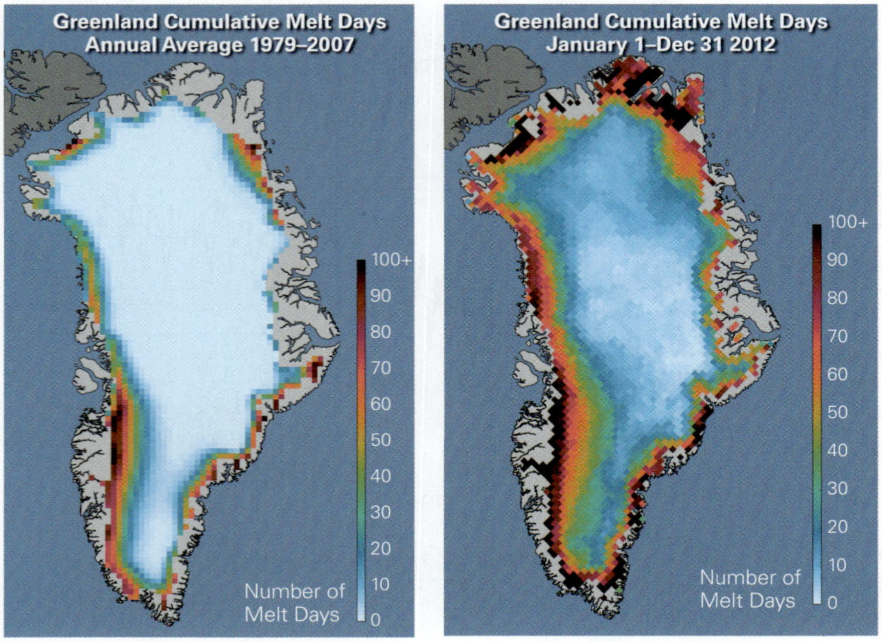

(c) The number of days during which melting of the Greenland Ice Sheet takes place has increased.

(d) The Muir Glacier in Alaska retreated 12 km between 1941 and 2004.

beginning of the industrial revolution. Studies of ice cores from the Antarctic ice sheet provide a CO_2 record back through the last 800,000 years, and demonstrate that during the alternating glacial and interglacial periods of the Late Pleistocene, CO_2 concentration varied between about 180 and 300 ppm (**Fig. 19.10c**). Thus, the increases that have happened since the beginning of the industrial revolution appear to exceed the range of natural fluctuations that occurred during the last 800,000 years. Have CO_2 concentrations ever been higher than they are today? Yes. During the Eocene climatic optimum, for example, which lasted from 47 to 55 Ma, there were times when CO_2 concentration spiked to values of over 1,000 ppm. (The climate was very different during the Eocene—for example, Antarctica was ice free.) Significantly, CH_4 concentrations have also demonstrably increased over the past two centuries (**Fig. 19.10d**), and they haven't been as high since the Eocene.

Observations of Contemporary Global Warming

The fundamental principle of the greenhouse effect requires that an increase in CO_2 concentration should cause atmospheric warming. But is warming actually taking place? During the past 30 years, researchers have published thousands of observations suggesting that it is. For example:

- Large ice shelves, such as the Larsen B Ice Shelf along the Antarctic Peninsula, and the Ayles Ice Shelf along Ellesmere Island in northernmost Canada, are breaking up (**Fig. 19.11a**).
- The area covered by sea ice in the Arctic Ocean has decreased substantially. Some estimates place the rate of ice-cover loss at about 3% per decade. This observation leads to the prediction that it may be possible to sail across the Arctic Ocean within decades (**Fig.19.11b**).
- The Greenland ice sheet has been melting at an accelerating pace (**Fig. 19.11c**). Studies suggest that the rate of ice loss has increased from 90 to 220 cubic km per year in the last 10 years and that, in places, the sheet is thinning by about 1 m per year.
- Most valley glaciers worldwide have been retreating rapidly, so that areas that were once ice covered are now ice free. The change is truly dramatic in many locations (**Fig. 19.11d**). For example, of the 150 glaciers that occupied valleys and cirques in Glacier National Park, Montana, when the park was founded in 1910, fewer than 25 remain, and according to models, the last glaciers there will disappear by mid-century.
- Taking into account both the melting of polar ice sheets and the melting of mountain glaciers, geologists estimate that, worldwide, glacial volumes are diminishing at a rate of about 400 cubic km every year.
- The area of permafrost in high latitudes has decreased substantially, and melt ponds have replaced frozen land across Siberia, Alaska, and arctic Canada.
- Average annual water vapor in the atmosphere has been increasing due to evaporation of warmer seas, possibly leading to more severe hurricanes.
- Biological phenomena that are sensitive to climate are being disrupted. For example, the time when sap in the maple trees of the northeastern United States starts to flow has changed; the mosquito line (the elevation at which mosquitoes can survive) has risen substantially; plant hardiness zones in North America have migrated northward (**Fig. 19.12**); and the average weight of polar bears has been decreasing, because the bears can no longer walk over pack ice to reach their hunting grounds in the sea.

Documenting warming by directly measuring temperatures over time can be challenging, because weather varies so significantly with location, even on a daily basis. To determine average temperatures, researchers must use statistical methods to analyze temperature measurements obtained at recording stations around the world. Results of such analyses support the concept that warming is happening, for they indicate that global mean atmospheric temperature has risen by almost 1°C during the last century and is higher now than it has been at any time during the past 2,000 years (**Fig. 19.13**). Although such a change may seem small, consider that the total temperature change between the last ice age and now was only 3° to 5°C.

SEE FOR YOURSELF...

RECEDING GLACIER, SWITZERLAND

Latitude
46°34′48.75″ N

Longitude
8°22′58.53″ E

From 4 km (~2.5 miles), look obliquely NNE.

A valley glacier near the town of Gletsch has receded. Bare land downslope from the toe of the glacier was covered by ice a century ago. Note that the lateral moraine lies like a bathtub ring high above the present-day ice surface, because the glacier has thinned.

FIGURE 19.12 Plant hardiness zones have been migrating northward.

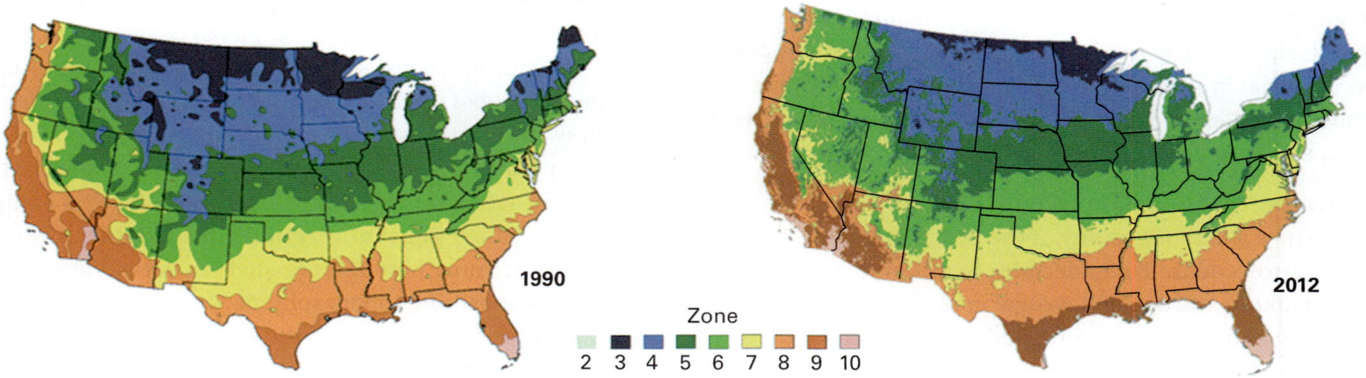

FIGURE 19.13 Measurements of global warming and global temperature anomalies.

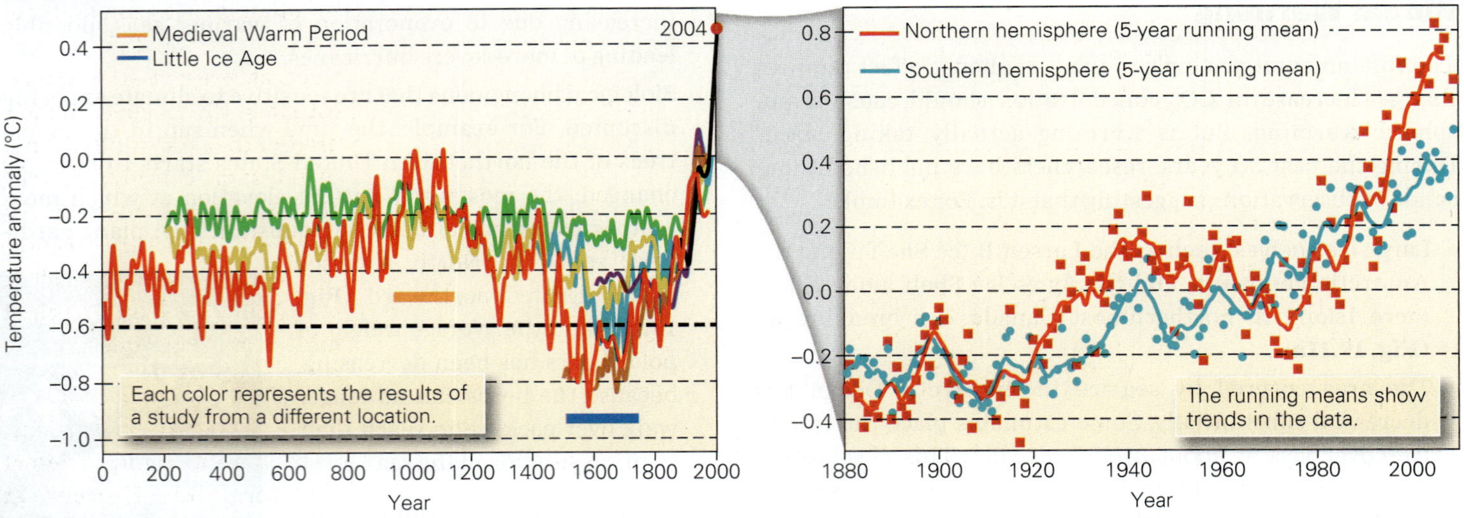

(a) Reconstructions of temperature during the past 2,000 years. Each color is a published estimate from a research team. The black line is the average of direct measurements.

(b) Graph showing the change in global average temperature since 1880 relative to a reference value. Note that both the northern and southern hemispheres show a temperature increase.

A small change in average temperature may have major consequences. Significantly, temperature change is not uniform around the world—some regions appear to be warming more than others, and some areas have been cooling. Because of the increase in atmospheric temperature, average measured values of near-surface ocean-water temperatures have also been rising (**Fig. 19.14**).

Interpretations and Potential Consequences of Climate Change

Climate data are not easy to obtain or interpret. Because climate studies can have major implications for global policy decisions, a group of leading scientists founded the *Intergovernmental Panel on Climate Change* (IPCC) in 1988. This panel, sponsored by the World Meteorological Organization and the United Nations, reviews published research on climate change and, every five years, summarizes the conclusions in an assessment report. The language describing the likelihood that global warming is happening, and that humans have contributed significantly to causing it, has become progressively less equivocal in successive versions of the report. The results of the Fifth Assessment Report, a synthesis for policy makers published in 2014, states:

> Warming of the climate system is unequivocal, and since the 1950s, many of the observed changes are unprecedented over decades to millennia. The atmosphere and ocean have warmed, the amounts of snow and ice have diminished, sea level has risen, and the concentrations of greenhouse gases have increased . . . . Anthropogenic greenhouse gas emissions have increased since the pre-industrial era, driven largely by economic and population growth, and are now higher than ever. This has led to atmospheric concentrations of carbon dioxide, methane and nitrous oxide that are unprecedented in at least the last 800,000 years. Their effects, together with those of other anthropogenic drivers, have been detected throughout the climate system and are extremely likely to have been the dominant cause of the observed warming since the mid-20th century.

In other words, the vast majority of climate researchers have concluded that global warming is real, and that the actions of people—burning fossil fuels, cutting down forests, paving over wetlands—have played a significant role in causing it. Other possible causes, such as changes in solar radiation or cosmic-ray flux, do not appear to be sufficient to cause observed warming.

If the report of the IPCC proves to be correct, society faces the challenge of either slowing global warming or dealing with its consequences. To understand the potential consequences of global warming, researchers study paleoclimate to see how the Earth's surface conditions changed with changing temperature in the past, for the record of the past may provide insight into the future. They also develop computer programs, called **general circulation models** (GCMs), that take into account how factors such as temperature, atmospheric composition, topography, ocean currents, and Earth's orbit affect the circulation of the atmosphere and, therefore, the distribution of climate belts. These models can provide

FIGURE 19.14 The difference in global temperature of shallow ocean water, for the period of 1850 to 2005, relative to the average temperature between 1961 and 2006.

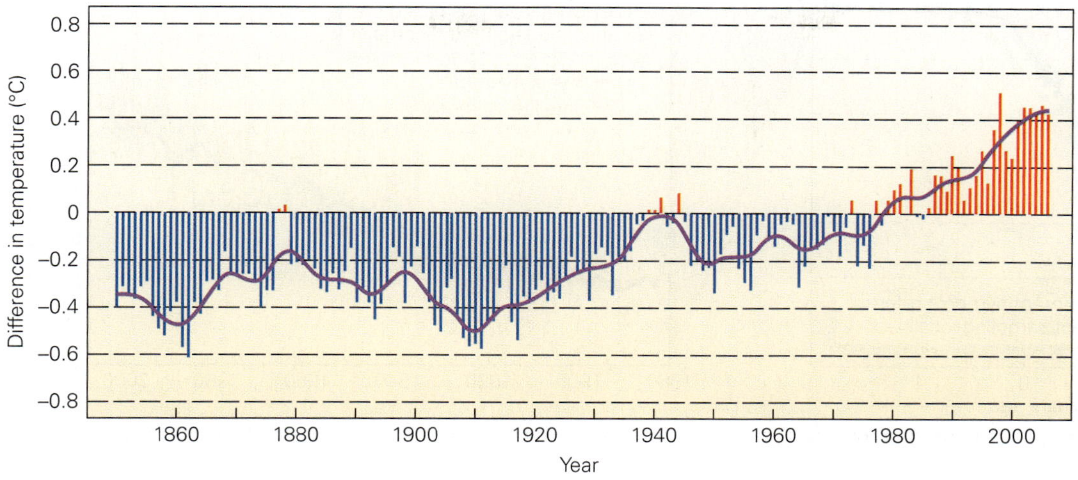

insight into when and why changes took place in the past, and whether they will happen in the future. Specifically, they can help to predict how rainfall, sea level, ice cover, and other physical phenomena may be influenced by the warming or cooling of the atmosphere. Not all researchers, however, agree on how to construct or interpret these models, so predictions from them remain controversial.

In a worst-case scenario, computer models predict that global warming will continue into the future at the present rate, so that by 2050—within the lifetime of many readers of this book—the average annual temperature will have increased in some parts of the world by 1.5° to 2.0°C. At these rates, by the end of the century, temperatures could be almost 4°C warmer (**Fig. 19.15a**), and by 2150, global temperatures may be 5° to 11°C warmer than at present—the warmest since the Eocene. Models predict that warming will not be the same everywhere—the greatest impact will be in the Arctic (**Fig. 19.15b**). According to some climate models, the following events might happen as a consequence:

- *Shift in climate belts*: As the climate overall warms, temperate and desert regions would occur at higher latitudes. Thus, regions that now host agriculture may become too dry to be farmable. The change in climate would also affect the amount and distribution of precipitation (rain and snow) that falls.
- *Ice retreat and snow-line rise*: Global warming will decrease the volume of water held in glacial reservoirs and will cause areas covered by sea ice to decrease. It will also cause snow-line elevations in mountainous areas to rise—this change will affect the amount of water held in the winter snowpack, and therefore, will decrease summer runoff.
- *Melting permafrost*: Warming climates will cause areas of unglaciated land that had remained frozen most of the

FIGURE 19.15 Model predictions of future global warming.

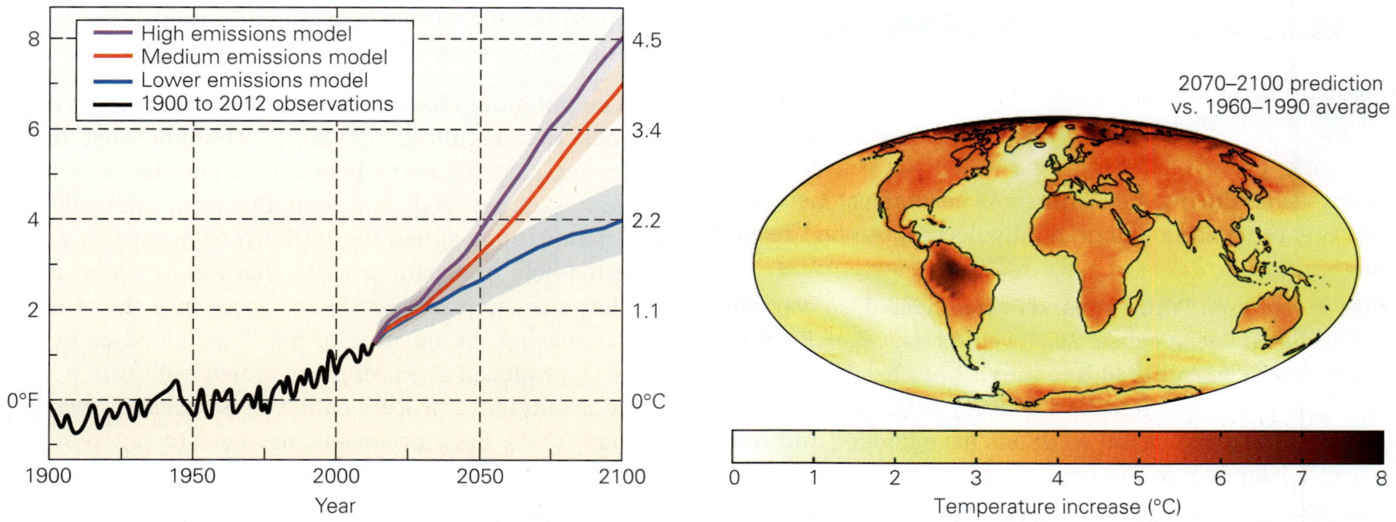

(a) Model calculations of global warming. Different models assume different rates of CO_2 emissions. All suggest significant global temperature increase by 2100.

(b) Global warming does not mean that all locations warm by the same amount. This map shows a model prediction of how temperature increase may vary with location for the time period 2070–2100.

FIGURE 19.16 Sea-level rise of the recent past and possible sea-level rise of the future.

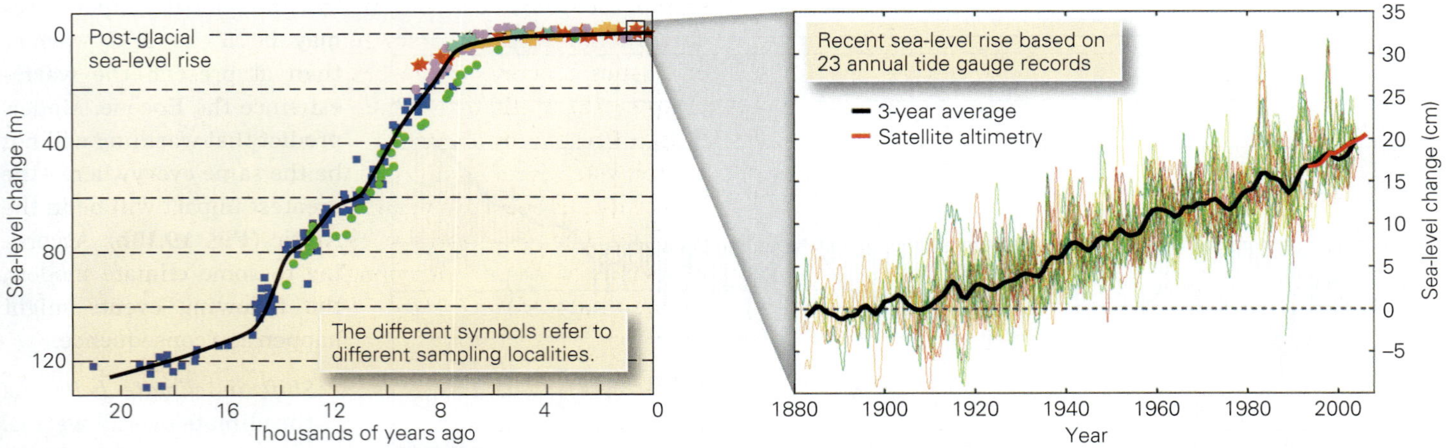

(a) Sea level rose rapidly after the last ice age due to melting of continental glaciers. It is continuing to rise slowly.

(b) Tide gauges document sea-level rise of the past 130 years.

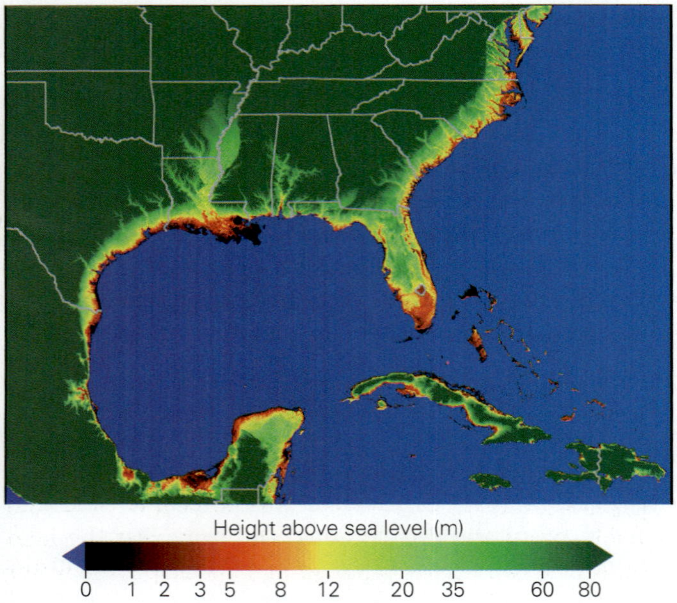

(c) The red and black areas of this map are so low that they could be flooded if sea level rose a few meters.

year to thaw during the summer. As permafrost melts, frozen organic matter will gradually decompose and release methane.

- *Rise in sea level*: Water expands when heated, so warming of the global ocean causes sea level to rise. Glacial melting due to global warming adds to the rise. Sea level has already changed noticeably in the last century (**Fig. 19.16a, b**), and as a result, some coastal wetlands have flooded and some oceanic islands have become submerged. The rate of further sea-level rise depends on the rate of global warming. Models suggest that by 2100, it may rise by an additional 20 to 60 cm. If seas rise a meter or two, the result would be to inundate regions of the world where 20% of the human population currently lives (**Fig. 19.16c**).
- *Stronger storms*: An increase in average ocean temperatures means that more of the ocean evaporates when a tropical depression passes over. This evaporation might nourish stronger hurricanes. Areas without deserts might receive increasing amounts of precipitation and, therefore, of flooding.
- *An increase in wildfires*: Warmer temperatures may decrease the moisture content in plants, and therefore, could lead to an increase in the frequency of wildfires.
- *An interruption of the oceanic heat conveyor*: Oceanic currents play a major role in transferring heat across latitudes. If global warming melts polar ice, the resulting freshwater would dilute surface ocean water at high latitudes. This water could not sink, and thus thermohaline circulation would be shut off (see Chapter 15), preventing the water from conveying heat.

The potential changes described above, along with other studies that estimate the large economic cost of global warming, have led many people to argue the issue needs to be addressed seriously and soon. But what can be done? The 160 nations that signed the 1997 Kyoto Accord, at a summit meeting held in Japan, propose that the first step would be to slow the input of greenhouse gases into the atmosphere by decreasing the burning of fossil fuels and/or by forcing the CO_2 produced at power plants down wells into pore space underground by a process called *carbon capture and sequestration (CCS)*. Such proposals, needless to say, remain very controversial, and even if industrialized countries succeed in in diminishing the release of greenhouse gases, the increased production of these gases by industrializing countries has, so far, more than made up the difference.

TAKE-HOME MESSAGE

Growing evidence indicates that greenhouse gases (CO_2 and CH_4) being introduced into the atmosphere by activities of society are causing global warming of both the air and sea that might not otherwise have happened. This change is associated with glacial melting, shifting climate belts, sea-level rise, and other phenomena.

QUICK QUESTION How do researchers develop predictions of future climate change?

19.6 Other Human Impacts on the Earth System

According to some researchers, perhaps only 20,000 people lived on Earth after the catastrophic eruption of the Toba volcano about 70,000 years ago, and by the dawn of civilization, 4000 B.C.E., the population was still, at most, a few tens of millions. But by the beginning of the 19th century, revolutions in industrial methods, agriculture, medicine, and hygiene had substantially lowered death rates and raised living standards, so that the human population began to grow at accelerating rates, reaching 1 billion in 1850. It took only 80 years for the population to double after that, reaching 2 billion in 1930. Now the doubling time is only 44 years—the population passed the 6 billion mark just before the year 2000 and surpassed the 7.3 billion mark in 2015 (**Fig. 19.17**).

As the population grows and the standard of living improves, per capita usage of resources increases. We use land for agriculture and grazing, forests for wood, rock and dirt for construction, oil and coal for energy or plastics, and ores for metals (see Chapter 12). Without a doubt, our usage of resources has affected the Earth System profoundly, and thus humanity has become a major agent of global change. Here, we examine some of these anthropogenic (human-induced) impacts.

FIGURE 19.17 Population now doubles about every 44 years. The Black Death pandemic caused an abrupt drop that lasted for a few decades.

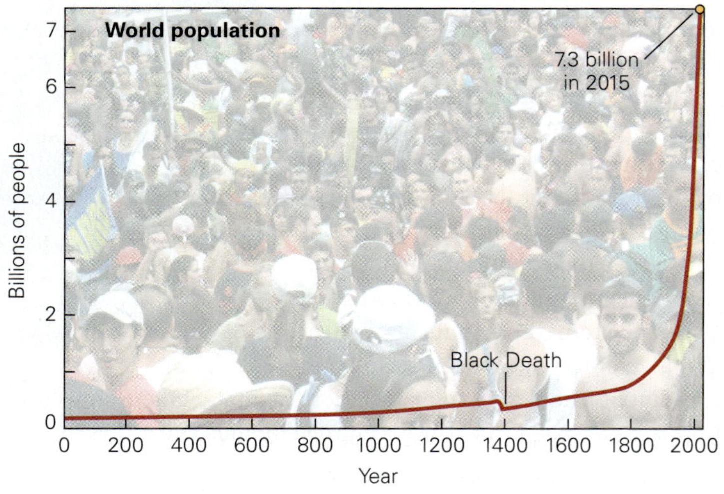

SEE FOR YOURSELF...

FIELDS AND VILLAGES, CHINA

Latitude
35°6′46.35″ N

Longitude
114°30′7.81″ E

Zoom to 4 km (~2.5 miles) and look down.

This landscape has been completely altered by human society. What was once forest or grassland has been divided into cultivated rectangles. In villages, houses and roads cover the surface. These changes affect both the carbon cycle and the hydrologic cycle.

The Modification of Landscapes

Every time we pick up a shovel and move a pile of rock or soil, we redistribute a portion of the Earth's crust. In the last century, the pace of Earth movement has accelerated, for now shovels in mines can move 300 cubic meters of rock in a single scoop, trucks can carry 200 tons of rock in a single load, and tankers can transport 500 million liters (~3 million barrels) of oil during a single journey. In North America, human activity now moves more sediment each year than the continent's rivers do. The extraction of rock during mining, building levees and dams along rivers, or of seawalls along the coast, and construction of highways and cities all involve redistribution of Earth materials (**Fig. 19.18**). In addition, people clear and plow fields, drain and fill wetlands, and pave over the land surface. All these activities change the landscape, the water table, and the supply of sediment.

The Modification of Ecosystems

In undisturbed areas, the *ecosystem* (an interconnected network of organisms and the physical environment in which they live) of a region is the product of evolution for an extended period of time. The ecosystem's flora and fauna include species that have adapted to living together in a particular climate and on the substrate available. Human-caused deforestation, overgrazing, agriculture, and urbanization disrupt ecosystems and lead to a decrease in biodiversity.

Archaeological studies have found that the earliest example of humans modifying an ecosystem occurred in the Stone Age, when hunters played a major role in causing mass extinctions of mammoths and other large mammals. Today,

FIGURE 19.18 Excavation, agriculture, and construction modify topography, drainage, infiltration, and ecology.

Humans move vast amounts of rock.
Agriculture eliminates diverse ecosystems.
Urbanization changes the water table.

SEE FOR YOURSELF...

URBANIZING A DESERT, ARIZONA

Latitude
33°33′15.42″ N

Longitude
111°50′28.92″ W

Zoom to 9 km (~6 miles) and look down.

The growing population of Phoenix causes intense change in a desert. The central portion of the image shows a network of dry washes in the original desert. To the north, we see new houses and a water canal, and to the south, farmlands and a golf course. Water usage in the area has increased dramatically with development.

less than 5% of Europe and the United States retain their original habitats. Worldwide, tropical rainforests cover less than half the area that they covered before the dawn of civilization, and they are disappearing at a rate of about 1.8% per year (**Fig. 19.19a, b**). Much of this loss comes from slash-and-burn agriculture, in which farmers and ranchers destroy forest to create open land for farming and grazing (**Fig. 19.19c**).

Some changes humans make to the land have permanent consequences in a human time frame. The heavy rainfall of tropical regions removes nutrients from the soil of the fields produced by deforestation, making the soil useless in just a few years. Overgrazing by domesticated animals can remove vegetation so completely that some grasslands have undergone desertification. And urbanization replaces the natural land surface with concrete or asphalt, a process that completely destroys an ecosystem and radically changes the amount of rain that infiltrates the land surface to become groundwater.

Pollution

The environment has always contained contaminants such as soot, dust, and the by-products of organisms. But when human populations grew, urbanization, industrial and agricultural activity, the production of electricity, and modern modes of transportation greatly increased both the quantity and diversity of contaminants that entered the air, surface water, and groundwater. These contaminants, or **pollution**, include both natural and synthetic materials in liquid, solid, and gaseous form. They have become a problem because society emits them too fast for the Earth System to naturally absorb or modify—small quantities of sewage can be absorbed by clay minerals in the soil or destroyed by bacterial metabolism, but large quantities overwhelm natural controls and can accumulate in destructive concentrations. Pollution of the Earth System represents a type of global change because it results in a redistribution and reformulation of materials. Some key problems associated with this change include the following:

› *Water contamination*: Humans dump a great variety of chemicals into surface water and groundwater. Examples

SEE FOR YOURSELF...

DEFORESTATION, AMAZON

Latitude
4°4′22.81″ S

Longitude
54°54′43.06″ W
Look down from 100 km (~60 miles).

A patchwork of fields cuts into the Amazon rainforest, south of the Amazon River in Brazil. Note the pattern of cutting—the fields border a grid of logging roads bulldozed across the landscape. Uncut forest remains in the upper left.

FIGURE 19.19 The area of forests has been shrinking. For example, tropical rainforests are being logged or burned.

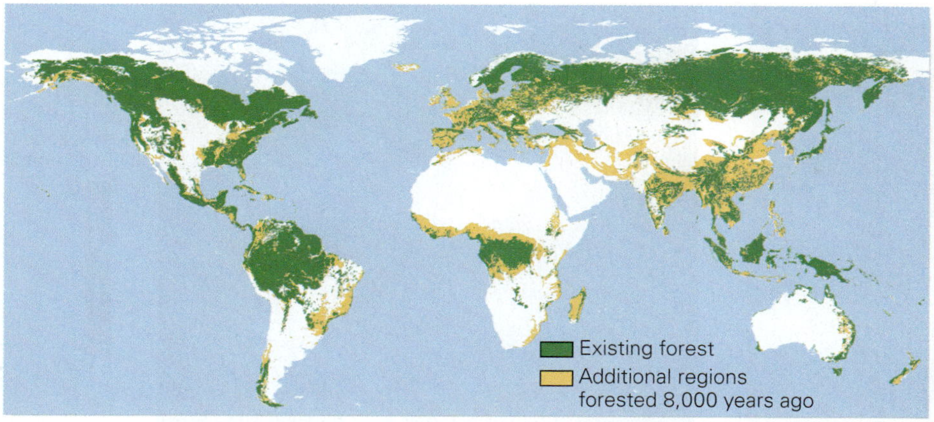

(a) Remaining forests today are much smaller than forests of 8,000 years ago.

(c) Part of the Amazon rainforest's destruction is due to slash-and-burn agriculture.

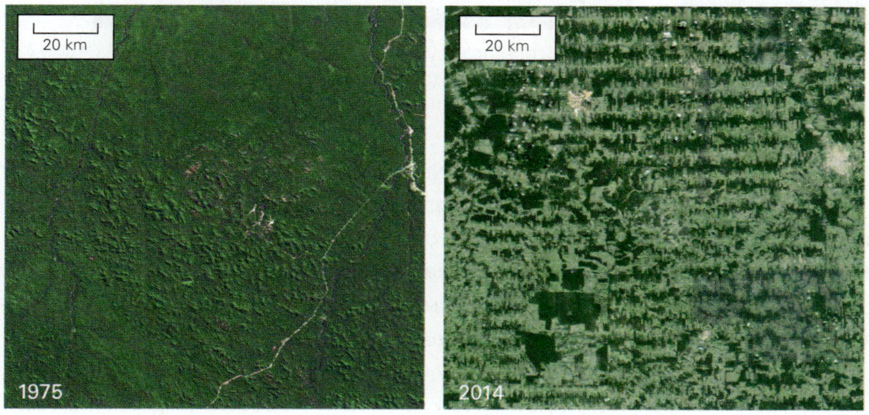

(b) Comparison of satellite imagery highlights areas that have undergone deforestation, such as this one near Ariquemes, Brazil.

include gasoline, other organic chemicals, radioactive waste, acids, fertilizers—the list could go on for pages. These chemicals affect biodiversity.

- *Destruction of habitat*: Agriculture, grazing, deforestation, urban sprawl, road construction, and other ways in which people change the surface of the land have resulted in the widespread destruction of habitats in which ecosystems thrived.
- *Smog*: The term was originally coined to refer to the dank, dark air that resulted when smoke from the burning of coal mixed with fog in London and other industrial cities. Another kind of smog, called *photochemical smog*, is the ozone-rich brown haze that blankets cities when exhaust from cars and trucks reacts with air in the presence of sunlight (**Fig. 19.20a**).
- *Acid runoff and acid rain:* Dissolution of sulfide-containing minerals in ores or coal by groundwater or stream water makes the water into *acid mine runoff*. When rain passes through air that contains sulfur-containing aerosols (emitted from power plants), the water dissolves the sulfur and yields **acid rain**. Wind can carry aerosols far from a power plant, so acid rain can damage a broad region.
- *Radioactive materials*: Nuclear weapons, nuclear energy, and medical waste transfer radioactive materials from rock to Earth's surface environment.
- *Ozone depletion*: When emitted into the atmosphere, human-produced chemicals, most notably chlorofluorocarbons (CFCs), react with ozone in the stratosphere. This reaction, which happens most rapidly on the surfaces of tiny ice crystals in polar stratospheric clouds, destroys ozone molecules, and yields an *ozone hole* over high-latitude regions, particularly during the spring (**Fig. 19.20b**). Note that the "hole" is not an area without any ozone, but is an area where atmospheric ozone has been decreased substantially. Ozone holes have dangerous consequences, for they affect the ability of the atmosphere to shield the Earth's surface from harmful ultraviolet radiation.

TAKE-HOME MESSAGE

Human activities impact the landscape and ecosystems significantly by moving rock and soil, by paving the surface, and by changing the character of land cover. Society also introduces contaminants into the environment at rates that cannot be accommodated by the Earth System, leading to pollution of land, air, surface water, and groundwater.

QUICK QUESTION What is the ozone hole, and why did it form?

FIGURE 19.20 Consequences of adding pollutants to the atmosphere.

(a) Photochemical smog hazes the view in Beijing, on an otherwise sunny day.

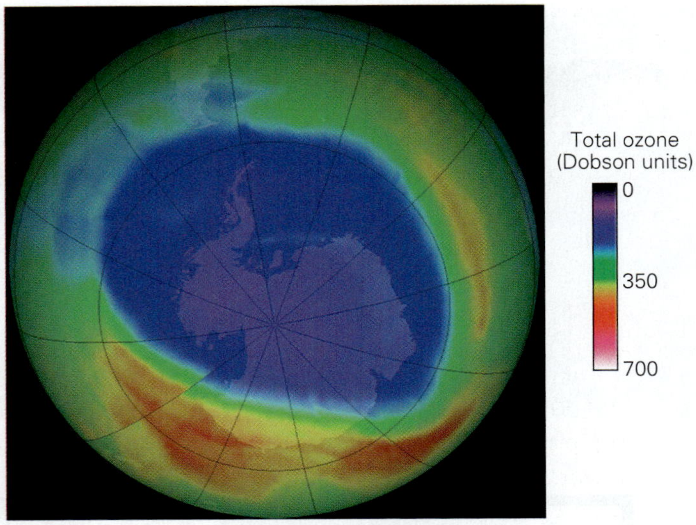

(b) The ozone hole over Antarctica in 2014. A Dobson unit is a measure of the amount of ozone in the atmosphere; red is more, purple is less.

19.7 The Future of the Earth

Most of the discussion in this book has focused on the past, for the geologic record preserved in rocks tells us about earlier times. Let's now bring this book to a close by facing in the opposite direction and speculating what the world might look like in geologic time to come.

In the geologic near term of centuries to millennia, the future of the world depends largely on human activities. Whether the Earth System undergoes a major disruption and shifts to a new equilibrium, or whether society achieves *sustainable growth* (an ability to prosper within the constraints of the Earth System), will depend on our own foresight and ingenuity. Projecting a few thousand years further into the future, we might see that the Earth has returned to ice-age conditions, with glaciers advancing over the locations of what are now major cities and the continental shelf becoming dry land. Alternatively, the ice age may be over for good, because of global warming—no one really knows for sure.

If we project millions of years into the future, the map of the planet will change significantly because of the continuing activity of plate tectonics. For example, during the next 50 million years or so, the Atlantic Ocean will probably become bigger, the Pacific Ocean will shrink, and the western part of California will migrate northward. Eventually, Australia will crush against the southern margin of Asia, and the islands of Indonesia may be flattened in between.

What will the record of our time look like to geologic observers tens of millions of years in the future? Because the world has changed so much, due to the activities of humans, some researchers refer to the last few millennia as the *Anthropocene*. Not only will strata deposited during the Anthropocene contain an array of unusual chemicals and trace fossils (from bottles to buildings), but sadly, it may record an abrupt decrease in biodiversity. Biologists suggest that dozens of species are becoming extinct every day, a rate that may be 1,000 times the rate of species loss prior to the advent of civilization. Causes include habitat destruction, poaching, introduction of invasive species (organisms from one part of the world that thrive in a different part of the world at the expense of native organisms), climate change, and environmental contamination. In fact, some biologists worry that the Anthropocene could be the time of the *sixth extinction*, a mass extinction as dramatic as the mass-extinction events that marked the end of the Paleozoic and the end of the Mesozoic, and thus might delineate the boundary between the end of the Cenozoic and a geologic era yet to come.

Predicting the map of the Earth hundreds of millions of years into the future is hard, because we don't know for sure where new subduction zones will develop. Perhaps subduction of the Pacific Ocean will lead to a collision of the Americas with Asia, to produce a supercontinent (called Amasia). Eventually a subduction zone will form on one side of the Atlantic Ocean, and the ocean will be consumed. As a consequence, the eastern margin of the Americas will likely collide with the western margin of Europe and Africa. The sites of major cities—New York, Miami, Rio de Janeiro, Buenos Aires, and London—will be incorporated into a collisional mountain belt, and likely will be subjected to metamorphism and igneous intrusion before being uplifted and eroded. Shallow seas may once again cover the interiors of continents and then later retreat, and glaciers may once again cover the continents—it happened in the past, so it could happen again. And if the past is the key to the future, we *Homo sapiens* might not be around to watch our cities enter the rock cycle, for biological evolution may have introduced new species to the biosphere, and there is no way to predict what these species will be like.

And what of the end of the Earth? Geologic catastrophes resulting from asteroid and comet collisions will undoubtedly occur in the future as they have in the past. Concern about these have led observatories to scan space to detect incoming objects, and space agencies to think about how to deflect objects that might be on a collision course. The explosion of a 20-m-diameter asteroid 27 km above Siberia in 2013 sent out waves that blew glass out of windows, leading to the injury of 1500 people. If a 100-m-diameter asteroid struck Earth, it would destroy everything within a blast ring almost 24 km in diameter—in other words, it could wipe a major city off the map. The meteorite associated with mass extinction at the end of the Mesozoic was about 10 km in diameter. We can't predict when such an object will strike next, but unless it can be diverted, Earth will endure radical readjustment of surface conditions in the future.

Bad as a collision with a meteorite might be, it's not likely that such a collision will destroy our planet. Rather, astronomers predict that our planet has reached only middle age and has 5 billion years yet to go. The inevitable end of the Earth won't occur until the Sun begins to run out of fuel. When this happens, thermal pressure caused by fusion reactions will no longer be able to prevent the Sun from collapsing inward due to the immense gravitational pull of its mass. Were the Sun a few times larger than it is, this collapse would trigger a supernova explosion. But stars the size of the Sun won't explode. Rather, the thermal energy generated when the Sun's interior collapses inward will heat the gases of its outer layers sufficiently to cause them to expand. As a result, the Sun will become a **red giant**, a huge star whose radius exceeds the present orbit of the Earth. As the Sun expands, its solar wind will increase, and by ejecting mass into space, its total mass will decrease. This change will allow the Earth's orbit to increase in radius—but alas, our planet won't escape. As the Sun's surface approaches and the Earth's orbit no longer lies in the habitable zone, the oceans and atmosphere will boil away, and then the solid planet itself will start to melt and finally vaporize. Finally, the Sun will engulf the Earth's orbit (**Fig. 19.21**), and when that happens, our planet's remaining atoms will mix with the gases of the Sun. When the red giant phase finishes, the remnants of the Sun will shrink to become a white dwarf, leaving behind an expanding ring of gas. When this ultimate global change has taken place, the atoms that once formed Earth and all its inhabitants throughout geologic time will drift into a nebula, perhaps to be incorporated in a future solar system, where the cycle of planetary formation and evolution could begin anew.

TAKE-HOME MESSAGE

In the near term, Earth's surface will be affected by the actions of human society, which appear to be playing a major role in decreasing biodiversity and could lead to a geologically significant mass extinction. Our planet may endure glaciations within the next few millennia. Over longer time scales, the map of the Earth will change due to plate interactions and sea-level change. The end of the Earth will likely happen in the far distant future, when the Sun becomes a red giant.

QUICK QUESTION What might happen to the atoms that make up the Earth long after the red-giant stage of Solar System evolution?

FIGURE 19.21 In about 5 billion years, the Sun will become a red giant.

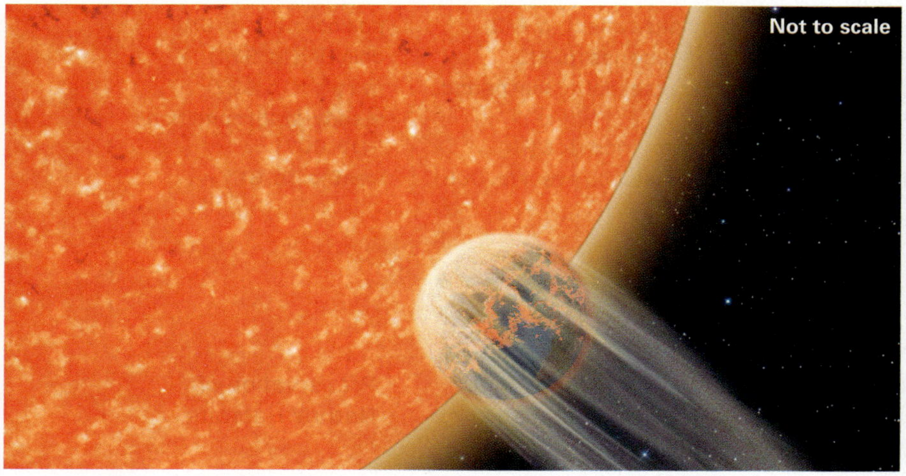

(a) As the surface of the red giant approaches Earth, the planet will evaporate like a giant comet.

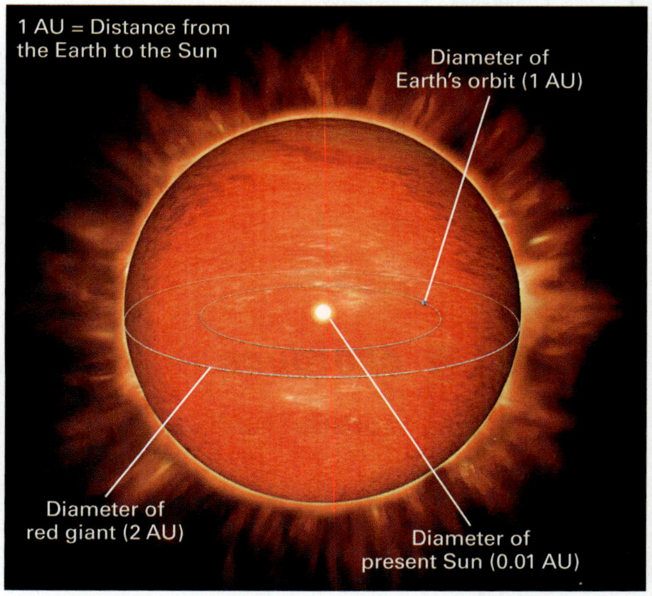

(b) At full size, the red giant will have a diameter twice that of Earth's orbit, so Earth's atoms will become part of the star.

Chapter 19 Review

Chapter Summary

- We refer to the global interconnecting web of physical and biological phenomena on Earth as the Earth System. Global change involves the transformations or modifications of physical and biological components of the Earth System through time. Unidirectional change results in transformations that never repeat, whereas cyclic change involves repetition of the same steps over and over.
- Examples of unidirectional change include the gradual evolution of the solid Earth from a homogeneous collection of planetesimals to a layered planet, the formation of the oceans, the gradual change in the composition of the atmosphere, and the evolution of life.
- Examples of physical cycles that take place on Earth include the supercontinent cycle, the sea-level cycle, and the rock cycle.
- A biogeochemical cycle involves the passage of a chemical among nonliving and living reservoirs. Examples include the hydrologic cycle and the carbon cycle. Global change occurs when factors change the relative proportions of the chemical in different reservoirs.
- Studies of long-term climate change show that, at times in the past, the Earth experienced greenhouse (warmer) periods; at other times, there were icehouse (cooler) periods. Factors leading to long-term climate change include the positions of continents, volcanic activity, the uplift of land, and the addition or removal of CO_2, an important greenhouse gas.
- Short-term climate change can be seen in the near-term record of the last million years, and can be studied by examining micropaleontology, oxygen-isotope ratios, bubbles in ice, and growth rings in trees.
- During only the past 15,000 years, we see that the climate has warmed and cooled a few times. Causes of short-term climate change include the Milankovitch cycle, fluctuations in solar radiation and cosmic rays, changes in reflectivity, and changes in ocean currents.
- Mass extinction, a catastrophic change in biodiversity, may be caused by the impact of a comet or asteroid or by intense volcanic activity.
- The addition of CO_2 and CH_4 to the atmosphere appears to be causing global warming, which could shift climate belts and lead to a rise in sea level. Sources for the added CO_2 include fossil fuel burning, cement production, and deforestation.
- During the last two centuries, humans have changed landscapes; modified ecosystems; and added pollutants to the land, air, and water at rates faster than the Earth System can process.
- In the future, in addition to climate change, the Earth will witness a continued rearrangement of continents resulting from plate tectonics and will likely suffer the impact of asteroids and comets. The end of the Earth may come in about 5 billion years, when the Sun runs out of fuel and becomes a red giant.

Guide Terms

acid rain (p. 625)
albedo (p. 616)
biogeochemical cycle (p. 607)
carbon cycle (p. 610)
climate (p. 611)
climate change (p. 611)
feedback mechanism (p. 613)
general circulation model (GCM) (p. 620)
global change (p. 604)
global cooling (p. 611)
global warming (p. 611)
greenhouse effect (p. 613)
greenhouse gas (p. 613)
greenhouse (hothouse) period (p. 611)
icehouse period (p. 611)
mass-extinction event (p. 605)
paleoclimate (p. 611)
pollution (p. 624)
red giant (p. 627)
steady-state condition (p. 607)
weather (p. 611)

GEOTOURS *THIS CHAPTER'S GEOTOUR EXERCISE (S) FEATURES:*

- Effects of Global Warming
- Effects of Desforestation
- Water Use in Arid Regions
- Preservation of Natural Habitats

Review Questions

1. Why do we use the term "Earth System" to describe the components of processes operating on this planet?
2. How have the Earth's crust and atmosphere changed since they first formed?
3. What processes control the rise and fall of sea level on Earth?
4. How does carbon cycle through the various Earth systems?
5. Contrast icehouse and greenhouse conditions.
6. What are the possible causes of long-term climatic change?
7. How do paleoclimatologists study ancient near-term climate change?
8. What factors explain short-term climatic change?
9. Give some examples of events that cause catastrophic change.
10. What is the ozone hole, and how does it affect us?
11. Describe how contemporary global warming takes place, and how society may play a role in causing it.
12. What effects might global warming have on the Earth System?
13. Give some examples of how humans have changed the Earth.
14. What are some likely scenarios for the long-term future of the Earth?

On Further Thought

15. If global warming continues, how will the distribution of grain crops change? Might this affect national economies? Why?
16. Currently, tropical rainforests are being cut down at a rate of about 1.8% per year. At this rate, how many more years will the forests survive? In the eastern United States, the proportion of land with forest cover today has increased over the past century. In fact, most of the farmland that existed in New York State in 1850 is forestland today. Why?
17. Using the library or the Web, examine the change in the nature of world fisheries that has taken place in the last 50 years. Is the world's fish biomass sustainable if these patterns continue? What has happened to whale populations during the past 50 years?

Online Resources

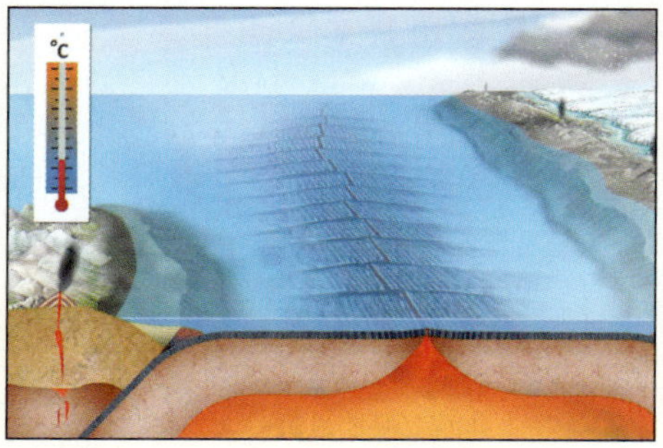

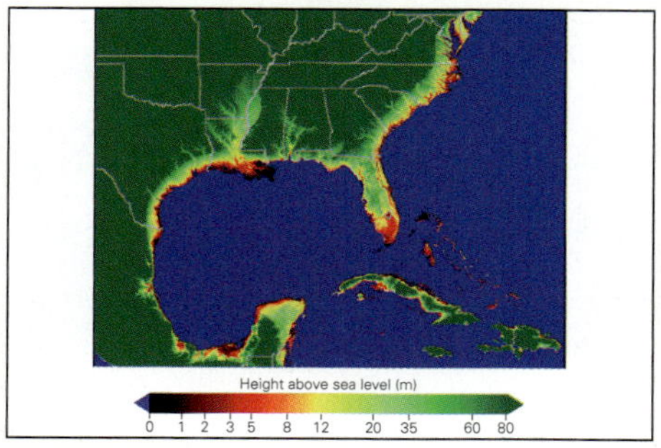

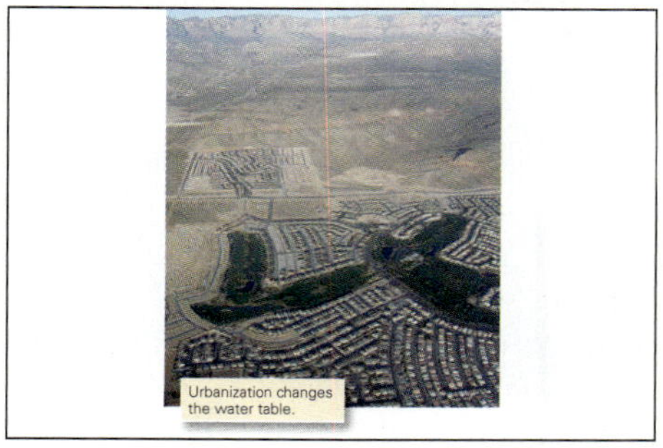

Animations

This chapter features animations simulating global sea change over time.

Assessment

This chapter includes visual exercises on geologic eras, and questions on the causes of and impacts of global change.

Additional Charts

Metric Conversion Chart

Length

1 kilometer (km) = 0.6214 mile (mi)
1 meter (m) = 1.094 yards = 3.281 feet
1 centimeter (cm) = 0.3937 inch
1 millimeter (mm) = 0.0394 inch
1 mile (mi) = 1.609 kilometers (km)
1 yard = 0.9144 meter (m)
1 foot = 0.3048 meter (m)
1 inch = 2.54 centimeters (cm)

Area

1 square kilometer (km^2) = 0.386 square mile (mi^2)
1 square meter (m^2) = 1.196 square yards (yd^2)
= 10.764 square feet (ft^2)
1 square centimeter (cm^2) = 0.155 square inch (in^2)
1 square mile (mi^2) = 2.59 square kilometers (km^2)
1 square yard (yd^2) = 0.836 square meter (m^2)
1 square foot (ft^2) = 0.0929 square meter (m^2)
1 square inch (in^2) = 6.4516 square centimeters (cm^2)

Volume

1 cubic kilometer (km^3) = 0.24 cubic mile (mi^3)
1 cubic meter (m^3) = 264.2 gallons
= 35.314 cubic feet (ft^3)
1 liter (l) = 1.057 quarts
= 33.815 fluid ounces
1 cubic centimeter (cm^3) = 0.0610 cubic inch (in^3)
1 cubic mile (mi^3) = 4.168 cubic kilometers (km^3)
1 cubic yard (yd^3) = 0.7646 cubic meter (m^3)
1 cubic foot (ft^3) = 0.0283 cubic meter (m^3)
1 cubic inch (in^3) = 16.39 cubic centimeters (cm^3)

Mass

1 metric ton = 2,205 pounds
1 kilogram (kg) = 2.205 pounds
1 gram (g) = 0.03527 ounce
1 pound (lb) = 0.4536 kilogram (kg)
1 ounce (oz) = 28.35 grams (g)

Pressure

1 kilogram per
square centimeter (kg/cm^2)* = 0.96784 atmosphere (atm)
= 0.98066 bar
= 9.8067×10^4 pascals (Pa)
1 bar = 0.1 megapascals (Mpa)
= 1.0×10^5 pascals (Pa)
= 29.53 inches of mercury (in a barometer)
= 0.98692 atmosphere (atm)
= 1.02 kilograms per square centimeter (kg/cm^2)
1 pascal (Pa) = 1 $kg/m/s^2$
1 pound per square inch = 0.06895 bars
= 6.895×10^3 pascals (Pa)
= 0.0703 kilogram per square centimeter

Temperature

To change from Fahrenheit (F) to Celsius (C):

$$°C = \frac{(°F - 32°)}{1.8}$$

To change from Celsius (C) to Fahrenheit (F):

$$°F = (°C \times 1.8) + 32°$$

To change from Celsius (C) to Kelvin (K):

$$K = °C + 273.15$$

To change from Fahrenheit (F) to Kelvin (K):

$$K = \frac{(°F - 32°)}{1.8} + 273.15$$

*Note: Because kilograms are a measure of mass whereas pounds are a unit of weight, pressure units incorporating kilograms assume a given gravitational constant (g) for Earth. In reality, the gravitational for Earth varies slightly with location.

Symbol | Atomic number | Name | Atomic weight: He 2 Helium 4.002

Alkali metals — Transition elements (metals) — Nonmetals — Inert gases

H 1 Hydrogen 1.007																	He 2 Helium 4.002
Li 3 Lithium 6.941	Be 4 Beryllium 9.0121											B 5 Boron 10.811	C 6 Carbon 12.011	N 7 Nitrogen 14.006	O 8 Oxygen 15.999	F 9 Fluorine 18.998	Ne 10 Neon 20.179
Na 11 Sodium 22.989	Mg 12 Magnesium 24.305											Al 13 Aluminum 26.981	Si 14 Silicon 28.085	P 15 Phosphorus 30.973	S 16 Sulfur 32.066	Cl 17 Chlorine 35.452	Ar 18 Argon 39.948
K 19 Potassium 39.098	Ca 20 Calcium 40.078	Sc 21 Scandium 44.955	Ti 22 Titanium 47.88	V 23 Vanadium 50.941	Cr 24 Chromium 51.996	Mn 25 Manganese 54.938	Fe 26 Iron 55.847	Co 27 Cobalt 58.933	Ni 28 Nickel 58.693	Cu 29 Copper 63.546	Zn 30 Zinc 65.39	Ga 31 Gallium 69.723	Ge 32 Germanium 72.61	As 33 Arsenic 74.921	Se 34 Selenium 78.96	Br 35 Bromine 79.904	Kr 36 Krypton 83.80
Rb 37 Rubidium 85.467	Sr 38 Strontium 87.62	Y 39 Yttrium 88.905	Zr 40 Zirconium 91.224	Nb 41 Niobium 92.906	Mo 42 Molybdenum 95.94	Tc 43 Technetium 98.907	Ru 44 Ruthenium 101.07	Rh 45 Rhodium 102.905	Pd 46 Palladium 106.42	Ag 47 Silver 107.868	Cd 48 Cadmium 112.411	In 49 Indium 114.82	Sn 50 Tin 118.710	Sb 51 Antimony 121.757	Te 52 Tellurium 127.60	I 53 Iodine 126.904	Xe 54 Xenon 131.29
Cs 55 Cesium 132.905	Ba 56 Barium 137.327	La 57 Lanthanum 138.905	Hf 72 Hafnium 178.49	Ta 73 Tantalum 180.947	W 74 Tungsten 183.85	Re 75 Rhenium 186.207	Os 76 Osmium 190.2	Ir 77 Iridium 192.22	Pt 78 Platinum 195.08	Au 79 Gold 196.966	Hg 80 Mercury 200.59	Tl 81 Thallium 204.383	Pb 82 Lead 207.2	Bi 83 Bismuth 208.980	Po 84 Polonium 208.982	At 85 Astatine 209.987	Rn 86 Radon 222.017
Fr 87 Francium 223.019	Ra 88 Radium 226.025	Ac 89 Actinium 227.027															

Ce 58 Cerium 140.115	Pr 59 Praseodymium 140.907	Nd 60 Neodymium 144.24	Pm 61 Promethium 144.912	Sm 62 Samarium 150.36	Eu 63 Europium 151.965	Gd 64 Gadolinium 157.25	Tb 65 Terbium 158.925	Dy 66 Dysprosium 162.50	Ho 67 Holmium 164.930	Er 68 Erbium 167.26	Tm 69 Thulium 168.934	Yb 70 Ytterbium 173.04	Lu 71 Lutetium 174.967

Th 90 Thorium 232.038	Pa 91 Protactinium 231.035	U 92 Uranium 238.028	Np 93 Neptunium 237.048	Pu 94 Plutonium 244.064	Am 95 Americium 243.061	Cm 96 Curium 247.070	Bk 97 Berkelium 247.070	Cf 98 Californium 251.079	Es 99 Einsteinium 252.083	Fm 100 Fermium 257.095	Md 101 Mendelevium 258.10	No 102 Nobelium 259.100	Lr 103 Lawrencium 262.11

The modern periodic table of the elements. Each column groups elements with related properties. For example, inert gases are listed in the column on the right. Metals are found in the central and left parts of the chart.

Glossary

a'a A lava flow with a rubbly surface.
abandoned meander A meander that dries out after it was cut off.
ablation The removal of ice at the toe of a glacier by melting, sublimation (the evaporation of ice into water vapor), and/or calving.
abrasion The process in which one material (such as sand-laden water) grinds away at another (such as a stream channel's floor and walls).
absolute age Numerical age (the age specified in years).
absolute plate velocity The movement of a plate relative to a fixed point in the mantle.
absolute zero The lowest temperature possible (−273.15°C); at absolute zero, vibrations and movements of atoms in a material cease.
abyssal plain A broad, relatively flat region of the ocean that lies at least 4.5 km below sea level.
Acadian orogeny A convergent mountain-building event that occurred around 400 million years ago, during which continental slivers accreted to the eastern edge of the North American continent.
accreted terrane A block of crust that collided with a continent at a convergent margin and stayed attached to the continent.
accretion A general term for the attachment of a block of crust to the margin of a larger block during orogeny.
accretionary coast A coastline that receives more sediment than erodes away.
accretionary disk The disk-shaped body of gas, ice, and dust onto which matter falls; it ultimately evolves into a solar system.
accretionary lapilli Hailstone-like clumps of wet ash that fall from a volcanic eruptive cloud.
accretionary orogen An orogen formed by the attachment of numerous buoyant slivers of crust to an older, larger continental block.
accretionary prism A wedge-shaped mass of sediment and rock scraped off the top of a downgoing plate and accreted onto the overriding plate at a convergent plate margin.
acid mine runoff A dilute solution of sulfuric acid, produced when sulfur-bearing minerals in mines react with rainwater, that flows out of a mine.
acid rain Precipitation in which air pollutants react with water to make a weak acid that then falls from the sky.
active continental margin A continental margin that coincides with a plate boundary.
active fault A fault that has moved recently or is likely to move in the future.
active sand The top layer of beach sand, which moves daily because of wave action.
active volcano A volcano that has erupted within the past few centuries and will likely erupt again.
adiabatic cooling The cooling of a body of air or matter without the addition or subtraction of thermal energy (heat).
adiabatic heating The warming of a body of air or matter without the addition or subtraction of heat.
advection A process of heat transfer in which heat is carried into a solid by a liquid or gas moving through fractures or pores in the solid.
aerosols Tiny solid particles or liquid droplets that remain suspended in the atmosphere for a long time.
aftershocks The series of smaller earthquakes that follow a major earthquake.
agent of erosion A physical entity (wind, moving water, flowing ice) that grinds and removes material from the Earth's surface.
air The mixture of gases that make up the Earth's atmosphere.
air-fall tuff Tuff formed when ash settles gently from the air.
air mass A body of air, about 1,500 km across, that has recognizable physical characteristics.
air pressure The push that air exerts on its surroundings.
albedo The reflectivity of a surface.
Alleghenian orogeny The orogenic event that occurred about 270 million years ago when Africa collided with North America.
alloy A metal containing more than one type of metal atom.
alluvial fan A gently sloping apron of sediment dropped by an ephemeral stream at the base of a mountain in arid or semiarid regions.
alluvium Sorted sediment deposited by a stream.
alluvium-filled valley A valley whose floor fills with sediment.
Alpine-Himalayan chain The largest orogenic belt on Earth today, formed by collisions of the former Gondwana continents with the southern margins of Europe and Asia.
amber Hardened (fossilized) ancient sap or resin.
amphibolite facies A set of metamorphic mineral assemblages formed under intermediate pressures and temperatures.
amplitude The height of a wave from crest to trough.
Ancestral Rockies The late Paleozoic uplifts of the Rocky Mountain region; they eroded away long before the present Rocky Mountains formed.
angiosperm A flowering plant.
angle of repose The angle of the steepest slope that a pile of uncemented material can attain without collapsing from the pull of gravity.
angularity The degree to which grains have sharp or rounded edges or corners.
angular unconformity An unconformity in which the strata below were tilted or folded before the unconformity developed; strata below the unconformity therefore have a different tilt than strata above.
anhedral grains Crystalline mineral grains without well-formed crystal faces.
anion A negatively charged ion.
annual probability The likelihood that a flood of a given size or larger will happen at a specified locality during any given year.
Antarctic bottom water mass The mass of cold, dense water that sinks along the coast of Antarctica.
antecedent stream A stream that cuts across an uplifted mountain range; the stream must have existed before the range uplifted and must then have been able to downcut as fast as the land was rising.

anthracite coal Shiny black coal formed at temperatures between 200°C and 300°C. A high-rank coal.
Anthropocene An informal name for the most recent interval of geologic time, during which the significant impact of human society on the Earth System will likely be preserved in the geologic record.
anticline A fold with an arch-like shape in which the limbs dip away from the hinge.
anticyclone The clockwise flow of air around a high-pressure mass.
anticyclonic flow A circulation of air around a high-pressure region in the atmosphere; it rotates clockwise in the northern hemisphere.
Antler orogeny The Late Devonian mountain-building event in which slices of deep-marine strata were pushed eastward, up and over the shallow-water strata on the western coast of North America.
anvil cloud A large cumulonimbus cloud that spreads laterally at the tropopause to form a broad, flat top.
aphanitic A textural term for fine-grained igneous rock.
apparent polar-wander path A path on the globe along which a magnetic pole appears to have wandered over time; in fact, the continents drift, while the magnetic pole stays fairly fixed.
aquiclude Sediment or rock that transmits no water.
aquifer Sediment or rock that transmits water easily.
aquitard Sediment or rock that does not transmit water easily and therefore retards the motion of the water.
archaea A kingdom of "old bacteria," now commonly found in extreme environments like hot springs. (Also called archaeobacteria.)
Archean Eon The middle Precambrian eon (4.0–2.5 Ga).
Archimedes' principle The mass of the water displaced by a block of material equals the mass of the whole block of material.
arête A residual knife-edge ridge of rock that separates two adjacent cirques.
argillaceous sedimentary rock Sedimentary rock that contains abundant clay.
arkose A clastic sedimentary rock containing both quartz and feldspar grains.
arroyo The channel of an ephemeral stream; dry wash; wadi.
artesian spring A location where the ground surface intersects a natural fracture (joint) that taps a confined aquifer in which the pressure can drive the water to the surface.
artesian well A well in which water rises on its own.
artificial levee A man-made retaining wall to hold back a river from flooding.
ash *See* Volcanic ash.
ash fall Ash that falls to the ground out of an ash cloud.
ash flow An avalanche of ash that tumbles down the side of an explosively erupting volcano.
assimilation The process of magma contamination in which blocks of wall rock fall into a magma chamber and dissolve.
asteroid One of the fragments of solid material, left over from planet formation or produced by collision of planetesimals, that resides between the orbits of Mars and Jupiter.
asthenosphere The layer of the mantle that lies between 100–150 km and 350 km deep; the asthenosphere is relatively soft and can flow when acted on by force.
atm A unit of air pressure that approximates the pressure exerted by the atmosphere at sea level.
atmosphere A layer of gases that surrounds a planet.
atoll A coral reef that develops around a circular reef surrounding a lagoon.
atom The smallest piece of an element that has the properties of the element; it consists of a nucleus surrounded by an electron cloud.
atomic mass The amount of matter in an atom; roughly, it is the sum of the number of protons plus the number of neutrons in the nucleus.
atomic number The number of protons in the nucleus of a given element.
atomic weight The number of protons plus the number of neutrons in the nucleus of a given element. (Also known as atomic mass.)
aurora australis The same phenomenon as the aurora borealis, but in the southern hemisphere.
aurora borealis A ghostly curtain of varicolored light that appears across the night sky in the northern hemisphere when charged particles from the Sun interact with the ions in the ionosphere.
avalanche A turbulent cloud of debris mixed with air that rushes down a steep hill slope at high velocity; the debris can be rock and/or snow.
avalanche chute A downslope hillside pathway along which avalanches repeatedly fall, consequently clearing the pathway of mature trees.
avulsion The process in which a river overflows a natural levee and begins to flow in a new direction.
axial surface The imaginary surface that encompasses the hinges of successive layers of a fold.
axial trough A narrow depression that runs along a mid-ocean ridge axis.
axis An imaginary line around which an object spins.
backscattered light Atmospheric scattered sunlight that returns to space.
backshore zone The zone of beach that extends from a small step cut by high-tide swash to the front of the dunes or cliffs that lie farther inshore.
backswamp The low marshy region between the bluffs and the natural levees of a floodplain.
backwash The gravity-driven flow of water back down the slope of a beach.
bacteria A type of tiny prokaryotic single-celled organism.
bajada An elongate wedge of sediment formed by the overlap of several alluvial fans emerging from adjacent valleys.
Baltica A Paleozoic continent that included crust that is now part of today's Europe.
banded-iron formation (BIF) Iron-rich sedimentary layers consisting of alternating gray beds of iron oxide and red beds of iron-rich chert.
bar (1) A sheet or elongate lens or mound of alluvium; (2) a unit of air-pressure measurement approximately equal to 1 atm.
barchan dune A crescent-shaped dune whose tips point downwind.
barrier island An offshore sand bar that rises above the mean high-water level, forming an island.
barrier reef A coral reef that develops offshore, separated from the coast by a lagoon.
basal sliding The phenomenon in which meltwater accumulates at the base of a glacier, so that the mass of the glacier slides on a layer of water or on a slurry of water and sediment.
basalt A fine-grained, mafic, igneous rock.
base level The lowest elevation a stream channel's floor can reach at a given locality.
basement Older igneous and metamorphic rocks making up the Earth's crust beneath sedimentary cover.
basement uplift Uplift of basement rock by faults that penetrate deep into the continental crust.
base metals Metals that are mined but not considered precious. Examples include copper, lead, zinc, and tin.
basin A fold or depression shaped like a right-side-up bowl.
Basin and Range Province A broad, Cenozoic continental rift that has affected a portion of the western United States in Nevada, Utah, and Arizona; in this province, tilted fault blocks form ranges, and alluvium-filled valleys are basins.
batholith A vast composite, intrusive, igneous rock body up to several hundred km long and 100 km wide, formed by the intrusion of numerous plutons in the same region.
bathymetric map A map illustrating the shape of the ocean floor.

bathymetric profile A cross section showing ocean depth plotted against location.
bathymetry Variation in depth.
bauxite A residual mineral deposit rich in aluminum.
baymouth bar A sandspit that grows across the opening of a bay.
beach A gently sloping fringe of sediment along the shore.
beach drift The gradual migration of sand along a beach.
beach erosion The removal of beach sand caused by wave action and longshore currents.
beach face A steeply concave part of the foreshore zone formed where the swash of the waves actively scours the sand.
beach nourishment When humans add sand to a beach, to replace sand that has been eroded away.
beach profile A profile showing the trace of a beach's surface, as it would appear on a vertical surface.
bearing (structural geology) The compass orientation of a linear feature.
bed A distinct layer of sediment or sedimentary rock, typically reflecting continuous deposition during a relatively short time interval.
bedding Layering or stratification in sedimentary rocks.
bed load Large particles, such as sand, pebbles, or cobbles, that bounce or roll along a streambed.
bedrock Rock still attached to the Earth's crust.
Bergeron process Precipitation involving the growth of ice crystals in a cloud at the expense of water droplets.
berm A horizontal or landward-sloping terrace in the backshore zone of a beach that receives sediment during a storm.
Big Bang A cataclysmic explosion that scientists suggest represents the formation of the Universe; before this event, all matter and all energy were packed into one volumeless point.
biochemical sedimentary rock Sedimentary rock formed from material (such as shells) produced by living organisms.
biodiversity The number of different species that exist at a given time.
biofuel Gas or liquid fuel made from plant material (biomass). Examples of biofuel include alcohol (from fermented sugar), biodiesel from vegetable oil, and wood.
biogenic minerals Substances that meet the definition of a mineral and are produced naturally by organisms (e.g., calcite in shells).
biogeochemical cycle The exchange of chemicals between living and nonliving reservoirs in the Earth System.
biomass The amount of organic material in a specified volume.
bioremediation The injection of oxygen and nutrients into a contaminated aquifer to foster the growth of bacteria that will ingest or break down contaminants.
biosphere The region of the Earth and atmosphere inhabited by life; this region stretches from a few km below the Earth's surface to a few km above.
bioturbation The mixing of sediment by burrowing animals such as clams and worms.
bituminous coal Dull, black intermediate-rank coal formed at temperatures between 100°C and 200°C.
black-lung disease Lung disease contracted by miners from the inhalation of too much coal dust.
black smoker The cloud of suspended minerals formed where hot water spews out of a vent along a mid-ocean ridge; the dissolved sulfide components of the hot water instantly precipitate when the water mixes with seawater and cools.
blind fault A fault that does not intersect the ground surface.
block Large, angular pyroclastic fragments consisting of volcanic rock, broken up during the eruption.
blocking temperature The temperature below which isotopes in a mineral are no longer free to move, so the radiometric clock starts.
blocky lava Lava that is so viscous that it breaks into boulder-like blocks as it moves; typically, such lavas are andesitic or rhyolitic.
blowout A deep, bowl-like depression scoured out of desert terrain by a turbulent vortex of wind.
blue shift The phenomenon in which a source of light moving toward you appears to have a higher frequency.
body fossil A relict of an organism's body, preserved in rock.
body waves Seismic waves that pass through the interior of the Earth.
bog A wetland dominated by moss and shrubs.
bolide A solid extraterrestrial object such as a meteorite, comet, or asteroid that explodes in the atmosphere.
bornhardt An inselberg with a loaf geometry, like that of Uluru (Ayers Rock) in central Australia.
Bowen's reaction series The sequence in which different silicate minerals crystallize during the progressive cooling of a melt.
braided stream A sediment-choked stream consisting of entwined subchannels.
breaker A water wave in which water at the top of the wave curves over the base of the wave.
breakwater An offshore wall, built parallel or at an angle to the beach, that prevents the full force of waves from reaching a harbor.
breccia Coarse sedimentary rock consisting of angular fragments; or rock broken into angular fragments by faulting.
breeder reactor A nuclear reactor that produces its own fuel.
brine Water that is not fresh but is less salty than seawater; brine may be found in estuaries.
brittle deformation The cracking and fracturing of a material subjected to stress.
brittle-ductile transition (brittle-plastic transition) The depth above which materials behave brittlely and below which materials behave ductilely (plastically); this transition typically lies between a depth of 10 and 15 km in continental crustal rock, and 60 m deep in glacial ice.
buoyancy The upward force acting on an object immersed or floating in fluid; the tendency of an object to float when placed in a fluid.
burial metamorphism Metamorphism due only to the consequences of very deep burial.
butte A medium-sized, flat-topped hill in an arid region.
caldera A large circular depression with steep walls and a fairly flat floor, formed after an eruption as the center of the volcano collapses into the drained magma chamber below.
caliche A solid mass created where calcite cements the soil together. (Also called calcrete.)
calorie A unit of energy approximately equal to 4.2 joules; 1 calorie can raise the temperature of 1 gm of water by 1°C.
calving The breaking off of chunks of ice at the edge of a glacier.
Cambrian explosion The remarkable diversification of life, indicated by the fossil record, that occurred at the beginning of the Cambrian Period.
Canadian Shield A broad, low-lying region of exposed Precambrian rock in the Canadian interior.
canyon A trough or valley with steeply sloping walls, cut into the land by a stream.
capacity (of a stream) The total quantity of sediment a stream can carry.
capillary fringe The thin subsurface layer in which water molecules seep up from the water table by capillary action to fill pores.
carbon cycle The movement of carbon among various reservoirs in the Earth System; this carbon can occur in the form of CO_2, $CaCO_3$, as pure carbon, or in a great variety of organic molecules.
carbonate rocks Rocks containing calcite and/or dolomite.
carbon-14 A radioactive isotope of the element carbon; the ratio of C14 to C12 can provide an isotopic date of organic carbon.
carbon-14 dating A radiometric dating process that can tell us the age of organic material containing carbon originally extracted from the atmosphere.

carbon sequestration The process of extracting carbon dioxide from sources (e.g., power plants) and sending it back underground to keep it out of the atmosphere and diminish the greenhouse effect.
cast Sediment that preserves the shape of a shell it once filled before the shell dissolved or mechanically weathered away.
catabatic winds Strong winds that form at the margin of a glacier where the warmer air above ice-free land rises and the cold, denser air from above the glaciers rushes in to take its place.
catastrophic change Change that takes place either instantaneously or rapidly in geologic time.
catchment *See* Drainage network.
cation A positively charged ion.
Celsius scale A metric-system measure of temperature in which the difference between the freezing point (0°C) and boiling point (100°C) of water is divided into 100 units; equivalent to centigrade scale.
cement Mineral material that precipitates from water and fills the spaces between grains, holding the grains together.
cementation The phase of lithification in which cement, consisting of minerals that precipitate from groundwater, partially or completely fills the spaces between clasts and attaches each grain to its neighbor.
Cenozoic Era The most recent era of the Phanerozoic Eon, lasting from 66 Ma up until the present.
chain reaction A self-perpetuating process in a nuclear reaction, whereby neutrons released during the fission trigger more fission.
chalk Very fine-grained limestone consisting of weakly cemented plankton shells.
change of state The process in which a material changes from one phase (liquid, gas, or solid) to another.
channel A trough dug into the ground surface by flowing water.
channeled scablands A barren, soil-free landscape in eastern Washington, scoured clean by a flood unleashed when a large glacial lake drained.
chatter marks Wedge-shaped indentations left on rock surfaces by glacial plucking.
chemical A material consisting of a distinct element or compound.
chemical bond The invisible link that holds together atoms in a molecule and/or in a crystal.
chemical formula The "recipe" that specifies the elements and their proportions in a compound.
chemical fossil Distinctive molecules or molecular fragments, formed from the remains of living organisms, that can be preserved in rock.
chemical reaction Interactions among atoms and/or molecules involving breaking or forming chemical bonds.
chemical sedimentary rocks Sedimentary rocks made up of minerals that precipitate directly from water solution.
chemical weathering The process in which chemical reactions alter or destroy minerals when rock comes in contact with water solutions and/or air.
chert A sedimentary rock composed of very fine-grained silica (cryptocrystalline quartz).
Chicxulub crater A circular excavation buried beneath younger sediment on the Yucatán Peninsula; geologists suggest that a meteorite landed there 65 Ma.
chimney (1) A conduit in a magma chamber in the shape of a long vertical pipe through which magma rises and erupts at the surface; (2) an isolated column of strata in an arid region.
chron The time interval between successive magnetic reversals.
cinder cone A subaerial volcano consisting of a cone-shaped pile of tephra whose slope approaches the angle of repose for tephra.
cinders Fragments of glassy rock ejected from a volcano.
cirque A bowl-shaped depression carved by a glacier on the side of a mountain.
cirrus cloud A wispy cloud that tapers into delicate, feather-like curls.
clast A fragment or grain produced by the physical or chemical weathering of a pre-existing rock.
clastic (detrital) sedimentary rock Sedimentary rock consisting of cemented-together detritus derived from the weathering of preexisting rock.
cleavage (1) The tendency of a mineral to break along preferred planes; (2) a type of foliation in low-grade metamorphic rock.
cleavage planes A series of surfaces on a crystal that form parallel to the weakest bonds holding the atoms of the crystal together.
cliff (scarp) retreat The change in the position of a cliff face caused by erosion.
climate The average weather conditions, along with the range of conditions, of a region over a year.
climate change The modification of average atmospheric temperature, oceanic temperature, and other weather characteristics in a given region over time.
climate-change model A computer-generated model designed to provide insight into how climate changed in the past and may change in the future, and what the consequences of climate change may be.
closure temperature The temperature at which parent and daughter isotopes can no longer escape from a mineral, so the ratio of parents to daughters can be used for isotopic dating.
cloud A mist of tiny water droplets in the sky.
coal A black, organic rock consisting of greater than 50% carbon; it forms from the buried and altered remains of plant material.
coalbed methane Natural gas created during the formation of coal, that gets trapped within the coal.
coal gasification The process of producing relatively clean-burning gases from solid coal.
coal rank A measurement of the carbon content of coal; higher-rank coal forms at higher temperatures.
coal reserve The quantities of discovered, but not yet mined, coal in sedimentary rock of the continents.
coal swamp A swamp whose oxygen-poor water allows thick piles of woody debris to accumulate; this debris transforms into coal upon deep burial.
coast The belt of land bordering the sea.
coastal plain Low-relief regions of land adjacent to the coast.
coastal wetland A flat-lying coastal area that floods during high tide and drains during low tide, and hosts salt-resistant plants.
cold front The boundary at which a cold air mass pushes underneath a warm air mass.
collision The process of two buoyant pieces of lithosphere converging and squashing together.
collisional orogen A mountain belt formed when two relatively buoyant blocks of crust converge.
color The characteristic of a material due to the spectrum of light emitted or reflected by the material, as perceived by eyes or instruments.
columnar jointing A type of fracturing that yields roughly hexagonal columns of basalt; columnar joints form when a dike, sill, or lava flow cools.
comet A ball of ice and dust, probably remaining from the formation of the Solar System, that orbits the Sun.
compaction The phase of lithification in which the pressure of the overburden on the buried rock squeezes out water and air that was trapped between clasts, and the clasts press tightly together.
competence (of a stream) The maximum particle size a stream can carry.
composite volcano *See* Stratovolcano.
compositional banding A type of metamorphic foliation, found in gneiss, defined by alternating bands of light and dark minerals.
compound A material composed of two or more elements that cannot be separated mechanically; the smallest piece is a molecule.
compressibility The degree to which a material's volume changes in response to squashing.

compression A push or squeezing felt by a body.
compressional waves Waves in which particles of material move back and forth parallel to the direction in which the wave itself moves.
concentration The proportion of one substance (the solute) dissolved within another (the solvent).
conchoidal fractures Smoothly curving, clamshell-shaped surfaces along which materials with no cleavage planes tend to break.
condensation The process of gas molecules linking together to form a liquid.
condensation nuclei Preexisting solid or liquid particles, such as aerosols, onto which water condenses during cloud formation.
conduction A process of heat transfer involving progressive migration of thermal energy from cooler to warmer regions in a material, without the physical flow of the material itself.
cone of depression The downward-pointing, cone-shaped surface of the water table in a location where the water table is experiencing drawdown because of pumping at a well.
confined aquifer An aquifer that is separated from the Earth's surface by an overlying aquitard.
conglomerate Very coarse-grained sedimentary rock consisting of rounded clasts.
consuming boundary *See* Convergent plate boundary.
contact The boundary surface between two rock bodies (as between two stratigraphic formations, between an igneous intrusion and adjacent rock, between two igneous rock bodies, or between rocks juxtaposed by a fault).
contact metamorphism *See* Thermal metamorphism.
contaminant plume A cloud of contaminated groundwater that moves away from the source of the contamination.
continental crust The crust beneath the continents.
continental divide A highland separating drainage that flows into one ocean from drainage that flows into another.
continental-drift The idea that continents have moved and are still moving slowly across the Earth's surface.
continental glacier A vast sheet of ice that spreads over thousands of square km of continental crust.
continental-interior desert An inland desert that develops because by the time air masses reach the continental interior, they have lost all of their moisture.
continental lithosphere Lithosphere topped by continental crust; this lithosphere reaches a thickness of 150 km.
continental margin A continent's coastline.
continental rifting The process by which a continent stretches and splits along a belt; if it is successful, rifting separates a larger continent into two smaller continents separated by a divergent boundary.
continental rise The sloping sea floor that extends from the lower part of the continental slope to the abyssal plain.
continental shelf A broad, shallowly submerged fringe of a continent; ocean-water depth over the continental shelf is generally less than 200 meters; the widest continental shelves occur over passive margins.
continental slope The slope at the edge of a continental shelf, leading down to the deep sea floor.
continental volcanic arc A long, curving chain of subaerial volcanoes on the margin of a continent adjacent to a convergent plate boundary.
contour lines Lines on a map along which a parameter has a constant value; for example, all points along a contour line on a topographic map are at the same elevation.
control rod Rods that absorb neutrons in a nuclear reactor and thus decrease the number of collisions between neutrons and radioactive atoms.
convection Heat transfer that results when warmer, less dense material rises while cooler, denser material sinks.
convection cell A distinct flow configuration for a volume of material that is moving during convective heat transport; simplistically, the material rises when warm and sinks when cool, and thus follows a loop-like path.
conventional reserve A hydrocarbon reserve that can be extracted simply by pumping from a reservoir rock.
convergence zone A place where two surface air flows meet so that air has to rise.
convergent margin *See* Convergent plate boundary.
convergent plate boundary A boundary at which two plates move toward each other so that one plate sinks (subducts) beneath the other; only oceanic lithosphere can subduct.
coral reef A mound of coral and coral debris forming a region of shallow water.
core The dense, iron-rich center of the Earth.
core-mantle boundary An interface 2,900 km below the Earth's surface separating the mantle and core.
Coriolis effect The deflection of objects, winds, and currents on the surface of the Earth owing to the planet's rotation.
cornice A huge, overhanging drift of snow built up by strong winds at the crest of a mountain ridge.
correlation The process of defining the age relations between the strata at one locality and the strata at another.
cosmic rays Nuclei of hydrogen and other elements that bombard the Earth from deep space.
cosmology The study of the overall structure of the Universe.
country rock (wall rock) The preexisting rock into which magma intrudes.
covalent bonding The attachment of one atom to another that develops when the atoms share electrons; one type of chemical bond.
crater (1) A circular depression at the top of a volcanic mound; (2) a depression formed by the impact of a meteorite.
craton A long-lived block of durable continental crust commonly found in the stable interior of a continent.
cratonic platform A province in the interior of a continent in which Phanerozoic strata bury most of the underlying Precambrian rock.
creep The gradual downslope movement of regolith.
crevasse A large crack that develops by brittle deformation in the top 60 m of a glacier.
critical mass A sufficiently dense and large mass of radioactive atoms in which a chain reaction happens so quickly that the mass explodes.
cross bed Internal laminations in a bed, inclined at an angle to the main bedding; cross beds are a relict of the slip face of dunes or ripples.
cross-cutting relations A determination of which geologic feature cuts across another, or is cut by the other; these relations form a basis for determining relative ages of the features.
cross section A diagram depicting the geometry of materials underground as they would appear on an imaginary vertical slice through the Earth.
crude oil Oil extracted directly from the ground.
crust The rock that makes up the outermost layer of the Earth.
crustal root Low-density crustal rock that protrudes downward beneath a mountain range.
crustal thickening The process by which the continental crust increases in thickness, becoming up to 70 km thick (vs. normal thickness of about 35–40 km); it can occur during continental collision.
cryosphere The ice components of the Earth System, including glaciers, snow, sea ice, and permafrost.
crystal A single, continuous piece of a mineral bounded by flat surfaces that formed naturally as the mineral grew.
crystal face The flat surfaces of a crystal, formed during the crystal's growth.
crystal form The geometric shape of a crystal, defined by the arrangement of crystal faces.
crystal habit The general shape of a crystal or cluster of crystals that grew unimpeded.

crystal lattice The orderly framework within which the atoms or ions of a mineral are fixed.
crystal structure The arrangement of atoms in a crystal.
crystalline rock A rock that consists of minerals that grew when a melt solidified, and eventually interlock like pieces of a jigsaw puzzle.
Crystalline solid Containing a crystal lattice.
cuesta An asymmetric ridge formed by tilted layers of rock, with a steep cliff on one side cutting across the layers and a gentle slope on the other side; the gentle slope is parallel to the layering.
cumulonimbus cloud A rain-producing, puffy cloud.
cumulus cloud A puffy, cotton-ball-shaped cloud.
current (1) A well-defined stream of ocean water; (2) the moving flow of water in a stream.
cut bank The outside bank of the channel wall of a meander, which is continually undergoing erosion.
cutoff A straight reach in a stream that develops when erosion eats through a meander neck.
cyanobacteria Blue-green algae; a type of archaea.
cycle A series of interrelated events or steps that occur in succession and can be repeated, perhaps indefinitely.
cyclone (1) The counterclockwise flow of air around a low-pressure mass; (2) the equivalent of a hurricane in the Indian Ocean.
cyclonic flow A circulation of air around a low-pressure region in the atmosphere; it rotates counterclockwise in the northern hemisphere.
cyclothem A repeated interval within a sedimentary sequence that contains a specific succession of sedimentary beds.
Darcy's law A mathematical equation stating that a volume of water, passing through a specified area of material at a given time, depends on the material's permeability and hydraulic gradient.
daughter isotope The decay product of radioactive decay.
day The time it takes for the Earth to spin once on its axis.
debris avalanche A chaotic cloud of debris (rock fragments, sediment, dust), suspended in air, that moves down a slope.
debris flow A downslope movement of mud mixed with larger rock fragments.
debris slide A sudden downslope movement of material consisting only of regolith.
decompression melting The kind of melting that occurs when hot mantle rock rises to shallower depths in the Earth so that pressure decreases while the temperature remains unchanged.
deep current An ocean current at a depth greater than 100 m.
deep-focus earthquake An earthquake that occurs at a depth between 300 and 670 km; below 670 km, earthquakes do not happen.
deflation The process of lowering the land surface by wind abrasion.
deformation A change in the shape, position, or orientation of a material, by bending, breaking, or flowing.
dehydration Loss of water.
delamination (plate tectonics) The process by which dense lithospheric mantle separates from the base of a plate and sinks into the mantle.
delta A wedge of sediment formed at a river mouth when the running water of the stream enters standing water, the current slows, the stream loses competence, and sediment settles out.
delta plain The low, swampy land on the surface of a delta.
delta-plain flood A flood in which water submerges a delta plain.
dendritic network A drainage network whose interconnecting streams resemble the pattern of branches connecting to a deciduous tree.
dendrochronologist A scientist who analyzes tree rings to determine the geologic age of features.
density Mass per unit volume.
denudation The removal of rock and regolith from the Earth's surface.
deposition The process by which sediment settles out of a transporting medium.
depositional environment A setting in which sediments accumulate; its character (fluvial, deltaic, reef, glacial, etc.) reflects local conditions.
depositional landform A landform resulting from the deposition of sediment where the medium carrying the sediment evaporates, slows down, or melts.
desert A region so arid that it contains no permanent streams, except for those that bring water in from elsewhere, and has very sparse vegetation cover.
desertification The process of transforming nondesert areas into desert.
desert pavement A mosaic-like stone surface forming the ground in a desert.
desert varnish A dark, rusty-brown coating of iron oxide and magnesium oxide that accumulates on the surface of the rock.
detachment fault A nearly horizontal fault at the base of a fault system.
detritus The chunks and smaller grains of rock broken off outcrops by physical weathering.
dewpoint temperature The temperature at which air becomes saturated so that dew can form.
diagenesis All of the physical, chemical, and biological processes that transform sediment into sedimentary rock and that alter the rock after the rock has formed.
differential stress A condition causing a material to experience a push or pull in one direction of a greater magnitude than the push or pull in another direction; in some cases, differential stress can result in shearing.
differential weathering What happens when different rocks in an outcrop undergo weathering at different rates.
differentiation (of a planet) A process early in a planet's history during which dense iron alloy melted and sank downward to form the core, leaving less-dense mantle behind.
diffraction The splitting of light into many tiny beams that interfere with one another.
digital elevation map (DEM) A computer-produced portrayal of elevation differences commonly using shading to simulate shadows; the data used to produce the map assigns elevations to each point on the map.
dike A tabular (wall-shaped) intrusion of rock that cuts across the layering of country rock.
dimension stone An intact block of granite or marble to be used for architectural purposes.
dip The angle of a plane's slope as measured in a vertical plane perpendicular to the strike.
dipole A magnetic field with a north and south pole, like that of a bar magnet.
dipole field (for Earth) The part of the Earth's magnetic field, caused by the flow of liquid iron alloy in the outer core, that can be represented by an imaginary bar magnet with a north and south pole.
dip-slip fault A fault in which sliding occurs up or down the slope (dip) of the fault.
dip slope A hill slope underlain by bedding parallel to the slope.
directional drilling The process of controlling the trajectory of a drill bit to make sure that the drill hole goes exactly where desired.
disappearing stream A stream that intersects a crack or sinkhole leading to an underground cavern, so that the water disappears into the subsurface and becomes an underground stream.
discharge The volume of water in a conduit or channel passing a point in 1 second.
discharge area A location where groundwater flows back up to the surface and may emerge at springs.
disconformity An unconformity parallel to the two sedimentary sequences it separates.
displacement (offset) The amount of movement or slip across a fault plane.

disseminated deposit A hydrothermal ore deposit in which ore minerals are dispersed throughout a body of rock.
dissolution A process during which materials dissolve in water.
dissolved load Ions dissolved in a stream's water.
distillation column A vertical pipe in which crude oil is separated into several components.
distributaries The fan of small streams formed where a river spreads out over its delta.
divergence zone A place where sinking air separates into two flows that move in opposite directions.
divergent plate boundary A boundary at which two lithosphere plates move apart from each other; they are marked by mid-ocean ridges.
diversification The development of many different species.
DNA (deoxyribonucleic acid) The complex molecule, shaped like a double helix, containing the code that guides the growth and development of an organism.
doldrums A belt with very slow winds along the equator.
dolostone A name sometimes used for a carbonate rock containing a high proportion of dolomite.
domain The broadest divisions of life (Archaea, Bacteria, Eukaryota).
dome Folded or arched layers with the shape of an overturned bowl.
Doppler effect The phenomenon in which the frequency of wave energy appears to change when a moving source of wave energy passes an observer.
dormant volcano A volcano that has not erupted for hundreds to thousands of years but does have the potential to erupt again in the future.
downcutting The process in which water flowing through a channel cuts into the substrate and deepens the channel relative to its surroundings.
downdraft Downward-moving air.
downgoing plate (slab) A lithosphere plate that has been subducted at a convergent margin.
downslope force The component of the force of gravity acting in the downslope direction.
downslope movement The tumbling or sliding of rock and sediment from higher elevations to lower ones.
downwelling zone A place where near-surface water sinks.
drag fold A fold that develops in layers of rock adjacent to a fault during or just before slip.
drainage divide A highland or ridge that separates one watershed from another.
drainage network (basin) An array of interconnecting streams that together drain an area.
drainage reversal When the overall direction of flow in a drainage network becomes the opposite of what it once had been.
drawdown The phenomenon in which the water table around a well drops because the users are pumping water out of the well faster than it flows in from the surrounding aquifer.
drilling mud A slurry of water mixed with clay that oil drillers use to cool a drill bit and flush rock cuttings up and out of the hole.
dripstone Limestone (travertine in a cave) formed by the precipitation of calcium carbonate out of groundwater.
dropstone A rock that drops to the sea floor once the iceberg that was carrying the rock melts.
drumlin A streamlined, elongate hill formed when a glacier overrides glacial till.
dry-bottom (polar) glacier A glacier so cold that its base remains frozen to the substrate.
dry wash The channel of an ephemeral stream when empty of water.
dry well (1) A well that does not supply water because the well has been drilled into an aquitard or into rock that lies above the water table; (2) a well that does not yield oil, even though it has been drilled into an anticipated reservoir.
ductile (plastic) deformation The bending and flowing of a material (without cracking and breaking) subjected to stress.
dune A pile of sand generally formed by deposition from the wind.
dust storm An event in which strong winds hit unvegetated land, strip off the topsoil, and send it skyward to form rolling dark clouds that block out the Sun.
dynamic metamorphism Metamorphism that occurs as a consequence of shearing alone, with no change in temperature or pressure.
dynamo A power plant generator in which water or wind power spins an electrical conductor around a permanent magnet.
dynamothermal metamorphism Metamorphism that involves heat, pressure, and shearing.
earth materials Any of the substances (minerals, rocks, metals, sediments, soils) composing the solid Earth.
earthquake A vibration caused by the sudden breaking or frictional sliding of rock in the Earth.
earthquake belt A relatively narrow, distinct belt of earthquakes that defines the position of a plate boundary.
earthquake engineering The design of buildings that can withstand shaking.
earthquake early warning system A communications network that provides an alert within microseconds after the first earthquake waves arrive at a seismograph near the epicenter, but before damaging vibrations reach population centers.
earthquake zoning The determination of where land is relatively stable and where it might collapse because of seismicity.
EarthScope A project, funded by the U.S. National Science Foundation, to study the Earth's interior, primarily using a large array of seismometers.
Earth System The global interconnecting web of physical and biological phenomena involving the solid Earth, the hydrosphere, and the atmosphere.
ebb tide The falling tide.
eccentricity cycle The cycle of the gradual change of the Earth's orbit from a more circular to a more elliptical shape; the cycle takes around 100,000 years.
ecliptic The plane defined by a planet's orbit.
ecosystem An environment and its inhabitants.
eddy An isolated, ring-shaped current of water.
Ediacaran fauna Multicellular invertebrate organisms that lived perhaps as early as 620 Ma and certainly by 565 Ma. They were named for a region in southern Australia.
effusive eruption An eruption that yields mostly lava, not ash.
Ekman spiral The change in flow direction of water with depth, caused by the Coriolis effect.
Ekman transport The overall movement of a mass of water, resulting from the Eckman spiral, in a direction 90° to the wind direction.
elastic-rebound theory The concept that earthquakes happen because stress builds up, causing rock adjacent to a fault to bend elastically until breaking and slip on a fault occurs; the slip relaxes the elastic bending and decreases stress.
elastic strain A change in shape of a material; the change disappears instantly when stress is removed.
electromagnet An electrical device that produces a magnetic field.
electron A negatively charged subatomic particle that orbits the nucleus of an atom; electrons are about 0.0005 the size of a proton.
electron microprobe A laboratory instrument that can focus a beam of electrons on a small part of a mineral grain in order to create a signal that defines its chemical composition.
element A material consisting entirely of one kind of atom; elements cannot be subdivided or changed by chemical reactions.
El Niño The flow of warm water eastward from the Pacific Ocean that reverses the upwelling of cold water along the western coast of South America and causes significant global changes in weather patterns.
embayment A low area of coastal land.

emergent coast A coast where the land is rising relative to sea level or sea level is falling relative to the land.
end moraine (terminal moraine) A low, sinuous ridge of till that develops when the terminus (toe) of a glacier stalls in one position for a while.
energy The capacity to do work.
energy density The amount of energy that can be produced per unit mass of a fuel.
energy grid The configuration of power-generating sources, power lines, and transformers that distributes electrical power nationally.
energy resource Something that can be used to produce work; in a geologic context, a material (such as oil, coal, wind, flowing water) that can be used to produce energy.
eon The largest subdivision of geologic time.
epeirogenic movement The gradual uplift or subsidence of a broad region of the Earth's surface.
epeirogeny An event of epeirogenic movement; the term is usually used in reference to the formation of broad mid-continent domes and basins.
ephemeral (intermittent) stream A stream whose bed lies above the water table, so that the stream flows only when the rate at which water enters the stream from rainfall or meltwater exceeds the rate at which water infiltrates the ground below.
epicenter The point on the surface of the Earth directly above the focus of an earthquake.
epicontinental sea A shallow sea overlying a continent.
equipotential surface An imaginary surface on which all points have the same potential energy.
epoch An interval of geologic time representing the largest subdivision of a period.
equant A term for a grain that has the same dimensions in all directions.
equatorial low The area of low pressure that develops over the equator because of the intertropical convergence zone.
equilibrium line (of a glacier) The boundary between the zone of accumulation and the zone of ablation.
equinox One of two days out of the year (September 22 and March 21) in which the Sun is directly overhead at noon at the equator.
era An interval of geologic time representing the largest subdivision of the Phanerozoic Eon.
erg Sand seas formed by the accumulation of dunes in a desert.
erosion The grinding away and removal of Earth's surface materials by moving water, air, or ice.
erosional coast A coastline where sediment is not accumulating and wave action grinds away at the shore.
erosional landform A landform that results from the breakdown and removal of rock or sediment.
erratic A boulder or cobble that was picked up by a glacier and deposited hundreds of kilometers away from the outcrop from which it detached.
eruptive style The character of a particular volcanic eruption; geologists name styles based on typical examples (e.g., Hawaiian, Strombolian).
esker A ridge of sorted sand and gravel that snakes across a ground moraine; the sediment of an esker was deposited in subglacial meltwater tunnels.
estuary An inlet in which seawater and river water mix; created when a coastal valley is flooded because of either rising sea level or land subsidence.
Eubacteria The kingdom of "true bacteria."
euhedral crystal A crystal whose faces are well formed and whose shape reflects crystal form.
eukaryote An organism whose cells contain a nucleus; all plants and animals consist of eukaryotic cells.
eukaryotic cell A cell with a complex internal structure, capable of building multicellular organisms.
eustatic sea-level change A global rising or falling of the ocean surface.
evaporate To change from liquid to vapor.
evaporite Thick salt deposits that form as a consequence of precipitation from saline water.
evapotranspiration The sum of evaporation from bodies of water and the ground surface and transpiration from plants and animals.
evolution (life evolution) The change in populations of life that take place over time, leading to the modification of a given species, and/or to the appearance of new species or extinction of existing species.
exfoliation The process by which an outcrop of rock splits apart into onion-like sheets along joints that lie parallel to the ground surface.
exhumation The process (involving uplift and erosion) that returns deeply buried rocks to the surface.
exotic terrane A block of land that collided with a continent along a convergent margin and attached to the continent; the term exotic implies that the land was not originally part of the continent to which it is now attached.
expanding Universe theory The theory that the whole Universe must be expanding because galaxies in every direction seem to be moving away from us.
explosive eruptions Violent volcanic eruptions that produce clouds and avalanches of pyroclastic debris.
external energy Energy from the Sun that drives components of the Earth System; it contributes to the circulation and flow of the atmosphere and hydrosphere, and serves a major role in driving erosion.
external process A geomorphologic process—such as downslope movement, erosion, or deposition—that is the consequence of gravity or of the interaction between the solid Earth and its fluid envelope (air and water). Energy for these processes comes from gravity and sunlight.
extinction The death of the last members of a species so that there are no parents to pass on their genetic traits to offspring.
extinct volcano A volcano that was active in the past but has now shut off entirely and will not erupt in the future.
extraordinary fossil A rare fossilized relict, or trace, of the soft part of an organism.
extratropical cyclone (wave cyclone, mid-latitude cyclone) A large, rotating storm system, in mid-latitudes, associated with a regional-scale low-pressure zone.
extrusive igneous rock Rock that forms by the freezing of lava above ground, after it flows or explodes out (extrudes) onto the surface and comes into contact with the atmosphere or ocean.
eye The relative calm in the center of a hurricane.
eye wall A rotating vertical cylinder of clouds surrounding the eye of a hurricane.
facet (of a gem) The ground and polished surface of a gem, produced by a gem cutter using a grinding lap.
facies (1) Sedimentary: a group of rocks and primary structures indicative of a given depositional environment; (2) metamorphic: a set of metamorphic mineral assemblages formed under a given range of pressures and temperatures.
Fahrenheit scale An English-system measure of temperature in which the difference between the freezing point (32°F) and the boiling point (212°F) is divided into 180 units.
failure surface A weak surface that forms the base of a landslide.
faint young Sun paradox The apparent contradiction implied by the fact that much of the Earth's surface temperature has remained above the melting point of water during the past 4 Ga, even though calculations indicate that the Sun produced much less energy when young.
fault A fracture on which one body of rock slides past another.
fault-block mountains An outdated term for a narrow, elongate range of mountains that develops in a continental-rift setting as normal faulting drops down blocks of crust or tilts blocks.

fault breccia Fragmented rock in which angular fragments were formed by brittle fault movement; fault breccia occurs along a fault.
fault creep Gradual movement along a fault that occurs in the absence of an earthquake.
fault gouge Pulverized rock consisting of fine powder that lies along fault surfaces; gouge forms by crushing and grinding.
faulting Slip events along a fault.
fault scarp A small step on the ground surface where one side of a fault has moved vertically with respect to the other.
fault system A grouping of numerous related faults.
fault trace (fault line) The intersection between a fault and the ground surface.
feedback mechanism A condition that arises when the consequence of a phenomenon influences the phenomenon itself.
felsic An adjective used in reference to igneous rocks that are rich in elements forming feldspar and quartz.
Ferrel cells The name given to the middle-latitude convection cells in the atmosphere.
fetch The distance across a body of water along which a wind blows to build waves.
field force A push or pull that applies across a distance (i.e., without contact between objects); examples are gravity and magnetism.
fine-grained A textural term for rock consisting of many fine grains or clasts.
firn Compacted granular ice (derived from snow) that forms where snow is deeply buried; if buried more deeply, firn turns into glacial ice.
fission A nuclear reaction during which the nucleus of a large atom splits to form two nuclei of smaller atoms; the process also releases neutrons and energy.
fission track A line of damage formed in the crystal lattice of a mineral by the impact of an atomic particle ejected during the decay of a radioactive isotope.
fissure A conduit in a magma chamber in the shape of a long crack through which magma rises and erupts at the surface.
fjord A deep, glacially carved, U-shaped valley flooded by rising sea level.
flank eruption An eruption that occurs when a secondary chimney, or fissure, breaks through the flank of a volcano.
flash flood A flood that occurs during unusually intense rainfall or as the result of a dam collapse, during which the floodwaters rise very fast.
flexing The process of folding in which a succession of rock layers bends and slip occurs between the layers.
flocculation The clumping together of clay suspended in river water into bunches that are large enough to settle out.
flood An event during which the volume of water in a stream becomes so great that it covers areas outside the stream's normal channel.
flood basalt Vast sheets of basalt that spread from a volcanic vent over an extensive surface of land; they may form where a rift develops above a continental hot spot, and where lava is particularly hot and has low viscosity.
flood-hazard map A representation of a portion of the Earth's surface that is designed to show how the danger of flooding varies with location.
floodplain The flat land on either side of a stream that becomes covered with water during a flood.
floodplain flood A flood during which a floodplain is submerged.
flood stage The stage when water reaches the top of a stream channel.
flood tide The rising tide.
floodway A mapped region likely to be flooded, in which people avoid constructing buildings.
flow fold A fold that forms when the rock is so soft that it behaves like weak plastic.
flowstone A sheet of limestone that forms along the wall of a cave when groundwater flows along the surface of the wall.
fluvial deposit Sediment deposited in a stream channel, along a stream bank, or on a floodplain.
flux Flow.
flux melting The transformation of hot solid to liquid that occurs when a volatile material injects into the solid.
focus The location where a fault slips during an earthquake (hypocenter).
fog A cloud that forms at ground level.
fold A bend or wrinkle of rock layers or foliation; folds form as a consequence of ductile deformation.
fold axis An imaginary line that, when moved parallel to itself, can trace out the shape of a folded surface.
fold-thrust belt An assemblage of folds and related thrust faults that develop above a detachment fault.
foliation Layering formed as a consequence of the alignment of mineral grains, or of compositional banding in a metamorphic rock.
foraminifera Microscopic plankton with calcitic shells, components of some limestones.
foreland sedimentary basin A basin located under the plains adjacent to a mountain front, which develops as the weight of the mountains pushes the crust down, creating a depression that traps sediment.
foreshocks The series of smaller earthquakes that precede a major earthquake.
foreshore zone The zone of beach regularly covered and uncovered by rising and falling tides.
formation *See* Stratigraphic formation.
fossil The remnant, or trace, of an ancient living organism that has been preserved in rock or sediment.
fossil assemblage A group of fossil species found in a specific sequence of sedimentary rock.
fossil correlation A determination of the stratigraphic relation between two sedimentary rock units, reached by studying fossils.
fossil fuel An energy resource such as oil or coal that comes from organisms that lived long ago and thus stores solar energy that reached the Earth then.
fossiliferous limestone Limestone consisting of abundant fossil shells and shell fragments.
fossilization The process of forming a fossil.
fossil succession The sequence of assemblages of fossil species preserved in the stratigraphic record.
fractional crystallization The process by which a magma becomes progressively more silicic as it cools, because early-formed crystals settle out.
fracture zone A narrow band of vertical fractures in the ocean floor; fracture zones lie roughly at right angles to a mid-ocean ridge, and the actively slipping part of a fracture zone is a transform fault.
fragmental igneous rock A rock consisting of igneous chunks and/or shards that are packed together, welded together, or cemented together after having solidified.
frequency The number of waves that pass a point in a given time interval.
fresh rock Rock whose mineral grains have their original composition and shape.
friction Resistance to sliding on a surface.
fringing reef A coral reef that forms directly along the coast.
front The boundary between two air masses.
frost wedging The process in which water trapped in a joint freezes, forces the joint open, and may cause the joint to grow.
fuel A transportable substance that can serve a supply of energy.
fuel rod A metal tube that holds the nuclear fuel in a nuclear reactor.
Fujita scale A scale that distinguishes among tornadoes on the basis of wind speed, path dimensions, and possible damage.
Ga Billions of years ago (abbreviation).
gabbro A coarse-grained, intrusive, mafic igneous rock.
Gaia The term used for the Earth System, with the implication that it resembles a complex living entity.

galaxy An immense system of hundreds of billions of stars.
gem A finished (cut and polished) gemstone ready to be set in jewelry.
gemstone A mineral that has special value because it is rare and people consider it beautiful.
gene An individual component of the DNA code that guides the growth and development of an organism.
general circulation model A numerical calculation that simulates the flow of the atmosphere and resulting phenomena, due to changes in atmospheric temperature and other parameters.
genetics The study of genes and how they transmit information.
geocentric Universe concept An ancient Greek idea suggesting that the Earth sat motionless in the center of the Universe while stars and other planets and the Sun orbited around it.
geochronology The science of dating geologic events in years.
geode A cavity in which euhedral crystals precipitate out of water solutions passing through a rock.
geographical pole The locations (north and south) where the Earth's rotational axis intersects the planet's surface.
geoid The imaginary shape a global ocean surface would take if impacted by only the Earth's gravitational pull and rotation.
geologic column A composite stratigraphic chart that represents the entirety of the Earth's history.
geologic cross section A representation of geologic features underground as they would appear crossing a vertical slice through the Earth down to a specified depth.
geologic history The sequence of geologic events that has taken place in a region.
geologic map A map showing the distribution of rock units and structures across a region.
geologic structures Features formed by the deformation of rock.
geologic time The span of time since the formation of the Earth.
geologic time scale A scale that describes the intervals of geologic time.
geology The study of the Earth, including our planet's composition, behavior, and history.
geophysics The study of the Earth using quantitative analysis of earthquake waves, magnetic fields, and gravity, and/or the production of mathematical models of the behavior of the Earth's interior.
geotherm The change in temperature with depth in the Earth.
geothermal energy Heat and electricity produced by using the internal heat of the Earth.
geothermal gradient The rate of change in temperature with depth.
geothermal region A region of current or recent volcanism in which magma or very hot rock heats up groundwater, which may discharge at the surface in the form of hot springs and/or geysers.
geyser A fountain of steam and hot water that erupts periodically from a vent in the ground in a geothermal region.
giant planets The four outer, or Jovian, planets of our Solar System, which are significantly larger than the rest of the planets and consist largely of gas and/or ice.
glacial abrasion The process by which clasts embedded in the base of a glacier grind away at the substrate as the glacier flows.
glacial advance The forward movement of a glacier's toe when the supply of snow exceeds the rate of ablation.
glacial drift Sediment deposited in glacial environments.
glacial incorporation The process by which flowing ice surrounds and incorporates debris.
glacial marine Sediment consisting of ice-rafted clasts mixed with marine sediment.
glacial outwash Coarse sediment deposited on a glacial outwash plain by meltwater streams.
glacially polished surface A polished rock surface created by the glacial abrasion of the underlying substrate.
glacial plucking (glacial quarrying) The process by which a glacier breaks off and carries away fragments of bedrock.
glacial rebound The process by which the surface of a continent rises back up after an overlying continental ice sheet melts away and the weight of the ice is removed.
glacial retreat The movement of a glacier's toe back toward the glacier's origin; glacial retreat occurs if the rate of ablation exceeds the rate of supply.
glacial striation Grooves or scratches cut into bedrock when clasts embedded in the moving glacier act like the teeth of a giant rasp.
glacial subsidence The sinking of the surface of a continent caused by the weight of an overlying glacial ice sheet.
glacial till Sediment transported by flowing ice and deposited beneath a glacier or at its toe.
glaciation (glacial period) A portion of an ice age during which huge glaciers grew and covered substantial areas of the continents.
glacier A river or sheet of ice that slowly flows across the land surface and lasts all year long.
glass A solid in which atoms are not arranged in an orderly pattern.
glassy igneous rock Igneous rock consisting entirely of glass, or of tiny crystals surrounded by a glass matrix.
glide horizon The surface along which a slump slips.
global change The transformations or modifications of both physical and biological components of the Earth System through time.
global circulation The movement of volumes of air in paths that ultimately take it around the planet.
global climate change Transformations or modifications in Earth's climate over time.
global cooling A fall in the average atmospheric temperature.
global positioning system (GPS) A satellite system people can use to measure rates of movement of the Earth's crust relative to one another, or simply to locate their position on the Earth's surface.
global warming A rise in the average atmospheric temperature.
gneiss A compositionally banded metamorphic rock typically composed of alternating dark- and light-colored layers.
Gondwana A supercontinent that consisted of today's South America, Africa, Antarctica, India, and Australia. (Also called Gondwanaland.)
graben A down-dropped crustal block bounded on either side by a normal fault dipping toward the basin.
grade (of an ore) The concentration of a useful metal in an ore—the higher the concentration, the higher the grade.
graded bed A layer of sediment, deposited by a turbidity current, in which grain size varies from coarse at the bottom to fine at the top.
graded stream A stream that has attained an equilibrium longitudinal profile in which the sediment input into an area equals sediment removal.
gradualism The theory that evolution happens at a constant, slow rate.
grain A fragment of a mineral crystal or of a rock.
grain rotation The process by which rigid, inequant mineral grains distributed through a soft matrix may rotate into parallelism as the rock changes shape owing to differential stress.
granite A coarse-grained, intrusive, silicic igneous rock.
granulite facies A set of metamorphic mineral assemblages formed at very high pressures and temperatures.
gravitational potential energy The energy stored in an object because of its position in a gravitational field; the energy is released when the object falls to a lower elevation.
gravitational spreading A process of lateral spreading that occurs in a material because of the weakness of the material; gravitational spreading causes continental glaciers to grow and mountain belts to undergo orogenic collapse.
gravity The attractive force that one mass exerts on another; the magnitude depends on the size of the objects and the distance between them.
gravity anomaly A deviation between the observed pull of gravity and the geoid at a particular location.

graywacke An informal term used for sedimentary rock consisting of sand-sized up to small-pebble-sized grains of quartz and rock fragments all mixed together in a muddy matrix; typically, graywacke occurs at the base of a graded bed.

great oxygenation event The time in Earth's history, about 2.4 Ga, when the concentration of oxygen in the atmosphere increased dramatically.

greenhouse conditions (greenhouse period) Relatively warm global climate leading to the rising of sea level for an interval of geologic time.

greenhouse effect The trapping of heat in the Earth's atmosphere by carbon dioxide and other greenhouse gases, which absorb infrared radiation; somewhat analogous to the effect of glass in a greenhouse.

greenhouse gases Atmospheric gases, such as carbon dioxide and methane, that regulate the Earth's atmospheric temperature by absorbing infrared radiation.

greenschist facies A set of metamorphic mineral assemblages formed under relatively low pressures and temperatures.

greenstone A low-grade metamorphic rock formed from basalt; if foliated, the rock is called greenschist.

Greenwich mean time (GMT) The time at the astronomical observatory in Greenwich, England; time in all other time zones is set in relation to GMT.

Grenville orogeny The orogeny that occurred about 1 billion years ago and yielded the belt of deformed and metamorphosed rocks that underlie the eastern fifth of the North American continent.

groin A concrete or stone wall built perpendicular to a shoreline in order to prevent beach drift from removing sand.

ground moraine A thin, hummocky layer of till left behind on the land surface during a rapid glacial recession.

groundwater Water that resides under the surface of the Earth, mostly in pores or cracks of rock or sediment.

groundwater contamination Addition of chemicals or microbes (e.g., from agricultural and industrial activities, and landfills or septic tanks) to the groundwater supply.

groundwater depletion The removal of groundwater at a rate exceeding the natural resupply of groundwater in a region.

group A succession of stratigraphic formations that have been lumped together, making a single, thicker stratigraphic entity.

growth ring A rhythmic layering that develops in trees, travertine deposits, and shelly organisms as a consequence of seasonal changes.

gusher A fountain of oil formed when underground pressure causes the oil to rise on its own out of a drilled hole.

guyot A seamount that had a coral reef growing on top of it, so that it is now flat-crested.

gymnosperm A plant whose seeds are "naked," not surrounded by a fruit.

gyre A large, circular flow pattern of ocean surface currents.

habitable zone (astronomy) The region in the Solar System where the intensity of radiation is sufficient to allow water to exist in liquid form on the surface of a planet.

Hadean Eon The oldest of the Precambrian eons; the time between Earth's origin and the formation of the first rocks that have been preserved.

Hadley cells The name given to the low-latitude convection cells in the atmosphere.

hail Falling ice balls from the sky, formed when ice crystallizes in turbulent storm clouds.

hail streak An approximately 2-by-10-km stretch of ground, elongate in the direction of a storm, onto which hail has fallen.

half-graben A wedge-shaped basin in cross section that develops as the hanging-wall block above a normal fault slides down and rotates; the basin develops between the fault surface and the top surface of the rotated block.

half-life The time it takes for half of a group of a radioactive element's isotopes to decay.

halocline The boundary in the ocean between surface-water and deep-water salinities.

hamada Barren, rocky highlands in a desert.

hand specimen A piece of rock, roughly the size of a fist, collected in order to examine.

hanging valley A glacially carved tributary valley whose floor lies at a higher elevation than the floor of the trunk valley.

hanging wall The rock or sediment above an inclined fault plane.

hardness (of a mineral) A measure of the relative ability of a mineral to resist scratching; it represents the resistance of bonds in the crystal structure from being broken.

hard water Groundwater that contains dissolved calcium and magnesium, usually after passing through limestone or dolomite.

head (1) The elevation of the water table above a reference horizon; (2) the edge of ice at the origin of a glacier.

headland A place where a hill or cliff protrudes into the sea.

head scarp The distinct step along the upslope edge of a slump where the regolith detached.

headward erosion The process by which a stream channel lengthens up its slope as the flow of water increases.

headwaters The beginning point of a stream.

heat Thermal energy resulting from the movement of molecules.

heat capacity A measure of the amount of heat that must be added to a material to change its temperature.

heat flow The rate at which heat rises from the Earth's interior up to the surface.

heat-transfer melting Melting that results from the transfer of heat from a hotter magma to a cooler rock.

heliocentric Universe concept An idea proposed by Greek philosophers around 250 B.C.E. suggesting that all heavenly objects including the Earth orbited the Sun.

heliosphere A bubble-like region in space in which solar wind has blown away most interstellar atoms.

Hercynian orogen The late Paleozoic orogen that affected parts of Europe; a continuation of the Alleghenian orogen.

heterosphere A term for the upper portion of the atmosphere, in which gases separate into distinct layers on the basis of composition.

hiatus The interval of time between deposition of the youngest rock below an unconformity and deposition of the oldest rock above the unconformity.

high-altitude westerlies Westerly winds at the top of the troposphere.

high-grade metamorphic rocks Rocks that metamorphose under relatively high temperatures.

high-level waste Nuclear waste containing greater than 1 million times the safe level of radioactivity.

hinge The portion of a fold where curvature is greatest.

hogback A steep-sided ridge of steeply dipping strata.

Holocene The period of geologic time since the last glaciation.

Holocene climatic maximum The period from 5,000 to 6,000 years ago, when Holocene temperatures reached a peak.

homosphere The lower part of the atmosphere, in which the gases have stirred into a homogenous mixture.

hoodoo The local name for the brightly colored shale and sandstone chimneys found in Bryce Canyon National Park in Utah.

horn A pointed mountain peak surrounded by at least three cirques.

hornfels Rock that undergoes metamorphism simply because of a change in temperature, without being subjected to differential stress.

horse latitudes The region of the subtropical high in which winds are weak.

horst The high block between two grabens.

hot spot A location at the base of the lithosphere, at the top of a mantle plume, where temperatures can cause melting.

hot-spot track A chain of now-dead volcanoes transported off the hot spot by the movement of a lithosphere plate.

hot-spot volcano An isolated volcano not caused by movement at a plate boundary, but rather by the melting of a mantle plume.
hot spring A spring that emits water ranging in temperature from about 30°C to 104°C.
Hubbert's Peak The point on a production graph showing a resource plotted against time, at which the rate of production levels off and starts to decrease.
hummocky surface An irregular and lumpy ground surface.
hurricane A huge rotating storm, resembling a giant spiral in map view, in which sustained winds blow over 119 km per hour.
hurricane track The path a hurricane follows.
hyaloclastite A rubbly extrusive rock consisting of glassy debris formed in a submarine or sub-ice eruption.
hydration The absorption of water into the crystal structure of minerals; a type of chemical weathering.
hydraulic conductivity The coefficient K in Darcy's law; hydraulic conductivity takes into account the permeability of the sediment or rock as well as the fluid's viscosity.
hydraulic gradient The slope of the water table.
hydraulic head The potential energy available to drive the flow of a given volume of groundwater at a location; it can be measured as an elevation above a reference.
hydrocarbon A chain-like or ring-like molecule made of hydrogen and carbon atoms; petroleum and natural gas are hydrocarbons.
hydrocarbon generation A process in which oil shale warms to temperatures of greater than about 90°C so kerogen molecules transform into oil and natural gas molecules.
hydrocarbon reserve A known supply of oil and gas held underground.
hydrocarbon system The association of source rock, migration pathway, reservoir rock, seal, and trap geometry that leads to the occurrence of a hydrocarbon reserve.
hydrofracturing ("fracking") A process by which drillers generate new fractures or open preexisting ones underground, by pumping a high-pressure fluid into a portion of the drill hole, in order to increase the permeability of surrounding hydrocarbon-bearing rocks.
hydrogen bond The attraction of a hydrogen atom to a negatively charged atom or molecule (e.g., hydrogen bonds attach water molecules to each other).
hydrologic cycle The continual passage of water from reservoir to reservoir in the Earth System.
hydrolysis The process in which water chemically reacts with minerals and breaks them down.
hydrosphere The Earth's water, including surface water (lakes, rivers, and oceans), groundwater, and liquid water in the atmosphere.
hydrothermal deposit An accumulation of ore minerals precipitated from hot-water solutions circulating through a magma or through the rocks surrounding an igneous intrusion.
hydrothermal metamorphism When very hot water passes through the crust and causes metamorphism of rock.
hypocenter (focus) The place within the Earth where earthquake energy originates; commonly, the hypocenter is the place on a fault where slip took place.
hypsometric curve A graph that plots surface elevation on the vertical axis and the percentage of the Earth's surface on the horizontal axis.
ice age An interval of time in which the climate was colder than it is today, glaciers occasionally advanced to cover large areas of the continents, and mountain glaciers grew; an ice age can include many glacials and interglacials.
iceberg A large block of ice that calves off the front of a glacier and drops into the sea.
icehouse period A period of time when the Earth's temperature was cooler than it is today and ice ages could occur.
ice-margin lake A meltwater lake formed along the edge of a glacier.
ice-rafted sediment Sediment carried out to sea by icebergs.
ice sheet A vast glacier that covers the landscape.
ice shelf A broad, flat region of ice along the edge of a continent formed where a continental glacier flowed into the sea.
ice stream A portion of a glacier that travels much more quickly than adjacent portions of the glacier.
ice tongue The portion of a valley glacier that has flowed out into the sea.
igneous rock Rock that forms when hot molten rock (magma or lava) cools and freezes solid.
ignimbrite Rock formed when deposits of pyroclastic flows solidify.
inactive fault A fault that last moved in the distant past and probably won't move again in the near future, yet is still recognizable because of displacement across the fault plane.
inactive sand The sand along a coast that is buried beneath a layer of active sand and moves only during severe storms or not at all.
incised meander A meander that lies at the bottom of a steep-walled canyon.
index fossil A fossil species that is widespread and survived for a relatively short duration, and can therefore be used to correlate strata.
index minerals Minerals that serve as good indicators of metamorphic grade.
induced seismicity Seismic events caused by the actions of people (e.g., filling a reservoir that lies over a fault with water).
industrial minerals Minerals that serve as the raw materials for manufacturing chemicals, concrete, and wallboard, among other products.
inequant A term for a mineral grain whose length and width are not the same.
inertia The tendency of an object at rest to remain at rest.
infiltrate Seep down into.
injection well A well in which a liquid is pumped down into the ground under pressure so that it passes from the well back into the pore space of the rock or regolith.
inner core The inner section of the core, extending from 5,155 km deep to the Earth's center at 6,371 km and consisting of solid iron alloy.
inselberg An isolated mountain or hill in a desert landscape created by progressive cliff retreat, so that the hill is surrounded by a pediment or an alluvial fan.
insolation Exposure to the Sun's rays.
intensity (seismology) A measure of the relative size of an earthquake (the severity of ground shaking) at a location, as determined by examining the amount of damage caused.
interglacial A period of time between two glaciations.
interior basin A basin with no outlet to the sea.
interlocking texture The texture of crystalline rocks in which mineral grains fit together like pieces of a jigsaw puzzle.
internal energy Energy that comes from the Earth's internal heat; it drives plate tectonics, and therefore, earthquakes, volcanoes, and mountain building.
internal process A process in the Earth System, such as plate motion, mountain building, or volcanism, ultimately caused by Earth's internal heat.
intertidal zone The area of coastal land across which the tide rises and falls.
intertropical convergence zone The equatorial convergence zone in the atmosphere.
intraplate earthquakes Earthquakes that occur away from plate boundaries.
intrusive contact The boundary between country rock and an intrusive igneous rock.
intrusive igneous rock Rock formed by the freezing of magma underground.
ion A version of an atom that has lost or gained electrons, relative to an electrically neutral version, so that it has a net electrical charge.
ionic bond The attachment of one atom to another that happens when one atom transfers electrons to another; one type of chemical bond.

ionosphere The interval of Earth's atmosphere, at an elevation between 50 and 400 km, containing abundant positive ions.
iron catastrophe The proposed event very early in Earth history when the Earth partly melted and molten iron sank to the center to form the core.
isobar A line on a map along which the air has a specified pressure.
isograd (1) A line on a pressure-temperature graph along which all points are taken to be at the same metamorphic grade; (2) a line on a map making the first appearance of a metamorphic index mineral.
isostasy (isostatic equilibrium) The condition that exists when the buoyancy force pushing lithosphere up equals the gravitational force pulling lithosphere down.
isostatic compensation The process in which the surface of the crust slowly rises or falls to reestablish isostatic equilibrium after a geologic event changes the density or thickness of the lithosphere.
isotherm Lines on a map or cross section along which the temperature is constant.
isotopes Different versions of a given element that have the same atomic number but different atomic weights.
isotopic dating (equivalent to radiometric dating) A method to determine a rock's numerical age by measuring the ratio of parent to daughter isotopes in minerals.
jet stream A fast-moving current of air that flows at high elevations.
jetty A man-made wall that protects the entrance to a harbor.
joints Naturally formed cracks in rocks.
joint set A group of systematic joints.
Jovian A term used to describe the outer gassy, Jupiter-like planets (gas-giant planets).
kame A stratified sequence of lateral-moraine sediment that's sorted by water flowing along the edge of a glacier.
karst landscape A region underlain by caves in limestone bedrock; the collapse of the caves creates a landscape of sinkholes separated by higher topography, or of limestone spires separated by low areas.
Kelvin (K) scale A measure of temperature in which 0 K is absolute zero and the freezing point of water is 273.15 K; divisions in the Kelvin scale have the same value as those in the Celsius scale.
kerogen The waxy molecules into which the organic material in shale transforms on reaching about 100°C. At higher temperatures, kerogen transforms into oil.
kettle hole A circular depression in the ground made when a block of ice calves off the toe of a glacier, becomes buried by till, and later melts.
kingdom A high-level taxonomic division of life. Eukaryota are divided into six kingdoms, including Plantae, Animalia, and Fungi.
knob-and-kettle topography A land surface with many kettle holes separated by round hills of glacial till.
K-T boundary event The mass extinction that happened at the end of the Cretaceous Period, 66 million years ago, possibly due to the collision of an asteroid with the Earth.
Kuiper Belt A diffuse ring of icy objects, remnants of Solar System formation, that orbit our Sun outside the orbit of Neptune.
laccolith A blister-shaped igneous intrusion that forms when magma injects between layers underground in a manner that pushes overlying layers upward to form a dome.
lag deposit The coarse sediment left behind in a desert after wind erosion removes the finer sediment.
lagoon A body of shallow seawater separated from the open ocean by a barrier island.
lahar A thick slurry formed when volcanic ash and debris mix with water, either in rivers or from rain or melting snow and ice on the flank of a volcano.
landform An individual feature of the landscape, characterized by a distinctive shape.
landscape The overall shape of the Earth's surface in a region.
landslide A sudden movement of rock and debris down a nonvertical slope.
landslide-potential map A map on which regions are ranked according to the likelihood that a mass movement will occur.
land subsidence Sinking elevation of the ground surface; the process may occur over an aquifer that is slowly draining and decreasing in volume because of pore collapse.
La Niña Years in which the El Niño event is not strong.
lapilli Any pyroclastic particle that is 2 to 64 mm in diameter (i.e., marble-sized); the particles can consist of frozen lava clots, pumice fragments, or ash clumps.
Laramide orogeny The mountain-building event that lasted from about 80 Ma to 40 Ma, in western North America; in the United States, it formed the Rocky Mountains as a result of basement uplift and the warping of the younger overlying strata into large monoclines.
large igneous province (LIP) A region in which huge volumes of lava and/or ash erupted over a relatively short interval of geologic time.
latent heat of condensation The heat released during condensation, which comes only from a change in state.
lateral moraine A strip of debris along the side margins of a glacier.
laterite soil A hard, brick-red, soil formed from iron-rich rock in a tropical environment; it consists primarily of insoluble iron and aluminum oxide and hydroxide and forms due to extreme leaching.
Laurentia A continent in the early Paleozoic Era composed of today's North America and Greenland.
Laurentide ice sheet An ice sheet that spread over northeastern Canada during the Pleistocene ice age(s).
lava Molten rock that has flowed out onto the Earth's surface.
lava dome A dome-like mass of rhyolitic lava that accumulates above the eruption vent.
lava flows Sheets or mounds of lava that flow onto the ground surface or sea floor in molten form and then solidify.
lava fountain A column of molten lava spraying upward under pressure from a volcanic vent.
lava lake A large pool of lava produced around a vent when lava fountains spew forth large amounts of lava in a short period of time.
lava tube The empty space left when a lava tunnel drains; this happens when the surface of a lava flow solidifies while the inner part of the flow continues to stream downslope.
leach To dissolve and carry away.
leader A conductive path stretching from a cloud toward the ground, along which electrons leak from the base of the cloud, and which provides the start for a lightning flash to the ground.
lightning bolt (lightning flash, lightning stroke) A giant spark or pulse of current that jumps across a gap of charge separation.
light year The distance that light travels in one Earth year (about 6 trillion miles or 9.5 trillion km).
lignite Low-rank coal that consists of 50% carbon.
limb The side of a fold, showing less curvature than at the hinge.
limestone Sedimentary rock composed of calcite.
liquefaction The process by which saturated, unconsolidated sediments are transformed into a substance that acts like a liquid as a result of ground shaking.
liquidus The lowest temperature at which all the components of a material have melted and transformed into liquid.
liquification The process by which wet sediment becomes a slurry; liquification may be triggered by earthquake vibrations.
lithification The transformation of loose sediment into solid rock through compaction and cementation.
lithologic correlation A correlation based on similarities in rock type.
lithosphere The relatively rigid, nonflowable, outer 100- to 150-km-thick layer of the Earth, constituting the crust and the top part of the mantle.
lithosphere plate One of many distinct pieces of the lithosphere (Earth's relatively rigid shell) that are separated from one another by breaks (plate boundaries).

little ice age A period of cooler temperatures, between 1500 and 1800 C.E., during which many glaciers advanced.
loam A type of soil consisting of roughly equal parts of sand, silt, and clay; it tends to be good for growth of crops.
local base level A base level upstream from a drainage network's mouth.
lodgment till A flat layer of till smeared out over the ground when a glacier overrides an end moraine as it advances.
loess Layers of fine-grained sediments deposited from the wind; large deposits of loess formed from fine-grained glacial sediment blown off outwash plains.
longitudinal (seif) dune A dune formed when there is abundant sand and a strong, steady wind, and whose axis lies parallel to the wind direction.
longitudinal profile A cross-sectional image showing the variation in elevation along the length of a river.
longshore current The flow of water parallel to the shore just off a coast, because of the diagonal movement of waves toward the shore.
longshore drift The movement of sediment laterally along a beach; it occurs when waves wash up a beach diagonally.
lower mantle The deepest section of the mantle, stretching from 670 km down to the core-mantle boundary.
low-grade metamorphic rocks Rocks that underwent metamorphism at relatively low temperatures.
low-velocity zone The asthenosphere underlying oceanic lithosphere in which seismic waves travel more slowly, probably because rock has partially melted.
luster The way a mineral surface scatters light.
L-waves (Love waves) Surface seismic waves that cause the ground to ripple back and forth, creating a snake-like movement.
Ma Millions of years ago (abbreviation).
macrofossil A fossil large enough to be seen with the naked eye.
mafic A term used in reference to magmas or igneous rocks that are relatively poor in silica and rich in iron and magnesium.
mafic magma A magma with 45 to 52% silica (i.e., a magma with a high proportion of magnesium and iron oxide).
magma Molten rock beneath the Earth's surface.
magma chamber A space below ground filled with magma.
magma contamination The process in which flowing magma incorporates components of the country rock through which it passes.
magmatic deposit An ore deposit formed when sulfide ore minerals accumulate at the bottom of a magma chamber.
magnetic anomaly The difference between the expected strength of the Earth's magnetic field at a certain location and the actual measured strength of the field at that location.
magnetic declination The angle between the direction a compass needle points at a given location and the direction of true north.
magnetic dipole An imaginary vector that points from the north magnetic pole to the south magnetic pole of a magnetic field.
magnetic field The region affected by the force emanating from a magnet.
magnetic field lines The trajectories along which magnetic particles would align, or charged particles would flow, if placed in a magnetic field.
magnetic force The push or pull exerted by a magnet.
magnetic inclination The angle between a magnetic needle free to pivot on a horizontal axis and a horizontal plane parallel to the Earth's surface.
magnetic poles The ends of a magnetic dipole; all magnetic dipoles have a north pole and a south pole.
magnetic reversal The change of the Earth's magnetic polarity; when a reversal occurs, the field flips from normal to reversed polarity, or vice versa.
magnetic-reversal chronology The history of magnetic reversals through geologic time.
magnetism An attractive or repulsive field force generated by permanent magnets or by an electrical current.
magnetization The degree to which a material can exert a magnetic force.
magnetometer An instrument that measures the strength of the Earth's magnetic field.
magnetosphere The region protected from the electrically charged particles of the solar winds by Earth's magnetic field.
magnetostratigraphy The comparison of the pattern of magnetic reversals in a sequence of strata, with a reference column showing the succession of reversals through time.
magnitude (of an earthquake) The number that represents the maximum amplitude of ground motion that would be measured by a seismometer placed at a specified distance from the epicenter.
mainshock (earthquake) The largest energy-generating event in a cluster of related earthquakes; the mainshock is usually an order of magnitude larger than aftershocks.
manganese nodules Lumpy accumulations of manganese-oxide minerals precipitated onto the sea floor.
mantle The thick layer of rock below the Earth's crust and above the core.
mantle plume A column of very hot rock rising up through the mantle.
marble A metamorphic rock composed of calcite and transformed from a protolith of limestone.
mare The broad, darker areas on the Moon's surface; they consist of flood basalts that erupted over 3 billion years ago and spread out across the Moon's lowlands.
marginal sea A small ocean basin created when sea-floor spreading occurs behind an island arc.
marine magnetic anomaly The difference between the *expected* strength of the Earth's main dipole field at a certain location on the sea floor and the *actual* measured strength of the magnetic field at that location.
maritime tropical air mass A mass of air that originates over tropical or subtropical oceanic regions.
marker bed A particularly unique layer that provides a definitive basis for correlation.
marsh A wetland dominated by grasses.
mass The amount of matter in an object; mass differs from weight in that its value does not depend on the strength of gravity.
mass-extinction event A time when vast numbers of species abruptly vanish.
mass movement (mass wasting) The gravitationally caused downslope transport of rock, regolith, snow, or ice.
matter The material substance of the universe; it consists of atoms and has mass.
matrix Finer-grained material surrounding larger grains in a rock.
meander A snake-like curve along a stream's course.
meandering stream A reach of stream containing many meanders (snake-like curves).
meander neck A narrow isthmus of land separating two adjacent meanders.
mean sea level The average level between the high and low tide over a year at a given point.
mechanical force A push, pull, or shear applied by one object on another; it can be applied only if the objects are in contact.
mechanical weathering *See* Physical weathering.
medial moraine A strip of sediment in the interior of a glacier, parallel to the flow direction of the glacier, formed by the lateral moraines of two merging glaciers.
Medieval Warm Period A period of high temperatures in the Middle Ages.
melt Molten (liquid) rock.
meltdown The melting of the fuel rods in a nuclear reactor that occurs if the rate of fission becomes too fast and the fuel rods become too hot.
melting curve The line defining the range of temperatures and pressures at which a rock melts.

melting temperature The temperature at which the thermal vibration of the atoms or ions in the lattice of a mineral is sufficient to break the chemical bonds holding them to the lattice, so a material transforms into a liquid.
meltwater lake A lake fed by glacial meltwater.
mesa A large, flat-topped hill (with a surface area of several square km) in an arid region.
mesopause The boundary that marks the top of the mesosphere of Earth's atmosphere.
mesosphere The cooler layer of atmosphere overlying the stratosphere.
Mesozoic Era The middle of the three Phanerozoic eras; it lasted from 252 Ma to 66 Ma.
metaconglomerate A metamorphic rock produced by metamorphism of a conglomerate; typically, it contains flattened pebbles and cobbles.
metal A solid composed almost entirely of atoms of metallic elements; it is generally opaque, shiny, smooth, malleable, and can conduct electricity.
metallic bond A chemical bond in which the outer atoms are attached to each other in such a way that electrons flow easily from atom to atom.
metamorphic aureole The region around a pluton, stretching tens to hundreds of meters out, in which heat transferred into the country rock and metamorphosed the country rock.
metamorphic facies A set of metamorphic mineral assemblages indicative of metamorphism under a specific range of pressures and temperatures.
metamorphic foliation A fabric defined by parallel surfaces or layers that develop in a rock as a result of metamorphism; schistocity and gneissic layering are examples.
metamorphic grade A representation of the intensity of metamorphism, meaning the amount or degree of metamorphic change.
metamorphic mineral New minerals that grow in place within a solid rock under metamorphic temperatures and pressures.
metamorphic mineral assemblage A group of minerals that form in a rock as a result of metamorphism.
metamorphic rock Rock that forms when preexisting rock changes into new rock as a result of an increase in pressure and temperature and/or shearing under elevated temperatures; metamorphism occurs without the rock first becoming a melt or a sediment.
metamorphic texture A distinctive arrangement of mineral grains produced by metamorphism.
metamorphic zone The region between two metamorphic isograds, typically named after an index mineral found within the region.
metamorphism The process by which one kind of rock transforms into a different kind of rock.
metasomatism The process by which a rock's overall chemical composition changes during metamorphism because of reactions with hot water that bring in or remove elements.
meteor A streak of bright, glowing gas created as a meteoroid vaporizes in the atmosphere due to friction.
meteoric water Water that falls to Earth from the atmosphere as either rain or snow.
meteorite A piece of rock or metal alloy that fell from space and landed on Earth.
micrite Limestone consisting of lime mud (i.e., very fine-grained limestone).
microfossil A fossil that can be seen only with a microscope or an electron microscope.
mid-ocean ridge A 2-km-high submarine mountain belt that forms along a divergent oceanic plate boundary.
migmatite A rock formed when gneiss is heated high enough so that it begins to partially melt, creating layers, or lenses, of new igneous rock that mix with layers of the relict gneiss.
migration (hydrocarbons) The movement of hydrocarbons from source rocks to reservoir rocks.
Milankovitch cycles Climate cycles that occur over tens to hundreds of thousands of years because of changes in Earth's orbit and tilt.
mine A site at which ore is extracted from the ground.
mineral A homogenous, naturally occurring, solid inorganic substance with a definable chemical composition and an internal structure characterized by an orderly arrangement of atoms, ions, or molecules in a lattice. Most minerals are inorganic.
mineral classes Groups of minerals distinguished from each other on the basis of chemical composition.
mineralogist A geoscientist specializing in the study of minerals.
mineral resources The minerals extracted from the Earth's upper crust for practical purposes.
Mississippi Valley-type (MVT) ore An ore deposit, typically in dolostone, containing lead- and zinc-bearing minerals that precipitated from groundwater that had moved up from several km depth in the upper crust; such deposits occur in the upper Mississippi Valley.
mixture A material consisting of two or more substances that can be separated mechanically (i.e., without chemical reactions).
Modified Mercalli Intensity scale An earthquake characterization scale based on the amount of damage that the earthquake causes.
Moho The seismic-velocity discontinuity that defines the boundary between the Earth's crust and mantle. Named for Andrija Mohorovičić.
Mohs hardness scale A list of ten minerals in a sequence of relative hardness, with which other minerals can be compared.
mold A cavity in sedimentary rock left behind when a shell that once filled the space weathers out.
molecule The smallest piece of a compound that has the properties of the compound; it consists of two or more atoms attached by chemical bonds.
Moment magnitude scale (*Mw*) A modern measure of earthquake size that reflects the amount of energy an earthquake produces, the size of rupture, and amount of displacement.
monocline A fold in the land surface whose shape resembles that of a carpet draped over a stair step.
monsoon A seasonal reversal in wind direction that causes a shift from a very dry season to a very rainy season in some regions of the world.
moon A sizable solid body locked in orbit around a planet.
moraine A sediment pile composed of till deposited by a glacier.
morphology (fossil morphology) The shape of features visible on a fossil.
mountain belts Chains or ranges of mountains.
mountain building The process of causing a belt of land to undergo significant uplift, usually in association with deformation.
mountain front The boundary between a mountain range and adjacent plains.
mountain (alpine) glacier A glacier that exists in or adjacent to a mountainous region.
mountain ice cap A mound of ice that submerges peaks and ridges at the crest of a mountain range.
mouth The outlet of a stream where it discharges into another stream, a lake, or a sea.
mudflow A downslope movement of mud at slow to moderate speed.
mud pot A viscuous slurry that forms in a geothermal region when hot water or steam rises into soils rich in volcanic ash and clay.
mudstone Very fine-grained sedimentary rock that will not easily split into sheets.
mylonite Rock formed during dynamic metamorphism and characterized by foliation that lies roughly parallel to the fault (shear zone) involved in the shearing process; mylonites have very fine grains formed by the nonbrittle subdivision of larger grains.
native metal A naturally occurring pure mass of a single metal in an ore deposit.
natural arch An arch that forms when erosion along joints leaves narrow walls of rock; when the lower part of the wall erodes while the upper part remains, an arch results.

natural hazard A natural feature of the environment that can cause injury to living organisms and/or damage to buildings and the landscape.

natural levees A pair of low ridges that appear on either side of a stream and develop as a result of the accumulation of sediment deposited naturally during flooding.

natural selection The process by which the fittest organisms survive to pass on their characteristics to the next generation.

neap tide An especially low tide that occurs when the angle between the direction of the Moon and the direction of the Sun is 90°.

nebula A cloud of gas or dust in space.

nebular theory The concept that planets grow out of rings of gas, dust, and ice surrounding a newborn star.

negative anomaly An area where the magnetic field strength is less than expected.

negative feedback Feedback that slows a process down or reverses it.

neocrystallization The growth of new crystals, not in the protolith, during metamorphism.

neutron A subatomic particle, in the nucleus of an atom, that has a neutral charge.

Nevadan orogeny A convergent-margin mountain-building event that took place in western North America during the Late Jurassic Period.

nonconformity A type of unconformity at which sedimentary rocks overlie basement (older intrusive igneous rocks and/or metamorphic rocks).

nonflowing artesian well An artesian well in which water rises on its own up to a level that lies below the ground surface.

nonfoliated metamorphic rock Rock containing minerals that recrystallized during metamorphism but has no foliation.

nonmetallic mineral resources Mineral resources that do not contain metals; examples include building stone, gravel, sand, gypsum, phosphate, and salt.

nonplunging fold A fold with a horizontal hinge.

nonrenewable resource A resource that nature will take a long time (hundreds to millions of years) to replenish or may never replenish.

nonsystematic joints Short cracks in rocks that occur in a range of orientations and are randomly placed and oriented.

nor'easter A large, midlatitude North American cyclone; when it reaches the east coast, it produces strong winds that come out of the northeast.

normal fault A fault in which the hanging-wall block moves down the slope of the fault.

normal force The component of the gravitational force acting perpendicular to a slope.

normal polarity Polarity in which the paleomagnetic dipole has the same orientation as it does today.

normal stress The push or pull that is perpendicular to a surface.

North Atlantic deep-water mass The mass of cold, dense water that sinks in the north polar regions.

northeast tradewinds Surface winds that come out of the northeast and occur in the region between the equator and 30°N.

nuclear bond The force that attaches subatomic particles to each other within the nucleus of an atom.

nuclear fuel Pellets of concentrated uranium oxide or a comparable radioactive material that can provide energy in a nuclear reactor.

nuclear fusion The process by which the nuclei of atoms fuse together, thereby creating new, larger atoms.

nuclear reaction A process that results in changing the nucleus of an atom by breaking or forming nuclear bonds.

nuclear reactor The part of a nuclear power plant where the fission reactions occur.

nuclear waste The radioactive material produced as a byproduct in a nuclear plant that must be disposed of carefully due to its dangerous radioactivity.

nucleus The central ball of an atom that consists of protons and neutrons (except for hydrogen, whose nuclei contains only a proton).

nuée ardente *See* Pyroclastic flow.

numerical age (in older literature, "absolute age") The age of a geologic feature given in years.

oasis A verdant region surrounded by desert, occurring at a place where natural springs provide water at the surface.

oblique-slip fault A fault in which sliding occurs diagonally along the fault plane.

obsidian An igneous rock consisting of a solid mass of volcanic glass.

occluded front A front that no longer intersects the ground surface.

oceanic crust The crust beneath the oceans; composed of gabbro and basalt, overlain by sediment.

oceanic plateau A region of oceanic floor that is higher than surrounding areas; such regions have particularly thick oceanic crust and are relicts of submarine large igneous provinces.

oceanic lithosphere Lithosphere topped by oceanic crust; it reaches a thickness of 100 km.

offshore bar A narrow ridge of sand that forms off the shore of a beach; some offshore bars rise above sea level, and separate a lagoon on one side from the open ocean on the other.

oil (geology) A liquid hydrocarbon that can be burned as a fuel.

Oil Age The period of human history, including our own, so named because the economy depends on oil.

oil field A region containing a significant amount of accessible oil underground.

oil reserve The known supply of oil held underground.

oil seep A location at the surface of the Earth where hydrocarbons are being released from underground naturally.

oil shale Shale containing kerogen.

oil trap A geologic configuration that keeps oil underground in the reservoir rock and prevents it from rising to the surface.

oil window The narrow range of temperatures under which oil can form in a source rock.

olistotrome A large, submarine slump block, buried and preserved.

Oort Cloud A cloud of icy objects, left over from Solar System formation, that orbit the Sun in a region outside of the heliosphere.

ophiolite A slice of oceanic crust that has been thrust onto continental crust.

ordinary well A well whose base penetrates below the water table and can thus provide water.

ore Rock containing native metals or a concentrated accumulation of ore minerals.

ore deposit An economically significant accumulation of ore.

ore minerals Minerals that have metal in high concentrations and in a form that can be easily extracted.

organic carbon Carbon that has been incorporated in an organism.

organic chemical A carbon-containing compound that occurs in living organisms, or that resembles such compounds; it consists of carbon atoms bonded to hydrogen atoms along with varying amounts of oxygen, nitrogen, and other chemicals.

organic coast A coast along which living organisms control landforms along the shore.

organic sedimentary rock Sedimentary rock (such as coal) formed from carbon-rich relicts of organisms.

organic shale Lithified, muddy, organic-rich ooze that contains the raw materials from which hydrocarbons eventually form.

original horizontality A geologic principle stating that strata, when first deposited, occur in roughly horizontal beds.

orogen (orogenic belt) A linear range of mountains.

orogenic collapse The process in which mountains begin to collapse under their own weight and spread out laterally.

orogeny A mountain-building event.

orographic barrier A landform that diverts air flow upward or laterally.

outcrop An exposure of bedrock.

outer core The section of the core, between 2,900 and 5,150 km deep, that consists of liquid iron alloy.
outwash plain A broad area of gravel and sandbars deposited by a braided stream network, fed by the melt water of a glacier.
overburden The weight of overlying rock on rock buried deeper in the Earth's crust.
overriding plate (slab) The plate at a subduction zone that overrides the downgoing plate.
oversaturated solution A solution that contains so much solute (dissolved ions) that precipitation begins.
oversized stream valley A large valley with a small stream running through it; the valley formed earlier when the flow was greater.
oxbow lake A meander that has been cut off yet remains filled with water.
oxidation reaction A reaction in which an element loses electrons; an example is the reaction of iron with air to form rust.
ozone O_3, an atmospheric gas that absorbs harmful ultraviolet radiation from the Sun.
ozone hole An area of the atmosphere, over polar regions, from which ozone has been depleted.
pahoehoe A lava flow with a surface texture of smooth, glassy, rope-like ridges.
paleoclimate The past climate of the Earth.
paleomagnetism The record of ancient magnetism preserved in rock.
paleontology The study of life history through the examination of the fossil record and its relation to the stratigraphic record.
paleontologist A geologist or biologist who studies the fossil record to reconstruct the history of life on Earth.
paleopole The supposed position of the Earth's magnetic pole in the past, with respect to a particular continent.
paleosol Ancient soil preserved in the stratigraphic record.
Paleozoic Era The oldest era of the Phanerozoic Eon (541–252 Ma).
Pangaea A supercontinent that assembled at the end of the Paleozoic Era.
Pannotia A supercontinent that may have existed sometime between 800 Ma and 600 Ma.
parabolic dunes Dunes formed when strong winds break through transverse dunes to make new dunes whose ends point upwind.
parallax The apparent movement of an object seen from two different points not on a straight line from the object (e.g., from your two different eyes).
parallax method A trigonometric method used to determine the distance from the Earth to a nearby star.
parent isotope A radioactive isotope that undergoes decay.
partial melting The melting in a rock of the minerals with the lowest melting temperatures, while other minerals remain solid.
passive margin A continental margin that is not a plate boundary.
passive-margin basin A thick accumulation of sediment along a tectonically inactive coast, formed over crust that stretched and thinned when the margin first began.
patterned ground A polar landscape in which the ground splits into pentagonal or hexagonal shapes.
pause An elevation in the atmosphere where temperature stops decreasing and starts increasing, or vice versa.
peat Compacted and partially decayed vegetation accumulating beneath a swamp.
pedalfer soil A temperate-climate soil characterized by well-defined soil horizons and an organic A-horizon.
pediment The broad, nearly horizontal bedrock surface at the base of a retreating desert cliff.
pedocal soil Thin soil, formed in arid climates. It contains very little organic matter, but significant precipitated calcite.
pegmatite A coarse-grained igneous rock containing crystals of up to tens of centimeters across and occurring in dike-shaped intrusions.
pelagic sediment Microscopic plankton shells and fine flakes of clay that settle out and accumulate on the deep-ocean floor.
Pelé's hair Droplets of basaltic lava that mold into long, glassy strands as they fall.
Pelé's tears Droplets of basaltic lava that mold into tear-shaped, glassy beads as they fall.
peneplain A nearly flat surface that lies at an elevation close to sea level; thought to be the product of long-term erosion.
perched water table A quantity of groundwater that lies above the regional water table because an underlying lens of impermeable rock or sediment prevents the water from sinking down to the regional water table.
percolation The process by which groundwater meanders through tiny, crooked channels in the surrounding material.
peridotite A coarse-grained ultramafic rock.
periglacial environment A region with widespread permafrost but without a blanket of snow or ice.
period An interval of geologic time representing a subdivision of a geologic era.
permafrost Permanently frozen ground.
permanent magnet A special material that behaves magnetically for a long time all by itself.
permanent stream A stream that flows year-round because its bed lies below the water table, or because more water is supplied from upstream than can infiltrate the ground.
permeability The degree to which a material allows fluids to pass through it via an interconnected network of pores and cracks.
permineralization The fossilization process in which plant material becomes transformed into rock by the precipitation of silica from groundwater.
petrified A term used by geologists to describe plant material that has transformed into rock by permineralization.
petroglyph Drawings formed by chipping into the desert varnish of rocks to reveal the lighter rock beneath.
petroleum *See* Oil.
phaneritic A textural term used to describe coarse-grained igneous rock.
Phanerozoic Eon The most recent eon, an interval of time from 542 Ma to the present.
phenocryst A large crystal surrounded by a finer-grained matrix in an igneous rock.
photochemical smog Brown haze that blankets a city when exhaust from cars and trucks reacts in the presence of sunlight.
photomicrograph A photograph of a thin section of rock, as seen through a microscope.
photosynthesis The process during which chlorophyll-containing plants remove carbon dioxide from the atmosphere, form tissues, and expel oxygen back to the atmosphere.
phreatomagmatic eruption An explosive eruption that occurs when water enters the magma chamber and turns into steam.
phyllite A fine-grained metamorphic rock with a foliation caused by the preferred orientation of very fine-grained mica.
phyllitic luster A silk-like sheen characteristic of phyllite, a result of the rock's fine-grained mica.
phylogenetic tree A chart representing the ideas of paleontologists showing which groups of organisms radiated from which ancestors.
phylogeny A representation of life evolution showing how populations of the present relate to populations of the past.
physical weathering The process in which intact rock breaks into smaller grains or chunks.
piedmont glacier A fan or lobe of ice that forms where a valley glacier emerges from a valley and spreads out into the adjacent plain.
pillow basalt Glass-encrusted basalt blobs that form when magma extrudes on the sea floor and cools very quickly.
placer deposit Concentrations of metal grains in stream sediment that develop when rocks containing native metals erode and create a mixture of sand grains and metal fragments; the moving water of the stream carries away lighter mineral grains.

planet An object that orbits a star, is roughly spherical, and has cleared its neighborhood of other objects.
planetesimal Tiny, solid pieces of rock and metal that collect in a planetary nebula and eventually accumulate to form a planet.
plankton Tiny plants and animals that float in sea or lake water.
plastic deformation The deformational process in which mineral grains behave like plastic and, when compressed or sheared, become flattened or elongate without cracking or breaking.
plate One of about 20 distinct pieces of the relatively rigid lithosphere.
plate boundary The border between two adjacent lithosphere plates.
plate-boundary earthquakes The earthquakes that occur along and define plate boundaries.
plate-boundary volcano A volcanic arc or mid-ocean ridge volcano, formed as a consequence of movement along a plate boundary.
plate interior A region away from the plate boundaries that consequently experiences few earthquakes.
plate tectonics *See* Theory of plate tectonics.
playa The flat, typically salty lake bed that remains when all the water evaporates in drier times; forms in desert regions.
Pleistocene Epoch The period of time from about 2 Ma to 14,000 years ago, during which the Earth experienced an ice age.
Pleistocene Ice Age The most recent ice age in Earth history. It began at about 2.6 Ma, and has included several advances and retreats.
plunge (structural geology) The angle that a line makes with respect to horizontal, as measured in a vertical plane.
plunge pool A depression at the base of a waterfall scoured by the energy of the falling water.
plunging fold A fold with a tilted hinge.
pluton An irregular or blob-shaped intrusion; can range in size from tens of m across to tens of km across.
pluvial lake A lake formed to the south of a continental glacier as a result of enhanced rainfall during an ice age.
point bar A wedge-shaped deposit of sediment on the inside bank of a meander.
polar cell A high-latitude convection cell in the atmosphere.
polar easterlies Prevailing winds that come from the east and flow from the polar high to the subpolar low.
polar front The convergence zone in the atmosphere at latitude 60°.
polar glacier *See* Dry-bottom glacier.
polar high The zone of high pressure in polar regions created by the sinking of air in the polar cells.
polarity The orientation of a magnetic dipole.
polarity chron The time interval between polarity reversals of Earth's magnetic field.
polarity subchron The time interval between magnetic reversals if the interval is of short duration (less than 200,000 years long).
polarized light A beam of filtered light waves that all vibrate in the same plane.
polar wander The phenomenon of the progressive changing through time of the position of the Earth's magnetic poles relative to a location on a continent; significant polar wander probably doesn't occur—in fact, poles seem to remain fairly fixed, while continents move.
polar-wander path The curving line representing the apparent progressive change in the position of the Earth's magnetic pole, relative to a locality X, assuming that the position of X on Earth has been fixed through time (in fact, poles stay fixed while continents move).
pollen Tiny grains involved in plant reproduction.
pollution Natural and synthetic contaminant materials introduced to the Earth's environment by the activities of humans.
polymorphs Two minerals that have the same chemical composition but a different crystal lattice structure.
pore A small, open space within sediment or rock.
pore collapse The closer packing of grains that occurs when groundwater is extracted from pores, thus eliminating the support holding the grains apart.
porosity The total volume of empty space (pore space) in a material, usually expressed as a percentage.
porphyritic A textural term for igneous rock that has phenocrysts distributed throughout a finer matrix.
Portland cement Cement made by mechanically mixing limestone, sandstone, and shale in just the right proportions, before heating in a kiln, to provide the correct chemical makeup of cement.
positive anomaly An area where the magnetic field strength is stronger than expected.
positive-feedback mechanism A mechanism that enhances the process that causes the mechanism in the first place.
post-glacial rebound The slow rise of land, after the weight of an overlying glacier has been removed, due to melting.
potentiometric surface The elevation to which water in an artesian system would rise if unimpeded; where there are flowing artesian wells, the potentiometric surface lies above ground.
pothole A bowl-shaped depression carved into the floor of a stream by a long-lived whirlpool carrying sand or gravel.
Precambrian Period The interval of geologic time between Earth's formation about 4.57 Ga and the beginning of the Phanerozoic Eon 542 Ma.
precession The gradual conical path traced out by Earth's spinning axis; simply put, it is the "wobble" of the axis.
precious metals Metals (like gold, silver, and platinum) that have high value.
Precipitate (chemistry, n.) A solid substance formed when atoms attach and settle out of a solution, or attach to the walls of the container holding the solution; (chemistry, v.) the action of forming a solid substance from a solution; (meteorology, v.) the dropping of snow or rain from the sky.
precipitation (1) The process by which atoms dissolved in a solution come together and form a solid; (2) rainfall or snow.
preferred orientation The metamorphic texture that exists where platy grains lie parallel to one another and/or elongate grains align in the same direction.
preservation potential The likelihood that an organism will be preserved in the fossil record.
pressure Force per unit area, or the "push" acting on a material in cases where the push (compressional stretch) is the same in all directions.
pressure gradient The rate of pressure change over a given horizontal distance.
pressure solution The process of dissolution at points of contact, between grains, where compression is greatest, producing ions that then precipitate elsewhere, where compression is less.
prevailing winds Surface winds that generally flow in the same direction for long time periods.
primary porosity The space that remains between solid grains or crystals immediately after sediment accumulates or rock forms.
principal aquifer The geologic unit that serves as the primary source of groundwater in a region.
principle of baked contacts When an igneous intrusion "bakes" (metamorphoses) surrounding rock, the rock that has been baked must be older than the intrusion.
principle of cross-cutting relations If one geologic feature cuts across another, the feature that has been cut is older.
principle of fossil succession In a stratigraphic sequence, different species of fossil organisms appear in a definite order; once a fossil species disappears in a sequence of strata, it never reappears higher in the sequence.
principle of inclusions If a rock contains fragments of another rock, the fragments must be older than the rock containing them.
principle of original continuity Sedimentary layers, before erosion, formed fairly continuous sheets over a region.
principle of original horizontality Layers of sediment, when originally deposited, are fairly horizontal.

principle of superposition In a sequence of sedimentary rock layers, each layer must be younger than the one below, for a layer of sediment cannot accumulate unless there is already a substrate on which it can collect.
principle of uniformitariansim The physical processes we observe today also operated in the past in the same way, and at comparable rates.
product (chemistry, n.) materials produced or formed by a chemical reaction.
prograde metamorphism Metamorphism that occurs as temperatures and pressures are increasing.
prokaryote An organism whose cells do not contain a nucleus; archaea and bacteria consist of prokaryotic cells.
Proterozoic Eon The most recent of the Precambrian eons (2,500–541 Ma).
protocontinent A block of crust composed of volcanic arcs and hotspot volcanoes sutured together.
protolith The original rock from which a metamorphic rock formed.
proton A positively charged subatomic particle in the nucleus of an atom.
protoplanet A body that grows by the accumulation of planetesimals but has not yet become big enough to be called a planet.
protoplanetary disk The portion of the accretionary disk outside of the proto-Sun; the matter within it ultimately becomes planets, moons, and other objects.
protoplanetary nebula A ring of gas and dust that surrounded the newborn Sun, from which the planets were formed.
protostar A dense body of gas that is collapsing inward because of gravitational forces and that may eventually become a star.
pumice A glassy igneous rock that forms from felsic frothy lava and contains abundant (over 50%) pore space.
pumice lapilli Marble-sized chunks consisting of frothy, siliceous igneous rock that fall from a volcanic eruptive cloud.
punctuated equilibrium The hypothesis that evolution takes place in fits and starts; evolution occurs very slowly for quite a while and then, during a relatively short period, takes place very rapidly.
P-waves Compressional seismic waves that move through the body of the Earth.
P-wave shadow zone A band between 103° and 143° from an earthquake epicenter, as measured along the circumference of the Earth, inside which P-waves do not arrive at seismograph stations.
pycnocline The boundary between layers of water of different densities.
pyroclastic debris Fragmented material that sprayed out of a volcano and landed on the ground or sea floor in solid form.
pyroclastic flow A fast-moving avalanche that occurs when hot volcanic ash and debris mix with air and flow down the side of a volcano.
pyroclastic rock Rock made from fragments that were blown out of a volcano during an explosion and were then packed or welded together.
quarry A site at which stone is extracted from the ground.
quartzite A metamorphic rock composed of quartz and transformed from a protolith of quartz sandstone.
quenching A sudden cooling of molten material to form a solid.
quick clay Clay that behaves like a solid when still (because of surface tension holding the water-coated clay flakes together) but that flows like a liquid when shaken.
radial network A drainage network in which the streams flow outward from a cone-shaped mountain and define a pattern resembling spokes on a wheel.
radiation (physics) Electromagnetic energy traveling away from a source through a medium or space.
radioactive decay The process by which a radioactive atom undergoes fission or releases particles, thereby being transformed into a new element.
radioactive isotope An unstable isotope of a given element.
radiometric dating The science of dating geologic events in years by measuring the ratio of parent radioactive atoms to daughter product atoms.
rain band A spiraling arm of a hurricane radiating outward from the eye.
rain shadow The inland side of a mountain range, which is arid because the mountains block rain clouds from reaching the area.
range (for fossils) The interval of a sequence of strata in which a specific fossil species appears.
rapids A reach of a stream in which water becomes particularly turbulent; as a consequence, waves develop on the surface of the stream.
rare earth element One of a group of 17 elements including the lanthanides, scandium, and yttrium; they are essential in the production of high-tech devices.
reach A specified segment of a stream's path.
reactant (chemistry) The starting materials of a chemical reaction.
recessional moraine The end moraine that forms when a glacier stalls for a while as it recedes.
recharge area A location where water enters the ground and infiltrates down to the water table.
recrystallization The process in which ions or atoms in minerals rearrange to form new minerals.
rectangular network A drainage network in which the streams join each other at right angles because of a rectangular grid of fractures that breaks up the ground and localizes channels.
recurrence interval The average time between successive geologic events.
red giant A very large, red star, formed when a star the size of our Sun starts to die and undergoes an immense expansion.
red shift The phenomenon in which a source of light moving away from you very rapidly shifts to a lower frequency; that is, toward the red end of the spectrum.
reef bleaching The death and loss of color of a coral reef.
reflected ray A ray that bounces off a boundary between two different materials.
reflection Energy waves bouncing off of a boundary.
refracted ray A ray that bends as it passes through a boundary between two different materials.
refraction The bending of a ray as it passes through a boundary between two different materials.
refractory materials (refractories) Substances that have a relatively high melting point and tend to exist in solid form.
reg A vast stony plain in a desert.
regional metamorphism *See also* Dynamothermal metamorphism; metamorphism of a broad region, usually the result of deep burial during an orogeny.
regolith Any kind of unconsolidated debris that covers bedrock.
regression The seaward migration of a shoreline caused by a lowering of sea level.
relative age The age of one geologic feature with respect to another.
relative humidity The ratio between the measured water content of air and the maximum possible amount of water the air can hold at a given condition.
relative plate velocity The movement of one lithosphere plate with respect to another.
relief The difference in elevation between adjacent high and low regions on the land surface.
renewable resource A resource that can be replaced by nature within a short time span relative to a human life span.
reservoir rock Rock with high porosity and permeability, so it can contain an abundant amount of easily accessible oil.
residence time The average length of time that a substance stays in a particular reservoir.
residual mineral deposit Soils in which the residuum left behind after leaching by rainwater is so concentrated in metals that the soil itself becomes an ore deposit.

resonance (seismology) A situation that arises when earthquake waves of a particular frequency cause particularly large-amplitude movements because energy input happens at just the right time.
resurgent dome The new mound, or cone, of igneous rock that grows within a caldera as an eruption begins anew.
retrograde metamorphism Metamorphism that occurs as pressures and temperatures are decreasing; for retrograde metamorphism to occur, water must be added.
return stroke An upward-flowing electric current from the ground that carries positive charges up to a cloud during a lightning flash.
reversed polarity Polarity in which the paleomagnetic dipole points north.
reverse fault A steeply dipping fault on which the hanging-wall block slides up.
rhythmic layering Banding in sediments, shells, trees, corals, or ice that repeats periodically; it may be correlated to annual cycles.
Richter scale A scale that defines earthquakes on the basis of the amplitude of the largest ground motion recorded on a seismogram.
ridge axis The crest of a mid-ocean ridge; the ridge axis defines the position of a divergent plate boundary.
ridge-push force A process in which gravity causes the elevated lithosphere at a mid-ocean ridge axis to push on the lithosphere that lies farther from the axis, making it move away.
rift A linear belt along which continental lithosphere stretches and pulls apart.
right-lateral strike-slip fault A strike-slip fault in which the block on the opposite fault plane from a fixed spot moves to the right of that spot.
rip current A strong, localized seaward flow of water perpendicular to a beach.
ripple mark Relatively small elongated ridges that form on a sedimentary bed surface at right angles to the direction of current flow.
riprap Loose boulders or concrete piled together along a beach to absorb wave energy before it strikes a cliff face.
roche moutonnée A glacially eroded hill that becomes elongate in the direction of flow and asymmetric; glacial rasping smoothes the upstream part of the hill into a gentle slope, while glacial plucking erodes the downstream edge into a steep slope.
rock A coherent, naturally occurring solid, consisting of an aggregate of minerals or a mass of glass.
rock burst A sudden explosion of rock off the ceiling or wall of an underground mine.
rock composition The chemical makeup of rock, typically represented by the proportions of different minerals.
rock cycle The succession of events that results in the transformation of Earth materials from one rock type to another, then another, and so on.
rockfall A mass of rock that separates from a cliff, typically along a joint, and then free-falls downslope.
rock flour Fine-grained sediment produced by glacial abrasion of the substrate over which a glacier flows.
rock glacier A slow-moving mixture of rock fragments and ice.
rockslide A sudden downslope movement of rock.
rocky coast An area of coast where bedrock rises directly from the sea, so beaches are absent.
Rodinia A proposed Precambrian supercontinent that existed around 1 billion years ago.
rogue wave Waves that are two to five times the size of most of the large waves passing a locality in a given time interval.
rotational axis The imaginary line through the center of the Earth around which the Earth spins.
running water Water that flows down the surface of sloping land in response to the pull of gravity.
R-waves (Rayleigh waves) Surface seismic waves that cause the ground to ripple up and down, like water waves in a pond.
sabkah A region of formerly flooded coastal desert in which stranded seawater has left a salt crust over a mire of mud that is rich in organic material.
salinity The degree of concentration of salt in water.
saltation The movement of a sediment in which grains bounce along their substrate, knocking other grains into the water column (or air) in the process.
salt dome A rising bulbous dome of salt that bends up the adjacent layers of sedimentary rock.
salt wedging The process in arid climates by which dissolved salt in groundwater crystallizes and grows in open pore spaces in rocks and pushes apart the surrounding grains.
sand dune A relatively large ridge of sand built up by a current of wind (or water); cross bedding typically occurs within the dune.
sandspit An area where the beach stretches out into open water across the mouth of a bay or estuary.
sandstone Coarse-grained sedimentary rock consisting almost entirely of quartz.
sand volcano (sand blow) A small mound of sand produced when sand layers below the ground surface liquify as a result of seismic shaking, causing the sand to erupt onto the Earth's surface through cracks or holes in overlying clay layers.
saprolite A layer of rotten rock created by chemical weathering in warm, wet climates.
Sargasso Sea The center of North Atlantic Gyre, named for the tropical seaweed sargassum, which accumulates in its relatively non-circulating waters.
saturated solution Water that carries as many dissolved ions as possible under given environmental conditions.
saturated zone The region below the water table where pore space is filled with water.
scattering The dispersal of energy that occurs when light interacts with particles in the atmosphere.
schist A medium-to-coarse-grained metamorphic rock that possesses schistosity.
schistosity Foliation caused by the preferred orientation of large mica flakes.
scientific method A sequence of steps for systematically analyzing scientific problems in a way that leads to verifiable results.
scoria A glassy, mafic, igneous rock containing abundant air-filled holes.
scoria cone An accumulation of lapilli-sized or larger fragments formed from a volcanic eruption that spatters clots of basaltic lava. (Also called cinder cone.)
scouring A process by which running water removes loose fragments of sediment from a streambed.
sea arch An arch of land protruding into the sea and connected to the mainland by a narrow bridge.
sea-floor spreading The gradual widening of an ocean basin as new oceanic crust forms at a mid-ocean ridge axis and then moves away from the axis.
sea ice Ice formed by the freezing of the surface of the sea.
seal A relatively impermeable rock, such as shale, salt, or unfractured limestone, that lies above a reservoir rock and stops the oil from rising further.
seam A sedimentary bed of coal interlayered with other sedimentary rocks.
seamount An isolated submarine mountain.
seasonal floods Floods that appear almost every year during seasons when rainfall is heavy or when winter snows start to melt.
seasonal well A well that provides water only during the rainy season when the water table rises below the base of the well.
sea stack An isolated tower of land just offshore, disconnected from the mainland by the collapse of a sea arch.
seawall A wall of riprap built on the landward side of a backshore zone in order to protect shore cliffs from erosion.

second The basic unit of time measurement, now defined as the time it takes for the magnetic field of a cesium atom to flip polarity 9,192,631,770 times, as measured by an atomic clock.
secondary enrichment The process by which a new ore deposit forms from metals that were dissolved and carried away from preexisting ore minerals.
secondary porosity New pore space in rocks, created some time after a rock first forms.
secondary recovery technique A process used to extract the quantities of oil that will not come out of a reservoir rock with just simple pumping.
sediment An accumulation of loose mineral grains, such as boulders, pebbles, sand, silt, or mud, that are not cemented together.
sediment liquefaction When pressure in the water in the pores push sediment grains apart so that they become surrounded by water and no longer rest against each other, and the sediment becomes able to flow like a liquid.
sedimentary basin A depression, created as a consequence of subsidence, that fills with sediment.
sedimentary rock Rock that forms either by the cementing together of fragments broken off preexisting rock or by the precipitation of mineral crystals out of water solutions at or near the Earth's surface.
sedimentary sequence A grouping of sedimentary units bounded on top and bottom by regional unconformities.
sedimentary structure A distinctive shape or form (examples—bed, ripple mark, cross bed) formed during deposition of sediment.
sediment budget The proportion of sand supplied to sand removed from a depositional setting.
sediment load The total volume of sediment carried by a stream.
sediment maturity The degree to which a sediment has evolved from a crushed-up version of the original rock into a sediment that has lost its easily weathered minerals and become well sorted and rounded.
sediment sorting The segregation of sediment by size.
seep A place where oil-filled reservoir rock intersects the ground surface, or where fractures connect a reservoir to the ground surface, so that oil flows out onto the ground on its own.
seiche Rhythmic movement in a body of water caused by ground motion.
seismic belts (seismic zones) The relatively narrow strips of crust on Earth under which most earthquakes occur.
seismicity Earthquake activity.
seismic ray The changing position of an imaginary point on a wave front as the front moves through rock.
seismic-reflection profile A cross-sectional view of the crust made by measuring the reflection of artificial seismic waves off boundaries between different layers of rock in the crust.
seismic retrofitting The strengthening of an already existing structure (building, bridge, etc.) so that it can withstand earthquake vibrations.
seismic tomography Analysis by sophisticated computers of global seismic data in order to create a three-dimensional image of variations in seismic-wave velocities within the Earth.
seismic velocity The speed at which seismic waves travel.
seismic-velocity discontinuity A boundary in the Earth at which seismic velocity changes abruptly.
seismic (earthquake) waves Waves of energy emitted at the focus of an earthquake.
seismogram The record of an earthquake produced by a seismograph.
seismologist A geophysicist who specializes in studying earthquakes.
seismometer (seismograph) An instrument that can record the ground motion from an earthquake.
semipermanent pressure cell A somewhat elliptical zone of high or low atmospheric pressure that lasts much of the year; it forms because high-pressure zones tend to be narrower over land than over sea.
Sevier orogeny A mountain-building event that affected western North America between about 150 Ma and 80 Ma, a result of convergent margin tectonism; a fold-thrust belt formed during this event.
shale Very fine-grained sedimentary rock that breaks into thin sheets.
shale gas Gas that comes directly from a source rock (organic shale).
shale oil Oil that is still residing in shale, a source rock.
shatter cones Small, cone-shaped fractures formed by the shock of a meteorite impact.
shear When one part of a material moves sideways, relative to another.
shear strain A change in shape of an object that involves the movement of one part of a rock body sideways past another part so that angular relationships within the body change.
shear stress A stress that moves one part of a material sideways past another part.
shear waves Seismic waves in which particles of material move back and forth perpendicular to the direction in which the wave itself moves.
shear zone A fault in which movement has occurred ductilely.
sheetwash A film of water less than a few mm thick that covers the ground surface during heavy rains.
shell (biology) A relatively hard, protective structure formed of minerals and surrounding the soft part of an invertebrate organism.
shield An older, interior region of a continent.
shield volcano A subaerial volcano with a broad, gentle dome, formed either from low-viscosity basaltic lava or from large pyroclastic sheets.
shocked quartz Grains of quartz that have been subjected to intense pressure such as occurs during a meteorite impact.
shock metamorphism The changes that can occur in a rock due to the passage of a shock wave, generally resulting from a meteorite impact.
shoreline The boundary between the water and land.
shortening The process during which a body of rock or a region of crust becomes shorter.
short-term climate change Climate change that takes place over hundreds to thousands of years.
Sierran arc A large continental volcanic arc along western North America that was initiated at the end of the Jurassic Period and lasted until about 80 million years ago.
silica SiO_2.
silicate rock Rock composed of silicate minerals.
silicates (silicate minerals) Minerals built from silicon-oxygen tetrahedra arranged in chains, sheets, or 3-D networks; they make up most of the Earth's crust and mantle.
siliceous sedimentary rock Sedimentary rock that contains abundant quartz.
silicon-oxygen tetrahedron The SiO_4^{4-} anionic group, in which four oxygen atoms surround a single silicon atom, thereby defining the corners of a tetrahedron.
silicic Rich in silica with relatively little iron and magnesium.
sill A nearly horizontal tabletop-shaped tabular intrusion that occurs between the layers of country rock.
siltstone Fine-grained sedimentary rock generally composed of very small quartz grains.
sinkhole A circular depression in the land that forms when an underground cavern collapses.
slab-pull force The force that downgoing plates (or slabs) apply to oceanic lithosphere at a convergent margin.

slate Fine-grained, low-grade metamorphic rock, formed by the metamorphism of shale.
slaty cleavage The foliation typical of slate, and reflective of the preferred orientation of slate's clay minerals, that allows slate to be split into thin sheets.
slickensides The polished surface of a fault caused by slip on the fault; lineated slickensides also have grooves that indicate the direction of fault movement.
slip face The leeward slope of a dune; sand that builds up at the crest of the dune slides down this face; slip faces are preserved as cross beds within sandstone layers.
slip lineations Linear marks on a fault surface created during movement on the fault; some slip lineations are defined by grooves, some by aligned mineral fibers.
slope failure The downslope movement of material on an unstable slope.
slumping Downslope movement in which a mass of regolith detaches from its substrate along a spoon-shaped, sliding surface and slips downward semicoherently.
smelting The heating of a metal-containing rock to high temperatures in a fire so that the rock will decompose to yield metal plus a nonmetallic residue (slag).
snottite A long gob of bacteria that slowly drips from the ceiling of a cave.
snow avalanche A cloud or mass of snow and ice, mixed with air and sometimes water, that moves down a slope.
snowball Earth A model proposing that, at times during Earth history, glaciers covered all land, and the entire ocean surface froze.
snow line The boundary above which snow remains all year.
soda straw A hollow stalactite in which calcite precipitates around the outside of a drip.
soil Sediment that has undergone changes at the surface of the Earth, including reaction with rainwater and the addition of organic material.
soil erosion The removal of soil by wind and runoff.
soil horizon Distinct zones within a soil, distinguished from each other by factors such as chemical composition and organic content.
soil moisture Underground water that wets the surface of the mineral grains and organic material making up soil, but lies above the water table.
soil profile A vertical sequence of distinct zones of soil.
solar cell A device that can produce electricity directly from incoming solar energy.
Solar System Our Sun and all the materials that orbit it (including planets, moons, asteroids, Kuiper Belt objects, and Oort Cloud objects).
solar wind A stream of particles with enough energy to escape from the Sun's gravity and flow outward into space.
solid-state diffusion The slow movement of atoms or ions through a solid.
solidus The highest temperature at which all the components of a material are solid; at the solidus temperature, the material begins to melt.
solifluction The type of creep characteristic of tundra regions; during the summer, the uppermost layer of permafrost melts, and the soggy, weak layer of ground then flows slowly downslope in overlapping sheets.
solstice A day on which the polar ends of the terminator (the boundary between the day hemisphere and the night hemisphere) lie 23.5° away from the associated geographic poles.
solution A material containing dissolved ions.
Sonoma orogeny A convergent-margin mountain-building event that took place on the western coast of North America in the Late Permian and Early Triassic periods.
sorting (1) The range of clast sizes in a collection of sediment; (2) the degree to which sediment has been separated by flowing currents into different-sized fractions.
source rock A rock (organic-rich shale) containing the raw materials from which hydrocarbons eventually form.
southeast tradewinds Tradewinds in the southern hemisphere, which start flowing northward, deflect to the west, and end up flowing from southeast to northwest.
southern oscillation The movement of atmospheric pressure cells back and forth across the Pacific Ocean, in association with El Niño.
specific gravity A number representing the density of a mineral, as specified by the ratio between the weight of a volume of the mineral and the weight of an equal volume of water.
speleothem A formation that grows in a limestone cave by the accumulation of travertine precipitated from water solutions dripping in a cave or flowing down the wall of a cave.
sphericity The measure of the degree to which a clast approaches the shape of a sphere.
spreading boundary *See* Divergent plate boundary.
spreading rate The rate at which sea floor moves away from a mid-ocean ridge axis, as measured with respect to the sea floor on the opposite side of the axis.
spring A natural outlet from which groundwater flows up onto the ground surface.
spring tide An especially high tide that occurs when the Sun is on the same side of the Earth as the Moon.
stable air Air that does not have a tendency to rise rapidly.
stable slope A slope on which downward sliding is unlikely.
stalactite An icicle-like cone that grows from the ceiling of a cave as dripping water precipitates limestone.
stalagmite An upward-pointing cone of limestone that grows when drips of water hit the floor of a cave.
standing wave A wave whose crest and trough remain in place as water moves through the wave.
star An object in the Universe in which fusion reactions occur pervasively, producing vast amounts of energy; our Sun is a star.
star dune A constantly changing dune formed by frequent shifts in wind direction; it consists of overlapping crescent dunes pointing in many different directions.
steady state condition The condition when proportions of a chemical in different reservoirs remain fairly constant even though there is a constant flux (flow) of the chemical among the reservoirs.
stellar nucleosynthesis The production of new, larger atoms by fusion reactions in stars; the process generates more massive elements that were not produced by the Big Bang.
stellar wind The stream of atoms emitted from a star into space.
stick-slip behavior Stop-start movement along a fault plane caused by friction, which prevents movement until stress builds up sufficiently.
stone rings Ridges of cobbles between adjacent bulges of permafrost ground.
stoping A process by which magma intrudes; blocks of wall rock break off and then sink into the magma.
storm An episode of severe weather in which winds, precipitation, and in some cases lightning become strong enough to be bothersome and even dangerous.
storm-center velocity A storm's (hurricane's) velocity along its track.
storm surge Excess seawater driven landward by wind during a storm; the low atmospheric pressure beneath the storm allows sea level to rise locally, increasing the surge.
strain The change in shape of an object in response to deformation (i.e., as a result of the application of a stress).
strata A succession of several layers or beds together.
strategic mineral A mineral containing elements of importance to technology (particularly to the military).
stratified drift Glacial sediment that has been redistributed and stratified by flowing water.
stratigraphic column A cross-section diagram of a sequence of strata summarizing information about the sequence.

stratigraphic formation A recognizable layer of a specific sedimentary rock type or set of rock types, deposited during a certain time interval, that can be traced over a broad region.
stratigraphic group Several adjacent stratigraphic formations in a succession.
stratigraphic sequence An interval of strata deposited during periods of relatively high sea level, and bounded above and below by regional unconformities.
stratopause The temperature pause that marks the top of the stratosphere.
stratosphere The stable, stratified layer of atmosphere directly above the troposphere.
stratovolcano A large, cone-shaped subaerial volcano consisting of alternating layers of lava and tephra.
stratus cloud A thin, sheet-like, stable cloud.
streak The color of the powder produced by pulverizing a mineral on an unglazed ceramic plate.
stream A ribbon of water that flows in a channel.
streambed The floor of a stream.
stream capacity The total quantity of sediment a stream carries.
stream capture (stream piracy) The situation in which headward erosion causes one stream to intersect the course of another, previously independent stream, so that the intersected stream starts to flow down the channel of the first stream.
stream competence The maximum particle size that a stream can carry.
stream gradient The slope of a stream's channel in the downstream direction.
stream piracy A process that happens when headward erosion by one stream causes the stream to intersect the course of another stream and capture its flow.
stream rejuvenation The renewed downcutting of a stream into a floodplain or peneplain, caused by a relative drop of the base level.
stream terrace When a stream downcuts through the alluvium of a floodplain so that a new, lower floodplain develops and the original floodplain becomes a step-like platform.
stress The push, pull, or shear that a material feels when subjected to a force; formally, the force applied per unit area over which the force acts.
stretching The process during which a layer of rock or a region of crust becomes longer.
striations Linear scratches in rock.
strike (structural geology) The compass orientation of a horizontal line on a plane.
strike-slip fault A fault in which one block slides horizontally past another (and therefore parallel to the strike line), so there is no relative vertical motion.
strip mining The scraping off of all soil and sedimentary rock above a coal seam in order to gain access to the seam.
stromatolite Layered mounds of sediment formed by cyanobacteria; cyanobacteria secrete a mucous-like substance to which sediment sticks, and as each layer of cyanobacteria gets buried by sediment, it colonizes the surface of the new sediment, building a mound upward.
structural control The condition in which geologic structures, such as faults, affect the distribution and drainage of water or the shape of the land surface.
subaerial Pertaining to land regions above sea level (i.e., under air).
subduction The process by which one oceanic plate bends and sinks down into the asthenosphere beneath another plate.
subduction zone The region along a convergent boundary where one plate sinks beneath another.
sublimation The evaporation of ice directly into vapor without first forming a liquid.
submarine canyon A narrow, steep canyon that dissects a continental shelf and slope.
submarine fan A wedge-shaped accumulation of sediment at the base of a submarine slope; fans usually accumulate at the mouth of a submarine canyon.
submarine slump The underwater downslope movement of a semi-coherent block of sediment along a weak mud detachment.
submergent coast A coast at which the land is sinking relative to sea level.
subpolar low The rise of air where the surface flow of a polar cell converges with the surface flow of a Ferrel cell, creating a low-pressure zone in the atmosphere.
subsidence The vertical sinking of the Earth's surface in a region, relative to a reference plane.
substrate A general term for material just below the ground surface.
subtropical high (subtropical divergence zone) A belt of high pressure in the atmosphere at 30° latitude formed where the Hadley cell converges with the Ferrel cell, causing cool, dense air to sink.
subtropics Desert climate regions that lie on either side of the equatorial tropics between the lines of 20° and 30° north or south of the equator.
summit eruption An eruption that occurs in the summit crater of a volcano.
sunspot cycle The cyclic appearance of large numbers of sunspots (black spots thought to be magnetic storms on the Sun's surface) every 9 to 11.5 years.
supercontinent cycle The process of change during which supercontinents develop and later break apart, forming pieces that may merge once again in geologic time to make yet another supercontinent.
supernova A short-lived, very bright object in space that results from the cataclysmic explosion marking the death of a very large star; the explosion ejects large quantities of matter into space to form new nebulae.
superplume A huge mantle plume.
superposed stream A stream whose geometry has been laid down on a rock structure and is not controlled by the structure.
superposition A geologic principle stating that, in a sequence of upright strata, younger strata lie above older strata, because younger strata must be deposited over older strata.
superrotation The faster rotation of the core, relative to the rest of the Earth.
supervolcano A volcano that erupts a vast amount (more than 1,000 cubic km) of volcanic material during a single event; none have erupted during recorded human history.
surface current An ocean current in the top 100 m of water.
surface load (bed load) Sediment that rolls and bounce along the ground (under the air) or along a stream bed (under water).
surface water Liquid or seasonally frozen water that resides at the surface of the Earth in oceans, lakes, streams, and marshes.
surface waves Seismic waves that travel along the Earth's surface.
surface westerlies The prevailing surface winds in North America and Europe, which come out of the west or southwest.
surf zone A region of the shore in which breakers crash onto the shore.
surge (glacial) A pulse of rapid flow in a glacier.
suspended load Tiny solid grains carried along by a stream without settling to the floor of the channel.
sustainable growth The ability of society to prosper without depleting the supply of natural resources, and without destroying the environment.
suture The surface across which two separate blocks of crust are attached, after ocean floor between them has been subducted.
swamp A wetland dominated by trees.
swash The upward surge of water that flows up a beach slope when breakers crash onto the shore.
S-waves Seismic shear waves that pass through the body of the Earth.

S-wave shadow zone A band between 103° and 180° from the epicenter of an earthquake inside of which S-waves do not arrive at seismograph stations.
swelling clay Clay possessing a mineral structure that allows it to absorb water between its layers and thus swell to several times its original size.
symmetry The condition in which the shape of one part of an object is a mirror image of the other part.
syncline A trough-shaped fold whose limbs dip toward the hinge.
systematic joints Long planar cracks that occur fairly regularly throughout a rock body.
tabular intrusions Sheet intrusions that are planar and of roughly uniform thickness.
tachylite A basaltic volcanic glass.
Taconic orogeny A convergent mountain-building event that took place around 400 million years ago, in which a volcanic island arc collided with eastern North America.
tailings pile A pile of waste rock from a mine.
talus A sloping apron of fallen rock along the base of a cliff.
tar Hydrocarbons that exist in solid form at room temperature.
tarn A lake that forms at the base of a cirque on a glacially eroded mountain.
tar sand Sandstone reservoir rock in which less viscous oil and gas molecules have either escaped or been eaten by microbes, so that only tar remains.
taxonomy The study and classification of the relationships among different forms of life.
tectonic foliation A planar fabric, such as cleavage, schistocity, or gneissic banding, that develops in rocks; caused by compression or shearing during deformation (e.g., during mountain building).
temperature A measure of the hotness or coldness of a material.
tension A stress that pulls on a material and could lead to stretching.
tephra Unconsolidated accumulations of pyroclastic grains.
terminal moraine The end moraine at the farthest limit of glaciation.
terminator The boundary between the half of the Earth that has daylight and the half experiencing night.
terrace The elevated surface of an older floodplain into which a younger floodplain had cut down.
terrestrial planets Planets that are of comparable size and character to the Earth and consist of a metallic core surrounded by a rock mantle.
thalweg The deepest part of a stream's channel.
theory A scientific idea supported by an abundance of evidence that has passed many tests and failed none.
theory of plate tectonics The theory that the outer layer of the Earth (the lithosphere) consists of separate plates that move with respect to one another.
thermal energy The total kinetic energy in a material due to the vibration and movement of atoms in the material.
thermal metamorphism Metamorphism caused by heat conducted into country rock from an igneous intrusion.
thermocline A boundary between layers of water with differing temperatures.
thermohaline circulation The rising and sinking of water driven by contrasts in water density, which is due in turn to differences in temperature and salinity; this circulation involves both surface and deep-water currents in the ocean.
thermophilic An adjective meaning "heat loving." Usually used in reference to bacteria or archaea that thrive in hot water deep underground, in geothermal springs on land, or in black smokers on the sea floor.
thermosphere The outermost layer of the atmosphere, containing very little gas.
thin section A 3/100-mm-thick slice of rock that can be examined with a petrographic microscope.
thin-skinned deformation A distinctive style of deformation characterized by displacement on faults that terminate at depth along a subhorizontal detachment fault.
thrust fault A gently dipping reverse fault; the hanging-wall block moves up the slope of the fault.
tidal bore A visible wall of water that moves toward shore with the rising tide in quiet waters.
tidal flat A broad, nearly horizontal plain of mud and silt, exposed or nearly exposed at low tide but totally submerged at high tide.
tidal power Energy produced by the daily rise and fall of the tides; people can utilize this energy, for example, by damming a bay or estuary, so that water passes through turbines when the tide changes.
tidal range The difference in sea level between high tide and low tide at a given point.
tide The daily rising or falling of sea level at a given point on the Earth.
tide-generating force The force, caused in part by the gravitational attraction of the Sun and Moon and in part by the centrifugal force created by the Earth's spin, that generates tides.
tidewater glacier A glacier that has entered the sea along a coast.
till A mixture of unsorted mud, sand, pebbles, and larger rocks deposited by glaciers.
tillite A rock formed from hardened ancient glacial deposits and consisting of larger clasts distributed through a matrix of sandstone and mudstone.
toe (terminus) The leading edge or margin of a glacier.
tombolo A narrow ridge of sand that links a sea stack to the mainland.
topographical map A map that uses contour lines to represent variations in elevation.
topography Variations in elevation.
topsoil The top soil horizons, which are typically dark and nutrient-rich.
tornado A near-vertical, funnel-shaped cloud in which air rotates extremely rapidly around the axis of the funnel.
tornado swarm Dozens of tornadoes produced by the same storm.
tower karst A karst landscape in which steep-sided residual bedrock towers remain between sinkholes.
trace fossil Fossilized imprints or debris that an organism leaves behind while moving on or through sediment; examples include footprints, burrows, and fecal matter.
transform fault A fault marking a transform plate boundary; along mid-ocean ridges, transform faults are the actively slipping segment of a fracture zone between two ridge segments.
transform plate boundary A boundary at which one lithosphere plate slips laterally past another.
transgression The inland migration of shoreline resulting from a rise in sea level.
transition zone The middle portion of the mantle, from 400 to 670 km deep, in which there are several jumps in seismic velocity.
transpiration The release of moisture as a metabolic by-product.
transverse dune A simple, wave-like dune that appears when enough sand accumulates for the ground surface to be completely buried, but only moderate winds blow.
trap A subsurface configuration of seal rocks and structures that keep oil and/or gas underground, so it doesn't seep out at the surface.
travel time (seismic-wave travel time) The time for seismic waves to pass through the Earth from the focus of the earthquake to a seismometer.in that rock.
travel-time curve A graph that plots the time since an earthquake began on the vertical axis and the distance to the epicenter on the horizontal axis.
travertine A rock composed of crystalline calcium carbonate ($CaCO_3$) formed by chemical precipitation from groundwater that has seeped out at the ground surface.

trellis network A drainage system that develops across a landscape of parallel valleys and ridges so that major tributaries flow down the valleys and join a trunk stream that cuts through the ridge; the resulting map pattern resembles a garden trellis.
trench A deep, elongate trough bordering a volcanic arc; a trench defines the trace of a convergent plate boundary.
triangulation The method for determining the map location of a point from knowing the distance between that point and three other points; this method is used to locate earthquake epicenters.
tributary A smaller stream that flows into a larger stream.
triple junction A point where three lithosphere plate boundaries intersect.
tropical cyclone A large, intense spiral storm formed in tropical latitudes—a more general term for storms known regionally as hurricanes, typhoons, or cyclones.
tropical depression A tropical storm with winds reaching up to 61 km per hour; such storms develop from tropical disturbances, and may grow to become hurricanes.
tropical disturbance Cyclonic winds that develop in the tropics.
tropopause The temperature pause marking the top of the troposphere.
troposphere The lowest layer of the atmosphere, where air undergoes convection and where most wind and clouds develop.
truncated spur A spur (elongate ridge between two valleys) whose end was eroded off by a glacier.
trunk stream The single larger stream into which an array of tributaries flow.
tsunami A large wave along the sea surface triggered by an earthquake or large submarine slump.
tuff A pyroclastic igneous rock composed of volcanic ash and fragmented pumice, formed when accumulations of the debris cement together.
tundra A cold, treeless region of land at high latitudes, supporting only species of shrubs, moss, and lichen capable of living on permafrost.
turbidite A graded bed of sediment built up at the base of a submarine slope and deposited by turbidity currents.
turbidity current A submarine avalanche of sediment and water that speeds down a submarine slope.
turbulence The chaotic twisting, swirling motion in flowing fluid.
typhoon The equivalent of a hurricane in the western Pacific Ocean.
ultimate base level Sea level; the level below which a trunk stream cannot cut.
ultramafic A term used to describe igneous rocks or magmas that are rich in iron and magnesium and very poor in silica.
ultramafic magma A magma with over 45% silica (i.e., a magma with a very high proportion of magnesium and iron oxide).
unconfined aquifer An aquifer that intersects the surface of the Earth.
unconformity A boundary between two different rock sequences representing an interval of time during which new strata were not deposited and/or were eroded.
unconsolidated Consisting of unattached grains.
unconventional reserve An accumulation of hydrocarbons that are too viscous to flow, and/or that occur in impermeable rock, so that they cannot be pumped simply by drilling a well (examples—tar sand, oil shale, shale oil/gas).
undercutting Excavation at the base of a slope that results in the formation of an overhang.
underground mining The extraction of ore or coal by digging shafts and tunnels underground.
undersaturated A term used to describe a solution capable of holding more dissolved ions.
uniformitarianism A geologic principle stating that processes that can be observed today also happened in the past, at comparable rates, and can explain features preserved in the geologic record. Put simply, the present is the key to the past.
Universe All of space and all the matter and energy within it.
unsaturated zone The region of the subsurface above the water table.
unstable air Air that is significantly warmer than air above and has a tendency to rise quickly.
unstable ground Land capable of slumping or slipping downslope in the near future.
unstable slope A slope on which sliding will likely happen.
updraft Upward-moving air.
uplift (structural geology) The vertical rise of a region of land (or the upward displacement of the Earth's surface).
upper mantle The uppermost section of the mantle, reaching down to a depth of 400 km.
upwelling zone A place where deep water rises in the ocean, or where hot magma rises in the asthenosphere.
U-shaped valley A steep-walled valley shaped by glacial erosion into the form of a U.
vacuum Space that contains very little matter in a given volume (e.g., a region in which air has been removed).
valley A trough with sloping walls, cut into the land by a stream.
valley glacier A river of ice that flows down a mountain valley.
Van Allen radiation belts Belts of solar wind particles and cosmic rays that surround the Earth, trapped by Earth's magnetic field.
van der Waals bonding The relatively weak attachment of two elements or molecules due to their polarity and not due to covalent or ionic bonding.
varve A pair of thin layers of glacial lake-bed sediment, one consisting of silt brought in during the spring floods and the other of clay deposited during the winter when the lake's surface freezes over and the water is still.
vascular plant A plant with woody tissue and seeds and veins for transporting water and food.
vein A seam of minerals that forms when dissolved ions carried by water solutions precipitate in cracks.
vein deposit A hydrothermal deposit in which the ore minerals occur in veins that fill cracks in preexisting rocks.
velocity-versus-depth curve A graph that shows the variation in the velocity of seismic waves with increasing depth in the Earth.
ventifact (faceted rock) A desert rock whose surface has been faceted by the wind.
vesicles Open holes in igneous rock formed by the preservation of bubbles in magma as the magma cools into solid rock.
viscosity The resistance of material to flow.
volatiles (volatile materials) Elements or compounds such as H_2O and CO_2 that evaporate at relatively low temperatures and can exist in gaseous forms at the Earth's surface.
volatility A specification of the ease with which a material evaporates.
volcanic agglomerate An accumulation consisting dominantly of volcanic bombs and other relatively large chunks of igneous material.
volcanic arc A curving chain of active volcanoes formed adjacent to a convergent plate boundary.
volcanic ash Tiny glass shards formed when a fine spray of exploded lava freezes instantly upon contact with the atmosphere.
volcanic bomb A smooth, streamlined piece of solidified lava with a diameter > 64 mm.
volcanic breccia A pyroclastic igneous rock that consists of fragments of volcanic debris, which either fall through the air and accumulate, or form when solidfying lava breaks up during flow.
volcanic danger-assessment map A map delineating areas that lie in the path of potential lava flows, lahars, debris flows, or pyroclastic flows of an active volcano.
volcanic debris flow A mixture of water and pyroclastic debris that moves downslope like wet concrete.

volcanic gas Elements or compounds that bubble out of magma or lava in gaseous form.
volcanic island arc The volcanic island chain that forms on the edge of the overriding plate where one oceanic plate subducts beneath another oceanic plate.
volcaniclastic deposit An accumulation of large quantities of fragmental igneous material (including both pyroclastic debris, and water-transported debris).
volcanic eruption An event during which gas, pyroclastic debris, and/or lava are ejected from a volcanic vent.
volcaniclastic rock A material composed of cemented-together grains of volcanic material; it includes both pyroclastic rocks and rocks formed from accumulations of water-transported volcanic debris.
volcano (1) A vent from which melt from inside the Earth spews out onto the planet's surface; (2) a mountain formed by the accumulation of extrusive volcanic rock.
V-shaped valley A valley whose cross-sectional shape resembles the shape of a V; the valley probably has a river running down the point of the V.
Wadati-Benioff zone A sloping band of seismicity defined by intermediate- and deep-focus earthquakes that occur in the downgoing slab of a convergent plate boundary.
wadi The name used in the Middle East and North Africa for a dry wash.
warm front A front in which warm air rises slowly over cooler air in the atmosphere.
waste rock Rock dislodged by mining activity yet containing no ore minerals.
waterfall A place where water drops over an escarpment.
water gap An opening in a resistant ridge where a trunk river has cut through the ridge.
watershed The region that collects water that feeds into a given drainage network.
water table The boundary, approximately parallel to the Earth's surface, that separates substrate in which groundwater fills the pores from substrate in which air fills the pores.
wave A disturbance that transmits energy from one point to another in the form of periodic motions.
wave base The depth, approximately equal in distance to half a wavelength in a body of water, beneath which there is no wave movement.
wave-cut bench A platform of rock, cut by wave erosion, at the low-tide line that was left behind a retreating cliff.
wave-cut notch A notch in a coastal cliff cut out by wave erosion.
wave erosion The combined effects of the shattering, wedging, and abrading of a cliff face by waves and the sediment they carry.
wave front The boundary between the region through which a wave has passed and the region through which it has not yet passed.
wavelength The horizontal difference between two adjacent wave troughs or two adjacent crests.
wave refraction (ocean) The bending of waves as they approach a shore so that their crests make no more than a 5° angle with the shoreline.
weather The conditions of temperature, wind speed, humidity, air pressure, and storminess at a location at a given time.
weathered rock Rock that has reacted with air and/or water at or near the Earth's surface.
weathering The processes that break up and corrode solid rock, eventually transforming it into sediment.
weather system A specific set of weather conditions, reflecting the configuration of air movement in the atmosphere, that affects a region for a period of time.
welded tuff Tuff formed by the welding together of hot volcanic glass shards at the base of pyroclastic flows.
well A hole in the ground dug or drilled in order to obtain water.
Western Interior Seaway A north-south-trending seaway that ran down the middle of North America during the Late Cretaceous Period.
wet-bottom (temperate) glacier A glacier with a thin layer of water at its base, over which the glacier slides.
wetted perimeter The area in which water touches a stream channel's walls.
wind abrasion The grinding away at surfaces in a desert by wind-blown sand and dust.
wind erosion The process of removing sediment directly by the action of fast moving air.
wind gap An opening through a high ridge that developed earlier in geologic history by stream erosion, but that is now dry.
xenolith A relict of wall rock surrounded by intrusive rock when the intrusive rock freezes.
yardang A mushroom-like column with a resistant rock perched on an eroding column of softer rock; created by wind abrasion in deserts where a resistant rock overlies softer layers of rock.
yazoo stream A small tributary that runs parallel to the main river in a floodplain because the tributary is blocked from entering the main river by levees.
Younger Dryas An interval of cooler temperatures that took place 4,500 years ago during a general warming/glacier-retreat period.
zeolite facies The metamorphic facies just above diagenetic conditions, under which zeolite minerals form.
zone of ablation The area of a glacier in which ablation (melting, sublimation, calving) subtracts from the glacier.
zone of accumulation (1) The layer of regolith in which new minerals precipitate out of water passing through, thus leaving behind a load of fine clay; (2) the area of a glacier in which snowfall adds to the glacier.
zone of aeration *See* Unsaturated zone.
zone of leaching The layer of regolith in which water dissolves ions and picks up very fine clay; these materials are then carried downward by infiltrating water.

Photo Credits

Pages ii-iii: Stephen Marshak; **p. ix (top)**: Stephen Marshak; **(bottom)**: NASA; **p. x**: Images provided by Google Earth mapping services/NASA, © DigitalGlobe, © Terra Metrics, © GeoEye, © Europa Technologies, Copyright 2015; **p. xi (top)**: Mark Schneider/Visuals Unlimited/Corbis; **(center):** Photos 12 / Alamy; **(bottom)**: Stephen Marshak; **p. xii (top)**: Richard Roscoe/Stocktrek Images/Corbis; **(center and bottom)**: Stephen Marshak; **p. xiii (top and center)**: Stephen Marshak; **(bottom)**: AP Photo/Natacha Pisarenko; **p. xiv (top)**: USGS; **(center and bottom)**: Stephen Marshak; **p. xv (all)**: Stephen Marshak; **p. xvi (top)**: NOAA; **(bottom)**: Reuters/Newscom; **p. xvii (both)**: Stephen Marshak; **p. xviii (top)**: Emma Marshak; **(bottom)**: Stephen Marshak; **p. xix (both)**: Stephen Marshak; **p. xxviii**: Kurt Burmeister.

Prelude

Pages 1-2: Stephen Marshak; **p. 2 (both)**: Stephen Marshak; **p. 3 (all)**: Stephen Marshak; **p. 4**: AP Photo/Hurriyet; **p. 6**: Stephen Marshak.

Chapter 1

Pages 10-11: NASA; **p.12 (a)**: Peter Apian, *Cosmographia*, Antwerp, 1524; **(b)**: Picture Library/Alamy; **p. 13 (a)**: Copyright Miloslav Druckmuller; **(b)**: NASA; **(c)**: NASA/JPL-Caltech; **p. 15**: Moonrunner Design; **p. 19 (a)**: SOHO (ESA & NASA); **(b)**: J. Hester and P. Scowen/NASA; **p. 21**: Stephen Marshak; **p. 22:** © R. Pelisson, SaharaMet; **p. 26**: NASA / Wikimedia; **p. 27**: JPL/NASA; **p. 29**: NASA/Science Source; **p. 31 (from left to right)**: Susan E. Degginger / Alamy; Susan E. Degginger / Alamy; Susan E. Degginger / Alamy; The Natural History Museum/Alamy; The Natural History Museum/Alamy; **p. 35 (a)**: Fred Espenak / Science Source; **(b)**: © Tom Bean; **(c)**: Marli Miller/Visuals Unlimited, Inc.; **p. 31**: Images provided by Google Earth mapping services/NASA, © DigitalGlobe, © Terra Metrics, © GeoEye, © Europa Technologies, Copyright 2015; **p. 32**: APOD/NASA; **p. 34**: Images provided by Google Earth mapping services/NASA, © DigitalGlobe, © Terra Metrics, © GeoEye, © Europa Technologies, Copyright 2015; **p. 39**: NASA/JPL-Caltech.

Chapter 2

Pages 42-43: Images provided by Google Earth mapping services/NASA, © DigitalGlobe, © Terra Metrics, © GeoEye, © Europa Technologies, Copyright 2015; **p. 44 (a)**: Alfred Wegener Institute for Polar and Marine Research; **(b)**: Ron Blakey, Colorado Plateau Geosystems; **p. 45 (both)**: Images provided by Google Earth mapping services/NASA, © DigitalGlobe, © Terra Metrics, © GeoEye, © Europa Technologies, Copyright 2015; **p. 52 (a, c)**: Images provided by Google Earth mapping services/NASA, © DigitalGlobe, © Terra Metrics, © GeoEye, © Europa Technologies, Copyright 2015; **(b)**: NOAA; **p. 53**: Images provided by Google Earth mapping services/NASA, © DigitalGlobe, © Terra Metrics, © GeoEye, © Europa Technologies, Copyright 2015; **p. 62**: EOS TRANSACTIONS, AMERICAN GEOPHYSICAL UNION, VOL. 78, PAGE 265, JULY 1, 1997, D. Smith et al: Viewing the Morphology of the Mid-Atlantic Ridge from a New perspective; **p. 63 (top)**: NOAA / University of Washington; **(bottom)**: Images provided by Google Earth mapping services/NASA, © DigitalGlobe, © Terra Metrics, © GeoEye, © Europa Technologies, Copyright 2015; **p. 64**: J.R. Delaney and D.S. Kelley, University of Washington; **p. 66**: Images provided by Google Earth mapping services/NASA, © DigitalGlobe, © Terra Metrics, © GeoEye, © Europa Technologies, Copyright 2015; **p. 67**: Images provided by Google Earth mapping services/NASA, © DigitalGlobe, © Terra Metrics, © GeoEye, © Europa Technologies, Copyright 2015; **p. 69**: Kevin Schafer / Alamy; **p. 74**: Cindy Ebinger University of Rochester; **p. 75**: Images provided by Google Earth mapping services / NASA, © DigitalGlobe, © Terra Metrics, © GeoEye, © Europa Technologies, Copyright 2015.

Chapter 3

Pages 82-83: Mark Schneider/Visuals Unlimited/Corbis; **p. 85 (left)**: Ken Lucas/ Visuals Unlimited; **(right)**: Stephen Marshak; **p. 88 (left)**: Mark A. Schneider/Science Source; **(center)**: incamerastock / Alamy; **p. 89 (a)**: Erich Schrempp/Science Source; **(c)**: Charles O'Rear /Corbis; **(d)**: John A. Jaszczak, Michigan Technological University; **p. 90**: 1996 Jeff Scovil; **p. 92 (a-b,d,f)**: Richard P. Jacobs /JLM Visuals; **(c)**: Breck P. Kent/JLM Visuals; **(e top)**: Stephen Marshak; **(e bottom)**: Andrew Silver/USGS; **(g)**: Scientifica / Visuals Unlimited / Corbis; **p. 93 (a)**: Farbled/Dreamstime.com; **(c)**: Dennis Kunkel Microscopy, Inc./Visuals Unlimited/Corbis; **(d)**: Construction Photography / Alamy; **p. 94 (a)**: Richard P. Jacobs/JLM Visuals; **(b)**: 1992 Jeff Scovil; **(c)**: Marli Miller/ Visuals Unlimited, Inc.; **(d-e)**: Richard P. Jacobs/JLM Visuals; **(h, left)**: Arco Images GmbH / Alamy; **(h, right)**: Ann Bryant/Geology.com; **p. 97 (top)**: 1996 Smithsonian Institution, photo Chip Clark; **(bottom)**: Images provided by Google Earth mapping services/NASA, © DigitalGlobe, © Terra Metrics, © GeoEye, © Europa Technologies, Copyright 2015; **p. 98 (left)**: Michael Langford/ Gallo Images / Getty Images; **(right)**: Ken Lucas / Visuals Unlimited; **p. 99 (a)**: Albert Copley/Visuals Unlimited; **(inset)**: Stephen J. Krasemann / Science Source; **(right)**: Images provided by Google Earth mapping services/NASA, © DigitalGlobe, © Terra Metrics, © GeoEye, © Europa Technologies, Copyright 2015; **p. 101:** Stephen Marshak.

Interlude A

Pages 102-103: Photos 12 / Alamy; **p. 104 (a, left)**: Stephen Marshak; **(a, center)**: Courtesy David W. Houseknecht, USGS; **(b, left)**: sciencephotos/ Alamy; **(b, center)**: Courtesy of Kent Ratajeski, Dept. of Geology and Geophysics, U of Wisconsin, Madison; **p. 105 (all)**: Stephen Marshak; **p. 106 (all)**: Stephen Marshak; **p. 108 (a)**: Tom Bean; **(b)**: Stephen Marshak; **(bottom, all)**: Stephen Marshak; **p. 109 (c)**: Stephen Marshak; **(d)**: Scenics & Science / Alamy; **(bottom)**: Product photo courtesy of JEOL, USA; **p. 110**: Stephen Marshak.

Chapter 4

Pages 112-113: Stephen Marshak; **p. 114 (a)**: Google Earth; **(b)**: Liysa/ Pacific Stock / Agefotostock; **(c)**: J.D. Griggs/ U.S. Geological Survey; **(d)**: Stephen Marshak; **p. 115 (b-c)**: Stephen Marshak; **p. 121 (a)**: USGS; **(b-d)**: Stephen Marshak; **p. 122 (both)**: Stephen Marshak; **p. 124 (both)**:

Stephen Marshak; **p. 125**: Google Earth; **p. 126 (both)**: Stephen Marshak; **p. 127 (top left)**: Dr. Kent Ratajeski; **(top center)**: Omphacite. 2006. Wikimedia; *http://en.wikipedia.org/wiki/Public_domain*; **(top right)**: Dr. Matthew Genge; **(bottom, left to right)**: Stephen Marshak; Mark A. Schneider/Science Source; Doug Sokell/Visuals Unlimited; **p. 130 (counterclockwise from top left)**: geoz / Alamy; Mark A. Schneider/Science Source; Joyce Photographics/Science Source; Siim Sepp /Alamy; Wally Eberhart/Visuals Unlimited/Corbis; Stephen Marshak; **(bottom both)**: Stephen Marshak; **p. 133**: Stephen Marshak.

Chapter 5

Pages 136-137: Richard Roscoe/Stocktrek Images/Corbis; **p. 139 (a)**: Yale Center for British Art, Paul Mellon Collection / Bridgeman Images; **(b)**: Jack Repcheck; **(c-d)**: Stephen Marshak; **p. 140 (a)**: Robert Francis/Agefotostock; **(b,d-f)**: Stephen Marshak; **(c)**: USGS; **p. 141 (a)**: Thomas Hallstein / Alamy; **(b)**: Stephen Marshak; **(bottom)**: Marli Miller / Visuals Unlimited; **p. 142 (a)**: AP Photo; **(b)**: Stephen Marshak; **(c)**: Stephen Weaver; **p. 143 (a)**: AFP/Getty Images; **(b)**: Photo by Suzanne MacLachlan, British Ocean Sediment Core Research Facility, National Oceanography Centre, Southampton; **(c-d)**: Stephen Marshak; **(e)**: AFP/Getty Images; **(f)**: USGS; **(g)**: Anthony Phelps/Reuters/Corbis; **p. 144 (top, both)**: Stephen Marshak; **(right)**: Images provided by Google Earth mapping services/NASA, © DigitalGlobe, © Terra Metrics, © GeoEye, © 2015; **p. 145 (a)**: Sunshine Pics /Alamy; (b): USGS; **p. 146**: Marli Miller/Visuals Unlimited; **p. 147 (a)**: Marli Miller/Visuals Unlimited; **(b)**: © Tom Bean 1985; **(c)**: Robert Harding World Imagery / Alamy; **p. 148 (a)**: USGS; **(b)**: AFP/Getty Images; **(c)**: USGS; **p. 149 (b)**: © Tom Pfeiffer / *www.volcanodiscovery.com*; **(c)**: USGS; **(bottom)**: Corbis; **p. 151 (top)**: Gary Braasch /Corbis; **(bottom)**: Stephen Marshak; **p. 155 (top)**: Google Earth; **(bottom)**: Images provided by Google Earth mapping services/NASA, © DigitalGlobe, © Terra Metrics, © GeoEye, © Europa Technologies, Copyright 2015; **p. 157**: Stephen Marshak; **p. 158**: Stephen Marshak; **p. 159 (a)**: USGS; **(b)**: Vittoriano Rastelli/ Corbis; **(c)**: AP Photo; **(d)**: Stephen Marshak; **(e)**: Alberto Garcia / Corbis; **(f)**: Philippe Bourseiller/Getty Images; **(g)**: Photo: Magnus T. Gudmundsson, University of Iceland; **(h)**: USGS; **p. 161**: Peter Turnley/Corbis; **p. 162**: Stephen Marshak; **p. 164 (a)**: Vittoriano Rastelli/Corbis; **(b)**: Sigurgeir Jonasson/Frank Lane Picture Agency/Corbis; **p. 165 (a)**: Gail Mooney/ Corbis; **p. 166 (a)**: Julian Baum / Science Source; **(b-e)**: NASA /JPL; **p. 167:** Reuters/Landov.

Interlude B

Pages 170-171: Stephen Marshak; **p. 172 (all):** Stephen Marshak; **p. 173 (top right)**: Visuals Unlimited; **(bottom, both)**: Stephen Marshak; **p. 174 (all)**: Stephen Marshak; **p. 175**: Stephen Marshak; **p.176 (both)**: Stephen Marshak; **p. 177 (all)**: Stephen Marshak; **p. 179 (all)**: Stephen Marshak; **p. 180**: Stephen Marshak; **p. 182**: US Dept. of Agriculture / Natural Resources Conservation Services; **p. 184 (a)**: Jim Richardson/Corbis; **(b and bottom):** Stephen Marshak.

Chapter 6

Pages 186-187: Stephen Marshak; **p. 188**: Stephen Marshak; **p.189 (both)**: Stephen Marshak; **p. 192 (all but bottom right)**: Stephen Marshak; **(bottom right)**: Scottsdale Community College; **p. 193 (top left)**: Stephen Marshak; **(top): right**: E.R. Degginger / Color-Pic, Inc.; **(bottom both)**: Stephen Marshak; **p. 194 (all)**: Stephen Marshak; **p. 195 (top both)**: Stephen Marshak; **(bottom left)**: Visuals Unlimited; **(bottom right)**: Stephen Marshak; **p. 196 (a)**: Marli Miller; **(b)**: M.W. Schmidt; **p. 197 (a-c)**: Stephen Marshak; **(left bottom)**: Images provided by Google Earth mapping services/NASA, © DigitalGlobe, © Terra Metrics, © GeoEye, © Europa Technologies, Copyright 2015; **p. 198**: Stephen Marshak; **p. 199 (a)**: All Canada Photos / Alamy; **(inset)**: Stephen Marshak; **(b)**: 1980 Grand Canyon Natural History Association; **p. 200 (both):** Stephen Marshak; **p. 201 (a):** IMAGINA Photography / Alamy; **(b-c)**: Stephen Marshak; **p. 202 (top a)**: Stephen Marshak; **(c)**: Marli Miller/Visuals Unlimited; **(bottom both)**: Stephen Marshak; **p. 203**: Images provided by Google Earth mapping services/NASA, © DigitalGlobe, © Terra Metrics, © GeoEye, © Europa Technologies, Copyright 2015; **p. 204 (a)**: Emma Marshak; **(b)**: Stephen Marshak; **(c)**: Marli Miller/Visuals Unlimited; **(d-e)**: Stephen Marshak; **(f)**: Callan Bentley; **p. 205 (both)**: Images provided by Google Earth mapping services/NASA, © DigitalGlobe, © Terra Metrics, © GeoEye, © Europa Technologies, Copyright 2015; **p. 206:** Stephen Marshak; **p. 208 (top)**: Stephen Marshak; **(bottom)**: Corbis; **p. 209 (a)**: The Natural History Museum / Alamy Stock Photo; **(b)**: G.R. "Dick" Roberts ©/Natural Sciences Image Library; **p. 211**: Stephen Marshak.

Chapter 7

Pages 214-215: Stephen Marshak; **p. 216 (top left)**: Stephen Marshak; **(top right)**: Visuals Unlimited; **(bottom left)**: Corbis; **(inset)**: Stephen Marshak; **(bottom right)**: Kurt Freihauf; **p. 220 (inset)**: Emma Marshak; **(a)**: Corbis; **(b)**: Stephen Marshak; **p. 221 (a-b)**: Stephen Marshak; **(c)**: Visuals Unlimited / Corbis; **p. 222 (both)**: Stephen Marshak; **p. 224 (all)**: Stephen Marshak; **p. 227**: Images provided by Google Earth mapping services/NASA, © DigitalGlobe, © Terra Metrics, © GeoEye, © Europa Technologies, Copyright 2015; **p. 230 (all)**: Stephen Marshak; **p. 232 (both)**: Images provided by Google Earth mapping services/NASA, © DigitalGlobe, © Terra Metrics, © GeoEye, © Europa Technologies, Copyright 2015.

Interlude C

Pages 236; 242 (all): Stephen Marshak.

Chapter 8

Pages 244-245: AP Photo/Natacha Pisarenko; **p. 247 (a)**: AFP/Getty Images; **(b)**: JIJI Press / AFP / Getty Images; **(c)**: AP Photo/ Kyodo News; **p. 248 (both)**: Photo Courtesy of Paul "Kip" Otis-Diehl, USMC, 29 Palms CA; **p. 251**: Peltzer et al. (1999), Evidence of nonlinear elasticity of the Crust. Science, v. 286. Copyright ©1999, AAAS; **p.253**: Images provided by Google Earth mapping services/NASA, © DigitalGlobe, © Terra Metrics, © GeoEye, © Europa Technologies, Copyright 2015; **p. 264 (b)**: Corbis; **(c)**: George Hall /Corbis; **p. 265**: Patrick Robert/Corbis Sygma; **p. 267**: New Madrid earthquake woodcut from Deven's Our First Century (1877); **p. 268 (from top down)**: AP Photo; Pacific Press Service / Alamy; M. Celebi, U.S. Geographical Survey; Reuters; **p. 269 (a)**: Barry Lewis / Alamy; **(b)**: Bettmann/Corbis; **p. 270 (left a)**: NOAA / National Geophysical Data Center (NGDC); **(c)**: Photo by Martin Luff *http://www.flickr.com/photos/23934380@N06/5469769673* *https://creativecommons.org/licenses/by-sa/2.0/* ; **(right a)**: Karl V. Steinbrugge Collection, University of California, Berkeley; **p. 271 (a)**: National Geophysical Data Center (NGDC); **(bottom)**: Images provided by Google Earth mapping services/ NASA, © DigitalGlobe, © Terra Metrics, © GeoEye, © Europa Technologies, Copyright 2015; **p. 273 (a)**: Vasily V. Titov, Associate Director, Tsunami Inundation Mapping Efforts (TIME), NOAA/PMEL-UW/JISAO, USA; (b): AFP/ Getty images; **(c)**: Photo by David Rydevik. 2004. Wikimedia *http://en.wikipedia.org/wiki/Public_domain*; **(d)**: Ikonos images copyright Centre for Remote Imaging, Sensing and Processing, National University of Singapore and Space Imaging; **p. 274 (a)**: AP Photo/Kyodo News; **(b top)**: EPA/The Tokyo Electric Power Company /Landov; **(b bottom)**: Air Photo Service/Reuters /Landov; **p. 275 (a-b)**: Reuters/Eduardo Munoz; **(c)**: USGS; **p. 279**: NOAA/NOA Center for Tsunami Research.

Interlude D

Page 282: USGS; **p. 284**: George Resch/ Fundamental Photographs, NYC; **p. 290**: Matthew Fouch, Arizona State University; **p. 291 (a)**: Courtesy Gregory Mountain, Lamont-Doherty Earth Observatory; **(b)**: Courtesy of Greg Moore, University of Hawaii and Nathan Bangs, University of Texas; **(inset)**: Courtesy Sercel and CGG Veritas; **p. 292 (c, both)**: Figures provided courtesy F. Lemoine & J. Frawley, NASA Goddard Space Flight Center; **(bottom)**: USGS; **p. 294**: USGS.

Chapter 9

Pages 296-297: Stephen Marshak; **p. 298**: NOAA/ETOPO1382; **p. 300 (all)**: Stephen Marshak; **p. 301 (both)**: Stephen Marshak; **p. 304 (top)**:

Google Earth; **(a)**: Galen Rowell/ Corbis; **(b-c)**: Stephen Marshak; **p. 305**: Stephen Marshak; **p. 307 (a)**: Stephen Marshak; **(b)**: USGS; **(c)**: Lloyd Cluff/Corbis; **p. 308 (a-c)**: Stephen Marshak; **(bottom)**: Images provided by Google Earth mapping services/NASA, © DigitalGlobe, © Terra Metrics, © GeoEye, © Europa Technologies, Copyright 2015; **p. 310 (a-d)**: Stephen Marshak; **p. 310 (e)**: Landsat/USGS; **p. 311 (both)**: Stephen Marshak; **p. 312 (top)**: Images provided by Google Earth mapping services/ NASA, © DigitalGlobe, © Terra Metrics, © GeoEye, © Europa Technologies, Copyright 2015; **(b)**: Stephen Marshak; **p. 313**: Stephen Marshak; **p. 314 (c)**: Images provided by Google Earth mapping services/NASA, © DigitalGlobe, © Terra Metrics, © GeoEye, © Europa Technologies, Copyright 2015; **p. 315 (left)**: Images provided by Google Earth mapping services/NASA, © DigitalGlobe, © Terra Metrics, © GeoEye, © Europa Technologies, Copyright 2015; **(b)**: Stephen Marshak; **p. 320**: Stephen Marshak; **p. 321**: Stephen Marshak; **p. 323 (both)**: Stephen Marshak.

Interlude E

Page 326: Stephen Marshak; **p. 328 (a, c)**: Stephen Marshak; **(b)**: Richard T. Nowitz / Corbis; **p. 330 (a)**: Sovfoto/UIG via Getty Images; **(b)**: Dirk Wiersma/ Science Source; **(c, d, f)**: Stephen Marshak; **(e)**: Kevin Schafer/ Corbis; **p. 331 (a)**: Stephen Marshak; **(b)**: UCL Micropaleontology Collections, UCL Museums & Collections; **p. 332 (a)**: Courtesy of Senckenberg, Messel Research Department; **(b)**: Humboldt-Universität zu Berlin Museum für Naturkunde. Photo by W. Harre; **p. 334**: Illustration by Karen Carr and Karen Carr Studio, Inc. © Smithsonian Institution.

Chapter 10

Pages 338-339: Stephen Marshak; **p. 342 (all)**: Stephen Marshak; **p. 344 (both)**: Stephen Marshak; **p. 345**: Stephen Marshak; **p. 347**: Stephen Marshak; **p. 349**: Images provided by Google Earth mapping services/ NASA, © DigitalGlobe, © Terra Metrics, © GeoEye, © Europa Technologies, Copyright 2015; **p. 350 (c)**: Jenning, C.W., 1997, California Dept. of Mines & Geology/ USGS; (d): Paul Karabinos, Williams College and USGS; **p. 351**: Images provided by Google Earth mapping services/NASA, © DigitalGlobe, © Terra Metrics, © GeoEye, © Europa Technologies, Copyright 2015; **p. 353 (all)**: Stephen Marshak; **p. 357**: Stephen Marshak; **p. 358 (a)**: Stephen Marshak; **(c)**: Damon Runberg, USGS; **p. 361**: USGS National Cventer of EROS and NASA, Landsat Project Science Office; **p. 363**: Stephen Marshak.

Chapter 11

Pages 364-365: Stephen Marshak; **p. 367 (both)**: artwork copyright Don Dixon / *www.cosmographica.com*; **p. 369 (a)**: Courtesy of Dr. J. William Schopf/UCLA; **(b)**: Stephen Marshak; **p. 370**: USGS; p. 373 (a): De Agostini Picture Library Universal Images Group/Newscom; **p. 373 (b)**: Stephen Marshak; **(c)** : Courtesy of Dr. Paul Hoffman, Harvard University; **p. 374 (b-c)**: Ronald C. Blakey; Colorado Plateau Geosystems, Inc.; **(bottom)**: Tom McHugh/Science Source; **p. 375 (b)**: Ronald C. Blakey; Colorado Plateau Geosystems, Inc.; **(c)**: Edward B. Daeschler; **p. 376**: Ronald C. Blakey; Colorado Plateau Geosystems, Inc.; **p. 377 (left)**: Mackenzie, J. 2012. Hillshaded Digital Elevation Model of the Continental US. *Http://www.udel.edu/johnmack/data_library/usa_dem.png*; **(top right)**: E.R. Degginger/Color-Pic; **(bottom right)**: Stephen Marshak; **p. 378**: Ronald C. Blakey; Colorado Plateau Geosystems, Inc.; **p. 379 (top left)**: Richard Bizley; **(top right)**: Ronald C. Blakey; Colorado Plateau Geosystems, Inc.; **(left)**: Images provided by Google Earth mapping services/NASA, © DigitalGlobe, © Terra Metrics, © GeoEye, © Europa Technologies, Copyright 2015; **p. 380**: Ronald C. Blakey; Colorado Plateau Geosystems, Inc.; **p. 382 (a)**: NASA, JPL; **(inset)**: Ron Blakey, Colorado Plateau Geosystems, Inc.; **(b)**: Images provided by Google Earth mapping services/NASA/ © 2015; **(c)**: Image Credit: Virgil L. Shaprton, Lunar and Planetary Institute; **(right)**: Images provided by Google Earth mapping services/NASA, © DigitalGlobe, © Terra Metrics, © GeoEye, © Europa Technologies, Copyright 2015; **p. 385**: Jacques Descloitres, MODIS Team, NASA Visible Earth; **pp. 386-387 (all)**: Ron Blakey, Colorado Plateau Geosystems.

Chapter 12

Pages 390-391: Stephen Marshak; **p. 398**: Data courtesy of Fugro. Credit: Virtual Seismic Atlas; *http://www.seismicatlas.org*; **p. 399 (a-b,d)**: Stephen Marshak; (c): AP Photo/Franship/HO: **p. 400**: Images provided by Google Earth mapping services/NASA, © DigitalGlobe, © Terra Metrics, © GeoEye, © Europa Technologies, Copyright 2015; **p. 403 (a)**: Field Museum Library/ Getty Images; **(d)**: Department of Natural Resources, Alaska; **p. 405**: Andrew Harrer/ Bloomberg via Getty Images; **p. 407 (a)**: Stephen Marshak; **(c)**: Stephen Marshak; **(d)**: Cultura Creative (RF) / Alamy; **p. 408**: Ron Chapple / Dreamstime; **p. 410**: Images provided by Google Earth mapping services/NASA, © DigitalGlobe, © Terra Metrics, © GeoEye, © Europa Technologies, Copyright 2015; **p. 411**: G.R. 'Dick' Roberts © Natural Sciences Image Library; **p. 412 (all)**: Stephen Marshak; **p. 415 (a)**: AP Photo/ Stapleton; **(b)**: AFP/ Getty Images; **p. 416 (a)**: US Coast Guard/Handout/Corbis; **(b)**: NASA/GSFC, MODIS; **p. 417**: Stephen Marshak; **(inset)**: Layne Kennedy / Corbis; **p. 418 (both)**: Richard P. Jacobs/ JLM Visuals; **p. 420**: Stephen Marshak; **p. 421 (b)**: Stephen Marshak; **(bottom)**: Images provided by Google Earth mapping services/ NASA, © DigitalGlobe, © Terra Metrics, © GeoEye, © Europa Technologies, Copyright 2015; **p. 423 (a)**: Richard P. Jacobs/JLM Visuals; **(b-c)**: Stephen Marshak; **p. 425 (a)**: Doug Sokell/Visuals Unlimited; **(b)**: A.J. Copley/ Visuals Unlimited.

Interlude F

Page 428: NOAA; **p. 430 (all)**: Stephen Marshak; **p. 432 (a)**: G.R. 'Dick' Roberts © Natural Sciences Image Library; **(b)**: Julie Dermansky/Corbis; **p. 433 (all)**: Stephen Marshak; **p. 435 (a)**: Melba Photo Agency / Alamy; **(b)**: Dr. David Smith, NASA Goddard Space Flight Center/MOLA Science Team; **(c)**: NASA/USGS Flagstaff; **(d)**: NASA/Johns Hopkins University Applied Physics Laboratory/Southwest Research Institute; **p. 438 (a)**: ESA/DLR/FU Berlin; **(b)**: AP Photo/ NASA.

Chapter 13

Pages 440-441: Reuters/Newscom; **p. 442 (both)**: Lloyd Cluff/Corbis; **p. 443 (c)**: Stephen Marshak; **(d)**: Marli Miller/ Visuals Unlimited, Inc.; **p. 444 (a, c-d)**: Stephen Marshak; **(bottom both)**: Images provided by Google Earth mapping services/NASA, © DigitalGlobe, © Terra Metrics, © GeoEye, © Europa Technologies, Copyright 2015; **p. 445 (a)**: Shana Reis/ EPA/Landov; **(b)**: Cascades Volcano Observatory/USGS; **(c-d)**: Stephen Marshak; **(right)**: Images provided by Google Earth mapping services/ NASA, © DigitalGlobe, © Terra Metrics, © GeoEye, © Europa Technologies, Copyright 2015; **p. 446 (a)**: Bob Shuster, USGS; **(b)**: National Geographic Image Collection / Alamy; **p. 447**: Stephen Marshak; **p. 448 (a)**: APA / AFP Photos; **(b)**: Alaska Stock/ Alamy; **p. 449**: Images provided by Google Earth mapping services/NASA, © DigitalGlobe, © Terra Metrics, © GeoEye, © Europa Technologies, Copyright 2015; **(a-b)**: Stephen Marshak; **p. 450 (a)**: Jerome Neufeld and Stephen Morris, Nonlinear Physics, University of Toronto; **(b)**: USGS/Barry W. Eakins; **(c)**: USGS, Geologic Investigations Series I-2809 by Barry W. Eakins, Joel E. Robinson, Toshiya Kanamatsu, Jiro Naka, John R. Smith, Eiichi Takahashi, and David A. Clague; **p. 451**: Stephen Marshak; **p. 456:** Breck P. Kent/JLM Visuals.

Chapter 14

Pages 462-463: Stephen Marshak; **p. 464 (a)**: Helen H. Richardson/ The Denver Post / Getty Images; **(b)**: John Gibson/Getty Images; **(c)**: Andy Clark/Reuters/Corbis; **p. 466 (left)**: Stephen Marshak; **(right)**: Images provided by Google Earth mapping services/NASA, © DigitalGlobe, © Terra Metrics, © GeoEye, © Europa Technologies, Copyright 2015; **p. 468 (both)**: Stephen Marshak; **p. 470 (a)**: Stephen Marshak; **(b)**: © Ron Niebrugge; **p. 471 (a-b)**: Stephen Marshak; **(c)**: Christina Neal/Alaska Volcano Observatory/USGS; **p.473 (a-b)**: Stephen Marshak; **p. 474**: Images provided by Google Earth mapping services/NASA, © DigitalGlobe, © Terra Metrics, © GeoEye, © Europa Technologies, Copyright 2015; **p. 475 (all)**: Stephen Marshak; **p. 476 (a)**: Stephen Marshak; **(b)**: Marti Miller/

Visuals Unlimited; **p. 477 (both)**: Stephen Marshak; **p. 478**: NASA/GSFC/meti/ersdac/jaros, and US/Japan ASTER Science Team; **p. 480:** 1997 Tom Bean; **p. 483 (top a)** : NASA images created by Jesse Allen, Earth Observatory, using data provided courtesy of the Landsat Project Science Office- copyright 2008; **(top b)**: Missouri Department of Transportation; **(c)**: Stringer /India /Reuters /Landov ; **(bottom a)**: Reuters/ Yoray Cohen/ Eilat Rescue Unit/ Landov; **(bottom b)**: USGS; **p. 484**: Courtesy Johnstown Area Heritage Association; **p. 485 (b)**: AP Photo/ The News-Star, Margaret Croft; **(c)**: Stephen Marshak; **p. 486 (all)**: Stephen Marshak; **p. 487 (both)**: Stephen Marshak; **p. 489**: Photo courtesy of the Bureau of Reclamation.

Chapter 15

Pages 492-493: Stephen Marshak; **p.496 (both)**: NOAA; **p. 499 (both)**: NASA; **p. 500 (b)**: Stephen Marshak; **(c)**: http://en.wikipedia.org/wiki/File:MtStMichel_avion.jpg, pd; **p. 501**: Stephen Marshak; **p.502 (both)**: Stephen Marshak; **p. 504 (a-b, f)**: Stephen Marshak; **(e)**: NASA; **p. 505**: Images provided by Google Earth mapping services/NASA, © DigitalGlobe, © Terra Metrics, © GeoEye, © Europa Technologies, Copyright 2015; **p. 506 (a):** G. R. 'Dick' Roberts © Natural Sciences Image Library; **(b)**: Stephen Marshak; **(d left)**: Stephen Marshak; **(d right)**: Emma Marshak; **p. 507 (bottom)**: Cody Duncan/ Alamy; **(right)**: Images provided by Google Earth mapping services/NASA, © DigitalGlobe, © Terra Metrics, © GeoEye, © Europa Technologies, Copyright 2015**; p. 508 (top both)**: Stephen Marshak; **(inset)**: Steve Bloom Images/ Alamy; **(b)**: Stephen Marshak; **p. 509**: Images provided by Google Earth mapping services/NASA, © DigitalGlobe, © Terra Metrics, © GeoEye, © Europa Technologies, Copyright 2015; **p. 514 (both a)**: USGS; **(b)**: Stephen Marshak; **(bottom)**: Images provided by Google Earth mapping services/NASA, © DigitalGlobe, © Terra Metrics, © GeoEye, © Europa Technologies, Copyright 2015; **p. 515**: Stephen Marshak; **p. 516**: Photograph by Robert Simmon, ASA Earth Observatory and NASA/NOAA GOES Project Science team; **p. 518**: NASA; **p. 519 (d)**: Images provided by Google Earth mapping services/NASA, © DigitalGlobe, © Terra Metrics, © GeoEye, © Europa Technologies, Copyright 2014; **(e)**: AP Photo/Susan Walsh; **(f)**: New York Times Photos & Graphics; **(g)**: Vincent Laforet/ Pool/ Reuters/ Corbis; **(h)**: Stephen Marshak.

Chapter 16

Pages 522-523: Emma Marshak; **p. 524 (a)**: USGS; **(c)**: Images provided by Google Earth mapping services/NASA, © DigitalGlobe, © Terra Metrics, © GeoEye, © Europa Technologies, Copyright 2015; **(right)**: Images provided by Google Earth mapping services/NASA, © DigitalGlobe, © Terra Metrics, © GeoEye, © Europa Technologies, Copyright 2015; **p. 526 (b)**: Photo courtesy of Eric Prokacki and Jim Best, University of Illinois; **(c)**: Stephen Marshak; **p. 531**: Stephen Marshak; **p. 533**: Stephen Marshak; **p. 535 (b)**: Allan Tuchman; **(c-e)**: Stephen Marshak; **p. 541 (all)**: Stephen Marshak; **p. 542 (top)**: Lois Kent; **(bottom left)**: Paul F. Hudson, University of Texas; **(bottom right)**: ML Sinibaldi / Corbis; **p. 543 (left)**: Lois Kent; **(right)**: Ashley Cooper/Corbis; **p. 544 (left)**: George Steinmetz / Corbis; **(center)**: Stephen Marshak; **(right)**: Image provided by Google Earth mapping services/ DigitalGlobe, Terra Metrics, NASA, Europa Technologies, Copyright 2015; **p. 545**: Stephen Marshak.

Chapter 17

Pages 548-549: Stephen Marshak; **p. 550**: Stephen Marshak; **p. 551**: Stephen Marshak; **p. 552**: Professor Andrew Danderfer; **p. 553 (all)**: Stephen Marshak; **p. 554 (both)**: Stephen Marshak; **p. 555 (top)**: Scott Wood / Barcroft Media / Getty Images; **(center and bottom)**: Stephen Marshak; **p. 556 (top)**: Stephen Marshak; **(bottom)**: Images provided by Google Earth mapping services/NASA, © DigitalGlobe, © Terra Metrics, © GeoEye, © Europa Technologies, Copyright 2015; **p. 557 (all)**: Stephen Marshak; **p. 560 (both)**: Stephen Marshak; **p.561 (top)**: Images provided by Google Earth mapping services/NASA, © DigitalGlobe, © Terra Metrics, © GeoEye, © Europa Technologies, Copyright 2015; **(bottom)**: Stephen Marshak; **p. 562 (all)**: Stephen Marshak; **p. 563 (top):** Images provided by Google Earth mapping services/NASA, © DigitalGlobe, © Terra Metrics, © GeoEye, © Europa Technologies, Copyright 2015; **(bottom)**: Stephen Marshak; **p. 564 (top)**: Images provided by Google Earth mapping services/NASA, © DigitalGlobe, © Terra Metrics, © GeoEye, © Europa Technologies, Copyright 2015; **(bottom)**: Eye Ubiquitous / Newscom; **p. 565 (top)**: Image courtesy Jacques Descloitres MODIS Rapid Response Team; **(bottom)**: United States Department of Agriculture, pd.

Chapter 18

Pages 568-569: Stephen Marshak; **p. 570**: Tom Bean/ Corbis; **p. 571 (a)**: Shutterstock; **(both b)**: Stephen Marshak; **(c center)**: Emma Marshak; **(c bottom)**: Ted Spiegel/ National Geographic Creative; **p. 572 (a)**: NASA-JPL; **(b)**: NASA; **p. 573 (b-c)**: Stephen Marshak; **(d)**: SRTM Team NASA/JPL/NIMA; **p. 574**: Images provided by Google Earth mapping services/NASA, © DigitalGlobe, © Terra Metrics, © GeoEye, © Europa Technologies, Copyright 2015; **p. 575 (top)**: Stephen Marshak; **(bottom)**: National Geophysical Data Center/ NOAA; **p. 578 (b)**: Ralph A. Clevenger/Corbis; **(c)**: Stephen Marshak; **(d both)**: ESA; **p. 579 (a)**: Shutterstock; **(b-d)**: Stephen Marshak; **(bottom right)**: Images provided by Google Earth mapping services/NASA, © DigitalGlobe, © Terra Metrics, © GeoEye, © Europa Technologies, Copyright 2015; **p. 580 (a-c)**: Stephen Marshak; **(d)**: 1986 Keith S. Walklet / Quietworks; **p. 581 (top)**: Marli Miller/ Visuals Unlimited, Inc.; **p. 581 (bottom)**: Wolfgang Meier/zefa/Corbis; **p. 582 (both)**: Stephen Marshak; **p. 583 (a-e1)**: Stephen Marshak; **(e2)**: Kevin Schafer/ Alamy; **p. 585 (c)**: Glenn Oliver/ Visuals Unlimited; **(e)**: Stephen Marshak; **p. 588**: Tom Bean/Corbis; **p. 590**: Stephen Marshak; **p. 591**: Lynda Dredge/ Geological Survey of Canada; **p. 593**: Detail of mural by Charles R. Knight, American Museum of Natural History, #4950(5), Photo by Denis Finnin; **p. 598 (top)**: Hendrick Averkamp, *Winter Scene with Ice Skaters*, ca. 1600. Courtesy of Rijksmuseum, Amsterdam; **(c)**: Stephen Marshak; **p. 599 (a)**: James Balog/Aurora Photos/Corbis; **(b)**: Roger J. Braithwaite; **(c)**: Michael Melford/Getty Images; **(d)**: NASA.

Chapter 19

Page 600-601: Photo by Felix Andrews, 2005, *https://creativecommons.org/licenses/by-sa/3.0/deed.en*; **p. 602**: Stephen Marshak; **p. 606**: DiBgd, 2007, *https://creativecommons.org/licenses/by-sa/3.0/deed.en*; **p. 610 (all)**: Stephen Marshak; **p. 614 (top)**: ISM/ PhotoTakeUSA.com; **(bottom)**: Bob Sacha/Corbis; **(inset):** NOAA; **p. 615 (top left to right)**: Nick Cobbing/ Alamy; Core Repository Lab at Lamong-Doherty Geological Observatory of Columbia University; **(bottom both)**: Stephen Marshak; **(inset)**: NOAA; **p. 618 (a, both)**: NASA/Goddard Space Flight Center Scientific Visualization Studio; **(c both):** National Snow and Ice Data Center; **(d top)**: USGS; **(d bottom)**: USGS photograph by Bruce Molnia; **p. 619**: Images provided by Google Earth mapping services/NASA, © DigitalGlobe, © Terra Metrics, © GeoEye, © Europa Technologies, Copyright 2015; **p. 623 (left)**: Stephen Marshak; (right): Images provided by Google Earth mapping services/NASA, © DigitalGlobe, © Terra Metrics, © GeoEye, © Europa Technologies, Copyright 2015; **p. 624 (left)**: Courtesy P&H Mining Equipment; **(center)**: Richard Hamilton Smith/ Corbis; **(right)**: Stephen Marshak; **(bottom both)**: Images provided by Google Earth mapping services/NASA, © DigitalGlobe, © Terra Metrics, © GeoEye, © Europoa Technologies, Copyright 2015; **p. 625 (b)**: Images provided by Google Earth mapping services/NASA, © DigitalGlobe, © Terra Metrics, © GeoEye, © Europoa Technologies, Copyright 2015; **(c)**: Nigel Dickinson/ Alamy; **p. 626 (a)**: Stephen Marshak; **(b)**: NASA.

Index

Note: *Italicized* page numbers refer to illustrations, tables, and figures. **Bold** page numbers refer to Guide Terms.

B

C

E

F

H

I

J

K

L

M

N

O

P

Q

R

S

T

U

V

W